Objective PHYSICS

Chapter-wise MCQs

for

NTA JEE Main/ BITSAT/ NEET/ AIIMS

• **Corporate Office :** 45, 2nd Floor, Maharishi Dayanand Marg, Corner Market,
Malviya Nagar, New Delhi-110017
Tel. : 011-49842349 / 49842350

By

Sanjeev Kumar Jha

Siddharth Shankar Singh

Typeset by Disha DTP Team

DISHA PUBLICATION

ALL RIGHTS RESERVED

© Publisher

No part of this publication may be reproduced in any form without prior permission of the publisher. The author and the publisher do not take any legal responsibility for any errors or misrepresentations that might have crept in. We have tried and made our best efforts to provide accurate up-to-date information in this book.

For further information about the books from DISHA,

Log on to **www.dishapublication.com** or email to **info@dishapublication.com**

CONTENTS

1 Physical World, Units and Measurements

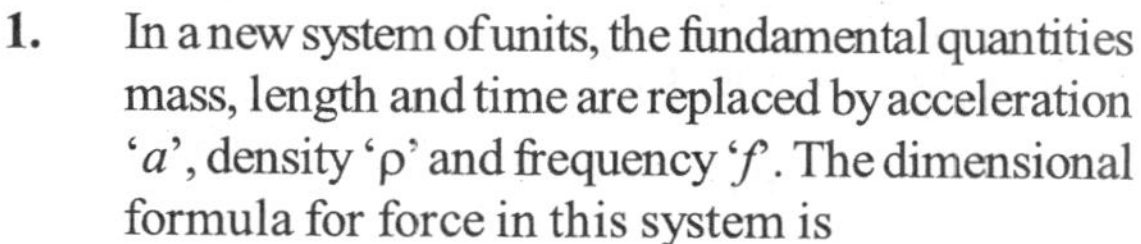

1. In a new system of units, the fundamental quantities mass, length and time are replaced by acceleration 'a', density 'ρ' and frequency 'f'. The dimensional formula for force in this system is
   (a) $[\rho a^4 f]$ (b) $[\rho a^4 f^{-6}]$
   (c) $[\rho^{-1} a^{-4} f^{6}]$ (d) $[\rho^{-1} a^{-4} f^{-1}]$

2. When two quantities are divided, the relative error in the result is given by
   (a) the product of the relative error in the individual quantities
   (b) the quotient of the relative error in the individual quantities
   (c) the difference of the relative error in the individual quantities
   (d) the sum of the relative error in the individual quantities

3. The momentum of an electron in an orbit is h/λ where h is a constant and λ is wavelength associated with it. The nuclear magneton of electron of charge e and mass m_e is given as $\mu_n = \frac{eh}{3672\pi m_e}$. The dimensions of μ_n are ($A \rightarrow$ current)
   Critical Thinking
   (a) $[ML^2A]$ (b) $[ML^3A]$
   (c) $[L^2A]$ (d) $[ML^2]$

4. If $Z = A^3$, then $\frac{\Delta Z}{Z}$ = ——
   (a) $\frac{\Delta A^3}{A}$ (b) $\left(\frac{\Delta A}{A}\right)^3$
   (c) $3\left(\frac{\Delta A}{A}\right)$ (d) $\left(\frac{\Delta A}{A}\right)^{1/3}$

5. The pitch and the number of circular scale divisions in a screw gauge with least count 0.02 mm are respectively
   (a) 1 mm and 100 (b) 0.5 mm and 50
   (c) 1 mm and 50 (d) 0.5 mm and 100

6. If L = 2. 331 cm, B = 2.1 cm, then L + B =
   (a) 4.4 cm (b) 4 cm
   (c) 4.43 cm (d) 4.431 cm

7. What is the correct number of significant figures in 0.0003026?
   (a) Four (b) Seven (c) Eight (d) Six

8. Number of significant figures in expression $\frac{4.327\text{ g}}{2.51\text{ cm}^3}$ is
   (a) 2 (b) 4 (c) 3 (d) 5

9. Error in the measurement of radius of a sphere is 1%. Then error in the measurement of volume is
   (a) 1% (b) 5% (c) 3% (d) 8%

10. In order to measure physical quantities in the sub-atomic world, the quantum theory often employs energy [E], angular momentum [J] and velocity [c] as fundamental dimensions instead of the usual mass, length and time. Then, the dimension of pressure in this theory is
    Toughnut
    (a) $\frac{[E]^4}{[J]^3[c]^3}$ (b) $\frac{[E]^2}{[J][c]}$
    (c) $\frac{[E]}{[J]^2[c]^2}$ (d) $\frac{[E]^3}{[J]^2[c]^2}$

11. In equation, $r = m^2 \sin \pi t$, where t represents time. If the unit of m is N, then the unit of r is
    (a) N (b) N^2 (c) Ns (d) N^2s

12. If $x = at + bt^2$, where x is the distance travelled by the body in kilometers while t is the time in seconds, then the unit of b is
    (a) km/s (b) kms (c) km/s^2 (d) kms^2

13. The electric field is given by $\vec{E} = \frac{A}{x^3}\hat{i} + By\hat{j} + Cz^2\hat{k}$. The SI units of A, B and C are respectively: [where x, y and z are in m]

    (a) $\frac{N\text{-}m^3}{C}$, V/m^2, N/m^2-C
    (b) V-m^2, V/m, N/m^2-C
    (c) V/m^2, V/m, N-C/m^2
    (d) V/m, N-m^3/C, N-C/m

14. If unit of length and force are increased 4 times. The unit of energy
(a) is increased by 4 times
(b) is increased by 16 times
(c) is increased by 8 times
(d) remains unchanged

15. The density of a material in CGS system of units is $4 g/cm^3$. In a system of units in which unit of length is 10 cm and unit of mass is 100 g, the value of density of material will be **Critical Thinking**
(a) 0.4 (b) 40 (c) 400 (d) 0.04

16. Young's modulus of steel is 1.9×10^{11} N/m^2. When expressed in CGS units of dyne/cm^2, it will be equal to (1N = 10^5 dyne, 1 m^2 = 10^4 cm^2)
(a) 1.9×10^{10} (b) 1.9×10^{11}
(c) 1.9×10^{12} (d) 1.9×10^{13}

17. The speed of light in vacuum, c, depends on two fundamental constants, the permeability of free space, μ_0 and the permittivity of free space, ε_0. The speed of light is given by $c = \frac{1}{\sqrt{\mu_0 \varepsilon_0}}$. The units of ε_0 are $N^{-1}C^2m^{-2}$. The units for μ_0 are
(a) $kg^{-1}m^{-1}C^2$ (b) $kg\, mC^{-2}$
(c) $kg\, ms^{-4}C^{-2}$ (d) $kg^{-1}s^{-3}C^{-2}$

18. Which of the following quantities has not been expressed in proper unit?
(a) torque, newton metre
(b) stress, newton metre^{-2}
(c) modulus of elasticity, newton metre^{-2}
(d) surface tension, newton metre^{-2}

19. E, m, J and G denote energy, mass, angular momentum and gravitational constant respectively, then the unit of $\frac{EJ^2}{m^5G^2}$ is
(a) newton (b) metre
(c) kilogram (d) unitless

20. In a certain system of units, 1 unit of time is 5 sec, 1 unit of mass is 20 kg and 1 unit of length is 10m. In this system, one unit of power will correspond to-
(a) 16 watts (b) $\frac{1}{16}$ watts
(c) 25 watts (d) $\frac{1}{25}$ watts

21. The moment of inertia of a body rotating about a given axis is 6.0 kg m^2 in the SI system. What is the value of the moment of inertia in a system of units in which the unit of length is 5 cm and the unit of mass is 10 g? **Toughnut**
(a) 2.4×10^3 (b) 2.4×10^5
(c) 6.0×10^3 (d) 6.0×10^5

22. The dimensional formula for relative density is
(a) $[ML^{-3}]$ (b) $[M^oL^{-3}]$
(c) $[M^oL^oT^{-1}]$ (d) $[M^oL^oT^o]$

23. The dimensions of speed and velocity are
(a) $[L^2T], [LT^{-1}]$ (b) $[LT^{-1}], [LT^{-2}]$
(c) $[LT], [LT]$ (d) $[LT^{-1}], [LT^{-1}]$

24. $[MLT^{-1}] + [MLT^{-1}] =$ .... **Tricky**
(a) $[M^oL^oT^o]$ (b) $[MLT^{-1}]$
(c) $2[MLT^{-1}]$ (d) None of these

25. The physical quantity that has no dimensions is
(a) strain (b) angular velocity
(c) angular momentum (d) linear momentum

26. Dimensions of specific heat are
(a) $[ML^2T^{-2}K]$ (b) $[ML^2T^{-2}K^{-1}]$
(c) $[ML^2T^2K^{-1}]$ (d) $[L^2T^{-2}K^{-1}]$

27. The dimensions of coefficient of self inductance are
(a) $[ML^2T^{-2}A^{-2}]$ (b) $[ML^2T^{-2}A^{-1}]$
(c) $[MLT^{-2}A^{-2}]$ (d) $[MLT^{-2}A^{-1}]$

28. The division of energy by time is X. The dimensional formula of X is same as that of
(a) momentum (b) power
(c) torque (d) electric field

29. Which of the following is a dimensional constant?
(a) Refractive index (b) Dielectric constant
(b) Relative density (d) Gravitational constant

30. Which one of the following represents the correct dimensions of the coefficient of viscosity?
(a) $[ML^{-1}T^{-1}]$ (b) $[MLT^{-1}]$
(c) $[ML^{-1}T^{-2}]$ (d) $[ML^{-2}T^{-2}]$

31. The ratio of the dimensions of Planck's constant and that of the moment of inertia is the dimensions of
(a) time (b) frequency
(c) angular momentum (d) velocity

32. Two quantities A and B have same dimensions which mathematical operation given below is physically meaningful? **Tricky**
(a) A/B (b) $A + B^2$ (c) $A^2 - B$ (d) $A = B^2$

33. The solar constant is defined as the energy incident per unit area per second. The dimensional formula for solar constant is
(a) $[M^0L^0T^0]$ (b) $[MLT^{-2}]$
(c) $[ML^2T^{-2}]$ (d) $[ML^0T^{-3}]$

34. Distance travelled by a particle at any instant 't' can be represented as S = A (t + B) + Ct^2. The dimensions of B are **Tricky**
(a) $[M^0L^1T^{-1}]$ (b) $[M^0L^0T^1]$
(c) $[M^0L^{-1}T^{-2}]$ (d) $[M^0L^2T^{-2}]$

35. If P,Q, R are physical quantities, having different dimensions, which of the following combinations can never be a meaningful quantity?
(a) (P – Q) / R (b) PQ – R
(c) PQ / R (d) $(PR - Q^2) / R$

36. Dimensions of 'resistance' are same as (where h is Planck's constant and e is charge)
(a) $\frac{h}{e}$ (b) $\frac{h^2}{e}$ (c) $\frac{h}{e^2}$ (d) $\frac{h^2}{e^2}$

37. If $v = \frac{a}{t} + bt^3$ where v = velocity and t is time
The dimensional formula of a and b are
(a) [T], $[T^{-3}]$ (b) [L], $[LT^{-4}]$
(c) $[T^{-3}]$, [T] (d) $[LT^{-4}]$, [L]

38. The displacement of a body at a particular second n is given by the expression $S_{nth} = u + \frac{a}{2}(2n-1)$. The dimensional formula of S_{nth} in this equation is 
(a) $[M^1L^0T^1]$ (b) $[M^0L^1T^0]$
(c) $[M^0L^1T^{-1}]$ (d) $[M^0L^0T^0]$

39. In the equation $P = \frac{RT}{V-b} e^{\frac{aV}{RT}}$
V = volume, P = pressure, R = universal gas constant, and T = temperature
The dimensional formula of a is same as that of
(a) V (b) P (c) T (d) R

40. Let Q denote the charge on the plate of a capacitor of capacitance C. The dimensional formula for $\frac{Q^2}{C}$ is
(a) $[L^2M^2T]$ (b) $[LMT^2]$
(c) $[L^2MT^{-2}]$ (d) $[L^2M^2T^2]$

41. Suppose the kinetic energy of a body oscillating with amplitude A and at a distance x is given by
$$K = \frac{Bx}{x^2 + A^2}$$

The dimensions of B are the same as that of
(a) work/time (b) work × distance
(c) work/distance (d) work × time

42. If the dimensions of a physical quantity are given by $M^a L^b T^c$, then the physical quantity will be
(a) velocity if a = 1, b = 0, c = – 1
(b) acceleration if a = 1, b = 1, c = – 2
(c) force if a = 0, b = – 1, c = – 2
(d) pressure if a = 1, b = – 1, c = – 2

43. Which one of the following represents the correct dimensions of the gravitational constant?
(a) $[M^{-1}L^3T^{-2}]$ (b) $[MLT^{-1}]$
(c) $[ML^{-1}T^{-2}]$ (d) $[ML^{-2}T^{-2}]$

44. Write the dimensions of a × b in the relation $E = \frac{b - x^2}{at}$, where E is the energy, x is the displacement and t is time
(a) ML^2T (b) $M^{-1}L^2T^1$
(c) ML^2T^{-2} (d) MLT^{-2}

45. The dimensions of the quantity $\vec{E} \times \vec{B}$ where $\vec{E}$ represents the electric field and $\vec{B}$ the magnetic field may be given as:
(a) $[MT^3]$ (b) $[M^2LT^{-5}A^{-2}]$
(c) $[M^2L^{-3}T^2A^{-1}]$ (d) $[MLT^{-2}A^{-2}]$

46. What are the dimensions of A/B in the relation $F = A\sqrt{x} + Bt^2$, where *F* is the force, *x* is the distance and *t* is time?
(a) ML^2T^{-2} (b) $L^{-1/2}T^2$ (c) $L^{-1/2}T^{-1}$ (d) LT^{-2}

47. A physical quantity x depends on quantities *y* and *z* as follows : $x = Ay + B \tan Cz$, where *A*, *B* and *C* are constants. Which of the following do not have the same dimensions :
(a) *x* and *B* (b) *C* and z^{-1}
(c) *y* and *B*/*A* (d) *x* and *A*

48. A quantity *X* is given by $\varepsilon_0 L \frac{\Delta V}{\Delta t}$ where ϵ_0 is the permittivity of the free space, *L* is a length, ΔV is a potential difference and Δt is a time interval. The dimensional formula for *X* is the same as that of **Critical Thinking**
(a) resistance (b) charge
(c) voltage (d) current

49. The thrust developed by a rocket-motor is given by $F = mv + A(P_1 - P_2)$ where m is the mass of the gas ejected per unit time, v is velocity of the gas, A is area of cross-section of the nozzle, P_1 and P_2 are the pressures of the exhaust gas and surrounding atmosphere. The formula is dimensionally
(a) correct
(b) wrong
(c) sometimes wrong, sometimes correct
(d) Data is not adequate

50. A, B, C and D are four different physical quantities having different dimensions. None of them is dimensionless. But we know that the equation AD = C ln (BD) holds true. Then which of the combination is not a meaningful quantity?

(a) $\frac{C}{BD}-\frac{AD^2}{C}$ (b) $A^2-B^2C^2$

(c) $\frac{A}{B}-C$ (d) $\frac{(A-C)}{D}$

51. If electronic charge e, electron mass m, speed of light in vacuum c and Planck's constant h are taken as fundamental quantities, the permeability of vacuum μ_0 can be expressed in units of **Critical Thinking**

(a) $\left(\frac{h}{me^2}\right)$ (b) $\left(\frac{hc}{me^2}\right)$

(c) $\left(\frac{h}{ce^2}\right)$ (d) $\left(\frac{mc^2}{he^2}\right)$

52. A boy recalls the relation almost correctly but forgets where to put the constant *c* (speed of light). He writes; $m=\frac{m_0}{\sqrt{1-v^2}}$, where m and m_0 stand for masses and *v* for speed. Right place of *c* is **Tricky**

(a) $m=\frac{cm_0}{\sqrt{1-v^2}}$ (b) $m=\frac{m_0}{c\sqrt{1-v^2}}$

(c) $m=\frac{m_0}{\sqrt{c^2-v^2}}$ (d) $m=\frac{m_0}{\sqrt{1-\frac{v^2}{c^2}}}$

53. When a small sphere moves at low speed through a fluid, the viscous force *F*, opposing the motion is experimentally found to depend upon the radius *r*, the velocity *v* of the sphere and the viscosity η of the fluid. Expression for force is **Toughnut**

(a) $4\pi\eta rv^2$ (b) $4\pi\eta r^2v$

(c) $2\pi\eta r^2v$ (d) $6\pi\eta rv$

54. With the usual notations, the following equation $S_t=u+\frac{1}{2}a(2t-1)$ is

(a) only numerically correct
(b) only dimensionally correct
(c) both numerically and dimensionally correct
(d) neither numerically nor dimensionally correct

55. Dimensionally wavelength is equivalent to

(a) $\frac{E\sqrt{LC}}{B}$ (b) $\frac{E}{B\sqrt{LC}}$

(c) $\frac{B\sqrt{LC}}{E}$ (d) $\frac{B}{E\sqrt{LC}}$

56. If velocity (V), force (F) and energy (E) are taken as fundamental units, then dimensional formula for mass will be

(a) $V^{-2}F^0E^3$ (b) V^0FE^2

(c) $VF^{-2}E^0$ (d) $V^{-2}F^0E$

57. The period of a body under SHM is represented by $T = P^aD^bS^c$; where P is pressure, D is density and S is surface tension. The value of a, b and c are.

(a) $-\frac{3}{2},\frac{1}{2},1$ (b) $-1,-2,3$

(c) $\frac{1}{2},-\frac{3}{2},\frac{1}{2}$ (d) $1,2,\frac{1}{3}$

58. A physical quantity of the dimensions of length that can be formed out of c, G and $\frac{e^2}{4\pi\varepsilon_0}$ is [c is velocity of light, G is universal constant of gravitation and e is charge] **Toughnut**

(a) $c^2\left[G\frac{e^2}{4\pi\varepsilon_0}\right]^{1/2}$ (b) $\frac{1}{c^2}\left[\frac{e^2}{G4\pi\varepsilon_0}\right]^{1/2}$

(c) $\frac{1}{c}G\frac{e^2}{4\pi\varepsilon_0}$ (d) $\frac{1}{c^2}\left[G\frac{e^2}{4\pi\varepsilon_0}\right]^{1/2}$

59. Assuming that the mass m of the largest stone that can be moved by a flowing river depends upon the velocity v of the water, its density ρ and the acceleration due to gravity g. Then m is directly proportional to :

(a) v^3 (b) v^4 (c) v^5 (d) v^6

60. A spherical body of mass m and radius r is allowed to fall in a medium of viscosity η. The time in which the velocity of the body increases from zero to 0.63 times the terminal velocity (v) is called time constant (τ). Dimensionally τ can be represented by :

(a) $\frac{mr^2}{6\pi\eta}$ (b) $\sqrt{\left(\frac{6\pi mr\eta}{g^2}\right)}$

(c) $\frac{m}{6\pi\eta rv}$ (d) None of these

61. If x and R stands for distance. Then which of the following is dimensionally same as $\int\frac{Rdx}{x^2}$?

(a) Rx^2 (b) $2xR$

(c) $\frac{R}{x}$ (d) $-\frac{R^2}{x}$

62. The unit of the coefficient of viscosity in S.I. system is
(a) m/kg-s (b) m-s/kg^2
(c) kg/m-s^2 (d) kg/m-s

63. The unit of the Stefan-Boltzmann's constant is
(a) W/m^2K^4 (b) W/m^2
(c) W/m^2K (d) W/m^2K^2

64. Unit of specific resistance is
(a) ohm/m^2 (b) ohm m^3
(c) ohm – m (d) ohm/m

65. Unit of magnetic moment is
(a) ampere–$metre^2$ (b) ampere–metre
(c) weber–$metre^2$ (d) weber/metre

66. Surface tension of a liquid is 70 dyne/cm. Its value in SI is
(a) 70 N/m (b) 7×10^{-2} N/m
(c) 7×10^2 N/m (d) 7×10^3 N/m

67. Which of the following represents incorrect unit?
(a) Pressure = N/m^2 (b) Surface tension = N/m
(c) Energy = kg-m/sec (d) $\frac{\text{Stress}}{\text{Strain}}$ = N/m^2

68. Which one of the following is not measured in units of energy ? **Tricky**
(a) Couple × angle
(b) Moment of inertia × (angular velocity)^2
(c) Force × distance
(d) Impulse × time

69. The numerical values of young's modulus in S.I. unit is β. What is its numerical value in cgs system?
(a) β (b) 10β (c) β/10 (d) 100 β

70. In the eqn. $\left(P+\frac{a}{V^2}\right)(V-b)$ = constant, the unit of a is **Tricky**
(a) dyne × cm^5 (b) dyne × cm^4
(c) dyne/cm^3 (d) dyne × cm^2

71. The value of resistance is 10.845 Ω and the value of current is 3.23 A. The potential difference is 35.02935 volt. Its value in significant number would be
(a) 35 V (b) 35.0 V
(c) 35.03 V (d) 35.029 V

72. Which of the following statements is/are correct?
I. 345.726 has six significant figures.
II. 0.004289 has seven singificant figures.
III. 125000 has three significant figures.
IV. 9.0042 has five significant figures.
(a) I only (b) II only
(c) I, III and IV (d) II, III and IV

73. Resistance R = V/I, here V = (100 ± 5)V and I = (10 ± 0.2) A. Find percentage error in R.
(a) 5% (b) 2% (c) 7% (d) 3%

74. A wire has a mass 0.3 ± 0.003 g, radius 0.5 ± 0.005 mm and length 6 ± 0.06 cm. The maximum percentage error in the measurement of its density is
(a) 1 (b) 2 (c) 3 (d) 4

75. The pitch of the screw gauge is 0.5 mm. Its circular scale contains 50 divisions. The least count of the screw gauge is
(a) 0.001 mm (b) 0.01 mm
(c) 0.02 mm (d) 0.025 mm

76. In a vernier callipers N division of vernier coincide with $(N-1)$ divisions of main scale in which length of a division is 1 mm. The least count of the instrument in cm is
(a) N (b) $N-1$ (c) $\frac{1}{10N}$ (d) $(1/N)-1$

77. One centimetre on the main scale of a vernier callipers is divided into 10 equal parts. If 10 divisions of vernier coincide with 8 small divisions of the main scale, the least count of vernier callipers is **Critical Thinking**
(a) 0.01 cm (b) 0.02 cm
(c) 0.05 cm (d) 0.005 cm

78. Relative density of a metal may be found with the help of spring balance. In air the spring balance reads (5.00 ± 0.05) N and in water it reads (4.00 ± 0.05) N. Then, the relative density along with the maximum permissible percentage error would be
(a) (5.00 ± 0.05) (b) (5.00 ± 11%)
(c) (5.00 ± 0.10) (d) (5.00 ± 6%)

79. The refractive index of water measured by the relation $\mu = \frac{\text{real depth}}{\text{apparent depth}}$ is found to have values of 1.34, 1.38, 1.32 and 1.36; the mean value of refractive index with percentage error is **Critical Thinking**
(a) 1.35 ± 1.48 % (b) 1.35 ± 0 %
(c) 1.36 ± 6 % (d) 1.36 ± 0 %

80. In an experiment four quantities a, b, c and d are measured with percentage error 1%, 2%, 3% and 4% respectively. Quantity P is calculated as follows
$P = \frac{a^3b^2}{cd}$ % error in P is
(a) 10% (b) 7% (c) 4% (d) 14%

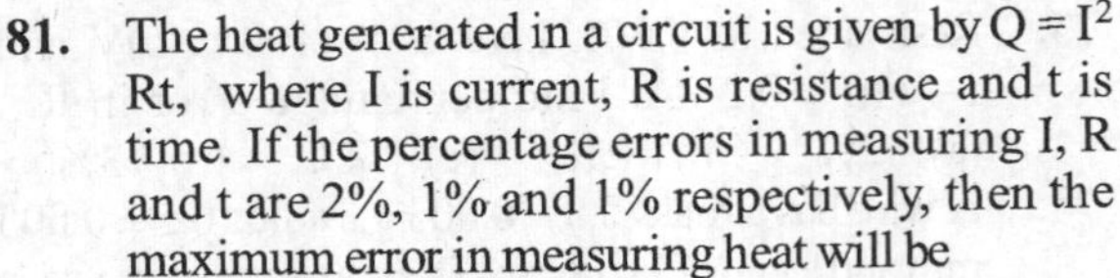

81. The heat generated in a circuit is given by $Q = I^2 Rt$, where I is current, R is resistance and t is time. If the percentage errors in measuring I, R and t are 2%, 1% and 1% respectively, then the maximum error in measuring heat will be

(a) 2% (b) 3% (c) 4% (d) 6%

82. The pressure on a square plate is measured by measuring the force on the plate and length of the sides of the plate by using the formula $P = \frac{F}{\ell^2}$. If the maximum errors in the measurement of force and length are 6% and 3% respectively, then the maximum error in the measurement of pressure is **Critical Thinking**

(a) 1% (b) 2% (c) 12% (d) 10%

83. An atomic clock has an accuracy of 1 part is 10^{10}. If two such clocks are operated with precision, then after running for 2500 years these will record a difference of nearly.

(a) 1 sec (b) 8 sec (c) 5 sec (d) 10 sec

84. Resistance of a given wire is obtained by measuring the current flowing in it and the voltage difference applied across it. If the percentage errors in the measurement of the current and the voltage difference are 3% each, then error in the value of resistance of the wire is **Critical Thinking**

(a) 6% (b) zero
(c) 1% (d) 3%

85. Which of the following is most accurate?

(a) A screw gauge of least count 0.001 mm
(b) A screw gauge having pitch 1 mm and 50 divisions on circular scale
(c) A vernier callipers of least count 0.01 mm
(d) Vernier callipers having 20 divisions on the sliding scale (vernier scale) coinciding 19 divisions on the main millimetre scale

86. A thin copper wire of length l metre increases in length by 2% when heated through 10°C. What is the percentage increase in area when a square copper sheet of length l metre is heated through 10°C?

(a) 4% (b) 8%
(c) 16% (d) None of these

87. A physical quantity ζ is calulated using the formula $\zeta = \frac{1}{10} xy^2 / z^{1/3}$, where x, y and z are experimentally measured quantities. If the fractional error in the measurement of x, y and z are 2 %, 1% and 3% respectively, then the fractional error in ζ will be **Critical Thinking**

(a) 0.5% (b) 5% (c) 6% (d) 7%

88. In a simple pendulum experiment, the maximum percentage error in the measurement of length is 2% and that in the observation of the time-period is 3%. Then the maximum percentage error in determination of the acceleration due to gravity g is

(a) 5% (b) 6% (c) 1% (d) 8%

89. Area of a square is (100 ± 2) m^2. Its side is

(a) (10 ± 1)m (b) (10 ± 0.1)m
(c) $(10 \pm \sqrt{2})$m (d) $10 \pm \sqrt{2}\%$

90. The values of kinetic energy K and potential energy U are measured as follows : K = 100.0 ± 2.0 J, U = 200.0 ± 1.0 J. Then the percentage error in the measurement of mechanical energy is –

(a) 2.5% (b) 1% (c) 0.5% (d) 1.5%

91. You measure two quantities as $A = 1.0\,\text{m} \pm 0.2\,\text{m}$, $B = 2.0\,\text{m} \pm 0.2\,\text{m}$. We should report correct value for $\sqrt{AB}$ as

(a) 1.4 m ± 0.4 m (b) 1.41 m ± 0.15 m
(c) 1.4 m ± 0.3 m (d) 1.4 m ± 0.2 m

92. The resistance of a metal is given by $R = \frac{V}{I}$, where V is potential difference and I is the current. In a circuit the potential difference across resistance is V = (8 ± 0.5) V and current in resistance, I = (2 ± 0.2) A. What is the value of resistance with its percentage error ?

(a) $4\Omega \pm 16.25\%$ (b) $(4 \pm 0.7)\Omega$
(c) $4\Omega \pm 0.7\%$ (d) $4\Omega \pm 7\%$

93. The density of a sphere is measured by measuring its mass and diameter. If, it is known that the maximum percentage errors in the measurement are 2% and 3%, then find the maximum percentage error in the measurement of density?

(a) 15% (b) 18% (c) 9% (d) 11%

94. The length of a cylinder is measured with a metre rod having least count 0.1 cm. Its diameter is measured with vernier callipers having least count 0.01 cm. Given that length is 5.0 cm and radius is 2.00 cm. The percentage error in the calculated value of the volume will be :

(a) 11% (b) 2% (c) 3% (d) 4%

95. Intensity observed in an interference pattern is $I = I_0 \sin^2 \theta$. At $\theta = 30°$ intensity $I = 5 \pm 0.0020$ W/m^2. Find percentage error in angle if $I_0 = 20$ W/m^2.

Toughnut

(a) $\frac{4}{\pi}\sqrt{3} \times 10^{-2}\%$ (b) $\frac{2}{\pi}\sqrt{3} \times 10^{-2}\%$

(c) $\frac{1}{\pi}\sqrt{3} \times 10^{-2}\%$ (d) $\frac{3}{\pi}\sqrt{3} \times 10^{-2}\%$

96. The resistance R of a wire is given by the relation $R = \frac{\rho \ell}{\pi r^2}$. Percentage error in the measurement of ρ, ℓ and r is 1%, 2% and 3% respectively. Then the percentage error in the measurement of R is

(a) 6 (b) 9 (c) 8 (d) 10

97. A physical quantity A is related to four observable quantities a, b, c and d as follows, $A = \frac{a^2 b^3}{c\sqrt{d}}$ and the percentage errors of measurement in a, b, c and d are 1%, 3%, 2% and 2% respectively. What is the percentage error in the quantity A?

(a) 12% (b) 7% (c) 5% (d) 14%

98. A physical quantity Q is related to four observables x, y, z and t by the relation $Q = \frac{x^{2/5} z^3}{y\sqrt{t}}$. The percentage errors of measurement in x, y, z and t are 2.5%, 2%, 0.5% and 1% respectively. The percentage error in Q will be

(a) 5% (b) 4.5% (c) 8% (d) 7.75%

99. The frequency (f) of a wire oscillating with a length ℓ, in p loops, under a tension T is given by $f = \frac{p}{2\ell}\sqrt{\frac{T}{\mu}}$ where μ = linear density of the wire. If the error made in determining length, tension and linear density be 1%, –2% and 4%, then find the percentage error in the calculated frequency.

Toughnut

(a) –4% (b) –2% (c) –1% (d) –5%

100. The current voltage relation of a diode is given by $I = \left(e^{1000V/T} - 1\right)$ mA, where the applied voltage V is in volts and the temperature T is in degree kelvin. If a student makes an error measuring ± 0.01 V while measuring the current of 5 mA at 300 K, what will be the error in the value of current in mA?

(a) 0.2 mA (b) 0.02 mA

(c) 0.5 mA (d) 0.05 mA

Answer KEYs

1	(b)	11	(b)	21	(b)	31	(b)	41	(b)	51	(c)	61	(c)	71	(b)	81	(d)	91	(d)
2	(d)	12	(c)	22	(d)	32	(a)	42	(d)	52	(d)	62	(d)	72	(c)	82	(c)	92	(a)
3	(c)	13	(a)	23	(d)	33	(d)	43	(a)	53	(d)	63	(a)	73	(c)	83	(b)	93	(d)
4	(c)	14	(b)	24	(b)	34	(b)	44	(b)	54	(a)	64	(c)	74	(d)	84	(a)	94	(c)
5	(c)	15	(b)	25	(a)	35	(a)	45	(b)	55	(a)	65	(a)	75	(b)	85	(a)	95	(a)
6	(a)	16	(c)	26	(d)	36	(c)	46	(b)	56	(d)	66	(b)	76	(c)	86	(a)	96	(b)
7	(a)	17	(b)	27	(a)	37	(b)	47	(d)	57	(a)	67	(c)	77	(b)	87	(b)	97	(d)
8	(c)	18	(d)	28	(b)	38	(c)	48	(d)	58	(d)	68	(d)	78	(b)	88	(d)	98	(a)
9	(c)	19	(d)	29	(d)	39	(b)	49	(a)	59	(d)	69	(b)	79	(a)	89	(a)	99	(a)
10	(a)	20	(b)	30	(a)	40	(c)	50	(d)	60	(d)	70	(b)	80	(d)	90	(b)	100	(a)

Hints & Solutions

1. (b) $F = ma = \rho$ volume 'a'
To write volume in terms of 'a' and 'f'
Volume $= L^3 = \left(\frac{L}{T^2}\right)^3 T^6 = a^3 f^{-6}$
$\therefore F = \rho a^4 f^{-6}$

2. (d) If $z = \frac{A}{B}$ then
$\frac{\Delta z}{z} = \frac{\Delta A}{A} + \frac{\Delta B}{B}$

3. (c) $\mu_n = \frac{ATML^2T^{-1}}{M} = L^2 A$

4. (c) as $z = A^3$, so $\frac{\Delta z}{z} = 3\frac{\Delta A}{A}$

5. (c) Least count of a screw gauge
$= \frac{\text{Pitch}}{\text{Number of circular scale divisions}}$
$= \frac{1\text{ mm}}{50} = 0.02\text{ mm}$
Therefore the pitch and no. of circular scale divisions are 1mm and 50 respectively.

6. (a) $L + B = 2.331 + 2.1 \cong 4.4$ cm
Since minimum significant figure is 2.

7. (a) If number is < 1, then zero(s) on right of decimal but left of first non-zero digit not significant.

8. (c) In multiplication or division the final result should return as many significant figures as there are in the original number with the least significant figures.

9. (c) $V = \frac{4}{3}\pi r^3$;
$\frac{\Delta V}{V} \times 100 = 3\left(\frac{\Delta r}{r}\right) \times 100 = 3 \times 1\% = 3\%$

10. (a) $[E] = [ML^2T^{-2}]$ ..... (i)
$[J] = [ML^2T^{-1}]$ ..... (ii)
$[C] = [LT^{-1}]$ ..... (iii)
Solving (i), (ii) and (iii) we get,
$\left[\frac{E}{C^2}\right] = [M]$, $\left[\frac{JC}{E}\right] = [L]$ and $\left[\frac{J}{E}\right] = [T]$
Now, [Pressure] $= [ML^{-1}T^{-2}]$
$= \left[\frac{E}{C^2}\right] \times \left[\frac{E}{JC}\right] \times \left[\frac{E^2}{J^2}\right] = \frac{[E]^4}{[J^3][C^3]}$

11. (b) Trigonometric ratio are a number and hence dimensionless

12. (c) $[x] = [bt^2]$. Hence $[b] = [x/t^2] = $ km/s^2

13. (a) Unit of Electric field (E) = (N/C) = (V/m)
So, unit of $A = E \times x^3 = \frac{N\text{-}m^3}{C} = (V\text{-}m^2)$
Unit of $B = E/y = \frac{N}{mc} = \left(\frac{V}{m^2}\right)$
Unit of $C = E/z^2 = \frac{N}{m^2 c} = \left(\frac{V}{m^3}\right)$

14. (b) The work done = force × displacement
$\therefore$ unit, $u_1 = Fs$
and $u_2 = 4F \times 4s = 16u$.

15. (b) In CGS system, $d = 4\frac{\text{g}}{\text{cm}^3}$
The unit of mass is 100g and unit of length is 10 cm, so
density $= \frac{4\left(\frac{100\text{g}}{100}\right)}{\left(\frac{10}{10}\text{cm}\right)^3} = \frac{\left(\frac{4}{100}\right)}{\left(\frac{1}{10}\right)^3}\frac{(100\text{g})}{(10\text{cm})^3}$
$= \frac{4}{100} \times (10)^3 \cdot \frac{100\text{g}}{(10\text{cm})^3} = 40$ unit

16. (c) It is given that Young's modulus (Y) is,
$Y = 1.9 \times 10^{11}$ N/m^2
1N $= 10^5$ dyne
So, $Y = 1.9 \times 10^{11} \times 10^5$ dyne/m^2
Convert meter to centimeter $\because$ 1m = 100 cm
$Y = 1.9 \times 10^{11} \times 10^5$ dyne/$(100)^2$ cm^2
$= 1.9 \times 10^{12}$ dyne/cm^2

17. (b) We have, $c = \frac{1}{\sqrt{\mu_0 \epsilon_0}} \Rightarrow \mu_0 = \frac{1}{\epsilon_0 c^2}$
$\therefore$ units of $\mu_0 = \frac{1}{\text{units of } \epsilon_0 c^2}$
$= \frac{1}{N^{-1}C^2m^{-2}(ms^{-1})^2} = Ns^2C^{-2}$
$= (\text{kgms}^{-2})s^2C^{-2} = \text{kgm } C^{-2}$

18. (d) Surface tension
$S = \frac{\text{Force}}{\text{Length}} = \frac{N}{m} = Nm^{-1}$

19. (d) Unitless quantity.

20. (b) 1 sec* = 5 sec, 1kg* = 20 kg, 1m* = 10m

1 watt = 1 kg m^2 sec^{-3}

$$= \frac{1}{20}\text{kg}*\left(\frac{1}{10}\text{m}*\right)^2\left(\frac{1}{5}\text{sec}*\right)^{-3}$$

$$= \frac{1}{20}\times\frac{1}{100}\times 125\text{ kg}*\text{m}*^2\text{ sec}*^{-3} = \frac{1}{16}\text{watt}*$$

21. (b) In new system, 1g* = 10g, 1cm* = 5cm.

$I = 6\times100\times g^*(20\,cm^*)^2 = 6\times100\times400\ g\times cm^{*2}$

$= 2.4\times10^5\ g^*\ cm^{*2}$

22. (d) Relative density $= \dfrac{\text{Density of Substance}}{\text{Density of water}}$

Hence no dimension.

23. (d) speed $= \dfrac{\text{distance}}{\text{time}}$ and velocity $= \dfrac{\text{displacement}}{\text{time}}$

24. (b) Only same dimensional quantity can be added or subtracted.

25. (a) Strain $= \dfrac{\text{Change in length}}{\text{Original length}}$

Hence no dimension.

26. (d) $s = \dfrac{Q}{m\theta} = \dfrac{ML^2T^{-2}}{MK} = [L^2T^{-2}K^{-1}]$

27. (a) Energy stored in an inductor, $U = \dfrac{1}{2}LI^2$

$\Rightarrow L = \dfrac{2U}{I^2}$

$\therefore [L] = \dfrac{[ML^2T^{-2}]}{[A]^2} = [ML^2T^{-2}A^{-2}]$

28. (b) Power $= \dfrac{\text{Energy}}{\text{time}}$

29. (d) Except Gravitational constant other are ratios of same dimensional quantity

30. (a) From stokes law

$F = 6\pi\eta rv \Rightarrow \eta = \dfrac{F}{6\pi rv}$

$\therefore \eta = \dfrac{MLT^{-2}}{[L][LT^{-1}]} \Rightarrow \eta = [ML^{-1}T^{-1}]$

31. (b) $\dfrac{\text{Planck's constant}}{\text{Moment of inertia}} = \dfrac{\frac{2\pi I\omega}{n}}{I}$

$\left[\text{As } \dfrac{nh}{2\pi} = I\omega\ \right]$

$= \dfrac{2\pi I(2\pi f)}{nI} = \left(\dfrac{4\pi^2}{n}.f\right) = [T^{-1}]$

32. (a) $[A] \neq [B^2]$ so, cannot be added or subtracted

33. (d) Energy incident per unit area per second

$= \dfrac{\text{Energy}}{\text{area}\times\text{second}} = \dfrac{ML^2T^{-2}}{L^2T} = MT^{-3}$

34. (b) In $S = A(t+B) + Ct^2$; B is added to time t. Therefore, dimensions of B are those of time.

35. (a) $[P] \neq [Q]$ so, P–Q is not meaningful

36. (c) $\dfrac{h}{e^2} = \dfrac{ML^2T^{-1}}{(AT)^2} = ML^2T^{-3}A^{-2}$

= Resistance (ohm)

37. (b) Dimensionally $v = \dfrac{a}{t} \Rightarrow a = LT^{-1}\times T = L$

also dimensionally

$v = bt^3 \Rightarrow b = \dfrac{LT^{-1}}{T^3} = LT^{-4}$

38. (c) $[S_n] = [M^oL^1T^o]$

39. (b) $[a] = \dfrac{[RT]}{[V]} = [P]$

40. (c) We know that $\dfrac{Q^2}{2C}$ is energy of capacitor, so it represent the dimension of energy $= [ML^2T^{-2}]$.

41. (b) From $K = \dfrac{Bx}{x^2+A^2} = \dfrac{Bx}{x^2} = \dfrac{B}{x}$

$\therefore$ B = K × x = K.E. × distance

= work × distance.

42. (d) Pressure $= \dfrac{MLT^{-2}}{L^2} = [ML^{-1}T^{-2}]$

$\Rightarrow a = 1, b = -1, c = -2.$

43. (a) $F = \dfrac{GMm}{R^2}$

$\therefore G = \dfrac{FR^2}{Mm} \Rightarrow G = [M^{-1}L^3T^{-2}]$

44. (b) Here, b and $x^2 = L^2$ have same dime nsions

Also, $a = \dfrac{x^2}{E\times t} = \dfrac{L^2}{(ML^2T^{-2})T} = M^{-1}T^1$

$a\times b = [M^{-1}L^2T^1]$

45. (b) $[\vec{E}\times\vec{B}] = [\vec{E}]\times[B]$

$[MLT^{-3}A^{-1}]\times[ML^{-2}A^{-1}] = [MLT^{-5}A^{-2}]$

46. (b) $[F] = [A]\times[x]^{1/2} = [B][t]^2$

$\Rightarrow \dfrac{[A]}{[B]} = \dfrac{[t]^2}{[x]^{1/2}} = [L^{-1/2}T^2]$

47. (d) $[x] = [A]\times[y] = [B] \Rightarrow [x] \neq [A]$

48. (d) Dimensionally $\varepsilon_0 L = C$
where C = capacitance
Now the given expression reduces to $C\Delta V/\Delta t$.
Dimensionally $C\Delta V = q$ where q is charge.
Again the given expression reduces to $q/\Delta t$.
Dimensionally $\frac{q}{\Delta t} = I$ where I is current

49. (a) Use principle of homogeneity.

50. (d) Dimension of A $\neq$ dimension of (C)
Hence A – C is not possible.

51. (c) Let μ_0 related with e, m, c and h as follows.
$\mu_0 = ke^a m^b c^c h^d$
$[MLT^{-2}A^{-2}] = [AT]^a [M]^b [LT^{-1}]^c [ML^2T^{-1}]^d$
$= [M^{b+d} L^{c+2d} T^{a-c-d} A^a]$
On comparing both sides we get
$a = -2,\ b = 0,\ c = -1,\ d = 1$
$\therefore\ [\mu_0] = \left[\frac{h}{ce^2}\right]$

52. (d) In, $1 - v^2$, v^2 should be dimensionsless, so it should be $1 - \frac{v^2}{c^2}$.

53. (d) We can thus say that the viscous force (F) is the function of radius (r), velocity (v) and viscosity (η).
or $F = f(\eta, r, v)$ or $F = k\eta^x r^y v^z$ ......(1)
Where k is a constant.
Now, dimensions of the constituents are
$\therefore\ [MLT^{-2}] = [ML^{-1}T^{-1}]^x [L]^y [LT^{-1}]^z$
$= [M^x L^{-x+y+z} T^{-x-z}]$
Equating the exponents of similar quantities of both sides we get, $x = 1$; $-x + y + z = 1$ and $-x - z = -2$
Solving for x, y & z, we get $x = y = z = 1$
Equation (1) becomes $F = k\eta rv$
Experimentally, it was found that
$k = 6\pi$ or $F = 6\pi\eta rv$, which is the famous Stokes' law.

54. (a) On LHS, S_t is the distance while on RHS, u is the speed, so it is not dimensionally correct. It is numerically correct only.

55. (a) $\frac{E}{B}$ = velocity = m/s and
$\sqrt{LC}$ = time period = s
$\therefore\ \frac{E\sqrt{LC}}{B} = m$ = wavelength

56. (d) Let (M) = $V^a F^b E^c$
Putting the dimensions of V, F and E, we have
$(M) = (LT^{-1})^a \times (MLT^{-2})^b \times (ML^2T^{-2})^c$
or $M^1 = M^{b+c} L^{a+b+2c} T^{-a-2b-2c}$
Equating the powers of dimensions, we have
$b + c = 1$
$a + b + 2c = 0;\ -a - 2b - 2c = 0$
which give $a = -2$, $b = 0$ and $c = 1$.
Therefore (M) = $(V^{-2} F^0 E)$.

57. (a) $T = [ML^{-1}T^{-2}]^a [L^{-3}M]^b [MT^{-2}]^c$
$\Rightarrow b + a + c = 0,\ -a - 3b = 0,\ -2a - 2c = 1$
$\Rightarrow a = -\frac{3}{2},\ b = \frac{1}{2},\ c = 1$

58. (d) Let dimensions of length is related as,
$L = [c]^x [G]^y \left[\frac{e^2}{4\pi\varepsilon_0}\right]^z \Rightarrow \frac{e^2}{4\pi\varepsilon_0} = ML^3T^{-2}$
$L = [LT^{-1}]^x [M^{-1}L^3T^{-2}]^y [ML^3T^{-2}]^z$
$[L] = [L^{x+3y+3z} M^{-y+z} T^{-x-2y-2z}]$
Comparing both sides
$z = y = \frac{1}{2},\ x = -2$ Hence, L =
$c^{-2}\left[G \cdot \frac{e^2}{4\pi\varepsilon_0}\right]^{1/2}$

59. (d) $m \propto v^x \rho^y g^z$
$M \equiv [LT^{-1}]^x [ML^{-3}]^y [LT^{-2}]^z$
$M^1L^0T^0 = M^y L^{x-3y+z} T^{-x-2z}$
$y = 1$, $x - 3y + z = 0,\ -x - 2z = 0$
Solving above equations
$x - 3 - \frac{x}{2} = 0 \Rightarrow \frac{x}{2} = 3 \Rightarrow x = 6$

60. (d) $F = 6\pi\eta rv$
$\Rightarrow mg = 6\pi\eta rv$ (dimensionally)
Check options dimensionally

61. (c) $\left[\int \frac{Rdx}{x^2}\right] = \frac{[R][x]}{[x^2]} = \frac{[R]}{[x]}$

62. (d) $F = 6\pi\eta vr \Rightarrow \eta = F/6\pi vr$
$\therefore$ Unit of coefficient of viscosity in S.I. system kg/m-s.

63. (a) $E = \sigma AT^4$
E is energy dissipated per second.
$\sigma = \frac{E}{AT^4} = \frac{\text{Watt}}{m^2K^4}$

64. (c) $R = \rho\frac{l}{A}\ \therefore\ \rho = \frac{RA}{l} = \frac{\Omega m^2}{m} = \Omega m$

65. (a) M = Pole strength × length
= amp – metre × metre = amp – metre^2

66. (b) $S = 70\dfrac{\text{dyne}}{\text{cm}} = \dfrac{70\times10^{-5}\ \text{N}}{10^{-2}\ \text{m}}$
$= 70\times10^{-3}$ N/m.

67. (c) Unit of energy will be kg-m^2/sec^2

68. (d) Unit of impulse × time = (kg ms^{-2} × s) × s = kgm

69. (b) $\dfrac{\text{N}}{\text{m}^2} = \dfrac{10^5\ \text{dynes}}{10^4\ \text{cm}^2} = 10$ dynes / cm^2 = 10 β

70. (b) As $\dfrac{a}{V^2} = P$
$\therefore\ a = PV^2 = \dfrac{\text{dyne}}{\text{cm}^2}(\text{cm}^3)^2 = \text{dyne}\times\text{cm}^4$

71. (b) The significant number in the potential, $V = iR$; should be the minimum of either i or R. So corresponding to $i = 3.23\ A$, we have only three significant numbers in $V = 35.02935\ V$. Thus the result is $V = 35.0\ V$.

72. (c) 0.004289 has four significant figures

73. (c) $R = \dfrac{V}{I}$ so, $\dfrac{\Delta R}{R}\times100 = \dfrac{\Delta V}{V}\times100 + \dfrac{\Delta I}{I}\times100$
$\Rightarrow \dfrac{\Delta R}{R}\times100 = \dfrac{5}{100}\times100 + \dfrac{0.2}{10}\times100 = 7\%$

74. (d) Density, $\rho = \dfrac{M}{V} = \dfrac{M}{\pi r^2\ell}$
$\therefore\ \dfrac{\Delta\rho}{\rho}\times100 = \left[\dfrac{\Delta M}{M} + \dfrac{2\Delta r}{r} + \dfrac{\Delta\ell}{\ell}\right]\times100 = 4$

75. (b) Least count $= \dfrac{0.5}{50} = 0.01$ mm

76. (c) $L.C. = \dfrac{\text{value of 1 division of main scale}}{\text{number of division on main scale}}$
$= \dfrac{1}{N}\ \text{mm} = \dfrac{1}{10N}\ \text{cm}$

77. (b) The value of 1 division of main scale $= \dfrac{1}{10}$
= 0.1 cm
The value of 1 division of vernier scale
$= \dfrac{8\times0.1}{10} = 0.08$ cm
Thus L.C. = 0.1 – 0.08 = 0.02 cm

78. (b)

79. (a) The mean value of refractive index,
$\mu = \dfrac{1.34+1.38+1.32+1.36}{4} = 1.35$ and
$\Delta\mu = \dfrac{|(1.35-1.34)|+|(1.35-1.38)|+|(1.35-1.32)|+|(1.35-1.36)|}{4}$
$= 0.02$
Thus $\dfrac{\Delta\mu}{\mu}\times100 = \dfrac{0.02}{1.35}\times100 = 1.48$

80. (d) $P = \dfrac{a^3b^2}{cd}$, $\dfrac{\Delta P}{P}\times100\% = 3\dfrac{\Delta a}{a}\times100\%$
$+\ 2\dfrac{\Delta b}{b}\times100\% + \dfrac{\Delta c}{c}\times100\% + \dfrac{\Delta d}{d}\times100\%.$
$= 3\times1\% + 2\times2\% + 3\% + 4\% = 14\%$

81. (d) $\dfrac{\Delta Q}{Q}\times100 = \dfrac{2\Delta I}{I}\times100 + \dfrac{\Delta R}{R}\times100 + \dfrac{\Delta t}{t}\times100$
$= 2\times2\% + 1\% + 1\% = 6\%.$

82. (c) $\dfrac{\Delta P}{P}\times100 = \dfrac{\Delta F}{F}\times100 + 2\dfrac{\Delta\ell}{\ell}\times100$
$= 6\% + 2\times3\% = 12\%$

83. (b)

84. (a) From Ohm's law,
$\Rightarrow R = \dfrac{V}{I} \Rightarrow \ln R = \ln V - \ln I$
$\Rightarrow \dfrac{\Delta R}{R} = \dfrac{\Delta V}{V} + \dfrac{\Delta I}{I} = 3\% + 3\% = 6\%$

85. (a)

86. (a) Since percentage increase in length = 2 %
Hence, percentage increase in area of square sheet
= 2 × 2% = 4%

87. (b) $\dfrac{\Delta\xi}{\xi}\times100 = \left(\dfrac{\Delta x}{x} + \dfrac{2\Delta y}{y} + \dfrac{1}{3}\dfrac{\Delta z}{z}\right)\times100$
$= 2 + 2\times1 + \dfrac{1}{3}\times3 = 5\%$

88. (d) As we know, time period of a simple pendulum
$T = 2\pi\sqrt{\dfrac{L}{g}} \Rightarrow g = \dfrac{4\pi^2L}{T^2}$
The maximum percentage error in g
$\dfrac{\Delta g}{g}\times100 = \dfrac{\Delta L}{L}\times100 + 2\left(\dfrac{\Delta T}{T}\times100\right)$
$= 2\% + 2(3\%) = 8\%$

89. (a) Area = (Length)^2
or length = (Area)$^{1/2}$ = $(100\pm2)^{1/2}$
$= (100)^{1/2} \pm \dfrac{1}{2}\times2 = (10\pm1)$m

90. (b) $M_E = K + U = 300\pm3J = E\pm\Delta E$ ∴
$\dfrac{\Delta E}{E}\times100 = 1\%$

91. (d) As given that,
$A = 1.0\ \text{m}\pm0.2\ \text{m}$, $B = 2.0\ \text{m}\pm0.2\ \text{m}$
So, $X = \sqrt{AB} = \sqrt{(1.0)(2.0)} = 1.414$ m
Rounding off upto two significant digit
$X = 1.4\ \text{m} = (r)$ (Let)

$\frac{\Delta x}{x}=\frac{1}{2}\left[\frac{\Delta A}{A}+\frac{\Delta B}{B}\right]=\frac{1}{2}\left[\frac{0.2}{1.0}+\frac{0.2}{2.0}\right]=0.212$

Rounding off upto one significant digit

$\Delta x=0.2\text{ m}=\Delta r$ (Let)

So, correct value of

$\sqrt{AB}=r+\Delta r=(1.4\pm0.2)\text{ m}$

92. **(a)** Actual $R=\frac{8}{2}=4$

% error in R = % error in V + % error in I

$=\frac{0.5}{8}\times100\%+\frac{0.2}{2}\times100\%=16.25\%$

$R=4\Omega\pm16.25\%$

93. **(d)** Let m and d be the mass and diameter of the sphere, then the density ρ of the sphere is given by

$$\rho=\frac{\text{mass}}{\text{volume}}=\frac{m}{\frac{4}{3}\pi\left(\frac{d}{2}\right)^3}=\frac{6m}{\pi d^3}$$

or $\left(\frac{d\rho}{\rho}\times100\right)_{max}=\left|\frac{dm}{m}\times100\right|+\left|\frac{3d(d)}{d}\times100\right|$

$=2+3(3)=11\%$

94. **(c)** $V=\pi r^2\ell$ $\quad\frac{\Delta V}{V}\%=2\frac{\Delta r}{r}\%+\frac{\Delta\ell}{\ell}\%$

$=2\times\frac{0.01}{2}\times100\%+\frac{0.1}{5}\times100\%=3\%$

95. **(a)** $I=I_0\sin^2\theta\Rightarrow\theta=\sin^{-1}\sqrt{\frac{I}{I_0}}$

$$d\theta=\frac{1}{2}\frac{\sqrt{I_0}}{\sqrt{I_0-I}}\frac{dI}{\sqrt{I_0}\sqrt{I}}$$

$$\frac{d\theta}{\theta}=\frac{1}{2}\frac{dI}{\sqrt{I(I_0-I)}\sin^{-1}\sqrt{I/I_0}}$$

$\therefore$ % error $=\frac{4}{\pi}\sqrt{3}\times10^{-2}\%$

96. **(b)** Given $R=\frac{\rho\ell}{\pi r^2}$, then

$\frac{\Delta R}{R}\times100$

$=\frac{\Delta\rho}{\rho}\times100+\frac{\Delta\ell}{\ell}\times100+2\frac{\Delta r}{r}\times100$

$=1\%+2\%+2\times3\%=9\%$

97. **(d)** Percentage error in A

$=\left(2\times1+3\times3+1\times2+\frac{1}{2}\times2\right)\%=14\%$

98. **(a)** Given: $Q=x^{2/5}y^{-1}t^{-1/2}z^3$

$\therefore\ \frac{\Delta Q}{Q}\times100$

$=\frac{2}{5}\frac{\Delta x}{x}\times100+\frac{\Delta y}{y}\times100$

$+\frac{1}{2}\frac{\Delta t}{t}\times100+3\frac{\Delta z}{z}\times100$

$=\frac{2}{5}\times2.5+2+\frac{1}{2}\times1+3\times0.5=5\%$

99. **(a)** Given $f=\frac{p}{2\ell}\sqrt{\frac{T}{\mu}}$ Taking log of both sides

$\log f=\log\left(\frac{p}{2}\right)-\log\ell+\frac{1}{2}\log T-\frac{1}{2}\log\mu$

Differentiating partially on both sides,

$\frac{df}{f}=0-\frac{d\ell}{\ell}+\frac{1}{2}\frac{dT}{T}-\frac{1}{2}\frac{d\mu}{\mu}$

or

$\frac{df}{f}\times100=\left(-\frac{d\ell}{\ell}\times100\right)+\left(\frac{1}{2}\frac{dT}{T}\times100\right)-\left(\frac{1}{2}\frac{d\mu}{\mu}\times100\right)$

$=(-1)+\frac{1}{2}(-2)-\frac{1}{2}(4)=-1-1-2=-4\%$

100. **(a)** The current voltage relation of diode is

$I=(e^{1000\,V/T}-1)$ mA (given)

When, $I=5mA$, $e^{1000\,V/T}=6mA$

Also, $dI=(e^{1000\,V/T})\times\frac{1000}{T}$ (By exponential function)

$=(6\,mA)\times\frac{1000}{300}\times(0.01)=0.2\text{ mA}$

2 Motion in a Straight Line

1. A body starts from rest and travels 's' m in 2nd second, then acceleration is
(a) $(2s)\ m/s^2$ (b) $(3s)\ m/s^2$
(c) $\left(\frac{2}{3}s\right)\ m/s^2$ (d) $\left(\frac{3}{2}s\right)\ m/s^2$

2. The velocity-time graph of a body is shown in fig. The ratio of average acceleration during the intervals OA and AB is
(a) 1
(b) $\frac{1}{2}$
(c) $\frac{1}{3}$
(d) 3

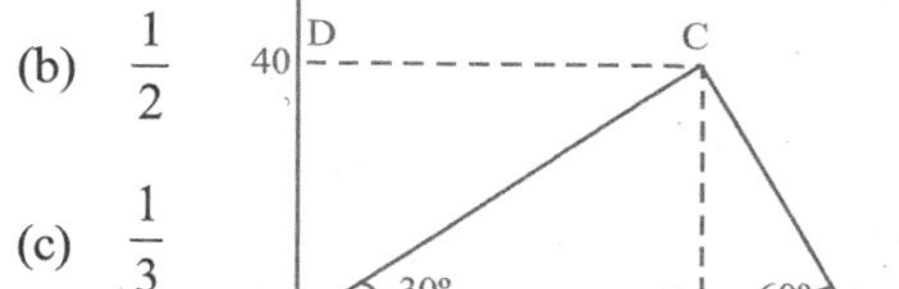

3. A bullet fired into a wooden block loses half of its velocity after penetrating 40 cm. It comes to rest after penetrating a further distance of
(a) $\frac{22}{3}$ cm (b) $\frac{40}{3}$ cm (c) $\frac{20}{3}$ cm (d) $\frac{22}{5}$ cm

4. A body covers 26, 28, 30, 32 meters in 10th, 11th, 12th and 13th seconds respectively. The body starts

Critical Thinking

(a) from rest and moves with uniform velocity
(b) from rest and moves with uniform acceleration
(c) with an initial velocity and moves with uniform acceleration
(d) with an initial velocity and moves with uniform velocity

5. The displacement x of a particle at the instant when its velocity is v is given by $v = \sqrt{3x + 16}$. Its acceleration and initial velocity are
(a) 1.5 units, 4 units (b) 3 units, 4 units
(c) 16 units, 1.6 units (d) 16 units, 3 units

6. The distance time graph of a particle at time t makes angles 45° with the time axis. After one second, it makes angle 60° with the time axis. What is the acceleration of the particle?
(a) $\sqrt{3}-1$ (b) $\sqrt{3}+1$ (c) $\sqrt{3}$ (d) 1

7. A particle experiences constant acceleration for 20 seconds after starting from rest. If it travels a distance s_1 in the first 10 seconds and distance s_2 in the next 10 seconds, then
(a) $s_2 = s_1$ (b) $s_2 = 2\ s_1$
(c) $s_2 = 3\ s_1$ (d) $s_2 = 4\ s_1$

8. In 1.0 s, a particle goes from point *A* to point *B*, moving in a semicircle of radius 1.0 m (see Figure). The magnitude of the average velocity is
(a) 3.14 m/s
(b) 2.0 m/s
(c) 1.0 m/s
(d) Zero

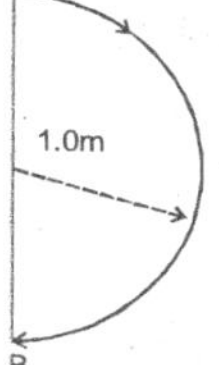

9. If distance covered by a particle is zero, what can you say about its displacement?
(a) It may or may not be zero
(b) It cannot be zero
(c) It is negative (d) It must be zero

10. A man of height h walks in a straight path towards a lamp post of height H with velocity v. Then velocity of the edge of the shadow on the ground will be

(a) $\frac{hv}{H+h}$ (b) $\frac{Hv}{H-h}$
(c) $\frac{H+h}{Hv}$ (d) $\frac{(H-h)}{Hh}$

11. The displacement-time graphs of two particles A and B are straight lines making angles of 30° and 60° respectively with the time axis. If the velocity of A is v_A and that of B is v_B, the value of v_A/v_B is
(a) 1/2 (b) $1/\sqrt{3}$ (c) $\sqrt{3}$ (d) 1/3

12. An athlete completes one round of a circular track of radius R in 40 sec. What will be his displacement at the end of 3 min. 20 sec ?
(a) Zero (b) 2R (c) $2\pi R$ (d) $7\pi R$

13. A body moves in straight line with velocity v_1 for $1/3^{rd}$ time and for remaining time with v_2. Find average velocity.
(a) $\frac{v_1}{3}+\frac{2v_2}{3}$ (b) $\frac{v_1}{3}+\frac{v_2}{3}$
(c) $\frac{2v_1}{3}+\frac{v_2}{3}$ (d) $v_1+\frac{2v_2}{3}$

14. A particle moves in straight line with velocity 6 m/s and 3 m/s for time intervals which are in ratio 1:2. Find average velocity.

(a) 2 m/s (b) 3 m/s (c) 4 m/s (d) 5 m/s

15. A particle moves from (2,3) m to (4,1) m. The magnitude of displacement is

(a) 2 m (b) $2\sqrt{3}$ m (c) $2\sqrt{2}$ m (d) $3\sqrt{2}$ m

16. Which of the following is not possible for a body in uniform motion?

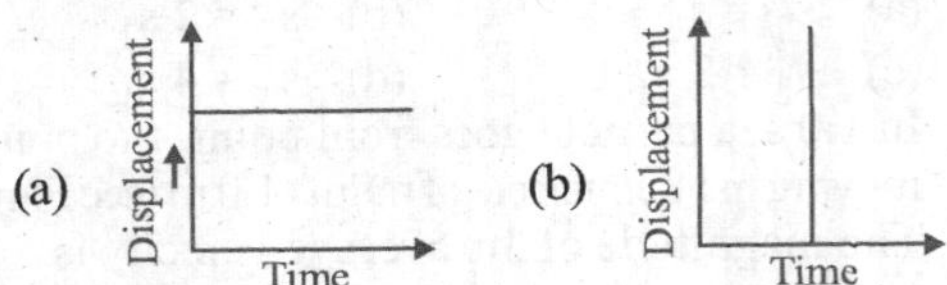

(c) Both (a) & (b) (d) None of these

17. A man leaves his house for a cycle ride. He comes back to his house after half-an-hour after covering a distance of one km. What is his average velocity for the ride?

(a) zero (b) 2 km h^{-1}

(c) 10 km s^{-1} (d) $\frac{1}{2}$ km s^{-1}

18. A point traversed half of the distance with a velocity v_0. The half of remaining part of the distance was covered with velocity v_1 & second half of remaining part by v_2 velocity. The mean velocity of the point, averaged over the whole time of motion is

(a) $\frac{v_0+v_1+v_2}{3}$ (b) $\frac{2v_0+v_1+v_2}{3}$

(c) $\frac{v_0+.2v_1+2v_2}{3}$ (d) $\frac{2v_0(v_1+v_2)}{(2v_0+v_1+v_2)}$

19. If a car covers 2/5th of the total distance with v_1 speed and 3/5th distance with v_2 then average speed is

(a) $\frac{1}{2}\sqrt{v_1v_2}$ (b) $\frac{v_1+v_2}{2}$

(c) $\frac{2v_1v_2}{v_1+v_2}$ (d) $\frac{5v_1v_2}{3v_1+2v_2}$

20. A person moves 30 m north and then 20 m towards east and finally $30\sqrt{2}$ m in south-west direction. The displacement of the person from the origin will be

(a) 10 m along north (b) 10 m along south

(c) 10 m along west (d) zero

21. A bus travelling the first one third distance at a speed of 10 km/h, the next one third at 20 km/h and the last one-third at 60 km/h. The average speed of the bus is

(a) 9 km/h (b) 16 km/h

(c) 18 km/h (d) 48 km/h

22. A bird flies with a speed of 10 km/h and a car moves with uniform speed of 8 km/h. Both start from B towards A (BA=40km) at the same instant. The bird having reached A, flies back immediately to meet the approaching car. As soon as it reaches the car, it flies back to A. The bird repeats this till both the car and the bird reach A simultaneously. The total distance flown by the bird is

Critical Thinking

(a) 80 km (b) 40 km (c) 50 km (d) 30 km

23. A car moves with a speed of 60 km/hr from point A to point B and then with the speed of 40 km/hr from point B to point C. Further it moves to a point D with a speed equal to its average speed between A and C. Points A, B, C and D are collinear and equidistant. The average speed of the car between A and D is

(a) 30 km/hr (b) 50 km/hr

(c) 48 km/hr (d) 60 km/hr

24. Three elephants A, B and C are moving along a straight line with constant speed in same direction as shown in figure. Speed of A is 5 m/s and speed of C is 10 m/s. Initially separation between A and B is 'd' and between B and C is also d. When 'B' catches 'C' separation between A and C becomes 3d. Then the speed of B will be

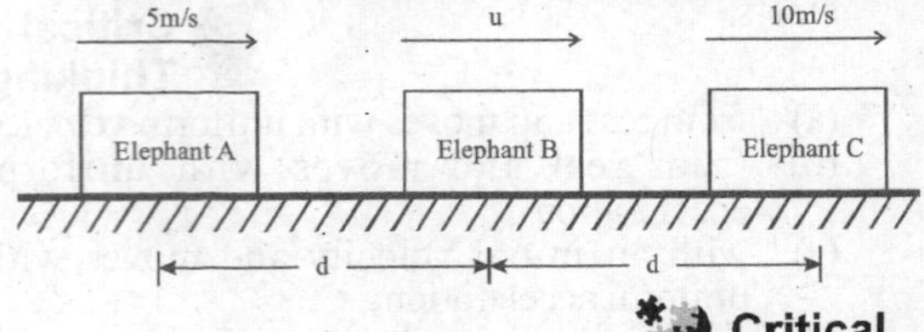

Critical Thinking

(a) 7.5 m/s (b) 15 m/s (c) 20 m/s (d) 5 m/s

25. If a body travels with constant acceleration, which of the following quantities remains constant?

(a) Displacement (b) Velocity

(c) Time (d) None of these

26. A bus starts moving with acceleration 2 m/s^2. A cyclist 96 m behind the bus starts simultaneously towards the bus at 20 m/s. After what time will he be able to overtake the bus?

(a) 4 sec (b) 8 sec (c) 18 sec (d) 16 sec

27. The velocity time graph of the motion of the body is as shown below

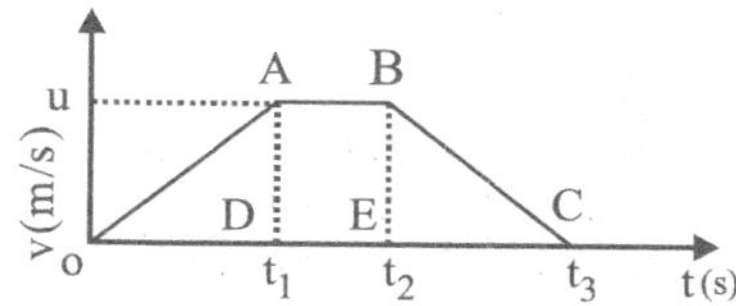

The total distance travelled by the body during the motion is equal to ____.

(a) $\frac{1}{2}$ (AD + BE) × OC

(b) $\frac{1}{2}$ (OA + BC) × OC

(c) $\frac{1}{2}$ (OC + AB) × AD

(d) $\frac{1}{2}$ (OA + AB) × BC

28. Stopping distance of a moving vehicle is directly proportional to
(a) square of the initial velocity
(b) square of the initial acceleration
(c) the initial velocity
(d) the initial acceleration

29. Which of the following graphs gives the equation $x = v_0 t + \frac{1}{2} at^2$ **Tricky**

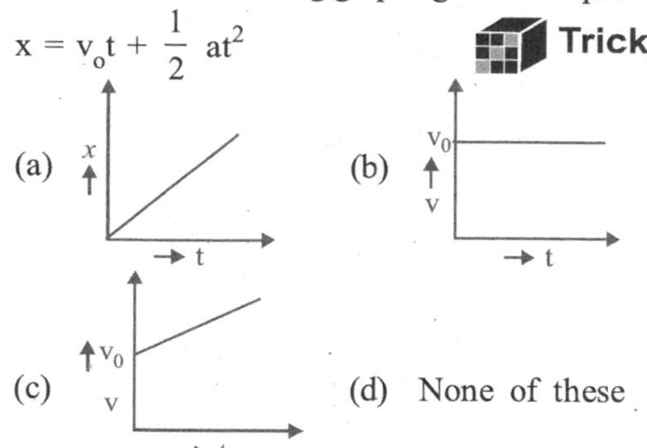

(d) None of these

30. The displacement of a particle is represented by the following equation: $S = 3t^3 + 7t^2 + 5t + 8$ where 5 is in meter and t in second. The acceleration of the particle at t = 15 is
(a) 14 m/s^2 (b) 18 m/s^2
(c) 32 m/s^2 (d) zero

31. If a train travelling at 20 m/s is to be brought to rest in a distance of 200 m, then its retardation should be
(a) 1 m/s^2 (b) 2 m/s^2 (c) 10 m/s^2 (d) 20 m/s^2

32. The distance travelled by a particle starting from rest and moving with an acceleration $\frac{4}{3} \text{ms}^{-2}$, in the third second is:
(a) 6m (b) 4m (c) $\frac{10}{3}$m (d) $\frac{19}{3}$m

33. If a car at rest accelerates uniformly to a speed of 144 km/h in 20 s, it covers a distance of
(a) 2880m (b) 1440m (c) 400m (d) 20m

34. The displacement x of a particle along a straight line at time t is given by : $x = a_0 + \frac{a_1 t}{2} + \frac{a_2}{3} t^2$. The acceleration of the particle is
(a) $\frac{a_2}{3}$ (b) $\frac{2a_2}{3}$ (c) $\frac{a_1}{2}$ (d) $a_0 + \frac{a_2}{3}$

35. The dependence of velocity of a body with time is given by the equation $v = 20 + 0.1t^2$. The body is in
(a) uniform retardation
(b) uniform acceleration
(c) non-uniform acceleration
(d) zero acceleration.

36. A car accelerates from rest at a constant rate α for some time, after which it decelerates at a constant rate β and comes to rest. If the total time elapsed is t, then the maximum velocity acquired by the car is
(a) $\left(\frac{\alpha^2 + \beta^2}{\alpha\beta}\right)t$ (b) $\left(\frac{\alpha^2 - \beta^2}{\alpha\beta}\right)$
(c) $\frac{(\alpha + \beta)t}{\alpha\beta}$ (d) $\frac{\alpha\beta t}{\alpha + \beta}$

37. The deceleration experienced by a moving motorboat after its engine is cut off, is given by $\frac{dv}{dt} = -KV^3$ where K is constant. If V_0 is the magnitude of the velocity at cut-off, the magnitude of the velocity at a time t after the cut-off is
(a) $\frac{V_0}{\sqrt{(2V_0^2 Kt + 1)}}$ (b) $V_0 e^{-Kt}$
(c) $V_0/2$ (d) V_0

38. The displacement of a particle as a function of time is shown in figure. It indicates that

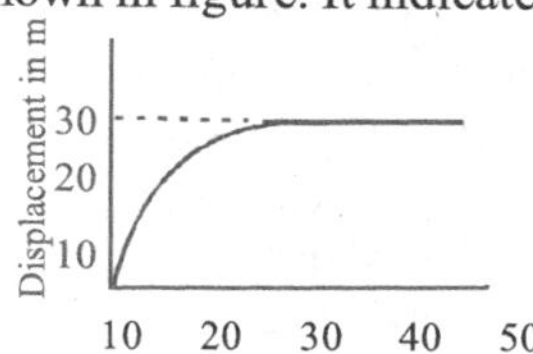

(a) the velocity of the particle is constant throughout
(b) the acceleration of the particle is constant throughout
(c) the particle starts with a constant velocity and is accelerated
(d) the motion is retarded and finally the particle stops

39. A particle moves along a straight line OX. At a time t (in second) the distance x (in metre) of the particle from O is given by $x = 40 + 12t - t^3$. How long would the particle travel before coming to rest?
(a) 24m (b) 40m (c) 56m (d) 16m

40. A particle moves a distance x in time t according to equation $x = (t + 5)^{-1}$. The acceleration of particle is proportional to
(a) $(\text{velocity})^{3/2}$ (b) $(\text{distance})^2$
(c) $(\text{distance})^{-2}$ (d) $(\text{velocity})^{2/3}$

41. A particle is moving eastwards with a velocity of 5 ms^{-1}. In 10 seconds the velocity changes to 5 ms^{-1} northwards. The average acceleration in this time is
(a) $\frac{1}{2} ms^{-2}$ towards north
(b) $\frac{1}{\sqrt{2}} ms^{-2}$ towards north - east
(c) $\frac{1}{\sqrt{2}} ms^{-2}$ towards north - west
(d) zero

42. It is given that $t = px^2 + qx$, where x is displacement and t is time. The acceleration of particle at origin is
(a) $-\frac{2p}{q^3}$ (b) $-\frac{2q}{p^3}$ (c) $\frac{2p}{q^3}$ (d) $\frac{2q}{p^3}$

43. An object, moving with a speed of 6.25 m/s, is decelerated at a rate given by: $\frac{dv}{dt} = -2.5\sqrt{v}$ where v is the instantaneous speed. The time taken by the object, to come to rest, would be
(a) 2 s (b) 4 s (c) 8 s (d) 1 s

44. A bike accelerates from rest at a constant rate 5 m/s^2 for some time after which it decelerates at a constant rate 3 m/s^2 to come to rest. If the total time elapsed is 8 second, the maximum velocity acquired by the bike is given by
(a) 5 m/s (b) 10 m/s (c) 12 m/s (d) 15 m/s

45. A metro train starts from rest and in 5 s achieves 108 km/h. After that it moves with constant velocity and comes to rest after travelling 45 m with uniform retardation. If total distance travelled is 395 m, find total time of travelling.
Critical Thinking
(a) 12.2 s (b) 15.3 s (c) 9 s (d) 17.2 s

46. A car, starting from rest, accelerates at the rate f through a distance S, then continues at constant speed for time t and then decelerates at the rate $\frac{f}{2}$ to come to rest. If the total distance traversed is 15 S, then
(a) $S = \frac{1}{6} ft^2$ (b) $S = ft$
(c) $S = \frac{1}{4} ft^2$ (d) $S = \frac{1}{72} ft^2$

47. A particle starting with certain initial velocity and uniform acceleration covers a distance of 12 m in first 3 seconds and a distance of 30 m in next 3 seconds. The initial velocity of the particle is
(a) 3 ms^{-1} (b) 2.5 ms^{-1}
(c) 2 ms^{-1} (d) 1 ms^{-1}

48. The velocity of an object moving rectilinearly is given as a function of time by $v = 4t - 3t^2$, where v is in m/s and t is in seconds. The average velocity of particle between $t = 0$ to $t = 2$ seconds is
(a) 0 (b) –2m/s (c) –4m/s (d) 8 m/s

49. A body starts from rest and is uniformly accelerated for 30 s. The distance travelled in the first 10 s is x_1, next 10 s is x_2 and the last 10 s is x_3. Then $x_1 : x_2 : x_3$ is the same as: Tricky
(a) 1 : 2 : 4 (b) 1 : 2 : 5 (c) 1 : 3 : 5 (d) 1 : 3 : 9

50. The acceleration of a particle, starting from rest, varies with time according to the relation $a = -s\omega^2 \sin \omega t$. The displacement of this particle at a time t will be
(a) $s \sin \omega t$ (b) $s \omega \cos \omega t$
(c) $s \omega \sin \omega t$ (d) $-\frac{1}{2}(s\omega^2 \sin \omega t)t^2$

51. A particle is moving along a straight line path according to the relation $s^2 = at^2 + 2bt + c$
s represents the distance travelled in t seconds and a, b, c are constants. Then the acceleration of the particle varies as
(a) s^{-3} (b) $s^{3/2}$ (c) $s^{-2/3}$ (d) s^2

52. A car of mass 1000 kg is moving at a speed of 30 m/s. Brakes are applied to bring the car to rest. If the deceleration is 5 m/s^2 the car comes to stop after travelling d m in t s. Then
(a) $d = 150, t = 5$ (b) $d = 120, t = 8$
(c) $d = 180, t = 6$ (d) $d = 90, t = 6$

53. A goods train accelerating uniformly on a straight railway track, approaches an electric pole standing on the side of track. Its engine passes the pole with velocity u and the guard's room passes with velocity v. The middle wagon of the train passes the pole with a velocity. Toughnut
(a) $\frac{u+v}{2}$ (b) $\frac{1}{2}\sqrt{u^2 + v^2}$
(c) $\sqrt{uv}$ (d) $\sqrt{\left(\frac{u^2 + v^2}{2}\right)}$

54. Starting from rest a particle moves in a straight line with acceleration $a = (25 - t^2)^{1/2}$ m/s^2 for $0 \le t \le 5$s,

$a = \frac{3\pi}{8}$ m/s^2 for t > 5s. The velocity of particle at t = 7s is:

(a) 11 m/s (b) 22 m/s (c) 33 m/s (d) 44 m/s

55. A car accelerates from rest with a constant acceleration α on a straight road. After gaining a velocity v_1 the car moves with the same velocity for some-time. Then the car decelerated to rest with a retardation β. If the total distance covered by the car is equal to S, the total time taken for its motion is

(a) $\frac{S}{v} + \frac{v}{2}\left(\frac{1}{\alpha} + \frac{1}{\beta}\right)$ (b) $\frac{S}{v} + \frac{v}{\alpha} + \frac{v}{\beta}$

(c) $\left(\frac{v}{\alpha} + \frac{v}{\beta}\right)$ (d) $\frac{S}{v} - \frac{v}{2}\left(\frac{v}{\alpha} + \frac{v}{\beta}\right)$

56. A truck has to carry a load in the shortest time from one station to another station situated at a distance L from the first. It can start up or slowdown at the same acceleration or deceleration a. What maximum velocity must the truck attain to satisfy this condition? Toughnut

(a) $\sqrt{La}$ (b) $\sqrt{2La}$ (c) $\sqrt{3La}$ (d) $\sqrt{5La}$

57. Two trains, each 40 m long are travelling in opposite direction with equal velocity 20 m/s. The time of crossing is

(a) 1s (b) 2s (c) 3s (d) Zero

58. The graph shown below represent Tricky

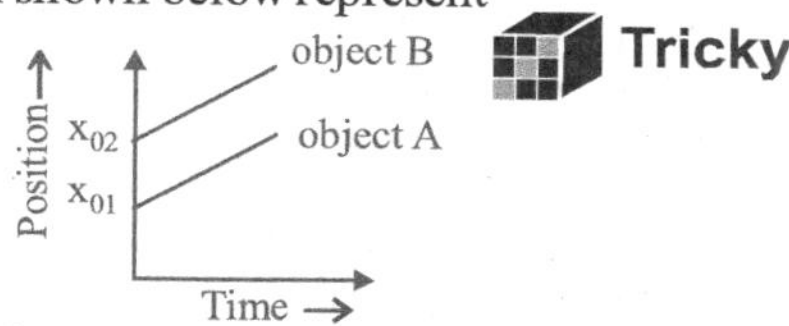

(a) A and B are moving with same velocity in opposite directions
(b) velocity of B is more than A in same direction
(c) velocity of A is more than B in same direction
(d) velocity of A and B is equal in same direction

59. Two cars A and B approach each other at the same speed, then what will be the velocity of A if velocity of B is 8 m/s?

(a) 16 m/s (b) 8 m/s (c) – 8 m/s
(d) Can't be determined.

60. A train of 150 m length is going towards north direction at a speed of 10 ms^{-1}. A parrot flies at a speed of 5 ms^{-1} towards south direction parallel to the railway track. The time taken by the parrot to cross the train is equal to Critical Thinking

(a) 12 s (b) 8 s (c) 15 s (d) 10 s

61. A boat takes 2 hours to travel 8 km and back in still water lake. With water velocity of 4 km h^{-1}, the time taken for going upstream of 8 km and coming back is

(a) 160 minutes (b) 80 minutes
(c) 100 minutes (d) 120 minutes

62. Two trains are each 50 m long moving parallel towards each other at speeds 10 m/s and 15 m/s respectively. After what time will they pass each other?

(a) $5\sqrt{\frac{2}{3}}$ sec (b) 4 sec
(c) 2 sec (d) 6 sec

63. A ship A is moving Westwards with a speed of 10 km h^{-1} and a ship B 100 km South of A, is moving Northwards with a speed of 10 km h^{-1}. The time after which the distance between them becomes shortest, is

(a) 5 h (b) $5\sqrt{2}$ h (c) $10\sqrt{2}$ h (d) 0 h

64. A bus is moving with a velocity of 10 ms^{-1} on a straight road. A scootorist wishes to overtake the bus in one minute. If the bus is at a distance of 1.2 km ahead, then the velocity with which he has to chase the bus is

(a) 20 ms^{-1} (b) 25 ms^{-1}
(c) 60 ms^{-1} (d) 30 ms^{-1}

65. Two particles P and Q simultaneously start moving from point A with velocities 15 m/s and 20 m/s respectively. The two particles move with acceleration equal in magnitude but opposite in direction. When P overtakes Q at B then its velocity is 30 m/s The velocity of Q at point B will be

(a) 30 m/s (b) 5 m/s (c) 20 m/s (d) 15 m/s

66. A body A begins to move with initial velocity 2 m/sec and continues to move at a constant acceleration a. $\Delta t = 10$ seconds after the body A begins to move a body B departs from the same point with an initial velocity 12 m/sec and moves with the same acceleration a. What is the maximum acceleration a at which the body B can overtake A ?

(a) 1 m/s^2 (b) 2 m/s^2
(c) 1/2 m/s^2 (d) 3 m/s^2

67. Two particles start moving from rest from the same point along the same straight line. The first moves with constant velocity v and the second with constant acceleration a. During the time that elapse before the second catches the first, the greatest distance between the particles is

(a) $\frac{v^2}{a}$ (b) $\frac{v^2}{2a}$ (c) $\frac{2v^2}{a}$ (d) $\frac{v^2}{4a}$

68. A man travelling in a car with a maximum constant speed of 20m/s watches the friend start off at a distance 100m ahead on a motor cycle with constant acceleration 'a'. The maximum value of 'a' for which the man in the car can reach his friend is

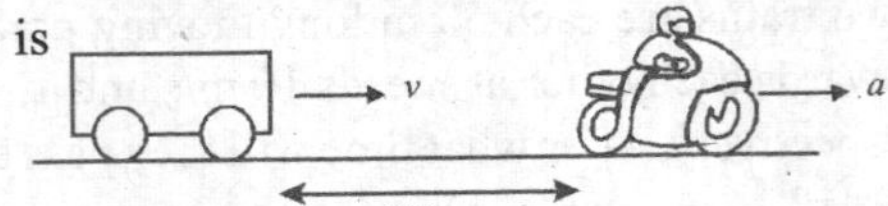

(a) 2 m/s^2 (b) 1 m/s^2 (c) 4 m/s^2 (d) 8 ms^{-2}

69. A body is thrown vertically upwards. If air resistance is to be taken into account, then the time during which the body rises is [assume no air resistance close to earth] **Tricky**

(a) equal to the time of fall
(b) less than the time of fall
(c) greater than the time of fall
(d) twice the time of fall

70. A body is thrown upwards and reaches half of its maximum height. At that position

(a) its acceleration is minimum
(b) its velocity is maximum
(c) its velocity is zero
(d) its acceleration is constant

71. A ball thrown vertically upwards after reaching a maximum height h, returns to the starting point after a time of 10 s. Its displacement is

(a) h (b) $2h$ (c) $10h$ (d) zero

72. The ball is projected up from ground with speed 30 m/sec. What is the average velocity for time 0 to 4 sec?

(a) 10 m/sec (b) 20 m/sec
(c) 15 m/sec (d) zero

73. A body is projected vertically upwards. If t_1 and t_2 be the times at which it is at height h above the projection while ascending and descending respectively, then h is

(a) $\frac{1}{2}gt_1t_2$ (b) gt_1t_2 (c) $2gt_1t_2$ (d) $2hg$

74. Two balls A and B of same mass are thrown from the top of the building. A thrown upward with velocity v and B, thrown down with velocity v, then

(a) velocity A is more than B at the ground
(b) velocity of B is more than A at the ground
(c) both A &B strike the ground with same velocity
(d) None of these

75. A rocket is fired upward from the earth's surface such that it creates an acceleration of 19.6 ms^{-2}. If after 5 s, its engine is switched off, the maximum height of the rocket from earth's surface would be

(a) 980m (b) 735m (c) 490m (d) 245m

76. A man throws balls with same speed vertically upwards one after the other at an interval of 2 sec. What should be the speed of throw so that more than two balls are in air at anytime? **Critical Thinking**

(a) Only with speed 19.6 m/s
(b) More than 19.6 m/s
(c) At least 9.8 m/s
(d) Any speed less then 19.6 m/s.

77. A stone is dropped from a rising balloon at a height of 76 m above the ground and reaches the ground in 6s. What was the velocity of the balloon when the stone was dropped? Take g = 10 m/s^2.

(a) $\left(\frac{52}{3}\right)$ m/s upward

(b) $\left(\frac{52}{3}\right)$ m/s downward

(c) 3 m/s (d) 9.8 m/s

78. Let A, B, C, D be points on a vertical line such that AB = BC = CD. If a body is released from position A, the times of descent through AB, BC and CD are in the ratio. **Toughnut**

(a) $1:\sqrt{3}-\sqrt{2}:\sqrt{3}+\sqrt{2}$
(b) $1:\sqrt{2}-1:\sqrt{3}-\sqrt{2}$
(c) $1:\sqrt{2}-1:\sqrt{3}$ (d) $1:\sqrt{2}:\sqrt{3}-1$

79. If two balls of masses m_1 and m_2($m_1 = 2m_2$) are dropped from the same height, then the ratio of the time taken by them to reach the ground will be

(a) $m_1 : m_2$ (b) $2m_2 : m_1$ (c) 1 : 1 (d) 1 : 2

80. A boy standing at the top of a tower of 20m height drops a stone. Assuming g = 10 ms^{-2}, the velocity with which it hits the ground is

(a) 10.0 m/s (b) 20.0 m/s (c) 40.0 m/s (d) 5.0 m/s

81. What will be the ratio of the distances moved by a freely falling body from rest on 4th and 5th seconds of journey ?

(a) 4:5 (b) 7:9 (c) 16:25 (d) 1:1

82. From a balloon moving upwards with a velocity of 12 ms^{-1}, a packet is released when it is at a height of 65 m from the ground. The time taken by it to reach the ground is (g = 10 ms^{-2})

(a) 5 s (b) 8 s (c) 4 s (d) 7 s

83. A ball dropped from a point A falls down vertically to C, through the midpoint B. The descending time from A to B and that from A to C are in the ratio

Tricky

(a) 1 : 1 (b) 1 : 2 (c) 1 : 3 (d) $1:\sqrt{2}$

84. A ball is dropped from a high rise platform at t = 0 starting from rest. After 6 seconds another ball is thrown downwards from the same platform with a speed v. The two balls meet at t = 18s. What is the value of v? (take $g = 10\ m/s^2$)

(a) 75 m/s (b) 55 m/s (c) 40 m/s (d) 60 m/s

85. A stone falls freely under gravity. It covers distances h_1, h_2 and h_3 in the first 5 seconds, the next 5 seconds and the next 5 seconds respectively. The relation between h_1, h_2 and h_3 is

Toughnut

(a) $h_1 = \frac{h_2}{3} = \frac{h_3}{5}$ (b) $h_2 = 3h_1$ and $h_3 = 3h_2$

(c) $h_1 = h_2 = h_3$ (d) $h_1 = 2h_2 = 3h_3$

86. From a building two balls *A* and *B* are thrown such that *A* is thrown upwards and *B* downwards (both vertically). If T_A and T_B are their respective time of flights then

(a) $T_A > T_B$ (b) $T_A = T_B$ (c) $T_A < T_B$

(d) their time of flights depend on their masses.

87. A ball is released from the top of tower of height h metre. It takes T second to reach the ground. What is the position in (m) from the ground of the ball in T/3 second ?

(a) $\frac{h}{9}$ (b) $\frac{7h}{9}$ (c) $\frac{8h}{9}$ (d) $\frac{17h}{18}$

88. A stone is dropped into a well in which the level of water is h below the top of the well. If v is velocity of sound, the time T after which the splash is heard is given by

Critical Thinking

(a) $T = 2h/v$ (b) $T = \sqrt{\left(\frac{2h}{g}\right)} + \frac{h}{v}$

(c) $T = \sqrt{\left(\frac{2h}{v}\right)} + \frac{h}{g}$ (d) $T = \sqrt{\left(\frac{h}{2g}\right)} + \frac{2h}{v}$

89. A ball is thrown vertically upwards. It was observed, at a height h twice with a time interval Δt. The initial velocity of the ball is-

Toughnut

(a) $\sqrt{8gh + g^2(\Delta t)^2}$ (b) $\sqrt{8gh + \left(\frac{g\Delta t}{2}\right)^2}$

(c) $\frac{1}{2}\sqrt{8gh + g^2(\Delta t)^2}$ (d) $\sqrt{8gh + 4g^2(\Delta t)^2}$

90. The balls are released from the top of a tower of height H at regular interval of time. When first ball reaches at the ground, the nth ball is to be just released and $\left(\frac{n+1}{2}\right)^{th}$ ball is at same distance 'h' from top of the tower. The value of h is.

Toughnut

(a) $\frac{2}{3}H$ (b) $\frac{3}{4}H$ (c) $\frac{4}{5}H$ (d) $\frac{5H}{6}$

91. A ball is dropped from the top of a tower of height 100 m and at the same time another ball is projected vertically upwards from ground with a velocity $25\ ms^{-1}$. Then the distance from the top of the tower, at which the two balls meet is

(a) 68.4m (b) 48.4m (c) 18.4m (d) 78.4m

92. A body thrown vertically so as to reach its maximum height in t second. The total time from the time of projection to reach a point at half of its maximum height while returning (in sec) is

(a) $\sqrt{2}t$ (b) $\left(1+\frac{1}{\sqrt{2}}\right)t$ (c) $\frac{3t}{2}$ (d) $\frac{t}{\sqrt{2}}$

93. A body dropped from top of a tower fall through 40 m during the last two seconds of its fall. The height of tower is ($g = 10\ m/s^2$)

(a) 60m (b) 45m (c) 80m (d) 50m

94. From a pole of height 10 m, a stone is thrown vertically upwards with a speed 5 m/s. The time taken by the stone, to hit the ground, is n times that taken by it to reach the highest point of its path. The value of n is [take $g = 10\ m/s^2$]

(a) 2 (b) 3 (c) 4 (d) 5

95. A body *A* is thrown vertically upward with the initial velocity v_1. Another body *B* is dropped from a height *h*. Find how the distance *x* between the bodies depends on the time *t* if the bodies begin to move simultaneously.

Critical Thinking

(a) $x = h - v_1 t$ (b) $x = (h - v_1)t$

(c) $x = h - \frac{v_1}{t}$ (d) $x = \frac{h}{t} - v_1$

96. A body is thrown upwords. If air resistance causing deceleration of $5\ m/s^2$, then ratio of time of ascent to time of descent is [take $g = 10\ m/s^2$]

(a) $\sqrt{\frac{1}{2}}$ (b) $\sqrt{\frac{1}{2.5}}$ (c) $\sqrt{\frac{1}{3}}$ (d) $\sqrt{\frac{1}{5}}$

97. A particle when thrown, moves such that it passes from same height at 2 and 10 seconds, then this height h is :

(a) 5g (b) g (c) 8g (d) 10g

98. A hunter tries to hunt a monkey with a small, very poisonous arrow, blown from a pipe with initial speed v_0. The monkey is hanging on a branch of a tree at height H above the ground. The hunter is at a distance L from the bottom of the tree. The monkey sees the arrow leaving the blow pipe and immediately loses the grip on the tree, falling freely down with zero initial velocity. The minimum initial speed v_0 of the arrow for hunter to succeed while monkey is in air is **Tricky**

(a) $\sqrt{\dfrac{g(H^2+L^2)}{2H}}$ (b) $\sqrt{\dfrac{gH^2}{\sqrt{H^2+L^2}}}$

(c) $\sqrt{\dfrac{g\sqrt{H^2+L^2}}{H}}$ (d) $\sqrt{\dfrac{2gH^2}{\sqrt{H^2+L^2}}}$

99. A ball is dropped from the top of a building. The ball takes 0.5 s to fall past the 3 m length of a window some distance from the top of the building. If the velocities of the ball at the top and at the bottom of the window are v_T and v_B respectively, then (take g = 10 m/s²) **Toughnut**

(a) $v_T + v_B = 12\ ms^{-1}$ (b) $v_T - v_B = 4.9\ ms^{-1}$

(c) $v_B\, v_T = 1\ ms^{-1}$ (d) $v_B / v_T = 1\ ms^{-1}$

100. From a building two balls A and B are thrown such that A is thrown upwards and B downwards (both vertically). If v_A and v_B are their respective velocities on reaching the ground, then

(a) $v_B > v_A$ (b) $v_A = v_B$

(c) $v_A > v_B$

(d) their velocities depend on their masses.

Answer KEYs

1	(c)	11	(d)	21	(c)	31	(a)	41	(c)	51	(a)	61	(a)	71	(d)	81	(b)	91	(d)
2	(c)	12	(a)	22	(c)	32	(c)	42	(a)	52	(d)	62	(b)	72	(a)	82	(a)	92	(b)
3	(b)	13	(a)	23	(c)	33	(c)	43	(a)	53	(d)	63	(a)	73	(a)	83	(d)	93	(b)
4	(c)	14	(c)	24	(b)	34	(b)	44	(d)	54	(b)	64	(d)	74	(c)	84	(a)	94	(c)
5	(a)	15	(c)	25	(d)	35	(c)	45	(d)	55	(a)	65	(b)	75	(b)	85	(a)	95	(a)
6	(a)	16	(c)	26	(b)	36	(d)	46	(d)	56	(a)	66	(a)	76	(b)	86	(a)	96	(c)
7	(c)	17	(a)	27	(c)	37	(d)	47	(d)	57	(b)	67	(b)	77	(a)	87	(c)	97	(d)
8	(b)	18	(d)	28	(a)	38	(d)	48	(a)	58	(d)	68	(a)	78	(b)	88	(b)	98	(a)
9	(d)	19	(d)	29	(c)	39	(c)	49	(c)	59	(c)	69	(b)	79	(c)	89	(c)	99	(b)
10	(b)	20	(c)	30	(c)	40	(a)	50	(a)	60	(d)	70	(d)	80	(b)	90	(b)	100	(b)

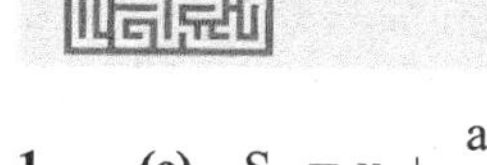

Hints & Solutions

1. (c) $S_n = u + \frac{a}{2}(2n - 1)$

or, $s = \frac{a}{2}(2 \times 2 - 1) \Rightarrow a = \frac{2}{3}\ \text{sm/s}^2$

2. (c) During OA, acceleration = tan 30° = $\frac{1}{\sqrt{3}}$ m/s²
During AB, acceleration = –tan60°
$= -\sqrt{3}\ \text{m/s}^2$

Required ratio $= \frac{1/\sqrt{3}}{\sqrt{3}} = \frac{1}{3}$

3. (b) For first part of penetration, by equation of motion

$\left(\frac{u}{2}\right)^2 - (u)^2 = 2aS$ or $a = -\frac{3u^2}{8S}$ ...(i)

For latter part of penetration

$(0)^2 - \left(\frac{u}{2}\right)^2 = 2aS'$, $S' = -\frac{u^2}{8a}$

$S' = -\frac{u^2}{8}\left(\frac{8S}{-3u^2}\right)$ (Using (i))

$S' = \frac{S}{3}$ or $S' = \frac{40}{3}$ cm

4. (c) The distance covered in n^{th} second is
$S_n = u + \frac{1}{2}(2n - 1)a$
where u is initial velocity & a is acceleration

then $26 = u + \frac{19a}{2}$ ....(i)

$28 = u + \frac{21a}{2}$ ....(ii)

$30 = u + \frac{23a}{2}$ ....(iii)

$32 = u + \frac{25a}{2}$ ....(iv)

From eqs. (i) and (ii) we get u = 7m/sec, a=2m/sec²
∴ The body starts with initial velocity u =7m/sec and moves with uniform acceleration a = 2m/sec²

5. (a) $v = \sqrt{3x + 16} \Rightarrow v^2 = 3x + 16$
$\Rightarrow v^2 - 16 = 3x$
Comparing with $v^2 - u^2 = 2aS$, we get, u = 4 units, 2a = 3 or a = 1.5 units

6. (a) Velocity at time t is tan 45° = 1. Velocity at time (t = 1) is $\tan 60° = \sqrt{3}$. Acceleration is change in velocity in one second = $\sqrt{3} - 1$.

7. (c) Let a be the constant acceleration of the particle. Then $s = ut + \frac{1}{2}at^2$ or

$s_1 = 0 + \frac{1}{2} \times a \times (10)^2 = 50a$

and $s_2 = \left[0 + \frac{1}{2}a(20)^2\right] - 50a = 150a$

$\therefore s_2 = 3s_1$

8. (b) $|\text{Average velocity}| = \frac{|\text{displacement}|}{\text{time}}$

$= \frac{2r}{t} = 2 \times \frac{1}{1} = 2\ \text{m/s}.$

9. (d) Distance covered by a particle is zero only when it is at rest. Therefore, its displacement must be zero.

10. (b) as ΔLOS and ΔMPS are similar so,

$\frac{H}{x} = \frac{h}{x - y} \Rightarrow (H - h)x = Hy$

$\Rightarrow (H - h)\frac{dx}{dt} = H\frac{dy}{dt} = Hv$

$\Rightarrow V_s = \frac{Hv}{H - h}$

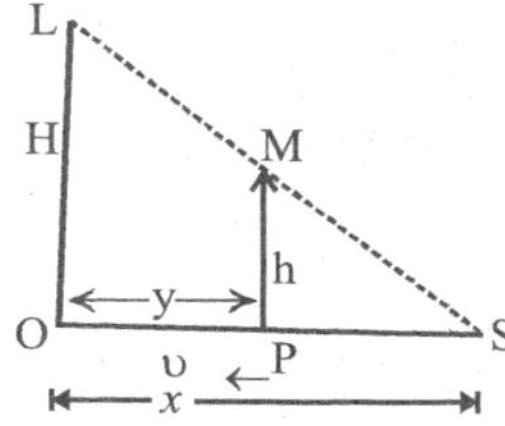

11. (d) $v_A = \tan 30°$ and $v_B = \tan 60°$

$\therefore \frac{v_A}{v_B} = \frac{\tan 30°}{\tan 60°} = \frac{1/\sqrt{3}}{\sqrt{3}} = \frac{1}{3}$

12. (a) Total time of motion is 3 min 20 sec = 20 sec. As time period of circular motion is 40 sec so in 20 sec athlete will complete 5 revolution i.e., he will be at starting point i.e., displacemet = zero.

13. (a) $v_{av} = \frac{v_1 t/3 + v_2(2t/3)}{t} = \frac{v_1}{3} + \frac{2v_2}{3}$

14. (c)

15. (c) Displacement

$= \sqrt{(4-2)^2 + (3-1)^2} = \sqrt{4+4} = 2\sqrt{2}\text{m}$

16. (c)

17. (a) Since displacement is zero.

18. (d) Let the total distance be d. Then for first half distance, time $= \frac{d}{2v_0}$, next distance. $= v_1 t$ and last half distance $= v_2 t$

$\therefore \; v_1 t + v_2 t = \frac{d}{2}; \; t = \frac{d}{2(v_1 + v_2)}$

Now average speed

$t = \frac{d}{\frac{d}{2v_0} + \frac{d}{2(v_1+v_2)} + \frac{d}{2(v_1+v_2)}}$

$= \frac{2v_0(v_1+v_2)}{(v_1+v_2)+2v_0}$

19. (d) Average speed$= \frac{\text{Total distance travelled}}{\text{Total time taken}}$

$= \frac{x}{\frac{2x/5}{v_1} + \frac{3x/5}{v_2}} = \frac{5v_1v_2}{3v_1 + 2v_2}$

20. (c) From figure,

$\overrightarrow{OA} = 0\hat{i} + 30\hat{j}$,

$\overrightarrow{AB} = 20\hat{i} + 0\hat{j}$

$\overrightarrow{BC} = -30\sqrt{2}\cos 45°\hat{i} - 30\sqrt{2}\sin 45°\hat{j}$

$= -30\hat{i} - 30\hat{j}$

$\therefore$ Net displacement,

$\overrightarrow{OC} = \overrightarrow{OA} + \overrightarrow{AB} + \overrightarrow{BC} = -10\hat{i} + 0\hat{j}$

$|\overrightarrow{OC}| = 10\text{m}$.

21. (c) Average speed

$= \frac{s}{\frac{s/3}{10} + \frac{s/3}{20} + \frac{s/3}{60}} = \frac{s}{s/18} = 18\text{km/h}$

22. (c)

$v_L = 8$ km/h, $s = v_0 \times t \Rightarrow t = \frac{40}{8} = 5\text{h}$.

Total distance flown by the bird $= 10 \times 5 = 50$ km.

23. (c) Let the points *A*, *B*, *C* and *D* be separated by 1 km. Then

$t_{AB} = \frac{1}{60} hr, \; t_{BC} = \frac{1}{40} hr$

$\therefore \langle v_{AC} \rangle = \frac{1+1}{\frac{1}{60} + \frac{1}{40}} = 48 \text{ km/hr} \Rightarrow t_{CD} \frac{1}{48} hr$.

Now $\langle v_{AD} \rangle = \frac{1+1+1}{\frac{1}{60} + \frac{1}{40} + \frac{1}{48}} = 48$ km/hr

24. (b) B catches C in time t then $t = \frac{d}{u-10}$.

Separation by this time has increased by 'd' between A and C hence

$(10-5) \times \frac{d}{(u-10)} = d, \; u = 15\text{ m/s}$

25. (d)

26. (b) At the highest point v = 0.

27. (c) distance travelled = area □OABC

$= \frac{1}{2} \times (OC + AB) \times AD$

28. (a) Let s be the distance travelled by the vehicle before it stops.

Final velocity $\upsilon = 0$, initial velocity $= u$

Using equation of motion $v^2 - u^2 = 2aS$

$0^2 - u^2 = 2aS$

Stopping distance, $S = -\frac{u^2}{2a}$

29. (c) given $x = V_0 t + \frac{1}{2} at^2$

$\frac{dx}{dt} = V_0 + \frac{1}{2} \times a \times 2t = V_0 + at$

$\Rightarrow V = V_0 + at$

so, option (c) represents the correct graph.

30. (c)

31. (a) $v^2 - u^2 = 2as$

$\Rightarrow a = \frac{u^2}{2as} = \frac{(20)^2}{2 \times 200} = 1 \text{ m/s}^2$

32. (c) Distance travelled in the nth second is given by

$t_n = u + \frac{a}{2}(2n-1)$

put $u = 0$, $a = \frac{4}{3}\text{ms}^{-2}, n = 3$

$\therefore \; d = 0 + \frac{4}{3 \times 2}(2 \times 3 - 1) = \frac{4}{6} \times 5 = \frac{10}{3}\text{m}$

33. (c) Initial velocity of car $(u) = 0$

Final velocity of car $(v) = 144$ km/hr $= 40$ m/s

Time taken = 20 s

We know that, $v = u + at$

$40 = a \times 20 \Rightarrow a = 2\ \text{m/s}^2$

Also, $v^2 - u^2 = 2as \Rightarrow s = \dfrac{v^2 - u^2}{2a}$

$\Rightarrow s = \dfrac{(40)^2 - (0)^2}{2 \times 2} = \dfrac{1600}{4} = 400$ m.

34. **(b)** $\dfrac{dx}{dt} = \dfrac{a_1}{2} + \dfrac{2a_2 t}{3} \Rightarrow \dfrac{d^2x}{dt^2} = \dfrac{2a_2}{3}$
acceleration required

35. **(c)** On differentiating, acceleration = 0.2t
$\Rightarrow$ a = f(t)

36. **(d)**

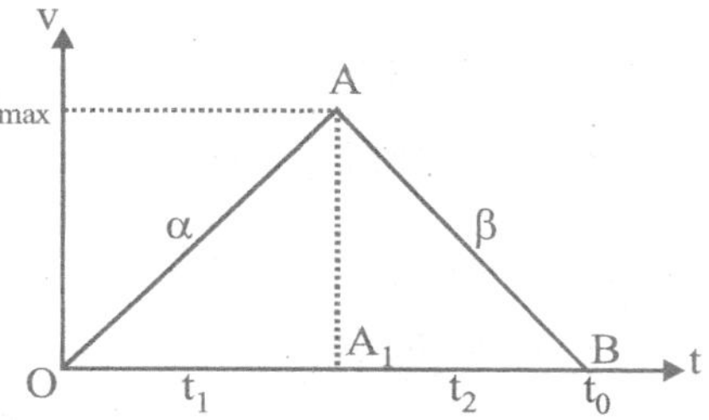

In fig., $AA_1 = v_{max.} = \alpha t_1 = \beta t_2$

But $t = t_1 + t_2 = \dfrac{v_{max}}{\alpha} + \dfrac{v_{max}}{\beta}$

$= v_{max}\left(\dfrac{1}{\alpha} + \dfrac{1}{\beta}\right) = v_{max}\left(\dfrac{\alpha+\beta}{\alpha\beta}\right)$

or, $v_{max} = t\left(\dfrac{\alpha\beta}{\alpha+\beta}\right)$

37. **(d)**

38. **(d)** From displacement-time graph, it is clear that in equal intervals of time displacements are not equal infact, decreases and after 40s displacement constant i.e. the particle stops.

39. **(c)** When particle comes to rest,

$V = 0 = \dfrac{dx}{dt} = \dfrac{d}{dt}(40 + 12t - t^3)$

$\Rightarrow \quad 12 - 3t^2 = 0$

$\Rightarrow \quad t^2 = \dfrac{12}{3} = 4 \ \therefore t = 2$ sec

Therefore distance travelled by particle before coming to rest,
$x = 40 + 12t - t^3 = 40 + 12 \times 2 - (2)^3 = 56$m

40. **(a)** $x = \dfrac{1}{t+5} \quad \therefore \ v = \dfrac{dx}{dt} = \dfrac{-1}{(t+5)^2}$

$\therefore \ a = \dfrac{d^2x}{dt^2} = \dfrac{2}{(t+5)^3} = 2x^3$

Now $\dfrac{1}{(t+5)} \propto v^{\frac{1}{2}}$

$\therefore \ \dfrac{1}{(t+5)^3} \propto v^{\frac{3}{2}} \propto a$

41. **(c)** Average acceleration = $\dfrac{\text{change in velocity}}{\text{time interval}}$

$= \dfrac{\Delta \vec{v}}{t}$

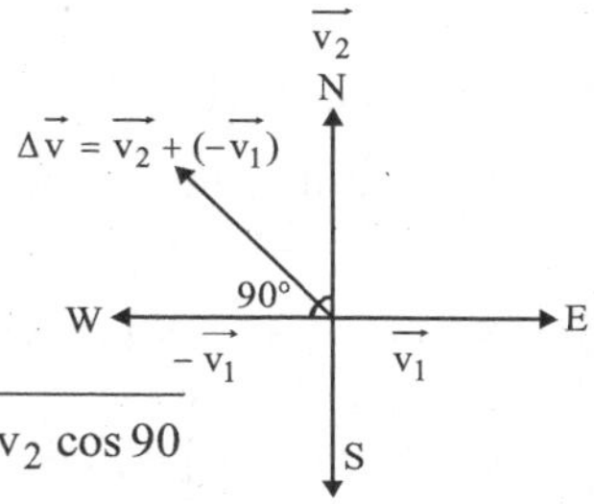

$\vec{v_1} = 5\hat{i}, \ \vec{v_2} = 5\hat{j}$

$\Delta \vec{v} = (\vec{v}_2 - \vec{v}_1)$

$= \sqrt{v_1^2 + v_2^2 + 2v_1v_2 \cos 90}$

$= \sqrt{5^2 + 5^2 + 0}$

[As $|v_1| = |v_2| = 5$ m/s] $= 5\sqrt{2}$ m/s

Avg. acc. $= \dfrac{\Delta \vec{v}}{t} = \dfrac{5\sqrt{2}}{10} = \dfrac{1}{\sqrt{2}}$ m/s^2

$\Rightarrow \tan\theta = \dfrac{5}{-5} = -1$

which means θ is in the second quadrant. (towards north-west)

42. **(a)** Differentiate two times and put $x = 0$.

43. **(a)** $\dfrac{dv}{dt} = -2.5\sqrt{v} \Rightarrow \dfrac{dv}{\sqrt{v}} = -2.5\, dt$

Integrating, $\displaystyle\int_{6.25}^{0} v^{-1/2} dv = -2.5\int_0^t dt$

$\Rightarrow \quad \left[\dfrac{v^{+1/2}}{(1/2)}\right]_{6.25}^{0} = -2.5[t]_0^t$

$\Rightarrow \quad -2(6.25)^{1/2} = -2.5t \ \Rightarrow \ t = 2$ sec

44. **(d)** $V_{max} = 0 + 5t_1$ ...(i)
$0 = V_{max} - 3t_2$ ...(ii)

$\Rightarrow t_1 + t_2 = 8 = V_{max}\left(\dfrac{1}{5} + \dfrac{1}{3}\right)$

$\Rightarrow V_{max} = \dfrac{8 \times 15}{8} = 15$m/s

45. **(d)** Given : u = 0, t = 5 sec, v = 108 km/hr = 30m/s
By eq^n of motion v = u + at

or $\quad a = \dfrac{v}{t} = \dfrac{30}{5} = 6\ \text{m/s}^2 \quad [\because u = 0]$

$S_1 = \dfrac{1}{2}at^2 = \dfrac{1}{2} \times 6 \times 5^2 = 75$ m

Distance travelled in first 5 sec is 75m.
Distance travelled with uniform speed of 30 m/s is S_2
$395 = S_1 + S_2 + S_3 \Rightarrow 395 = 75 + S_2 + 45$
$\Rightarrow S_2 = 275$ m

Time taken to travel 275 m $= \frac{275}{30} = 9.2$ sec

For retarding motion, we have
$0^2 - 30^2 = 2(-a) \times 45$, we get $a = 10$ m/s²

$S = ut + \frac{1}{2}at^2 \Rightarrow 45 = 30t + \frac{1}{2}(-10)t^2$
$\Rightarrow 45 = 30t - 5t^2$
On solving we get, $t = 3$ sec
Total time taken $= 5 + 9.2 + 3 = 17.2$ sec.

46. (d) Distance from A to B $= S = \frac{1}{2}ft_1^2$
Distance from B to C $= (ft_1)t$
Distance from C to D $= \frac{u^2}{2a} = \frac{(ft_1)^2}{2(f/2)}$
$= ft_1^2 = 2S$

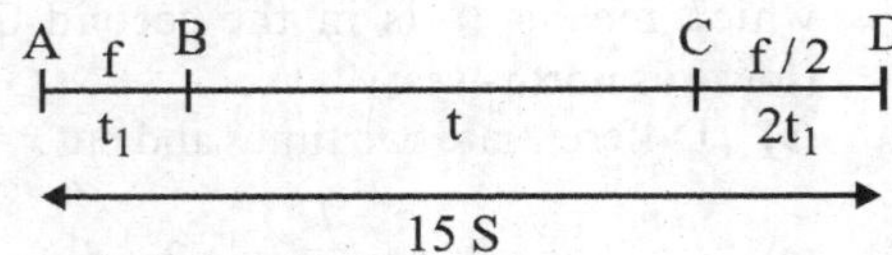

$\Rightarrow S + f t_1 t + 2S = 15S$
$\Rightarrow f t_1 t = 12S$ .... (i)
$\frac{1}{2} f t_1^2 = S$ .... (ii)
Dividing (i) by (ii), we get $t_1 = \frac{t}{6}$
$\Rightarrow S = \frac{1}{2}f\left(\frac{t}{6}\right)^2 = \frac{ft^2}{72}$

47. (d) Let u be the initial velocity that have to find and a be the uniform acceleration of the particle.
For $t = 3$s, distance travelled $S = 12$ m and for $t = 3 + 3 = 6$ s distance travelled $S' = 12 + 30 = 42$ m
From, $S = ut + 1/2\, at^2$
$12 = u \times 3 + \frac{1}{2} \times a \times 3^2$
or $24 = 6u + 9a$ ...(i)
Similarly, $42 = u \times 6 + \frac{1}{2} \times a \times 6^2$
or $42 = 6u + 18a$ ...(ii)
On solving, we get $u = 1$ m s⁻¹

48. (a) $v = 4t - 3t^2$
$\frac{ds}{dt} = 4t - 3t^2$; $\int_0^s ds = \int_0^t (4t - 3t^2)\, dt$
$s = \frac{4t^2}{2} - \frac{3t^3}{3}$; $s = 2t^2 - t^3$
$v_{avg} = \frac{s_{(2)} - s_{(0)}}{2 - 0}$ $[s_{(2)} = 8 - 8 = 0]$
$v_{avg} = \frac{0 - 0}{2} = 0$

49. (c) $x_1 = \frac{1}{2}a(10)^2 = 50a$
$x_2 = (10a)(10) + \frac{1}{2}a(10)^2 = 150a$
$x_3 = (20a)(10) + \frac{1}{2}a(10)^2 = 250a$
$\Rightarrow x_1 : x_2 : x_3 :: 1 : 3 : 5$

50. (a) $a = \frac{d^2x}{dt^2} = -s\omega^2 \sin \omega t$.
On integrating, $\frac{dx}{dt} = s\omega^2 \frac{\cos \omega t}{\omega} = s\omega \cos \omega t$
Again on integrating, we get
$x = s\omega \frac{\sin \omega t}{\omega} = s \sin \omega t$

51. (a) $s^2 = at^2 + 2bt + c$ $\therefore$ $2s\frac{ds}{dt} = 2at + 2b$
or $\frac{ds}{dt} = \frac{at + b}{s}$, again differentiating
$\frac{d^2s}{dt^2} = \frac{a.s - (at + b)\frac{ds}{dt}}{s^2} = \frac{as - (at + b)\left(\frac{at + b}{s}\right)}{s^2}$
$\therefore \frac{d^2s}{dt^2} = \frac{as^2 - (at + b)^2}{s^3}$ $\therefore a = \frac{d^2s}{dt^2} \propto s^{-3}$.

52. (d) Given: mass of car $m = 1000$ kg; $u = 30$ m/s; $v = 0$ m/s
retardation, $-a = 5$ m/s²
By equation, $v^2 - u^2 = 2as$
$0 - (30)^2 = -2 \times 5 \times d$
$\therefore d = \frac{900}{10} = 90$ m
and $a = \frac{v - u}{t}$ $\therefore t = \frac{v - u}{a} = \frac{0 - 30}{-5} = 6$s

53. (d) Let 'S' be the distance between two ends 'a' be the constant acceleration.
As we know $v^2 - u^2 = 2aS$ or, $aS = \frac{v^2 - u^2}{2}$.

Let v be velocity at mid point.

Therefore, $v_c^2 - u^2 = 2a\frac{S}{2} \Rightarrow v_c^2 = u^2 + aS$

$$v_c^2 = u^2 + \frac{v^2 - u^2}{2} \Rightarrow v_c = \sqrt{\frac{u^2 + v^2}{2}}$$

54. **(b)** $V_7 = V_5 + \frac{3\pi}{8} \times 2$

To calculate V_5 i.e. velocity at end of 5 sec.

$$a = (25 - t^2)^{1/2} \Rightarrow \frac{dv}{dt} = (25 - t^2)^{1/2}$$

$$\int dv = \int 5\,(25 - 25\sin^2\theta)^{1/2} \cos\theta d\theta \quad \begin{cases} t = 5\sin\theta \\ \frac{dt}{d\theta} = 5\cos\theta \\ t = 0, \theta = 0 \\ t = 5, \theta = \frac{\pi}{2} \end{cases}$$

$$= \int_0^{\pi/2} 5 \times 5\cos^2\theta d\theta = \frac{25\pi}{4}$$

$$V_7 = \frac{25\pi}{4} + \frac{3\pi}{4} = \frac{28\pi}{4} = 7\pi = 22\text{m/s}$$

55. **(a)** From $v = u + at$, we have

$v = 0 + \alpha t_1 \Rightarrow t_1 = v/\alpha$

$0 = v - \beta t_3 \Rightarrow t_3 = v/\beta.$

Now, $S = \frac{1}{2}vt_1 + vt_2 + \frac{1}{2}vt_3 = \frac{v^2}{2\alpha} + vt_2 + \frac{v^2}{2\beta}$

$$\Rightarrow t_2 = \frac{S}{v} - \frac{v}{2}\left(\frac{1}{\alpha} + \frac{1}{\beta}\right).$$

Hence, $t_1 + t_2 + t_3 = \frac{S}{v} + \frac{v}{2}\left(\frac{1}{\alpha} + \frac{1}{\beta}\right)$

56. **(a)** Let v be the maximum velocity attained and t the total time of journey. t' is the duration of acceleration and retardation. Then $v = 0 + at'$.

$$\therefore L = \frac{1}{2}at'^2 + v\,(t - 2t') + \frac{1}{2}at'^2$$

$$= a\left(\frac{v}{a}\right)^2 + v\left(t - \frac{2v}{a}\right)$$

$$= \frac{v^2}{a} + vt - \frac{2v^2}{a} = vt - \frac{v^2}{a}$$

$$t = \frac{L}{v} + \frac{v}{a} \Rightarrow \frac{dt}{dv} = -\frac{L}{v^2} + \frac{1}{a}$$

When t is minimum, $\frac{dt}{dv} = 0 \quad \therefore v_{max} = \sqrt{La}$

57. **(b)** Time $= \frac{\text{total length}}{\text{relative velocity}} = \frac{X + X}{20 + 20} = \frac{2X}{40}$ s $= 2$s

58. **(d)** As slope is same, so velocity of A and B will be equal

59. **(c)** Velocity of A is same as that of B in magnitude but opposite in direction.

60. **(d)** So by figure the velocity of parrot w.r.t. train is $= 5-(-10) = 15$m/sec so time taken to cross the train is

$$= \frac{\text{length of train}}{\text{relative velocity}} = \frac{150}{15} = 10 \text{ sec}$$

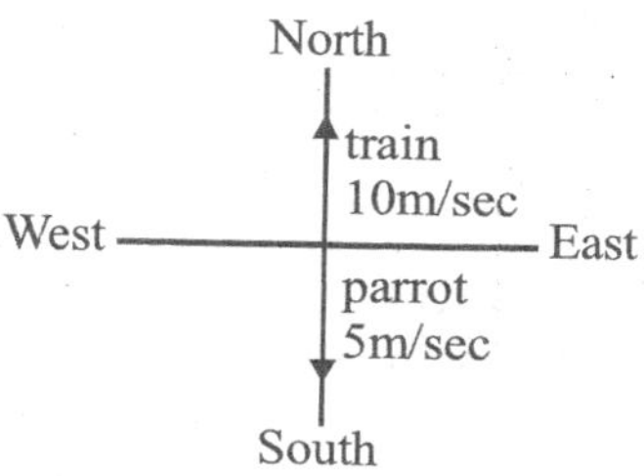

61. **(a)** Velocity of boat $= \frac{8+8}{2} = 8\,\text{km h}^{-1}$

Velocity of water $= 4\,\text{km h}^{-1}$

$$t = \frac{8}{8-4} + \frac{8}{8+4} = \frac{8}{3}\text{h} = 160 \text{ minute}$$

62. **(b)** Relative speed of each train with respect to each other be, $v = 10 + 15 = 25$ m/s

Here distance covered by each train = sum of their lengths $= 50 + 50 = 100$ m

$\therefore$ Required time $= \frac{100}{25} = 4$ sec .

63. **(a)** $\vec{V}_A = 10(-\hat{i})$

$\vec{V}_B = 10(\hat{j})$

$\vec{V}_{BA} = 10\hat{j} + 10\hat{i}$

$= 10\sqrt{2}$ km/h

Distance OB = 100 cos 45°

$= 50\sqrt{2}$ km

Time taken to each the shortest distance between

A and B $= \frac{OB}{V_{BA}} = \frac{50\sqrt{2}}{10\sqrt{2}} = 5$h

64. **(d)** Speed to cover 1200 m by scootarist

$v_r \times 60 = 1200 \Rightarrow v_r = 20$

speed to overtake bus

$v = v_r + 10 = 30$ m/s

65. (b) As magnitude of acceleration is same

$$\frac{30-15}{t}=\frac{20-V_Q}{t}$$

$15=20-V_Q \Rightarrow V_Q=5\text{m/s}$

66. (a) $a=\frac{v_0''-v_0'}{\Delta t}$

On substituting

$v'_0=2$ m/s, $v''_0=12$ m/s and $\Delta t=10$ sec,

we get, $a=1$ m/s^2.

67. (b) Initial relative velocity, $v_{AB}=v-0=v$.

Acceleration $a_{AB}=0-a=-a$.

For max. separation $v_{AB}=0$

$$0=v^2-2as \Rightarrow s=\frac{v^2}{2a}$$

68. (a) $v^2=u^2+2as$ $400>2a(100)$; $a<2$

69. (b) Let the initial velocity of ball be u $\therefore$ Time of rise

$$t_1=\frac{u}{g+a} \text{ and height reached } =\frac{u^2}{2(g+a)}$$

Time of fall t_2 is given by

$$\frac{1}{2}(g-a)t_2^2=\frac{u^2}{2(g+a)}$$

$$t_2=\frac{u}{\sqrt{(g+a)(g-a)}}=\frac{u}{(g+a)}\sqrt{\frac{g+a}{g-a}}$$

$\therefore$ $t_2>t_1$ because $\frac{1}{g+a}<\frac{1}{g-a}$

70. (d) During the complete journey acceleration (a = g) remains constant close to the earth provided no air resistance

71. (d) As ball returns to starting point so displacement is zero.

72. (a) We should know the displacement in this time.

$$\vec{S}=\vec{u}t+\frac{\vec{a}t^2}{2} \text{ (we take upward as positive)}$$

$$S=30\times4-10\times4\times\frac{4}{2}=40\,\text{m.}$$

The average velocity will be 10 m/sec.

73. (a) $h=ut_1-\frac{1}{2}gt_1^2$ Also $h=ut_2-\frac{1}{2}gt_2^2$

After simplify above equations, we get

$$h=\frac{1}{2}gt_1t_2.$$

74. (c) If h is the height of the building, then

$v_A^2=v^2+2gh$ and $v_B^2=(-v)^2+2gh$.

Clearly $v_A=v_B$.

75. (b) Velocity when the engine is switched off

$v=19.6\times5=98\,\text{ms}^{-1}$ $h_{max}=h_1+h_2$

where $h_1=\frac{1}{2}at^2$ & $h_2=\frac{v^2}{2a}$

$$h_{max}=\frac{1}{2}\times19.6\times5\times5+\frac{98\times98}{2\times9.8}=735\,\text{m}$$

76. (b) Height attained by balls in 2 sec is

$$=\frac{1}{2}\times9.8\times4=19.6\text{m}$$

the same distance will be covered in 2 second (for descent)

Time interval of throwing balls, remaining same. So, for two balls remaining in air, the time of ascent or descent must be greater than 2 seconds. Hence speed of balls must be greater than 19.6 m/sec.

77. (a) $S=ut+\frac{1}{2}at^2$

$$-76=4\times6-\frac{1}{2}\times10\times(6)^2 \Rightarrow u=\frac{52}{3}\text{ m/s}$$

78. (b) $S=AB=\frac{1}{2}gt_1^2$

$\Rightarrow$ $2S=AC=\frac{1}{2}g(t_1+t_2)^2$

and $3S=AD=\frac{1}{2}g(t_1+t_2+t_3)^2$

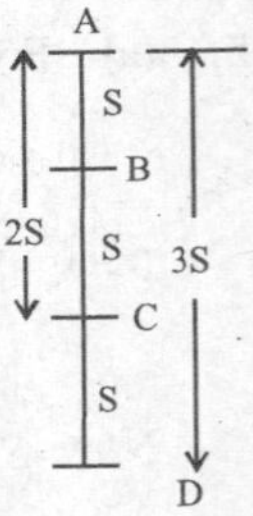

$$t_1=\sqrt{\frac{2S}{g}}$$

$$t_1+t_2=\sqrt{\frac{4S}{g}},\quad t_2=\sqrt{\frac{4S}{g}}-\sqrt{\frac{2S}{g}}$$

$$t_1+t_2+t_3=\sqrt{\frac{6S}{g}}\qquad t_3=\sqrt{\frac{6S}{g}}-\sqrt{\frac{4S}{g}}$$

$$t_1:t_2:t_3::1:(\sqrt{2}-\sqrt{1}):(\sqrt{3}-\sqrt{2})$$

79. (c) Time of flight is independent of mass

80. (b)

81. (b) $S_{n\text{th}}=u+\frac{a}{2}(2n-1)$

$$\text{so, }\frac{S_4}{S_5}=\frac{0+\frac{g}{2}(7)}{0+\frac{g}{2}(9)}=\frac{7}{9}$$

82. (a) $s=ut+\frac{1}{2}at^2$

$-65=12t-5t^2$ on solving we get, $t=5$s

83. (d) For A to B

$$S=\frac{1}{2}gt^2 \quad \text{...(i)}$$

For A to C

$$2S=\frac{1}{2}gt'^2 \quad \text{...(ii)}$$

Dividing (i) by (ii) we get $\frac{t}{t'} = \frac{1}{\sqrt{2}}$

84. (a) Clearly distance moved by 1st ball in 18s = distance moved by 2nd ball in 12s.
Now, distance moved in 18 s by 1st ball
$= \frac{1}{2} \times 10 \times 18^2 = 90 \times 18 = 1620$ m
Distance moved in 12 s by 2nd ball
$= ut + \frac{1}{2}gt^2 \quad \therefore \; 1620 = 12v + 5 \times 144$
$\Rightarrow v = 135 - 60 = 75 \text{ ms}^{-1}$

85. (a) $\because \quad h = \frac{1}{2}gt^2 \; \therefore \; h_1 = \frac{1}{2}g(5)^2 = 125$
$h_1 + h_2 = \frac{1}{2}g(10)^2 = 500 \Rightarrow h_2 = 375$
$\therefore \quad h_1 + h_2 + h_3 = \frac{1}{2}g(15)^2 = 1125$
$\Rightarrow \; h_3 = 625$
$h_2 = 3h_1,\; h_3 = 5h_1 \text{ or } h_1 = \frac{h_2}{3} = \frac{h_3}{5}$

86. (a)

87. (c) In $\frac{T}{3}$ sec, the distance travelled
$= \frac{1}{2}g\left(\frac{T}{3}\right)^2 = \frac{h}{9}$
$\therefore$ Position of the ball from the ground
$= h - \frac{h}{9} = \frac{8h}{9}$ m

88. (b) Time taken by the stone to reach the water level $t_1 = \sqrt{\frac{2h}{g}}$
Time taken by sound to come to the mouth of the well, $t_2 = \frac{h}{v}$
$\therefore$ Total time $t_1 + t_2 = \sqrt{\frac{2h}{g}} + \frac{h}{v}$

89. (c) $h = ut - \frac{1}{2}gt^2$
$\Rightarrow gt^2 - 2ut + 2h = 0$
solving for t we get
$t_1 + t_2 = 2u/g$
$t_1 \times t_2 = 2h/g$
so, $\Delta t = |t_1 - t_2| = (t_1 + t_2)^2 - 4t_1t_2$
putting value we get
$u = \frac{1}{2}\sqrt{8hg + g^2\Delta t^2}$

90. (b)

91. (d) Let the two balls Pand Q meet at height x m from the ground after time t s from the start. We have to find distance, $BC = (100 - x)$

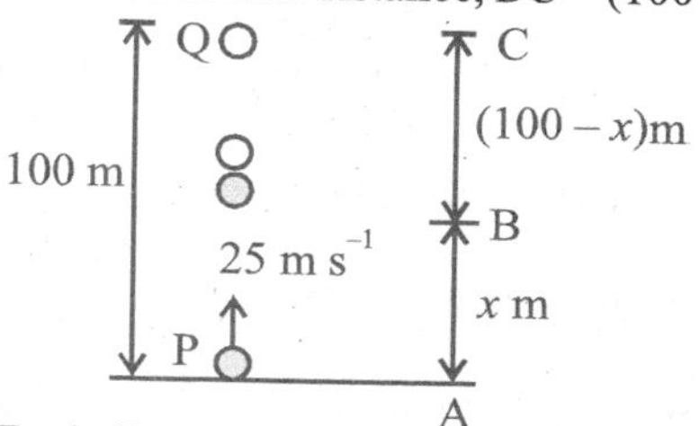

For ball P
$S = x$ m, $u = 25 \text{ m s}^{-1}$, $a = -g$
From $S = ut + \frac{1}{2}at^2$
$x = 25t - \frac{1}{2}gt^2$ .... (i)
For ball Q
$S = (100 - x)$ m, $u = 0$, $a = g$
$\therefore \; 100 - x = 0 + \frac{1}{2}gt^2$ .... (ii)
Adding eqns. (i) and (ii), we get
$100 = 25t$ or $t = 4$ s
From eqn. (i),
$x = 25 \times 4 - \frac{1}{2} \times 9.8 \times (4)^2 = 21.6$ m
Hence distance from the top of the tower
$= (100 - x)$ m $= (100 - 21.6$ m$) = 78.4$ m

92. (b)

93. (b) Let the body fall through the height of tower in t seconds. From, $D_n = u + \frac{a}{2}(2n - 1)$ we have, total distance travelled in last 2 second of fall is
$D = D_t + D_{(t-1)}$
$= \left[0 + \frac{g}{2}(2t-1)\right] + \left[0 + \frac{g}{2}\{2(t-1) - 1\}\right]$
$= \frac{g}{2}(2t-1) + \frac{g}{2}(2t-3) = \frac{g}{2}(4t-4)$
$= \frac{10}{2} \times 4(t-1)$
or, $40 = 20\,(t - 1)$ or $t = 2 + 1 = 3$s
Distance travelled in t second is
$s = ut + \frac{1}{2}at^2 = 0 + \frac{1}{2} \times 10 \times 3^2 = 45$m

94. (c) Given, H = 10 m, u = 5 m/s, g = 10 m/s^2
Speed on reaching ground
$v = \sqrt{u^2 + 2gh}$
Now, $v = u + at \Rightarrow \sqrt{u^2 + 2gh} = -u + gt$

Time taken to reach highest point is $t = \frac{u}{g}$,

$\Rightarrow \quad t = \frac{u + \sqrt{u^2 + 2gH}}{g} = \frac{nu}{g}$ (from question)

$\Rightarrow \quad 2gH = n(n-2)u^2$

$\Rightarrow \quad n(n-2) = \frac{2gH}{u^2} = \frac{2 \times 10 + 10}{5 \times 5} = 8 \Rightarrow n = 4$

95. (a) The distance travelled by the body A is h_1 given by

$v_1 t - \frac{gt^2}{2}$ and that travelled by the body B is

$h_2 = \frac{gt^2}{2}$

The distance between the bodies

$= x = h - (h_1 + h_2)$.

Since $h_1 + h_2 = v_1 t$, the relation sought is

$x = h - v_1 t$

96. (c) $\frac{\text{Time of ascent}}{\text{Time of descent}} = \frac{\left(\frac{u}{g+a}\right)}{\frac{u}{\sqrt{(g+a)(g-a)}}}$

$= \frac{\sqrt{(g+a)(g-a)}}{(g+a)} = \sqrt{\frac{g-a}{g+a}}$

$= \sqrt{\frac{10-5}{10+5}} = \sqrt{\frac{5}{15}} = \sqrt{\frac{1}{3}}$

97. (d)

B (v = 0)
4 sec 4 sec
(2 sec) A C
(t = 0) D

As the time taken from D to A = 2 sec.

and D → A → B → C = 10 sec (given).

As ball goes from B → C $(u = 0, t = 4\,\text{sec})$

$v_c = 0 + 4g.$

As it moves from C to D, $s = ut + \frac{1}{2}gt^2$

$s = 4g \times 2 + \frac{1}{2}g \times 4 = 10\text{ g}.$

98. (a) $t = \frac{s}{v} = \sqrt{H^2 + L^2} \div \sqrt{\frac{2H}{g}}$

99. (b) $v_B^2 = v_T^2 + 2\,(10)\,(3)$

$\Rightarrow \quad v_B^2 = v_T^2 + 60 \quad$ ..(i)

Also, $v_B = v_T + (10)\,(0.5)$ .. (ii)

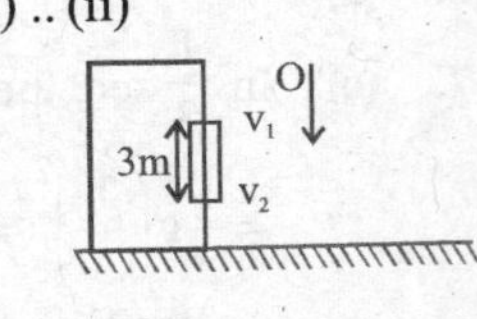

Solve eqs. (i) and (ii)

we get

$v_T - v_B = 4.9$ m/s

100. (b) Ball A is thrown upwards from the building. During its downward journey when it comes back to the point of throw, its speed is equal to the speed of throw. So, for the journey of both the balls from point A to B.

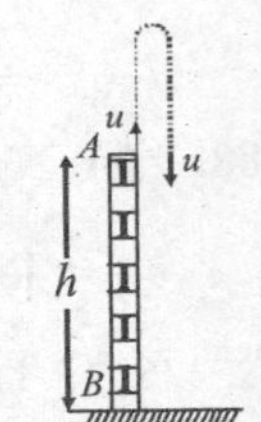

We can apply $v^2 - u^2 = 2gh$.

As u, g, h are same for both the balls, $v_A = v_B$

3 Motion in a Plane

1. If a unit vector is represented by $0.5\hat{i} + 0.8\hat{j} + c\hat{k}$, then the value of c is

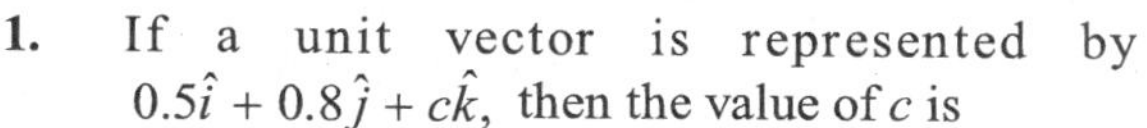

   (a) 1 (b) $\sqrt{0.8}$
   (c) $\sqrt{0.11}$ (d) $\sqrt{0.01}$

2. If the magnitudes of vectors $\vec{A}$, $\vec{B}$ and $\vec{C}$ are 12, 5 and 13 units respectively and $\vec{A} + \vec{B} = \vec{C}$, the angle between vectors A and B is :
   (a) 0 (b) π (c) $\frac{\pi}{2}$ (d) $\frac{\pi}{4}$

3. The velocity of a projectile at the initial point A is $(2\hat{i} + 3\hat{j})$ m/s its velocity (in m/s) at point B is
   (a) $-2\hat{i} + 3\hat{j}$
   (b) $2\hat{i} - 3\hat{j}$
   (c) $2\hat{i} + 3\hat{j}$
   (d) $-2\hat{i} - 3\hat{j}$

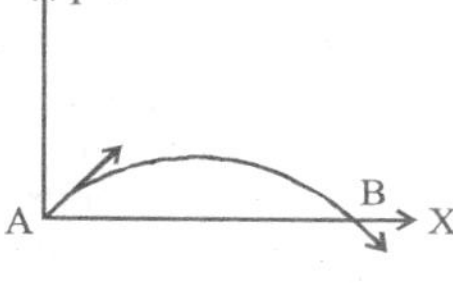

4. A body is projected from the ground with a velocity 50 m/s at an angle of 30°. It crosses a wall after 3 sec. How far beyond the wall the stone will strike the ground?
   [take g = 10 m/s^2]
   (a) $50\sqrt{3}$ (b) $70\sqrt{2}$ c) $15\sqrt{3}$ (d) $16\sqrt{2}$

5. A stone is projected horizontally with a 5 m/s from the top of a plane inclined at an angle 45° with the horizontal. How far from the point of projection will the particle strike the plane?
   (a) $5\sqrt{2}$ m (b) $11\sqrt{2}$ m
   (c) $12\sqrt{2}$ m (d) $15\sqrt{2}$ m

6. Two balls are projected at an angle θ and $(90° - \theta)$ to the horizontal with the same speed. The ratio of their maximum vertical heights is **Tricky**
   (a) 1 : 1 (b) $\tan\theta : 1$
   (c) $1 : \tan\theta$ (d) $\tan^2\theta : 1$

7. A body is thrown with a velocity of 9.8 ms^{-1} making an angle of 30° with the horizontal. It will hit the ground after a time
   (a) 3.0 s (b) 2.0 s (c) 1.5 s (d) 1 s

8. The velocity of projection of a body is increased by 2%. Other factors remaining unchanged, what will be the percentage change in the maximum height attained?
   (a) 1% (b) 2% (c) 4% (d) 8%

9. A cricket ball is hit with a velocity 25 ms^{-1}, 60° above the horizontal. How far above the ground, ball passes over a fielder 50 m from the bat (consider the ball is struck very close to the ground)? Take $\sqrt{3} = 1.7$ and g = 10 ms^{-2}
   (a) 6.8m (b) 7m (c) 5m (d) 10m

10. A stone is thrown from a point with a speed 5 m/s at an elevation angle of θ. From the same point and at the same instant, a person starts running with a constant speed 2.5 m/s to catch the stone. If the person will be able to catch the ball then, what should be the angle of projection θ?
    (a) 75° (b) 30° **Critical Thinking**
    (c) 60° (d) 45°

11. If the resultant of the vectors $3\hat{i} + 4\hat{j} + 5\hat{k}$ and $5\hat{i} + 3\hat{j} + 4\hat{k}$ makes an angle θ with x-axis, then $\cos\theta$ is
    (a) 0.07 (b) 0.574 (c) 0.111 (d) 0.123

12. Let $\vec{C} = \vec{A} + \vec{B}$ then
    (a) $|\vec{C}|$ is always greater than $|\vec{A}|$
    (b) it is possible to have $|\vec{C}| < |\vec{A}|$ and $|\vec{C}| < |\vec{B}|$
    (c) $\vec{C}$ is always equal to $\vec{A} + \vec{B}$
    (d) $\vec{C}$ is never equal to $\vec{A} + \vec{B}$

13. Vector $\vec{A}$ makes equal angle with x, y and z-axis. Value of its components in terms of magnitude of $\vec{A}$ will be
    (a) $\frac{A}{\sqrt{3}}$ (b) $\frac{A}{\sqrt{2}}$ (c) $\sqrt{3}A$ (d) $\frac{\sqrt{3}}{A}$

14. The vector that must be added to the vector $\hat{i} - 3\hat{j} + 2\hat{k}$ and $3\hat{i} - 6\hat{j} + 7\hat{k}$ so that the resultant vector is a unit vector along the y-axis, is
    (a) $4\hat{i} - 2\hat{j} + 5\hat{k}$ (b) $-4\hat{i} - 2\hat{j} + 5\hat{k}$
    (c) $3\hat{i} - 4\hat{j} + 5\hat{k}$ (d) null vector

15. It is found that $|A + B| = |A|$. This necessarily implies,
(a) $B = 0$
(b) A, B are antiparallel
(c) A, B are perpendicular
(d) $A.B \leq 0$

16. The resultant of two vectors $\vec{A}$ and $\vec{B}$ is perpendicular to the vector $\vec{A}$ and its magnitude is equal to half the magnitude of vector $\vec{B}$. The angle between $\vec{A}$ and $\vec{B}$ is
(a) 120° (b) 150° (c) 135° (d) 180°

17. If $\vec{A} = \vec{B} - \vec{C}$, then, the angle between $\vec{A}$ and $\vec{B}$ is
(a) $\tan^{-1}\dfrac{B^2 + A^2 - C^2}{2AB}$
(b) $\sin^{-1}\dfrac{B^2 + A^2 - C^2}{2AB}$
(c) $\cos^{-1}\dfrac{A^2 + B^2 - C^2}{2AB}$
(d) $\sec^{-1}\dfrac{A^2 + B^2 - C^2}{2AB}$

18. The resultant of vectors $\vec{P}$ and $\vec{Q}$ is $\vec{R}$. On reversing the direction of $\vec{Q}$, the resultant vector becomes $\vec{S}$. Then, correct relation is  **Tricky**
(a) $R^2 + S^2 = 2(P^2 + Q^2)$.
(b) $R^2 + S^2 = P^2 + Q^2$
(c) $R^2 + P^2 = S^2 + Q^2$
(d) $P^2 + S^2 = 2(Q^2 + R^2)$

19. The x and y components of $\vec{A}$ are 4 m and 6 m, respectively. The x and y components of $(\vec{A} + \vec{B})$ are 10 m and 9 m respectively. The magnitude of vector $\vec{B}$ is :
(a) 19m (b) $\sqrt{27}$ m (c) $\sqrt{45}$ m (d) $\sqrt{50}$ m

20. If the sum of two unit vectors is a unit vector, then the magnitude of their difference is
(a) 1 (b) $\sqrt{2}$ (c) $\sqrt{3}$ (d) 2

21. If $\vec{A} = 5\hat{i} + 7\hat{j} - 3\hat{k}$ and $\vec{B} = 2\hat{i} + 2\hat{j} - a\hat{k}$ are perpendicular vectors, the value of a is:
(a) −2 (b) 8 (c) −7 (d) −8

22. The condition for $\vec{A} + \vec{B}$ to be perpendicular to $\vec{A} - \vec{B}$ is that
(a) $|\vec{A}| = |\vec{B}|$ (b) $\vec{A} = \vec{B}$
(c) $\vec{B} = 0$ (d) $|\vec{A} + \vec{B}| = |\vec{A} - \vec{B}|$

23. Let $\vec{a}$ and $\vec{b}$ be two unit vectors. If the vectors $\vec{c} = \hat{a} + 2\hat{b}$ and $\vec{d} = 5\hat{a} - 4\hat{b}$ are perpendicular to each other, then the angle between $\hat{a}$ and $\hat{b}$ is :
(a) $\dfrac{\pi}{6}$ (b) $\dfrac{\pi}{2}$ (c) $\dfrac{\pi}{3}$ (d) $\dfrac{\pi}{4}$

24. A particle crossing the origin of co-ordinates at time t = 0, moves in the xy-plane with a constant acceleration a in the y-direction. If its equation of motion is $y = bx^2$ (b is a constant), its velocity component in the x-direction is
(a) $\sqrt{\dfrac{2b}{a}}$ (b) $\sqrt{\dfrac{a}{2b}}$ (c) $\sqrt{\dfrac{a}{b}}$ (d) $\sqrt{\dfrac{b}{a}}$

25. The position of particle is given by $\vec{r} = 2t^2\hat{i} + 3t\hat{j} + 4\hat{k}$, where t is in second and the coefficients have proper units for $\vec{r}$ to be in metre. The $\vec{a}(t)$ of the particle at t = 1 s is
(a) 4 m s^{-2} along y-direction
(b) 3 m s^{-2} along x-direction
(c) 4 m s^{-2} along x-direction
(d) 2 m s^{-2} along z-direction

26. A particle has an initial velocity $3\hat{i} + 4\hat{j}$ and an acceleration of $0.4\hat{i} + 0.3\hat{j}$. Its speed after 10 sec is
(a) $7\sqrt{2}$ units (b) 7 units
(c) 8.5 units (d) 10 units

27. A particle moves in the X-Y plane with a constant acceleration 1.5 m/s² in the direction making an angle of 37° with the X-axis. At $t = 0$ the particle is at the origin and its velocity is 8.0 m/s along the X-axis. Find the position of the particle at $t = 4.0$ s.
(a) (41.6m, 7.2 m) (b) (50.3 m, 8.2 m)
(c) (60.2 m, 8.2 m) (d) (11.2 m, 8 m)

28. The coordinates of a particle moving in x–y plane at any instant of time t are $x = 4t^2$; $y = 3t^2$. The speed of the particle at that instant is
(a) 10 t (b) 5 t (c) 3 t (d) 2 t

29. A bullet is dropped from the same height when another bullet is fired horizontally. They will hit the ground
(a) one after the other
(b) simultaneously
(c) depends on the observer
(d) None of these

30. If V_1 is velocity of a body projected from the point A and V_2 is the velocity of a body projected from point B which is vertically below the highest point C. if both the bodies collide, then
(a) $V_1 = \dfrac{1}{2}V_2$
(b) $V_2 = \dfrac{1}{2}V_1$
(c) $V_1 = V_2$
(d) Two bodies can't collide.

31. Which of the following is correct ?
(a) $\vec{A}.\vec{B} \neq \vec{B}.\vec{A}$
(b) $\vec{A}.(\vec{B}+\vec{C}) = A.B+\vec{C}$
(c) $\vec{A}\times\vec{B} \neq \vec{B}\times A$
(d) $\vec{A}.(\vec{B}+\vec{C}) \neq \vec{A}.\vec{B}+\vec{A}.\vec{C}$

32. If none of the vectors $\vec{\mathbf{A}}$, $\vec{\mathbf{B}}$ and $\vec{\mathbf{C}}$ are zero and if $\vec{\mathbf{A}}\times\vec{\mathbf{B}}=0$ and $\vec{\mathbf{B}}\times\vec{\mathbf{C}}=0$, the value of $\vec{\mathbf{A}}\times\vec{\mathbf{C}}$ is:
(a) unity (b) zero
(c) $\mathbf{B}^2$ (d) AC cos θ

33. If $|\vec{\mathbf{A}}\times\vec{\mathbf{B}}|=\sqrt{3}\vec{\mathbf{A}}.\vec{\mathbf{B}}$, then the value of $|\vec{\mathbf{A}}+\vec{\mathbf{B}}|$ is : **Tricky**
(a) $\left(A^2+B^2+\dfrac{AB}{\sqrt{3}}\right)^{1/2}$
(b) $A+B$
(c) $(A^2+B^2+\sqrt{3}AB)^{1/2}$
(d) $(A^2+B^2+AB)^{1/2}$

34. $\vec{a}=3\hat{i}-5\hat{j}$ and $\vec{b}=6\hat{i}+3\hat{j}$ are two vectors and $\vec{c}$ is a vector such that $\vec{c}=\vec{a}\times\vec{b}$ then $|\vec{a}|:|\vec{b}|:|\vec{c}|$
(a) $\sqrt{34}:\sqrt{45}:\sqrt{39}$ (b) $\sqrt{34}:\sqrt{45}:39$
(c) $34:39:45$ (d) $39:35:34$

35. If $|\vec{a}|=4, |\vec{b}|=2$ and the angle between $\vec{a}$ and $\vec{b}$ is $\pi/6$ then $(\vec{a}\times\vec{b})^2$ is equal to
(a) 48 (b) 16 (c) 4 (d) 2

36. If the vectors $(\hat{i}+\hat{j}+\hat{k})$ and $3\hat{i}$ form two sides of a triangle, the area of the triangle is :
(a) $\sqrt{3}$ (b) $2\sqrt{3}$ (c) $\dfrac{3}{\sqrt{2}}$ (d) $3\sqrt{2}$

37. The vector having magnitude equal to 3 and perpendicular to the two vectors $\vec{\mathbf{A}}=2\hat{i}+2\hat{j}+\hat{k}$ and $\vec{\mathbf{B}}=2\hat{i}-2\hat{j}+3\hat{k}$ is:
(a) $\pm(2\hat{i}-\hat{j}-2\hat{k})$ (b) $\pm(3\hat{i}+\hat{j}-2\hat{k})$
(c) $-(3\hat{i}+\hat{j}-3\hat{k})$ (d) $(3\hat{i}-\hat{j}-3\hat{k})$

38. Let two vectors $\vec{\mathbf{A}}=3\hat{i}+\hat{j}+2\hat{k}$ and $\vec{\mathbf{B}}=2\hat{i}-2\hat{j}+4\hat{k}$. Consider the unit vector perpendicular to both $\vec{\mathbf{A}}$ and $\vec{\mathbf{B}}$ is
(a) $\dfrac{\hat{i}-\hat{j}-\hat{k}}{\sqrt{3}}$ (b) $\dfrac{\hat{i}-\hat{j}-\hat{k}}{2\sqrt{3}}$
(c) $\dfrac{-\hat{i}-\hat{j}-\hat{k}}{\sqrt{3}}$ (d) $\dfrac{\hat{i}-\hat{j}-\hat{k}}{2\sqrt{3}}$

39. A projectile can have the same range R for two angles of projection. If t_1 and t_2 be the times of flight in two cases, then what is the product of two times of flight? **Tricky**
(a) $t_1t_2 \propto R$ (b) $t_1t_2 \propto R^2$
(c) $t_1t_2 \propto 1/R$ (d) $t_1t_2 \propto 1/R^2$

40. A particle of mass m is projected with a velocity υ making an angle of 30° with the horizontal. The magnitude of $(V_h \times h)$ of the projectile when the particle is at its maximum height *h*
(a) $\dfrac{\sqrt{3}}{2}\dfrac{\upsilon^2}{g}$ (b) zero
(c) $\dfrac{\upsilon^3}{\sqrt{2}g}$ (d) $\dfrac{\sqrt{3}}{16}\dfrac{\upsilon^3}{g}$

41. The range of a projectile is R when the angle of projection is 40°. For the same velocity of projection and range, the other possible angle of projection is
(a) 45° (b) 50° (c) 60° (d) 40°

42. If the angles of projection of a projectile with same initial velocity exceed or fall short of 45° by equal amounts , then the ratio of horizontal ranges is
(a) 1 : 2 (b) 1 : 3 (c) 1 : 4 (d) 1 : 1

43. The equation of a projectile is $y=\sqrt{3}x-\dfrac{gx^2}{2}$
The angle of projection is given by
(a) $\tan\theta=\dfrac{1}{\sqrt{3}}$ (b) $\tan\theta=\sqrt{3}$
(c) $\dfrac{\pi}{2}$ (d) zero.

44. A plane flying horizontally at a height of 1500 m with a velocity of 200 ms^{-1} passes directly overhead on antiaircraft gun. Then the angle with the horizontal at which the gun should be fired from the shell with a muzzle velocity of 400 ms^{-1} to hit the plane, is
(a) 90° (b) 60° (c) 30° (d) 45°

45. The equation of trajectory of projectile is given by $y=\dfrac{x}{\sqrt{3}}-\dfrac{gx^2}{20}$, where x and y are in metre.
The maximum range of the projectile is
(a) $\dfrac{8}{3}$m (b) $\dfrac{4}{3}$m (c) $\dfrac{3}{4}$m (d) $\dfrac{3}{8}$m

46. A bullet is fired with a speed of 1500 m/s in order to hit a target 100 m away. If g = 10 m/s². The gun should be aimed
(a) 15 cm above the target
(b) 10 cm above the target
(c) 2.2 cm above the target
(d) directly towards the target

47. A projectile is thrown in the upward direction making an angle of 60° with the horizontal direction with a velocity of 147 ms^{-1}. Then the time after which its inclination with the horizontal is 45°, is

(a) 15 s (b) 10.98 s
(c) 5.49 s (d) 2.745 s

48. Two pegs A and B thrown with speeds in the ratio 1 : 3 acquired the same heights. If A is thrown at an angle of 30° with the horizontal, the angle of projection of B will be

(a) 0° (b) $\sin^{-1}\left(\frac{1}{8}\right)$
(c) $\sin^{-1}\left(\frac{1}{6}\right)$ (d) $\sin^{-1}\left(\frac{1}{2}\right)$

49. A body projected at an angle with the horizontal has a range 300 m. If the time of flight is 6 s, then the horizontal component of velocity is

(a) $30\ m\ s^{-1}$ (b) $50\ m\ s^{-1}$
(c) $40\ m\ s^{-1}$ (d) $45\ m\ s^{-1}$

50. A boy can throw a stone up to a maximum height of 10 m. The maximum horizontal distance that the boy can throw the same stone up to will be

(a) $20\sqrt{2}$ m (b) 10m
(c) $10\sqrt{2}$ m (d) 20m

51. A projectile is given an initial velocity of $(\hat{i}+2\hat{j})$ m/s, where $\hat{i}$ is along the ground and $\hat{j}$ is along the vertical. If $g = 10\ m/s^2$, the equation of its trajectory is :

(a) $y = x - 5x^2$ (b) $y = 2x - 5x^2$
(c) $4y = 2x - 5x^2$ (d) $4y = 2x - 25x^2$

52. A body is projected vertically upwards with a velocity u, after time t another body is projected vertically upwards from the same point with a velocity v, where $v < u$. If they meet as soon as possible, then choose the correct option

(a) $t = \dfrac{u - v + \sqrt{u^2 + v^2}}{g}$

(b) $t = \dfrac{u - v + \sqrt{u^2 - v^2}}{g}$

(c) $t = \dfrac{u + v + \sqrt{u^2 - v^2}}{g}$

(d) $t = \dfrac{u - v + \sqrt{u^2 - v^2}}{2g}$

53. A particle is projected with a velocity v such that its range on the horizontal plane is twice the greatest height attained by it. The range of the projectile is (where g is acceleration due to gravity)

(a) $\dfrac{4v^2}{5g}$ (b) $\dfrac{4g}{5v^2}$ (c) $\dfrac{v^2}{g}$ (d) $\dfrac{4v^2}{\sqrt{5}g}$

54. A projectile is fired from the surface of the earth with a velocity of 5 ms^{-1} and angle θ with the horizontal. Another projectile fired from another planet with a velocity of 3 ms^{-1} at the same angle follows a trajectory which is identical with the trajectory of the projectile fired from the earth. The value of the acceleration due to gravity on the planet is (in ms^{-2}) given $g = 9.8\ m/s^2$

Critical Thinking

(a) 3.5 (b) 5.9 (c) 16.3 (d) 110.8

55. A projectile is thrown at an angle of 40° with the horizontal and its range is R_1. Another projectile is thrown at an angle 40° with the vertical and its range is R_2. What is the relation between R_1 and R_2?

(a) $R_1 = R_2$ (b) $R_1 = 2R_2$
(c) $2R_1 = R_2$ (d) $R_1 = 4R_2/5$

56. A particle is projected at an angle of elevation α and after t seconds it appears to have an angle of elevation β as seen from point of projection. The initial velocity will be

(a) $\dfrac{gt}{2\sin(\alpha - \beta)}$ (b) $\dfrac{gt\cos\beta}{2\sin(\alpha - \beta)}$

(c) $\dfrac{\sin(\alpha - \beta)}{2gt}$ (d) $\dfrac{2\sin(\alpha - \beta)}{gt\cos\beta}$

57. A projectile is thrown in the upward direction making an angle of 60° with the horizontal direction with a velocity of 147 ms^{-1}. Then the time after which its inclination with the horizontal is 45°, is

Critical Thinking

(a) $15\left(\sqrt{3}-1\right)$s (b) $15\left(\sqrt{3}+1\right)$s
(c) $7.5\left(\sqrt{3}-1\right)$s (d) $7.5\left(\sqrt{3}+1\right)$s

58. If a particle is projected with speed u from ground at an angle with horizontal ,then radius of curvature of a point where velocity vector is perpendicular to initial velocity vector is given by

(a) $\dfrac{u^2\cos^2\theta}{g}$ (b) $\dfrac{u^2\cot^2\theta}{g\sin\theta}$

(c) $\dfrac{u^2}{g}$ (d) $\dfrac{u^2\tan^2\theta}{g\cos\theta}$

59. A ball rolls off to the top of a staircase with a horizontal velocity u m/s. If the steps are h metre high and b metre wide, the ball will hit the edge of the nth step, if

(a) $n = \frac{2hu}{gb^2}$ (b) $n = \frac{2hu^2}{gb}$

(c) $n = \frac{2hu^2}{gb^2}$ (d) $n = \frac{hu^2}{gb^2}$

60. A body of mass m is projected horizontally with a velocity v from the top of a tower of height h and it reaches the ground at a distance x from the foot of the tower. If a second body of mass 2m is projected horizontally from the top of a tower of height 2h, it reaches the ground at a distance 2x from the foot of the tower. The horizontal velocity of the second body is

(a) v (b) $2v$ (c) $\sqrt{2v}$ (d) $v/2$

61. A bomb is dropped on an enemy post by an aeroplane flying horizontally with a velocity of 60 km h^{-1} and at a height of 490 m. At the time of dropping the bomb, how far the aeroplane should be from the enemy post so that the bomb may directly hit the target ? **Critical Thinking**

(a) $\frac{400}{3}$ m (b) $\frac{500}{3}$ m

(c) $\frac{1700}{3}$ m (d) 498 m.

62. A projectile of mass m is thrown with a velocity v making an angle 60° with the horizontal. Neglecting air resistance, the change in velocity from the departure A to its arrival at B, along the vertical direction is

(a) 2v (b) $\sqrt{3}$v (c) v (d) $\frac{v}{\sqrt{3}}$

63. A particle is projected at angle 37° with the incline plane in upward direction with speed 10 m/s. The angle of incline plane is given 53°. Then the maximum height attained by the particle from the incline plane will be

(a) 3m (b) 4m (c) 5m (d) zero

64. An aircraft moving with a speed of 250 m/s is at a height of 6000 m, just overhead of an anti aircraft gun. If the muzzle velocity is 500 m/s, the firing angle q should be:

(a) 30° (b) 45° (c) 60° (d) 75°

65. A particle is projected with a certain velocity at an angle α above the horizontal from the foot of an inclined plane of inclination 30°. If the particle strikes the plane normally then α is

(a) $30° + \tan^{-1}\left(\frac{\sqrt{3}}{2}\right)$ (b) $30° + \tan^{-1}\frac{1}{2}$

(c) $30° + \tan^{-1}1$ (d) 60°

66. A cricket ball thrown across a field is at heights h_1 and h_2 from point of projection at times t_1 and t_2 respectively after the throw. The ball is caught by a fielder at the same height as that of projection. The time of flight of the ball in this journey is **Critical Thinking**

(a) $\frac{h_1t_2^2 - h_2t_1^2}{h_1t_2 - h_2t_1}$ (b) $\frac{h_1t_2^2 + h_2t_1^2}{h_1t_2 + h_2t_1}$

(c) $\frac{h_1t_2}{h_2t_1 - h_1t_2}$ (d) None

67. You throw a ball with a $\vec{v} = (3\hat{i} + 4\hat{j})$ m/s towards a wall, where it hits at height h_1. Suppose that the launch velocity were, instead, $\vec{v} = (5\hat{i} + 4\hat{j})$ m/s and h_2 is height, then **Tricky**

(a) $h_2 = h_1$ (b) $h_2 < h_1$

(c) $h_2 > h_1$ (d) $h_2 \geq h_1$

68. If retardation produced by air resistance of projectile is one-tenth of acceleration due to gravity, the time to reach maximum height
(a) decreases by 11 percent
(b) increases by 11 percent
(c) decreases by 9 percent
(d) increases by 9 percent

69. For a stone thrown from a lower of unknown height, the maximum range for a projection speed of 10 m/s is obtained for a projection angle of 30°. The corresponding distance between the foot of the lower and the point of landing of the stone is

(a) 10m (b) 20m

(c) $(20/\sqrt{3})$m (d) $(10/\sqrt{3})$m

70. A jet plane flying at a constant velocity v at a height $h = 8$ km, is being tracked by a radar R located at O directly below the line of flight. If the angle θ is decreasing at the rate of 0.025 rad/s, the velocity of the plane when $\theta = 60°$ is :

(a) 1440 km/h (b) 960 km/h
(c) 1920 km/h (d) 480 km/h

71. An object is projected with a velocity of 20 m/s making an angle of 45° with horizontal. The equation for the trajectory is $h = Ax - Bx^2$ where h is height, x is horizontal distance, A and B are constants. The ratio $A : B$ is ($g = 10\ \text{ms}^{-2}$)

(a) 1 : 5 (b) 5 : 1 (c) 1 : 40 (d) 40 : 1

72. The position of a projectile launched from the origin at $t = 0$ is given by $\vec{r} = (40\hat{i} + 50\hat{j})$ m at t = 2s. If the projectile was launched at an angle θ from the horizontal, then θ is (take g = 10 ms^{-2})

(a) $\tan^{-1}\frac{2}{3}$ (b) $\tan^{-1}\frac{3}{2}$
(c) $\tan^{-1}\frac{7}{4}$ (d) $\tan^{-1}\frac{4}{5}$

73. A particle is projected from a tower as shown in figure, then the distance from the foot of the tower where it will strike the ground will be

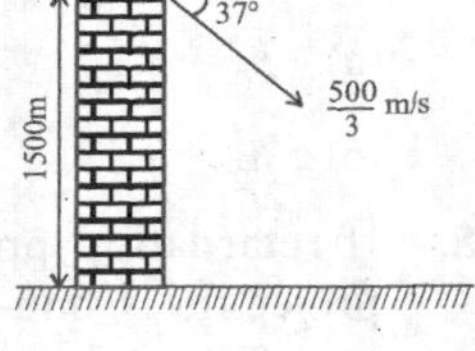

(a) 4000/3 m (b) 2000/m
(c) 1000/3 m (d) 2500/3 m

74. For an observer on trolley direction of projection of particle is shown in the figure, while for observer on ground ball rise vertically. The maximum height reached by ball from trolley is

(a) 10m (b) 15m (c) 20m (d) 5m

75. A particle P is projected from a point on the surface of smooth inclined plane (see figure). Simultaneously another particle Q is released on the smooth inclined plane from the same position. P and Q collide on the inclined plane after $t = 4$ second. The speed of projection of P is

(a) 5 m/s (b) 10 m/s (c) 15 m/s (d) 20 m/s

76. A large number of bullets are fired in all directions with the same speed v. What is the maximum area on the ground on which these bullets will spread?

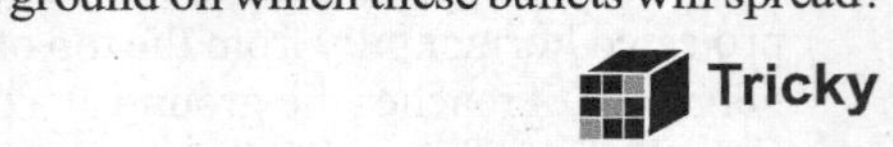

(a) $\frac{\pi v^2}{g}$ (b) $\frac{\pi v^4}{g^2}$
(c) $\pi^2 \frac{v^2}{g^2}$ (d) $\frac{\pi^2 v^4}{g^2}$

77. Two particles are projected simultaneously from the level ground as shown in figure. They may collide after a time :

(a) $\frac{x \sin\theta_2}{u_1}$ (b) $\frac{x\cos\theta_2}{u_2}$
(c) $\frac{x\sin\theta_2}{u_1 \sin(\theta_2 - \theta_1)}$ (d) $\frac{x\sin\theta_1}{u_2\sin(\theta_2 - \theta_1)}$

78. A body is thrown horizontally with a velocity $\sqrt{2gh}$ from the top of a tower of height h. It strikes the level ground through the foot of the tower at a distance x from the tower. The value of x is

(a) h (b) $h/2$ (c) $2h$ (d) $2h/3$

79. Three particles A, B and C are thrown from the top of a tower with the same speed. A is thrown up, B is thrown down and C is horizontally. They hit the ground with speeds v_A, v_B and v_C respectively then,

(a) $v_A = v_B = v_C$ (b) $v_A = v_B > v_C$
(c) $v_B > v_C > v_A$ (d) $v_A > v_B = v_C$

80. It was calculated that a shell when fired from a gun with a certain velocity and at an angle of elevation $5\pi/36$ rad should strike a given target. In actual practice, it was found that a hill just prevented the trajectory. At what angle (rad) of elevation should the gun be fired to hit the target

(a) $\frac{5\pi}{36}$ (b) $\frac{11\pi}{36}$ (c) $\frac{7\pi}{36}$ (d) $\frac{13\pi}{36}$

81. A boy is standing on a cart moving along x-axis with the speed of 10 m/s. When the cart reaches the origin he throws a stone in the horizontal x-y plane with the speed of 5 m/s with respect to himself at an angle θ with the x-axis. It is found that the stone hits a ball lying at rest at a point whose co-ordinates are $(\sqrt{3}\text{m}, 1\text{m})$. The value of θ is (gravitational effect is to be ignored)

(a) 30° (b) 60° (c) 90° (d) 120°

82. A projectile is fired with a velocity v at right angle to the slope which is inclined at an angle θ with the horizontal. The range of the projectile along the inclined plane is:

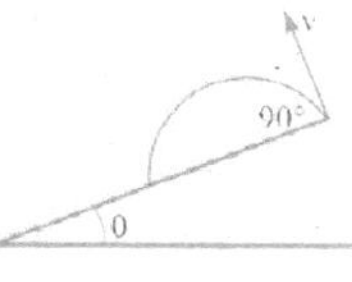

(a) $\frac{2v^2 \tan\theta}{g}$ (b) $\frac{v^2 \sec\theta}{g}$

(c) $\frac{2v^2 \tan\theta \sec\theta}{g}$ (d) $\frac{v^2 \sin\theta}{g}$

83. A boat is moving with a velocity $3\hat{i} + 4\hat{j}$ with respect to ground. The water in the river is moving with a velocity $-3\hat{i} - 4\hat{j}$ with respect to ground. The relative velocity of the boat with respect to water is

(a) $8\hat{j}$ (b) $-6\hat{i} - 8\hat{j}$ (c) $6\hat{i} + 8\hat{j}$ (d) $5\sqrt{2}$

84. A boat which has a speed of 5 km/hr in still water crosses a river of width 1 km along the shortest possible path in 15 minutes. The velocity of the river water in km/hr is

(a) 3 (b) 4 (c) $\sqrt{21}$ (d) 1

85. A swimmer wants to cross a river straight. He swim at 5 km/hr in still water. A river 1 km wide flows at the rate of 3 km/hr. Which of the following figure shows the correct direction for the swimmer along which he should strike?

($V_s \rightarrow$ velocity of swimmer, $V_r \rightarrow$ velocity of river, $V \rightarrow$ resultant velocity) **Tricky**

(a)

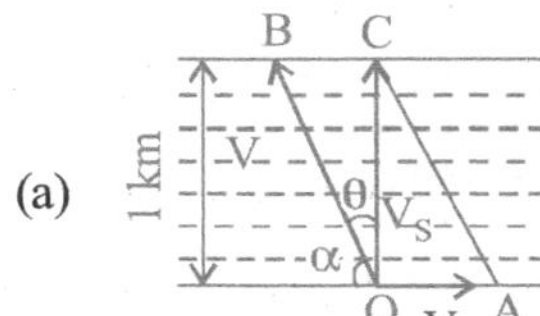

(b)

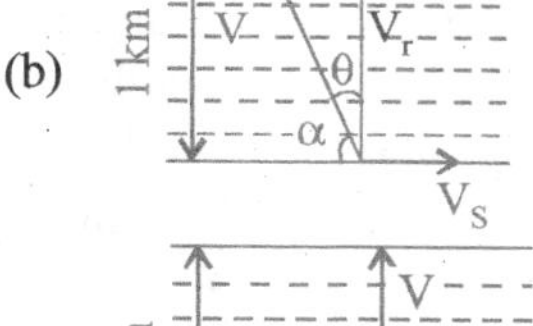

(c)

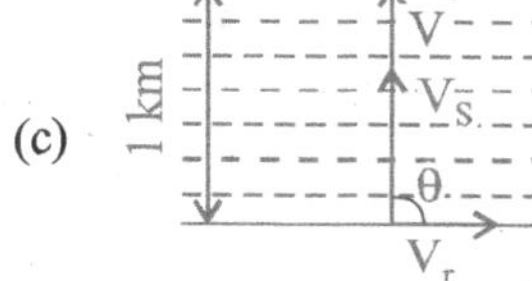

(d)

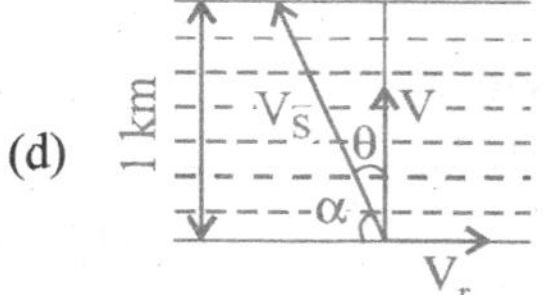

86. If V_r is the velocity of rain falling vertically and V_m is the velocity of a man walking on a level road, and θ is the angle with vertical at which he should hold the umbrella to protect himself, than the relative velocity of rain w.r.t. the man is given by:

(a) $V_{rm} = \sqrt{V_r^2 + V_m^2 + 2V_rV_m \cos\theta}$

(b) $V_{rm} = \sqrt{V_r^2 + V_m^2 - 2V_rV_m \cos\theta}$

(c) $V_{rm} = \sqrt{V_r^2 + V_m^2}$

(d) $V_{rm} = \sqrt{V_r^2 - V_m^2}$

87. An aircraft executes a horizontal loop of radius 1.00 km with a stedy speed of 900 km/h. The ratio of centripetal acceleration to acceleration due to gravity is [g = 9.8 m/s²]

(a) 6.38 (b) 9.98 (c) 11.33 (d) 12.13

88. A boat B is moving upstream with velocity 3 m/s with respect to ground. An observer standing on boat observes that a swimmer S is crossing the river perpendicular to the direction of motion of boat. If river flow velocity is 4 m/s and swimmer crosses the river of width 100 m in 50 sec, then

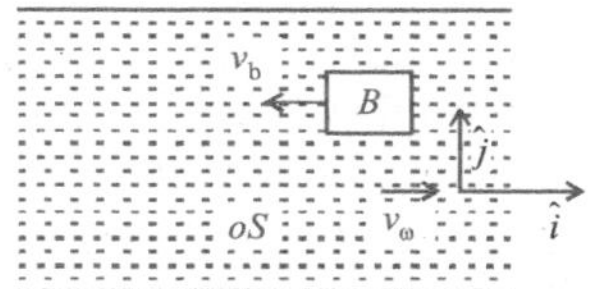

(a) velocity of swimmer w.r.t ground is $\sqrt{13}$ m/s
(b) drift of swimmer along river is zero
(c) drift of swimmer along river will be 50 m
(d) velocity of swimmer w.r.t ground is 2 m/s

89. Two particles A and B separated by a distance $2R$ are moving counter clockwise along the same circular path of radius R each with uniform speed v. At time $t = 0$, A is given a tangential acceleration of magnitude $a = \frac{72v^2}{25\pi R}$ then **Critical Thinking**

(a) the time lapse for the two bodies to collide is $\frac{6\pi R}{5v}$

(b) the angle covered by A is $11\pi/6$

(c) angular velocity of A is $\frac{11v}{5R}$

(d) radial acceleration of A is $289v^2/5R$

90. The length of second's hand in a watch is 1 cm. The change in velocity of its tip in 15 seconds is:

(a) zero (b) $\frac{\pi}{30\sqrt{2}}$ cm/s

(c) $\frac{\pi}{30}$ cm/s (d) $\frac{\pi\sqrt{2}}{30}$ cm/s

91. A stone tied to the end of a string of 1 m long is whirled in a horizontal circle with a constant speed. If the stone makes 22 revolution in 44 seconds, what is the magnitude and direction of acceleration of the stone?

(a) π^2 m s^{-2} and direction along the radius towards the centre.

(b) π^2 m s^{-2} and direction along the radius away from the centre.

(c) π^2 m s^{-2} and direction along the tangent to the circle.

(d) $\pi^2/4$ m s^{-2} and direction along the radius towards the centre.

92. A particle moves in a circle of radius 30 cm. Its linear speed is given by: $V = 2t$, where t in second and v in m/s. Find out its radial and tangential acceleration at t = 3 sec respectively.

(a) 220 m/sec^2, 50 m/sec^2

(b) 110 m/sec^2, 5 m/sec^2

(c) 120 m/sec^2, 2 m/sec^2

(d) 110 m/sec^2, 10 m/sec^2

93. Three points are located at the vertices of an equilateral triangle whose side equal to a. They all start moving simultaneously with velocity v constant in modulus, with first point heading continually for the second, the second for the third, and the third for the first. How soon will the points converge? **Tricky**

(a) $3v/2a$ (b) $2a/5v$ (c) $5v/3a$ (d) $2a/3v$

94. A man in a row boat must get from point A to point B on the opposite bank of the river (see figure). The distance BC = a. The width of the river AC = b. At what minimum speed u relative to the still water should the boat travel to reach the point B? The velocity of flow of the river is v_0.

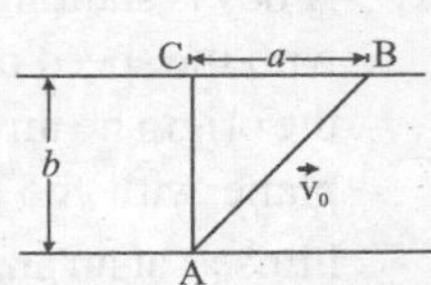

(a) $\sqrt{a^2 + b^2}/v_0$ (b) $\frac{v_0 b}{\sqrt{a^2 + b^2}}$

(c) $v_0 a/b$ (d) $v_0 b/a$

95. Rain, pouring down at an angle α with the vertical has a speed of 10 ms^{-1}. A girl runs against the rain with a speed of 8 ms^{-1} and sees that the rain makes an angle β with the vertical, then relation between α and β is

(a) $\tan\alpha = \frac{8 + 10\sin\beta}{10\cos\beta}$

(b) $\tan\beta = \frac{8 + 10\sin\alpha}{10\cos\alpha}$

(c) $\tan\alpha = \tan\beta$

(d) $\tan\alpha = \cot\beta$

96. A 2 m wide truck is moving with a uniform speed $v_0 = 8$ m/s along a straight horizontal road. A pedestrain starts to cross the road with a uniform speed v when the truck is 4 m away from him. The minimum value of v so that he can cross the road safely is **Critical Thinking**

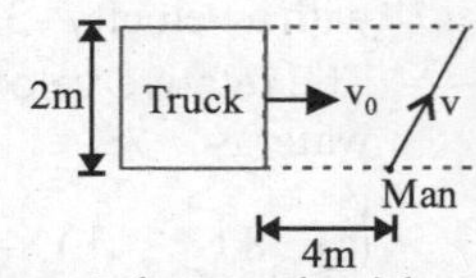

(a) 2.62 m/s (b) 4.6 m/s

(c) 3.57 m/s (d) 1.414 m/s

97. Two boats A and B, move away from a buoy anchored at the middle of a river along the mutually perpendicular straight lines : the boat A along the river and the boat B across the river. Having moved off an equal distance from the buoy the boat returned. What is the ratio of times of motion of boats $\frac{\tau_A}{\tau_B}$, if the velocity of each boat with respect to water is 1.2 times greater than the stream velocity. **Toughnut**

(a) 2.3 (b) 1.8 (c) 0.5 (d) 0.2

98. A balloon starts rising from the surface of the earth. The ascension rate is constant and equal to v_0. Due to the wind the balloon gathered the horizontal velocity component $v_x = ay$, where a is a constant and y is the height of ascent. The tangential, acceleration of the balloon is

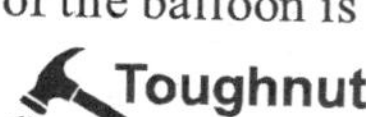

(a) a^2y/v_0 (b) $a^2y/\sqrt{1+(ay/v_0)^2}$

(c) $a^2y/\sqrt{1+v_0^2}$ (d) $a^2v_0/\sqrt{1+(2y/a)^2}$

99. The angle which the velocity vector of a projectile thrown with a velocity v at an angle θ to the horizontal will make with the horizontal after time t of its being thrown up is:

(a) θ

(b) $\tan^{-1}(\theta/t)$

(c) $\tan^{-1}\left(\frac{v\cos\theta}{v\sin\theta - gt}\right)$

(d) $\tan^{-1}\left(\frac{v\sin\theta - gt}{v\cos\theta}\right)$

100. Two identical particles are projected horizontally in opposite directions with a speed of 5 ms^{-1} each from the top of a tall tower as shown. Assuming $g = 10\ ms^{-2}$, the distance between them at the moment when their velocity vectors become mutually perpendicular is

(a) 2.5m

(b) 5m

(c) 10m

(d) 20m

Answer KEYs

1	(c)	11	(b)	21	(d)	31	(c)	41	(b)	51	(b)	61	(b)	71	(d)	81	(a)	91	(a)
2	(c)	12	(b)	22	(a)	32	(b)	42	(d)	52	(b)	62	(b)	72	(c)	82	(c)	92	(c)
3	(b)	13	(a)	23	(c)	33	(d)	43	(b)	53	(a)	63	(a)	73	(a)	83	(c)	93	(d)
4	(a)	14	(b)	24	(b)	34	(b)	44	(b)	54	(a)	64	(c)	74	(b)	84	(a)	94	(b)
5	(a)	15	(b)	25	(c)	35	(b)	45	(b)	55	(a)	65	(a)	75	(b)	85	(d)	95	(b)
6	(d)	16	(b)	26	(a)	36	(c)	46	(c)	56	(b)	66	(a)	76	(b)	86	(c)	96	(c)
7	(d)	17	(c)	27	(a)	37	(a)	47	(c)	57	(c)	67	(b)	77	(c)	87	(a)	97	(b)
8	(c)	18	(a)	28	(a)	38	(a)	48	(c)	58	(b)	68	(c)	78	(c)	88	(a)	98	(b)
9	(c)	19	(c)	29	(b)	39	(a)	49	(b)	59	(c)	69	(d)	79	(a)	89	(b)	99	(d)
10	(c)	20	(c)	30	(b)	40	(d)	50	(d)	60	(c)	70	(b)	80	(d)	90	(d)	100	(b)

Hints & Solutions

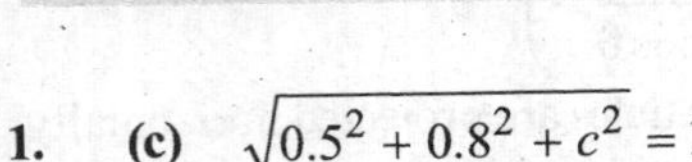

1. (c) $\sqrt{0.5^2 + 0.8^2 + c^2} = 1$

or $c = \sqrt{1 - 0.89} = \sqrt{0.11}$

2. (c) $13^2 = 12^2 + 5^2 + 2 \times 12 \times 5 \cos\theta$

$\therefore \quad \theta = 90°$

3. (b) At point B the direction of velocity component of the projectile along Y - axis reverses.

Hence, $\vec{V}_B = 2\hat{i} - 3\hat{j}$

4. (a) Required distance from the wall

$$= \frac{u^2 \sin 2\theta}{g} - u\cos\theta \times t = 50\sqrt{3}\text{ m}$$

5. (a) Range $= \dfrac{2u^2}{g\sin\theta} = 5\sqrt{2}$ m

6. (d) $\dfrac{H_1}{H_2} = \dfrac{u^2 \sin^2\theta / 2g}{u^2 \sin^2(90° - \theta)/2g} = \tan^2\theta$

7. (d) Time of flight $= \dfrac{2u\sin\theta}{g}$

$$= \frac{2 \times 9.8 \times \sin 30°}{9.8} = 2 \times \frac{1}{2} = 1\text{ sec.}$$

8. (c) We know that, $y_m = H = \dfrac{(u\sin\theta)^2}{2g}$

$$= \frac{u^2 \sin^2\theta}{2g}$$

$\therefore \dfrac{\Delta H}{H} = \dfrac{2\Delta u}{u}$. Given $\dfrac{\Delta u}{u} = 2\%$

$\therefore \dfrac{\Delta H}{H} = 2 \times 2 = 4\%$

9. (c) $y = x\tan\theta - \dfrac{1}{2}\dfrac{gx^2}{u^2\cos^2\theta}$

$$y = 50\tan 60° - \frac{10 \times 50 \times 50}{2 \times 25 \times 25 \times \cos^2 60°} = 5\text{m}$$

10. (c)

11. (b) The resultant vector

$$= (3\hat{i} + 4\hat{j} + 5\hat{k}) + (5\hat{i} + 3\hat{j} + 4\hat{k}) = 8\hat{i} + 7\hat{j} + 9\hat{k}$$

$$\cos\theta = \frac{8}{\sqrt{8^2 + 7^2 + 9^2}} = 0.574$$

12. (b) As $\vec{C} = \vec{A} + \vec{B}$ and $|\vec{C}| < |\vec{B}|$ $\therefore$ $|\vec{C}| < |\vec{A}|$

13. (a) Let the components of $\vec{A}$ make angles α, β, and γ with x, y and z-axis respectively then $\alpha = \beta = \gamma$ and

$\cos^2\alpha + \cos^2\beta + \cos^2\gamma = 1$

$\Rightarrow 3\cos^2\alpha = 1 \Rightarrow \cos\alpha = \dfrac{1}{\sqrt{3}}$

$\therefore A_x = A_y = A_z = A\cos\alpha = \dfrac{A}{\sqrt{3}}$

14. (b) Unit vector along y axis $= \hat{j}$, so the required vector

$= \hat{j} - [(\hat{i} - 3\hat{j} + 2\hat{k}) + (3\hat{i} + 6\hat{j} - 7\hat{k})]$

$= -4\hat{i} - 2\hat{j} + 5\hat{k}$

15. (b)

16. (b) $\dfrac{B}{2} = \sqrt{A^2 + B^2 + 2AB\cos\theta}$ ...... (i)

$\therefore \quad \tan 90° = \dfrac{B\sin\theta}{A + B\cos\theta} \Rightarrow A + B\cos\theta = 0$

$\therefore \quad \cos\theta = -\dfrac{A}{B}$

Hence, from (i) $\dfrac{B^2}{A} = A^2 + B^2 - 2A^2$

$\Rightarrow A = \sqrt{3}\dfrac{B}{2} \Rightarrow \cos\theta = -\dfrac{A}{B} = -\dfrac{\sqrt{3}}{2}$

$\therefore \theta = 150°$

17. (c) Given, $\vec{A} = \vec{B} - \vec{C}$

$\therefore \quad \vec{C} = \vec{B} - \vec{A}$

If θ is the angle between $\vec{A}$ and $\vec{B}$, then

$C^2 = B^2 + A^2 - 2AB\cos\theta$

$\therefore \cos\theta = \dfrac{B^2 + A^2 - C^2}{2AB}$

18. (a) We have $R^2 = P^2 + Q^2 + 2PQ\cos\theta$ ...(i)

and $S^2 = P^2 + Q^2 - 2PQ\cos\theta$ ...(ii)

Adding equations (i) and (ii), we get

$R^2 + S^2 = 2(P^2 + Q^2)$.

19. (c) $\vec{A} = 4\hat{i} + 6\hat{j}$. If B_x and B_y are the components of vector along x and y axes, then

$4 + B_x = 10, \quad \therefore \quad B_x = 6.$

Also $6 + B_y = 9, \quad \therefore \quad B_y = 3.$

Thus, $B = \sqrt{B_x^2 + B_y^2} = \sqrt{6^2 + 3^2} = \sqrt{45}$.

20. **(c)**

21. **(d)** $\vec{A}.\vec{B}=0$

or $(5\hat{i}+7\hat{j}-3\hat{k}).\ (2\hat{i}+2\hat{j}-a\hat{k})=0$

$\therefore$ $5\times2+7\times2+3a=0$ or $a=-8$

22. **(a)** $(\vec{A}+\vec{B}).(\vec{A}-\vec{B})=0$

or $\vec{A}.\vec{A}+\vec{B}.\vec{A}-\vec{A}.\vec{B}-\vec{B}.\vec{B}=0$

$\therefore$ $A=B.$

23. **(c)** Let $\vec{c}=\hat{a}+2\hat{b}$ and $\vec{d}=5\hat{a}-4\hat{b}$

Since $\vec{c}$ and $\vec{d}$ are perpendicular to each other

$\therefore$ $\vec{c}.\vec{d}=0 \Rightarrow \left(\hat{a}+2\hat{b}\right).\left(5\hat{a}-4\hat{b}\right)=0$

$\Rightarrow 5+6\hat{a}.\hat{b}-8=0$ $\qquad (\because \vec{a}.\vec{a}=1)$

$\Rightarrow \hat{a}.\hat{b}=\dfrac{1}{2} \Rightarrow \theta=\dfrac{\pi}{3}$

24. **(b)** $y = bx^2$. Differentiating w.r.t to t an both sides, we get

$$\frac{dy}{dx}=b2x\frac{dx}{dt} \Rightarrow v_y=2bxv_x$$

Again differentiating w.r.t to t on both sides we get

$$\frac{dv_y}{dt}=2bv_x\frac{dx}{dt}+2bx\frac{dv_x}{dt}=2bv_x^2+0$$

[$\dfrac{dv_x}{dt}=0$, because the particle has constant acceleration along y-direction]

Now, $\dfrac{dv_y}{dt}=a=2bv_x^2$;

$$v_x^2=\frac{a}{2b} \Rightarrow v_x=\sqrt{\frac{a}{2b}}$$

25. **(c)** $\vec{r}=2t^2\hat{i}+3t\hat{j}+4\hat{k}$

$$\therefore \vec{v}=\frac{d\vec{r}}{dt}=\frac{d}{dt}(2t^2\hat{i}+3t\hat{j}+4\hat{k})=4t\hat{i}+3\hat{j}$$

$$\vec{a}=\frac{d\vec{v}}{dt}=\frac{d}{dt}(4t\hat{i}+3\hat{j})=4\hat{i}$$

$\therefore \vec{a}=4\,ms^{-2}$ along x-direction

26. **(a)** $\vec{u}=3\hat{i}+4\hat{j} \Rightarrow ux=3$ and $uy=4$

$\vec{a}=0.4\hat{i}+0.3\hat{j} \Rightarrow ax=0.4$ and $ay=0.3$

So, $V_x = ux + a_xt = 3 + 0.4\times10 = 3 + 4 = 7$

$V_y = uy + a_yt = 4 + 0.3\times10 = 4 + 3 = 7$

$\vec{V}=7\hat{i}+7\hat{j}$

$\Rightarrow |\vec{V}|=\sqrt{7^2+7^2}=7\sqrt{2}$ units

27. **(a)**

28. **(a)** According to the question, at any instant t, $x=4t^2, y=3t^2$

$$\therefore v_x=\frac{dx}{dt}=\frac{d}{dt}(4t^2)=8t$$

and $v_y=\dfrac{dy}{dt}=\dfrac{d}{dt}(3t^2)=6t$

The speed of the particle at instant t.

$$v=\sqrt{v_x^2+v_y^2}=\sqrt{(8t)^2+(6t)^2}=10t$$

29. **(b)** In both the cases, the initial velocity in the vertical downward direction is zero. So they will hit the ground simultaneously.

30. **(b)** Two bodies will collide at the highest point if both cover the same vertical height in the same time.

So $\dfrac{V_1^2\sin^2 30°}{2g}=\dfrac{V_2^2}{2g}$

$\Rightarrow \dfrac{V_2}{V_1}=\sin 30°=\dfrac{1}{2}$ $\qquad \therefore V_2=\dfrac{1}{2}V_1$

31. **(c)** Scalar product is distributive over addition.

$\vec{i}\times\vec{j}=-\vec{j}\times\vec{i}$

32. **(b)** For the given conditions, vectors $\vec{A},\vec{B}$ and $\vec{C}$ are parallel, so $\vec{A}\times\vec{C}=AC\sin 0°=0$.

33. **(d)** $|\vec{A}\times\vec{B}|=\sqrt{3}.\vec{A}.\vec{B}$

or $AB\sin\theta=\sqrt{3}\ AB\cos\theta$

$\therefore$ $\tan\theta=\sqrt{3}$, or $\theta=60°$

Thus $|\vec{A}+\vec{B}|=\sqrt{A^2+B^2+2AB\cos 60°}$

$=\sqrt{A^2+B^2+AB}$.

34. **(b)** We have $\vec{a}\times\vec{b}=39\vec{k}=\vec{c}$

Also $|\vec{a}|=\sqrt{34}, |\vec{b}|=\sqrt{45}, |\vec{c}|=39$;

$\therefore |\vec{a}|:|\vec{b}|:|\vec{c}|=\sqrt{34}:\sqrt{45}:39$.

35. **(b)** We have, $\vec{a}.\vec{b}=|\vec{a}||\vec{b}|\cos\dfrac{\pi}{6}$

$=4\times2\times\dfrac{\sqrt{3}}{2}=4\sqrt{3}$.

Now, $(\vec{a}\times\vec{b})^2+(\vec{a}\cdot\vec{b})^2=a^2b^2$;

$\Rightarrow (\vec{a}\times\vec{b})^2+48=16\times4 \Rightarrow (\vec{a}\times\vec{b})^2=16$

36. **(c)** The area of the triangle

$$A=\frac{1}{2}|(\hat{i}+\hat{j}+\hat{k})\times(3\hat{i})|=\frac{3}{\sqrt{2}}.$$

37. (a) The required vector is,

$$= 3\frac{(\vec{A}\times\vec{B})}{|\vec{A}\times\vec{B}|} = 3\frac{[(2\hat{i}+2\hat{j}+\hat{k})\times(2\hat{i}-2\hat{j}+3\hat{k})]}{|\vec{A}\times\vec{B}|}$$

$$= 3\frac{(8\hat{i}-4\hat{j}-8\hat{k})}{\sqrt{8^2+4^2+8^2}} = 2\hat{i}-\hat{j}-2\hat{k}.$$

38. (a) Angle between $\vec{A}$ and $\vec{B}$ is given by

$$\cos\theta = \frac{\vec{A}\cdot\vec{B}}{AB} = \frac{3}{\sqrt{21}}$$

The unit vector perpendicular to $\vec{A}$ and $\vec{B}$ is given by

$$\hat{n} = \frac{\vec{A}\times\vec{B}}{|\vec{A}\times\vec{B}|} = \frac{(3\hat{i}+\hat{j}+2\hat{k})\times(2\hat{i}-2\hat{j}+4\hat{k})}{|(3\hat{i}+\hat{j}+2\hat{k})\times(2\hat{i}-2\hat{j}+4\hat{k})|}$$

$$= \frac{\hat{i}-\hat{j}-\hat{k}}{\sqrt{3}}$$

39. (a) $t_1 = \dfrac{2u\sin\theta}{g}$ and

$$t_2 = \frac{2u\sin(90-\theta)}{g} = \frac{2u\cos\theta}{g}$$

$$\therefore\ t_1 t_2 = \frac{4u^2\cos\theta\sin\theta}{g^2} = \frac{2}{g}\left[\frac{u^2\sin 2\theta}{g}\right]$$

$$= \frac{2}{g}R,$$

where R is the range. Hence $t_1 t_2 \propto R$

40. (d) $V_h = V\cos\theta$

where h is the maximum height

$$V_h \times h = (\upsilon\cos\theta)\left(\frac{\upsilon^2\sin^2\theta}{2g}\right)$$

$$= \frac{\upsilon^3\sin^2\theta\cos\theta}{2g} = \frac{\sqrt{3}\,\upsilon^3}{16g}$$

41. (b) Horizontal range $R = \dfrac{u^2\sin 2\theta}{g}$.

Range is same for angle of projection θ and (90° – θ).

42. (d) For complementary angles of projection (45° + α) and (45° – α) with same initial velocity u, range R is same.

$\theta_1 + \theta_2 = (45° + \alpha) + (45° - \alpha) = 90°$

43. (b) Comparing the given equation with

$$y = x\tan\theta - \frac{gx^2}{2u^2\cos^2\theta},$$ we get

$\tan\theta = \sqrt{3}$

44. (b) Horizontal distance covered should be same for the time of collision.

$400\cos\theta = 200$ or $\cos\theta = \dfrac{1}{2}$ or $\theta = 60°$

45. (b) Comparing the given equation with the equation of trajectory of a projectile,

$$y = x\tan\theta - \frac{gx^2}{2u^2\cos^2\theta}$$

we get, $\tan\theta = \dfrac{1}{\sqrt{3}} \Rightarrow \theta = 30°$

and $2u^2\cos^2\theta = 20 \Rightarrow u^2 = \dfrac{20}{2\cos^2\theta}$

$$= \frac{40}{3}$$

Now, $R_{max} = \dfrac{u^2}{g} = \dfrac{40}{3\times 10} = \dfrac{4}{3}$ m

46. (c) The bullet performs a horizontal journey of 100 cm with constant velocity of 1500 m/s. The bullet also performs a vertical journey of h with zero initial velocity and downward acceleration g.

∴ For horizontal journey, time (t) = $\dfrac{\text{Distance}}{\text{Velocity}}$

$$\therefore\ t = \frac{100}{1500} = \frac{1}{15}\text{ sec} \quad \ldots(1)$$

The bullet performs vertical journey for this time.

For vertical journey, $h = ut + \dfrac{1}{2}gt^2$

$$h = 0 + \frac{1}{2}\times 10\times\left(\frac{1}{15}\right)^2$$

or, $h = \dfrac{20}{9}$ cm = 2.2cm

47. (c) Velocity of projectile u = 147 ms^{-1}

angle of projection $\alpha = 60°$

Let, the time taken by the projectile from O to A be t where direction β = 45°. As horizontal component of velocity remains constant during the projectile motion.

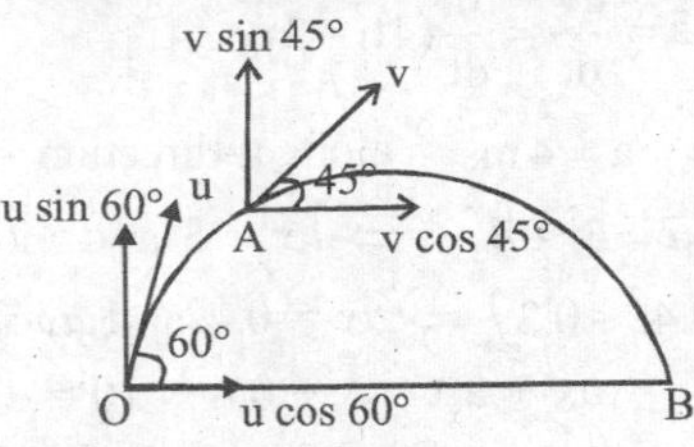

$\Rightarrow$ v cos 45° = u cos 60°

$$\Rightarrow\ v\times\frac{1}{\sqrt{2}} = 147\times\frac{1}{2} \Rightarrow v = \frac{147}{\sqrt{2}}\text{ms}^{-1}$$

For Vertical motion,

$v_y = u_y - gt$

$\Rightarrow \quad v\sin 45° = 45\sin 60° - 9.8\,t$

$\Rightarrow \quad \frac{147}{\sqrt{2}} \times \frac{1}{\sqrt{2}} = 147 \times \frac{\sqrt{3}}{2} - 9.8\,t$

$\Rightarrow \quad 9.8\,t = \frac{147}{2}(\sqrt{3}-1) \Rightarrow t = 5.49\text{ s}$

48. **(c)** max height, $H_A = \frac{u_A^2 \sin^2 30°}{2g}$;

$H_B = \frac{u_B^2 \sin^2 \theta}{2g}$

As we know, $H_A = H_B$

$\frac{u_A^2 \sin^2 30°}{2g} = \frac{u_B^2 \sin^2 \theta}{2g}$

$\Rightarrow \frac{\sin^2 \theta}{\sin^2 30°} = \frac{u_A^2}{u_B^2}$

$\sin^2 \theta = \left(\frac{u_A}{u_B}\right)^2 \sin^2 30° \Rightarrow \sin^2 \theta = \frac{1}{36}$

$\sin \theta = \frac{1}{6} \Rightarrow \theta = \sin^{-1}\left(\frac{1}{6}\right)$

49. **(b)** As we know, $R = u\cos\theta \times t$

Given, $R = 300$ m, $t = 6$ s

$\therefore \; u\cos\theta = \frac{R}{t} = \frac{300}{6} = 50\text{ms}^{-1}$

50. **(d)** $R = \frac{u^2 \sin^2 \theta}{g}$, $H = \frac{u^2 \sin^2 \theta}{2g}$

H_{max} at $2\theta = 90°$ So, $H_{max} = \frac{u^2}{2g}$

$\frac{u^2}{2g} = 10 \Rightarrow u^2 = 10g \times 2$

$R = \frac{u^2 \sin 2\theta}{g} \Rightarrow R_{max} = \frac{u^2}{g} = 20$ metre

51. **(b)** From equation, $\vec{v} = \hat{i} + 2\hat{j}$

$\Rightarrow \quad x = t \qquad \ldots$(i)

$y = 2t - \frac{1}{2}(10t^2) \qquad \ldots$(ii)

From (i) and (ii), $y = 2x - 5x^2$

52. **(b)** Let the two bodies meet each other at a height h after time T of the projection of second body. Then before meeting, the first body was in motion for time $(t+T)$ whereas the second body was in motion for time T.

The distance moved by the first body in time $(t+T)$

$= u(t+T) - \frac{1}{2}g(t+T)^2.$

And the distance moved by the second body in time $T = vT - \frac{1}{2}gT^2 = h$ (supposed above). ....(1)

$\because$ The two bodies meet each other,

$\therefore$ They are equidistant from the point of projection.

Hence, $u(t+T) - \frac{1}{2}g(t+T)^2 = vT - \frac{1}{2}gT^2$

or $gt^2 + 2t(gT-u) + 2(v-u)T = 0$.........(2)

Also from (1) we get, $h = vT - \frac{1}{2}gT^2$

$\therefore \quad \frac{dh}{dT} = v - gT$

$\therefore \quad h$ increases as T increases

$\therefore \quad T$ is minimum when h is minimum i.e., when

$\frac{dh}{dT} = 0$, i.e. when $v - gT = 0$ or $T = v/g$.

Substituting this value of T in (2), we get

$gt^2 + 2t(v-u) + 2(v-u)(v/g) = 0$

or $t = \frac{2g(u-v) + \sqrt{4g^2(u-v)^2 + 8vg^2(u-v)}}{2g^2}$

or $t = \frac{u - v + \sqrt{u^2 - v^2}}{g}$

neglecting the negative sign which gives negative value of t.

53. **(a)** We know,

$R = 4H\cot\theta \Rightarrow \cot\theta = \frac{1}{2}$

From triangle we can say that

$\sin\theta = \frac{2}{\sqrt{5}}$, $\cos\theta = \frac{1}{\sqrt{5}}$

$\sqrt{5}$ 2 θ 1

$\therefore$ Range of projectile $R = \frac{2v^2 \sin\theta\cos\theta}{g}$

$= \frac{2v^2}{g} \times \frac{2}{\sqrt{5}} \times \frac{1}{\sqrt{5}} = \frac{4v^2}{5g}$

54. **(a)** Horizontal range $= \frac{u^2 \sin 2\theta}{g}$ so $g \propto u^2$

or $\frac{g_{planet}}{g_{earth}} = \frac{(u_{planet})^2}{(u_{earth})^2}$

Therefore $g_{planet} = \left(\frac{3}{5}\right)^2 (9.8\ m/s^2)$

$= 3.5\ m/s^2$

55. **(a)** R is same for both θ and (90° – θ). If angle w.r.t. vertical is 40° then w.r.t. horizontal direction it will be 90° – 40° = 50°.

56. **(b)** $t = \frac{2u \sin(\alpha - \beta)}{g \cos\beta}$; $u = \frac{gt \cos\beta}{2 \sin(\alpha - \beta)}$

57. **(c)** At the two points of the trajectory during projectile motion, the horizontal component of the velocity is same. Then u cos 60° = v cos 45°.

$\Rightarrow 150 \times \frac{1}{2} = v \times \frac{1}{\sqrt{2}}$ or $\frac{150}{\sqrt{2}}$ m/s

Initially: $u_y = u \sin 60° = \frac{150\sqrt{3}}{2}$ m/s

Finally: $v_y = v \sin 45° = \frac{150}{\sqrt{2}} \times \frac{1}{\sqrt{2}} = \frac{150}{2}$ m/s

But $v_y = u_y + a_y t$ or $\frac{150}{2} = \frac{150\sqrt{3}}{2} - 10t$

$10t = \frac{150}{2}\left(\sqrt{3} - 1\right)$ or $t = 7.5\left(\sqrt{3} - 1\right) s$

58. **(b)** Horizontal components of velocity at O and P are equal.

$\therefore v \cos(90° - \theta) = u \cos\theta$

or $v \sin\theta = u \cos\theta$

or $v = u \cot\theta$

At P, $\frac{v_T^2}{R} = a_c$;

$\frac{u^2 \cot^2\theta}{g \sin\theta} = R$

59. **(c)** If the ball hits the nth step, then the horizontal distance traversed = nh. Here, the velocity along the horizontal direction = u. Initial velocity along the vertical direction = 0.

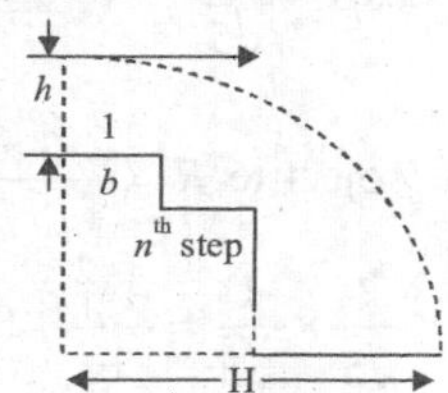

So $nb = ut$ ...(i)

$nh = 0 + \frac{1}{2}gt^2$ ...(ii)

From $t = \frac{nb}{v}$, putting in eq. (ii)

$nh = \frac{1}{2}g \times \left(\frac{nb}{u}\right)^2$ or $n = \frac{2hu^2}{gb^2}$

60. **(c)** $x = v\sqrt{\frac{2h}{g}}, 2x = v'\sqrt{\frac{2(2h)}{g}} = \sqrt{2}v$

61. **(b)** Time taken for vertical direction motion

$t = \sqrt{\frac{2h}{g}} = \sqrt{\frac{2 \times 490}{9.8}} = \sqrt{100} = 10\ s$

The same time is for horizontal direction.

$\therefore x = vt = \left(60 \times \frac{5}{18}\right) \times 10 = \frac{500}{3}$ m

62. **(b)**

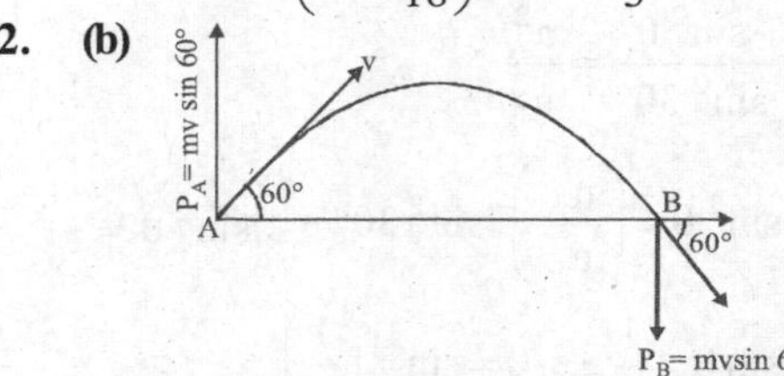

As the figure drawn above shows that at points A and B the vertical component of velocity is v sin 60° but their directions are opposite. Hence, change in velocity.

$\Delta v = v \sin 60° - (-v \sin 60°) = 2v \sin 60° = \sqrt{3}\,v$

63. **(a)** $H = \frac{u_\perp^2}{2g_\perp} \Rightarrow H = \frac{10 \sin 37° \times 10 \sin 37°}{2 \times g \cos 53°} = 3m$

64. **(c)** $500 \cos\theta = 250 \Rightarrow \cos\theta = \frac{1}{2}$ or $\theta = 60°$.

65. **(a)**

66. **(a)** $h_1 = u \sin\theta\, t_1 + \frac{1}{2}gt_1^2$;

$h_2 = u \sin\theta\, t_2 + \frac{1}{2}gt_2^2$

So, $\frac{t_1}{t_2} = \frac{h_1 + \frac{1}{2}gt_1^2}{h_2 + \frac{1}{2}gt_2^2}$

$\Rightarrow h_1 t_2 - h_2 t_1 = \frac{1}{2}g\left(t_1 t_2^2 - t_1^2 t_2\right)$

Time of flight $= 2u \sin\theta / g = \frac{h_1 t_2^2 - h_2 t_1^2}{h_1 t_2 - h_2 t_1}$

[Use above eqn to simplify]

67. (b) $x = 3t_1 = 5t_2 \Rightarrow t_1 = x/3$ and $t_2 = x/5$

Now $h_1 = 4t_1 - \frac{1}{2}gt_1^2 = \frac{4}{3}x - \frac{gx^2}{18}$.

and $h_2 = 4t_2 - \frac{1}{2}gt_2^2 = \frac{4x}{5} - \frac{gx^2}{10}$

Clearly, $h_2 < h_1$.

68. (c) The time to reach the maximum height without air resistance; $t_1 = \frac{u}{g} = \frac{u}{10}$,

and with air resistance, $t_2 = \frac{u}{10+1} = \frac{u}{11}$

$$= \frac{t_1 - t_2}{t_1} \times 100 = \left[\frac{\frac{u}{10} - \frac{u}{11}}{\frac{u}{10}}\right] \times 100$$

= 9 % (decreases)

69. (d) $\tan\theta = \frac{u^2}{Rg}$

$$\Rightarrow R = \frac{u^2}{g\tan\theta} = \frac{100}{10 \times \sqrt{3}} = \frac{10}{\sqrt{3}} \text{ m}$$

70. (b) $x = h\cot\theta$

$\therefore \frac{dx}{dt} = h\frac{d(\cot\theta)}{dt}$

$= h(-\text{cosec}^2\theta)\frac{d\theta}{dt}$

or $v = h\,\text{cosec}^2\theta\left(\frac{-d\theta}{dt}\right)$

$= (8 \times 10^3)\,\text{cosec}^2 60° \times 0.025 = 266.67$ m/s

$= 960$ km/h

71. (d) Given $h = Ax - Bx^2$, on comparing with

$y = x\tan\theta - \frac{gx^2}{2u^2\cos^2\theta}$, we get

$A = \tan\theta = \tan 45° = 1$,

and $B = \frac{g}{2u^2\cos^2\theta} = \frac{10}{2 \times 20^2 \times \cos^2 45°} = \frac{1}{40}$

$\therefore \frac{A}{B} = 40$.

72. (c) From question,

Horizontal velocity (initial), $u_x = \frac{40}{2} = 20$ m/s

Vertical velocity (initial), $50 = u_y t + \frac{1}{2}gt^2$

$\Rightarrow u_y \times 2 + \frac{1}{2}(-10) \times 4$; or, $u_y = \frac{70}{2} = 35$ m/s ;

$\therefore \tan\theta = \frac{u_y}{u_x} = \frac{35}{20} = \frac{7}{4} \Rightarrow$ Angle $\theta = \tan^{-1}\frac{7}{4}$

73. (a) $150 = 100t + \frac{1}{2}10t^2$; $t = 10$ sec

$x = \frac{400}{3}t = \frac{4000}{3}$ m

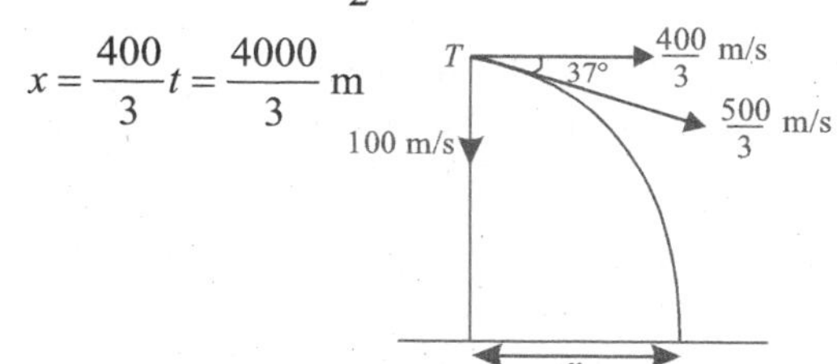

74. (b) $10 - v\cos 60° = 0 \Rightarrow v = 20$ m/s

$$H = \frac{v^2\sin^2 60°}{2g} = 15\text{m}$$

75. (b)

It can be observed from figure that P and Q shall collide if the initial component of velocity of P on inclined plane i.e., along incline $u_{II} = 0$ that is particle is projected perpendicular to incline. Time of flight on an inclined plane of inclination α is given by

$T = \frac{2u\sin(\theta - \alpha)}{g\cos\alpha}$,

$\Rightarrow 4 = \frac{2u\sin 90°}{10 \times \frac{1}{2}} \Rightarrow u = 10$ m/s

76. (b) Maximum possible horizontal range $= v^2/g$

Maximum possible area of the circle

$= \pi\left(\frac{v^2}{g}\right)^2 = \frac{\pi v^4}{g^2}$ $\left[\text{Here } r = \frac{v^2}{g}\right]$

77. (c) $x + u_2\cos\theta_2 t = u_1\cos\theta_1 t$

$\therefore t = \frac{x}{u_1\cos\theta_1 - u_2\cos\theta_2}$ ...(i)

Also $u_1\sin\theta_1 = u_2\sin\theta_2$ ...(ii)

After solving above equations, we get

$t = \frac{x\sin\theta_2}{u_1\sin(\theta_2 - \theta_1)}$.

78. (c) Using equation to trajectory

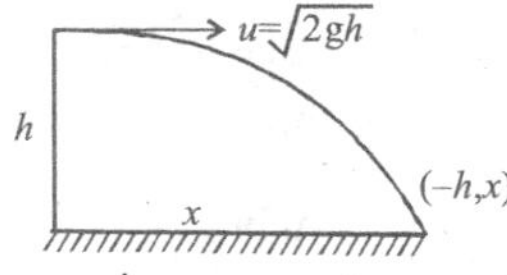

$$-h = x\tan(0°) - \frac{gx^2}{2(2gh)(\cos^2 0°)}$$

$$\Rightarrow x = 2h$$

79. (a) **For A:** It goes up with velocity u will it reaches its maximum height (i.e. velocity becomes zero) and comes back to O and attains velocity u.

Using $v^2 = u^2 + 2as \Rightarrow v_A = \sqrt{u^2 + 2gh}$

For B, going down with velocity u

$$\Rightarrow v_B = \sqrt{u^2 + 2gh}$$

For C, horizontal velocity remains same, i.e. u.

Vertical velocity $= \sqrt{0 + 2gh} = \sqrt{2gh}$

The resultant $v_C = \sqrt{v_x^2 + v_y^2} = \sqrt{u^2 + 2gh}$.

Hence $v_A = v_B = v_C$

80. (d) Ranges for complementary angles are same

$$\therefore \text{ Required angle} = \frac{\pi}{2} - \frac{5\pi}{36} = \frac{13\pi}{36}$$

81. (a) From figure,

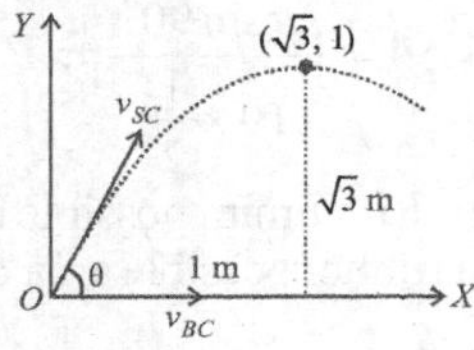

$$\tan\theta = \frac{\sqrt{3}}{1} = \sqrt{3} \Rightarrow \theta = 30°$$

82. (c) If t is the time of flight, then

$$0 = vt - \frac{1}{2}(g\cos\theta)t^2 \Rightarrow t = \frac{2v}{g}\, g\cos\theta.$$

$$R = 0 + \frac{1}{2}(g\sin\theta)t^2 = \frac{1}{2}g\sin\theta\left(\frac{2v}{g\cos\theta}\right)^2$$

$$= \frac{2v^2}{g}\tan\theta\sec\theta$$

83. (c) Relative velocity

$$= (3\hat{i} + 4\hat{j}) - (-3\hat{i} - 4\hat{j}) = 6\hat{i} + 8\hat{j}.$$

84. (a) Speed along the shortest path

$$= \frac{1}{15/60} = 4\text{km/hr}$$

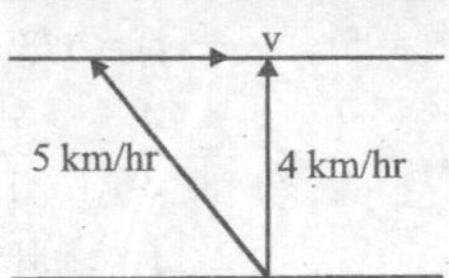

Speed of water $v = \sqrt{5^2 - 4^2} = 3$ km/hr

85. (d) The swimmer will cross straight if the resultant velocity of river flow and swimmer acts perpendicular to the direction of river flow. It will be so if the swimmer moves making an angle α with the upstream. i.e. goes along OB.

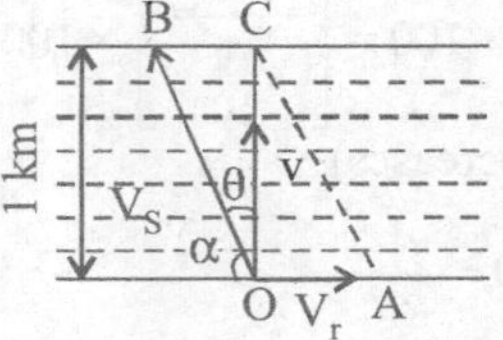

86. (c) According to pythagorus theorem

$$V_{rm} = \sqrt{V_r^2 + V_m^2}$$

87. (a) $a_c = \frac{v^2}{r} = \frac{(250)^2}{10^3} = 62.5 \text{ m/s}^2$

$$\Rightarrow a_c/g = \frac{62.5}{9.8} = 6.38$$

88. (a) $\vec{v}_{SB} = v\hat{j} = \vec{v}_s + 3\hat{i}$

$$\vec{v}_S = v\hat{j} - 3\hat{i} \text{ and } v = \frac{100}{50} = 2\text{m/s}$$

$$\therefore \left|\vec{v}_S\right| = \sqrt{v^2 + (3)^2} = \sqrt{2^2 + 9} = \sqrt{13} \text{ m/s}$$

Drift $= 50 \times 3 = 150$ m

89. (b) As when they collide

$$vt + \frac{1}{2}\left(\frac{72v^2}{25\pi R}\right)t^2 - \pi R = vt$$

$$\therefore t = \frac{5\pi R}{6v}$$

Now, angle covered by $A = \pi + \dfrac{vt}{R}$

Put t, $\therefore$ angle covered by A $= \dfrac{11\pi}{6}$

90. (d) $\Delta v = \sqrt{2}v = \sqrt{2}\omega r = \sqrt{2}\left(\dfrac{2\pi}{60}\right) \times 1 = \dfrac{\pi\sqrt{2}}{30}$ cm/s

91. (a) $a_r = \omega^2 R$

$a_r = (2\pi 2)^2 R = 4\pi^2 2^2 R$

$= 4\pi^2\left(\dfrac{22}{44}\right)^2 (1) \quad \left[\because v = \dfrac{22}{44}\right]$

$a_t = \dfrac{dv}{dt} = 0$

$a_{net} = a_r = \pi^2$ ms^{-2} and direction along the radius towards the centre.

92. (c) Given : r = 30 cm = 0.3 cm = 0.3 m and V = 2t

Radial acceleration at t = 3 sec

$$a_r = \frac{v^2}{r} = \frac{4t^2}{0.3} = \frac{4 \times (3)^2}{0.3} = 120 \text{ m/s}^2$$

and tangential acceleration

$$a_t = \frac{dv}{dt} = 2 \text{ m/s}^2$$

93. (d) The motions of the points are sketched in the figure. As they start moving simultaneously symmetrically, they will meet at the centroid of the triangle.

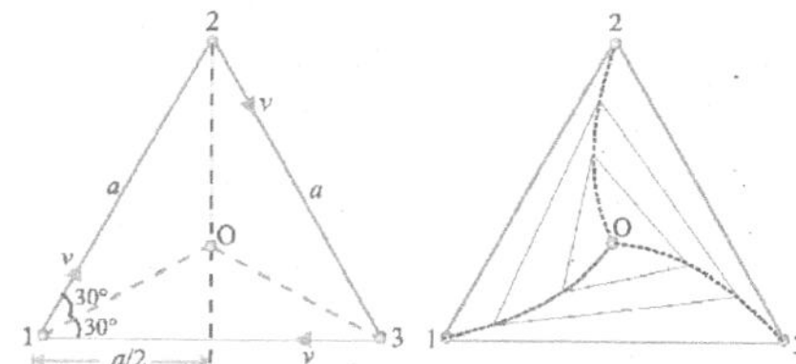

The velocity of any point towards centroid of triangle O

$= v\cos 30° = \dfrac{\sqrt{3}v}{2}$.

And its displacement $= \dfrac{a/2}{\cos 30°} = \dfrac{a}{\sqrt{3}}$.

Time taken to converge the points

$= \dfrac{\text{displacement}}{\text{velocity}} = \dfrac{2a}{3v}$.

94. (b) $\dfrac{v_0 b}{\sqrt{a^2 + b^2}}$

95. (b) $\tan\beta = \dfrac{v_{girl} + v_{rain}\sin\alpha}{v_{rain}\cos\alpha}$

$\Rightarrow \tan\beta = \dfrac{8 + 10\sin\alpha}{10\cos\alpha}$

96. (c) Let the man starts crossing the road at an angle θ as shown in figure. For safe crossing the condition is that the man must cross the road by the time the truck describes the distance $4 + AC$ or $4 + 2\cot\theta$.

$$\therefore \frac{4 + 2\cot\theta}{8} = \frac{2/\sin\theta}{v} \text{ or } v = \frac{8}{2\sin\theta + \cos\theta} \quad \text{....(i)}$$

For minimum v, $\dfrac{dv}{d\theta} = 0$

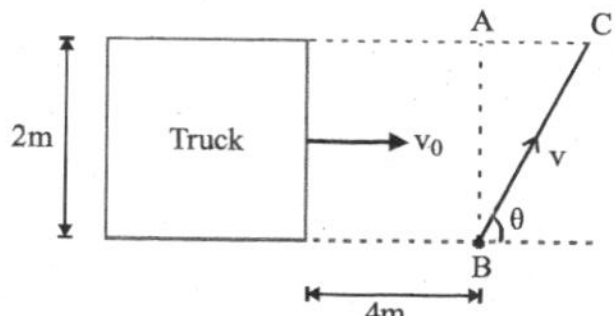

or $\dfrac{-8(2\cos\theta - \sin\theta)}{(2\sin\theta + \cos\theta)^2} = 0$

or $2\cos\theta - \sin\theta = 0$ or $\tan\theta = 2$

From equation (i),

$$v_{min} = \frac{8}{2\left(\frac{2}{\sqrt{5}}\right) + \frac{1}{\sqrt{5}}} = \frac{8}{\sqrt{5}} = 3.57 \text{ m/s}$$

97. (b) Suppose the stream velocity is $v_s = v$, then the velocity of each boat with respect to water is $v_b = 1.2\,v$. Let each boat travel a distance ℓ. Then for boat A, time of motion

$$\tau_A = \frac{\ell}{v_b + v_s} + \frac{\ell}{v_b - v_s}$$

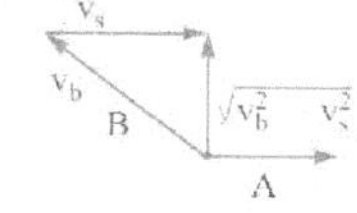

$$= \left[\frac{\ell}{1.2v + v} + \frac{\ell}{1.2v - v}\right] = \frac{60\ell}{11v} \quad \text{....(i)}$$

For the boat B, time of motion

$$\tau_B = \frac{\ell}{\sqrt{v_b^2 - v_s^2}} + \frac{\ell}{\sqrt{v_b^2 - v_s^2}} = \frac{2\ell}{\sqrt{v_b^2 - v_s^2}} \quad \text{....(ii)}$$

$$= \frac{2\ell}{\sqrt{(1.2v)^2 - v^2}} = \frac{3.01\ell}{v}$$

The ratio $\dfrac{\tau_A}{\tau_B} = \dfrac{(60\ell/11v)}{(3.01\ell/v)} \approx 1.8$

98. (b) Since velocity in vertical direction is constant,

$\therefore \quad a_y = \dfrac{dv_y}{dt} = 0$

The acceleration in horizontal direction,

$$a_x = \frac{dv_x}{dt} = \frac{d(av_0 t)}{dt} = av_0$$

$$a = \sqrt{a_x^2 + a_y^2} = \sqrt{(av_0)^2 + 0} = av_0$$

The total acceleration is av_0 and directed along horizontal direction.

Let θ is the angle that the resultant velocity makes with horizontal, then

Normal acceleration $a_n = a \sin\theta$ and tangential acceleration

$a_t = a \cos\theta$, we have $x = \frac{ay^2}{2v_0}$

or $y = \sqrt{\frac{2xv_0}{a}}$

Differentiating both si of equation (iii) w.r.t. x, we get

$$1 = \frac{a}{2v_0} \times 2y \times \frac{dy}{dx}$$

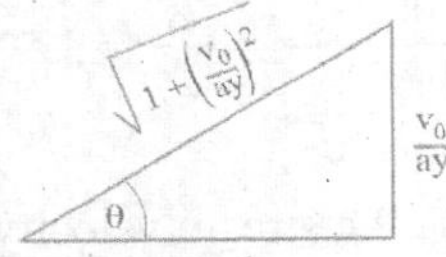

or $\frac{dy}{dx} = \frac{v_0}{ay} = \tan\theta$

Now $a_x = a \sin\theta = av_0 \times \frac{(v_0/ay)}{\sqrt{1+\left(\frac{v_0}{ay}\right)^2}}$

$$= \frac{av_0}{\sqrt{1+\left(\frac{ay}{v_0}\right)^2}}$$

$$a_t = a\cos\theta = av_0 \times \frac{1}{\sqrt{1+\left(\frac{v_0}{ay}\right)^2}} = a\,v_0$$

$$\frac{ay}{\sqrt{(ay)^2 + v_0^2}} = \frac{a^2 y}{\sqrt{1+\left(\frac{ay}{v_0}\right)^2}}$$

99. **(d)** Horizontally after time t

$v\cos\theta = v\cos\beta$ ..... (i)

[β = angle with horizontal after time t]

Vertically,

$v\sin\theta - gt = v\sin\beta$ ..... (ii)

Dividing eq^n (ii)/(i) we get

$$\tan\beta = \frac{v\sin\theta - gt}{v\cos\theta} \Rightarrow \beta = \tan^{-1}\left(\frac{v\sin\theta - gt}{v\cos\theta}\right)$$

100. **(b)** At a time t when velocity vector become mutually perpendicular

v cos 45° = 5 horizontal component

$$v = \frac{5}{\cos 45°} = 5\sqrt{2} \text{ m/s}$$

Vertically,

v sin 45° = gt

$$\Rightarrow t = \frac{v\sin 45°}{g} = \frac{5\sqrt{2} \times 1/\sqrt{2}}{10} = \frac{5}{10} = \frac{1}{2}$$

so, $OA = OB = v$ cos 45° × t = 5 × $\frac{1}{2}$ = 2.5

$\Rightarrow$ AB = 2.5 × 2 = 5 m

4 Laws of Motion

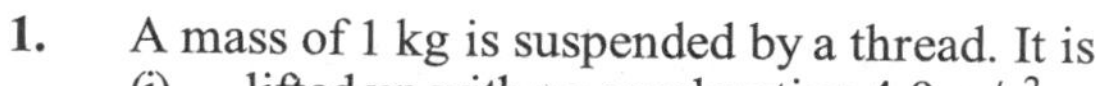

1. A mass of 1 kg is suspended by a thread. It is
(i) lifted up with an acceleration 4.9 m/s^2,
(ii) lowered with an acceleration 4.9 m/s^2.
The ratio of the tensions is
(a) 3 : 1 (b) 1 : 2 (c) 1 : 3 (d) 2 : 1

2. A monkey is decending from the branch of a tree with constant acceleration. If the breaking strength is 75% of the weight of the monkey, the minimum acceleration with which monkey can slide down without breaking the branch is

(a) g (b) $\frac{3g}{4}$ (c) $\frac{g}{4}$ (d) $\frac{g}{2}$

3. In the figure shown, the pulleys and strings are massless. The acceleration of the block of mass $4m$ just after the system is released from rest is ($\theta = \sin^{-1} 3/5$)

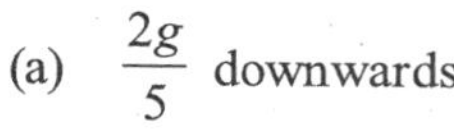

(a) $\frac{2g}{5}$ downwards
(b) $\frac{2g}{5}$ upwards

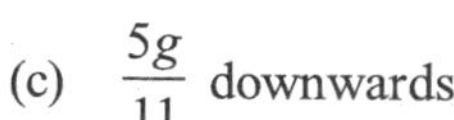

(c) $\frac{5g}{11}$ downwards (d) $\frac{5g}{11}$ upwards

4. Two blocks A and B of masses 3 m and m respectively are connected by a massless and inextensible string. The whole system is suspended by a massless spring as shown in figure.

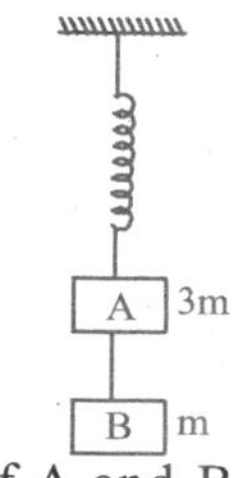

The magnitudes of acceleration of A and B immediately after the string is cut, are respectively:

Critical Thinking

(a) $\frac{g}{3}, g$ (b) g, g (c) $\frac{g}{3}, \frac{g}{3}$ (d) $g, \frac{g}{3}$

5. One end of string of length l is connected to a particle of mass 'm' and the other end is connected to a small peg on a smooth horizontal table. If the particle moves in circle with speed 'v' the net force on the particle (directed towards centre) will be (T represents the tension in the string) :-

(a) $T + \frac{mv^2}{l}$ (b) $T - \frac{mv^2}{l}$
(c) Zero (d) T

6. Three blocks A, B and C of masses 4 kg, 2 kg and 1 kg respectively, are in contact on a frictionless surface, as shown. If a force of 14 N is applied on the 4 kg block then the contact force between A and B is
(a) 6 N (b) 8 N
(c) 18 N (d) 2 N

7. A balloon with mass 'm' is descending down with an acceleration 'a' (where $a < g$). How much mass should be removed from it so that it starts moving up with an acceleration 'a'?
(a) $\frac{2ma}{g+a}$ (b) $\frac{2ma}{g-a}$ (c) $\frac{ma}{g+a}$ (d) $\frac{ma}{g-a}$

8. A bob is hanging over a pulley inside a car through a string. The second end of the string is in the hand of a person standing in the car.

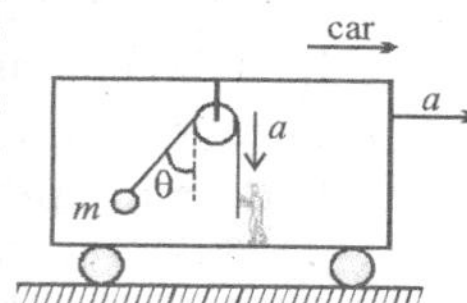

The car is moving with constant acceleration a directed horizontally as shown in figure. Other end of the string is pulled with constant acceleration a vertically. The tension in the string is equal to

Critical Thinking

(a) $m\sqrt{g^2 + a^2}$ (b) $m\sqrt{g^2 + a^2} - ma$
(c) $m\sqrt{g^2 + a^2} + ma$ (d) $m(g + a)$

9. Five persons A, B, C, D and E are pulling a cart of mass 100 kg on a smooth surface and cart is moving with acceleration 3 m/s^2 in east direction. When person A stops pulling, it moves with acceleration 1 m/s^2 in the west direction. When person B stops pulling, it moves with acceleration 24 m/s^2 in the north direction. The magnitude of acceleration of the cart when only A and B pull the cart keeping their directions same as the old directions, is
(a) 26 m/s^2 (b) $3\sqrt{71}$ m/s^2
(c) 30 m/s^2 (d) 25 m/s^2

10. A particle of mass 0.3 kg subject to a force $F = -kx$ with $k = 15$ N/m . What will be its initial acceleration if it is released from a point 20 cm away from the origin ?
(a) 15 m/s^2 (b) 3 m/s^2
(c) 10 m/s^2 (d) 5 m/s^2

11. If a stone is thrown out of an accelerated train, then acceleration of the stone at any instant depends on
(a) force acting on it at that instant
(b) acceleration of the train
(c) Both (a) & (b)
(d) None of these

12. A 600 kg rocket is set for a vertical firing. If the exhaust speed is 1000 ms^{-1} , the mass of the gas ejected per second to supply the thrust needed to overcome the weight of rocket is
(a) 117.6 kg s^{-1} (b) 58.6 kg s^{-1}
(c) 6 kg s^{-1} (d) 76.4 kg s^{-1}

13. An object of mass 10 kg moves at a constant speed of 10 ms^{-1}. A constant force, that acts for 4 sec on the object, gives it a speed of 2 ms^{-1} in opposite direction. The force acting on the object is
(a) –3 N (b) –30 N (c) 3 N (d) 30 N

14. A player stops a football weighting 0.5 kg which comes flying towards him with a velocity of 10m/s. If the impact lasts for 1/50th sec. and the ball bounces back with a velocity of 15 m/s, then the average force involved is
(a) 250 N (b) 1250 N (c) 500 N (d) 625 N

15. A truck accelerates on a horizontal road due to the force exerted by the
(a) road (b) engine (c) earth (d) driver

16. A block of mass 4 kg is suspended through two light spring balances A and B. Then A and B will read respectively :

(a) 4 kg and zero kg
(b) zero kg and 4 kg
(c) 4 kg and 4 kg
(d) 2 kg and 2 kg

17. A block of mass m is placed on a smooth wedge of inclination θ. The whole system is accelerated horizontally so that the block does not slip on the wedge. The force exerted by the wedge on the block (g is acceleration due to gravity) will be
(a) mg/cos θ (b) mg cos θ
(c) mg sin θ (d) mg

18. If n bullets each of mass m are fired with a velocity v per second from a machine gun, the force required to hold the gun in position is
(a) $(n+1)\,mv$ (b) $\frac{mv}{n^2}$
(c) $\frac{mv}{n}$ (d) mnv

19. A ball of mass 10 g moving perpendicular to the plane of the wall strikes it and rebounds in the same line with the same velocity. If the impulse experienced by the wall is 0.54 Ns, the velocity of the ball is
(a) 27 ms^{-1} (b) 3.7 ms^{-1}
(c) 54 ms^{-1} (d) 37 ms^{-1}

20. Two bodies of masses 1 kg and 2 kg moving with same velocities are stopped by the same force. Then the ratio of their stopping distances is
(a) 1 : 2 (b) 2 : 1 (c) $\sqrt{2}:1$ (d) $1:\sqrt{2}$

21. A hammer weighing 3 kg strikes the head of a nail with a speed of 2 ms^{-1} drives it by 1 cm into the wall. The impulse imparted to the wall is
(a) 6Ns (b) 3Ns (c) 2Ns (d) 12 Ns

22. The force required to stop a car of mass 800 kg, moving at a speed of 20 ms^{-1} over a distance of 25 m in 2.5 sec is
(a) 1200 N (b) 6400 N (c) 1600 N (d) 1800 N

23. A boy, sitting on the topmost birth in the compartment of a train which is just going to stop on the railway station, drops an apple aiming at the open hand of his brother situated vertically below his own hand at a distance of 2m. The apple will fall
(a) in the hand of his brother
(b) slightly away from the hand of his brother in the direction of motion of the train
(c) slightly away from the hand of his brother opposite to the direction of motion of the train
(d) None of the above

24. A shopper pushes a shopping cart of a store with a constant force of 75 N [forward]. The shopping cart exerts a force of 75 N [backward] on the shopper
(a) only if the velocity of the cart is constant.
(b) only if there is no friction between the cart and the floor.
(c) only if the velocity of the cart is increasing.
(d) under all circumstances assuming the system to be the shopper and cart.

25. A block is placed on a rough horizontal plane. A time dependent horizontal force F = kt acts on the block, where k is a positive constant. The acceleration - time graph of the block is :

Tricky

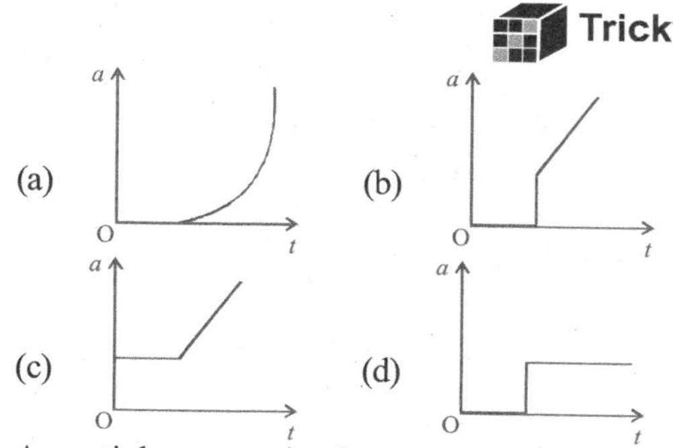

26. A particle moves in the *X-Y* plane under the influence of a force such that its linear momentum is $\vec{p}(t) = A\,[\hat{i}\cos(kt) - \hat{j}\sin(kt)]$, where *A* and *k* are constants. The angle between the force and the momentum is
(a) 0° (b) 30° (c) 45° (d) 90°

27. A stone is dropped from a height *h*. It hits the ground with a certain momentum *P*. If the same stone is dropped from a height 100% more than the previous height, the momentum when it hits the ground will change by :
(a) 68% (b) 41% (c) 200% (d) 100%

28. Sand is being dropped on a conveyor belt at the rate of M kg/s. The force (in N) necessary to keep the belt moving with a constant velocity of v m/s will be:
(a) Mv (b) 2 Mv (c) $\frac{Mv}{2}$ (d) $\frac{Mv}{3}$

29. If *a* cricketer catches a ball of mass 150 gm moving with a velocity of 20 m/s, then he experiences a force of (Time taken to complete the catch is 0.1 sec.)
(a) 300 N (b) 30 N (c) 3 N (d) 0.3 N

30. A bullet is fired from a gun. The force on the bullet is given by $F = 600 - 2 \times 10^5\, t$ where, F is in newton and t in second. The force on the bullet becomes zero as soon as it leaves the barrel. What is the average impulse imparted to the bullet?

Toughnut

(a) 1.8 N-s (b) zero (c) 9 N-s (d) 0.9 N-s

31. A 5000 kg rocket is set for vertical firing. The exhaust speed is 800 ms^{-1}. To give an initial upward acceleration of 20 ms^{-2}, the amount of gas ejected per second to supply the needed thrust will be (g = 10 ms^{-2})
(a) 127.5 kg s^{-1} (b) 187.5 kg s^{-1}
(c) 185.5 kg s^{-1} (d) 137.5 kg s^{-1}

32. A satellite in a force free space sweeps stationary interplanetary dust at a rate (dM/dt) = αv. The acceleration of satellite is
(a) $\frac{-2\alpha v^2}{M}$ (b) $\frac{-\alpha v^2}{M}$ (c) $\frac{-\alpha v^2}{2M}$ (d) $-\alpha v^2$

33. A particle of mass *m* is made to move with uniform speed *v* along the perimeter of a regular polygon of *n* sides, inscribed in a circle of radius *a*. The magnitude of impulse applied at each corner of the polygon is :
(a) $2\,m\,v\sin\frac{\pi}{n}$ (b) $m\,v\sin\frac{\pi}{n}$
(c) $m\,v\sin\frac{n}{\pi}$ (d) $2\,m\,v\sin\frac{n}{\pi}$

34. The force 'F' acting on a particle of mass 'm' is indicated by the force-time graph shown below. The change in momentum of the particle over the time interval from zero to 8 s is :

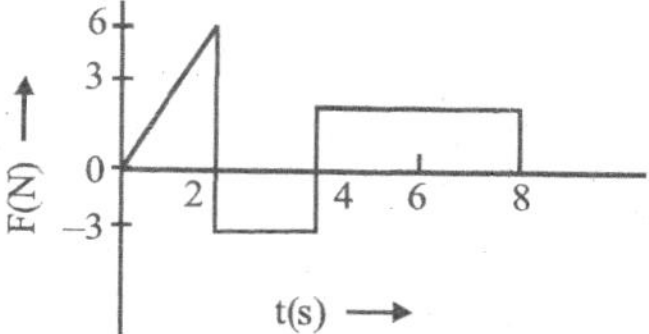

(a) 24 Ns (b) 20 Ns (c) 12 Ns (d) 6 Ns

35. For the given situation as shown in the figure, the value of θ to keep the system in equilibrium will be
(a) 30°
(b) 45°
(c) 0°
(d) 90°

T$_1$, 2 m, θ, 60N, 2 m, T$_2$, W = 60N

36. If the coefficient of friction between all surfaces is 0.5, then find the minimum force F to have equilibrium of system. (assume strings and pulleys are massless)

Critical Thinking

(a) $\frac{4000}{17}$ N
(b) $\frac{1000}{17}$ N
(c) $\frac{2000}{17}$ N
(d) $\frac{500}{17}$ N

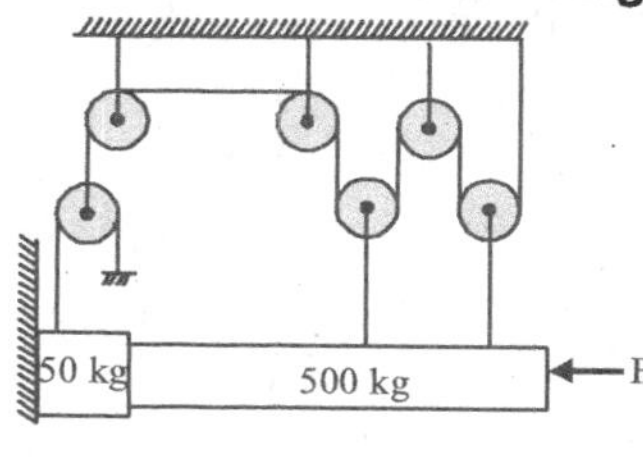

37. A person of mass 60 kg is inside a lift of mass 940 kg and presses the button on control panel. The lift starts moving upwards with an acceleration 1.0 m/s^2. If g = 10 ms^{-2}, the tension in the supporting cable is **Critical Thinking**

(a) 8600 N (b) 9680 N
(c) 11000 N (d) 1200 N

38. If two masses (M & m) are connected on a horizontal plane and a force is applied on the combination, then the tension T depends on

(a) the force applied on the system
(b) whether force is applied on M or m
(c) both (a) and (b)
(d) Can't be predicted.

39. A solid sphere of 2 kg is suspended from a horizontal beam by two supporting wires as shown in fig. Tension in each wire is approximately (g = 10 ms^{-2})

(a) 30 N
(b) 20 N
(c) 10 N
(d) 5 N

40. Three identical blocks of masses m = 2 kg are drawn by a force F = 10. 2 N with an acceleration of 0. 6 ms^{-2} on a frictionless surface, then what is the tension (in N) in the string between the blocks *B* and *C*?

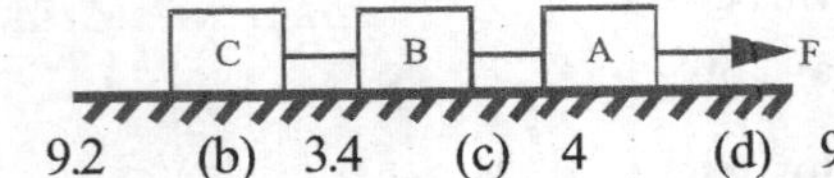

(a) 9.2 (b) 3.4 (c) 4 (d) 9.8

41. A 1 kg block and a 0.5 kg block move together on a horizontal frictionless surface. Each block exerts a force of 6 N on the other. The block move with a uniform acceleration of

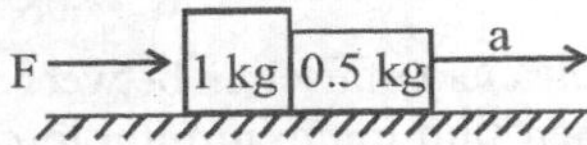

(a) 3 ms^{-2} (b) 6 ms^{-2}
(c) 9 ms^{-2} (d) 12 ms^{-2}

42. For the system shown in figure, the correct expression is

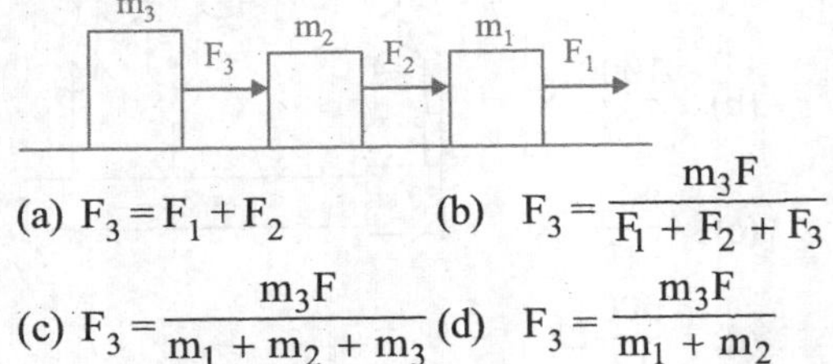

(a) $F_3 = F_1 + F_2$ (b) $F_3 = \frac{m_3 F}{F_1 + F_2 + F_3}$

(c) $F_3 = \frac{m_3 F}{m_1 + m_2 + m_3}$ (d) $F_3 = \frac{m_3 F}{m_1 + m_2}$

43. Which of the following is true about acceleration, a for the system?

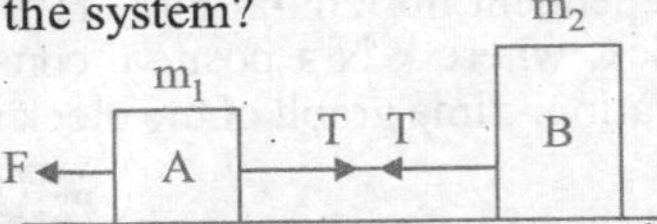

(a) Acceleration is more in A, when force is applied on A.
(b) Acceleration is more in B, when force is applied on B.
(c) Acceleration is same and does not depend on whether the force is applied on m_1 or m_2
(d) Acceleration depends on the tension in the string.

44. Two blocks of masses 2 kg and 4 kg are attached by an inextensible light string as shown in the figure. If a force of 120 N pulls the blocks vertically upward, the tension in the string is (take g = 10 ms^{-2})

(a) 20 N
(b) 15 N
(c) 35 N
(d) 40 N

45. A spring balance is attached to the ceiling of a lift. A man hangs his bag on the spring and the spring reads 49 N, when the lift is stationary. If the lift moves downward with an acceleration of 5 m/s^2, the reading of the spring balance will be

(a) 24 N (b) 74 N (c) 15 N (d) 49 N

46. The elevator shown in fig. is descending with an acceleration of 2 m/s^2. The mass of the block A = 0.5 kg. The force exerted by the block A on block B is

(a) 2 N
(b) 4 N
(c) 6 N
(d) 8 N

47. A block A of mass 7 kg is placed on a frictionless table. A thread tied to it passes over a frictionless pulley and carries a body B of mass 3 kg at the other end. The acceleration of the system is (given g = 10 ms^{-2})

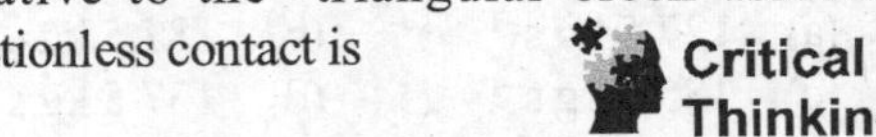

(a) 100 ms^{-2} (b) 3 ms^{-2}
(c) 10 ms^{-2} (d) 30 ms^{-2}

48. A triangular block of mass M with angles 30°, 60°, and 90° rests with its 30°–90° side on a horizontal table. A cubical block of mass m rests on the 60°–30° side. The acceleration which M must have relative to the table to keep m stationary relative to the triangular block assuming frictionless contact is **Critical Thinking**

(a) g (b) $\frac{g}{\sqrt{2}}$ (c) $\frac{g}{\sqrt{3}}$ (d) $\frac{g}{\sqrt{5}}$

49. An overweight acrobat, "weighing" in at 115 kg, wants to perform a single hand stand. He tries to cheat by resting one foot against a smooth frictionless vertical wall. The horizontal force there is 130 N. What is the magnitude of the force exerted by the floor on his hand? Answer in N.

(a) 1134 (b) 1257 (c) 997 (d) 1119

50. A monkey of mass 20 kg is holding a vertical rope. The rope will not break when a mass of 25 kg is suspended from it but will break if the mass exceeds 25 kg. What is the maximum acceleration with which the monkey can climb up along the rope ? ($g = 10\ m/s^2$)

(a) 2.5 m/s (b) 5 m/s^2 (c) 10 m/s^2 (d) 25 m/s^2

51. Two blocks $m_1 = 5$ gm and $m_2 = 10$ gm are hung vertically over a light frictionless pulley as shown here. What is the acceleration of the masses when they are left free?

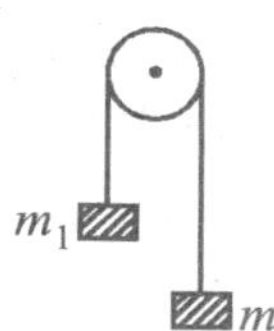

(a) g / 3 (b) g / 2 (c) g (d) g / 5

(where g is acceleration due to gravity)

52. In the figure (i) an extensible string is fixed at one end and the other end is pulled by a tension *T*. In figure (ii) another identical string is pulled by tension *T* at both the ends. The ratio of elongation in equilibrium of string in (i) to the elongation of string in (ii) is

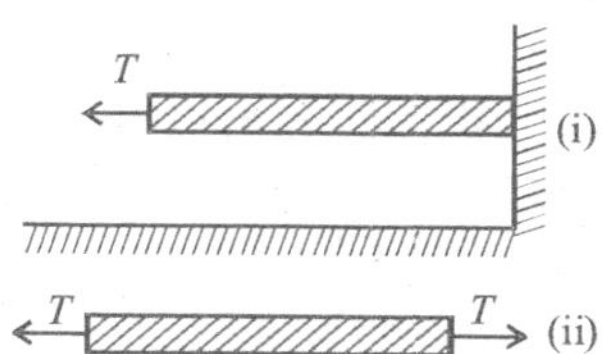

(a) 1 : 1 (b) 1 : 2 (c) 2 : 1 (d) 0

53. Two monkeys of masses 10 kg and 8 kg are moving along a vertical rope which is light and inextensible, the former climbing up with an acceleration of 2m/s^2 while the latter coming down with a uniform velocity of 2m/s. Find the tension (in newtons). **Tricky**

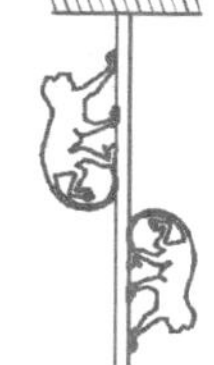

(a) 200 N (b) 150 N (c) 300 N (d) 100 N

54. A weight *W* is supported by two cables as shown. The tension in the cable making angle θ with horizontal will be the minimum when the value of θ is

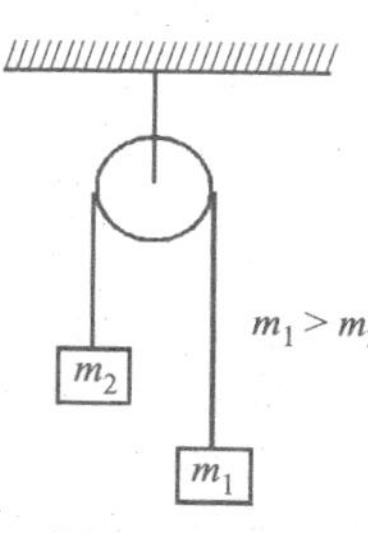

(a) 0 (b) 30°
(c) 60° (d) 45°

55. For the arrangement shown in the figure let '*a*' and *T* be the acceleration of the blocks and tension in the string respectively. The string and the pulley are frictionless and massless. Which of the graphs show the correct relationship between '*a*' and *T* for the system in which sum of the two masses m_1 and m_2 is constant? **Tricky**

$m_1 > m_2$

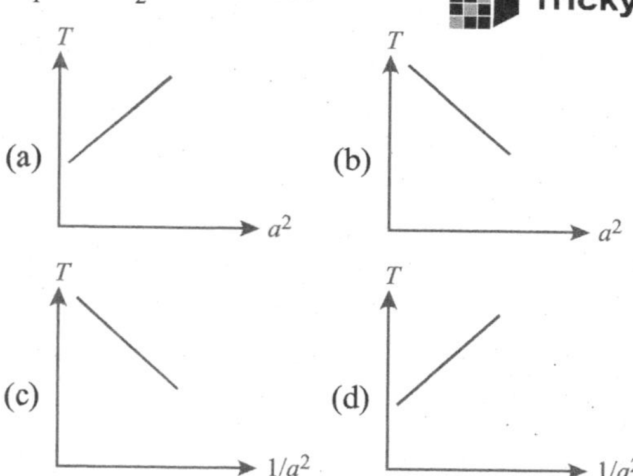

56. A body of mass 5 kg under the action of constant force $\vec{F} = F_x\hat{i} + F_y\hat{j}$ has velocity at t = 0 s as $\vec{v} = (6\hat{i} - 2\hat{j})$ m/s and at t = 10s as $\vec{v} = +6\hat{j}$ m/s. The force $\vec{F}$ is:

(a) $(-3\hat{i} + 4\hat{j})$ N (b) $\left(-\frac{3}{5}\hat{i} + \frac{4}{5}\hat{j}\right)$ N

(c) $(3\hat{i} - 4\hat{j})$ N (d) $\left(\frac{3}{5}\hat{i} - \frac{4}{5}\hat{j}\right)$ N

57. A pail filled with sand has a total mass of 60 kg. A crane is lowering it such that it has an initial downward acceleration of 1.5 m/s^2. A hole in the pail allows sand to leak out. If the force exerted by the crane on the pail does not change, what mass of sand must leak out before the downward acceleration decreases to zero? **Critical Thinking**

(a) 9.2 kg (b) 20 kg (c) 40 kg (d) 51 kg

58. One end of a massless rope, which passes over a massless and frictionless pulley *P* is tied to a hook *C* while the other end is free.

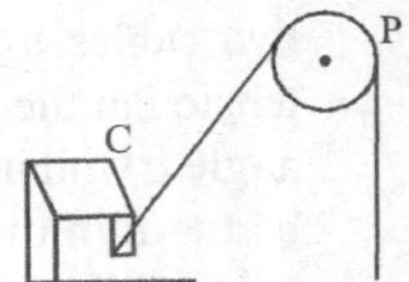

Maximum tension that the rope can bear is 360 N. With what value of maximum safe acceleration (in ms^{-2}) can a man of 60 kg climb on the rope?

(a) 16 (b) 6 (c) 4 (d) 8

59. A smooth ring P of mass *m* can slide on a fixed horizontal rod. A string tied to the ring passes over a fixed pulley and carries a block Q of mass ($m/2$) as shown in the figure.

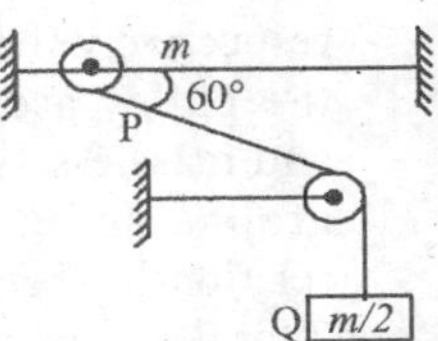

At an instant, the string between the ring and the pulley makes an angle 60° with the rod. The initial acceleration of the ring is **Critical Thinking**

(a) $\frac{2g}{3}$ (b) $\frac{2g}{6}$ (c) $\frac{2g}{9}$ (d) $\frac{g}{3}$

60. A horizontal uniform rope of length *L*, resting on a frictionless horizontal surface, is pulled at one end by force *F*. What is the tension in the rope at a distance *l* from the end where the force is applied?

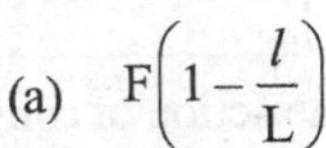

(a) $F\left(1-\frac{l}{L}\right)$ (b) $2F\left(1-\frac{l}{2L}\right)$

(c) $\frac{F}{L}$ (d) $\frac{F}{l}\left(1-\frac{l}{L}\right)$

61. In the figure acceleration of bodies A, B and C are shown with directions. Values *b* and *c* are w.r.t. ground whereas *a* is acceleration of block A w.r.t. wedge C.

Acceleration of block A w.r.t. ground is

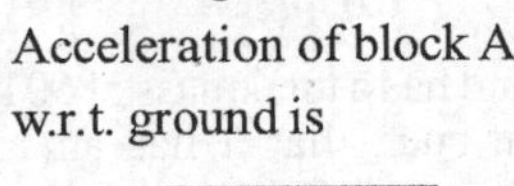

(a) $\sqrt{(b+c)^2+a^2}$

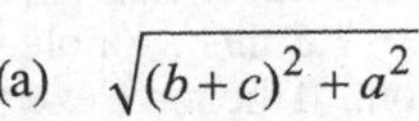

(b) $c-(a+b)\cos\theta$

(c) $\sqrt{(b+c)^2+c^2-2(b+c).c.\cos\theta}$

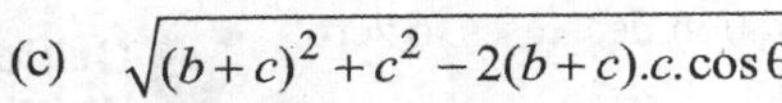

(d) $\sqrt{(b+c)^2+c^2+2(b+c).c.\cos\theta}$

62. Two blocks of masses m and M are connected by means of a metal wire of cross-sectional area A passing over a frictionless fixed pulley as shown in the figure. The system is then released. If M = 2 m, then the tension per unit crossectional area produced in the wire is

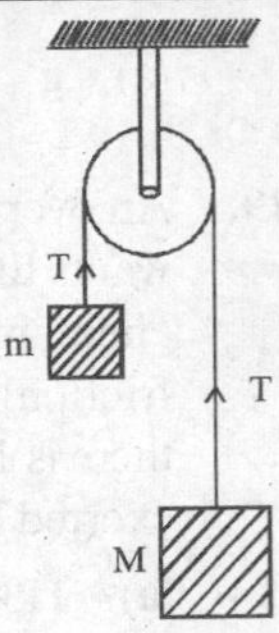

(a) $\frac{2mg}{3A}$ (b) $\frac{4mg}{3A}$ (c) $\frac{mg}{A}$ (d) $\frac{3mg}{4A}$

63. A ball of mass 0.2 kg is thrown vertically upwards by applying a force by hand. If the hand moves 0.2 m while applying the force and the ball goes upto 2 m height further, find the magnitude of the force. (Consider g = 10 m/s²).

(a) 4 N (b) 16 N (c) 20 N (d) 22 N

64. A uniform sphere of weight W and radius 5 cm is being held by a string as shown in the figure. The tension in the string will be : **Tricky**

(a) $12\frac{W}{5}$

(b) $5\frac{W}{12}$

(c) $13\frac{W}{5}$

(d) $13\frac{W}{12}$

8 cm

65. Two blocks of mass $M_1 = 20$ kg and $M_2 = 12$ kg are connected by a metal rod of mass 8 kg. The system is pulled vertically up by applying a force of 480 N as shown. The tension at the mid-point of the rod is :

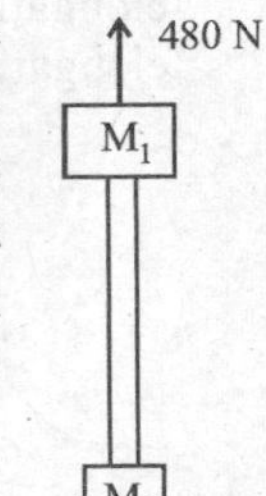

(a) 144 N (b) 96 N

(c) 240 N (d) 192 N

66. Two smooth cylindrical bars weighing *W* each lie next to each other in contact.

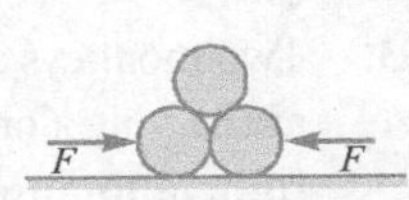

A similar third bar is placed over the two bars as shown in figure. Neglecting friction, the minimum horizontal force on each lower bar necessary to keep them together is

(a) $\frac{W}{2}$ (b) W (c) $\frac{W}{\sqrt{3}}$ (d) $\frac{W}{2\sqrt{3}}$

67. A string of negligible mass going over a clamped pulley of mass m supports a block of mass M as shown in the figure. The force on the pulley by the clamp is given by

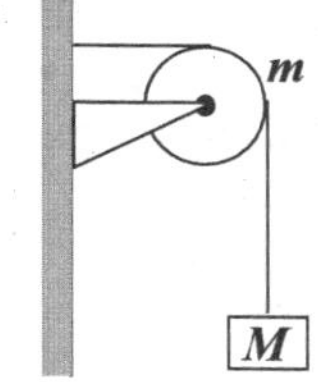

(a) $\sqrt{2}$ Mg
(b) $\sqrt{2}$ mg
(c) $\sqrt{(M+m)^2+m^2}\ g$
(d) $\sqrt{(M+m)^2+M^2}\ g$

68. The force required to just move a body up the inclined plane is double the force required to just prevent the body from sliding down the plane. The coefficient of friction is μ. The inclination θ of the plane is
(a) $\tan^{-1}\mu$ (b) $\tan^{-1}(\mu/2)$
(c) $\tan^{-1}2\mu$ (d) $\tan^{-1}3\mu$

69. A system shown in the figure. Assume that cylinder remains in contact with the wedge and block hence the velocity of cylinder is

Critical Thinking

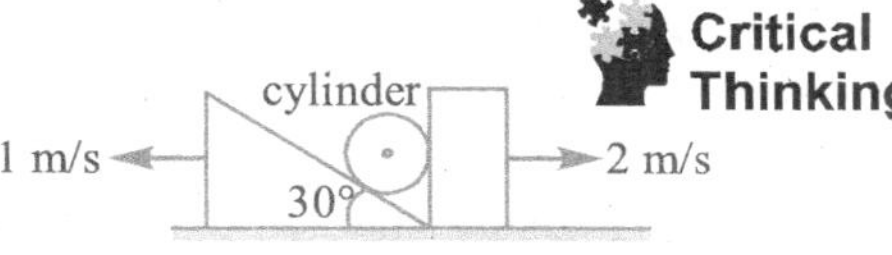

(a) $\frac{\sqrt{19-4\sqrt{3}}}{2}$ m/s (b) $\frac{\sqrt{13}}{2}$ m/s
(c) $\sqrt{3}$ m/s (d) $\sqrt{7}$ m/s

70. A conveyor belt is moving at a constant speed of 2m/s. A box is gently dropped on it. The coefficient of friction between them is $\mu = 0.5$. The distance that the box will move relative to belt before coming to rest on it taking $g = 10\ ms^{-2}$, is
(a) 1.2m (b) 0.6m (c) zero (d) 0.4m

71. A uniform rod AB of length $3r$ remains in equilibrium on a hemispherical bowl of radius r as shown in figure. Ignoring friction, the inclination of the rod θ with the horizontal is
(a) $\cos^{-1}(1/3)$
(b) $\sin^{-1}(1/3)$
(c) $\cos^{-1}(0.9)$
(d) $\sin^{-1}(0.9)$

72. The upper half of an inclined plane of inclination θ is perfectly smooth while lower half is rough. A block starting from rest at the top of the plane will again come to rest at the bottom, if the coefficient of friction between the block and lower half of the plane is given by

(a) $\mu = \frac{2}{\tan\theta}$ (b) $\mu = 2\tan\theta$
(c) $\mu = \tan\theta$ (d) $\mu = \frac{1}{\tan\theta}$

73. A 20 kg block B is suspended from a cord attached to a 40 kg cart A. Find the ratio of the acceleration of block in cases (i) and (ii) shown in the figure immediately after the system is released from rest. (neglect friction)

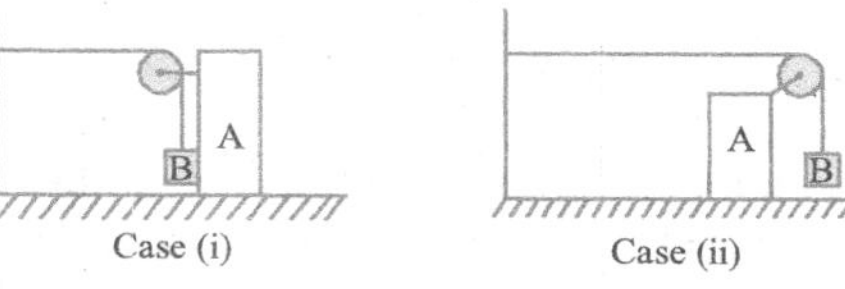

(a) $\frac{\sqrt{2}}{3}$ (b) $3\sqrt{2}$ (c) $\frac{3}{2}$ (d) $\frac{3}{2\sqrt{2}}$

74. In the diagram shown, friction is completely absent. If a force F has been applied on the wedge such that it moves with a constant velocity than value of normal reaction N′ is

Tricky

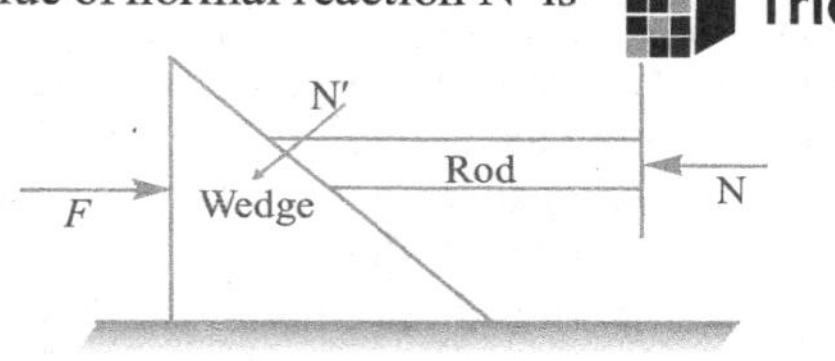

(a) $>F$ (b) $<F$
(c) $=F$ (d) cannot find

75. It is difficult to move a cycle with brakes on because
(a) rolling friction opposes motion on road
(b) sliding friction opposes motion on road
(c) rolling friction is more than sliding friction
(d) sliding friction is more than rolling friction

76. The acceleration of the system shown in the figure is given by the expression (coefficient of friction between m_1 and surface is μ)

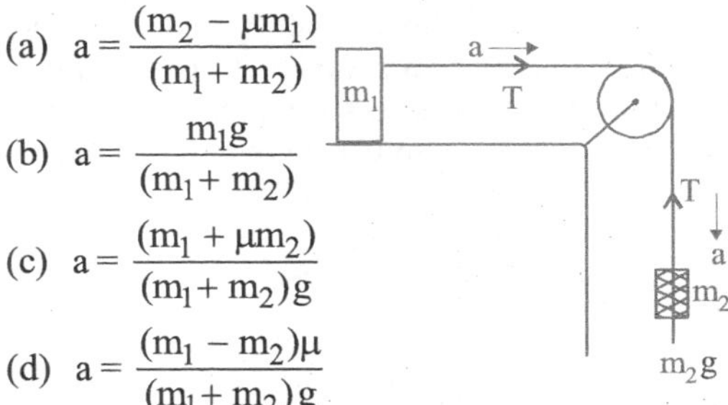

(a) $a = \frac{(m_2 - \mu m_1)}{(m_1 + m_2)}$

(b) $a = \frac{m_1 g}{(m_1 + m_2)}$

(c) $a = \frac{(m_1 + \mu m_2)}{(m_1 + m_2)g}$

(d) $a = \frac{(m_1 - m_2)\mu}{(m_1 + m_2)g}$

77. A horizontal force of 10 N is necessary to just hold a block stationary against a wall. The coefficient of friction between the block and the wall is 0.2. The weight of the block is

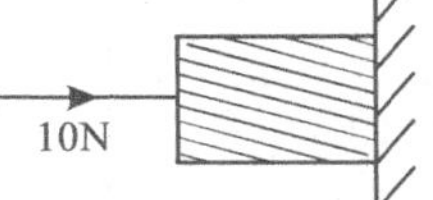

(a) 20 N (b) 50 N
(c) 100 N (d) 2 N

78. A block B is pushed momentarily along a horizontal surface with an initial velocity V. If μ is the coefficient of sliding friction between B and the surface, block B will come to rest after a time

(a) $\frac{g\mu}{V}$ (b) $\frac{g}{V}$

(c) $\frac{V}{g}$ (d) $\frac{V}{g(\mu)}$

79. A small mass slides down a fixed inclined plane of inclination θ with the horizontal. The coefficient of friction is $\mu = \mu_0 x$ where x is the distance through which the mass slides down and μ_0 is a constant. Then the speed is maximum after the mass covers a distance of

(a) $\frac{\cos\theta}{\mu_0}$ (b) $\frac{\sin\theta}{\mu_0}$ (c) $\frac{\tan\theta}{\mu_0}$ (d) $\frac{2\tan\theta}{\mu_0}$

80. Two identical smooth surfaced solid cylinders of radius r are placed touching along their lengths on a horizontal surface. A third cylinder of same material but twice the radius of that of the cylinders is placed lengthwise on them so that the system remains at rest. If all three cylinders have the same length, then minimum value of the coefficient of friction between smaller cylinders and the surface is:

Critical Thinking

(a) $\frac{1}{\sqrt{2}}$

(b) $\frac{1}{3}$

(c) $\frac{1}{3\sqrt{2}}$

(d) None of these

81. A block is placed on a rought inclined plane. The angle of the incline, θ, is slowly increased from the horizontal position. At a certain angle, the block starts to slide along the plane. The angle of the incline is increased further.

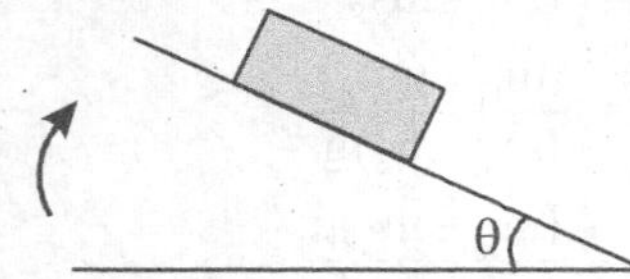

Consider the following graphs :

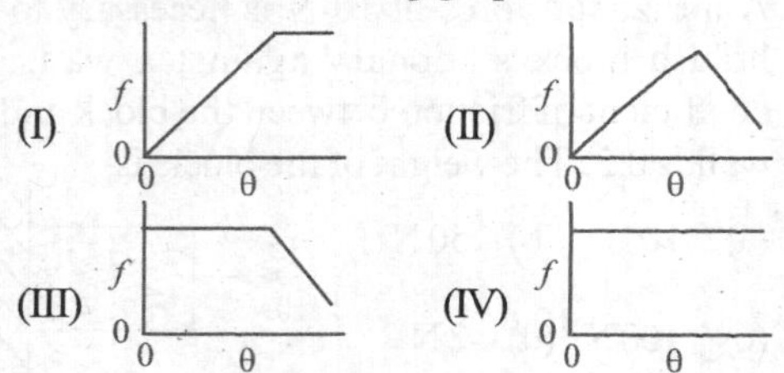

Which of the above graphs correctly depicts the variation of the fractional force, f, exerted by the plane on the block, as a function of θ? (Assume that the block does not topple.)

(A) I (B) II (C) III (D) IV

82. The minimum force required to start pushing a body up rough (frictional coefficient μ) inclined plane is F_1 while the minimum force needed to prevent it from sliding down is F_2. If the inclined plane makes an angle θ from the horizontal such that $\tan\theta = 2\mu$ then the ratio $\frac{F_1}{F_2}$ is

(a) 1 (b) 2 (c) 3 (d) 4

83. A body starts from rest on a long inclined plane of slope 45°. The coefficient of friction between the body and the plane varies as $\mu = 0.3\,x$, where x is distance travelled down the plane. The body will have maximum speed (for $g = 10\ \text{m/s}^2$) when $x =$

(a) 9.8m (b) 27m (c) 12m (d) 3.33m

84. A block A of mass 4 kg is placed on another block B of mass 5 kg, and the block B rests on a smooth horizontal table. If the minimum force that can be applied on A so that both the blocks move together is 12 N, the maximum force that can be applied to B for the blocks to move together will be:

(a) 30 N (b) 25 N (c) 27 N (d) 48 N

85. The two blocks, $m = 10$ kg and $M = 50$ kg are free to move as shown. The coefficient of static friction between the blocks is 0.5 and there is no friction between M and the ground. A minimum horizontal force F is applied to hold m against M that is equal to

Critical Thinking

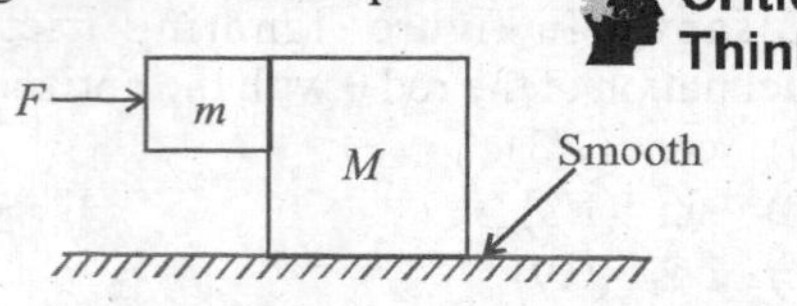

(a) 100 N (b) 50 N (c) 240 N (d) 180 N

86. A given object takes n times as much time to slide down a 45° rough incline as it takes to slide down a perfectly smooth 45° incline. The coefficient of friction between the object and the incline is

(a) $(1 - 1/n^2)$ (b) $1/(1 - n^2)$

(c) $\sqrt{(1 - 1/n^2)}$ (d) $1/\sqrt{(1 - n^2)}$

87. A block of mass $m = 2$ kg is placed on a plank of mass $M = 10$ kg which is placed on a smooth horizontal plane. The coefficient of friction between the block and the plank is $\mu = \frac{1}{3}$. If a horizontal force F is applied on the plank, then find the maximum value of F for which the block and the plank move together. (Take $g = 10$ m/s^2)

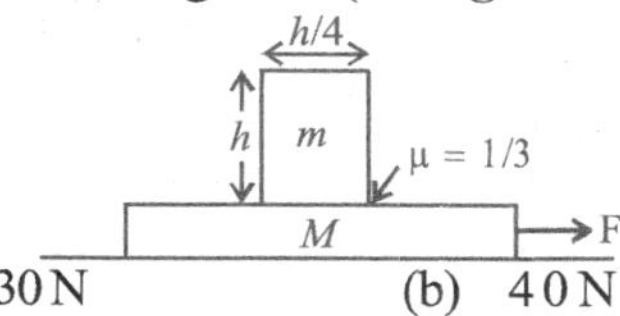

(a) 30 N (b) 40 N
(c) 120 N (d) None of the above

88. If μ be the coefficient of friction between the block and the cart, horizontal acceleration of the cart that is required to prevent block B from falling is:

(a) μ/g (b) g/μ
(c) g (d) $(\mu^2+1)g$

μ → Cart B a

μ = 0.2

89. An insect of mass m, starts moving on a rough inclined surface from point A. As the surface is very sticky, the coefficient of friction between the insect and the incline is $\mu = 1$. Assume that it can move in any direction, up the incline or down the incline then –

Critical Thinking

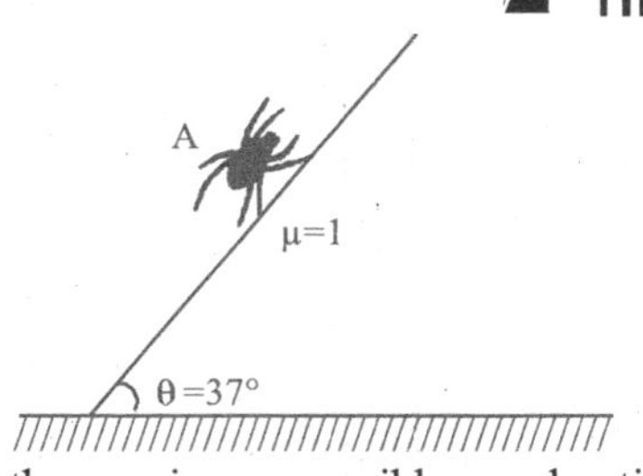

(a) the maximum possible acceleration of the insect can be 14 m/s^2
(b) the maximum possible acceleration of the insect can be 2 m/s^2
(c) the insect can move with a constant velocity
(d) the insect cannot move with a constant velocity

90. A particle rests on the top of a hemisphere of radius R. Find the smallest horizontal velocity that must be imparted to the particle if it is to leave the hemisphere without sliding down is

(a) $\sqrt{gR}$ (b) $\sqrt{2gR}$ (c) $\sqrt{3gR}$ (d) $\sqrt{5gR}$

91. A car is moving in a circular horizontal track of radius 10 m with a constant speed of 10 m/s. A bob is suspended from the roof of the car by a light wire of length 1.0 m. The angle made by the wire with the vertical is

(a) 0° (b) $\frac{\pi}{3}$ (c) $\frac{\pi}{6}$ (d) $\frac{\pi}{4}$

92. A particle of mass m rotates with a uniform angular speed ω. It is viewed from a frame rotating about the z-axis with a uniform angular velocity ω_0. The centrifugal force on the particle is:

(a) $m\omega^2 r$ (b) $m\omega_0^2 r$
(c) $m\left(\frac{\omega+\omega_0}{2}\right)a$ (d) zero

93. A bridge is in the from of a semi-circle of radius 40m. The greatest speed with which a motor cycle can cross the bridge without leaving the ground at the highest point is ($g = 10$ m s^{-2}) (frictional force is negligibly small)

(a) 40 m s^{-1} (b) 20 m s^{-1}
(c) 30 m s^{-1} (d) 15 m s^{-1}

94. A particle tied to a string describes a vertical circular motion of radius r continually. If it has a velocity $\sqrt{3\ gr}$ at the highest point, then the ratio of the respective tensions in the string holding it at the highest and lowest points is

(a) 4 : 3 (b) 5 : 4 (c) 1 : 4 (d) 3 : 2

95. An aircraft executes a horizontal loop with a speed of 150 m/s with its wings banked at an angle of 12°. The radius of the loop is : ($g = 10$ m/s^2)

(a) 10.6 km (b) 9.6 km (c) 7.4 km (d) 5.8 km

96. In the given figure, a smooth parabolic wire track lies in the xy-plane (vertical). The shape of track is defined by the equation $y = x^2$. A ring of mass m which can slide freely on the wire track, is placed at the position A (1,1). The track is rotated with constant angular speed ω such there is no relative slipping between the ring and the track. The value of ω is

Critical Thinking

(a) $\sqrt{g/2}$ (b) $\sqrt{g}$
(c) $\sqrt{2g}$ (d) $2\sqrt{g}$

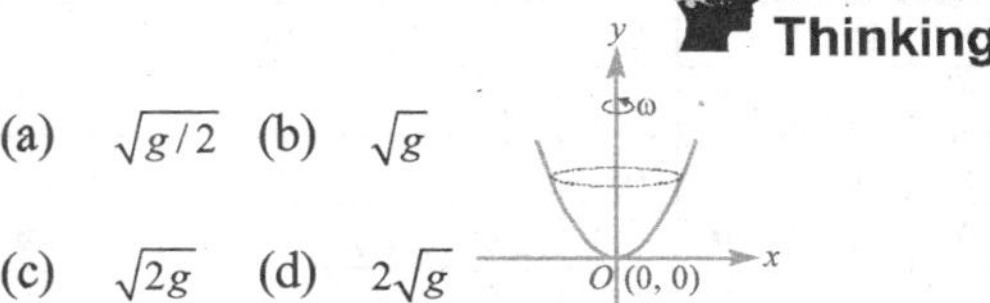

97. A conical pendulum of length 1 m makes an angle $\theta = 45°$ w.r.t. Z-axis and moves in a circle in the XY plane. The radius of the circle is 0.4 m and its centre is vertically below O. The speed of the pendulum, in its circular path, will be : (Take $g = 10$ ms^{-2})

(a) 0.4 m/s
(b) 4 m/s
(c) 0.2 m/s
(d) 2 m/s

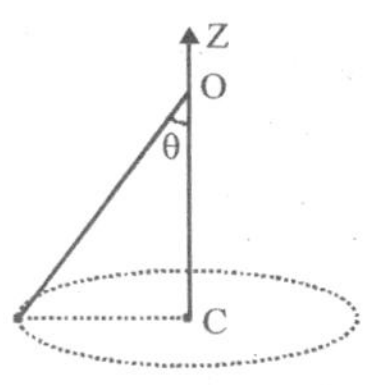

98. A car is negotiating a curved road of radius R. The road is banked at an angle θ. the coefficient of friction between the tyres of the car and the road is μ_s. The maximum safe velocity on this road is :

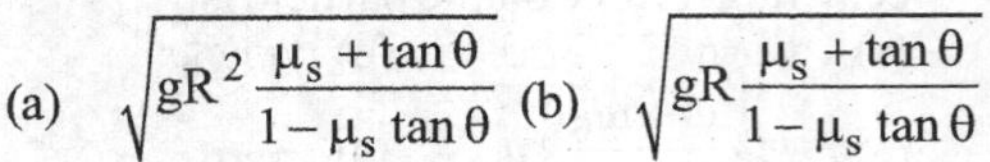

(a) $\sqrt{gR^2 \frac{\mu_s + \tan\theta}{1 - \mu_s \tan\theta}}$ (b) $\sqrt{gR \frac{\mu_s + \tan\theta}{1 - \mu_s \tan\theta}}$

(c) $\sqrt{\frac{g}{R} \frac{\mu_s + \tan\theta}{1 - \mu_2 \tan\theta}}$ (d) $\sqrt{\frac{g}{R^2} \frac{\mu_s + \tan\theta}{1 - \mu_s \tan\theta}}$

99. A particle is moving along a circular path in the *xy* plane (see figure). When it crosses the *x*-axis, it has an acceleration along the path of 1.5 m/s², and is moving with a speed of 10 m/s in the negative *y*-direction. The total acceleration of the particle is :

(a) 50 î - 1.5 ĵ m/s²

(b) - 50 î - 1.5 ĵ m/s²

(c) 10 î - 1.5 ĵ m/s²

(d) 1.5 î - 50 ĵ m/s²

100. Two particles of equal mass are connected to a rope *AB* of negligible mass such that one is at end *A* and other dividing the length of rope in the ratio 1 : 2 from *B*. The rope is rotated about end *B* in a horizontal plane. Ratio of tensions in the smaller part to the other is (ignore effect of gravity) **Critical Thinking**

(a) 4 : 3 (b) 1 : 4 (c) 1 : 2 (d) 1 : 3

Answer KEYs

1	(a)	11	(a)	21	(a)	31	(b)	41	(d)	51	(a)	61	(c)	71	(c)	81	(b)	91	(d)
2	(c)	12	(c)	22	(b)	32	(b)	42	(c)	52	(a)	62	(b)	72	(b)	82	(c)	92	(b)
3	(c)	13	(b)	23	(b)	33	(a)	43	(c)	53	(a)	63	(d)	73	(d)	83	(d)	93	(b)
4	(a)	14	(d)	24	(d)	34	(c)	44	(d)	54	(b)	64	(d)	74	(a)	84	(c)	94	(c)
5	(d)	15	(a)	25	(b)	35	(b)	45	(a)	55	(b)	65	(d)	75	(d)	85	(c)	95	(a)
6	(a)	16	(c)	26	(d)	36	(c)	46	(b)	56	(a)	66	(d)	76	(a)	86	(a)	96	(c)
7	(a)	17	(a)	27	(b)	37	(c)	47	(b)	57	(a)	67	(d)	77	(d)	87	(a)	97	(d)
8	(c)	18	(d)	28	(a)	38	(b)	48	(c)	58	(c)	68	(d)	78	(d)	88	(b)	98	(b)
9	(d)	19	(a)	29	(b)	39	(b)	49	(a)	59	(c)	69	(d)	79	(c)	89	(c)	99	(b)
10	(c)	20	(a)	30	(d)	40	(b)	50	(a)	60	(a)	70	(d)	80	(c)	90	(a)	100	(a)

Hints & Solutions

1. **(a)** In case (i) we have
$T_1-(1\times g)=1\times 4.9$
$\Rightarrow T_1=9.8+4.9=14.7\,\text{N}$
In case (ii), $1\times g-T_2=1\times 4.9$
$\Rightarrow T_2=9.8-4.9=4.9\,\text{N}$ $\quad\therefore \dfrac{T_1}{T_2}=\dfrac{14.7}{4.9}=\dfrac{3}{1}$

2. **(c)** Let T be the tension in the branch of a tree when monkey is descending with acceleration a. Then mg – $T = ma$; and T = 75% of weight of monkey
$\therefore ma = mg-\left(\dfrac{75}{100}\right)mg=\left(\dfrac{1}{4}\right)mg$ or $a=\dfrac{g}{4}$.

3. **(c)** The *FBD* of blocks is as shown
From Newton's second law
$4mg-2T\cos\theta=4mA$ .... (i)
and $T-mg=ma$ .... (ii)
$\cos\theta = 4/5$ and from constraint we get
$a=A\cos\theta$ .... (iii)

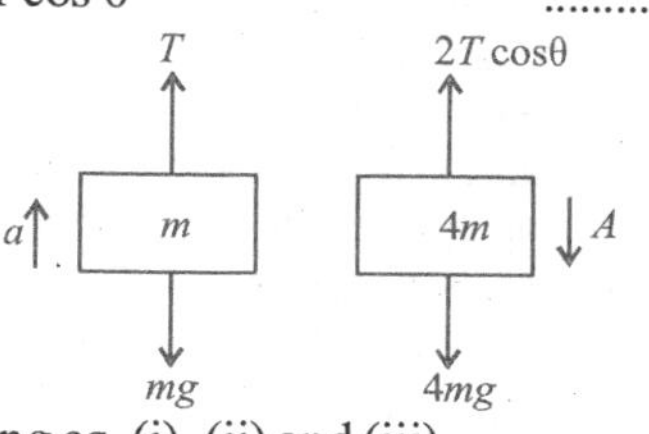

Solving eq. (i), (ii) and (iii)
we get acceleration of block of mass $4m$,

$A=\dfrac{5g}{11}$ downwards.

4. **(a)**

Before cutting the string
$kx=T+3\,mg$ ...(i)
$T=mg$ ...(ii)
$\Rightarrow\ kx=4mg$
After cutting the string
$T=0$

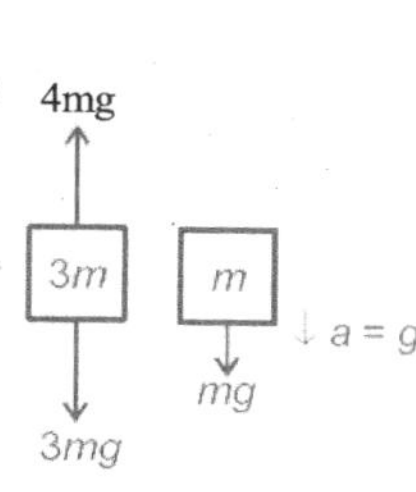

$a_A=\dfrac{4mg-3mg}{3m}$

$a_A=\dfrac{g}{3}\uparrow$ and $a_B=\dfrac{mg}{m}=g\downarrow$

5. **(d)** Net force on particle in uniform circular motion is centripetal force $\left(\dfrac{mv^2}{l}\right)$ which is provided by tension in string so the net force will be equal to tension i.e., T.

6. **(a)** Acceleration of system $a=\dfrac{F_{net}}{M_{total}}$
$=\dfrac{14}{4+2+1}=\dfrac{14}{7}=2\text{ m/s}^2$

The contact force between A and B
$=(m_B+m_C)\times a=(2+1)\times 2=6\text{N}$

7. **(a)** Let upthrust of air be F_a then
For downward motion of balloon
$F_a=mg-ma$
$mg-F_a=ma$
For upward motion $F_a-(m-\Delta m)g=(m-\Delta m)a$
Therefore $\Delta m=\dfrac{2ma}{g+a}$

8. **(c)**

(Force diagram in the frame of the car)
Applying Newton's law perpendicular to string
$mg\sin\theta = ma\cos\theta \quad\Rightarrow \tan\theta=\dfrac{a}{g}$
Applying Newton's law along string
$\Rightarrow T-m\sqrt{g^2+a^2}=ma$
or $T=m\sqrt{g^2+a^2}+ma$

9. **(d)** When all are pulling
$\vec{F}_{net}=100\times 3\hat{i}$ .... (i)
When A stops

$\vec{F}_{net} - \vec{F}_A = 100 \times 1(-\hat{i})$ ....(ii)

When B stops

$\vec{F}_{net} - \vec{F}_B = 100 \times 24\hat{j}$ ....(iii)

From these three we get

$\vec{F}_A + \vec{F}_B = 100(7\hat{i} - 24\hat{j})$

$\therefore$ Required $acc^n = \sqrt{7^2 + (-24)^2} = 25$ m/s²

10. (c) Mass (m) = 0.3 kg $\Rightarrow$ F = m.a = – 15 x

$a = -\frac{15}{0.3} x = \frac{-150}{3} x = -50x$

$a = -50 \times 0.2 = 10 \, m/s^2$

11. (a) Since force at a point at any instant is related to the acceleration at that point, at that instant and acceleration is determined only by the instantaneous force and not by any history of the motion of the particle. Therefore, the moment the stone is thrown out of an accelerated train, it has no horizontal force and acceleration, if air resistance is neglected.

12. (c) Thrust $= \frac{udM}{dt} = mg$

$\Rightarrow \frac{dM}{dt} = \frac{mg}{u} = \frac{600 \times 10}{1000} = 6 \, kg \, s^{-1}$

13. (b) Here $u = 10 \, ms^{-1}$, $v = -2 \, ms^{-1}$, $t = 4$ s, a = ?

Using $a = \frac{v-u}{t} = \frac{-2-10}{4} = -3 \, m/s^2$

$\therefore$ Force, F = ma = 10×(–3) = –30 N

14. (d) Here m = 0.5 kg ; u = – 10 m/s ; t = 1/50 s ; $v = +15 \, ms^{-1}$

Force = m (v– u)/t = 0.5 (10 + 15) × 50 = 625 N

15. (a) Due to reaction force by road

16. (c) In series each spring will have same force. Here it is 4 kg-wt.

17. (a) N = m a sin θ + mg cos θ ......(1)

also m g sin θ = m a cos θ ......(2)

from (2) a = g tan θ

$\therefore N = mg\frac{\sin^2\theta}{\cos\theta} + mg\cos\theta$,

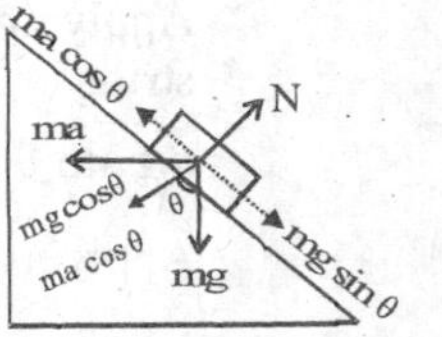

or $N = \frac{mg}{\cos\theta}$

18. (d) By Newton's second law of motion

$F = n(mv) = nmv$

19. (a) As the ball, m = 10 g = 0.01 kg rebounds after striking the wall

$\therefore$ Change in momentum $= mv - (-mv) = 2mv$

Inpulse = Change in momentum $= 2mv$

$\therefore v = \frac{\text{Impulse}}{2m} = \frac{0.54 \text{ N s}}{2 \times 0.01 \text{ kg}} = 27 \text{ m s}^{-1}$

20. (a) $S = \frac{V^2 - U^2}{2a}$ and a = F/m, so

$\frac{S_1}{S_2} = \frac{\left(\frac{0-U^2}{2F/m}\right)}{\left(\frac{0-U^2}{2F/2m}\right)} = 1:2$

21. (a) As we know, |impulse| = |change in momentum|

$= |p_2 - p_1| = |0 - mv_1| = |0 - 3 \times 2| = 6$ Ns

22. (b) As we know, $|a| = \left|\frac{v^2 - u^2}{2S}\right| = \frac{400}{50} = 8$ m/s².

So force required = F = ma = 800 × 8 = 6400 N

23. (b) The apple will fall slightly away from the hand of his brother in the direction of motion of the train due to inertia of motion. When train is just going to stop, the boy and his brother slows down with train but the apple which is in free fall continue to move with the same speed and therefore, falls slightly away from the hand in the direction of motion of the train.

24. (d) This is an example of Newton's Third Law of Motion applicable in all circumstances

25. (b) Graph (b) correctly depicts the acceleration-time graph of the block.

26. (d) $\vec{p}(t) = A[\hat{i}\cos(kt) - \hat{j}\sin(kt)]$

$\vec{F} = \frac{d\vec{p}}{dt} = Ak\,[-\hat{i}\sin(kt) - \hat{j}\cos(kt)]$

Here, $\vec{F}.\vec{P} = 0$ But $\vec{F}.\vec{p} = Fp\cos\theta$

$\therefore \cos\theta = 0 \Rightarrow \theta = 90°$.

27. (b) Momentum $P = mv = m\sqrt{2gh}$

($\because v^2 = u^2 + 2gh$; Here $u = 0$)

When stone hits the ground momentum

$P = m\sqrt{2gh}$

when same stone dropped from $2h$ (100% of initial) then momentum $P' = m\sqrt{2g(2h)} = \sqrt{2}P$

Which is changed by 41% of initial.

28. (a) $F = \frac{d(Mv)}{dt} = M\frac{dv}{dt} + v\frac{dM}{dt}$

$\therefore$ v is constant,

$\therefore F = v\frac{dM}{dt}$ But $\frac{dM}{dt} = Mkg/s$

$\therefore$ F = vM newton.

29. (b) Net force experienced = $\frac{\text{Total Impulse}}{\text{Time taken}}$

$= \frac{m\Delta v}{t} = 0.15 \times \frac{20}{0.1} = 30N$

30. (d) Given $F = 600 - (2 \times 10^5 t)$
The force is zero at time t, given by 0 $= 600 - 2 \times 10^5 t$

$\Rightarrow \quad t = \frac{600}{2 \times 10^5} = 3 \times 10^{-3}$ seconds

$\therefore$ Impulse $= \int_0^t F dt = \int_0^{3\times10^{-3}} (600 - 2\times10^5 t)\, dt$

$= \left[600t - \frac{2\times10^5 t^2}{2}\right]_0^{3\times10^{-3}}$

$= 600 \times 3 \times 10^{-3} - 10^5(3 \times 10^{-3})^2$
$= 1.8 - 0.9 = 0.9$ Ns

31. (b) Given : Mass of rocket (m) = 5000 kg
Exhaust speed (v) = 800 m/s
Acceleration of rocket (a) = 20 m/s^2
Gravitational acceleration (g) = 10 m/s^2
We know that upward force
$F = m(g + a) = 5000\,(10 + 20) = 5000 \times 30$
$= 150000$ N.
We also know that amount of gas ejected

$\left(\frac{dm}{dt}\right) = \frac{F}{v} = \frac{150000}{800} = 187.5 \text{ kg/s}^{-1}$

32. (b) Thrust on the satellite,

$F = \frac{-vdM}{dt} = -v(\alpha v) = -\alpha v^2$

Acceleration $= \frac{F}{M} = \frac{-\alpha v^2}{M}$

33. (a) $\vec{J} = m(\vec{v}_f - \vec{v}_i)$
$= 2\,mv \sin\theta/2 = 2mv \sin \pi/n.$

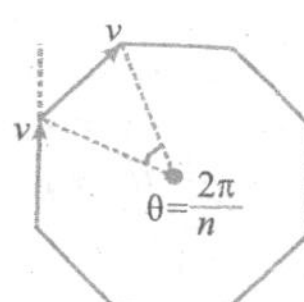

34. (c) Change in momentum,

$\Delta p = \int F dt$

= Area of F-t graph
= ar of Δ – ar of ▭ + ar of ▭

$= \frac{1}{2} \times 2 \times 6 - 3 \times 2 + 4 \times 3 = 12$ N-s

35. (b) In eqbm $T_1 \cos\theta$
$= T_2 = 60$N. ...(1)
$T_1 \sin\theta = 60$ N ...(2)
$\therefore \quad \tan\theta = 1 \Rightarrow \quad \theta = 45°.$

36. (c) $T = 2f + 500$ .... (i)
$8T = 5000 - f$ .... (ii)
$f = \frac{1}{2}F$ .... (iii)
From (i), (ii) and (iii)
$F = \frac{2000}{17}$N

37. (c)

Total mass = (60 + 940) kg = 1000 kg
Let T be the tension in the supporting cable, then
$T - 1000g = 1000 \times 1$
$\Rightarrow T = 1000 \times 11 = 11000$ N

38. (b) When force is applied on m_1 then $T = m_2a$ and when force is applied on m_2, then $T = m_1a$. Thus value of T is different for each case. And it depends on whether the force is applied on m_1, or m_2.

39. (b) $2\,T\cos 60° = mg$ or $T = mg = 2 \times 10 = 20$ N.

40. (b) $F = (m + m + m) \times a$
$\therefore a = \frac{10.2}{6} \text{m/s}^2$
$\therefore T_2 = ma = 2 \times \frac{10.2}{6} = 3.4$N

41. (d)

42. (c) Common acceleration of system is
$a = \frac{F}{m_1 + m_2 + m_3}$
$\therefore$ Force on m_3 is $F_3 = m_3 \times a$
$= \frac{m_3 F}{m_1 + m_2 + m_3}$

43. (c) $a = \frac{F}{m_1 + m_2}$ So the acceleration is same whether the force is applied on m_1 or m_2.

44. (d) Acceleration of the system

$$a = \frac{F - 4g - 2g}{4 + 2} = \frac{120 - 40 - 20}{6}$$

$= 10\ ms^{-2}$

From figure

$T - 2g = 2a$

$T = 2(a + g) = 2(10 + 10)$

$= 40$ N

FBD of block

45. (a) For the bag accelerating down

$mg - T = ma$

$\therefore T = m(g - a)$

$= \frac{49}{10}(10 - 5)$

$= 24.5$ N

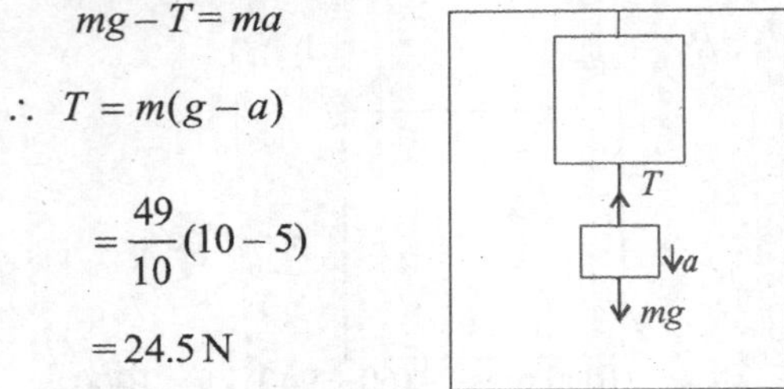

46. (b) $R = mg - ma = 0.5 \times 10 - 0.5 \times 2 = 5 - 1 = 4$ N

47. (b) $a = \frac{m_2}{m_1 + m_2} g = \frac{3}{7+3} \times 10 = 3\ m/s^2$

48. (c) $ma \cos 30° = mg \sin 30°$

$\therefore a = \frac{g}{\sqrt{3}}$

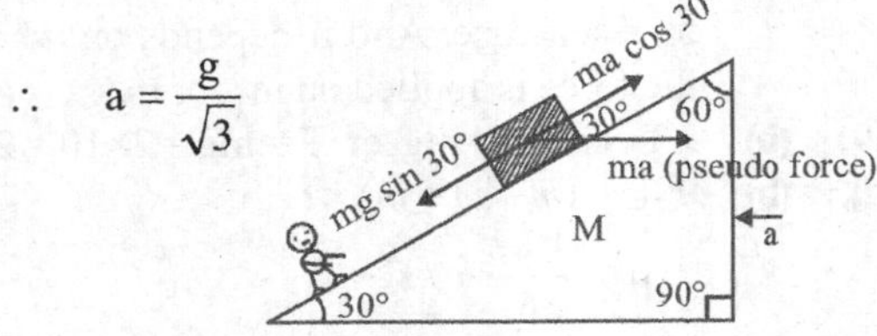

49. (a) The acrobat has a force acting on his hand that we resolve into two perpendicular components: the vertical one is the reaction to the weight (115×9.8 N $= 1127$ N) and the horizontal one balances the 130 N force from the wall. These two forces give a resultant force F of

$$F = \sqrt{1127^2 + 130^2} = 1134 \text{ N}$$

50. (a) T = Tension caused in string by monkey

$= m(g + a)$

$\therefore T \leq 25 \times 10 \Rightarrow 20(10 + a) \leq 250$

or, $10 + a\ 12.5 \Rightarrow a \leq 2.5$

51. (a) Let T be the tension in the string.

$\therefore 10g - T = 10a$ ....(i)

$T - 5g = 5a$ ....(ii)

Adding (i) and (ii),

$5g = 15a \Rightarrow a = \frac{g}{3}\ m/s^2$

52. (a) Tension in both string shall be same which can be observed by making FBD of string in figure (i)

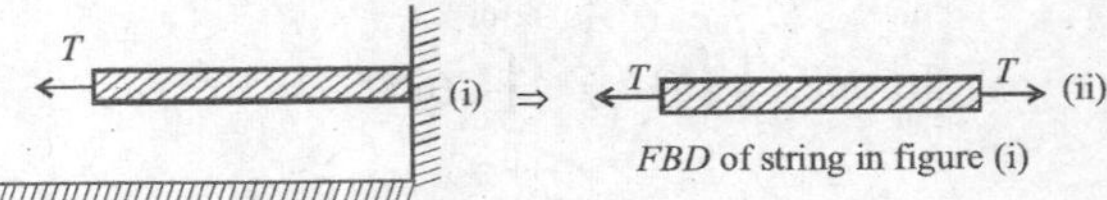

53. (a) $T = T_1 + T_2 = m_1(g + a) + m_2 g$

$= 10(10 + 2) + 8(10) = 120 + 80 = 200$N

54. (b) $T_2 \sin\theta + T_1 \sin 60° = W$

$T_1 \cos 60° = T_2 \cos\theta$

$T_2 \sin\theta + T_2 \cos\theta \cdot \tan 60° = W$

$T_2(\sin\theta + \sqrt{3}\cos\theta) = W$

$2T_2(\sin 30° \sin\theta + \cos 30° \cos\theta) = W$

$\cos(30° - \theta)$ is maximum

If $30° - \theta = 0; \theta = 30°$

55. (b) Let $m_1 + m_2 = m$ = constant and

$m_1 = x \Rightarrow m_2 = m - x$

$a = \frac{x - (m - x)}{m} g$

$T = \frac{2x(m - x)}{m} g \Rightarrow T = \frac{m}{2g}(g^2 - a^2)$

56. (a) From question, Mass of body, $m = 5$ kg

Velocity at $t = 0$, $u = (6\hat{i} - 2\hat{j})$ m/s

Velocity at $t = 10$s, $v = +\ 6\hat{j}$ m/s

Force, $F = ?$

$a = \frac{v - u}{t} = \frac{6\hat{j} - (6\hat{i} - 2\hat{j})}{10} = \frac{-3\hat{i} + 4\hat{j}}{5}\ m/s^2$

Force, $F = ma = 5 \times \frac{(-3\hat{i} + 4\hat{j})}{5} = (-3\hat{i} + 4\hat{j})N$

57. (a) Free body diagram at start

Free body diagram when $a = 0$

$m_1 g - F_c = m_1 a$; $F_c = m_1(g - a)$

$F_c - m_2 g = 0$; $F_c = m_2 g$

$m_1(g - a) = m_2 g$

$m_2 = \frac{m_1(g - a)}{g} = 50.8$ kg ;

$\Delta m = 9.2$ kg

58. (c) $mg - T = ma$

$\therefore a = g - \frac{T}{m} = 10 - \frac{360}{60} = 4 m/s^2$

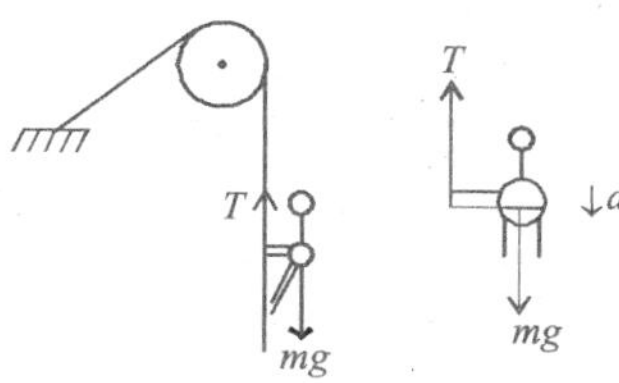

59. **(c)** $\frac{m}{2}g - T = \frac{m}{2}a$ ....(i)

$T\cos 60^\circ = \frac{\text{ma}}{\cos 60^\circ}$ ...(ii)

Solving (i) and (ii)

acceleration of ring $= \frac{2g}{9}$

60. **(a)** Let a be the acceleration of the rope and M be its total mass. Then

$T = \frac{M}{L}(L-\ell)a$ ... (i)

and $F - T = \frac{M}{L} \times \ell a$ ... (ii)

Dividing (i) and (ii)

$\frac{F-T}{T} = \frac{\ell}{L-\ell} \Rightarrow F(L-\ell) - T(L-\ell) = T\ell$

$\Rightarrow F(L-\ell) = T(L-\ell+\ell) = T \times L$

$\Rightarrow T = F\left(1 - \frac{\ell}{L}\right)$

61. **(c)** $a = b + c$

Net acceleration of A

$= \sqrt{a^2 + c^2 + 2ac\cos(\pi - \theta)}$

$= \sqrt{(b+c)^2 + c^2 - 2(b+c).c.\cos\theta}$

62. **(b)** Tension in the wire, $T = \left(\frac{2mM}{m+M}\right)g$

$= \frac{\text{Force/Tension}}{\text{Area}} = \frac{2mM}{A(m+M)}g$

$= \frac{2(m \times 2m)g}{A(m+2m)} = \frac{4m^2}{3mA}g = \frac{4mg}{3A}$

(M = 2 m given)

63. **(d)** Let the velocity of the ball just when it leaves the hand is u then applying, $v^2 - u^2 = 2as$ for upward journey

$\Rightarrow -u^2 = 2(-10) \times 2 \Rightarrow u^2 = 40$

Again applying $v^2 - u^2 = 2as$ for the upward journey of the ball, when the ball is in the hands of the thrower,

$v^2 - u^2 = 2as$

$\Rightarrow 40 - 0 = 2(a)\,0.2 \Rightarrow a = 100 \text{ m/s}^2$

$\therefore F = ma = 0.2 \times 100 = 20\ N$

$\Rightarrow N - mg = 20 \Rightarrow N = 20 + 2 = 22N$

64. **(d)** $PQ = \sqrt{OP^2 + OQ^2}$

$= \sqrt{13^2 + 5^2} = 12$

Tension in the string

$T = w\cos\theta = \frac{13}{12}W$

65. **(d)** Acceleration produced in upward direction

$a = \frac{F}{M_1 + M_2 + \text{Mass of metal rod}}$

$= \frac{480}{20+12+8} = 12 \text{ ms}^{-2}$

Tension at the mid point

$T = \left(M_2 + \frac{\text{Mass of rod}}{2}\right)a$

$= (12 + 4) \times 12 = 192$ N

66. **(d)** $2N\cos 30^\circ = W,$

$N = \frac{W}{\sqrt{3}}$

and $N \sin 30^\circ = F$

or $\frac{W}{\sqrt{3}} \times \frac{1}{2} = F$, or $F = \frac{W}{2\sqrt{3}}$.

67. **(d)** At equilibrium $T = Mg$

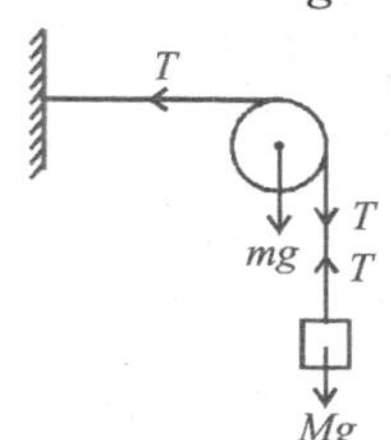

F.B.D. of pulley

$F_1 = (m+M)g$

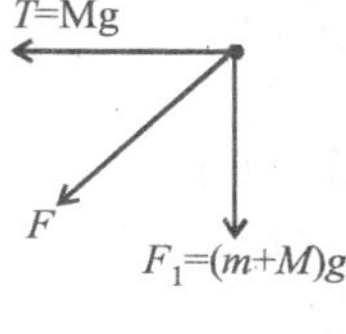

The resultant force on pulley is

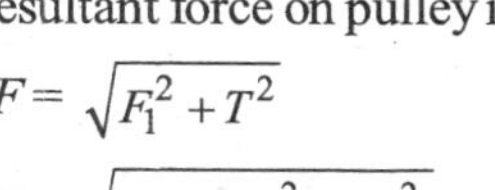

$F = \sqrt{F_1^2 + T^2}$

$F = \sqrt{(m+M)^2 + M^2}\,g$

68. **(d)** In case (a) In case (b)

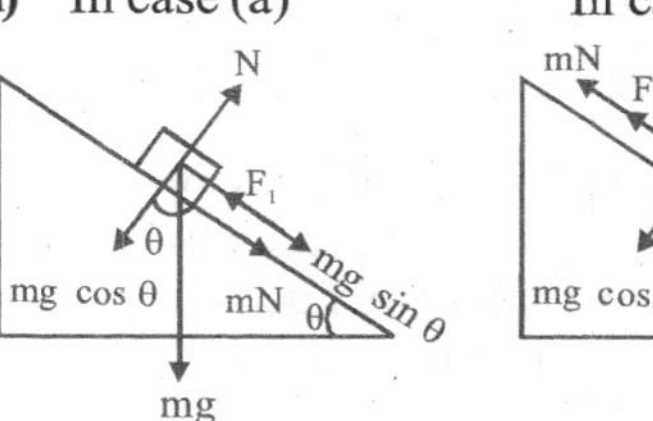

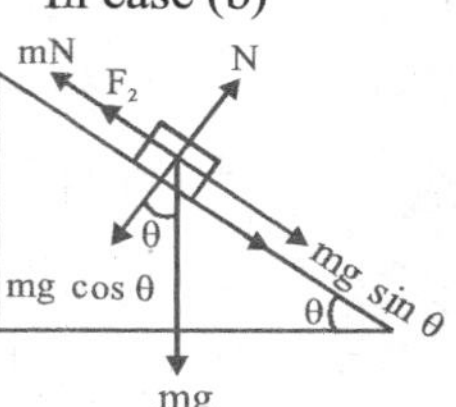

$mg \sin \theta = F_1 - \mu N$

$N = mg \cos \theta \therefore\ mg \sin \theta + \mu\, mg \cos \theta = F_1$

In second case (b)

$\mu N + F_2 = mg \sin \theta$

$\Rightarrow \mu\, mg \cos \theta - F_2 = mg \sin \theta$

or $F_2 = mg \sin \theta - \mu mg \cos \theta$ but $F_1 = 2F_2$

therefore $mg \sin \theta + \mu\, mg \cos \theta$

$= 2(mg \sin \theta - \mu\, mg \cos \theta)$

$mg \sin \theta = 3\, \mu\, mg \cos \theta$

or $\tan \theta = 3\mu$ or $\theta = \tan^{-1}(3\mu)$

69. (d) Supposing block is fixed and wedge moves with velocity 3m/s relative to block, and so

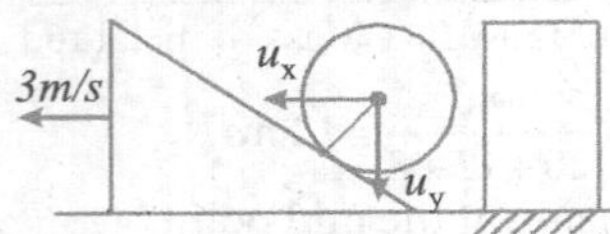

$3 \sin 30° = v_y \cos 30°, \quad \therefore\ v_y = \sqrt{3}$ m/s

and $u_x = \dfrac{1}{\sin 30°} = 2$m/s

$\therefore u = \sqrt{u_x^2 + u_y^2} = \sqrt{4+3} = \sqrt{7}$ m/s.

70. (d) Frictional force on the box $f = \mu mg$

$\therefore$ Acceleration in the box $a = \mu g = 5$ ms^{-2}

$v^2 = u^2 + 2as \Rightarrow 0 = 2^2 + 2 \times (5)\, s \Rightarrow s = -\dfrac{2}{5}$

w.r.t. belt $\Rightarrow$ distance $= 0.4$ m

71. (c) From the geometry

$AD = 2r;\ AC = 2r \cos \theta$

$CD = 2r \sin \theta$

$AG = 1.5r;\ GC = (2r \cos$

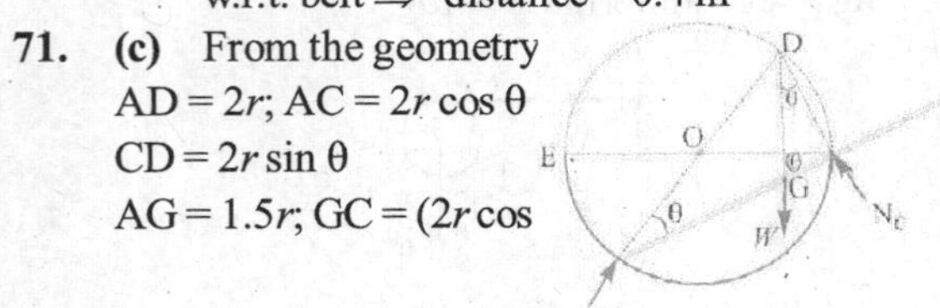

$\tan \theta = \dfrac{GC}{CD} = \dfrac{2r \cos \theta - 1.5r}{2r \sin \theta}$

After simplifying, we get, $\cos \theta = 0.9$.

72. (b)

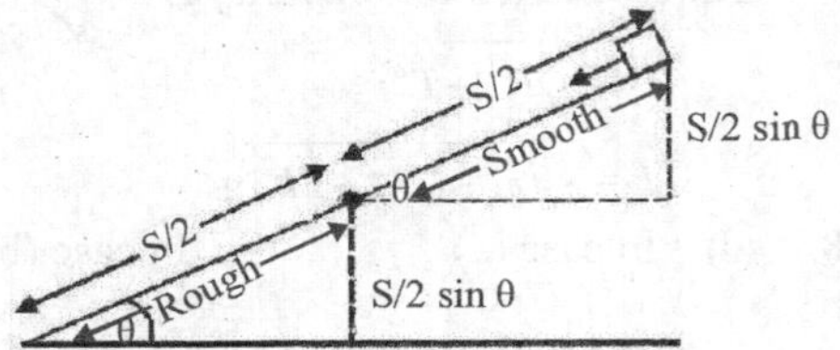

For upper half of inclined plane

$v^2 = u^2 + 2a\, S/2 = 2\,(g \sin \theta)\, S/2 = gS \sin \theta$

For lower half of inclined plane

$0 = u^2 + 2\,g\,(\sin \theta - \mu \cos \theta)\, S/2$

$\Rightarrow -gS \sin \theta = gS\,(\sin\theta - \mu \cos \theta)$

$\Rightarrow 2 \sin \theta = \mu \cos \theta$

$\Rightarrow \mu = \dfrac{2 \sin \theta}{\cos \theta} = 2 \tan \theta$

73. (d) **Case I :**

$T - N = 40\,a$

and $20g - T = 20\,a$

Also $N = 20\,a$

After simplifying, we get

$a = \dfrac{g}{4}$

Acceleration of block B, $= \sqrt{2}a = \dfrac{g}{2\sqrt{2}}$.

Case II :

$T = 40\,a$

and $20\,g - T = 20\,a$

After simplifying above equation, we get

$a = g/3$

Ratio $= \dfrac{g/2\sqrt{2}}{g/3} = \dfrac{3}{2\sqrt{2}}$.

74. (a) For the wedge,

$N' \sin \theta = F$

$\therefore\ N' = \dfrac{F}{\sin \theta}$. As $\sin\theta < 1$, and so $N' > F$

75. (d) When brakes are on, the wheels of the cycle will slide on the road instead of rolling there. It means the sliding friction will come into play instead of rolling friction. The value of sliding friction is more than that of rolling friction.

76. (a)

77. (d) For the block to remain stationary with the wall

$f = W$

$\mu N = W$

$0.2 \times 10 = W \Rightarrow W = 2N$

78. (d) Friction is the retarding force for the block

$F = ma = \mu R = \mu mg$

Therefore, from the first equation of motion

$v = u - at$

$0 = V - \mu g \times t \Rightarrow \dfrac{V}{\mu g} = t$

79. (c) Acceleration of mass at distance x

$a = g(\sin\theta - \mu_0 x\cos\theta)$

Speed is maximum, when $a = 0$

$g(\sin\theta - \mu_0 x\cos\theta) = 0$

$x = \dfrac{\tan\theta}{\mu_0}$

80. (c) $\sin\theta = \dfrac{r}{3r} = \dfrac{1}{3}, \cos\theta = \sqrt{8/9}$

If mass of smaller cylinder is m, then mass of bigger one will be 4m. For the equilibrium of upper cylinder,

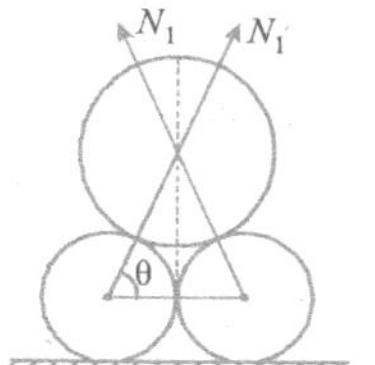

N_2 N_1 θ f mg

$2N_1\sin\theta = 4mg$

$\therefore \quad N_1 = \dfrac{2mg}{\sin\theta}$ ...(i)

Now for the equilibrium of lower cylinder,

$N_2 = N_1\sin\theta + mg$ ...(ii)

and $f = N_1\cos\theta$

or $\mu N_2 = N_1\cos\theta$ ...(iii)

After solving above equations, we get

$\mu = \dfrac{1}{3\sqrt{2}}$.

81. (b)

82. (c)

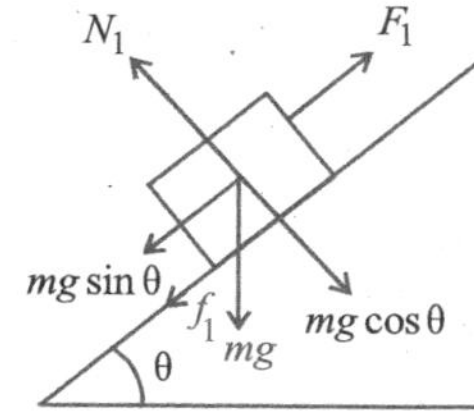

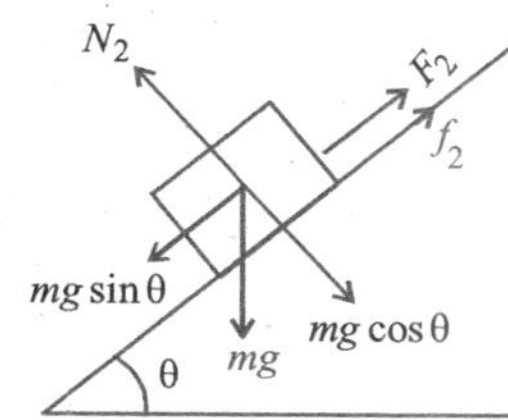

For the upward motion of the body

$mg\sin\theta + f_1 = F_1$

or, $F_1 = mg\sin\theta + \mu mg\cos\theta$

For the downward motion of the body,

$mg\sin\theta - f_2 = F_2$

or $F_2 = mg\sin\theta - \mu mg\cos\theta$

$\therefore \quad \dfrac{F_1}{F_2} = \dfrac{\sin\theta + \mu\cos\theta}{\sin\theta - \mu\cos\theta}$

$\Rightarrow \dfrac{\tan\theta + \mu}{\tan\theta - \mu} = \dfrac{2\mu + \mu}{2\mu - \mu} = \dfrac{3\mu}{\mu} = 3$

83. (d) When the body has maximum speed then

$\mu = 0.3x = \tan 45°$ $\therefore x = 3.33$ m

84. (c) Minimum force on A

= frictional force between the surfaces = 12 N

Therefore maximum acceleration

A 4 kg
B 5 kg
Smooth table

$a_{max} = \dfrac{12N}{4kg} = 3\text{m/s}^2$

Hence maximum force,

F_{max} = total mass × a_{max} = 9 × 3 = 27 N

85. (c) As m would slip in vertically downward direction, then

$mg = \mu N$

$\Rightarrow N = \dfrac{mg}{\mu} = \dfrac{100}{0.5} = 200$ Newton

Same normal force would accelerated M,

thus $a_M = \dfrac{200}{50} = 4\ \text{m/s}^2$

Taking $m + M$ as system

$F = (m + M)4 = 240$ N

86. (a)

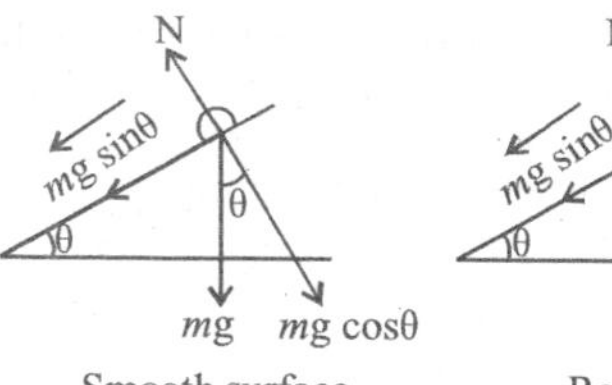

Smooth surface

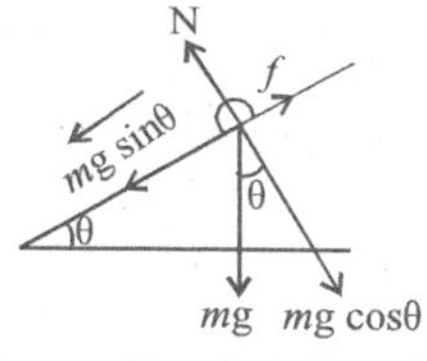

Rough surface

For smooth surface, $s = \dfrac{1}{2}g\sin\theta t_1^2$ ...(i)

For rough surface, $a = g(\sin\theta - \mu\cos\theta)\,t_2^2$

$\therefore s = \dfrac{1}{2}g(\sin\theta - \mu\cos\theta)t_2^2$ ...(ii)

From (i) and (ii),

$\therefore s = \dfrac{1}{2}g\sin\theta t_1^2 = \dfrac{1}{2}g(\sin\theta - \mu\cos\theta)t_2^2$

Given, $\theta = 45°$ $\therefore t_1^2 = (1-\mu)t_2^2$

Also, given that, $t_2 = nt_1$ $\therefore t_1^2 = (1-\mu)n^2t_1^2$

$\dfrac{1}{n^2} = 1 - \mu$ $\quad \therefore \mu = \left(1 - \dfrac{1}{n^2}\right)$

87. (a) For no slipping between m and M,

$F \le (M + m)g/3$

$F \le 40\,N$

For no toppling of m block

$F \le (M + m)g/4$

$F \le 30$ N

$\therefore F_{min} = 30N$

88. **(b)** $mg = \mu N = \mu ma \quad \therefore a = \dfrac{g}{\mu}$.

89. **(c)** $mg\sin\theta + mg\cos\theta = ma$

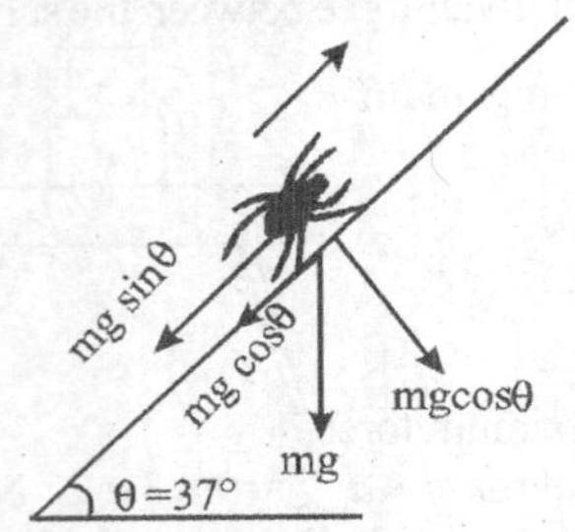

$$a = g\sin\theta + g\cos\theta = 10\left[\frac{3}{5}+\frac{4}{5}\right] = 14\ m/\sec^2$$

If $f_r = mg\sin\theta$

$= mg \times \dfrac{3}{5} = \dfrac{3mg}{5} < f_{r\max}$

$f_r < f_{r\max}$

$\dfrac{3mg}{5} < \dfrac{4mg}{5}$

Hence insect can move with constant velocity.

90. **(a)** The velocity should be such that the centripetal acceleration is equal to the acceleration due to gravity $\dfrac{v^2}{R} = g$

or $v = \sqrt{gR}$.

91. **(d)** Given; speed = 10 m/s; radius r = 10 m

Angle made by the wire with the vertical

$$\tan\theta = \frac{v^2}{rg} = \frac{10^2}{10\times 10} = 1 \Rightarrow \theta = 45° = \frac{\pi}{4}$$

92. **(b)** The centrifugal force on the particle = mass × acceleration of the frame

$= m \times \omega_0^2 r = m\omega_0^2 r$

93. **(b)** $v = \sqrt{gr} = \sqrt{10\times 40} = 20\ \text{m s}^{-1}$

94. **(c)** Tension at the highest point

$$T_{top} = \frac{mv^2}{r} - mg = 2mg \quad (\therefore v_{top} = \sqrt{3gr})$$

Tension at the lowest point

$T_{bottom} = 2mg + 6mg = 8mg$

$$\therefore \frac{T_{top}}{T_{bottom}} = \frac{2mg}{8mg} = \frac{1}{4}.$$

95. **(a)** $\tan\theta = \dfrac{v^2}{rg}$

$\tan 12° = (150)^2 / r \times 10$

$$\Rightarrow r = \frac{2250}{0.2125} = 10.6\ \text{km}$$

96. **(c)**

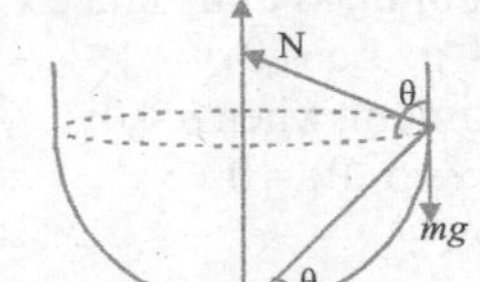

$N\cos\theta = mg$ and $N\sin\theta = m\omega^2 r$

$$\therefore \quad \tan\theta = \frac{\omega^2 r}{g} \quad \ldots\text{(i)}$$

Given $\quad y = x^2$

$$\therefore \quad \frac{dy}{dx} = 2x$$

or $\quad \tan\theta = 2\times 1 = 2 \quad \ldots\text{(ii)}$

From above equations, we get

$\omega = \sqrt{2g} \quad (r = 1\text{ m})$

97. **(d)** Given, $\theta = 45°$, r = 0.4 m, g = 10 m/s²

$T\sin\theta = \dfrac{mv^2}{r}$ ...... (i)

$T\cos\theta = mg$ ...... (ii)

From equation (i) & (ii) we have,

$\tan\theta = \dfrac{v^2}{rg}$

$v^2 = rg \quad \because \theta = 45°$

Hence, speed of the pendulum in its circular path,

$v = \sqrt{rg} = \sqrt{0.4\times 10} = 2$ m/s

98. **(b)** On a banked road, $\dfrac{V_{max}^2}{Rg} = \left(\dfrac{\mu_s + \tan\theta}{1 - \mu_s\tan\theta}\right)$

Maximum safe velocity of a car on the banked road

$$V_{max} = \sqrt{Rg\left[\frac{\mu_s + \tan\theta}{1-\mu_s\tan\theta}\right]}$$

99. **(b)** Centripetal acceleration

$a_n = \dfrac{v^2}{r} = \dfrac{10^2}{2} = 50\ \text{m/s}^2$, along negative x-axis.

so $\quad \vec{a} = -50\hat{i} - 1.5\hat{j}\ \text{m/s}^2$

100. **(a)** B T_2 T_1 T_1 λ 2λ

$T_1 = m\omega^2(3\ell)$

and $\quad T_2 - T_1 = m\omega^2\ell$

or $\quad T_2 = T_1 + m\omega^2\ell = m\omega^2(3\ell) + m\omega^2\ell$

$= 4m\omega^2\ell$

$$\therefore \quad \frac{T_2}{T_1} = \frac{4}{3}.$$

5 Work, Energy and Power

1. A ball whose kinetic energy is E, is projected at an angle of 45° to the horizontal. The kinetic energy of the ball at the highest point of its flight will be
   (a) E (b) $E/\sqrt{2}$ (c) E/2 (d) zero.

2. A load hangs from a travelling crane, moving horizontally with velocity v. If the load is not to swing more than 4 m horizontally, when the crane is stopped suddenly, what is the maximum allowable speed of the crane? **Critical Thinking**
   (a) 4.05 m/s
   (b) 4.00 m/s
   (c) 3.00 m/s
   (d) 3.50 m/s

3. A 10 kg block is pulled in the vertical plane along a frictionless surface in the form of an arc of a circle of radius 10 m. The applied force is of 200 N as shown in figure. If the block had started from rest at A, the velocity at B would be
   (a) 1.7 m/s
   (b) 17 m/s
   (c) 27 m/s
   (d) 34 m/s

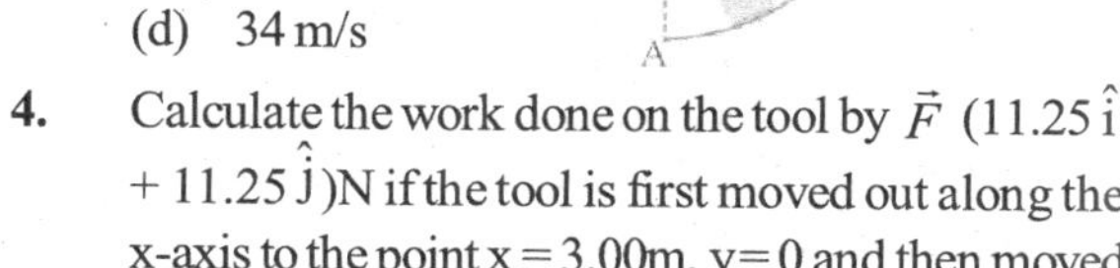

4. Calculate the work done on the tool by $\vec{F}$ $(11.25\hat{i} + 11.25\hat{j})$N if the tool is first moved out along the x-axis to the point x = 3.00m, y = 0 and then moved parallel to the y-axis to x = 3.00m, y = 3.00 m.
   (a) 67.5 J (b) 85 J (c) 102 J (d) 7.5 J

5. A force F acting on an object varies with distance x as shown here. The force is in N and x in m. The work done by the force in moving the object from x = 0 to x = 6 m is
   (a) 18.0 J (b) 13.5 J (c) 9.0 J (d) 4.5 J

6. A particle describe a horizontal circle of radius 0.5 m with uniform speed. The centripetal force acting is 10 N. The work done in describing a semicircle is
   (a) zero (b) 5 J (c) 5π J (d) 10π J

7. A force F_x acts on a particle such that its position x changes as shown in the figure. The work done by the particle as it moves from x = 0 to 20 m is
   (a) 37.5 J (b) 10 J
   (c) 45 J (d) 22.5 J

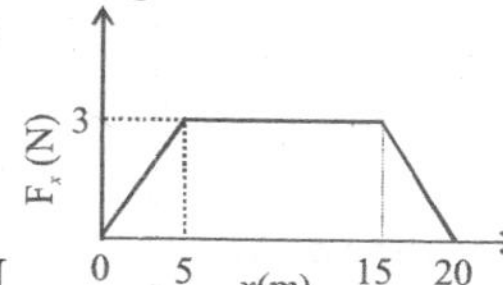

8. Two identical beads of m = 100 gram are connected by an inextensible massless string can slide along the two arms AC and BC of a rigid smooth wire frame in a vertical plane. If the system is released from rest, the kinetic energy of the first particle when they have moved by a distance of 0.1 m is $8x \times 10^{-3}$ J. Find the value of x. (g = 10 m/s²)
   (a) 8
   (b) 6
   (c) 9
   (d) 11

9. A cord is used to lower vertically a block of mass M, a distance d at a constant downward acceleration of g/4. The work done by the cord on the block is
   (a) $Mg\frac{d}{4}$ (b) $3Mg\frac{d}{4}$ (c) $-3Mg\frac{d}{4}$ (d) Mg d

10. A spring of spring constant 5×10^3 N/m is stretched initially by 5 cm from the unstretched position. Then the work required to stretch it further by another 5 cm is
    (a) 18.75 J (b) 25.00 J (c) 6.25 J (d) 12.50 J

11. A uniform force of $(3\hat{i} + \hat{j})$ newton acts on a particle of mass 2 kg. The particle is displaced from position $(2\hat{i} + \hat{k})$ meter to position $(4\hat{i} + 3\hat{j} - \hat{k})$ meter. The work done by the force on the particle is
    (a) 6 J (b) 13 J (c) 15 J (d) 9 J

12. The position of a particle of mass 4 g, acted upon by a constant force is given by $x = 4t^2 + t$, where x is in metre and t in second. The work done during the first 2 seconds is
    (a) 128 mJ (b) 512 mJ (c) 576 mJ (d) 144 mJ

13. A body moves a distance of 10 m along a straight line under the action of a force of 5 newtons. If the work done is 25 joules, the angle which the force makes with the direction of motion of body is
(a) 0º (b) 30º (c) 60º (d) 90º

14. A block of mass m is suspended by a light thread from a lift. The lift is moving upward with uniform velocity v. From the frame of lift, the work done by tension on the block in t seconds will be
(a) $-mgvt$
(b) 0
(c) $\frac{mgvt}{2}$
(d) $mgvt$

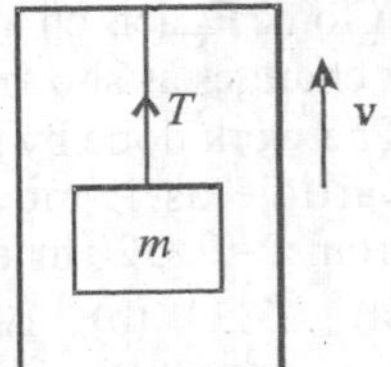

15. The work done by a force $\vec{F}=(-6x^3\hat{i})\,N$, in displacing a particle from $x=4\,m$ to $x=-2\,m$ is
(a) $360J$ (b) $240J$ (c) $-240J$ (d) $-360J$

16. Work done by static friction on an object
(a) may be positive (b) must be negative
(c) must be zero (d) None of these

17. A particle is placed at the origin and a force $F=kx$ is acting on it (where k is positive constant). If $U(0)=0$, the graph of $U(x)$ versus x will be (where U is the potential energy function) : **Tricky**

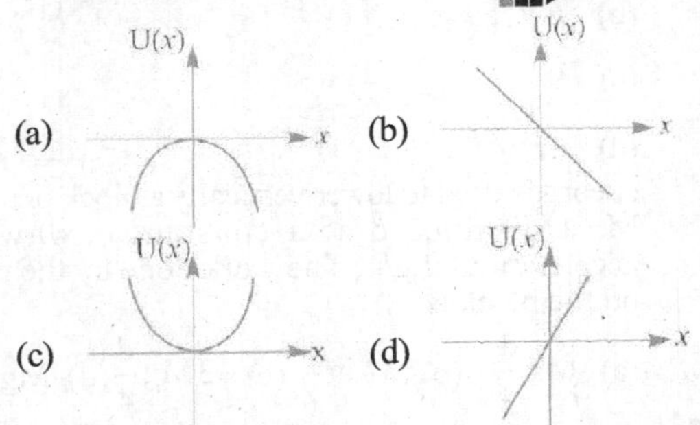

18. Calculate the K.E and P.E. of the ball half way up, when a ball of mass 0.1 kg is thrown vertically upwards with an initial speed of 20 ms^{-1}.
(a) 10 J, 20 J (b) 10 J, 10 J
(c) 15 J, 8 J (d) 8 J, 16 J

19. At time $t=0$s particle starts moving along the x-axis. If its kinetic energy increases uniformly with time t, the net force acting on it must be proportional to
(a) $\sqrt{t}$ (b) constant
(c) t (d) $\frac{1}{\sqrt{t}}$

20. Two bodies of masses 4 kg and 5 kg are moving with equal momentum. Then the ratio of their respective kinetic energies is
(a) 4 : 5 (b) 2 : 1 (c) 1 : 3 (d) 5 : 4

21. A rod of mass m and length ℓ is made to stand at an angle of 60° with the vertical. Potential energy of the rod in this position is
(a) $mg\ell$ (b) $\frac{mg\ell}{2}$ (c) $\frac{mg\ell}{3}$ (d) $\frac{mg\ell}{4}$

22. A 2 kg block slides on a horizontal floor with a speed of 4m/s. It strikes a uncompressed spring, and compresses it till the block is motionless. The kinetic friction force is 15N and spring constant is 10,000 N/m. The spring compresses by **Critical Thinking**
(a) 8.5 cm (b) 5.5 cm (c) 2.5 cm (d) 11.0 cm

23. The potential energy of a conservative system is given by $U = ay^2 - by$, where y represents the position of the particle and a as well as b are constants. What is the force acting on the system ?
(a) $-ay$ (b) $-by$ (c) $2ay-b$ (d) $b-2ay$

24. A block of mass $m = 0.1$ kg is connected to a spring of unknown spring constant k. It is compressed to a distance x from its equilibrium position and released from rest. After approaching half the distance $\left(\frac{x}{2}\right)$ from equilibrium position, it hits another block and comes to rest momentarily, while the other block moves with a velocity 3 ms^{-1}.
The total initial energy of the spring is
(a) 0.3 J (b) 0.6 J (c) 0.8 J (d) 1.5 J

25. A block of 1 kg is kept on a rough surface of an elevator moving up with constant velocity of 5m/s. In 10 seconds work done by normal reaction (the block does not slide on the inclined surface)

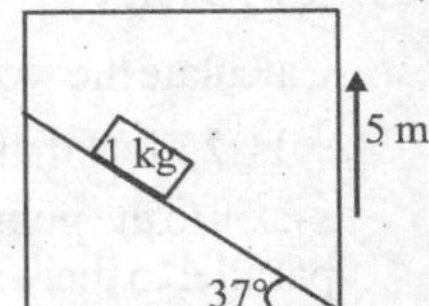

(i) from ground frame is 320 J
(ii) is equal to work done by friction force in elevator frame
(iii) is equal to work done by friction in ground frame

Of the three statements given above, the one that is true is given by the choice
(a) only (i) (b) (ii) and (iii)
(c) (i) and (ii) (d) only (iii)

26. 300 J of work is done in sliding a 2 kg block up an inclined plane of height 10 m. Taking g = 10 m/s^2, work done against friction is
(a) 100 J (b) zero (c) 1000 J (d) 200 J

27. A force acts on a 30 gm particle in such a way that the position of the particle as a function of time is given by $x = 3t - 4t^2 + t^3$, where x is in metres and t is in seconds. The work done during the first 4 seconds is
(a) 576 mJ (b) 450 mJ (c) 490 mJ (d) 530 mJ

28. A man starts walking from a point on the surface of earth (assumed smooth) and reaches diagonally opposite point.
What is the work done by him?
(a) Zero (b) Positive
(c) Negative (d) Nothing can be said

29. The components of a force acting on a particle are varying according to the graphs shown. When the particles move from (0, 5, 6) to (2, 10, 0) then the work done by this force is **Tricky**

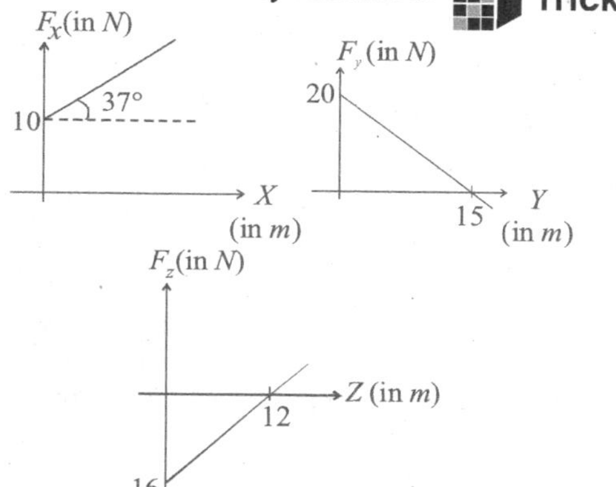

(a) 192 J (b) $\frac{400}{3}$ J
(c) $\frac{287}{2}$ J (d) None of these

30. A force $F = -K(y\hat{i} + x\hat{j})$ (where K is a positive constant) acts on a particle moving in the xy plane. Starting from the origin, the particle is taken along the positive x axis to the point $(a, 0)$, and then parallel to the y axis to the point (a, a), The total work done by the force F on the particle is
(a) $-2Ka^2$ (b) $2Ka^2$ (c) $-Ka^2$ (d) Ka^2

31. A block of mass 1 kg is pulled along the curve path ACB by a tangential force as shown in figure. The work done by the frictional force when the block moves from A to B is

μ = 0.2
C
A B
x = 10 m

(a) 5 J
(b) 10 J
(c) 20 J
(d) None of these

32. If stretch in a spring of force constant k is tripled then the ratio of elastic potential energy in the two cases will be
(a) 9 : 1 (b) 1 : 6 (c) 3 : 1 (d) 1 : 3

33. One man takes 1 min. to raise a box to a height of 1 metre and another man takes 1/2 min. to do so. The energy of the
(a) two is different (b) two is same
(c) first is more (d) second is more

34. If the momentum of a body is increased by 50%, then the percentage increase in its kinetic energy is
(a) 50% (b) 100% (c) 125% (d) 200%

35. A ball dropped from a height of 2m reaches to a height of 1.5m before hitting the ground. Then the percentage of potential energy lost is
(a) 25 (b) 30 (c) 50 (d) 100

36. A body starts from rest and acquires a velocity V in time T. The work done on the body in time t will be proportional to
(a) $\frac{V}{T}t$ (b) $\frac{V^2}{T}t^2$ (c) $\frac{V^2}{T^2}t$ (d) $\frac{V^2}{T^2}t^2$

37. Four particles given, have same momentum. Which has maximum kinetic energy
(a) Proton (b) Electron
(c) Deutron (d) α-particles

38. A running man has half the kinetic energy of that of a boy of half of his mass. The man speeds up by 1m/s so as to have same K.E. as that of the boy. The original speed of the man will be
(a) $\sqrt{2}$ m/s (b) $(\sqrt{2}-1)$ m/s
(c) $\frac{1}{(\sqrt{2}-1)}$ m/s (d) $\frac{1}{\sqrt{2}}$ m/s

39. A steel ball of mass 5g is thrown downward with velocity 10 m/s from height 19.5 m. It penetrates sand by 50 cm. The change in mechanical energy will be (g = 10 m/s^2)
(a) 1J (b) 1.25 J (c) 1.5 J (d) 1.75 J

40. The figure gives the potential energy function $U(x)$ for a system in which a particle is in one-dimensional motion.
In which region the magnitude of the force on the particle is greatest :

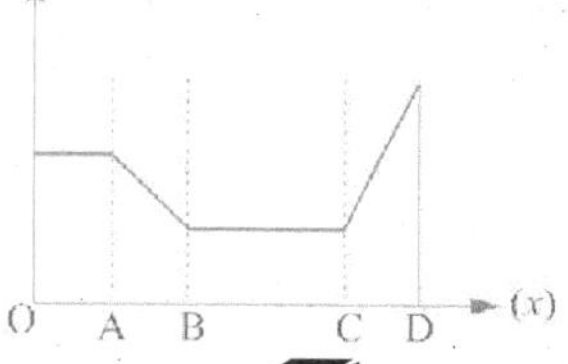

Tricky

(a) OA (b) AB (c) BC (d) CD

41. A ball is allowed to fall from a height of 10 m. If there is 40% loss of energy due to air friction, then velocity of the ball when it hit the ground is
(a) $\sqrt{190}$ m/s (b) $\sqrt{180}$ m/s
(c) $\sqrt{150}$ m/s (d) $\sqrt{120}$ m/s

42. A body falls freely under gravity. Its velocity is v when it has lost potential energy equal to U. What is the mass of the body?
(a) U^2/v^2 (b) $2U^2/v^2$ (c) $2U/v^2$ (d) U/v^2

43. When a body is projected vertically up from the ground with certain velocity, its potential energy and kinetic energy at a point A are in the ratio 2 : 3. If the same body is projected with double the previous velocity, then at the same point A the ratio of its potential energy to kinetic energy is
(a) 9 : 1 (b) 2 : 9 (c) 1 : 9 (d) 9 : 2

44. A uniform chain of length 2 m and mass 0.1 kg overhangs a smooth table with its two third part lying on the table. Find the kinetic energy of the chain as it completely slips- off the table.
(a) $\frac{8}{9}$ J (b) $\frac{12}{5}$ J (c) $\frac{3}{7}$ J (d) $\frac{11}{3}$ J

45. A 1 kg particle at a height of 8m has a speed of 10m/s down a fixed incline making an angle 53° with horizontal as shown in figure. It slides on a horizontal section of length 10 m at ground level and then up a fixed incline making an angle 37° with horizontal. All surfaces have $\mu_k = 0.5$. How far (in meters) from point O (bottom of right inclined plane), along the incline making an angle 37° with horizontal, does the particle first come to rest ? **Toughnut**

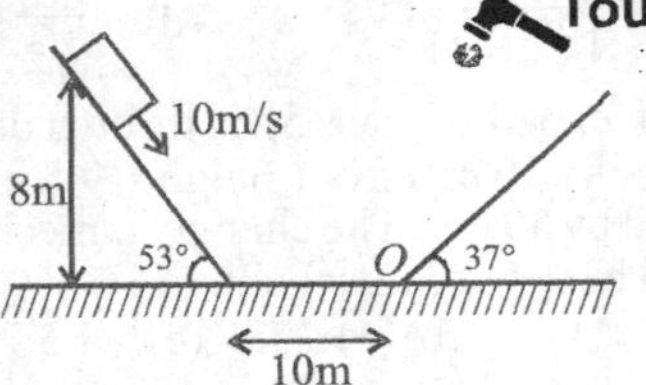

(a) 1m (b) 0.5m (c) 2m (d) 5m

46. A man places a chain of mass m and length L on a table slowly. Initially the lower end of the chain just touches the table. The man drops the chain when half of the chain is in vertical position. Then work done by the man in this process is
(a) $-mg\frac{L}{2}$ (b) $-\frac{mgL}{4}$ (c) $-\frac{3mgL}{8}$ (d) $-\frac{mgL}{8}$

47. A particle of mass 10 g moves along a circle of radius 6.4 cm with a constant tangential acceleration. What is the magnitude of this acceleration if the kinetic energy of the particle becomes equal to 8×10^{-4} J by the end of the second revolution after the beginning of the motion ?
(a) 0.1 m/s² (b) 0.15 m/s²
(c) 0.18 m/s² (d) 0.2 m/s²

48. A variable force P is maintained tangent to a frictionless cylindrical surface of radius a as shown in figure. By slowly varying this force, a block of weight W is moved and the spring to which it is stretched from position 1 to position 2. The work done by the force P is **Toughnut**

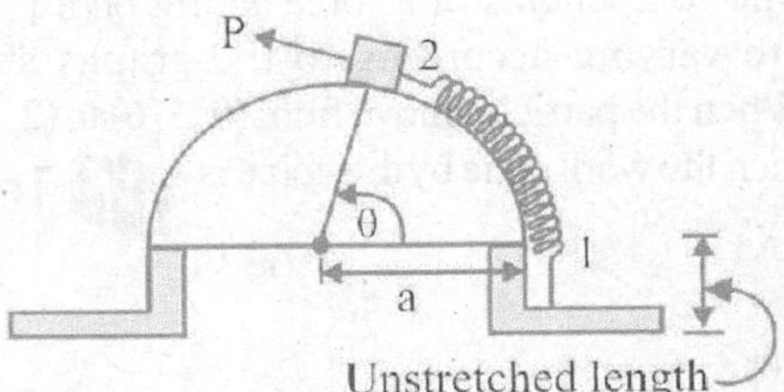

(a) $Wa\sin\theta$ (b) $\frac{1}{2}ka^2\theta^2$
(c) $Wa\sin\theta + ka^2\theta^2$ (d) $Wa\sin\theta + \frac{1}{2}ka^2\theta^2$

49. A straight bar, of mass 15 kg and length 2 m, at rest on a frictionless horizontal surface, receives an instantaneous impulse of 7.5 Ns perpendicular to the bar. If the impulse is applied at the mid point of the bar, the energy transfered is
(a) 3.2 J
(b) 1.9 J
(c) 3.8 J
(d) 2.5 J

50. A particle, which is constrained to move along the x-axis, is subjected to a force in the same direction which varies with the distance x of the particle from the origin as $F(x) = -kx + ax^3$. Here k and a are positive constants. For $x \geq 0$, the functional form of the potential energy $U(x)$ of the particle is

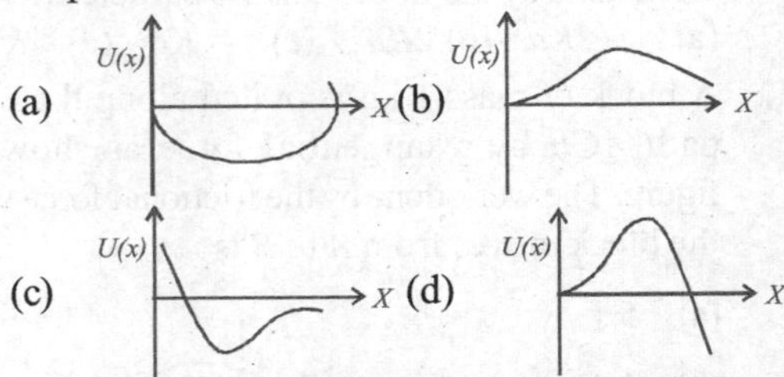

51. Two blocks of masses m and M are joined with an ideal spring of spring constant k and kept on a rough surface as shown. The spring is initially unstretched and the coefficient of friction

between the blocks and the horizontal surface is μ. What should be the maximum speed of the block of mass M such that the smaller block does not move?

Critical Thinking

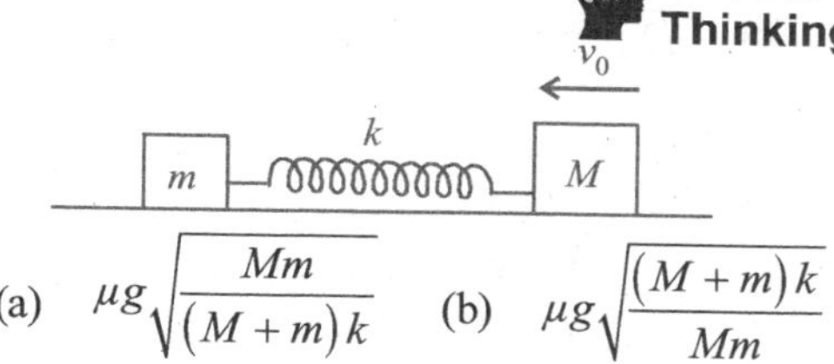

(a) $\mu g\sqrt{\frac{Mm}{(M+m)k}}$ (b) $\mu g\sqrt{\frac{(M+m)k}{Mm}}$

(c) $\mu g\sqrt{\frac{(2M+m)m}{kM}}$ (d) None of these

52. Two blocks of masses $m_1 = 10$ kg and $m_2 = 20$ kg are connected by a spring of stiffness $k = 200$N/m. The coefficient of friction between the blocks and the fixed horizontal surface is $\mu = 0.1$. Find the minimum constant horizontal force F (in newtons) to be applied to m_1 in order to slide the mass m_2. [Take $g = 10$ m/s²]

(a) $\mu m_1 g + \frac{\mu m_2 g}{2}$ (b) $\mu m_1 g + \mu m_2 g$

(c) $\mu m_1 g - \frac{\mu m_2 g}{2}$ (d) $\frac{\mu m_1 g + \mu m_2 g}{2}$

53. Two objects each of mass m are connected with a uniform rope having same mass m and length $\ell = 10/3\ m$ are shown in the figure. The rope can slide on a frictionless pulley which is fixed to the edge of the table. Initially 1/3 of the rope hangs vertically. Friction is negligible everywhere.

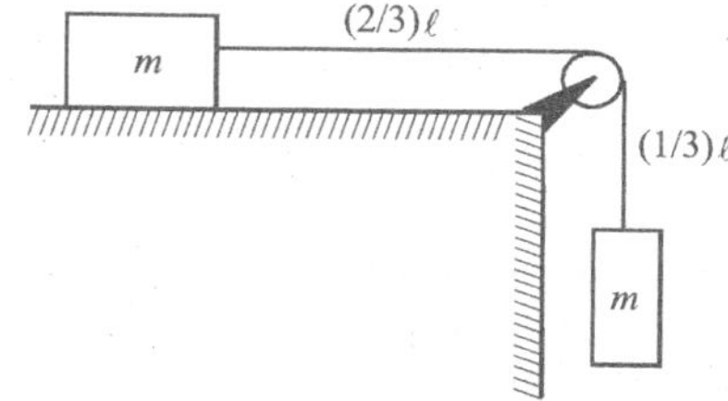

If the system is released from rest and object on the right hand side descend further by a distance of $\ell/3$, what would be the speed of the object (in m/s) ?

(a) 2.12 m/s (b) 3.33 m/s
(c) 0.33 m/s (d) 4.12 m/s

54. A bead of mass m is sliding down the fixed inclined rod without friction. It is connected to a point P on the horizontal surface with a light spring of spring constant k. The bead is initially released from rest and the spring is initially unstressed and vertical. The bead just stops at the bottom of the inclined rod. Find the angle which the inclined rod makes with horizontal.

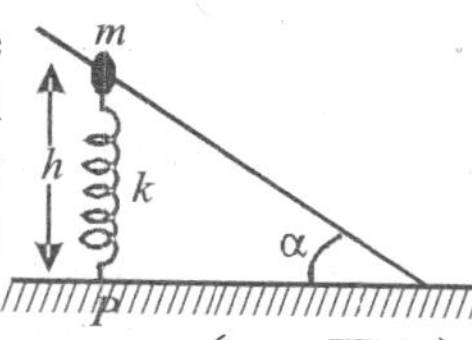

(a) $\cot^{-1}\left(1+\sqrt{\frac{2mg}{kh}}\right)$ (b) $\tan^{-1}\left(1+\sqrt{\frac{2mg}{kh}}\right)$

(c) $\cot^{-1}\left(1+\sqrt{\frac{mg}{kh}}\right)$ (d) $\tan^{-1}\left(1+\sqrt{\frac{mg}{kh}}\right)$

55. A body of mass m is accelerated uniformly from rest to a speed v in a time T. The instantaneous power delivered to the body as a function of time is given by

(a) $\frac{mv^2}{T^2}.t^2$ (b) $\frac{mv^2}{T^2}.t$

(c) $\frac{1}{2}\frac{mv^2}{T^2}.t^2$ (d) $\frac{1}{2}\frac{mv^2}{T^2}.t$

56. A car of mass m starts from rest and accelerates so that the instantaneous power delivered to the car has a constant magnitude p_o. The instantaneous velocity of this car is proportiourl to:

(a) $t^2 p_o$ (b) $t^{1/2}$ (c) $t^{-1/2}$ (d) $\frac{t}{\sqrt{m}}$

57. How much water, a pump of 2 kW can raise in one minute to a height of 10 m, take g = 10 m/s²?

(a) 1000 (b) 1200 (c) 100 (d) 2000

58. In figure, a block having a mass of 5 kg is released from rest, where the spring acting on the body are horizontal and have a tension of 50 N. The force constant of each spring is, k = 500 N/m. The velocity of the block after it has descended by 0.3 m.

Toughnut

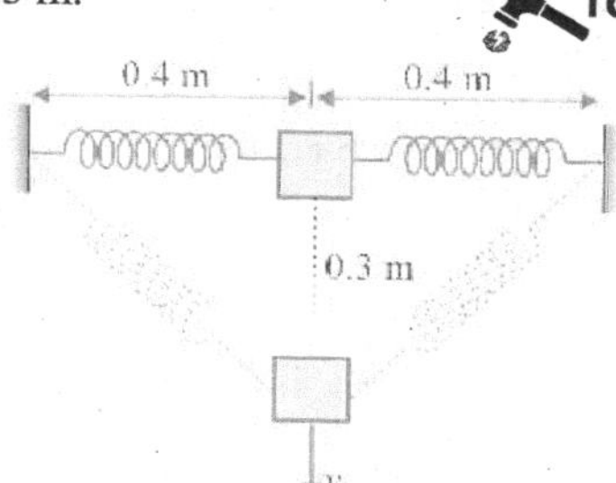

(a) 0 (b) 1 m/s (c) 2 m/s (d) none

59. The engine of a vehicle delivers constant power. If the vehicle is moving up the inclined plane then, its velocity,

(a) must remain constant
(b) must increase
(c) must decrease
(d) may increase, decrease or remain same.

60. A body is moved along a straight line by a machine delivering a constant power. The distance moved by the body in time 't' is proportional to **Tricky**

(a) $t^{3/4}$ (b) $t^{3/2}$ (c) $t^{1/4}$ (d) $t^{1/2}$

61. A body projected vertically from the earth reaches a height equal to earth's radius before returning to the earth. The power exerted by the gravitational force is greatest

(a) at the highest position of the body
(b) at the instant just before the body hits the earth
(c) it remains constant all through
(d) at the instant just after the body is projected

62. A body of mass 10 kg moves with a velocity v of 2 m/s along a circular path of radius 8 m. The power produced by the body will be

(a) 10 J/s (b) 98 J/s (c) 49 J/s (d) zero

63. A constant power P is applied to a car starting from rest. If v is the velocity of the car at time t, then

(a) $v \propto t$ (b) $v \propto \frac{1}{t}$

(c) $v \propto \sqrt{t}$ (d) $v \propto \frac{1}{\sqrt{t}}$

64. If two persons A and B take 2 seconds and 4 seconds respectively to lift an object to the same height h, then the ratio of their powers is

(a) 1 : 2 (b) 1 : 1 (c) 2 : 1 (d) 1 : 3

65. A 10 H.P. motor pumps out water from a well of depth 20 m and fills a water tank of volume 22380 litres at a height of 10 m from the ground. The running time of the motor to fill the empty water tank is ($g = 10ms^{-2}$)

(a) 5 minutes (b) 10 minutes
(c) 15 minutes (d) 20 minutes

66. If a machine gun fires n bullets per second each with kinetic energy K, then the power of the machine gun is

(a) nK^2 (b) $\frac{K}{n}$ (c) n^2K (d) nK

67. A ring of mass m can slide over a smooth vertical rod as shown in figure. The ring is connected to a spring of force constant $k = 4\,mg/R$, where $2R$ is the natural length of the spring. The other end of spring is fixed to the ground at a horizontal distance $2R$ from the base of the rod. If the mass is released at a height $1.5R$, then the velocity of the ring as it reaches the ground is **Critical Thinking**

(a) $\sqrt{gR}$
(b) $2\sqrt{gR}$
(c) $\sqrt{2gR}$
(d) $\sqrt{3gR}$

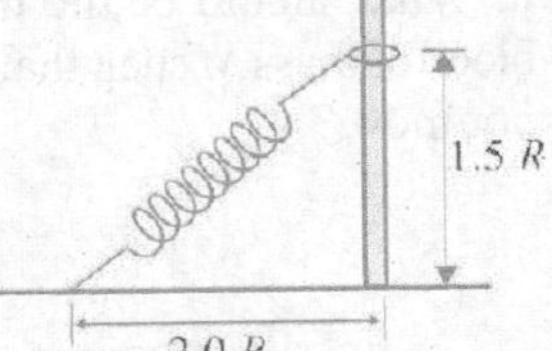

68. An engineer claims to have made an engine delivering 10kW power with fuel consumption of 1 g/s. The calorific value of fuel is 2 kcal/g. This claim is

(a) valid
(b) invalid
(c) depends on engine design
(d) dependent on load

69. A force applied by an engine of a train of mass 2.05×10^6 kg changes its velocity from 5m/s to 25 m/s in 5 minutes. The power of the engine is

(a) 1.025 MW (b) 2.05 MW
(c) 5 MW (d) 6 MW

70. A 10 m long iron chain of linear mass density 0.8 kg m^{-1} is hanging freely from a rigid support. If $g = 10\ ms^{-2}$, then the power required to left the chain upto the point of support in 10 second

(a) 10 W (b) 20W (c) 30 W (d) 40 W

71. The net power of all the forces acting on a particle (P) versus time curve is shown. Work done upon the particle from A to B **Tricky**

(a) increases
(b) decreases
(c) first increases then decreases
(d) first decreases then increases

72. A wind-powered generator converts wind energy into electrical energy. Assume that the generator converts a fixed fraction of the wind energy intercepted by its blades into electrical energy. For wind speed v, the electrical power output will be proportional to

(a) v (b) v^2 (c) v^3 (d) v^4

73. A body of mass 1 kg begins to move under the action of a time dependent force $\vec{F} = (2t\hat{i} + 3t^2\hat{j})$ N, where $\hat{i}$ and $\hat{j}$ are unit vectors alogn x and y axis. What power will be developed by the force at the time t?

(a) $(2t^2 + 3t^3)$W (b) $(2t^2 + 4t^4)$W
(c) $(2t^3 + 3t^4)$ W (d) $(2t^3 + 3t^5)$W

74. The energy content of gasoline is 3.6×10^7 J/L. A motor with an efficiency of 20% is needed at full output power of 45kW for 50.0 minutes. How many litres of gasoline are required to operate the motor for this amount of time ?

(a) 0.31 L (b) 0.38 L (c) 1.6 L (d) 19 L

75. An escalator has 2400 W of power available to move passengers from the first floor of a mall to the second, 6.0 m vertically. If the average mass of the passengers is 65 kg, what is the maximum number of passengers that can be carried to the second floor in 1.0 minute?

(a) 4

(b) 5

(c) 22

(d) 37

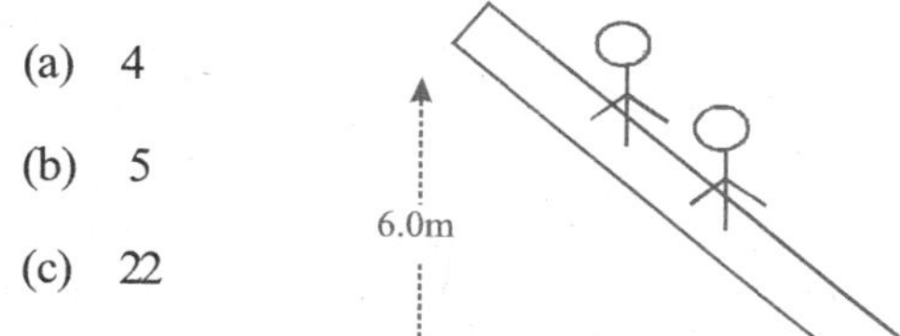

76. An iron ball of mass *m*, suspended by a light inextensible string of length ℓ from a fixed point *O*, is shifted by an angle θ_0 as shown so as to strike the vertical wall perpendicularly. The maximum angle made by the string with vertical after the first collision, if e is the coefficient of restitution, is

(a) $\sin^{-1}\{1-e^2(1-\cos\theta_0)\}$

(b) $\cos^{-1}\{1-e^2(1-\cos\theta_0)\}$

(c) $\tan^{-1}\{1-e^2(1-\cos\theta_0)\}$

(d) zero

77. Mass m_1 strikes m_2 which is at rest. The ratio of masses for which they will collide again (collision between ball and wall are elastic, coefficient of restitution between m_1 and m_2 is *e* and all the surfaces are smooth.) **Tricky**

(a) $<\frac{e}{2+e}$

(b) $>\frac{2e}{2+e}$

(c) $\geq\frac{e}{2(2+e)}$

(d) None of the above

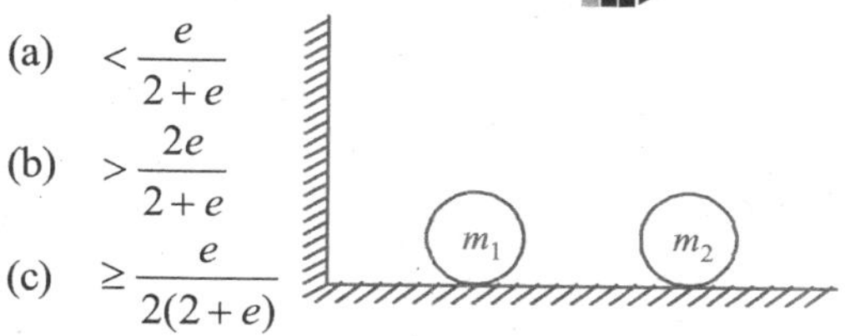

78. Three masses m, 2m and 3m are moving in x-y plane with speed 3u, 2u and u respectively as shown in figure. The three masses collide at the same point at P and stick together. The velocity of resulting mass will be

(a) $\frac{u}{12}\left(\hat{i}+\sqrt{3}\hat{j}\right)$

(b) $\frac{u}{12}\left(\hat{i}-\sqrt{3}\hat{j}\right)$

(c) $\frac{u}{12}\left(-\hat{i}+\sqrt{3}\hat{j}\right)$

(d) $\frac{u}{12}\left(-\hat{i}-\sqrt{3}\hat{j}\right)$

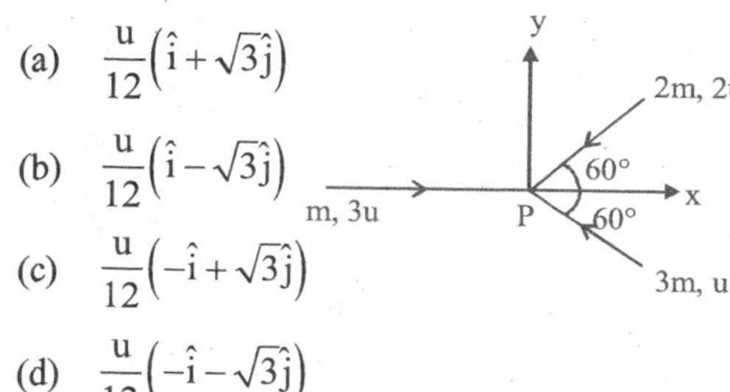

79. Two small bodies of masses 'm' and '2m' are placed in a fixed smooth horizontal circular hollow tube of mean radius 'r' as shown. The mass 'm' is moving with speed 'u' and the mass '2m' is stationary. After their first collision, the time elapsed for next collision is [coefficient of restitution e = 1/2]

(a) $\frac{2\pi r}{u}$ (b) $\frac{4\pi r}{u}$

(c) $\frac{3\pi r}{u}$ (d) $\frac{12\pi r}{u}$

80. A block lying on a smooth surface with spring connected to it is pulled by an external force as shown. Initially the velocity of ends A and B of the spring are 4 m/s and 2 m/s respectively. If the energy of the spring is increasing at the rate of 20 J/sec, then the stretch in the spring is **Critical Thinking**

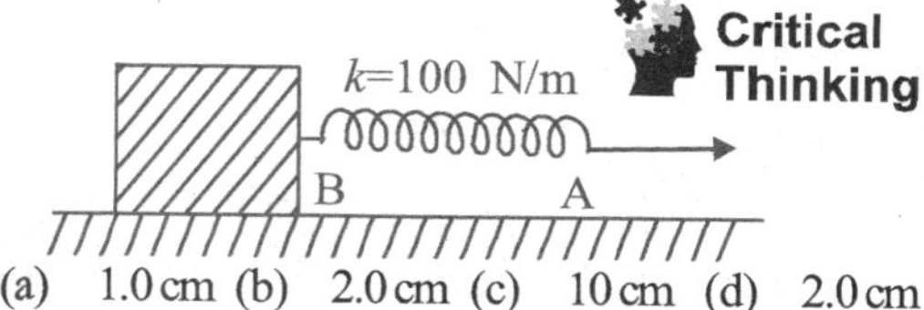

(a) 1.0 cm (b) 2.0 cm (c) 10 cm (d) 2.0 cm

81. A car of weight W is on an inclined road that rises by 100 m over a distance of 1 Km and applies a constant frictional force $\frac{W}{20}$ on the car. While moving uphill on the road at a speed of 10 ms^{-1}, the car needs power P. If it needs power $\frac{P}{2}$ while moving downhill at speed *v* then value of *v* is:

(a) 20 ms^{-1} (b) 5 ms^{-1} (c) 15 ms^{-1} (d) 10 ms^{-1}

82. A block of mass $m = 1$kg is moving with a constant acceleration $a = 1$ m/s^2 on a rough horizontal plane. The coefficient of friction between the block and plane is $\mu = 0.1$. The initial velocity of block is zero at $t = 0$. The power delivered by the external agent at a time $t = 2$ sec from the beginning is equal to (Take g = 10m/s^2)

(a) 1 watt (b) 2 watt (c) 3 watt (d) 4 watt

83. A ball of mass m hits the floor making an angle θ as shown in the figure. If e is the coefficient of restitution, then which relation is true, for the velocity component before and after collision?

Tricky

(a) $V^1 \sin\theta' = V\sin\theta$

(b) $V^1 \sin\theta' = -\sin\theta$

(c) $V^1 \cos\theta' = V\cos\theta$

(d) $V^1 \cos\theta' = -V\cos\theta$

84. A rubber ball is dropped from a height of 5m on a plane, where the acceleration due to gravity is not shown. On bouncing it rises to 1.8 m. The ball loses its velocity on bouncing by a factor of

(a) $\frac{16}{25}$ (b) $\frac{2}{5}$ (c) $\frac{3}{5}$ (d) $\frac{9}{25}$

85. A ball of mass m moving with a constant velocity strikes against a ball of same mass at rest. If e = coefficient of restitution, then what will be the ratio of velocity of two balls after collision?

(a) $\frac{1-e}{1+e}$ (b) $\frac{e-1}{e+1}$ (c) $\frac{1+e}{1-e}$ (d) $\frac{2+e}{e-1}$

86. The block of mass M moving on the frictionless horizontal surface collides with the spring of spring constant k and compresses it by length L. The maximum momentum of the block after collision is

(a) $\frac{kL^2}{2M}$ (b) $\sqrt{Mk}\,L$

(c) $\frac{ML^2}{k}$ (d) zero

87. An object of mass 2.0 kg makes an elastic collision with another object of mass M at rest and continues to move in the original direction but with one-fourth of its original speed. What is the value of M?

(a) 0.75 kg (b) 1.0 kg

(c) 1.2 kg (d) None of these

88. A bullet of mass 20g and moving with 600 m/s collides with a block of mass 4 kg hanging with the string. What is velocity of bullet when it comes out of block, if block rises to height 0.2 m after collision ?

Critical Thinking

(a) 200m/s (b) 150 m/s (c) 400 m/s (d) 300 m/s

89. A block of mass 0.50 kg is moving with a speed of 2.00 ms^{-1} on a smooth surface. It strikes another mass of 1.00 kg and then they move together as a single body. The energy loss during the collision is

(a) 0.16 J (b) 1.00 J (c) 0.67 J (d) 0.34 J

90. The heart of man pumps 5 litres of blood through the arteries per minute at a pressure of 150 mm of mercury. If the density of mercury be 13.6×10^3 kg/m^3 and g = 10m/s^2 then the power of heart in watt is :

(a) 2.35 (b) 3.0 (c) 1.50 (d) 1.70

91. An engine is hauling a train of mass M kg on a level track at a constant speed v m/s. The resistance due to friction is f N / kg. What extra power must the engine develop to maintain the speed up a gradient of h in s :

(a) $\frac{Mghv}{s}$ (b) $\frac{Mghs}{v}$ (c) $Mghvs$ (d) zero

92. An electric pump is used to fill an overhead tank of capacity 9m^3 kept at a height of 10m above the ground. If the pump takes 5 minutes to fill the tank by consuming 10 kW power the efficiency of the pump should be (g = 10 ms^{-2})

(a) 60% (b) 40% (c) 20% (d) 30%

93. Water falls from a height of 60 m at the rate of 15 kg/s to operate a turbine. The losses due to frictional force are 10% of energy. How much power is generated by the turbine?(g = 10 m/s^2)

(a) 8.1 kW (b) 10.2 kW

(c) 12.3 kW (d) 7.0 kW

94. A particle of mass m is moving in a circular path of constant radius r such that its centripetal acceleration a_c is varying with time t as $a_c = k^2rt^2$ where k is a constant. The power delivered to the particles by the force acting on it is

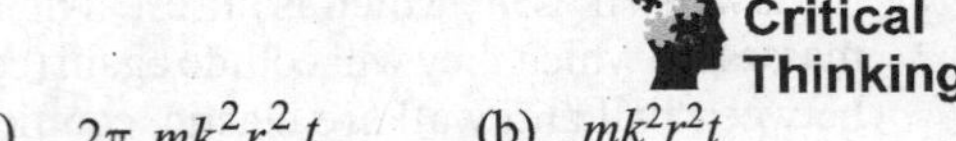

Critical Thinking

(a) $2\pi\, mk^2r^2\, t$ (b) mk^2r^2t

(c) $\frac{(mk^4r^2t^5)}{3}$ (d) zero

95. A uniform rope of linear mass density λ and length ℓ is coiled on a smooth horizontal surface. One end is pulled up with constant velocity v. Then the average power applied by the external agent in pulling the entire rope just off the horizontal surface is

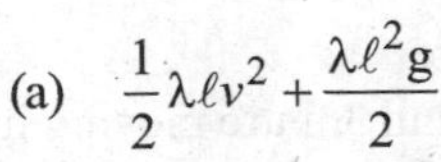

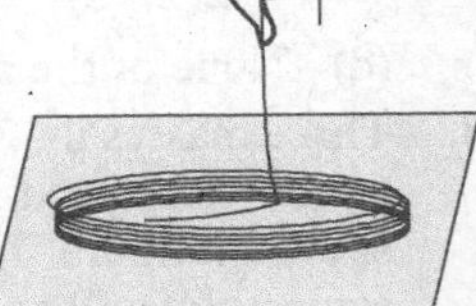

(a) $\frac{1}{2}\lambda\ell v^2 + \frac{\lambda\ell^2 g}{2}$

(b) $\lambda\ell gv$

(c) $\frac{1}{2}\lambda v^3 + \frac{\lambda \ell v g}{2}$ (d) $\lambda \ell v g + \frac{1}{2}\lambda v^3$

96. A particle falls from a height h on a fixed horizontal plane and rebounds. If e is the coefficient of restitution, the total distance travelled by the particle before it stops rebounding is Toughnut

(a) $\frac{h}{2}\frac{[1-e^2]}{[1+e^2]}$ (b) $\frac{h[1-e^2]}{[1+e^2]}$

(c) $\frac{h}{2}\frac{[1+e^2]}{[1-e^2]}$ (d) $\frac{h[1+e^2]}{[1-e^2]}$

97. A collision occurs between two identical balls of mass m each, moving with velocities $\vec{u}_1$ and $\vec{u}_2$. If the collision is head on and the energy lost in the collision is $\Delta E = \frac{3}{16}(\vec{u}_1 - \vec{u}_2)^2$ then the coefficient of restitution is

(a) 0.25 (b) 0.75 (c) 0.5 (d) 0.9

98. A massive disc of radius R is moved with a constant velocity u on a frictionless table. Another small disc collides with it elastically with a speed of $v_0 = 0.3$ m/s, the velocities of the discs being parallel. The distance d shown in the figure is equal to $R/2$, friction between the discs is negligible. For which u (in m/s) will the small disc move perpendicularly to its original motion after the collision ? Toughnut

(a) 0.1
(b) 0.5
(c) 1.0
(d) 0.01

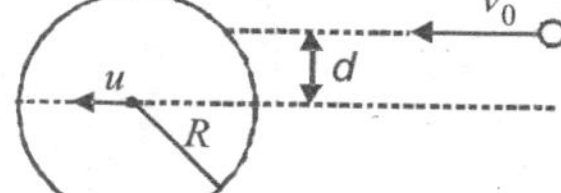

99. A shell is fired from a cannon with a velocity v (m/sec.) at an angle θ with the horizontal direction. At the highest point in its path it explodes into two pieces of equal mass. One of the pieces retraces its path to the cannon and the speed (in m/sec.) of the other piece immediately after the explosion is

Critical Thinking

(a) $3v \cos\theta$ (b) $2v \cos\theta$

(c) $\frac{3}{2}v \cos\theta$ (d) $\sqrt{\frac{3}{2}}v \cos\theta$

100. An object of mass m is projected vertically upwards with a speed of v_0. At the same moment another object of mass M, which is initially above the projected one, is dropped from a height of h. The two point like objects collide completely inelastically, and they stick to each other. Find kinetic energy (in J) of combined mass just before it hits the ground. (Given : $m = 1$ kg, $v_0 = 20$ m/s, $M = 3$ kg, $h = 20$m, $g = 10$ m/s^2). Toughnut

(a) 550 J (b) 650 J (c) 450 J (d) 250 J

Answer KEYs

1	(c)	11	(d)	21	(d)	31	(c)	41	(d)	51	(c)	61	(b)	71	(a)	81	(c)	91	(a)
2	(a)	12	(c)	22	(b)	32	(a)	42	(c)	52	(a)	62	(d)	72	(c)	82	(d)	92	(d)
3	(b)	13	(c)	23	(d)	33	(b)	43	(c)	53	(b)	63	(a)	73	(d)	83	(a)	93	(a)
4	(a)	14	(b)	24	(b)	34	(c)	44	(a)	54	(a)	64	(c)	74	(d)	84	(b)	94	(b)
5	(b)	15	(a)	25	(a)	35	(a)	45	(a)	55	(b)	65	(c)	75	(d)	85	(a)	95	(c)
6	(a)	16	(a)	26	(a)	36	(d)	46	(c)	56	(b)	66	(d)	76	(b)	86	(b)	96	(d)
7	(c)	17	(a)	27	(a)	37	(b)	47	(a)	57	(b)	67	(b)	77	(a)	87	(d)	97	(c)
8	(a)	18	(b)	28	(a)	38	(c)	48	(d)	58	(a)	68	(b)	78	(d)	88	(a)	98	(a)
9	(c)	19	(d)	29	(c)	39	(b)	49	(b)	59	(a)	69	(b)	79	(b)	89	(c)	99	(a)
10	(a)	20	(d)	30	(c)	40	(d)	50	(d)	60	(b)	70	(d)	80	(c)	90	(d)	100	(b)

Hints & Solutions

1. (c) Let u be the speed with which the ball of mass m is projected.
Then the kinetic energy (E) at the point of projection is
$E = \frac{1}{2}mu^2$ ...(i)
When the ball is at the highest point of its flight, the speed of the ball is $\frac{u}{\sqrt{2}}$ (Remember that the horizontal component of velocity does not change during a projectile motion).
∴ The kinetic energy at the highest point
$= \frac{1}{2}m\left(\frac{u}{\sqrt{2}}\right)^2 = \frac{1}{2}\frac{mu^2}{2} = \frac{E}{2}$ [From (i)]

2. (a) $y = 10 - \sqrt{10^2 - 4^2} = 0.84m$
Thus $\frac{1}{2}mv^2 = mgy$
or $v = \sqrt{2gy} = \sqrt{2\times 9.8\times 0.84}$
$= 4.05$ m/s.

3. (b) $W_{gravity} + W_F = \Delta K$
$-10\times g(10 - 10\cos 60°) + 200\times 10\sin 60°$
$= \frac{1}{2}\times 10\times v_f^2 - 0$ ∴ $v_f = 17$ m/s

4. (a) Net displacement is $(3\hat{i} + 3\hat{j})$
so work done $= \vec{F}.\Delta\vec{s} = 11.25(\hat{i}+\hat{j}).(3\hat{i}+3\hat{j})$
$= 33.75\times 2 = 67.50$ J

5. (b)
F(N) A B C x(m) 0 1 2 3 1 2 3 4 5 6 7
Work done = area under F-x graph
= area of trapezium OABC $= \frac{1}{2}(3+6)(3)$
$= 13.5$ J

6. (a) $W = F s\cos 90° =$ zero

7. (c) $W =$ area of $F - x$ graph
= area of Δ + area of rectangle + area of Δ
$= \frac{5\times 3}{2} + 10\times 3 + \frac{5\times 3}{2} = 45\ J$

8. (a) Let m be the mass of the beads.
By energy conservation
$mgh = \frac{1}{2}mv_1^2 + \frac{1}{2}mv_2^2$ ....(i)
$\frac{v_1}{v_2} = \frac{4}{3}$ ....(ii)
$v_1 = \frac{4\sqrt{2}}{5}, v_2 = \frac{3\sqrt{2}}{5}$;
$K.E. = \frac{1}{2}\times 100\times 10^{-3}\times\frac{16\times 2}{25} = 64\times 10^{-3}$ J
x = 8.

9. (c) As the cord is trying to hold the motion of the block, work done by the cord is negative.
$W = -M(g-a)d$
$= -M\left(g - \frac{g}{4}\right)d = \frac{-3Mgd}{4}$

10. (a) $W_1 = \frac{1}{2}\times 5\times 10^3(0.05)^2$
$\Rightarrow W_2 = \frac{1}{2}\times 5\times 10^3(0.10)^2$
∴ $\Delta W = \frac{1}{2}\times 5\times 10^3\times 0.15\times 0.05 = 18.75$J.

11. (d) Given : $\vec{F} = 3\hat{i} + \hat{j}$
$\vec{r_1} = (2\hat{i} + \hat{k}),\ \vec{r_2} = (4\hat{i} + 3\hat{j} - \hat{k})$
$\vec{r} = \vec{r_2} - \vec{r_1} = (4\hat{i} + 3\hat{j} - \hat{k}) - (2\hat{i} + \hat{k})$
or $\vec{r} = 2\hat{i} + 3\hat{j} - 2\hat{k}$
So work done by the given force $w = \vec{f}.\vec{r}$
$= (3\hat{i} + \hat{j}).(2\hat{i} + 3\hat{j} - 2\hat{k}) = 6 + 3 = 9$J

12. (c) here, $m = 4$, $g = 4\times 10^{-3}$ kg
$x = 4t^2 + t$ ∴ $\frac{dx}{dt} = 8t + 1$ $\frac{d^2x}{dt^2} = 8$
Work done, $W = \int f dx = \int m\frac{d^2x}{dt^2}\left(\frac{dx}{dt}\right)dt$
$= \int_0^2 (4\times 10^{-3})(8)(8t+1)dt$
$= 32\times 10^{-3}\int_0^2 (8t+1)dt = 32\times 10^{-3}\left[\frac{8t^2}{2} + t\right]_0^2$
$= 32\times 10^{-3}[4(2)^2 + 2 - 0] = 576$ mJ

13. (c) $W = F s \cos\theta$, $\cos\theta = \frac{W}{Fs} = \frac{25}{5\times10} = \frac{1}{2}$, $\theta = 60°$.

14. (b) In the frame of lift displacement is zero, so W = F.S = T.S = 0.

15. (a) $W = \int_{x_1}^{x_2} Fdx = \int_{4}^{-2} -6x^3 dx$

$$= -6\left|\frac{x^4}{4}\right|_4^{-2} = \frac{-6}{4}[(-2)^4 - (4)^2] = 360\text{ J}.$$

16. (a) Consider the blocks shown in the figure to be moving together due to friction between them.

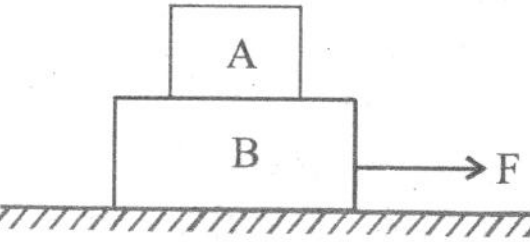

The free body diagrams of both the blocks are shown below.

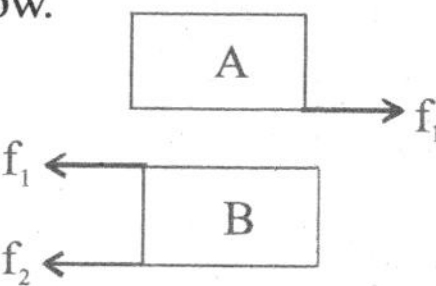

Work done by static friction on A is positive and on B is negative

17. (a) $U = -\int_0^x Fdx = -\int_0^x kxdx = -\frac{1}{2}kx^2$.

It is correctly drawn in (a)

18. (b) Total energy at the time of projection

$$= \frac{1}{2}mv^2 = \frac{1}{2}\times 0.1(20)^2 = 20\text{J}$$

Half way up, P.E. becomes half the P.E. at the top i.e. P.E. $= \frac{20}{2} = 10$J

$\therefore$ K.E. = 20 – 10 = 10J.

19. (d) Given, $\frac{dk}{dt}$ = constant

$$\Rightarrow k \propto t \Rightarrow v \propto \sqrt{t}$$

Also, $P = Fv = \frac{dk}{dt}$ = constant

$$\Rightarrow F \propto \frac{1}{v} \Rightarrow F \propto \frac{1}{\sqrt{t}}$$

20. (d) Kinetic energy of a body, $K = \frac{p^2}{2m}$

As $p_1 = p_2$ (Given)

$$\therefore \frac{K_1}{K_2} = \frac{m_2}{m_1} = \frac{5}{4}$$

21. (d) For any uniform rod, the mass is supposed to be concentrated at its centre.

$\therefore$ height of the mass from ground is, $h = (l/2)\sin 30°$

$\therefore$ Potential energy of the rod

$$= m\times g\times\frac{\ell}{2}\sin 30°$$

$$= m\times g\times\frac{\ell}{2}\times\frac{1}{2} = \frac{mg\ell}{4}$$

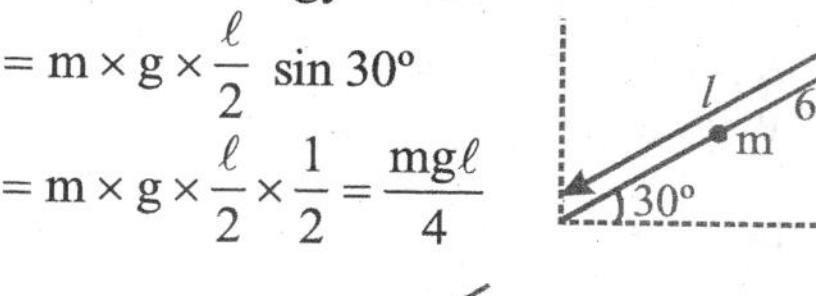

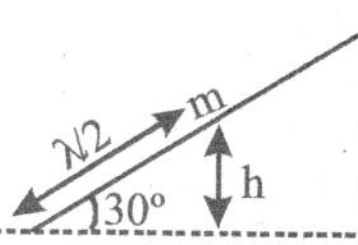

22. (b) Let the blow compress the spring by x before stopping.

Kinetic energy of the block = (P.E of compressed spring) + work done against function.

$$\frac{1}{2}\times 2\times(4)^2 = \frac{1}{2}\times 10{,}000\times x^2 + (+15)\times x$$

$$10{,}000x^2 + 30x - 32 = 0$$

$$\Rightarrow 5000x^2 + 15x - 16 = 0$$

$$\therefore x = -\frac{15\pm\sqrt{(15)^2 - 4\times(5000)(-16)}}{2\times 5000}$$

= 0.055m = 5.5cm.

23. (d) $F = -\frac{dU}{dy} = b - 2ay$

24. (b) Applying momentum conservation

$$m_1u_1 + m_2u_2 = m_1v_1 + m_2v_2$$

$$0.1u + m(0) = 0.1(0) + m(3)$$

$$0.1u = 3m$$

$$\frac{1}{2}0.1u^2 = \frac{1}{2}m(3)^2$$

Solving we get, $u = 3$

$$\frac{1}{2}kx^2 = \frac{1}{2}K\left(\frac{x}{2}\right)^2 + \frac{1}{2}(0.1)3^2$$

$$\Rightarrow \frac{3}{4}kx^2 = 0.9 \Rightarrow \frac{3}{2}\times\frac{1}{2}kx^2 = 0.9$$

$\therefore \frac{1}{2}Kx^2 = 0.6$ J (total initial energy of the spring)

25. (a) W.D. from ground frame = 6.4 × 50 = 320 J
From elevator frame, displacement = 0.

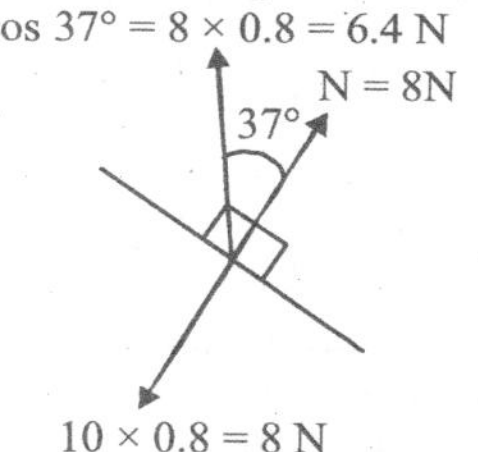

26. **(a)** Work done against gravity $= mg \sin\theta \times d$
$= 2 \times 10 \times 10 = 200$ J ($d \sin\theta = 10$)
Actual work done = 300 J
Work done against friction = 300 – 200 = 100 J

27. **(a)** $x = 3t - 4t^2 + t^3 \quad \frac{dx}{dt} = 3 - 8t + 3t^2$

Acceleration $= \frac{d^2x}{dt^2} = -8 + 6t$

Acceleration after 4 sec $= -8 + 6 \times 4 = 16\ \text{ms}^{-2}$
Displacement in 4 sec $= 3 \times 4 - 4 \times 4^2 + 4^3 = 12$ m
$\therefore$ Work = Force × displacement
= Mass × acc. × disp. $= 3 \times 10^{-3} \times 16 \times 12 = 576$ mJ

28. **(a)** As surface is smooth so work done against friction is zero. Also the displacement and force of gravity are perpendicular so work done against gravity is zero.

29. **(c)** From given graph :

$$\vec{F} = \left(\frac{3}{4}x + 10\right)\hat{i} + \left(20 - \frac{4}{3}y\right)\hat{j} + \left(\frac{4}{3}z - 16\right)\hat{k}$$

$$W = \int \vec{F} \cdot \vec{ds}$$

$$= \int_{(0,5,6)}^{(2,10,0)} \left(\frac{3}{4}x + 10\right)\hat{i} + \left(20 - \frac{4}{3}y\right)\hat{j} + \left(\frac{4}{3}z - 16\right)\hat{k}$$

$$[dx\,\hat{i} + dy\,\hat{j} + dz\,\hat{k}] = \frac{287}{2}\text{J}$$

Work done can also be found by finding area under these curves.

30. **(c)** The expression of work done by the variable force F on the particle is given by

$$W = \int \vec{F} \cdot \vec{d\ell}$$

In going from (0, 0) to (*a*, 0), the coordinate of *x* varies from 0 to '*a*', while that of *y* remains zero. Hence, the work done along this path is :

$$W_1 = \int_0^a (-Kx\hat{j}) \cdot dx\hat{i} = 0 \qquad [\because \hat{j} \cdot \hat{i} = 0]$$

In going from (*a*, 0) to (*a*, *a*) the coordinate of *x* remains constant (= *a*) while that of *y* changes from 0 to '*a*'. Hence, the work done along this path is

$$W_2 = \int_0^a [(-K(y\hat{i} + a\hat{j}) \cdot dy\,\hat{j}] = ka\int_0^a dy = -Ka^2$$

Hence, $W = W_1 + W_2 = -ka^2$

31. **(c)** Work done by friction

$$= \int \vec{F} \cdot \vec{ds} = \int_0^x \mu mg \cos\theta \frac{dx}{\cos\theta} = \mu mg\,x = 20\text{ J}$$

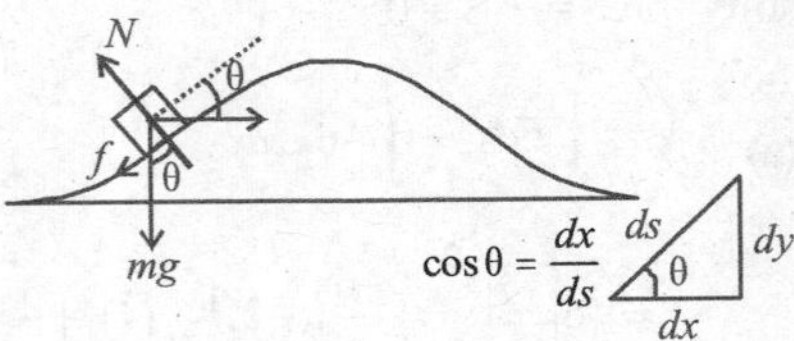

32. **(a)** For a given spring, $u = \frac{1}{2}kx^2$

$$\therefore \frac{u_2}{u_1} = \frac{\frac{1}{2}Kx_2^2}{\frac{1}{2}Kx_1^2} = \frac{(3x)^2}{x^2} = 9:1$$

33. **(b)** Energy required = mgh
In both cases, h is the same. Hence energy both is the same.

34. **(c)** Initial momentum $(p_1) = p$; Final momentum $(p_2) = 1.5$ p and initial kinetic energy $(K_1) = K$.

Kinetic energy $(K) = \frac{p^2}{2m} \propto p^2$

or, $\frac{K_1}{K_2} = \left(\frac{p_1}{p_2}\right)^2 = \left(\frac{p}{1.5p}\right)^2 = \frac{1}{2.25}$

or, $K_2 = 2.25\,K$.
Therefore, increase in kinetic energy is $2.25\,K - K = 1.25\,K$ or 125%.

35. **(a)** $U_1 = mgh_1$ and $U_2 = mgh_2$

$$\%\ \text{energy lost} = \frac{U_1 - U_2}{U_1} \times 100$$

$$= \frac{mgh_1 - mgh_2}{mgh_1} \times 100 = \left(\frac{h_1 - h_2}{h_1}\right) \times 100$$

$$= \frac{2 - 1.5}{2} \times 100 = 25\%$$

36. **(d)** Work done on the body is gain in the kinetic energy. Acceleration of the body is a = V/T.

Velocity acquired in time t is $v = at = \frac{V}{T}t$

K.E. acquired $\propto v^2$. That is work done $\propto \frac{V^2t^2}{T^2}$

37. **(b)** $E = \frac{P^2}{2m} \therefore E \propto \frac{1}{m}$ [If P = constant]
i.e. the lightest particle will possess maximum kinetic energy and in the given option mass of electron is minimum.

38. **(c)** Let m = mass of boy, M = mass of man
v = velocity of boy, V = velocity of man

$$\frac{1}{2}MV^2 = \frac{1}{2}\left[\frac{1}{2}mv^2\right] \qquad ...(i)$$

$$\frac{1}{2}M(V+1)^2 = 1\left[\frac{1}{2}mv^2\right] \quad ...(ii)$$

Putting $m = \frac{M}{2}$ and solving $V = \frac{1}{\sqrt{2}-1}$

39. (b) $v^2 = u^2 + 2gh = (10)^2 + 2\times10\times19.5 = 490$

K.E. at the ground

$= \frac{1}{2}mv^2 = \frac{1}{2}\times\frac{5}{1000}\times490 = \frac{49}{40}$ J

P.E. = mgh $= \frac{5}{1000}\times10\times\left(\frac{-50}{100}\right) = -\frac{1}{40}$ J

$\therefore$ Change in energy $= \frac{49}{40} - \left(-\frac{1}{40}\right)$

$= \frac{50}{40} = 1.25$J

40. (d) $|F| = \frac{dU}{dx}$, which is greatest in the reagion CD.

41. (d)

42. (c) $U = (1/2)Mv^2$

43. (c) Let E be the total energy then

$\frac{P.E}{K.E} = \frac{mgh}{E-mgh} = \frac{2}{3} \Rightarrow E = \frac{5}{2}$ mgh

When velocity is double then inital energy becomes $4E$.

So, $\frac{mgh}{4E-mgh} = NL = \frac{mgh}{10mgh-mgh}$

On solving we get $\frac{P.E}{K.E} = \frac{1}{9}$.

44. (a) $U_1 = \int_0^{\ell/3} -\frac{m}{\ell}gxdx = -\frac{1}{18}mg\ell;$

$U_2 = \int_0^{\ell} -\frac{m}{\ell}gxdx = -\frac{1}{2}mg\ell$

loss in P.E. $= U_1 - U_2 = \frac{4}{9}$mgl

$= \frac{4}{9}\times0.1\times10\times2 = \frac{8}{9}$J = Final K.E.

45. (a) Let x be the distance travelled by the particle along the 37° incline, after which it comes to rest. From work-energy theorem, Work done by gravity on block on incline 53° + work done by friction on block on incline 53° + work done by friction on ground level + work done by gravity on block on incline 37° + work done by friction on block on incline 37° = change in K.E.

$mg(8) - \mu mg\cos53°(10) - \mu mg(10) - \mu mg\cos37°(x)$

$- mg\sin37°(x) = 0 - \frac{1}{2}(1)(10)^2$

On solving we get $x = 1$m

46. (c) The work done by man is negative of magnitude of decreases in potential energy of chain

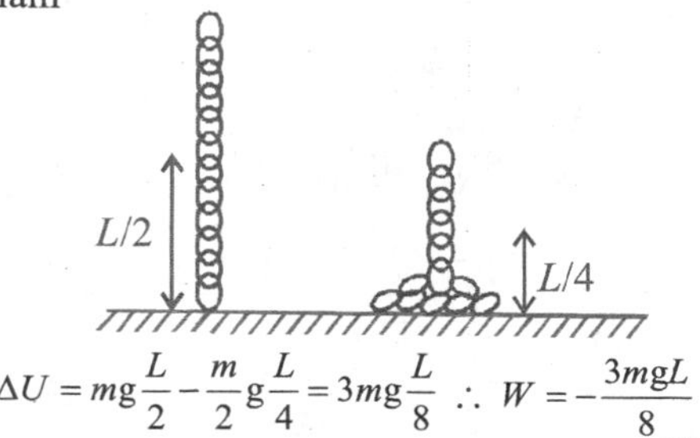

$\Delta U = mg\frac{L}{2} - \frac{m}{2}g\frac{L}{4} = 3mg\frac{L}{8} \therefore W = -\frac{3mgL}{8}$

47. (a) Given: Mass of particle, M = 10g = $\frac{10}{1000}$kg

radius of circle R = 6.4 cm

Kinetic energy E of particle = 8×10^{-4}J

acceleration a_t = ?

$\frac{1}{2}mv^2 = E \Rightarrow \frac{1}{2}\left(\frac{10}{1000}\right)v^2 = 8\times10^{-4}$

$\Rightarrow v^2 = 16\times10^{-2} \Rightarrow v = 4\times10^{-1} = 0.4$ m/s

Now, using

$v^2 = u^2 + 2a_t s \quad (s = 4\pi R)$

$(0.4)^2 = 0^2 + 2a_t\left(4\times\frac{22}{7}\times\frac{6.4}{100}\right)$

$\Rightarrow a_t = (0.4)^2\times\frac{7\times100}{8\times22\times6.4} = 0.1$ m/s^2

48. (d) $W = mgh + \frac{1}{2}kx^2 = mg(a\sin\theta) + \frac{1}{2}k(a\theta)^2$

$= Wa\sin\theta + \frac{1}{2}ka^2\theta^2.$

49. (b) $J = m(v_f - v_i)$ or $7.5 = 15(v_f - 0)$

$\Rightarrow v_f = 0.5$ m/s

$\therefore \quad K = \frac{1}{2}mv^2 = \frac{1}{2}\times15\times0.5^2 = 1.9$J

50. (d) $dU_{(x)} = -Fdx$

$\therefore \quad U_x = -\int_0^x Fdx = \frac{kx^2}{2} - \frac{ax^4}{4}$

$U = 0$ at $x = 0$ and at $x = \sqrt{\frac{2k}{a}}$; $\Rightarrow$ we have potential energy zero twice (out of which one is at origin). Also, when we put $x = 0$ in the given function, we get $F = 0$. But $F = -\frac{dU}{dx}$

$\Rightarrow$ At $x = 0$; $\frac{dU}{dx} = 0$ i.e. the slope of the graph should be zero. These characteristics are represented by (d).

51. (c) For the smaller block to move $kx_0 = \mu mg$ and from work energy theorem

$$-\mu Mgx_0 - \frac{1}{2}k\,x_0^2 = -\frac{1}{2}M\,v_0^2$$

$$+\mu Mg\left(\frac{\mu mg}{k}\right) + \frac{1}{2}k\left(\frac{\mu mg}{k}\right)^2 = \frac{1}{2}Mv^2$$

$$v = \mu g\sqrt{\frac{(2M+m)m}{kM}}$$

52. (a) $W_F + W_{Sp} + W_{\text{fric}} = \Delta k$

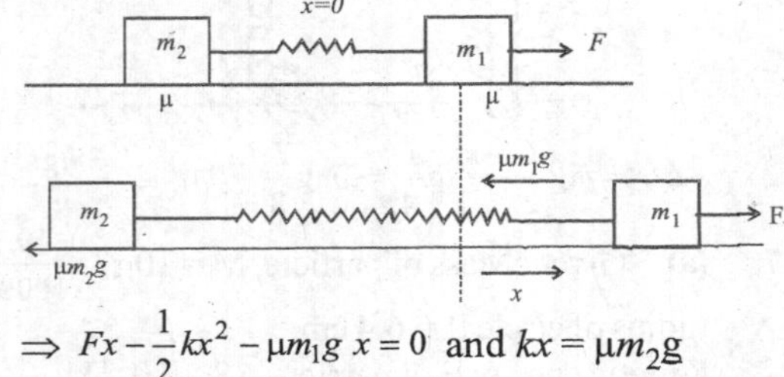

$$\Rightarrow Fx - \frac{1}{2}kx^2 - \mu m_1 g\,x = 0 \text{ and } kx = \mu m_2 g$$

$$\Rightarrow F - \frac{1}{2}\mu m_2 g - \mu m_1 g = 0 \Rightarrow F = \mu m_1 g + \frac{\mu m_2 g}{2}$$

53. (b)

Initial Final

2m/3 m/3

m/3

m

m

By using mechanical energy conservation

$$v = \frac{10}{3} = 3.33 \text{ m/s}$$

54. (a) Loss in gravitational P.E. = gain in spring P.E.

$$mgh = \frac{1}{2}k\,(h\cot\alpha - h)^2 \text{ or } (\cot\alpha - 1) = \sqrt{\frac{2mg}{kh}}$$

or $\cot\alpha = 1 + \sqrt{\frac{2mg}{kh}}$

55. (b) u = 0; v = u + aT; v = aT

Instantaneous power = F × v = m. a. at = $m.a^2.t$

$\therefore$ Instantaneous power = $m\frac{v^2}{T^2}t$

56. (b) Constant power of car $P_0 = F.V = ma.v$

$$P_0 = m\frac{dv}{dt}.v$$

$P_0 dt = mvdv$ Integrating

$$P_0.t = \frac{mv^2}{2} \qquad v = \sqrt{\frac{2P_0 t}{m}}$$

$\because$ P_0, m and 2 are constant $\quad \therefore\ v \propto \sqrt{t}$

57. (b) $P = \frac{W}{t}$. Here, P = 2kW = 2000 W.

W = Mgh = M × 10 × 10 = 100 M

and t = 60 s.

This gives, M = 1200 kg

Its volume = 1200 litre as 1 litre of water contains 1 kg of its mass.

58. (a) $x_i = \frac{50}{500} = 0.1\text{m},\ \ell_i = 0.4 - 0.1 = 0.3\text{m}$

$$x_f = \sqrt{0.4^2 + 0.3^2} - 0.3 = 0.2\text{m}$$

$$2\times\frac{1}{2}kx_i^2 + mg\times 0.3 = 2\times\frac{1}{2}kx_f^2 + \frac{1}{2}mv^2 \Rightarrow v = 0$$

59. (a)

60. (b) We know that F × v = Power

$\therefore\ F \times v = c$ where c = constant

$$\therefore\ m\frac{dv}{dt}\times v = c \qquad \left(\because F = ma = \frac{mdv}{dt}\right)$$

$$\therefore\ m\int_0^v vdv = c\int_0^t dt \qquad \therefore\ \frac{1}{2}mv^2 = ct$$

$$\therefore\ v = \sqrt{\frac{2c}{m}}\times t^{1/2}$$

$$\therefore\ \frac{dx}{dt} = \sqrt{\frac{2c}{m}}\times t^{1/2} \qquad \text{where } v = \frac{dx}{dt}$$

$$\therefore\ \int_0^x dx = \sqrt{\frac{2c}{m}}\times\int_0^t t^{1/2}dt$$

$$x = \sqrt{\frac{2c}{m}}\times\frac{2t^{3/2}}{3} \Rightarrow x \propto t^{3/2}$$

61. (b) Power exerted by a force is given by P = F.v

When the body is just above the earth's surface, its velocity is greatest. At this instant, gravitational force is also maximum. Hence, the power exerted by the gravitational force is greatest at the instant just before the body hits the earth.

62. (d) The power of body is given by = $\vec{F}.\vec{v}$ as the body is moving in circular path, centripetal force and velocity are at 90°, or power = 0.

63. (a) Power, P = F.v = $m\frac{dv}{dt}.v$

As P is constant, $\frac{v}{dt}$ = constant

$\Rightarrow v \propto dt \Rightarrow v \propto t$

64. **(c)** Power $= \dfrac{\text{work done}}{\text{time}}$

Therefore power of A, $P_A = \dfrac{mgh}{t_A}$

and power of B, $P_B = \dfrac{mgh}{t_B}$

$\therefore \quad \dfrac{P_A}{P_B} = \dfrac{t_B}{t_A} = \dfrac{4}{2} = 2:1$

65. **(c)** Volume of water to raise = 22380 *l*
$= 22380 \times 10^{-3} m^3$

$$P = \frac{mgh}{t} = \frac{V\rho gh}{t} \Rightarrow t = \frac{V\rho gh}{P}$$

$$t = \frac{22380 \times 10^{-3} \times 10^3 \times 10 \times 10}{10 \times 746} = 15 \text{ min}$$

66. **(d)** Power $= \dfrac{\text{total work done}}{\text{time}}$

$$= \frac{\frac{1}{2}Mv^2}{t} = \frac{1}{2}(mv^2)n\left(\because \frac{M}{t} = mn\right)$$

$$= kn\left[\because K.E.\ K = \frac{1}{2}mv^2\right]$$

67. **(b)** Initial length of the spring,

$$\ell_i = \sqrt{(2R)^2 + (1.5R)^2} = 2.5R$$

$\therefore \quad x_i = 2.5R - 2R = 0.5R.$

Now $\frac{1}{2}kx_i^2 + mg(1.5R) = \frac{1}{2}mv^2$

or $\frac{1}{2}k(0.5R)^2 + mg(1.5R)$

$$= \frac{1}{2}mv^2 \qquad \left[k = \frac{4mg}{R}\right] \qquad \therefore \quad v = 2\sqrt{gR}$$

68. **(b)** P = 10 kW = 10000 W
Fuel consumption = 1g/s
Calorific value = 2 kcal/g
$\therefore$ Energy produced = 2 kcal/s
Input power = 2 kcal/s = 2000 cal/s
= 2000 × 4.18 J/s = 8.4 k W
$\therefore$ This claim is invalid.

69. **(b)** Power $= \dfrac{\text{Work done}}{\text{Time}} = \dfrac{\frac{1}{2}m\left(v^2 - u^2\right)}{t}$

$$P = \frac{1}{2} \times \frac{2.05 \times 10^6 \times \left[(25)^2 - (5^2)\right]}{5 \times 60}$$

$P = 2.05 \times 10^6$ W = 2.05 MW

70. **(d)** m = 10 × 0.8kg = 8kg
height of iron chain = 5m

$$P = \frac{mgh}{t} = \frac{8 \times 10 \times 5}{10} W = 40W$$

71. **(a)** Area under graph increases. Hence work done upon the particle from *A* to *B* increases.

72. **(c)** Number of particles string the blades/time $\propto$ velocity of wind
K.E. of particle $\propto$ (Velocity of wind)^2
$\therefore$ Power output $\propto$ (No. of particles striking/ time) × (K.E. of particle)
$\therefore$ Power output $\propto v^3$

73. **(d)** Given force $\vec{F} = 2t\hat{i} + 3t^2\hat{j}$
According to Newton's second law of motion,

$$m\frac{d\vec{v}}{dt} = 2t\hat{i} + 3t^2\hat{j} \qquad (m = 1\text{ kg})$$

$$\Rightarrow \int_0^{\vec{v}} d\vec{v} = \int_0^t \left(2t\hat{i} + 3t^2\hat{j}\right)dt \Rightarrow \vec{v} = t^2\hat{i} + t^3\hat{j}$$

Power P = $\vec{F}\cdot\vec{v}\ (2t\hat{i} + 3t^2\hat{j})\cdot(t^2\hat{i} + t^3\hat{j})$
$= (2t^3 + 3t^5)W$

74. **(d)** $E = Pt = 45 \times 10^3 \times 50 \times 60 = 1.35 \times 10^8$

$$V = \frac{1.35 \times 10^8}{3.6 \times 10^7} = 3.75\text{ L}$$

Since the motor is 20% efficient, five times as much gasoline is needed.

75. **(d)** $Pt = nmgh$
2400 (60) = n (65) (9.8) (6) ; n = 37

76. **(b)** Speed just before collision
$= \sqrt{2g\ell(1 - \cos\theta_0)}$

Speed just after collision = $e\sqrt{2g\ell(1 - \cos\theta_0)}$
From conservation of energy,

$$\frac{1}{2}me^2[2g\ell(1 - \cos\theta_0)] = mg\ell(1 - \cos\theta)$$

$\therefore \ \theta = \cos^{-1}\{1 - e^2(1 - \cos\theta_0)\}$

77. **(a)** $m_1v = -m_1v_1 + m_2v_2,$

$e = \dfrac{2v_1}{v}$ for limiting condition

$$\frac{m_1}{m_2} < \frac{e}{2 + e}$$

78. **(d)** From the law of conservation of momentum we know that,
$m_1u_1 + m_2u_2 + ... = m_1v_1 + m_2v_2 + ...$
Given $m_1 = m$, $m_2 = 2m$ and $m_3 = 3m$
and $u_1 = 3u$, $u_2 = 2u$ and $u_3 = u$
Let the velocity when they stick = $\vec{v}$

Then, according to question,

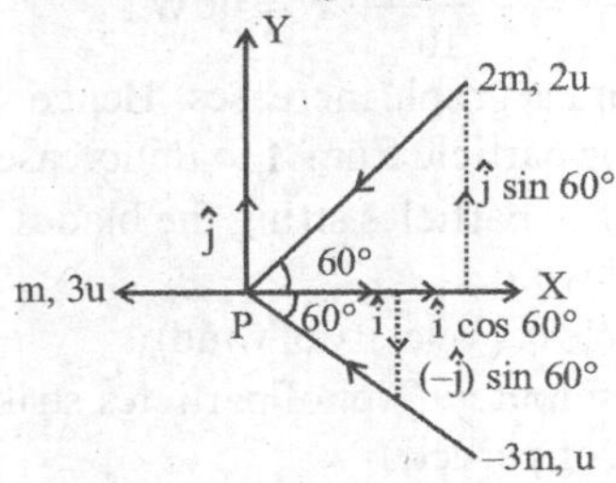

$$m \times 3u\,(\hat{i}) + 2m \times 2u\,(-\hat{i}\cos 60^\circ - \hat{j}\sin 60^\circ)$$

$$+3\,m \times u\,(-\hat{i}\cos 60^\circ + \hat{j}\sin 60^\circ) = (m+2m+3m)\,\vec{v}$$

$$\Rightarrow 3mu\hat{i} - 4mu\frac{\hat{i}}{2} - 4mu\left(\frac{\sqrt{3}}{2}\hat{j}\right) - 3mu\frac{\hat{i}}{2}$$

$$+3mu\left(\frac{\sqrt{3}}{2}\hat{j}\right) = 6m\,\vec{v}$$

$$\Rightarrow \quad mu\hat{i} - \frac{3}{2}mu\hat{i} - \frac{\sqrt{3}}{2}mu\hat{j} = 6m\,\vec{v}$$

$$\Rightarrow \quad -\frac{1}{2}mu\hat{i} - \frac{\sqrt{3}}{2}mu\hat{j} = 6m\,\vec{v}$$

$$\Rightarrow \vec{v} = \frac{u}{12}(-\hat{i} - \sqrt{3}\hat{j})$$

79. **(b)** If just after collision, relative velocity = v then

$\frac{v}{u} = \frac{1}{2} \qquad \therefore \omega_{rel} = \frac{v}{r} = \frac{u}{2r}$

$\therefore$ time between 1st and 2nd collision, $t = \frac{2\pi}{\omega_{rel}}$

$= \frac{4\pi r}{u}$

80. **(c)** Let x_A and x_B be the position of ends A and B at time t from the block, then stretched length of the spring will be

$\ell_2 = x_A - x_B$

and so the stretch

$\Delta\ell = \ell_2 - \ell_1 = (x_A - x_B) - \ell_1$ (ℓ_1 natural length of the spring)

So, $U = \frac{1}{2}k\Delta\ell^2 = \frac{1}{2}k[(x_A - x_B) - \ell_1]^2$

$$P = \frac{dU}{dt} = \frac{1}{2}k \cdot 2(x_A - x_B - \ell_1)\left(\frac{dx_A}{dt} - \frac{dx_B}{dt}\right)$$

$P = F(v_A - v_B) \qquad F = \frac{P}{v_A - v_B}$

$$\Delta\ell = \frac{F}{k} = \frac{P}{(v_A - v_B)k} = \frac{20}{(4-2)\times 100}$$

$\Delta\ell = 0.1$ m $= 10$ cm

81. **(c)** While moving downhill power

$$P = \left(w\sin\theta + \frac{w}{20}\right)10$$

$$P = \left(\frac{w}{10} + \frac{w}{20}\right)10 = \frac{3w}{2}$$

$$\frac{P}{2} = \frac{3w}{4} = \left(\frac{w}{10} - \frac{w}{20}\right)V$$

$$\frac{3}{4} = \frac{v}{20} \Rightarrow v = 15 \text{ m/s}$$

w cosθ

$\tan\theta = \frac{1}{10}$

$\therefore$ Speed of car while moving downhill v = 15 m/s.

82. **(d)** $m = 1$ kg, $a = 1$ m/s²

$\mu = 0.1$

a, F_{ext}, $f = 1N$

$f = \mu mg = 0.1 \times 1 \times 10 = 1N$

$F_{ext} = ma + f \qquad F_{ext} = 1 + 1 = 2N$

$v = u + at = 0 + 1 \times 2 = 2$ m/sec.

$\therefore \quad P = F_{ext} \cdot v = 2 \times 2 = 4$ watt.

83. **(a)** As the floor exerts a force on the ball along the normal, & no force parallel to the surface, therefore the velocity component along the parallel to the floor remains constant. Hence $V\sin\theta = V^1 \sin\theta^1$.

84. **(b)** According to principle of conservation of energy

Potential energy = kinetic energy

$$\Rightarrow mgh = \frac{1}{2}mv^2 \Rightarrow v = \sqrt{2gh}$$

If h_1 and h_2 are initial and final heights, then

$$\Rightarrow v_1 = \sqrt{2gh_1},\ v_2 = \sqrt{2gh_2}$$

Loss in velocity,

$$\Delta v = v_1 - v_2 = \sqrt{2gh_1} - \sqrt{2gh_2}$$

$\therefore$ fractional loss in velocity

$$= \frac{\Delta v}{v_1} = \frac{\sqrt{2gh_1} - \sqrt{2gh_2}}{\sqrt{2gh_1}} = 1 - \sqrt{\frac{h_2}{h_1}}$$

$$= 1 - \sqrt{\frac{1.8}{5}} = 1 - \sqrt{0.36} = 1 - 0.6 = 0.4 = \frac{2}{5}$$

85. **(a)** As $u_2 = 0$ and $m_1 = m_2$, therefore from $m_1u_1 + m_2u_2 = m_1v_1 + m_2v_2$

we get $u_1 = v_1 + v_2$

$$\text{Also, } e = \frac{v_2 - v_1}{u_1} = \frac{v_2 - v_1}{v_2 + v_1} = \frac{1 - v_1/v_2}{1 + v_1/v_2},$$

which gives $\frac{v_1}{v_2} = \frac{1-e}{1+e}$

86. **(b)** $\frac{1}{2}Mv^2 = \frac{1}{2}kL^2$

$$\Rightarrow v = \sqrt{\frac{k}{M}}\,.L$$

M

$$\text{Momentum} = M \times v = M \times \sqrt{\frac{k}{M}}.L$$
$$= \sqrt{kM}.L$$

87. (d) For the object of mass 2.0 kg.
$$\frac{\Delta k}{k} = \frac{k - k/4}{k} = \frac{3}{4}$$
Kinetic energy transferred
$$\frac{\Delta k}{k} = \frac{4m_1m_2}{(m_1+m_2)^2}$$
Here, $m_1 = 2.0$ kg, $m_2 = M$
$$\therefore \frac{3}{4} = \frac{4 \times 2M}{(2+M)^2} \Rightarrow M = \frac{2}{3}\text{kg or 6kg}$$

88. (a) Initial, $K.E. = \frac{1}{2}mv^2 = \frac{1}{2} \times \frac{20}{1000} \times 600 \times 600$
$= 3600$ J
Change in K.E. = P.E.
$$\frac{1}{2}m(v^2 - v'^2) = mgh$$
$$\Rightarrow 3600 - \frac{1}{2} \times \frac{20}{1000} \times {v_1}^2 = 4 \times 10 \times 80$$
$$\Rightarrow v_1 = 200\,m/s$$

89. (c) Initial kinetic energy of the system
$$K.E_i = \frac{1}{2}mu^2 + \frac{1}{2}M(0)^2 = \frac{1}{2} \times 0.5 \times 2 \times 2 + 0 = 1J$$
For collision, applying conservation of linear momentum
$$m \times u = (m+M) \times v$$
$$\therefore\ 0.5 \times 2 = (0.5+1) \times v \Rightarrow v = \frac{2}{3}m/s$$
Final kinetic energy of the system is
$$K.E_f = \frac{1}{2}(m+M)v^2 = \frac{1}{2}(0.5+1) \times \frac{2}{3} \times \frac{2}{3} = \frac{1}{3}J$$
$\therefore$ Energy loss during collision
$$= \left(1 - \frac{1}{3}\right)J = 0.67J$$

90. (d) Power $\vec{F}\cdot\vec{V} = PA\vec{V} = \rho ghAV$
$$\left[\because P = \frac{F}{A} \text{ and } P = \rho gh\right]$$
$$= 13.6 \times 10^3 \times 10 \times 150 \times 10^{-3} \times 0.5 \times 10^{-3}/60$$
$$= \frac{102}{60} = 1.70 \text{ watt}$$

91. (a) Extra power, $P = mg\sin\theta \times v$
$$= Mg \times \frac{h}{s} \times v = \frac{Mghv}{s}.$$

92. (d) $P_{out} = \frac{mgh}{t} = \frac{9000 \times 10 \times 10}{5 \times 60} = 3000\,W$
$$P_{in} = 10 \times 10^3\text{ W}.$$
$$\therefore\ \eta = \frac{P_{out}}{P_{in}} \times 100 = \frac{3000}{10 \times 10^3} \times 100 = 30\%$$

93. (a) Given, $h = 60$m, $g = 10\,ms^{-2}$,
Rate of flow of water = 15 kg/s
$\therefore$ Power of the falling water
$= 15\,kgs^{-1} \times 10\,ms^{-2} \times 60\,m = 900$ watt.
Loss in energy due to friction
$$= 9000 \times \frac{10}{100} = 900 \text{ watt.}$$
$\therefore$ Power generated by the turbine
$= (9000 - 900)$ watt $= 8100$ watt $= 8.1$ kW

94. (b) The centripetal acceleration
$$a_c = k^2rt^2 \text{ or } \frac{v^2}{r} = k^2rt^2 \quad \therefore \quad v = krt$$
So, tangential acceleration, $a_t = \frac{dv}{dt} = kr$
Work is done by tangential force.
Power $= F_t \cdot v \cdot \cos 0° = (ma_t)(krt) = (mkr)(krt)$
$= mk^2r^2t$

95. (c) Initial KE. = 0, Initial P.E. = 0
When the rope is just pulled off the table,
final K.E. $= \frac{1}{2}(\lambda\ell)v^2$, final P.E. $= (\lambda\ell)g\ell/2$
time taken $= t = \ell/v$
$$\text{Average power} = \frac{\text{net change in energy}}{\text{time}}$$
$$= \frac{\frac{1}{2}\lambda\ell v^2 + \lambda\ell g\ell/2}{\ell/v} = \frac{1}{2}\lambda v^3 + \frac{\lambda\ell vg}{2}$$

96. (d) The velocity of particle after falling through height h
$$u = \sqrt{2gh} \qquad \text{...(i)}$$
After first rebounding, the velocity of ball is eu and after attaining maximum height it will come to the ground with same velocity eu. So, after second rebounding its velocity will be e^2u.
Similarly, after third fourth ... etc reboundings its velocities will be e^2u, e^4u, ... etc.
Since, it first rebounds with velocity eu so if it attains height h then from
$$v^2 = u^2 - 2gh$$
$$\therefore\quad 0 = e^2u^2 - 2gh_1$$
$$\text{or}\quad h_1 = \frac{e^2u^2}{2g} = \frac{e^2 2gh}{2g} = e^2h \text{ [from Eq. (i)]}$$

The same height the ball travels while approaching ground. Now, it rebounds with velocity e^2u so if it attains a height h_2 then

$$0 = e^4u^2 - 2gh_2$$

or $h_2 = e^4h$

The similar process will follow for further reboundings

Hence, the total distance travelled by the practice before it stops rebounding.

$$= h + 2h_1 + 2h_2 + ...\infty \quad = h + 2e^2h + 2e^4h + ..\infty$$

$$= h + 2e^2h(1 + e^2 + e^4 + ...\infty) \ = h + 2e^2h\left(\frac{1}{1-e^2}\right)$$

$$= h\left(1 + \frac{2e^2}{1-e^2}\right) = \left(\frac{1+e^2}{1-e^2}\right)h$$

97. **(c)** Energy lost in the collision is

$$\Delta E = \frac{1}{2}mu_{rel}^2\left(1 - e^2\right)$$

$$\therefore \frac{3}{16}\left(\vec{u}_1 - \vec{u}_2\right)^2 = \frac{1}{2}mu_{rel}^2(1 - e^2) \ \Rightarrow e = 0.5$$

98. **(a)** $v_{0y} = v_0 \sin\theta$, this component does not change

$v_{0x} = v_0 \cos\theta$ and $u_x \cos\theta$

For elastic collision relative velocity before collision is equal to relative velocity after collision along x-axis

$v_0 \cos\theta - u\cos\theta = u\cos\theta + v_x$

if $v_x \cos\theta = v_{0y} \sin\theta$

By solving $u = 0.1$

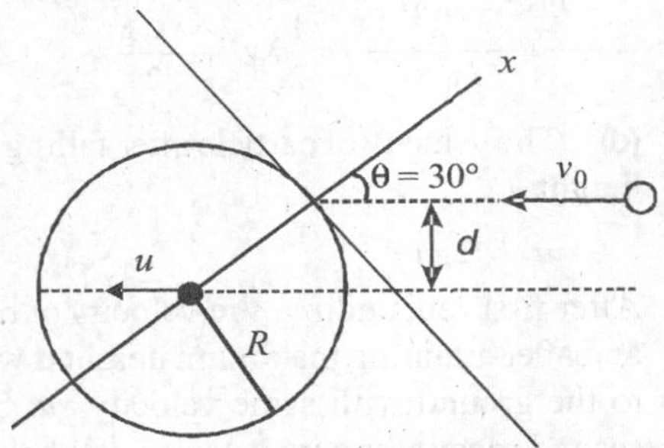

99. **(a)** As one piece retraces its path, the speed of this piece just after explosion should be $v\cos\theta$

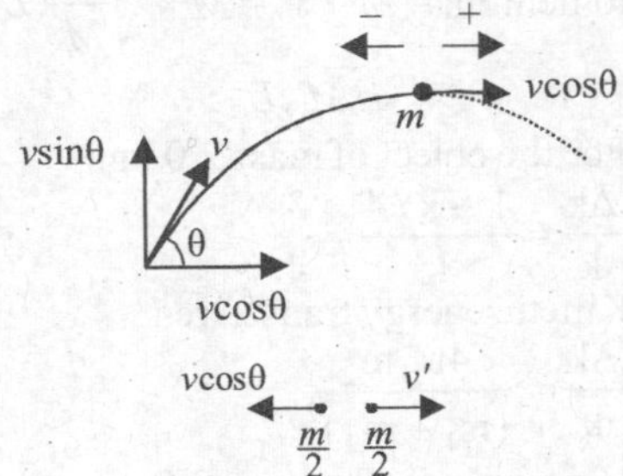

(At highest point just after explosion)

Applying conservation of linear momentum at the highest point;

$$m\,(v\cos\theta\,) = \frac{m}{2}\times v' - \frac{m}{2}\times v\cos\theta$$

$$3\,v\cos\theta = v'$$

100. **(b)** Using relative velocity, time of flight before collision will be

$$t = \frac{20}{20} = 15$$

T O W E R

B $h' = 1/2 \times 10 \times 1^2$ =5m

$h - h' = 15\text{m}$

$gt = 10\text{m/s}$

$20 - 10 \times 1 = 10\text{m/s}$

15

By *COM* at the time of collision

$3 \times 10 - 1 \times 10 = 4 \times v$

$2 \times 10 = 4 \times v;\ 5 = v$

$v = 5$ m/s

For 1-D motion

$v^2 = u^2 + 2as = 5^2 + 2 \times 10 \times 15 = 25 + 300 = 325$

$K = 650$ J

6 System of Particles and Rotational Motion

1. The mass per unit length of a non-uniform rod of length L varies as $m = \lambda x$ where λ is constant. The centre of mass of the rod will be at :

   (a) $\frac{2}{3}L$ (b) $\frac{3}{2}L$ (c) $\frac{1}{2}L$ (d) $\frac{4}{3}L$

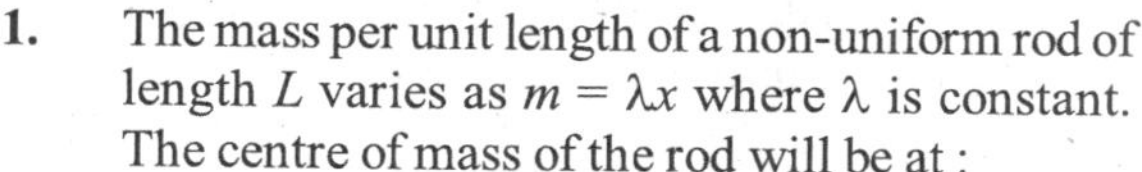

2. A pulley fixed to the ceiling carries a string with blocks of mass m and 3 m attached to its ends. The masses of string and pulley are negligible. When the system is released, its centre of mass moves with what acceleration ?

   (a) 0 (b) $-g/4$ (c) $g/2$ (d) $-g/2$

3. Three identical spheres, each of mass 1 kg are kept as shown in figure, touching each other, with their centres on a straight line. If their centres are marked P, Q, R respectively, the distance of centre of mass of the system from P is

   (a) $\frac{PQ+PR+QR}{3}$ (b) $\frac{PQ+PR}{3}$

   (c) $\frac{PQ+QR}{3}$ (d) $\frac{PR+QR}{3}$

4. Two spheres A and B of masses m and $2m$ and radii $2R$ and R respectively are placed in contact as shown. The COM of the system lies

   (a) inside A
   (b) inside B
   (c) at the point of contact
   (d) None of these

5. A ball rolls without slipping. The radius of gyration of the ball about an axis passing through its centre of mass is K. If radius of the ball be R, then the fraction of total energy associated with its rotational energy will be

   (a) $\frac{K^2}{R^2}$ (b) $\frac{K^2}{K^2+R^2}$

   (c) $\frac{R^2}{K^2+R^2}$ (d) $\frac{K^2+R^2}{R^2}$

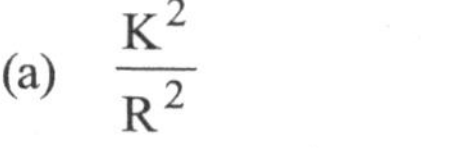

6. A hollow smooth uniform sphere A of mass m rolls without sliding on a smooth horizontal surface. It collides head on elastically with another stationary smooth solid sphere B of the same mass m and same radius. The ratio of kinetic energy of B to that of A just after the collision is

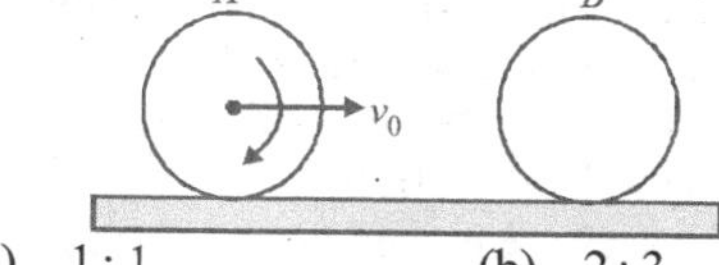

   (a) 1 : 1 (b) 2 : 3
   (c) 3 : 2 (d) None of these

7. A solid sphere, disc and solid cylinder all of the same mass and made of the same material are allowed to roll down (from rest) on an inclined plane, then **Tricky**

   (a) solid sphere reaches the bottom first
   (b) solid sphere reaches the bottom last
   (c) disc will reach the bottom first
   (d) all reach the bottom at the same time

8. A solid cylinder of mass m & radius R rolls down inclined plane without slipping. The speed of its C.M. when it reaches the bottom is

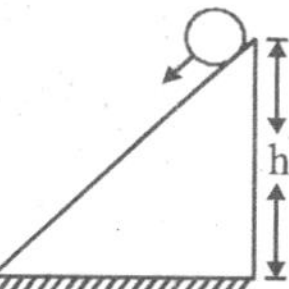

Critical Thinking

   (a) $\sqrt{2gh}$ (b) $\sqrt{4gh/3}$
   (c) $\sqrt{3/4gh}$ (d) $\sqrt{4gh}$

9. The least coefficient of friction for an inclined plane inclined at angle α with horizontal in order that a solid cylinder will roll down without slipping is

   (a) $\frac{2}{3}\tan\alpha$ (b) $\frac{2}{7}\tan\alpha$
   (c) $\tan\alpha$ (d) $\frac{5}{7}\tan\alpha$

10. An annular ring with inner and outer radii R_1 and R_2 is rolling without slipping with a uniform angular speed. The ratio of the forces experienced by the two particles situated on the inner and outer parts of the ring, $\frac{F_1}{F_2}$ is

   (a) $\left(\frac{R_1}{R_2}\right)^2$ (b) $\frac{R_2}{R_1}$ (c) $\frac{R_1}{R_2}$ (d) 1

11. There are some passengers inside a stationary railway compartment. The centre of mass of the compartment itself (without the passengers) is C_1, while the centre of mass of the 'compartment plus passengers' system is C_2. If the passengers move about inside the compartment then
(a) both C_1 and C_2 will move with respect to the ground
(b) neither C_1 nor C_2 will be stationary with respect to the ground
(c) C_1 will move but C_2 will be stationary with respect to the ground
(d) C_2 will move but C_1 will be stationary with respect to the ground

12. Three masses are placed on the x-axis : 300 g at origin, 500 g at $x = 40$ cm and 400 g at $x = 70$ cm. The distance of the centre of mass from the origin is
(a) 40 cm (b) 45 cm (c) 50 cm (d) 30 cm

13. A body A of mass M while falling vertically downwards under gravity breaks into two parts; a body B of mass $\frac{1}{3}$ M and a body C of mass $\frac{2}{3}$ M. The centre of mass of bodies B and C taken together shifts compared to that of body A
(a) does not shift
(b) depends on height of breaking
(c) towards body B
(d) towards body C

14. A system consists of three particles, each of mass m and located at (1, 1), (2, 2) and (3, 3). The co-ordinates of the centre of mass are
(a) (1, 1) (b) (2, 2) (c) (3, 3) (d) (6, 6)

15. Two bodies of masses 2 kg and 4 kg are moving with velocities 2 m/s and 10 m/s respectively along same direction. Then the velocity of their centre of mass will be
(a) 8.1 m/s (b) 7.3 m/s
(c) 6.4 m/s (d) 5.3 m/s

16. Four particles of masses m_1, m_2, m_3 and m_4 are placed at the vertices A, B, C and D as respectively of a square shown. The COM of the system will lie at diagonal AC if

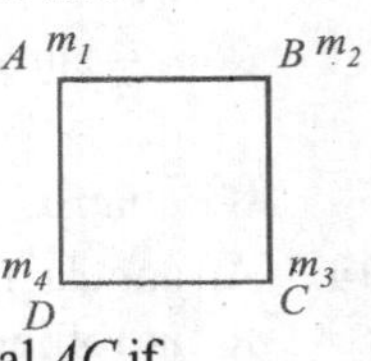

(a) $m_1 = m_3$ (b) $m_2 = m_4$
(c) $m_1 = m_2$ (d) $m_3 = m_4$

17. In the figure shown ABC is a uniform wire. If centre of mass of wire lies vertically below point A, then $\frac{BC}{AB}$ is close to :

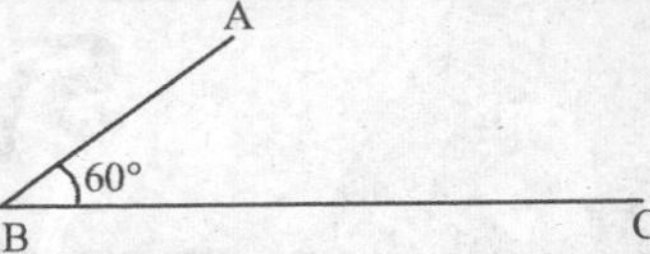

(a) 1.85
(b) 1.5
(c) 1.37

(d) 3

18. A circular disc of radius R is removed from a bigger circular disc of radius 2R such that the circumferences of the discs coincide. The centre of mass of the new disc is α / R form the centre of the bigger disc. The value of α is
(a) 1/4 (b) 1/3 (c) 1/2 (d) 1/6

19. The centre of mass of three bodies each of mass 1 kg located at the points (0, 0), (3, 0) and (0, 4) in the XY plane is
(a) $\left(\frac{4}{3}, 1\right)$ (b) $\left(\frac{1}{3}, \frac{2}{3}\right)$ (c) $\left(\frac{1}{2}, \frac{1}{2}\right)$ (d) $\left(1, \frac{4}{3}\right)$

20. Three identical rods are hinged at point A as shown. The angle made by rod AB with vertical is

Tricky

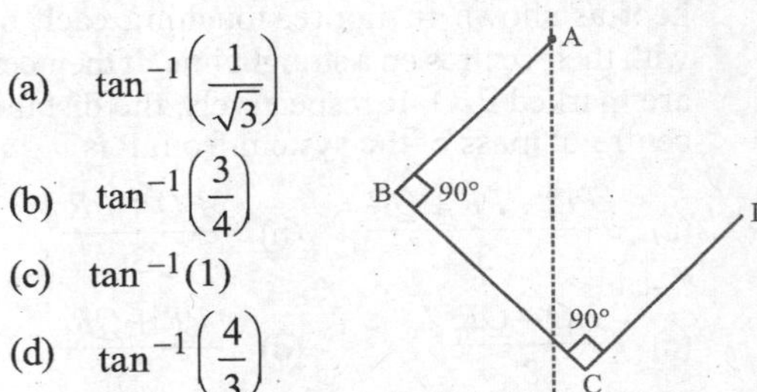

(a) $\tan^{-1}\left(\frac{1}{\sqrt{3}}\right)$
(b) $\tan^{-1}\left(\frac{3}{4}\right)$
(c) $\tan^{-1}(1)$
(d) $\tan^{-1}\left(\frac{4}{3}\right)$

21. A spool is pulled horizontally by two equal and opposite forces as shown in fig. Which of the following statements are correct?
(a) The centre of mass moves towards left.
(b) The centre of mass moves towards right
(c) The centre of mass remains stationary
(d) The net force about the centre of mass of the spool is zero

22. A metal sheet 14 cm × 2 cm of uniform thickness is cut into two pieces of width 2 cm. The two pieces are joined and laid along XY plane as shown. The centre of mass has the coordinates

Critical Thinking

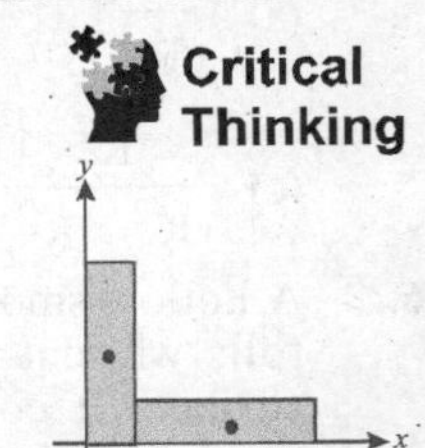

(a) (1, 1)
(b) (7/2, 7/2)
(c) (13/4, 9/4)
(d) (12/7, 8/7)

23. A rod PQ of mass M and length L is hinged at end P. The rod is kept horizontal by a massless string tied to point Q as shown in figure. When string is cut, the initial angular acceleration of the rod is Tricky

(a) g /L
(b) 2g/L
(c) $\frac{2g}{3L}$
(d) $\frac{3g}{2L}$

24. Two objects P and Q initially at rest move towards each other under mutual force of attraction. At the instant when the velocity of P is v and that of Q is 2v, the velocity of centre of mass of the system is
(a) v (b) 3v (c) 2v (d) zero

25. A uniform thin rod AB of length L has linear mass density $\mu(x) = a + \frac{bx}{L}$, where x is measured from A. If the CM of the rod lies at a distance of $\left(\frac{7}{12}\right)$L from A, then a and b are related as :
(a) $a=2b$ (b) $2a=b$
(c) $a=b$ (d) $3a=2b$

26. (n – 1) equal point masses each of mass m are placed at the vertices of a regular n-polygon. The vacant vertex has a position vector **a** with respect to the centre of the polygon. The position vector of centre of mass is
(a) $-\frac{1}{(n-a)}a$ (b) $\frac{a}{(n+1)}$
(c) $\frac{a}{n}$ (d) $\frac{n}{a+1}$

27. A right circular cone of base diameter 3 cm and height 6cm is cut from a solid cylinder of diameter 5cm and height 12cm. Find the position of *CG* of rest of the body.
(a) 2.1 cm (b) 6.3 cm (c) 8.2 cm (d) 5.3 cm

28. A solid sphere of radius *R* is placed on a smooth horizontal surface. A horizontal force *F* is applied at height h from the lowest point. For the maximum acceleration of the centre of mass,
(a) $h=R$ (b) $h=2R$ (c) $h=0$
(d) The acceleration will be same whatever h may be

29. Two persons of masses 55 kg and 65 kg respectively, are at the opposite ends of a boat. The length of the boat is 3.0 m and weighs 100 kg. The 55 kg man walks up to the 65 kg man and sits with him. If the boat is in still water the centre of mass of the system shifts by : Tricky
(a) 3.0m (b) 2.3m (c) zero (d) 0.75m

30. A wheel rotates with a constant acceleration of 2.0 radian/sec^2. If the wheel starts from rest, the number of revolutions it makes in the first ten seconds will be approximately
(a) 8 (b) 16 (c) 24 (d) 32

31. When a ceiling fan is switched off, its angular velocity falls to half while it makes 36 rotations. How many more rotations will it make before coming to rest?
(a) 24 (b) 36 (c) 18 (d) 12

32. A mass is revolving in a circle which is in the plane of paper. The direction of angular acceleration is
(a) upward the radius
(b) towards the radius
(c) tangential
(d) at right angle to angular velocity

33. In a bicycle, the radius of rear wheel is twice the radius of front wheel. It r_f and r_r are the radii and v_f and v_r are the speeds of topmost points of wheels then
(a) $v_r=2v_f$ (b) $v_f=2v_r$
(c) $v_f=v_r$ (d) $v_f=4v_r$

34. A particle moving in a circular path has an angular momentum of L. If the frequency of rotation is halved, then its angular momentum becomes
(a) $\frac{L}{2}$ (b) L (c) $\frac{L}{3}$ (d) $\frac{L}{4}$

35. The instantaneous angular position of a point on a rotating wheel is given by the equation $\theta(t) = 2t^3 - 6t^2$. The torque on the wheel becomes zero at
(a) t = 1s (b) t = 0.5 s
(c) t = 0.25 s (d) t = 2s

36. A particle of mass 2 kg is on a smooth horizontal table and moves in a circular path of radius 0.6 m. The height of the table from the ground is 0.8 m. If the angular speed of the particle is 12 rad s^{-1}, the magnitude of its angular momentum about a point on the ground right under the centre of the circle is
(a) 14.4 kg m^{2}s^{-1} (b) 8.64 kg m^{2}s^{-1}
(c) 20.16 kg m^{2}s^{-1} (d) 11.52 kg m^{2}s^{-1}

37. If the angular momentum of a particle of mass m rotating along a circular path of radius r with uniform speed is L, the centripetal force acting on the particle is

(a) $\frac{L^2}{mr^2}$ (b) $\frac{L^2}{mr}$ (c) $\frac{L}{mr}$ (d) $\frac{L^2 m}{r}$

38. Two particles which are initially at rest, move towards each other under the action of their internal attraction. If their speeds are v and $2v$ at any instant, then the speed of centre of mass of the system will be:

(a) $2v$ (b) zero (c) 1.5 (d) v

39. If the linear density (mass per unit length) of a rod of length 3m is proportional to x, where x is the distance from one end of the rod, the distance of the centre of gravity of the rod from this end is

Toughnut

(a) 2.5m (b) 1m (c) 1.5m (d) 2m

40. With O as the origin of the coordinate axis, the X and Y-coordinates of the centre of mass of the system of particles shown in the figure may be given as. Here m and $2m$ represent the masses of the particles.

(a) $\left(-\frac{b}{2}, 0\right)$

(b) $\left(-\frac{b}{2}, b\right)$

(c) $\left(-\frac{b}{3}, b\right)$

(d) $\left(-\frac{2}{5}b, b\right)$

41. A thin bar of length L has a mass per unit length λ, that increases linearly with distance from one end. If its total mass is M and its mass per unit length at the lighter end is λ_O, then the distance of the centre of mass from the lighter end is:

(a) $\frac{L}{2} - \frac{\lambda_o L^2}{4M}$ (b) $\frac{L}{3} + \frac{\lambda_o L^2}{8M}$

(c) $\frac{L}{3} + \frac{\lambda_o L^2}{4M}$ (d) $\frac{2L}{3} - \frac{\lambda_o L^2}{6M}$

42. Two particles, each of mass 2 kg are put at (2m,0) and (0, 2m), as shown in figure. Now 1 kg mass of particle A is put on to the particle B. The change in x-coordinate of centre of mass of the system is

(a) 0.5m

(b) 1m

(c) 1.5m

(d) None of these

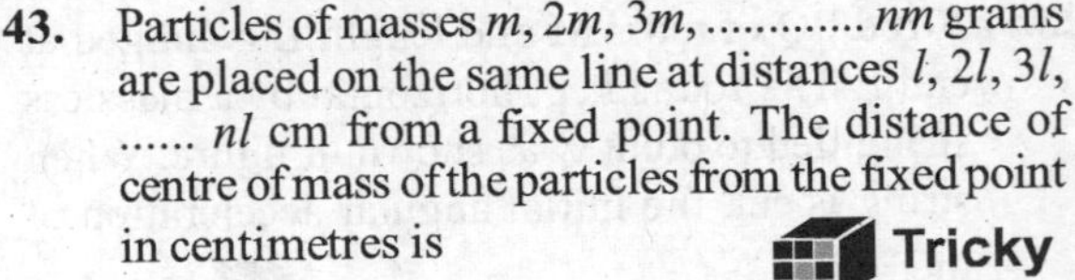

43. Particles of masses m, $2m$, $3m$, ... nm grams are placed on the same line at distances l, $2l$, $3l$, ... nl cm from a fixed point. The distance of centre of mass of the particles from the fixed point in centimetres is

Tricky

(a) $\frac{(2n+1)l}{3}$ (b) $\frac{1}{n+1}$

(c) $\frac{n(n^2+1)l}{2}$ (d) $\frac{2l}{n(n^2+1)}$

44. A wheel of radius 0.1m (wheel A) is attached by a non-stretching belt to a wheel of radius 0.2m (wheel B). The belt does not slip. By the time wheel B turns through 1 revolution, wheel A will rotate through

Tricky

(a) $\frac{1}{2}$ revolution

(b) 1 revolution

(c) 2 revolution (d) 4 revolution

45. The moment of inertia of a uniform circular disc of radius 'R' and mass 'M' about an axis passing from the edge of the disc and normal to the disc is

(a) MR^2 (b) $\frac{1}{2}MR^2$ (c) $\frac{3}{2}MR^2$ (d) $\frac{7}{2}MR^2$

46. Moment of inertia does not depend upon

(a) distribution of mass

(b) axis of rotation

(c) point of application of force

(d) None of these

47. Two identical discs of mass m and radius r are arranged as shown in the figure. If α is the angular acceleration of the lower disc and a_{cm} is acceleration of centre of mass of the lower disc, then relation between a_{cm}, α and r is

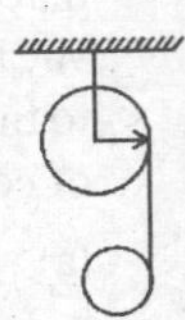

Tricky

(a) $a_{cm} = \alpha/r$ (b) $a_{cm} = 2\alpha r$

(c) $a_{cm} = \alpha r$ (d) None of these

48. A wheel having angular momentum 2π kg-m²/s about its vertical axis, rotates at the rate of 60 rpm about this axis, The torque which can stop the wheel's rotation in 30 sec would be

(a) $\frac{\pi}{18}$Nm (b) $\frac{2\pi}{15}$Nm (c) $\frac{\pi}{12}$Nm (d) $\frac{\pi}{15}$Nm

49. A wheel of radius 1m rolls forward half a revolution on a horizontal ground. The magnitude of the displacement of the point of the wheel initially in contact with the ground is

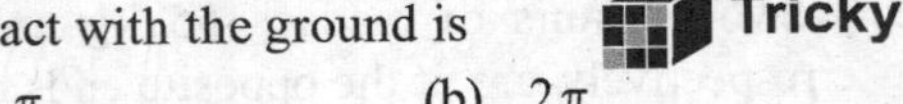

Tricky

(a) π (b) 2π

(c) $\sqrt{2}\pi$ (d) $\sqrt{\pi^2+4}$

50. A sphere of mass 'm' is given some angular velocity about a horizontal axis through its centre and gently placed on a plank of mass 'm'. The coefficient of friction between the two is μ. The plank rests on a smooth horizontal surface. The initial acceleration of the plank is

(a) zero (b) $(7/5)\mu g$ (c) μg (d) $2\mu g$

51. Moment of inertia of a circular wire of mass M and radius R about its diameter is

(a) $MR^2/2$ (b) MR^2
(c) $2MR^2$ (d) $MR^2/4$

52. Moment of inertia of a hollow cylinder of mass M and radius r about its own axis is

(a) $\frac{2}{3}Mr^2$ (b) $\frac{2}{5}Mr^2$

(c) $\frac{1}{3}Mr^2$ (d) Mr^2

53. Linear acceleration of cylinder of mass m_2 is a_2. Then angular acceleration α_2 is (given that there is no slipping).

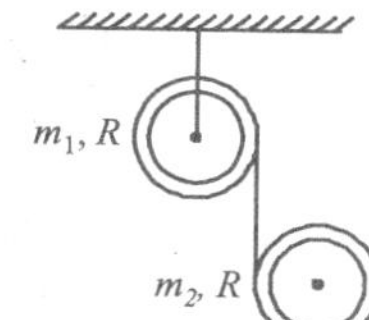

(a) $\frac{a_2}{R}$ (b) $\frac{(a_2+g)}{R}$

(c) $\frac{2(a_2+g)}{R}$ (d) None of these

54. Which of the following has the highest moment of inertia when each of them has the same mass and the same outer radius

(a) a ring about its axis, perpendicular to the plane of the ring
(b) a disc about its axis, perpendicular to the plane of the ring
(c) a solid sphere about one of its diameters
(d) a spherical shell about one of its diameters

55. Radius of gyration of a body depends upon

(a) axis of rotation
(b) translational motion
(c) shape of the body
(d) area of the body

56. A and B are moving in 2 circular orbits with angular velocity 2ω and ω respectively. Their positions are as shown at $t=0$. Find the time when they will meet for the first time.

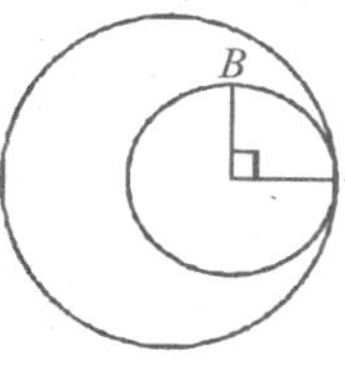

Toughnut

(a) $\frac{\pi}{2\omega}$ (b) $\frac{3\pi}{2\omega}$ (c) $\frac{\pi}{\omega}$
(d) they will never meet

57. If I_{xy} is the moment of inertia of a ring about a tangent in the plane of the ring and $I_{x'y'}$ is the moment of inertia of a ring about a tangent perpendicular to the plane of the ring then

(a) $I_{xy}=I_{x'y'}$ (b) $I_{xy}=\frac{1}{2}I_{x'y'}$

(c) $I_{x'y'}=\frac{3}{4}I_{xy}$ (d) $I_{xy}=\frac{3}{4}I_{x'y'}$

58. The moment of inertia of a uniform circular disc (figure) is maximum about an axis perpendicular to the disc and passing through

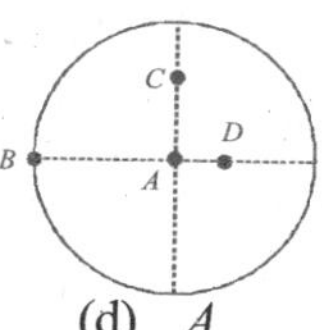

(a) B (b) C (c) D (d) A

59. One solid sphere A and another hollow sphere B are of same mass and same outer radii, Their moments of inertia about their diameters are respectively I_A and I_B, such that

(a) $I_A = I_B$ (b) $I_A > I_B$
(c) $I_A < I_B$ (d) $I_A / I_B = \rho_A = \rho_B$

Here ρ_A and ρ_B represent their densities.

60. A wheel having moment of inertia 2 kg-m^2 about its vertical axis, rotates at the rate of 60 rpm about this axis, The torque which can stop the wheel's rotation in one minute would be

(a) 1.12 Nm (b) 0.83 Nm
(c) 0.55 Nm (d) 0.21 Nm

61. Three particles, each of mass m gram, are situated at the vertices of an equilateral triangle ABC of side ℓ cm (as shown in the figure). The moment of inertia of the system about a line AX perpendicular to AB and in the plane of ABC, in gram-cm^2 units will be

(a) $\frac{3}{2}m\ell^2$

(b) $\frac{3}{4}m\ell^2$

(c) $2m\ell^2$

(d) $\frac{5}{4}m\ell^2$

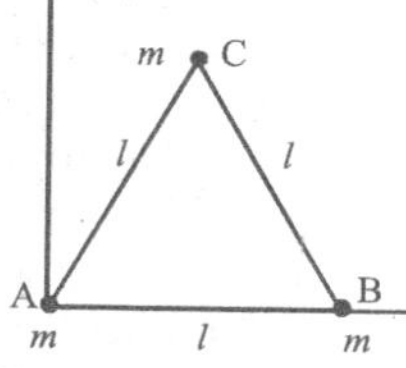

62. A round disc of moment of inertia I_2 about its axis perpendicular to its plane and passing through its centre is placed over another disc of moment of inertia I_1 rotating with an angular velocity ω about the same axis. The final angular velocity of the combination of discs is

(a) $\frac{(I_1+I_2)\omega}{I_1}$ (b) $\frac{I_2\omega}{I_1+I_2}$

(c) ω (d) $\frac{I_1\omega}{I_1+I_2}$

63. The ratio of the radii of gyration of a circular disc about a tangential axis in the plane of the disc and of a circular ring of the same radius about a tangential axis in the plane of the ring is
(a) $1:\sqrt{2}$ (b) $1:3$ (c) $2:1$ (d) $\sqrt{5}:\sqrt{6}$

64. The moment of inertia of a thin uniform rod of mass M and length L about an axis passing through its midpoint and perpendicular to its length is I_0. Its moment of inertia about an axis passing through one of its ends and perpendicular to its length is
(a) $I_0+ML^2/2$ (b) $I_0+ML^2/4$
(c) I_0+2ML^2 (d) I_0+ML^2

65. The moment of inertia of the rectangular plate ABCD, (AB = 2 BC) is minimum along the axis

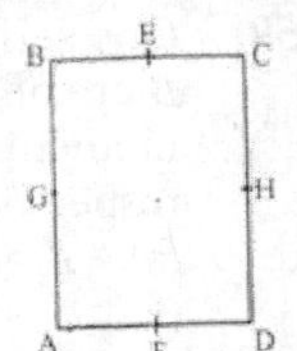

(a) GH
(b) EF
(c) BC
(d) AC

66. Four point masses, each of value m, are placed at the corners of a square ABCD of side ℓ. The moment of inertia of this system about an axis passing through A and parallel to BD is
(a) $2m\ell^2$ (b) $\sqrt{3}m\ell^2$ (c) $3m\ell^2$ (d) $m\ell^2$

67. Consider a uniform square plate of side 'a' and mass 'm'. The moment of inertia of this plate about an axis perpendicular to its plane and passing through one of its corners is
(a) $\frac{5}{6}ma^2$ (b) $\frac{1}{12}ma^2$ (c) $\frac{7}{12}ma^2$ (d) $\frac{2}{3}ma^2$

68. Point masses 1, 2, 3 and 4 kg are lying at the points (0, 0, 0), (2, 0, 0), (0, 3, 0) and (–2, –2, 0) respectively. The moment of inertia of this system about X-axis will be
(a) 43 kg m^2 (b) 34 kg m^2
(c) 27 kg m^2 (d) 72 kg m^2

69. The moment of inertia of a circular disc of mass M and radius R about an axis passing through the centre of mass is I_0. The moment of inertia of another circular disc of same mass and thickness but half the density about the same axis is
(a) $\frac{I_0}{8}$ (b) $\frac{I_0}{4}$ (c) $8I_0$ (d) $2I_0$

70. A hollow sphere of mass 2 kg is kept on a rough horizontal surface. A force of 10 N is applied at the centre of the sphere as shown in the figure. Find the minimum value of μ so that the sphere starts pure rolling. (Take g = 10m/s²)

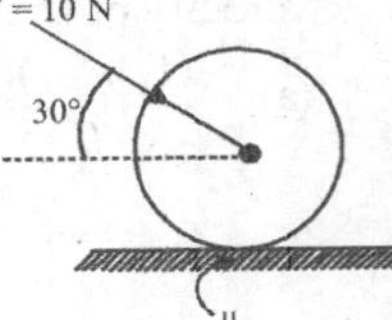

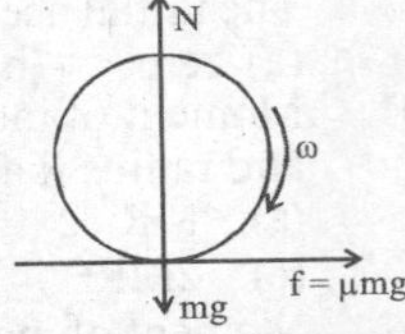

(a) $\sqrt{3}\times0.16$ (b) $\sqrt{3}\times0.08$
(c) $\sqrt{3}\times0.1$ (d) Data insufficient

71. A sphere is spinned in a clockwise direction by angular velocity $\frac{V_0}{R}$ and then it is released to a rough surface. The time elapsed by the sphere, till it starts pure rolling (coefficient of friction is μ) is [Radius of sphere is R].

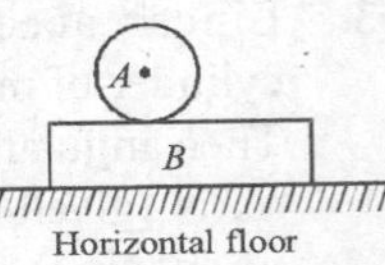

(a) $\frac{2V_0}{7\mu g}$ (b) $\frac{2V_0}{5\mu g}$ (c) $\frac{3V_0}{5\mu g}$ (d) $\frac{5V_0}{7\mu g}$

72. A cylinder *A* rolls without slipping on a plank *B*. The velocities of center of the cylinder and that of the plank are 4m/s and 2m/s respectively in same direction, with respect to the ground. Find the angular velocity of the cylinder (in rad/s) if its radius is 1m.

A
B
Horizontal floor

(a) 2 rad/sec (b) 4 rad/sec
(c) 6 rad/sec (d) 10 rad/sec

73. A thick-walled hollow sphere has outside radius R_0. It rolls down an incline without slipping and its speed at the bottom is v_0. Now the incline is waxed, so that it is practically frictionless and the sphere is observed to slide down (without any rolling). Its speed at the bottom is observed to be $5v_0/4$. The radius of gyration of the hollow sphere about an axis through its centre is
(a) $3R_0/2$ (b) $3R_0/4$ (c) $9R_0/16$ (d) $3R_0$

74. Consider a thin uniform square sheet made of a rigid material. If its side is '*a*' mass *m* and moment of inertia I about one of its diagonals, then
(a) $I>\frac{ma^2}{12}$ (b) $\frac{ma^2}{24}<I<\frac{ma^2}{12}$
(c) $I=\frac{ma^2}{24}$ (d) $I=\frac{ma^2}{12}$

75. From a solid sphere of mass M and radius R a cube of maximum possible volume is cut. Moment of inertia of cube about an axis passing through its center and perpendicular to one of its faces is

(a) $\frac{4MR^2}{9\sqrt{3}\pi}$ (b) $\frac{4MR^2}{3\sqrt{3}\pi}$ (c) $\frac{MR^2}{32\sqrt{2}\pi}$ (d) $\frac{MR^2}{16\sqrt{2}\pi}$

76. Two discs rotating about their respective axis of rotation with angular speeds 2 rads^{-1} and 5 rads^{-1} are brought into contact such that their axes of rotation coincide. Now, the angular speed of the system becomes 4 rads^{-1}. If the moment of inertia of the second disc is 1×10^{-3} kg m^2, then the moment of inertia of the first disc (in kg m^2) is

(a) 0.25×10^{-3} (b) 1.5×10^{-3}
(c) 1.25×10^{-3} (d) 0.5×10^{-3}

77. Two bodies have their moments of inertia I and 2I respectively about their axis of rotation. If their kinetic energies of rotation are equal, their angular momenta will be in the ratio

(a) 2 : 1 (b) 1 : 2 (c) $\sqrt{2}:1$ (d) $1:\sqrt{2}$

78. M.I of a circular loop of radius R about the axis in figure is

(a) MR^2
(b) $(3/4) MR^2$
(c) $MR^2/2$
(d) $2MR^2$

79. A rod of mass *m* and length *l* is bent in to shape of *L*. Its moment of inertia about the axis shown in figure

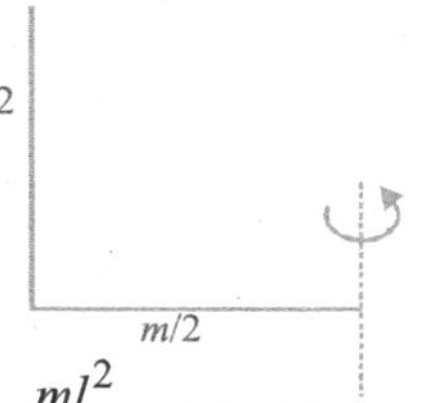

(a) $\frac{ml^2}{6}$ (b) $\frac{ml^2}{3}$ (c) $\frac{ml^2}{2}$ (d) None

80. The moment of inertia of a hollow thick spherical shell of mass M and its inner radius R_1 and outer radius R_2 about its diameter is

(a) $\frac{2M}{5}\frac{(R_2^5 - R_1^5)}{(R_2^3 - R_1^3)}$ (b) $\frac{2M}{3}\frac{(R_2^5 - R_1^5)}{(R_2^3 - R_1^3)}$

(c) $\frac{4M}{5}\frac{(R_2^5 - R_1^5)}{(R_2^3 - R_1^3)}$ (d) $\frac{4M}{3}\frac{(R_2^5 - R_1^5)}{(R_2^3 - R_1^3)}$

81. The moment of inertia of a uniform semicircular wire of mass m and radius r, about an axis passing through its centre of mass and perpendicular to its plane is $mr^2\left(1-\frac{k}{\pi^2}\right)$ then find the value of k.

(a) 2 (b) 3 (c) 4 (d) 5

82. A solid cylinder of mass 50 kg and radius 0.5 m is free to rotate about the horizontal axis. A massless string is wound round the cylinder with one end attached to it and other hanging freely. Tension in the string required to produce an angular acceleration of 2 revolutions s^{-2} is

(a) 25 N (b) 50 N (c) 78.5 N (d) 157 N

83. A child with mass m is standing at the edge of a playground merry - go - round (A large uniform disc which rotates in horizontal plane about a fixed vertical axis in parks) with moment of inertia I, radius R, and initial angular velocity w as shown in the figure. The child jumps off the edge of the merry-go-round with a velocity v with respect to the ground in direction tangent to periphery of the disc as shown. The new angular veloity of the merry-go-round is :

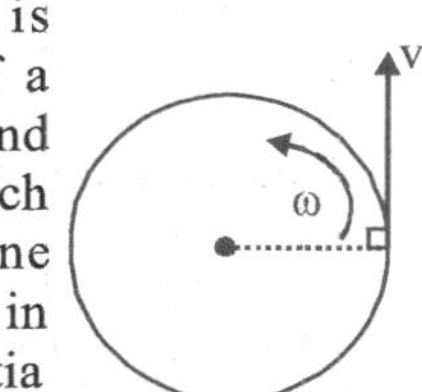

(a) $\sqrt{\frac{I\omega^2 - mv^2}{I}}$

(b) $\sqrt{\frac{(I + mR^2)\omega^2 - mv^2}{I}}$

(c) $\frac{I\omega - mvR}{I}$

(d) $\frac{(I + mR^2)\omega - mvR}{I}$

84. Two circular discs of same mass and thickness are made up of two different metals with densities d_X and d_Y ($d_X > d_Y$). Their moments of inertia about the axes passing through their centers of gravity and perpendicular to their planes are I_X and I_Y. Which one is correct?

(a) $I_x > I_y$ (b) $I_x < I_y$
(c) $I_x = I_y$ (d) None of these

85. There is a flat uniform triangular plate *ABC* such that *AB* = 4 cm, *BC* = 3 cm and angle *ABC* = 90°. The moment of inertia of the plate about *AB*, *BC* and *CA* as axis is respectively I_1, I_2 and I_3. Which one of the following is true?

Tricky

(a) $I_3 > I_2$ (b) $I_2 > I_1$ (c) $I_3 > I_1$ (d) $I_1 > I_2$

86. The moment of inertia of a body about a given axis is 1.2 kg m^2. Initially, the body is at rest. In order to produce a rotational kinetic energy of 1500 joule, an angular acceleration of 25 radian/sec^2 must be applied about that axis for a duration of

(a) 4 seconds (b) 2 seconds
(c) 8 seconds (d) 10 seconds

87. A fly wheel rotating about a fixed axis has a kinetic energy of 360 joule when its angular speed is 30 radian/sec. The moment of inertia of the wheel about the axis of rotation is

(a) 0.6 kg/m^2 (b) 0.15 kg m^2
(c) 0.8 kg m^2 (d) 0.75 kg m^2

88. A bullet of mass 10 g and speed 500 m/s is fired into a door and gets embedded exactly at the centre of the door. The door is 1.0 m wide and weighs 12 kg. It is hinged at one end and rotates about a vertical axis practically without friction. The angular speed of the door just after the bullet embeds into it will be :

(a) 6.25 rad/sec (b) 0.625 rad/sec
(c) 3.35 rad/sec (d) 0.335 rad/sec

89. A thin wire of length L and uniform linear mass density ρ is bent into a circular loop with centre at O as shown. The moment of inertia of the loop about the axis XX' is

(a) $\frac{\rho L^3}{8\pi^2}$ (b) $\frac{\rho L^3}{16\pi^2}$ (c) $\frac{5\rho L^3}{16\pi^2}$ (d) $\frac{3\rho L^3}{8\pi^2}$

90. A particle of mass m is attached to a thin uniform rod of length a and mass 4 m. The distance of the particle from the centre of mass of the rod is $a/4$. The moment of inertia of the combination about an axis passing through O normal to the rod is

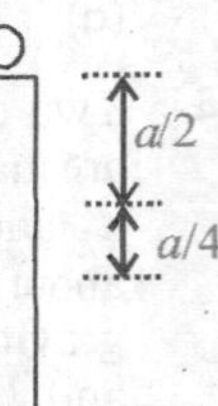

(a) $\frac{64}{48}ma^2$ (b) $\frac{91}{48}ma^2$ (c) $\frac{27}{48}ma^2$ (d) $\frac{51}{48}ma^2$

91. From a disc of radius R and mass M, a circular hole of diameter R, whose rim passes through the centre is cut. What is the radius of gyration of the remaining part of the disc about a perpendicular axis, passing through the centre ?

(a) $\frac{13}{2}R$ (b) $\sqrt{\frac{13}{16}}R$ (c) $\frac{11}{3}R$ (d) $\sqrt{\frac{13}{9}}R$

92. Four identical thin rods each of mass M and length l, form a square frame. Moment of inertia of this frame about an axis through the centre of the square and perpendicular to its plane is :

(a) $\frac{2}{3}Ml^2$ (b) $\frac{13}{3}Ml^2$ (c) $\frac{1}{3}Ml^2$ (d) $\frac{4}{3}Ml^2$

93. From a circular disc of radius R and mass $9M$, a small disc of radius $R/3$ is removed from the disc. The moment of inertia of the remaining disc about an axis perpendicular to the plane of the disc and passing through O is

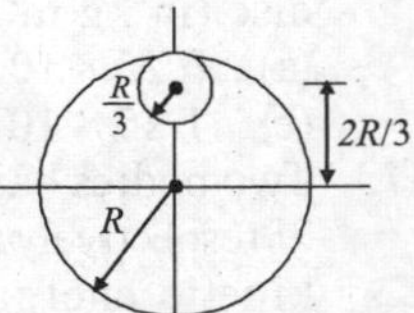

(a) $4MR^2$ (b) $\frac{40}{9}MR^2$
(c) $10MR^2$ (d) $\frac{37}{9}MR^2$

94. A horizontal turn table in the form of a disc of radius r carries a gun at G and rotates with angular velocity ω_0 about a vertical axis passing through the centre O. The increase in angular velocity of the system if the gun fires a bullet of mass m with a tangential velocity v with respect to the gun is (moment of inertia of gun + table about O is I_0)

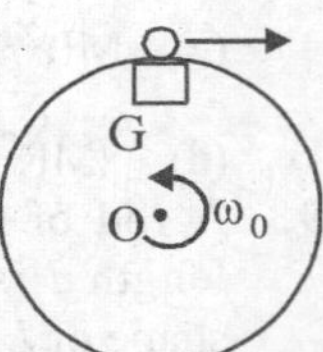

(a) $\frac{mvr}{I_0 + mr^2}$ (b) $\frac{2mvr}{I_0}$
(c) $\frac{v}{2r}$ (d) $\frac{mvr}{2I_0}$

95. A tangential force of 20 N is applied on a cylinder of mass 4 kg and moment of inertia 0.02 kg m^2 about its own axis. If the cylinder rolls without slipping, then linear acceleration of its centre of mass will be

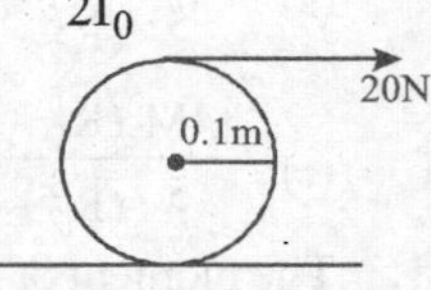

(a) 6.7 m/s^2 (b) 10 m/s^2
(c) 3.3 m/s^2 (d) None of these

96. A toy car rolls down the inclined plane as shown in the fig. It loops at the bottom. What is the relation between H and h? Tricky

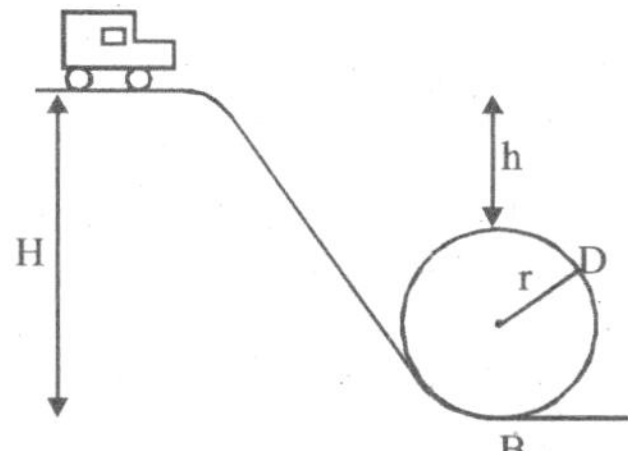

(a) $\frac{H}{h} = 2$ (b) $\frac{H}{h} = 3$ (c) $\frac{H}{h} = 4$ (d) $\frac{H}{h} = 5$

97. A thin ring, a disk and an annular cylinder, of same mass M, are released from a point 3.6 m from the ground up an inclined plane of 30° degree inclination. The ring and the disk have the same radius R. Times taken by the ring and disk to reach the ground are in the ratio,

(a) $\sqrt{2} : \sqrt{1.5}$ (b) $\sqrt{1.4} : \sqrt{1.5}$

(c) $\sqrt{1.5} : \sqrt{1.5}$ (d) $\sqrt{1.5} : \sqrt{1.6}$

98. A rolling body is kept on a plank B. There is sufficient friction between A and B and no friction between B and the inclined plane.

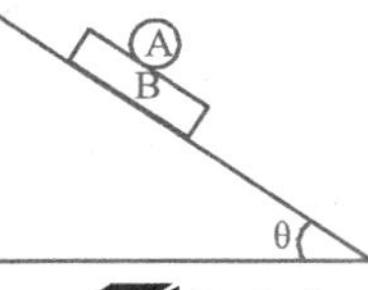

Then body :

(a) A rolls

(b) A does not experience any friction

(c) A and B has equal acceleration and unequal velocities

(d) A rolls depending upon the angle of inclination θ

99. In the figure shown mass of both, the spherical body and block is m. Moment of inertia of the spherical body about centre of mass is $2mR^2$. The spherical body rolls on the horizontal surface. There is no slipping at any surfaces in contact. The ratio of kinetic energy of the spherical body to that of block is

(a) 3/4 (b) 1/3 (c) 2/3 (d) 1/2

100. The free end of a thread wound on a bobbin is passed round a nail A hammered into the wall. The thread is pulled at a constant velocity. Assuming pure rolling of bobbin, find the velocity v_0 of the centre of the bobbin at the instant when the thread forms an angle α with the vertical.

(a) $\frac{vR}{R\sin\alpha - r}$ (b) $\frac{vR}{R\sin\alpha + r}$

(c) $\frac{2vR}{R\sin\alpha + r}$ (d) $\frac{v}{R\sin\alpha + r}$

Answer KEYs

1	(a)	11	(c)	21	(b)	31	(d)	41	(c)	51	(a)	61	(d)	71	(a)	81	(c)	91	(b)
2	(c)	12	(a)	22	(c)	32	(a)	42	(a)	52	(d)	62	(d)	72	(a)	82	(d)	92	(d)
3	(b)	13	(a)	23	(d)	33	(c)	43	(a)	53	(c)	63	(d)	73	(b)	83	(d)	93	(a)
4	(c)	14	(b)	24	(d)	34	(a)	44	(c)	54	(a)	64	(b)	74	(d)	84	(b)	94	(a)
5	(b)	15	(b)	25	(b)	35	(a)	45	(c)	55	(a)	65	(b)	75	(a)	85	(b)	95	(a)
6	(c)	16	(b)	26	(a)	36	(a)	46	(c)	56	(d)	66	(c)	76	(d)	86	(b)	96	(d)
7	(a)	17	(c)	27	(b)	37	(a)	47	(b)	57	(d)	67	(d)	77	(d)	87	(c)	97	(a)
8	(b)	18	(b)	28	(d)	38	(b)	48	(d)	58	(a)	68	(a)	78	(b)	88	(b)	98	(b)
9	(c)	19	(d)	29	(c)	39	(d)	49	(d)	59	(c)	69	(d)	79	(a)	89	(d)	99	(c)
10	(c)	20	(b)	30	(b)	40	(c)	50	(c)	60	(d)	70	(b)	80	(a)	90	(b)	100	(a)

Hints & Solutions

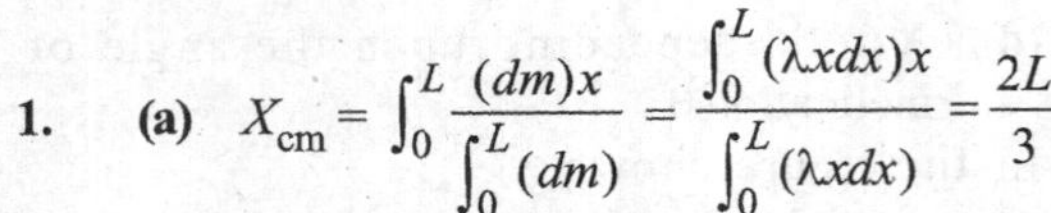

1. (a) $X_{cm} = \int_0^L \frac{(dm)x}{\int_0^L (dm)} = \frac{\int_0^L (\lambda x dx)x}{\int_0^L (\lambda x dx)} = \frac{2L}{3}$

2. (c) When the system is released, heavier mass move downward and the lighter one upward. Thus, centre of mass will move towards the heavier mass with acceleration

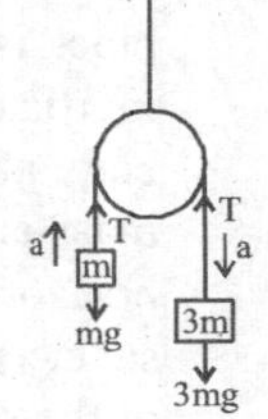

$a = \left(\frac{3m - m}{3m + m}\right) g = \frac{g}{2}$

3. (b) $x_{cm} = \frac{1 \times 0 + 1 \times PQ + 1 \times PR}{1 + 1 + 1} = \frac{PQ + PR}{3}$

and $y_{cm} = 0$

4. (c) Let O(0, 0) be the centre of sphere A then,

$x_{cm} = \frac{m \times 0 + 2m \times 3R}{m + 2m} = 2R$

= at the point of contact

5. (b) $\frac{\text{Rotational KE}}{\text{Total KE}} = \frac{\frac{1}{2}mv^2\left(\frac{K^2}{R^2}\right)}{\frac{1}{2}mv^2\left(1 + \frac{K^2}{R^2}\right)}$

$= \frac{K^2}{K^2 + R^2}$

6. (c) After collision velocity of COM of *A* becomes zero and that of *B* becomes equal to initial velocity of COM of *A*. But angular velocity of *A* remains unchanged as the two spheres are smooth.

7. (a)

8. (b) By energy conservation

$(K.E)_i + (P.E)_i = (K.E)_f + (P.E)_f$

$(K.E)_i = 0, (P.E)_i = mgh, (P.E)_f = 0$

$(K.E)_f = ½I\omega^2 + ½mv_{cm}^2$

Where I (moment of inertia) = $½mR^2$ (for solid cylinder)

so $mgh = ½(½mR^2)\left(\frac{v_{cm}^2}{R^2}\right) + ½mv_{cm}^2$

$\Rightarrow v_{cm} = \sqrt{4gh/3}$

9. (c) $\mu = \frac{F}{R} = \frac{mg \sin \alpha}{mg \cos \alpha} = \tan \alpha$

10. (c) $a_1 = \frac{v_1^2}{R_1} = \frac{\omega^2 R_1^2}{R_1} = \omega^2 R_1$

$a_2 = \frac{v_2^2}{R_2} = \omega^2 R_2$

Taking particle masses equal

$\frac{F_1}{F_2} = \frac{ma_1}{ma_2} = \frac{a_1}{a_2} = \frac{R_1}{R_2}$

11. (c)

12. (a) $X_{cm} = \frac{m_1x_1 + m_2x_2 + m_3x_3}{m_1 + m_2 + m_3}$

$X_{cm} = \frac{300 \times (0) + 500(40) + 400 \times 70}{300 + 500 + 400}$

$X_{cm} = \frac{500 \times 40 + 400 \times 70}{1200} = 40 \text{ cm}$

13. (a) Does not shift as no external force acts. The centre of mass of the system continues its original path. It is only the internal forces which comes into play while breaking.

14. (b) The coordinates of C.M of three particle are

$x = \frac{m_1x_1 + m_2x_2 + m_3x_3}{m_1 + m_2 + m_3}$

& $y = \frac{m_1y_1 + m_2y_2 + m_3y_3}{m_1 + m_2 + m_3}$

here $m_1 = m_2 = m_3 = m$

so $x = \frac{(x_1 + x_2 + x_3)m}{m + m + m} = 2$,

$y = \frac{(y_1 + y_2 + y_3)m}{m + m + m} = 2$

so coordinates of C.M. of three particle are (2,2)

15. (b) $\vec{v}_{cm} = \frac{m_1\vec{v}_1 + m_2\vec{v}_2}{m_1 + m_2} = \frac{2 \times 2 + 4 \times 10}{2 + 4} = 7.3 \text{ m/s}$

16. (b) Possible when $m_2 = m_4$

17. (c) Centre of mass

$x_{cm} = \frac{x}{2} = \frac{(\rho x)\left(\frac{x}{2}\right)\frac{1}{2} + \rho y\left(\frac{y}{2}\right)}{\rho(x + y)} \Rightarrow \frac{1}{2} + \frac{y}{x} = \frac{y^2}{x^2}$

$\therefore \frac{BC}{AB} = \frac{y}{x}$

$= \frac{1 + \sqrt{3}}{2} = 1.37$

A
x
B (0,0)
y
C

18. (b)

19. (d) $X_{CM} = \dfrac{m_1x_1 + m_2x_2 + m_3x_3}{m_1 + m_2 + m_3}$

$= \dfrac{1\times 0 + 1\times 3 + 1\times 0}{1+1+1} = 1$

$Y_{CM} = \dfrac{m_1y_1 + m_2y_2 + m_3y_3}{m_1 + m_2 + m_3}$

$= \dfrac{1\times 0 + 1\times 0 + 1\times 4}{1+1+1} = \dfrac{4}{3}$

Therefore the coordinates of centre of mass are $\left(1, \dfrac{4}{3}\right)$.

20. (b) Vertical line from hinge A must pass through C.M. of rod system.

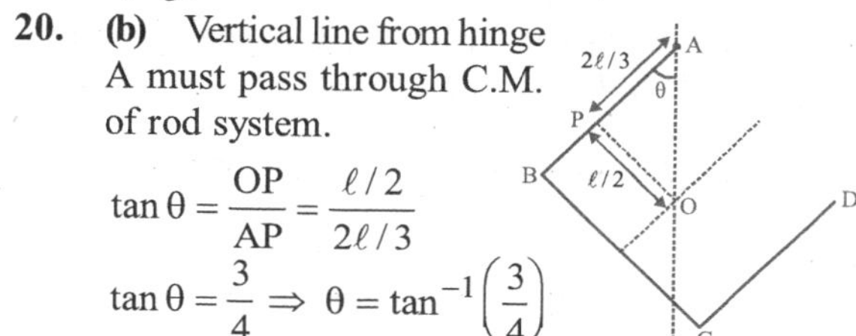

$\tan\theta = \dfrac{OP}{AP} = \dfrac{\ell/2}{2\ell/3}$

$\tan\theta = \dfrac{3}{4} \Rightarrow \theta = \tan^{-1}\left(\dfrac{3}{4}\right)$

21. (b) Friction force will act towards right.

22. (c)

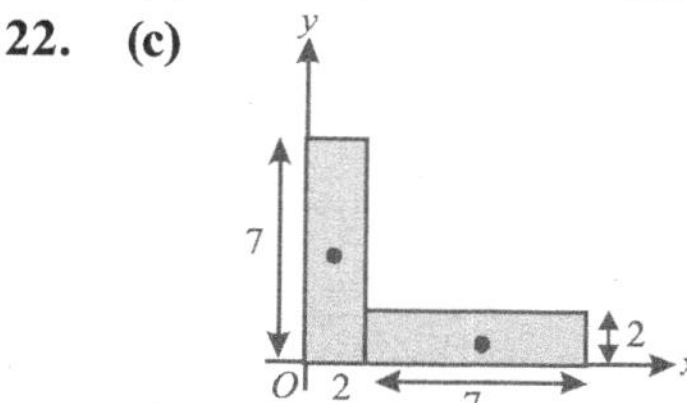

$X_{cm} = \dfrac{m\times 5.5 + 1\times m}{2m} = \dfrac{6.5}{2} = \dfrac{13}{4}$ cm.

$Y_{cm} = \dfrac{m\times 3.5 + 1\times m}{2m} = \dfrac{4.5}{2} = \dfrac{9}{4}$ cm.

23. (d)

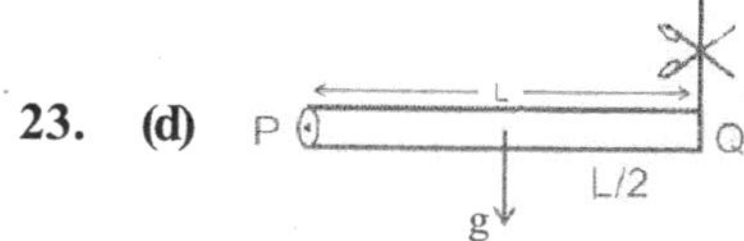

angular acceleration $\alpha = \dfrac{3g}{2L}$

24. (d) Centre of mass is at rest $\therefore V_{CM} = 0$

25. (b)

26. (a) Let the C.M. be 'b'. Then,

$\dfrac{(n-1)mb + ma}{mn} = 0 \Rightarrow b = -\dfrac{1}{n-1}a$

27. (b)

28. (d) As friction is absent at the point of contact,

$\text{Acceleration} = \dfrac{\text{Force}}{\text{Mass}}$

It is independent of h

29. (c)

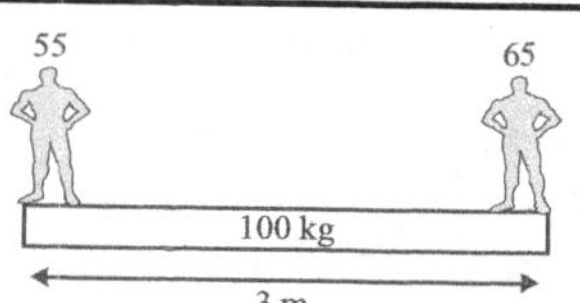

There is no external force so centre of mass of the system will not shift.

30. (b) $\theta = \omega_0 t + \dfrac{1}{2}\alpha t^2 \Rightarrow \theta = 100\,\text{rad}$

$\therefore$ Number of revolutions $= \dfrac{100}{2\pi} = 16$ (approx.)

31. (d)

32. (a) According to right hand thumb rule.

33. (c) The velocity of top point of the wheel is twice that of centre of mass. And the speed of centre of mass is same for both the wheels. $[v = \omega r]$

34. (a) Angular momentum of particle is given by:

$L = mr^2\omega = 2\pi mr^2 f \quad [\because W = 2\pi f]$

If frequency is halved then,

$L' = mr^2\dfrac{\omega}{2} = \pi mr^2 f \; \therefore L' = \dfrac{L}{2}$

35. (a) When angular acceleration (α) is zero then torque on the wheel becomes zero.

$\theta(t) = 2t^3 - 6t^2 \Rightarrow \dfrac{d\theta}{dt} = 6t^2 - 12t$

$\Rightarrow \alpha = \dfrac{d^2\theta}{dt^2} = 12t - 12 = 0 \;\therefore\; t = 1$ sec.

36. (a) Angular momentum,

$L_0 = mvr \sin 90°$

$= 2 \times 0.6 \times 12 \times 1 \times 1$

[As $V = r\omega$, Sin 90° = 1]

So, $L_0 = 14.4$ kgm^2/s

0.6m
O
0.8m
1 m
O'

37. (a) From the relation, angular momentum,

$L = mvr \qquad v = \dfrac{L}{mr}$

Centripetal force acting on the particle

$F = \dfrac{mv^2}{r} = \dfrac{m\left(\dfrac{L}{mr}\right)^2}{r} = \dfrac{L^2}{mr^3}$

38. (b) If no external force acts on a system of particles, the centre of mass remains at rest. So, speed of centre of mass is zero.

39. (d) Consider an element of length dx at a distance x from end A.

Here, mass per unit length λ of rod

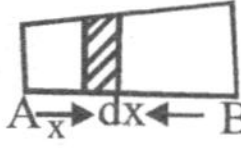

$\lambda \propto x \Rightarrow \lambda = kx$

$\therefore\; dm = \lambda dx = kx\, dx$

Position of centre of gravity of rod from end A.

$$x_{CG} = \frac{\int_0^L x\,dm}{\int_0^L dm}$$

$$\therefore\ x_{CG} = \frac{\int_0^3 x(kx\,dx)}{\int_0^3 kx\,dx} = \frac{\left[\frac{x^3}{3}\right]_0^3}{\left[\frac{x^2}{2}\right]_0^3} = \frac{\frac{(3)^3}{3}}{\frac{(3)^3}{2}} = 2m$$

40. (c) $\bar{x} = \frac{m(-2b)+2m(-b)+m\times 0+2m(b)}{m+2m+m+2m} = \frac{-b}{3}$

and $\bar{y} = +b.$

41. (c)

42. (a) $x_i = \frac{2\times 2+2\times 0}{2+2} = 1\text{m}$. and $x_f = \frac{1\times 2+3\times 0}{1+3} = 0.5\,\text{m}$.

$\therefore\ \Delta x = 0.5\,\text{m}$

43. (a) $x_{cm} = \frac{m\times \ell+2m\times 2\ell+3m\times 3\ell+...+nm\times n\ell}{m+2m+3m+..+nm}$

$$= \frac{(1+4+9+..+n^2)\ell}{1+2+..+n}$$

$$= \frac{\ell n(n+1)(2n+1)/6}{n(n+1)/2} = \frac{(2n+1)\ell}{3}$$

44. (c) Let speed of belt be v

Angular speeds of wheels

$$\omega_B = \frac{v}{2\pi R_B},\ \omega_A = \frac{v}{2\pi R_A};\ \frac{\omega_A}{\omega_B} = \frac{R_B}{R_A} = 2$$

45. (c) M.I. of a uniform circular disc of radius 'R' and mass 'M' about an axis passing through C.M. and normal to the disc is

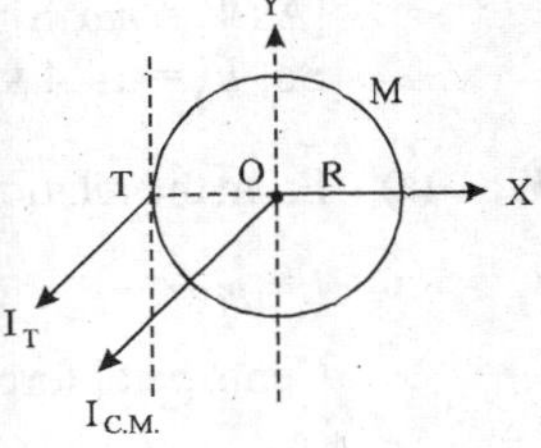

$$I_{C.M.} = \frac{1}{2}MR^2$$

From parallel axis theorem

$$I_T = I_{C.M.} + MR^2 = \frac{1}{2}MR^2 + MR^2 = \frac{3}{2}MR^2$$

46. (c)

47. (b) $Tr = \frac{mr^2}{2}\alpha_1$ ....(1)

$Tr = \frac{mr^2}{2}\alpha$ .... (2)

$\alpha_1 = \alpha$ .... (3)

Acceleration of point b = acceleration of point a

$r\alpha_1 = a_{cm} - r\alpha$ .... (4)

Hence, $2r\alpha = a_{cm}$

48. (d) $\tau \times \Delta t = L_0$ $\{\because$ since $L_f = 0\}$

or $\tau \times 30 = 2\pi$

$$\tau = \frac{\pi}{15}\text{N}-\text{m}$$

49. (d)

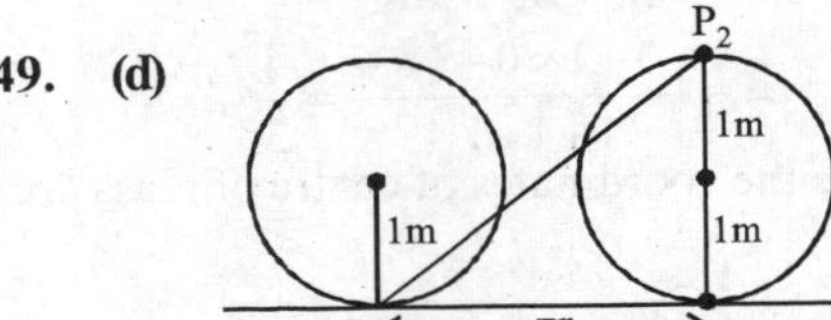

Linear distance moved by wheel in half revolution $= \pi r$. Point P_1 after half revolution reaches at P_2 vertically 2m above the ground.

$\therefore$ Displacement P_1P_2

$= \sqrt{\pi^2 r^2 + 2^2} = \sqrt{\pi^2 + 4}$ $[\because r = 1m]$

50. (c) FBD for sphere and block

$$a_1 = \frac{f_r}{m} = \frac{\mu mg}{m},\ a_2 = \frac{f_r}{m} = \frac{\mu mg}{m}$$

$$\vec{a}_1 = \mu g\hat{i},\ \vec{a}_2 = -\mu g\hat{i}$$

51. (a) Perp axis theorem $I_x + I_y = I_z$

$\Rightarrow\ 2I = MR^2$ So $I = \frac{MR^2}{2}$

52. (d) Moment of inertia of a hollow cylinder of mass M and radius r about its own axis is Mr^2.

53. (c) $m_2 g - T = m_2 a_2$ ... (i)

$TR = \frac{m_2 R^2}{2}\alpha_2$ ... (ii)

$\alpha_1 R = a_2 - \alpha_2 R$ ... (iii)

$TR = \left(\frac{m_1 R^2}{2}\right)\alpha_1$ ... (iv)

$$\alpha_2 = \frac{2T}{m_2 R} = \frac{2}{m_2 R}.(m_2 a_2 + m_2 g)$$

$$= \frac{2(a_2 + g)}{R}$$

54. (a)

55. (a) Radius of gyration of a body depends on the axis of rotation.

56. (d) **Case 1 :** When they rotate in same sense

$2m\pi = 2\omega t$

$\frac{3\pi}{2} + 2n\pi = \omega t$; $2m\pi = 2\left(\frac{3\pi}{2} + 2n\pi\right)$

$2m = 3 + 4n$; $m = \frac{3}{2} + 2n \Rightarrow m - 2n = \frac{3}{2}$

Not possible for m and n being integer.

Case 2 : When they rotate in opposite sense

$2m\pi = 2\omega t$

$\frac{\pi}{2} + 2n\pi = \omega t$; $2m\pi = 2\left(\frac{\pi}{2} + 2n\pi\right)$

$2m\pi = \pi + 4n\pi$; $2m - 4n = 1$

Not possible for m and n integer.

57. (d) I_{xy}, moment of inertia of a ring about its tangent in the plane of ring $I_{x'y} = \frac{3}{2}MR^2$

Moment of inertia about a tangent perpendicular to the plane of ring $I_{xy} = 2MR^2$

$\therefore\ I_{xy} = \frac{3}{4}(2MR^2) = \frac{3}{2}MR^2$ or $I_{xy} = \frac{3}{4}I_{x^1y^1}$

58. (a) According to parallel axis theorem of the moment of Inertia

$I = I_{cm} + md^2$

d is maximum for point B so I_{max} about B.

59. (c) In a hollow sphere, the mass is distributed away from the axis of rotation. So, its moment of inertia is greater than that of a solid sphere.

60. (d) $\tau \times \Delta t = I\omega$

or $\tau \times 60 = 2 \times 2 \times 60\pi/60 \Rightarrow \tau = 0.21$ N – m

$\left(\because\ f = 60\text{rpm} \therefore\ \omega = 2\pi f = 2\pi \times \frac{60}{60}\right)$

61. (d) $I_{AX} = m(AB)^2 + m(OC)^2$

$= m\ell^2 + m(\ell\cos 60°)^2$

$= m\ell^2 + m\ell^2/4 = 5/4\ m\ell^2$

62. (d) Angular momentum will be conserved

$I_1\omega = I_1\omega' + I_2\omega' \Rightarrow \omega' = \frac{I_1\omega}{I_1 + I_2}$

63. (d) Circular disc ...(1)

$I_{y_1} = \frac{MR^2}{4}$

$\therefore I'_{y_1} = \frac{MR^2}{4} + MR^2 = \frac{5}{4}MR^2$

Circular ring ...(2)

$I_{y_2} = \frac{MR^2}{2}$

$\therefore I'_{y_2} = \frac{MR^2}{2} + MR^2 = \frac{3}{2}MR^2$

$I'_{y_1} = MK_1^2,\ I'_{y_2} = MK_2^2$

$\therefore \frac{K_1^2}{K_2^2} = \frac{I'_{y_1}}{I'_{y_2}} \Rightarrow K_1 : K_2 = \sqrt{5} : \sqrt{6}$

64. (b) By theorem of parallel axes,

$I = I_{cm} + Md^2$

$I = I_0 + M(L/2)^2 = I_0 + ML^2/4$

65. (b) The distribution of mass about axis EF is minimum so radius of gyration is minimum and therefore moment of inertia is minimum about EF.

66. (c)

$I_{nn'}$ = M.I due to the point mass at B + M.I due to the point mass at D + M.I due to the point mass at C.

$= 2 \times m\left(\frac{\ell}{\sqrt{2}}\right)^2 + m(\sqrt{2}\ell)^2 = m\ell^2 + 2m\ell^2 = 3m\ell^2$

67. (d) $I_{nn'} = \frac{1}{12}M(a^2 + a^2) = \frac{Ma^2}{6}$

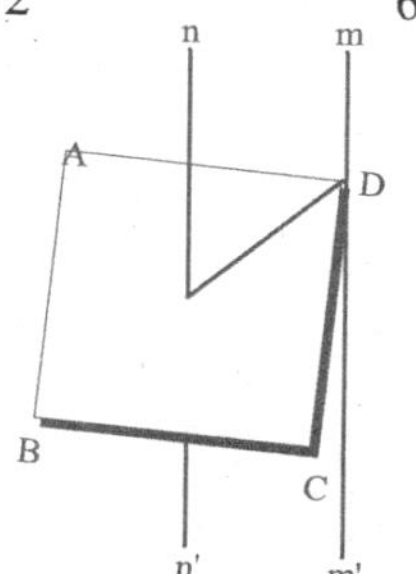

Also, $DO = \frac{DB}{2} = \frac{\sqrt{2}a}{2} = \frac{a}{\sqrt{2}}$

According to parallel axis theorem

$$I_{mm'} = I_{nn'} + M\left(\frac{a}{\sqrt{2}}\right)^2 = \frac{Ma^2}{6} + \frac{Ma^2}{2}$$

$$= \frac{Ma^2 + 3Ma^2}{6} = \frac{2}{3}Ma^2$$

68. **(a)** Moment of inertia of the whole system about the axis of rotation will be equal to the sum of the moments of inertia of all the particles.

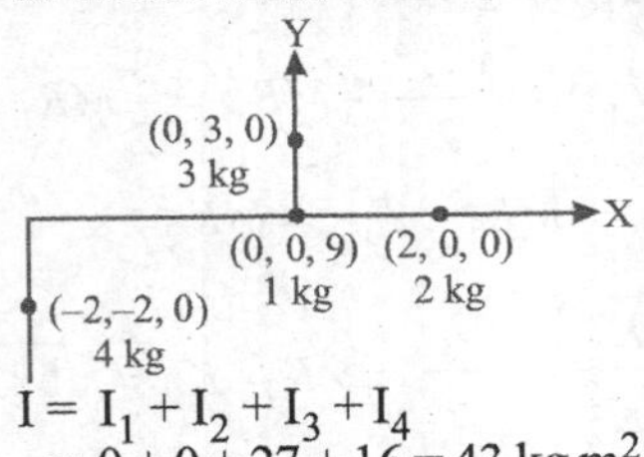

$\therefore \quad I = I_1 + I_2 + I_3 + I_4$
$= 0 + 0 + 27 + 16 = 43$ kg m^2

69. **(d)**

70. **(b)** $10\cos 30° - f = 2a$ ... (i)

$\tau = I\alpha$

$\Rightarrow fr = \frac{2}{3} \times 2 \times r^2 \times \alpha$ ... (ii)

(where r is radius of sphere)

From (i) and (ii), we get

$F_y = 10 \sin 30°$, $F_x = 10 \cos 30°$, α, a, f, N, 20 N

$f = 2\sqrt{3}$ newton,

$N = 20 + 10 \sin 30° = 25$

$f = \mu N \Rightarrow \mu = \frac{f}{N} = \frac{2\sqrt{3}}{25} = 0.08 \times \sqrt{3}$

71. **(a)**

72. **(a)** For rolling, $v_A = 2$m/s

or $4 - 1.\omega = 2$

or $\omega = 2$ rad/sec

(clockwise)

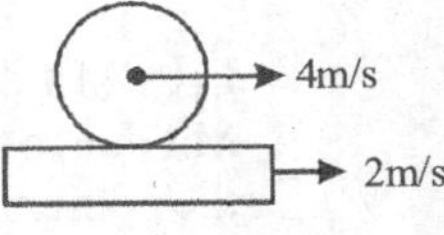

73. **(b)**

74. **(d)** For a thin uniform square sheet

$$I_1 = I_2 = I_3 = \frac{ma^2}{12}$$

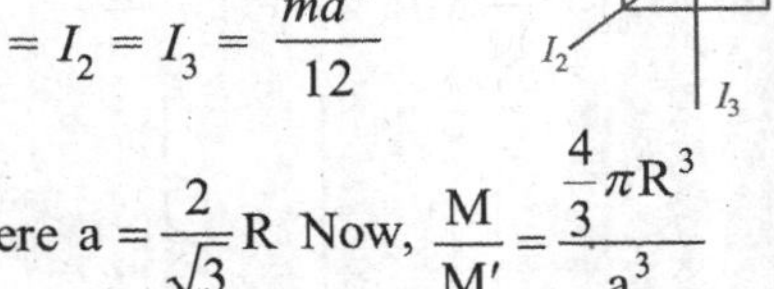

75. **(a)** Here $a = \frac{2}{\sqrt{3}}R$ Now, $\frac{M}{M'} = \frac{\frac{4}{3}\pi R^3}{a^3}$

$$= \frac{\frac{4}{3}\pi R^3}{\left(\frac{2}{\sqrt{3}}R\right)^3} = \frac{\sqrt{3}}{2}\pi.$$

$$M' = \frac{2M}{\sqrt{3}\pi}$$

Moment of inertia of the cube about the given axis, $I = \frac{M'a^2}{6}$

$$= \frac{\frac{2M}{\sqrt{3}\pi} \times \left(\frac{2}{\sqrt{3}}R\right)^2}{6} = \frac{4MR^2}{9\sqrt{3}\pi}$$

76. **(d)**

77. **(d)** $K = \frac{L^2}{2I} \Rightarrow L^2 = 2KI \Rightarrow L = \sqrt{2KI}$

$$\frac{L_1}{L_2} = \sqrt{\frac{K_1}{K_2} \cdot \frac{I_1}{I_2}} = \sqrt{\frac{K}{K} \cdot \frac{I}{2I}} = \frac{1}{\sqrt{2}}$$

$\Rightarrow L_1 : L_2 = 1 : \sqrt{2}$

78. **(b)** Use theorem of parallel axes.

79. **(a)** $I = \frac{\frac{m}{2}\left(\frac{\ell}{2}\right)^2}{3} + \frac{m}{2}\left(\frac{\ell}{2}\right)^2 = \frac{m\ell^2}{6}$

80. **(a)** $\rho = \frac{M}{\frac{4}{3}\pi(R_2^3 - R_1^3)}$

$$I_{shell} = \frac{2}{5}M_2R_2^2 - \frac{2}{5}M_1R_1^2 \quad .... (1)$$

$M_2 = \rho \times \frac{4}{3}\pi R_2^3$

$$= \frac{MR_2^3}{(R_2^3 - R_1^3)}; \; M_1 = \frac{MR_1^3}{R_2^3 - R_1^3}$$

R_2, R_1

Putting values of M_1 and M_2 in eq. (1),

$$I_{shell} = \frac{2M}{5}\frac{(R_2^5 - R_1^5)}{(R_2^3 - R_1^3)}$$

81. **(c)** Moment of inertia about z-axis,

$I_z = mr^2$ (about centre of mass)

Applying parallel axes theorem,

$I_z = I_{cm} + mk^2$

$I_{cm} = I_z - m\left(\frac{2}{\pi}r\right)^2$

$= mr^2 - \frac{m4r^2}{\pi^2} = mr^2\left(1 - \frac{4}{\pi^2}\right)$ i.e., $k = 4$

m, $(2/\pi)r$, r, z

82. **(d)** Here $\alpha = 2$ revolutions/s^2 $= 4\pi$ rad/s^2 (given)

$$I_{cylinder} = \frac{1}{2}MR^2 = \frac{1}{2}(50)(0.5)^2$$

$$= \frac{25}{4} \text{ Kg-m}^2$$

As $\tau = I\alpha$ so $TR = I\alpha$

$$\Rightarrow T = \frac{I\alpha}{R} = \frac{\left(\frac{25}{4}\right)(4\pi)}{(0.5)} N = 50\,\pi N = 157\,N$$

83. **(d)** Let the angular velocity of disc after child jumps off. be ω'

∴ From conservation of angular momentum

$(I + mR^2)\,\omega = mvR + I\omega$

$$\therefore \quad \omega' = \frac{\left(I + mR^2\right)\omega - mvR}{I}$$

84. **(b)**

85. **(b)** Moment of Inertia depend upon mass and distribution of masses as $I = \Sigma mr^2$.

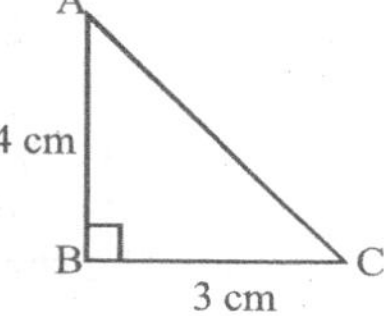

Further, as the distance of masses is more, more is the moment of Inertia.

If we choose BC as axis. Distance is maximum. Hence, Moment of Inertia is maximum.

$\therefore I_2 > I_1, I_2 > I_3$

86. **(b)** $I = 1.2$ kg m^2, $E_r = 1500$ J, $\alpha = 25$ rad/sec^2, $\omega_1 = 0$, $t = ?$

As $E_r = \frac{1}{2}I\omega^2$, $\omega = \sqrt{\frac{2E_r}{I}}$

$$= \sqrt{\frac{2 \times 1500}{1.2}} = 50 \text{ rad/sec}$$

From $\omega_2 = \omega_1 + \alpha t$

$50 = 0 + 25\,t$, $t = 2$ seconds

87. **(c)** $E_r = \frac{1}{2}I\omega^2 \Rightarrow$

$$I = \frac{2E_r}{\omega^2} = \frac{2 \times 360}{30 \times 30} = 0.8 \text{ kg m}^2$$

88. **(b)** Let x = distance between hinge and centre of the door = L/2 = $\frac{1}{2}$ = 0.5 m

As per conservation of angular momentum

$mvx = I\omega$

$$\Rightarrow \omega = \frac{mvx}{I} = \frac{0.01 \times 500 \times 0.5}{\frac{1}{3} \times 12 \times (1)^2} = 0.625 \text{ rad/s}$$

89. **(d)**

90. **(b)** Moment of inertia

$$= m\left(\frac{3a}{4}\right)^2 + m_1\frac{a^2}{3}$$

For the centre of rod

$$\left(\frac{m_1a^2}{12} + \frac{m_1a^2}{4}\right) = \frac{m_1a^2}{3}$$

$\therefore \quad m_1 = 4m$

$$\text{Total } I = m\left(\frac{3a}{4}\right)^2 + \frac{4ma^2}{3}$$

$$= \frac{9ma^2}{16} + \frac{4ma^2}{3}$$

$$= \frac{(27 + 64)}{48}ma^2 = \frac{91}{48}ma^2$$

91. **(b)**

92. **(d)** Moment of inertia of a thin rod of length l about an axis passing through centre and perpendicular to the rod = $\frac{1}{12}Ml^2$.

Thus moment of inertia of the frame.

$$\frac{ml^2}{12} + \frac{ml^2}{4} = \frac{4ml^2}{12} = \frac{ml^2}{3}$$

$$\text{Total M.I.} = 4 \times \frac{ml^2}{3}$$

93. **(a)**

94. **(a)** Given that I_0 is the moment of inertia of table and gun and m the mass of bullet.

Initial angular momentum of system about centre:

$L_i = (I_0 + mr^2)\,\omega_0$ ...(i)

Let w be the nagular velocity of table after the bullet is fired.

Final angular momentum

$Lf = I_0\omega - m\,(v - r\omega)r$ ...(ii)

Where $(v - r\omega)$ is absolute velocity of bullet to the right.

Equations (i) and (ii); we get

$$(\omega - \omega_0) = \frac{mvr}{I_0 + mr^2}$$

95. **(a)** Let friction force = f

F + f = ma ...(i)

(F – f) R = Iα ...(ii)

From eq^{n}s. (i) and (ii),

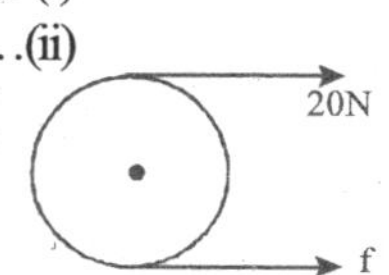

$$2F = ma + \frac{I\alpha}{R}$$

Use $\alpha = \frac{a}{R}$ (for pure rolling)

$$2F = ma + \frac{Ia}{R^2}$$

$$40 = 4a + \frac{0.02a}{(0.1)^2};\ a = \frac{40}{6} = 6.7\ m/s^2$$

96. (d) Velocity at the bottom and top of the circle is $\sqrt{5gr}$ and $\sqrt{gr}$. Therefore (1/2)M(5gr)=MgH and (1/2) M (gr) = Mgh.

97. (a) $t = \frac{1}{\sin\theta}\sqrt{\frac{2h}{g}\left[1+\frac{K^2}{R^2}\right]} \Rightarrow t \propto \left(1+\frac{K^2}{R^2}\right)$

so, $t_1 : t_2 = \sqrt{2} : \sqrt{1.5}$

98. (b) Both the sphere and the plank will slide down with same acceleration g sin θ.

99. (c) Let v be the linear velocity of centre of mass of the spherical body and w its angular velocity about centre of mass. Then $\omega = \frac{v}{2R}$

KE of spherical body $K_1 = \frac{1}{2}mv^2 + \frac{1}{2}I\omega^2$

$$K_1 = \frac{1}{2}mv^2 + \frac{1}{2}(2mR)^2\left(\frac{v^2}{4R^2}\right) = \frac{3}{4}mv^2 \quad ...(i)$$

Speed of the block will be

$$v' = (\omega)(3R) = 3R\omega = (3R)\left(\frac{v}{2R}\right) = \frac{3}{2}v$$

∴ KE of block $K_2 = \frac{1}{2}mv^2$

$$= \frac{1}{2}m\left(\frac{3}{2}v\right)^2 = \frac{9}{8}mv^2 \quad .....(ii)$$

From equations (i) and (ii), $\frac{K_1}{K_2} = \frac{2}{3}$

100. (a) When the thread is pulled, the bobbin rolls to the right.

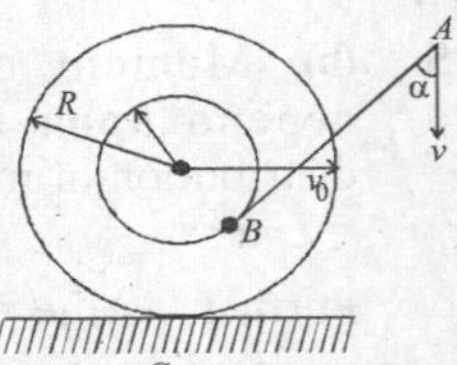

Resultant velocity of point B along the thread is $v = v_0 \sin\alpha - \omega r$, where $v_0 \sin\alpha$ is the component of translational velocity along the thread and ωr linear velocity due to rotation. As the bobbin rolls without slipping, $v_0 = \omega R$. Solving the obtained equations, we get $v_0 = \frac{vR}{R\sin\alpha - r}$

7 Gravitation

1. Let ω be the angular velocity of the earth's rotation about its axis. Assume that the acceleration due to gravity on the earth's surface has the same value at the equator and the poles. An object weighed at the equator gives the same reading as a reading taken at a depth d below earth's surface at a pole (d << R). The value of d is

(a) $\frac{\omega^2 R^2}{g}$ (b) $\frac{\omega^2 R^2}{2g}$ (c) $\frac{2\omega^2 R^2}{g}$ (d) $\frac{\sqrt{Rg}}{g}$

2. In a gravitational field, at a point where the gravitational potential is zero
   (a) the gravitational field is necessarily zero
   (b) the gravitational field is not necesarily zero
   (c) any value between one and infinite
   (d) None of these

3. The gravitational field strength due to a solid sphere (mass M, radius R) varies with distance r from centre as **Tricky**

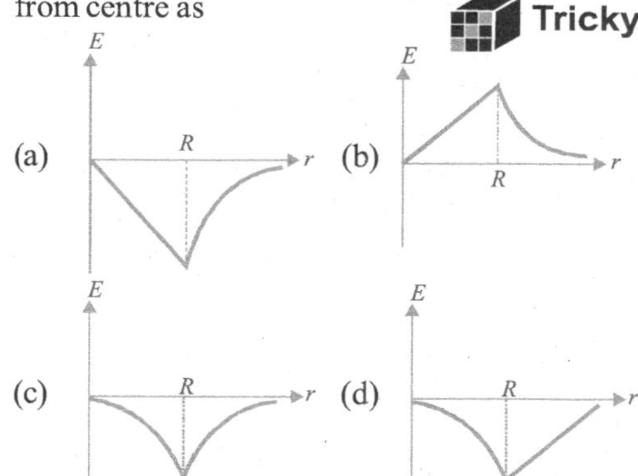

4. A particle of mass M is situated at the centre of a spherical shell of same mass and radius a. The gravitational potential at a point situated at $\frac{a}{2}$ distance from the centre, will be:

(a) $-\frac{3GM}{a}$ (b) $-\frac{2GM}{a}$ (c) $-\frac{GM}{a}$ (d) $-\frac{4GM}{a}$

5. Assuming the radius of the earth as R, the change in gravitational potential energy of a body of mass m, when it is taken from the earth's surface to a height $3R$ above its surface, is

(a) $3\ mgR$ (b) $\frac{3}{4} mgR$ (c) $1\ mgR$ (d) $\frac{3}{2} mgR$

6. The figure shows a planet in elliptical orbit around the sun S. Where is the kinetic energy of the planet maximum ?

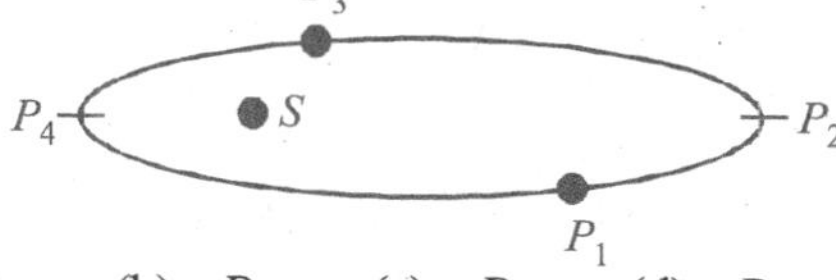

(a) P_1 (b) P_2 (c) P_3 (d) P_4

7. A planet revolves about the sun in elliptical orbit. The areal velocity $\left(\frac{dA}{dt}\right)$ of the planet is $4.0 \times 10^{16}\ m^2/s$. The least distance between planet and the sun is 2×10^{12} m. Then the maximum speed of the planet in km/s is **Critical Thinking**

(a) 10 (b) 20 (c) 30 (d) 40

8. In planetary motion the areal velocity of position vector of a planet depends on angular velocity (ω) and the distance of the planet from sun (r). If so, the correct relation for areal velocity is

(a) $\frac{dA}{dt} \propto \omega r$ (b) $\frac{dA}{dt} \propto \omega^2 r$

(c) $\frac{dA}{dt} \propto \omega r^2$ (d) $\frac{dA}{dt} \propto \sqrt{\omega r}$

9. The period of moon's rotation around the earth is nearly 29 days. If moon's mass were 2 fold its present value and all other things remain unchanged, the period of moon's rotation would be nearly

(a) $29\sqrt{2}$ days (b) $29/\sqrt{2}$ days
(c) 29×2 days (d) 29 days

10. Which of the following graphs represents the motion of a planet moving about the sun ?

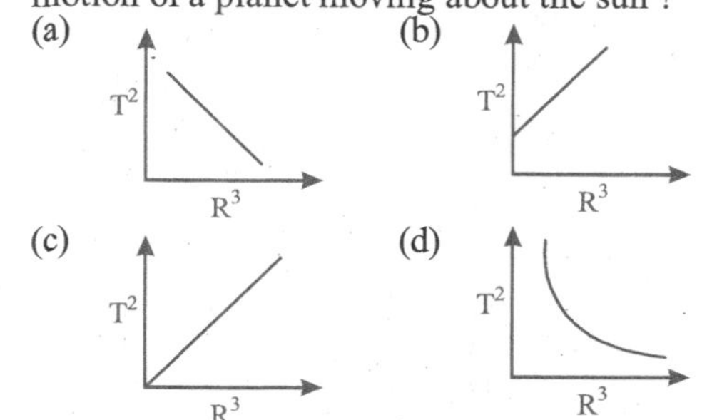

11. If three equal masses m are placed at the three vertices of an equilateral triangle of side $1/m$ then what force acts on a particle of mass 2m placed at the centroid?
(a) Gm^2 (b) $2Gm^2$ (c) Zero (d) $-Gm^2$

12. Two particles of equal mass 'm' go around a circle of radius R under the action of their mutual gravitational attraction. The speed of each particle with respect to their centre of mass is
(a) $\sqrt{\frac{Gm}{4R}}$ (b) $\sqrt{\frac{Gm}{3R}}$ (c) $\sqrt{\frac{Gm}{2R}}$ (d) $\sqrt{\frac{Gm}{R}}$

13. If suddenly the gravitational force of attraction between Earth and a body revolving around it becomes zero, then the body will
(a) continue to move in its orbit with same velocity
(b) move tangentially to the original orbit with same velocity
(c) become stationary in its orbit
(d) move towards the earth

14. A straight rod of length L extends from $x = a$ to $x = L + a$. Find the gravitational force it exerts on a point mass m at $x = 0$ if the linear density of rod $\mu = A + Bx^2$.
Critical Thinking
(a) $Gm\left[\frac{A}{a} + BL\right]$
(b) $Gm\left[A\left(\frac{1}{a} - \frac{1}{a+L}\right) + BL\right]$
(c) $Gm\left[BL + \frac{A}{a+L}\right]$ (d) $Gm\left[BL - \frac{A}{a}\right]$

15. Two concentric uniform shells of mass M_1 and M_2 are as shown in the figure. A particle of mass m is located just within the shell M_2 on its inner surface. Gravitational force on 'm' due to M_1 and M_2 will be
Tricky
(a) zero
(b) $\frac{GM_1m}{b^2}$
(c) $\frac{G(M_1+M_2)m}{b^2}$
(d) None of these

16. The figure shows elliptical orbit of a planet m about the sun S. The shaded area SCD is twice the shaded area SAB. If t_1 is the time for the planet to move from C to D and t_2 is the time to move from A to B then

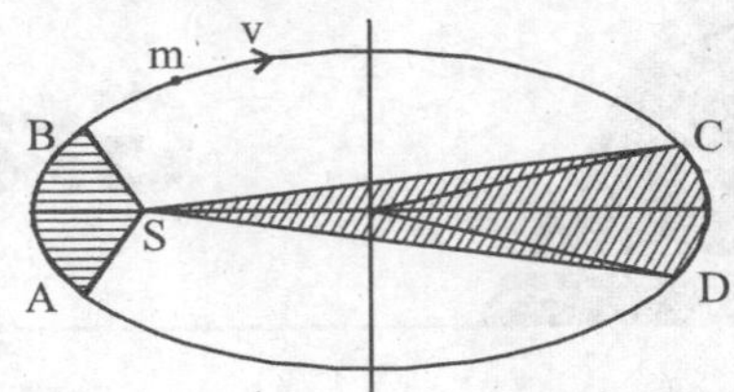

(a) $t_1 = 4t_2$ (b) $t_1 = 2t_2$
(c) $t_1 = t_2$ (d) $t_1 > t_2$

17. If the distance of earth is halved from the sun, then the no. of days in a year will be
(a) 365 (b) 182.5 (c) 730 (d) 129

18. If the earth is at one-fourth of its present distance from the sun, the duration of the year will be
(a) half the present year
(b) one-eighth the present year
(c) one-sixth the present year
(d) one-tenth the present year

19. The distance of Neptune and Saturn from the sun is nearly 10^{13} and 10^{12} meter respectively. Assuming that they move in circular orbits, their periodic times will be in the ratio
(a) 10 (b) 100 (c) $10\sqrt{10}$ (d) 1000

20. A satellite A of mass m is at a distance of r from the surface of the earth. Another satellite B of mass 2m is at a distance of 2r from the earth's centre. Their time periods are in the ratio of
(a) 1 : 2 (b) 1 : 16 (c) 1 : 32 (d) $1 : 2\sqrt{2}$

21. The time taken by the earth to travel over half its orbit, remote from the sun, separated by the minor axis is about 2 days more than half the year, then the eccentricity of the orbit is
(a) 1/30 (b) 1/60 (c) 1/15 (d) 1/70

22. For a particle inside a uniform spherical shell, the gravitational force on the particle is
(a) infinite (b) zero
(c) $\frac{-Gm_1m_2}{r^2}$ (d) $\frac{Gm_1m_2}{r^2}$

23. Two spheres of masses m and M are situated in air and the gravitational force between them is F. The space around the masses is now filled with a liquid of specific gravity 3. The gravitational force will now be
(a) $\frac{F}{9}$ (b) $3F$ (c) F (d) $\frac{F}{3}$

24. A mass M is split into two parts m and $(M - m)$, which are then separated by a certain distance. What ratio of m/M maximizes the gravitational force between the two parts?
(a) $\frac{1}{3}$ (b) $\frac{1}{2}$ (c) $\frac{1}{4}$ (d) $\frac{1}{5}$

25. Two stars of mass m_1 and m_2 are parts of a binary system. The radii of their orbits are r_1 and r_2 respectively, measured from the C.M. of the system. The magnitude of gravitational force m_1 exerts on m_2 is

(a) $\frac{m_1 m_2 G}{(r_1+r_2)^2}$ (b) $\frac{m_1 G}{(r_1+r_2)^2}$ (c) $\frac{m_2 G}{(r_1+r_2)^2}$ (d) $\frac{(m_1+m_2)}{(r_1+r_2)^2}$

26. Two bodies of masses 4 kg and 9 kg are separated by a distance of 60 cm. A 1 kg mass is placed in between these two masses. If the net force on 1 kg is zero, then its distance from 4 kg mass is

(a) 26 cm (b) 30 cm (c) 28 cm (d) 24 cm

27. If masses of two point objects is doubled and distance between them is tripled, then gravitational force of attraction between them will nearly

(a) increase by 225% (b) decrease by 44%
(c) decrease by 56% (d) increase by 125%

28. A central particle M is surrounded by a square array of other particles, separated by either distance d or distance $d/2$ along the perimeter of the square. The magnitude of the gravitational force on the central particle due to the other particles is

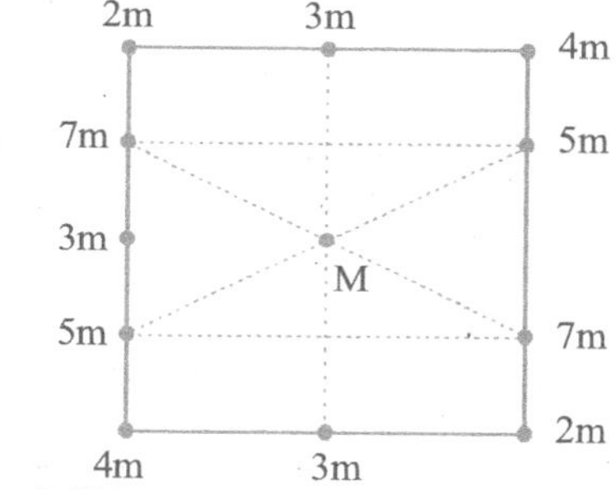

(a) $\frac{9\ GMm}{d^2}$ (b) $\frac{5\ GMm}{d^2}$ (c) $\frac{3\ GMm}{d^2}$ (d) $\frac{GMm}{d^2}$

29. Three particles P, Q and R are placed as per given figure. Masses of P, Q and R are $\sqrt{3}$m, $\sqrt{3}$m and m respectively. The gravitational force on a fourth particle S of mass m is equal to **Critical Thinking**

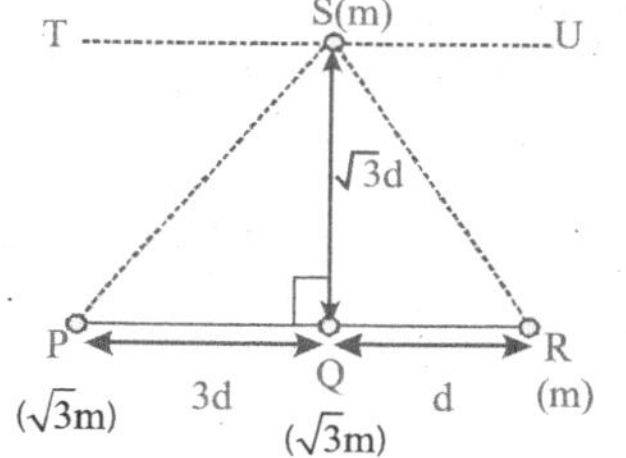

(a) $\frac{\sqrt{3}GM^2}{2d^2}$ in ST direction only

(b) $\frac{\sqrt{3}GM^2}{2d^2}$ in SQ direction and $\frac{\sqrt{3}GM^2}{2d^2}$ in SU direction

(c) $\frac{\sqrt{3}GM^2}{2d^2}$ in SQ direction only

(d) $\frac{\sqrt{3}GM^2}{2d^2}$ in SQ direction and $\frac{\sqrt{3}GM^2}{2d^2}$ in ST direction

30. The acceleration due to gravity on the planet A is 9 times the acceleration due to gravity on planet B. A man jumps to a height of 2m on the surface of A. What is the height of jump by the same person on the planet B?

(a) $\frac{2}{3}$m (b) $\frac{2}{9}$m (c) 18m (d) 6m

31. Assume that the acceleration due to gravity on the surface of the moon is 0.2 times the acceleration due to gravity on the surface of the earth. If R_e is the maximum range of a projectile on the earth's surface, what is the maximum range on the surface of the moon for the same velocity of projection **Critical Thinking**

(a) $0.2R_e$ (b) $2R_e$ (c) $0.5R_e$ (d) $5R_e$

32. The height of the point vertically above the earth's surface, at which acceleration due to gravity becomes 1% of its value at the earth's surface is (Radius of the earth = R)

(a) 8R (b) 9R (c) 10R (d) 20R

33. The speed of earth's rotation about its axis is ω. Its speed is increases to x times to make the effective acceleration due to gravity equal to zero at the equator. Then x is :

(a) 1 (b) 8.5 (c) 17 (d) 34

34. A particle is suspended from a spring and it stretches the spring by 1 cm on the surface of earth. By how much amount the same particle will stretch the same spring at a place 800 km above the surface of earth.

(a) 1.59 cm (b) 2.38 cm (c) 0.79 cm (d) 1.38 cm

35. The density of a newly discovered planet is twice that of earth. The acceleration due to gravity at the surface of the planet is equal to that at the surface of the earth. If the radius of the earth is R, the radius of the planet would be **Critical Thinking**

(a) 1/2R (b) 2R (c) 4R (d) 1/4R

36. Imagine a new planet having the same density as that of earth but it is 3 times bigger than the earth in size. If the acceleration due to gravity on the surface of earth is g and that on the surface of the new planet is g', then

(a) g' = g/9 (b) g' = 27g (c) g' = 9g (d) g' = 3g

37. If the earth were to rotates faster than its present speed, the weight of an object will
(a) increase at the equator but remain unchanged at the poles
(b) decrease at the equator but remain unchanged at the poles
(c) remain unchanged at the equator but decrease at the poles
(d) remain unchanged at the equator but increase at the poles

38. Four similar particles of mass m are orbiting in a circle of radius r in the same angular direction because of their mutual gravitational attractive force. Then, velocity of a particle is given by

Toughnut

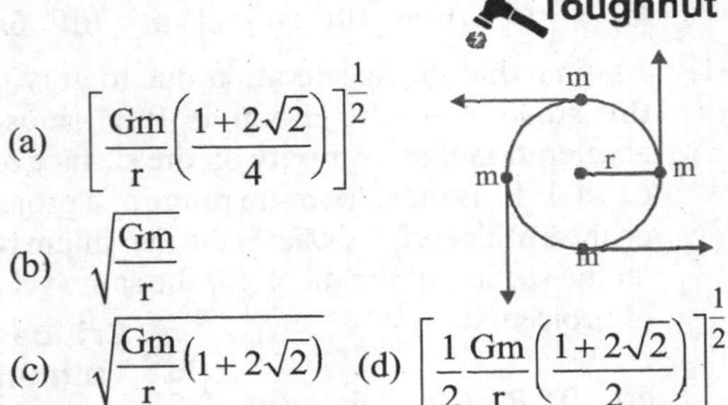

(a) $\left[\frac{Gm}{r}\left(\frac{1+2\sqrt{2}}{4}\right)\right]^{\frac{1}{2}}$
(b) $\sqrt{\frac{Gm}{r}}$
(c) $\sqrt{\frac{Gm}{r}(1+2\sqrt{2})}$ (d) $\left[\frac{1}{2}\frac{Gm}{r}\left(\frac{1+2\sqrt{2}}{2}\right)\right]^{\frac{1}{2}}$

39. Consider two solid uniform spherical objects of the same density ρ. One has radius R and the other has radius 2R. They are in outer space where the gravitational field from other objects are negligible. If they are arranged with their surface touching, what is the contact force between the objects due to their traditional attraction?

Critical Thinking

(a) $G\pi^2R^4$ (b) $\frac{128}{81}G\pi^2R^4\rho^2$
(c) $\frac{128}{81}G\pi^2$ (d) $\frac{128}{87}\pi^2R^4G$

40. Two spherical bodies of mass *M* and 5*M* & radii *R* & 2*R* respectively are released in free space with initial separation between their centres equal to 12 *R*. If they attract each other due to gravitational force only, then the distance covered by the smaller body just before collision is
(a) 2.5*R* (b) 4.5*R* (c) 7.5*R* (d) 1.5*R*

41. From a sphere of mass M and radius R, a smaller sphere of radius $\frac{R}{2}$ is carved out such that the cavity made in the original sphere is between its centre and the periphery (See figure). For the configuration in the figure where the distance between the centre of the original sphere and the removed sphere is 3R, the gravitational force between the two sphere is

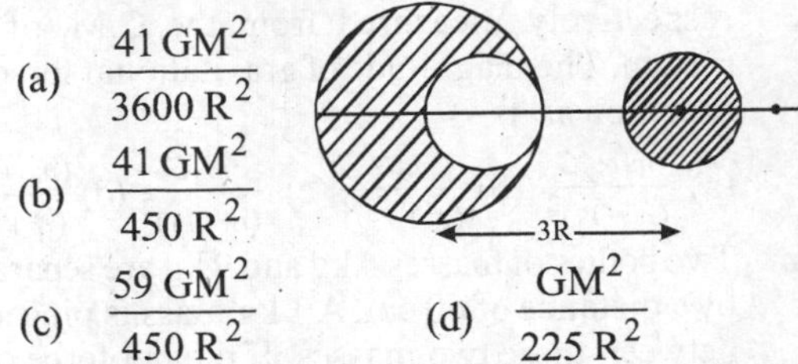

(a) $\frac{41\,GM^2}{3600\,R^2}$
(b) $\frac{41\,GM^2}{450\,R^2}$
(c) $\frac{59\,GM^2}{450\,R^2}$ (d) $\frac{GM^2}{225\,R^2}$

42. The value of acceleration due to gravity on moving from equator to poles will
(a) decrease (b) increase
(c) remain same (d) become half

43. The weight of a body at the centre of the earth is
(a) zero
(b) infinite
(c) same as on the surface of earth
(d) None of these

44. Which of the following graphs shows the correct variation of acceleration due to gravity with the height above the earth's surface? Tricky

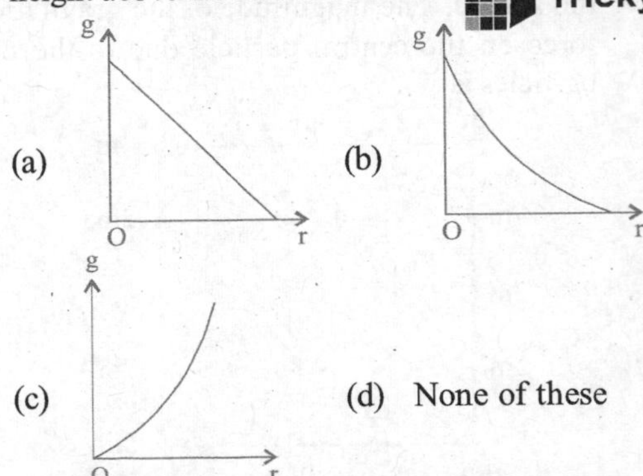

(d) None of these

45. At what height from the ground will the value of g be the same as that in 10 km deep mine below the surface of earth?
(a) 20 km (b) 10 km (c) 15 km (d) 5 km

46. The ratio of radii of earth to another planet is 2/3 and the ratio of their mean densities is 4/5. If an astronaut can jump to a maximum height of 1.5 m on the earth, with the same effort, the maximum height he can jump on the planet is
(a) 1 m (b) 0.8 m (c) 0.5 m (d) 1.25 m

47. The value of 'g' at a particular point is 9.8 m/s^2. Suppose the earth suddenly shrinks uniformly to half its present size without losing any mass. The value of 'g' at the same point (assuming that the distance of the point from the centre of the earth does not shrink) will now be
(a) 4.9 m/sec^2 (b) 3.1 m/sec^2
(c) 9.8 m/sec^2 (d) 19.6 m/sec^2

48. R is the radius of the earth and ω is its angular velocity and g_p is the value of g at the poles. The effective value of g at the latitude $\lambda = 60°$ will be equal to

(a) $g_p - \frac{1}{4}R\omega^2$ (b) $g_p - \frac{3}{4}R\omega^2$

(c) $g_p - R\omega^2$ (d) $g_p + \frac{1}{4}R\omega^2$

49. What should be the velocity of rotation of earth due to rotation about its own axis so that the weight of a person becomes $\frac{2}{3}$ of the present weight at the equator. Equatorial radius of the earth is R

(a) $\left(\frac{2g}{3R}\right)^{\frac{1}{2}}$ (b) $\left(\frac{g}{3R}\right)^{\frac{1}{2}}$ (c) $\left(\frac{g}{7R}\right)^{\frac{1}{2}}$ (d) $\left(\frac{g}{5R}\right)^{\frac{1}{2}}$

50. If the radius of the earth were to shrink by 1%, with its mass remaining the same, the acceleration due to gravity on the earth's surface would

(a) decrease by 1% (b) decrease by 2%

(c) increase by 1% (d) increase by 2%

51. While approaching a planet circling a distant star, an astronaut determines the radius of a planet is half of that of the earth. After landing on its surface, he finds its acceleration due to gravity is twice as that of the earth. The ratio of the mass of planet and that of the earth.

(a) 1 : 2 (b) 2 : 3 (c) 3 : 4 (d) 4 : 5

52. The depth d at which the value of acceleration due to gravity becomes $\frac{1}{n}$ times the value at the surface of the earth, is [R = radius of the earth]

(a) $\frac{R}{n}$ (b) $R\left(\frac{n-1}{n}\right)$ (c) $\frac{R}{n^2}$ (d) $R\left(\frac{n}{n+1}\right)$

53. The radii of two planets are respectively R_1 and R_2 and their densities are respectively ρ_1 and ρ_2. The ratio of the accelerations due to gravity at their surfaces is

Tricky

(a) $g_1 : g_2 = \frac{\rho_1}{R_1^2} : \frac{\rho_2}{R_2^2}$

(b) $g_1 : g_2 = R_1R_2 : \rho_1\rho_2$

(c) $g_1 : g_2 = R_1R_2 : R_2\rho_1$

(d) $g_1 : g_2 = R_1\rho_1 : R_2\rho_2$

54. The change in the value of acceleration of earth towards sun, when the moon comes from the position of solar eclipse to the position on the other side of earth in line with sun is:
(mass of the moon = 7.36×10^{22} kg, radius of the moon's orbit = 3.8×10^8 m).

(a) 6.73×10^{-5} m/s^2 (b) 6.73×10^{-3} m/s^2

(c) 6.73×10^{-2} m/s^2 (d) 6.73×10^{-4} m/s^2

55. Which graph correctly presents the variation of acceleration due to gravity with the distance from the centre of the earth (radius of the earth $= R_E$)?

Tricky

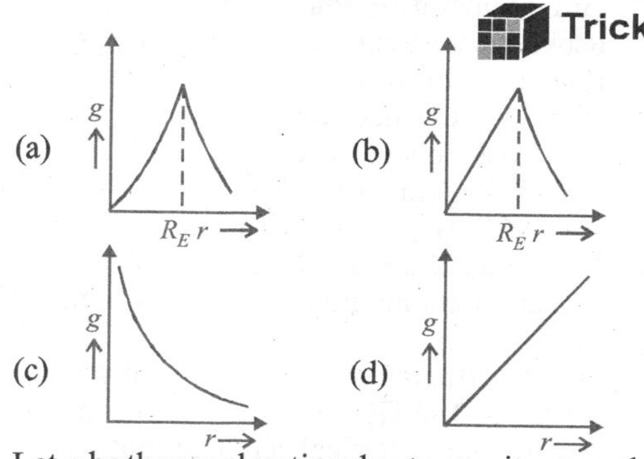

56. Let g be the acceleration due to gravity at earth's surface and K be the rotational kinetic energy of the earth. Suppose the earth's radius decreases by 2% keeping all other quantities same, then

(a) g decreases by 2% and K decreases by 4%

(b) g decreases by 4% and K increases by 2%

(c) g increases by 4% and K decreases by 4%

(d) g decreases by 4% and K increases by 4%

57. Two concentric spherical shells are as shown in figure. Choose the correct statement given below:

Tricky

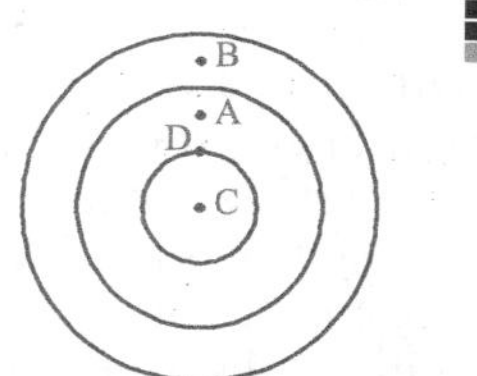

(a) Potential at A is greater than B

(b) Gravitational field at A is less than B

(c) As one moves from C to D then potential remains constant

(d) As one moves from D to A then gravitational field decreases

58. The potential energy of a rock, having mass m and rotating at a height of 3.2×10^6 m from the earth surface, is

(a) $-6\ mgR_e$ (b) $-0.67\ mgR_e$

(c) $-0.99\ mgR_e$ (d) $-0.33\ mgR_e$

59. Intensity of the gravitational field inside the hollow spherical shell is

(a) variable (b) minimum

(c) maximum (d) zero

60. If 'g' is the acceleration due to gravity on the earth's surface, the gain in the potential energy of an object of mass 'm' raised from the surface of the earth to a height equal to the radius 'R' of the earth is

(a) $\frac{1}{4}mgR$ (b) $\frac{1}{2}mgR$ (c) $2\,mgR$ (d) mgR

61. Radius of moon is 1/4 times that of earth and mass is 1/81 times that of earth. The point at which gravitational field due to earth becomes equal and opposite to that of moon, is (Distance between centres of earth and moon is 60R, where R is radius of earth)
(a) 5.75 R from centre of moon
(b) 16 R from surface of moon
(c) 53 R from centre of earth
(d) 54 R from centre of earth

62. Let V and E denote the gravitational potential and gravitational field at a point. It is possible to have
(a) V = 0 and E = 0 (b) V = 0 and E $\neq$ 0
(c) V $\neq$ 0 and E = 0 (d) All of the above

63. A uniform spherical shell gradually shrinks maintaining its shape. The gravitational potential at the centre
(a) increases (b) decreases
(c) remains constant (d) cannot say

64. The gravitational potential at the centre of a square of side 'a' and four equal masses (m each) placed at the corners of a square is
(a) Zero (b) $4\sqrt{2}\dfrac{Gm}{a}$
(c) $-4\sqrt{2}\dfrac{Gm}{a}$ (d) $-4\sqrt{2}\dfrac{Gm^2}{a}$

65. Three identical stars, each of mass M, form an equilateral triangle (stars are positioned at the corners) that rotates around the centre of the triangle. The system is isolated and edge length of the triangle is L. The amount of work done, that is required to dismantle the system is :
(a) $\dfrac{3GM^2}{L}$ (b) $\dfrac{3GM^2}{2L}$ (c) $\dfrac{3GM^2}{4L}$ (d) $\dfrac{GM^2}{2L}$

66. Inside a uniform sphere of density ρ there is a spherical cavity whose centre is at a distance ℓ from the centre of the sphere. Find the strength *F* of the gravitational field inside the cavity at the point *P*. **Critical Thinking**
(a) $\dfrac{4}{3}G\pi\rho\vec{\ell}$
(b) $\dfrac{1}{3}G\pi\rho\vec{\ell}$
(c) $\dfrac{2}{3}G\pi\rho\vec{\ell}$
(d) $\dfrac{1}{2}G\pi\rho\vec{\ell}$

67. The work done required to increase the separation distance from x_1 to $x_1 + d$ between two masses m_1 and m_2 is
(a) $\dfrac{Gm_1m_2d}{x_1(x_1+d)}$ (b) $\dfrac{Gm_1m_2x_1}{(x_1+d)d}$
(c) $\dfrac{-Gm_1m_2x_1}{(x_1+d)}$ (d) none of these

68. Gravitational field at the centre of a semicircle formed by a thin wire AB of mass m and length ℓ is **Toughnut**

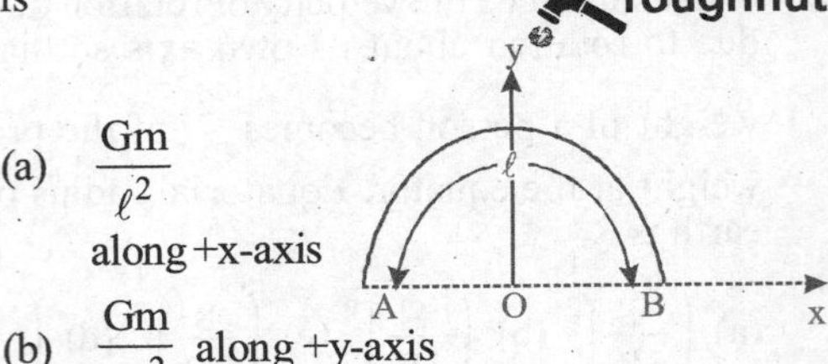

(a) $\dfrac{Gm}{\ell^2}$ along +x-axis
(b) $\dfrac{Gm}{\pi\ell^2}$ along +y-axis
(c) $\dfrac{2\pi Gm}{\ell^2}$ along +x-axis
(d) $\dfrac{2\pi Gm}{\ell^2}$ along +y-axis

69. In a certain region of space, gravitational field is given by $I = -(K/r)$. Taking the reference point to be at $r = r_0$ with $V = V_0$, find the potential.
(a) $K\log\dfrac{r}{r_0} + V_0$ (b) $K\log\dfrac{r_0}{r} + V_0$
(c) $K\log\dfrac{r}{r_0} - V_0$ (d) $\log\dfrac{r_0}{r} - V_0 r$

70. A point *P* lies on the axis of a fixed ring of mass *M* and radius *R*, at a distance 2*R* from its centre *O*. A small particle starts from *P* and reaches *O* under gravitational attraction only. Its speed at *O* will be **Toughnut**

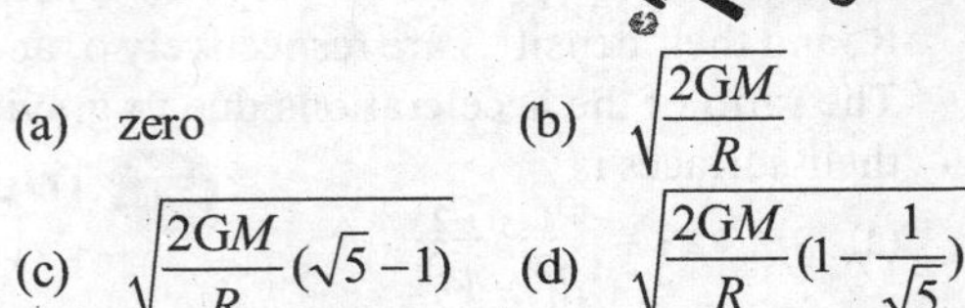

(a) zero (b) $\sqrt{\dfrac{2GM}{R}}$
(c) $\sqrt{\dfrac{2GM}{R}(\sqrt{5}-1)}$ (d) $\sqrt{\dfrac{2GM}{R}\left(1-\dfrac{1}{\sqrt{5}}\right)}$

71. The gravitational field, due to the 'left over part' of a uniform sphere (from which a part as shown, has been 'removed out'), at a very far off point, P, located as shown, would be (nearly) :

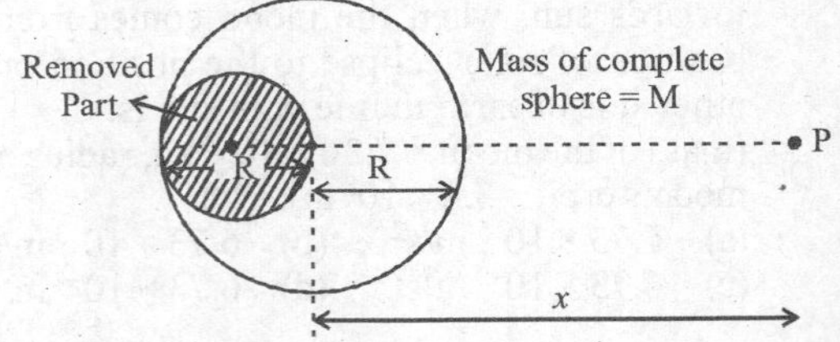

(a) $\frac{5}{6}\frac{GM}{x^2}$ (b) $\frac{8}{9}\frac{GM}{x^2}$ (c) $\frac{7}{8}\frac{GM}{x^2}$ (d) $\frac{6}{7}\frac{GM}{x^2}$

72. A space vehicle approaching a planet has a speed v, when it is very far from the planet. At that moment tangent of its trajectory would miss the centre of the planet by distance R. If the planet has mass M and radius r, what is the smallest value of R in order that the resulting orbit of the space vehicle will just miss the surface of the planet? **Toughnut**

(a) $\frac{r}{v}\left[v^2+\frac{2GM}{r}\right]^{\frac{1}{2}}$ (b) $vr\left[1+\frac{2GM}{r}\right]$

(c) $\frac{r}{v}\left[v^2+\frac{2GM}{r}\right]$ (d) $\frac{2GMv}{r}$

73. A skylab of mass m kg is first launched from the surface of the earth in a circular orbit of radius 2R (from the centre of the earth) and then it is shifted form this circular orbit of radius 3R. The minimum energy required to place the lab in the first orbit and to shift the lab from first orbit to the second orbit are

(a) $\frac{3}{4}mgR, \frac{mgR}{6}$ (b) $\frac{3}{4}mgR, \frac{mgR}{12}$

(c) mgR, mgR (d) 2mgR, mgR

74. With what minimum speed should m be projected from point C in presence of two fixed masses M each at A and B as shown in the figure such that mass m should escape the gravitational attraction of A and B **Tricky**

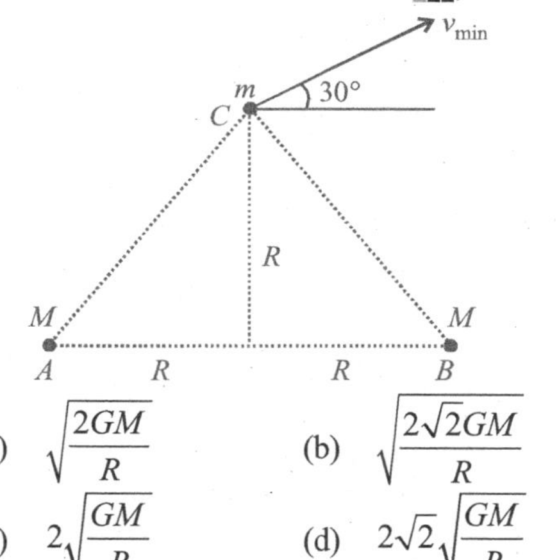

(a) $\sqrt{\frac{2GM}{R}}$ (b) $\sqrt{\frac{2\sqrt{2}GM}{R}}$

(c) $2\sqrt{\frac{GM}{R}}$ (d) $2\sqrt{2}\sqrt{\frac{GM}{R}}$

75. An artificial satellite is first taken to a height equal to half the radius of earth. Assume that it is at rest on the earth's surface initially and that it is at rest at this height. Let E_1 be the energy required. It is then given the appropriate orbital speed such that it goes in a circular orbit at that height. Let E_2 be the energy required. The ratio $\frac{E_1}{E_2}$ is

(a) 4 : 1 (b) 3 : 1 (c) 1 : 1 (d) 1 : 2

76. A satellite is launched in the equatorial plane in such a way that it can transmit signals upto 60° latitude on the earth. The angular velocity of the satellite is **Critical Thinking**

(a) $\sqrt{\frac{GM}{8R^3}}$ (b) $\sqrt{\frac{GM}{2R^3}}$ (c) $\sqrt{\frac{GM}{4R^3}}$ (d) $\sqrt{\frac{3\sqrt{3}GM}{8R^3}}$

77. A satellite of mass m revolves around the earth of radius R at a height x from its surface. If g is the acceleration due to gravity on the surface of the earth, the orbital speed of the satellite is

(a) $\frac{gR^2}{R+x}$ (b) $\frac{gR}{R-x}$ (c) gx (d) $\left(\frac{gR^2}{R+x}\right)^{1/2}$

78. For a satellite orbiting in an orbit, close to the surface of earth, to escape, what is the percentage increase in the kinetic energy required?

(a) 41% (b) 61% (c) 81% (d) 98%

79. The radius of the earth is reduced by 4%. The mass of the earth remains unchanged. What will be the change in escape velocity?

(a) Increased by 2% (b) Decreased by 4%

(c) Increased by 6% (d) Decreased by 8%

80. A satellite of mass m is orbiting the earth in a circular orbit of radius R. It starts losing energy due to small air resistance at the rate of C J/s. Find the time taken for the satellite to reach the earth.

(a) $\frac{GMm}{C}\left[\frac{1}{R}-\frac{1}{r}\right]$ (b) $\frac{GMm}{2C}\left[\frac{1}{R}+\frac{1}{r}\right]$

(c) $\frac{GMm}{2C}\left[\frac{1}{R}-\frac{1}{r}\right]$ (d) $\frac{2GMm}{C}\left[\frac{1}{R}+\frac{1}{r}\right]$

81. Two spheres each of mass M are situated at a distance $2d$ (see figure). A particle of mass m (m << M) is taken along the path shown in figure. The work done in the process from A to B is

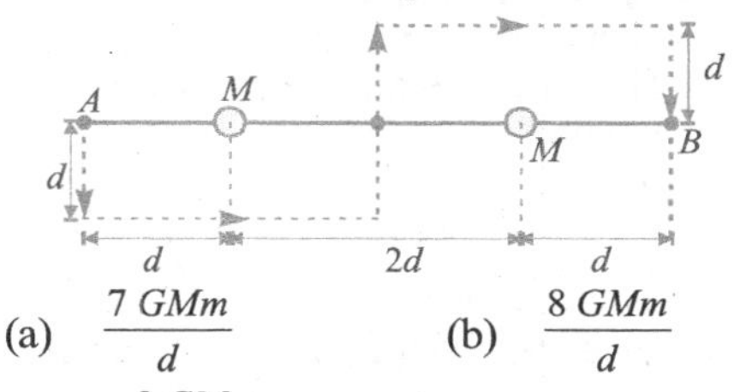

(a) $\frac{7\,GMm}{d}$ (b) $\frac{8\,GMm}{d}$

(c) $-\frac{8\,GMm}{d}$ (d) zero

82. An asteriod of mass m is approaching earth initially at a distance of $10R_e$ with speed v_i. It hits the earth with a speed v_f (R_e and M_e are radius and mass of earth), then **Critical Thinking**

(a) $v_f^2 = v_i^2 + \frac{2Gm}{M_e R}\left(1 - \frac{1}{10}\right)$

(b) $v_f^2 = v_i^2 + \frac{2GM_e}{R_e}\left(1 + \frac{1}{10}\right)$

(c) $v_f^2 = v_i^2 + \frac{2GM_e}{R_e}\left(1 - \frac{1}{10}\right)$

(d) $v_f^2 = v_i^2 + \frac{2Gm}{R_e}\left(1 - \frac{1}{10}\right)$

83. In older times, people used to think that the earth was flat. Imagine that the earth is indeed not a sphere of radius R, but an infinite plate of thickness H. What value of H is needed to allow the same gravitational acceleration to be experienced as on the surface of the actual earth? (Assume that the earth's density is uniform and equal in the two models **Toughnut**

(a) $2R/3$ (b) $4R/3$ (c) $8R/3$ (d) $R/3$

84. A shell is fired vertically from the earth with speed $\frac{V_{esc}}{N}$, where N is some number greater than one and V_{esc} is escape speed for the earth. Neglecting the rotation of the earth and air resistance, the maximum altitude attained by the shell will be (R_E is radius of the earth) **Critical Thinking**

(a) $\frac{N^2 R_E}{N^2 - 1}$ (b) $\frac{NR_E}{N^2 - 1}$ (c) $\frac{R_E}{N^2 - 1}$ (d) $\frac{R_E}{N^2}$

85. A spherical uniform planet is rotating about its axis. The velocity of a point on its equator is v. Due to the rotation of planet about its axis the acceleration due to gravity g at equator is 1/2 of g at poles. The escape velocity of a particle on the pole of planet in terms of v is

(a) $v_e = 2v$ (b) $v_e = v$
(c) $v_e = v/2$ (d) $v_e = \sqrt{3}v$

86. The acceleration due to gravity on the surface of the moon is 1/6 that on the surface of earth and the diameter of the moon is one-fourth that of earth. The ratio of escape velocities on earth and moon will be

(a) $\frac{\sqrt{6}}{2}$ (b) $\sqrt{24}$ (c) 3 (d) $\frac{\sqrt{3}}{2}$

87. Three equal masses (each m) are placed at the corners of an equilateral triangle of side 'a'. Then the escape velocity of an object from the circumcentre P of triangle is :

A, B, C, P

(a) $\sqrt{\frac{2\sqrt{3}\,Gm}{a}}$

(b) $\sqrt{\frac{\sqrt{3}\,Gm}{a}}$

(c) $\sqrt{\frac{6\sqrt{3}\,Gm}{a}}$ (d) $\sqrt{\frac{3\sqrt{3}\,Gm}{a}}$

88. A geo-stationary satellite orbits around the earth in a circular orbit of radius 36,000km. Then, the time period of a spy satellite orbiting a few hundred km above the earth's surface (R_{earth} = 6,400km) will approximately be

(a) 1/2 hr (b) 1 hr (c) 2 hr (d) 4 hr

89. A satellite can be in a geostationary orbit around earth at a distance r from the centre. If the angular velocity of earth about its axis doubles, a satellite can now be in a geostationary orbit around earth if its distance from the centre is

(a) $\frac{r}{2}$ (b) $\frac{r}{2\sqrt{2}}$ (c) $\frac{r}{(4)^{1/3}}$ (d) $\frac{r}{(2)^{1/3}}$

90. The mass M of a planet-earth is uniformly distributed over a spherical volume of radius R. Calculate the energy needed to disassemble the planet against the gravitational pull amongst its constituent particles.
Given : $MR = 2.5 \times 10^{31}$ kg-m and $g = 10$ m/s^2

(a) 3.0×10^{32} J (b) 2.5×10^{32} J
(c) 0.5×10^{28} J (d) 1.5×10^{32} J

91. Two bodies of masses M_1 and M_2 are placed at a distance d apart. What is the potential at the position where the gravitational field due to them is zero?

(a) $-\frac{G}{d}\left(M_1 + M_2 + 2\sqrt{M_1}\sqrt{M_2}\right)$

(b) $-\frac{G}{d}\left(M_1 + M_2 - 2\sqrt{M_1}\sqrt{M_2}\right)$

(c) $-\frac{G}{d}\left(2M_1 + M_2 + 2\sqrt{M_1}\sqrt{M_2}\right)$

(d) $-\frac{G}{2d}\left(M_1 + M_2 + 2\sqrt{M_1}\sqrt{M_2}\right)$

92. At what height from the surface of earth the gravitational potential and the value of g are -5.4×10^7 J kg^{-1} and 6.0 ms^{-2} respectively?
Take the radius of earth as 6400 km :

(a) 2600 km (b) 1600 km
(c) 1400 km (d) 2000 km

93. Which one of the following plots represents the variation of gravitational field on a particle with distance r due to a thin spherical shell of radius R? (r is measured from the centre of the spherical shell) **Tricky**

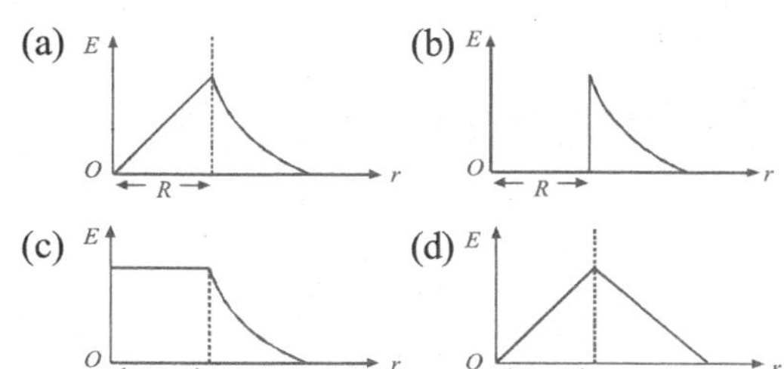

94. The gravitational potential of two homogeneous spherical shells A and B of same surface density at their respective centres are in the ratio 3 : 4. If the two shells coalesce into single one such that surface charge density remains same, then the ratio of potential at an internal point of the view shell to shell A is equal to **Toughnut**

(a) 3 : 2 (b) 4 : 3 (c) 5 : 3 (d) 5 : 6

95. The gravitational field in a region is given by $\vec{g} = 5\text{N/kg}\,\hat{i} + 12\text{N/kg}\,\hat{j}$. The change in the gravitational potential energy of a particle of mass 1 kg when it is taken from the origin to a point (7 m, – 3 m) is:

(a) 71 J (b) $13\sqrt{58}$J (c) –71 J (d) 1 J

96. The gravitational field due to a mass distribution is $E = K/x^3$ in the x-direction. (K is a constatnt). Taking the gravitational potential to be zero at infinity, its value at a distance x is

(a) K/x (b) $K/2x$ (c) K/x^2 (d) $K/2x^2$

97. Infinite number of masses, each 1 kg are placed along the x-axis at $x = \pm 1m, \pm 2m, \pm 4m, \pm 8m, \pm 16m$ ..... the magnitude of the resultant gravitational potential in terms of gravitational constant G at the orgin ($x = 0$) is

(a) $G/2$ (b) G (c) $2G$ (d) $4G$

98. Two rings each of radius 'a' are coaxial and the distance between their centres is a. The masses of the rings are M_1 and M_2. The work done in transporting a particle of a small mass m from centre C_1 to C_2 is : **Critical Thinking**

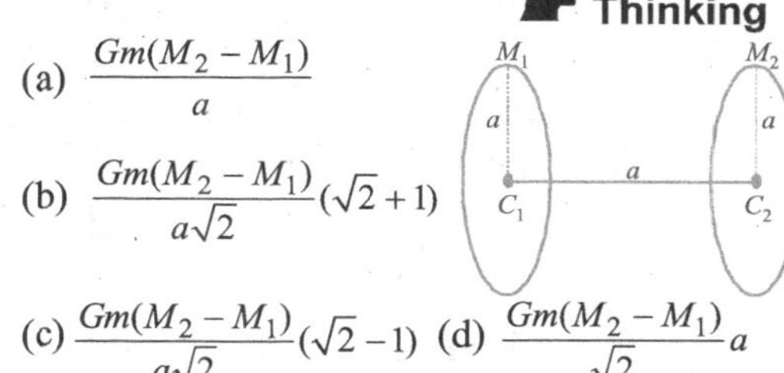

(a) $\dfrac{Gm(M_2 - M_1)}{a}$

(b) $\dfrac{Gm(M_2 - M_1)}{a\sqrt{2}}(\sqrt{2}+1)$

(c) $\dfrac{Gm(M_2 - M_1)}{a\sqrt{2}}(\sqrt{2}-1)$ (d) $\dfrac{Gm(M_2 - M_1)}{\sqrt{2}}a$

99. A system consists of two stars of equal masses that revolve in a circular orbit about a centre of mass midway between them. Orbital speed of each star is v and period is T. Find the mass M of each star (G is gravitational constant) **Critical Thinking**

(a) $\dfrac{2Gv^3}{\pi T}$ (b) $\dfrac{v^3T}{\pi G}$ (c) $\dfrac{v^3T}{2\pi G}$ (d) $\dfrac{2Tv^3}{\pi G}$

100. A uniform ring of mass m and radius r is placed directly above a uniform sphere of mass M and of equal radius. The centre of the ring is directly above the centre of the sphere at a distance $r\sqrt{3}$ as shown in the figure.

The gravitational field due to the ring at a distance $\sqrt{3}$r is

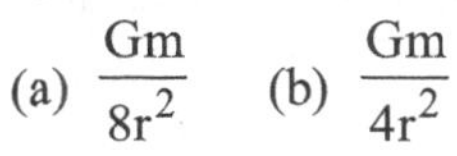

(a) $\dfrac{Gm}{8r^2}$ (b) $\dfrac{Gm}{4r^2}$

(c) $\sqrt{3}\dfrac{Gm}{8r^2}$ (d) $\dfrac{Gm}{8r^3\sqrt{3}}$

Answer KEYs

1	(a)	**11**	(c)	**21**	(b)	**31**	(d)	**41**	(a)	**51**	(a)	**61**	(d)	**71**	(c)	**81**	(d)	**91**	(a)
2	(a)	**12**	(a)	**22**	(b)	**32**	(b)	**42**	(b)	**52**	(b)	**62**	(d)	**72**	(a)	**82**	(c)	**92**	(a)
3	(a)	**13**	(b)	**23**	(c)	**33**	(c)	**43**	(a)	**53**	(d)	**63**	(a)	**73**	(b)	**83**	(a)	**93**	(b)
4	(a)	**14**	(b)	**24**	(b)	**34**	(c)	**44**	(b)	**54**	(a)	**64**	(c)	**74**	(b)	**84**	(c)	**94**	(c)
5	(b)	**15**	(b)	**25**	(a)	**35**	(a)	**45**	(d)	**55**	(b)	**65**	(b)	**75**	(c)	**85**	(a)	**95**	(d)
6	(d)	**16**	(b)	**26**	(d)	**36**	(d)	**46**	(b)	**56**	(c)	**66**	(a)	**76**	(a)	**86**	(b)	**96**	(d)
7	(d)	**17**	(d)	**27**	(c)	**37**	(b)	**47**	(c)	**57**	(c)	**67**	(a)	**77**	(d)	**87**	(c)	**97**	(c)
8	(c)	**18**	(b)	**28**	(c)	**38**	(a)	**48**	(a)	**58**	(b)	**68**	(d)	**78**	(a)	**88**	(c)	**98**	(c)
9	(d)	**19**	(c)	**29**	(c)	**39**	(b)	**49**	(b)	**59**	(d)	**69**	(a)	**79**	(a)	**89**	(c)	**99**	(d)
10	(c)	**20**	(d)	**30**	(c)	**40**	(c)	**50**	(d)	**60**	(b)	**70**	(d)	**80**	(c)	**90**	(d)	**100**	(c)

Hints & Solutions

1. (a) $g\left(1-\frac{d}{R}\right)=g-\omega^2R$; $d=\frac{\omega^2R^2}{g}$

2. (a) $E=-\frac{dV}{dx}\Rightarrow$ if $V=0$, E must be zero.

3. (a) $\vec{g}'=-\frac{g\vec{r}}{R}$ for $r\le R$ and $g'=\frac{g}{(1+r/R)^2}$ for $r\ge R$
so option (a) is correct.

4. (a) Potential at the given point
= Potential at the point due to the shell
+ Potential due to the particle
$=-\frac{GM}{a}-\frac{2GM}{a}=-\frac{3GM}{a}$

5. (b) Gravitational potential energy (GPE) on the surface of earth,
$E_1=-\frac{GMm}{R}$
GPE at $3R$, $E_2=-\frac{GMm}{(R+3R)}=-\frac{GMm}{4R}$
$\therefore$ Change in GPE
$=E_2-E_1=-\frac{GMm}{4R}+\frac{GMm}{R}=\frac{3GMm}{4R}$
$=\frac{3gR^2m}{4R}\left(\because g=\frac{GM}{R^2}\right)=\frac{3}{4}mgR$

6. (d) Angular momentum of the planet about S is conserved. So, mvr = constant.
v is maximum when r is minimum.
So, v is maximum at point P_4. Hence K.E. is maximum at P_4.

7. (d) $\frac{L}{2m}=\frac{dA}{dt}$ (L = angular momentum)
$\frac{mv_{max}r_{min}}{2m}=\frac{dA}{dt}$; $v_{max}=\frac{2\,dA/dt}{r_{min}}=40$

8. (c) $\frac{dA}{dt}=\frac{L}{2m}\Rightarrow\frac{dA}{dt}\propto vr\propto\omega r^2$

9. (d) Time period does not depend upon the mass of satellite

10. (c) 11. (c)

12. (a) Here, centripetal force will be given by the gravitational force between the two particles.
$\frac{Gm^2}{(2R)^2}=m\omega^2R$
$\Rightarrow\frac{Gm}{4R^3}=\omega^2\Rightarrow\omega=\sqrt{\frac{Gm}{4R^3}}$
If the velocity of the two particles with respect to the centre of gravity is v then $v=\omega R$
$v=\sqrt{\frac{Gm}{4R^3}}\times R=\sqrt{\frac{Gm}{4R}}$

13. (b) Body will move tangentially to the original orbit with same velocity.

14. (b) $\because dF=\frac{Gm(\mu dx)}{x^2}$

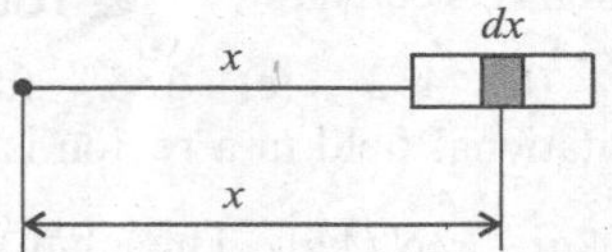

$F=Gm\int(A+Bx)\frac{dx}{x}$
$F=Gm\left[A\left(\frac{1}{a}-\frac{1}{a+L}\right)+BL\right]$

15. (b) Gravitational force will be due to M_1 only.

16. (b) According to Kepler's law, the areal velocity of a planet around the sun always remains constant.
SCD : A_1-t_1 (areal velocity constant)
SAB : A_2-t_2
$\frac{A_1}{t_1}=\frac{A_2}{t_2}$, $t_1=t_2\cdot\frac{A_1}{A_2}$, (given $A_1=2A_2$)
$=t_2\cdot\frac{2A_2}{A_2}$ $\therefore$ $t_1=2t_2$

17. (d) From Kepler's law of periods,
$T_2=T_1\left(\frac{R_2}{R_1}\right)^{3/2}=365\left(\frac{R/2}{R}\right)^{3/2}$
$=365\times\frac{1}{2\sqrt{2}}=129$ days.

18. (b) $\frac{T_1^2}{T_2^2}=\frac{r_1^3}{r_2^3}=(4)^3$ $\therefore T_2=T_1/8.$

19. (c) $T^2\propto R^3$ (According to Kepler's law)
$T_1^2\propto(10^{13})^3$ and $T_2^2\propto(10^{12})^3$
$\therefore\frac{T_1^2}{T_2^2}=(10)^3$ or $\frac{T_1}{T_2}=10\sqrt{10}$

20. (d) Time period does not depend on the mass. Also, $T^2\propto r^3$.

21. **(b)**

22. **(b)** Various regions of spherical shell attract the point mass inside it in various directions. These forces cancel each other completely. Therefore the gravitational force on the particle is zero.

23. **(c)** Gravitational force is independent of medium. Hence, this will remain same.

24. **(b)** $F = G\frac{m(M-m)}{r^2}$

For maximum value of F, $dF/dm = 0$, and so $m = M/2$.

25. **(a)** $F = \frac{Gm_1m_2}{r^2} = \frac{Gm_1m_2}{(r_1+r_2)^2}$.

26. **(d)**

A x C (60–x) B

4kg 1kg 9kg

60cm

$$\therefore \ G\frac{4\times 1}{x^2} = G\frac{9\times 1}{(60-x)^2}$$

$$\text{or } \frac{2}{3} = \frac{x}{(60-x)} \Rightarrow x = 24 \text{ cm}$$

27. **(c)** % change

$$= \frac{\left[\frac{G(2m_1)(2m_2)}{(3r)^2} - \frac{Gm_1m_2}{r^2}\right]}{\frac{Gm_1m_2}{r^2}} \times 100$$

$$= \frac{\frac{4}{9}-1}{1} \times 100 = -56\%,$$

–ve sign indicates that force of attraction decreases.

28. **(c)** $F = \frac{GM(3m)}{d^2} = \frac{3GMm}{d^2}$.

29. **(c)** In horizontal direction

$$\text{Net force} = \frac{G\sqrt{3}mm}{12d^2}\cos 30° - \frac{Gm^2}{4d^2}\cos 60°$$

$$= \frac{Gm^2}{8d^2} - \frac{Gm^2}{8d^2} = 0$$

In vertical direction,

Net force

$$= \frac{G\sqrt{3}m^2}{12d^2}\cos 60° + \frac{G\sqrt{3}m^2}{3d^2} + \frac{Gm^2}{4d}\cos 30°$$

$$= \frac{\sqrt{3}Gm^2}{24d^2} + \frac{\sqrt{3}Gm^2}{3d^2} + \frac{\sqrt{3}Gm^2}{8d^2}$$

$$= \frac{\sqrt{3}Gm^2}{d^2}\left[\frac{1+8+3}{24}\right] = \frac{\sqrt{3}Gm^2}{2d^2} \text{ along SQ.}$$

30. **(c)** Applying conservation of total mechanical energy principle

$$\frac{1}{2}mv^2 = mg_Ah_A = mg_Bh_B$$

$$\Rightarrow g_Ah_A = g_Bh_B$$

$$\Rightarrow h_B = \left(\frac{g_A}{g_B}\right)h_A = 9\times 2 = 18 \text{ m}$$

31. **(d)** Range of projectile $R = \frac{u^2 \sin 2\theta}{g}$

if u and θ are constant then $R \propto \frac{1}{g}$

$$\frac{R_m}{R_e} = \frac{g_e}{g_m} \Rightarrow \frac{R_m}{R_e} = \frac{1}{0.2}$$

$$\Rightarrow R_m = \frac{R_e}{0.2} \Rightarrow R_m = 5R_e$$

32. **(b)** $\frac{g'}{g} = \left(\frac{R}{R+h}\right)^2 \Rightarrow \frac{1}{100} = \left(\frac{R}{R+h}\right)^2$

$\Rightarrow h = 9R$

33. **(c)** $g' = g - \omega'^2 R$, for g' to be zero,

$\omega' = \sqrt{g/R} \simeq 17\omega$

34. **(c)** $x = \frac{mg}{k} = \frac{GMm}{R^2 k} \Rightarrow x \propto \frac{1}{R^2} \Rightarrow x_2$

$$= \left(\frac{6400}{7200}\right)^2 x_1 = 0.79 \text{ cm}$$

35. **(a)** $g = \frac{GM}{R^2}$ also $M = d\times\frac{4}{3}\pi R^3$

$\therefore g = \frac{4}{3}d\pi R$ at the surface of planet

$$g_p = \frac{4}{3}(2d)\pi R', \ g_e = \frac{4}{3}(d)\pi R$$

$g_e = g_p \Rightarrow dR = 2d\,R' \Rightarrow R' = R/2$

36. **(d)** We know that

$$g = \frac{GM}{R^2} = \frac{G\left(\frac{4}{3}\pi R^3\right)\rho}{R^2} = \frac{4}{3}\pi GR\rho$$

$$\frac{g'}{g} = \frac{R'}{R} = \frac{3R}{R} = 3 \qquad \therefore g' = 3g$$

37. **(b)** At the equator, $g' = g - R\omega^2$

As ω increases, g' decreases and hence weight decreases.

At the pole, g' = g.

So weight remain unchanged.

38. (a)

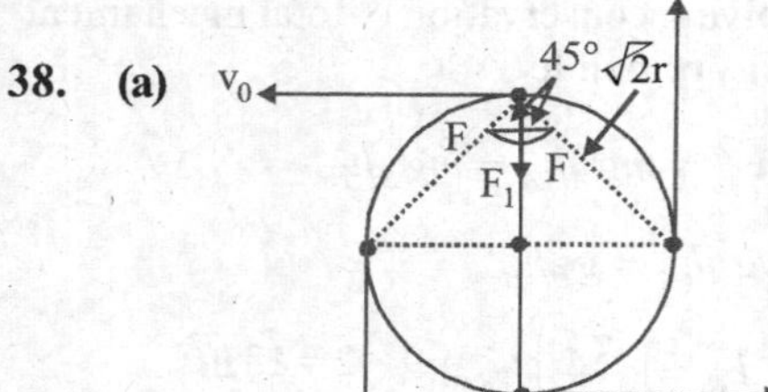

Centripetal force = net gravitational force

$$\Rightarrow \frac{mv_0^2}{r} = 2F\cos 45° + F_1 = \frac{2Gm^2}{(\sqrt{2})^2}\frac{1}{\sqrt{2}} + \frac{Gm^2}{4r^2}$$

$$\frac{mv_0^2}{r} = \frac{Gm^2}{4r^2}[2\sqrt{2}+1]$$

$$\Rightarrow v_0 = \left[\frac{Gm(2\sqrt{2}+1)}{4r}\right]^{\frac{1}{2}}$$

39. (b) $m_1 = \frac{4}{3}\pi R^3\rho,\ m_2 = \frac{4}{3}\pi(2R)^3\rho$, distance between the centres of the two spherical objects, r = 3R.

$$F = \frac{GM_1m_2}{r^2}$$

$$= G\left(\frac{4}{3}\pi R^3\rho\right)\left(8\times\frac{4}{3}\pi R^3\rho\right)\times\frac{1}{(3R)^2}$$

$$= \frac{128}{81}G\pi^2R^4\rho^2$$

40. (c)

41. (a) Volume of removed sphere

$$V_{remo} = \frac{4}{3}\pi\left(\frac{R}{2}\right)^3 = \frac{4}{3}\pi R^3\left(\frac{1}{8}\right)$$

Volume of the sphere (remaining)

$$V_{remain} = \frac{4}{3}\pi R^3 - \frac{4}{3}\pi R^3\left(\frac{1}{8}\right) = \frac{4}{3}\pi R^3\left(\frac{7}{8}\right)$$

Therefore mass of sphere carved and remaining sphere are at respectively $\frac{1}{8}$M and $\frac{7}{8}$M.

Therefore, gravitational force between these two sphere,

$$F = \frac{GMm}{r^2} = \frac{G\frac{7M}{8}\times\frac{1}{8}M}{(3R)^2} = \frac{7}{64\times 9}\frac{GM^2}{R^2}$$

$$\simeq \frac{41}{3600}\frac{GM^2}{R^2}$$

42. (b) Acceleration due to gravity (g) is given by

$$g = \frac{GM}{R^2} \Rightarrow g \propto \frac{1}{R^2}$$

As one moves from the equator to the poles, the radius of the earth decreases, hence g increases.

43. (a) The weight (= mg) of the body at the centre of the earth is zero, because the value of g at centre is zero.

44. (b) Acceleration due to gravity with height h varies as $g \propto \frac{1}{r^2}$

(when r = R + h). Thus variation of g and r is a parabolic curve.

45. (d)

46. (b) K.E. of the person = P.E. at the maximum height

$$\frac{1}{2}mv^2 = m\frac{GM}{R^2}h.$$

$$\therefore\ m\frac{GM}{R^2}.h = mG\frac{M'}{R'^2}h'$$

Given $\rho = \frac{4}{5}p', \frac{R}{R'} = \frac{2}{3}$

$$\therefore\quad mG.\frac{4}{3}\frac{\pi R^3\rho}{R^2}.h = mG.\frac{4}{3}\frac{\pi R'^3\rho'}{R'^2}.h'$$

$$\Rightarrow \frac{2}{3}R'.\frac{4}{5}\rho'1.5 = R'\rho'h' \ \therefore\ h' = 0.8\text{ m}.$$

47. (c) $g = \frac{GM}{r^2}$. Since M and r are constant, so $g = 9.8\ m/s^2$

48. (a) $g = g_p - R\omega^2\cos^2\lambda$

$$= g_p - \omega^2 R\cos^2 60° = g_p - \frac{1}{4}R\omega^2$$

49. (b)

50. (d) As $g = \frac{GM}{R^2}$

when radius is reduced by 1% then

$$R' = R - \frac{R}{100} = .99R,$$

$$\text{so } g' = \frac{GM}{(.99R)^2} = 1.02\frac{GM}{R^2}$$

$\Rightarrow$ % increase = 2%

51. (a) $\frac{g_p}{g_e} = \frac{2}{1}; \frac{R_p}{R_e} = \frac{1}{2}$

So, $M_p/M_e = 1 : 2$

52. (b) $g' = g\left(1 - \frac{d}{R}\right) \Rightarrow \frac{g}{n} = g\left(1 - \frac{d}{R}\right)$

$\Rightarrow d = \left(\frac{n-1}{n}\right)R$

53. (d) $g = \frac{4}{3}\pi\rho GR \quad \therefore \frac{g_1}{g_2} = \frac{R_1\rho_1}{R_2\rho_2}$

54. (a)

55. (b) Acceleration due to gravity,

$g \propto r$ (if $r < R_E$) and

$g \propto \frac{1}{r^2}$ (if $r > R_E$)

56. (c)

57. (c) Inside a shell $V = -\frac{GM}{R}$ = constant

and $E = 0$

Outside the shell, $V = -\frac{GM}{r}$ and $E = \frac{GM}{r^2}$

As r increases, V increases and E decreases.

58. (b) Mass of the rock = m and height of rock from earth $(h) = R_e/2 = 3.2 \times 10^6$ m

We know that gravitational potential energy of the rock rotating at height

$h = -\frac{GM_e m}{R_e + h} = -\frac{2}{3}mgR_e$

(where, $GM_e = gR_e^2$ and $h = R_e$)

59. (d) At a point inside a spherical shell, the value of gravitational intensity, I = 0.

If V = 0 then gravitational field is necessarily zero.

60. (b)

61. (d) $E_{earth} = E_{moon}$

$\Rightarrow \frac{GM}{x^2} = \frac{GM/81}{(60R - x)^2}$

$\Rightarrow \frac{1}{x} = \frac{1}{9(60R - x)}$

$\Rightarrow x = 54$ R from centre of earth.

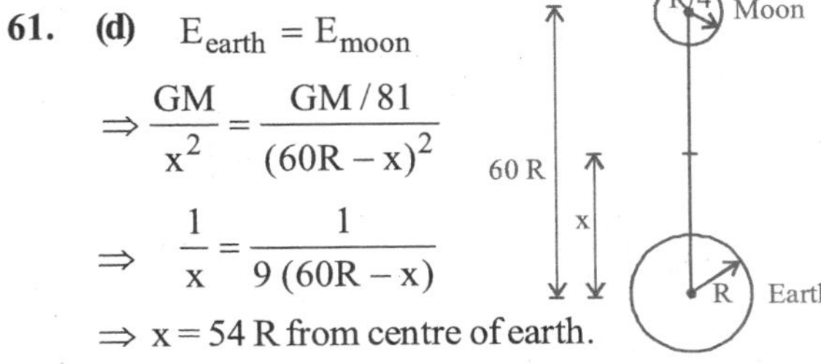

62. (d)

63. (a) The gravitational potential at the centre of uniform spherical shell is equal to the gravitational potential at the surface of shell i.e.,

$V = \frac{-GM}{a}$, where a is radius of spherical shell

Now, if the shell shrinks then its radius decrease then density increases, but mass is constant. so from above expression if a decreases, then V increases.

64. (c) 65. (b)

66. (a) For calculation of gravitational field intensity inside the cavity.

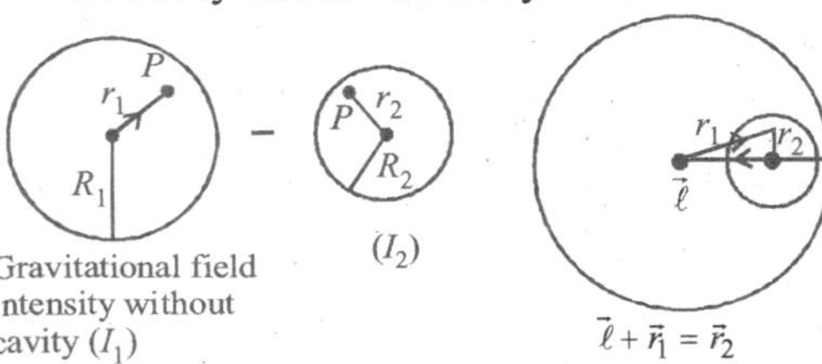

Gravitational field intensity without cavity (I_1)

(I_2)

$\vec{\ell} + \vec{r}_1 = \vec{r}_2$

$\vec{I}_1 = \frac{G\left(\frac{4}{3}\pi R_1^3\right)\rho(-\vec{r}_1)}{R_1^3}$,

$\vec{I}_2 = \frac{G\left(\frac{4}{3}\pi R_2^3\right)\rho(-\vec{r}_2)}{R_2^3}$

$\vec{I} = \vec{I}_1 - \vec{I}_2$ ($\vec{I}$ – intensity inside the cavity)

$= \frac{4}{3}G\pi\rho[-\vec{r}_1 + \vec{r}_2] = \frac{4}{3}G\pi\rho\vec{\ell}$

67. (a)

68. (d) Let mass per unit light of wire, $\lambda = \frac{m}{\ell}$

and $\pi r = \ell$, $r = \frac{\ell}{\pi}$.

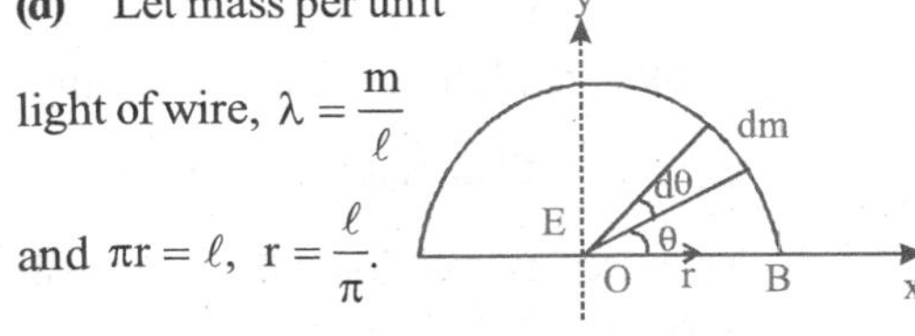

mass of element, $dm = \lambda r d\theta$ then $dE = \frac{Gdm}{r^2}$

$\int_0^{\pi} dE = \int_0^{\pi} \frac{G\lambda r d\theta}{r^2}(\hat{i}\cos\theta + \hat{j}\sin\theta)$

$E = \frac{G\lambda}{r}\left[\int_0^{\pi}\hat{i}\cos\theta + \int_0^{\pi}\hat{j}\sin\theta\right]$

$= \frac{2G\lambda}{r}\hat{j} = \frac{2GM}{\ell r}\hat{j} = \frac{2Gm\pi}{\ell^2}\hat{j}$ (along y-axis)

69. **(a)** We know that intensity is negative gradient of potential,
i.e., $I = -(dV/dr)$ and as here $I = -(K/r)$, so

$$\frac{dV}{dr} = \frac{K}{r}, \text{ i.e., } \int dV = K\int \frac{dr}{r}$$

or $V - V_0 = K \log \frac{r}{r_0}$

so $V = K \log \frac{r}{r_0} + V_0$

70. **(d)** Gravitational potential at P, $V_p = \frac{-GM}{\sqrt{5}R}$

Gravitational potential at O, $V_O = \frac{-GM}{R}$

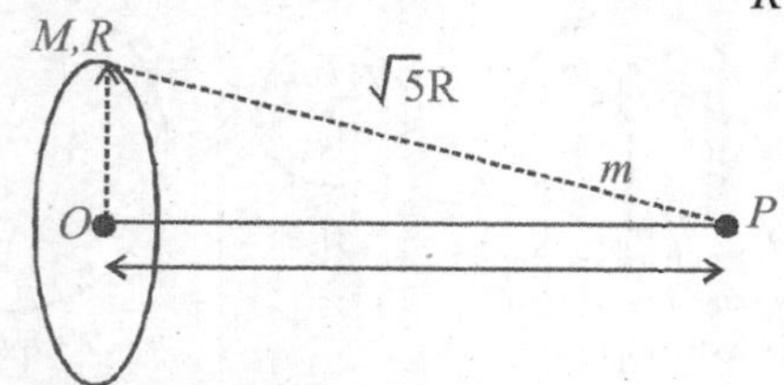

By work energy theorem,

$$W = \Delta K \Rightarrow m[V_P - V_O] = \frac{1}{2}mv^2$$

$$m\left[\frac{GM}{R} - \frac{GM}{\sqrt{5}R}\right] = \frac{1}{2}mv^2 \text{ or } V = \sqrt{\frac{2GM}{R}\left(1 - \frac{1}{\sqrt{5}}\right)}$$

71. **(c)** Let mass of smaller sphere (which has to be removed) is m

Radius $= \frac{R}{2}$ (from figure)

$$\frac{M}{\frac{4}{3}\pi R^3} = \frac{m}{\frac{4}{3}\pi\left(\frac{R}{2}\right)^3} \Rightarrow m = \frac{M}{8}$$

Mass of the left over part of the sphere

$$M' = M - \frac{M}{8} = \frac{7}{8}M$$

Therefore gravitational field due to the left over part of the sphere

$$= \frac{GM'}{x^2} = \frac{7}{8}\frac{GM}{x^2}$$

72. **(a)** From the principle of conserving angular momentum, we have

mvR = mv'r ...(i)

[v' = speed when spaceship is just touching the plane]

From conserving of energy, we have

$$\frac{1}{2}mv^2 = \frac{1}{2}mv'^2 - \frac{GMm}{r} \quad \text{...(ii)}$$

Solving Eqs. (i) and (ii), we get

$$R = \frac{r}{v}\left[v^2 + \frac{2GM}{r}\right]^{1/2}$$

73. **(b)**

74. **(b)** $\frac{1}{2}mv_{min}^2 - \frac{GMm}{R\sqrt{2}} \times 2 = 0$

$$\Rightarrow v_{min} = \sqrt{\frac{2GM\sqrt{2}}{R}}$$

75. **(c)** $E_1 = -\frac{GM_e m}{R_e + \frac{R_e}{2}} - \left(-\frac{GM_e m}{R_e}\right) = \frac{GM_e m}{3R_e}$

$$E_2 = \frac{1}{2}mv_0^2 = \frac{1}{2}m.\frac{GM_e}{R + \frac{R_e}{2}} = \frac{GM_e m}{3R_e}$$

$$\therefore \frac{E_1}{E_2} = 1:1$$

76. **(a)** In ΔAOB: $\cos 60° = \frac{R}{OB} \Rightarrow OB = 2R$

Here gravitational force will provide the required centripetal force.

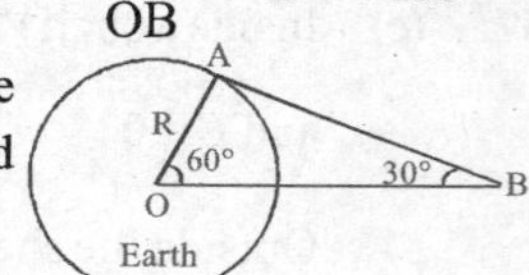

Hence, $\frac{GMm}{(OB)^2} = m(OB)\omega^2$

$$\Rightarrow \omega = \sqrt{\frac{GM}{(OB)^3}} = \sqrt{\frac{GM}{(2R)^3}} \Rightarrow \omega = \sqrt{\frac{GM}{8R^3}}$$

77. **(d)** Gravitational force provides the necessary centripetal force.

$$\therefore \frac{mv^2}{(R+x)} = \frac{GmM}{(R+x)^2} \text{ also } g = \frac{GM}{R^2}$$

$$\therefore \frac{mv^2}{(R+x)} = m\left(\frac{GM}{R^2}\right)\frac{R^2}{(R+x)^2} \quad \frac{n!}{r!(n-r)!}$$

$$\therefore \frac{mv^2}{(R+x)} = mg\frac{R^2}{(R+x)^2}$$

$$\therefore v^2 = \frac{gR^2}{R+x} \Rightarrow v = \left(\frac{gR^2}{R+x}\right)^{1/2}$$

78. **(a)** Orbital velocity for close to earth orbits $= \sqrt{gR}$. Escape velocity required $= \sqrt{2gR}$.

% Increase required $= \frac{v_e - v_0}{v_0} \times 100$

$$= \frac{\sqrt{gR}(\sqrt{2}-1)}{\sqrt{gR}} \times 100 = 41\%$$

79. (a) Escape velocity $= v = \sqrt{\dfrac{2GM}{R}}$

$$\Rightarrow \quad v^2 = \frac{2GM}{R} \qquad \text{.......(i)}$$

$$v^2 = (2GM)R^{-1}$$

Differentiating both sides, we get,

$$2v\frac{dv}{dR} = -\frac{2GM}{R^2} \Rightarrow v\frac{dv}{dR} = \frac{GM}{R^2} \qquad \text{......(ii)}$$

Dividing (ii) by (i), $\dfrac{1}{v}\dfrac{dv}{dR} = -\dfrac{1}{2R}$

$$\Rightarrow \quad \left|\frac{dv}{v}\right| \times 100 = \frac{1}{2} \times 4\% = 2\%$$

$\therefore$ If the radius decreases by 4%, escape velocity will increase by 2%.

80. (c) $E = -\dfrac{GMm}{2r}$

$$-\frac{dE}{dt} = \frac{GMm}{2r}\frac{1}{r^2}\frac{dr}{dt}$$

$$\int_0^t dt = -\frac{GMm}{2C}\int_r^R \frac{dr}{r^2}; \quad t = \frac{GMm}{2C}\left[\frac{1}{R} - \frac{1}{r}\right]$$

81. (d) $W_{AB} = m(V_B - V_A)$; But $V_A = V_B$ and so $W_{AB} = 0$.

82. (c) $-\dfrac{GM_e m}{10R_e} + \dfrac{1}{2}mv_i^2 = -\dfrac{GM_e m}{R_e} + \dfrac{1}{2}mv_f^2$

$$\therefore \quad v_f^2 = v_i^2 + \frac{2GM_e}{R_e}\left(1 - \frac{1}{10}\right).$$

83. (a)

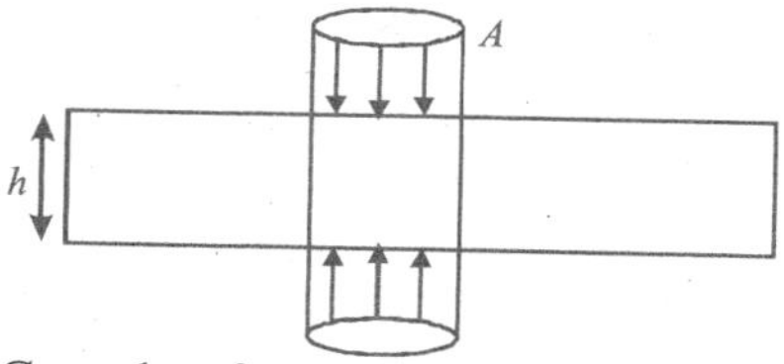

Gauss law for gravitation

$$\int \vec{g}\cdot d\vec{s} = -m_{in}\cdot 4\pi G; \quad g = \frac{GM}{R^2}$$

$$2 \times \frac{GM}{R^2} \times A = \frac{M}{\frac{4}{3}\pi R^3}(h \times A) \times 4\pi G$$

$$\Rightarrow h = \frac{2R}{3}$$

84. (c) By conservation of energy

$$-\frac{GMm}{R_E} + \frac{1}{2}\frac{m}{N^2}\frac{GM}{2R_E} = -\frac{GMm}{H}$$

$$\Rightarrow H = \frac{N^2 R_E}{N^2 - 1}$$

$$\text{Altitude} = H - R = \frac{R_E}{N^2 - 1}$$

85. (a) $v = \omega R$

$g = g_0 - \omega^2 R$ [g = at equator, g_0 = at poles]

$$\frac{g_0}{2} = g_0 - \omega^2 R; \qquad \omega^2 R = \frac{g_0}{2};$$

$$v^2 = \frac{g_0 R}{2}$$

$$v_e = \sqrt{2g_0 R} = \sqrt{4v^2} = 2v$$

86. (b) $v_e = \sqrt{2g_e R_e}$; $v_m = \sqrt{2g_m R_m}$

$$\frac{v_e}{v_m} = \sqrt{\frac{g_e}{g_m}\frac{R_e}{R_m}} = \sqrt{\frac{g_e}{g_e/6}\frac{R_e}{R_e/4}} = \sqrt{24}$$

87. (c) $\dfrac{-Gmm'}{r} \times 3 + \dfrac{1}{2}m'v_e^2 = 0$

or $\dfrac{-3Gm}{\left(\frac{a}{2}/\cos 30°\right)} + \dfrac{1}{2}v_e^2 = 0 \qquad v_e = \sqrt{\dfrac{6\sqrt{3}\,Gm}{a}}$.

88. (c) A satellite revolving near the earth's surface has a time period of 84.6 min.

We know that as the height increases, the time period increases. Thus the time period of the spy satellite should be slightly greater than 84.6 minutes.

$\therefore \quad T_s = 2$ hr

89. (c) Angular speed of earth = angular speed of geostationary satellite.

If ω is double, time period become half using

$$T^2 \propto r'^3 \qquad r' = \frac{r}{4^{1/3}}$$

90. (d)

91. (a) Let the gravitational field be zero at a point distant x from M_1.

$$\frac{GM_1}{x^2} = \frac{GM_2}{(d-x)^2}; \quad \frac{x}{d-x} = \sqrt{\frac{M_1}{M_2}}$$

$$x\sqrt{M_2} = \sqrt{M_1}d - x\sqrt{M_1}$$

$$x\left[\sqrt{M_1} + \sqrt{M_2}\right] = \sqrt{M_1}d$$

$$x = \frac{d\sqrt{M_1}}{\sqrt{M_1} + \sqrt{M_2}}, \quad d - x = \frac{d\sqrt{M_2}}{\sqrt{M_1} + \sqrt{M_2}}$$

Potential at this point due to both the masses will be

$$-\frac{GM_1}{x}-\frac{GM_2}{(d-x)}$$

$$=-G\left[\frac{M_1\left(\sqrt{M_1}+\sqrt{M_2}\right)}{d\sqrt{M_1}}+\frac{M_2\left(\sqrt{M_1}+\sqrt{M_2}\right)}{d\sqrt{M_2}}\right]$$

$$=-\frac{G}{d}\left(\sqrt{M_1}+\sqrt{M_2}\right)^2$$

$$=-\frac{G}{d}\left(M_1+M_2+2\sqrt{M_1}\sqrt{M_2}\right)$$

92. (a) As we know, gravitational potential (v) and acceleration due to gravity (g) with height

$$V=\frac{-GM}{R+h}=-5.4\times10^7 \quad ...(i)$$

and $g=\frac{GM}{(R+h)^2}=6 \quad ...(ii)$

Dividing (i) by (ii)

$$\frac{\frac{-GM}{R+h}}{\frac{GM}{(R+h)^2}}=\frac{-5.4\times10^7}{6}\Rightarrow\frac{5.4\times10^7}{(R+h)}=6$$

$\Rightarrow R+h=9000$ km so, $h=2600$ km

93. (b) The Gravitational field due to a thin spherical shell of radius R at distance r.

$$E=\frac{GM}{r^2}\ (\text{If } r>R)$$

For $r=R$ i.e. on the surface of the shell

$$E=\frac{GM}{R^2}$$

For $r<R$ i.e. inside the shell

$E=0$

94. (c) $M_A=\sigma4\pi R_A^2, M_B=\sigma4\pi R_B^2,$

where σ is surface density

$$V_A=\frac{-GM_A}{R_A},\ V_B=\frac{-GM_B}{R_B}$$

$$\frac{V_A}{V_B}=\frac{M_A}{M_B}\frac{R_B}{R_A}=\frac{\sigma4\pi R_A^2}{\sigma4\pi R_B^2}\frac{R_B}{R_A}=\frac{R_A}{R_B}$$

Given $\frac{V_A}{V_B}=\frac{R_A}{R_B}=\frac{3}{4}$ then $R_B=\frac{4}{3}R_A$

for new shell of mass M and radius R

$$M=M_A+M_B=\sigma4\pi R_A^2+\sigma4\pi R_B^2$$

$$\sigma4\pi R^2=\sigma4\pi(R_A^2+R_B^2)$$

then

$$\frac{V}{V_A}=\frac{M}{R}\frac{R_A}{R_B}=\frac{\sigma4\pi(R_A^2+R_B^2)}{(R_A^2+R_B^2)^{1/2}}=\frac{R_A}{\sigma4\pi R_A^2}$$

$$=\frac{\sqrt{R_A^2+R_B^2}}{R_A}=\frac{5}{3}$$

95. (d) Gravitational field, $I=\left(5\hat{i}+12\hat{j}\right)$ N/kg

$$I=-\frac{dv}{dr}$$

$$v=-\left[\int_0^x I_x dx+\int_0^y I_y dy\right]=-\left[I_x.x+I_y.y\right]$$

$$=-\left[5(7-0)+12(-3-0)\right]$$

$$=-\left[35+(-36)\right]=1\ \text{J/kg}$$

i.e., change in gravitational potential 1 J/kg. Hence change in gravitational potential energy 1 J

96. (d) $V=-\int_\infty^x E\,dx=\int_\infty^x Kx^{-3}\,dx=\frac{K}{2x^2}.$

97. (c) $x=0\quad 1\quad 2\quad 4\quad 8$

$$V=-\frac{GM}{r}=-G\left(\frac{1}{1}+\frac{1}{2}+\frac{1}{4}+\frac{1}{8}+......\infty\right)$$

$$=-G\left(\frac{1}{1-1/2}\right)=-2G$$

98. (c) $W=m(V_2-V_1)$

when, $V_1=-\left[\frac{GM_1}{a}+\frac{GM_2}{\sqrt{2}a}\right],$

$$V_2=-\left[\frac{GM_2}{a}+\frac{GM_1}{\sqrt{2}a}\right]$$

$$\therefore\quad W=\frac{Gm(M_2-M_1)}{a\sqrt{2}}(\sqrt{2}-1).$$

99. (d) $\frac{Mv^2}{R}=\frac{GM^2}{4R^2}\Rightarrow M=\frac{4Rv^2}{G}$

$$v=\frac{2\pi R}{T}\qquad R=\frac{vT}{2\pi}\qquad M=\frac{v^3T2}{\pi G}$$

100. (c) The gravitational field due to the ring at a distance $\sqrt{3}r$ is given by

$$E=\frac{Gm(\sqrt{3}r)}{[r^2+(\sqrt{3}r)^2]^{3/2}}\Rightarrow E=\frac{\sqrt{3}Gm}{8r^2}$$

8 Mechanical Properties of Solids

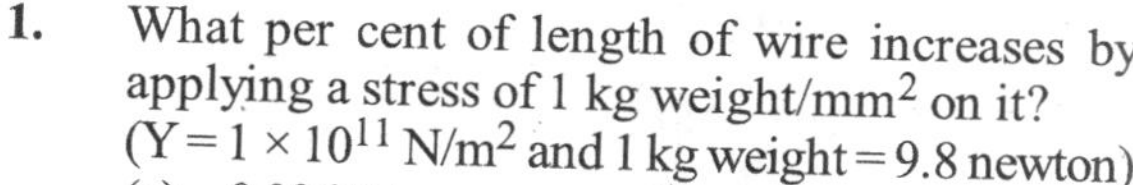

1. What per cent of length of wire increases by applying a stress of 1 kg weight/mm^2 on it? ($Y = 1 \times 10^{11}$ N/m^2 and 1 kg weight = 9.8 newton)
(a) 0.0067% (b) 0.0098%
(c) 0.0088% (d) 0.0078%

2. An elevator cable is to have a maximum stress of 7×10^7 N/m^2 to allow for appropriate safety factors. Its maximum upward acceleratiom is 1.5 m/s^2. If the cable has to support the total weight of 2000 kg of a loaded elevator, the area of cross–section of the cable should be

Critical Thinking

(a) 3.28 cm^2 (b) 2.38 cm^2
(c) 0.328 cm^2 (d) 8.23 cm^2

3. To break a wire, a force of 10^6 N/m^2 is required. If the density of the material is 3×10^3 kg/m^3, then the length of the wire which will break by its own weight will be
(a) 34m (b) 30m (c) 300m (d) 3m

4. A rubber cord catapult has cross-sectional area 25 mm^2 and initial length of rubber cord is 10 cm. It is stretched to 5 cm and then released to project a missile of mass 5 gm. Taking $Y_{rubber} = 5 \times 10^8$ N/m^2. Velocity of projected missile is
(a) 20 ms^{-1} (b) 100 ms^{-1}
(c) 250 ms^{-1} (d) 200 ms^{-1}

5. Which of the following affects the elasticity of a substance?
(a) Change in temperature
(b) Hammering and annealing
(c) Impurity in substance
(d) All of the above

6. A metal wire of length L is suspended vertically from a rigid support when the body of mass M is attached to the lower end of the wire, the elongation of the wire is l. The elastic potential energy stored in the wire is
(a) mgl (b) $mgl/2$ (c) $mgl/3$ (d) $mgl/4$

7. A body of mass 10 kg is attached to a wire of radius 3 cm. It's breaking stress is 4.8×10^7 Nm^{-2}, the area of cross-section of the wire is 10^{-6} m^2. What is the maximum angular velocity with which it can be rotated in the horizontal circle?
(a) 1 rad sec^{-1} (b) 2 rad sec^{-1}
(c) 4 rad sec^{-1} (d) 8 rad sec^{-1}

8. For the same cross-sectional area and for a given load, the ratio of depressions for the beam of a square cross-section and circular cross-section is

Critical Thinking

(a) $3:\pi$ (b) $\pi:3$ (c) $1:\pi$ (d) $\pi:1$

9. The length of a metal is ℓ_1 when the tension in it is T_1 and is ℓ_2 when the tension is T_2. The original length of the wire is
(a) $\frac{\ell_1 + \ell_2}{2}$ (b) $\frac{\ell_1 T_2 + \ell_2 T_1}{T_1 + T_2}$
(c) $\frac{\ell_1 T_2 - \ell_2 T_1}{T_2 - T_1}$ (d) $\sqrt{T_1 T_2 \ell_1 \ell_2}$

10. A light rod of length $2m$ suspended from the ceiling horizontally by means of two vertical wires of equal length. A weight W is hung from a light rod as shown in figure. The rod hung by means of a steel wire of cross-sectional area $A_1 = 0.1$ cm^2 and brass wire of cross-sectional area $A_2 = 0.2$ cm^2. To have equal stress in both wires, $T_1/T_2 =$

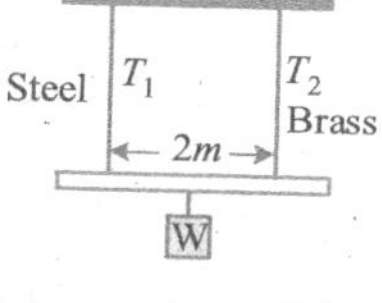

(a) 1/3 (b) 1/4
(c) 4/3 (d) 1/2

11. For an equal stretching force F, the young's modulus (Y_s) for steel and rubber (Y_r) are related as
(a) $Y_s = Y_r$ (b) $Y_s < Y_r$ (c) $Y_s > Y_r$ (d) $Y_s \geq Y_r$

12. which of the following is correct for young's modulus of elasticity (γ)? [where r = radius of cross section of wire, l = length of wire]
(a) $\gamma \propto r^2$ (b) $\gamma \propto l^3$ (c) $\gamma \propto 1/r^2$ (d) $\gamma \propto l^2$

13. If the length of a wire is reduced to half, then it can hold the
(a) half load (b) same load
(c) double load (d) one fourth load

14. A 2 m long rod of radius 1 cm which is fixed from one end is given a force of 8 N. The longitudinal strain developed will be [take $\gamma = 2.5 \times 10^{11}$ N/m^2]
(a) 10^{-8} (b) 10^{-6} (c) 10^{-5} (d) 10^{-4}

15. A vertical metal cylinder of radius 2 cm and length 2 m is fixed at the lower end and a load of 100 kg is put on it. Find the strain. [Young's modulus of the metal = 2×10^{11} N/m^2]
(a) 4×10^{-6} (b) 3×10^{-8}
(c) 2×10^{-9} (d) 6×10^{-8}

16. The graph given is a stress-strain curve for

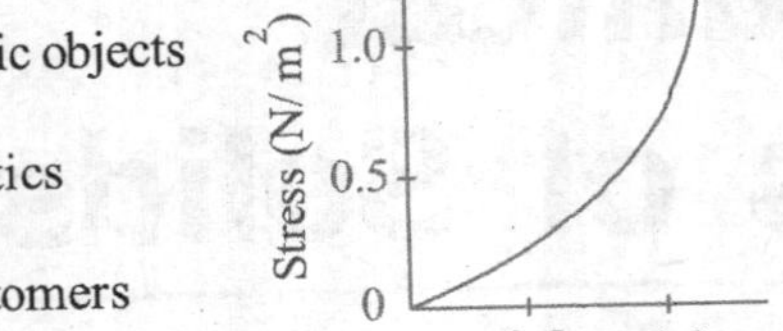

(a) elastic objects

(b) plastics

(c) elastomers

(d) None of these

17. If the ratio of radii of two wires of same material is 3 : 1 and ratio of their lengths is 5 : 1, then the ratio of the normal forces that will produce the same extension in the length of two wires is

(a) 2:1 (b) 4:1 (c) 1:4 (d) 1:1

18. An iron bar of length ℓ cm and cross section A cm^2 is pulled by a force of F dynes from ends so as to produce an elongation $\Delta\ell$ cm. Which of the following statement is correct ?

(a) Elongation is inversely proportional to length

(b) Elongation is directly proportional to cross section A

(c) Elongation is inversely proportional to cross-section

(d) Elongation is directly proportional to Young's modulus

19. Two wires of equal lengths are made of the same material. Wire A has a diameter that is twice as that of wire B. If identical weights are suspended from the ends of these wires, the increase in length is **Tricky**

(a) four times for wire A as for wire B

(b) twice for wire A as for wire B

(c) half for wire A as for wire B

(d) one-fourth for wire A as for wire B

20. A beam of metal supported at the two edges is loaded at the centre. The depression at the centre is proportional to **Tricky**

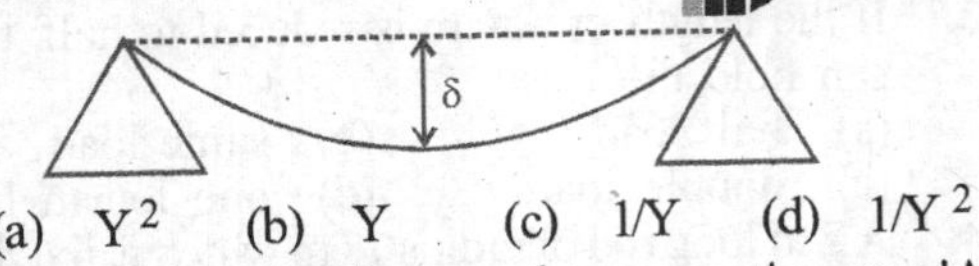

(a) Y^2 (b) Y (c) 1/Y (d) $1/Y^2$

21. A steel ring of radius r and cross-section area 'A' is fitted on to a wooden disc of radius R(R > r). If Young's modulus be E, then the force with which the steel ring is expanded is

(a) $AE\frac{R}{r}$ (b) $AE\left(\frac{R-r}{r}\right)$

(c) $\frac{E}{A}\left(\frac{R-r}{A}\right)$ (d) $\frac{Er}{AR}$

22. The adjacent graph shows the extension (Δl) of a wire of length 1m suspended from the top of a roof at one end with a load W connected to the other end. if the cross-sectional area of the wire is $10^{-6}m^2$, calculate the Young's modulus of the material of the wire

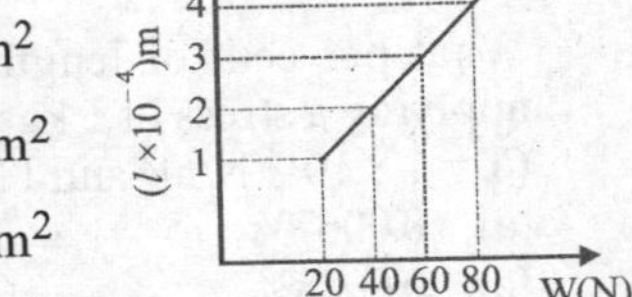

(a) 2×10^{11} N/m^2

(b) 2×10^{-11} N/m^2

(c) 2×10^{-12} N/m^2

(d) 2×10^{-13} N/m^2

23. Two persons pull a rope towards themselves. Each person exerts a force of 100 N on the rope. Find the Young's modulus of the material of the rope if it extends in length by 1 cm. Original length of the rope = 2 m and the aea of cross-section = 2 cm^2. **Critical Thinking**

(a) 10^8 N/m^2 (b) 10^7 N/m^2

(c) 10^6 N/m^2 (d) 10^5 N/m^2

24. If stress/strain is x in eastic region and y in the region of yield, then

(a) x=y (b) x>y (c) x<y (d) x=2y

25. A metallic rod breaks when strain produced is 0.2%. The Young's modulus of the material of the rod is 7×10^9 N/m^2. What should be its area of cross-section to support a load of 10^4 N ?

(a) 7.1×10^{-8} m^2 (b) 7.1×10^{-6} m^2

(c) 7.1×10^{-4} m^2 (d) 7.1×10^{-2} m^2

26. A steel wire of original length 1 m and cross-sectional area 4.00 mm^2 is clamped at the two ends so that it lies horizontally and without tension. If a load of 2.16 kg is suspended from the middle point of the wire, what would be its vertical depression?
Y of the steel - 2.0×10^{11}N/m^2. Take g = 10 m/s^2

(a) 1.5 cm (b) 2.8 cm (c) 3.2 cm (d) 4.1 cm

27. Two wires are made of the same material and have the same volume. However first wire has cross-sectional area A and second wire has cross-sectional area $5A$. If the length of first wire increases by Δl on applying force f, how much force is needed to stretch second wire by the same amount?

(a) $14f$ (b) $6f$ (c) $25f$ (d) $9f$

28. One end of uniform wire of length L and of weight W is attached rigidly to a point in the roof and a weight W_1 is suspended from its lower end. If s is the area of cross section of the wire, the stress in the wire at a height $\left(\frac{3L}{4}\right)$ from its lower end is

(a) $\frac{W_1}{s}$ (b) $\left[W_1 + \frac{W}{4}\right]s$

(c) $\left[W_1 + \frac{3W}{4}\right]/s$ (d) $\frac{W_1 + W}{s}$

29. The following four wires are made of the same material. Which of these will have the largest extension when the same tension is applied ?

(a) length = 50 cm, diameter = 0.5 mm
(b) length = 100 cm, diameter = 1 mm
(c) length = 200 cm, diameter = 2 mm
(d) length = 300 cm, diameter = 3 mm.

30. The stress versus strain graphs for wires of two materials A and B are as shown in the figure. If Y_A and Y_B are the Young's moduli of the materials, then

(a) $Y_B = 2\,Y_A$
(b) $Y_A = Y_B$
(c) $Y_B = 3\,Y_A$
(d) $Y_A = 3\,Y_B$

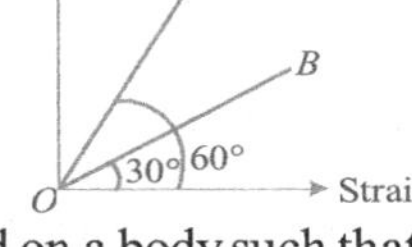

31. When forces are applied on a body such that it is still in static equilibrium, then the extent to which the body gets deformed, depends on

(a) nature of the material
(b) magnitude of deforming force
(c) Both (a) & (b)
(d) None of these

32. What is the minimum diameter of a brass rod if it is to support a 400N load without exceeding the elastic limit ? Assume that the stress for the elastic limit is 379 MPa.

(a) 1.16 mm (b) 2.32 mm
(c) 0.16 mm (d) 1.35 mm

33. The elastic limit of steel is 8×10^8 N/m^2 and its Young's modulus 2×10^{11} N/m^2. Find the maximum elongation of a half-meter steel wire that can be given without exceeding the elastic limit.

(a) 2mm (b) 4mm (c) 5mm (d) 6mm

34. A steel wire and a copper wire of equal length and equal cross-sectional area are joined end to end and the combination is subjected to a tension. Find the ratio of the stresses developed in the two wires and Y of steel = 2×10^{11} N/m^2. Y of copper = 1.3×10^{11} N/m^2.

(a) 1 (b) 3 (c) 5 (d) 7

35. Two wires A and B of same material and of equal length with the radii in the ratio 1 : 2 are subjected to identical loads. If the length of A increases by 8 mm, then the increase in length of B is

(a) 2 mm (b) 4 mm (c) 8 mm (d) 16 mm

36. The length of elastic string, obeying Hooke's law is ℓ_1 metres when the tension 4N and ℓ_2 metres when the tension is 5N. The length in metres when the tension is 9N is –

(a) $5\ell_1 - 4\ell_2$ (b) $5\ell_2 - 4\ell_1$
(c) $9\ell_1 - 8\ell_2$ (d) $9\ell_2 - 8\ell_1$

37. A steel wire of length l and cross section area A is stretched by 1 cm under a given load. When the same load is applied to another steel wire of double its length and half of its cross section area, the amount of stretching (extension) is

(a) 0.5 cm (b) 2 cm (c) 4 cm (d) 1.5 cm

38. Two wires are made of the same material and have the same volume. However wire 1 has cross-sectional area A and wire 2 has cross-sectional area $9A$. If the length of wire 1 increases by Δx on applying force F, how much force is needed to stretch wire 2 by the same amount?

(a) $16F$ (b) $25F$ (c) $81F$ (d) $64F$

39. The force exerted by a special compression device is given as function of compression x as $F_x(x) = kx(x - \ell)$ for $0 \le x \le \ell$, where ℓ is maximum possible compression and k is a constant. The force exerted by the device under compression is maximum when compression is –

(a) 0 (b) $\ell/4$ (c) $\ell/\sqrt{2}$ (d) $\ell/2$

40. A force of 6×10^6 Nm^{-2} is required for breaking a material. Then density ρ of the material is 3×10^3 kg m^{-3}. If the wire is to break under its own weight, the length of the wire made of that material should be (take g = 10 ms^{-2})

(a) 20 m (b) 200 m (c) 100 m (d) 2000 m

41. A steel wire of cross-sectional area 3×10^{-6} m^2 can with stand a maximum strain of 10^{-3}. Young's modulus of steel is 2×10^{11} N/m^2. The maximum mass the wire can hold is:

(a) 40 kg (b) 60 kg (c) 80 kg (d) 100 kg

42. A structural steel rod has a radius of 10 mm and length of 1.0 m. A 100 kN force stretches it along its length. Young's modulus of structural steel is 2×10^{11} Nm^{-2}. The percentage strain is about

(a) 0.16% (b) 0.32% (c) 0.08% (d) 0.24%

43. A thick rope of density ρ and length L is hung from a rigid support. The Young's modulus of the material of rope is Y. The increase in length of the rope due to its own weight is
(a) $(1/4)\rho g L^2/Y$ (b) $(1/2)\rho g L^2/Y$
(c) $\rho g L^2/Y$ (d) $\rho g L/Y$

44. If the ratio of lengths, radii and Young's moduli of steel and brass wires in the figure are a, b and c respectively, then the corresponding ratio of increase in their lengths is : **Critical Thinking**

(a) $\frac{3c}{2ab^2}$

(b) $\frac{2a^2c}{b}$

(c) $\frac{3a}{2b^2c}$

(d) $\frac{2ac}{b^2}$

Steel
M
Brass
2M

45. What per cent of length of wire increases by applying a stress of 1 kg weight/mm² on it? ($Y = 1 \times 10^{11}$ N/m² and 1 kg weight = 9.8 newton)
(a) 0.0067% (b) 0.0098%
(c) 0.0088% (d) 0.0078%

46. A metal wire of length L_1 and area of cross-section A is attached to a rigid support. Another metal wire of length L_2 and of the same cross-sectional area is attached to the free end of the first wire. A body of mass M is then suspended from the free end of the second wire. If Y_1 and Y_2 are the young's moduli of the wires respectively, the effective force constant of the system of two wires is **Toughnut**

(a) $\frac{(Y_1Y_2)A}{2(Y_1L_2+Y_2L_1)}$ (b) $\frac{(Y_1Y_2)A}{(L_1L_2)^{1/2}}$

(c) $\frac{(Y_1Y_2)A}{Y_1L_2+Y_2L_1}$ (d) $\frac{(Y_1Y_2)^{1/2}A}{(L_2L_1)^{1/2}}$

47. A steel wire 1.5 m long and of radius 1 mm is attached with a load 3 kg at one end the other end of the wire is fixed. It is whirled in a vertical circle with a frequency 2 Hz. Find the elongation of the wire when the weight is at the lowest position. [$Y = 2 \times 10^{11}$ N/m², $g = 10$ ms^{-2}]
(a) 1.77×10^{-3} m (b) 7.17×10^{-3} m
(c) 3.17×10^{-7} m (d) 1.37×10^{-7} m

48. A platform is suspended by four wires at its corners. The wires are 3m long and have a diameter of 2.0mm. Young's modulus for the material of the wires is 1,80,000 *MPa*. How far will the platform drop (due to elongation of the wires) if a 50 kg load is placed at the centre of the platform? **Toughnut**
(a) 0.25 mm (b) 0.65 mm (c) 1.65 mm (d) 0.35 mm

49. A force of 10^3 newton, stretches the length of a hanging wire by 1 millimetre. The force required to stretch a wire of same material and length but having four times the diameter by 1 millimetre is
(a) 4×10^3 N (b) 16×10^3 N
(c) $\frac{1}{4} \times 10^3$ N (d) $\frac{1}{16} \times 10^3$ N

50. A copper wire of length 1.0 m and a steel wire of length 0.5 m having equal cross-sectional areas are joined end to end. The composite wire is stretched by a certain load which stretches the copper wire by 1 mm. If the Young's modulii of copper and steel are respectively 1.0×10^{11} Nm^{-2} and 2.0×10^{11} Nm^{-2}, the total extension of the composite wire is :
(a) 1.75 mm (b) 2.0 mm (c) 1.50 mm (d) 1.25 mm

51. The diagram shows stress v/s strain curve for the materials A and B. From the curves we infer that :
(a) A is brittle but B is ductile
(b) A is ductile and B is brittle
(c) Both A and B are ductile
(d) Both A and B are brittle

Stress
Strain
A
B

52. A square frame of ABCD consisting of five steel bars of cross section area 400 mm² and joined by pivot is subjected to action of two forces P = 40 kN in the direction of the diagonal as shown. Find change in angle at A if Young's modulus $Y = 2 \times 10^5$ N/min.
(a) $\frac{1}{2000}$ rad
(b) $\frac{1}{1000}$ rad
(c) $\frac{\sqrt{2}}{1000}$ rad
(d) none

A a B a D a C 40 kN

53. A uniformly tapering conical wire is made from a material of Young's modulus Y and has a normal, unextended length L. The radii, at the upper and lower ends of this conical wire, have values R and 3R, respectively. The upper end of the wire is fixed to a rigid support and a mass M is suspended from its lower end. The equilibrium extended length, of this wire, would equal : **Toughnut**

(a) $L\left(1+\frac{2}{9}\frac{Mg}{\pi YR^2}\right)$ (b) $L\left(1+\frac{1}{9}\frac{Mg}{\pi YR^2}\right)$

(c) $L\left(1+\frac{1}{3}\frac{Mg}{\pi YR^2}\right)$ (d) $L\left(1+\frac{2}{3}\frac{Mg}{\pi YR^2}\right)$

54. Which of the following is the correct relation? Y = Young's modulus & G = modulus of rigidity?
(a) Y < G (b) Y > G
(c) Y = G (d) None of these

55. The ratio of shearing stress to the corresponding shearing strain is called
(a) bulk modulus (b) Young's modulus
(c) modulus of rigidity (d) None of these

56. Two rods A and B of the same material and length have their radii r_1 and r_2 respectively. When they are rigidly fixed at one end and twisted by the same couple applied at the other end, the ratio $\left(\frac{\text{Angle of twist at the end of A}}{\text{Angle of twist at the end of B}}\right)$ is
(a) r_1^2/r_2^2 (b) r_1^3/r_2^3 (c) r_2^4/r_1^4 (d) r_1^4/r_2^4

57. A metallic wire of length 2.0 m is elongated by 2.0 mm. Area of cross-section of the wire is 4.0 mm^2. The elastic potential energy stored in the wire in elongated condition is [young's modulus of the metallic wire is = 2×10^{11} N/m^2]
(a) 8.23 (b) 0.83 (c) 6.23 (d) 0.63

58. When a pressure of 100 atmosphere is applied on a spherical ball, then its volume reduces to 0.01%. The bulk modulus of the material of the rubber in dyne/cm^2 is
(a) 10×10^{12} (b) 100×10^{12}
(c) 1×10^{12} (d) 10×10^{12}

59. A uniform cube is subjected to volume compression. If each side is decreased by 1%, then bulk strain is
(a) 0.01 (b) 0.06 (c) 0.02 (d) 0.03

60. When a 4 kg mass is hung vertically on a light spring that obeys Hooke's law, the spring stretches by 2 cms. The work required to be done by an external agent in stretching this spring by 5 cms will be (g = 9.8 m/sec^2)
(a) 4.900 joule (b) 2.450 joule
(c) 0.495 joule (d) 0.245 joule

61. Consider four steel wires of dimensions given below (d = diameter and l = length) :
(A) l = 1m, d = 1mm (B) l = 2m, d = 2 mm
(C) l = 2m, d = 1mm (D) l = 1m, d = 2 mm
If same force is applied to all the wires then the elastic potential energy stored will be maximum in wire:
Tricky
(a) A (b) B (c) C (d) D

62. If in a wire of Young's modulus Y, longitudinal strain X is produced, then the value of potential energy stored in its unit volume will be
(a) YX^2 (b) $2YX^2$ (c) $Y^2X/2$ (d) $YX^2/2$

63. The Poisson's ratio of a material is 0.5. If a force is applied to a wire of this material, there is a decrease in the cross-sectional area by 4%. The percentage increase in the length is :
(a) 1% (b) 2% (c) 2.5% (d) 4%

64. A material has poisson's ratio 0.50. If a uniform rod of it suffers a longitudinal strain of 2×10^{-3}, then the percentage change in volume is
(a) 0.6 (b) 0.4 (c) 0.2 (d) Zero

65. The system is rotated with angular speed ω_0 (see figure). What is the ratio of energy stored in each wire?
Critical Thinking
ω_0 2l;A;Y l;A;Y m 2m
(a) 31:9 (b) 50:9
(c) 47:9 (d) 8:9

66. When a force is applied on a wire of uniform cross-section area 3×10^{-6} m^2 and length 4m, the increase in length is 1 mm. Energy stored in it will be ($Y = 2 \times 10^{11}$ N/m^2)
(a) 6250 J (b) 0.177 J (c) 0.075 J (d) 0.150 J

67. A metal rod of Young's modulus 2×10^{10} N m^{-2} undergoes an elastic strain of 0.06%. The energy per unit volume stored in J m^{-3} is
(a) 3600 (b) 7200 (c) 10800 (d) 14400

68. When the load on a wire is increasing slowly from 2 kg to 4 kg, the elongation increases from 0.6 mm to 1 mm. The work done during this extension of the wire is (g = 10 m/s^2)
(a) 9×10^{-3} J (b) 12×10^{-3} J
(c) 14×10^{-3} J (d) 16×10^{-3} J

69. A solid cube is subjected to a pressure of 5×10^5 Nm^{-2}. Each side of the cube is shortened by 1 %. Find x if 1.67×10^x N/m^2 be the bulk modulus of elasticity of the cube.
(a) 7 (b) 11 (c) 12 (d) 15

70. Two parallel and opposite forces, each of magnitude 4000 N are applied tangentially to the upper and lower faces of a cubical metal block 25 cm on a side. The angle of Shear is [shear modulas of metal is 80 G Pa]
(a) 8×10^{-7}rad (b) 7×10^{-7}rad
(c) 6×10^{-6}rad (d) 5×10^{-5}rad

71. Two Metal strips are riveted together at their ends by four rivets, each of diameter α = 6 mm. The maximum tension that can be exerted by the riveted strip (if the Shearing stress on the rivet is not to exceed 6.9×10^7 Pa) is?
(a) 6.8×10^2N (b) 7.8×10^3N
(c) 8.28×10^4N (c) 9.1×10^3N

72. Two cylinders A and B of the same material have same length, their radii being in the ratio 1 : 2 respectively. The two are joined end to end as shown. One end of cylinder A is rigidly clamped while free end of cylinder B is twisted through an angle θ. The angle of twist of cylinder A is

Critical Thinking

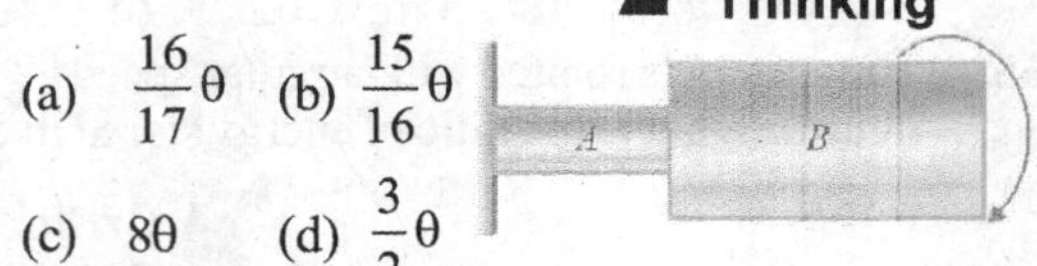

(a) $\frac{16}{17}\theta$ (b) $\frac{15}{16}\theta$

(c) 8θ (d) $\frac{3}{2}\theta$

73. The pressure in an explosion chamber is 345 MPa. What would be the percent change in volume of a piece of copper subjected to this pressure ? The bulk modulus for copper is 138 Gpa (= 138×10^9 Pa)

Critical Thinking

(a) 0.1% (b) 0.5% (c) 0.25% (d) 0.2%

74. The bulk modulus of a spherical object is 'B'. If it is subjected to uniform pressure 'p', the fractional decrease in radius is

(a) $\frac{B}{3p}$ (b) $\frac{3p}{B}$ (c) $\frac{p}{3B}$ (d) $\frac{p}{B}$

75. The Young's modulus of the material of a wire is 2×10^{10} Nm^{-2}. If the elongation strain is 1%, then the energy stored in the wire per unit volume in Jm^{-3} is

(a) 10^6 (b) 10^8 (c) 2×10^6 (d) 2×10^8

76. A 5 metre long wire is fixed to the ceiling. A weight of 10 kg is hung at the lower end and is 1 metre above the floor. The wire was elongated by 1 mm. The energy stored in the wire due to stretching is

(a) zero (b) 0.05 joule

(c) 100 joule (d) 500 joule

77. A wire suspended vertically from one of its ends is stretched by attaching a weight of 200N to the lower end. The weight stretches the wire by 1 mm. Then the elastic energy stored in the wire is

(a) 0.2 J (b) 10 J (c) 20 J (d) 0.1 J

78. Two, spring P and Q of force constants k_p and $k_Q \left(k_Q = \frac{k_p}{2} \right)$ are stretched by applying forces of equal magnitude. If the energy stored in Q is E, then the energy stored in P is

(a) E (b) 2E (c) E/2 (d) E/4

79. A spherical ball contracts in volume by 0.02% when subjected to a pressure of 100 atmosphere. Assuming one atmosphere = 10^5 N m^{-2}, the bulk modulus of the material of the ball is

(a) 0.02×10^5 N/m^2 (b) 0.02×10^7 N/m^2

(c) 50×10^7 N/m^2 (d) 50×10^9 N/m^2

80. A circular tube of mean radius 8 cm and thickness 0.04 cm is melted up and recast into a solid rod of the same length. The ratio of the torsional rigidities of the circular tube and the solid rod is

Toughnut

(a) $\frac{(8.02)^4 - (7.98)^4}{(0.8)^4}$ (b) $\frac{(8.02)^2 - (7.98)^2}{(0.8)^2}$

(c) $\frac{(0.8)^2}{(8.02)^4 - (7.98)^4}$ (d) $\frac{(0.8)^2}{(8.02)^3 - (7.98)^2}$

Answer KEYs

1	(b)	9	(c)	17	(d)	25	(c)	33	(a)	41	(d)	49	(b)	57	(b)	65	(b)	73	(c)
2	(a)	10	(d)	18	(c)	26	(a)	34	(a)	42	(a)	50	(d)	58	(c)	66	(c)	74	(c)
3	(a)	11	(c)	19	(d)	27	(c)	35	(a)	43	(b)	51	(b)	59	(d)	67	(a)	75	(a)
4	(c)	12	(c)	20	(c)	28	(c)	36	(b)	44	(c)	52	(b)	60	(b)	68	(c)	76	(b)
5	(d)	13	(b)	21	(b)	29	(a)	37	(c)	45	(b)	53	(c)	61	(c)	69	(a)	77	(d)
6	(b)	14	(a)	22	(a)	30	(d)	38	(c)	46	(c)	54	(b)	62	(d)	70	(a)	78	(c)
7	(c)	15	(a)	23	(a)	31	(c)	39	(d)	47	(a)	55	(c)	63	(d)	71	(b)	79	(d)
8	(a)	16	(c)	24	(b)	32	(a)	40	(b)	48	(b)	56	(c)	64	(b)	72	(a)	80	(a)

Hints & Solutions

1. **(b)** Stress $= 1$ kg wt/mm^2 $= 9.8$ N/mm^2 $= 9.8 \times 10^6$ N/m^2.

$Y = 1 \times 10^{11} N/m^2$, $\dfrac{\Delta\ell}{\ell} \times 100 = ?$

$$Y = \frac{\text{Stress}}{\text{Strain}} = \frac{\text{Stress}}{\Delta\ell/\ell}$$

$$\therefore \frac{\Delta\ell}{\ell} = \frac{\text{Stress}}{Y} = \frac{9.8 \times 10^6}{1 \times 10^{11}}$$

$$\frac{\Delta\ell}{\ell} \times 100 = 9.8 \times 10^{-11} \times 100 \times 10^6$$

$= 9.8 \times 10^{-3} = 0.0098\%$

2. **(a)** Given, the breaking strength of cable $f_u = 7 \times 10^7$ N/m^2

The force carried by the cable,

$F = m(g + a) = 2000(9.8 + 1.5) = 22600$ N

The area of cross-section, $A = \dfrac{F}{f_u} = \dfrac{22600}{7 \times 10^7}$
$= 3.28 \times 10^{-4}$ m^2.

3. **(a)** $L = \dfrac{S}{dg} = \dfrac{10^6}{3 \times 10^3 \times 10} = \dfrac{100}{3} = 34\,m$

4. **(c)** Young's modulus of rubber, Y_{rubber}

$= \dfrac{F}{A} \times \dfrac{\ell}{\Delta\ell} \Rightarrow F = YA.\dfrac{\Delta\ell}{\ell}$

On putting the values from question,

$$F = \frac{5 \times 10^8 \times 25 \times 10^{-6} \times 5 \times 10^{-2}}{10 \times 10^{-2}} = 6250\text{ N}$$

kinetic energy = potential energy of rubber

$$\frac{1}{2}mv^2 = \frac{1}{2}F\Delta\ell$$

$$v = \sqrt{\frac{F\Delta\ell}{m}} = \sqrt{\frac{6250 \times 5 \times 10^{-2}}{5 \times 10^{-3}}} = \sqrt{62500}$$

$= 250$ m/s

5. **(d)** The elasticity of a material depends upon the temperature of the material. Hammering & annealing reduces elastic property of a substance.

6. **(b)**

7. **(c)** Given that $F/A = 4.8 \times 10^7$ Nm^{-2}

$\therefore$ $F = 4.8 \times 10^7 \times A$

or $\dfrac{mv^2}{r} = 4.8 \times 10^7 \times 10^{-6} = 48$

or $\dfrac{mr^2\omega^2}{r} = 48$ or $\omega^2 = \dfrac{48}{mr}$

$$\omega = \sqrt{\left(\frac{48}{10 \times 0.3}\right)} = \sqrt{16} = 4 \text{ rad/sec}$$

8. **(a)** $\delta = \dfrac{W\ell^3}{3YI}$, where W = load, ℓ = length of beam and I is geometrical moment of inertia for rectangular beam,

$I = \dfrac{bd^3}{12}$ where b = breadth and d = depth

For square beam $b = d$ $\therefore$ $I_1 = \dfrac{b^4}{12}$

For a beam of circular cross-section,

$$I_2 = \left(\frac{\pi r^4}{4}\right)$$

$$\therefore \quad \delta_1 = \frac{W\ell^3 \times 12}{3Yb^4} = \frac{4W\ell^3}{Yb^4}$$

(for sq. cross section)

and $\delta_2 = \dfrac{W\ell^3}{3Y(\pi r^4/4)} = \dfrac{4W\ell^3}{3Y(\pi r^4)}$

(for circular cross-section)

Now $\dfrac{\delta_1}{\delta_2} = \dfrac{3\pi r^4}{b^4} = \dfrac{3\pi r^4}{(\pi r^2)^2} = \dfrac{3}{\pi}$

($\because b^2 = \pi r^2$ i.e., they have same cross-sectional area)

9. **(c)** If ℓ is the original length of wire, then change in length of first wire,

$\Delta\ell_1 = (\ell_1 - \ell)$

Change in length of second wire,

$\Delta\ell_2 = (\ell_2 - \ell)$

Now, $Y = \dfrac{T_1}{A} \times \dfrac{\ell}{\Delta\ell_1} = \dfrac{T_2}{A} \times \dfrac{\ell}{\Delta\ell_2}$

or $\dfrac{T_1}{\Delta\ell_1} = \dfrac{T_2}{\Delta\ell_2}$ or $\dfrac{T_1}{\ell_1 - \ell} = \dfrac{T_2}{\ell_2 - \ell}$

or $T_1\ell_2 - T_1\ell = T_2\ell_1 - \ell T_2$

or $\ell = \dfrac{T_2\ell_1 - T_1\ell_2}{T_2 - T_1}$

10. **(d)** For stress to be equal, $\dfrac{T_1}{A_1} = \dfrac{T_2}{A_2}$

$\therefore$ $\dfrac{T_1}{T_2} = \dfrac{A_1}{A_2} = \dfrac{1}{2}$

11. **(c)** Young's modulus of steel is highest.

12. **(c)** $Y = \dfrac{F}{\pi r^2} \times \dfrac{\ell}{\Delta\ell} \Rightarrow Y \propto \dfrac{1}{r^2}$

13. **(b)** Load holding capacity does not depends on length

14. (a) strain $= \frac{F}{\pi r^2 Y} = 10^{-8}$

15. (a) Strain $= \frac{F/A}{Y}$ & F = mg

$A = \pi r^2$

16. (c) The given graph does not obey Hooke's law. and there is no well defined plastic region. So the graph represents elastomers.

17. (d) $Y = \frac{F\ell}{A\Delta\ell}; \Rightarrow F = \frac{YA\Delta\ell}{\ell}$

$\frac{F_1}{F_2} = \frac{A_1}{A_2} \times \frac{l_2}{l_1} = \frac{r_1^2}{r_2^2} \times \frac{l_2}{l_1} = \frac{4}{1} \times \frac{1}{4} = 1:1.$

18. (c) According to Hooke's law

Stress $\propto$ strain i.e., $\frac{F}{A} \propto \frac{\Delta l}{l}$

$\Rightarrow$ For same F & l, $\Delta l \propto \frac{1}{A}$

19. (d) $l = \frac{FL}{AY} \Rightarrow l \propto \frac{1}{r^2}$ (F, L and Y are same)

$\frac{l_A}{l_B} = \left(\frac{r_B}{r_A}\right)^2 = \left(\frac{r_B}{2r_B}\right)^2 = \frac{1}{4} \Rightarrow l_A = 4l_B$

or $l_B = \frac{l_A}{4}$

20. (c)

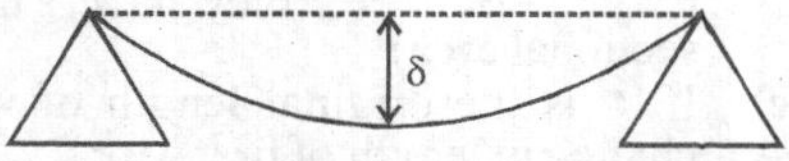

For a beam, the depression at the centre is given by, $\delta = \left(\frac{f\,L}{4Ybd^3}\right)$

[f, L, b, d are constants for a particular beam]

i.e. $\delta \propto \frac{1}{Y}$

21. (b) Initial length (circumference) of the ring $= 2\pi r$

Final length (circumference) of the ring $= 2\pi R$

Change in length $= 2\pi R - 2\pi r$

strain $= \frac{2\pi(R-r)}{2\pi r} = \frac{R-r}{r}$

Young's modulus $E = \frac{F/A}{l/L} = \frac{F/A}{(R-r)/r}$

$\therefore F = AE\left(\frac{R-r}{r}\right)$

22. (a) From the graph $l = 10^{-4}$m, $F = 20$ N

$A = 10^{-6}\text{m}^2$, L = 1m

$\therefore Y = \frac{FL}{Al} = \frac{20 \times 1}{10^{-6} \times 10^{-4}}$

$= 20 \times 10^{10} = 2 \times 10^{11}\text{N/m}^2$

23. (a) $Y = \frac{F}{A} \times \frac{L}{\Delta L} = \frac{100}{2 \times 10^{-4}} \times \frac{2}{1 \times 10^{-2}} = 10^8\text{ N/m}^2$

24. (b)

25. (c) Maximum possible strain = 0.2/100

$\therefore A = \frac{F}{Y \times \text{strain}} = \frac{10^4 \times 100}{(7 \times 10^9) \times 0.2}$

$= 7.1 \times 10^{-4}\text{ m}^2$

26. (a)

27. (c) Volume i.e. $(A \times l)$ must be constant

28. (c) Force

= weight suspended + weight of $\frac{3L}{4}$ of wire

$= W_1 + \frac{3W}{4}$

stress $= \frac{\text{force}}{\text{area}}$

29. (a) $Y = \frac{T/A}{\Delta\ell/\ell} \Rightarrow \Delta\ell = \frac{T \times \ell}{A \times Y} = \frac{T}{Y} \times \frac{\ell}{A}$

Here, $\frac{T}{Y}$ is constant. Therefore, $\Delta\ell \propto \frac{\ell}{A}$.

$\frac{\ell}{A}$ is largest in the first case.

30. (d)

31. (c) The deformation of a body on application of a force depends on the nature of the material and the magnitude of the applied force.

32. (a) To find the minimum diameter, and hence minimum cross-sectional area, we assume that the force $F = 400$ N brings us to the elastic limit. Then from the stress, $F/A = 379 \times 10^6$ Pa, we get

$A = \frac{400\text{N}}{379 \times 10^6\text{ Pa}} = 1.0554 \times 10^{-6}\text{m}^2.$

Then, $A = \frac{\pi D^2}{4}$

$D^2 = \frac{4A}{\pi} = \frac{4\,(1.0554 \times 10^{-6}\text{m}^2)}{\pi}$

$= 1.344 \times 10^{-6}$

and $D = \sqrt{1.344 \times 10^{-6}\text{m}^2}$

$= 1.16 \times 10^{-3}$ m = 1.16 mm

33. (a) $Y = \left(\frac{F}{A}\right) \times \frac{L}{\Delta L}$

$\Rightarrow \Delta L \left(\frac{F}{A}\right)\frac{L}{Y} = \frac{8 \times 10^8 \times 0.5}{2 \times 10^{11}} = 2\text{ mm}$

34. **(a)** Given $l_{steel} = l_{copper}$ and $A_{steel} = A_{copper}$

so, $\dfrac{\text{stress of steel}}{\text{stress of copper}} = \dfrac{F_{steel}}{A_{steel}} \times \dfrac{A_{copper}}{V_{copper}}$

$= \dfrac{F_{steel}}{F_{copper}} = 1$

35. **(a)** Ratio of radii $r_1 : r_2 = 1:2$

Ratio of area, $A_1 : A_2 = \pi r_1^2 : \pi r_2^2$

$A_1 : A_2 = 1 : 4$

Now, $\text{Stress}_1 : \text{Stress}_2 = 4 : 1$

So, $\text{Strain}_1 : \text{Strain}_2 = 4 : 1$

$\therefore \dfrac{l_1}{l_2} = \dfrac{4}{1} \Rightarrow 4l_2 = l_1 = 8$

$\therefore l_2 = 2$ mm

Increase in length of B is 2 mm.

36. **(b)** Let ℓ_0 be the unstretched length and ℓ_3 be the length under a tension of 9N. Then

$$Y = \frac{4\ell_0}{A(\ell_1 - \ell_0)} = \frac{5\ell_0}{A(\ell_2 - \ell_0)} = \frac{9\ell_0}{A(\ell_3 - \ell_0)}$$

These give

$$\frac{4}{\ell_1 - \ell_0} = \frac{5}{\ell_2 - \ell_0} \Rightarrow \ell_0 = 5\ell_1 - 4\ell_2$$

Further, $\dfrac{4}{\ell_1 - \ell_0} = \dfrac{9}{\ell_2 - \ell_0}$

Substituting the value of ℓ_0 and solving, we get $\ell_3 = 5\ell_2 - 4\ell_1$

37. **(c)** Young's modulus of elasticity is

$Y = \dfrac{F/A}{\Delta L / L}$ $\quad \therefore \Delta L = \dfrac{FL}{AY}$

So, $\Delta L \propto \dfrac{L}{A}$

$\therefore \dfrac{\Delta L_2}{\Delta L_1} = \dfrac{L_2}{L_1} \times \dfrac{A_1}{A_2} = \dfrac{2}{1} \times \dfrac{2}{1} = 4$

$\Delta L_2 = 4 \times \Delta L_1 = 4 \times 1 = 4$ cm

38. **(c)**

39. **(d)** $\dfrac{dF}{dx} = 0$ at $\ell/2$

40. **(b)** Breaking stress $= \dfrac{\text{Force}}{\text{area}}$

The breaking force will be its own weight.

$F = mg = V\rho g = \text{area} \times \ell\rho g$

Breaking stress $= 6 \times 10^6 = \dfrac{\text{area} \times \ell \times \rho g}{\text{area}}$

or $\ell = \dfrac{6 \times 10^6}{3 \times 10^3 \times 10} = 200$ m.

41. **(d)** $m = \dfrac{Y \times \text{strain} \times A}{2}$

$= \dfrac{2 \times 10^{11} \times 10^{-3} \times 3 \times 10^{-6}}{2} = 100$ kg

42. **(a)** Given: $F = 100$ kN $= 10^5$ N

$Y = 2 \times 10^{11}$ Nm^{-2}

$\ell_0 = 1.0$ m

radius $r = 10$ mm $= 10^{-2}$ m

From formula, $Y = \dfrac{\text{Stress}}{\text{Strain}}$

$\Rightarrow \text{Strain} = \dfrac{\text{Stress}}{Y} = \dfrac{F}{AY}$

$= \dfrac{10^5}{\pi r^2 Y} = \dfrac{10^5}{3.14 \times 10^{-4} \times 2 \times 10^{11}} = \dfrac{1}{628}$

Therefore % strain $= \dfrac{1}{628} \times 100 = 0.16\%$

43. **(b)** $\Delta\ell = \dfrac{F(L/2)}{AY} = \dfrac{(AL\rho g)(L/2)}{AY}$

$= \left(\dfrac{1}{2}\right) \rho g L^2 / Y$

44. **(c)** According to questions,

$\dfrac{\ell_s}{\ell_b} = a, \dfrac{r_s}{r_b} = b, \dfrac{y_s}{y_b} = c, \dfrac{\Delta\ell s}{\Delta\ell_b} = ?$

As, $y = \dfrac{F\ell}{A\Delta\ell} \Rightarrow \Delta\ell = \dfrac{F\ell}{Ay}$

$\Delta\ell_s = \dfrac{3mg\ell_s}{\pi r_s^2 . y_s}$ $\quad [\because F_s = (M + 2M)g]$

$\Delta\ell_b = \dfrac{2Mg\ell_b}{\pi r_b^2 . y_b}$ $\quad [\because F_b = 2Mg]$

$\therefore \dfrac{\Delta\ell_s}{\Delta\ell_b} = \dfrac{\dfrac{3Mg\ell_s}{\pi r_s^2 . y_s}}{\dfrac{2Mg.\ell_b}{\pi r_b^2 . y_b}} = \dfrac{3a}{2b^2 C}$

45. **(b)** Stress = 1 kg wt/mm^2 = 9.8 N/mm^2
= 9.8×10^6 N/m^2.

$Y = 1 \times 10^{11}$ N/m^2, $\dfrac{\Delta\ell}{\ell} \times 100 = ?$

$Y = \dfrac{\text{Stress}}{\text{Strain}} = \dfrac{\text{Stress}}{\Delta\ell/\ell}$

$\therefore \frac{\Delta \ell}{\ell} = \frac{\text{Stress}}{Y} = \frac{9.8 \times 10^6}{1 \times 10^{11}}$

$\frac{\Delta \ell}{\ell} \times 100 = 9.8 \times 10^{-11} \times 100 \times 10^6$

$= 9.8 \times 10^{-3} = 0.0098\ \%$

46. **(c)** Using the usual expression for the Young's modulus, the force constant for the wire can be written as $k = \frac{F}{\Delta l} = \frac{YA}{L}$ where the symbols have their usual meanings. Now the two wires together will have an effective force constant $\left[\frac{k_1 k_2}{k_1 + k_2}\right]$. Substituting the corresponding lengths and the Young's moduli we get the answer.

47. **(a)**

48. **(b)** $\Delta L = \frac{LF}{AY}$, where $L = 3\text{m}$,

$A = \pi(1.0 \times 10^{-3}\ \text{m})^2 = 3.14 \times 10^{-6}\ \text{m}^2$,

and since each wire supports one-quarter of the load,

$F = \frac{(50\ \text{kg})(9.8\ \text{m/s}^2)}{4} = 123\ \text{N}$

$\Delta L = \frac{(3\text{m})(123\text{N})}{(3.14 \times 10^{-6}\ \text{m}^2)(1.8 \times 10^{11}\ \text{N/m}^2)}$,

$= 65 \times 10^{-5}$ m or 0.65 mm

49. **(b)** $F = Y \times A \times \frac{l}{L} \Rightarrow F \propto r^2$ (Y, l and and L are constant)

If diameter is made four times then force required will be 16 times, i.e., 16×10^3 N

50. **(d)** $Y_c \times (\Delta L_c / L_c) = Y_s \times (\Delta L_s / L_s)$

$\Rightarrow 1 \times 10^{11} \times \left(\frac{1 \times 10^{-3}}{1}\right) = 2 \times 10^{11} \times \left(\frac{\Delta L_s}{0.5}\right)$

$\therefore \Delta L_s = \frac{0.5 \times 10^{-3}}{2} = 0.25$ mm

Therefore, total extension of the composite wire

$= \Delta L_c + \Delta L_s = 1\ \text{mm} + 0.25\ \text{mm} = 1.25\ \text{mm}$

51. **(b)** For brittle material, there is no yield point.

52. **(b)**

53. **(c)** Consider a small element dx of radius r,

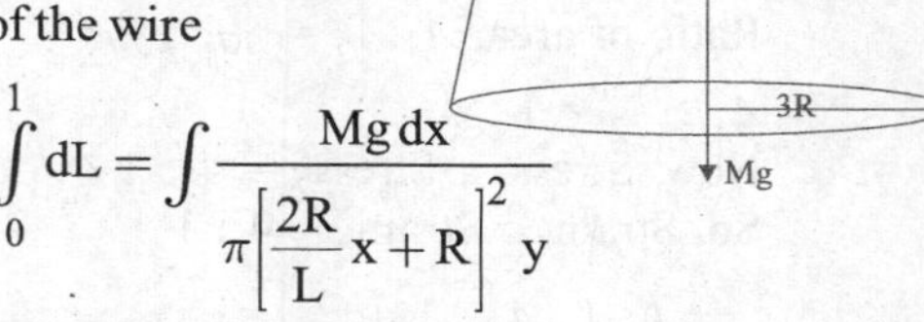

$r = \frac{2R}{L}x + R$

At equilibrium change in length of the wire

$\int_0^1 dL = \int \frac{Mg\,dx}{\pi\left[\frac{2R}{L}x + R\right]^2 y}$

Taking limit from 0 to L

$\Delta L = \frac{Mg}{\pi y}\left[\frac{-1}{\left[\frac{2Rx}{L} + R\right]_0^L} \times \frac{L}{2R}\right] = \frac{MgL}{3\pi R^2 y}$

The equilibrium extended length of wire $= L + \Delta L$

$= L + \frac{MgL}{3\pi R^2 Y} = L\left(1 + \frac{1}{3}\frac{Mg}{\pi Y R^2}\right)$

54. **(b)** Young's modulus $Y = 2G(1 - \sigma)$
G = modulus of rigidity and
σ = poissons's ratio

55. **(c)**

56. **(c)** Couple per unit angle of twist, $C = \frac{\pi \eta r^4}{2\ell}$

$\therefore$ Couple $\tau = C\theta = \frac{\pi \eta r^4 \theta}{2\ell}$

Here η, ℓ, C & τ are same. So, $r^4\theta$ = constant

$\therefore \frac{\theta_1}{\theta_2} = \left(\frac{r_2^4}{r_1^4}\right)$

57. **(b)**

58. **(c)** Bulk modulus $K = \frac{P}{\frac{\Delta V}{V}} = \frac{100}{0.01/100}$

$= 10^6$ atm $= 10^{11}$ N/m^2 $= 10^{12}$ dyne/cm^2

59. **(d)** If side of the cube is L. then $V = L^3$

$\Rightarrow \frac{dV}{V} = 3\frac{dL}{L}$

$\therefore$ % change in volume = 3 × (% change in length)
= 3 × 1% = 3%

$\therefore$ Bulk strain $\frac{\Delta V}{V} = 0.03$

60. **(b)** $K = \frac{F}{x} = \frac{4 \times 9.8}{2 \times 10^{-2}} = 19.6 \times 10^2$

Work done $= \frac{1}{2} 19.6 \times 10^2 \times (0.05)^2 = 2.45\,J$

61. (c) We have, $U = \frac{F^2}{2k}$

where $k = \frac{Yl}{A} = \frac{Yl}{\frac{1}{4}\pi d^2}$ $\quad \therefore \quad U \propto \frac{d^2}{l}$

62. (d) $\text{P.E.} = \frac{1}{2} \times \text{stress} \times \text{strain} \times \text{volume}$

or $(\text{P.E./volume}) = \frac{1}{2} \times (Y \times \text{strain})(\text{strain})$

$= \frac{1}{2} Y (\text{strain})^2 = \frac{1}{2} Y X^2$

63. (d) $\frac{\Delta r / r}{\Delta l / l} = 0.5 = \frac{1}{2}, \frac{\Delta r}{r} = \frac{1}{2}\frac{\Delta l}{l}$

64. (b) $\frac{dV}{V} = (1+2\sigma)\frac{dL}{L}$

$\frac{dV}{V} = 2 \times 2 \times 10^{-3} = 4 \times 10^{-3}$ $\quad \left[\because \sigma = 0.5 = \frac{1}{2}\right]$

$\therefore$ Percentage change in volume $= 4 \times 10^{-1}$ $= 0.4\%$

65. (b) $T_1 - T_2 = m\omega_0^2 (2l)$

$T_2 = m\omega_0^2 (3l)$

$\therefore T_1 = 5 m\omega_0^2 l.$

ω_0 $2l$ T_1 l T_2 Ⓐ AY m Ⓑ AY m

$\frac{\text{Energy}(A)}{\text{Energy}(B)} = \frac{(\text{stress})^2_A \text{ volume A}}{(\text{stress})^2_B \text{ volume B}}$

$= \frac{(5)^2}{(3)^2} \times \left(\frac{A \times 2l}{A \times l}\right) = \frac{50}{9}$

66. (c) $e = \frac{\Delta Y}{\ell} = \frac{1 \times 10^{-3}}{4}$

$u = \frac{e^2 Y}{2} = \left(\frac{10^{-3}}{4}\right)^2 \times \frac{2 \times 10^{11}}{2} = 0.075 \text{ J}.$

67. (a) $\text{U/volume} = \frac{1}{2} \text{Y} \times \text{strain}^2 = 3600 \text{ J m}^{-3}$

$[\text{Strain} = 0.06 \times 10^{-2}]$

68. (c) $\Delta w = \frac{1}{2}[F_2 \ell_2 - F_1 \ell_1]$

$= \frac{1}{2}\ [4 \times 10 \times 1 \times 10^{-3} - 2 \times 10 \times 0.6 \times 10^{-3}$ $= 14 \times 10^{-3} \text{ J}]$

69. (a) Volumetric strain, $\frac{\Delta V}{V} = 3 \times \frac{\Delta \ell}{\ell}$

$= 3 \times 0.01 = 0.03$

Bulk modulus, $B = \frac{\Delta P}{(\Delta V / V)} = \frac{5 \times 10^5}{0.03}$

$= 1.67 \times 10^7 \text{ N/m}^2$

70. (a) Shear stress $= \frac{F}{A}$

$\tan\theta = \frac{\text{shear stress}}{\text{shear modulus}}$

$\Rightarrow \tan\theta = \frac{4000}{(0.25)^2 \times 8 \times 10^{10}}$ $\quad \tan\theta = 8 \times 10^{-7}$

for small angle $\tan\theta = \theta = 8 \times 10^{-7}$ rad

71. (b)

72. (a) Let θ_1 and θ_2 are the angle of twist produced in cylinders A and B respectively.

Given, $\theta_1 + \theta_2 = \theta$ ...(i)

On being in series, the torque acts at their free ends are equal.

We have $\tau = \frac{\pi\eta r^4 \theta}{2\ell}$

$\therefore \frac{\pi\eta r^4 \theta_1}{2\ell} = \frac{\pi\eta (2r)^4 \theta_2}{2\ell} \Rightarrow \theta_1 = 16\theta_2$ ...(ii)

From (i) and (ii), we have

$\theta_1 = \frac{16}{17}\theta$

73. (c) The bulk modulus is defined as

$B = -\frac{\Delta p}{\Delta V / V}$, where the minus sign is inserted because ΔV is negative when Δp is positive.

$100\left|\frac{\Delta V}{V}\right| = 100\frac{\Delta p}{B} = 100\frac{345 \times 10^6}{138 \times 10^9} = 0.25\%$

74. (c) Bulk modulus is given by

$B = \frac{p}{\left(\frac{\Delta V}{V}\right)}$ or $\frac{\Delta V}{V} = \frac{p}{B}$

$3\frac{\Delta R}{R} = \frac{p}{B}$ (here, $\frac{\Delta R}{R}$ = fractional decreases in radius)

$\Rightarrow \frac{\Delta R}{R} = \frac{p}{3B}$

75. (a) $Y = 2 \times 10^{10} \text{ Nm}^{-2}; \frac{\Delta l}{l} = 0.01$

Work done per unit volume

$= \frac{1}{2} \times \text{stress} \times \text{strain} = \frac{1}{2} \times Y \times \text{strain}^2$

$= \frac{1}{2} Y \left(\frac{\Delta l}{l}\right)^2$

$\therefore \quad U = \frac{1}{2} \times 2 \times 10^{10} \times \left(\frac{1}{100}\right)^2 = U = 10^6 \text{ Jm}^{-3}$

76. **(b)** $W=\frac{1}{2}\times F\times l=\frac{1}{2}mgl$

$=\frac{1}{2}\times 10\times 10\times 1\times 10^{-3}=0.05J$

77. **(d)** Elastic energy $=\frac{1}{2}\times F\times x$

$F=200$ N, $x=1$ mm $=10^{-3}$ m

$\therefore\ E=\frac{1}{2}\times 200\times 1\times 10^{-3}=0.1$ J

78. **(c)** Here, $k_Q=\frac{k_p}{2}$

According to Hooke's law

$\therefore\quad F_p=-k_p x_p$

$F_Q=-k_Q x_Q \Rightarrow \frac{F_P}{F_Q}=\frac{k_P}{k_Q}\frac{x_P}{x_Q}$

$F_p=F_Q$ [Given]

$\therefore\quad \frac{x_P}{x_Q}=\frac{k_Q}{k_p}$ ...(i)

Energy stored in a spring is $U=\frac{1}{2}kx^2$

$\therefore\ \frac{U_P}{U_Q}=\frac{k_P x_P^2}{k_Q x_Q^2}=\frac{k_P}{k_Q}\times\frac{k_Q^2}{k_P^2}=\frac{1}{2}\quad\left[\because k_Q=\frac{k_P}{2}\right]$

$\Rightarrow\ U_P=\frac{U_Q}{2}=\frac{E}{2}\qquad [\because U_Q=E]$

79. **(d)** $B=-\frac{\Delta P}{\Delta V/V}=\frac{100\times 10^5}{(0.02/100)}$

$=50\times 10^9$ N/m²s

80. **(a)** $C_1=\frac{\pi\eta(r_2^4-r_1^4)}{2\ell},\ C_2=\frac{\pi\eta r^4}{2\ell}$

Initial volume = Final volume

$\therefore\ \pi[r_2^2-r_1^2]\ell\rho=\pi r^2\ell\rho$

$\Rightarrow r^2=r_2^2-r_1^2\Rightarrow r^2=(r_2+r_1)(r_2-r_1)$

$\Rightarrow r^2=(8.02+7.98)(8.02-7.98)$

$\Rightarrow r^2=16\times 0.04=0.64$ cm $\Rightarrow r=0.8$ cm

$\therefore\ \frac{C_1}{C_2}=\frac{r_2^4-r_1^4}{r^4}=\frac{[8.02]^4-[7.98]^4}{[0.8]^4}$

9 Mechanical Properties of Fluids

1. The total weight of a piece of wood is 6 kg. In the floating state in water its $\frac{1}{3}$ part remains inside the water. On this floating piece of wood what maximum weight is to be put such that the whole of the piece of wood is to be drowned in the water?
(a) 15 kg (b) 14 kg (c) 10 kg (d) 12 kg

2. A vessel contains oil (density = 0.8 gm/cm^3) over mercury (density = 13.6 gm/cm^3). A homogeneous sphere floats with half of its volume immersed in mercury and the other half in oil. The density of the material of the sphere in gm/cm^3 is
(a) 3.3 (b) 6.4 (c) 7.2 (d) 12.8

3. In a hydraulic lift, compressed air exerts a force F_1 on a small piston having a radius of 5 cm. This pressure is transmitted to a second piston of radius 15 cm. If the mass of the load to be lifted is 1350 kg, find the value of F_1 ? The pressure necessary to accomplish this task is
(a) 1.4×10^5 Pa (b) 2×10^5 Pa
(c) 1.9×10^5 Pa (d) 1.9 Pa

4. The force acting on a window of area 50 cm × 50 cm of a submarine at a depth of 2000 m in an ocean, interior of which is maintained at sea level atmospheric pressure is
(Density of sea water = 10^3 kgm^{-3}, g = 10 ms^{-2})
(a) 10^6 N (b) 5×10^5 N
(c) 25×10^6 N (d) 5×10^6 N

5. A concrete sphere of radius *R* has a cavity of radius *r* which is packed with sawdust. The specific gravities of concrete and sawdust are respectively 2.4 and 0.3 for this sphere to float with its entire volume submerged under water. Ratio of mass of concrete to mass of sawdust will be
(a) 8 (b) 4 (c) 3 (d) zero

6. A cubical box is to be constructed with iron sheets 1 mm in thickness. What can be the minimum value of the external edge so that the cube does not sink in water?
[δ_{iron} = 8000 kg/m^3, δ_{water} = 1000 kg/m^3]
(a) 4.8 cm (b) 5.8 cm (c) 6.9 cm (d) 7.3 cm

7. The pressure at the bottom of a tank containing a liquid does not depend on
(a) acceleration due to gravity
(b) height of the liquid column
(c) area of the bottom surface
(d) nature of the liquid

8. A body B is capable of remaining stationary inside a liquid at the position shown in Fig. (a). If the whole system is gently placed on smooth inclined plane (Fig (b)) and is allowed to slide down, then ($0 < \theta < 90°$). The body will

Tricky

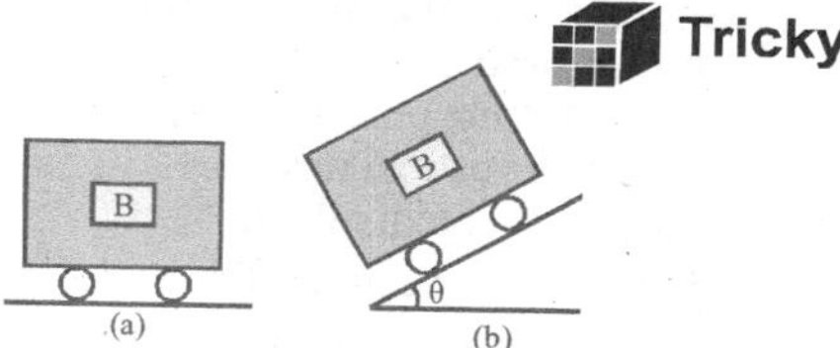

(a) move up (relative to liquid)
(b) move down (relative to liquid)
(c) remain stationary (relative to liquid)
(d) move up for some inclination θ and will move down for another inclination θ

9. A hollow sphere of mass M = 50 kg and radius $r = \left(\frac{3}{40\pi}\right)^{1/3}$ m is immersed in a tank of water (density $\rho_w = 10^3$ kg/m^3). The sphere is tied to the bottom of a tank by two wires A and B as shown. Tension in wire A is (g = 10 m/s^2)

Critical Thinking

(a) $125\sqrt{2}$N
(b) 125 N
(c) $250\sqrt{2}$N
(d) 250 N

10. In rising from the bottom of a lake, to the top, the temperature of an air bubble remains unchanged, but its diameter gets doubled. If *h* is the barometric height (expressed in m of mercury of relative density ρ) at the surface of the lake, the depth of the lake is
(a) 8ρhm (b) 7ρhm (c) 9ρhm (d) 12ρhm

11. A hemispherical portion of radius R is removed from the bottom of a cylinder of radius R. The volume of the remaining cylinder is V and its mass M. It is suspended by a string in a liquid of density ρ where it stays vertical. The upper surface of the cylinder is at a depth h below the liquid surface. The force on the bottom of the cylinder by the liquid is

(a) Mg
(b) $Mg - V\rho g$
(c) $Mg + \pi R^2 h\rho g$
(d) $\rho g(V + \pi R^2 h)$

12. A thin tube sealed at both ends is 100 cm long. It lies horizontally, the middle 20 cm containing mercury and two equal ends containing air at standard atmospheric pressure. If the tube is now turned to a vertical position, by what amount will the mercury be displaced ?
(Given : cross-section of the tube can be assumed to be uniform)

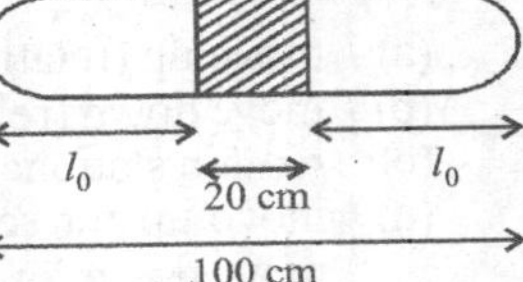

(a) 2.95 cm
(b) 5.18 cm
(c) 8.65 cm
(d) 0.0 cm

13. A right circular cone of density ρ, floats just immersed with its vertex downwards in a vessel containing two liquids of densities σ_1 and σ_2 respectively, the planes of separation of the two liquids cuts off from the axis of the cone a fraction z of its length. Find z. Toughnut

(a) $\left(\frac{\rho+\sigma_2}{\sigma_1+\sigma_2}\right)^{1/3}$ (b) $\left(\frac{\rho-\sigma_2}{\sigma_1-\sigma_2}\right)^{1/3}$
(c) $\left(\frac{\rho-\sigma_2}{\sigma_1+\sigma_2}\right)^{1/2}$ (d) $\left(\frac{\rho-\sigma_2}{\sigma_1-\sigma_2}\right)$

14. A sphere of solid material of specific gravity 8 has a concentric spherical cavity and just sinks in water. The ratio of radius of cavity to that of outer radius of the sphere must be

(a) $\frac{7^{1/3}}{2}$ (b) $\frac{5^{1/3}}{2}$ (c) $\frac{9^{1/3}}{2}$ (d) $\frac{3^{1/3}}{2}$

15. A vessel containing water is given a constant acceleration 'a' towards the right along a straight horizontal path. Which of the following diagrams represents the surface of the liquid ? Tricky

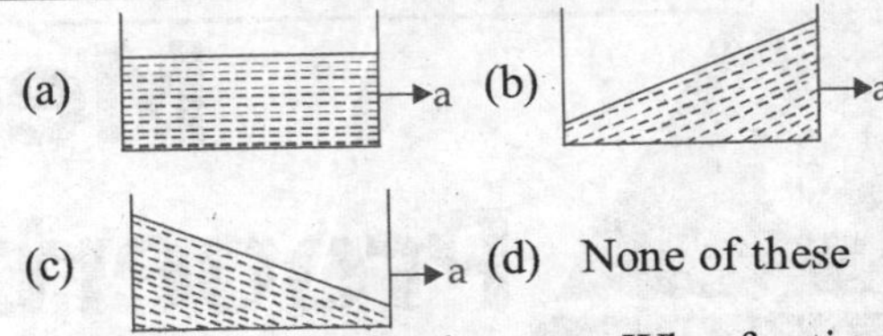

(d) None of these

16. An iceberg is floating in ocean. What fraction of its volume is above the water ? (Given : density of ice = 900kg/m^3 and density of ocean water = 1030 kg/m^3)

(a) $\frac{90}{103}$ (b) $\frac{13}{103}$ (c) $\frac{10}{103}$ (d) $\frac{1}{103}$

17. A cone full of water, is placed on its side on a horizontal table, the thrust on its base is x times the weight of the contained fluid, where 2α is the vertical angle of the cone. Find the value of x.
(a) $3 \cos \alpha$ (b) $3 \sin \alpha$ (c) $2 \sin \alpha$ (d) $2 \cos \alpha$

18. What is the absolute pressure of the gas above the liquid surface in the tank shown in fig. Density of oil = 820 kg/m^3, density of mercury = 13.6×10^3 kg/m^3. Given 1 atmospheric pressure = 1.01×10^5 N/m^2. Critical Thinking

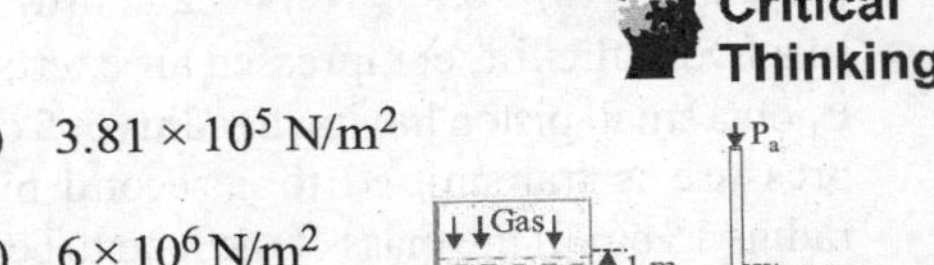

(a) 3.81×10^5 N/m^2
(b) 6×10^6 N/m^2
(c) 5×10^7 N/m^2
(d) 4.6×10^2 N/m^2

19. Figure shows a U-tube of uniform cross-sectional area A, accelerated with acceleration a as shown. If d is the separation between the limbs, then what is the difference in the levels of the liquid in the U-tube is

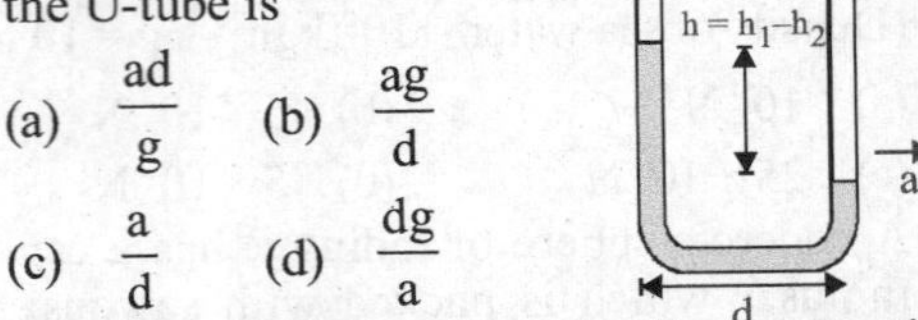

(a) $\frac{ad}{g}$ (b) $\frac{ag}{d}$
(c) $\frac{a}{d}$ (d) $\frac{dg}{a}$

20. The two thigh bones, each of cross-sectional area 10 cm^2 support the upper part of a human body of mass 40 kg. Estimate the average pressure sustained by the bones. Take g = 10 m/s^2
(a) 2×10^5 N/m^2 (b) 5×10^4 N/m^2
(c) 2×10^7 N/m^2 (d) 3×10^6 N/m^2

21. A hemispherical bowl just floats without sinking in a liquid of density 1.2×10^3 kg/m^3. If outer diameter and the density of the bowl are 1 m and 2×10^4 kg/m^3 respectively then the inner diameter of the bowl will be Critical Thinking

(a) 0.94m (b) 0.97m (c) 0.98m (d) 0.99m

22. A uniform rod of density ρ is placed in a wide tank containing a liquid of density $\rho_0 (\rho_0 > \rho)$. The depth of liquid in the tank is half the length of the rod. The rod is in equilibrium, with its lower end resting on the bottom of the tank. In this position the rod makes an angle θ with the horizontal, then: Toughnut

(a) $\sin\theta = \frac{1}{2}\sqrt{\rho_0/\rho}$ (b) $\sin\theta = \frac{1}{2}\cdot\frac{\rho_0}{\rho}$

(c) $\sin\theta = \sqrt{\rho/\rho_0}$ (d) $\sin\theta = \rho_0/\rho$

23. A solid sphere of density $\eta\,(>1)$ times lighter than water is suspended in a water tank by a string tied to its base as shown in fig. If the mass of the sphere is m, then the tension in the string is given by

(a) $\left(\frac{\eta-1}{\eta}\right)mg$

(b) ηmg

(c) $\frac{mg}{\eta-1}$

(d) $(\eta-1)mg$

24. A boy can reduce the pressure in his lungs to 750 mm of mercury. Using a straw he can drink water from a glass upto the maximum depth of (atmospheric pressure = 760 mm of mercury; density of mercury = 13.6 gcm^{-3})
(a) 13.6 cm (b) 9.8 cm (c) 10 cm (d) 76 cm

25. A wooden block, with a coin placed on its top, floats in water as shown in fig. the distance l and h are shown there. After some time the coin falls into the water. Then Tricky

(a) ℓ decreases and h increases
(b) ℓ increases and h decreases
(c) both ℓ and h increases
(d) both ℓ and h decreases

Coin

26. The density ρ of water of bulk modulus B at a depth y in the ocean is related to the density at surface ρ_0 by the relation

(a) $\rho = \rho_0\left[1 - \frac{\rho_0 gy}{B}\right]$ (b) $\rho = \rho_0\left[1 + \frac{\rho_0 gy}{B}\right]$

(c) $\rho = \rho_0\left[1 + \frac{B}{\rho_0 hgy}\right]$ (d) $\rho = \rho_0\left[1 - \frac{B}{\rho_0 hgy}\right]$

27. Two non-mixing liquids of densities ρ and nρ (n > 1) are put in a container. The height of each liquid is h. A solid cylinder of length L and density d is put in this container. The cylinder floats with its axis vertical and length pL(p < 1) in the denser liquid. The density d is equal to :
(a) $\{1+(n+1)p\}\rho$ (b) $\{2+(n+1)p\}\rho$
(c) $\{2+(n-1)p\}\rho$ (d) $\{1+(n-1)p\}\rho$

28. Two wooden blocks A and B float in a liquid of density ρ_L as shown. The distance L and H are shown. After some time, block B falls into the liquid, so that L decreases and H increases. If density of block B is ρ_B, find the correct option.

(a) $\rho_L = \rho_B$

(b) $\rho_L > \rho_B$

(c) $\rho_L < \rho_B$

(d) unpredicatbale

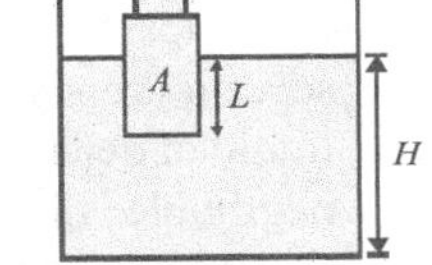

29. A homogeneous solid cylinder of length L ($L < H/2$), cross-sectional area $A/5$ is immersed such that it floats with its axis vertical at the liquid-liquid interface with length $L/4$ in the denser liquid as shown in the figure. The lower density liquid is open to atmosphere having pressure P_0. Then density D of solid is given by

(a) $\frac{5}{4}d$

(b) $\frac{d}{4}$

(c) $4d$

(d) $\frac{d}{5}$

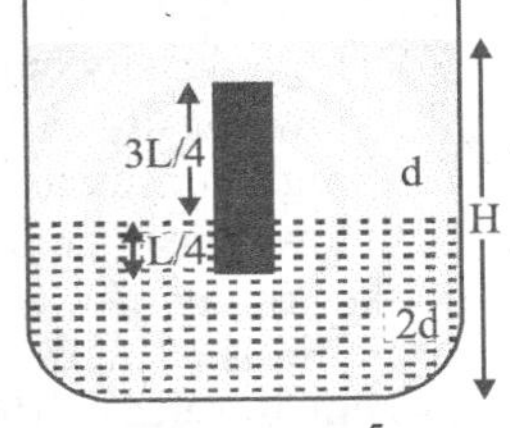

30. The compressibility of water is 4×10^{-5} per unit atmospheric pressure. The decrease in volume of 100 cm^3 of water under a pressure of 100 atmosphere will be
(a) 0.4 cm^3 (b) 4×10^{-5} cm^3
(c) 0.025 cm^3 (d) 0.004 cm^3

31. A hollow wooden cylinder of height h, inner radius R and outer radius $2R$ is placed in a cylindrical container of radius $3R$. When water is poured into the container, the minimum height H of the container for which cylinder can float inside freely is

(a) $\frac{h\rho_{water}}{\rho_{water}+\rho_{wood}}$

(b) $\frac{h\rho_{wood}}{\rho_{water}}$ (c) h (d) $\frac{h^2}{R}$

32. A balloon of volume V, contains a gas whose density is to that of the air at the earth's surface as 1 : 15. If the envelope of the balloon be of weight w but of negligible volume, find the acceleration with which it will begin to ascend.

(a) $\left(\frac{7Vg\sigma - w}{Vg\sigma + w}\right)\times g$ (b) $\left(\frac{2Vg\sigma - w}{Vg\sigma + w}\right)\times g$

(c) $\left(\frac{14Vg\sigma - w}{Vg\sigma + w}\right)\times g$ (d) $\left(\frac{14Vg\sigma + w}{Vg\sigma - w}\right)\times g$

33. A uniform wooden stick of length L, cross-section area A and density d is immersed in a liquid of density $4d$. A small body of mass m and negligible volume is attached at the lower end of the rod so that the stick floats vertically in stable equilibrium then

Critical Thinking

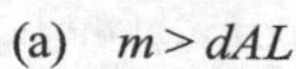
(a) $m > dAL$

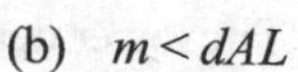
(b) $m < dAL$

(c) $m < dAL/2$

(d) $m < dAL/4$

34. Streamline flow is more likely for liquids with

(a) high density and low viscosity
(b) low density and high viscosity
(c) high density and high viscosity
(d) low density and low viscosity

35. In the arrangement as shown, $m_B = 3m$, density of liquid is r and density of block B is 2ρ. The system is released from rest so that block B moves up when in liquid and moves down when out of liquid with the same acceleration. Find the mass of block A.

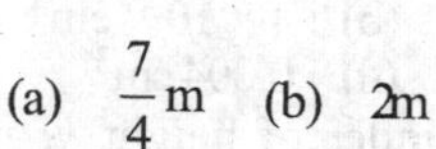
(a) $\frac{7}{4}$m (b) 2m

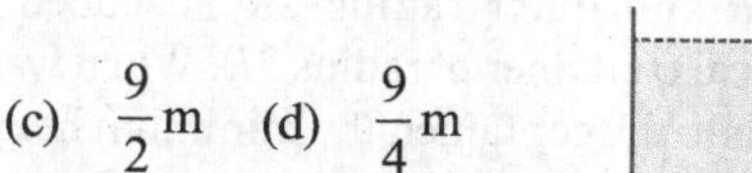
(c) $\frac{9}{2}$m (d) $\frac{9}{4}$m

36. In Bernoulli's theorem which of the following is conserved?

(a) Mass (b) Linear momentum
(c) Energy (d) Angular momentum

37. Water flows through a frictionless tube with a varying cross-section as shown in fig. Pressure P at points along the axis is represented by

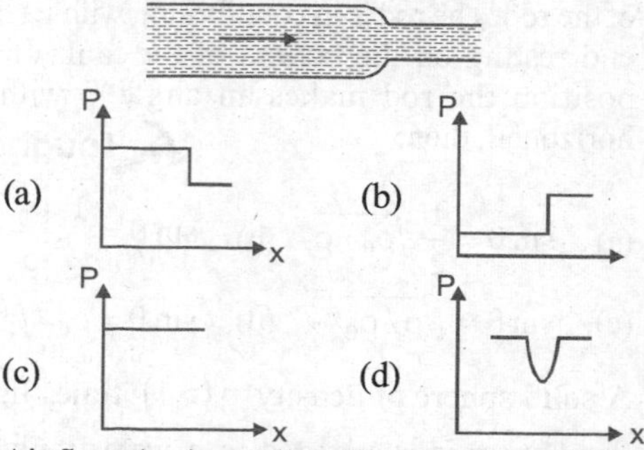

38. Air flows horizontally with a speed v = 106 km/hr. A house has plane roof of area A = 20m^2. The magnitude of aerodynamic lift of the roof is

(a) 1.127×10^4 N (b) 5.0×10^4 N
(c) 1.127×10^5 N (d) 3.127×10^4 N

39. A wind with speed 40 m/s blows parallel to the roof of a house. The area of the roof is 250 m^2. Assuming that the pressure inside the house is atmospheric pressure, the force exerted by the wind on the roof and the direction of the force will be (ρ_{air} = 1.2 kg/m^3)

(a) 4.8×10^5 N, upwards
(b) 2.4×10^5 N, upwards
(c) 2.4×10^5 N, downwards
(d) 4.8×10^5 N, downwards

40. In the figure shown the velocity and pressure of theliquid at the cross section (2) are given by (If P_0 is the atmospheric pressure).

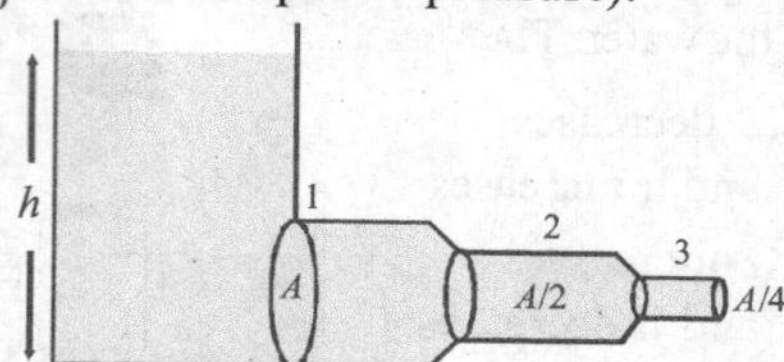

(a) $\sqrt{2hg}, P_0+\frac{\rho hg}{2}$ (b) $\sqrt{hg}, P_0+\frac{\rho hg}{2}$

(c) $\sqrt{\frac{hg}{2}}, P_0+\frac{3\rho hg}{4}$ (d) $\sqrt{\frac{hg}{2}}, P_0-\frac{\rho hg}{4}$

41. A cylinderical vessel contains a liquid of density ρ filled upto a height h. The uper surface of the liquid is in contact with a pistion of mass m and area of cross-section A. A small hole is drilled at the bottom of the vessel. (Neglect the viscous effects) The speed with which the liquid comes out of the hoe is :

(a) $\sqrt{2gh}$ (b) $\sqrt{2g\left(h+\frac{m}{\rho A}\right)}$

(c) $\sqrt{g\left(h+\frac{m}{\rho A}\right)}$ (d) $\sqrt{g\left(h+\frac{2m}{\rho A}\right)}$

42. A wide vessel with a small hole at the bottom is filled with water (density ρ_1, height h_1) and kerosene (density ρ_2, height h_2). Neglecting viscosity effects, the speed with which water flows out is :

(a) $[2g(h_1+h_2)]^{1/2}$
(b) $[2g(h_1\rho_1+h_2\rho_2)]^{1/2}$
(c) $[2g(h_1+h_2(\rho_2/\rho_1))]^{1/2}$
(d) $[2g\,(h_1+h_2(\rho_1/\rho_2))]^{1/2}$

43. The cylindrical tube of a spray pump has radius, R, one end of which has n fine holes, each of radius r. If the speed of the liquid in the tube is V, the speed of the ejection of the liquid through the holes is :

(a) $\frac{VR^2}{nr^2}$ (b) $\frac{VR^2}{n^3r^2}$ (c) $\frac{V^2R}{nr}$ (d) $\frac{VR^2}{n^2r^2}$

44. A fluid is in streamline flow across a horizontal pipe of variable area of cross section. For this which of the following statements is correct?

(a) The velocity is minimum at the narrowest part of the pipe and the pressure is minimum at the widest part of the pipe
(b) The velocity is maximum at the narrowest part of the pipe and pressure is maximum at the widest part of the pipe
(c) Velocity and pressure both are maximum at the narrowest part of the pipe
(d) Velocity and pressure both are maximum at the widest part of the pipe

45. Oil is filled in a cylindrical container upto height 4m. A small hole of area 'p' is punched in the wall of the container at a height 1.52m from the bottom. The cross sectional area of the container is Q. If $\frac{p}{q}=0.1$ then v is (where v is the velocity of oil coming out of the hole)

(a) $5\sqrt{2}$ (b) $6\sqrt{3}$ (c) $8\sqrt{2}$ (d) $7\sqrt{5}$

46. A broad vessel with water stands on a smooth surface. The level of the water in the vessel is h. The vessel together with the water weighs G. The side wall of the vessel has at the bottom a plugged hole (with rounded edges) with an area A. At what coefficient of friction between the bottom and the surface will the vessel begin to move if the plug is removed ?

(a) $\frac{\rho ghA}{G}$ (b) $\frac{2\rho ghA}{G}$

(c) $\frac{\rho ghA}{2G}$ (d) $\frac{2\rho ghA}{3G}$

47. For a fluid which is flowing steadily, the level in the vertical tubes is best represented by

Tricky

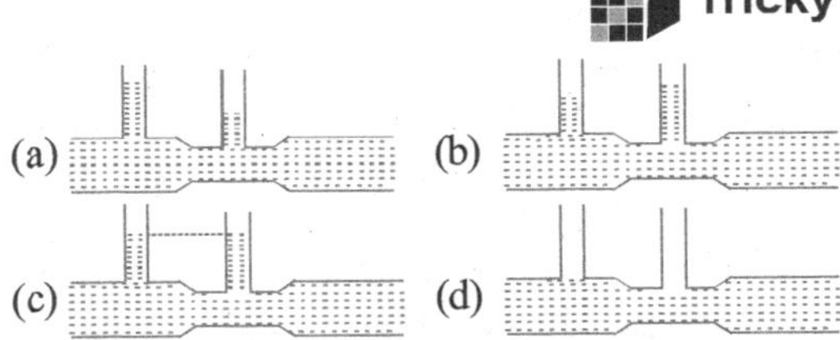

48. For the arrangement shown in the figure, find the time interval in seconds after which the water jet ceases to cross the wall. Area of the cross section of the tank $A=\sqrt{5}\text{m}^2$ and area of the orifice $a=4\text{ cm}^2$. [Assume that the container remaining fixed]

Toughnut

(a) 1000s
(b) 2000s
(c) 1500s
(d) 500s

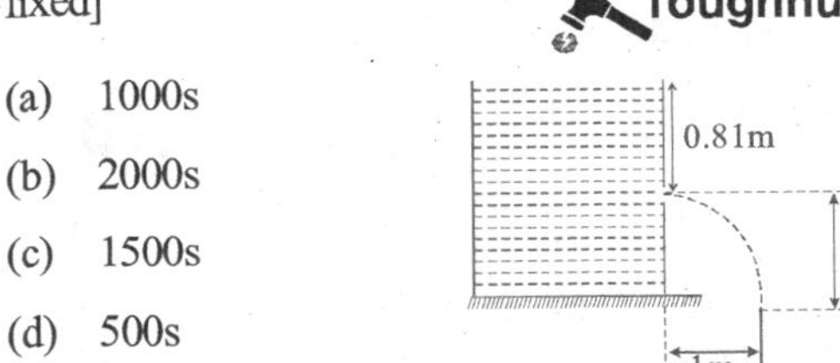

49. Air of density 1.2 kg m^{-3} is blowing across the horizontal wings of an aeroplane in such a way that its speeds above and below the wings are 150 ms^{-1} and 100 ms^{-1}, respectively. The pressure difference between the upper and lower sides of the wings, is

(a) 60 Nm^{-2} (b) 180 Nm^{-2}
(c) 7500 Nm^{-2} (d) 12500 Nm^{-2}

50. In the figure shown, a light container is kept on a horizontal rough surface of coefficient of friction $\mu=\frac{Sh}{V}$. A very small hole of area S is made at depth h. Water of volume V is filled in the container. The friction is not sufficient to keep the container at rest. The acceleration of the container initially is

Toughnut

(a) $\frac{V}{Sh}g$
(b) g
(c) zero
(d) $\frac{Sh}{V}g$

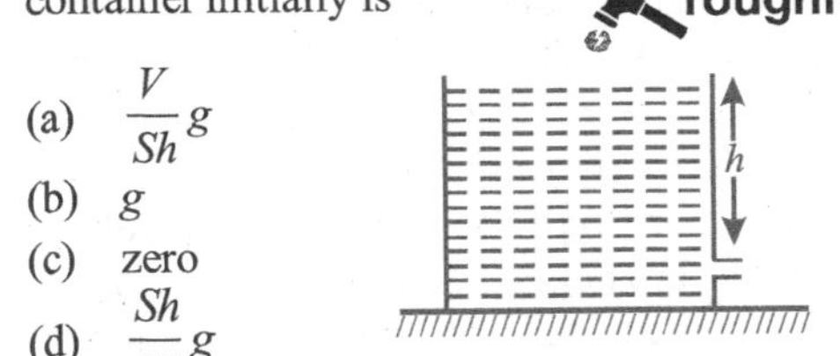

51. An incompressible liquid flows through a horizontal tube as shown in the figure. Then the velocity 'v' of the fluid is

(a) 3.0 m/s
(b) 1.5 m/s
(c) 1.0 m/s
(d) 2.25 m/s

52. Figure shows a liquid flowing through a tube at the rate of 0.1 m^3/s. The tube is branched into two semicircular tubes of cross - sectional area A/3 and 2A/3. The velocity of liquid at Q is (the cross-section of the main tube is $A = 10^{-2}$ m^2 and $v_P = 20$ m/s)

(a) 5 m/s
(b) 30 m/s
(c) 35 m/s
(d) None of these

53. A large tank filled with water to a height 'h' is to be emptied through a small hole at the bottom. The ratio of time taken for the level of water to fall from h to $\frac{h}{2}$ and $\frac{h}{2}$ to zero is **Tricky**

(a) $\sqrt{2}$ (b) $\frac{1}{\sqrt{2}}$ (c) $\sqrt{2}-1$ (d) $\frac{1}{\sqrt{2}-1}$

54. A square box of water has a small hole located in one of the bottom corner. When the box is full and sitting on a level surface, complete opening of the hole results in a flow of water with a speed v_0, as shown in figure. When the box is half empty, it is tilted by 45° so that the hole is at the lowest point. Now the water will flow out with a speed of

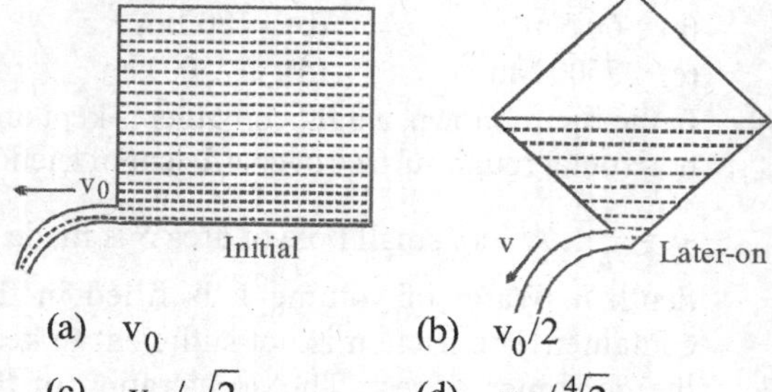

(a) v_0 (b) $v_0/2$
(c) $v_0/\sqrt{2}$ (d) $v_0/\sqrt[4]{2}$

55. A horizontal tube has different cross sections at points A and B. The area of cross section are a_1 and a_2, respectively, and pressures at these points are $p_1 = \rho g h_1$ and $p_2 = \rho g h_2$, where ρ is the density of liquid flowing in the tube and h_1 and h_2 are heights of liquid columns in vertical tubes connected at A and B. If $h_1 - h_2 = h$, then the flow rate of the liquid in the horizontal tube is **Critical Thinking**

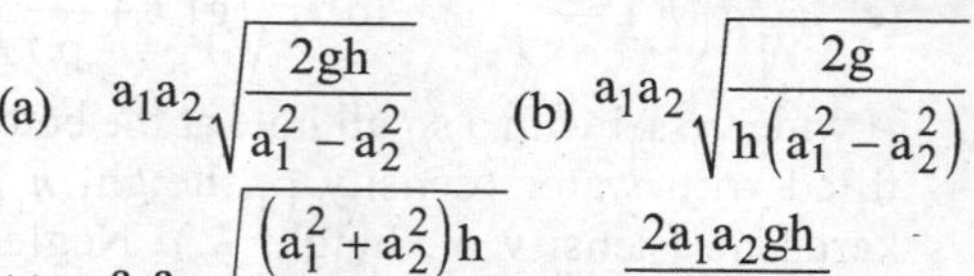

(a) $a_1a_2\sqrt{\frac{2gh}{a_1^2-a_2^2}}$ (b) $a_1a_2\sqrt{\frac{2g}{h(a_1^2-a_2^2)}}$

(c) $a_1a_2\sqrt{\frac{(a_1^2+a_2^2)h}{2g(a_1^2-a_2^2)}}$ (d) $\frac{2a_1a_2gh}{\sqrt{a_1^2-a_2^2}}$

56. Water is filled in a cylindrical container to a height of 3m. The ratio of the cross-sectional area of the orifice and the beaker is 0.1. The square of the speed of the liquid coming out from the orifice is ($g = 10$ m/s^2)

(a) 50 m^2/s^2
(b) 50.5 m^2/s^2
(c) 51 m^2/s^2
(d) 52 m^2/s^2

57. Water is flowing through a horizontal tube having cross-sectional areas of its two ends being A and A' such that the ratio A/A' is 5. If the pressure difference of water between the two ends is 3×10^5 N m^{-2}, the velocity of water with which it enters the tube will be (neglect gravity effects)

(a) 5 m s^{-1} (b) 10 m s^{-1}
(c) 25 m s^{-1} (d) $50\sqrt{10}$ ms^{-1}

58. There are two identical small holes P and Q of area of cross-section a on the opposite sides of a tank containing a liquid of density ρ. The difference in height between the holes is h. Tank is resting on a smooth horizontal surface. Horizontal force which will has to be applied on the tank to keep it in equilibrium is

(a) $gh\rho a$ (b) $\frac{2gh}{\rho a}$

(c) $2\rho agh$ (d) $\frac{\rho gh}{a}$

59. In a cylindrical water tank, there are two small holes A and B on the wall at a depth of h_1, from the surface of water and at a height of h_2 from the bottom of water tank. Surface of water is at height of h_2 from the bottom of water tank. Surface of water is at heigh H from the bottom of water tank. Water coming out from both holes strikes the ground at the same point S. Find the ratio of h_1 and h_2

(a) Depends on H

(b) 1 : 1

(c) 2 : 2

(d) 1 : 2

60. Equal volumes of two immiscible liquids of densities ρ and 2ρ are filled in a vessel as shown in figure. Two small holes are punched at depth $h/2$ and $3h/2$ from the surface of lighter liquid. If v_1 and v_2 are the velocities of a flux at these two holes, then v_1/v_2 is

(a) $\frac{1}{2\sqrt{2}}$

(b) $\frac{1}{2}$

(c) $\frac{1}{4}$

(d) $\frac{1}{\sqrt{2}}$

61. A tank with a small hole at the bottom has been filled with water and kerosene (specific gravity 0.8). The height of water is 3m and that of kerosene 2m. When the hole is opened the velocity of fluid coming out from it is nearly: (take g =10 ms^{-2} and density of water = 10^3 kg m^{-3})

(a) 10.7 ms^{-1} (b) 9.8 ms^{-1}

(c) 8.5 ms^{-1} (d) 7.6 ms^{-1}

62. If it takes 5 minutes to fill a 15 litre bucket from a water tap of diameter $\frac{2}{\sqrt{\pi}}$ cm then the Reynold's number for the flow is close to : (density of water = 10^3 kg/m^3 and viscosity of water = 10^{-3} Pa.s)

(a) 1100 (b) 11,000 (c) 550 (d) 5500

63. Water is flowing on a horizontal fixed surface, such that its flow velocity varies with y (vertical direction) as

$v = k\left(\frac{2y^2}{a^2} - \frac{y^3}{a^3}\right)$. If coefficient of viscosity for water is η, what will be shear stress between layers of water at $y = a$.

Critical Thinking

(a) $\frac{\eta k}{a}$ (b) $\frac{\eta}{ka}$

(c) $\frac{\eta a}{k}$ (d) None of these

64. A container filled with viscous liquid is moving vertically downwards with constant speed $3v_0$. At the instant shown, a sphere of radius r is moving vertically downwards (in liquid) has speed v_0. The coefficient of viscosity is η. There is no relative motion between the liquid and the container. Then at the shown instant, the magnitude of viscous force acting on sphere is

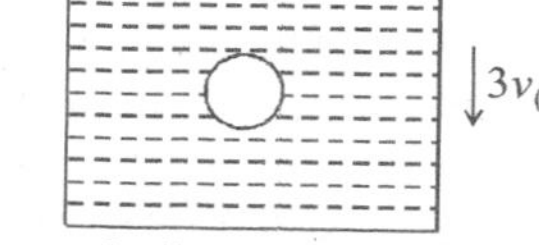

(a) $6\pi\eta r v_0$

(b) $12\pi\eta r v_0$

(c) $18\pi\eta r v_0$

(d) $24\pi\eta r v_0$

65. As the temperature of a liquid is raised, the coefficient of viscosity

(a) decreases

(b) increases

(c) remains the same

(d) may increase or decrease depending on the nature of liquid

66. After terminal velocity is reached, the acceleration of a body falling through a fluid is

(a) equal to g (b) zero

(c) less than g (d) greater than g

67. An air bubble of radius 1 cm rises with terminal velocity 0.21 cm/s in liquid column. If the density of liquid is 1.47 × 10^3 kg/m^3. Then the value of coefficient of viscosity of liquid ignoring the density of air, will be

Critical Thinking

(a) 1.71 × 10^4 poise (b) 1.82 × 10^4 poise

(c) 1.78 × 10^4 poise (d) 1.52 × 10^4 poise

68. The relative velocity of two parallel layers of water is 8 cm/sec. If the perpendicular distance between the layers is 0.1 cm. Then velocity gradient will be

(a) 80/sec (b) 60 /sec (c) 50/sec (d) 40/sec

69. If a ball of steel (density ρ = 7.8 g cm^{-3}) attains a terminal velocity of 10 cms^{-1} when falling in a tank of water (coefficient of viscosity η_{water} = 8.5 × 10^{-4} $Pa-s$) then its terminal velocity in glycerine ($\rho = 12\ gcm^{-3}$, $\eta = 13.2\ Pa-s$) would be nearly

(a) 1.6 × 10^{-5} cms^{-1} (b) 6.25 × 10^{-4} cms^{-1}

(c) 6.45 × 10^{-4} cms^{-1} (d) 1.5 × 10^{-5} cms^{-1}

70. What is the velocity v of a metallic ball of radius r falling in a tank of liquid at the instant when its acceleration is one -half that of a freely falling body ? (The densities of metal and of liquid are ρ and σ respectively, and the viscosity of the liquid is η).

(a) $\frac{r^2 g}{9\eta}(\rho - 2\sigma)$ (b) $\frac{r^2 g}{9\eta}(2\rho - \sigma)$

(c) $\frac{r^2 g}{9\eta}(\rho - \sigma)$ (d) $\frac{2r^2 g}{9\eta}(\rho - \sigma)$

71. A spherical ball of iron of radius 2 mm is falling through a column of glycerine. If densities of glycerine and iron are respectively 1.3×10^3 kg/m^3 and 8×10^3 kg/m^3. η for glycerine = 0.83 Nm^{-2} sec, then the terminal velocity is
(a) 0.7 m/s (b) 0.07 m/s
(c) 0.007 m/s (d) 0.0007 m/s

72. A small spherical ball falling through a viscous medium of negligible density has terminal velocity v. Another ball of the same mass but of radius twice that of the earlier falling through the same viscous medium will have terminal velocity
(a) v (b) v/4 (c) v/2 (d) 2v

73. The velocity of water in a river is 18 km/hr near the surface. If the river is 5 m deep, find the shearing stress between the horizontal layers of water. The co-efficient of viscosity of water = 10^{-2} poise.

(a) 10^{-1} N/m^2 (b) 10^{-2} N/m^2
(c) 10^{-3} N/m^2 (d) 10^{-4} N/m^2

74. A rain drop of radius 0.3mm falling vertically downwards in air has a terminal velocity of 1m/s. The viscosity of air is 18×10^{-5} poise. The viscous force on the drop is
(a) 101.73×10^{-4} dyne (b) 101.73×10^{-5} dyne
(c) 16.95×10^{-5} dyne (d) 16.95×10^{-4} dyne

75. The average mass of rain drops is 3.0×10^{-5} kg and their average terminal velocity is 9 m/s. Calculate the energy transferred by rain to each square metre of the surface at a place which receives 100 cm of rain in a year.

Critical Thinking

(a) 3.5×10^5 J (b) 4.05×10^4 J
(c) 3.0×10^5 J (d) 9.0×10^4 J

76. Which of the following is the velocity time graph of a small spherical body falling through a long columns of a viscous liquid? Tricky

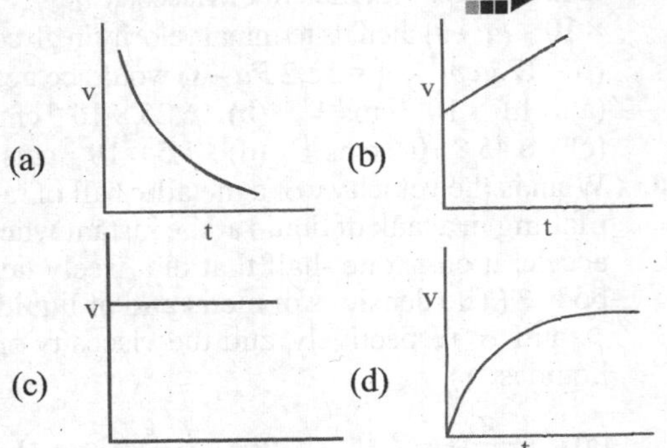

77. A certain number of spherical drops of a liquid of radius 'r' coalesce to form a single drop of radius 'R' and volume 'V'. If 'T' is the surface tension of the liquid, then :
(a) energy $= 4VT\left(\frac{1}{r}-\frac{1}{R}\right)$ is released
(b) energy $= 3VT\left(\frac{1}{r}+\frac{1}{R}\right)$ is absorbed
(c) energy $= 3VT\left(\frac{1}{r}-\frac{1}{R}\right)$ is released
(d) energy is neither released nor absorbed

78. A soap film of surface tension 3×10^{-2} formed in a rectangular frame cam support a straw as shown in Fig. If g = 10 ms^{-12}, the mass of the straw is

10 cm
Soap film
Straw

(a) 0.006 g
(b) 0.06 g
(c) 0.6 g
(d) 6 g

79. Two soap bubbles of radii a and b combine to form a single bubble of radius c. If P is the external pressure, then the surface tension of the soap solution is
(a) $\frac{P(c^3+a^3+b^3)}{4(a^2+b^2-c^2)}$ (b) $\frac{P(c^3-a^3-b^3)}{4(a^2+b^2-c^2)}$
(c) $Pc^3 - 4a^2 - 4b^2$ (d) $Pc^2 - 2a^2 - 3b^2$

80. Drops of liquid of density ρ are floating half immersed in a liquid of density σ. If the surface tension of liquid is T, the radius of the drop will be
(a) $\sqrt{\frac{3T}{g(3\rho-\sigma)}}$ (b) $\sqrt{\frac{6T}{g(2\rho-\sigma)}}$
(c) $\sqrt{\frac{3T}{g(2\rho-\sigma)}}$ (d) $\sqrt{\frac{3T}{g(4\rho-3\sigma)}}$

81. A soap bubble of radius R is surrounded by another soap bubble of radius $2R$, as shown. Take surface tension = S. Then the pressure inside the smaller soap bubble, in excess of the atmospheric pressure, will be
(a) $4S/R$
(b) $3S/R$
(c) $6S/R$
(d) None of these

Atmosphere

82. Two spherical bubbles are in contact with each other internally as shown. The radius of curvature of the common surface is R, then Tricky

(a) $R > R_1$

(b) $R_1 > R > R_2$

(c) $R < R_2$

(d) $R = R_1$

83. Water of density ρ in a clean aquarium forms a meniscus, as illustrated in the figure. Calculate the difference in height h between the centre and the edge of the meniscus. The surface tension of water is γ. **Toughnut**

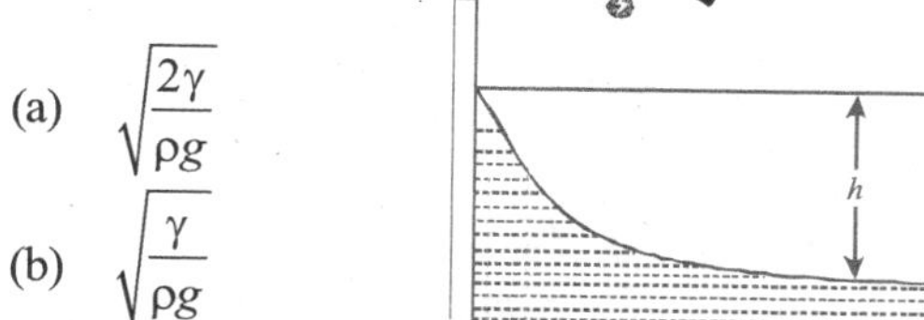

(a) $\sqrt{\frac{2\gamma}{\rho g}}$

(b) $\sqrt{\frac{\gamma}{\rho g}}$

(c) $\frac{1}{2}\sqrt{\frac{\gamma}{\rho g}}$ (d) $2\sqrt{\frac{\gamma}{\rho g}}$

84. The lower end of a capillary tube of radius 2.00mm is dipped 10.00cm below the surface of water in a beaker. Calculate the pressure within a bubble blown at its end in water, in excess of atmospheric pressure.
[Surface tension of water 72×10^{-3} N/m]

(a) 718 Nm^{-2} (b) 912 Nm^{-2}
(c) 1160 Nm^{-2} (d) 1052 Nm^{-2}

85. A water film is formed between two straight parallel wires of 10 cm length 0.5 cm apart. If the distance between wires is increased by 1 mm. What will be the work done ?
(surface tension of water = 72 dyne/cm)

(a) 36 erg (b) 288 erg (c) 144 erg (d) 72 erg

86. A vertical capillary tube with inside diameter 0.5mm is submerged into water so that the length of its part protruding over the surface of water is equal to 2.5mm. Find the radius of curvature of the meniscus. **Toughnut**

(a) 0.3 mm (b) 0.6 mm
(c) 0.9 mm (d) 1.2 mm

87. A large number of droplets, each of radius, r coalesce to form a bigger drop of radius, R. An engineer designs a machine so that the energy released in this process is converted into the kinetic energy of the drop. Velocity of the drop is (T = surface tension, ρ = density) **Critical Thinking**

(a) $\left[\frac{T}{\rho}\left(\frac{1}{r}-\frac{1}{R}\right)\right]^{1/2}$ (b) $\left[\frac{6T}{\rho}\left(\frac{1}{r}-\frac{1}{R}\right)\right]^{1/2}$

(c) $\left[\frac{3T}{\rho}\left(\frac{1}{r}-\frac{1}{R}\right)\right]^{1/2}$ (d) $\left[\frac{2T}{\rho}\left(\frac{1}{r}-\frac{1}{R}\right)\right]^{1/2}$

88. An air bubble of radius 0.1 cm is in a liquid having surface tension 0.06 N/m and density 10^3 kg/m^3. The pressure inside the bubble is 1100 Nm^{-2} greater than the atmospheric pressure. At what depth is the bubble below the surface of the liquid? (g = 9.8 ms^{-2}) **Toughnut**

(a) 0.1 m (b) 0.15 m (c) 0.20 m (d) 0.25 m

89. Radius of a capillary is 2×10^{-3} m. A liquid of weight 6.28×10^{-4} N may remain in the capillary then the surface tension of liquid will be

(a) 5×10^{-3} N/m (b) 5×10^{-2} N/m
(c) 5 N/m (d) 50 N/m

90. Two capillary of length L and 2L and of radius R and 2R are connected in series. The net rate of flow of fluid through them will be (given rate to the flow through single capillary, $X = \frac{\pi PR^4}{8\eta L}$)

(a) $\frac{8}{9}X$ (b) $\frac{9}{8}X$ (c) $\frac{5}{7}X$ (d) $\frac{7}{5}X$

Answer KEYs

1	(d)	10	(b)	19	(a)	28	(b)	37	(a)	46	(b)	55	(a)	64	(b)	73	(b)	82	(c)
2	(c)	11	(d)	20	(a)	29	(a)	38	(a)	47	(a)	56	(a)	65	(a)	74	(a)	83	(a)
3	(c)	12	(b)	21	(c)	30	(a)	39	(b)	48	(a)	57	(a)	66	(b)	75	(b)	84	(d)
4	(d)	13	(b)	22	(a)	31	(b)	40	(c)	49	(c)	58	(c)	67	(d)	76	(d)	85	(c)
5	(b)	14	(a)	23	(d)	32	(c)	41	(b)	50	(d)	59	(b)	68	(a)	77	(c)	86	(b)
6	(a)	15	(c)	24	(a)	33	(a)	42	(c)	51	(c)	60	(d)	69	(b)	78	(c)	87	(b)
7	(c)	16	(b)	25	(d)	34	(b)	43	(a)	52	(a)	61	(b)	70	(c)	79	(b)	88	(a)
8	(d)	17	(b)	26	(b)	35	(d)	44	(b)	53	(c)	62	(d)	71	(b)	80	(c)	89	(b)
9	(c)	18	(a)	27	(d)	36	(c)	45	(a)	54	(d)	63	(a)	72	(c)	81	(c)	90	(a)

Hints & Solutions

1. **(d)** Weight of submerged part of the block

$W = \frac{1}{3} v \text{ (Density of water) } g \quad ...(i)$

Excess weight, = weight of water having $\frac{2}{3}$ volume of the block.

$W' = \frac{2}{3} v \text{ (Density of water) } g \quad ...(ii)$

Dividing (ii) by (i), $\frac{W'}{W} = \frac{2/3}{1/3}$

$\therefore W' = 2W \Rightarrow W' = 2 \times 6 = 12$ kg

2. **(c)**

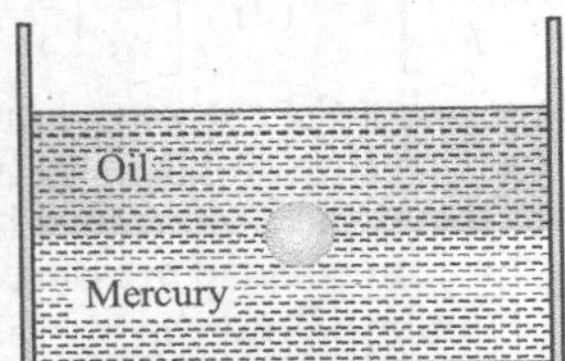

As the sphere floats in the liquid. Therefore its weight will be equal to the upthrust force on it

Weight of sphere $= \frac{4}{3}\pi R^3 \rho g \quad \ldots (i)$

Upthrust due to oil and mercury

$= \frac{2}{3}\pi R^3 \times \sigma_{oil} g + \frac{2}{3}\pi R^3 \sigma_{Hg} g \ldots (ii)$

Equating (i) and (ii)

$\frac{4}{3}\pi R^3 \rho g = \frac{2}{3}\pi R^3 0.8g + \frac{2}{3}\pi R^3 + 13.6g$

$\Rightarrow 2\rho = 0.8 + 13.6 = 14.4 \Rightarrow \rho = 7.2$

3. **(c)** Since pressure is transmitted undiminished throughout the fluid (Pascal's law)

$F_1 = \frac{A_1}{A_2} F_2 = \frac{\pi (5 \times 5)}{\pi (15 \times 15)} (1350 \times 9.81)$

$\approx 1.5 \times 10^3$ N

The air pressure that will produce this force is

$P = \frac{F_1}{A_1} = \frac{1.5 \times 10^3}{\pi (5 \times 10^{-2} \text{ m})^2} = 1.9 \times 10^5$ Pa

4. **(d)** The pressure outside the submarine is

$P = P_a + \rho g h$

Pressure inside the submarine is P_a.

Net pressure acting on the window is

$P_g = P - P_a = \rho g h$

$= 10^{-3}$ kgm$^{-3} \times 10$ ms$^{-2} \times 2000$ m

$= 2 \times 10^7$ Pa

Area of window is

$A = 50 \text{ cm} \times 50 \text{ cm} = 2500 \times 10^{-4} \text{ m}^2$

Force on the window is

$F = P_g A = 2 \times 10^7 \text{ Pa} \times 2500 \times 10^{-4} \text{ m}^2$

$= 5 \times 10^6$ N

5. **(b)**

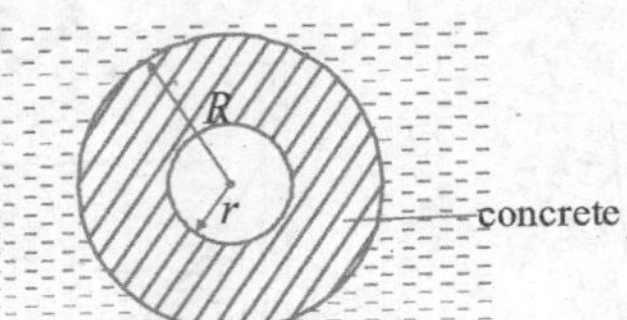

$(m_{concrete} + m_{sawdust})g = V\rho_w g$

$2.4 \times \frac{4}{3}\pi(R^3 - r^3) + 0.3 \times \frac{4}{3}\pi r^3 = \frac{4}{3}\pi R^3 \times 1$

or $R^3 = \frac{3}{2} r^3$

$\therefore \frac{m_{concrete}}{m_{sawdust}} = \frac{\rho_{concrete} \frac{4}{3}\pi(R^3 - r^3)}{\rho_{sawdust} \frac{4}{3}\pi r^3}$

$= \frac{2.4}{0.3} \frac{\left(\frac{3}{2}r^3 - r^3\right)}{r^3} = 4$

6. **(a)**

7. **(c)** $P = h\rho g$ i.e. pressure does not depend upon the area of bottom surface.

8. **(d)** Due to acceleration down the plane, both body B and liquid will received the same component of acceleration down the plane.

9. **(c)** $T_1 = T_2, Mg + \sqrt{2}T_1$

$= \frac{4}{3}\pi r^3 \rho_w g \; T_1$

$= 250\sqrt{2}$N

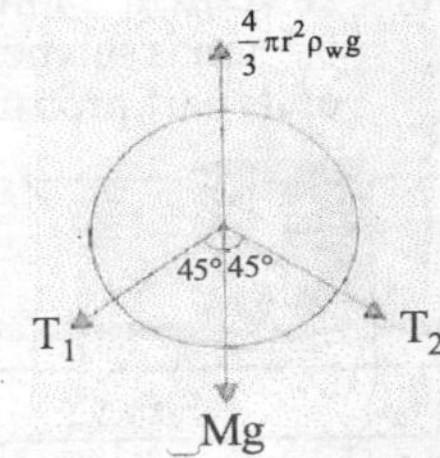

10. **(b)**

11. **(d)** According to Archimedes principle

Upthrust = Wt. of fluid displaced

$F_{bottom} - F_{top} = V\rho g$

$\therefore F_{bottom} = F_{top} + V\rho g$

$= P_1 \times A + V\rho g$

$= (h\rho g) \times (\pi R^2) + V\rho g$

$= \rho g [\pi R^2 h + V]$

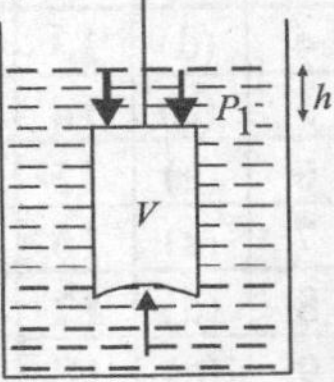

12. (b)

13. (b) VAB is the given cone. Let its height be h and semi-vertical angle α. Let the base AB of the cone be in the surface. CD is the surface of separation of two liquids, O and O' are the centres of the base AB and surface of separation CD.

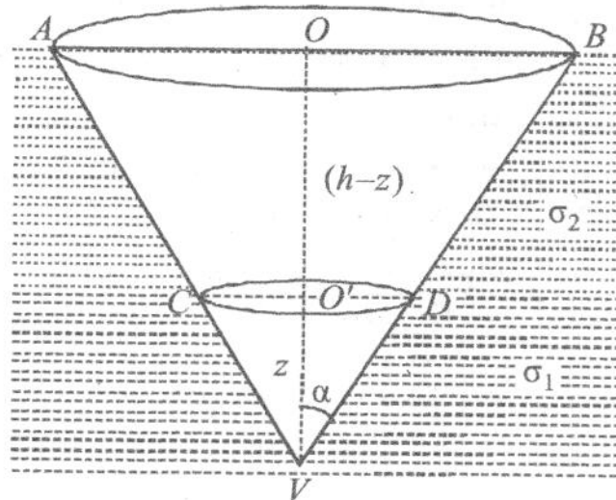

∴ For equilibrium,
weight of the cone = (weight of liquid of density σ_1 displaced) + (weight of liquid of density σ_2 displaced)

or $\frac{1}{3}\pi h^3 \tan^2 \alpha \rho g = \frac{1}{3}\pi z^3 \tan^2 \alpha \sigma_1 g$

$+\frac{1}{3}\pi (h^3 - z^3) \tan^2 \alpha . \sigma_2 g$

or $h^3 \rho = z^3 \sigma_1 + (h^3 - z^3) \sigma_2$

or $h^3 (\rho - \sigma_2) = z^3 (\sigma_1 - \sigma_2)$

or $z = h \left(\frac{\rho - \sigma_2}{\sigma_1 - \sigma_2}\right)^{1/3}$.

14. (a) Let ρ be the density of the material. ρ_0 be the density of water when the sphere has just started sinking, the weight of the sphere = weight of water displaced (approx).

$\Rightarrow \frac{4}{3}\pi (R^3 - r^3) \rho g = \frac{4}{3}\pi R^3 \rho_0 g$

$\Rightarrow (R^3 - r^3)\rho = R^3 \rho 0 \Rightarrow \frac{(R^3 - r^3)}{R^3} = \frac{\rho^0}{\rho}$

$\Rightarrow \frac{r}{R} = \frac{(7)^{1/3}}{2}$

15. (c) Let us consider a small dotted segment of thickness dx for observation.

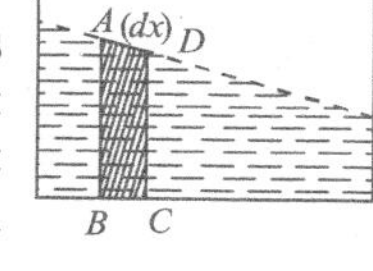

Since, this segment is accelerated towards right, a net force is acting in this segment towards right from the liquid towards the left of $ABCD$. According to Newton's third law, the segment $ABCD$ will also apply a force on the previous section creating a pressure on it which makes the liquid rise.

16. (b) Let V be the volume of iceberg and let x be the fraction of volume above water.
Using law of floatation, weight of floating body = weight of liquid displaced by part of the floating body inside the liquid.
Therefore, $V\rho_{ice} g = (1 - x) V\rho_{water} g$.
Using the value of ρ_{ice} and ρ_{water}, we get x = (13/103).

17. (b)

18. (a)

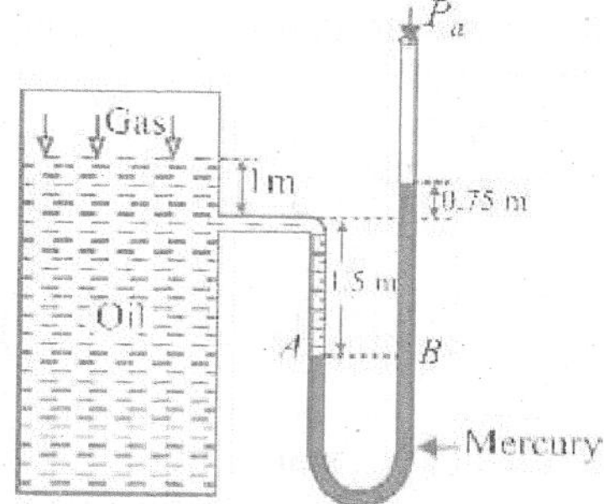

Suppose P_{gas} is the pressure of the gas on the oil. As the points A and B are at the same level in the mercury columns, so

$P_A = P_B$

or $P_{gas} + \rho_{oil}\, gh_{oil} = P_a + \rho_{Hg}\, g\, h_{Hg}$

or $P_{gas} + 820 \times 9.8 \times (1 + 1.50)$
$= P_a + 13.6 \times 10^3 \times 9.8 \times (1.5 + 0.75)$

or $P_{gas} + 20.09 \times 10^3 = P_a + 299.88 \times 10^3$

∴ $P_{gas} - Pa = 299.88 \times 10^3 - 20.09 \times 10^3$

or $[P_{gas}]_{gauge} = 279.8 \times 10^3\ N/m^2 = 2.8 \times 10^5\ N/m^2$

Absolute pressure of gas

$[P_{gas}]_{absolute} = [P_{gas}]_{gauge} + P_a$
$= 2.8 \times 10^5 + 1.01 \times 10^5 = 3.81 \times 10^5\ N/m^2$

19. (a) Mass of liquid in horizontal portion of U-tube = $Ad\rho$
Pseudo force on this mass = $Ad\rho a$
Force due to pressure difference in the two limbs
= $(h_1 \rho g - h_2 \rho g) A$
Equating both the forces
$(h_1 - h_2)\, \rho g A = Ad\rho a$

$\Rightarrow (h_1 - h_2) = \frac{Ad\rho a}{\rho g A} = \frac{ad}{g}$

20. (a) Total cross-sectional area of the thigh bones
$A = 2(10 \times 10^{-4}) = 2 \times 10^{-3}\ m^2$
Force acting on the bones = $mg = 40 \times 10$ = 400 N

∴ $P_{av} = \frac{F}{A} = \frac{400}{2 \times 10^{-3}} = 2 \times 10^5\ N/m^2$

21. **(c)** Weight of the bowl = mg

$$= V\rho g = \frac{4}{3}\pi\left[\left(\frac{D}{2}\right)^3 - \left(\frac{d}{2}\right)^3\right]\rho g$$

Where D = Outer diameter

d = Inner diameter, ρ = Density of bowl

Weight of the liquid diplaced by the bowl

$$= V\sigma g = \frac{4}{3}\pi\left(\frac{D}{2}\right)^3 \sigma g$$

where σ is the density of the liquid

For the floation

$$\frac{4}{3}\pi\left(\frac{D}{2}\right)^3 \sigma g = \frac{4}{3}\pi\left[\left(\frac{D}{2}\right)^3 - \left(\frac{d}{2}\right)^3\right]\rho g$$

$$\Rightarrow \left(\frac{1}{2}\right)^3 \times 1.2\times10^3 = \left[\left(\frac{1}{2}\right)^3 - \left(\frac{d}{2}\right)^3\right] 2\times10^4$$

By solving we get d = 0.98 m.

22. **(a)** Let L = PQ = length of rod

$\therefore$ SP = SQ = $\frac{L}{2}$

Weight of rod,

W = Alρg. acting

At point S

And force of buoyancy, $F_B = Al\rho_0 g$. [l = PR]

Which acts at mid-point of PR.

For rotational equilibrium.

$$Al\rho_0 g \times \frac{\ell}{2}\cos\theta = AL\rho g \times \frac{L}{2}\cos\theta$$

$$\Rightarrow \frac{l^2}{L^2} = \frac{\rho}{\rho_0} \Rightarrow \frac{l}{L} = \sqrt{\frac{\rho}{\rho_0}}$$

From figure, $\sin\theta = \frac{h}{l} = \frac{L}{2l} = \frac{1}{2}\sqrt{\frac{\rho_0}{\rho}}$

23. **(d)** $T + mg = F_b$

$\therefore$ $T = F_b - mg$

$= V\rho_w g - mg$

$= \frac{m}{(\rho_w/\eta)}\rho_w g - mg$

$= (\eta - 1)mg$.

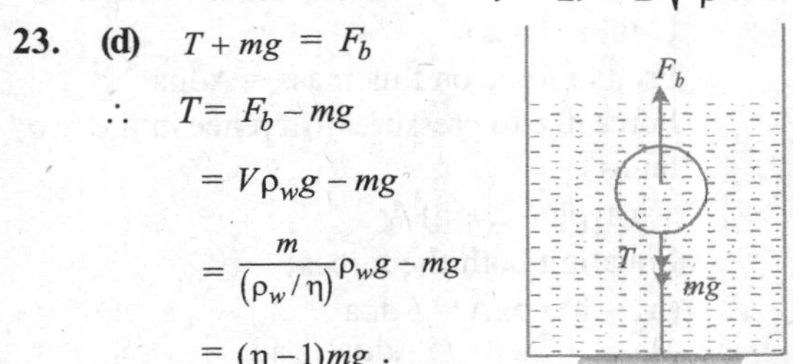

24. **(a)** Pressure difference between lungs and atmosphere

= (760 – 750) mm of Hg

= 10 mm of Hg = 1 cm of Hg

Let the boy can suck water from depth *h*. Then

Pressure difference = $h\rho_{water}g$ = 1 cm of Hg

or, $h \times 1\text{g cm}^{-3} \times 980 \text{ cm s}^{-2}$

$= 1 \text{ cm} \times 13.6\text{g cm}^{-3} \times 980 \text{ cm s}^{-2}$

$\therefore$ h = 13.6 cm

The boy can suck water from the depth of 13.6 cm

25. **(d)** As the block moves up with the fall of coil, *l* decreases, similarly *h* will also decrease because when the coin is in water, it displaces water equal to its own volume only.

26. **(b)** Bulk modulus,

$$B = -V_0\frac{\Delta p}{\Delta V} \Rightarrow \Delta V = -V_0\frac{\Delta p}{B}$$

$$\Rightarrow V = V_0\left[1 - \frac{\Delta p}{B}\right]$$

$$\therefore \text{Density, } \rho = \rho_0\left[1 - \frac{\Delta p}{B}\right]^{-1} = \rho_0\left[1 + \frac{\Delta p}{B}\right]$$

where, $\Delta p = p - p_0 = h\rho_0 g$

= pressure difference between depth and surface of ocean

$$\therefore \rho = \rho_0\left[1 + \frac{\rho_0 g y}{B}\right] \text{ (As h = y)}$$

27. **(d)** As we know,

Pressure P = Vdg

Here, LAdg = (pL) A(nρ)g + (1 – p)LAρg

$\Rightarrow$ d = (1 – p)ρ + pn ρ = [1 + (n – 1)p]ρ

28. **(b)** As block B falls, L will decrease.

Block B will displace, volume of liquid (V_1) Equal to its own volume when it is in the liquid.

$$V_1 = V_B = \frac{M_B}{\rho_B} \quad \text{...(i)}$$

When block B is on block A, it will displace the volume of liquid (V_2) whose weight is equal to the weight of the block B.

$V_2\rho_L g = M_B g = (V_1\rho_B)g$ or $V_2\rho L = V_1\rho_B$.

Since H increases, $V_1 > V_2$

So, $\rho_L > \rho_B$.

29. **(a)** Weight of cylinder = Upthrust due to upper liquid + Upthrust due to lower liquid.

$$D\left(\frac{A}{5} \times L \times g\right) = d\left(\frac{A}{5}\right)\left(\frac{3}{4}L\right)g + 2d\left(\frac{A}{5}\right)\left(\frac{L}{4}\right) \times g$$

$$\therefore \quad D = \frac{5d}{4}$$

30. **(a)**

31. **(b)** $mg = v\rho_{wood}g = (3\pi R^2 h)\rho_{wood}g$

$(3\pi R^2 h)\rho_{wood}g = (3\pi R^2 H)\rho_w g$

$$\frac{h\rho_{wood}}{\rho_{water}} = H$$

32. (c) Let σ be the density of the gas, then that of the air is 15σ. Then the weight of the balloon = weight of the gas + weight of the envelope $= Vg\sigma + w$.
If f be the required acceleration of the balloon acting vertically upward and then from "mass × acceleration = forces acting in the sense of acceleration" we get

$$\frac{(Vg\sigma + w)}{g} \times a = \text{force of buoyance} - \text{wt. of the balloon with gas} = V15\sigma g - (Vg\sigma + w)$$

$$\text{or } a = \left(\frac{14Vg\sigma - w}{Vg\sigma + w}\right) \times g$$

33. (a) For equilibrium
$4dA\ell g = (dAL + m)g$

and $\frac{\ell}{2} > Y_{cm}$ (for rotational equilibrium)

$$Y_{cm} = \frac{m \times 0 + dAL(L/2)}{m + dAL} = \frac{dAL^2}{2(m + dAL)}$$

$$\ell > \frac{AL^2d}{(ALd + m)} \Rightarrow \frac{m + dAL}{4dA} > \frac{AL^2d}{(ALd + m)}$$

$m > ALd$

34. (b)

35. (d) Acceleration when block B is in the liquid,

$$a_1 = \frac{m_A g - (m_B g - \text{upthrust})}{(m_A + 3m)}$$

$$= \frac{m_A g - \left(3mg - \frac{3m}{2\rho}\rho g\right)}{(m_A + 3m)} (\uparrow)$$

Acceleration when block B is outside if the liquid,

$$a_2 = \frac{3mg - m_A g}{(m_A + 3m)} (\downarrow)$$

Given $a_1 = a_2$, we get

$$m_A g - \frac{3}{2}mg = 3mg - m_A g$$

$$\Rightarrow 2m_A g - \frac{9}{2}mg \Rightarrow m_A = \left(\frac{9}{4}\right)m$$

36. (c) In Bernoulli's theorem only law of conservation of energy is obeyed.

37. (a)

38. (a) Air flows just above the roof and there is no air flow just below the roof inside the room. Therefore $v_1 = 0$ and $v_2 = v$. Applying Bernaulli's theorem at the points inside and outside the roof, we obtain.

$(1/2)\rho v_1^2 + \rho g h_1 + P_1$
$= (1/2)\rho v_2^2 + \rho g h_2 + P_2$.

Since $h_1 = h_2 = h$, $v_1 = 0$ and $v_2 = v_1$

$P_1 = P_2 + 1/2\rho v^2 \Rightarrow P_1 - P_2 = \Delta P = 1/2\rho v^2$.

Since the area of the roof is A, the aerodynamic lift exerted on it $= F = (\Delta P)A$

$\Rightarrow F = 1/2\rho A v^2$

where ρ = density of air $= 1.3$ kg/m^3
$A = 20$ m^2, $v = 29.44$ m/sec.

$\Rightarrow F = \{1/2 \times 1.3 \times 20 \times (29.44)^2\}$ N
$= 1.127 \times 10^4$ N.

39. (b) According to Bernoulli's theorem,

$P + \frac{1}{2}\rho v^2 = P_0 + 0$

So, $\Delta P = \frac{1}{2}\rho v^2$

$F = \Delta PA = \frac{1}{2}\rho v^2 A$

$= \frac{1}{2} \times 1.2 \times 40 \times 40 \times 250 = 2.4 \times 10^5$ N (upwards)

40. (c)

41. (b) $h\rho g + \frac{mg}{A} = \frac{1}{2}\rho v^2$

42. (c) $P_a + \frac{1}{2}\rho_1 v_1^2 + 0$

$= Pa + \frac{1}{2}\rho_2 v_2^2 + (\rho_1 g h_1 + \rho_2 g h_2)$

As $v_2 << v_1$, $\therefore v_1 = \sqrt{2g(h_1 + h_2\frac{\rho_2}{\rho_1})}$

43. (a) Inflow rate of volume of the liquid = Outflow rate of volume of the liquid

$$\pi R^2 V = n\pi r^2(v) \Rightarrow v = \frac{\pi R^2 V}{n\pi r^2} = \frac{VR^2}{nr^2}$$

44. (b) According to Bernoulli's theorem,

$P + \frac{1}{2}\rho v^2$ = constant and Av = constant

If A is minimum, v is maximum, P is minimum.

45. (a) **46.** (b)

47. (a) From continuity equation, velocity at cross-section (1) is more than that at cross-section (2). Hence $P_1 < P_2$.

48. (a) Let x be the height of the water level from the hole $v = \sqrt{2gx} = \frac{\ell}{t} = \frac{1}{t}$ ...(i)

For vertical motion $x = \frac{1}{2}gt^2$ ...(ii)

Now solving (i) & (ii) we get
$\Rightarrow x = 0.25$m (i.e. level goes down from 0.81m to 0.25m.)

Using equation of continuity

$\sqrt{5}.\frac{dx}{dt} = \sqrt{2gx}.4 \times 10^{-4}$ On solving $t = 1000$s

49. (c) Pressure difference

$$P_2 - P_1 = \frac{1}{2}\rho\left(v_2^2 - v_1^2\right)$$

$$= \frac{1}{2} \times 1.2\left((150)^2 - (100)^2\right) = 7500\ \text{Nm}^{-2}$$

50. (d) Let the density of water be ρ, then the force by escaping liquid on container $= \rho S(\sqrt{2gh})^2$

∴ Acceleration of container

$$a = \frac{2\rho Sgh - \mu\rho Vg}{\rho V} = \left(\frac{2Sh}{V} - \mu\right) g$$

Now, $\mu = \frac{Sh}{V}$ $\quad$ ∴ $a = \frac{Sh}{V} g$

51. (c) $A_1 v_1 = A_2 v_2 + A_3 v_3$

$A \times 3 = A \times 1.5 + 1.5A \times v_3 \Rightarrow v_3 = 1$ m/s

52. (a) From the equation of continuity,

$$\text{Rate} = r = \frac{A}{3} V_P + \left(A - \frac{A}{3}\right) V_Q$$

$$\therefore \quad V_P + 2V_Q = \frac{3r}{A} = \frac{3 \times 0.1}{10^{-2}} = 30\,\text{m/s}$$

As $V_P = 20$ m/s, so $V_Q = 5$ m/s.

53. (c) $t = k(\sqrt{h_1} - \sqrt{h_2})$

$\therefore\ t_1 = k(\sqrt{h} - \sqrt{h/2})$ and $t_2 = k\left(\sqrt{h/2} - 0\right)$

$\therefore\ \frac{t_1}{t_2} = \sqrt{2} - 1$

54. (d) $v_0 = \sqrt{2gh} \Rightarrow v_2 = \sqrt{2g\frac{h}{\sqrt{2}}} = \frac{v_0}{\sqrt[4]{2}}$

55. (a) Let a_1 and a_2 be cross sectional areas and v_1 and v_2 the velocities of liquid flow at A and B. If p_1 and p_2 are pressures of liquid recorded by manometer, then

$$p_1 + \frac{1}{2}\rho v_1^2 = p_2 + \frac{1}{2}\rho v_2^2$$

$$\Rightarrow p_1 - p_2 = \frac{1}{2}\rho v_1^2\left(\frac{v_2^2}{v_1^2} - 1\right)$$

By equation of continuity, $a_1 v_1 = a_2 v_2$

Also, $p_1 - p_2 = h\rho g$

Substituting these values, we have

$$v_1 = \sqrt{\frac{2gh}{\frac{a_1^2}{a_2^2} - 1}}$$

Rate of flow of liquid is $a_1 v_1 = a_1 a_2 \sqrt{\frac{2gh}{a_1^2 - a_2^2}}$

56. (a) If v_2 is the required speed then

$$A_1 v_1 = A_2 v_2$$

or $\quad v_1 = \frac{A_2}{A_1} v_2 = 0.1 v_2 \quad \ldots \text{(i)}$

From Bernoulli's equation, we have

$$P_a + \frac{1}{2}\rho v_1^2 + \rho g(3 - 0.525) = P_a + \frac{1}{2}\rho v_2^2 + 0 \quad \ldots \text{(ii)}$$

After solving above equations, we get

$$v_2^2 = 50\ \text{m}^2/\text{s}^2$$

57. (a) According to Bernoulli's theorem

$$P_1 + \frac{1}{2}\rho v_1^2 = P_2 + \frac{1}{2}\rho v_2^2 \quad \ldots \text{(i)}$$

From question,

$$P_1 - P_2 = 3 \times 10^5, \frac{A_1}{A_2} = 5$$

According to equation of continuity

$$A_1 v_1 = A_2 v_2 \text{ or, } \frac{A_1}{A_2} = \frac{v_2}{v_1} = 5 \Rightarrow v_2 = 5v_1$$

From equation (i)

$$P_1 - P_2 = \frac{1}{2}\rho\left(v_2^2 - v_1^2\right)$$

$$\text{or } 3 \times 10^5 = \frac{1}{2} \times 1000\left(25 v_1^2 - v_1^2\right)$$

$$\Rightarrow \quad 600 = 6v_1 \times 4v_1 \Rightarrow v_1^2 = 25$$

$$\therefore \quad v_1 = 5\,\text{m/s}$$

58. (c) Force $F = \rho Q v = \rho a v^2 = \rho a \times 2gh$

Thus net force

$= F_2 - F_1$

$= \rho a v_2^2 - \rho a v_1^2$

$= \rho a(v_2^2 - v_1^2)$

$= \rho a[2g(y + h) - 2gy]$

$= \rho a \times 2gh$

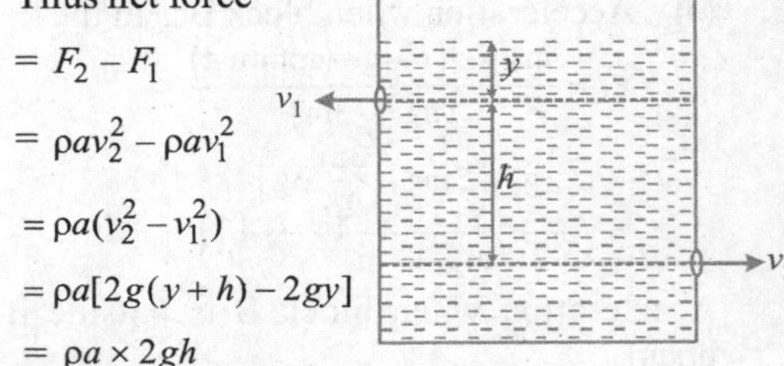

59. (b) Range is same for both holes

$\therefore\ 2\sqrt{(H - h_1)h_1} = 2\sqrt{(H - h_2)h_2}$

Squaring both sides,

$4(H - h_1)\,h_1 = 4(H - h_2)\,h_2$

$Hh_1 - h_1^2 = Hh_2 - h_2^2$

On solving we get,

$H = h_1 + h_2$ (not possible) and $h_1 - h_2 = 0$

Hence, the ratio of $\frac{h_1}{h_2} = 1{:}1$.

60. (d) $v_1 = \sqrt{2g(h/2)}$.

For v_2 we can replace the liquid of density ρ and height h to a liquid of density 2ρ and height $h/2$.

Thus $v_2 = \sqrt{2gh} = \sqrt{2}v$.

61. (b) According to Toricelli's theorem,
Velocity of efflex,
$V_{eff} = \sqrt{2gh} = \sqrt{2\times 9.8\times 5} \cong 9.8\,ms^{-1}$

62. (d) Given: Diameter of water tap $= \frac{2}{\sqrt{\pi}}$ cm

$\therefore$ Radius, $r = \frac{1}{\sqrt{\pi}}\times 10^{-2}$ m $\quad \frac{dm}{dt} = \rho AV$

$$\frac{15}{5\times 60} = 10^3 \times \pi\left(\frac{1}{\sqrt{\pi}}\right)^2 \times 10^{-4} V \Rightarrow V = 0.05 \text{ m/s}$$

Reynold's number, $R_e = \frac{\rho V r}{n}$

$$= \frac{10^3 \times 0.5 \times \frac{2}{\sqrt{\pi}} 10^{-2}}{10^{-3}} \cong 5500$$

63. (a) Newton's law of viscosity, $F = \eta A \frac{dv}{dy}$

$$\text{Stress} = \frac{F}{A} = \eta\left(\frac{dv}{dy}\right) = \eta k\left(\frac{4y}{a^2} - \frac{3y^2}{a^3}\right)$$

At $y = a$, stress $= \eta k\left(\frac{4}{a} - \frac{3}{a}\right) = \frac{\eta k}{a}$

64. (b) Relative to liquid, the velocity of sphere is $2v_0$ upwards.
$\therefore$ Viscous force on sphere
$= 6\pi\eta\, r\, 2v_0$ downwards
$= 12\pi\eta\, r\, v_0$ downwards

65. (a) As the temperature rises the atoms of the liquid become more mobile and the coefficient of viscosity falls.

66. (b) When terminal velocity is reached then body moves with constant velocity hence, acceleration is zero.

67. (d) Using the formula of the terminal velocity of a body falling through a viscous medium,

$$V = \frac{2r^2(\rho - \sigma)g}{9\eta} \Rightarrow \eta = \frac{2r^2(\rho - \sigma)g}{9v}$$

Where ρ is the density of material of body and σ is the density of medium.
In case of the air bubble $\rho = 1$ and $\sigma = 1.47 \times 10^3$ kg/ms and the air bubble rises up.

$$\eta = \frac{2r^2\sigma g}{9V} = \frac{2\times(10^{-2})^2 \times 1.47\times 10^3 \times 9.8}{9\times 0.21\times 10^{-2}}$$

$= 1.52 \times 10^3$ decapoise $= 1.52 \times 10^4$ Poise

68. (a) dv = 8 cm/s and dx = 0.1 cm

Velocity gradient $= \frac{dv}{dx} = \frac{8}{0.1} = 80/\text{s}$.

69. (b) $v \propto \frac{\rho - \rho_0}{\eta}$

$$\therefore \frac{v_2}{v_1} = \frac{\rho - \rho_{02}}{\rho - \rho_{01}} \times \frac{\eta_1}{\eta_2} = \frac{7.8 - 1.2}{7.8 - 1} \times \frac{8.5\times 10^{-4} \times 10}{13.2}$$

$= 6.25 \times 10^{-4}\, cms^{-1}$

70. (c)

71. (b) Terminal velocity, $v_0 = \frac{2r^2(\rho - \rho_0)g}{9\eta}$

$$= \frac{2\times(2\times 10^{-3})^2 \times (8 - 1.3)\times 10^3 \times 9.8}{9\times 0.83}$$

$= 0.07\, ms^{-1}$

72. (c) If ρ is the density of the ball and ρ' that of the another ball, m for the balls are the same, but $r' = 2r$
$\therefore mg = 6\pi r\eta v$ (by Stoke's law)
or, $6\pi r\eta v = 6\pi 2r\eta v'$ $\quad$ So, $v' = \frac{v}{2}$

73. (b) $\eta = 10^{-2}$ poise
$v = 18$ km/h $= \frac{18000}{3600} = 5$ m/s $\quad l = 5$ m
Strain rate $= \frac{v}{l}$
Coefficient of viscosity, $\eta = \frac{\text{shearing stress}}{\text{strain rate}}$
$\therefore$ Shearing stress $= \eta \times$ strain rate
$= 10^{-2} \times \frac{5}{5} = 10^{-2}\, Nm^{-2}$

74. (a) Viscous force $= 6\pi\eta rv$
$= 6\pi \times 18 \times 10^{-5} \times 0.03 \times 100$
$= 101.73 \times 10^{-4}$ dyne

75. (b) Total volume of rain drops, received 100 cm in a year by area 1 m^2
$= 1m^2 \times \frac{100}{100} m = 1\, m^3$
As we know, density of water,
$d = 10^3$ kg/m^3
Therefore, mass of this volume of water
$M = d \times v = 10^3 \times 1 = 10^3$ kg
Average terminal velocity of rain drop
$v = 9$ m/s (given)
Therefore, energy transferred by rain,
$E = \frac{1}{2} mv^2 = \frac{1}{2} \times 10^3 \times (9)^2 = \frac{1}{2} \times 10^3 \times 81$
$= 4.05 \times 10^4$ J

76. (d) Initially velocity increases with respect to time & becomes constant after some time, this constant velocity is called terminal velocity.

77. (c) As surface area decreases so energy is released.
Energy released $= 4\pi R^2 T[n^{1/3} - 1]$ (where $R = n^{1/3} r$)

$$= 4\pi R^3 T\left[\frac{1}{r} - \frac{1}{R}\right] = 3VT\left[\frac{1}{r} - \frac{1}{R}\right]$$

78. **(c)** $m \times 10 = 2 \times 3 \times 10^{-2} \times \frac{10}{100}$

or $m = 6 \times 10^{-4}\,kg = 6 \times 10^{-4} \times 10^3\,g = 0.6\,g$

79. **(b)** $\left(P + \frac{4\sigma}{a}\right)\left(\frac{4}{3}\pi a^3\right) + \left(P + \frac{4\sigma}{b}\right)\left(\frac{4}{3}\pi b^3\right)$

$= \left(P + \frac{4\sigma}{c}\right)\left(\frac{4}{3}\pi c^3\right)$

or $P\,[a^3 + b^3 - c^3] = 4\sigma\,[c^2 - a^2 - b^2]$

or $\sigma = \frac{P\left(c^3 - a^3 - b^3\right)}{4\left(a^2 + b^2 - c^2\right)}$

80. **(c)** Balancing the forces acting on the drop, we get

$\frac{4}{3}\pi r^3 \rho g = 2\pi r T + \frac{1}{2}.\frac{4}{3}\pi r^3 \sigma \Rightarrow r = \sqrt{\frac{3T}{(2\rho - \sigma)g}}$

81. **(c)** $P_1 - P_0 = \frac{4S}{2R}$ ...... (i)

$P_2 - P_1 = \frac{4S}{R}$ ...... (ii)

Add (i) and (ii), $P_2 - P_0 = \frac{6S}{R}$

$\bullet P_2$ $\bullet P_1$ $\bullet P_0$

82. **(c)** Pressure difference is largest between atmosphere and smaller bubbles. Hence radius of curvature (R) is smallest.

83. **(a)**

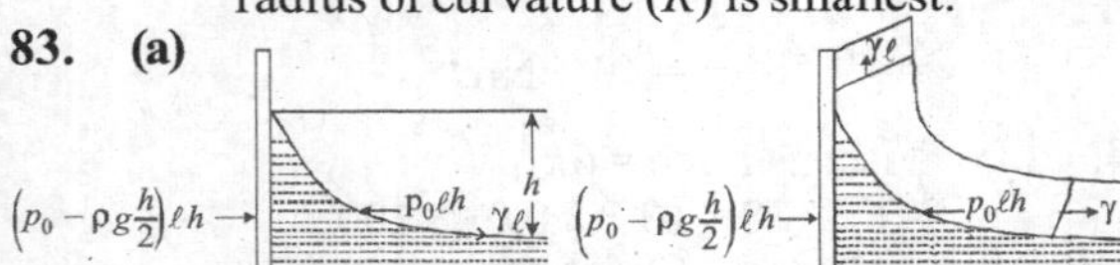

Balancing forces in horizontal direction

$\left(P_0 - \rho g \frac{h}{2}\right)\ell h + \gamma \ell = P_0 \ell h \Rightarrow h = \sqrt{\frac{2\gamma}{\rho g}}$

84. **(d)** The pressure at a point just outside the bubble

$P_0 = P_{atm} + h\rho g = P_{atm} + 10 \times 10^{-2} \times 10^3 \times 9.8$

$= P_{atm} + 980\,Nm^{-2}$

Now excess pressure within the bubble compared to a point just outside

$= \frac{2T}{R} = \frac{2 \times 72 \times 10^{-3}}{2 \times 10^{-3}} = 72\,Nm^{-2}$

$\therefore$ Inside pressure

= excess pressure + outside pressure

$= P_{atm} + 980 + 72 = P_{atm} + 1052\,Nm^{-2}$

$\therefore$ Pressure within the bubble in excess of atmospheric pressure $= 1052\,Nm^{-2}$.

85. **(c)** Work done = Surface tension × increase in area of the film

$W = S \times \Delta A$

Increase in area = Final area – initial area

$= 10 \times (0.5 + 0.1) - 10 \times 0.5 = 1\,cm^2$

$\therefore$ $W = 72 \times 2 \times 1 = 144$ erg

[$\because$ There are 2 free surfaces; $\therefore$ $\Delta A = 2 \times 1$].

86. **(b)** Let the liquid rise to a height h.

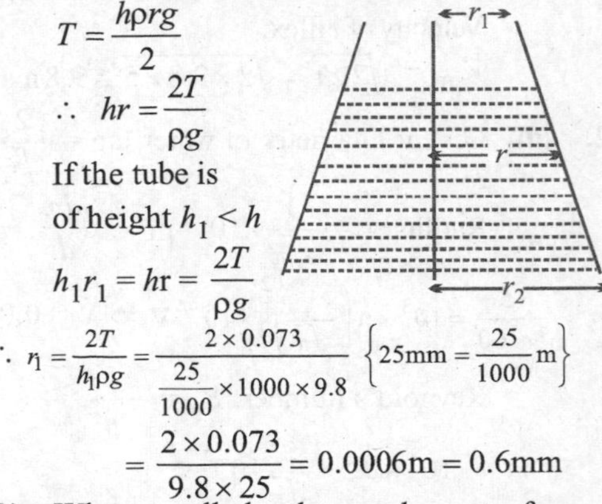

$T = \frac{h\rho r g}{2}$

$\therefore$ $hr = \frac{2T}{\rho g}$

If the tube is of height $h_1 < h$

$h_1 r_1 = hr = \frac{2T}{\rho g}$

$\therefore r_1 = \frac{2T}{h_1 \rho g} = \frac{2 \times 0.073}{\frac{25}{1000} \times 1000 \times 9.8}$ $\left\{25mm = \frac{25}{1000}m\right\}$

$= \frac{2 \times 0.073}{9.8 \times 25} = 0.0006m = 0.6mm$

87. **(b)** When small droplets coelesce to form a bigger drop, energy released in this process is given by,

$4\pi R^3 T\left[\frac{1}{r} - \frac{1}{R}\right]$

According to question

$\frac{1}{2}mv^2 = 4\pi R^3 T\left[\frac{1}{r} - \frac{1}{R}\right]$

$\Rightarrow \frac{1}{2}\left[\frac{4}{3}\pi R^3 \rho\right]v^2 = 4\pi R^3 T\left[\frac{1}{r} - \frac{1}{R}\right]$

$\Rightarrow V^2 = \frac{6T}{\rho}\left[\frac{1}{r} - \frac{1}{R}\right] \Rightarrow V = \left[\frac{6T}{\rho}\left(\frac{1}{r} - \frac{1}{R}\right)\right]^{\frac{1}{2}}$

88. **(a)** Let depth of bubble below the liquid surface, $h = ?$

As we know, $\rho_{Excess} = h\rho g + \frac{2s}{r}$

$\Rightarrow$ $1100 = h \times 10^3 \times 9.8 + \frac{2 \times 6 \times 10^{-2}}{10^{-3}}$

$\Rightarrow 1100 = 9800\,h + 120 \Rightarrow 9800h = 1100 - 120$

$\Rightarrow$ $h = \frac{980}{9800} = 0.1\,m$

89. **(b)** $W = mg = \rho(\pi r^2 h)g$

or $W \simeq \rho(\pi r^2) \times \left(\frac{2T}{\rho r g}\right) g$ or $W = 2\pi r T$

$\therefore T = \frac{W}{2\pi r} = \frac{6.28 \times 10^4}{2\pi \times 2 \times 10^{-3}} = 5 \times 10^{-2}\,N/m.$

90. **(a)** Fluid resistance is given by $R = \frac{8\eta L}{\pi r^4}$

When two capillary tubes of same size are joined in series, then equivalent fluid resistance is

$R_S = \frac{8\eta L}{\pi R^4} + \frac{8\eta \times 2L}{\pi (2R)^4} = \left(\frac{8\eta L}{\pi R^4}\right) \times \frac{9}{8}$

Rate of flow $= \frac{P}{R_S} = \frac{\pi P R^4}{8\eta L} \times \frac{8}{9} = \frac{8}{9}X$ $\left[\text{as } X = \frac{\pi P R^4}{8\eta L}\right]$

10 Thermal Properties of Matter

1. If two rods A and B of equal length L, and different areas of cross-section A_1 and A_2 have one end each at temperature T_1 and T_2, have equal rates of flow of heat, then

   (a) $A_1 = A_2$ (b) $\frac{A_1}{A_2} = \frac{K_1}{K_2}$

   (c) $\frac{A_1}{A_2} = \frac{K_2}{K_1}$ (d) $K_1 = K_2$

2. 4200 J of work is required for
   (a) increasing the temperature of 10 g of water through 10°C
   (b) increasing the temperature of 100 g of water through 10°C
   (c) increasing the temperature of 1 kg of water through 10°C
   (d) increasing the temperature of 500 g of water through 10°C

3. The latent heat of vaporization of a substance is always
   (a) greater than its latent heat of fusion
   (b) greater than its latent heat of sublimation
   (c) equal to its latent heat of sublimation
   (d) less than its latent heat of fusion

4. Six identical conducting rods are joined as shown in figure. Points A and D are maintained at 200°C and 20°C respectively. The temperature of junction B will be **Tricky**

   (a) 120°C (b) 100°C (c) 140°C (d) 80°C

5. 500 g of water and 100 g of ice at 0°C are in a calorimeter whose water equivalent is 40 g. 10 g of steam at 100°C is added to it. Then water in the calorimeter is : (Latent heat of ice = 80 cal/g, Latent heat of steam = 540 cal/ g)

   (a) 580 g (b) 590 g (c) 600 g (d) 610 g

6. The temperature of the two outer surfaces of a composite slab, consisting of two materials having coefficients of thermal conductivity K and $2K$ and thickness x and $4x$, respectively, are T_2 and $T_1 (T_2 > T_1)$. The rate of heat transfer through the slab, in a steady state is $\left(\frac{A(T_2 - T_1)K}{x}\right) f$, with f equal to **Critical Thinking**

   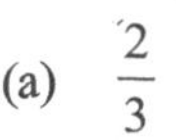

   (a) $\frac{2}{3}$

   (b) $\frac{1}{2}$

   (c) 1

   (d) $\frac{1}{3}$

7. A block of ice at –10°C is slowly heated and converted to steam at 100°C. Which of the following curves represents the phenomenon qualitatively **Tricky**

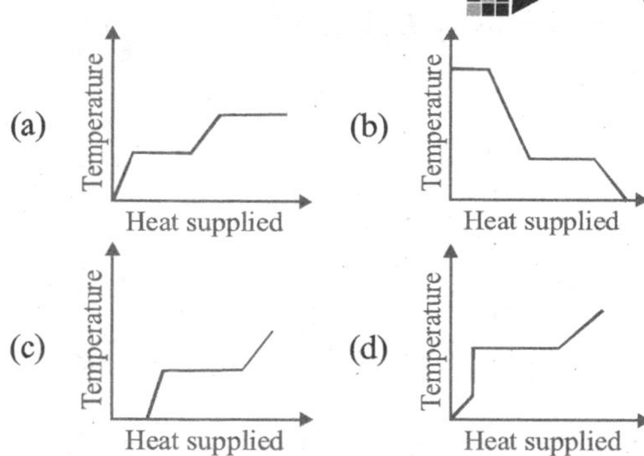

8. Assuming no heat losses, the heat released by the condensation of x g of steam at 100°C can be used to convert y g of ice at 0°C into water at 100°C, the ratio $x : y$ is :

   (a) 1 : 1 (b) 1 : 2 (c) 1 : 3 (d) 3 : 1

9. Which of the following will expand the most for same rise in temperature?
   (a) Aluminium
   (b) Glass
   (c) Wood
   (d) All will expand same

10. If α, β and γ are coefficient of linear, area and volume expansion respectively, then

    (a) $\gamma = 3\alpha$ (b) $\alpha = 3\gamma$

    (c) $\beta = 3\alpha$ (d) $\gamma = 3\beta$

11. An iron tyre is to be fitted on to a wooden wheel 1m in diameter. The diameter of tyre is 6 mm smaller than that of wheel. The tyre should be heated so that its temperature increases by a minimum of (the coefficient of cubical expansion of iron is 3.6×10^{-5}/°C)
(a) 167°C (b) 334°C (c) 500°C (d) 1000°C

12. A metal sheet with a circular hole is heated. The hole **Tricky**
(a) gets larger
(b) gets smaller
(c) remains of the same size
(d) gets deformed

13. The coefficient of apparent expansion of mercury in a glass vessel is 153×10^{-6}/°C and in a steel vessel is 144×10^{-6}/°C. If α for steel is 12×10^{-6}/°C, then that of glass is
(a) 9×10^{-6}/°C (b) 6×10^{-6}/°C
(c) 36×10^{-6}/°C (d) 27×10^{-6}/°C

14. What is the equivalent thermal conductivity of the rods in figure given below, if the length of each cylinder be ℓ and area of cylinder having thermal conductivities K_1 and K_3 be A while that of the middle cylinder having thermal conductivity K_2 be 2A? **Critical Thinking**

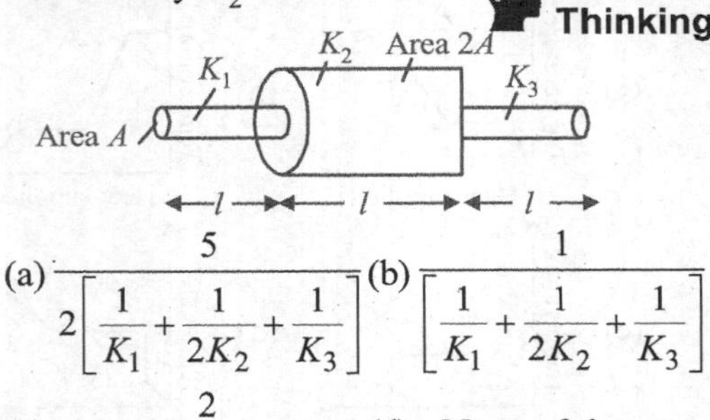

(a) $\dfrac{5}{2\left[\dfrac{1}{K_1}+\dfrac{1}{2K_2}+\dfrac{1}{K_3}\right]}$ (b) $\dfrac{1}{\left[\dfrac{1}{K_1}+\dfrac{1}{2K_2}+\dfrac{1}{K_3}\right]}$

(c) $\dfrac{2}{5\left[\dfrac{1}{K_1}+\dfrac{1}{2K_2}+\dfrac{1}{K_3}\right]}$ (d) None of these

15. A piece of metal weighs 45g in air and 25g in a liquid of density 1.5×10^3 kg – m^{-3} kept at 30°C. When the temperature of the liquid is raised to 40°C, the metal piece weighs 27g. The density of liquid at 40°C, is 1.25×10^3 kg –m^{-3}. The coefficient of linear expansion of metal is
(a) 1.3×10^{-3}/°C (b) 5.2×10^{-3}/°C
(c) 2.6×10^{-3}/°C (d) 0.26×10^{-3}/°C

16. On a new scale of temperature (which is linear) and called the W scale, the freezing and boiling points of water are 39° W and 239° W respectively. What will be the temperature on the new scale, corresponding to a temperature of 39° C on the Celsius scale?
(a) 78° W (b) 117° W (c) 200° W (d) 139° W

17. A solid ball of metal has a spherical cavity inside it. The ball is heated. The volume of cavity will **Tricky**
(a) decrease (b) increase
(c) remain unchanged
(d) have its shape changed

18. On a linear temperature scale Y, water freezes at –160° Y and boils at – 50° Y. On this Y scale, a temperature of 340 K would be read as : (water freezes at 273 K and boils at 373 K)
(a) –73.7° Y (b) –233.7° Y
(c) –86.3° Y (d) –106.3° Y

19. A glass flask is filled up to a mark with 50 cc of mercury at 18°C. If the flask and contents are heated to 38°C, how much mercury will be above the mark? (α for glass is 9×10^{-6}/°C and coefficient of real expansion of mercury is 80×10^{-6}/°C)
(a) 0.85 cc (b) 0.46 cc
(c) 0.153 cc (d) 0.05 cc

20. A bar of iron is 10 cm at 20°C. At 19°C it will be (α of iron = 11×10^{-6}/°C)
(a) 11×10^{-6} cm longer (b) 11×10^{-6} cm shorter
(c) 11×10^{-5} cm shorter (d) 11×10^{-5} cm longer

21. Two marks on a glass rod 10 cm apart are found to increase their distance by 0.08 mm when the rod is heated from 0°C to 100°C. A flask made of the same glass as that of rod measures a volume of 1000 cc at 0°C. The volume it measures at 100°C in cc is
(a) 1002.4 (b) 1004.2 (c) 1006.4 (d) 1008.2

22. One end of a uniform rod of length 1m is placed in boiling water while its other end is placed in melting ice. A point P on the rod is maintained at a constant temperature of 800°C. The mass of steam produced per second is equal to the mass of ice melted per second. If specific latent heat of steam is 7 times the specific latent heat of ice, the distance of P from the steam chamber must be **Critical Thinking**
(a) (1/7) m (b) (1/8) m
(c) (1/9) m (d) (1/10) m

23. Which of the following circular rods, (given radius r and length l) each made of the same material and whose ends are maintained at the same temperature will conduct most heat?
(a) $r = 2r_0; l = 2l_0$ (b) $r = 2r_0; l = l_0$
(c) $r = r_0; l = 2l_0$ (d) $r = r_0; l = l_0$

24. A beaker contains 200 gm of water. The heat capacity of the beaker is equal to that of 20 gm of water. The initial temperatue of water in the beaker is 20°C. If 440 gm of hot water at 92°C is poured in it, the final temperature, neglecting radiation loss, will be nearest to

(a) 58°C (b) 68°C (c) 73°C (d) 78°C

25. In a vertical U-tube containing a liquid, the two arms are maintained at different temperatures t_1 and t_2. The liquid columns in the two arms have heights l_1 and l_2 respectively. The coefficient of volume expansion of the liquid is equal to **Tricky**

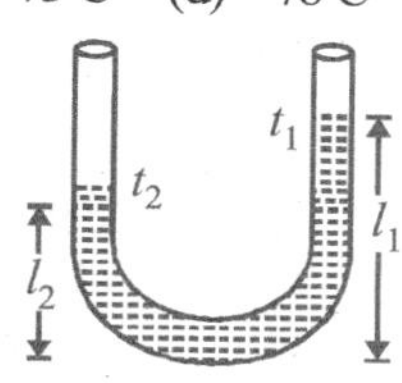

(a) $\frac{l_1 - l_2}{l_2 t_1 - l_1 t_2}$ (b) $\frac{l_1 - l_2}{l_1 t_1 - l_2 t_2}$

(c) $\frac{l_1 + l_2}{l_2 t_1 + l_1 t_2}$ (d) $\frac{l_1 + l_2}{l_1 t_1 + l_2 t_2}$

26. The coefficient of linear expansion for a certain metal varies with temperature as $\alpha(T)$. If L_0 is the initial length of the metal and the temperature of metal is changed from T_0 to T ($T_0 > T$), then

(a) $L = L_0 \int_{T_0}^{T} \alpha(T)\,dT$

(b) $L = L_0 \left[1 + \int_{T_0}^{T} \alpha(T)\,dT\right]$

(c) $L = L_0 \left[1 - \int_{T_0}^{T} \alpha(T)\,dT\right]$

(d) $L > L_0$

27. The length of a metallic rod is 5 m at 0°C and becomes 5.01 m, on heating upto 100°C. The linear expansion of the metal will be

(a) 2.33×10^{-5}/°C (b) 6.0×10^{-5}/°C

(c) 4.0×10^{-5}/°C (d) 2.0×10^{-5}/°C

28. A pendulum clock is 5 seconds fast at temperature of 15°C and 10 seconds slow at a temperature of 30°C. At what temperature does it give the correct time? (take time interval = 24 hours)

(a) 18°C (b) 20°C (c) 22°C (d) 25°C

29. A pendulum clock loses 12 s a day if the temperature is 40°C and gains 4 s a day if the temperature is 20° C. The temperature at which the clock will show correct time, and the co-efficient of linear expansion (α) of the metal of the pendulum shaft are respectively :

(a) 30°C; $\alpha = 1.85 \times 10^{-3}$/°C

(b) 55°C; $\alpha = 1.85 \times 10^{-2}$/°C

(c) 25°C; $\alpha = 1.85 \times 10^{-5}$/°C

(d) 60°C; $\alpha = 1.85 \times 10^{-4}$/°C

30. The apparent coefficient of expansion of a liquid when heated in a copper vessel is C and that when heated in a silver vessel is S. If A is the linear coefficient of expansion of copper, then the linear coefficient of expansion of silver is

(a) $\frac{C + S - 3A}{3}$ (b) $\frac{C + 3A - S}{3}$

(c) $\frac{S + 3A - C}{3}$ (d) $\frac{C + S + 3A}{3}$

31. Coefficient of linear expansion of brass and steel rods are α_1 and α_2. Lengths of brass and steel rods are ℓ_1 and ℓ_2 respectively. If $(\ell_2 - \ell_1)$ is maintained same at all temperatures, which one of the following relations holds good?

(a) $3\alpha_1\ell_2 = \alpha_2\ell_{1_2}$ (b) $4\alpha_1\ell_2 = \alpha_2\ell_1$

(c) $2\alpha_1\ell_2 = \alpha_2\ell_1$ (d) $\alpha_1\ell_1 = \alpha_2\ell_2$

32. If H_C, H_K and H_F are heat required to raise the temperature of one gram of water by one degree in Celcius, Kelvin and Fahrenheit temperature scales respectively then

(a) $H_K > H_C > H_F$ (b) $H_F > H_C > H_K$

(c) $H_K = H_C > H_F$ (d) $H_K = H_C = H_F$

33. A mass of 50g of water in a closed vessel, with surroundings at a constant temperature takes 2 minutes to cool from 30°C to 25°C. A mass of 100g of another liquid in an identical vessel with identical surroundings takes the same time to cool from 30° C to 25° C. The specific heat of the liquid is : (The water equivalent of the vessel is 30g.)

(a) 2.0 kcal/kg (b) 7 kcal/kg

(c) 3 kcal/kg (d) 0.5 kcal/kg

34. The coefficient of linear expansion of glass is α_g per °C and the cubical expansion of mercury is γ_m per °C. The volume of the bulb of a mercury thermometer at 0°C is V_0 and cross section of the capillary is A_0. What is the length of mercury column in capillary at T°C, if the mercury just fills the bulb at 0°C?

(a) $\frac{V_0 T(\gamma_m + 3\alpha_g)}{A_0(1 + 2\alpha_g T)}$ (b) $\frac{V_0 T(\gamma_m - 3\alpha_g)}{A_0(1 + 2\alpha_g T)}$

(c) $\frac{V_0 T(\gamma_m + 2\alpha_g)}{A_0(1 + 3\alpha_g T)}$ (d) $\frac{V_0 T(\gamma_m - 2\alpha_g)}{A_0(1 + 3\alpha_g T)}$

35. The density of water at 4°C is 1000.0 kg/m³ and at 100°C it is 958.4 kg/m³. The cubic expansivity of water between these temperatures is

(a) 4.5×10^{-3}/K (b) 5.4×10^{-5}/K

(c) 4.5×10^{-4}/K (d) 5.4×10^{-6}/K

36. A rod PQ of length l is pivoted at an end P and freely rotated in a horizontal plane at an angular speed ω about a vertical axis passing through P. If coefficient of linear expansion of material of rod is α, find the percentage change in its angular velocity if temperature of system is increased by ΔT is
(a) $(\alpha\Delta T\times 100)\%$ (b) $(2\alpha\Delta T\times 100)\%$
(c) $(3\alpha\Delta T\times 100)\times 100\%$
(d) $(4\alpha\Delta T\times 100)\%$

37. A steel rod of length 1m is heated from 25°C to 75°C keeping its length constant. The longitudinal strain developed in the rod is (Given: Coefficient of linear expansion of steel $= 12\times 10^{-6}$/°C)
(a) 6×10^{-6} (b) -6×10^{-5}
(c) -6×10^{-4} (d) zero

38. A steel rail of length 5 m and area of cross-section 40 cm^2 is prevented from expanding along its length while the temperature rises by 10°C. If coefficient of linear expansion and Young's modulus of steel are 1.2×10^{-5} K^{-1} and 2×10^{11} Nm^{-2} respectively, the force developed in the rail is approximately: **Toughnut**
(a) 2×10^7 N (b) 1×10^5 N
(c) 2×10^9 N (d) 3×10^{-5} N

39. When a block of iron floats in mercury at 0°C, fraction k_1 of its volume is submerged, while at the temperature 60 °C, a fraction k_2 is seen to be submerged. If the coefficient of volume expansion of iron is γ_{Fe} and that of mercury is γ_{Hg}, then the ratio k_1/k_2 can be expressed as
(a) $\frac{1+60\gamma_{Fe}}{1+60\gamma_{hg}}$ (b) $\frac{1-60\gamma_{Fe}}{1+60\gamma_{Hg}}$
(c) $\frac{1+60\gamma_{Fe}}{1-60\gamma_{Hg}}$ (d) $\frac{1+60\gamma_{Hg}}{1+60\gamma_{Fe}}$

40. The density of steel at 0 °C is 8.0 g/cc. At what temperature is density lesser by 0.1 %? Co-efficient of linear expansion of steel is 10^{-5} /°C.
(a) 37.2°C (b) 33.3°C (c) 55.4°C (d) 40.2°C

41. A thin steel ring of inner diameter 40 cm and cross-sectional area 1 mm^2, is heated until it easily slides on a rigid cylinder of diameter 40.05 cm. [For steel, $\alpha = 10^{-5}$ /°C, Y = 200 GPa]
When the ring cools down, the tension in the ring will be: **Critical Thinking**
(a) 1000 N (b) 500 N (c) 250 N (d) 100 N

42. Consider two identical iron spheres, one which lie on a thermally insulating plate, while the other hangs from an insulatory thread. Equal amount of heat is supplied to the two spheres **Tricky**

A B

(a) temperature of A will be greater than B
(b) temperature of B will be greater than A
(c) their temperature will be equal
(d) can't be predicted

43. A horizontal tube, open at both ends, contains a column of liquid. The length of this liquid column does not change with temperature. Let γ = coefficient of volume expansion of the liquid and α = coefficient of linear expansion of the material of the tube, then
(a) γ = α (b) γ = 2α (c) γ = 3α (d) γ = 0

44. A cylindrical steel alloy is inserted into a circular hole of diameter 2.60 m in a brass plate. When the plug and the plate are at a temperature 20°C, the diameter of the plug is 0.010 mm smaller than that of the hole. The temperature at which the plug will just fit in is (given $\alpha_{steel} = 11\times 10^{-6}$/°C and $\alpha_{brass} = 19\times 10^{-6}$/°C)
(a) – 20° C (b) – 48 °C
(c) 10 °C (d) None of these

45. The ratio of the coefficient of volume expansion of a glass container to that of a viscous liquid kept inside the container is 1 : 4. What fraction of the inner volume of the container should the liquid occupy so that the volume of the remaining vacant space will be same at all temperatures?
(a) 2 : 5 (b) 1 : 4 (c) 1 : 64 (d) 1 : 8

46. An iron tyre is to be fitted on to a wooden wheel 1m in diameter. The diameter of tyre is 6 mm smaller than that of wheel. The tyre should be heated so that its temperature increases by a minimum of (the coefficient of cubical expansion of iron is 3.6×10^{-5}/°C) **Toughnut**
(a) 167°C (b) 334°C (c) 500°C (d) 1000°C

47. Mass of water which absorbs or emits the same amount of heat as is done by the body for the same rise or fall in temperature is known as
(a) thermal capacity of the body
(b) specific heat capacity of the body
(c) latent heat capacity of the body
(d) water equivalent of the body

48. A lead bullet strikes against a steel plate with a velocity 200 m s^{-1}. If the impact is perfectly inelastic and the heat produced is equally shared between the bullet and the target, then the rise in temperature of the bullet is (specific heat capacity of lead = 125 J kg^{-1} K^{-1})
(a) 80°C (b) 60°C (c) 160°C (d) 40°C

49. Certain amount of heat is given to 100 g of copper to increase its temperature by 21° C. If the same amount of heat is given to 50 g of water, then the rise in its temperature is (Specific heat capacity of copper = 400 J kg^{-1} K^{-1} and that for water = 4200 J kg^{-1} K^{-1})
(a) 4° C (b) 5.25° C(c) 8° C (d) 6° C

50. A hammer of mass 1 kg having speed of 50 m/s, hit a iron nail of mass 200 gm. If specific heat of iron is 0.105 cal/gm°C and half the energy is converted into heat. the raise in temperature of nail is
(a) 7.1°C (b) 9.2°C (c) 10.5°C (d) 12.1°C

51. In an energy recycling process, 100 g of steam at 100°C becomes water at 100°C which converts y g of ice at 0°C into water at 100°C. The numeric value of y is
(a) 100 (b) 200 (c) 300 (d) 400

52. In a water-fall the water falls from a height of 100 m. If the entire K.E. of water is converted into heat, the rise in temperature of water will be
(a) 0.23°C (b) 0.46°C (c) 2.3°C (d) 0.023°C

53. A student takes 50gm wax (specific heat = 0.6 kcal/kg°C) and heats it till it boils. The graph between temperature and time is as follows. Heat supplied to the wax per minute and boiling point are respectively

Critical Thinking

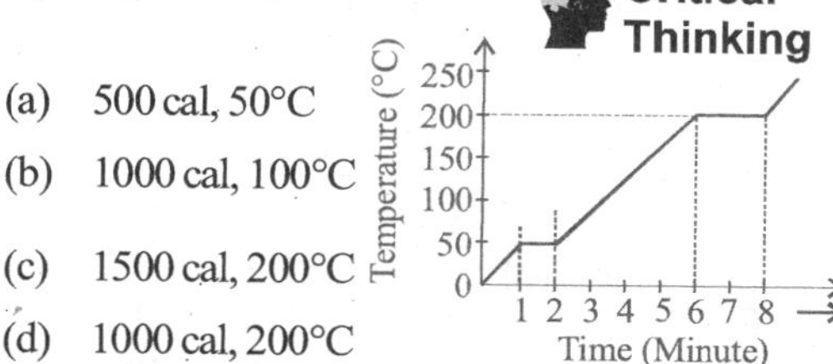

(a) 500 cal, 50°C
(b) 1000 cal, 100°C
(c) 1500 cal, 200°C
(d) 1000 cal, 200°C

54. 19 g of water at 30°C and 5 g of ice at – 20°C are mixed together in a calorimeter. What is the final temperature of the mixture? Given specific heat of ice = 0.5 cal $g^{-1}(°C)^{-1}$ and latent heat of fusion of ice = 80 cal g^{-1}
(a) 0°C (b) –5°C (c) 5°C (d) 10°C

55. The heat (Q) supplied to a solid, which is otherwise thermally isolated from its surroundings, is plotted as a function of its absolute temperature, θ. It is found that they are related by the equation.
$Q = a\theta^2 + b\theta^4$. ($a$, b are constants).
The heat capacity of the solid is given by

(a) $a\frac{\theta^3}{3} + b\frac{\theta^5}{5}$ (b) $a\theta + b\theta^3$

(c) $a\frac{\theta}{3} + b\frac{\theta^3}{5}$ (d) $2a\theta + 4b\theta^3$

56. A 2 kg copper block is heated to 500°C and then it is placed on a large block of ice at 0°C. If the specific heat capacity of copper is 400 J/kg°C and latent heat of fusion of water is 3.5 × 10^5 J/kg, the amount of ice that can melt is
(a) (7/8) kg (b) (7/5) kg (c) (8/7) kg (d) (5/7) kg

57. Two spheres of different materials one with triple the radius and one-fifth wall thickness of the other are filled with ice. If the time taken for complete melting of ice in the larger sphere is 30 minute and for smaller one is 20 minute, the ratio of thermal conductivities of the materials of larger spheres to that of smaller sphere is

Critical Thinking

(a) 1/8 (b) 3/4 (c) 2/3 (d) 1/2

58. Five rods of same dimensions are arranged as shown in the figure. They have thermal conductivities K_1, K_2, K_3, K_4 and K_5. When points A and B are maintained at different temperatures, no heat flows through the central rod if

Tricky

(a) $K_1 = K_4$ and $K_2 = K_3$
(b) $K_1K_4 = K_2K_3$
(c) $K_1K_2 = K_3K_4$
(d) $\frac{K_1}{K_4} = \frac{K_2}{K_3}$

59. Ice starts forming in a lake with water at 0°C when the atomspheric temperature is –10°C. If the time taken for the first 1 cm of ice to be formed is 7 hours, then the time taken for the thickness of ice to change from 1 cm to 2 cm is
(a) 7 hours (b) 14 hours
(c) 21 hours (d) 3.5 hours

60. A kettle with 3 litre water at 27°C is heated by operating coil heater of power 2 kW. The heat is lost to the atmosphere at constant rate 130J/sec, when its lid is open. In how much time will water heated to 97°C with the lid open? (specific heat of water = 4.2 kJ/kg)
(a) 472 sec (b) 693 sec (c) 912 sec (d) 1101 sec

61. The two ends of a metal rod are maintained at temperatures 100°C and 110°C. The rate of heat flow in the rod is found to be 4.0 J/s. If the ends are maintained at temperatures 200°C and 210°C, the rate of heat flow will be
(a) 16.8 J/s (b) 8.0 J/s (c) 4.0 J/s (d) 44.0 J/s

62. One end of a thermally insulated rod is kept at a temperature T_1 and the other at l_2. The rod is composed of two sections of length l_1 and l_2 and thermal conductivities K_1 and K_2 respectively. The temperature at the interface of the two section is

Tricky

(a) $\dfrac{(K_1l_1T_1+K_2l_2T_2)}{(K_1l_1+K_2l_2)}$

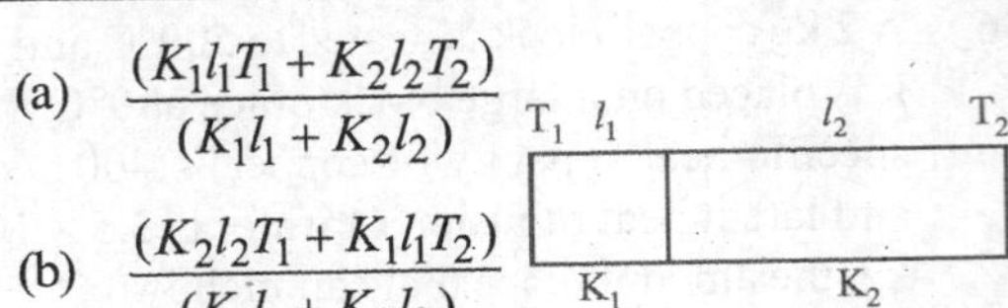

(b) $\dfrac{(K_2l_2T_1+K_1l_1T_2)}{(K_1l_1+K_2l_2)}$

(c) $\dfrac{(K_2l_1T_1+K_1l_2T_2)}{(K_2l_1+K_1l_2)}$ (d) $\dfrac{(K_1l_2T_1+K_2l_1T_2)}{(K_1l_2+K_2l_1)}$

63. A black body is at a temperature of 2880 K. The energy of radiation emitted by this object with wavelength between 499 nm and 500 nm is U_1, between 999 nm and 1000 nm is U_2 and between 1499 nm and 1500 nm is U_3. Wien's constant $b = 2.88 \times 10^6$ nm-K. Then

(a) $U_1 = 0$ (b) $U_2 = 0$
(c) $U_1 = U_2$ (d) $U_2 > U_1$

64. A cylindrical rod of aluminium is of length 20 cms and radius 2 cms. The two ends are maintained at temperatures of 0ºC and 50ºC [the coefficient of thermal conductivity is $\dfrac{0.5 \text{ cal}}{\text{cm} \times \text{sec} \times {}^\circ\text{C}}$]. Then the thermal resistance of the rod in $\dfrac{\text{cal}}{\text{sec} \times {}^\circ\text{C}}$ is

(a) 318 (b) 31.8 (c) 3.18 (d) 0.318

65. A metal ball of surface area 200 square cm, temperature 527ºC is surrounded by a vessel at 27ºC. If the emissivity of the metal is 0.4, then the rate of loss of heat from the ball is approximately

$\left[\sigma = 5.67 \times 10^{-8} \dfrac{\text{joule}}{\text{m}^2 \times \text{sec} \times \text{K}^2}\right]$ **Toughnut**

(a) 108 joule (b) 168 joule
(c) 182 joule (d) 192 joule

66. Two rods of same length and transfer a given amount of heat 12 second, when they are joined as shown in figure (i).

But when they are joined as shown in figure (ii), then they will transfer same heat in same conditions in

(a) 24 s
(b) 13 s
(c) 15 s
(d) 48 s

l
Fig. (i)

l l
Fig. (ii)

67. The rectangular surface of area 8 cm × 4 cm of a black body at temperature 127°C emits energy E per second. If the length and breadth are reduced to half of the initial value and the temperature is raised to 327°C, the rate of emission of energy becomes

(a) $\dfrac{3}{8}$E (b) $\dfrac{81}{16}$E (c) $\dfrac{9}{16}$E (d) $\dfrac{81}{64}$E

68. A tube shaped like a parallelogram is filled with liquid is kept in a vertical plane. Tube is heated at mid point of ad. The heat flows only by convection. Then **Tricky**

(a) $T_d = T_a > T_b = T_c$
(b) $T_a > T_b > T_c > T_d$
(c) $T_a > T_b > T_d = T_c$
(d) $T_d > T_a > T_b > T_c$

a b d c

69. In the plot of temperature versus time showing changes in the state of ice on heating, which part represents constant temperature? **Tricky**

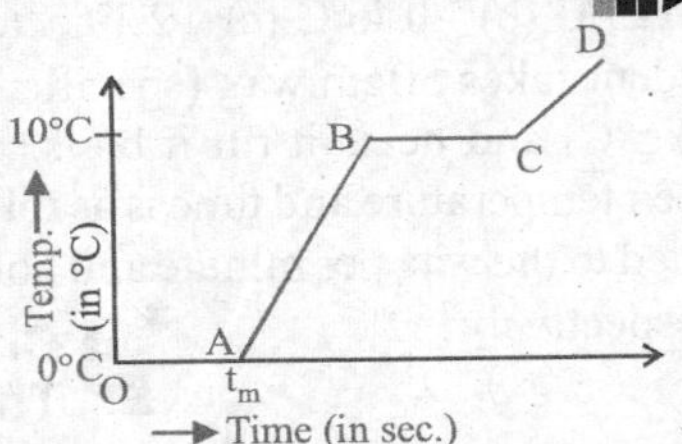

(a) OA (b) AB
(c) CD (d) All of these

70. A piece of ice falls from a height h so that it melts completely. Only one-quarter of the heat produced is absorbed by the ice and all energy of ice gets converted into heat during its fall. The value of h is : [Latent heat of ice is 3.4×10^5 J/kg and g = 10 N/kg]

(a) 34 km (b) 544 km (c) 136 km (d) 68 km

71. 12 identical rods made of same material are arranged in the form of a cube. The temperature of *P* and *R* are maintained at 90°C and 30°C respectively. Then the temperature of point *V*, when steady state is reached is **Tricky**

(a) 65°C
(b) 60°C
(c) 20°C
(d) 50°C

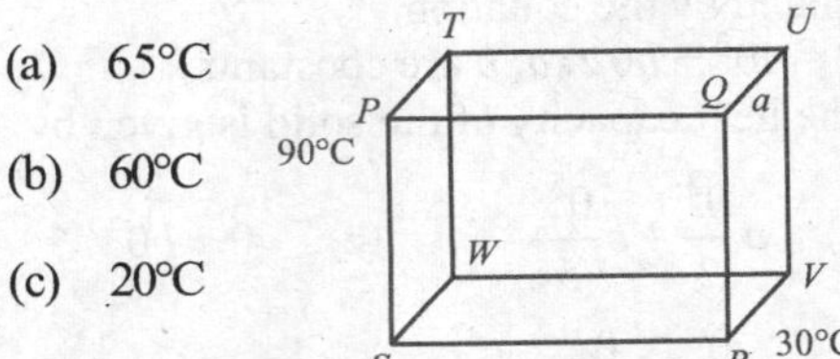

72. A sphere and a cube of same material and same volume are heated upto same temperature and allowed to cool in the same surroundings. The ratio of the amounts of radiations emitted will be

(a) 1 : 1 (b) $\frac{4\pi}{3}:1$

(c) $\left(\frac{\pi}{6}\right)^{1/3}:1$ (d) $\frac{1}{2}\left(\frac{4\pi}{3}\right)^{2/3}:1$

73. Four identical rods of same material are joined end to end to form a square. If the temperature difference between the ends of a diagonal is 100°C, then the temperture difference between the ends of other diagonal will be

(a) 0°C (b) $\frac{100}{l}°C$ (c) $\frac{100}{2l}°C$ (d) 100°C

(where l is the length of each rod)

74. The figure shows a system of two concentric spheres of radii r_1 and r_2 are kept at temperatures T_1 and T_2, respectively. The radial rate of flow of heat in a substance between the two concentric spheres is proportional to **Tricky**

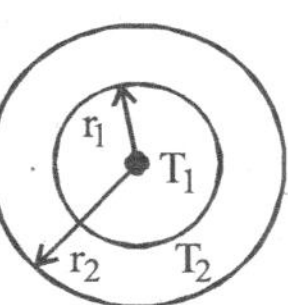

(a) $In\left(\frac{r_2}{r_1}\right)$ (b) $\frac{(r_2 - r_1)}{(r_1 r_2)}$

(c) $(r_2 - r_1)$ (d) $\frac{r_1 r_2}{(r_2 - r_1)}$

75. 300 gm of water at 25°C is added to 100 g of ice at 0°C. The final temperature of the mixture is

(a) $-\frac{5}{3}°C$ (b) $-\frac{5}{2}°C$ (c) 5°C (d) 0°C

76. A slab of stone of area 0.36 m^2 and thickness 0.1 m is exposed on the lower surface to steam at 100°C. A block of ice at 0°C rests on the upper surface of the slab. In one hour 4.8 kg of ice is melted. The thermal conductivity of slab is : (Given latent heat of fusion of ice = 3.36×10^5 J kg^{-1}.):

(a) 1.24 J/m°C (b) 1.29 J/m°C
(c) 2.05 J/m°C (d) 1.02 J/m°C

77. A partition wall has two layers of different materials A and B in contact with each other. They have the same thickness but the thermal conductivity of layer A is twice that of layer B. At steady state the temperature difference across the layer B is 50 K, then the corresponding difference across the layer A is

(a) 50 K (b) 12.5 K (c) 25 K (d) 60 K

78. The tempertaure of equal masses of three different liquids A, B and C are 12°C, 19°C and 28°C respectively. The temperature when A and B are mixed is 16°C and when B and C are mixed is 23°C. The temperature when A and C are mixed is **Critical Thinking**

(a) 18.2°C (b) 22°C
(c) 20.2°C (d) 25.2°C

79. 2 kg of ice at -20^0C is mixed with 5kg of water at 20^0C in an insulating vessel having a negligible heat capacity. Calculate the final mass of water remaining in the container. It is given that the specific heats of water & ice are 1kcal/kg/^{0}C & 0.5 kcal/kg/^{0}C while the latent heat of fusion of ice is 80 kcal/kg

(a) 7 kg (b) 6 kg (c) 4 kg (d) 2 kg

80. An experiment takes 10 minutes to raise the temperature of water in a container from 0°C to 100°C and another 55 minutes to convert it totally into steam by a heater supplying heat at a uniform rate. Neglecting the specific heat of the container and taking specific heat of water to be 1 cal/g°C, the heat of vapourization according to this experiment will come out to be :

(a) 560 cal/ g (b) 550 cal/ g
(c) 540 cal/ g (d) 530 cal/ g

81. The room heater can maintain only 16°C in the room when the temperature outside is –20°C. It is not warm and comfortable, that is why the electric stove with power of 1kW is also plugged in. Together these two devices maintain the room temperature of 22°C. Determine the thermal power of the heater. **Critical Thinking**

(a) 3 kW (b) 4 kW (c) 5 kW (d) 6 kW

82. Three liquids of equal volumes are thoroughly mixed. If their specific heats are s_1, s_2, s_3 and their temperatures θ_1, θ_2, θ_3 and their masses m_1, m_2, m_3 respectively, then the final temperature of the mixture is

(a) $\frac{s_1\theta_1 + s_2\theta_2 + s_3\theta_3}{m_1s_1 + m_2s_2 + m_3s_3}$

(b) $\frac{m_1s_1\theta_1 + m_2s_2\theta_2 + m_3s_3\theta_3}{m_1s_1 + m_2s_2 + m_3s_3}$

(c) $\frac{m_1s_1\theta_1 + m_2s_2\theta_2 + m_3s_3\theta_3}{m_1\theta_1 + m_2\theta_2 + m_3\theta_3}$

(d) $\frac{m_1\theta_1 + m_2\theta_2 + m_3\theta_3}{s_1\theta_1 + s_2\theta_2 + s_3\theta_3}$

83. A body cools from 50.0°C to 49.9°C in 5s. How long will it take to cool from 40.0°C to 39.9°C? Assume the temperature of surroundings to be 30.0°C and Newton's law of cooling to be valid

(a) 2.5 s (b) 10 s (c) 20 s (d) 5 s

84. A container contains hot water at 100°C. If in time T_1 temperature falls to 80° C and the time T_2 temperature falls to 60°C form 80°C, then

(a) $T_1 = T_2$ (b) $T_1 > T_2$
(c) $T_1 < T_2$ (d) None of these

85. According to Newton's law of cooling, the rate of cooling of a body is proportional to $(\Delta\theta)^n$, where $\Delta\theta$ is the difference of the temperature of the body and the surroundings, and n is equal to

(a) two (b) three (c) four (d) one

86. A body cools in a surrounding which is at a constant temperature of θ_0. Assume that it obeys Newton's law of cooling. Its temperature θ is plotted against time t. Tangents are drawn to the curve at the points $P(\theta = \theta_2)$ and $Q(\theta = \theta_1)$. These tangents meet the time axis at angle of ϕ_2 and ϕ_1, as shown, then

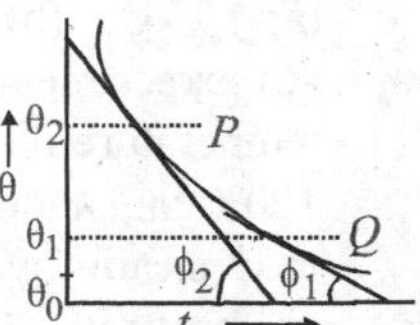

(a) $\dfrac{\tan\phi_2}{\tan\phi_1} = \dfrac{\theta_1 - \theta_0}{\theta_2 - \theta_0}$ (b) $\dfrac{\tan\phi_2}{\tan\phi_1} = \dfrac{\theta_2 - \theta_0}{\theta_1 - \theta_0}$

(c) $\dfrac{\tan\phi_1}{\tan\phi_2} = \dfrac{\theta_1}{\theta_2}$ (d) $\dfrac{\tan\phi_1}{\tan\phi_2} = \dfrac{\theta_2}{\theta_1}$

87. Consider two hot bodies B_1 and B_2 which have temperatures 100°C and 80°C respectively at t = 0. The temperature of the surroundings is 40°C. The ratio of the respective rates of cooling R_1 and R_2 of these two bodies at t = 0 will be

(a) $R_1 : R_2 = 3 : 2$ (b) $R_1 : R_2 = 5 : 4$
(c) $R_1 : R_2 = 2 : 3$ (d) $R_1 : R_2 = 4 : 5$

88 A cube, a pyramid (with faces identical) and a sphere (all of them hollow) are made from the same material and have equal mass and bound equal volume. They are heated to the same temperature and then left to cool. After some time,

1. sphere will have the highest temperature
2. pryamid will have the highest temperature.
3. cube will have the lowest temperature.
4. sphere will have the lowest temperature.
5. pyramid will have the lowest temperature

Correct option will be : Tricky

(a) 2 and 3 (b) 3 and 1
(c) 1 and 5 (d) 2 and 4

89. A sphere of density ρ, specific heat capacity c and radius r is hung by a thermally insulating thread in an enclosure which is kept at a lower temperature than the sphere. The temperature of the sphere starts to drop at a rate which depends upon the temperature difference between the sphere and the enclosure and the nature of the surface of sphere and is proportional to

Critical Thinking

(a) $\dfrac{c}{r^3\rho}$ (b) $\dfrac{1}{r^3\rho c}$ (c) $3r^3\rho c$ (d) $\dfrac{1}{r\rho c}$

90. A hot body, obeying Newton's law of cooling is cooling down from its peak value 80°C to an ambient temperature of 30°C. It takes 5 minutes in cooling down from 80°C to 40°C. How much time will it take to cool down from 62°C to 32°C? (Given In 2 = 0.693, In 5 = 1.609)

(a) 3.75 minutes (b) 8.6 minutes
(c) 9.6 minutes (d) 6.5 minutes

91. Which of the given graphs proves Newton's law of cooling? Tricky

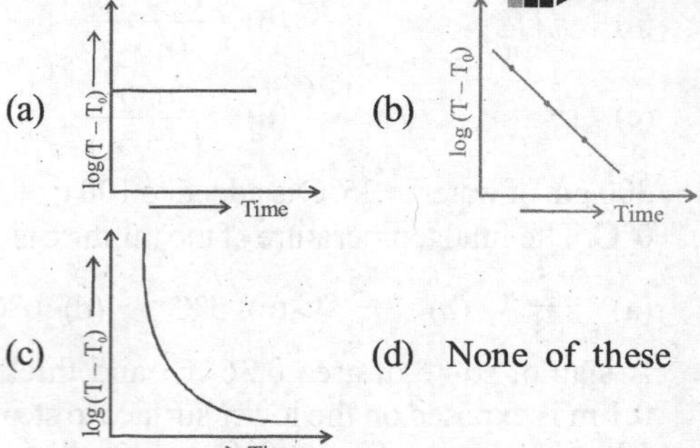

(d) None of these

92. A body cools in a surrounding of constant temperature 30°C. Its heat capacity is 2J/kg °C. Initial temperature of the body is 40°C. Assume Newton's law of cooling is valid. The body cools to 38°C in 10 minutes. In further 10 minutes it will cool from 38°C to

(a) 36°C (b) 36.4°C (c) 37°C (d) 37.5°C

93. A 5.0 g bullet (specific heat of material of bullet = 128 J/kg°C) moving with a velocity of 200 m/s enters a sand bag and stops. If the entire kinetic energy of the bullet is changed into heat energy that is added to the bullet, then the rise in the temperature of the bullet is

Critical Thinking

(a) 312.5°C (b) 156°C (c) 500°C (d) 624°C

94. A system S receives heat continuously from an electrical heater of power 10 W. The temperature of S becomes constant at 50°C when the surrounding temperature is 20°C. After the heater is switched off, S cools from 35.1°C to 34.9°C in 1 minute. The heat capacity of S is **Toughnut**

(a) 100 J/°C (b) 300 J/°C
(c) 750 J/°C (d) 1500 J/°C

95. An ideal Black-body at room temperature is thrown into a furnace. It is observed that **Tricky**

(a) initially it is the darkest body and at later times the brightest
(b) it is the darkest body at all times
(c) it cannot be distinguished at all times
(d) initially it is the darkest body and at later times it cannot be distinguished

96. A black body is at 727° C. It emits energy at a rate which is proportional to

(a) $(1000)^4$ (b) $(1000)^2$ (c) $(727)^4$ (d) $(727)^2$

97. A spherical black body with a radius of 12 cm radiates 450 W power at 500 K. If the radius were halved and the temperature doubled, the power radiated in watt would be

(a) 225 (b) 450 (c) 900 (d) 1800

98. Three rods of identical cross-sectional area and made from the same metal from the sides of an isosceles traingle *ABC*, right-angled at *B*. The points *A* and *B* are maintained at temperatures *T* and $(\sqrt{2})\,T$ respectively. In the steady state, the temperature of the point *C* is T_c. Assuming that only heat conduction takes place, T_c / T is **Critical Thinking**

(a) $\dfrac{1}{2(\sqrt{2}-1)}$ (b) $\dfrac{3}{\sqrt{2}+1}$

(c) $\dfrac{1}{\sqrt{3}(\sqrt{2}-1)}$ (d) $\dfrac{1}{\sqrt{2}+1}$

99. Two rods A and B of different materials are welded together as shown in figure. Their thermal conductivities are K_1 and K_2. The thermal conductivity of the composite rod will be :

(a) $\dfrac{3(K_1+K_2)}{2}$

(b) K_1+K_2

(c) $2(K_1+K_2)$

(d) $\dfrac{K_1+K_2}{2}$

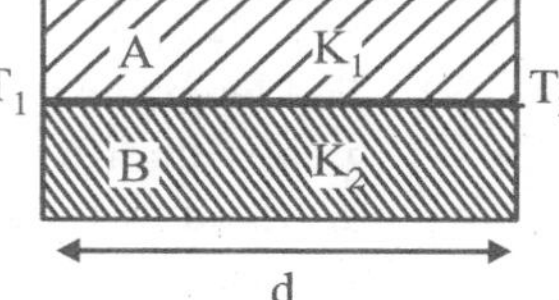

100. A large cylindrical rod of length *L* is made by joining two identical rods of copper and steel of length $\left(\dfrac{L}{2}\right)$ each. The rods are completely insulated from the surroundings. If the free end of copper rod is maintained at 100°C and that of steel at 0°C then the temperature of junction is (Thermal conductivity of copper is 9 times that of steel) **Toughnut**

(a) 90°C (b) 50°C (c) 10°C (d) 67°C

Answer KEYs

1	(c)	**11**	(c)	**21**	(a)	**31**	(d)	**41**	(c)	**51**	(c)	**61**	(c)	**71**	(d)	**81**	(d)	**91**	(b)
2	(b)	**12**	(a)	**22**	(c)	**32**	(c)	**42**	(b)	**52**	(a)	**62**	(d)	**72**	(c)	**82**	(b)	**92**	(b)
3	(a)	**13**	(a)	**23**	(b)	**33**	(d)	**43**	(b)	**53**	(c)	**63**	(d)	**73**	(a)	**83**	(b)	**93**	(b)
4	(c)	**14**	(a)	**24**	(b)	**34**	(b)	**44**	(b)	**54**	(c)	**64**	(d)	**74**	(d)	**84**	(c)	**94**	(d)
5	(b)	**15**	(c)	**25**	(a)	**35**	(c)	**45**	(b)	**55**	(a)	**65**	(c)	**75**	(d)	**85**	(d)	**95**	(a)
6	(d)	**16**	(b)	**26**	(b)	**36**	(b)	**46**	(c)	**56**	(c)	**66**	(d)	**76**	(a)	**86**	(b)	**96**	(a)
7	(a)	**17**	(b)	**27**	(d)	**37**	(c)	**47**	(d)	**57**	(d)	**67**	(d)	**77**	(c)	**87**	(a)	**97**	(d)
8	(c)	**18**	(c)	**28**	(c)	**38**	(b)	**48**	(a)	**58**	(b)	**68**	(a)	**78**	(c)	**88**	(c)	**98**	(b)
9	(a)	**19**	(c)	**29**	(c)	**39**	(a)	**49**	(a)	**59**	(c)	**69**	(a)	**79**	(b)	**89**	(d)	**99**	(d)
10	(a)	**20**	(c)	**30**	(b)	**40**	(b)	**50**	(a)	**60**	(a)	**70**	(c)	**80**	(b)	**90**	(b)	**100**	(a)

Hints & Solutions

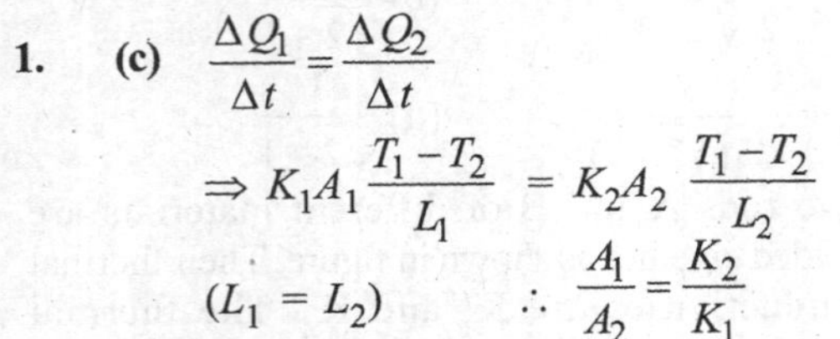

1. (c) $\frac{\Delta Q_1}{\Delta t} = \frac{\Delta Q_2}{\Delta t}$

$\Rightarrow K_1 A_1 \frac{T_1 - T_2}{L_1} = K_2 A_2 \frac{T_1 - T_2}{L_2}$

$(L_1 = L_2)$ $\therefore \frac{A_1}{A_2} = \frac{K_2}{K_1}$

2. (b) Work done to raise the temperature of 100 gm water through 10°C is
$W = JQ = 4.2 \times (100 \times 10^{-3} \times 1000 \times 10)$
$= 4200$ J

3. (a) The latent heat of vaporization is always greater than latent heat of fusion because in liquid to vapour phase change there is a large increase in volume. Hence more heat is required as compared to solid to liquid phase change.

4. (c) Temperature difference between the end points A and D = 200 – 20 = 180°C
As the resistances for the three parts are equal, the temperature difference must be distribuited equally in the three parts (= 180/3 = 60°C)
∴ Temperature of B = 200°C – 60° = 140°C.

5. (b) As 1g of steam at 100°C melts 8g of ice at 0°C.
10 g of steam will melt 8× 10 g of ice at 0°C
Water in calorimeter = 500 + 80 + 10g = 590g

6. (d) The thermal resistance is given by

$$\frac{x}{KA} + \frac{4x}{2KA} = \frac{x}{KA} + \frac{2x}{KA} = \frac{3x}{KA}$$

$$\therefore \frac{dQ}{dt} = \frac{\Delta T}{\frac{3x}{KA}} = \frac{(T_2 - T_1)KA}{3x}$$

$$= \frac{1}{3}\left\{\frac{A(T_2 - T_1)K}{x}\right\} \qquad \therefore f = \frac{1}{3}$$

7. (a) Initially, on heating temperature rises from – 100°C to 0°C. then ice melts and temperature does not rise. After the whole ice has melted, temperature beigins to rise until it reaches 100°C. then it becomes constant, as at the boiling point will not rise.

8. (c) The heat lost in condensation = $x \times 540$ cal.
$\therefore \; x \times 540 = y \times 80 + y \times 1 \times (100 - 0)$
or $\frac{x}{y} = \frac{1}{3}$.

9. (a) Among glass, wood and metals, metals expand more for same rise in temperature.

10. (a) $V + \Delta V = (L + \Delta L)^3 = (L + \alpha L \Delta T)^3$
$= L^3 + (1 + 3\alpha\Delta T + 3\alpha^2\Delta T^2 + \alpha^3\Delta T^3)$
$\Rightarrow$ α^2 and α^3 terms are neglected.
$\therefore \; V(1 + \gamma\Delta T) = V(1 + 3\alpha\Delta T)$
$1 + \gamma\Delta T = 1 + 3\alpha\Delta T \qquad \therefore \gamma = 3\alpha.$

11. (c) Initial diameter of tyre = (1000 – 6) *mm* = 994 *mm*, so initial radius of tyre

$$R = \frac{994}{2} = 497\,mm$$

and change in diameter ΔD = 6 *mm* so $\Delta R = \frac{6}{2} = 3\,mm$
After increasing temperature by Δθ tyre will fit onto wheel
Increment in the length (circumference) of the iron tyre
$\Delta L = L \times \alpha \times \Delta\theta = L \times \frac{\gamma}{3} \times \Delta\theta \quad [\text{As } \alpha = \frac{\gamma}{3}]$

$$2\pi\Delta R = 2\pi R\left(\frac{\gamma}{3}\right)\Delta\theta \Rightarrow \Delta\theta$$

$$= \frac{3}{\gamma}\frac{\Delta R}{R} = \frac{3 \times 3}{3.6 \times 10^{-5} \times 497} \Rightarrow \Delta\theta \simeq 500°C$$

12. (a) The area of circular hole increases when we heat the metal sheet & expansion of metal sheet will be independent of shape & size of the hole.

13. (a) $\gamma_{real} = \gamma_{app.} + \gamma_{vessel}$
So $(\gamma_{app.} + \gamma_{vessel})_{glass} = (\gamma_{app} + \gamma_{vessel})_{steel}$
$\Rightarrow \; 153 \times 10^{-6} + (\gamma_{vessel})_{glass}$
$= 144 \times 10^{-6} + (\gamma_{vessel})_{steel}$
Further, $(\gamma_{vessel})_{steel} = 3\alpha = 3 \times (12 \times 10^{-6})$
$= 36 \times 10^{-6}/°C$
$\Rightarrow \; 153 \times 10^{-6} + (\gamma_{vessel})_{glass}$
$= 144 \times 10^{-6} + 36 \times 10^{-6}$
$\Rightarrow \; (\gamma_{vessel})_{glass} = 3\alpha = 27 \times 10^{-6}\,/°C$
$\Rightarrow \; \alpha = 9 \times 10^{-6}\,/°C$

14. (a) $\frac{\ell}{K_1 A} + \frac{\ell}{2K_2 A} + \frac{\ell}{K_3 A} = \frac{\ell}{KA} + \frac{\ell}{2KA} + \frac{\ell}{KA}$

$$\frac{5}{2K}=\left(\frac{1}{K_1}+\frac{1}{2K_2}+\frac{1}{K_3}\right)$$

$$K_{eq}=\frac{5}{2\left(\frac{1}{K_1}+\frac{1}{2K_2}+\frac{1}{K_3}\right)}$$

15. (c) $V_1=\frac{(45-25)g}{d_1}$ and $V_2=\frac{(45-27)g}{d_2}$

$\therefore \quad \frac{V_1}{V_2}=\frac{20}{18}\times\frac{d_2}{d_1}$

or $\quad \frac{V_1}{V_1(1+r_{metal}\times 10)}=\frac{20}{18}\times\left[\frac{d_1}{(1+r_\ell\times 10)d_1}\right]$

After simplifying, we get

$\alpha=\frac{\gamma_{metal}}{3}=2.6\times10^{-3}/°C$

16. (b)

17. (b) The volume of cavity inside the solid ball increases when it is heated.

18. (c) $\frac{\text{Reading on any scale}-\text{LFP}}{\text{UFP}-\text{LFP}}$

= constant for all scales

$$\frac{340-273}{373-273}=\frac{°y-(-160)}{-50-(-160)}\Rightarrow\frac{67}{100}=\frac{y+160}{110}$$

$\therefore \quad y=-86.3°$ Y

19. (c) Due to volume expansion of both mercury and flask, the change in volume of mercury relative to flask is given by

$\Delta V=V_0\left[\gamma_L-\gamma_g\right]\Delta\theta=V\left[\gamma_L-3\alpha_g\right]\Delta\theta$

$=50\,[\,180\times10^{-6}-3\times9\times10^{-6}](38-18)$

$=0.153$ cc

20. (c) $L=L_0(1+\alpha\Delta\theta)\Rightarrow\frac{L_1}{L_2}=\frac{1+\alpha(\Delta\theta)_1}{1+\alpha(\Delta\theta)_2}$

$\Rightarrow\quad\frac{10}{L_2}=\frac{1+11\times10^{-6}\times20}{1+11\times10^{-6}\times19}\Rightarrow L_2=9.99989$

$\Rightarrow$ Length is shorten by

$10-9.99989=0.00011=11\times10^{-5}$ cm.

21. (a) $T_1=0°C,\ T_2=100°C$

$\gamma=\frac{V_2-V_1}{100V_1};\alpha=\frac{\ell_2-\ell_1}{100\ell_1}$

$\ell_1=10$ cm; $\ell_2-\ell_1=0.08$ mm $=0.008$ cm

$\alpha=\frac{0.008}{10\times100}=8\times10^{-6}/°C;$

$\gamma=3\alpha=24\times10^{-6}/°C.$

$\therefore \quad 24\times10^{-6}=\frac{V_2-1000}{1000\times100}$

$V_2-1000=24\times10^{-6}\times10^5=2.4$

$\therefore \quad V_2=1002.4$ cc

22. (c) $Q_1=kA\left(\frac{800-100}{x}\right)t=mL_{steam}$

$Q_2=kA\left(\frac{800-0}{1-x}\right)t=mL_{ice}$

Dividing, $\left(\frac{700}{x}\right)\left(\frac{1-x}{800}\right)=7$

or $1-x=8x$ or $x=\frac{1}{9}$m

23. (b)

24. (b) Let the final temperature be T.

Then $200\times1\times(T-20)+20\times(T-20)$

$=440\,(92-T)$

Solving it, we get $T=68°C$.

25. (a) Suppose, height of liquid in each arm before rising the temperature is l.

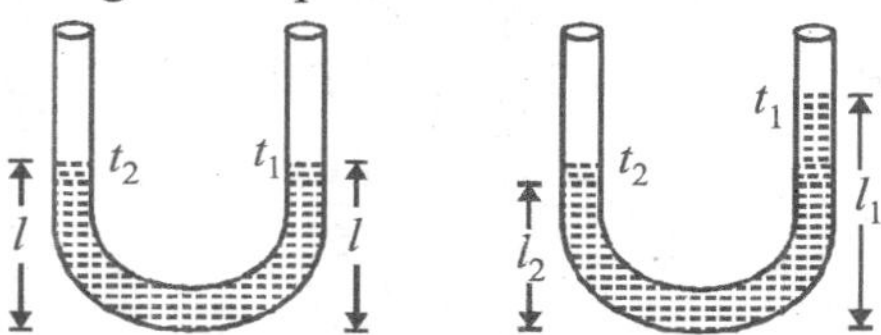

With temperature rise height of liquid in each arm increases *i.e.* $l_1>l$ and $l_2>l$

Also $l=\frac{l_1}{1+\gamma t_1}=\frac{l_2}{1+\gamma t_2}$

$\Rightarrow l_1+\gamma l_1t_2=l_2+\gamma l_2t_1\Rightarrow\gamma=\frac{l_1-l_2}{l_2t_1-l_1t_2}.$

26. (b)

27. (d) $\ell=5$m $\quad t_1=0°C$

$\ell_2=5.01$m $\quad t_2=100°C$

$\alpha=\frac{\ell_2-\ell_1}{\ell_1(t_2-t_1)}=\frac{5.01-5}{5\times100}=2\times10^{-5}/°C.$

28. (c) $\Delta t=\frac{1}{2}\alpha\,\Delta T\times t$

$\therefore \quad 5=\frac{1}{2}\alpha\,(T-15)\times86400$

and $10=\frac{1}{2}\alpha(30-T)\times86400$

29. (c) Time lost/gained per day $=\frac{1}{2}\alpha\Delta\theta\times86400$ second

$12=\frac{1}{2}\alpha(40-\theta)\times86400$ .... (i)

$4=\frac{1}{2}\alpha(\theta-20)\times86400$ ....(ii)

On dividing we get, $3=\frac{40-\theta}{\theta-20}$

$3\theta-60=40-\theta$

$4\theta=100 \Rightarrow \theta=25°C$

30. (b) $\gamma_r=\gamma_a+\gamma_v$ where γ_r = coefficient of real expansion, γ_a = coefficient of apparent expansion and γ_v = coefficient of expansion of vessel.

For copper: $\gamma_r=C+3\alpha_{Cu}=C+3A$

For silver, $\gamma_r=S+3\alpha_{Ag}$

$\Rightarrow C+3A=S+3\alpha_{Ag} \Rightarrow \alpha_{Ag}=\frac{C-S+3A}{3}$

31. (d) From question, $(\ell_2-\ell_1)$ is maintained same at all temperatures hence change in length for both rods should be same

i.e., $\Delta\ell_1=\Delta\ell_2$

As we know, coefficient of linear expansion,

$\alpha=\frac{\Delta\ell}{\ell_0\Delta T}$

$\ell_1\alpha_1\Delta T=\ell_2\alpha_2\Delta T \Rightarrow \ell_1\alpha_1=\ell_2\alpha_2$

32. (c) $\frac{C}{5}=\frac{F-3^2}{9} \Rightarrow \frac{\Delta C}{5}=\frac{\Delta F}{9}$

$\Delta C=\frac{5}{9}\Delta F \Rightarrow \Delta C=\Delta K$

33. (d) $\frac{(m_W C_W+W)\Delta T}{t}=\frac{(m_\ell C_\ell+W)\Delta T}{t}$

(W = water equivalent of the vessel)

or, $m_W C_W=m_\ell C_\ell$

$\therefore$ Specific heat of liquid, $C=\frac{m_W C_W}{m_\ell}$

$=\frac{50\times1}{100}=0.5$ kcal / kg

34. (b)

35. (c) $V=V_0(1+\alpha\Delta\theta)$ where $\Delta\theta$ is the difference in temperature, α is the coefficient of volume expansion.

i.e., $\frac{V-V_0}{V_0}\cdot\frac{1}{\Delta\theta}=\alpha$

$$\alpha=\frac{\frac{m}{d_2}-\frac{m}{d_1}}{m/d_1}\cdot\frac{1}{\Delta\theta}=\frac{(d_1-d_2)}{(d_1d_2)}\times\frac{d_1}{\Delta\theta}$$

$$=\frac{(d_1-d_2)}{d_2}\cdot\frac{1}{\Delta\theta}=\frac{1000-958.4}{958.4\times96}$$

$=4.5\times10^{-4}$ per degree K.

36. (b)

37. (c) Strain developed

$\varepsilon=\alpha\Delta T=(12\times10^{-6})(50)=6\times10^{-4}$

Strain will be negative, as the rod is in a compressed state.

38. (b) Young's modulus

$=\frac{\text{Thermal stress}}{\text{Strain}}=\frac{F/A}{\Delta L/L}$

$Y=\frac{F}{A.\alpha.\Delta\theta}\left(\because \frac{\Delta L}{L}=\alpha\Delta\theta\right)$

Force developed in the rail $F=YA\alpha\Delta t$

$=2\times10^{11}\times40\times10^{-4}\times1.2\times10^{-5}\times10$

$=9.6\times10^4=1\times10^5$ N

39. (a)

40. (b) $\rho=\frac{\rho_0}{1+\gamma\Delta\theta}\cong\rho_0(1-\gamma\Delta\theta)$; $\frac{\Delta\rho}{\rho}=-\gamma\Delta\theta$

$\frac{1}{100}=3\times10^{-5}\Delta\theta$; $\Delta\theta=\frac{100}{3}=33.3°C$

41. (c) $\frac{\Delta\ell}{\ell}=\frac{\pi\times0.05}{\pi\times40}=\frac{1}{800}$

$T=Y\frac{\Delta\ell}{\ell}\times A$

$=200\times10^9\times\frac{1}{800}\times1\times10^{-6}=250$ N

42. (b) Temperature of B will be higher because, due to expansion centre of mass B will come down same heat is supplied but in B, Potential energy is decreased therefore internal energy gain will be more.

43. (b) Let A_0 and At be the areas of cross-section of the tube at temperature 0 and t respectively,

l = length of the liquid column (constant)

V_0 and V_t be the volumes of the liquid at temperature 0 and t respectively,

$V_0=\ell A_0$ $\quad V_t=\ell A_t$

$V_t=V_0(1+\gamma t)\,A_t=A_0(1+2\alpha t)$

$\therefore V_t=\ell A_0(1+2\alpha t)=V_0(1+\gamma t)$

$=\ell A_0(1+\gamma t)$ or $\gamma=2\alpha$.

44. (b) $2.60+2.60\times\alpha_{steel}\Delta T=2.60\,\alpha_{brass}\Delta T$

45. (b) When there is no change in liquid level in vessel then $\gamma'_{real}=\gamma'_{vessel}$

Change in volume in liquid relative to vessel

$\Delta V_{app}=V\gamma'_{app}\Delta\theta=V(\gamma'_{real}-\gamma'_{vessel})$

46. **(c)** Initial diameter of tyre = (1000 – 6) *mm* = 994 *mm*, so initial radius of tyre

$R = \frac{994}{2} = 497 mm$

and change in diameter $\Delta D = 6\ mm$

so $\Delta R = \frac{6}{2} = 3\ mm$

After increasing temperature by $\Delta\theta$ tyre will fit onto wheel

Increment in the length (circumference) of the iron tyre

$\Delta L = L \times \alpha \times \Delta\theta = L \times \frac{\gamma}{3} \times \Delta\theta \quad [\text{As } \alpha = \frac{\gamma}{3}]$

$2\pi\Delta R = 2\pi R\left(\frac{\gamma}{3}\right)\Delta\theta \Rightarrow \Delta\theta$

$= \frac{3\ \Delta R}{\gamma\ R} = \frac{3 \times 3}{3.6 \times 10^{-5} \times 497} \Rightarrow \Delta\theta \simeq 500°C$

47. **(d)** Water equivalent is the quantity of water whose thermal capacity is same as the heat capacity of the body.

48. **(a)** Heat produced $= ms\Delta T = \frac{1}{2}\left(\frac{1}{2}mv^2\right)$

$\Rightarrow \Delta T = \frac{v^2}{4s} = \frac{(200)^2}{4 \times 125} = \frac{4 \times 10^4}{4 \times 125} = 80°C$

49. **(a)**

50. **(a)** $W = JQ \Rightarrow \frac{1}{2}\left(\frac{1}{2}Mv^2\right) = J(m.c.\Delta\theta)$

$\frac{1}{4} \times 1 \times (50)^2 = 4.2[200 \times 0.105 \times \Delta\theta]$

$\Rightarrow \Delta\theta = 7.1°C$

51. **(c)** Specific heat of water = 4200 J/kg-K

Latent heat of fusion $= 3.36 \times 10^5$ J/kg

Latent heat of vapourisation

$= 22.68 \times 10^5$ J/kg

$x \times 10^{-3} \times 22.68 \times 10^5$ J

$= y \times 10^{-3} \times 3.36 \times 10^5 \text{ J} + y \times 10^{-3} \times 4200 \times 100$

$\therefore \frac{x}{y} = \frac{7.56}{22.68} = \frac{1}{3}$

52. **(a)** $\Delta\theta = 0.0023h = 0.0023 \times 100 = 0.23°C$

53. **(c)** Since specific heat = 0.6 *kcal*/*g* × °C = 0.6 *cal*/*g* × °C

From graph it is clear that in a minute, the temperature is raised from 0°C to 50°C.

⇒ Heat required for a minute = 50 × 0.6 × 50 = 1500 *cal*.

Also from graph, Boiling point of wax is 200°C.

54. **(c)** **55.** **(a)**

56. **(c)** Let x kg of ice can melt

Using law of Calorimetry,

Heat lost by copper = Heat gained by ice

$2 \times 400 \times (500 - 0) = x \times 3.5 \times 10^5$

or $x = \frac{2 \times 400 \times 500}{3.5 \times 10^5} = \frac{8}{7}$ kg

57. **(d)** Let K_1 and K_2 be the thermal conductivities of larger and smaller sphere.

For bigger sphere,

$\frac{K_1 4\pi (3r)^2 \times 100}{t/4} = \frac{\frac{4}{3}\pi(3r)^3 \rho L}{30 \times 60}$

For smaller sphere,

$\frac{K_2 \times 4\pi r^2 \times 100}{t} = \frac{\frac{4}{3}\pi r^3 \rho L}{20 \times 60} \quad \therefore \frac{K_1}{K_2} = \frac{1}{2}$

58. **(b)**

59. **(c)** Since no heat flows through the central rod, temperature at the ends of the central rod will be same. So, it will form a wheatstone like network.

So, $\frac{K_1}{K_2} = \frac{K_3}{K_4} \Rightarrow k_1k_4 = k_2k_3$

60. **(a)** By the law of conservation of energy, energy given by heater must be equal to the sum of energy gained by water and energy lost from the lid.

$Pt = ms\ \Delta\theta$ + energy lost

$2000t = 3 \times 4.2 \times 10^3 \times (97 - 27) + 130t$

$\Rightarrow t = 472$ sec

61. **(c)** As the temperature difference $\Delta T = 10°C$ as well as the thermal resistance is same for both the cases, so thermal current or rate of heat flow will also be same for both the cases.

62. **(d)** **63.** **(d)**

64. **(d)** Thermal resistance $R = l/KA$

Where l = 20cm. & A(cylindrical rod)

$= \pi r^2 = 40\pi cm^2$

So $R = \frac{20}{0.5 \times 40\pi} = 0.318 \frac{cal}{sec \times° C}$

65. **(c)** According to Stefan's Law, the rate of loss of heat is $\frac{Q}{t} = \sigma A(T_1^4 - T_2^4) \times e$

here $\sigma = 5.67 \times 10^{-8} J/m^2 \times sec.K^2$,

$T_1 = 527 + 273 = 800K$,

$T_2 = 27 + 273 = 300K$ & $A = 200 \times 10^{-4} m^2$

So. $\dfrac{Q}{t} = 5.67\times10^{-8}\times2\times10^{-2}$

$[(800)^4-(300)^4]\times0.4$

$\cong 182$ joule

66. (d) $t \propto \dfrac{\ell}{A},\ t' \propto \dfrac{2\ell}{A/2} \Rightarrow t' = 4t = 4\times12 = 48$

67. (d) $E = \sigma\times \text{area}\times T^4$; T increases by a factor $\dfrac{3}{2}$. Area increases by a factor $\dfrac{1}{4}$.

$$\Rightarrow E' = \frac{1}{4}\times\left(\frac{3}{2}\right)^4\times E = \frac{81}{64}E$$

68. (a)

69. (a) OA refers to change of state from ice to water without change of temperature.

70. (c) According to question only one-quarter of the heat produced by falling piece of ice is absorbed in the melting of ice.

i.e., $\dfrac{mgh}{4} = mL$

$$\Rightarrow h = \frac{4L}{g} = \frac{4\times3.4\times10^5}{10} = 136 \text{ km}.$$

71. (d) $\dfrac{90-V}{2R_T} = \dfrac{V-30}{R_T} \Rightarrow V = 50°C$

72. (c) $Q = \sigma A t\,(T^4 - T_0^4)$

If T, T_0, σ and t are same for both bodies, then

$$\frac{Q_{sphere}}{Q_{cube}} = \frac{A_{sphere}}{A_{cube}} = \frac{4\pi r^2}{6a^2} \quad ...(i)$$

But according to problem,

Volume of sphere = Volume of cube

$$\Rightarrow \frac{4}{3}\pi r^3 = a^3 \Rightarrow a = r\left(\frac{4}{3}\pi\right)^{1/3}$$

Substituting the value of *a* in equation (i), we get

$$\frac{Q_{sphere}}{Q_{cube}} = \frac{4\pi r^2}{6a^2} = \frac{4\pi r^2}{6\left\{\left(\frac{4}{3}\pi\right)^{1/3} r\right\}^2} = \left(\frac{\pi}{6}\right)^{1/3} : 1$$

73. (a) Suppose temperature difference between A and B is 100°C and $\theta_A > \theta_B$

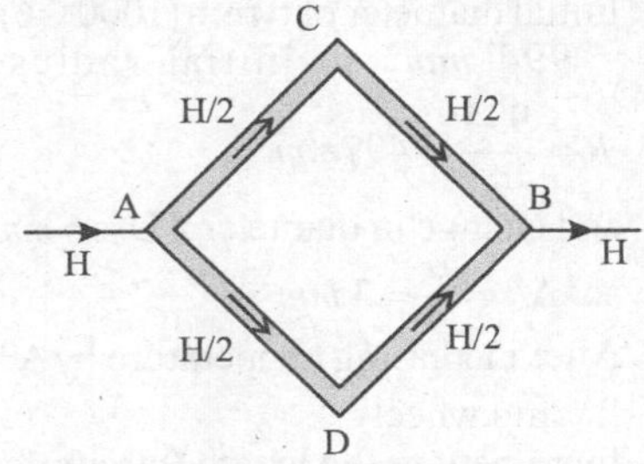

Heat current will flow from A to B via path ACB and ADB. Since all the rod are identical, so, $(\Delta\theta)_{AC} = (\Delta\theta)_{AD}$

(Because heat current $H = \dfrac{\Delta\theta}{R}$; here R = same for all)

$\Rightarrow \theta_A - \theta_C = \theta_A - \theta_D \Rightarrow \theta_C = \theta_D$

i.e., temperature difference between C and D will be zero.

74. (d)

75. (d) $\theta_{mix} = \dfrac{m_W\theta_W - \dfrac{m_iL_i}{S_W}}{m_i + m_W}$

$$= \frac{300\times25 - \dfrac{100\times80}{1}}{100+300} = -1.25°C$$

Which is not possible. Hence $\theta_{mix} = 0°C$

76. (a)

100°C (steam)

0.1 m $A = 0.36\ m^2$

ice 0°C

$$\frac{dQ}{dt} = \frac{KA(100-0)}{1} = m\frac{dL}{dt},$$

$$\frac{K\times100\times0.36}{0.1} = \frac{4.8\times3.36\times10^5}{60\times60}$$

$K = 1.24$ J/m/s/°C

77. (c) Let T be temperature of the junction

Here, $K_A = 2K_B$,

$T - T_B = 50$ K

At the steady state,

$H_A = H_B$

T_A T T_B
A B
L L

$$\therefore \frac{K_AA(T_A - T)}{L} = \frac{K_BA(T - T_B)}{L}$$

$$2K_B(T_A - T) = K_B(T - T_B)$$

$$T_A - T = \frac{T - T_B}{2} = \frac{50K}{2} = 25K$$

78. (c) Heat gain = heat lost

$C_A(16-12) = C_B(19-16) \Rightarrow \frac{C_A}{C_B} = \frac{3}{4}$

and $C_B(23-19) = C_c(28-23) \Rightarrow \frac{C_B}{C_C} = \frac{5}{4}$

$\Rightarrow \frac{C_A}{C_C} = \frac{15}{16}$ ...(i)

If θ is the temperature when A and C are mixed then,

$C_A(\theta - 12) = C_C(28-\theta)$

$\Rightarrow \frac{C_A}{C_C} = \frac{28-\theta}{\theta-12}$ ...(ii)

On solving equations (i) and (ii) $\theta = 20.2°C$

79. (b)

80. (b) As $Pt = mC\Delta T$

So, $P \times 10 \times 60 = mC\ 100$ ...(i)

and $P \times 55 \times 60 = mL$ ...(ii)

Dividing equation (i) by (ii) we get

$\frac{10}{55} = \frac{C \times 100}{L}$

$\therefore L = 550$ cal./g.

81. (d) Rate of heat loss with only room heater

$P_h = \frac{\Delta Q}{\Delta t} = C(16+20)$, C = constant

while for both heater and stove it is

$P_h + P_s = \left(\frac{\Delta Q}{\Delta t}\right)' = C(22+20)$

$\frac{P_h}{P_h + P_s} = \frac{36}{42} \Rightarrow 7P_h = 6P_h + 6P_s$

$\Rightarrow P_h = 6P_s = 6\text{kW}$

82. (b) Let lost temperature is T°C then

$m_1s_1T_1 + m_2s_2T_2 + m_3s_3T_3$

= released energy

$= m_1s_1T + m_2s_2T + m_3s_3T$ = energy taken

$\Rightarrow T = \frac{m_1s_1T_1 + m_2s_2T_2 + m_3s_3T_3}{m_1s_1 + m_2s_2 + m_3s_3}$

83. (b) $\frac{50-49.9}{5} = K\left(\frac{50+49.9}{2} - 30\right)$ ....(i)

$\frac{40-39.9}{t} = K\left[\frac{40+39.9}{2} - 30\right]$ ....(ii)

From equations (i) and (ii), we get $t \approx 10$ s.

84. (c) Rate of colling $= \frac{-d\theta}{dt} \propto \left(\frac{\theta_1+\theta_2}{2} - \theta_0\right)$

In second case average temperature will be less hence rate of colling will be less. Therefore time taken will be more than 4 minutes.

85. (d)

86. (b) For θ-t plot, rate of cooling $= \frac{dQ}{dt}$ = slope of the curve.

AT P, $\frac{dQ}{dt} = |\tan(180° - \phi_2)|$

$= \tan\phi_2 = k(\theta_2 - \theta_1)$

where k = constant.

At Q,

$\frac{dQ}{dt} = |\tan(180° - \varphi_1)| = \tan\varphi_1 = k(\theta_1 - \theta_0)$

$\therefore \frac{\tan\phi_2}{\tan\phi_1} = \frac{\theta_2 - \theta_0}{\theta_1 - \theta_0}$

87. (a) Initially at $t = 0$

Rate of cooling (R) ∝ Fall in temperature of body $(\theta - \theta_0)$

$\Rightarrow \frac{R_1}{R_2} = \frac{\theta_1 - \theta_0}{\theta_2 - \theta_0} = \frac{100-40}{80-40} = \frac{3}{2}$

88. (c)

89. (d) $\left[\frac{-dT}{dr}\right] = \frac{A}{mc}[T - T_0] = \frac{4\pi r^2}{\rho \times \frac{4}{3}\pi r^3 c}[T - T_0]$

$\therefore \left[\frac{-dT}{dr}\right] \propto \frac{1}{\rho rc}$

90. (b)

91. (b) When hot water temperature (T) and surrounding temparature (T_0) readings are noted, and $\log(T - T_0)$ is plotted versus time, we get a straight line having a negative slope; as a proof of Newton's law of cooling.

92. (b) We have $\theta - \theta_s = (\theta_0 - \theta_s)\,e^{-kt}$ ...(1)

where θ_0 = initial temperature of body = 40°C

θ = temperature of body after time t

θ_s = temperature of surrounding

Since body cools from 40°C to 38°C in 10 min, we have

$38° - 30° = (40° - 30°)\,e^{-10k}$ .... (2)

Let after 10 min, the body temp. be θ.

$\theta - 30° = (38° - 30°)\,e^{-10k}$ ..... (3)

Dividing equ. (2) by equ. (3) gives,

$\frac{8°}{\theta - 30°} = \frac{10°}{8°} \Rightarrow \theta - 30° = 6.4° \Rightarrow \theta = 36.4°$

93. (b) $\frac{1}{2}mv^2 = mC\Delta T$

$\therefore \quad \Delta T = \frac{v^2}{2C} = \frac{200^2}{2 \times 128} = 156°C$

94. (d) Rate of loss of heat $\propto$ difference in temperature with the surroundings.

At 50°C, $\frac{dQ}{dt} = k(50-20) = 10$,

where k = constant

$\therefore \quad k = 1/3$

At an angle temperature of 35°C,

$\frac{dQ}{dt} = \frac{1}{3}(35-20) J/s = 5 J/s$

Heat lost in 1 minutes

$= \frac{dQ}{dt} \times 60 J = 5 \times 60 J = 300 J = Q$

Fall in temperature = 0.2°C = $\Delta\theta$.

$Q = c\Delta\theta$.

Heat capacity

$= c = \frac{Q}{d\theta} = \frac{300J}{0.2°C} = 1500 J/°C$.

95. (a) According to Kirchoff's law, good absorbers are good emitters as well.

At high temperature (in the furnace), since it absorbs more energy, it emits more radiations as well and hence is the brightest.

96. (a) According to Stefan's law,

$E \propto T^4$

$\propto (t+273)^4 K \qquad [727°C = (727+273)K]$

$\propto (727+273)^4 K$

$\propto (1000)^4 K$

97. (d) The energy radiated per second by a black body is given by Stefan's Law

$\frac{E}{t} = \sigma T^4 \times A$, where A is the surface area.

$\frac{E}{t} = \sigma T^4 \times 4\pi r^2$ ($\because$ For a sphere, $A = 4\pi r^2$)

Dividing (ii) and (i), we get

$$\frac{E/t}{450} = \frac{(1000)^4(0.06)^2}{(500)^4(0.12)^2} = \frac{2^4}{2^2} = 4$$

$$\Rightarrow \frac{E}{t} = 450 \times 4 = 1800 \text{ W}$$

98. (b) Heat flow from B to A, A to C and C to B (for steady state condition, $\Delta Q/\Delta t$ is same)

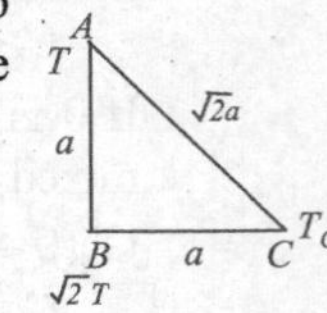

Where $\frac{\Delta Q}{\Delta t} = \frac{kA\Delta T}{\ell}$

For sides AC and CB $\left(\frac{\Delta T}{\sqrt{2}a}\right)_{AC} = \left(\frac{\Delta T}{a}\right)_{CB}$

$$\Rightarrow \frac{T - T_c}{\sqrt{2}a} = \frac{T_c - \sqrt{2}T}{a}$$

$$\Rightarrow T - T_c = \sqrt{2}T_c - 2T$$

$$\Rightarrow 3T = T_c(\sqrt{2}+1) \Rightarrow \frac{T_c}{T} = \frac{3}{\sqrt{2}+1}$$

99. (d) Heat current $H = H_1 + H_2$

$$= \frac{K_1A(T_1 - T_2)}{d} + \frac{K_2A(T_1 - T_2)}{d}$$

$$\frac{K_{EQ}2A(T_1 - T_2)}{d} = \frac{A(T_1 - T_2)}{d}[K_1 + K_2]$$

Hence equivalent thermal conductivities for two rods of equal area is given by

$$K_{EQ} = \frac{k_1 + k_2}{2}$$

100. (a) From formula temperature of junction;

$$\theta = \frac{K_{copper}\theta_{copper}l_{steel} + K_{steel}\theta_{steel}l_{copper}}{K_{copper}l_{steel} + K_{steel}l_{copper}}$$

$$= \frac{9k \times 100 \times \frac{L}{2} + k \times 0 \times \frac{L}{2}}{9k \times \frac{L}{2} + k \times \frac{L}{2}} = \frac{\frac{900}{2}kL}{\frac{10kL}{2}} = 90°C$$

11 Thermodynamics

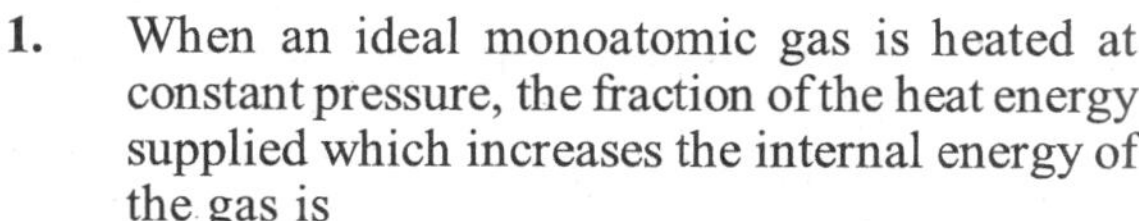

1. When an ideal monoatomic gas is heated at constant pressure, the fraction of the heat energy supplied which increases the internal energy of the gas is
(a) 2/5 (b) 3/5 (c) 3/7 (d) 5/7

2. The internal energy change in a system that has absorbed 2 Kcal of heat and done 500 J of work is
(a) 8900 J (b) 6400 J (c) 5400 J (d) 7900 J

3. In a given process on an ideal gas, $dW = 0$ and $dQ < 0$. Then for the gas **Tricky**
(a) the temperature will decrease
(b) the volume will increase
(c) the pressure will remain constant
(d) the temperature will increase

4. Unit mass of a liquid with volume V_1 is completely changed into a gas of volume V_2 at a constant external pressure P and temperature T. If the latent heat of evaporation for the given mass is L, then the increase in the internal energy of the system is
(a) Zero (b) $P(V_2 - V_1)$
(c) $L - P(V_2 - V_1)$ (d) L

5. Two kg of water is converted into steam by boiling at atmospheric pressure. The volume changes from $2 \times 10^{-3} m^3$ to $3.34\ m^3$. The work done by the system is about
(a) – 340 kJ (b) –170 kJ (c) 170 kJ (d) 340 kJ

6. A gas is compressed from a volume of $2 m^3$ to a volume of $1 m^3$ at a constant pressure of 100 N/m^2. Then it is heated at constant volume by supplying 150 J of energy. As a result, the internal energy of the gas: **Critical Thinking**
(a) increases by 250 J (b) decreases by 250 J
(c) increases by 50 J (d) decreases by 50 J

7. The internal energy of an ideal gas depends upon
(a) specific volume (b) pressure
(c) temperature (d) density

8. Thermodynamics is concerned in part with transformations between
(a) different forms of heat energy
(b) internal energy at various temperatures
(c) one form of mechanical energy into other forms
(d) heat, internal energy and mechnical work

9. 110 joules of heat is added to a gaseous system, whose internal energy is 40J; then the amount of external work done is
(a) 150 J (b) 70 J (c) 110 J (d) 40 J

10. The change in internal energy of a thermo-dynamical system which has absorbed 2 kcal of heat and done 400 J of work is (1 cal = 4.2 J)
(a) 2 kJ (b) 8 kJ (c) 3.5 kJ (d) 5.5 kJ

11. If the amount of heat given to a system is 35 J and the amount of work done on the system is 15 J, then the change in internal energy of the system is
(a) –50 J (b) 20 J (c) 30 J (d) 50 J

12. A perfect gas goes from a state A to another state B by absorbing 8×10^5 J of heat and doing 6.5×10^5 J of external work. It is now transferred between the same two states in another process in which it absorbs 10^5 J of heat. In the second process **Critical Thinking**
(a) work done by gas is 10^5 J
(b) work done on gas is 10^5 J
(c) work done by gas is 0.5×10^5 J
(d) work done on the gas is 0.5×10^5 J

13. Some of the thermodynamic parameters are state variables while some are process variables. Some grouping of the parameters are given. Choose the correct one
(a) State variables : Temperature, no. of moles
Process variables : Internal energy, work done by the gas
(b) State variables : Volume, temperature
Process variables : Internal energy, work done by the gas
(c) State variables : work done by the gas, heat rejected by the gas.
Process variables : Temperature, volume
(d) State variables : Internal energy, volume
Process variables : Work done by the gas, heat absorbed by the gas

14. An insulated container of gas has two chambers separated by an insulating partition. One of the chambers has volume V_1 and contains ideal gas at pressure P_1 and temperature T_1. The other chamber has volume V_2 and contains ideal gas at pressure P_2 and temperature T_2. If the partition is removed without doing any work on the gas, the final equilibrium temperature of the gas in the container will be **Critical Thinking**

(a) $\frac{T_1T_2(P_1V_1+P_2V_2)}{P_1V_1T_2+P_2V_2T_1}$ (b) $\frac{P_1V_1T_1+P_2V_2T_2}{P_1V_1+P_2V_2}$

(c) $\frac{P_1V_1T_2+P_2V_2T_1}{P_1V_1+P_2V_2}$ (d) $\frac{T_1T_2(P_1V_1+P_2V_2)}{P_1V_1T_1+P_2V_2T_2}$

15. A cube of side 5 cm made of iron and having a mass of 1500 g is heated from 25° C to 400°C. The specific heat for iron is 0.12 cal/g°C and the coefficient of volume expansion is 3.5×10^{-5}/°C, the change in the internal energy of the cube is (atm pressure = 1×10^5 N/m^2)
(a) 320 kJ (b) 282 kJ (c) 141 kJ (d) 423 kJ

16. 4 kg of oxygen gas is heated so as to raise its temperature from 20 to 120°C. If the heating is done at constant pressure, the external work done by the gas is (C_p = 0.219 cal/g°C and C_v = 0.157 cal/g°C)
(a) 628 kJ (b) 104 kJ (c) 366 kJ (d) 206 kJ

17. Two cylinders *A* and *B* fitted with pistons contain equal amounts of an ideal diatomic gas at 300 K. The piston of *A* is free to move while that of *B* is held fixed. The same amount of heat is given to the gas in each cylinder. If the rise in temperature of the gas in *A* is 30 K, then the rise in temperature of the gas in *B* is **Critical Thinking**
(a) 30 K (b) 18 K (c) 50 K (d) 42 K

18. The specific heat of a gas at constant pressure is greater than the specific heat of the same gas at constant volume because
(a) work is done in the expansion of the gas at constant pressure.
(b) work is done in the expansion of the gas at constant volume.
(c) the attraction between the molecules increases at constant pressure.
(d) the molecular attraction increases at constant volume.

19. The specific heat of a gas in an isothermal process is
(a) infinite (b) zero
(c) negative (d) remains constant

20. The work done in an adiabatic change in a particular gas depends only upon
(a) change in volume
(b) change in temperature
(c) change in pressure
(d) None of these

21. Which of the following holds good for an isochoric process?
(a) No work is done on the gas
(b) No work is done by the gas
(c) Both (a) and (b)
(d) None of these

22. The slopes of isothermal and adiabatic curves are related as
(a) isothermal curve slope = adiabatic curve slope
(b) isothermal curve slope = $\gamma \times$ adiabatic curve slope
(c) adiabatic curve slope = $\gamma \times$ isothermal curve slope
(d) adiabatic curve slope = $\frac{1}{2} \times$ isothermal curve slope

23. 2 k mol of hydrogen at NTP expands isobarically to twice its initial volume. The change in its internal energy is (C_v = 10 kJ/kg.K and atm pressure = 1×10^5 N/m^2)
(a) 10.9 MJ (b) 9.10 MJ
(c) 109 MJ (d) 1.09 MJ

24. When a system is taken from state i to state f along the path iaf, it is found that Q = 50 cal and W = 20 cal. Along the path ibf Q = 36 cal. W along the path ibf is

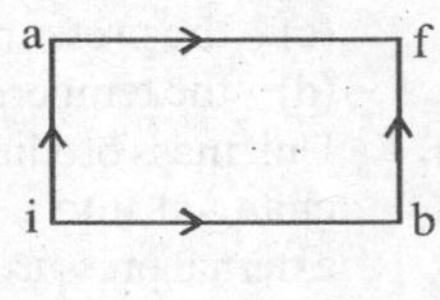

(a) 14 cal (b) 6 cal (c) 16 cal (d) 66 cal

25. Choose the incorrect statement related to an isobaric process.
(a) $\frac{V}{T}$ = constant
(b) $W = P\Delta V$
(c) Heat given to a system is used up in raising the temperature only.
(d) $\Delta Q > W$

26. A thermodynamic system undergoes cyclic process ABCDA as shown in fig. The work done by the system in the cycle is **Tricky**
(a) P_0V_0
(b) $2P_0V_0$
(c) $\frac{P_0V_0}{2}$
(d) Zero

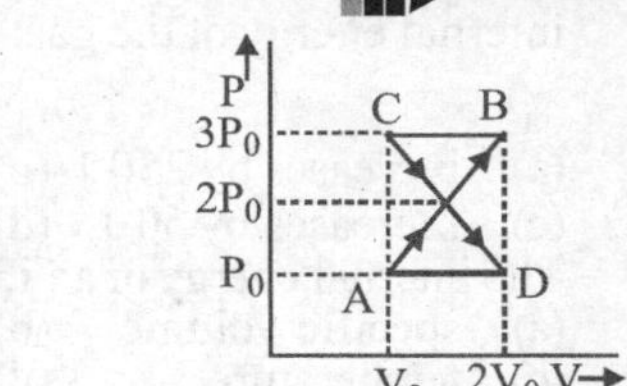

27. What will be the final pressure if an ideal gas in a cylinder is compressed adiabatically to $\frac{1}{3}$rd of its volume?

(a) Final pressure will be three times less than initial pressure.
(b) Final pressure will be three times more than initial pressure.
(c) Change in pressure will be more than three times the initial pressure.
(d) Change in pressure will be less than three times the initial pressure.

28. In the P-V diagram, I is the initial state and F is the final state. The gas goes from I ot F by (i) IAF, (ii) IBF, (iii) ICF. The heat absorbed by the gas is

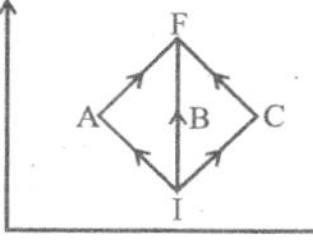

(a) the same in all three processes
(b) the same in (i) and (ii)
(c) greater in (i) than in (ii)
(d) the same in (i) and (iii)

29. A closed system undergoes a process $1 \rightarrow 2$ for which the values $W_{1\text{-}2}$ and $Q_{1\text{-}2}$ are 50 kJ and – 20 kJ respectively. If the system is returned to state 1 and $Q_{2\rightarrow 1}$ is + 10 kJ the work done $W_{2\rightarrow 1}$ is
(a) 40 kJ (b) 50 kJ (c) –60 kJ (d) –50 kJ

30. During an adiabatic process of an ideal gas, if P is proportional to $\frac{1}{V^{1.5}}$, then the ratio of specific heat capacities at constant pressure to that at constant volume for the gas is
(a) 1.5 (b) 0.25 (c) 0.75 (d) 0.4

31. The relation between U, P and V for an ideal gas in an adiabatic process is given by relation $U = a + bPV$. Find the value of adiabatic exponent (γ) of this gas.
(a) $\frac{b+1}{b}$ (b) $\frac{b+1}{a}$ (c) $\frac{a+1}{b}$ (d) $\frac{a}{a+b}$

32. An ideal gas is initially at P_1, V_1 is expanded to P_2, V_2 and then compressed adiabatically to the same volume V_1 and pressure P_3. If W is the net work done by the gas in complete process which of the following is true?

Tricky

(a) $W > 0 ; P_3 > P_1$ (b) $W < 0 ; P_3 > P_1$
(c) $W > 0 ; P_3 < P_1$ (d) $W < 0 ; P_3 < P_1$

33. For an ideal gas graph is shown for three processes. Process 1, 2 and 3 are respectively.
(a) Isobaric, adiabatic, isochoric
(b) Adiabatic, isobaric, isochoric
(c) Isochoric, adiabatic, isobaric
(d) Isochoric, isobaric, adiabatic

34. During an adiabatic process an object does 100J of work and its temperature decreases by 5K. During another process it does 25J of work and its temperature decreases by 5K. Its heat capacity for 2nd process is
(a) 20 J/K (b) 24 J/K (c) 15 J/K (d) 100 J/K

35. Heat energy absorbed by a system in going through a cyclic process shown in the given figure is
(a) $10^7 \pi$ J
(b) $10^4 \pi$ J
(c) $10^2 \pi$ J
(d) $10^{-3} \pi$ J

36. The pressure inside a tyre is 4 times that of atmosphere. If the tyre bursts suddenly at temperature 300 K, what will be the new temperature?
(a) $300(4)^{7/2}$ (b) $300(4)^{2/7}$
(c) $300(2)^{7/2}$ (d) $300(4)^{-2/7}$

37. A monatomic ideal gas, initially at temperature T_1, is enclosed in a cylinder fitted with a frictionless piston. The gas is allowed to expand adiabatically to a temperature T_2 by releasing the piston suddenly. If L_1 and L_2 are the length of the gas column before and after expansion respectively, then $\frac{T_1}{T_2}$ is given by
(a) $\left(\frac{L_1}{L_2}\right)^{2/3}$ (b) $\frac{L_1}{L_2}$ (c) $\frac{L_2}{L_1}$ (d) $\left(\frac{L_2}{L_1}\right)^{2/3}$

38. One mole of an ideal gas is taken from state A to state B by three different processes, (i) ACB (ii) ADB (iii) AEB as shown in the P-V diagram. The heat absorbed by the gas is–

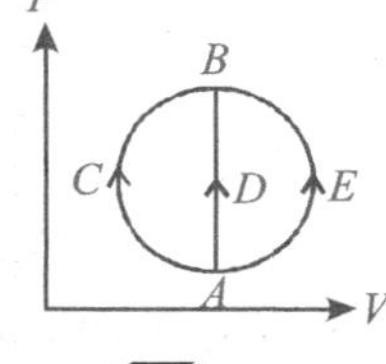

Tricky

(a) greater in process (ii) than in (i)
(b) the least in process (ii)
(c) the same in (i) and (iii)
(d) less in (iii) than in (ii)

39. During an isothermal expansion, a confined ideal gas does –150 J of work against its surroundings. This implies that
(a) 150 J heat has been removed from the gas
(b) 300 J of heat has been added to the gas
(c) no heat is transferred because the process is isothermal
(d) 250 J of heat has been added to the gas

40. A mass of diatomic gas ($\gamma = 1.4$) at a pressure of 2 atmospheres is compressed adiabatically so that its temperature rises from 27°C to 927°C. The pressure of the gas in final state is

(a) 28 atm (b) 68.7 atm (c) 256 atm (d) 8 atm

41. One mole of an ideal gas goes from an initial state A to final state B via two processes : It first undergoes isothermal expansion from volume V to $3V$ and then its volume is reduced from $3V$ to V at constant pressure. The correct P-V diagram representing the two processes is : **Tricky**

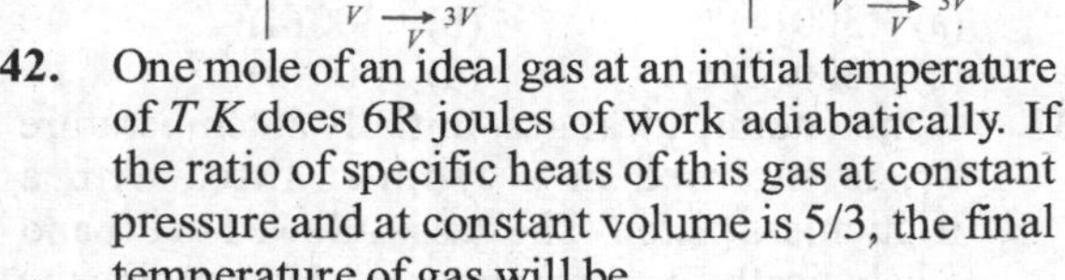

42. One mole of an ideal gas at an initial temperature of T K does 6R joules of work adiabatically. If the ratio of specific heats of this gas at constant pressure and at constant volume is 5/3, the final temperature of gas will be

(a) $(T-4)\,K$ (b) $(T+2.4)\,K$
(c) $(T-2.4)\,K$ (d) $(T+4)\,K$

43. In pressure-volume diagram, the isochoric, isothermal, isobaric and iso-entropic parts respectively, are **Tricky**

(a) BA, AD, DC,CB
(b) DC, CB, BA, AD
(c) AB, BC, CD, DA
(d) CD, DA, AB, BC

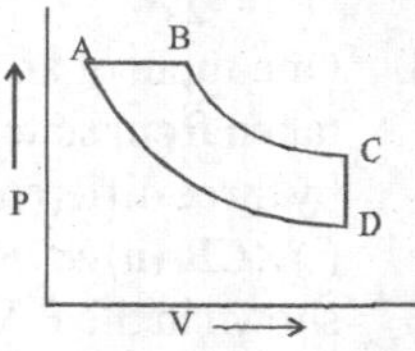

44. A certain diatomic gas has the same specific heats as an ideal gas but a slightly different equation of state : $PV = R(T + \alpha T^2)$, $\alpha = 0.001\ K^{-1}$. The temperature of the gas is raised from $T_1 = 300$ K to T_2 at constant pressure. It is found that work done on the gas is 70% higher than what would be on an ideal gas. Choose the correct statement(s).

(a) $T_2 = 400$ K, internal energy increases by 250R per mole.
(b) $T_2 = 400$ K, internal energy increases by 350R per mole.
(c) Total heat absorbed in the process is 450R per mole.
(d) Total heat absorbed in the process is 520R per mole.

45. Starting with the same initial conditions, an ideal gas expands from volume V_1 to V_2 in three different ways. The work done by the gas is W_1 if the process is purely isothermal, W_2 if purely isobaric and W_3 if purely adiabatic. Then

(a) $W_2 > W_1 > W_3$
(b) $W_2 > W_3 > W_1$
(c) $W_1 > W_2 > W_3$
(d) $W_1 > W_3 > W_2$

46. Which of the following graphs correctly represents the variation of $\beta = -\dfrac{dV/dP}{V}$ with P for an ideal gas at constant temperature? **Tricky**

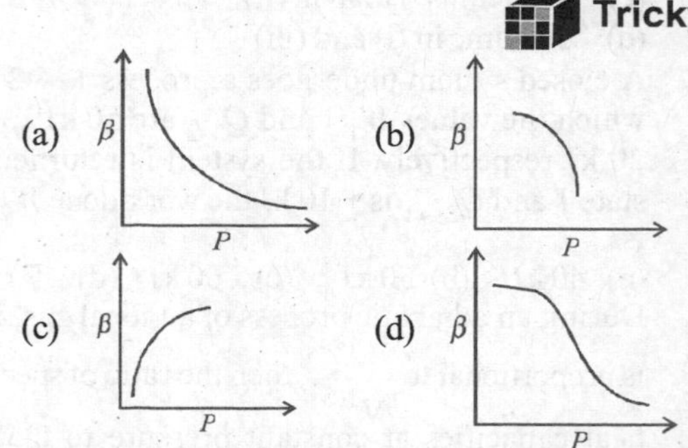

47. There are two processes ABC and DEF. In which of the process is the amount of work done by the gas greater? **Critical Thinking**

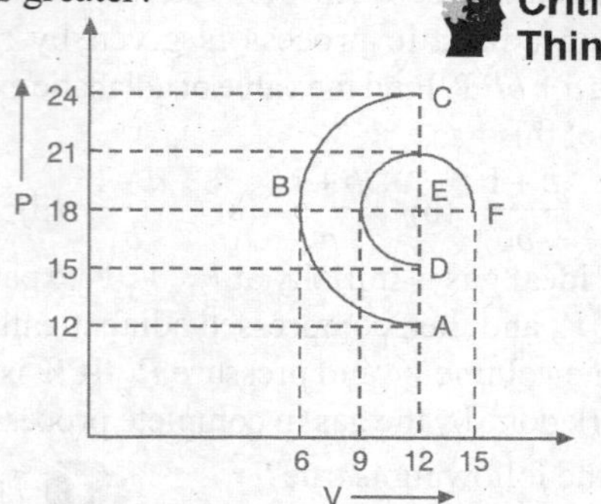

(a) ABC
(b) DEF
(c) Equal in both processes
(d) It cannot be predicted

48. In a thermodynamic process, pressure of a fixed mass of gas is changed in such a manner that the gas molecules gives out 30 joules of heat and 10 J of work is done on the gas. If the initial internal energy of the gas was 40J, then final internal energy will be

(a) zero (b) 80 J (c) 20 J (d) –20 J

49. Calculate the work done when 1 mole of a perfect gas is compressed adiabatically. The initial pressure and volume of the gas are 10^5 N/m^2 and 6 litre respectively. The final volume of the gas is 2 litres. Molar specific heat of the gas at constant volume is 3R/2. [Given $(3)^{5/3} = 6.19$]

Toughnut

(a) –957 J (b) +957 J (c) – 805 J (d) + 805 J

50. If a given mass of an ideal gas followed a relation VT = constant during a process, then which of the following graphs correctly represents the process ?

Tricky

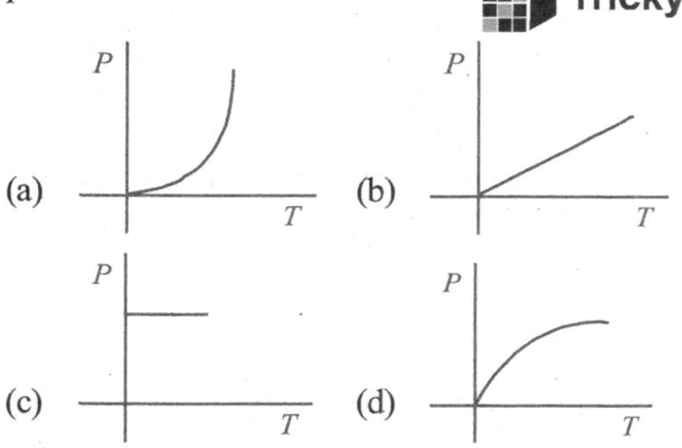

51. The state of an ideal gas is changed through an isothermal process at temperature T_0 as shown in figure. The work done by gas in going from state B to C is double the work done by gas in going from state A to B. If the pressure in the state B is $P_0/2$ then the pressure of the gas in state C is

Critical Thinking

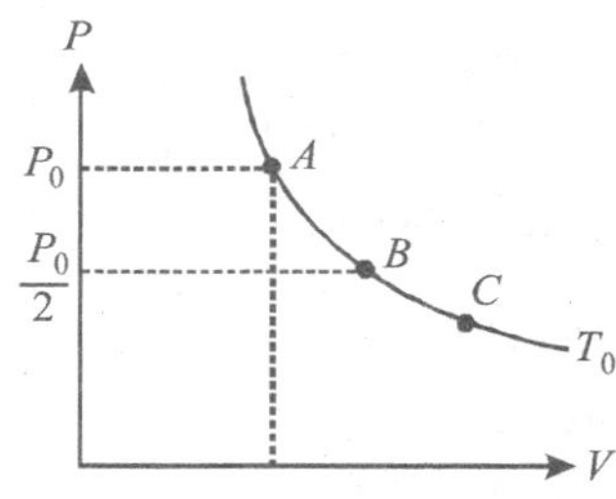

(a) $P_0/2$ (b) $P_0/4$ (c) $P_0/6$ (d) $P_0/8$

52. Certain perfect gas is found to obey $PV^{3/2}$ = constant during an adiabatic process. If such a gas at initial temperature T is adiabatically compressed to half the initial volume, its final temperature will be

(a) $\sqrt{2}$ T (b) 2T (c) $2\sqrt{2}$ T (d) 4T

53. Figure shows the variation of internal energy (U) with the pressure (P) of 2.0 mole gas in cyclic process abcda. The temperature of gas at c and d are 300 K and 500 K. Calculate the heat absorbed by the gas during the process.

Toughnut

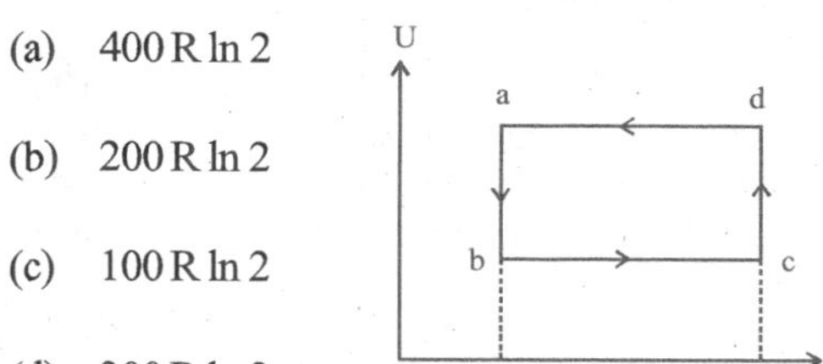

(a) 400 R ln 2
(b) 200 R ln 2
(c) 100 R ln 2
(d) 300 R ln 2

54. A one mole sample of an ideal gas is carried around the thermodynamic cycle shown in the figure. The cycle consists of three steps : (i) an isothermal expansion (a → b), (ii) an isobaric compression (b → c), and (iii) a constant volume increases in pressure (c → a). If T_a = 400 K, P_a = 4 atm and $P_b = P_c$ = 1 atm, the work done by the gas per cycle is

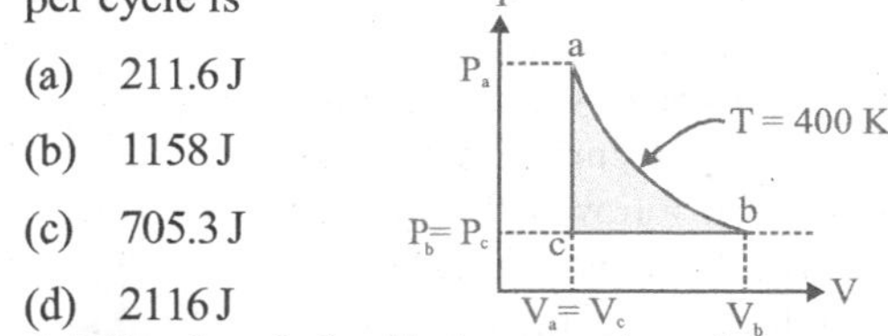

(a) 211.6 J
(b) 1158 J
(c) 705.3 J
(d) 2116 J

55. A 500 ml sealed cylinder contains nitrogen at a pressure of 1 atm. A tiny glass tube lies at the bottom of the cylinder. Its volume is 0.50 ml and it contains hydrogen at a pressure of 4.5 atm. The glass tube is broken so that hydrogen also fills the cylinder. The new pressure in the cylinder is (1 atm = 1 × 10^5 N/m^2)

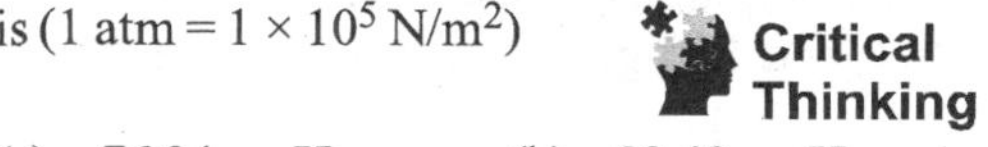

(a) 76.34 cm Hg (b) 82.40 cm Hg
(c) 94.24 cm Hg (d) 104.34 cm Hg

56. In a reversible cyclic process of a gaseous system
(a) ΔQ = ΔU (b) ΔU = ΔW
(c) ΔW = 0 (d) ΔU = 0

57. Which of the following processes is irreversible?
(a) Transfer of heat by radiation
(b) Adiabatic changes performed slowly
(c) Extremely slow extension of a spring
(d) Isothermal changes performed slowly

58. A Carnot engine works between a source and a sink maintained at constant temperatures T_1 and T_2. For efficiency to be the greatest
(a) T_1 and T_2 should be high
(b) T_1 and T_2 should be low
(c) T_1 should be low and T_2 should be high
(d) T_1 should be high and T_2 should be low

59. The temperature of source and sink of a heat engine are 127°C and 27°C respectively. An inventor claims its efficiency to be 26%, then:
(a) it is impossible
(b) it is possible with high probability
(c) it is possible with low probability
(d) data are insufficient.

60. A steam engine takes 0.1 kg of steam at 100°C per minute and cools it down to 20°C. What is the heat rejected by the steam engine per second? [Latent heat of vaporization of steam = 540 cal g^{-1}]
(a) 6.2×10^4 cal/sec (b) 1.03×10^3 cal/sec
(c) 62×10^3 cal/sec (d) None of these

61. If an air conditioner is put in the middle of a room and started working
(a) the room can be cooled slightly
(b) the temperature of the room will not change
(c) the room will become slightly warmer
(d) the same temperature will be attained in the room as by putting it on the window in the standard position

62. The pressure-volume diagram shows six curved paths that can be followed by the gas (connected by vertical paths). Which two of them should be part of a closed cycle if the net work done by the gas is to be its maximum positive value?

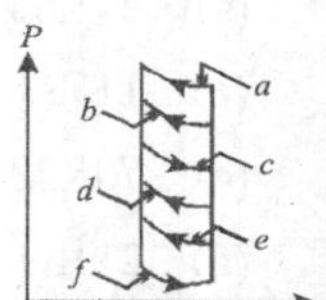

(a) *af* (b) *ae* (c) *ac* (d) *ce*

63. In the following *P-V* diagram two adiabatics cut two isothermals at temperatures T_1 and T_2 (fig). The value of $\frac{V_a}{V_d}$ will be

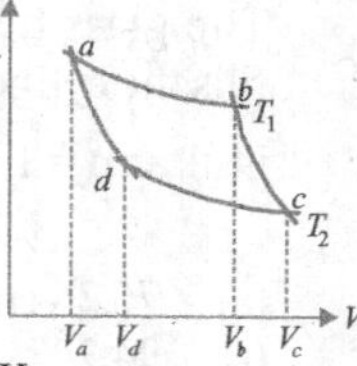

(a) $\frac{V_b}{V_c}$ (b) $\frac{V_c}{V_b}$ (c) $\frac{V_d}{V_a}$ (d) $V_b V_c$

64. A diatomic ideal gas undergoes a thermodynamic change according to the P-V diagram shown in the figure. The total heat given to the gas is nearly (use ln 2 = 0.7)

Critical Thinking

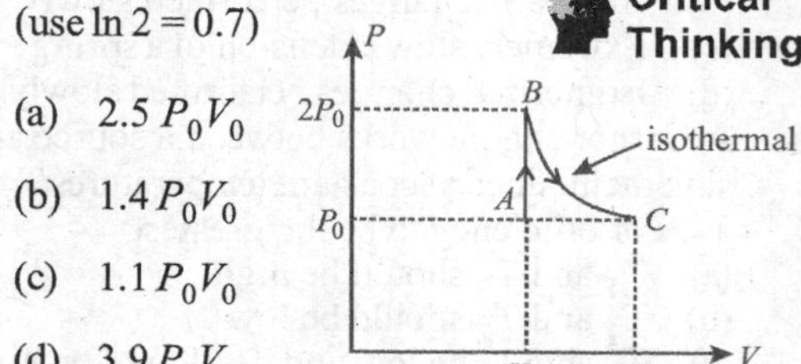

(a) $2.5\,P_0V_0$
(b) $1.4\,P_0V_0$
(c) $1.1\,P_0V_0$
(d) $3.9\,P_0V_0$

65. In *P-V* diagram shown in figure *ABC* is a semicircle. The work done in the process *ABC* is

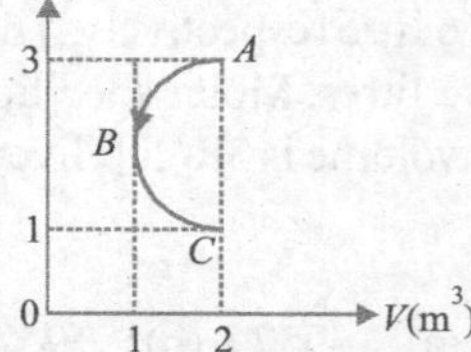

(a) 4 J
(b) $\frac{-\pi}{2}$ J
(c) $\frac{\pi}{2}$ J
(d) zero

66. On *P-V* coordinates, the slope of an isothermal curve of a gas at a pressure *P* = 1MPa and volume *V* = 0.0025 m³ is equal to –400 MPa/m³. If C_p / C_v = 1.4, the slope of the adiabatic curve passing through this point is :

Critical Thinking

(a) –56 MPa/m³ (b) –400 MPa/m³
(c) –560 MPa/m³ (d) None of these

67. A refrigerator with coefficient of performance $\frac{1}{3}$ releases 200 J of heat to a hot reservoir. Then the work done on the working substance is
(a) $\frac{100}{3}$ J (b) 100 J (c) $\frac{200}{3}$ J (d) 150 J

68. A Carnot engine takes 3×10^6 cal. of heat from a reservoir at 627°C, and gives it to a sink at 27°C. The work done by the engine is
(a) 4.2×10^6 J (b) 8.4×10^6 J
(c) 16.8×10^6 J (d) zero

69. A diatomic ideal gas is used in a car engine as the working substance. If during the adiabatic expansion part of the cycle, volume of the gas increases from *V* to 32 V, the efficiency of the engine is

Toughnut

(a) 0.5 (b) 0.75 (c) 0.99 (d) 0.25

70. A Carnot engine operating between temperatures T_1 and T_2 has efficiency $\frac{1}{6}$. When T_2 is lowered by 62 K its efficiency increases to $\frac{1}{3}$. Then T_1 and T_2 are, respectively
(a) 372 K and 330 K (b) 330 K and 268 K
(c) 310 K and 248 K (d) 372 K and 310 K

71. If the co-efficient of performance of a refrigerator is 5 and operates at the room temperature 27°C, the temperature inside the refrigerator is
(a) 240 K (b) 250 K (c) 230 K (d) 260 K

72. The temperature-entropy diagram of a reversible engine cycle is given in the figure. Its efficiency is

(a) $\frac{1}{4}$ (b) $\frac{1}{2}$

(c) $\frac{2}{3}$ (d) $\frac{1}{3}$

73. A Carnot engine whose efficiency is 50% has an exhaust temperature of 500 K. If the efficiency is to be 60% with the same intake temperature, the exhaust temperature must be (in K)
(a) 800 (b) 200 (c) 400 (d) 600

74. For an ideal gas four processes are marked as 1, 2, 3 and 4 on P-V diagram as shown in figure. The amount of heat supplied to the gas in the process 1, 2, 3 and 4 are Q_1, Q_2, Q_3 and Q_4 respectively, then correct order of heat supplied to the gas is – [AB is process-1, AC is process-2, AD is adiabatic process-3 and AE is process-4] **Tricky**

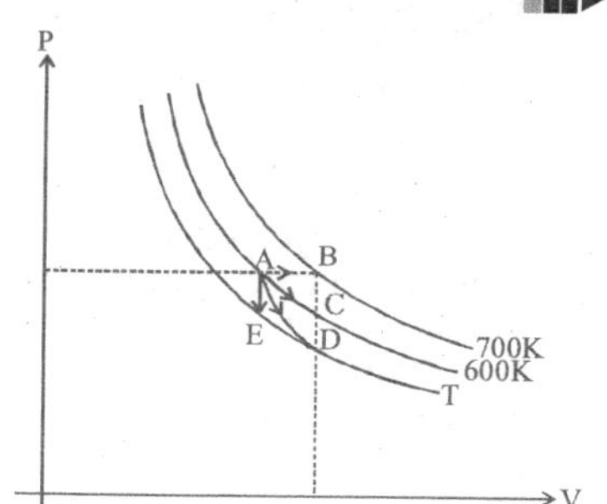

(a) $Q_1 > Q_2 > Q_3 > Q_4$ (b) $Q_1 > Q_2 > Q_4 > Q_3$
(c) $Q_1 > Q_4 > Q_2 > Q_3$ (d) $Q_1 < Q_2 < Q_3 < Q_4$

75. An ideal gas can be expanded from an initial state to a certain volume through two different processes,
(A) $PV^2 = K$ and (B) $P = KV^2$, where K is a positive constant. Then, choose the correct option from the following.
(a) Final temperature in (A) will be greater than in (B)
(b) Final temperature in (B) will be greater than in (A)
(c) Work done by the gas in both the processes would be equal
(d) Total heat given to the gas in (A) is greater than in (B)

76. Suppose 0.5 mole of an ideal gas undergoes an isothermal expansion as energy is added to its heat Q.
Graph shows the final volume V_f versus Q.

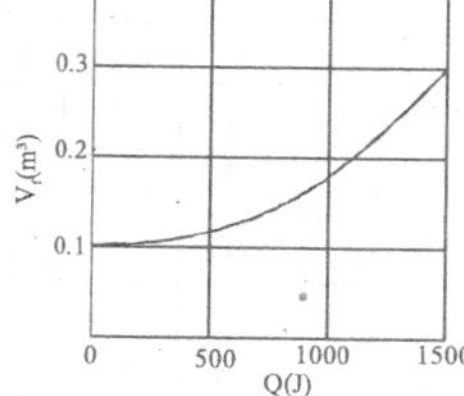

The temperature of the gas is (use ln 9 = 2 and $R = \frac{25}{3}$ J/mol-K)
(a) 360 K (b) 293 K (c) 386 K (d) 412 K

77. A gas is expanded from volume V_0 to $2V_0$ under three different processes. Process 1 is isobaric process, process 2 is isothermal and process 3 is adiabatic.

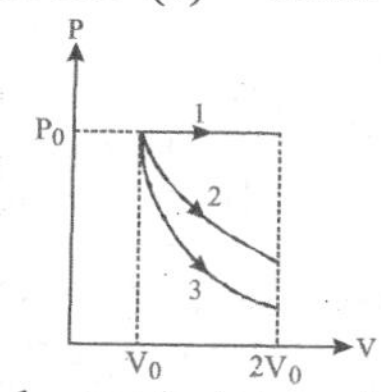

Let ΔU_1, ΔU_2 and ΔU_3 be the change in internal energy of the gas is these three processes. Then–

(a) $\Delta U_1 > \Delta U_2 > \Delta U_3$ (b) $\Delta U_1 < \Delta U_2 < \Delta U_3$
(c) $\Delta U_2 < \Delta U_1 < \Delta U_3$ (d) $\Delta U_2 < \Delta U_3 < \Delta U_1$

78. The *P-V* diagram of a gas system undergoing cyclic process is shown here. The work done during isobaric compression is

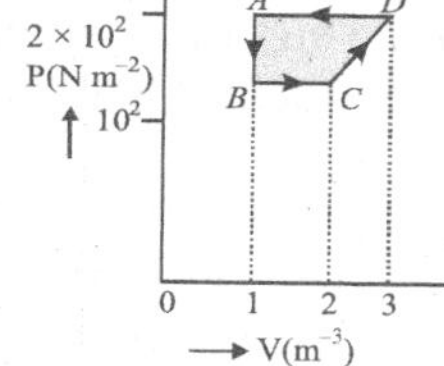

(a) 100 J (b) 200 J
(c) 600 J (d) 400 J

79. 1 gm of water at a pressure of 1.01×10^5 Pa is converted into steam without any change of temperature. The volume of 1 g of steam is 1671 cc and the latent heat of evaporation is 540 cal. The change in internal energy due to evaporation of 1 gm of water is
(a) ≈167 cal (b) ≈500 cal
(c) 540 cal (d) 581 cal

80. A system changes from the state (P_1, V_1) to (P_2, V_2) as shown in the figure. What is the work done by the system? **Tricky**

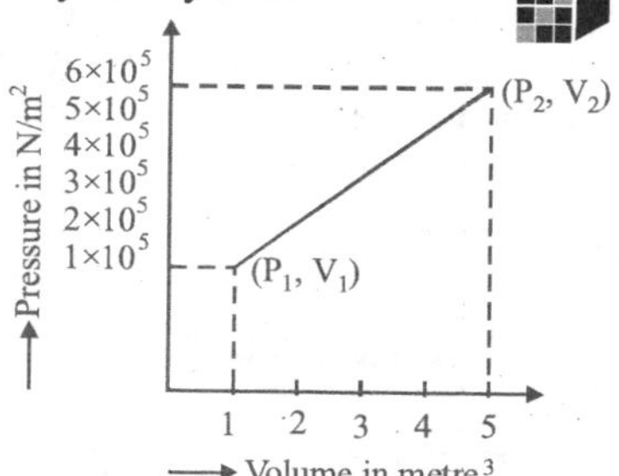

(a) 7.5×10^5 joule (b) 7.5×10^5 erg
(c) 12×10^5 joule (d) 6×10^5 joule

81. When the state of a gas adiabatically changed from an equilibrium state A to another equilibrium state B an amount of work done on the stystem is 35 J. If the gas is taken from state A to B via process in which the net heat absorbed by the system is 12 cal, then the net work done by the system is (1 cal = 4.19 J)
(a) 13.2 J (b) 15.4 J (c) 12.6 J (d) 16.8 J

82. A Carnot engine is working between 127°C and 27°C. The increase in efficiency will be maximum when the temperature of

(a) the source is increased by 50°C
(b) the sink is decreased by 50°C
(c) source is increased by 25°C and that of sink is decreased by 25°C
(d) both source and sink are decreased by 25°C each.

83. When 1 kg of ice at 0°C melts to water at 0°C, the resulting change in its entropy, taking latent heat of ice to be 80 cal/°C, is

(a) 273 cal/K (b) 253 cal/K
(c) 263 cal/K (d) 293 cal/K

84. If the energy input to a Carnot engine is thrice the work it performs then, the fraction of energy rejected to the sink is

(a) $\frac{1}{3}$ (b) $\frac{1}{4}$ (c) $\frac{2}{5}$ (d) $\frac{2}{3}$

85. The efficiency of an ideal gas with adiabatic exponent 'γ' for the shown cyclic process would be **Critical Thinking**

(a) $\frac{(2\ln 2-1)}{\gamma/(\gamma-1)}$

(b) $\frac{(1-2\ln 2)}{\gamma/(\gamma-1)}$

(c) $\frac{(2\ln 2+1)}{\gamma/(\gamma-1)}$

(d) $\frac{(2\ln 2-1)}{\gamma/(\gamma+1)}$

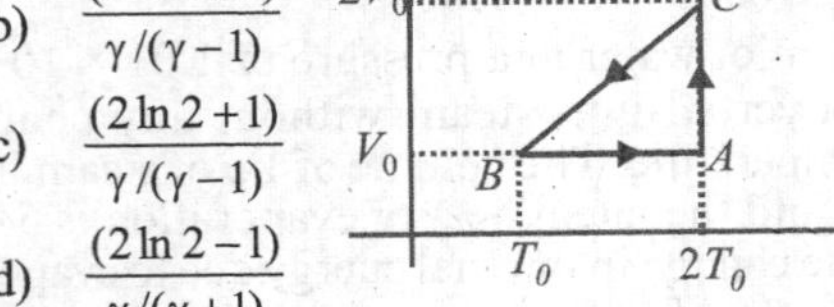

86. In a Carnot engine, the temperature of reservoir is 927°C and that of sink is 27°C. If the work done by the engine when it transfers heat from reservoir to sink is 12.6×10^6J, the quantity of heat absorbed by the engine from the reservoir is

(a) 16.8×10^6 J (b) 4×10^6 J
(c) 7.6×10^6 J (d) 4.2×10^6 J

87. A monoatomic ideal gas goes through a process $p = p_0 - \alpha V$ where p_0 and α are positive constants and V is its volume. At what volume will the entropy of gas be maximum? **Toughnut**

(a) $\frac{5p_0}{6\alpha}$ (b) $\frac{p_0}{2\alpha}$ (c) $\frac{p_0}{4\alpha}$ (d) $\frac{5p_0}{8\alpha}$

88. An ideal gas has temperature T_1 at the initial state i shown in the P-V diagram. The gas has a higher temperature T_2 at the final states a and b, which it can reach the paths shown. The change in entropy: **Tricky**

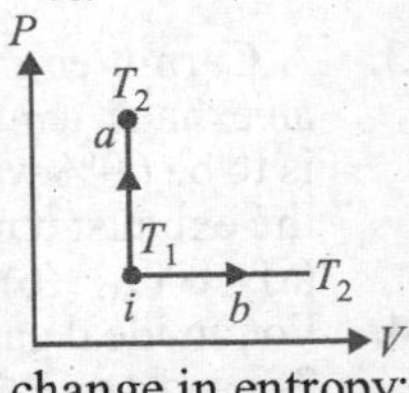

(a) greatest in a
(b) greatest in b
(c) same in a and b
(d) nothing can be said

89. In Carnot engine efficiency is 40% at hot reservoir temperature T. For efficiency 50% what will be temperature of hot reservoir?

(a) $\frac{T}{5}$ (b) $\frac{2T}{5}$ (c) 6T (d) $\frac{6T}{5}$

90. An ideal gas is subjected to cyclic process involving four thermodynamic states, the amounts of heat (Q) and work (W) involved in each of these states

$Q_1 = 6000$ J; $Q_2 = -5500$ J; $Q_3 = -3000$ J; $Q_4 = +3500$ J

$W_1 = 2500$ J; $W_2 = -1000$ J; $W_3 = -1200$ J; $W_4 = x$J

The ratio of the net work done by the gas to the total heat absorbed by the gas is η. The values of x and η respectively are

(a) 500; 7.5% (b) 700; 10.5%
(c) 1000; 21% (d) 1500; 15%

Answer KEYs

1	(b)	10	(b)	19	(a)	28	(c)	37	(d)	46	(a)	55	(a)	64	(d)	73	(c)	82	(b)
2	(d)	11	(d)	20	(b)	29	(c)	38	(d)	47	(b)	56	(d)	65	(c)	74	(b)	83	(d)
3	(a)	12	(d)	21	(c)	30	(a)	39	(a)	48	(c)	57	(a)	66	(c)	75	(b)	84	(d)
4	(c)	13	(d)	22	(c)	31	(a)	40	(c)	49	(a)	58	(d)	67	(d)	76	(a)	85	(a)
5	(d)	14	(a)	23	(a)	32	(b)	41	(d)	50	(a)	59	(a)	68	(b)	77	(a)	86	(a)
6	(a)	15	(b)	24	(b)	33	(d)	42	(a)	51	(d)	60	(b)	69	(b)	78	(d)	87	(d)
7	(c)	16	(b)	25	(c)	34	(c)	43	(d)	52	(a)	61	(c)	70	(d)	79	(b)	88	(b)
8	(d)	17	(d)	26	(d)	35	(c)	44	(b)	53	(a)	62	(b)	71	(b)	80	(c)	89	(d)
9	(b)	18	(a)	27	(c)	36	(d)	45	(a)	54	(d)	63	(a)	72	(d)	81	(b)	90	(b)

Hints & Solutions

1. (b) By first law, $\Delta Q = \Delta U + \Delta W$

$$= \frac{f}{2}nRT + P\int dV = \frac{f}{2}nRT + PV$$

$$\Delta Q = \frac{f}{2}nRT + nRT = \frac{3}{2}nRT + nRT = \frac{5}{2}nRT$$

$$\therefore \text{ Fraction} = \frac{(3/2)\,nRT}{(5/2)\,nRT} = \frac{3}{5}$$

2. (d) $\Delta Q = \Delta U + \Delta W \Rightarrow 2 \times 10^3 \times 4.2 = \Delta U + 500$
$\Rightarrow \Delta U = 7900\ J$

3. (a) From the first law of thermodynamics
$dQ = dU + dW$
Here $dW = 0$ (given)
$\therefore\ dQ = dU$
Now since $dQ < 0$ (given)
$\therefore\ dQ$ is negative
$\Rightarrow dU = -\text{ve} \Rightarrow dU$ decreases.
$\Rightarrow$ Temperature decreases.

4. (c) $Q = mL = 1 \times L = L;\ W = P(V_2 - V_1)$
Now $Q = \Delta U + W$
or $L = \Delta U + P(V_2 - V_1)$
$\therefore\ \Delta U = L - P(V_2 - V_1)$

5. (d) $W = P(\Delta V) = 1 \times 10^5 \times (3.34 - 2 \times 10^{-3})$
$= 340 \times 10^3$J.

6. (a) As we know,
$\Delta Q = \Delta U + \Delta W$ (Ist law of thermodynamics)
$\Rightarrow \Delta Q = \Delta U + P\Delta V$
or $150 = \Delta U + 100(1-2) = \Delta U - 100$
$\therefore\ \Delta U = 150 + 100 = 250$ J
Thus the internal energy of the gas increases by 250 J

7. (c) Internal energy of an ideal gas depends only on the temperature.

8. (d) Thermodynamics concerned with ΔH, ΔU and ΔW.

9. (b) $\Delta Q = \Delta U + \Delta W$
$\Rightarrow \Delta W = \Delta Q - \Delta U = 110 - 40 = 70$ J

10. (b) According to first law of thermodynamics
$Q = \Delta U + W$
Given : $Q = 2$ kcal $= 2000 \times 4.2 = 8400$ J
$W = 400$ J
$\therefore\ \Delta U = Q - W = 8400 - 400 = 8000$ J

11. (d) According to first law of thermodynamics
$\Delta Q = \Delta U + \Delta W$
$\Delta U = \Delta Q - \Delta W$
$\Delta Q = 35$J, $\Delta W = -15$J
$\therefore\ \Delta U = 35\text{J} - (-15\text{J}) = 50$J

12. (d) $dU = dQ - dW = (8 \times 10^5 - 6.5 \times 10^5)$
$= 1.5 \times 10^5$ J

$dW = dQ - dU = 10^5 - 1.5 \times 10^5$
$= -0.5 \times 10^5\ J$
– ve sign indicates that work done on the gas is 0.5×10^5 J.

13. (d) Internal energy and volume depend upon states.

14. (a) Here $Q = 0$ and $W = 0$. Therefore from first law of thermodynamics $\Delta U = Q + W = 0$
$\therefore$ Internal energy of the system with partition = Internal energy of the system without partition.
$n_1C_vT_1 + n_2C_vT_2 = (n_1 + n_2)C_vT$

$$\therefore\ T = \frac{n_1T_1 + n_2T_2}{n_1 + n_2}$$

But $n_1 = \dfrac{P_1V_1}{RT_1}$ and $n_2 = \dfrac{P_2V_2}{RT_2}$

$$\therefore\ T = \frac{\dfrac{P_1V_1}{RT_1} \times T_1 + \dfrac{P_2V_2}{RT_2} \times T_2}{\dfrac{P_1V_1}{RT_1} + \dfrac{P_2V_2}{RT_2}} = \frac{T_1T_2\,(P_1V_1 + P_2V_2)}{P_1V_1T_2 + P_2V_2T_1}$$

15. (b) $Q = mC\Delta T = 1.5 \times 0.12 \times 4200 \times (400 - 25)$
$= 2.83 \times 10^5$ J
$W = P(\Delta V) = P(V\gamma\Delta T)$
$= 10^5 \times (5 \times 10^{-2})^3 \times 3.5 \times 10^{-5} \times 375 = 0.164$ J
Thus $Q = \Delta U + W$
or $2.83 \times 10^5 = \Delta U + 0.164$; $\Delta U = 282$ kJ

16. (b) $W = P(\Delta V) = m(C_p - C_v)\Delta T$
$= 4(0.219 - 0.157) \times 4200 \times (120 - 20)$
$\simeq 104 \times 10^3$J

17. (d) $Q = nC_p \times 30 = n \times \dfrac{7R}{2} \times 30$
and $Q = n \times \dfrac{5R}{2} \times \Delta T$

$$\therefore\ n \times \frac{7R}{2} \times 30 = n \times \frac{5R}{2} \times \Delta T \text{ or } \Delta T = 42\text{ K}.$$

18. (a) $C_p - C_v$ = work done

19. (a) In isothermal process temperature remains constnat.
i.e., $\Delta T = 0$. Hence according to
$$C = \frac{Q}{m\Delta T} \Rightarrow C_{iso} = \infty$$

20. (b) In adiabatic process, no heat is taken or given by the system i.e.,
$\Delta Q = 0 \Rightarrow \Delta U = -\Delta W$
If ΔW is negative (work done on system), then ΔU increases & temperature increases and vice-versa. So work done in adiabatic change in particular gas (ideal gas) depends on change in temperature.

21. (c) In an isochoric process, no work is done on or by the gas. V is constant

22. (c) $\dfrac{\text{Slope of adiabatic curve}}{\text{Slope of isothermal curve}}$
$$= \frac{(dP/dV)_{adi}}{(dP/dV)_{iso}} = +\gamma$$
So slope to adiabatic curve is $\gamma \left(= \dfrac{C_P}{C_V}\right)$ times of isothermal curve, as clear also from figure.

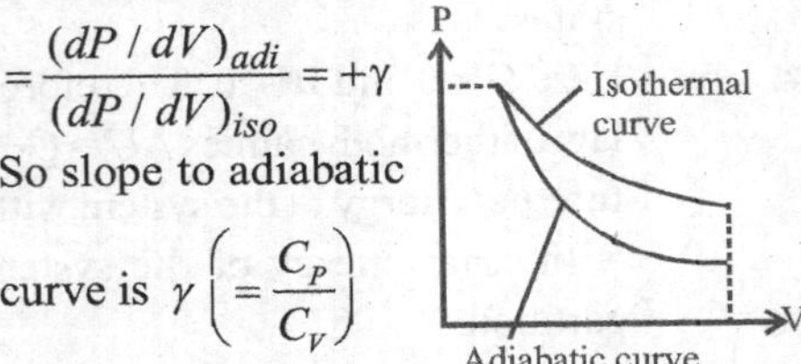

23. (a) $\Delta U = nC_v\Delta T = 2 \times 10^3 \times 20 \times 273 = 10.9\text{MJ}.$

24. (b) For path iaf, Q = 50 cal W = 20 cal
By first law of thermodynamics,
$\Delta U = Q - W = 50 - 20 = 30$ cal.
For path ibf
Q' = 36 cal
W' = ?
or, W' = Q' − ΔU'
Since, the change in internal energy does not depend on the path, therefore
ΔU' = 30 cal
∴ W' = Q' − ΔU' = 36 − 30 = 6 cal.

25. (c)

26. (d) Work done by the system in the cycle = Area under P-V curve and V-axis
$$= \frac{1}{2}(2P_0 - P_0)(2V_0 - V_0) + \left[-\left(\frac{1}{2}\right)(3P_0 - 2P_0)(2V_0 - V_0)\right]$$
$$= \frac{P_0V_0}{2} - \frac{P_0V_0}{2} = 0$$

27. (c) $P_1V_1^{\gamma} = P_2V_2^{\gamma}$ (Adiabatic change)
$$P_2 = P_1\left(\frac{V_1}{V_2}\right)^{\gamma} = P_1\left(\frac{V_1}{V_1/3}\right)^{\gamma} = P_2(3)^{\gamma}$$

28. (c) Heat absorbed in a thermodynamic process is given by $\Delta Q = \Delta U + \Delta W$.
Here ΔU is same for all the three processes as it depends only on initial and final states.
But $\Delta W_I = +$ Ve, $\Delta W_{II} = 0$, $\Delta W_{III} = -$ ve
$\therefore \quad \Delta Q_I > \Delta Q_{II}$

29. (c) $-20 = \Delta U + 50 \Rightarrow \Delta U = -70.$
From $2 \to 1$, $\Delta U = 70$ kJ.
Now $10 = 70 + W \Rightarrow W = -60$ kJ.

30. (a) As $P \propto \dfrac{1}{V^{1.5}}$, So $PV^{1.5}$ = constant
$\therefore \gamma = 1.5$ ($\because$ Process is adiabatic)
As we know, $\dfrac{C_P}{C_V} = \gamma \qquad \therefore \dfrac{C_P}{C_V} = 1.5$

31. (a) $U = a + bPV$ ......(1)
In adiabatic change,
$$dU = -dW = \frac{nR}{\gamma - 1}(T_2 - T_1) = \frac{nR}{\gamma - 1}(dT)$$
$$\Rightarrow U = \int dU = \frac{nR}{\gamma - 1}\int dT$$
or $U = \left(\dfrac{nR}{\gamma - 1}\right)T + a = \dfrac{PV}{\gamma - 1} + a$ ......(2)
where a is the constant of integration.
Comparing (1) and (2), we get

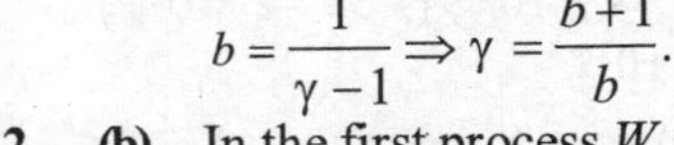

$$b = \frac{1}{\gamma - 1} \Rightarrow \gamma = \frac{b+1}{b}.$$

32. (b) In the first process W is + ve as ΔV is positive, in the second process W is − ve as ΔV is − ve and area under the curve of second process is more
∴ Net Work < 0 and also $P_3 > P_1$.

33. (d) Isochoric process dV = 0
W = 0 proceess 1
Isobaric : $W = P\Delta V = nR\Delta T$
Adiabatic $|W| = \dfrac{nR\Delta T}{\gamma - 1} \quad 0 < \gamma - 1 < 1$
As workdone in case of adiabatic process is more so process 3 is adiabatic and process 2 is isobaric.

34. (c) For adiabatic process, dU = − 100 J which remains same for other processes also.

Let C be the heat capacity of 2nd process then

$-(C)5 = dU + dW = -100 + 25 = -75$

$\therefore$ C = 15 J/K

35. (c) As $\Delta U = 0$ in a cyclic process,

$\Delta Q = \Delta W =$ area of circle $= \pi r^2$

or $\Delta W = 10^2 \pi J$

36. (d) Under adiabatic change

$$\frac{T_2}{T_1} = \left(\frac{P_1}{P_2}\right)^{\frac{1-\gamma}{\gamma}} \text{ or } T_2 = T_1 (P_1/P_2)^{\frac{1-\gamma}{\gamma}}$$

$\therefore\ T_2 = 300(4/1)^{\frac{1-(7/5)}{(7/5)}}$; $\gamma = 1.4 = 7/5$ for air

or $T_2 = 300(4)^{-2/7}$

37. (d) Here $TV^{\gamma-1} =$ constant

As $\gamma = \frac{5}{3}$, hence $TV^{2/3} =$ constant

Now $T_1 L_1^{2/3} = T_2 L_2^{2/3}$ $(\because V \propto L)$;

Hence, $\frac{T_1}{T_2} = \left(\frac{L_2}{L_1}\right)^{2/3}$

38. (d) Heat absorbed by gas in three processes is given by

$Q_{ACB} = \Delta U + W_{ACB}$

$Q_{ADB} = \Delta U$

$Q_{AEB} = \Delta U + W_{AEB}$

The change in internal energy in all the three cases is same and W_{ACB} is +ve, W_{AEB} is –ve.

Hence $Q_{ACB} > Q_{ADB} > Q_{AEB}$

39. (a) If a process is expansion then work done is positive so answer will be (a).

40. (c) $T_1 = 273 + 27 = 300K$

$T_2 = 273 + 927 = 1200K$

For adiabatic process,

$P^{1-\gamma} T^{\gamma} =$ constant $\Rightarrow P_1^{1-\gamma} T_1^{\gamma} = P_2^{1-\gamma} T_2^{\gamma}$

$$\Rightarrow \left(\frac{P_2}{P_1}\right)^{1-\gamma} = \left(\frac{T_1}{T_2}\right)^{\gamma} \Rightarrow \left(\frac{P_1}{P_2}\right)^{1-\gamma} = \left(\frac{T_2}{T_1}\right)^{\gamma}$$

$$\left(\frac{P_1}{P_2}\right)^{1-1.4} = \left(\frac{1200}{300}\right)^{1.4} \Rightarrow \left(\frac{P_1}{P_2}\right)^{-0.4} = (4)^{1.4}$$

$$\left(\frac{P_2}{P_1}\right)^{0.4} = 4^{1.4} \text{ or, } P_2 = P_1 4^{\left(\frac{1.4}{0.4}\right)} = P_1 4^{\left(\frac{7}{2}\right)}$$

$= P_1(2^7) = 2 \times 128 = 256$ atm

41. (d) 1st process is isothermal expansion which is only correct shown in option (d)

2nd process is isobaric compression which is correctly shown in option (d)

42. (a) $T_1 = T$, $W = 6R$ joules, $\gamma = \frac{5}{3}$

$$W = \frac{P_1V_1 - P_2V_2}{\gamma - 1} = \frac{nRT_1 - nRT_2}{\gamma - 1}$$

$$= \frac{nR(T_1 - T_2)}{\gamma - 1}$$

$n = 1, T_1 = T \Rightarrow \frac{R(T - T_2)}{5/3 - 1} = 6R$

$\Rightarrow T_2 = (T - 4)K$

43. (d) From C to D, V is constant. So process is isochoric. From D to A, the curve represents constant temperature. So the process is isothermal.

From A to B, pressure is constant . So, the process is isobaric.

BC represents constant entropy.

44. (b)

45. (a) Work done is equal to area under the curve on PV diagram.

46. (a) $PV =$ constant. Differentiating,

$$\frac{PdV}{dP} = -V; \beta = -\left(\frac{1}{V}\right)\left(\frac{dV}{dP}\right) = \left(\frac{1}{P}\right) \Rightarrow \beta \times P = 1$$

$\therefore$ Graph between β and P will be a rectangular hyperbola.

47. (b) $W_{ABC} = \frac{\pi r^2}{2} = \frac{\pi(6)^2}{2} = 18\pi$

$W_{DEF} = \frac{\pi \times 3^2}{2} + \frac{\pi \times 3^2}{4} + (15 - 12) \times 18$

$= 6.75\pi + 54$

$W_{DEF} > W_{ABC}$

48. (c) Given that $dQ = -30$ J and $dW = -10$ J

$E_i = 40$ J and let final internal energy $= E_f$

So, $dQ = E_f - E_i + dW$

$-30 = E_f - 40 - 10$

$E_f = 20$ J

49. (a) For an adiabatic change $PV^{\gamma} =$ constant

$P_1V_1^{\gamma} = P_2V_2^{\gamma}$

As molar specific heat of gas at constant volume

$$C_v = \frac{3}{2}R$$

$$C_P = C_V + R = \frac{3}{2}R + R = \frac{5}{2}R;$$

$$\gamma = \frac{C_P}{C_V} = \frac{(5/2)R}{(3/2)R} = \frac{5}{3}$$

$\therefore$ From eq^n. (1)

$$P_2 = \left(\frac{V_1}{V_2}\right)^{\gamma} P_1 = \left(\frac{6}{2}\right)^{5/3} \times 10^5 \text{ N/m}^2$$

$= (3)^{5/3} \times 10^5 = 6.19 \times 10^5$ N/m^2

Work done

$$= \frac{1}{1-(5/3)} [6.19 \times 10^5 \times 2 \times 10^{-3} - 10^{-5} \times 6 \times 10^{-3}]$$

$$= -\left[\frac{2 \times 10^2 \times 3}{2}(6.19-3)\right]$$

$= -3 \times 10^2 \times 3.19 = -957$ joules

[–ve sign shows external work done on the gas]

50. (a) VT = constant

$\therefore PV = nRT$

$$V = \left[\frac{nRT}{P}\right] \quad \therefore \frac{nRT}{P} \cdot T = \text{constant}$$

$\Rightarrow P \propto T^2$.

51. (d) Work done by gas in going isothermally from state A to B is

$$\Delta W_{AB} = nRT \ln \frac{P_A}{P_B} = nRT \ln 2 \quad \text{...(i)}$$

Work done by gas in going isothermally from state B to C is

$$\Delta W_{BC} = nRT \ln \frac{P_B}{P_C} = nRT \frac{P_0}{2P_C} \quad \text{...(ii)}$$

It is given that $\Delta W_{BC} = 2\Delta W_{AB}$

$$\ln \frac{P_0}{2P_C} = \ln(2)^2 \quad \therefore P_C = \frac{P_0}{8}$$

52. (a) $PV^{3/2}$ = constant

$PV = nRT$

$\Rightarrow V^{1/2} T$ = const.

$$\Rightarrow \left(\frac{V}{2}\right)^{1/2} T' = \text{const.}$$

$$\Rightarrow T' = T\sqrt{2}$$

53. (a) Change in internal energy for cyclic process (ΔU) = 0.

For process a $\to$ b, (P-constant)

$W_{a \to b} = P\Delta V$

$= nR\Delta T = -400\ R$

For process b $\to$ c, (T-constant)

$W_{b \to c} = -2R(300) \ln 2$

For process c $\to$ d, (P-constant)

$W_{c \to d} = +400R$

For process d $\to$ a, (T-constant)

$W_{d \to a} = +2R(500) \ln 2$

Net work (ΔW) = $W_{a \to b} + W_{b \to c} + W_{c \to d} + W_{d \to a}$

$\Delta W = 400R \ln 2$

$\therefore$ dQ = dU + dW, first law of thermodynamics

$\therefore$ dQ = 400 R ln 2.

54. (d)

55. (a) For tiny glass tube,

$P_1 V_1 = P_2 V_2$

$$\therefore P_2 = \frac{P_1 V_1}{V_2} = \frac{4.5 \times 0.50}{500} = 0.0045 \text{ atm.}$$

Thus P = 1atm + 0.0045 atm

= 1.0045 atm = 76.34 cm of Hg.

56. (d) In reversible cyclic Process

$\Delta U = 0$

57. (a) Slow isothermal expansion or compression of an ideal gas is reversible process, while the other given processes are irreversible in nature.

58. (d) $\eta = 1 - \frac{T_2}{T_1}$ So for η be high T_1 must be high and T_2 must be low.

59. (a) $\eta = 1 - \frac{300}{400} = \frac{100}{400} = \frac{1}{4}$

$\eta = \frac{1}{4} \times 100 = 25\%$

Hence, it is not possible to have efficiency more than 25%.

60. (b)

61. (c) The working of an air coditioner is similar to the working of a refrigerator. An air conditioner removes heat from the room, does some work and rejects the heat to the surroundings. As air conditioner is put in the middle of the room then due to continuous, external work the room will become slightly warmer.

62. (b) Work done in expansion is positive and in compression it is negative.

63. (a) We know that $\frac{V_a}{V_b} = \frac{V_d}{V_c} \Rightarrow \frac{V_a}{V_d} = \frac{V_b}{V_c}$

64. (d) $Q_{AB} = \Delta U_{AB} + W_{AB}$

$W_{AB} = 0$

$$\Delta U_{AB} = \frac{f}{2} nR\Delta T \Rightarrow \frac{f}{2}(\Delta PV)$$

$$\Delta U_{AB} = \frac{5}{2}(\Delta PV) \Rightarrow Q_{AB} = 2.5\, P_0 V_0$$

Process *BC*:

$Q_{BC} = \Delta U_{BC} + W_{BC} = 0 + 2P_0V_0 \ln 2 = 1.4\, P_0V_0$

$Q_{net} = Q_{AB} + Q_{BC} = 3.9\, P_0V_0$

65. (c) $W=\frac{\pi r_1 r_2}{2}=\frac{\pi\times1\times1}{2}=\pi/2$ J

66. (c) Slope of adiabatic curve
$=\gamma\times$ slope of isothermal curve
$=1.4\times(-400)=-560$ MPa/m^3

67. (d) The coefficient of performance of a refrigerator is given by
$$\alpha=\frac{Q_2}{W}=\frac{Q_2}{Q_1-Q_2}$$
Substituting the given values, we get
$$\frac{1}{3}=\frac{Q_2}{200-Q_2}$$
$\Rightarrow 200-Q_2=3Q_2 \Rightarrow 4Q_2=200$
or $Q_2=\frac{200}{4}\text{J}=50\text{J}$
$\therefore\ W=Q_1-Q_2=200\text{ J}-50\text{ J}=150\text{ J}$

68. (b) $\eta=\frac{(627+273)-(273+27)}{627+273}$
$=\frac{900-300}{900}=\frac{600}{900}=\frac{2}{3}$
work $=(\eta)\times$ Heat $=\frac{2}{3}\times3\times10^6\times4.2$ J
$=8.4\times10^6$ J

69. (b) The efficiency of cycle is $\eta=1-\frac{T_2}{T_1}$
For adiabatic process $TV^{\gamma-1}=$ constant
For diatomic gas $\gamma=\frac{7}{5}$
$T_1V_1^{\gamma-1}=T_2V_2^{\gamma-1}\qquad T_1=T_2\left(\frac{V_2}{V_1}\right)^{\gamma-1}$
$T_1=T_2(32)^{\frac{7}{5}-1}=T_2(2^5)^{2/5}=T_2\times4$
$T_1=4T_2$
$\therefore\ \eta=\left(1-\frac{1}{4}\right)=\frac{3}{4}=0.75$

70. (d) $\eta_1=1-\frac{T_2}{T_1}\Rightarrow\frac{1}{6}=1-\frac{T_2}{T_1}\Rightarrow\frac{T_2}{T_1}=\frac{5}{6}\ \ldots$ (i)
$\eta_2=1-\frac{T_2-62}{T_1}\Rightarrow\frac{1}{3}=1-\frac{T_2-62}{T_1}\quad\ldots$(ii)
On solving Eqs. (i) and (ii)
$T_1=372$ K and $T_2=310$ K

71. (b) Here, Coefficient of performance $(\beta)=5$
$T_1=27°C=(27+273)K=300$ K
As, $\beta=\frac{T_2}{T_1-T_2}\Rightarrow5=\frac{T_2}{300-T_2}$
or $1500-5T_2=T_2$ or $6T_2=1500$
$\therefore\ T_2=\frac{1500}{6}=250$ K

72. (d)

$Q_1=T_0S_0+\frac{1}{2}T_0S_0=\frac{3}{2}T_0S_0$
$Q_2=T_0(2S_0-S_0)=T_0S_0$ and $Q_3=0$
$\eta=\frac{W}{Q_1}=\frac{Q_1-Q_2}{Q_1}$
$=1-\frac{Q_2}{Q_1}=1-\frac{T_0S_0}{\frac{3}{2}T_0S_0}=\frac{1}{3}$

73. (c) $\eta=1-\frac{T_2}{T_1}$ or $\frac{50}{100}=1-\frac{500}{T_1}\Rightarrow T_1=1000\,K$
Also, $\frac{60}{100}=1-\frac{T_2}{1000}\Rightarrow T_2=400$ K

74. (b) In process-1 heat supplied = area under AB curve + n × c_v × 100 (isobaric process)
In process-2 heat supplied = area under AC curve (isothermal process)
In process-3 heat supplied = 0 (adiabatic process)
In process-4 heat supplied = n × c_v (T – 600) (isobaric process)

75. (b) Process (A) $\frac{nRT}{V}V^2=K$; TV = constant;
$T\propto\frac{1}{V}$
Process (B) $\frac{nRT}{V}=KV^2$; $\frac{T}{V^3}=$ constant;
$T\propto V^3$

76. (a) $Q=W=nRT\ln\frac{V_f}{V_i}$
$T=\frac{Q}{nR\ln(V_f/V_i)}=\frac{1500}{0.5\times25/3\times\ln3}$
$\Rightarrow\ T=\frac{1500}{0.5\times25/3\times1}=360\,K$

77. (a) Since volume is same in all three process therefore temperature will be least having least pressure.

78. (d) Isobaric compression is represented by curve AO
Work done = area under AD
$=2\times10^2\times(3-1)=4\times10^2=400$ J.

79. **(b)** $dW = P\Delta V = 1.01 \times 10^5[1671-1] \times 10^{-6}$ Joule

$= \frac{1.01 \times 167}{4.2}$ cal. $= 40$ cal. nearly

$\Delta Q = mL = 1 \times 540,$

$\Delta Q = \Delta W + \Delta U$

or $\Delta U = 540 - 40 = 500$ cal.

80. **(c)** $W = W_1 + W_2$

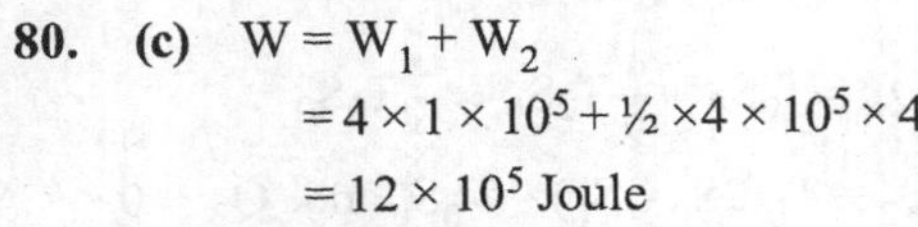

$= 4 \times 1 \times 10^5 + ½ \times 4 \times 10^5 \times 4$

$= 12 \times 10^5$ Joule

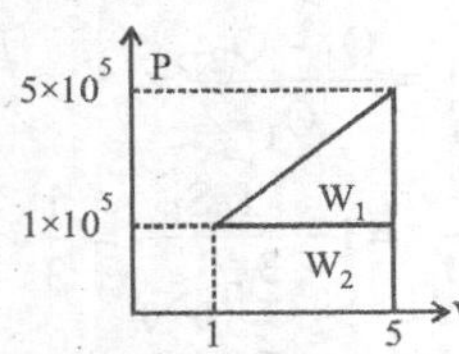

81. **(b)** In the first-case adiabatic change,

$\Delta Q = 0, \Delta W = -35$ J

From 1st law of thermodynamics,

$\Delta Q = \Delta U + \Delta W,$

or $0 = \Delta U - 35$

$\therefore \Delta U = 35$ J

In the second case

$\Delta Q = 12$ cal $= 12 \times 4.2$ J $= 50.4$ J

$\Delta W = \Delta Q - \Delta U = 50.4 - 35 = 15.4$ J

82. **(b)** $\frac{T_1 - T_2}{T_1}$ is maximum in case (b).

83. **(d)** Change in entropy is given by

$$dS = \frac{dQ}{T} \text{ or } \Delta S = \frac{\Delta Q}{T} = \frac{mL_f}{273}$$

$$\Delta S = \frac{1000 \times 80}{273} = 293 \text{ cal/K.}$$

84. **(d)** Efficiency $\eta = \frac{W_{output}}{Heat_{input}} = \frac{w}{3w} = \frac{1}{3}$

$$\eta = 1 - \frac{Q_2}{Q_1} = \frac{1}{3} \quad \therefore \quad \frac{Q_2}{Q_1} = \frac{2}{3}$$

85. **(a)** $W_{AB} = 0,\ W_{BC} = P\Delta V = nR\Delta T = -nRT_0$

$$W_{CA} = nRT\ell n\frac{V_f}{V_i} = nR(2T_0)\ell n2$$

$$Q_{BC} = nC_p\Delta T = \left(\frac{nR\gamma}{\gamma-1}\right)T_0$$

$$\text{Efficiency, } \eta = \frac{W}{Q} = \left[\frac{2\ell n2 - 1}{\gamma/(\gamma-1)}\right]$$

86. **(a)** As we know $\eta = \frac{W}{Q_1} = 1 - \frac{T_2}{T_1}$

$$\Rightarrow \eta = 1 - \frac{300\text{ K}}{1200\text{K}} = \frac{3}{4}$$

$$\frac{3}{4} = \frac{W}{Q_1} \Rightarrow Q_1 = W \times \frac{4}{3} \Rightarrow Q_1 = 12.6 \times 10^6 \times \frac{4}{3}$$

$Q_1 = 16.8 \times 10^6$J.

87. **(d)** $ds = nC_v dT + PdV = 0$

$$nR\frac{dT}{dV} + (p_0 - \alpha V) = 0$$

$$pV = nRT$$

$$p_0V - \alpha V^2 = nRT$$

$$p_0 - 2\alpha V = nR\frac{dT}{dV}$$

$$-(p_0 - \alpha V)(\gamma - 1) = p_0 - 2\alpha V$$

$$-p_0(\gamma-1) + \alpha(\gamma-1)V = p_0 - 2\alpha V$$

$$p_0V = \alpha V(\gamma+1)$$

$$V = \frac{p_0\gamma}{\alpha(\gamma+1)}$$

$$V = \frac{p_0 \times \frac{5}{3}}{\alpha\left(\frac{5}{3}+1\right)} = \frac{5p_0}{8\alpha}$$

88. **(b)** $Q_a = \Delta U + 0 = \Delta U$ and $Q_b = \Delta U + P\Delta V$

As $Q_b > Q_a$,

$\therefore$ Change in entropy is greater in case (b).

89. **(d)** $\eta = 1 - \frac{T_2}{T_1}$

$T_1 = T$ (Temperature of hot reservoir)

For $\eta = 40\%$,

$$\frac{40}{100} = 1 - \frac{T_2}{T} \Rightarrow \frac{T_2}{T} = \frac{3}{5} \Rightarrow T_2 = \frac{3}{5}T$$

For $\eta = 50\%$,

$$\frac{50}{100} = 1 - \frac{\frac{3}{5}T}{T_1} \Rightarrow T_1 = \frac{6T}{5}$$

90. **(b)** $Q = Q_1 + Q_2 + Q_3 + Q_4$

$= 6000 - 5500 - 3000 + 3500 = +1000$ J

$W = W_1 + W_2 + W_3 + W_4$

$= 2500 - 1000 - 1200 + x = +300 + x$

In cyclic process, $\Delta U = 0$

Now, $Q = \Delta U + W$

or $1000 = 0 + (300 + x)$

$\therefore x = 700$ J

$$\eta = \frac{W}{Q_1 + Q_4}$$

$$= \frac{1000}{6000 + 3500} = 10.5\%$$

12 Kinetic Theory

1. The velocity of sound in air is 332 m s^{-1} at NTP. Find the rms speed of air molecules at NTP. ($\gamma = 1.41$)
   (a) 484 ms^{-1} (b) 418 ms^{-1}
   (c) 248 ms^{-1} (d) 382 ms^{-1}
2. The density of a gas is 6×10^{-2} kg/m^3 and the root mean square velocity of the gas molecules is 500 m/s. The pressure exerted by the gas on the walls of the vessel is
   (a) $5 \times 10^3 \text{ N/m}^2$ (b) $1.2 \times 10^{-4} \text{ N/m}^2$
   (c) $0.83 \times 10^{-4} \text{ N/m}^2$ (d) 30 N/m^2
3. Four molecules have speeds 2 km/sec, 3 km/sec, 4 km/sec and 5 km/sec. The root mean square speed of these molecules (in km/sec) is
   (a) $\sqrt{54/4}$ (b) $\sqrt{54/2}$ (c) 3.5 (d) $3\sqrt{3}$
4. The absolute gas temperature at which the root mean square speed of helium molecules exceeds their most probable speed by 200 m/s is
   (a) 110.2 K (b) 90.2 K (c) 190.2 K (d) 100.2 K
5. N (< 100) molecules of a gas have velocities 1, 2, 3, ... N km/s respectively. Then ratio of rms speed and average speed is **Tricky**
   (a) 1 (b) $\frac{\sqrt{(2N+1)(N+1)}}{6N}$
   (c) $\frac{\sqrt{(2N+1)(N+1)}}{6}$ (d) $2\sqrt{\frac{2N+1}{6(N+1)}}$
6. If the molecules in a tank of hydrogen have the same RMS speed as the molecules in another tank of oxygen, we may be sure that
   (a) the pressures are the same
   (b) the hydrogen is at the higher temperature
   (c) the temperatures are the same
   (d) the oxygen is at the higher temperature
7. H_2, O_2, N_2 and He are enclosed in identical containers under the similar conditions of pressure and temperature. The gases will have
   (a) same R.M.S. speed (b) same $\frac{K.E.}{kg}$
   (c) different $\frac{K.E.}{mole}$ (d) same $\frac{K.E.}{vol}$
8. When do real gases approach the ideal gas behaviour?
   (a) At low pressure and low temperature
   (b) At low pressure and high temperature
   (c) At high pressure and high temperature
   (d) At high pressure and low temperature
9. According to the kinetic theory of gases, the pressure exerted by a gas on the wall of the container is measured as
   (a) rate of change of momentum imparted to the walls per second per unit area.
   (b) momentum imparted to the walls per unit area
   (c) change of momentum imparted to the walls per unit volume.
   (d) change in momentum per unit volume
10. In kinetic theory of gases, which of the following statement regarding elastic collisions of the molecules is wrong?
    (a) Kinetic energy is lost in collisions
    (b) Kinetic energy remains constant in collision
    (c) Momentum is conserved in collision
    (d) Pressure of the gas remains constant in collisions
11. At room temperature a diatomic gas is found to have an r.m.s. speed of 1930 ms^{-1}. The gas is:

    (a) H_2 (b) Cl_2 (c) O_2 (d) F_2
12. Gases exert pressure on the walls of containing vessel because the gas molecules
    (a) possess momentum
    (b) collide with each other
    (c) have finite volume
    (d) obey gas laws
13. Consider a gas with density ρ and $\bar{c}$ as the root mean square velocity of its molecules contained in a volume. If the system moves as whole with velocity, then the pressure exerted by the gas is
    (a) $\frac{1}{3}\rho\bar{c}^2$ (b) $\frac{1}{3}\rho(c+v)^2$
    (c) $\frac{1}{3}\rho(\bar{c}-v)^2$ (d) $\frac{1}{3}\rho(c^{-2}-v)^2$
14. The pressure of a gas is raised from 27°C to 927°C. The root mean square speed is
    (a) $\sqrt{(927/27)}$ times the earlier value
    (b) remain the same
    (c) gets halved (d) get doubled

15. Why does the pressure of an ideal gas increase when it is heated at constant volume ?
(a) The gas molecules expand
(b) The molecules move at the same speed, but hit the walls more often
(c) The molecules move faster and hit the walls more often
(d) The number of molecules of gas increases

16. Maxwell's velocity distribution curve is given for two different temperature. For the given curves.

(a) $T_1 > T_2$
(b) $T_1 < T_2$
(c) $T_1 \leq T_2$
(d) $T_1 = T_2$

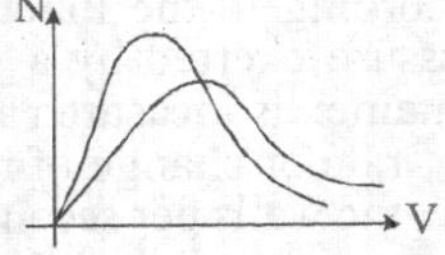

17. Helium gas is filled in a closed vessel (having negligible thermal expansion coefficient) when it is heated from 300 K to 600 K, then average kinetic energy of helium atom will be
(a) $\sqrt{2}$ times (b) 2 times
(c) unchanged (d) half

18. The temperature of an ideal gas is increased from 120 K to 480 K. If at 120 K the root-mean-square velocity of the gas molecules is v, at 480 K it becomes
(a) 4v (b) 2v (c) v/2 (d) v/4

19. Figure shows a parabolic graph between T and 1/V for a mixture of a gas undergoing an adiabatic process. What is the ratio of V_{rms} of molecules and speed of sound in mixture?

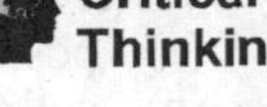

(a)  $\sqrt{3/2}$
(b) $\sqrt{2}$
(c) $\sqrt{2/3}$
(d) $\sqrt{3}$

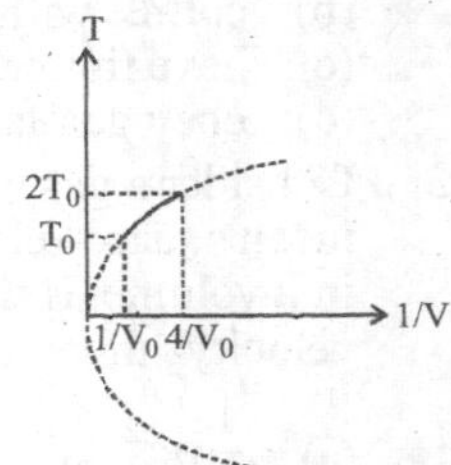

20. A nitrogen molecule has some rms speed at 0°C on the surface of the earth. With this speed, it goes straight up. If there is no collisions with other molecules, the molecule will rise up to a height of

Critical Thinking

(a) 8.2 km (b) 12.4 km
(c) 10.6 km (d) 15.2 km

21. At constant volume, temperature is increased then
(a) collision on walls will be less
(b) number of collisions per unit time will increase
(c) collisions will be in straight lines
(d) collisions will not change

22. At what temperature is the r.m.s velocity of a hydrogen molecule equal to that of an oxygen molecule at 47°C?
(a) 80 K (b) –73 K (c) 3 K (d) 20 K

23. Modern vacuum pumps can evacuate a vessel down to a pressure of 4.0×10^{-15} atm. at room temperature (300 K). Taking $R = 8.0\ JK^{-1}\ mole^{-1}$, 1 atm = 10^5 Pa and $N_{Avogadro} = 6 \times 10^{23}\ mole^{-1}$, the mean distance between molecules of gas in an evacuated vessel will be of the order of:
(a) 0.2 μm (b) 0.2 mm (c) 0.2 cm (d) 0.2 nm

24. The quantity of gas in a closed vessel is halved and the velocities of its molecules are doubled. The final pressure of the gas will be
(a) *P* (b) 2*P* (c) *P*/2 (d) 4*P*

25. Pressure versus temeprature graph of an ideal gas of equal number of moles of different volumes are plotted as shown in figure. Choose the correct alternative

Tricky

(a) $V_1 = V_2, V_3 = V_4$ and $V_2 > V_3$
(b) $V_1 = V_2, V_3 = V_4$ and $V_2 < V_3$
(c) $V_1 = V_2 = V_3 = V_4$
(d) $V_4 > V_3 > V_2 > V_1$

26. Consider a collection of a large number of dust particles each with speed *v*. The direction of velocity is randomly distributed in the collection. What is the magnitude of the relative velocity between a pairs in the collection ?
(a) $\frac{3v}{\pi}$ (b) $\frac{4v}{\pi}$ (c) $\frac{2v}{\pi}$ (d) $\frac{v}{\pi}$

27. Relation between pressure (*P*) and energy (*E*) of a gas is
(a) $P = \frac{2}{3}E$ (b) $P = \frac{1}{3}E$
(c) $P = \frac{1}{2}E$ (d) $P = 3E$

28. Five gas molecules chosen at random are found to have speeds of 500, 600, 700, 800 and 900 m/s
(a) the root mean square speed and the average speed are the same.
(b) the root mean square speed is 14 m/s higher than the average speed.
(c) the root mean square speed is 14 m/s lower than the average speed.
(d) the root mean square speed is $\sqrt{14}$ m/s higher than the average speed.

29. Two containers A and B are partly filled with water and closed. The volume of A is twice that of B and it contains half the amount of water in B. If both are at the same temperature, the water vapour in the containers will have pressure in the ratio of
(a) 1 : 2 (b) 1 : 1 (c) 2 : 1 (d) 4 : 1

30. A gas mixture consists of molecules of type 1, 2 and 3, with molar masses $m_1 > m_2 > m_3$. v_{rms} and $\bar{K}$ are the r.m.s. speed and average kinetic energy of the gases. Which of the following is true?
(a) $(v_{rms})_1 < (v_{rms})_2 < (v_{rms})_3$ and $(\bar{K})_1 = (\bar{K})_2 = (\bar{K})_3$
(b) $(v_{rms})_1 = (v_{rms})_2 = (v_{rms})_3$ and $(\bar{K})_1 = (\bar{K})_2 > (\bar{K})_3$
(c) $(v_{rms})_1 > (v_{rms})_2 > (v_{rms})_3$ and $(\bar{K})_1 < (\bar{K})_2 > (\bar{K})_3$
(d) $(v_{rms})_1 > (v_{rms})_2 > (v_{rms})_3$ and $(\bar{K})_1 < (\bar{K})_2 < (\bar{K})_3$

31. One gram mole of nitrogen at 27°C and 1 atm pressure is contained in a vessel and the molecules are moving with their rms speed. The number of collisions per second which the molecules make with an area of $1 m^2$ on the vessel's wall is
(a) 2×10^{27} (b) 2×10^{20}
(c) 2×10^{10} (d) 2×10^{24}

32. In an ideal gas at temperature T, the average force that a molecule applies on the walls of a closed container depends on T as T^q. A good estimate for q is:
(a) $\frac{1}{2}$ (b) 2 (c) 1 (d) $\frac{1}{4}$

33. Three closed vessels A, B and C are at the same temperature T and contain gases which obey the Maxwellian distribution of velocities. Vessel A contain only O_2, B only N_2 and C a mixture of equal quantities of O_2 and N_2. If the average speed of the O_2 molecules in vessel A is v_1 that of the N_2 molecules in vessel B is v_2, the average speed of the O_2 molecules in vessel C is **Critical Thinking**
(a) $\frac{v_1 + v_2}{2}$ (b) v_1
(c) $(v_1 . v_2)^{\frac{1}{2}}$ (d) $\sqrt{\frac{3kT}{M}}$

34. The molecules of a given mass of a gas have r.m.s. velocity of 200 ms^{-1} at 27°C and 1.0×10^5 Nm^{-2} pressure. When the temperature and pressure of the gas are respectively, 127°C and 0.05×10^5 Nm^{-2}, the r.m.s. velocity of its molecules in ms^{-1} is :
(a) $100\sqrt{2}$ (b) $\frac{400}{\sqrt{3}}$ (c) $\frac{100\sqrt{2}}{3}$ (d) $\frac{100}{3}$

35. N molecules each of mass m of a gas *A* and 2N molecules each of mass 2m of gas *B* are contained in the same vessel which is maintained at temperature T. The mean square velocity of molecules of *B* type is v^2 and the mean square rectangular component of the velocity of *A* type is denoted by ω^2. Then ω^2/v^2 is **Toughnut**
(a) 2 (b) 1 (c) 1/3 (d) 2/3

36. The density (ρ) versus pressuure (P) of a given mass of an ideal gas is shown at two temperatures T_1 and T_2
Then relation between T_1 and T_2 may be **Tricky**

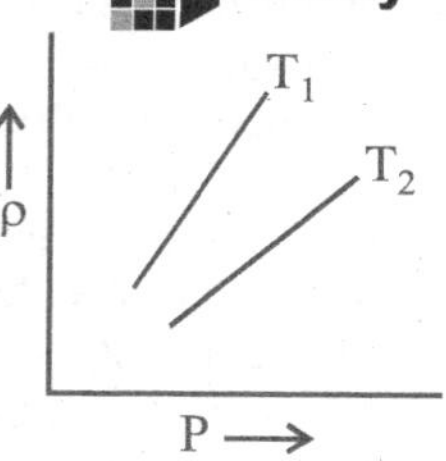

(a) $T_1 > T_2$
(b) $T_2 > T_1$
(c) $T_1 = T_2$
(d) All the three are possible

37. Boyle' law is applicable for an
(a) adiabatic process.
(b) isothermal process.
(c) isobaric process.
(d) isochoric process

38. What will be the ratio of number of molecules of a monoatomic and a diatomic gas in a vessel, if the ratio of their partial pressures is 5 : 3 ?
(a) 5 : 1 (b) 3 : 1 (c) 5 : 3 (d) 3 : 5

39. A balloon contains 1500 m^3 of helium at 27°C and 4 atmospheric pressure. The volume of helium at – 3°C temperature and 2 atmospheric pressure will,
(a) 1500 m^3 (b) 1700 m^3
(c) 1900 m^3 (d) 2700 m^3

40. The figure shows graph of pressure and volume of a gas at two different temperatures T_1 and T_2. Which of the following inferences is correct?

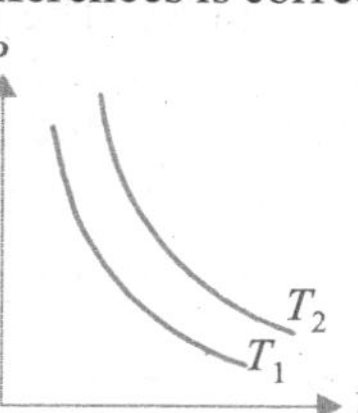

(a) $T_1 > T_2$
(b) $T_1 = T_2$
(c) $T_1 < T_2$
(d) None of these

41. The average translational kinetic energy of O_2 (relative molar mass 32) molecules at a particular temperature is 0.048 eV. The translational kinetic energy of N_2 (relative molar mass 28) molecules in eV at the same temperature is
(a) 0.0015 (b) 0.003 (c) 0.048 (d) 0.768

42. At 10° C the value of the density of a fixed mass of an ideal gas divided by its pressure is x. At 110°C this ratio is:
(a) x (b) $\frac{383}{283}x$ (c) $\frac{10}{110}x$ (d) $\frac{283}{383}x$

43. Two vessels separately contain two ideal gases A and B at the same temperature. The pressure of A being twice that of B. Under such conditions, the density of A is found to be 1.5 times the density of B. The ratio of molecular weight of A and B is :
(a) $\frac{3}{4}$ (b) 2 (c) $\frac{1}{2}$ (d) $\frac{2}{3}$

44. A perfect gas at 27°C is heated at constant pressure so as to double its volume. The final temperature of the gas will be, close to
(a) 327°C (b) 200°C (c) 54°C (d) 300°C

45. The average translational energy and the rms speed of molecules in a sample of oxygen gas at 300 K are 6.21×10^{-21} J and 484 m/s respectively. The corresponding values at 600 K are nearly (assuming ideal gas behavior) Critical Thinking
(a) 12.42×10^{-21} J, 968 m/s
(b) 8.78×10^{-21} J, 684 m/s
(c) 6.21×10^{-21} J, 968 m/s
(d) 12.42×10^{-21} J, 684 m/s

46. The temperature of an air bubble while rising from bottom to surface of a lake remains constant but its diameter is doubled if the pressure on the surface is equal to h meter of mercury column and relative density of mercury is ρ then the depth of lake in metre is Toughnut
(a) $2\rho h$ (b) $4\rho h$ (c) $8\rho h$ (d) $7\rho h$

47. Three containers of the same volume contain three different gases. The masses of the molecules are m_1, m_2 and m_3 and the number of molecules in their respective containers are N_1, N_2 and N_3. The gas pressure in the containers are P_1, P_2 and P_3 respectively. All the gases are now mixed and put in one of these containers. The pressure P of the mixture will be
(a) $P<(P_1+P_2+P_3)$ (b) $P=\frac{P_1+P_2+P_3}{3}$
(c) $P=P_1+P_2+P_3$ (d) $P>(P_1+P_2+P_3)$

48. Temperature of a body is only on manifestation of the mean
(a) total mechanical energy of a molecule of the body
(b) potential energy of a molecule of the body
(c) rotational kinetic energy of a molecule of the body
(d) translational kinetic energy of a molecule of the body

49. The mean free path of molecules of a gas, (radius 'r') is inversely proportional to :
(a) r^3 (b) r^2 (c) r (d) $\sqrt{r}$

50. A closed hollow insulated cylinder is filled with gas at 0°C and also contains an insulated piston of negligible weight and negligible thickness at the middle point. The gas on one side of the piston is heated to 100°C. If the piston moves 5 cm, the length of the hollow cylinder is
(a) 13.65 cm (b) 27.3 cm
(c) 38.6 cm (d) 64.6 cm

51. An air bubble of volume v_0 is released by a fish at a depth h in a lake. The bubble rises to the surface. Assume constant temperature and standard atmospheric pressure above the lake. The volume of the bubble just before touching the surface will be (density of water is ρ)

(a) v_0 (b) $v_0(\rho gh/p)$
(c) $\frac{v_0}{\left(1+\frac{\rho gh}{p}\right)}$ (d) $v_0\left(1+\frac{\rho gh}{p}\right)$

52. One kg of a diatomic gas is at a pressure of $8 \times 10^4 N/m^2$. The density of the gas is $4 kg/m^3$. What is the energy of the gas due to its thermal motion?
(a) 5×10^4 J (b) 6×10^4 J
(c) 7×10^4 J (d) 3×10^4 J

53. The maximum attainable temperature of ideal gas in the process $P = P_0 - \alpha V^2$ where P_0 and α are +ve constants.
(a) $\frac{2P_0}{3nR}\left(\frac{P_0}{3\alpha}\right)^{1/2}$ (b) $\frac{P_0}{2nR}\left(\frac{2P_0}{3\alpha}\right)^{1/2}$
(c) $\frac{2nR}{P_0}\left(\frac{2P_0}{3\alpha}\right)^{1/2}$ (d) $\frac{2P_0}{nR}\left(\frac{P_0}{2\alpha}\right)^{1/2}$

54. In the given (V – T) diagram, what is the relation between pressure P_1 and P_2 ?

(a) $P_2 > P_1$
(b) $P_2 < P_1$
(c) Cannot be predicted
(d) $P_2 = P_1$

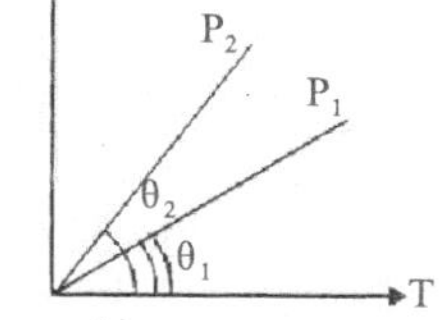

55. One mole of an ideal gas undergoes a process

$$P = \frac{P_0}{1+\left(\frac{V_0}{V}\right)^2}$$

Here P_0 and V_0 are constant. Change in temperature of the gas when volume is changed from $V = V_0$ to $V = 2V_0$

(a) $\frac{-2P_0V_0}{5R}$ (b) $\frac{11P_0V_0}{10R}$ (c) $\frac{-5P_0V_0}{4R}$ (d) P_0V_0

56. The internal energy of monatomic and diatomic gases are respectively due to
(a) linear motion and rolling motion
(b) rolling motion and linear motion
(c) linear motion and rotatory motion
(d) rotatory motion and linear motion

57. The equation of state for 5 g of oxygen at a pressure P and temperature T, when occupying a volume V, will be
(a) PV = (5/16) RT (b) PV = (5/32) RT
(c) PV = 5 RT (d) PV = (5/2) RT
where R is the gas constant.

58. A gas at 27°C temperature and 30 atmospheric pressure is allowed to expand to the atmospheric pressure. If the volume becomes 10 times its initial volume, then the final temperature becomes
(a) 100°C (b) 173°C (c) 273°C (d) –173°C

59. Air is pumped into an automobile tube upto a pressure of 200 kPa in the morning when the air temperature is 22°C. During the day, temperature rises to 42°C and the tube expands by 2%. The pressure of the air in the tube at this temperature, will be approximately

Critical Thinking

(a) 212 kPa (b) 209 kPa
(c) 206 kPa (d) 200 kPa

60. Which of the following will have maximum total kinetic energy at temperature 300K ?
(a) $1\ kg\ H_2$ (b) $1\ kg\ He$
(c) $\frac{1}{2} kg\ H_2 + \frac{1}{2} kg\ He$ (d) $\frac{1}{4} kg\ H_2 + \frac{3}{4} kg\ He$

61. One litre of oxygen at a pressure of 1 atm, and 2 litres of nitrogen at a pressure of 0.5 atm. are introduced in the vessel of 1 litre capacity, without any change in temperature. The total pressure would be

Critical Thinking

(a) 1.5 atm. (b) 0.5 atm. (c) 2.0 atm. (d) 1.0 atm.

62. The degree of freedom of a molecule of a triatomic gas is
(a) 2 (b) 4 (c) 6 (d) 8

63. A polyatomic gas with n degrees of freedom has a mean energy per molecule given by

(a) $\frac{nkT}{N}$ (b) $\frac{nkT}{2N}$ (c) $\frac{nkT}{2}$ (d) $\frac{3kT}{2}$

64. A vessel contains a mixture of one mole of oxygen and two moles of nitrogen at 300 K. The ratio of the average rotational kinetic energy per O_2 molecule to that per N_2 molecule is
(a) 1 : 1 (b) 1 : 2 (c) 2 : 1
(d) depends on the moments of inertia of the two molecules

65. Two different masses m and $3m$ of an ideal gas are heated separately in a vessel of constant volume, the pressure P and absolute temperature T, graphs for these two cases are shown in the figure as A and B. The ratio of slopes of curves B to A is
(a) 3 : 1
(b) 1 : 3
(c) 9 : 1
(d) 1 : 9

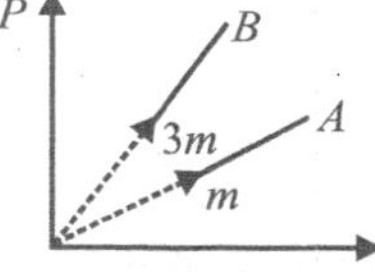

66. Two gases occupy two containers A and B the gas in A, of volume $0.10 m^3$, exerts a pressure of 1.40 MPa and that in B of volume $0.15 m^3$ exerts a pressure 0.7 MPa. The two containers are united by a tube of negligible volume and the gases are allowed to intermingle. Then if the temperature remains constant, the final pressure in the container will be (in MPa)

Toughnut

(a) 0.70 (b) 0.98 (c) 1.40 (d) 210

67. Two thermally insulated vessels 1 and 2 are filled with air at temperatures (T_1, T_2), volume (V_1, V_2) and pressure (P_1, P_2) respectively. If the valve joining the two vessels is opened, the temperature inside the vessel at equilibrium will be
(a) $T_1 + T_2$ (b) $(T_1 + T_2)/2$
(c) $\frac{T_1T_2(P_1V_1 + P_2V_2)}{P_1V_1T_2 + P_2V_2T_1}$ (d) $\frac{T_1T_2(P_1V_1 + P_2V_2)}{P_1V_1T_1 + P_2V_2T_2}$

68. The equation of state of a gas is given by $\left(P+\frac{aT^2}{V}\right)V^c=(RT+b\ ,)$where a, b, c and R are constants. The isotherms can be represented by $P=AV^m-BV^n$, where A and B depend only on temperature and

(a) $m=-c$ and $n=-1$ (b) $m=c$ and $n=-1$
(c) $m=-c$ and $n=1$ (d) $m=c$ and $n=-1$

69. An ideal gas is expanding such that PT^2 = constant. The coefficient of volume expansion of the gas is

(a) 1/T (b) 2/T (c) 3/T (d) 4/T

70. A non-linear triatomic gas is filled inside a vessel. If 'α' fraction of moles dissociate into individual atoms, then average degree of freedom for the mixture is : (neglect vibrational degrees of freedom) **Tricky**

(a) $\frac{3\alpha+6}{\alpha+1}$ (b) $\frac{\alpha+6}{2\alpha+1}$ (c) $\frac{3\alpha+6}{\alpha+2}$ (d) $\frac{3\alpha+6}{2\alpha+1}$

71. The amount of heat energy required to raise the temperature of 1g of Helium at NTP, from T_1K to T_2K is

(a) $\frac{3}{2}N_ak_B(T_2-T_1)$ (b) $\frac{3}{4}N_ak_B(T_2-T_1)$

(c) $\frac{3}{4}N_ak_B\frac{T_2}{T_1}$ (d) $\frac{3}{8}N_ak_B(T_2-T_1)$

72. The specific heat of a gas

(a) has only two values c_p and c_v
(b) has a unique value at a given temperature
(c) can have any value between 0 and ∞
(d) depends upon the mass of the gas

73. The specific heats at constant pressure is greater than that of the same gas at constant volume because **Tricky**

(a) at constant pressure work is done in expanding the gas
(b) at constant volume work is done in expanding the gas
(c) the molecular attraction increases more at constant pressure
(d) the molecular vibration increases more at constant pressure

74. During an adiabatic process, the pressure of a gas is found to be proportional to the cube of its absolute temperature. The ratio $\frac{C_p}{C_v}=\gamma$ for the gas is

(a) 2 (b) 3/2 (c) 5/3 (d) 4/3

75. A mixture of n_1 moles of monoatomic gas and n_2 moles of diatomic gas has $\frac{C_p}{C_v}=\gamma=1.5$ then

(a) $n_1=n_2$ (b) $2n_1=n_2$
(c) $n_1=2n_2$ (d) $2n_1=3n_2$

76. Two moles of ideal helium gas are in a rubber balloon at 30°C. The balloon is fully expandable and can be assumed to require no energy in its expansion. The temperature of the gas in the balloon is slowly changed to 35°C. The amount of heat required in raising the temperature is nearly (take $R=8.31$ J/mol.K) **Critical Thinking**

(a) 62 J (b) 104 J (c) 124 J (d) 208 J

77. Two cylinders A and B fitted with pistons contain equal amounts of an ideal diatomic gas at 300 K. The piston of A is free to move, while that B is held fixed. The same amount of heat is given to the gas in each cylinder. If the rise in temperature of the gas in A is 30 K, then the rise in temperature of the gas in B is **Critical Thinking**

(a) 30 K (b) 18 K (c) 50 K (d) 42 K

78. When an ideal diatomic gas is heated at constant pressure, the fraction of the heat energy supplied which increases the internal energy of the gas is

(a) 2/5 (b) 3/5 (c) 3/7 (d) 5/7

79. If one mole of a monatomic gas $\left(\gamma=\frac{5}{3}\right)$ is mixed with one mole of a diatomic gas $\left(\gamma=\frac{7}{5}\right)$, the value of γ for mixture is

(a) 1.40 (b) 1.50 (c) 1.53 (d) 3.07

80. 4.0 g of a gas occupies 22.4 litres at NTP. The specific heat capacity of the gas at constant volume is $5.0JK^{-1}$. If the speed of sound in this gas at NTP is 952 ms^{-1}, then the heat capacity at constant pressure is (Take gas constant R = 8.3 $JK^{-1}\ mol^{-1}$)

(a) 7.5 $JK^{-1}\ mol^{-1}$ (b) 7.0 $JK^{-1}\ mol^{-1}$
(c) 8.5 $JK^{-1}\ mol^{-1}$ (d) 8.0 $JK^{-1}\ mol^{-1}$

81. The molar specific heats of an ideal gas at constant pressure and volume are denoted by C_p and C_v, respectively. If $\gamma=\frac{C_p}{C_v}$ and R is the universal gas constant, then C_v is equal to

(a) $\frac{R}{(\gamma-1)}$ (b) $\frac{(\gamma-1)}{R}$

(c) γR (d) $\frac{1+\gamma}{1-\gamma}$

82. The specific heat of Ar at constant volume is 0.075 kg^{-1} K^{-1}. Calculate the atomic weight (R = 2 cal mol^{-1} K^{-1})
(a) 40 (b) 40.4 (c) 40.2 (d) 40.80

83. A mixture of ideal gases N_2 and He are taken in the mass ratio of 14 : 1 respectively. Molar heat capacity of the mixture at constant pressure is
(a) 6R/19 (b) 13R/6 (c) 6R/13 (d) 19R/6

84. The ratio C_P/C_V for a gas mixture consisting of 8g of helium and 16 g of oxygen is Tricky
(a) 24.2/15 (b) 15/23 (c) 27/17 (d) 17/27

85. The molar specific heat at constant pressure of an ideal gas is (7/2) R. The ratio of specific heat at constant pressure to that at constant volume is
(a) 8/7 (b) 5/7 (c) 9/7 (d) 7/5

86. If for a gas, $\frac{R}{C_v} = 0.67$, the gas is made up of molecules which are
(a) diatomic
(b) mixture of diatomic and polyatomic molecules
(c) monoatomic
(d) polyatomic

87. At very high temperatures vibrational degrees also becomes active. At such temperatures an ideal diatomic gas has a molar specific heat at constant pressure, C_p is
(a) 3R/2 (b) 5R/2 (c) 6R/2 (d) 9R/2

88. For hydrogen gas, $C_p - C_v$ = a and for oxygen gas, $C_p - C_v = b$, so the relation between a and b is given by
(a) $a = 16\,b$ (b) $16\,b = a$
(c) $a = 4\,b$ (d) $a = b$

89. The value of $C_p - C_v$ is 1.00R for a gas sample in state A and is 1.06R in state B. Let p_A, p_B denote the pressure and T_A, T_B denote the temperature of the states A and B respectively. Then most likely
(a) $p_A < p_B$ and $T_A > T_B$ (b) $p_A > p_B$ and $T_A < T_B$
(c) $p_A = p_B$ and $T_A < T_B$ (d) $p_A > p_B$ and $T_A = T_B$

90. A monoatomic ideal gas is taken through a reversible process whose equation is given by :

$$p = kV^{-\frac{1}{2}}$$

where p is the pressure and V is the volume of the gas.
The molar heat capacity of the gas in the above process, is
(a) $C_P + 2R$ (b) $C_V - 2R$
(c) $C_V + 2R$ (d) $C_P + 2R$

91. One mole of a gas occupies 22.4 lit at N.T.P. Calculate the difference between two molar specific heats of the gas.
J = 4200 J/kcal.
(a) 1.979 k cal/kmol K (b) 2.378 k cal/kmol K
(c) 4.569 kcal/kmol K (d) 3.028 k cal/ kmol K

92. For a gas, difference between two specific heats is 5000 J/mole°C. If the ratio of specific heat is 1.6, the two specific heats are in J/mole-°C * Critical Thinking
(a) $C_P = 1.33 \times 10^4$, $C_V = 2.66 \times 10^4$
(b) $C_P = 13.3 \times 10^4$, $C_V = 8.33 \times 10^3$
(c) $C_P = 1.33 \times 10^4$, $C_V = 8.33 \times 10^3$
(d) $C_P = 2.6 \times 10^4$, $C_V = 8.33 \times 10^4$

93. The molar specific heat at constant pressure of an ideal gas is (9/2)R. The ratio of specific heat at constant pressure to that at constant volume is Tricky
(a) 1.58 (b) 1.82 (c) 1.28 (d) 1.44

94. For a certain gas the ratio of specific heats is given to be $\gamma = 1.5$. For this gas
(a) $C_V = 3R/J$ (b) $C_P = 3R/J$
(c) $C_P = 5R/J$ (d) $C_V = 5R/J$

95. One mole of a diatomic gas is taken through the process $PV^n = k$, where n and k are constant. If the heat capacity of gas is negative, then the value of n may be Tricky
(a) $\frac{5}{7}$ (b) $-\frac{5}{7}$ (c) $\frac{9}{7}$ (d) $-\frac{9}{7}$

96. The relation between internal energy U, pressure P and volume V of an ideal gas in an adiabatic process is
$U = 2 + 3PV$. What is the value of the ratio of the molar specific heats $\left(\frac{C_P}{C_V}\right)$?
(a) 2/3 (b) 4/3 (c) 3/2 (d) 1

97. Find the value of $\gamma = \frac{C_P}{C_V}$ for a mixture consisting of n_1 moles of a monoatomic gas and n_2 moles of a gas of diatomic molecules:
(a) $\frac{n_1}{n_2}$ (b) $\frac{5n_1 + 7n_2}{3n_1 + 5n_2}$
(c) $\frac{3n_1 + 5n_2}{5n_1 + 7n_2}$ (d) $\frac{7n_2 + 3n_1}{5n_1 + 3n_2}$

98. Graph of specific heat at constant volume for a monatomic gas is

(a) 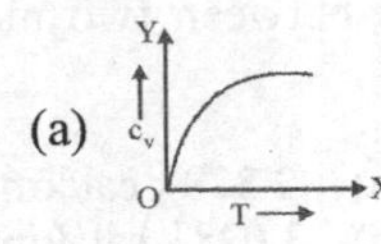

(b) 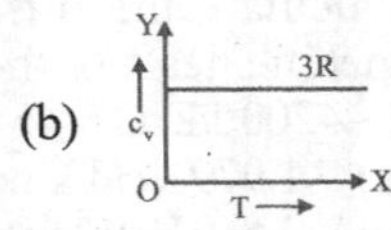

(c) 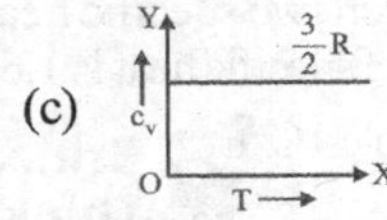

(d)

99. The K.E. of one mole of an ideal gas is E = (3/2) RT. Then C_p will be

(a) 0.5 R (b) 0.1 R (c) 1.5 R (d) 2.5 R

100. For a gas sample with N_0 number of molecules, function N(V) is given by :

$$N(V) = \frac{dN}{dV} = \left[\frac{3N_0}{V_0^3}\right]V^2 \text{ for } 0 \le V \le V_0 \text{ and}$$

$N(V) = 0$ for

$V > V_0$ where dN is number of molecules in speed range V to $V + dV$. The rms speed of the gas molecule is -

(a) $\sqrt{\frac{2}{5}}V_0$ (b) $\sqrt{\frac{3}{5}}V_0$

(c) $\sqrt{2V_0}$ (d) $\sqrt{3V_0}$

Answer KEYs

1	(a)	11	(a)	21	(b)	31	(a)	41	(c)	51	(d)	61	(c)	71	(d)	81	(a)	91	(a)
2	(a)	12	(a)	22	(d)	32	(c)	42	(d)	52	(a)	62	(c)	72	(c)	82	(a)	92	(c)
3	(a)	13	(a)	23	(b)	33	(b)	43	(a)	53	(a)	63	(c)	73	(a)	83	(d)	93	(c)
4	(c)	14	(d)	24	(b)	34	(b)	44	(a)	54	(b)	64	(a)	74	(b)	84	(a)	94	(b)
5	(d)	15	(c)	25	(a)	35	(d)	45	(d)	55	(b)	65	(a)	75	(a)	85	(d)	95	(c)
6	(d)	16	(b)	26	(b)	36	(b)	46	(d)	56	(a)	66	(b)	76	(d)	86	(c)	96	(b)
7	(d)	17	(b)	27	(a)	37	(b)	47	(c)	57	(b)	67	(c)	77	(d)	87	(d)	97	(b)
8	(b)	18	(b)	28	(b)	38	(c)	48	(d)	58	(d)	68	(a)	78	(d)	88	(d)	98	(c)
9	(a)	19	(b)	29	(b)	39	(d)	49	(b)	59	(b)	69	(c)	79	(b)	89	(a)	99	(d)
10	(c)	20	(b)	30	(a)	40	(c)	50	(d)	60	(a)	70	(d)	80	(d)	90	(c)	100	(b)

Hints & Solutions

1. (a) Velocity of sound in a gas is $v\sqrt{\frac{\gamma P}{\rho}}$;

rms speed $C=\sqrt{\frac{3P}{\rho}}$

Dividing $\frac{C}{v}=\sqrt{\frac{3}{\gamma}}\Rightarrow C=v\sqrt{\frac{3}{\gamma}}$

$=332\times\sqrt{\frac{3}{1.41}}=484$ m/s

2. (a) $P=\frac{1}{3}\rho\upsilon^2=\frac{1}{3}\times(6\times10^{-2})\times(500)^2$

$=5\times10^3\ N/m^2$

3. (a) $v_{rms}=\left[\frac{(2)^2+(3)^2+(4)^2+(5)^2}{4}\right]^{1/2}$

$=\sqrt{\left[\frac{54}{4}\right]}$

4. (c) If T be the required temperature, then,

$v_{rms}-v_p=200\ m/s$

or $\sqrt{\frac{3RT}{M_{He}}}-\sqrt{\frac{2RT}{M_{He}}}=200\ m/s$

or $\sqrt{T}\,[\sqrt{3}-\sqrt{2}]=\sqrt{\frac{M_{He}}{R}}(200\ m/s)$

Substituting $M_{He}=4\times10^{-3}$ kg, R = 8.314 J/mol-K,

T = 190.2 K.

5. (d) $V_{rms}=\sqrt{\frac{1^2+2^2+......+N^2}{N}}$

$=\sqrt{\frac{N(N+1)(2N+1)}{6N}}$

$V_{mean}=\frac{1+2+.....N}{N}=\frac{N(N+1)}{2N}$

6. (d) $v_{rms}=\sqrt{\frac{3RT}{M}}$

Same V_{rms}, implies $\frac{T}{M}$ same.

Hence, $T\propto M$

M for oxygen is more than that of hydrogen, therefore, oxygen is at higher temperature than hydrogen.

7. (d) $\frac{K.E.}{vol}=\frac{3}{2}P$. Here P is constant

8. (b)

9. (a) P = rate of change of momentum per unit area.

10. (c) In elastic collision kinetic energy is conserved.

11. (a) $\because\quad C=\sqrt{\frac{3RT}{M}}$

$(1930)^2=\frac{3\times8.314\times300}{M}$

$M=\frac{3\times8.314\times300}{1930\times1930}\approx2\times10^{-3}$ kg

The gas is H_2.

12. (a)

13. (a) Pressure of the gas will not be affected by motion of the system, hence by

$v_{rms}=\sqrt{\frac{3P}{\rho}}\Rightarrow\bar{c}^2=\frac{3P}{\rho}\Rightarrow P=\frac{1}{3}p\bar{c}^2$

14. (d) $c_{rms}\propto\sqrt{T}$

As temperature increases from 300 K to 1200 K that is four times, so, c_{rms} will be doubled.

15. (c) As temperature increases, velocity of molecules increases and hence the pressure.

16. (b) Higher is the temperature greater is the most probable velocity.

17. (b) Average K.E. $=\frac{1}{2}mc^2=\frac{3}{2}kT\propto T$

$\therefore\quad\frac{(Av.K.E.)_{600K}}{(Av.K.E.)_{300K}}=\frac{600}{300}=2$

$\therefore$ At 600K it will be 2 times than that at 300K.

18. (b) $\frac{(V_{rms})_1}{(V_{rms})_2}=\sqrt{\frac{T_1}{T_2}}\Rightarrow\frac{V}{(V_{rms})_2}=\sqrt{\frac{120}{480}}$

$\Rightarrow\quad\frac{V}{(V_{rms})_2}=\frac{1}{2}\Rightarrow(V_{rms})_2=2V$

19. (b) From graph, T^2V = const. .....(i)

As we know that $TV^{\gamma-1}$ = const

$\Rightarrow VT^{\frac{1}{\gamma-1}}$ = cons. ....(ii)

On comparing (1) and (2), we get

$\Rightarrow\gamma=3/2$

Also $v_{rms}=\sqrt{\frac{3P}{\rho}}$ and $v_{sound}=\sqrt{\frac{P\gamma}{\rho}}$

$\Rightarrow \dfrac{v_{rms}}{v_{sound}} = \sqrt{\dfrac{3}{\gamma}} = \sqrt{2}$

20. **(b)** $v_{rms} = \sqrt{\dfrac{3RT}{M}} = \sqrt{\dfrac{3 \times 8.31 \times 273}{28 \times 10^{-3}}} = 493$ m/s

Thus $\dfrac{1}{2} m v_{rms}^2 = \dfrac{mgh}{1 + \dfrac{h}{R}}$. After substituting the values and simplifying, we get $h = 12.4$ km

21. **(b)** As the temperature increases, the average velocity increases. So, the collisions are faster.

22. **(d)** $v_{rms} = \sqrt{\dfrac{8RT}{\pi M}}$

For v_{rms} to be equal $\dfrac{T_{H_2}}{M_{H_2}} = \dfrac{T_{O_2}}{M_{O_2}}$

Here $M_{H_2} = 2;\ M_{O_2} = 32$;

$T_{O_2} = 47 + 273 = 320$ K

$\therefore \dfrac{T_{H_2}}{2} = \dfrac{320}{32} \Rightarrow T_{H_2} = 20$ K

23. **(b)**

24. **(b)** $v_{rms} = \sqrt{\dfrac{3P}{\rho}} = \sqrt{\dfrac{3PV}{m}} \Rightarrow P = \dfrac{m\,(v_{rms})^2}{3V}$

If P' be the final pressure then

$\dfrac{P'}{P} = \left[\dfrac{\dfrac{\dfrac{m}{2}(2v_{rms})^2}{3V}}{\dfrac{m(v_{rms})^2}{3V}}\right] = 2 \Rightarrow P' = 2P.$

25. **(a)** From ideal gas equation $PV = \mu RT$

$\Rightarrow$ Slope of $P.T$ curve $\dfrac{P}{T} = \dfrac{\mu R}{V}$

$\Rightarrow$ Slope $\propto \dfrac{1}{V}$

It means line of smaller slope represent greater volume of gas.

For the given problem figure

Point 1 and 2 are on the same line so they will represent same volume i.e. $V_1 = V_2$

Similarly point 3 and 4 are on the same line so they will represent same volume i.e. $V_3 = V_4$

Also slope of line 1-2 is less than 3-4.

Hence $(V_1 = V_2) > (V_3 = V_4)$

26. **(b)** Figure shows the particles each moving with same spped v but the different directions. Consider any two particles having angle θ between directions of their velocities.

Then, $\vec{v}_{rel} = \vec{v}_B - \vec{v}_A$

i.e., $v_{rel} = \sqrt{v^2 + v^2 - 2vv\cos\theta}$

$\Rightarrow v_{rel} = \sqrt{2v^2(1 - \cos\theta)} = 2v\sin(\theta/2)$

So averaging v_{rel} over all pairs

$$\bar{v}_{rel} = \frac{\int_0^{2\pi} v_{rel}\, d\theta}{\int_0^{2\pi} d\theta} = \frac{\int_0^{2\pi} 2v\sin(\theta/2)}{\int_0^{2\pi} d\theta}$$

$$= \frac{2v \times 2[-\cos(\theta/2]_0^{2\pi}}{2\pi}$$

$\Rightarrow \bar{v}_{rel} = (4v/\pi) > v$ [as $4/\pi > 1$]

27. **(a)** $P = \dfrac{1}{3} Nmc^2 = \dfrac{2}{3} \times \left(\dfrac{1}{2} Nm\right) c^2 = \dfrac{2}{3} K.E$

28. **(b)** $v_{av} = \left[\dfrac{500 + 600 + 700 + 800 + 900}{5}\right] = 700$ m/s

and

$v_{rms} = \sqrt{\dfrac{500^2 + 600^2 + 700^2 + 800^2 + 900^2}{5}} = 714$ m/s

Thus v_{rms} is greater than average speed by 14 m/s.

29. **(b)** Vapour pressure does not depend on the amount of substance. It depends on the temperature alone.

30. **(a)** $v_{rms} \propto \dfrac{1}{\sqrt{M}} \Rightarrow (v_{rms})_1 < (v_{rms})_2 < (v_{rms})_3$

also in mixture temperature of each gas will be same, hence kinetic energy also remains same.

31. **(a)** Number of collisions can be calculated as

$n(2mv_{rms}) = PA$

$\therefore n = \left[\dfrac{PA}{2mv_{rms}}\right];\ v_{rms} = \sqrt{\dfrac{3kT}{m}}$

32. **(c)** Pressure, $P = \dfrac{1}{3}\dfrac{mN}{V} V_{rms}^2$

or, $P = \dfrac{(mN)T}{V}$

If the gas mass and temperature are constant then

$P \propto (V_{rms})^2 \propto T$

So, force $\propto (V_{rms})^2 \propto T$

i.e., Value of $q = 1$

33. **(b)** All three vessels are at same temperature. According to Maxwell's distribution of speed, average speed of molecules of a gas $v \propto \sqrt{T}$.

$\therefore$ The velocity of oxygen molecules will be same in A as well as C where M is the mass of an oxygen molecule.

34. (b) Here $v_1 = 200$ m/s;
temperature $T_1 = 27°C = 27 + 273 = 300$ k
temperature $T_2 = 127°C = 127 + 273 = 400$ k, V = ?
R.M.S. Velocity, $V \propto \sqrt{T}$

$$\Rightarrow \frac{v}{200} = \sqrt{\frac{400}{300}} \Rightarrow v = \frac{400}{\sqrt{3}} \text{ m/s}$$

35. (d) Mean kinetic energy of the two types of molecules should be equal. The mean square velocity of A type molecules = $\omega^2 + \omega^2 + \omega^2 = 3\omega^2$

Therefore, $\frac{1}{2}m(3\omega^2) = \frac{1}{2}(2m)v^2$

This gives $\omega^2 / v^2 = 2/3$

36. (b) According to ideal gas equation
PV = nRT

$$PV = \frac{m}{M}RT, \ P = \frac{\rho}{M}RT \text{ or } \frac{\rho}{P} = \frac{M}{RT}$$

or $\frac{\rho}{P} \propto \frac{1}{T}$

Here, $\frac{\rho}{P}$ represent the slope of graph

Hence $T_2 > T_1$

37. (b) According to boyle's law, at constant temperature.

$P \propto \frac{1}{V}$ or $P_1V_1 = P_2V_2$

38. (c)

39. (d) $\frac{P_1V_1}{T_1} = \frac{P_2V_2}{T_2}$

$$\therefore V_2 = \frac{P_1V_1}{T_1} \times \frac{T_2}{P_2} = \frac{4 \times 1500 \times 270}{300 \times 2} = 2700 \text{ m}^3$$

40. (c) With increase in temperature, the product (PV) increases.

41. (c) Average translational kinetic energy of an ideal gas molecule is $\frac{3}{2}$kT which depends on temperature only. Therefore, if temperature is same, translational kinetic energy of O_2 and N_2 both will be equal.

42. (d) Let the mass of the gas be m.
At a fixed temperature and pressure, volume is fixed.

Density of the gas, $\rho = \frac{m}{V}$

Now $\frac{\rho}{P} = \frac{m}{PV} = \frac{m}{nRT}$

$\Rightarrow \frac{m}{nRT} = x$ (By question)

$\Rightarrow xT = \text{constant} \Rightarrow x_1T_1 = x_2T_2$

$$\Rightarrow x_2 \Rightarrow \frac{x_1T_1}{T_2} = \frac{283}{383}x \quad \left[\begin{array}{l} \therefore \\ T_1 = 283K \\ T_2 = 383K \end{array}\right]$$

43. (a) From PV = nRT

$$P_A = \frac{\rho_A M_A}{RT} \text{ and } P_B = \frac{\rho_B M_B}{RT}$$

From question,

$$\frac{P_A}{P_B} = \frac{\rho_A}{\rho_B}\frac{M_A}{M_B} = 2\frac{M_A}{M_B} = \frac{3}{2}$$

So, $\frac{M_A}{M_B} = \frac{3}{4}$

44. (a) Given, $V_1 = V$, $V_2 = 2V$
$T_1 = 27° + 273 = 300$ K so, $T_2 = ?$
From charle's law

$$\frac{V_1}{T_1} = \frac{V_2}{T_2} \text{ or, } \frac{V}{300} = \frac{2V}{T_2}$$

($\because$ Pressure is constant)

$\therefore$ $T_2 = 600$ K $= 600 - 273 = 327°C$

45. (d) The formula for average kinetic energy is

$$\overline{K.E.} = \frac{3}{2}kT \quad \therefore \frac{\overline{(K.E.)}_{600K}}{\overline{(K.E.)}_{300K}} = \frac{600}{300}$$

$$\Rightarrow \overline{(K.E.)}_{600K} = 2 \times 6.21 \times 10^{-21} \text{ J} = 12.42 \times 10^{-21} \text{ J}$$

Also the formula for r.m.s. velocity is

$$C_{rms} = \sqrt{\frac{3KT}{m}} \quad \therefore \frac{(C_{rms})_{600K}}{(C_{rms})_{300K}} = \sqrt{\frac{600}{300}}$$

$$\Rightarrow (C_{rms})_{600K} = \sqrt{2} \times 484 = 684 \text{ m/s}$$

46. (d) From Boyle's law (T = constant)

$P_1V_1 = P_2V_2$

$$\therefore (Hd_{waer} + h\,d_{mercury})\,g\,(\frac{4}{3}\pi r^3) = hd_{mercury}\,g(\frac{4}{3}\pi(2r)^3)$$

$$\Rightarrow Hd_{water} = 8hd_{mercury} - hd_{mercury}$$

$$\Rightarrow H = 7h\frac{d_{mercury}}{d_{water}}$$

$\therefore$ $H = 7h\rho$

47. (c) According to Dalton's law of partial pressures, we have $P = P_1 + P_2 + P_3$

48. (d) Mean translational $K.E.$ of any molecule whether monoatomic, diatomic or polyatomic, at a given temperature T is $\frac{3}{2}kT$.

49. (b) Mean free path $\lambda_m = \frac{1}{\sqrt{2}\pi d^2 n}$

where d = diameter of molecule and d = 2r

$\therefore \quad \lambda_m \propto \frac{1}{r^2}$

50. (d) Using Charle's law, we have $\frac{V}{T}$ = constant

$\Rightarrow \frac{\frac{l}{2}+5}{373} = \frac{\frac{l}{2}-5}{273}$

As the piston moves 5 cm, the length of one side will be $\left(\frac{l}{2}+5\right)$ and other side $\left(\frac{l}{2}-5\right)$. On solving this equaiton, we get l = 64.6 cm.

51. (d) As the bubble rises the pressure gets reduced for constant temperature, if P is the standard atmospheric pressure, then

$(P + \rho gh)V_0 = PV$

or $V = V_0\left(1 + \frac{\rho gh}{P}\right)$

52. (a) Volume $= \frac{\text{mass}}{\text{density}} = \frac{1}{4}\text{ m}^3$

$K.E = \frac{5}{2}PV = \frac{5}{2} \times 8 \times 10^4 \times \frac{1}{4} = 5 \times 10^4\ J$

Alternatively:

$K.E = \frac{5}{2}nRT = \frac{5}{2}\frac{m}{M}RT = \frac{5}{2}\frac{m}{M} \times \frac{PM}{d}$

$[\because PM = dRT]$

$= \frac{5}{2}\frac{mP}{d} = \frac{5}{2} \times \frac{1 \times 8 \times 10^4}{4} = 5 \times 10^4$ J

53. (a)

54. (b) $P_1 > P_2$

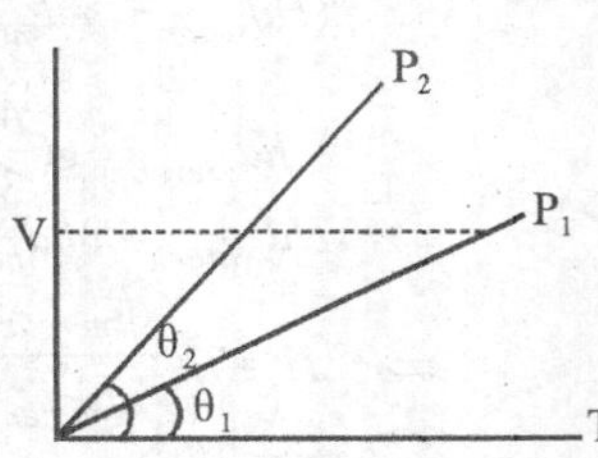

As V = constant $\Rightarrow P \propto T$

Hence from V–T graph $P_1 > P_2$

55. (b)

56. (a) The internal energy of a monatomic gas is due to linear motion only and that of the diatomic gas is due to rolling (Linear rotatory) motion.

57. (b) $n = \frac{5}{32} \quad \therefore PV = \frac{5}{32}RT \quad [\because PV = nRT]$

58. (d)

59. (b) $\frac{P_1V_1}{T_1} = \frac{P_2V_2}{T_2}$ Here, $P_1 = 200$kPa

$T_1 = 22°C = 295$ K $\quad T_2 = 42°C = 315$K

$V_2 = V_1 + \frac{2}{100}V_1 = 1.02V_1$

$\therefore \quad P_2 = \frac{200 \times 315V_1}{295 \times 1.02V_1} = 209.37$kPa

60. (a) Total Kinetic Energy $= U = \frac{f}{2}nRT$

In case of H_2 degree of freedom is greatest and number of moles n is highest.

So this is the case of maximum kinetic energy.

61. (c)

62. (c) No. of degree of freedom $= 3K - N$ where K is no. of atom and N is the number of relations between atoms. For triatomic gas,

$K = 3, N = {}^3C_2 = 3$

No. of degree of freedom $= 3(3) - 3 = 6$

63. (c) According to law of equipartition of energy, the energy per degree of freedom is $\frac{1}{2}kT$.

For a polyatomic gas with n degrees of freedom, the mean energy per molecule $= \frac{1}{2}nkT$

64. (a)

65. (a) $PV = \frac{m}{M}RT; \therefore \frac{P}{T} = Cm$

or $\quad \frac{\text{slope of } B}{\text{slope of } A} = \frac{m_B}{m_A} = \frac{3m}{m} = 3$

66. (b) We know that

$P_AV_A = nRT, P_BV_B = n_BRT$ and

$P_f(V_A + V_B) = (n_A + n_B)RT$

$P_f(V_A+V_B)=P_AV_A+P_BV_B$

$\therefore \quad P_f=\left(\frac{P_AV_A+P_BV_B}{V_A+V_B}\right)$

$=\frac{1.4\times0.1+0.7\times0.15}{0.1+0.15}$ MPa

$=0.98$MPa

67. (c) $\frac{P(V_1+V_2)}{T}=\frac{P_1V_1}{T_1}+\frac{P_2V_2}{T_2}\ \ldots$(i)

Also $\quad P(V_1+V_2) = P_1V_1+P_2V_2 \ldots$(ii)

After solving above equations, we get

$T=\left[\frac{(P_1V_1+P_2V_2)T_1T_2}{P_1V_1T_2+P_2V_2T_1}\right]$

68. (a) $\left(P+\frac{aT^2}{V}\right)V^c=(RT+b)$

$\Rightarrow P=(RT+b)V^{-c}-(aT^2)V^{-1}$

On comparing with equation, $P=AV^m-BV^n$, $m=-c$ and $n=-1$.

69. (c)

70. (d) $f_{\max}=\frac{(n\alpha)(3)(3)+(n-n\alpha)(6)}{(n\alpha)(3)+(n-n\alpha)}=\frac{3\alpha+6}{2\alpha+1}$

71. (d) From first law of thermodynamics

$\Delta Q=\Delta U+\Delta W=\frac{3}{2}\cdot\frac{1}{4}R(T_2-T_1)+0$

$=\frac{3}{8}N_aK_B(T_2-T_1)\ [\because K=\frac{R}{N}]$

72. (c) $C=\frac{Q}{m\Delta T}$; If $\Delta T=0$, $C=\infty$ and if $Q=0$, then $C=0$

73. (a) Work done is to be done in expanding the gas at constant pressure.

74. (b) $p^{1-\gamma}T^{\gamma}$ = constant

or $\quad p\propto T^{[\gamma/(\gamma-1)]}$ $\qquad$ Here, $p\propto T^3$

or $\quad \frac{\gamma}{\gamma-1}=3$ or $\qquad \gamma=\frac{3}{2}$

75. (a)

76. (d) The heat is supplied at constant pressure. Therefore,

$Q=n\,C_p\,\Delta t$

$=2\left[\frac{5}{2}R\right]\times\Delta t=2\times\frac{5}{2}\times8.31\times5=208$ J

$\left(\because C_p=\frac{5}{2}R \text{ for mono-atomic gas}\right)$

77. (d) A is free to move, therefore heat will be supplied at constant pressure

$\therefore \quad \Delta Q_A=nC_p\Delta T_A \qquad \ldots$(i)

B is held fixed, therefore heat will be supplied at constant volume.

$\therefore \quad \Delta Q_B=nC_v\Delta T_B \qquad \ldots$(ii)

But $\Delta Q_A=\Delta Q_B$ $\qquad$ (given)

$\therefore \quad nC_p\Delta T_A=nC_v\Delta T_B$

$\therefore \ \Delta T_B=\left(\frac{C_p}{C_v}\right)\Delta T_A$

$=\gamma(\Delta T_A)$ $\qquad$ [$\gamma=1.4$ (diatomic)]

$=(1.4)(30\,\text{K})$

$\therefore \quad \Delta T_B=42$ K

78. (d) $\Delta Q=nC_p\Delta T$ at constant pressure

$\Delta U=nC_V\Delta T$

Fraction $=\frac{\Delta U}{\Delta Q}=\frac{C_V}{C_P}=\frac{1}{\gamma}=\frac{5}{7}$ for diatomic gas.

79. (b)

80. (d) Molar mass of the gas = 4g/mol

Speed of sound

$V=\sqrt{\frac{\gamma RT}{m}}\Rightarrow 952=\sqrt{\frac{\gamma\times3.3\times273}{4\times10^{-3}}}$

$\Rightarrow\gamma=1.6=\frac{16}{10}=\frac{8}{5}$

Also, $\gamma=\frac{C_P}{C_V}=\frac{8}{5}$

So, $C_P=\frac{8\times5}{5}=8\text{JK}^{-1}\text{mol}^{-1}$

$[C_V=5.0\ \text{JK}^{-1}$ given$]$

81. (a) $C_p-C_v=R\Rightarrow C_p=C_v+R$

$\because \gamma=\frac{C_p}{C_v}=\frac{C_v+R}{C_v}=\frac{C_v}{C_v}+\frac{R}{C_v}$

$\Rightarrow\gamma=1+\frac{R}{C_v}\Rightarrow\frac{R}{C_v}=\gamma-1\Rightarrow C_v$

$=\frac{R}{\gamma-1}$

82. (a) $M=\frac{C_V}{S_V}=\frac{\frac{3\times2}{2}}{0.075}=40\ \therefore\ C_V=\frac{3}{2}R$

83. (d) $\frac{M_{N_2}}{M_{H_e}}=\frac{14}{1}M_{N_2}=14m$ and $M_{H_e}=m$

$C_p=\frac{n_1C_{p_1}+n_2C_{p_2}}{n_1+n_2}=\frac{\frac{14m}{28}\times\frac{7}{2}R+\frac{m}{4}\times\frac{5}{2}R}{\frac{14m}{28}+\frac{m}{4}}$

$$=\frac{2\times\frac{7}{2}R+\frac{5}{2}R}{2+1}=\frac{7R+\frac{5}{2}R}{3}=\frac{19R}{6}$$

84. **(a)** γ for He $= 5/3$; γ for $O_2 = 7/5$

8 g of He is equal to 2 moles of He and 16 g of $O_2 = \frac{1}{2}$ mole of O_2 gas. Total one has 2½ moles.

$\therefore$ The weighted average is

$$=\left(2\times\frac{5}{3}+\frac{7}{5}\times\frac{1}{2}\right)\times\frac{2}{5}=\left\{\frac{10}{3}+\frac{7}{10}\right\}\times\frac{2}{5}$$

$$=\frac{121}{30}\times\frac{2}{5}=\frac{24.2}{15}$$

85. **(d)** $C_P=\frac{7}{2}R;\ C_V=C_P-R=\frac{7}{2}R-R=\frac{5}{2}R$

$$\frac{C_P}{C_V}=\frac{7/2\,R}{5/2\,R}=\frac{7}{5}$$

86. **(c)** Since $\frac{R}{C_V}=0.67\Rightarrow C_V=\frac{3}{2}R,$ hence gas is monoatomic.

87. **(d)** At high temperature for diatomic gas $f=7$

$$C_p=\frac{fR}{2}+R=\frac{9R}{2}$$

88. **(d)** Both are diatomic gases and $C_p-C_v=R$ for all gases.

89. **(a)** For ideal gas $C_p-C_v=R$

If $C_p-C_v=1.06\,R$,

then gas will be real gas. Thus pressure is high and temperature is low for real gas.

90. **(c)**

91. **(a)** $V=22.4$ litre $=22.4\times10^{-3}\,m^3, J=4200$ J/kcal

by ideal gas equation for one mole of a gas,

$$R=\frac{PV}{T}=\frac{1.013\times10^5\times22.4\times10^{-3}}{273}$$

$$C_p-C_v=\frac{R}{J}=\frac{1.013\times10^5\times22.4}{273\times4200}=1.979$$

kcal/kmol K

92. **(c)** Given

$C_P-C_V=5000\ J/mole\,°C$ ....(i)

$\frac{C_P}{C_V}=1.6$ ....(ii)

From Equation (i) & (ii),

$$\Rightarrow\frac{C_P}{C_V}-\frac{C_V}{C_V}=\frac{5000}{C_V}\Rightarrow1.6-1=\frac{5000}{C_V}$$

$$\Rightarrow C_V=\frac{5000}{0.6}=8.33\times10^3$$

Hence $C_P=1.6, C_V = 1.6\times8.33\times10^3$

$=1.33\times10^4$

93. **(c)** $C_p=\left(\frac{f}{2}+1\right)R=\frac{9}{2}R$

$\Rightarrow f=7$

$$\frac{C_p}{C_v}=1+\frac{2}{f}=1+\frac{2}{7}=1.28$$

94. **(b)**

95. **(c)** $C=\frac{R}{\gamma-1}+\frac{R}{1-n}$ for a polytropic process

$\Rightarrow C=\frac{5R}{2}+\frac{R}{1-n}$ $(\because\ r=\frac{7}{5}$ for a diatomic gas)

For C to be negative, the possible value of n will be $\frac{9}{7}$.

96. **(b)** **97.** **(b)**

98. **(c)** For a monatomic gas

$C_V=\frac{3}{2}R$

So correct graph is

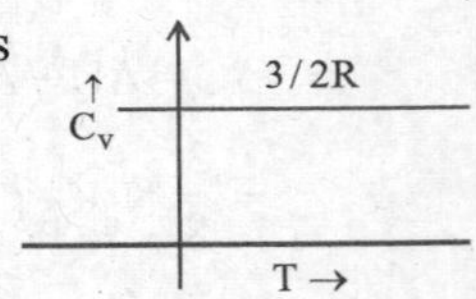

99. **(d)** $C_v=dE/dT=\frac{3}{2}R$

$$C_p=\frac{3}{2}R+R=\frac{5}{2}R=2.5R$$

100. **(b)** $V_{rms}^2=\langle V^2\rangle=\frac{V_1^2+V_2^2+V_3^2+......}{N}$

$$=\frac{\int V^2dN}{\int dN}\text{ here }\frac{dN}{dV}=N(V)$$

$$V_{rms}^2=\frac{1}{N}\int_0^\infty N(V)V^2dV$$

$$=\frac{1}{N}\int_0^{V_0}\left[\frac{3N}{V_0^3}V^2\right]V^2\,dV=\frac{3}{5}V_0^2$$

$$\Rightarrow V_{rms}=\sqrt{\frac{3}{5}}V_0$$

13 Oscillations

1. A particle performs SHM in a straight line. In the first second, starting from rest, it travels a distance a and in the next second it travels a distance b in the same direction. The amplitude of the SHM is

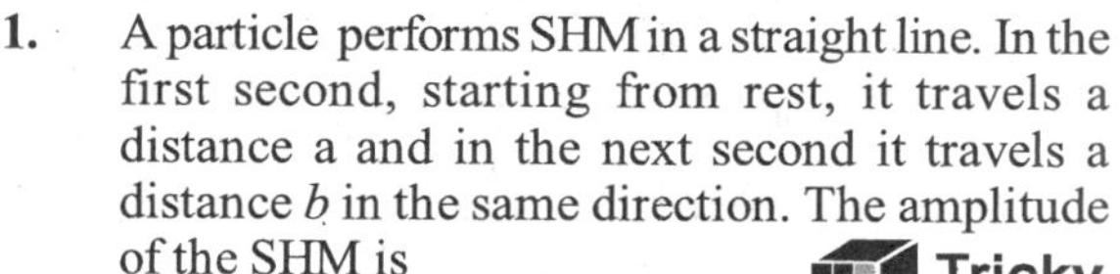

Tricky

(a) $a-b$ (b) $\frac{2a-b}{3}$
(c) $\frac{2a^2}{3a-b}$ (d) None of these

2. A simple harmonic oscillator of angular frequency 2 rad s^{-1} is acted upon by an external force F = sin t N. If the oscillator is at rest in its equilibrium position at t = 0, its position at later times is given by:

Tricky

(a) $\sin t+\frac{1}{2}\cos 2t$ (b) $\cos t-\frac{1}{2}\sin 2t$
(c) $\sin t-\frac{1}{2}\sin 2t$ (d) $\sin t+\frac{1}{2}\sin 2t$

3. Part of a simple harmonic motion is graphed in the figure, where y is the displacement from the mean position. The correct equation describing this S.H. M. is

(a) $y=4\cos(0.6t)$
(b) $y=2\sin\left(\frac{10}{3}t-\frac{\pi}{2}\right)$
(c) $y=4\sin\left(\frac{10}{3}t+\frac{\pi}{2}\right)$ (d) $y=2\cos\left(\frac{10}{3}t+\frac{\pi}{2}\right)$

4. If a is the amplitude of SHM, then K.E. is equal to the P.E. at .... distance from the mean position.
(a) $\frac{a}{\sqrt{2}}$ (b) $\frac{a}{2}$ (c) $\frac{a}{4}$ (d) a

5. Which of the following is true about total mechanical energy of SHM ?
(a) It is zero at mean position.
(b) It is zero at extreme position.
(c) It is always zero.
(d) It is never zero.

6. In S.H.M. the ratio of kinetic energy at mean position to the potential energy when the displacement is half of the amplitude is
(a) $\frac{4}{1}$ (b) $\frac{2}{3}$ (c) $\frac{4}{3}$ (d) $\frac{1}{2}$

7. Starting from the origin a body oscillates simple harmonically with a period of 2 s. After what time will its kinetic energy be 75% of the total energy?

Critical Thinking

(a) $\frac{1}{6}$s (b) $\frac{1}{4}$s (c) $\frac{1}{3}$s (d) $\frac{1}{12}$s

8. For a particle executing SHM the displacement x is given by $x = A\cos\omega t$. Identify the graph which represents the variation of potential energy (P.E.) as a function of time t and displacement x.

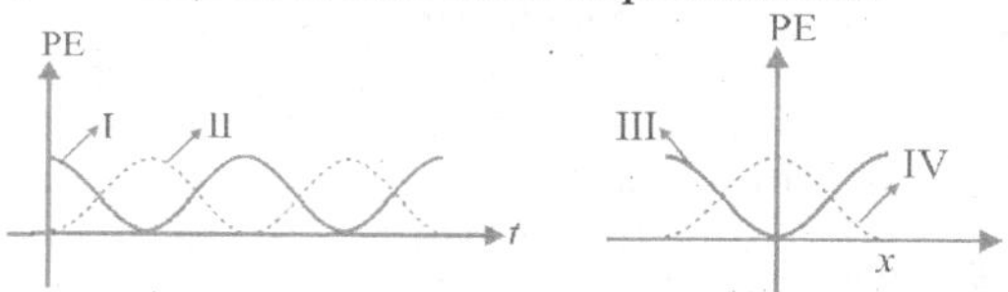

(a) I, III (b) II, IV (c) II, III (d) I, IV

9. The total energy of the particle executing simple harmonic motion of amplitude A is 100 J. At a distance of 0.707 A from the mean position, its kinetic energy is
(a) 25 J (b) 50 J (c) 100 J (d) 12.5 J

10. When the displacement of a particle executing simple harmonic motion is half of its amplitude, the ratio of its kinetic energy to potential energy is
(a) 1 : 3 (b) 2 : 1 (c) 3 : 1 (d) 1 : 2

11. The phase difference between the instantaneous velocity and acceleration of a particle executing simple harmonic motion is
(a) π (b) 0.707π (c) zero (d) 0.5π

12. A particle of mass 1 kg is moving in S.H.M. with an amplitude 0.02 and a frequency of 60 Hz. The maximum force acting on the particle is

Critical Thinking

(a) $144\,\pi^2$ (b) $188\,\pi^2$
(c) $288\,\pi^2$ (d) None of these

13. The displacement of a particle in simple harmonic motion in one time period is [A = amplitude]
(a) A (b) 2 A (c) 4 A (d) Zero

14. Suppose a tunnel is dug along a diameter of the earth. A particle is dropped from a point, a distance h directly above the tunnel, the motion of the particle is

Tricky

(a) simple harmonic (b) parabolic
(c) oscillatory (d) non-periodic

15. The acceleration of a particle undergoing SHM is graphed in figure. At point 2 the velocity of the particle is
(a) zero
(b) negative
(c) positive
(d) None of these

16. A particle starts with S.H.M. from the mean position as shown in figure below. Its amplitude is A and its time period is T. At one time, its speed is half that of the maximum speed. What is the displacement at that time?

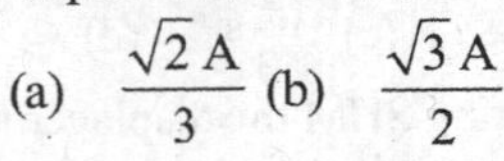

(a) $\frac{\sqrt{2}\,A}{3}$ (b) $\frac{\sqrt{3}\,A}{2}$
(c) $\frac{2A}{\sqrt{3}}$ (d) $\frac{3A}{\sqrt{2}}$

17. A body executing linear simple harmonic motion has a velocity of 3 m/s when its displacement is 4 cm and a velocity of 4 m/s when its displacement is 3 cm. What is the amplitude of oscillation?
(a) 5 cm (b) 7.5 cm (c) 10 cm (d) 12.5 cm

18. The ratio of maximum acceleration to maximum velocity in a simple harmonic motion is 10 s^{-1}. At, t = 0 the displacement is 5 m. The initial phase is $\frac{\pi}{4}$. What is the maximum acceleration?
(a) 500 m/s^2 (b) $500\sqrt{2}$ m/s^2
(c) 750 m/s^2 (d) $750\sqrt{2}$ m/s^2

19. The displacement of a particle in SHM is $x = 10\sin\left(2t - \frac{\pi}{6}\right)$ metre. When its displacement is 6 m, the velocity of the particle (in m s^{-1}) is
(a) 8 (b) 24 (c) 16 (d) 10

20. Two particles are executing simple harmonic motion of the same amplitude A and frequency ω along the *x*-axis. Their mean position is separated by distance $X_0 (X_0 > A)$. If the maximum separation between them is $(X_0 + A)$, the phase difference between their motion is **Critical Thinking**
(a) $\frac{\pi}{3}$ (b) $\frac{\pi}{4}$ (c) $\frac{\pi}{6}$ (d) $\frac{\pi}{2}$

21. A graph of the square of the velocity against the square of the acceleration of a given simple harmonic motion is **Tricky**

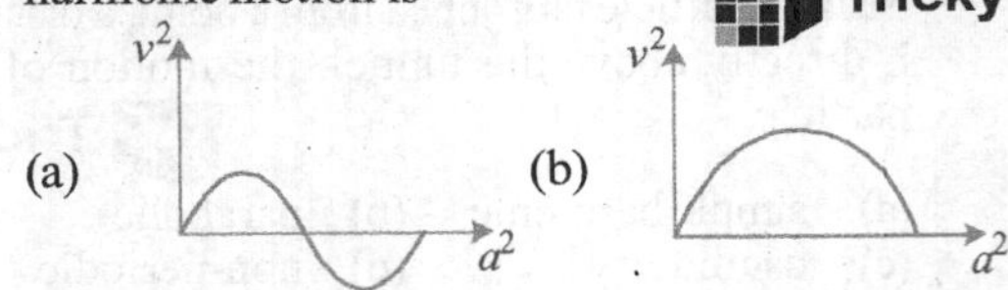

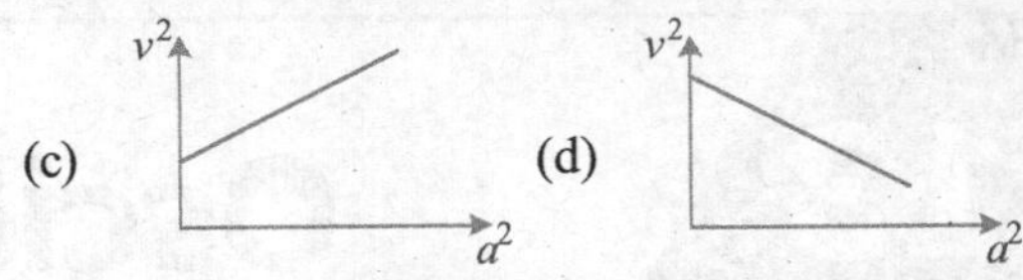

22. For a particle moving according to the equation $x = a\cos\pi t$, the displacement in 3 s is
(a) 0 (b) 0.5a (c) 1.5a (d) 2a

23. A body oscillates with SHM according to the equation (in SI units), $x = 5\cos\left(2\pi t - \frac{\pi}{4}\right)$. Its instantaneous displacement at $t = 1$ second is
(a) $\frac{\sqrt{2}}{5}$ m (b) $\frac{1}{\sqrt{3}}$ m (c) $\frac{5}{\sqrt{2}}$ m (d) $\frac{1}{2}$ m

24. A point mass oscillates along the x-axis according to the law $x = x_0\cos(\omega t - \pi/4)$. If the acceleration of the particle is written as $a = A\cos(\omega t + \delta)$, then
(a) $A = x_0\omega^2, \delta = 3\pi/4$ (b) $A = x_0, \delta = -\pi/4$
(c) $A = x_0\omega^2, \delta = \pi/4$ (d) $A = x_0\omega^2, \delta = -\pi/4$

25. The displacement vs time of a particle executing SHM is shown in figure. **Tricky**
The initial phase ϕ is
(a) $-\pi < \phi < -\frac{\pi}{2}$
(b) $\pi < \phi < \frac{3\pi}{2}$
(c) $-\frac{3\pi}{2} < \phi < -\pi$
(d) $\frac{\pi}{2} < \phi < \pi$

26. The amplitude of a particle executing SHM is 4 cm. At the mean position the speed of the particle is 16 cm/sec. The distance of the particle from the mean position at which the speed of the particle becomes $8\sqrt{3}$ cm/s, will be
(a) $2\sqrt{3}$ cm (b) $\sqrt{3}$ cm (c) 1 cm (d) 2 cm

27. Two simple harmonic motions are represented by the equations $y_1 = 0.1\sin\left(100\pi t + \frac{\pi}{3}\right)$ and $y_2 = 0.1\cos\pi t$.
The phase difference of the velocity of particle 1 with respect to the velocity of particle 2 is
(a) $\frac{\pi}{3}$ (b) $\frac{-\pi}{6}$ (c) $\frac{\pi}{6}$ (d) $\frac{-\pi}{3}$

28. Two particles are oscillating along two close parallel straight lines side by side, with the same frequency and amplitudes. They pass each other, moving in opposite directions when their displacement is half of the amplitude. The mean positions of the two particles lie on a straight line perpendicular to the paths of the two particles. The phase difference is **Toughnut**
(a) 0 (b) 2π/3 (c) π (d) π/6

29. A particle execuits SHM with time period 8 s. Initially, it is at its mean position. The ratio of distance travelled by it in the 1st second to that in the 2nd second is

(a) $\sqrt{2}:1$ (b) $1:(\sqrt{2}-1)$

(c) $(\sqrt{2}+1):\sqrt{2}$ (d) $(\sqrt{2}-1):1$

30. The *x-t* graph of a particle undergoing simple harmonic motion is shown below. The acceleration of the particle at $t = 4/3$ s is **Critical Thinking**

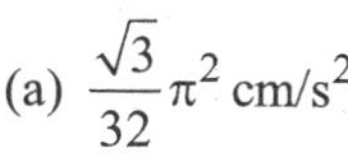

(a) $\frac{\sqrt{3}}{32}\pi^2\ \text{cm/s}^2$

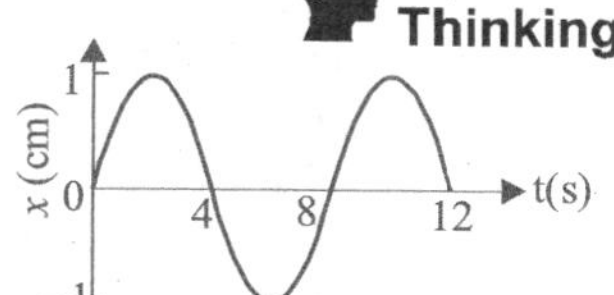

(b) $\frac{-\pi^2}{32}\ \text{cm/s}^2$

(c) $\frac{\pi^2}{32}\ \text{cm/s}^2$

(d) $-\frac{\sqrt{3}}{32}\pi^2\text{cm/s}^2$

31. A particle executing SHM has velocities u and v and acceleration a and b in two of its positions. Find the distance between these two positions.

(a) $\frac{u^2-v^2}{a+b}$ (b) $\frac{v^2+u^2}{a-b}$

(c) $\frac{v^2+u^2}{a+b}$ (d) $\frac{v^2-u^2}{a-b}$

32. A particle executing harmonic motion is having velocities v_1 and v_2 at distances is x_1 and x_2 from the equilibrium position. The amplitude of the motion is

(a) $\sqrt{\frac{v_1^2x_2-v_2^2x_1}{v_1^2+v_2^2}}$ (b) $\sqrt{\frac{v_1^2x_1^2-v_2^2x_2^2}{v_1^2+v_2^2}}$

(c) $\sqrt{\frac{v_1^2x_2^2-v_2^2x_1^2}{v_1^2-v_2^2}}$ (d) $\sqrt{\frac{v_1^2x_2^2+v_2^2x_1^2}{v_1^2+v_2^2}}$

33. A coin is placed on a horizontal platform which undergoes vertical simple harmonic motion of angular frequency ω. The amplitude of oscillation is gradually increased. The coin will leave contact with the platform for the first time **Critical Thinking**

(a) at the mean position of the platform

(b) for an amplitude of $\frac{g}{\omega^2}$

(c) for an amplitude of $\frac{g^2}{\omega^2}$

(d) at the highest position of the platform

34. Two simple harmonic motions of angular frequency 100 and 1000 rad s^{-1} have the same displacement amplitude. The ratio of their maximum accelerations is:

(a) 1 : 10 (b) $1:10^2$ (c) $1:10^3$ (d) $1:10^4$

35. Two particles execute SHM on same straight line with same mean position, same time period 6 second and same amplitude 5cm. Both the particles start SHM from their mean position (in same direction) with a time gap of 1 second. The maximum separation between the two particles during their motion is **Critical Thinking**

(a) 2 cm (b) 3 cm (c) 4 cm (d) 5 cm

36. A particle is executing simple harmonic motion with amplitude A. When the ratio of its kinetic energy to the potential energy is $\frac{1}{4}$, its displacement from its mean position is

(a) $\frac{2}{\sqrt{5}}A$ (b) $\frac{\sqrt{3}}{2}A$ (c) $\frac{3}{4}A$ (d) $\frac{1}{4}A$

37. A mass of 4 kg suspended from a spring of force constant 800 N m^{-1} executes simple harmonic oscillations. If the total energy of the oscillator is 4 J, the maximum acceleration (in m s^{-2}) of the mass is

(a) 5 (b) 15 (c) 45 (d) 20

38. A body executes simple harmonic motion. The potential energy (P.E.), the kinetic energy (K.E.) and total energy (T.E.) are measured as a function of displacement *x*. Which of the following statement is true?

(a) P.E. is maximum when $x = 0$.

(b) K.E. is maximum when $x = 0$.

(c) T.E. is zero when $x = 0$.

(d) K.E. is maximum when x is maximum.

39. If $<E>$ and $<U>$ denote the average kinetic and the average potential energies respectively of mass describing a simple harmonic motion, over one period, then the correct relation is

(a) $<E> = <U>$ (b) $<E> = 2<U>$

(c) $<E> = -2<U>$ (d) $<E> = -<U>$

40. What do you conclude from the graph about the frequency of KE, PE and SHM ?

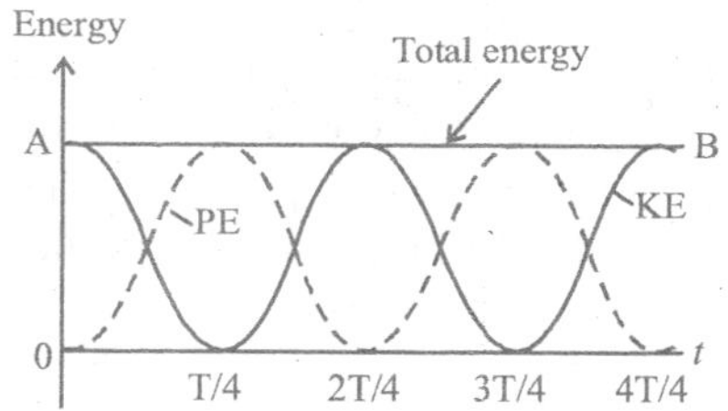

(a) Frequency of KE and PE is double the frequency of SHM.
(b) Frequency of KE and PE is four times the frequency SHM.
(c) Frequency of PE is double the frequency of K.E.
(d) Frequency of KE and PE is equal to the frequency of SHM.

41. A particle of mass m executes simple harmonic motion with amplitude a and frequency ν. The average kinetic energy during its motion from the position of equilibrium to the end is **Critical Thinking**

(a) $2\pi^2 ma^2\nu^2$ (b) $\pi^2 ma^2\nu^2$
(c) $\frac{1}{4}ma^2\nu^2$ (d) $4\pi^2 ma^2\nu^2$

42. A particle of mass m is executing oscillations about the origin on the X-axis with amplitude A. Its potential energy is given as $U(x) = \alpha x^4$ where α is a positive constant. The x coordinate of the particle where the potential energy is one third of the kinetic energy is

(a) $\pm A/2$ (b) $\pm A/\sqrt{2}$
(c) $\pm A/3$ (d) $\pm A/\sqrt{3}$

43. A particle of mass 10 gm is describing S.H. M. along a straight line with period of 2 sec and amplitude of 10 cm. Its kinetic energy when it is at 5 cm from its equilibrium position is

(a) $37.5\pi^2$ ergs (b) $3.75\pi^2$ ergs
(c) $375\pi^2$ ergs (d) $0.375\pi^2$ erg

44. A particle of mass 1 kg is placed in a potential field. Its potential energy is given by $U = 10x^2 + 5$. The frequency of oscillations of the particle is given by **Critical Thinking**

(a) $(\sqrt{10})$ (b) $(\sqrt{5})$ (c) $\left(\sqrt{\frac{10}{\pi}}\right)$ (d $\left(\frac{\sqrt{5}}{\pi}\right)$

45. A body is executing simple harmonic motion. At a displacement x from mean position, its potential energy is $E_1 = 2$J and at a displacement y from mean position, its potential energy is $E_2 = 8$J. The potential energy E at a displacement $(x + y)$ from mean position is

(a) 10J (b) 14J (c) 18J (d) 4J

46. The particle executing simple harmonic motion has a kinetic energy $K_0 \cos^2 \omega t$. The maximum values of the potential energy and the total energy are respectively

(a) $K_0/2$ and K_0 (b) K_0 and $2K_0$
(c) K_0 and K_0 (d) 0 and $2K_0$.

47. A particle of mass m oscillates with simple harmonic motion between points x_1 and x_2, the equilibrium position being O. Its potential energy is plotted. It will be as given below in the graph. **Tricky**

(a) X_1 O X_2 (b) X_1 O X_2
(c) X_1 O X_2 (d) X_1 O X_2

48. A linear harmonic oscillator of force constant 2×10^6 N/m and amplitude 0.01 m has a total mechanical energy of 160 J. Its

(a) potential energy is 160 J
(b) potential energy is 100 J
(c) potential energy is zero
(d) potential energy is 120 J

49. The angular velocity and the amplitude of a simple pendulum is ω and a respectively. At a displacement x from the mean position if its kinetic energy is T and potential energy is V, then the ratio of T to V is

(a) $\frac{(a^2 - x^2\omega^2)}{x^2\omega^2}$ (b) $\frac{x^2\omega^2}{(a^2 - x^2\omega^2)}$
(c) $\frac{(a^2 - x^2)}{x^2}$ (d) $\frac{x^2}{(a^2 - x^2)}$

50. A simple pendulum oscillates in air with time period T and amplitude A. As the time passes

(a) T and A both decrease
(b) T increases and A is constant
(c) T remains same and A decreases
(d) T decreases and A is constant

51. If the maximum velocity and acceleration of a particle executing SHM are equal in magnitude, the time period will be

(a) 1.57 sec (b) 3.14 sec
(c) 6.28 sec (d) 12.56 sec

52. A particle executes simple harmonic motion between x = – A and x = + A. The time taken for it to go from O to A/2 is T_1 and T_1 and to go from A/2 to A is T_2. Then **Tricky**

(a) $T_1 < T_2$ (b) $T_1 > T_2$
(c) $T_1 = T_2$ (d) $T_1 = 2T_2$

53. If a particle takes 0.5 sec to reach position of minimum velocity from previous such position, then

(a) T = 6 sec, ν = 1/6 Hz (b) T = 2 sec, ν = 1 Hz
(c) T = 3 sec, ν = 3 Hz (d) T = 1 sec, ν = 1 Hz

54. A simple harmonic motion has an amplitude A and time period T. The time required by it to travel from x = A to x = A/2 is

(a) T/6 (b) T/4 (c) T/3 (d) T/2

55. Frequency of oscillation is proportional to

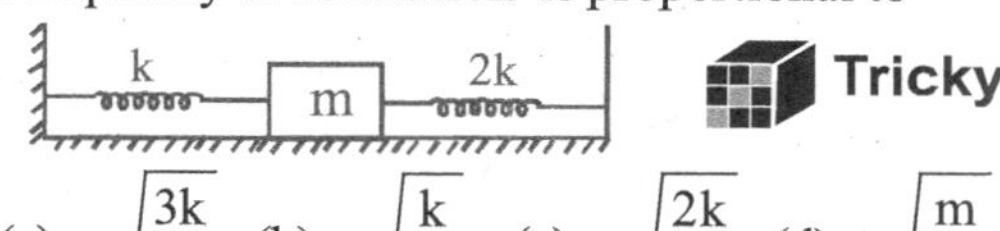

Tricky

(a) $\sqrt{\frac{3k}{m}}$ (b) $\sqrt{\frac{k}{m}}$ (c) $\sqrt{\frac{2k}{m}}$ (d) $\sqrt{\frac{m}{3k}}$

56. The bob of a simple pendulum of mass m and total energy E will have maximum linear momentum equal to

(a) $\sqrt{\frac{2E}{m}}$ (b) $\sqrt{2mE}$ (c) 2mE (d) mE^2

57. A particle executing simple harmonic motion covers a distance equal to half of its amplitude in one second. Then the time period of the simple harmonic motion is

(a) 4 s (b) 6 s (c) 8 s (d) 12 s

58. The maximum velocity of a particle, executing simple harmonic motion with an amplitude 7 mm, is 4.4 m/s. The period of oscillation is

(a) 0.01 s (b) 10 s (c) 0.1 s (d) 100 s

59. A body is executing S.H.M. When its displacement from the mean position are 4 cm and 5 cm, it has velocities 10 cm s^{-1} and 8 cm s^{-1} respectively. Its periodic time is

(a) $\pi/2$ s (b) π s (c) $3\pi/2$ s (d) 2π s

60. The displacement of an object attached to a spring and executing simple harmonic motion is given by $x = 2 \times 10^{-2} \cos \pi t$ metre. The time at which the maximum speed first occurs is

(a) 0.25 s (b) 0.5 s (c) 0.75 s (d) 0.125 s

61. A particle at the end of a spring executes S.H.M with a period t_1, while the corresponding period for another spring is t_2. If the period of oscillation with the two springs in series is T then

Tricky

(a) $T^{-1} = t_1^{-1} + t_2^{-1}$ (b) $T^2 = t_1^2 + t_2^2$

(c) $T = t_1 + t_2$ (d) $T^{-2} = t_1^{-2} + t_2^{-2}$

62. A particle is executing SHM along a straight line. Its velocities at distances x_1 and x_2 from the mean position are V_1 and V_2, respectively. Its time period is

(a) $2\pi\sqrt{\frac{x_2^2 - x_1^2}{V_1^2 - V_2^2}}$ (b) $2\pi\sqrt{\frac{V_1^2 + V_2^2}{x_1^2 + x_2^2}}$

(c) $2\pi\sqrt{\frac{V_1^2 - V_2^2}{x_1^2 - x_2^2}}$ (d) $2\pi\sqrt{\frac{x_1^2 - x_2^2}{V_1^2 - V_2^2}}$

63. A simple harmonic wave having an amplitude a and time period T is represented by the equation $y = 5\sin\pi(t+4)m$. Then the value of amplitude (a) in (m) and time period (T) in second are

(a) a = 10, T = 2 (b) a = 5, T = 1

(c) a = 10, T = 1 (d) a = 5, T = 2

64. What will be the force constant of the spring system shown in figure?

Tricky

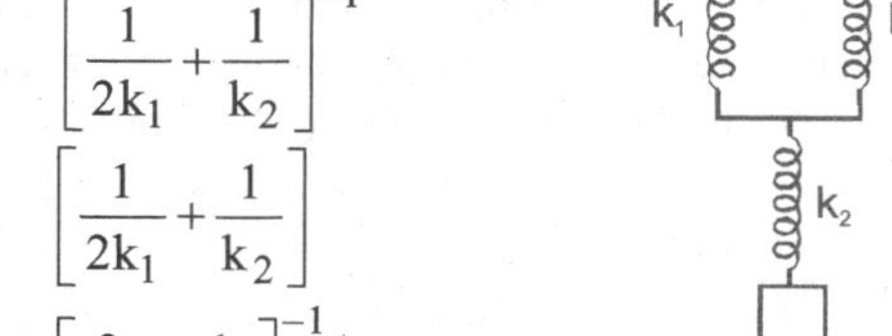

(a) $\frac{k_1}{2} + k_2$

(b) $\left[\frac{1}{2k_1} + \frac{1}{k_2}\right]^{-1}$

(c) $\left[\frac{1}{2k_1} + \frac{1}{k_2}\right]$

(d) $\left[\frac{2}{k_1} + \frac{1}{k_2}\right]^{-1}$

65. A particle of mass m oscillates with a potential energy $U = U_0 + \alpha x^2$, where U_0 and α are constants and x is the displacement of particle from equilibrium position. The time period of oscillation is

(a) $2\pi\sqrt{\frac{m}{\alpha}}$ (b) $2\pi\sqrt{\frac{m}{2\alpha}}$ (c) $\pi\sqrt{\frac{2m}{\alpha}}$ (d) $2\pi\sqrt{\frac{m}{\alpha^2}}$

66. The average value of potential energy for one complete oscillation cycle of a body of mass *m* executing S.H.M. having angular frequency ω and maximum amplitude *a* is given by:

(a) $\frac{1}{2}m\omega^2 a^2$ (b) $\frac{1}{4}m\omega^2 a^2$

(c) $\frac{1}{8}m\omega^2 a^2$ (d) $\frac{1}{16}m\omega^2 a^2$

67. A wall clock uses a vertical spring-mass system to measure the time. Each time the mass reaches an extreme position, the clock advances by a second. The clock gives correct time at the equator. If the clock is taken to the poles it will

Tricky

(a) run slow (b) run fast

(c) stop working (d) give correct time

68. The height of liquid column in a U tube is 0.3 m. If the liquid in one of the limbs is depressed and then released, then the time period of the liquid column will be

(a) 1.1 sec (b) 19 sec (c) 0.11 sec (d) 2 sec

69. A block rests on a horizontal table which is executing SHM in the horizontal plane with an amplitude 'a'. If the coefficient of friction is 'μ', then the block just starts to slip when the frequency of oscillation is

(a) $\frac{1}{2\pi}\sqrt{\frac{\mu g}{a}}$ (b) $\sqrt{\frac{\mu g}{a}}$ (c) $2\pi\sqrt{\frac{a}{\mu g}}$ (d) $\sqrt{\frac{a}{\mu g}}$

70. On earth, a body suspended on a spring of negligible mass causes extension L and undergoes oscillations along length of the spring with frequency f. On the Moon, the same quantities are L/n and f' respectively. The ratio f'/f is

(a) n (b) $\frac{1}{n}$ (c) $n^{-1/2}$ (d) 1

71. A second's pendulum is placed in a space laboratory orbiting around the earth at a height 3 R from the earth's surface where R is earth's radius. The time period of the pendulum will be

(a) zero (b) $2\sqrt{3}$ (c) 4 sec (d) infinite

72. A bent tube of uniform cross-section area A has a non-viscous liquid of density ρ. The mass of liquid in the tube is m. The time period of oscillation of the liquid is **Toughnut**

(a) $2\pi\sqrt{\frac{m}{\rho g A}}$

(b) $2\pi\sqrt{\frac{m}{2\rho g A}}$

(c) $2\pi\sqrt{\frac{2m}{\rho g A}}$

(d) None of these

73. A pendulum made of a uniform wire of cross sectional area A has time period T. When an additional mass M is added to its bob, the time period changes to T_M. If the Young's modulus of the material of the wire is Y then $\frac{1}{Y}$ is equal to:

(*g = gravitational acceleration*) **Critical Thinking**

(a) $\left[1-\left(\frac{T_M}{T}\right)^2\right]\frac{A}{Mg}$ (b) $\left[1-\left(\frac{T}{T_M}\right)^2\right]\frac{A}{Mg}$

(c) $\left[\left(\frac{T_M}{T}\right)^2-1\right]\frac{A}{Mg}$ (d) $\left[\left(\frac{T_M}{T}\right)^2-1\right]\frac{Mg}{A}$

74. Masses M_A and M_B hanging from the ends of strings of lengths L_A and L_B are executing simple harmonic motions. If their frequencies are $f_A = 2f_B$, then

(a) $L_A = 2L_B$ and $M_A = M_B/2$

(b) $L_A = 4L_B$ regardless of masses

(c) $L_A = L_B/4$ regardless of masses

(d) $L_A = 2L_B$ and $M_A = 2M_B$

75. A simple pendulum has time period T_1. The point of suspension is now moved upward according to the relation $y = Kt^2$, ($K = 1$ m/s²) where y is the vertical displacement. The time period now becomes T_2. The ratio of $\frac{T_1^2}{T_2^2}$ is ($g = 10$ m/s²)

(a) 5/6 (b) 6/5 (c) 1 (d) 4/5

76. A cylindrical block of wood (density = 650 kg m^{-3}), of base area 30cm² and height 54 cm, floats in a liquid of density 900 kg m^{-3}. The block is depressed slightly and then released. The time period of the resulting oscillations of the block would be equal to that of a simple pendulum of length (nearly) **Critical Thinking**

(a) 52 cm (b) 65 cm (c) 39 cm (d) 26 cm

77. A body executes simple harmonic motion under the action of a force F_1 with a time period $\frac{4}{5}$ s. If the force is changed to F_2, it executes S.H.M. with time period $\frac{3}{5}$ s. If both the forces F_1 and F_2 act simultaneously in the same direction on the body, its time period in second is

(a) $\frac{12}{25}$ (b) $\frac{7}{5}$ (c) $\frac{24}{25}$ (d) $\frac{5}{7}$

78. If the differential equation for a simple harmonic motion is $\frac{d^2y^2}{dt^2} + 2y = 0$, the time-period of the motion is

(a) $\pi\sqrt{2}s$ (b) $\frac{\sqrt{2}s}{\pi}$ (c) $\frac{\pi}{\sqrt{2}}s$ (d) $2\pi s$

79. Four massless springs whose force constants are $2k$, $2k$, k and $2k$ respectively are attached to a mass M kept on a frictionless plane (as shown in figure). If the mass M is displaced in the horizontal direction, then the frequency of the system is **Tricky**

(a) $\frac{1}{2\pi}\sqrt{\frac{k}{4M}}$

(b) $\frac{1}{2\pi}\sqrt{\frac{4k}{M}}$

(c) $\frac{1}{2\pi}\sqrt{\frac{k}{7M}}$

(d) $\frac{1}{2\pi}\sqrt{\frac{7k}{M}}$

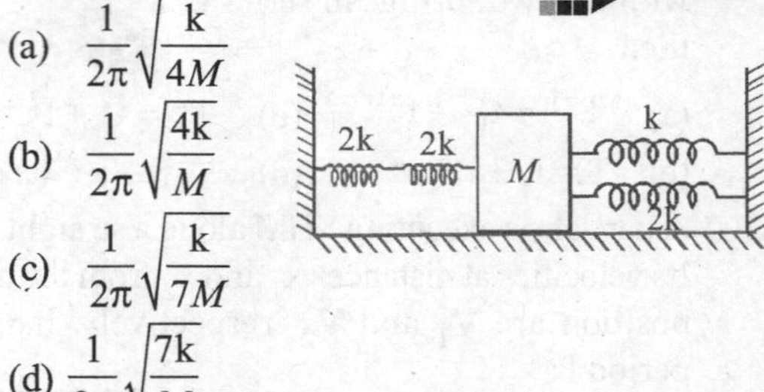

80. The bob of a simple pendulum executes simple harmonic motion in water with a period t, while the period of oscillation of the bob is t_0 in air. Neglecting frictional force of water and given that the density of the bob is $(4/3)\times 1000$ kg/m³. What relationship between t and t_0 is true?

(a) $t = t_0$ (b) $t = t_0/2$ (c) $t = 2t_0$ (d) $t = 4t_0$

81. A U-tube is of non uniform cross-section. The area of cross-sections of two sides of tube are A and 2A (see fig.). It contains non-viscous liquid of mass m. The liquid is displaced slightly and free to oscillate. Its time period of oscillations is

Toughnut

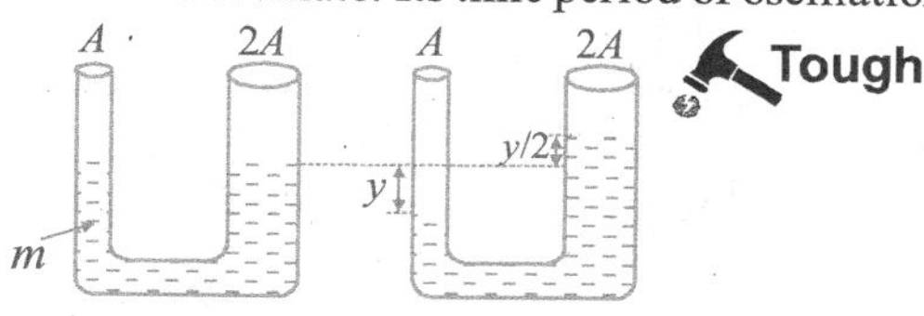

(a) $T = 2\pi\sqrt{\frac{m}{3\rho g A}}$ (b) $T = 2\pi\sqrt{\frac{m}{2\rho g A}}$

(c) $T = 2\pi\sqrt{\frac{m}{\rho g A}}$ (d) None of these

82. A small ball of density $4\rho_0$ is released from rest just below the surface of a liquid. The density of liquid increases with depth as $\rho = \rho_0(1 + ay)$ where $a = 2m^{-1}$ is a constant. Find the time period of its oscillation. (Neglect the viscosity effects).

(a) $\frac{2\pi}{\sqrt{5}}$ sec (b) $\frac{\pi}{\sqrt{5}}$ sec

(c) $\frac{\pi}{2\sqrt{5}}$ sec (d) $\frac{3\pi}{2\sqrt{5}}$ sec

83. A rod of mass M and length L is hinged at its centre of mass so that it can rotate in a vertical plane. Two springs each of stiffness k are connected at its ends, as shown in the figure. The time period of SHM is

Toughnut

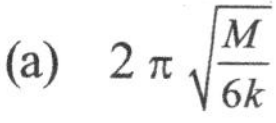

(a) $2\pi\sqrt{\frac{M}{6k}}$

(b) $2\pi\sqrt{\frac{M}{3k}}$

(c) $2\pi\sqrt{\frac{ML}{k}}$

(d) $\pi\sqrt{\frac{M}{6k}}$

84. Disregarding gravity, find the period of oscillation of the particle connected with four springs as shown in the figure.
(Given: $\theta = 45°, \beta = 30°$)

Critical Thinking

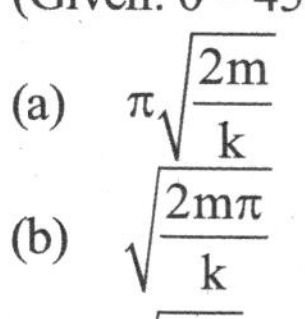

(a) $\pi\sqrt{\frac{2m}{k}}$

(b) $\sqrt{\frac{2m\pi}{k}}$

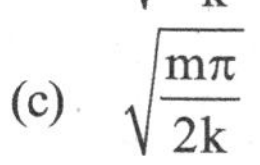

(c) $\sqrt{\frac{m\pi}{2k}}$

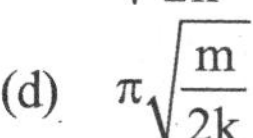

(d) $\pi\sqrt{\frac{m}{2k}}$

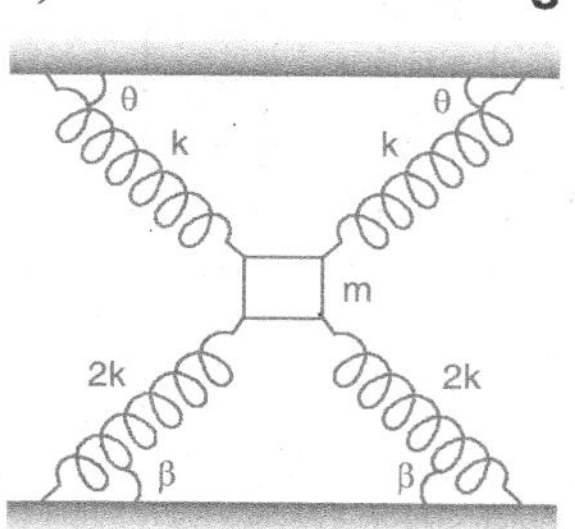

85. Two springs, each of spring constant $k = 100$ N/m, are attached to a block of mass 2 kg as shown in the figure. The block can slide smoothly along a horizontal platform clamped to the opposite walls of the trolley of mass 5kg. If the block is displaced by x cm. and released, the period of oscillation in seconds is

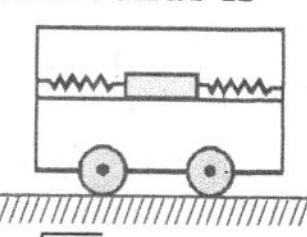

(a) $T = 2\pi\sqrt{\frac{1}{20}}$ (b) $T = 2\pi\sqrt{\frac{7}{1000}}$

(c) $T = 2\pi\sqrt{\frac{1}{140}}$ (d) $T = 2\pi\frac{49}{100}$

86. If a body oscillates at the angular frequency ω_d of the driving force, then the oscillations are called

(a) free oscillations
(b) coupled oscillations
(c) forced oscillations
(d) maintained oscillations

87. In case of a forced vibration, the resonance wave becomes very sharp when the

(a) quality factor is small
(b) damping force is small
(c) restoring force is small
(d) applied periodic force is small

88. When an oscillator completes 100 oscillations its amplitude reduces to $\frac{1}{3}$ of its initial value. What will be its amplitude, when it completes 200 oscillations?

(a) $\frac{1}{8}$ (b) $\frac{2}{3}$ (c) $\frac{1}{6}$ (d) $\frac{1}{9}$

89. A forced oscillator is acted upon by a force $F = F_0 \sin \omega t$. The amplitude of oscillation is given by $\frac{55}{\sqrt{2\omega^2 - 36\omega + 9}}$. The resonant angular frequency is

(a) 2 unit (b) 9 unit (c) 18 unit (d) 36 unit

90. A particle, with restoring force proportional to displacement and resisting force proportional to velocity is subjected to a force F sin ωt. If the amplitude of the particle is maximum for $\omega = \omega_1$ and the energy of the particle is maximum for $\omega = \omega_2$, then

(a) $\omega_1 = \omega_0$ and $\omega_2 \neq \omega_0$
(b) $\omega_1 = \omega_0$ and $\omega_2 = \omega_0$
(c) $\omega_1 \neq \omega_0$ and $\omega_2 = \omega_0$
(d) $\omega_1 \neq \omega_0$ and $\omega_2 \neq \omega_0$

91. A particle of mass m is attached to a spring (of spring constant k) and has a natural angular frequency ω_0. An external force $F(t)$ proportional to $\cos \omega t (\omega \neq \omega_0)$ is applied to the oscillator. The displacement of the oscillator will be proportional to **Critical Thinking**

(a) $\dfrac{1}{m(\omega_0^2 + \omega^2)}$ (b) $\dfrac{1}{m(\omega_0^2 - \omega^2)}$

(c) $\dfrac{m}{\omega_0^2 - \omega^2}$ (d) $\dfrac{m}{(\omega_0^2 + \omega^2)}$

92. What is the amplitude of simple harmonic motion at resonance in the ideal case of zero damping?

(a) Zero (b) -1 (c) 1 (d) Infinite

93. If a simple pendulum has significant amplitude (upto a factor of $1/e$ of original) only in the period between $t = 0s$ to $t = \tau s$, then τ may be called the average life of the pendulum. when the spherical bob of the pendulum suffers a retardation (due to viscous drag) proportional to its velocity with b as the constant of proportionality, the average life time of the pendulum is (assuming damping the small) in seconds **Critical Thinking**

(a) $\dfrac{0.693}{b}$ (b) b (c) $\dfrac{1}{b}$ (d) $\dfrac{2}{b}$

94. A pendulum with time period of 1s is losing energy. At certain time its energy is 45 J. If after completing 15 oscillations, its energy has become 15 J, its damping constant (in s^{-1}) is :

(a) $\dfrac{1}{2}$ (b) $\dfrac{1}{30}\ln 3$ (c) 2 (d) $\dfrac{1}{15}\ln 3$

95. In damped oscillations, the amplitude of oscillations is reduced to one-third of its inital value a_0 at the end of 100 oscillations. When the oscillator completes 200 oscillations, its amplitude must be

(a) $a_0/2$ (b) $a_0/4$ (c) $a_0/6$ (d) $a_0/9$

96. A block connected to a spring oscillates vertically. A damping force F_d, acts on the block by the surrounding medium. Given as $F_d = -bV$ b is a positive constant which depends on :

(a) viscosity of the medium
(b) size of the block
(c) shape of the block
(d) All of these

97. The amplitude of a damped oscillator becomes $\left(\dfrac{1}{3}\right)^{rd}$ in 2 seconds. If its amplitude after 6 seconds is $\dfrac{1}{n}$ times the original amplitude, the value of n is **Critical Thinking**

(a) 3^2 (b) 3^3 (c) $\sqrt[3]{3}$ (d) 2^3

98. In case of sustained forced oscillations the amplitude of oscillations

(a) decreases linearly
(b) decreases sinusoidally
(c) decreases exponentially
(d) always remains constant

99. The amplitude of velocity of a particle is given by, $V_m = V_0/(a\omega^2 - b\omega + c)$ where V_0, a, b and c are positive :
The condition for a single resonant frequency is

(a) $b^2 < 4ac$ (b) $b^2 = 4ac$
(c) $b^2 = 5ac$ (d) $b^2 = 7ac$

100. The time period of simple pendulum of length 1m suspended in a car that is moving with constant speed 36 km/hr around a circle of radius 10 m is:

(a) $\pi\sqrt{2}$ sec (b) $\dfrac{\pi}{10}\sqrt{2}$ sec

(c) $\dfrac{\pi}{10\sqrt{2}}$ sec (d) $\dfrac{5\pi}{\sqrt{2}}$ sec

Answer KEYs

1	(c)	11	(d)	21	(d)	31	(a)	41	(b)	51	(c)	61	(b)	71	(d)	81	(a)	91	(b)
2	(c)	12	(c)	22	(d)	32	(c)	42	(b)	52	(a)	62	(a)	72	(a)	82	(a)	92	(d)
3	(b)	13	(d)	23	(c)	33	(b)	43	(c)	53	(d)	63	(d)	73	(c)	83	(a)	93	(d)
4	(a)	14	(c)	24	(a)	34	(b)	44	(d)	54	(a)	64	(b)	74	(c)	84	(a)	94	(d)
5	(d)	15	(a)	25	(a)	35	(d)	45	(c)	55	(a)	65	(b)	75	(b)	85	(c)	95	(d)
6	(a)	16	(b)	26	(d)	36	(a)	46	(c)	56	(b)	66	(b)	76	(c)	86	(c)	96	(d)
7	(a)	17	(a)	27	(b)	37	(d)	47	(b)	57	(d)	67	(d)	77	(a)	87	(b)	97	(b)
8	(a)	18	(b)	28	(b)	38	(b)	48	(b)	58	(a)	68	(a)	78	(a)	88	(d)	98	(a)
9	(b)	19	(c)	29	(b)	39	(a)	49	(c)	59	(b)	69	(a)	79	(b)	89	(b)	99	(b)
10	(c)	20	(a)	30	(d)	40	(a)	50	(c)	60	(b)	70	(d)	80	(c)	90	(c)	100	(b)

Hints & Solutions

1. (c) When particle starts from mean position,

$x = A\cos\omega t$

$\therefore\ (A-a) = A\cos(\omega\times 1)$

or $(A-a) = A\cos\omega$ ...(i)

Also $A-(a+b) = A\cos(\omega\times 2)$

or $A-(a+b) = A\cos 2\omega$ ...(ii)

After simplifying above equations, we get

$A = \left[\frac{2a^2}{3a-b}\right]$

2. (c) As we know,

$F = ma \Rightarrow a \propto F$ or, $a \propto \sin t$

$\Rightarrow \frac{dv}{dt} \propto \sin t \Rightarrow \int_0^0 dV \propto \int_0^t \sin t\, dt$

$V \propto -\cos t + 1$

$\int_0^x dx = \int_0^t (-\cos t + 1)\, dt$

$x = \sin t - \frac{1}{2}\sin 2t$

3. (b) From the figure, $\frac{T}{2} = 0.3\pi$,

$\therefore\ T = 0.6\pi$ and $\phi_0 = -\frac{\pi}{2}$.

Thus $y = A\sin(\omega t + \phi_0) = 2\sin\left(\frac{2\pi}{0.6\pi}t + -\frac{\pi}{2}\right)$

$= 2\sin\left(\frac{10t}{3} - \frac{\pi}{2}\right)$.

4. (a) If displacement of particle is y, then

$KE = \frac{1}{2}m\omega^2(a^2 - y^2)$ & $P.E. = \frac{1}{2}m\omega^2 y^2$

If KE = PE then $\frac{1}{2}m\omega^2 y^2$

$= \frac{1}{2}m\omega^2 a^2 - \frac{1}{2}m\omega^2 y^2$

$\Rightarrow 2y^2 = a^2 \quad \therefore y = \frac{a}{\sqrt{2}}$

5. (d) Total mechanical energy is constant throughout the motion and equals

$\frac{1}{2}m\omega^2 A^2$.

6. (a)

7. (a) $K.E. = \frac{1}{2}ma^2\omega^2\cos^2\omega t$

and $T.E. = \frac{1}{2}ma^2\omega^2$

Given K.E. = 0.75 T.E.

$\Rightarrow 0.75 = \cos^2\omega t \Rightarrow \omega t = \frac{\pi}{6}$

$\Rightarrow t = \frac{\pi}{6\times\omega} \Rightarrow t = \frac{\pi\times 2}{6\times 2\pi} \Rightarrow t = \frac{1}{6}\ s$

8. (a) In $x = A\cos\omega t$, the particle starts oscillating from extreme position. So at $t = 0$, its potential energy is maximum.

9. (b)

10. (c) Given, $x = \frac{A}{2}$

$\therefore$ from $x = A\sin\omega t \Rightarrow \omega t = 30°$

And, $\frac{KE}{PE} = \cot^2\omega t = \left(\sqrt{3}\right)^2 = 3$

11. (d) Let $y = A\sin\omega t$

$v_{inst} = \frac{dy}{dt} = A\omega\cos\omega t = A\omega\sin(\omega t + \pi/2)$

Acceleration $= -A\omega^2\sin\omega t = A\omega^2\sin(\pi + \omega t)$

$\therefore \phi = \frac{\pi}{2} = 0.5\pi$

12. (c) Max. force = mass × max. acceleration

$= m\,4\pi^2 v^2 a = 1\times 4\times\pi^2\times(60)^2\times 0.02$

$= 288\,\pi^2$

13. (d) As seen from figure after one time period the bob return to its equilibrium position, so diplacement of the particle is zero, but distance covered by the particle in one time period is 4A (where A is amplitude of bob, when it does S.H.M.)

14. (c) When a particle is dropped from a height h above the centre of tunnel.

(i) It will oscillate, through the earth to a height h on both sides.

(ii) The motion of particle is periodic.

(iii) The motion of particle will not be S.H.M.

15. (a) At point 2, the acceleration of the particle is maximum, which is at the extreme position. At extreme position, the velocity of the particle will be zero.

16. (b) $v = \omega[A^2 - x^2]^{1/2} \Rightarrow x = \left[A^2 - \frac{v^2}{\omega^2}\right]^{1/2}$

Given that $v = \frac{v_{max}}{2} = \frac{A\omega}{2}$.

so, $\left[A^2 - \frac{A^2\omega^2}{4\omega^2}\right]^{1/2} = \frac{\sqrt{3}}{2}.A$

17. (a) Velocity in SHM is given by

$v = \omega\sqrt{a^2 - y^2}$

At $y = 4$ cm = 0.04 m, $v = 3$ m/s

$\therefore\ 3 = \omega\sqrt{a^2 - (0.04)^2}$ ....(1)

At $y = 3$ cm = 0.03 m, $v = 4$ m/s

$\therefore\ 4 = \omega\sqrt{a^2 - (0.03)^2}$ ....(2)

Dividing (2) by (1), we get a = 0.05 = 5 cm

18. (b) Given that, $\frac{A_{max}}{v_{max}} = 10$ i.e., $\omega = 10\ s^{-1}$

Displacement is given by

$x = a \sin(\omega t + \pi/4)$

at $t = 0$, $x = 5$; $5 = a \sin 45° \Rightarrow a = 5\sqrt{2}$

$A_{max} = a\omega^2 = 500\sqrt{2}\ m/s^2$

19. (c) Given, $x = 10 \sin\left(2t - \frac{\pi}{6}\right)$

A = 10 and ω = 2 Hz

$\therefore\ \upsilon = \omega\sqrt{A^2 - d^2} = 2\sqrt{(10)^2 - (6)^2}$

$= 2\sqrt{100 - 36} = 2 \times 8 = 16 m\ s^{-1}$

20. (a)

21. (d) $v^2 = \omega^2(A^2 - x^2)$ ...(i)

and $a^2 = (\omega^2 x^2) = \omega^4 x^2$ ... (ii)

From above equations, we have

$v^2 = -\frac{a^2}{\omega^2} + \omega^2 A^2 \Rightarrow y = mx + c$

It represents straight line with negative slope.

22. (d) Initially $t = 0$

$x = a \cos\pi t = -a \cos 0° = a$

Finally at $t = 3$

$x = a \cos 3\pi = -3a$

Total displacement = 2a

23. (c) Given $t = 1$s

$\therefore\ x = 5 \cos\left(2\pi + \frac{\pi}{4}\right) = 5 \cos\frac{\pi}{4} = \frac{5}{\sqrt{2}}$ m

i.e., displacement at $t = 1$s is $\frac{5}{\sqrt{2}}$ m

24. (a)

25. (a) For $x = (-A)$, we have

$-A = A \sin(\omega \times 0 + \phi_0)$ or $\phi_0 = -\frac{\pi}{2}$.

So for $x < (-A)$, $\phi_0 < (-\pi/2)$.

26. (d) At mean position velocity is maximum

i.e., $v_{max} = \omega a \Rightarrow \omega = \frac{v_{max}}{a} = \frac{16}{4} = 4$

$\therefore v = \omega\sqrt{a^2 - y^2} \Rightarrow 8\sqrt{3} = 4\sqrt{4^2 - y^2}$

$\Rightarrow 192 = 16(16 - y^2) \Rightarrow 12 = 16 - y^2$

$\Rightarrow y = 2$cm.

27. (b) $v_1 = \frac{dy_1}{dt} = 0.1 \times 100\pi \cos\left(100\pi t + \frac{\pi}{3}\right)$

$v_2 = \frac{dy_2}{dt} = -0.1\pi \sin \pi t$

$= 0.1\pi \cos\left(\pi t + \frac{\pi}{2}\right)$

$\therefore$ Phase diff. $= \phi_1 - \phi_2 = \frac{\pi}{3} - \frac{\pi}{2} = \frac{2\pi - 3\pi}{6} = \frac{\pi}{6}$

28. (b) Equation of S.H.M. is given by

$x = A \sin(\omega t + \delta)$

$(\omega t + \delta)$ is called phase.

When $x = \frac{A}{2}$, then $\sin(\omega t + \delta) = \frac{1}{2}$

$\Rightarrow \omega t + \delta = \frac{\pi}{6}$

or $\phi_1 = \frac{\pi}{6}$

1 A
2 –A x = A/2
x=0

For second particle,

$\phi_2 = \pi - \frac{\pi}{6} = \frac{5\pi}{6}$

$\therefore \phi = \phi_2 - \phi_1 = \frac{4\pi}{6} = \frac{2\pi}{3}$

29. (b)

30. (d) From the graph it is clear that the amplitude is 1 cm and the time period is 8 second. Therefore the equation for the S.H.M. is

$x = a \sin\left(\frac{2\pi}{T}\right) \times t = 1 \sin\left(\frac{2\pi}{8}\right)t = \sin\frac{\pi}{4}t$

The velocity (v) of the particle at any instant of time 't' is

$v = \frac{dx}{dt} = \frac{d}{dt}\left[\sin\left(\frac{\pi}{4}\right)t\right] = \frac{\pi}{4}\cos\left(\frac{\pi}{4}\right)t$

The acceleration of the particle is

$\frac{d^2x}{dt^2} = -\left(\frac{\pi}{4}\right)^2 \sin\left(\frac{\pi}{4}\right)t$

At $t = \frac{4}{3}s$ we get

$\frac{d^2x}{dt^2} = -\left(\frac{\pi}{4}\right)^2 \sin\frac{\pi}{4} \times \frac{4}{3} = \frac{-\sqrt{3}\pi^2}{32}$ cm/s^2

31. (a)

32. (c) $v_1 = \omega\sqrt{a^2 - x_1^2}$, $v_2 = \omega\sqrt{a^2 - x_2^2}$

We get, $a = \sqrt{\dfrac{v_1^2 x_2^2 - v_2^2 x_1^2}{v_1^2 - v_2^2}}$

33. (b) For block A to move in S.H.M.

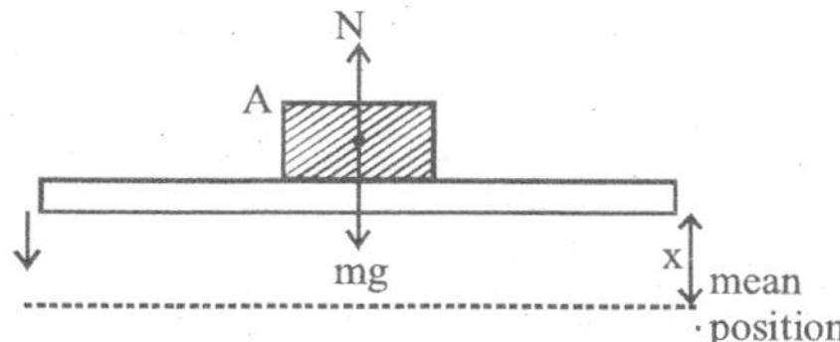

$mg - N = m\omega^2 x$

where x is the distance from mean position

For block to leave contact N = 0

$\Rightarrow mg = m\omega^2 x \Rightarrow x = \dfrac{g}{\omega^2}$

34. (b)

35. (d) Figure shows the mapping of the two SHMs with circular motions having phase difference

$\phi = \omega t = \dfrac{2\pi}{6} \times 1 = \dfrac{\pi}{3}$ rad

The maximum separation between the two particles is

$S_{max} = 2A\sin\dfrac{\pi}{6}$

or $S_{max} = 2 \times 5 \times \dfrac{1}{2} = 5$cm.

36. (a) $\therefore \dfrac{\frac{1}{2}m\omega^2(A^2 - x^2)}{\frac{1}{2}m\omega^2 x^2} = \dfrac{1}{4} \Rightarrow \dfrac{A^2 - x^2}{x^2} = \dfrac{1}{4}$

$4A^2 - 4x^2 = x^2 \Rightarrow x^2 = \dfrac{4}{5}A^2 \Rightarrow x = \dfrac{2}{\sqrt{5}}A.$

37. (d)

38. (b) $K = \dfrac{1}{2}m\omega^2(A^2 - x^2) \therefore K_{max} = \dfrac{1}{2}m\omega^2 A^2$, at $x = 0$.

39. (a) $E_{av} = U_{av} = \dfrac{1}{4}m\omega^2 A^2$

40. (a) KE and PE completes two vibration in a time during which SHM completes one vibration. Thus frequency of PE or KE is double than that of SHM.

41. (b) The kinetic energy of a particle executing S.H.M. is given by

$K = \dfrac{1}{2}ma^2\omega^2\sin^2\omega t$

$= \left\langle \dfrac{1}{2}m\omega^2 a^2 \sin^2\omega t \right\rangle = \dfrac{1}{2}m\omega^2 a^2 \langle \sin^2\omega t \rangle$

$= \dfrac{1}{2}m\omega^2 a^2 \left(\dfrac{1}{2}\right)$

$= \dfrac{1}{4}m\omega^2 a^2 \quad \left(\because \langle \sin^2\theta \rangle = \dfrac{1}{2}\right)$

$= \dfrac{1}{4}ma^2(2\pi\nu)^2 \quad (\because \omega = 2\pi\nu)$

or, $\langle K \rangle = \pi^2 m a^2 \nu^2$

42. (b)

43. (c) Kinetic energy $K = \dfrac{1}{2}m\omega^2(a^2 - y^2)$

$= \dfrac{1}{2} \times 10 \times \left(\dfrac{2\pi}{2}\right)^2 [10^2 - 5^2] = 375\,\pi^2$ *ergs*

44. (d) $F = -\dfrac{dU}{dx} = -20x$

$a = \dfrac{F}{M} = \dfrac{-20x}{1} = -20x$

$-20x = -\omega^2(x)$

$\therefore \omega = \sqrt{20} = 2\sqrt{5}$

$f = \dfrac{\omega}{2\pi} = \dfrac{2\sqrt{5}}{2\pi} = \left(\dfrac{\sqrt{5}}{\pi}\right)$

45. (c) $E_1 = \dfrac{1}{2}kx^2$, $E_2 = \dfrac{1}{2}ky^2$,

$E = \dfrac{1}{2}k(x+y)^2 = \dfrac{1}{2}kx^2 + \dfrac{1}{2}ky^2 + kxy$

$= E_1 + E_2 + 2\sqrt{E_1E_2} = 2 + 8 + 2\sqrt{16} = 18$J

46. (c)

47. (b) Potential energy $\propto$ (Amplitude)^2

$\propto$ (Displacement)^2

P.E. is maximum at maximum distance.

P.E. is zero at equilibrium point.

P.E. curve is parabolic.

48. (b) Force constant $k = 2 \times 10^6$ N/m

Amplitude $(x) = 0.01$ m

Potential Energy $= \dfrac{1}{2}kx^2$

$= \dfrac{1}{2} \times (2 \times 10^6) \times (0.01)^2 = 100$ J

49. (c) P.E., $V = \dfrac{1}{2}m\omega^2 x^2$ and

K.E., $T = \dfrac{1}{2}m\omega^2(a^2 - x^2) \therefore \dfrac{T}{V} = \dfrac{a^2 - x^2}{x^2}$

50. (c) When a simple pendulum oscillates in air continously then its time period remains same whereas amplitude decreases.

51. (c) **52.** (a)

53. (d) $\frac{T}{2} = 0.5$ sec $\Rightarrow T = 1$ sec $\quad \nu = \frac{1}{T} = 1$ Hz

54. (a) At extreme position

$y = A\sin(\omega t + \frac{\pi}{2}) = A\cos\omega t$

Now $y = A/2$ then $A/2 = A\cos\omega t$

$\Rightarrow \omega t = \cos^{-1}(1/2) \quad \frac{2\pi}{T}t = \frac{\pi}{3} \Rightarrow t = T/6$

55. (a) $k_{eq} = k + 2k = 3k$

56. (b) $p_{max} = \sqrt{2mE_{max}}$

57. (d)

58. (a) Maximum velocity, $v_{max} = a\omega$

$v_{max} = a \times \frac{2\pi}{T}$

$\Rightarrow T = \frac{2\pi a}{v_{max}} = \frac{2 \times 3.14 \times 7 \times 10^{-3}}{4.4} \approx 0.01$ s

59. (b) Using $v^2 = \omega^2(a^2 - y^2)$ we have

$10^2 = \omega^2(a^2 - 4^2)$

and $8^2 = \omega^2(a^2 - 5^2)$;

so $10^2 - 8^2 = \omega^2(5^2 - 4^2) = (3\omega^2)$ or $6 = 3\omega$

or $\omega = 2$ or $T = 2\pi/\omega = 2\pi/2 = \pi$ s.

60. (b)

61. (b) $t_1 = 2\pi\sqrt{\frac{m}{k_1}}, t_2 = 2\pi\sqrt{\frac{m}{k_2}}$

when springs are in series then,

$k_{eff} = \frac{k_1 k_2}{k_1 + k_2} \quad \therefore T = 2\pi\sqrt{\frac{m(k_1 + k_2)}{k_1 k_2}}$

$\therefore T = 2\pi\sqrt{\frac{m}{k_2} + \frac{m}{k_1}} = 2\pi\sqrt{\frac{t_2^2}{(2\pi)^2} + \frac{t_1^2}{(2\pi)^2}}$

$\Rightarrow T^2 = t_1^2 + t_2^2$

62. (a) As we know, for particle undergoing SHM,

$V = \omega\sqrt{A^2 - X^2}$; $\quad V_1^2 = \omega^2(A^2 - x_1^2)$

$V_2^2 = \omega^2(A^2 - x_2^2)$

Substructing we get,

$\frac{V_1^2}{\omega^2} + x_1^2 = \frac{V_2^2}{\omega^2} + x_2^2$

$w = \sqrt{\frac{V_1^2 - V_2^2}{x_2^2 - x_1^2}} \Rightarrow T = 2\pi\sqrt{\frac{x_2^2 - x_1^2}{V_1^2 - V_2^2}}$

63. (d) $y = 5\sin(\pi t + 4\pi)$, comparing it with standard equation

$y = a\sin(\omega t + \phi) = a\sin\left(\frac{2\pi t}{T} + \phi\right)$

$a = 5m$ and $\frac{2\pi t}{T} = \pi t \Rightarrow T = 2\,\text{sec}$.

64. (b) Two springs of force constants k_1 and k_1 are in parallel. Hence

$k' = k_1 + k_1 = 2k_1$

The third spring k_2 is in series with spring of force constant k'.

$\therefore \frac{1}{k} = \left[\frac{1}{2k_1} + \frac{1}{k_2}\right]$ or $k = \left[\frac{1}{2k_1} + \frac{1}{k_2}\right]^{-1}$

65. (b) $U = U_0 + \alpha x^2 \Rightarrow F = -\frac{dU}{dx} = -2\alpha x$

$a = \frac{F}{m} = -\frac{2\alpha}{m}x$

$\Rightarrow \omega^2 = \frac{2\alpha}{m} \Rightarrow T = \frac{2\pi}{\omega} = 2\pi\sqrt{\frac{m}{2\alpha}}$

66. (b) $U_{average} = \frac{1}{T}\int_0^T U dt$

$= \frac{1}{T}\int_0^T \frac{1}{2}\omega^2 a^2 \sin^2(\omega t + \phi)\,dt = \frac{1}{4}m\omega^2 a^2$

67. (d) If it gives correct time at equator, it will give correct time at poles also because the time period of spring-mass system is independent of g.

68. (a)

69. (a) For the block is about to slip, $\mu g = \omega^2 a$

$\omega = \sqrt{\frac{\mu g}{a}} \Rightarrow \nu = \frac{1}{2\pi}\sqrt{\frac{\mu g}{a}}$

70. (d) Oscillations along spring length are independent of gravitation.

71. (d) The second pendulum placed in a space laboratory orbiting around the earth is in a weightlessness state. Hence $g = 0$ so $T = \infty$

72. (a) The restoring force

$F = -(PA)$

$= -[\rho g(2y\sin 30°)]A$

$= \rho g A(-y)$

$\therefore \quad a = \frac{F}{m} = \frac{\rho g A}{m}$ Thus $T = 2\pi\sqrt{\frac{m}{\rho g A}}$

73. (c) As we know, time period, $T = 2\pi\sqrt{\frac{\ell}{g}}$

When additional mass M is added then

$T_M = 2\pi\sqrt{\frac{\ell + \Delta\ell}{g}}$

$\frac{T_M}{T} = \sqrt{\frac{\ell + \Delta\ell}{\ell}}$ or $\left(\frac{T_M}{T}\right)^2 = \frac{\ell + \Delta\ell}{\ell}$

or, $\left(\frac{T_M}{T}\right)^2 = 1 + \frac{Mg}{Ay}$ $\left[\because \Delta\ell = \frac{Mg\ell}{Ay}\right]$

$\therefore \frac{1}{y} = \left[\left(\frac{T_M}{T}\right)^2 - 1\right]\frac{A}{Mg}$

74. (c) $f_A = \frac{1}{2\pi}\sqrt{\frac{g}{L_A}}$ and $f_B = \frac{f_A}{2} = \frac{1}{2\pi}\sqrt{\frac{g}{L_B}}$

$\therefore \frac{f_A}{f_{A/2}} = \frac{1}{2\pi}\sqrt{\frac{g}{L_A}} \times 2\pi\sqrt{\frac{L_B}{g}} \Rightarrow 2 = \sqrt{\frac{L_B}{L_A}}$

$\Rightarrow 4 = \frac{L_B}{L_A}$, regardless of mass.

75. (b)

76. (c) h = Length of block immerged in water

mg = F_B

$\rho A l\,g = \rho_B A h g$

$650 \times A \times 54 \times 10^{-2} g = 900 \times A \times hg$

$\Rightarrow h = 0.39\ m = 39$ cm.

77. (a) **78. (a)**

79. (b) Springs on the left of the block are in series, hence their equivalent spring constant is

$K_1 = \frac{(2k)(2k)}{2k + 2k} = k$

Springs on the right of the block are in parallel, hence their equivalent spring constant is

$k_2 = k + 2k = 3k$

Now again both K_1 and K_2 are in parallel

$\therefore K_{eq} = k_1 + k_2 = k + 3k = 4k$

Hence, frequency is

$f = \frac{1}{2\pi}\sqrt{\frac{K_{eq}}{M}} = \frac{1}{2\pi}\sqrt{\frac{4k}{M}}$

80. (c) $t_0 = 2\pi\sqrt{\frac{\ell}{g}}$.

The restoring force in a liquid

$F = -(mg - V\rho_e g)\sin\theta$

$= -\left(mg - \frac{m}{\left(\frac{4}{3}\times 1000\right)} \times 1000g\right)\left(\frac{x}{\ell}\right)$

or $a = \left(g - \frac{3g}{4}\right)\left(\frac{-x}{\ell}\right) = \frac{g}{4}\left(\frac{-x}{\ell}\right)$

$\therefore t = 2\pi\sqrt{\frac{\ell}{(g/4)}} = 2t_0$.

81. (a) Suppose the liquid in left side limb is displaced slightly by y, the liquid in right limb will increase by $y/2$. The restoring force

$F = -PA = -\rho g\left(\frac{3y}{2}\right) \times 2A = 3\rho g A(-y)$.

$a = \frac{F}{m} = 3\rho g A(-y)/m$

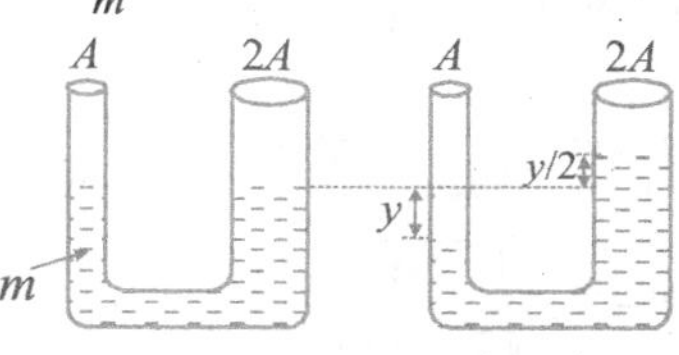

On comparing with, $a = -\omega^2 y$, we get

$\omega = \sqrt{\frac{3\rho g A}{m}}$ and $T = 2\pi\sqrt{\frac{m}{3\rho g A}}$

82. (a) In equilibrium position

$mg = F_B = \frac{m}{4\rho_0}\rho_0(1 + ay_0)g \Rightarrow 4 = 1 + ay_0$

$y_0 = \frac{3}{2} = 1.5$

Now displace it downward by Δy

$\therefore F = mg - \frac{m}{4\rho_0}\rho_0[1 + a(y_0 + \Delta y)]g$

$= mg - \frac{mg}{4}[1 + ay_0 + a\Delta y]$

$F = -\frac{mga}{4}\Delta y$; Acceleration $= -\left(\frac{ag}{4}\right)\Delta y$

$\therefore T = \frac{2\pi}{\omega} = 2\pi\sqrt{\frac{4}{ag}} = \frac{2\pi}{\sqrt{5}}$ sec.

83. (a) The restoring torque (for small θ)

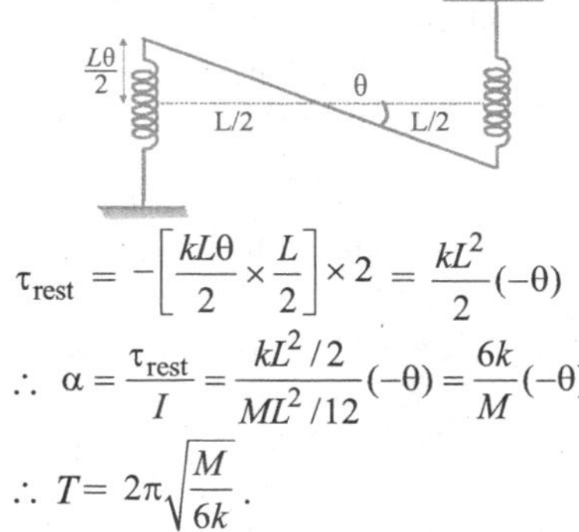

$\tau_{rest} = -\left[\frac{kL\theta}{2} \times \frac{L}{2}\right] \times 2 = \frac{kL^2}{2}(-\theta)$

$\therefore \alpha = \frac{\tau_{rest}}{I} = \frac{kL^2/2}{ML^2/12}(-\theta) = \frac{6k}{M}(-\theta)$

$\therefore T = 2\pi\sqrt{\frac{M}{6k}}$.

84. (a) $2k\sin^2\theta = k_1$ or $k_1 = 2k\sin^2\theta$

and $k_2 = 2(2k)\sin^2\beta$

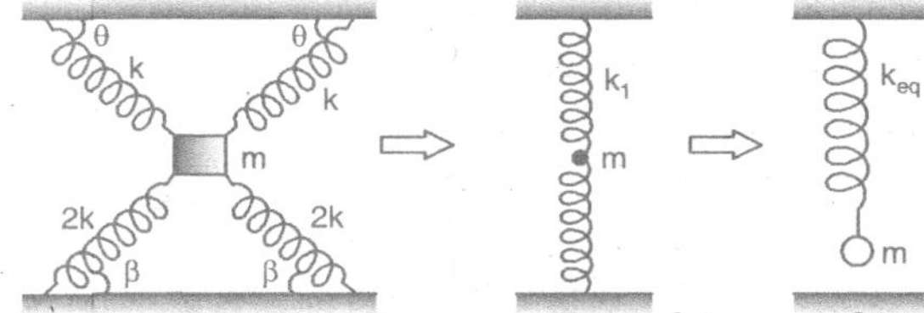

Then $k_{eq} = k_1 + k_2 = 2k[\sin^2\theta + 2\sin^2\beta]$

$= 2k[\sin^2 45° + 2\sin^2 30°] = 2k\left(\frac{1}{2} + \frac{1}{2}\right) = 2k$

Then $T = 2\pi\sqrt{\dfrac{m}{k_{eq}}} = 2\pi\sqrt{\dfrac{m}{2k}} = \pi\sqrt{\dfrac{2m}{k}}$

85. (c) $T = 2\pi\sqrt{\dfrac{\mu}{2k}} \quad \because \mu = \dfrac{m_1 m_2}{m_1 + m_2}$

$T = 2\pi\sqrt{\dfrac{10}{7\times 200}} = \dfrac{2\times 5}{7} = \dfrac{10}{7} = 2\pi\sqrt{\dfrac{1}{140}}$

86. (c) In forced oscillations, the body oscillates at the angular frequency of the driving force.

87. (b) The resonance wave becomes very sharp when damping force is small.

88. (d)

89. (b) At resonance, amplitude of oscillation is maximum

$\Rightarrow 2\omega^2 - 36\omega + 9$ is minimum

$\Rightarrow 4\omega - 36 = 0$ (derivative is zero) $\Rightarrow \omega = 9$

90. (c) At maximum energy of the particle, velocity resonance takes place, which occurs when frequency of external periodic force is equal to natural frequency of undamped vibrations, i.e. $\omega_2 = \omega_0$.

Further amplitude resonance takes place at a frequency of external force which is less than the frequency of undamped natural vibrations, i.e. $\omega_1 \neq \omega_0$.

91. (b) $x = A\sin(\omega t + \phi)$

where $A = \dfrac{F_0}{m\sqrt{(\omega_0^2 - \omega^2)^2}} = \dfrac{F_0}{m(\omega_0^2 - \omega^2)}$

Here damping effect is considered to be zero

$\therefore x \propto \dfrac{1}{m(\omega_0^2 - \omega^2)}$

92. (d)

93. (d) For damped harmonic motion,

$ma = -kx - mbv$ or $ma + mbv + kx = 0$

Solution to above equation is

$x = A_0 e^{-\frac{bt}{2}} \sin\omega t$; with $\omega^2 = \dfrac{k}{m} - \dfrac{b^2}{4}$

where amplitude drops exponentially with time

i.e., $A_\tau = A_0 e^{-\frac{b\tau}{2}}$

Average time τ is that duration when amplitude drops by 63%, i.e., becomes A_0/e.

Thus, $A_\tau = \dfrac{A_0}{e} = A_0 e^{-\frac{b\tau}{2}}$

or $\dfrac{b\tau}{2} = 1$ or $\tau = \dfrac{2}{b}$

94. (d) As we know, $E = E_0 e^{-\frac{bt}{m}}$

$15 = 45e^{-\frac{b15}{m}} \Rightarrow \dfrac{1}{3} = e^{-\frac{b15}{m}}$

[As no. of oscillations = 15 so t = 15 sec]

Taking log on both sides

$\dfrac{b}{m} = \dfrac{1}{15}\ln 3$

95. (d)

96. (d) $F = -bV$, b depends on all the three i.e, shape and size of he block and viscosity of the medium.

97. (b) Amplitude of a damped oscillator at any instant t is given by

$A = A_0 e^{-bt/2m}$

where A_0 is the original amplitude

From question,

When $t = 2$ s, $A = \dfrac{A_0}{3} \therefore \dfrac{A_0}{3} = A_0 e^{-2b/2m}$

or, $\dfrac{1}{3} = e^{-b/m}$ ...(i)

When $t = 6$ s, $A = \dfrac{A_0}{n} \therefore \dfrac{A_0}{n} = A_0 e^{-6b/2m}$

or, $\dfrac{1}{n} = e^{-3b/m} = (e^{-b/m})^3$ or, $\dfrac{1}{n} = \left(\dfrac{1}{3}\right)^3$

$\therefore n = 3^3$ (Using eq. (i))

98. (a) In case of sustained force oscillations the amplitude of oscillations decreases linearly.

99. (b) $V_m = V_0/(a\omega^2 - b\omega + c)$

If there is a single resonant frequency, then this equation should be satisfied for only one that particular resonant frequency, hence $a\omega^2 - b\omega + c = 0$ will have equal roots therefore;

$D = 0 \Rightarrow (-b)^2 - 4ac = 0 \Rightarrow b^2 = 4ac$

100. (b) Given, $V = 36\times\dfrac{5}{18}$ m/s = 10 m/s,

$g = 10$ m/s², $r = 10$ m, $l = 1$ m

$T = 2\pi\sqrt{\dfrac{\ell}{\sqrt{g^2 + \left(\dfrac{v^2}{r}\right)^2}}} = 2\pi\sqrt{\dfrac{1}{\sqrt{10^2 + \left(\dfrac{100}{10}\right)^2}}}$

$= \dfrac{\pi}{10}\sqrt{2}$ sec

14 Waves

1. When two sound waves travel in the same direction in a medium, the displacements of a particle located at 'x' at time 't' is given by :
   $y_1 = 0.05 \cos(0.50\,\pi x - 100\,\pi t)$
   $y_2 = 0.05 \cos(0.46\,\pi x - 92\,\pi t)$
   where y_1, y_2 and x are in meters and t in seconds. The speed of sound in the medium is :
   (a) 92 m/s (b) 200 m/s
   (c) 100 m/s (d) 332 m/s
2. The extension in a string, obeying Hooke's law, is x. The speed of sound in the stretched string is v. If the extension in the string is increased to $1.5x$, the speed of sound will be
   (a) 1.22v (b) 0.61v (c) 1.50v (d) 0.75v
3. A wave travelling on a string is described by $y(x, t) = 0.005 \sin(80.0x - 3.0t)$
   The period of the wave is
   (a) 3.00 s (b) 2.09 s (c) 0.48 s (d) 0.05 s
4. A closed organ pipe (closed at one end) is excited to support the third overtone. It is found that air in the pipe has **Tricky**
   (a) three nodes and three antinodes
   (b) three nodes and four antinodes
   (c) four nodes and three antinodes
   (d) four nodes and four antinodes
5. If there are six loops for 1 m length in transverse mode of Melde's experiment., the no. of loops in longitudinal mode under otherwise identical conditions would be
   (a) 3 (b) 6 (c) 12 (d) 8
6. If n_1, n_2 and n_3 are the fundamental frequencies of three segments into which a string is divided, then the original fundamental frequency n of the string is given by **Critical Thinking**
   (a) $\frac{1}{n} = \frac{1}{n_1} + \frac{1}{n_2} + \frac{1}{n_3}$
   (b) $\frac{1}{\sqrt{n}} = \frac{1}{\sqrt{n_1}} + \frac{1}{\sqrt{n_2}} + \frac{1}{\sqrt{n_3}}$
   (c) $\sqrt{n} = \sqrt{n_1} + \sqrt{n_2} + \sqrt{n_3}$
   (d) $n = n_1 + n_2 + n_3$
7. When a sound wave goes from one medium to another, the quantity that remains unchanged is
   (a) frequency (b) amplitude
   (c) wavelength (d) speed
8. Of the following, the equation of plane progressive wave is
   (a) $y = r \sin \omega t$
   (b) $y = r \sin(\omega t - kx)$
   (c) $y = \frac{a}{\sqrt{r}} \sin(\omega t - kx)$
   (d) $y = \frac{a}{r} \sin(\omega t - kr)$
9. Consider the three waves z_1, z_2 and z_3 as
   $z_1 = A \sin(kx - \omega t)$; $z_2 = A \sin(kx + \omega t)$
   $z_3 = A \sin(ky - \omega t)$
   Which of the following represents a standing wave? **Tricky**
   (a) $z_1 + z_2$ (b) $z_2 + z_3$
   (c) $z_3 + z_1$ (d) $z_1 + z_2 + z_3$
10. Two waves are represented by the equations $y_1 = a \sin(\omega t + kx + 0.57)\,m$ and $y_2 = a \cos(\omega t + kx)\,m$, where x is in meter and t in sec. The phase difference between them is
    (a) 1.0 radian (b) 1.25 radian
    (c) 1.57 radian (d) 0.57 radian
11. A transverse wave is represented by $y = A \sin(\omega t - kx)$. For what value of the wavelength is the wave velocity equal to the maximum particle velocity?
    (a) $\frac{\pi A}{2}$ (b) πA (c) $2\pi A$ (d) A
12. The displacement y of a particle in a medium can be expressed as,
    $y = 10^{-6} \sin\left(100t + 20x + \frac{\pi}{4}\right) m$ where t is in second and x in meter. The speed of the wave is
    (a) 20 m/s (b) 5 m/s
    (c) 2000 m/s (d) 5π m/s
13. A progressive sound wave of frequency 500 Hz is travelling through air with a speed of 350 ms^{-1}. A compression maximum appears at a place at a given instant. The minimum time interval after which the rarefraction maximum occurs at the same point, is

(a) 200 s (b) $\frac{1}{250}$ s
(c) $\frac{1}{500}$ s (d) $\frac{1}{1000}$ s

14. At t = 0, the shape of a travelling pulse is given by

$$y(x, 0) = \frac{4 \times 10^{-3}}{8 - (x)^{-2}}$$

where x and y are in metres. The wave function for the travelling pulse if the velocity of propagation is 5 m/s in the x direction is given by **Tricky**

(a) $y(x, t) = \frac{4 \times 10^{-3}}{8 - (x^2 - 5t)}$

(b) $y(x, t) = \frac{4 \times 10^{-3}}{8 - (x - 5t)^2}$

(c) $y(x, t) = \frac{4 \times 10^{-3}}{8 - (x + 5t)^2}$

(d) $y(x, t) = \frac{4 \times 10^{-3}}{8 - (x^2 + 5t)}$

15. A plane progressive simple harmonic sound wave of angular frequency 680 rad/s moves with speed 340 m/s in the direction which makes equal angle with each x, y and z-axis. The phase difference $(\phi_1 - \phi_2)$ between the oscillations of the particle in the medium located at the positions $(\sqrt{3}, \sqrt{3}, \sqrt{3})$ and $(2\sqrt{3}, 2\sqrt{3}, 2\sqrt{3})$ is (assume $\cos\theta > 0$)
(a) 2 radian (b) 3 radian
(c) 4 radian (d) 6 radian

16. The amplitude of a wave disturbance propagating in the positive x–direction is given by $y = \frac{1}{1+x^2}$ at $t = 0$ and $y = \frac{1}{2+x^2-2x}$ at $t = 2s$, where x and y are in meter. Assuming that the shape of the wave disturbance does not change during the propagation, the speed of the wave is
(a) 0.5 m/s (b) 1 m/s (c) 1.5 m/s (d) 2 m/s

17. A small piece of cork in a ripple tank oscillates up and down as ripple pass it. If the ripples travelling at 0.3 m/s have a wavelength of 1.5π cm and the cork vibrates with an amplitude of 5 mm, the maximum velocity of the cork is:
(a) 20 cm/sec (b) 20 m/sec
(c) 0.02 m/sec (d) 200 cm/sec

18. The wave described by $y = 0.25 \sin(10\pi x - 2\pi t)$, where x and y are in meters and t in seconds, is a wave travelling along the: **Tricky**
(a) –ve x direction with frequency 1 Hz.
(b) +ve x direction with frequency π Hz and wavelength $\lambda = 0.2$ m.
(c) +ve x direction with frequency 1 Hz and wavelength $\lambda = 0.2$ m
(d) –ve x direction with amplitude 0.25 m and wavelength $\lambda = 0.2$ m

19. When a longitudinal wave propagates through a medium, the particles of the medium execute simple harmonic oscillations about their mean positions. These oscillations of a particle are characterised by an invariant
(a) kinetic energy
(b) potential energy
(c) sum of kinetic energy and potential energy
(d) difference between kinetic energy and potential energy

20. In a transverse wave the distance between a crest and neighbouring trough at the same instant is 4.0 cm and the distance between a crest and trough at the same place is 1.0 cm. The next crest appears at the same place after a time interval of 0.4s. The maximum speed of the vibrating particles in the medium is :
(a) $\frac{3\pi}{2}$ cm/s (b) $\frac{5\pi}{2}$ cm/s
(c) $\frac{\pi}{2}$ cm/s (d) 2π cm/s

21. An earthquake generates both transverse (*S*) and longitudinal (*P*) sound waves in the earth. The speed of S waves in about 4.5 km/s and that of *P* waves is about 8.0 km/s. A seismograph records *P* and *S* waves from an earthquake. The first *P* wave arrives 4.0 min. before the first *S* wave. The epicenter of the earthquake is located at a distance about **Critical Thinking**
(a) 25 km (b) 250 km
(c) 2500 km (d) 5000 km

22. A progressive wave is travelling in a medium such that frequency of oscillation and displacement amplitude of the particles of the medium are *f* and *A* respectively. The ratio of their acceleration amplitude and velocity amplitude is
(a) $2\pi f$ (b) πf (c) $\frac{2\pi f}{A}$ (d) $\frac{\pi f}{A}$

23. In a plane progressive harmonic wave particle speed is always less than the wave speed if
(a) amplitude of wave is less than $\frac{\lambda}{2\pi}$
(b) amplitude of wave is greater than $\frac{\lambda}{2\pi}$
(c) amplitude of wave is less than λ
(d) amplitude of wave is greater than $\frac{\lambda}{\pi}$

24. Equation of a progressive wave is given by
$y = 4\sin\left[\pi\left(\frac{t}{5}-\frac{x}{9}\right)+\frac{\pi}{6}\right]$
Then which of the following is correct?
(a) $v = 5$ cm (b) $\lambda = 18$ cm
(c) $a = 0.04$ cm (d) $f = 50$ Hz

25. A transverse sinusoidal wave moves along a string in the positive x-direction at a speed of 10 cm/s. The wavelength of the wave is 0.5 m and its amplitude is 10 cm. At a particular time t, the snap–shot of the wave is shown in figure. The velocity of point P when its displacement is 5 cm is

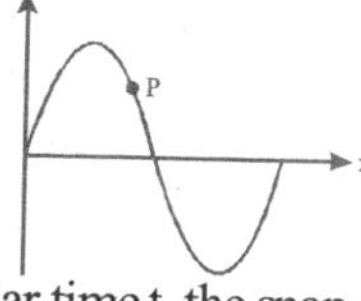

(a) $\frac{\sqrt{3}\pi}{50}\hat{j}$ m/s (b) $-\frac{\sqrt{3}\pi}{50}\hat{j}$ m/s
(c) $\frac{\sqrt{3}\pi}{50}\hat{i}$ m/s (d) $-\frac{\sqrt{3}\pi}{50}\hat{i}$ m/s

26. Sound waves are travelling in a medium whose adiabatic elasticity is E and isothermal elasticity E′. The velocity of sound waves is proportional to
(a) E′ (b) $\sqrt{E}$ (c) $\sqrt{E'}$ (d) $\frac{E}{E'}$

27. What is the effect of humidity on sound waves when humidity increases?
(a) Speed of sound waves is more
(b) Speed of sound waves is less
(c) Speed of sound waves remains same
(d) Speed of sound waves becomes zero

28. The bulk modulus of a liquid of density 8000 kg m^{-3} is 2×10^9 N m^{-2}. The speed of sound in that liquid is (in m s^{-1})
(a) 200 (b) 250 (c) 100 (d) 500

29. The velocity of sound in hydrogen is 1224 m/s. Its velocity in a mixture of hydrogen and oxygen containing 4 parts by volume of hydrogen and 1 part oxygen is

Critical Thinking

(a) 1224 m/s (b) 612 m/s
(c) 2448 m/s (d) 306 m/s

30. When a sound wave of frequency 300 Hz passes through a medium, the maximum displacement of a particle of the medium is 0.1 cm. The maximum velocity of the particle is equal to
(a) 60 πcms^{-1} (b) 30 πcms^{-1}
(c) 30 cms^{-1} (d) 60 cms^{-1}

31. Velocity of sound waves in air is 330 m/s. For a particular sound wave in air, a path difference of 40 cm is equivalent to phase difference of 1.6π. The frequency of this wave is
(a) 165 Hz (b) 150 Hz
(c) 660 Hz (d) 330 Hz

32. A hospital uses an ultrasonic scanner to locate tumours in a tissue. The operating frequency of the scanner is 4.2 MHz. The speed of sound in a tissue is 1.7 km/s. The wavelength of sound in tissue is close to
(a) 4×10^{-4} m (b) 8×10^{-4} m
(c) 4×10^{-3} m (d) 8×10^{-3} m

33. A person speaking normally produces a sound intensity of 40 dB at a distance of 1 m. If the threshold intensity for reasonable audibility is 20 dB, the maximum distance at which he can be heard clearly is
(a) 4m (b) 5m (c) 10m (d) 20m

34. The fundamental frequency of a closed end organ pipe is n. Its length is doubled and radius is halved. Its frequency will become nearly
(a) n/2 (b) n/3 (c) n (d) 2n

35. The fundamental frequency of an organ pipe is 512 Hz. If its length is increased, then frequency will
(a) decrease
(b) increase
(c) remains same
(d) cannot be predicted

36. Tube *A* has both ends open while tube *B* has one end closed, otherwise they are identical. The ratio of fundamental frequency of tube *A* and *B* is

Tricky

(a) 1 : 2 (b) 1 : 4 (c) 2 : 1 (d) 4 : 1

37. If the length of a stretched string is shortened by 40% and the tension increased by 44% then the ratio of the final and initial fundamental frequencies is

Critical Thinking

(a) 3 : 4 (b) 4 : 3 (c) 1 : 3 (d) 2 : 1

38. The number of possible natural oscillation of air column in a pipe closed at one end of length 85 cm whose frequencies lie below 1250 Hz are : (velocity of sound = 340 ms^{-1})
(a) 4 (b) 5 (c) 7 (d) 6

39. The fundamental frequency of a closed organ pipe of length 20 cm is equal to the second overtone of an organ pipe open at both the ends. The length of organ pipe open at both the ends is
(a) 100 cm (b) 120 cm (c) 140 cm (d) 80 cm

40. Two strings A and B, made of same material, are stretched by same tension. The radius of string A is double of radius of B. A transverse wave travels on A with speed v_A and on B with speed v_B. The ratio v_A / v_B is
(a) 1/2 (b) 2 (c) 1/4 (d) 4

41. In the sonometer experiment, a tuning fork of frequency 256 Hz is in resonance with 0.4 m length of the wire when the iron load attached to free end of wire is 2 kg. If the load is immersed in water, the length of the wire in resonance would be (specific gravity of iron = 8) Toughnut

(a) 0.37m (b) 0.43m (c) 0.31m (d) 0.2m

42. Two identical piano wires kept under the same tension T have a fundamental frequency of 600 Hz. The fractional increase in the tension of one of the wires which will lead to occurrence of 6 beats/s when both the wires oscillate together would be

(a) 0.02 (b) 0.03 (c) 0.04 (d) 0.01

43. The length of the wire between two ends of a sonometer is 100 cm. What should be the positions of two bridges below the wire so that the three segments of the wire have their fundamental frequencies in the ratio of 1 : 3 : 5? Tricky

(a) $\frac{1500}{23}$ cm, $\frac{2000}{23}$ cm (b) $\frac{1500}{23}$ cm, $\frac{500}{23}$ cm

(c) $\frac{1500}{23}$ cm, $\frac{300}{23}$ cm (d) $\frac{300}{23}$ cm, $\frac{1500}{23}$ cm

44. An organ pipe P_1 closed at one end vibrating in its first overtone and another pipe P_2 open at both ends vibrating in third overtone are in resonance with a given tuning fork. The ratio of the length of P_1 to that of P_2 is

(a) 8/3 (b) 3/8 (c) 1/2 (d) 1/3

45. Length of a sonometer wire between two fixed ends is 110 cm. If the fundamental frequencies are in the ratio of 1 : 2 : 3, then what is the ratio of lengths of these segments of the wire?

(a) 3 : 2 : 1 (b) 6 : 3 : 2
(c) 6 : 2 : 3 (d) 2 : 3 : 6

46. The equation of a wave on a string of linear mass density 0.04 kg m^{-1} is given by

$$y = 0.02(m)\sin\left[2\pi\left(\frac{t}{0.04(s)} - \frac{x}{0.50(m)}\right)\right]$$

The tension in the string is

(a) 4.0 N (b) 12.5 N (c) 0.5 N (d) 6.25 N

47. A sonometer wire of length 1.5 m is made of steel. The tension in it produces an elastic strain of 1%. What is the fundamental frequency of steel if density and elasticity of steel are 7.7×10^3 kg/m^3 and 2.2×10^{11} N/m^2 respectively?

(a) 188.5 Hz (b) 178.2 Hz
(c) 200.5 Hz (d) 770 Hz

48. A sonometer wire supports a 4 kg load and vibrates in fundamental mode with a tuning fork of frequency 416 Hz. The length of the wire between the bridges is now doubled. In order to maintain fundamental mode, the load should be changed to

(a) 1 kg (b) 2 kg (c) 4 kg (d) 16 kg

49. Two pulses in a stretched string whose centres are initially 8 cm apart are moving towards each other as shown in the figure. The speed of each pulse is 2 cm/s. After 2 s, the total energy of the pulses will be Tricky

(a) Zero
(b) Purely kinetic
(c) Purely potential
(d) Partly kinetic and partly potential

8 cm

50. A stretched wire 60 cm long is vibrating with its fundamental frequency of 256 Hz. If the length of the wire is decreased to 15 cm and the tension remains the same. Then the fundamental freuqency of the vibration of the wire will be

(a) 1024 (b) 572 (c) 256 (d) 64

51. A string is stretched between fixed points separated by 75.0 cm. It is observed to have resonant frequencies of 420 Hz and 315 Hz. There are no other resonant frequencies between these two. Then, the lowest resonant frequency for this string is

(a) 105 Hz (b) 1.05 Hz
(c) 1050 Hz (d) 10.5 Hz

52. The fundamental frequency of an open organ pipe is 300 Hz. The first overtone of this pipe has same frequency as first overtone of a closed organ pipe. If speed of sound is 330 m/s, then the length of closed organ pipe is Critical Thinking

(a) 41 cm (b) 37 cm (c) 31 cm (d) 80 cm

53. In a standing wave formed as a result of reflection from a surface, the ratio of the amplitude at an antinode to that at node is x. The fraction of energy that is reflected is

(a) $\left[\frac{x-1}{x}\right]^2$ (b) $\left[\frac{x}{x+1}\right]^2$

(c) $\left[\frac{x-1}{x+1}\right]^2$ (d) $\left[\frac{1}{x}\right]^2$

54. String 1 is connected with string 2. The mass per unit length in string 1 is μ_1 and the mass per unit length in string 2 is $4\mu_1$. The tension in the strings is T. A travelling wave is coming from the left. What fraction of the energy in the incident wave goes into string 2 ?

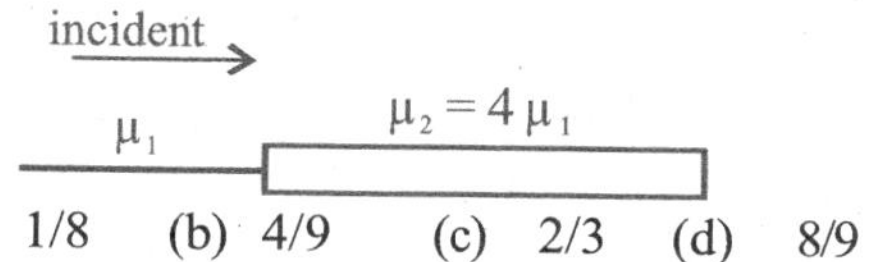

(a) 1/8 (b) 4/9 (c) 2/3 (d) 8/9

55. A uniform rope of length L and mass m_1 hangs vertically from a rigid support. A block of mass m_2 is attached to the free end of the rope. A transverse pulse of wavelength λ_1 is produced at the lower end of the rope. The wavelength of the pulse when it reaches the top of the rope is λ_2 the ratio λ_2/λ_1 is **Critical Thinking**

(a) $\sqrt{\dfrac{m_1}{m_2}}$ (b) $\sqrt{\dfrac{m_1+m_2}{m_2}}$

(c) $\sqrt{\dfrac{m_2}{m_1}}$ (d) $\sqrt{\dfrac{m_1+m_2}{m_1}}$

56. A composite string is made up by joining two strings of different masses per unit length μ and 4μ. The composite string is under the same tension. A transverse wave pulse $Y = (6\text{ mm})\sin(5t + 40x)$, where '$t$' is in seconds and '$x$' is in metres, is sent along the lighter string towards the joint. The joint is at $x = 0$. The equation of the wave pulse reflected from the joint is **Critical Thinking**

(a) $Y = (2\text{ mm})\sin(5t - 40x)$
(b) $Y = (4\text{ mm})\sin(40x - 5t)$
(c) $Y = -(2\text{ mm})\sin(5t - 40x)$
(d) $Y = (2\text{ mm})\sin(5t - 10x)$

57. A pipe of length ℓ_1, closed at one end is kept in a chamber of gas of density ρ_1. A second pipe open at both ends is placed in a second chamber of gas of density ρ_2. The compressibility of both the gases is equal. Calculate the length of the second pipe if frequency of first overtone in both the cases is equal

(a) $\dfrac{4}{3}\ell_1\sqrt{\dfrac{\rho_2}{\rho_1}}$ (b) $\dfrac{4}{3}\ell_1\sqrt{\dfrac{\rho_1}{\rho_2}}$

(c) $\ell_1\sqrt{\dfrac{\rho_2}{\rho_1}}$ (d) $\ell_1\sqrt{\dfrac{\rho_1}{\rho_2}}$

58. Three sound waves of equal amplitudes have frequencies $(\nu - 1)$, ν, $(\nu + 1)$. They superpose to give beats. The number of beats produced per second will be :

(a) 3 (b) 2 (c) 1 (d) 4

59. Two tones of frequencies n_1 and n_2 are sounded together. The beats can be heard distinctly when **Tricky**

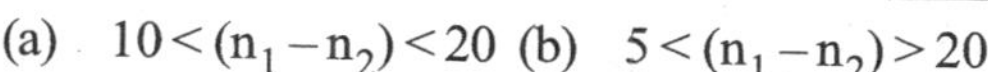
(a) $10 < (n_1 - n_2) < 20$ (b) $5 < (n_1 - n_2) > 20$
(c) $5 < (n_1 - n_2) < 20$ (d) $0 < (n_1 - n_2) < 10$

60. Two travelling waves $y_1 = A\sin[k(x - ct)]$ and $y_2 = A\sin[k(x + ct)]$ are superimposed on string. The distance between adjacent nodes is

(a) ct/π (b) $ct/2\pi$ (c) $\pi/2k$ (d) π/k

61. Two factories are sounding their sirens at 800 Hz. A man goes from one factory to other at a speed of 2m/s. The velocity of sound is 320 m/s. The number of beats heard by the person in one second will be:

(a) 2 (b) 4 (c) 8 (d) 10

62. Two sound sources S_2 and S_1 emit pure sinusoidal coherent waves in phase. If the speed of sound is 340 m/s, then find out the frequencies for which constructive interference occurs at P. **Critical Thinking**

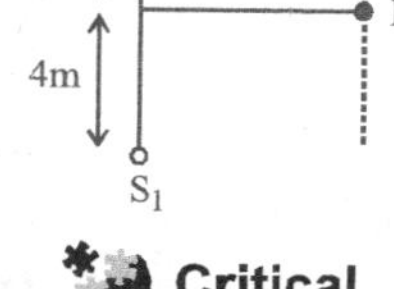

(a) 170Hz (b) 340 Hz
(c) 510Hz (d) All of these

63. A tuning fork arrangement (pair) produces 4 beats/sec with one fork of frequency 288 cps. A little wax is placed on the unknown fork and it then produces 2 beats/sec. The frequency of the unknown fork is

(a) 286 cps (b) 292 cps
(c) 294 cps (d) 288 cps

64. A tuning fork of frequency 512 Hz makes 4 beats per second with the vibrating string of a piano. The beat frequency decreases to 2 beats per sec when the tension in the piano string is slightly increased. The frequency of the piano string before increasing the tension was

(a) 510 *Hz* (b) 514 *Hz*
(c) 516 *Hz* (d) 508 *Hz*

65. Two vibrating tuning forks produce progressive waves given by $Y_1 = 4\sin 500\pi t$ and $Y_2 = 2\sin 506\pi t$. Number of beats produced per minute is

(a) 360 (b) 180 (c) 60 (d) 3

66. The wavelength of two waves are 50 and 51 cm respectively. If the temperature of the room is 20°C then what will be the number of beats produced per second by these waves, when the speed of sound at 0°C is 332 m/s?

(a) 24 (b) 14
(c) 10 (d) none of these

67. In the figure shown the wave speed is v. The velocity of car is v_0. The beat frequency for the observer will be **Tricky**

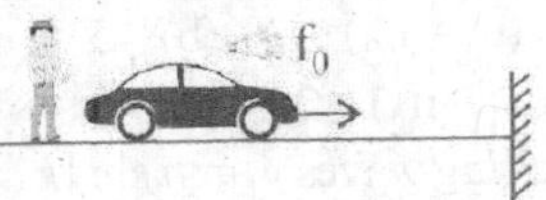

(a) $\frac{2f_0vv_0}{v^2+v_0^2}$ (b) $\frac{2f_0v^2}{v^2-v_0^2}$

(c) $\frac{2f_0vv_0}{v^2-v_0^2}$ (d) $\frac{f_0vv_0}{v^2-v_0^2}$

68. An open pipe is in resonance in 2nd harmonic with frequency f_1. Now one end of the tube is closed and frequency is increased to f_2 such that the resonance again occurs in nth harmonic. Choose the correct option **Critical Thinking**

(a) $n=3,\ f_2=\frac{3}{4}f_1$ (b) $n=3,\ f_2=\frac{5}{4}f_1$

(c) $n=5,\ f_2=\frac{3}{4}f_1$ (d) $n=5,\ f_2=\frac{5}{4}f_1$

69. Two sources of sound placed close to each other are emitting progressive waves given by $y_1 = 4 \sin 600\pi t$ and $y_2 = 5 \sin 608\pi t$. An observer located near these two sources of sound will hear:
(a) 4 beats per second with intensity ratio 25 : 16 between waxing and waning.
(b) 8 beats per second with intensity ratio 25 : 16 between waxing and waning
(c) 8 beats per second with intensity ratio 81 : 1 between waxing and waning
(d) 4 beats per second with intensity ratio 81 : 1 between waxing and waning

70. If wind blows from a stationary sounding object to a stationary listener, then the apparent frequency n' and actual frequency n are related as **Tricky**
(a) $n' \geq n$ (b) $n' < n$ (c) $n' = n$ (d) $n' > n$

71. A car is moving towards a high cliff. The car driver sounds a horn of frequency f. The reflected sound heard by the driver has as frequency $2f$. If v be the velocity of sound, then the velocity of the car, in the same velocity units, will be
(a) v/2 (b) v/√2 (c) v/3 (d) v/4

72. Choose the false statement(s) from the following.
I. Change in frequency due to Doppler effect will be positive if the distance between source and listener increases.
II. Change in frequency due to Doppler effect will be negative if the distance between source and listener
(a) I only (b) II only
(c) I and II (d) None of these

73. Two trains are moving towards each other with speeds of 20m/s and 15 m/s relative to the ground. The first train sounds a whistle of frequency 600 Hz. The frequency of the whistle heard by a passenger in the second train before the train meets, is (the speed of sound in air is 340 m/s)
(a) 600 Hz (b) 585 Hz
(c) 645 Hz (d) 666 Hz

74. Two trains move towards each other with the same speed. The speed of sound is 340 m/s. If the height of the tone of the whistle of one of them heard on the other changes 9/8 times, then the speed of each train should be
(a) 20 m/s (b) 2 m/s
(c) 200 m/s (d) 2000 m/s

75. A source producing sound of frequency 170 Hz is approaching a stationary observer with a velocity 17 ms^{-1}. The apparent change in the wavelength of sound heard by the observer is (speed of sound in air = 340 ms^{-1})
(a) 0.1 m (b) 0.2 m (c) 0.4 m (d) 0.5 m

76. A man is watching two trains, one leaving and the other coming in with equal speeds of 4m/sec. If they sound their whistles, each of frequency 240 Hz, the number of beats heard by the man (velocity of sound in air = 320 m/sec) will be equal to **Critical Thinking**
(a) 6 (b) 3 (c) 0 (d) 12

77. A sound source emits frequency of 180 Hz when moving towards a rigid wall with speed 5 m/s and an observer is moving away from wall with speed 5 m/s. Both source and observer moves on a straight line which is perpendicular to the wall. The number of beats per second heard by the observer will be [Speed of sound = 355 m/s]
(a) 5 beats/s (b) 10 beats/s
(c) 6 beats/s (d) 8 beats/s

78. A police car moving at 22 m/s, chases a motorcyclist. The policeman sounds his horn at 176 Hz, while both of them move towards a stationary siren of frequency 165 Hz. The speed of the motorcycle, if it is given that he does not observe any beats is

Police car Motorcycle

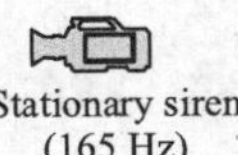

(a) 33 m/s (b) 22 m/s (c) zero (d) 11 m/s

79. In the figure shown a source of sound of frequency 510 Hz moves with constant velocity $v_s = 20$ m/s in the direction shown. The wind is blowing at a constant velocity $v_w = 20$ m/s towards an observer who is at rest at point B. Corresponding to the sound emitted by the source at initial position A, the frequency detected by the observer is equal to (speed of sound relative to air = 330 m/s)

(a) 510 Hz

(b) 500 Hz

(c) 525 Hz

(d) 550 Hz

80. A train has just completed a U-curve in a track which is a semicircle. The engine at the forward end of the semicircular part of the track while the last carriage is at the rear end of the semicircular track. The driver blows a whistle of frequency 200 Hz. Velocity of sound is 340 m/s. Then the apparent frequency as observed by a passenger in the middle of a train when the speed of the train is 30 m/s is

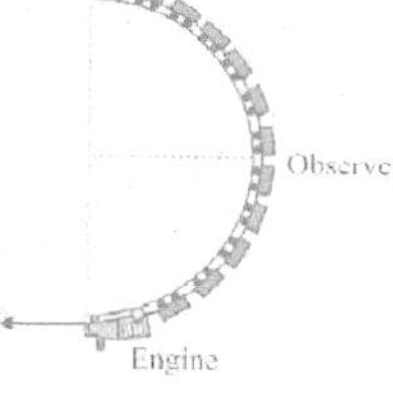

Critical Thinking

(a) 209 Hz (b) 288 Hz
(c) 200 Hz (d) 181 Hz

81. A source of sound attached to the bob of a simple pendulum execute SHM. The difference between the apparent frequency of sound as received by an observer during its approach and recession at the mean position of the SHM motion is 2% of the natural frequency of the source. The velocity of the source at the mean position is (velocity of sound in the air is 340 m/s)
[Assume velocity of sound source << velocity of sound in air]

Toughnut

(a) 1.4 m/s (b) 3.4 m/s
(c) 1.7 m/s (d) 2.1 m/s

82. A source of sound S emitting waves of frequency 100 Hz and an observor O are located at some distance from each other. The source is moving with a speed of 19.4 ms^{-1} at an angle of 60° with the source observer line as shown in the figure. The observor is at rest. The apparent frequency observed by the observer is (velocity of sound in air 330 ms^{-1})

Critical Thinking

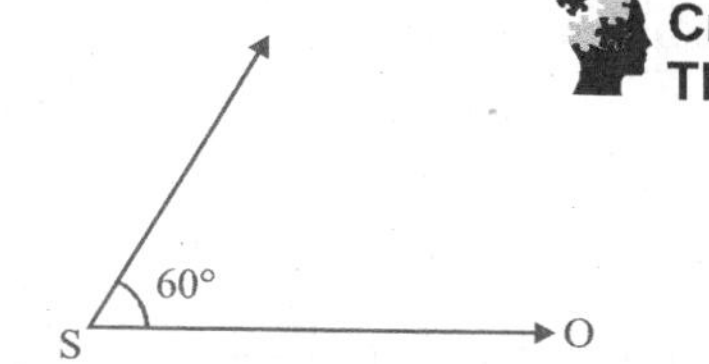

(a) 103 Hz (b) 106 Hz
(c) 97 Hz (d) 100 Hz

83. A speeding motorcyclist sees trafic jam ahead of him. He slows down to 36 km/hour. He finds that traffic has eased and a car moving ahead of him at 18 km/hour is honking at a frequency of 1392 Hz. If the speeds of sound is 343 m/s, the frequency of the honk as heard by him will be :

(a) 1332 Hz (b) 1372 Hz
(c) 1412 Hz (d) 1464 Hz

84. An observer moves towards a stationary source of sound with a speed 1/8 th of the speed of sound. The wavelength and frequency of the sound emitted are λ and f respectively. The apparent frequency and wavelength recorded by the observer are respectively.

(a) 0.4f, 0.4λ (b) 1.6f, 1.6λ
(c) 1.4f, λ (d) 1.1f, λ

85. A parachutist jumps from the top of a very high tower with a siren of frequency 800 Hz on his back. Assume his initial velocity to be zero. After falling freely for 12s, he observes that the frequency of sound heard by him reflected from level ground below him is differing by 700Hz w.r.t. the original frequency. What was the height of tower. Velocity of sound in air is 330 m/s, and $g = 10\ m/s^2$.

Toughnut

(a) 511.5 m. (b) 1057.5 m.
(c) 757.5 m. (d) 1215.5 m.

Answer KEYs

1	(b)	**10**	(a)	**19**	(c)	**28**	(d)	**37**	(d)	**46**	(d)	**55**	(b)	**64**	(d)	**73**	(d)	**82**	(a)
2	(a)	**11**	(c)	**20**	(b)	**29**	(b)	**38**	(d)	**47**	(b)	**56**	(c)	**65**	(b)	**74**	(a)	**83**	(c)
3	(b)	**12**	(b)	**21**	(c)	**30**	(a)	**39**	(b)	**48**	(d)	**57**	(b)	**66**	(b)	**75**	(a)	**84**	(d)
4	(d)	**13**	(d)	**22**	(a)	**31**	(c)	**40**	(a)	**49**	(b)	**58**	(b)	**67**	(c)	**76**	(a)	**85**	(b)
5	(a)	**14**	(b)	**23**	(a)	**32**	(a)	**41**	(a)	**50**	(a)	**59**	(d)	**68**	(d)	**77**	(a)		
6	(a)	**15**	(d)	**24**	(b)	**33**	(c)	**42**	(a)	**51**	(a)	**60**	(d)	**69**	(d)	**78**	(b)		
7	(a)	**16**	(a)	**25**	(a)	**34**	(a)	**43**	(a)	**52**	(a)	**61**	(d)	**70**	(c)	**79**	(c)		
8	(b)	**17**	(a)	**26**	(b)	**35**	(a)	**44**	(b)	**53**	(c)	**62**	(d)	**71**	(c)	**80**	(c)		
9	(a)	**18**	(c)	**27**	(a)	**36**	(c)	**45**	(b)	**54**	(d)	**63**	(b)	**72**	(c)	**81**	(b)		

Hints & Solutions

1. (b) Standard equation

$$y(x,t) = A\cos\left(\frac{\omega}{V}x - \omega t\right)$$

From any of the displacement equation Say y_1

$\frac{\omega}{V} = 0.50\pi$ and $\omega = 100\pi$

$\therefore \frac{100\pi}{V} = 0.5\pi$

$\therefore V = \frac{100\pi}{0.5\pi} = 200$ m/s

2. (a) According to Hooke's law $F_R \propto x$ [Restoring force $F_R = T$, tension of spring] Velocity of sound by a stretched string

$v = \sqrt{\frac{T}{m}}$ where m is the mass per unit length

Hence $v \propto \sqrt{T}$ or, $\frac{v}{v'} = \sqrt{\frac{T}{T'}}$

or $v' = v\sqrt{\frac{T'}{T}} = v\sqrt{\frac{1.5x}{x}} = 1.22$ v

3. (b) Time period, $T = \frac{2\pi}{\omega}$; from given eqn.

$\omega = 3.0\ s^{-1}$

or $T = \frac{2\pi}{3} = 2.09$ s

4. (d) Third overtone has a frequency 7 n, which means $L = \frac{7\lambda}{4}$ = three full loops + one half loop, which would make four nodes and four antinodes.

5. (a) No. of loops in longitudinal mode $= \frac{6}{2} = 3$

6. (a) Total length of string $\ell = \ell_1 + \ell_2 + \ell_3$ (As string is divided into three segments)

But frequency $\propto \frac{1}{\text{length}}$ $\left(\because f = \frac{1}{2\ell}\sqrt{\frac{T}{m}}\right)$

so $\frac{1}{n} = \frac{1}{n_1} + \frac{1}{n_2} + \frac{1}{n_3}$

7. (a) Frequency of wave is a function of the source of waves. Therefore, it remains unchanged.

8. (b) The position of such a wave changes in two dimensional plane with time.

9. (a) Standing waves are produced when two waves propagate in opposite direction. As z_1 & z_2 are propagating in +ve x-axis & –ve x-axis

so, $z_1 + z_2$ will represent a standing wave.

10. (a) Here, $y_1 = a\sin(\omega t + kx + 0.57)$

and $y_2 = a\cos(\omega t + kx)$

$= a\sin\left[\frac{\pi}{2} + (\omega t + kx)\right]$

Phase difference, $\Delta\phi = \phi_2 - \phi_1$

$= \frac{\pi}{2} - 0.57$

$= \frac{3.14}{2} - 0.57 = 1.57 - 0.57 = 1$ radian

11. (c) $y = A\sin(\omega t - kx)$

Particle velocity,

$v_p = \frac{dy}{dt} = A\omega\cos(\omega t - kx)$

$\therefore v_{p\,max} = A\omega$

wave velocity $= \frac{\omega}{k}$

$\therefore A\omega = \frac{\omega}{k}$

i. e., $A = \frac{1}{k}$ But $k = \frac{2\pi}{\lambda}$

$\therefore \lambda = 2\pi A$

12. (b) From equation, $\omega = 100$

$\therefore \frac{2\pi}{T} = 100 \Rightarrow \nu = \frac{100}{2\pi}$

$\frac{2\pi}{\lambda} = 20 \Rightarrow \lambda = \frac{2\pi}{20}$

$v = \lambda\nu = \frac{2\pi}{20} \times \frac{100}{2\pi} = 5\text{m/s}$

13. (d) $T = \frac{1}{500}$s; $t = \frac{T}{2} = \frac{1}{1000}$ s

14. (b) $y(x, t) = f(x - vt)$

$y = (x, 0) = \frac{4\times10^{-3}}{8 - x^2}$

For a travelling wave in the x-direction

$$y(x, t) = \frac{4 \times 10^{-3}}{8 - (x - 5t)^2}$$

15. (d) $y(\vec{r}, t) = A \sin(\omega t - \vec{k} \cdot \vec{r})$

$$\hat{k} = \frac{2\pi}{\lambda}(\cos\alpha \hat{i} + \cos\beta \hat{j} + \cos\gamma \hat{k}) = \frac{k}{\sqrt{3}}(\hat{i} + \hat{j} + \hat{k})$$

$\phi_1 - \phi_2 = 2\,(3) = 6$ rad.

16. (a) At $t = 0$, $y = \frac{1}{1 + x^2}$ or $x = \sqrt{\frac{1 - y}{y}} = x_1$

At $t = 2s$, $y = \frac{1}{2 + x^2 - 2x} = \frac{1}{1 + (x - 1)^2}$

or $(x - 1)^2 = \frac{1 - y}{y}$ or $x = 1 + \sqrt{\frac{1 - y}{y}} = x_2$

∴ Speed of the wave

$$v = \frac{\Delta x}{\Delta t} = \frac{x_2 - x_1}{t_2 - t_1} = \frac{1}{2 - 0} = 0.5\,\text{m/s}$$

17. (a)

18. (c) $y = 0.25 \sin(10\pi x - 2\pi t)$

Comparing this equation with the standard wave equation

$y = a \sin(kx - \omega t)$

We get, $k = 10\pi \Rightarrow \frac{2\pi}{\lambda} = 10\pi \Rightarrow \lambda = 0.2$ m

And $\omega = 2\pi$ or, $2\pi v = 2\pi \Rightarrow v = 1$ Hz.

The sign inside the bracket is negative, hence the wave travels in + ve x- direction.

19. (c) The sum of their KE and PE is a constant.

20. (b)

21. (c) If x be the distance of epicentre from the seismograph, then

$\frac{x}{v_s} - \frac{x}{v_p} = 4 \times 60$ or $\frac{x}{4.5} - \frac{x}{8} = 4 \times 60$ on simplifying, we get

$x = 2500$ km

22. (a) $v_{max} = (2\pi f)A$

$$a_{max} = (2\pi f)^2 A \quad \frac{a_{max}}{v_{max}} = (2\pi f)$$

23. (a) $v_P < v$

or $\omega A < f\lambda$ or $A < \frac{f\lambda}{\omega}$ or $\frac{\lambda}{2\pi}$

24. (b) The standard equation of a progressive wave is

$$y = a \sin\left[2\pi\left(\frac{t}{T} - \frac{x}{\lambda}\right) + \phi\right]$$

The given equation can be written as

$$y = 4 \sin\left[2\pi\left(\frac{t}{10} - \frac{x}{18}\right) + \frac{\pi}{6}\right]$$

∴ $a = 4$ cm, $T = 10$ s, $\lambda = 18$ cm and $\phi = \pi/6$

Hence, (b) is correct.

25. (a) Since the wave is sinusoidal moving in positive x-axis the point will move parallel to y-axis therefore options (c) and (d) are ruled out. As the wave moves forward in positive X-direction, the point should move upwards i.e. in the positive Y-direction. Therefore correct option is a.

Alternate solution-1 :

Equation of a wave moving in positive x-axis is given as $y = A \sin(\omega t - \phi)$ or $v_P = A\omega \cos(\omega t - \phi)$

Here $y = 5$ cm, $A = 10$ cm,

∴ $5 = 10 \sin(\omega t - \phi) \Rightarrow \omega t - \phi = 30°$

Substituting this value in the equation of velocity we get $v_P = 0.10 \times \omega \cos 30°$

Now $v = \nu\lambda$ ∴ $\nu = \frac{v}{\lambda} = \frac{0.10}{0.5} = 0.2$

∴ $\omega = 2\pi\nu = 2\pi \times 0.2 = 0.4\pi$

$\Rightarrow v_P = 0.1 \times 0.4\pi \times \frac{\sqrt{3}}{2} = \frac{\sqrt{3}}{50}\pi$.

It has to be in positive y direction.

26. (b) $v = \sqrt{\frac{\gamma P}{\rho}} = \sqrt{\frac{E}{\rho}}$

27. (a) Velocity of sound $= \sqrt{\frac{\gamma RT}{M}}$

When water vapour are present in air, average molecular weight of air decreases and hence velocity increases.

28. (d) The speed of sound in liquid,

$$v = \sqrt{\frac{k}{\rho}} = \sqrt{\frac{2 \times 10^9}{8000}} = \sqrt{\frac{1}{4} \times 10^6}$$

$$v = \frac{1}{2} \times 10^3 = 500\,\text{m s}^{-1}.$$

29. (b) If $\rho_H = 1$, then $\rho_{mix} = \frac{4 \times 1 + 1 \times 16}{(4 + 1)} = 4$

$$\frac{v_{mix}}{v_H} = \sqrt{\frac{\rho_H}{\rho_{mix}}} = \sqrt{\frac{1}{4}} = \frac{1}{2}$$

$$v_{mix} = \frac{v_H}{2} = \frac{1224}{2} = 612\ \text{m/s}$$

30. (a) $V_{max} = a\omega = 0.1 \times 2\pi \times 300 = 60\pi \text{ cm s}^{-1}$

31. (c) From $\Delta x = \frac{\lambda}{2\pi}\Delta\phi$,

$$\lambda = 2\pi\frac{\Delta x}{\Delta\phi} = \frac{2\pi(0.4)}{1.6\pi} = 0.5 \text{ m}$$

$$\nu = \frac{v}{\lambda} = \frac{330}{0.5} = 660 \text{ Hz}$$

32. (a) Frequency (n) = 4.2 MHz = 4.2×10^6 Hz and speed of sound (v) = 1.7 km/s = 1.7×10^3 m/s. Wave length of sound in tissue

$$(\lambda) = \frac{v}{n} = \frac{1.7 \times 10^3}{4.2 \times 10^6} = 4 \times 10^{-4} \text{ m}.$$

33. (c) $40 = 10\log\frac{I_1}{I_0}$, $\therefore \frac{I_1}{I_0} = 10^4$

and $20 = 10\log\frac{I_2}{I_0}$, $\therefore \frac{I_2}{I_0} = 10^2$ $\therefore \frac{I_1}{I_2} = 100.$

Also $\frac{I_1}{I_2} = 100 = \frac{r_2^2}{r_1^2} = \frac{r_2^2}{1^2}$ or $r_2 = 10$ m.

34. (a) Frequency does not depend upon radius. As length is doubled, fundamental frequency becomes half.

35. (a)

36. (c) $\frac{\lambda_A}{\lambda_B} = \frac{1}{2} \Rightarrow \frac{n_A}{n_B} = \frac{2}{1}$

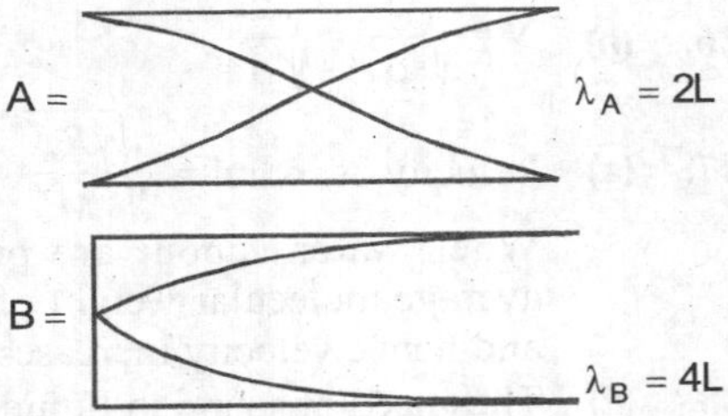

A = $\lambda_A = 2L$

B = $\lambda_B = 4L$

37. (d) $n = \frac{1}{2\ell}\sqrt{\frac{T}{m}}$ $\therefore \frac{n_2}{n_1} = \frac{\ell_1}{\ell_2}\sqrt{\frac{T_2}{T_1}}$

$$= \frac{\ell_1}{\left[\ell_1 - \frac{40}{100}\ell_1\right]}\frac{\sqrt{T_1 + \left(\frac{44}{100}\right)T_1}}{T_1}$$

$$= \frac{100}{60} \times \frac{12}{10} = 2:1$$

38. (d) In case of closed organ pipe frequency,

$$f_n = (2n+1)\frac{v}{4l}$$

for $n = 0, f_0 = 100$ Hz
$n = 1, f_1 = 300$ Hz
$n = 2, f_2 = 500$ Hz
$n = 3, f_3 = 700$ Hz
$n = 4, f_4 = 900$ Hz
$n = 5, f_5 = 1100$ Hz
$n = 6, f_6 = 1300$ Hz

Hence possible natural oscillation whose frequencies < 1250 Hz = 6(n = 0, 1, 2, 3, 4, 5)

39. (b) Fundamental frequency of closed organ pipe

$$V_c = \frac{V}{4l_c}$$

Fundamental frequency of open organ pipe

$$V_0 = \frac{V}{2l_0}$$

Second overtone frequency of open organ pipe

$$= \frac{3V}{2l_0}$$

From question,

$$\frac{V}{4l_c} = \frac{3V}{2l_0}$$

$\Rightarrow l_0 = 6l_c = 6 \times 20 = 120$ cm

40. (a) $\frac{v_A}{v_B} = \frac{D_B}{D_A} = \frac{1}{2}$

41. (a) $\sqrt{T/l}$ = constant; Tension decreases by a factor (8 – 1) / 8, length decreases by square root of this i.e. 0.77.

42. (a) For fundamental mode,

$$f = \frac{1}{2\ell}\sqrt{\frac{T}{\mu}}$$

Taking logarithm on both sides, we get

$$\log f = \log\left(\frac{1}{2\ell}\right) + \log\left(\sqrt{\frac{T}{\mu}}\right)$$

$$= \log\left(\frac{1}{2\ell}\right) + \frac{1}{2}\log\left(\frac{T}{\mu}\right)$$

or $\log f = \log\left(\frac{1}{2\ell}\right) + \frac{1}{2}[\log T - \log\mu]$

Differentiating both sides, we get

$$\frac{df}{f} = \frac{1}{2}\frac{dT}{T} \quad \text{(as } \ell \text{ and } \mu \text{ are constants)}$$

$$\Rightarrow \frac{dT}{T} = 2 \times \frac{df}{f}$$

Here df = 6
f = 600 Hz

$$\therefore \frac{dT}{T} = \frac{2\times 6}{600} = 0.02$$

43. (a) From formula, $f = \frac{1}{x}\sqrt{\frac{T}{m}} \Rightarrow \frac{1}{f} \propto l$

$$\therefore l_1 : l_2 : l_3 = \frac{1}{f_1} : \frac{1}{f_2} : \frac{1}{f_3} = f_2f_3 : f_1f_3 : f_1f_2$$

[Given: $f_1 : f_2 : f_3 = 1:3:5$]

$= 15:5:3$

Therefore the positions of two bridges below the wire are

$$\frac{15\times 100}{15+5+3}\text{ cm and } \frac{15\times 100 + 5\times 100}{15+5+3}\text{ cm}$$

i.e., $\frac{1500}{23}$ cm, $\frac{2000}{23}$ cm

44. (b) $3\times\frac{v}{4l_c} = 4\times\frac{v}{2l_0}$ or $\frac{l_c}{l_0} = \frac{3v}{4}\times\frac{2}{4v} = \frac{3}{8}$

45. (b) Frequency $v = \frac{1}{2L}\sqrt{\frac{T}{m}} \quad \because V \propto \frac{1}{l}$

$$\therefore \quad l_1 : l_2 : l_3 = \frac{1}{1} : \frac{1}{2} : \frac{1}{3} = 6:3:2$$

46. (d) $T = \mu v^2 = \mu\frac{\omega^2}{k^2} = 0.04\frac{(2\pi/0.004)^2}{(2\pi/0.50)^2}$

6.25 N

47. (b) Fundamental frequency,

$$f = \frac{v}{2\ell} = \frac{1}{2\ell}\sqrt{\frac{T}{\mu}} = \frac{1}{2\ell}\sqrt{\frac{T}{A\rho}}$$

$$\left[\because v = \sqrt{\frac{T}{\mu}} \text{ and } \mu = \frac{m}{\ell}\right]$$

Also, $Y = \frac{T\ell}{A\Delta\ell} \Rightarrow \frac{T}{A} = \frac{Y\Delta\ell}{\ell}$

$$\Rightarrow f = \frac{1}{2\ell}\sqrt{\frac{\gamma\Delta\ell}{\ell\rho}} \quad \text{....(i)}$$

Putting the value of $\ell, \frac{\Delta\ell}{\ell}, \rho$ and γ in eq^n.

(i) we get,

$$f = \sqrt{\frac{2}{7}}\times\frac{10^3}{3}$$

or, $f \approx 178.2$ Hz

48. (d) Load supported by sonometer wire = 4 kg

Tension in sonometer wire = 4 g

If μ = mass per unit length

then frequency $\upsilon = \frac{1}{2l}\sqrt{\frac{T}{\mu}}$

$$\Rightarrow 416 = \frac{1}{2l}\sqrt{\frac{4g}{\mu}}$$

When length is doubled, i.e., $l' = 2l$

Let new load = L

As, $\upsilon' = \upsilon$

$$\therefore \frac{1}{2l'}\sqrt{\frac{Lg}{\mu}} = \frac{1}{2l}\sqrt{\frac{4g}{\mu}}$$

$$\Rightarrow \frac{1}{4l}\sqrt{\frac{Lg}{\mu}} = \frac{1}{2l}\sqrt{\frac{4g}{\mu}}$$

$$\Rightarrow \sqrt{L} = 2\times 2 \Rightarrow L = 16\text{ kg}$$

49. (b) After 2 s, the each wave travels a distance = $2\times 2 = 4$ m.

The wave shape is shown in figure.

Thus energy is purely kinetic.

50. (a) $L_0 = 60$ cm $\qquad v_0 = 256$ Hz.

$$v = \frac{1}{2L}\sqrt{\frac{T}{m}} \qquad \therefore \quad v \propto \frac{1}{L}$$

$$\frac{v_1}{v_0} = \frac{L_0}{L_1}$$

$$\Rightarrow v_1 = v_0\frac{L_0}{L_1} = 256\times\frac{60}{15} = 1024\text{ Hz.}$$

51. (a) Given $\frac{nv}{2\ell} = 315$ and $(n+1)\frac{v}{2\ell} = 420$

$$\Rightarrow \frac{n+1}{n} = \frac{420}{315} \Rightarrow n = 3$$

Hence $3\times\frac{v}{2\ell} = 315 \Rightarrow \frac{v}{2\ell} = 105$ Hz

The lowest resonant frequency is when n = 1

Therefore lowest resonant frequency = 105 Hz.

52. (a) For open pipe, $n = \frac{v}{2\ell}$, where n_0 is the fundamental frequency of open pipe.

$$\therefore \ell = \frac{v}{2n} = \frac{330}{2\times 300} = \frac{11}{20}$$

As freq. of 1st overtone of open pipe = freq. of 1st overtone of closed pipe

$$\therefore 2\frac{v}{2\ell} = 3\frac{v}{4\ell'}$$

$$\Rightarrow \ell' = \frac{3\ell}{4} = \frac{3}{4}\times\frac{11}{20} = 41.25\text{ cm}$$

53. (c) $\frac{A_1 + A_2}{A_1 - A_2} = x$; $\frac{A_2}{A_1} = \frac{x-1}{x+1}$; Energy $\propto A^2$

$\Rightarrow \left(\frac{x-1}{x+1}\right)^2$

54. (d) $\frac{E_r}{E_i} = \left(\frac{A_r}{A_i}\right)^2 = \left(\frac{V_2 - V_1}{V_1 + V_2}\right)^2 = 1/9$

Therefore, $\frac{E_t}{E_i} = 8/9$

55. (b) From figure, tension $T_1 = m_2 g$

$T_2 = (m_1 + m_2)g$

As we know

Velocity $\propto \sqrt{T}$ So,

$\lambda \propto \sqrt{T}$

$\Rightarrow \frac{\lambda_1}{\lambda_2} = \frac{\sqrt{T_1}}{\sqrt{T_2}}$

$\Rightarrow \frac{\lambda_2}{\lambda_1} = \sqrt{\frac{m_1 + m_2}{m_2}}$

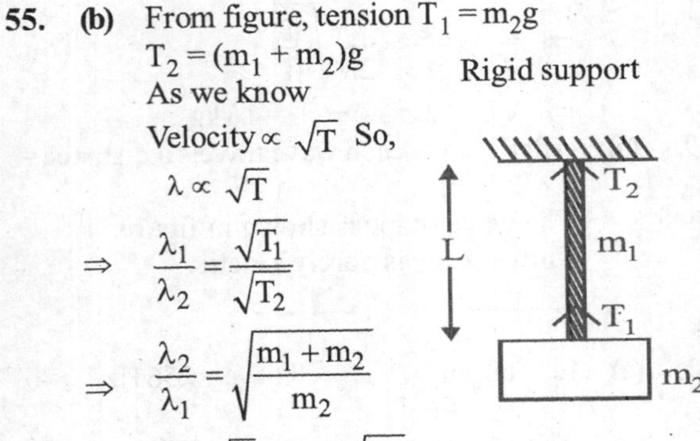

56. (c) $v_1 = \sqrt{\frac{T}{\mu}}$; $v_2 = \sqrt{\frac{T}{4\mu}}$ 4μ μ

$v_2 < v_1 \Rightarrow$ 2nd medium is denser

∴ The wave reflected from the denser medium has phase change of π.

$\Rightarrow A_r = \frac{v_2 - v_1}{v_2 + v_1} \times 6 = \frac{\frac{v_1}{2} - v_1}{\frac{v_2}{2} + v_1} \times 6 = -2\text{mm}$

$\Rightarrow$ eq^n of reflected wave pulse is

$Y = -(2\text{mm}) \sin(5t - 40x)$

57. (b) Frequency of first overtone in closed pipe,

$\nu = \frac{3v}{4\ell_1}\sqrt{\frac{P}{\rho_1}}$ ...(i)

Frequency of first overtone in open pipe,

$\nu' = \frac{1}{\ell_2}\sqrt{\frac{P}{\rho_2}}$ ...(ii)

From equation (i) and (ii)

$\Rightarrow \ell_2 = \frac{4}{3}\ell_1\sqrt{\frac{\rho_1}{\rho_2}}$

58. (b) Maximum number of beats

$= (\nu + 1) - (\nu - 1) = 2$

59. (d) As number of beats/sec = diff. in frequencies has to be less than 10, therefore $0 < (n_1 - n_2) < 10$

60. (d) $y = A\sin(kx - kct) + A\sin(kx + kct)$

$= 2A\sin\left(\frac{kx - kct + kx + kct}{2}\right)\cdot\cos\left(\frac{kx - kct - kx - kct}{2}\right)$

$= 2A\sin(kct).\cos kx.$

Thus $\frac{2\pi}{\lambda} = k$, ∴ $\lambda = \frac{2\pi}{k}$

The distance between adjacent nodes

$= \frac{\lambda}{2} = \frac{\pi}{k}$

61. (d) Frequency of sound produced by siren, $f = 800$ Hz

Speed of observer, $u = 2$ m/s

Velocity of sound, $v = 320$ m/s

No. of beats heard per second = ?

No. of extra waves received by the observer per second $= \pm 4\lambda$

∴ No. of beats/ sec

$= \frac{2}{\lambda} - \left(-\frac{2}{\lambda}\right) = \frac{4}{\lambda}$

$= \frac{2\times 2}{\frac{320}{800}} = \frac{2\times 2\times 800}{320} = 10$ $\left(\because \lambda = \frac{V}{f}\right)$

62. (d) Path difference,

$\Delta x = S_2P - S_1P = \sqrt{(2\sqrt{10})^2 + 3^2} - \sqrt{4^2 + 3^2}$

$= 7 - 5 = 2$m

For constructive interference,

$\Delta x = n\lambda$, where n = 1, 2, 3, ...

$\Rightarrow f = \frac{nv}{\Delta x} = \frac{1\times 340}{2}, \frac{2\times 340}{2}, \frac{3\times 340}{2}, ...$

= 170 Hz, 340 Hz, 510 Hz .....

63. (b) A tuning fork produces 4 beats/sec with another tuning fork of frequency 288 cps. From this information we can conclude that the frequency of unknown fork is 288 + 4 cps or 288 – 4 cps i.e. 292 cps or 284 cps. When a little wax is placed on the unknown fork, it produces 2 beats/sec. When a little wax is placed on the unknown fork, its frequency decreases and simultaneously the beat frequency decreases confirming that the frequency of the unknown fork is 292 cps.

64. (d) The frequency of the piano string $= 512 \pm 4 = 516$ or 508. When the tension is increased, beat frequency decreases to 2, it means that frequency of the string is 508 as frequency of string increases with tension.

65. (b) Equation of progressive wave is given by $Y = A\sin 2\pi ft$

Given $Y_1 = 4\sin 500\pi t$ and $Y_2 = 2\sin 506\pi t$.
Comparing the given equations with equation of progressive wave, we get
$2f_1 = 500 \Rightarrow f_1 = 250$
$2f_2 = 506 \Rightarrow f_2 = 253$
Beats $= f_2 - f_1 = 253 - 250 = 3$ beats/sec
$= 3 \times 60 = 180$ beats/minute.

66. (b) $\lambda_1 = 50$ cm. $\lambda_2 = 51$ cm.

$$v \propto \sqrt{T} \Rightarrow \frac{v_1}{v_2} = \sqrt{\frac{T_2}{T_1}} = \sqrt{\frac{273+20}{273}}$$

$\Rightarrow v_2 = 319.23.$

$$\nu_1 = \frac{v_2}{\lambda_1} = \frac{319.23}{0.50} = 640 \text{ Hz.}$$

$$\nu_2 = \frac{v_2}{\lambda_2} = \frac{319.23}{51 \times 10^{-2}} = 625.94 = 626 \text{ Hz.}$$

No. of beats $= \nu_2 - \nu_1 = 14$ Hz

67. (c) $$f_2 = \frac{f_0 v}{v + v_0}$$

The wave which reaches wall f_1 is reflected.

$$f_1 = \frac{f_0 v}{v - v_0}$$

The reflected frequency is f_1 as the wall is at rest.

$$Beats = f_1 - f_2 = \frac{f_0 v}{v - v_0} - \frac{f_0 v}{v + v_0} = \frac{2 f_0 v v_0}{v^2 - v_0^2}$$

68. (d)

n^{th} harmonic

$\lambda = \ell$

$$\therefore f_1 = \frac{v}{\lambda} = \frac{v}{\ell} \quad \text{... (i)}$$

$$\lambda = \frac{4\ell}{n}$$

$$\therefore f_2 = \frac{v}{\lambda} = \frac{nv}{4\ell} \quad \text{... (ii)}$$

Here n is a odd number. From (i) and (ii)

$$f_2 = \frac{n}{4} f_1$$

For first resonance, $n = 5, f_2 = \frac{5}{4} f_1$

69. (d) $2\pi f_1 = 600\pi$
$f_1 = 300$ ... (i)
$2\pi f_2 = 608\pi$
$f_2 = 304$ ... (ii)
$|f_1 - f_2| = 4$ beats

$$\frac{I_{max}}{I_{m.n}} = \frac{(A_1 + A_2)^2}{(A_1 + A_2)^2} = \frac{(5+4)^2}{(5-4)^2} = \frac{81}{1}, \text{ where } A_1,$$

A_2 are amplitudes of given two sound wave.

70. (c) Frequency is not affected by blowing of wind so long as source and listener are stationary.

71. (c) Let f' be the frequency of sound heard by cliff.

$$\therefore f' = \frac{vf}{v - v_c} \quad \text{......(1)}$$

Now for the reflected wave cliff. acts as a source

$$\therefore 2f' = \frac{f'(v + v_c)}{v} \quad \text{......(2)}$$

$$2f = \frac{(v + v_c)f}{v - v_c} \Rightarrow 2v - 2v_c = v + v_c$$

or $\frac{v}{3} = v_c$

72. (c) Change in frequency has nothing to do with distance between source and listener.

73. (d) $$\nu' = \nu\left(\frac{v + v_D}{v - v_S}\right)$$

Here, $\nu = 600$ Hz, $v_D = 15$ m/s
$v_s = 20$ m/s, $v = 340$ m/s

$$\therefore \quad \nu' = 600\left(\frac{355}{320}\right) \approx 666 \text{ Hz}$$

74. (a) Here, $\nu' = \frac{9}{8}\nu$

Source and observer are moving in opposite direction, therefore, apparent frequency

$$\nu' = \nu \times \frac{(v+u)}{(v-u)} \Rightarrow \frac{9}{8}\nu = \nu \times \frac{340+u}{340-u}$$

$$\Rightarrow u = \frac{340}{17} = 20 \text{ m/sec.}$$

75. (a) $$\lambda = \frac{v}{n} = \frac{340}{170} = 2m, \quad n' = \frac{340}{340-17} \times 170$$

$n' = 178.9$ *Hz*

Now $$\lambda' = \frac{v}{n'} = \frac{340}{178.9} = 1.9$$

$\Rightarrow \lambda - \lambda' = 2 - 1.9 = 0.1$

76. (a)

S S
v_s v_s
Receding train
Approaching train

Frequency of sound heard by the man from approaching train.

$$n_a = n\left(\frac{v}{v-v_s}\right) = 240\left(\frac{320}{320-4}\right) = 243\,Hz$$

Frequency of sound heard by the man from receding train $n_r = n\left(\frac{v}{v+v_s}\right)$

$$= 240\left(\frac{320}{320+4}\right) = 237\ Hz$$

Hence, number of beats heard by man per sec

$= n_a - n_r = 243 - 237 = 6$

77. **(a)** Frequency heard by observer directly coming from source $= \frac{355-5}{355+5} \times 180 = 175$ Hz. $f_2 \rightarrow$ frequency heard by observer after reflection

$$= \left[\frac{355}{355-5} \times\right]\left[\frac{355-5}{355}\right] \times 180 = 180\ \text{Hz}$$

$f_2 - f_1 = 5$ Hz

78. **(b)** $176\left(\frac{v-v_0}{v-22}\right) = 165\frac{v+v_0}{v}$

Here $v = 330$ m/s, after simplifying, we get $v_0 = 22$ m/s.

79. **(c)** Apparent frequency

$$n' = n\frac{(u+v_w)}{(u+v_w - v_s\cos 60°)}$$

$$= \frac{510\,(330+20)}{330+20-20\cos 60°}$$

$$= 510 \times \frac{350}{340} = 525\ \text{Hz}$$

80. **(c)** The line of motion of engine is perpendicular to line of motion of the observer at that instant, so $f' = f = 200$ Hz.

81. **(b)** $f_1 = f_0\left(\frac{V_0}{V_0 - V}\right) f_2 = f_0\left(\frac{V_0}{V_0+V}\right)$

$$f_1 - f_2 = f_0 V_0\left(\frac{1}{V_0 - V} - \frac{1}{V_0+V}\right)$$

$$= f_0 V_0\left(\frac{V_0 + V - V_0 + V}{V_0^2 - V^2}\right)$$

$$= f_0 V_0 \times \frac{2V}{V_0^2} = f_0\frac{2V}{V_0}$$

Given $\frac{2Vf_0}{V_0} = 0.02 \times f_0 \Rightarrow V = 0.01\,V_0 = 3.4$ m/s

82. **(a)** Here, original frequency of sound, $f_0 = 100$ Hz

Speed of source $V_s = 19.4\cos 60° = 9.7$

From Doppler's formula

19.4
60°
19.4 cos 60° = 9.7

$$f = f_0\left(\frac{V - V_0}{V - V_s}\right)$$

$$f = 100\left(\frac{V-0}{V-(+9.7)}\right)$$

$$f = 100\frac{V}{V\left(1-\frac{9.7}{V}\right)} = \frac{100}{\left(1-\frac{9.7}{330}\right)} = 103\text{Hz}$$

Apparent frequency f = 103 Hz

83. **(c)** According to Doppler's effect

Apparent frequency

$$n' = n\left(\frac{v+v_0}{v+v_s}\right) = 1392\left(\frac{343+10}{343+5}\right)$$

$= 1412$ Hz

84. **(d)** $f_{apparent} = \left(\frac{u+u/5}{u}\right) f = \frac{6}{5}f = 1.2f$

Wavelength remains constant (unchanged) in this case.

85. **(b)** Let the sound observed by the parachutist at $t_0 = 12$s be produced at t_1s. Velocity of source at the instant of sound $= gt_1$ and velocity of observer at the instant of observing same sound $= gt_0$. Hence the relation between apparent frequency f' and original frequency f will be $f' = f\left(\frac{v+gt_0}{v-gt_1}\right)$.

Here $f = 800$ Hz, $g = 10$ m/s^2,

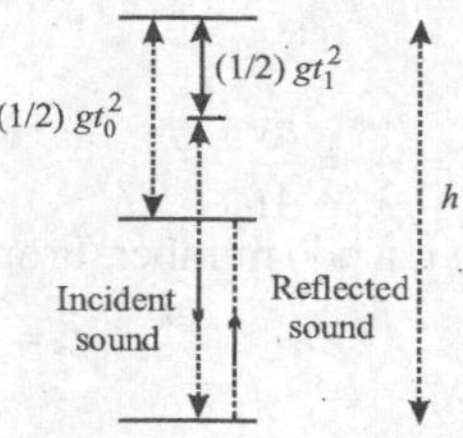

$v = 330$ m/s, $t_0 = 12$s and

$f' = 800 + 700 = 1500$ Hz

Putting these, we get, $t_1 = 9$s

Now the distance travelled by sound in $(t_0 - t_1)$ sec is

$$v\,(t_0 - t_1) = \left(h - \frac{1}{2}gt_0^2\right) + \left(h - \frac{1}{2}gt_1^2\right)$$

Putting the values, we get, $h = 1057.5$m.

15 Electric Charges and Fields

1. Two balls of same mass and carrying equal charge are hung from a fixed support of length l. At electrostatic equilibrium, assuming that angles made by each thread is small, the separation, x between the balls is proportional to :
(a) l (b) l^2 (c) $l^{2/3}$ (d) $l^{1/3}$

2. The force of repulsion between two electrons at a certain distance is F. The force between two protons separated by the same distance is (m_p = 1836 m_e)
(a) 2F (b) F (c) 1836F (d) $\frac{F}{1836}$

3. The force between two small charged spheres having charges of 1×10^{-7} C and 2×10^{-7} C placed 20 cm apart in air is
(a) 4.5×10^{-2} N (b) 4.5×10^{-3} N
(c) 5.4×10^{-2} N (d) 5.4×10^{-3} N

4. Two charge q and $-3q$ are placed fixed on x–axis separated by distance d. Where should a third charge $2q$ be placed such that it will not experience any force? **Tricky**

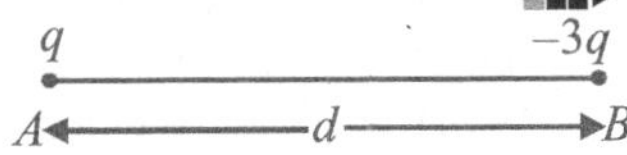

(a) $\frac{d-\sqrt{2}\,d}{2}$ (b) $\frac{d+\sqrt{3}\,d}{2}$

(c) $\frac{d+3d}{2}$ (d) $\frac{d-\sqrt{5}\,d}{2}$

5. Two insulated charged metalic sphere P and Q have their centres separated by a distance of 60 cm. The radii of P and Q are negligible compared to the distance of separation. The mutual force of electrostatic repulsion if the charge on each is 3.2×10^{-7} C is
(a) $5.2 \times 10^{-4}\,N$ (b) $2.5 \times 10^{-3}\,N$
(c) $1.5 \times 10^{-3} N$ (d) $3.5 \times 10^{-4}\,N$

6. If a charge q is placed at the centre of the line joining two equal charges Q such that the system is in equilibrium then the value of q is
(a) Q/2 (b) –Q/2 (c) Q/4 (d) –Q/4

7. Two positive ions, each carrying a charge q, are separated by a distance d. If F is the force of repulsion between the ions, the number of electrons missing from each ion will be (e being the charge of an electron)

(a) $\frac{4\pi\varepsilon_0 Fd^2}{e^2}$ (b) $\sqrt{\frac{4\pi\varepsilon_0 Fe^2}{d^2}}$

(c) $\sqrt{\frac{4\pi\varepsilon_0 Fd^2}{e^2}}$ (d) $\frac{4\pi\varepsilon_0 Fd^2}{q^2}$

8. A solid conducting sphere of radius a has a net positive charge 2Q. A conducting spherical shell of inner radius b and outer radius c is concentric with the solid sphere and has a net charge – Q. The surface charge density on the inner and outer surfaces of the spherical shell will be **Tricky**

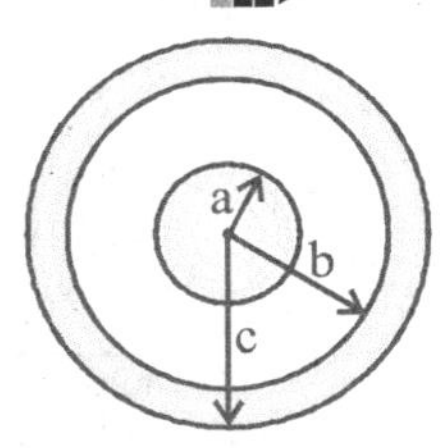

(a) $-\frac{2Q}{4\pi b^2}, \frac{Q}{4\pi c^2}$

(b) $-\frac{Q}{4\pi b^2}, \frac{Q}{4\pi c^2}$

(c) $0, \frac{Q}{4\pi c^2}$

(d) None of the above

9. Three charges $+q$, $+2q$ and $+4q$ are connected by strings as shown in the figure. What is ratio of tensions in the strings AB and BC ?

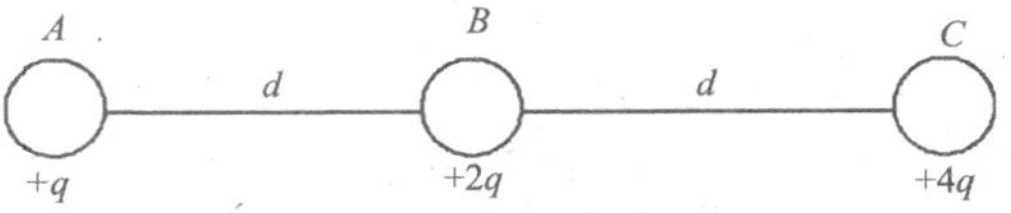

(a) 1 : 2 (b) 1 : 3 (c) 2 : 1 (d) 3 : 1

10. 1 C charge is equivalent to charge on how much number of protons?
(a) 6×10^{18} (b) 7×10^{19}
(c) 8×10^{20} (d) 9×10^{21}

11. The figure shows a charge $+q$ at point P held in equilibrium in air with the help of four $+q$ charges situated at the vertices of a square. The net electrostatic force on q is given by

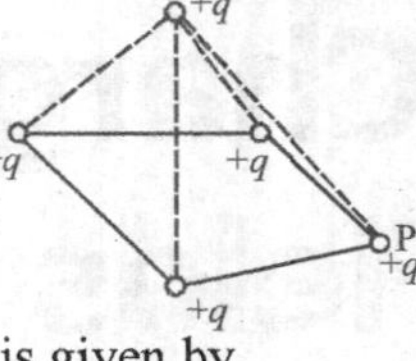

(a) Newton's
(b) Coulomb's law
(c) Principle of superposition
(d) Net electric flux out the position of $+q$.

12. The metal knob of a gold leaf electroscope is touched with a positively charged rod. When it is taken away the leaves stay separated. Now the metal knob is touched by negatively charged rod. The separation between the leaves

Tricky

(a) increases
(b) decreases
(c) remains same
(d) first increases then decreases.

13. Which of the following graphs shows the correct variation of force when the distance r between two charges varies ?

Tricky

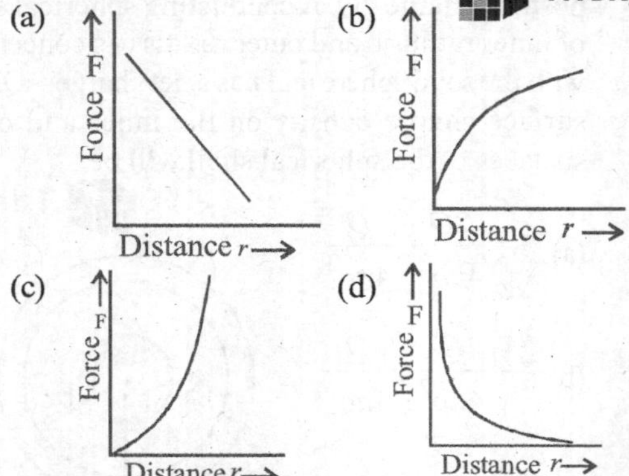

14. The electric charge required to expand a soap bubble to twice its dimension is

(a) $8\pi\sqrt{\epsilon_0\, r^3(7Pr+12T)}$
(b) $8\pi\sqrt{\epsilon_0\, r^2(7Pr+12T)}$
(c) $8\pi\sqrt{\epsilon_0\, r^3(6Pr+12T)}$
(d) $8\pi\sqrt{\epsilon_0\, r^3(Pr+T)}$

15. A large nonconducting sheet M is given a uniform charge density. Two uncharged small metal rods A and B are placed near the sheet as shown in figure. Then

M A B

(a) M attracts A
(b) M attracts B
(c) A attracts B
(d) All of these

16. Three charges $-q_1$, $+q_2$ and $-q_3$ are place as shown in the figure. The x-component of the force on $-q_1$ is proportional to

Critical Thinking

(a) $\frac{q_2}{b^2}-\frac{q_3}{a^2}\cos\theta$

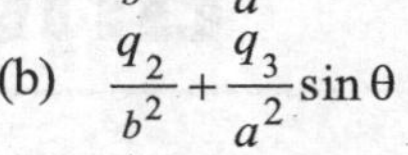

(b) $\frac{q_2}{b^2}+\frac{q_3}{a^2}\sin\theta$

(c) $\frac{q_2}{b^2}+\frac{q_3}{a^2}\cos\theta$

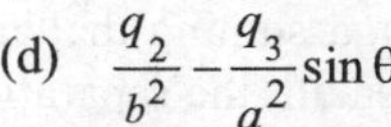

(d) $\frac{q_2}{b^2}-\frac{q_3}{a^2}\sin\theta$

Y, $-q_3$, a, θ, b, $-q_1$, $+q_2$, X

17. A charged ball B hangs from a silk thread S, which makes an angle θ with a large charged conducting sheet P, as shown in the figure. The surface charge density σ of the sheet is proportional to

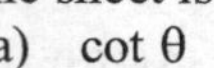

(a) cot θ
(b) cos θ
(c) tan θ
(d) sin θ

18. Two particles A and B having equal charges are placed at a distance d apart. A third charged particle placed on the perpendicular bisection of AB at distance x. The third particle experiences maximum force when

(a) $x=\frac{d}{\sqrt{2}}$
(b) $x=\frac{d}{2}$
(c) $x=\frac{d}{2\sqrt{2}}$
(d) $x=\frac{d}{3\sqrt{2}}$

19. Among two discs A and B, first have radius 10 cm and charge 10^{-6} μC and second have radius 30 cm and charge 10^{-5}C. When they are touched, charge on both q_A and q_B respectively will, be

(a) $q_A = 2.75\,\mu C$, $q_B = 3.15\,\mu C$
(b) $q_A = 1.09\,\mu C$, $q_B = 1.53\,\mu C$
(c) $q_A = q_B = 5.5\,\mu C$
(d) None of these

20. Two pith balls carrying equal charges are suspended from a common point by strings of equal length. The equilibrium separation between them is r. Now the strings are rigidly clamped at half the height. The equilibrium separation between the balls now become

Critical Thinking

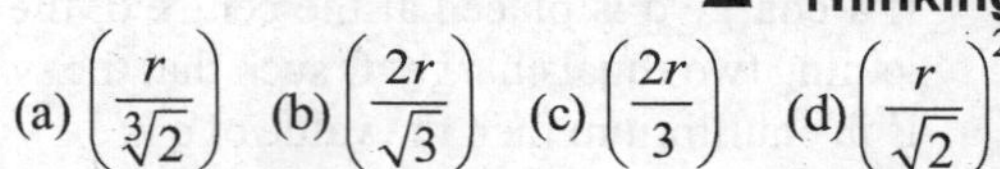

(a) $\left(\frac{r}{\sqrt[3]{2}}\right)$ (b) $\left(\frac{2r}{\sqrt{3}}\right)$ (c) $\left(\frac{2r}{3}\right)$ (d) $\left(\frac{r}{\sqrt{2}}\right)^2$

21. Two equal point charges each of 3μC are separated by a certain distance in metres. If they are located at $(\hat{i}+\hat{j}+\hat{k})$ and $(2\hat{i}+3\hat{j}+\hat{k})$, then the electrostatic force between them is
(a) 9×10^{3} N (b) 16×10^{-3} N
(c) 10^{-3} N (d) 9×10^{-2} N

22. In fig., two equal positive point charges $q_1=q_2=2.0$ μC interact with a third point charge $Q=4.0$ μC. The magnitude, as well as direction, of the net force on Q is
(a) 0.23 N in the +x-direction
(b) 0.46 N in the +x-direction
(c) 0.23 N in the –x-direction
(d) 0.46 N in the –x-direction

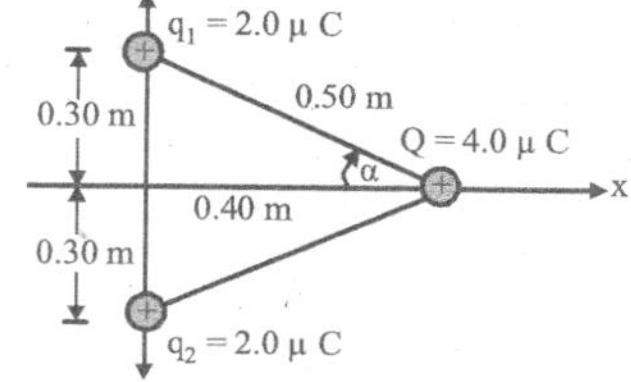

23. Force between two identical charges placed at a distance of r in vacuum is F. Now a slab of dielectric of dielectric contrant 4 is inserted between these two charges. If the thickness of the slab is r/2, then the force between the charges will become
(a) F (b) $\frac{3}{5}$F (c) $\frac{4}{9}$F (d) $\frac{F}{2}$

24. A total charge Q is broken in two parts Q_1 and Q_2 and they are placed at a distance R from each other. The maximum force of repulsion between them will occur. when
(a) $Q_2=\frac{Q}{R},\ Q_1=Q-\frac{Q}{R}$
(b) $Q_2=\frac{Q}{4},\ Q_1=Q-\frac{2Q}{3}$
(c) $Q_2=\frac{Q}{4},\ Q_1=\frac{3Q}{4}$
(d) $Q_1=\frac{Q}{2},\ Q_2=\frac{Q}{2}$

25. Two particle of equal mass m and charge q are placed at a distance of 16 cm. They do not experience any force. The value of $\frac{q}{m}$ is
(a) 1 (b) $\sqrt{\frac{\pi\varepsilon_0}{G}}$ (c) $\sqrt{\frac{G}{4\pi\varepsilon_0}}$ (d) $\sqrt{4\pi\varepsilon_0 G}$

26 A uniformly charged conducting sphere of 4.4 m diameter has a surface charge density of 60 μC m^{-2}. The charge on the sphere is
(a) $7.3\times10^{-3}C$ (b) 3.7×10^{-6} C
(c) 7.3×10^{-6} C (d) 3.7×10^{-3} C

27. Three identical spheres, each having a charge q and radius R, are kept in such a way that each touches the other two. The magnitude of the electric force on any sphere due to the other two is **Critical Thinking**
(a) $\frac{1}{4\pi\varepsilon_0}\left(\frac{q}{R}\right)^2$ (b) $\frac{\sqrt{3}}{4\pi\varepsilon_0}\left(\frac{q}{R}\right)^2$
(c) $\frac{\sqrt{3}}{16\pi\varepsilon_0}\left(\frac{q}{R}\right)^2$ (d) $\frac{\sqrt{5}}{16\pi\varepsilon_0}\left(\frac{q}{R}\right)^2$

28. Two identical blocks are kept on a frictionless horizontal table connected by a spring of stiffness k and of original length ℓ_0. A total charge Q is distributed on the block such that maximum elongation of spring at equilibrium is equal to x. Value of Q is **Critical Thinking**
(a) $2\ell_0\sqrt{4\pi\varepsilon_0 k(\ell_0+x)}$
(b) $2x\sqrt{4\pi\varepsilon_0 k(\ell_0+x)}$
(c) $2(\ell_0+x)\sqrt{4\pi\varepsilon_0 kx}$
(d) $(\ell_0+x)\sqrt{4\pi\varepsilon_0 kx}$

29. Two small balls having the same mass and charge and located on the same vertical at heights h_1 and h_2 are thrown in the same direction along the horizontal at the same velocity v. The first ball touches the ground at a distance ℓ from the initial vertical. At what height H_2 will the second ball be at this instant ? The air drag and the effect of the charges induced on the ground should be neglected.
(a) $h_1+h_2-g\left(\frac{\ell}{v}\right)^2$ (b) $h_1-h_2-g\left(\frac{\ell}{v}\right)^2$
(c) $h_1+h_2-g\left(\frac{\ell}{v}\right)^{1/2}$ (d) $\frac{h_1+h_2}{2}-g\left(\frac{\ell}{v}\right)^2$

30. Two identical beads each have a mass m and charge q. When placed in a hemispherical bowl of radius R with frictionless, nonconductive walls, the beads move, and at equilibrium the distance between them is R (Fig.). Determine the charge on each bead. **Toughnut**

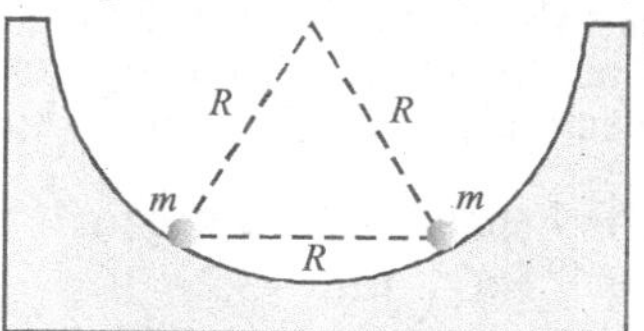

(a) $R\left(\frac{mg}{k_e\sqrt{3}}\right)^{1/2}$ (b) $R\left(\frac{mg}{k_e\sqrt{2}}\right)^{1/2}$
(c) $R\left(\frac{mg}{k_e 2\sqrt{3}}\right)^{1/2}$ (d) $R\left(\frac{2mg}{k_e\sqrt{3}}\right)^{1/2}$

31. If E_a be the electric field strength of a short dipole at a point on its axial line and E_e that on the equatorial line at the same distance, then
(a) $E_e=2E_a$ (b) $E_a=2E_e$
(c) $E_a=E_e$ (d) None of these

32. When an electric dipole $\vec{P}$ is placed in a uniform electric field $\vec{E}$ then at what angle between $\vec{P}$ and $\vec{E}$ the value of torque will be maximum?
(a) 90° (b) 0° (c) 180° (d) 45°

33. Which of the following graphs shows the correct variation in magnitude of torque on an electric dipole rotated in a uniform electric field from stable equillibrium to unstable equillibrium? **Tricky**

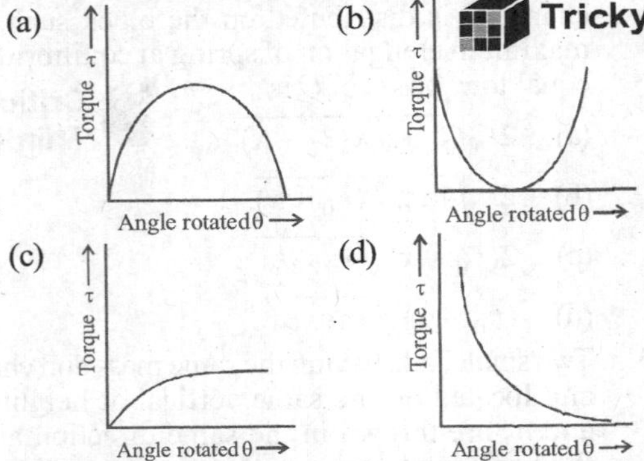

34. A rod of length 2.4 *m* and radius 4.6 *mm* carries a negative charge of 4.2×10^{-7} C spread uniformly over it surface. The electric field near the mid–point of the rod, at a point on its surface is
(a) -8.6×10^5 N C^{-1} (b) 8.6×10^4 N C^{-1}
(c) -6.7×10^5 N C^{-1} (d) 6.7×10^4 N C^{-1}

35. If electric field in a region is radially outward with magnitude E = Ar, the charge contained in a sphere of radius r centred at the origin is
(a) $\frac{1}{4\pi\varepsilon_o}Ar^3$ (b) $4\pi\varepsilon_o Ar^3$
(c) $\frac{1}{4\pi\varepsilon_o}\frac{A}{r^3}$ (d) $\frac{4\pi\varepsilon_o A}{r^3}$

36. The electric field intensity just sufficient to balance the earth's gravitational attraction on an electron will be: (given mass and charge of an electron respectively are 9.1×10^{-31} kg and 1.6×10^{-19} C.)
(a) -5.6×10^{-11} N/C (b) -4.8×10^{-15} N/C
(c) -1.6×10^{-19} N/C (d) -3.2×10^{-19} N/C

37. The insulation property of air breaks down when the electric field is 3×10^6 Vm^{-1}. The maximum charge that can be given to a sphere of diameter 5 m is approximately
(a) 2×10^{-2} C (b) 2×10^{-3} C
(c) 2×10^{-4} C (d) 2×10^{-5} C

38. A hollow insulated conduction sphere is given a positive charge of 10 μC. What will be the electric field at the centre of the sphere if its radius is 2 m?
(a) Zero (b) 5 μCm^{-2}
(c) 20 μCm^{-2} (d) 8 μCm^{-2}

39. The number of electric lines of force that radiate outwards from one coulomb of charge in vacuum is
(a) 1.13×10^{11} (b) 1.13×10^{10}
(c) 0.61×10^{11} (d) 0.61×10^9

40. The electric intensity due to a dipole of length 10 cm and having a charge of 500 μC, at a point on the axis at a distance 20 cm from one of the charges in air, is
(a) 6.25×10^7 N/C (b) 9.28×10^7 N/C
(c) 13.1×10^{11} N/C (d) 20.5×10^7 N/C

41. ABC is an equilateral triangle. Charges +q are placed at each corner as shown in fig. The electric intensity at centre O will be **Tricky**
(a) $\frac{1}{4\pi\epsilon_o}\frac{q}{r}$
(b) $\frac{1}{4\pi\epsilon_o}\frac{q}{r^2}$
(c) $\frac{1}{4\pi\epsilon_o}\frac{3q}{r^2}$
(d) zero

+q A, r, O, r, r, +q B, +q C

42. Intensity of an electric field (*E*) depends on distance *r*, due to a dipole, is related as
(a) $E \propto \frac{1}{r}$ (b) $E \propto \frac{1}{r^2}$ (c) $E \propto \frac{1}{r^3}$ (d) $E \propto \frac{1}{r^4}$

43. If the dipole of moment 2.57×10^{-17} cm is placed into an electric field of magnitude 3.0×10^4 N/C such that the fields lines are aligned at 30° with the line joining P to the dipole, what torque acts on the dipole?
(a) 7.7×10^{-13} Nm (b) 3.855×10^{-13} Nm
(c) 3.855×10^{-15} Nm (d) 7.7×10^{-15} Nm

44. An electric dipole is placed at an angle of 30° with an electric field of intensity 2×10^5 NC^{-1}, It experiences a torque of 4 Nm. Calculate the charge on the dipole if the dipole length is 2 cm.
(a) 8mC (b) 4mC (c) 8 μC (d) 2mC

45. On decreasing the distance between the two charges of a dipole which is perpendicular to electric field and decreasing the angle between the dipole and electric field, the torque on the dipole
(a) increases (b) decreases
(c) remains same (d) cannot be predicted

46. Let $\rho(r) = \frac{Q}{\pi R^4} r$ be the charge density distribution for a solid sphere of radius *R* and total charge *Q*. For a point '*P*' inside the sphere at distance r_1 from the centre of the sphere, the magnitude of electric field is :
(a) $\frac{Q}{4\pi\epsilon_0 r_1^2}$ (b) $\frac{Qr_1^2}{4\pi\epsilon_0 R^4}$ (c) $\frac{Qr_1^2}{3\pi\epsilon_0 R^4}$ (d) 0

47. In the figure the electric lines on the right have twice the separation of those on the left. If a charge particle takes time t to move a distance x in left region, then it will take time to travel the same distance in the right side region is :

(a) $\frac{t}{2}$ (b) t

(c) $\sqrt{2}\,t$ (d) $2t$

48. n identical point charges are kept symmetrically on the periphery of the circle $x^2 + y^2 = R^2$ in xy plane. The resultant electric field at (0, 0, R) is E_1 and at (0, 0, 2R) is E_2. The ratio of $\frac{E_1}{E_2}$ is

(a) $\frac{5\sqrt{5}}{4\sqrt{2}}$ (b) $\frac{5}{2}$ (c) $\frac{5}{4}$ (d) $\frac{5\sqrt{5}}{2\sqrt{2}}$

49. Two point dipoles of dipole moment $\vec{p}_1$ and $\vec{p}_2$ are at a distance x from each other and $\vec{p}_1 \parallel \vec{p}_2$. The force between the dipoles is: **Tricky**

(a) $\frac{1}{4\pi\varepsilon_0}\frac{4p_1p_2}{x^4}$ (b) $\frac{1}{4\pi\varepsilon_0}\frac{3p_1p_2}{x^4}$

(c) $\frac{1}{4\pi\varepsilon_0}\frac{6p_1p_2}{x^4}$ (d) $\frac{1}{4\pi\varepsilon_0}\frac{8p_1p_2}{x^4}$

50. Two identical electric dipoles are arranged on x-axis as shown in figure. Electric field at the origin will be

(a) Zero (b) $\frac{kp\sqrt{2}}{r^3}\hat{j}$

(c) $\frac{-kp\sqrt{2}}{r^3}\hat{j}$

(d) $\frac{-kp}{r^3}\hat{i} - \frac{kp}{r^3}\hat{j}$

51. A liquied drop having 6 excess electrons is kept stationary under a uniform electric field of 25.5 KVm^{-1}. The density of liquid is 1.26×10^3 kg m^{-3}. The radius of the drop is (neglect buoyany)

(a) 4.3×10^{-7} m (b) 7.3×10^{-7} m

(c) 0.078×10^{-7} m (d) 3.4×10^{-7} m

52. An electric dipole, consisting of two opposite charges of 2×10^{-6} C each separated by a distance 3 cm is placed in an electric field of 2×10^5 N/C. Torque acting on the dipole is

(a) $12 \times 10^{-1} N-m$ (b) $12 \times 10^{-2} N-m$

(c) $12 \times 10^{-3} N-m$ (d) $12 \times 10^{-4} N-m$

53. Point charge q moves from point P to point S along the path PQRS (as shown in fig.) in a uniform electric field E pointing co-parallel to the positive direction of X-axis.

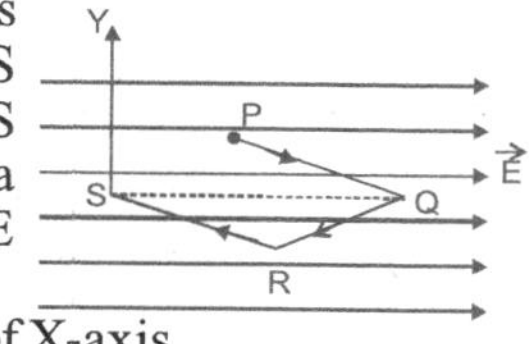

The coordinates of the points P, Q, R and S are (a, b, 0), (2 a, 0, 0), (a, –b, 0) and (0, 0, 0) respectively.

The workdone by the field in the above case is given by the expression **Tricky**

(a) q E A (b) – q E A

(c) q E A $\sqrt{2}$ (d) $qE\sqrt{[(2a)^2 + b^2]}$

54. A pendulum bob of mass m carrying a charge q is at rest with its string making an angle θ with the vertical in a uniform horizontal electric field E. The tension in the string is

(a) $\frac{mg}{\sin\theta}$ and $\frac{qE}{\cos\theta}$ (b) $\frac{mg}{\cos\theta}$ and $\frac{qE}{\sin\theta}$

(c) $\frac{qE}{mg}$ (d) $\frac{mg}{qE}$

55. Two very long line charges of uniform charge density $+\lambda$ and $-\lambda$ are placed along same line with the separation between the nearest ends being $2a$, as shown in figure. The electric field intensity at point O is **Critical Thinking**

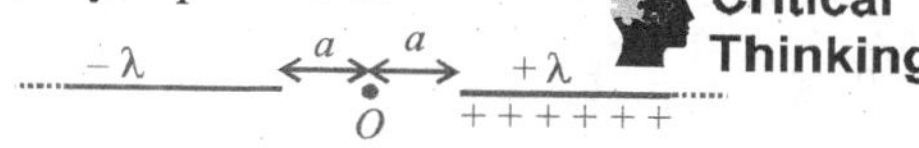

(a) $\frac{\lambda}{2\pi\varepsilon_0 a}$ (b) 0

(c) $\frac{\lambda}{\pi\varepsilon_0 a}$ (d) $\frac{\lambda}{4\pi\varepsilon_0 a}$

56. Figure shows an electric quadrupole, with quadrupole moment ($Q = 2q\ell^2$). The electric field at a distance from its centre at the axis of the quadrupole is given by

(a) $\left(\frac{1}{4\pi\epsilon_0}\right)\frac{Q}{r^4}$ (b) $\left(\frac{1}{4\pi\epsilon_0}\right)\frac{2Q}{r^4}$

(c) $\left(\frac{1}{4\pi\epsilon_0}\right)\frac{3Q}{r^4}$ (d) None of these

57. A spherical portion has been removed from a solid sphere having a charge distributed uniformly in its volume as shown in the figure. The electric field inside the emptied space is **Tricky**

(a) zero everywhere

(b) non-zero and uniform

(c) non-uniform

(d) zero only at its centre

58. Find the electric field vector at P (a, a, a) due to three infinitely long lines of charges along the x, y and z-axes, respectively. The charge density, i.e., charge per unit length of each wire is λ.

(a) $\frac{\lambda}{3\pi\varepsilon_0 a}(\hat{i}+\hat{j}+\hat{k})$

(b) $\frac{\lambda}{2\pi\varepsilon_0 a}(\hat{i}+\hat{j}+\hat{k})$

(c) $\frac{\lambda}{2\sqrt{2}\pi\varepsilon_0 a}(\hat{i}+\hat{j}+\hat{k})$

(d) $\frac{\sqrt{2}\lambda}{\pi\varepsilon_0 a}(\hat{i}+\hat{j}+\hat{k})$

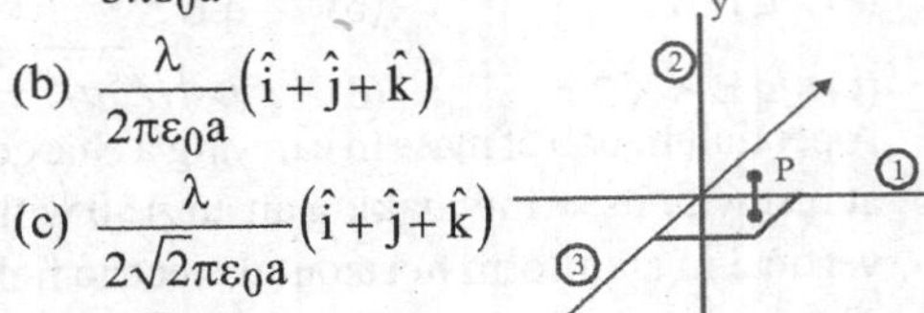

59. A particle of charge q and mass m moves rectilinearly under the action of electric field $E = A - Bx$, where A and B are positive constants and x is distance from the point where particle was initially at rest then the distance traveled by the particle before coming to rest and acceleration of particle at that moment are respectively : **Critical Thinking**

(a) $\frac{2A}{B}, 0$ (b) $0, -\frac{qA}{m}$

(c) $\frac{2A}{B}, -\frac{qA}{m}$ (d) $\frac{-2A}{B}, -\frac{qA}{m}$

60. A ring of charge with radius 0.5 m has 0.002 π m gap. If the ring carries a charge of +1 C, the electric field at the centre is **Tricky**

(a) 7.5×10^7 NC^{-1}

(b) 7.2×10^7 NC^{-1}

(c) 6.2×10^7 NC^{-1}

(d) 6.5×10^7 NC^{-1}

61. A point charge 50 μC is located in the x-y plane at a point whose position vector is $\vec{r} = (2\hat{i} + 3\hat{j})\, m$. Then electric field at the point whose position vector is $\vec{r} = (8\hat{i} - 5\hat{j})\, m$ (in vector form) will be

(a) $90(-3\hat{i} + 4\hat{j})$ V/m

(b) $900(3\hat{i} - 4\hat{j})$ V/m

(c) $90(3\hat{i} - 4\hat{j})$ V/m

(d) $900(-3\hat{i} + 4\hat{j})$ V/m

62. A thin glassrod is bent into a semicircle of radius r. A charge +Q is uniformly distributed along the upper half, and a charge –Q is uniformly distributed along the lower half, as shown in fig. The electric field E at P, the centre of the semicircle, is **Toughnut**

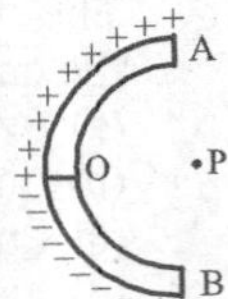

(a) $\frac{Q}{\pi^2\varepsilon_0 r^2}$ (b) $\frac{2Q}{\pi^2\varepsilon_0 r^2}$ (c) $\frac{4Q}{\pi^2\varepsilon_0 r^2}$ (d) $\frac{Q}{4\pi^2\varepsilon_0 r^2}$

63. A charge is situated at a certain distance from an electric dipole in the end-on position experiences a force F. If the distance of the charge is doubled, the force acting on the charge will be

(a) F/4 (b) F/8 (c) 2F (d) F/2

64. The thickness of a flat sheet of metal foil is d, and its area is S. A charge q is located at a distance ℓ from the centre of the sheet such that $d << \sqrt{S} << \ell$. Determine the force F with which the sheet is attracted to the charge q, assuming that the straight line connecting the charge to the centre of the sheet is perpendicular to the surface of the sheet. (approximately) **Critical Thinking**

(a) $\frac{q^2 Sd}{8\pi^2\varepsilon_0\ell^5}$ (b) $\frac{q^2 Sd}{4\pi^2\varepsilon_0\ell^5}$

(c) $\frac{q^2 Sd}{6\pi^2\varepsilon_0\ell^5}$ (d) $\frac{2q^2 Sd}{3\pi^2\varepsilon_0\ell^5}$

65. A thin conducting ring of radius R is given a charge + Q. The electric field at the centre O of the ring due to the charge on the part AKB of the ring is E. The electric field at the centre due to the charge on the part ACDB of the ring is

(a) E along KO (b) E along OK

(c) E along KO (d) 3 E along OK

66. A circular wire loop of radius x carries a total charge q distributed uniformly over its length. A small length dl of the wire is cut off. Find the value of m if $\frac{qdl}{m\pi^2\varepsilon_0 a^3}$ be the electric field at the centre due to remaining wire.

(a) 5 (b) 6 (c) 3 (d) 8

67. Three identical positive charges Q are arranged at the vertices of an equilateral triangle. The side of the triangle is a. Find the intensity of the field at the vertex of a regular tetrahedron of which the triangle is the base.

(a) $\sqrt{6}\frac{KQ}{a^2}$ (b) $\sqrt{2}\frac{KQ}{a^2}$

(c) $\sqrt{3}\frac{KQ}{a^2}$ (d) None of these

68. The electric field intensity at the centre of a uniformly charged hemispherical shell is E_0. Now two portions of the hemisphere are cut from either side, and the remaining portion is shown in Fig. If $\alpha = \beta = \pi/3$, then the electric field intensity at the centre due to the remaining portion is **Tricky**

(a) $E_0/3$ (b) $E_0/6$ (c) $E_0/2$
(d) information insufficient

69. Consider a uniform spherical charge distribution of radius R_1 centred at the origin O. In this distribution, a spherical cavity of radius R_2, centred at P with distance $OP = a = R_1 - R_2$ (see figure) is made.

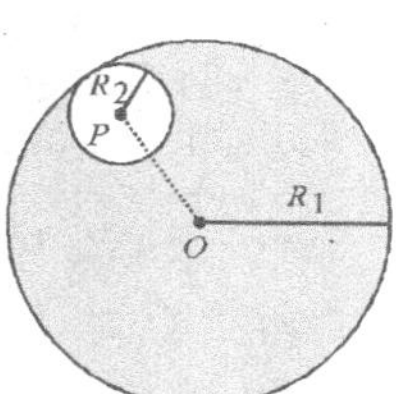

If the electric field inside the cavity at position $\vec{r}$ is $\vec{E}(\vec{r})$, then the correct statement is

(a) $\vec{E}$ is uniform, its magnitude is independent of R_2 but its direction depends on $\vec{r}$

(b) $\vec{E}$ is uniform, its magnitude depends on R_2 and its direction depends on $\vec{r}$

(c) $\vec{E}$ is uniform, its magnitude is independent of a but its direction depends on $\vec{a}$

(d) $\vec{E}$ is uniform and both its magnitude and direction depend on $\vec{a}$

70. A thin spherical shell of radius R has charge Q spread uniformly over its surface. Which of the following graphs most closely represents the electric field $E(r)$ produced by the shell in the range $0 \le r < \infty$, where r is the distance from the centre of the shell? **Tricky**

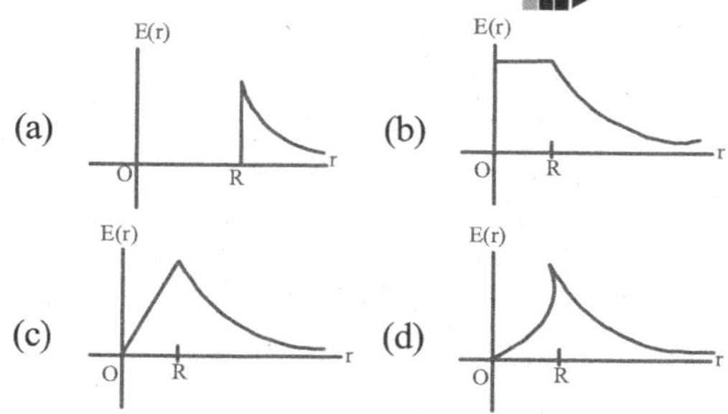

71. Find the force experienced by a semicircular rod having a charge q as shown in Fig. Radius of the wire is R, and the line of charge with linear charge density λ passes through its centre and is perpendicular to the plane of wire. **Toughnut**

(a) $\dfrac{\lambda q}{2\pi^2 \varepsilon_0 R}$ (b) $\dfrac{\lambda q}{\pi^2 \varepsilon_0 R}$

(c) $\dfrac{\lambda q}{4\pi^2 \varepsilon_0 R}$ (d) $\dfrac{\lambda q}{4\pi \varepsilon_0 R}$

72. A particle of charge $-q$ and mass m moves in a circle of radius r around an infinitely long line charge of linear charge density $+\lambda$. Then time period will be **Tricky**

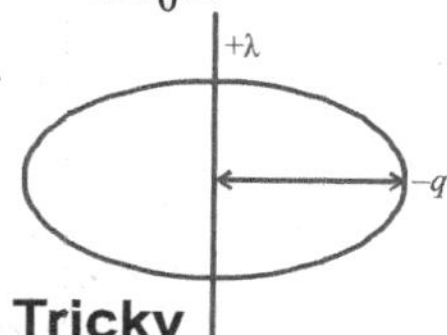

(a) $T = 2\pi r\sqrt{\dfrac{m}{2k\lambda q}}$ (b) $T^2 = \dfrac{4\pi^2 m}{2k\lambda q} r^3$

(c) $T = \dfrac{1}{2\pi r}\sqrt{\dfrac{2k\lambda q}{m}}$ (d) $T = \dfrac{1}{2\pi r}\sqrt{\dfrac{m}{2k\lambda q}}$

73. An electric dipole of moment $\vec{P}$ is placed in a uniform electric field $\vec{E}$ such that $\vec{P}$ points along $\vec{E}$. If the dipole is slightly rotated about an axis perpendicular to the plane containing $\vec{E}$ and $\vec{P}$ and passing through the centre of the dipole, the dipole executes simple harmonic motion. Consider I to be the moment of inertia of the dipole about the axis of rotation. What is the time period of such oscillation?

(a) $\sqrt{(pE/I)}$ (b) $2\pi\sqrt{(I/pE)}$

(c) $2\pi\sqrt{(I/2pE)}$ (d) None of these

74. Two point dipoles $p\hat{k}$ and $\dfrac{p}{2}\hat{k}$ are located at (0, 0, 0) and (1m, 0, 2m) respectively. The resultant electric field due to the two dipoles at the point (1m, 0, 0) is

(a) $\dfrac{9p}{32\pi \epsilon_0}\hat{k}$ (b) $\dfrac{-7p}{32\pi \epsilon_0}\hat{k}$

(c) $\dfrac{7p}{32\pi \epsilon_0}\hat{k}$ (d) $\dfrac{6p}{\epsilon_0}\hat{k}$

75. The dipole moment of the given charge distribution is

(a) $-\dfrac{4Rq}{\pi}\hat{i}$ (b) $\dfrac{4Rq}{\pi}\hat{i}$

(c) $-\dfrac{2Rq}{\pi}\hat{i}$ (d) $\dfrac{2Rq}{\pi}\hat{i}$

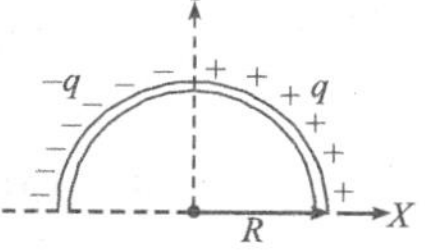

76. The total electric flux emanating from a closed surface enclosing an α-particle is (e-electronic charge)

(a) $\frac{2e}{\varepsilon_0}$ (b) $\frac{e}{\varepsilon_0}$ (c) $e\varepsilon_0$ (d) $\frac{\varepsilon_0 e}{4}$

77. A point charge $+Q$ is positioned at the center of the base of a square pyramid as shown. The flux through one of the four identical upper faces of the pyramid is **Tricky**

(a) $\frac{Q}{16\varepsilon_0}$

(b) $\frac{Q}{4\varepsilon_0}$

(c) $\frac{Q}{8\varepsilon_0}$

(d) 0

78. Electric flux over a surface in an electric field may be

(a) positive (b) negative
(c) zero (d) All of these

79. If the electric flux entering and leaving an enclosed surface respectively is ϕ_1 and ϕ_2, the electric charge inside the surface will be

(a) $(\phi_2+\phi_2)\times\varepsilon_o$ (b) $(\phi_2-\phi_2)\times\varepsilon_o$
(c) $(\phi_1+\phi_2)\times\varepsilon_o$ (d) $(\phi_2-\phi_1)\times\varepsilon_o$

80. Electric charges are distributed in a small volume. The flux of the electric field through a spherical surface of radius 1 m surrounding the total charge is 100 V-m. The flux over the concentric sphere of radius 2 m will be

(a) 25 V-m (b) 50 V-m
(c) 100 V-m (d) 200 V-m

81. A loop of diameter d is rotated in a uniform electric field until the position of maximum electric flux is found. The flux in this position is measured to be ϕ. What is the electric field strength ?

(a) $\frac{4\phi}{\pi d^2}$ (b) $\frac{2\phi}{\pi d^2}$ (c) $\frac{\phi}{\pi d^2}$ (d) $\frac{\pi\phi d^2}{4}$

82. Consider an electric field $\vec{E}=E_0\hat{x}$ where E_0 is a constant. The flux through the shaded area (as shown in the figure) due to this field is

(a) $2E_0a^2$ (b) $\sqrt{2}E_0a^2$

(c) E_0a^2 (d) $\frac{E_0a^2}{\sqrt{2}}$

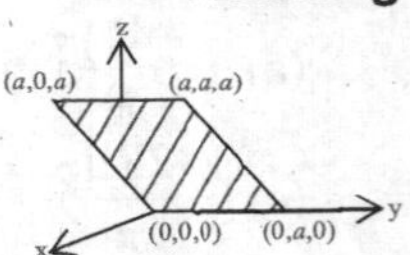

83. The inward and outward electric flux for a closed surface in units of N-m²/C are respectively 8×10^3 and 4×10^3. Then the total charge inside the surface is [where ε_0 = permittivity constant]

(a) $4\times10^3 C$ (b) $-4\times10^3 C$

(c) $\frac{(-4\times10^3)}{\varepsilon}C$ (d) $-4\times10^3\varepsilon_0 C$

84. The electric field in a region of space is given by, $\vec{E}=E_o\hat{i}+2E_o\hat{j}$ where E_o = 100 N/C. The flux of the field through a circular surface of radius 0.02 m parallel to the Y-Z plane is nearly:

(a) 0.125 Nm²/C (b) 0.02 Nm²/C
(c) 0.005 Nm²/C (d) 3.14 Nm²/C

85. A solid sphere of radius R has a charge Q distributed in its volume with a charge density $\rho=kr^a$, where k and a are constants and r is the distance from its centre. If the electric field at $r=\frac{R}{2}$ is $\frac{1}{8}$ times that at $r=R$, the value of a is.

(a) 3 (b) 5
(c) 2 (d) both (a) and (b)

86. A disc of radius $a/4$ having a uniformly distributed charge 6C is placed in the x-y plane with its centre at $(-a/2, 0, 0)$. A rod of length a carrying a uniformly distributed charge 8C is placed on the x-axis from $x=a/4$ to $x=5a/4$. Two point charges −7C and 3 C are placed at $(a/4, -a/4, 0)$ and $(-3a/4, 3a/4, 0)$, respectively. Consider a cubical surface formed by six surfaces $x=\pm a/2$, $y=\pm a/2$, $z=\pm a/2$. The electric flux through this cubical surface is

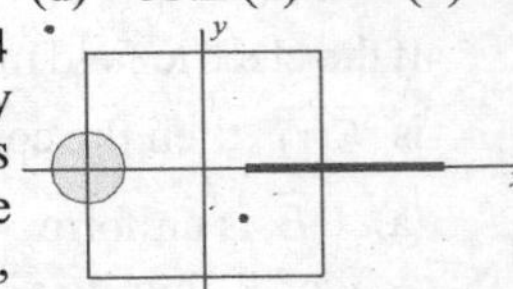

(a) $\frac{-2C}{\varepsilon_0}$ (b) $\frac{2C}{\varepsilon_0}$ (c) $\frac{10C}{\varepsilon_0}$ (d) $\frac{12C}{\varepsilon_0}$

87. A system consists of a uniform charged sphere of radius R and a surrounding medium filled by a charge with the volume density $\rho=\frac{\alpha}{r}$, where α is a positive constant and r is the distance from the centre of the charge. The charge of the sphere for which the electric field intensity E outside the sphere is independent of r is – **Toughnut**

(a) $\pi R^2\alpha$ (b) $4\pi R^2\alpha$
(c) $2\pi R^2\alpha$ (d) $3\pi R^2\alpha/4$

88. A charged particle q is placed at the centre O of cube of length L ($A\ B\ C\ D\ E\ F\ G\ H$). Another same charge q is placed at a distance L from O. Then the electric flux through $ABCD$ is

(a) $q/4\pi\epsilon_0 L$
(b) zero
(c) $q/2\pi\epsilon_0 L$
(d) $q/3\pi\epsilon_0 L$

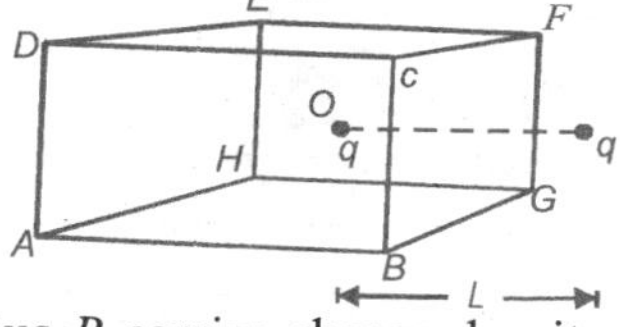

89. A sphere of radius R carries charge density ρ proportional to the square of the distance from the centre such that $\rho = CR^2$, where C is a positive constant. At a distance $R/2$ from the centre, the magnitude of the electric field is

(a) $\frac{CR^3}{20\epsilon_0}$ (b) $\frac{CR^3}{10\epsilon_0}$ (c) $\frac{CR^3}{5\epsilon_0}$ (d) $\frac{CR^3}{40\epsilon_0}$

90. Figure shows a uniformaly charged hemisphere of radius R. It has a volume charge density ρ. If the electric field at a point 2R, above its centre is E, then what is the electric field at the point 2R below its centre? Critical Thinking

(a) $\rho R/6\varepsilon_0 + E$ (b) $\rho R/12\varepsilon_0 - E$
(c) $-\rho R/6\varepsilon_0 + E$ (d) $\rho R/12\varepsilon_0 + E$

91. A uniformly charged and infinitely long line having a linear charge density λ is placed at a normal distance y from a point O. Consider an imaginary sphere of radius R with O as centre and $R > y$. Electric flux through the surface of the sphere is

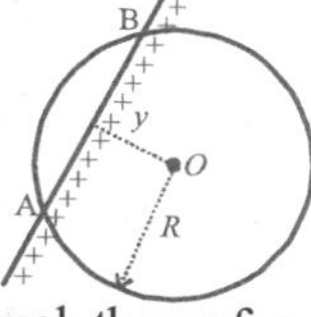

Tricky

(a) zero (b) $\frac{2\lambda R}{\varepsilon_0}$

(c) $\frac{2\lambda\sqrt{R^2 - y^2}}{\varepsilon_0}$ (d) $\frac{\lambda\sqrt{R^2 + y^2}}{\varepsilon_0}$

92. Flux passing through the shaded surface of a sphere when a point charge q is placed at the center is (radius of the sphere is R)

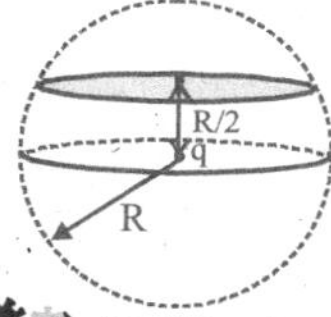

(a) q/ε_0 (b) $q/2\varepsilon_0$
(c) $q/4\varepsilon_0$ (d) zero

Critical Thinking

93. In the figure the net electric flux through the area A is $\phi = \vec{E} \cdot \vec{A}$ when the system is in air. On immersing the system in water the net electric flux through the area

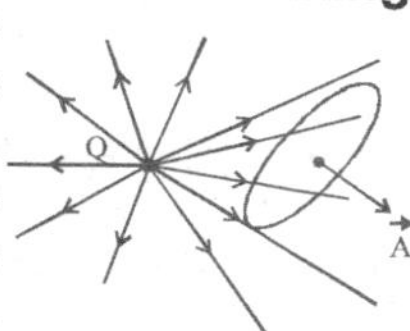

(a) becomes zero (b) remains same
(c) increases (d) decreases

94. A surface has the area vector $\vec{A} = (2\hat{i} + 3\hat{j})m^2$. The flux of an electric field through it if the field is $\vec{E} = 4\hat{i}\frac{V}{m}$:

(a) 8V-m (b) 12V-m (c) 20V-m (d) zero

95. For a given surface the Gauss's law is stated as $\oint \vec{E} \cdot dA = 0$. From this we can conclude that
(a) E is necessarily zero on the surface
(b) E is perpendicular to the surface at every point
(c) the total flux through the surface is zero
(d) the flux is only going out of the surface

96. A charge q is placed at the centre of the open end of a cylindrical vessel. The flux of the electric field through the surface of the vessel is

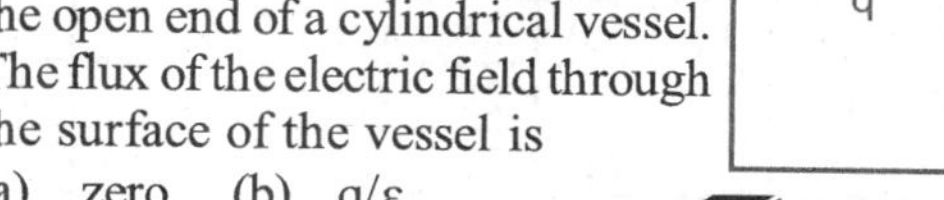

(a) zero (b) q/ε_o
(c) $q/2\varepsilon_o$ (d) $2q/\varepsilon_o$

97. The surface density on the copper sphere is σ. The electric field strength on the surface of the sphere is
(a) σ (b) $\sigma/2$ (c) $\sigma/2\varepsilon_o$ (d) Q/ε_o

98. An electric dipole is put in north-south direction in a sphere filled with water. Which statement is correct?
(a) Electric flux is coming towards sphere.
(b) Electric flux is coming out of sphere.
(c) Electric flux entering into sphere and leaving the sphere are same.
(d) Water does not permit electric flux to enter into sphere.

99. The magnitude of the average electric field normally present in the atmosphere just above the surface of the Earth is about 150 N/C, directed inward towards the center of the Earth. This gives the total net surface charge carried by the Earth to be: [Given $\varepsilon_o = 8.85 \times 10^{-12}\ C^2/N\text{-}m^2$, $R_E = 6.37 \times 10^6$ m] Critical Thinking

(a) +670 kC (b) −670 kC
(c) −680 kC (d) +680 kC

100. One-fourth of a sphere of radius R is removed as shown in Fig. An electric field E exists parallel to the xy plane. Find the flux through the remaining curved part. Toughnut

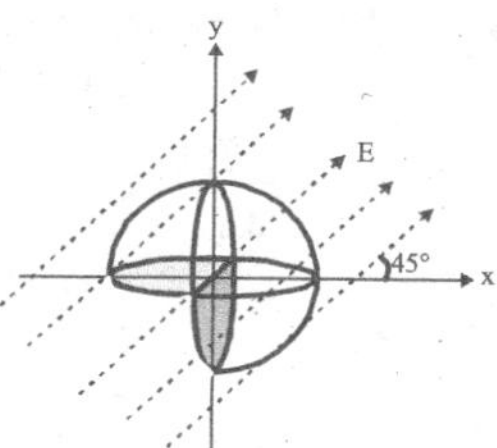

(a) $\pi R^2 E$ (b) $\sqrt{2}\pi R^2 E$
(c) $\pi R^2 E/\sqrt{2}$ (d) None of these

Answer KEYs

1	(d)	11	(c)	21	(b)	31	(b)	41	(d)	51	(b)	61	(b)	71	(b)	81	(a)	91	(c)
2	(b)	12	(b)	22	(b)	32	(a)	42	(c)	52	(c)	62	(a)	72	(a)	82	(c)	92	(c)
3	(b)	13	(d)	23	(c)	33	(a)	43	(b)	53	(b)	63	(b)	73	(b)	83	(d)	93	(d)
4	(b)	14	(a)	24	(d)	34	(c)	44	(d)	54	(b)	64	(a)	74	(b)	84	(a)	94	(a)
5	(b)	15	(d)	25	(d)	35	(b)	45	(b)	55	(a)	65	(b)	75	(b)	85	(c)	95	(c)
6	(d)	16	(b)	26	(d)	36	(a)	46	(b)	56	(c)	66	(d)	76	(a)	86	(a)	96	(a)
7	(c)	17	(c)	27	(c)	37	(b)	47	(c)	57	(b)	67	(a)	77	(c)	87	(c)	97	(d)
8	(a)	18	(c)	28	(c)	38	(a)	48	(a)	58	(b)	68	(c)	78	(d)	88	(b)	98	(c)
9	(b)	19	(c)	29	(a)	39	(a)	49	(b)	59	(c)	69	(d)	79	(d)	89	(d)	99	(c)
10	(a)	20	(a)	30	(a)	40	(a)	50	(c)	60	(b)	70	(a)	80	(c)	90	(b)	100	(c)

Hints & Solutions

1. **(d)** In equilibrium, $F_e = T\sin\theta$

$mg = T\cos\theta$

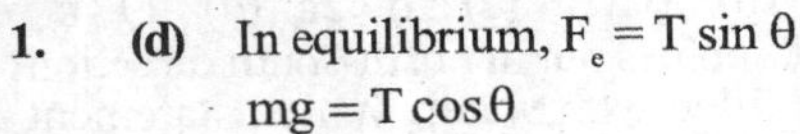

$$\tan\theta = \frac{F_e}{mg} = \frac{q^2}{4\pi\epsilon_0 x^2 \times mg}$$

also $\tan\theta \approx \sin = \dfrac{x/2}{l}$

Hence, $$\frac{x}{2l} = \frac{q^2}{4\pi\epsilon_0 x^2 \times mg}$$

$$\Rightarrow x^3 = \frac{2q^2 l}{4\pi\epsilon_0 mg} \quad \therefore x = \left(\frac{q^2 l}{2\pi\epsilon_0 mg}\right)^{1/3}$$

$\Rightarrow x \propto l^{1/3}$

2. **(b)** Electrostatic force is given by

$$F = \frac{1}{4\pi\varepsilon_0}\frac{q_1 q_2}{r^2}$$

Here, charge and distance are same. So, force between two protons will be same

3. **(b)** Here, $q_1 = 1 \times 10^{-7}$C, q_2 and 2×10^{-7} C, $r = 20$ cm $= 20 \times 10^{-2}$ m

$$F = \frac{q_1 q_2}{4\pi\varepsilon_0 r^2} = \frac{9\times10^9 \times 1\times10^{-7} \times 2\times10^{-7}}{(20\times10^{-2})^2}$$

$= 4.5 \times 10^{-3}$ N

4. **(b)** $2q$ at P, q at A, $-3q$ at B; $PA = \ell$, $AB = d$

Let a charge $2q$ be placed at P, at a distance I from A where charge q is placed, as shown in figure.

The charge $2q$ will not experience any force, when force, when force of repulsion on it due to q is balanced by force of attraction on it due to $-3q$ at B where AB $= d$

$$\text{or} \quad \frac{(2q)(q)}{4\pi\varepsilon_0 \ell^2} = \frac{(2q)(-3q)}{4\pi\varepsilon_0(\ell+d)^2}$$

$(\ell + d)^2 = 3\ell^2$ or $2\ell^2 - 2\ell d - d^2 = 0$

$$\therefore \quad \ell = \frac{2d \pm \sqrt{4d^2 + 2d^2}}{4} = \frac{d}{2} \pm \frac{\sqrt{3}d}{2}$$

5. **(b)** Here $q_1 = q_2 = 3.2 \times 10^{-7}$ C, $r = 60$ cm $= 0.6$ m

Electrostatic force, $F = \dfrac{1}{4\pi\varepsilon_0}\dfrac{q_1 q_2}{r^2}$

$$= \frac{9\times10^9(3.2\times10^{-7})^2}{(0.6)^2} = 2.56 \times 10^{-3}\ N$$

6. **(d)**

7. **(c)** Let n be the number of electrons missing.

$$F = \frac{1}{4\pi\varepsilon_0}\cdot\frac{q^2}{d^2} \Rightarrow q = \sqrt{4\pi\varepsilon_0 d^2 F} = ne$$

$$\therefore \quad n = \sqrt{\frac{4\pi\varepsilon_0 F d^2}{e^2}}$$

8. **(a)** Surface charge density $(\sigma) = \dfrac{\text{Charge}}{\text{Surface area}}$

So $\sigma_{\text{inner}} = \dfrac{-2Q}{4\pi b^2}$

and $\sigma_{\text{Outer}} = \dfrac{Q}{4\pi c^2}$

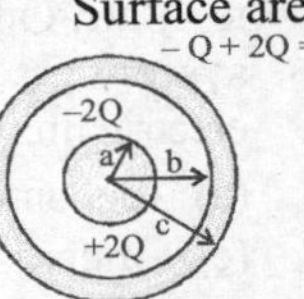

9. (b) $T_{AB} = \frac{kq\cdot 2q}{d^2} + \frac{kq\cdot 4q}{(2d)^2} = \frac{3kq^2}{d}$

$T_{BC} = \frac{k\cdot 4q\cdot 2q}{d^2} + \frac{kq\cdot 4q}{(2d)^2} = \frac{9kq^2}{d^2}$

10. (a) Number of protons

$= \frac{1C}{1.66\times10^{-19}C} \approx 6\times10^{18}$

11. (c)

12. (b) On touching the metal knob with a positively charged rod some electrons from the gold leaves get transferred to the rod making gold leaves positively charged and they get separated. When a negatively charged rod is touched with metal knob some negative charge flows to the gold leaves lessening the positive charge there and the separation between the leaves decreases.

13. (d) From Coulomb's law $F = \frac{Kq_1q_2}{r^2}$ i.e., $F \propto \frac{1}{r^2}$ which is correctly shown by graph (d).

14. (a)

15. (d) Each induce other oppositely near by.

16. (b) Force on charge q_1 due to q_2 is

$F_{12} = k\frac{q_1q_2}{b^2}$

Force on charge q_1 due to q_3 is

$F_{13} = k\frac{q_1q_3}{a^2}$

The X - component of the force (F_x) on q_1 is $F_{12} + F_{13}\sin\theta$

$\therefore\ F_x = k\frac{q_1q_2}{b^2} + k\frac{q_1q_2}{a^2}\sin\theta$

$\therefore\ F_x \propto \frac{q_2}{b^2} + \frac{q_3}{a^2}\sin\theta$

17. (c) $T\sin\theta = \frac{\sigma}{\varepsilon_0 K}.q$ ... (i)

$T\cos\theta = mg$ .. (ii)

Dividing (i) by (ii),

$\tan\theta = \frac{\sigma q}{\varepsilon_0 K.mg}$ $\therefore \sigma \propto \tan\theta$

18. (c)

19. (c) The charge on disc A is 10^{-6} μC. The charge on disc B is 10×10^{-6} μC. The total charge on both $= 11$μC. When touched, this charge will be distributed equally i.e. 5.5 μC on each disc.

20. (a)

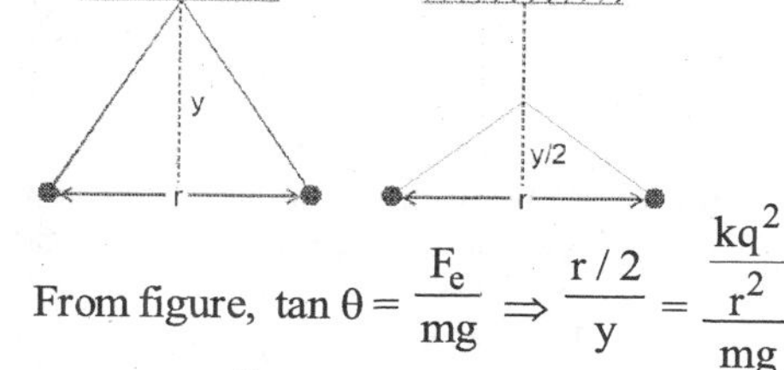

From figure, $\tan\theta = \frac{F_e}{mg} \Rightarrow \frac{r/2}{y} = \frac{\frac{kq^2}{r^2}}{mg}$

$[\because F = \frac{kq^2}{r^2}$ from coulomb's law$]$

$\Rightarrow r^3 \propto y \Rightarrow r'^3 \propto \frac{y}{2} \Rightarrow \frac{r'}{r} = \frac{1}{2^{1/3}} \Rightarrow r' = \frac{r}{\sqrt[3]{2}}$

21. (b)

22. (b) $F_{net} = 2|F_{31}|\cos\alpha$

$= 2\times\frac{1}{4\pi\varepsilon_0}\times\frac{2\times4\times10^{-12}}{(0.5)^2}\times\frac{4}{5} = 0.46\ N$

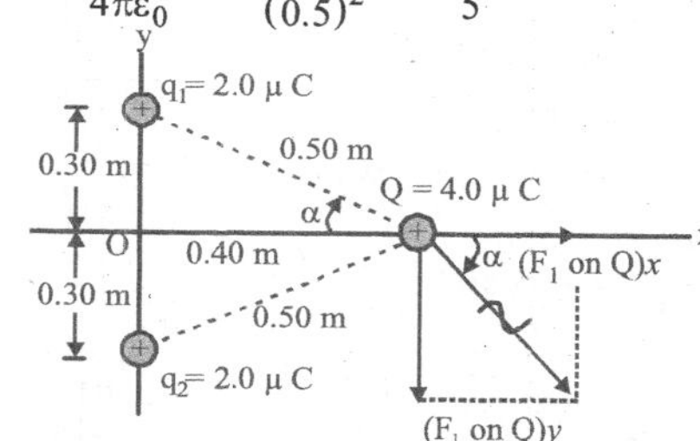

23. (c) 24. (d)

25. (d) They will not experience any force if $|\vec{F}_G| = |\vec{F}_e|$

$\Rightarrow G\frac{m^2}{(16\times10^{-2})^2} = \frac{1}{4\pi\varepsilon_0}.\frac{q^2}{(16\times10^{-2})^2}$

$\Rightarrow \frac{q}{m} = \sqrt{4\pi\varepsilon_0 G}$

26. (d) Here, $D = 2r = 4.4\ m$, or $= r = 2.2\ m$

$\sigma = 60\ \mu C\ m^{-2}$

Charge on the sphere, $q = \sigma \times 4\pi r^2$

$= 60\times10^{-6}\times4\times\frac{22}{7}\times(2.2)^2 = 3.7\times10^{-3}$ C

27. (c) For external points, a charged sphere behaves as if the whole of its charge is concentrated at its centre.

Force on A due to B,

$F_{AB} = \frac{1}{4\pi\varepsilon_0}\frac{q^2}{(2R)^2}$

$= \frac{1}{4\pi\varepsilon_0}\frac{q^2}{4R^2}$ along $\overrightarrow{BA}$

And force on A due to C,

$F_{AC} = \frac{1}{4\pi\varepsilon_0}\frac{q^2}{(2R)^2} = \frac{1}{4\pi\varepsilon_0}\frac{q^2}{4R^2}$ along $\overrightarrow{CA}$

Now as angle between BA and CA is 60° and $|F_{AB}| = |F_{AC}| = F$

$\therefore F_A = \sqrt{F^2 + F^2 + 2FF\cos 60} = \sqrt{3}F$

$= \frac{1}{4\pi\varepsilon_0}\frac{\sqrt{3}}{4}\left(\frac{q}{R}\right)^2$

28. (c) For maximum elongation charges on the blocks must be equal to $Q/2$ on each block.

$\therefore \frac{1}{4\pi\varepsilon_0}\frac{\frac{Q}{2}\frac{Q}{2}}{(\ell_0 + x)^2} = kx \quad Q = 2(\ell_0 + x)\sqrt{4\pi\varepsilon_0 kx}.$

29. (a)

30. (a) The bowl exerts a normal force N on each bead, directed along the radial line or at 60.0° above the horizontal. Consider the free-body diagram of the bead on the left with the electric force F_e applied:

$\Sigma F_y = N\sin 60° - mg = 0$

$\Rightarrow N = \frac{mg}{\sin 60°}$

$\Sigma F_x = -F_e + N\cos 60° = 0$

$\Rightarrow \frac{k_e q^2}{R^2} = N\cos 60° = \frac{mg}{\tan 60°} = \frac{mg}{\sqrt{3}}$

$k_e = \frac{1}{4\pi\varepsilon_0} \approx 9.0\times10^9\,\text{Nm}^2/\text{C}^2 \quad \therefore q = R\left(\frac{mg}{k_e\sqrt{3}}\right)^{1/2}$

31. (b) We have $E_a = \frac{2kp}{r^3}$ and $E_e = \frac{kp}{r^3}$;

$\therefore E_a = 2E_e$

32. (a) $\vec{\tau} = \vec{P}\times\vec{E} = PE\sin\theta$

For $q = 90°$, $\vec{\tau}_{max} = PE$

33. (a)

34. (c) Here, $\ell = 2.4\ m$, $r = 4.6\ mm = 4.6\times10^{-3}\ m$

$q = -4.2\times10^{-7}$ C

Linear charge density, $\lambda = \frac{q}{\ell}$

$= \frac{-4.2\times10^{-7}}{2.4} = -1.75\times10^{-7}\ \text{C m}^{-1}$

Electric field, $E = \frac{\lambda}{2\pi\varepsilon_0 r}$

$= \frac{-1.75\times10^{-7}}{2\times3.14\times8.854\times10^{-12}\times4.6\times10^{-3}}$

$= -6.7\times10^5\ \text{N C}^{-1}$

35. (b) $E = \frac{q}{4\pi\varepsilon_o r^2} \Rightarrow Ar = \frac{q}{4\pi\epsilon_0 r^2}$

$\Rightarrow q = 4\pi\varepsilon_o Ar^3$

36. (a) $-eE = mg$

$\vec{E} = -\frac{9.1\times10^{-31}\times10}{1.6\times10^{-19}} = -5.6\times10^{-11}\,\text{N/C}$

37. (b) $E = \frac{kQ}{r^2} \Rightarrow Q = \frac{E\times r^2}{k} = \frac{3\times10^6\times(2.5)^2}{9\times10^9}$

$= 2\times10^{-3}$ C

38. (a)

39. (a) Here, $q = 1$ C, $\varepsilon_0 = 8.85\times10^{-12}\ C^2N^{-1}m^{-2}$

Number of lines of force = Electric force

$= \frac{q}{\varepsilon_0} = \frac{1}{8.85\times10^{-12}} = 1.13\times10^{11}$

40. (a)

41. (d) Unit positive charge at O will be repelled equally by three charges at the three corners of triangle. By symmetry, resultant $\vec{E}$ at O would be zero.

42. (c) Intensity of electric field due to a Dipole

$E = \frac{p}{4\pi\varepsilon_0 r^3}\sqrt{3\cos^2\theta + 1} \Rightarrow E \propto \frac{1}{r^3}$

43. (b) $\tau = \left(2.57\times10^{-17}\,\text{Cm}\right)\left(3.0\times10^4\frac{\text{N}}{\text{C}}\right)\left(\frac{1}{2}\right)$

$= 3.855\times10^{-13}$ Nm.

44. (d) Torque, $\vec{\tau} = \vec{p}\times\vec{E} = pE\sin\theta$

$4 = p\times2\times10^5\times\sin 30°$

or, $p = \frac{4}{2\times10^5\times\sin 30°} = 4\times10^{-5}$ Cm

Dipole moment, $p = q\times l$

$q = \frac{p}{l} = \frac{4\times10^{-5}}{0.02} = 2\times10^{-3}\,\text{C} = 2\text{mC}$

45. (b) Since $\tau = pE\sin\theta$ on decreasing the distance between the two charges, and on decreasing angle θ between the dipole and electric field, $\sin\theta$ decreases therefore torque decreases.

46. (b)

47. (c) $x = \frac{1}{2}at^2 = \frac{1}{2}\left(\frac{Eq}{m}\right)t^2$ and $x = \frac{1}{2}\left(\frac{Eq}{2m}\right)t'^2$

$\therefore t' = \sqrt{2}t.$

48. (a) $E_1 = \frac{nkQR}{(R^2+R^2)^{3/2}}, E_2 = \frac{nkQ(2R)}{(R^2+(2R)^2)^{3/2}}$

$\therefore = \frac{E_1}{E_2} = \frac{5\sqrt{5}}{4\sqrt{2}}$

49. (b) Force of interaction

$F = \frac{1}{4\pi\epsilon_0}.\frac{3p_1p_2}{x^4}$

50. (c)

51. (b) $F = qE = mg\ (q = 6e = 6 \times 1.6 \times 10^{-19})$

Density (d)

$$\frac{\text{mass}}{\text{volume}} = \frac{m}{\frac{4}{3}\pi r^3} \text{ or } r^3 = \frac{m}{\frac{4}{3}\pi d}$$

Putting the value of v and m (= 2E/g) and solving we get $r = 7.8 \times 10^{-7}$ m

52. (c) Charges (q) = 2×10^{-6} C, Distance (d) = 3 cm = 3×10^{-2} m and electric field (E) = 2×10^5 N/C. Torque $(\tau) = q.d.$

$E = (2 \times 10^{-6}) \times (3 \times 10^{-2}) \times (2 \times 10^5)$

$= 12 \times 10^{-3}$ N–m.

53. (b) $w = \vec{F}.\vec{s} = Eq\hat{i}.(-a\hat{i} - b\hat{j}) = -Eqa$

54. (b) In equilibrium,

$T\cos\theta = mg$ ....(1)

$T\sin\theta = qE$ ....(2)

From (1), $T = \dfrac{mg}{\cos\theta}$

From (2), $T = \dfrac{qE}{\sin\theta}$

55. (a) The field at O due to small element dx is

$$dE = \frac{1}{4\pi\varepsilon_0}\cdot\frac{\lambda dx}{x^2}$$

Hence, due to one wire,

$$E_1 = \int_a^{\infty}\frac{1}{4\pi\varepsilon_0}\cdot\frac{\lambda dx}{x^2}\quad E_1 = \frac{1}{4\pi\varepsilon_0}\cdot\frac{\lambda}{a} \text{ towards left.}$$

Electric field at O due to other wire,

$$E_2 = \frac{1}{4\pi\varepsilon_0}\cdot\frac{\lambda}{a} \text{ towards left}$$

$\therefore$ Net field at O is $E = 2\times\dfrac{1}{4\pi\varepsilon_0}\dfrac{\lambda}{a} = \dfrac{\lambda}{2\pi\varepsilon_0 a}$

56. (c) $$E = \frac{1}{4\pi\epsilon_0}\left[\frac{2p}{(r-\ell/2)^3} - \frac{2p}{(r+\ell/2)^3}\right] = \frac{1}{4\pi\epsilon_0}\frac{3Q}{r^4}.$$

57. (b) **58.** (b)

59. (c) $F = qE = q(A - Bx)$

$$ma = q(A - Bx) \Rightarrow a = \frac{q}{m}(A - Bx) \quad \text{...(i)}$$

$$\frac{vdv}{dx} = \frac{q}{m}(A - Bx);\ vdv = \frac{q}{m}(A - Bx)dx$$

$$\int_0^0 vdv = \frac{q}{m}\int_0^x (A - Bx)dx;\ Ax - \frac{Bx^2}{2} = 0$$

$$x = 0, x = \frac{2A}{B} \quad \text{...(ii)}$$

From eqs. (i) and (ii)

$$\frac{q}{m}(A - Bx) = \frac{q}{m}\left(A - B\times\frac{2A}{B}\right) = \frac{-qA}{m}.$$

60. (b) Charge on the element opposite to the gap is

$$dq = \frac{Q}{2\pi r}(0.002\pi)$$
$$= \frac{1}{2\pi(0.5)}\times\frac{2\pi}{1000} = 2\times10^{-3}C$$
$$E = \frac{9\times10^9\times2\times10^{-3}}{(0.5)^2} = 7.2\times10^7\ NC^{-1}$$

61. (b) $$\vec{E} = \frac{kQ}{r^3}\vec{r} = \frac{9\times10^9\times50\times10^{-6}}{|\vec{r}-\vec{r}_0|^3}\times(\vec{r}-\vec{r}_0)$$

where $\vec{r} - \vec{r}_0 = (8\hat{i} - 5\hat{j}) - (2\hat{i} + 3\hat{j}) = 6\hat{i} - 8\hat{j}$

$\vec{E} = 900\,(3\hat{i} - 4\hat{j})$ V/m

62. (a) Take PO as the x-axis and PA as the y-axis. Consider two elements EF and E'F' of width $d\theta$ at angular distance θ above and below PO, respectively.

The magnitude of the fields at P due to either element is

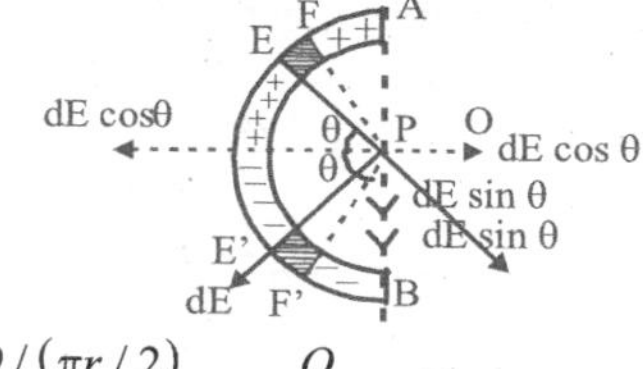

$$dE = \frac{1}{4\pi\varepsilon_0}\frac{rd\theta\times Q/(\pi r/2)}{r^2} = \frac{Q}{2\pi^2\varepsilon_0 r^2}d\theta$$

Resolving the fields, we find that the components along PO sum up to zero, and hence the resultant field is along PB. Therefore, field at P due to pair of elements is $2dE\sin\theta$

$$E = \int_0^{\pi/2} 2d\,E\sin\theta$$

$$= 2\int_0^{\pi/2}\frac{Q}{2\pi\varepsilon_0 r^2}\sin\theta d\theta = \frac{Q}{\pi^2\varepsilon_0 r^2}$$

63. (b) $E = \dfrac{1}{4\pi\varepsilon_0}\dfrac{2p}{r^3};\ E \propto \dfrac{1}{r^3} \Rightarrow F \propto \dfrac{1}{r^3}$

Hence, the force will become F/8.

64. (a) Since the sheet is metallic, the charges must be redistributed over its surface so that the field in the bulk of the sheet is zero. In the first approximation, we can assume that this distribution is uniform and has density $-\sigma$ and σ over the upper and the lower surface respectively of the sheet. According to the superposition principle, we obtain the condition for the absence of the field in the bulk of the sheet :

$$\frac{q}{4\pi\varepsilon_0\ell^2} - \frac{\sigma}{\varepsilon_0} = 0$$

Let us now take into consideration the non uniformity of the field produced by the point charge since it is the single cause of the force F of interaction. The upper surface of the sheet must be attracted with a force $\frac{\sigma Sq}{4\pi\varepsilon_0\ell^2}$, while the lower surface must be repelled with a force $\frac{\sigma Sq}{4\pi\varepsilon_0(\ell+d)^2}$.

Consequently, the force of attraction of the sheet to the charge is

$$F = \frac{\sigma Sq}{4\pi\varepsilon_0\ell^2}\left[1-\frac{1}{(1+d/\ell)^2}\right] \approx \frac{q^2 Sd}{8\pi^2\varepsilon_0\ell^5}$$

65. (b) **66.** (d) **67.** (a)

68. (c) The magnitude of electric field intensity due to each part of the hemispherical surface at the centre 'O' is same.

Suppose it is E.

$$E+\frac{E}{2}+\frac{E}{2}=E_0$$

or $2E = E_0$ or $E=\frac{E_0}{2}$

69. (d)

70. (a) The electric field inside a thin spherical shell of radius R has charge Q spread uniformly over its surface is zero.

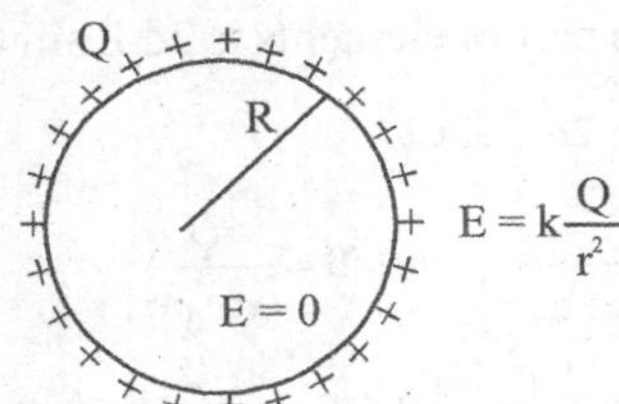

Outside the shell the electric field is $E = k\frac{Q}{r^2}$.

These characteristics are represented by graph (a).

71. (b) $F_{net} = \int dq\, E\cos\theta$

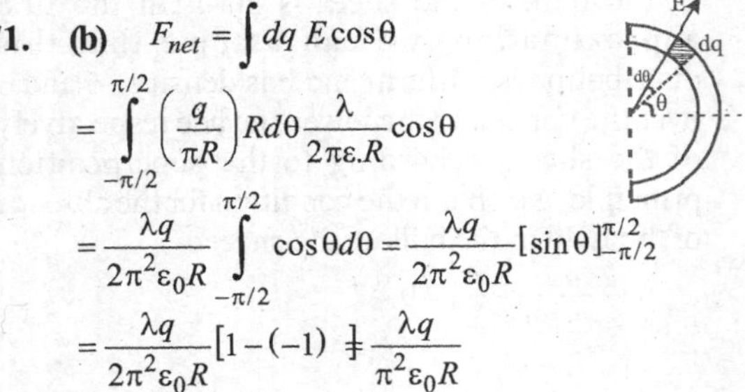

$$= \int_{-\pi/2}^{\pi/2}\left(\frac{q}{\pi R}\right)Rd\theta\frac{\lambda}{2\pi\varepsilon.R}\cos\theta$$

$$= \frac{\lambda q}{2\pi^2\varepsilon_0 R}\int_{-\pi/2}^{\pi/2}\cos\theta d\theta = \frac{\lambda q}{2\pi^2\varepsilon_0 R}[\sin\theta]_{-\pi/2}^{\pi/2}$$

$$= \frac{\lambda q}{2\pi^2\varepsilon_0 R}[1-(-1)] = \frac{\lambda q}{\pi^2\varepsilon_0 R}$$

72. (a) We have centripetal force equation

$$q\left(\frac{2k\lambda}{r}\right) = \frac{mv^2}{r} \quad \text{so } v = \sqrt{\frac{2kq\lambda}{m}}$$

Now, $T = \frac{2\pi r}{v} = 2\pi r\sqrt{\frac{m}{2kq\lambda}}$ where $k = \frac{1}{4\pi\varepsilon_0}$

73. (b)

74. (b) The given point is at axis of $\frac{\vec{p}}{2}$ dipole and at equatorial line of $\vec{p}$ dipole so that field at given point is $\vec{E}_1+\vec{E}_2$

$$\vec{E}_1 = \frac{2K(p/2)}{2^3} = \frac{Kp}{8}(+\hat{k}) \qquad \vec{E}_2 = \frac{Kp}{1}(-\hat{k})$$

$$\vec{E}_1+\vec{E}_2 = -\frac{7}{8}Kp(-\hat{k}) = -\frac{7p}{32\pi\,\varepsilon_0}\hat{k}$$

75. (b)

76. (a) According to Gauss's law total electric flux through a closed surface is $\frac{1}{\varepsilon_0}$ times the total charge inside that surface.

Electric flux, $\phi_E = \frac{q}{\varepsilon_0}$

Charge on α-particle $= 2e \quad \therefore \phi_E = \frac{2e}{\varepsilon_0}$

77. (c) Flux going in pyramid $= \frac{Q}{2\varepsilon_0}$.

which is divided equally among all 4 faces.

$\therefore$ Flux through one face $= \frac{Q}{8\varepsilon_0}$

78. (d)

79. (d) $\Delta\phi = \frac{\theta_{in}}{\varepsilon_0} \Rightarrow Qin\ (\phi = \phi_2 - \phi)\varepsilon_0$

80. (c) Flux does not depend on the size and shape of the close surface, and so, it remains same.

81. (a) $\phi = EA\cos 0° = E\times\frac{\pi d^2}{4}, \quad \therefore E = \frac{4\phi}{\pi d^2}$.

82. (c) $\int \overline{E.dA} = \phi$

$$\Rightarrow \phi = E_0\left(a\times a\sqrt{2}\right)\cos 45° = E_0 a^2$$

83. (d) By Gauss's law $\phi = \frac{1}{\varepsilon_0}(Q_{enclosed})$

$$\Rightarrow Q_{enclosed} = \phi\varepsilon_0 = (-8\times10^3 + 4\times10^3)\varepsilon_0$$
$$= -4\times10^3\,\varepsilon_0 \text{ coulomb.}$$

84. (a)

85. **(c)** Using Gauss's law, we have

$$\oint \vec{E}\cdot d\vec{A} = \frac{1}{\epsilon_0}\int(\rho dv) = \frac{1}{\epsilon_0}\int_0^R kr^a \times 4\pi r^2 dr$$

or $E \times 4\pi r^2 = \left(\frac{4\pi k}{\epsilon_0}\right)\frac{R^{(a+3)}}{(a+3)}$ $\therefore$ $E_1 = \frac{kR^{(a+1)}}{\epsilon_0 (a+3)}$

For $r = \frac{R}{2}\cdot E_2 = \frac{k\left(\frac{R}{2}\right)^{a+1}}{\epsilon_0 (a+3)}$

Given, $E_2 = \frac{E_1}{8}$ or $\frac{k\left(\frac{R}{2}\right)^{a+1}}{\epsilon_0 (a+3)} = \frac{1}{8}\frac{kR^{(a+1)}}{\epsilon_0 (a+3)}$

$\therefore$ $\frac{1}{2^{a+1}} = \frac{1}{8}$ or $a = 2$.

86. **(a)** From the figure it is clear that the charge enclosed in the cubical surface is $3C + 2C - 7C = -2C$. Therefore the electric flux through the cube is

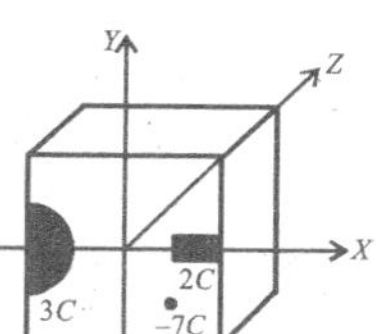

$$\phi = \frac{q_{in}}{\varepsilon_0} = \frac{-2C}{\varepsilon_0}$$

87. **(c)** $\oint E_p dS = \frac{Q+Q'}{\varepsilon_0}$

where Q' is the charge outside the sphere

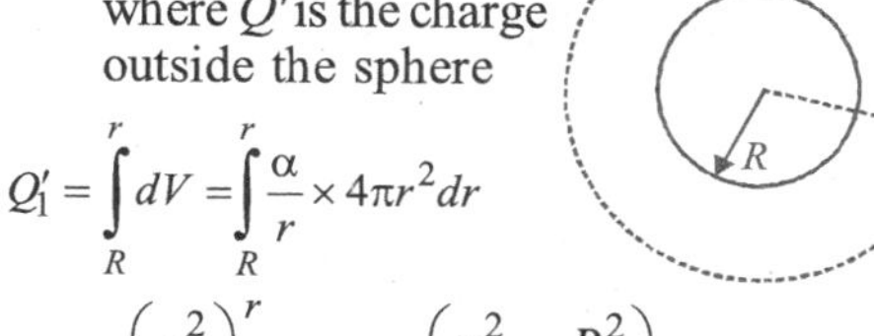

$$Q_1' = \int_R^r dV = \int_R^r \frac{\alpha}{r}\times 4\pi r^2 dr$$

$$= 4\pi\alpha\left(\frac{r^2}{2}\right)_R^r = 4\pi\alpha\left(\frac{r^2}{2} - \frac{R^2}{2}\right) = 2\pi\alpha(r^2 - R^2)$$

$$E_p \times 4\pi r^2 = \frac{Q + 2\pi\alpha(r^2 - R^2)}{\varepsilon_0}$$

$$E_p = \frac{Q}{4\pi r^2\varepsilon_0} + \frac{\alpha}{2\varepsilon_0} - \frac{\alpha R^2}{2r^2\varepsilon_0}$$

E is independent of r if $\frac{Q}{4\pi r^2\varepsilon_0} - \frac{\alpha R^2}{2r^2\varepsilon_0} = 0$

$Q = 2\pi R^2\alpha$

88. **(b)**

89. **(d)** For, $r = R/2$

Using Gauss's law, we have

$$\oint \vec{E}\cdot d\vec{A} = \frac{q_{in}}{\epsilon_0} \text{ or } E \times 4\pi r^2 = \int_0^{R/2}\frac{\rho 4\pi r^2 dr}{\epsilon_0}$$

or $E = \frac{Cr^3}{5\epsilon_0} = \frac{CR^3}{40\epsilon_0}$.

90. **(b)** Let us complete the sphere. Electric field due to lower part at A is equal to electric field due to upper part at B = E (given) Electric field due to lower part at

B = electric field due to full sphere

– electric field due to upper part

$$= \frac{kQ}{(2R)^2} - E$$

$$= \frac{1}{4\pi\varepsilon_0}\frac{\rho(4/3)\pi R^3}{4R^2} - E$$

$$= \frac{\rho R}{12\varepsilon_0} - E$$

91. **(c)** Charge on length

$AB = 2\sqrt{R^2 - y^2}\times\lambda$

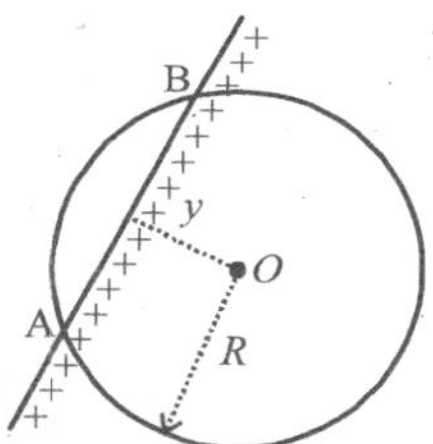

Electric flux

$$= \oint \vec{E}\cdot\vec{ds} = \frac{2\lambda\sqrt{R^2 - y^2}}{\varepsilon_0}$$

92. **(c)** $\alpha = 60°$. Solid angle subtended by BCD is

$\omega = 2\pi(1 \cos\alpha) = \pi$

Solid angle subtended by ABDE is

$\omega_{(ABCDE)} - \omega_{(BCD)} = 2\pi - \pi = \pi$

Hence, flux through ABDE is

$$\phi = \frac{q}{\varepsilon_0}\frac{\pi}{4\pi} = \frac{q}{4\varepsilon_0}$$

93. **(d)** Since electric field $\vec{E}$ decreases inside water, therefore flux $\phi = \vec{E}.\vec{A}$ also decreases.

94. **(a)** $\phi = \vec{E}.\vec{A} = 4\hat{i}.(2\hat{i} + 3\hat{j}) = 8$ V-m

95. **(c)** $\oint \vec{E}.d\vec{A} = 0$, represents charge inside close surface is zero. Electric field as any point on the surface may be zero.

96. **(a)** The flux is zero according to Gauss' Law because it is a open surface which enclosed a charge q.

97. **(d)** According to Gauss's theorem,

$$E\oint ds = \frac{q}{\epsilon_0}\left[\text{Here} \oint ds = 4\pi R^2\right]$$

$\therefore E = \frac{q/4\pi R^2}{\epsilon_0}$ $[\because q/4\pi R^2 = \sigma]$

or $E = \sigma/\varepsilon_o$

98. (c) If electric dipole, the flux coming out from positive charge is equal to the flux coming in at negative charge *i.e.* total charge on sphere = 0. From Gauss law, total flux passing through the sphere = 0.

99. (c) Given,

Electric field E = 150 N/C

Total surface charge carried by earth q = ?

According to Gauss's law.

$$\phi = \frac{q}{\epsilon_0} = EA$$

or, $q = \epsilon_0 EA = \epsilon_0 E\pi r^2 = 8.85 \times 10^{-12} \times 150 \times (6.37 \times 10^6)^2$.

$\simeq 680$ KC

As electric field directed inward hence

q = – 680 KC

100. (c) $\phi_{plain} + \phi_{curve} = 0$ or $\phi_{plain} = -\phi_{curve}$

$$\vec{A}_1 = -\frac{\pi R^2}{2}\hat{\imath}, \vec{A}_2 = -\frac{\pi R^2}{2}\hat{\jmath}$$

$$\vec{E} = E\cos 45°\hat{\imath} + E\sin 45°\,\hat{\jmath}$$

$$= \frac{E}{\sqrt{2}}\hat{\imath} + \frac{E}{\sqrt{2}}\hat{\jmath} \text{ and}$$

$$\phi = \vec{E}\cdot\left(\vec{A}_1 + \vec{A}_2\right)$$

$$= \frac{-E}{\sqrt{2}}\frac{\pi R^2}{2} - \frac{E}{\sqrt{2}}\frac{\pi R^2}{2} = \frac{-\pi R^2 E}{\sqrt{2}}$$

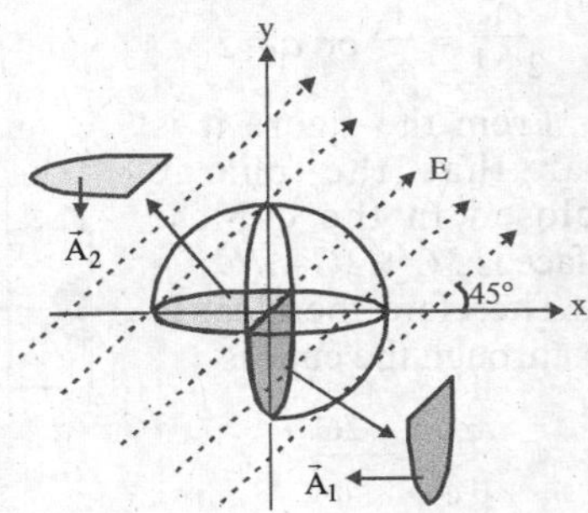

This is the flux entering. So flux is $\frac{\pi R^2 E}{\sqrt{2}}$

16 Electrostatic Potential and Capacitance

1. Charges $+q$ and $-q$ are placed at points A and B respectively which are a distance $2L$ apart, C is the midpoint between A and B. The work done in moving a charge $+Q$ along the semicircle CRD is

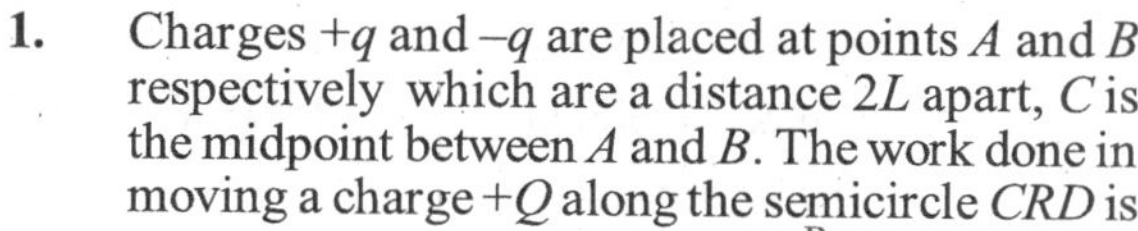

(a) $\frac{qQ}{2\pi\varepsilon_0 L}$

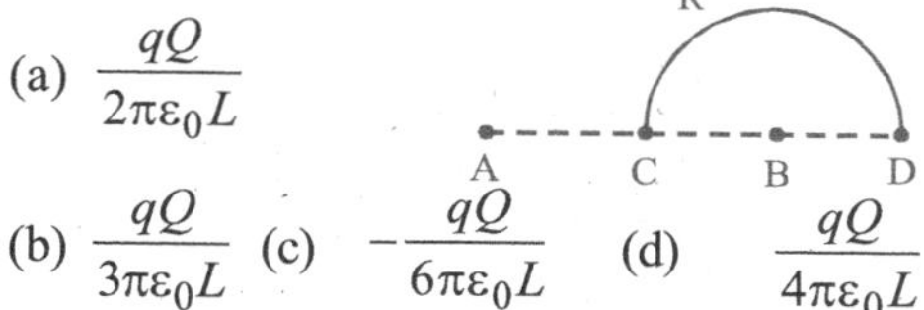

(b) $\frac{qQ}{3\pi\varepsilon_0 L}$ (c) $-\frac{qQ}{6\pi\varepsilon_0 L}$ (d) $\frac{qQ}{4\pi\varepsilon_0 L}$

2. Four identical particles each of mass m and charge q are kept at the four corners of a square of length L. The final velocity of these particles after setting them free will be

(a) $\left[\frac{Kq^2}{mL}(5.4)\right]^{1/2}$ (b) $\left[\frac{Kq^2}{mL}(1.35)\right]^{1/2}$

(c) $\left[\frac{Kq^2}{mL}(2.7)\right]^{1/2}$ (d) Zero

3. There is an infinite straight chain of alternating charges q and $-$q. The distance between the two neighbouring charges is equal to a. Find the interaction energy of any charge with all the other charges.

Tricky

(a) $-\frac{2q^2}{4\pi\varepsilon_0 a}$ (b) $\frac{2q^2 \log_e 2}{4\pi\varepsilon_0 a}$

(c) $-\frac{2q^2 \log_e 2}{4\pi\varepsilon_0 a}$ (d) None of these

4. Three identical particles, each possessing the mass m and charge $+q$, are placed at the corners of an equilateral triangle with side r_0. The particles are simultaneously set free and start flying apart symmetrically due to Coulomb's repulsion foces. The work performed by Coulomb's forces acting on each particle until the particles fly from one another to a very large distance is (where $k = 1/4\pi\varepsilon_0$.)

Critical Thinking

(a) $\frac{3kq^2}{r_0}$ (b) $\frac{kq^2}{r_0}$ (c) $\frac{3kq^2}{2r_0}$ (d) $\frac{kq^2}{2r_0}$

5. Two charges q_1 and q_2 are placed 30 cm apart, as shown in the figure. A third charge q_3 is moved along the arc of a circle of radius 40 cm from C to D. The change in the potential energy of the system is $\frac{q_3}{4\pi\epsilon_0}k$, where k is

Critical Thinking

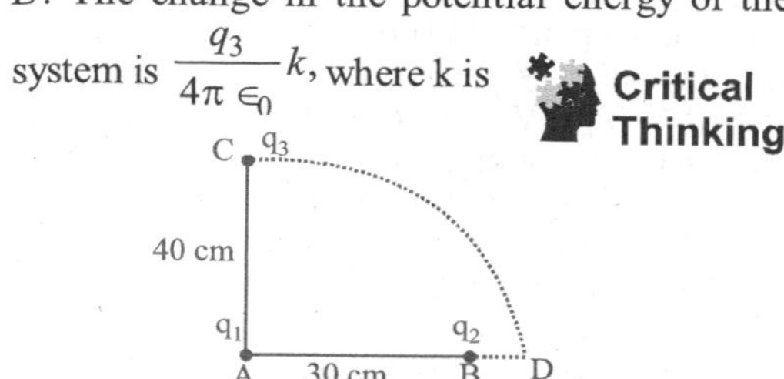

(a) $8q_1$ (b) $6q_1$ (c) $8q_2$ (d) $6q_2$

6. When air in a capacitor is replaced by a medium of dielectric constant K, the capacity

(a) decreases K times (b) increases K times
(c) increases K^2 times (d) remains constant

7. The work done in placing a charge of 8×10^{-18} coulomb on a condenser of capacity 100 micro-farad is

(a) 3.1×10^{-26} joule (b) 4×10^{-10} joule
(c) 32×10^{-32} joule (d) 16×10^{-32} joule

8. Consider the situation shown in the figure. The capacitor A has a charge q on it whereas B is uncharged. The charge appearing on the capacitor B a long time after the switch is closed is

(a) zero
(b) $q/2$
(c) q
(d) $2q$

9. A parallel plate capacitor is located horizontally such that one of the plates is submerged in a liquid while the other is above the liquid surface. When plates are charged the level of liquid

Tricky

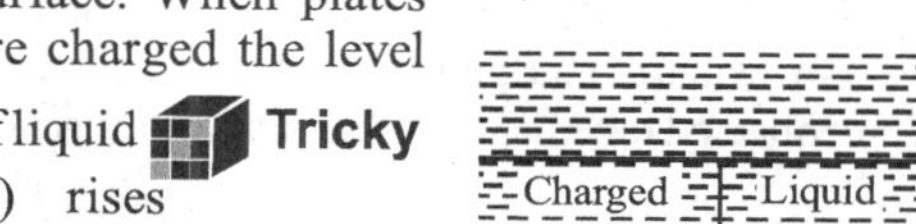

(a) rises
(b) falls
(c) remains unchanged
(d) may rise or fall depending on the of charge amount

10. What is the effective capacitance between points X and Y?

$C_1 = 8\mu F$, $C_3 = 8\mu F$, $C_5 = 10\mu F$, $C_2 = 8\mu F$, $C_4 = 8\mu F$; X, A, C, B, D, Y

(a) 8 μF
(b) 9 μF
(c) 10 μF
(d) 12 μF

11. A hollow metal sphere of radius 5 cm is charged such that the potential on its surface is 10 V. The potential at a distance of 2 cm from the centre of the sphere is

(a) zero (b) 10 V (c) 4 V (d) 10/3 V

12. A unit charge moves on an equipotential surface from a point A to point B, then

(a) $V_A - V_B = +ve$ (b) $V_A - V_B = 0$
(c) $V_A - V_B = -ve$ (d) it is stationary

13. A cube of a metal is given a positive charge Q. For this system, which of the following statements is true?

(a) Electric potential at the surface of the cube is zero
(b) Electric potential within the cube is zero
(c) Electric field is normal to the surface of the cube
(d) Electric field varies within the cube

14. In moving from A to B along an electric field line, the work done by the electric field on an electron is 6.4×10^{-19} J. If ϕ_1 and ϕ_2 are equipotential surfaces, then the potential difference $V_C - V_A$ is

Tricky

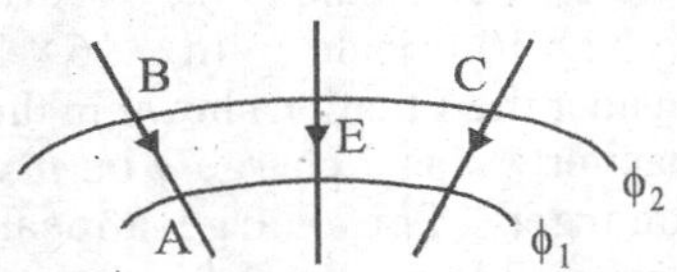

(a) –4 V (b) 4 V (c) zero (d) 6.4 V

15. A non-conducting ring of radius 0.5 m carries a total charge of 1.11×10^{-10} C distributed non-uniformly on its circumference producing an electric field E everywhere in space. The value of the integral $\int_{\ell=\infty}^{\ell=0} -E.d\ell$ ($\ell = 0$ being center of the ring) in volts is

(a) +2 (b) –1 (c) –2 (d) zero

16. Two equally charged spheres of radii a and b are connected together. What will be the ratio of electric field intensity on their surfaces?

Critical Thinking

(a) $\frac{a}{b}$ (b) $\frac{a^2}{b^2}$ (c) $\frac{b}{a}$ (d) $\frac{b^2}{a^2}$

17. If the electrostatic potential were given by $\phi = \phi_0 (x^2 + y^2 + z^2)$, where ϕ_0 is constant, then the charge density giving rise to the above potential would be.

(a) 0 (b) $-6\phi_0 \varepsilon_0$
(c) $-2\phi_0 \varepsilon_0$ (d) $-\frac{6\phi_0}{\varepsilon_0}$

18. A, B and C are three points in a uniform electric field. The electric potential is

Tricky

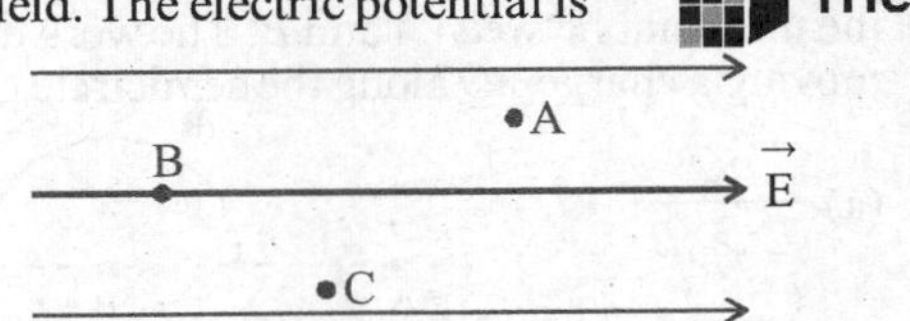

(a) maximum at B
(b) maximum at C
(c) same at all the three points A, B and C
(d) maximum at A

19. An electric charge 10^{-3} μ C is placed at the origin (0, 0) of X – Y co-ordinate system. Two points A and B are situated at $(\sqrt{2}, \sqrt{2})$ and (2, 0) respectively. The potential difference between the points A and B will be [in volt]

(a) 4.5 (b) 9 (c) zero (d) 2

20. Which of the following figure shows the correct equipotential surfaces of a system of two positive charges?

Tricky

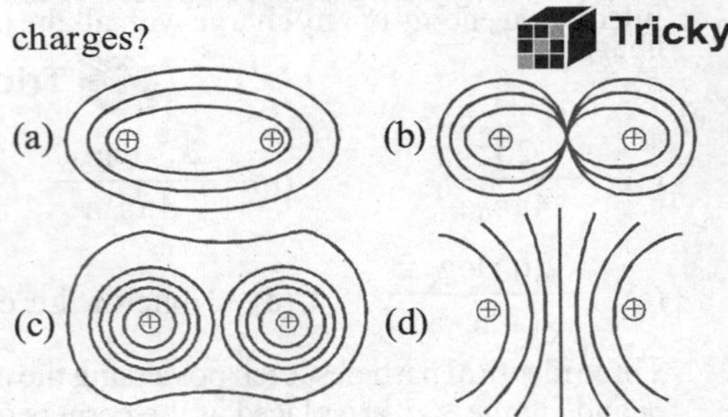

21. Equipotential surfaces are shown in figure. Then the electric field strength will be

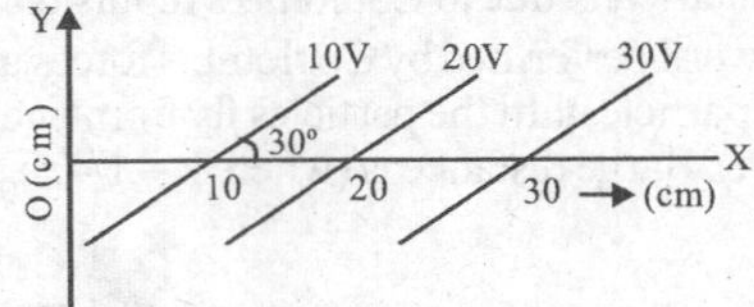

(a) 100 Vm^{-1} along X-axis
(b) 100 Vm^{-1} along Y-axis
(c) 200 Vm^{-1} at an angle 120° with X-axis
(d) 50 Vm^{-1} at an angle 120° with X-axis

22. Two concentric spheres of radii R and r have similar charges with equal surface charge densities (σ). What is the electric potential at their common centre ?

(a) σ/ε_0 (b) $\dfrac{\sigma}{\varepsilon_0}(R-r)$

(c) $\dfrac{\sigma}{\varepsilon_0}(R+r)$ (d) None of these

23. From a point charge, there is a fixed point A. At A, there is an electric field of 500 V/m and potential difference of 3000 V. Distance between point charge and A will be

(a) 6m (b) 12m (c) 16m (d) 24m

24. Two metal pieces having a potential difference of 800 V are 0.02 m apart horizontally. A particle of mass 1.96×10^{-15} kg is suspended in equilibrium between the plates. If e is the elementary charge, then charge on the particle is

(a) 8 (b) 6 (c) 0.1 (d) 3

25. A point charge of magnitude + 1 μC is fixed at (0, 0, 0). An isolated uncharged spherical conductor, is fixed with its center at (4, 0, 0). The potential and the induced electric field at the centre of the sphere is :

(a) 1.8×10^5 V and -5.625×10^6 V/m
(b) 0 V and 0 V/m
(c) 2.25×10^5 V and -5.625×10^6 V/m
(d) 2.25×10^5 V and 0 V/m

26. A plastic disc is charged on one side with a uniform surface charge density σ and then three quadrant of the disk are removed. The remaining quadrant is shown in figure, with $V = 0$ at infinity, the potential due to the remaining quadrant at point P is

(a) $\dfrac{\sigma}{2\varepsilon_0}[(r^2+R)^{1/2}-r]$

(b) $\dfrac{\sigma}{2\varepsilon_0}[R-r]$

(c) $\dfrac{\sigma}{8\varepsilon_0}[(r^2+R^2)^{1/2}-r]$

(d) None of these

27. A conducting sphere of radius R is given a charge Q. The electric potential and the electric field at the centre of the sphere respectively are: Tricky

(a) 0 and $\dfrac{Q}{4\pi\varepsilon_0 R^2}$

(b) $\dfrac{Q}{4\pi\varepsilon_0 R}$ and 0

(c) $\dfrac{Q}{4\pi\varepsilon_0 R}$ and $\dfrac{Q}{4\pi\varepsilon_0 R^2}$

(d) Both are 0

28. A conducting disc of radius R rotating about its axis with an angular velocity ω. Then the potential difference between the centre of the disc and its edge is (no magnetic field is present)

(a) zero (b) $\dfrac{m_e\omega^2R^2}{2e}$

(c) $\dfrac{m_e\omega R^3}{3e}$ (d) $\dfrac{em_e\omega R^2}{2}$

29. Four points a, b, c and d are set at equal distance from the centre of a dipole as shown in figure. The electrostatic potential V_a, V_b, V_c, and V_d would satisfy the following relation: Tricky

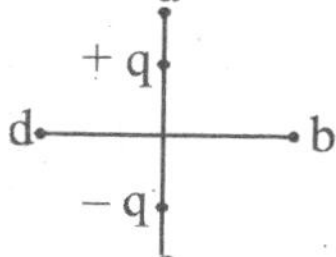

(a) $V_a > V_b > V_c > V_d$
(b) $V_a > V_b = V_d > V_c$
(c) $V_a > V_c = V_b = V_d$
(d) $V_b = V_d > V_a > V_c$

30. The potential at a point x (measured in μm) due to some charges situated on the x-axis is given by $V(x) = 20/(x^2-4)$ volt. The electric field E at x = 4 μm is given by

(a) (10/9) volt/μm and in the +ve x direction
(b) (5/3) volt/μm and in the –ve x direction
(c) (5/3) volt/μm and in the +ve x direction
(d) (10/9) volt/μm and in the –ve x direction

31. The expression $E = -\dfrac{dv}{dr}$ implies, that electric field is in that direction in which

(a) increase in potential is steepest.
(b) decrease in potential is steepest.
(c) change in potential is minimum.
(d) None of these

32. In a hollow spherical shell, potential (V) changes with respect to distance (s) from centre as

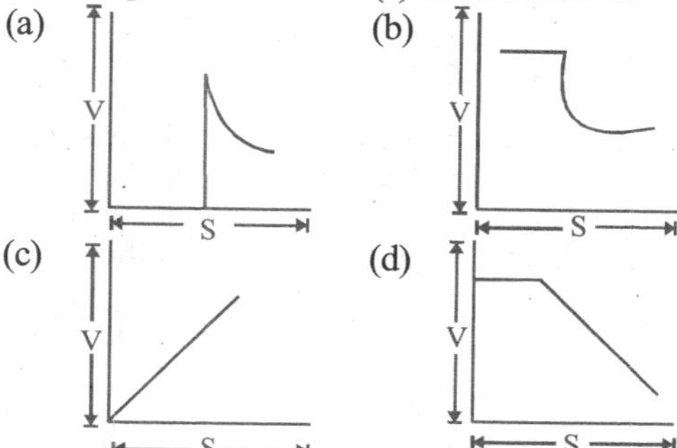

33. The 1000 small droplets of water each of radius r and charge Q, make a big drop of spherical shape. The potential of big drop is how many times the potential of one small droplet ? Critical Thinking

(a) 1 (b) 10 (c) 100 (d) 1000

34. The electric potential at a point (x, y) in the x – y plane is given by V = – kxy. The field intensity at a distance r from the origin varies as

(a) r^2 (b) r (c) $\frac{1}{r}$ (d) $\frac{1}{r^2}$

35. Four charges $q_1 = 2 \times 10^{-8}$ C, $q_2 = -2 \times 10^{-8}$ C, $q_3 = -3 \times 10^{-8}$ C, and $q_4 = 6 \times 10^{-8}$ C are placed at four corners of a square of side $\sqrt{2}$ m. What is the potential at the centre of the square?

(a) 270 V (b) 300 V (c) Zero (d) 100 V

36. A charge Q is distributed over two concentric hollow spheres of radii r and R ($R > r$) such that the surface densities are equal. The potential at the common centre is $\frac{1}{4\pi\varepsilon_0}$ times–

(a) $Q\left[\frac{r+R}{r^2+R^2}\right]$ (b) $\frac{Q}{2}\left(\frac{r+R}{r^2+R^2}\right)$

(c) $2Q\left(\frac{r+R}{r^2+R^2}\right)$ (d) zero

37. Three concentric charged metallic spherical shells A, B and C have radii a, b and c; charge densities σ, $-\sigma$ and σ and potentials V_A, V_B and V_C respectively. Then which of the following relations is correct? **Critical Thinking**

(a) $V_A = (a+b+c)\frac{\sigma}{\varepsilon_0}$

(b) $V_B = \left(\frac{a^2}{b} - b + c\right)\frac{\sigma}{\varepsilon_0}$

(c) $V_C = \left(\frac{a^2+b^2}{b} + c\right)\frac{\sigma}{\varepsilon_0}$

(d) $V_A = V_B = V_C = (a+b+c)\frac{\sigma}{\varepsilon_0}$

38. An electron having charge e and mass m starts from the lower plate of two metallic plates separated by a distance d. If the potential difference between the plates is V, the time taken by the electron to reach the upper plate is given by **Critical Thinking**

(a) $\sqrt{\frac{2md^2}{eV}}$ (b) $\sqrt{\frac{md^2}{eV}}$

(c) $\sqrt{\frac{md^2}{2eV}}$ (d) $\frac{2md^2}{eV}$

39. There is a uniform electrostatic field in a region. The potential at various points on a small sphere centred at P, in the region, is found to vary between in the limits 589.0 V to 589.8 V. What is the potential at a point on the sphere whose radius vector makes an angle of 60° with the direction of the field ?

(a) 589.5 V (b) 589.2 V (c) 589.4 V (d) 589.6 V

40. A charge of 3 coulomb moving in a uniform electric field experiences a force of 3000 newton. The potential difference between the two points situated in a field at a distance of 1 cm from each other will be :

(a) 100 (b) 5000 (c) 10 (d) 50

41. Three identical metallic uncharged spheres A, B and C each of radius a, are kept at the corners of an equilateral triangle of side d (d >> a) as shown in Fig. The fourth sphere (of radius a), which has a charge q, touches A and is then removed to a position far away. B is earthed and then the earth connection is removed. C is then earthed. The charge on C is **Critical Thinking**

(a) $\frac{qa}{2d}\left(\frac{2d-a}{2d}\right)$

(b) $\frac{qa}{2d}\left(\frac{2d-a}{d}\right)$

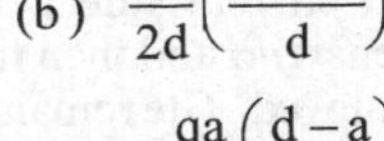

(c) $-\frac{qa}{2d}\left(\frac{d-a}{d}\right)$

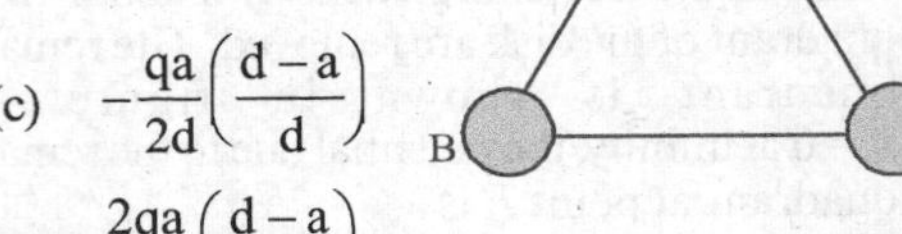

(d) $\frac{2qa}{d}\left(\frac{d-a}{2d}\right)$

42. Three charges $-q$, $+q$ and $+q$ are situated in X-Y plane at points (0, $-a$), (0, 0) and (0, a) respectively. The potential at a point distant r (r > a) in a direction making an angle θ from Y-axis will be

(a) $\frac{Kq}{r}\left(1-\frac{2a\cos\theta}{r}\right)$ (b) $\frac{2kq\cos\theta}{r^2}$

(c) $\frac{Kq}{r}$ (d) $\frac{Kq}{r}\left(1+\frac{2a\cos\theta}{r}\right)$

43. In a region, the potential is represented by V(x, y, z) = 6x – 8xy – 8y + 6yz, where V is in volts and x, y, z are in metres. The electric force experienced by a charge of 2 coulomb situated at point (1, 1, 1) is :

(a) $6\sqrt{5}$ N (b) 30 N (c) 24 N (d) $4\sqrt{35}$N

44. Figure shows a system of three concentric metal shells A, B and C with radii a, $2a$ and $3a$ respectively. Shell B is earthed and shell C is given a charge Q. Now if shell C is connected to shell A, then the final charge on the shell B, is equal to

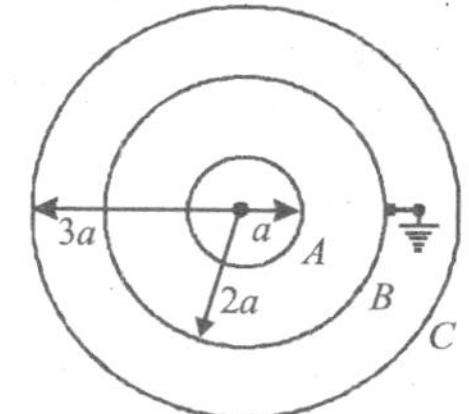

(a) $-4Q/13$ (b) $-8Q/11$ (c) $-5Q/3$ (d) $-3Q/7$

45. A charge $+q$ fixed at each of the points $x = x_0$, $x = 3x_0$, $x = 5x_0$, ... upto ∞ on X-axis and charge $-q$ is fixed on each of the points $x = 2x_0$, $x = 4x_0$, ... upto ∞. Here x_0 is a positive constant. Take the potential at a point due to a charge Q at a distance r form it to be $\frac{Q}{4\pi\varepsilon_0 r}$, then the potential at the origin due to above system of charges will be:

(a) zero (b) infinite

(c) $\frac{q}{8\pi\varepsilon_0 x_0 \log_e 2}$ (d) $\frac{q \log_e 2}{4\pi\varepsilon_0 x_0}$

46. Two small identical metal balls of radius r are at a distance a from each other and are charged, one with a potential V_1 and the other with a potential V_2. The charges on the balls are :

(a) $q_1 = V_1 a, q_2 = V_2 a$

(b) $q_1 = V_1 r, q_2 = V_2 r$

(c) $q_1 = \left(\frac{V_1 + V_2}{2}\right) a, q_2 = \left(\frac{V_1 + V_2}{2}\right) r$

(d) $q_1 = -\frac{r}{a}\left(rV_2 - aV_1\right), q_2 = -\frac{r}{a}\left(rV_1 - aV_2\right)$

47. A sphere of radius $2R$ has a uniform charge density ρ. The difference in the electric potential at $r = R$ and $r = 0$ from the centre is

(a) $\frac{-\rho R^2}{\epsilon_0}$ (b) $\frac{-2\rho R^2}{\epsilon_0}$ (c) $\frac{\rho R^2}{3\epsilon_0}$ (d) $\frac{-\rho R^2}{6\epsilon_0}$

48. The electric field intensity at all points in space is given by $\vec{E} = \sqrt{3}\hat{i} - \hat{j}$ volt/metre. The nature of equipotential lines in xy-plane is given by

Tricky

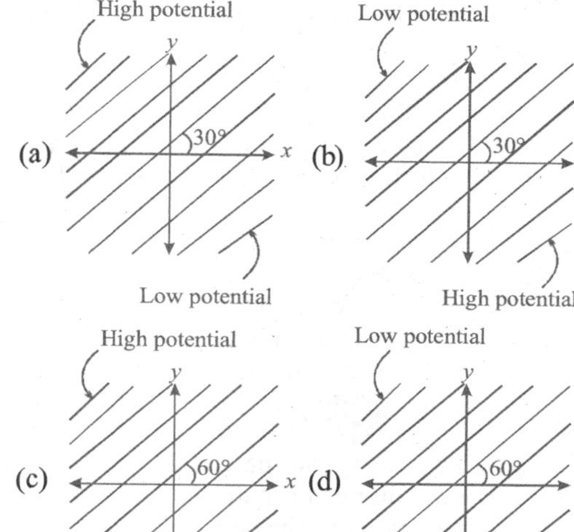

49. The electric potential at a point (x, y, z) is given by $V = -x^2y - xz^3 + 4$. The electric field $\vec{E}$ at that point is

(a) $\vec{E} = \hat{i}\,2xy + \hat{j}(x^2 + y^2) + \hat{k}(3xz - y^2)$

(b) $\vec{E} = \hat{i}\,z^3 + \hat{j}\,xyz + \hat{k}\,z^2$

(c) $\vec{E} = \hat{i}\,(2xy - z^3) + \hat{j}\,xy^2 + \hat{k}\,3z^2 x$

(d) $\vec{E} = \hat{i}\,(2xy + z^3) + \hat{j}\,x^2 + \hat{k}\,3xz^2$

50. An infinite non-conducting sheet has a surface charge density $\sigma = 0.1\ \mu C/m^2$ on one side. How far apart are equipotential surfaces whose potential differ by 50 volt ?

(a) 8.8 mm (b) 8.8 cm

(c) 8.8 μm (d) 8.8 pm

51. Two concentric conducting spherical shells of radii a_1 and a_2 ($a_2 > a_1$) are charged to potentials ϕ_1 and ϕ_2, respectively. Find the charge on the inner shell.

Tricky

(a) $q_1 = 4\pi\varepsilon_0\left(\frac{\phi_1 - \phi_2}{a_2 - a_1}\right)a_1a_2$

(b) $q_1 = 4\pi\varepsilon_0\left(\frac{\phi_1 + \phi_2}{a_2 + a_1}\right)a_1a_2$

(c) $q_1 = \pi\varepsilon_0\left(\frac{\phi_1 - \phi_2}{a_2 + a_1}\right)a_1a_2$

(d) $q_1 = 2\pi\varepsilon_0\left(\frac{\phi_1 + \phi_2}{a_2 - a_1}\right)a_1a_2$

52. The potential energy of a system of two charges is negative when
(a) both the charges are positive
(b) both the charges are negative
(c) one charge is positive and other is negative
(d) both the charges are separated by infinite distance

53. In the electric field of a point charge q, a certain charge is carried from point A to B, C, D and E. Then the work done is

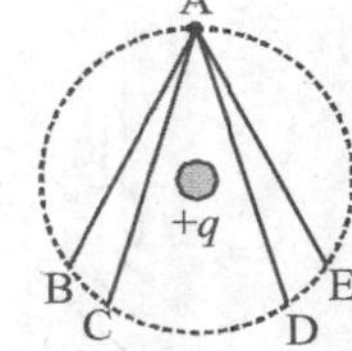

Tricky

(a) least along the path AB
(b) least along the path AD
(c) zero along all the paths AB, AC, AD and AE
(d) least along AE

54. Figure shows a system of three positive charges placed at the vertices of an equilateral triangle. To decrease the potential energy of the system,

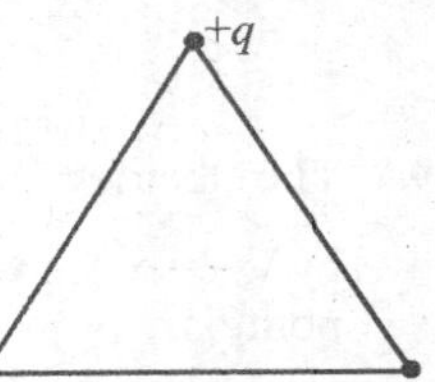

(a) a positive charge should be placed at centroid.
(b) a negative charge should be placed at centroid.
(c) distance between the charges should be decreased.
(d) it should be rotated by an angle of $\frac{\pi}{2}$ radian.

55. Two conducting spheres of radii R_1 and R_2 having charges Q_1 and Q_2 respectively are connected to each other. There is
(a) no change in the energy of the system
(b) an increase in the energy of the system
(c) always a decrease in the energy of the system
(d) a decrease in the energy of the system unless $Q_1R_2 = Q_2R_1$

56. A and B are two points in an electric field. If the work done in carrying 4.0C of electric charge from A to B is 16.0 J, the potential difference between A and B is
(a) zero (b) 2.0V (c) 4.0V (d) 16.0V

57. Two points P and Q are maintained at the potentials of 10 V and – 4 V, respectively. The work done in moving 100 electrons from P to Q is:
(a) 9.60×10^{-17} J (b) -2.24×10^{-16} J
(c) 2.24×10^{-16} J (d) -9.60×10^{-17} J

58. A ball of mass 1 g carrying a charge 10^{-8} C moves from a point A at potential 600 V to a point B at zero potential. The change in its K.E. is
(a) -6×10^{-6} erg (b) -6×10^{-6} J
(c) 6×10^{-6} J (d) 6×10^{-6} erg

59. As per the diagram, a point charge $+q$ is placed at the origin O. Work done in taking another point charge $-Q$ from the point A [coordinates $(0, a)$] to another point B [coordinates $(a, 0)$] along the straight path AB is:

Critical Thinking

(a) zero

(b) 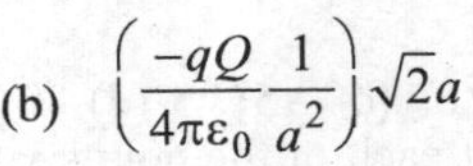 $\left(\frac{-qQ}{4\pi\varepsilon_0}\frac{1}{a^2}\right)\sqrt{2}a$

(c) 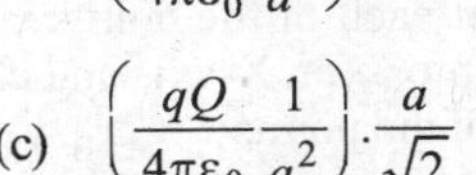 $\left(\frac{qQ}{4\pi\varepsilon_0}\frac{1}{a^2}\right)\cdot\frac{a}{\sqrt{2}}$

(d) 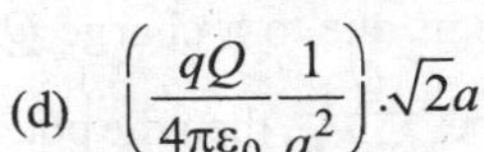 $\left(\frac{qQ}{4\pi\varepsilon_0}\frac{1}{a^2}\right)\cdot\sqrt{2}a$

60. Positive and negative point charges of equal magnitude are kept at $\left(0,0,\frac{a}{2}\right)$ and $\left(0,0,\frac{-a}{2}\right)$ respectively. The work done by the electric field when another positive point charge is moved from $(-a, 0, 0)$ to $(0, a, 0)$ is
(a) positive (b) negative (c) zero
(d) depends on the path connecting the initial and final positions

61. Four identical charges are placed at the four vertices of a square lying in YZ plane. A fifth charge is moved along X axis. The variation of potential energy (U) along X axis is correctly represented by

Tricky

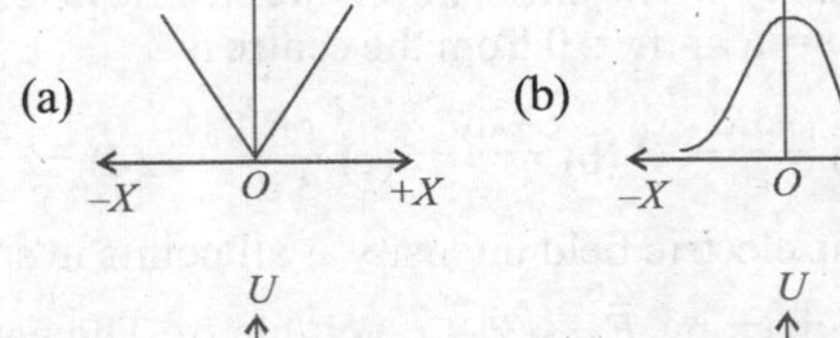

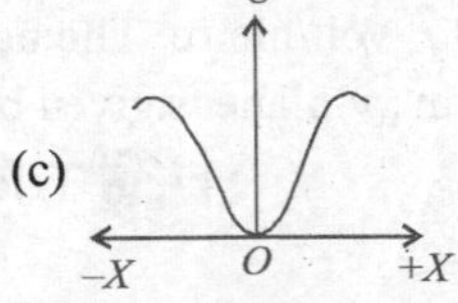

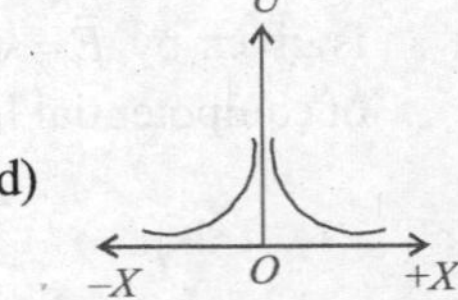

62. Each corner of a cube of side l has a negative charge, $-q$. The electrostatic potential energy of a charge q at the centre of the cube is

(a) $-\frac{4q^2}{\sqrt{2}\pi\varepsilon_0 l}$ (b) $\frac{\sqrt{3}q^2}{4\pi\varepsilon_0 l}$

(c) $\frac{4q^2}{\sqrt{2}\pi\varepsilon_0 l}$ (d) $-\frac{4q^2}{\sqrt{3}\pi\varepsilon_0 l}$

63. Two identical thin rings each of radius R meters are coaxially placed at a distance R meters apart. If Q_1 coulomb and Q_2 coulomb are respectively the charges uniformly spread on the two rings, the work done in moving a charge q from the centre of one ring to that of other is

(a) zero (b) $\frac{q(Q_1 - Q_2)(\sqrt{2}-1)}{\sqrt{2}.4\pi\varepsilon_0 R}$

(c) $\frac{q\sqrt{2}(Q_1 + Q_2)}{4\pi\varepsilon_0 R}$ (d) $\frac{q(Q_1 + Q_2)(\sqrt{2}+1)}{\sqrt{2}.4\pi\varepsilon_0 R}$

64. On decreasing the distance between the plates of a parallel plate capacitor, its capacitance
(a) remains unaffected
(b) decreases
(c) first increases then decreases.
(d) increases

65. If in a parallel plate capacitor, which is connected to a battery, we fill dielectrics in whole space of its plates, then which of the following increases?
(a) Q and V (b) V and E (c) E and C (d) Q and C

66. Two vertical metallic plates carrying equal and opposite charges are kept parallel to each other like a parallel plate capacitor. A small spherical metallic ball is suspended by a long insulated thread such that it hangs freely in the centre of the two metallic plates. The ball, which is uncharged, is taken slowly towards the positively charged plate and is made to touch that plate. Then the ball will
(a) stick to the positively charged plate
(b) come back to its original position and will remain there
(c) oscillate between the two plates touching each plate in turn
(d) oscillate between the two plates without touch them

67. Eight drops of mercury of equal radii possessing equal charges combine to form a big drop. Then the capacitance of bigger drop compared to each individual small drop is
(a) 8 times (b) 4 times
(c) 2 times (d) 32 times

68. A parallel plate condenser is filled with two dielectrics as shown. Area of each plate is A m^2 and the separation is t m.
The dielectric constants are k_1 and k_2 respectively. Its capacitance in farad will be

k_1 | k_2

(a) $\frac{\varepsilon_0 A}{t}(k_1 + k_2)$ (b) $\frac{\varepsilon_0 A}{t}.\frac{k_1 + k_2}{2}$

(c) $\frac{2\varepsilon_0 A}{t}(k_1 + k_2)$ (d) $\frac{\varepsilon_0 A}{t}.\frac{k_1 - k_2}{2}$

69. The resultant capacity of n condensers of capacitances $C_1, C_2 .... C_n$ connected in parallel is
(a) $C_p = C_1 + C_2 + ... + C_n$
(b) $C_p = C_1 - C_2 - C_3 .... - C_n$
(c) $\frac{1}{C_p} = \frac{1}{C_1} + \frac{1}{C_2} + ... + \frac{1}{C_n}$
(d) $C_p = \frac{1}{C_1} + \frac{1}{C_2} + ... + \frac{1}{C_n}$

70. If earth is assumed to be a conducting sphere having radius R = 6400 km, it's capacitance will be :
(a) 711 μF (b) 218 μF (c) 16 μF (d) 8 μF

71. A capacitor has two circular plates whose radius are 8cm and distance between them is 1mm. When mica (dielectric constant = 6) is placed between the plates, the capacitance of this capacitor and the energy stored when it is given potential of 150 volt respectively are

Critical Thinking

(a) 1.06×10^{-5} F, 1.2×10^{-9} J
(b) 1.068×10^{-9} F, 1.2×10^{-5} J
(c) 1.2×10^{-9} F, 1.068×10^{-5} J
(d) 1.6×10^{-9} F, 1.208×10^{-5} J

72. Three condensor each of capacitance 2F are put in series. The resultant capacitance is
(a) 6 F (b) 3/2 F (c) 2/3 F (d) 5 F

73. Two spherical conductors A_1 and A_2 of radii r_1 and r_2 ($r_2 > r_1$) are placed concentrically in air. A_1 is given a charge +Q while A_2 is earthed. Then the equivalent capacitance of the system is

Tricky

(a) $\frac{4\pi \epsilon_0 r_1 r_2}{r_2 - r_1}$
(b) $4\pi \epsilon_0 (r_1 + r_2)$
(c) $4\pi \epsilon_0 r_2$
(d) $4\pi \epsilon_0 r_1$

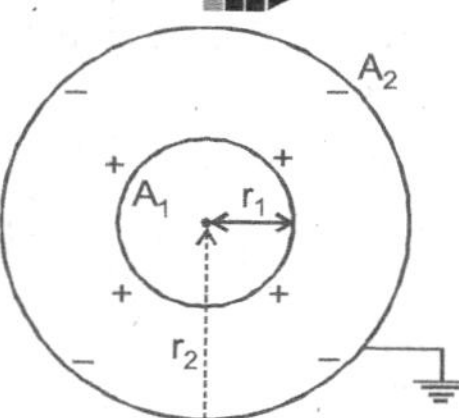

74. The capacitor, whose capacitance is 6, 6 and 3μF respectively are connected in series with 20 volt line. Find the charge on 3μF.

6μF 6μF 3μF

20μF

(a) 30 μc
(b) 60 μF
(c) 15 μF
(d) 90 μF

75. A parallel plate capacitor with air between the plates has a capacitance of 8 pF. Calculate the capacitance if the distance between the plates is reduced by half and the space between them is filled with a substance of dielectric constant. ($\varepsilon_r = 6$)

(a) 72 pF (b) 81 pF (c) 84 pF (d) 96 pF

76. A parallel plate air capacitor has a capacitance of 100 μF. The plates are at a distance *d* apart. If a slab of thickness *t* ($t < d$) and dielectric constant 5 is introduced between the parallel plates, then the capacitance will be

(a) 50 μF (b) 100 μF (c) 200 μF (d) 500 μF

77. A parallel plate capacitor having a separation between the plates d, plate area A and material with dielectric constant K has capacitance C_0. Now one-third of the material is replaced by another material with dielectric constant 2K, so that effectively there are two capacitors one with area $\frac{1}{3}$ A, dielectric constant 2K and another with area $\frac{2}{3}$ A and dielectric constant K. If the capacitance of this new capacitor is C then $\frac{C}{C_0}$ is

(a) 1 (b) $\frac{4}{3}$ (c) $\frac{2}{3}$ (d) $\frac{1}{3}$

78. A parallel plate capacitor with air between the plates is charged to a potential difference of 500V and then insulated. A plastic plate is inserted between the plates filling the whole gap. The potential difference between the plates now becomes 75V. The dielectric constant of plastic is

Critical Thinking

(a) 10/3 (b) 5 (c) 20/3 (d) 10

79. The effective capacitance of combination of equal capacitors between points A and B shown in figure is

(a) C (b) 2C
(c) 3C (d) $\frac{C}{2}$

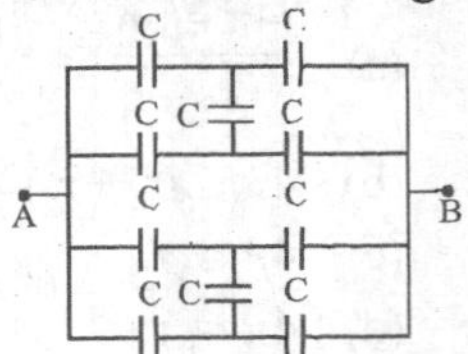

80. Three capacitors are connected in the arms of a triangle ABC as shown in figure 5 V is applied between A and B. The voltage between B and C is

C 2μF 3μF A B 2μF

(a) 2V (b) 1V
(c) 3V (d) 1.5V

81. To obtain 3 μF capacity from three capacitors of 2 μF each, they will be arranged. Tricky

(a) all the three in series
(b) all the three in parallel
(c) two capacitors in series and the third in parallel with the combinatioin of first two
(d) two capacitors in parallel and the third in series with the combinatioin of first two

82. The energy required to charge a parallel plate condenser of plate separation *d* and plate area of cross-section *A* such that the uniform electric field between the plates is *E*, is

(a) $\epsilon_0 E^2 Ad$ (b) $\frac{1}{2}\epsilon_0 E^2 Ad$
(c) $\frac{1}{2}\epsilon_0 E^2/Ad$ (d) $\epsilon_0 E^2/Ad$

83. In the circuit given below, the charge in μC, on the capacitor having capacitance 5 μF is

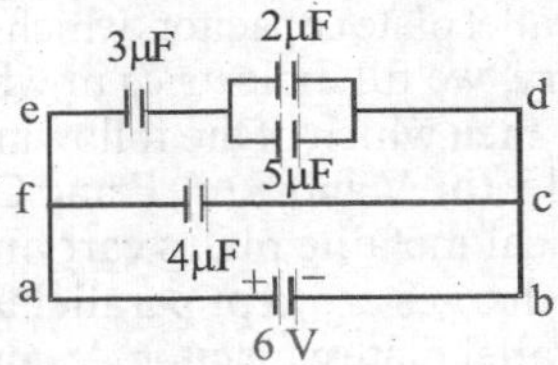

(a) 4.5 (b) 9 (c) 7 (d) 15

84. Capacitance (in F) of a spherical conductor with radius 1 m is

(a) 1.1×10^{-10} (b) 10^6
(c) 9×10^{-9} (d) 10^{-3}

85. The capacitance of a metallic sphere is 1μF, then it's radius is nearly

(a) 1.11 m (b) 10 m (c) 9 km (d) 1.11 cm

86. A network of six identical capacitors, each of value *C* is made as shown in the figure. Equivalent capacitance between points *A* and *B* is

(a) *C*/4
(b) 3*C*/4
(c) 4*C*/3
(d) 3*C*

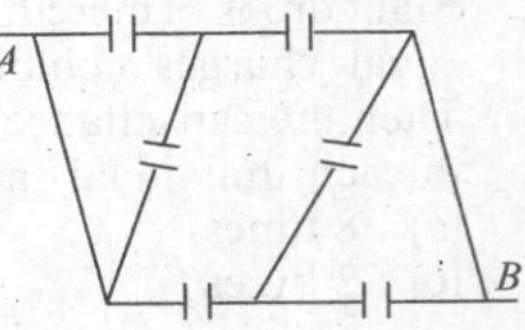

87. A series combination of n_1 capacitors, each of capacity C_1 is charged by source of potential difference 4 V. When another parallel combination of n_2 capacitors each of capacity C_2 is charged by a source of potential difference V, it has the same total energy stored in it as the first combination has. The value of C_2 in terms of C_1 is then

(a) $16\frac{n_2}{n_1}C_1$ (b) $\frac{2C_1}{n_1 n_2}$

(c) $2\frac{n_2}{n_1}C_1$ (d) $\frac{16C_1}{n_1 n_2}$

88. Two capacitors of capacitances C_1 and C_2 are connected in series, assume that $C_1 < C_2$. The equivalent capacitance of this arrangement is C, where

(a) $C < C_1/2$ (b) $C_1/2 < C < C_2/2$

(c) $C_1 < C < C_2$ (d) $C_2 < C < 2C_2$

89. Four identical square plates of side a are arranged as shown. The equivalent capacity between A and C **Tricky**

(a) $\frac{3\varepsilon_0 a^2}{2d}$ (b) $\frac{3\varepsilon_0 a^2}{5d}$

(c) $\frac{3\varepsilon_0 a^2}{3d}$ (d) $\frac{5\varepsilon_0 a^2}{3d}$.

90. Two parallel plate capacitors of capacitance C and 2C are connected in parallel and charged to a potential difference V. The battery is then disconnected, and the region between the plates of C is filled completely with a material of dielectric constant K. The common potential difference across the combination becomes

(a) $\frac{2V}{K+2}$ (b) $\frac{V}{K+2}$ (c) $\frac{3V}{K+3}$ (d) $\frac{3V}{K+2}$

91. Seven capacitors each of capacitance $2\mu F$ are to be connected in a configuration to obtain an effective capacitance of $\left(\frac{10}{11}\right)\mu F$. Which of the combination (s) shown in figure will achieve the desired result ?

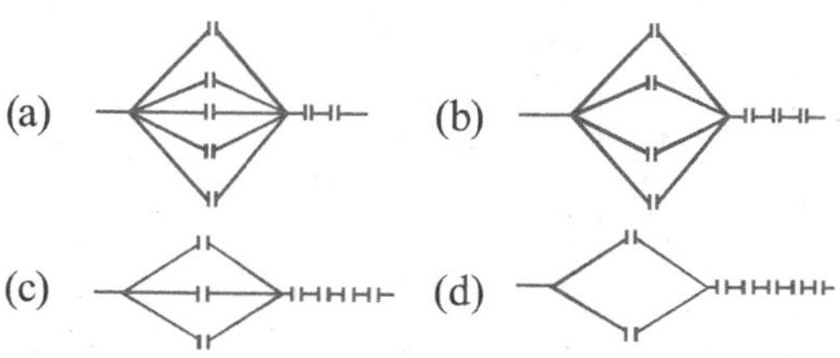

92. In figure, there is a four way key at the middle. If key is shown from situation BD to AD, then how much charge will flow through point O? **Tricky**

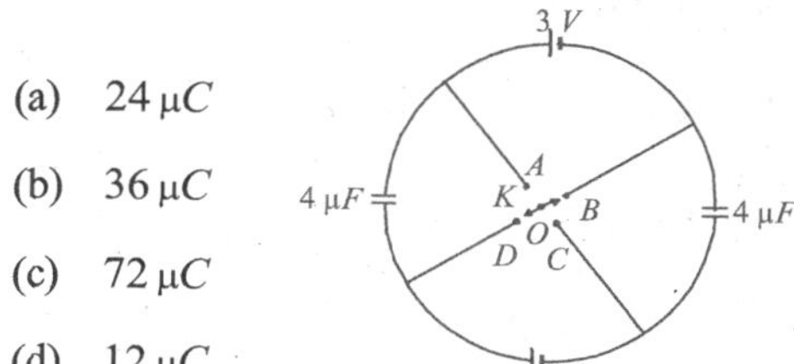

(a) 24 μC

(b) 36 μC

(c) 72 μC

(d) 12 μC

93. An uncharged paralle plate capacitor having a dielectric of dielectric constant K is connected to a similar air cored parallel plate capacitor charged to a potential V_0. The two share the charge, and the common potential becomes V. The dielectric constant K is

(a) $\frac{V_0}{V}-1$ (b) $\frac{V_0}{V}+1$ (c) $\frac{V}{V_0}-1$ (d) $\frac{V}{V_0}+1$

94. A capacitor of capacity C_1 is charged upto V volt and then connected to an uncharged capacitor of capacity C_2. Then final potential difference across each will be

(a) $\frac{C_2 V}{C_1+C_2}$ (b) $\left(1+\frac{C_2}{C_1}\right)V$

(c) $\frac{C_1 V}{C_1+C_2}$ (d) $\left(1-\frac{C_2}{C_1}\right)V$

95. A parallel plate capacitor of area 'A' plate separation 'd' is filled with two dielectrics as shown. What is the capacitance of the arrangement ? **Critical Thinking**

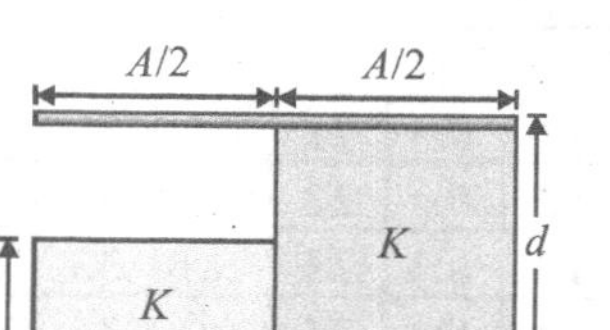

(a) $\frac{3K\varepsilon_0 A}{4d}$ (b) $\frac{4K\varepsilon_0 A}{3d}$

(c) $\frac{(K+1)\varepsilon_0 A}{2d}$ (d) $\frac{K(K+3)\varepsilon_0 A}{2(K+1)d}$

96. In the circuit shown, the effective capacitance between points X and Y is

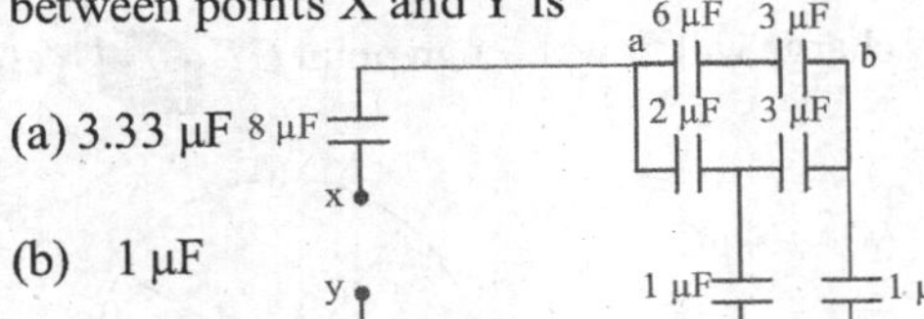

(a) 3.33 μF

(b) 1 μF

(c) 0.44 μF

(d) None of these

97. A combination of parallel plate capacitors is maintained at a certain potential difference.

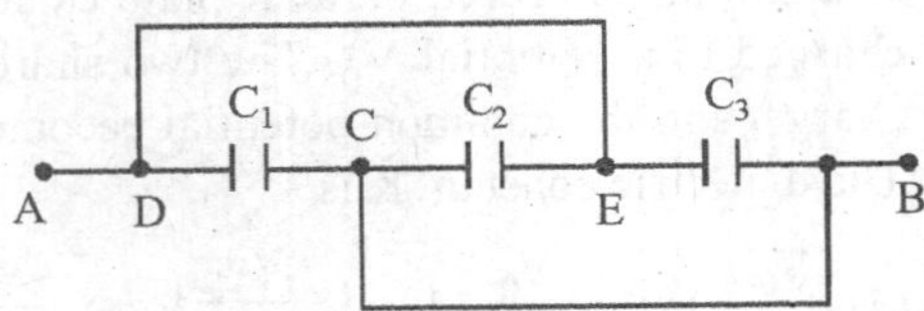

When a 3 mm thick slab is introduced between all the plates, in order to maintain the same potential difference, the distance between the plates is increased by 2.4 mm. **Toughnut**

Find the dielectric constant of the slab.

(a) 3 (b) 4 (c) 5 (d) 6

98. For the configuration of media of permittivities ε_0, ε and ε_0 between parallel plates each of area A, as shown in Fig. the equivalent capacitance is

(a) $\varepsilon_0 A/d$

(b) $\varepsilon\varepsilon_0 A/d$

(c) $\dfrac{\varepsilon\varepsilon_0 A}{d(\varepsilon+\varepsilon_0)}$

(d) $\dfrac{\varepsilon\varepsilon_0 A}{(2\varepsilon+\varepsilon_0)d}$

99. In a Van de Graaff generator, a spherical metal shell is to be 15×10^6 V electrode. The dielectric strength of the gas surrounding the electrode is $5 \times 10^7 V\, m^{-1}$. The minimum radius of the spherical shell required is

(a) 1 *m* (b) 2 *m* (c) 1.5 *m* (d) 3 *m*

100. A unit positive point charge of mass *m* is projected with a velocity *V* inside the tunnel as shown. The tunnel has been made inside a uniformly charged nonconducting sphere (charge density ρ). The minimum velocity with which the point charge should be projected such that it can reach the opposite end of the tunnel is equal to **Toughnut**

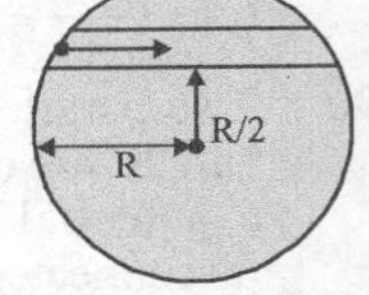

(a) $[\rho R^2/4m\varepsilon_0]^{1/2}$

(b) $[\rho R^2/24m\varepsilon_0]^{1/2}$

(c) $[\rho R^2/6m\varepsilon_0]^{1/2}$

(d) zero because the initial and the final points are at same potential

Answer KEYs

1	(c)	11	(b)	21	(c)	31	(b)	41	(c)	51	(a)	61	(b)	71	(b)	81	(c)	91	(a)
2	(c)	12	(b)	22	(c)	32	(b)	42	(d)	52	(c)	62	(d)	72	(c)	82	(a)	92	(c)
3	(c)	13	(d)	23	(a)	33	(c)	43	(d)	53	(c)	63	(b)	73	(a)	83	(b)	93	(a)
4	(b)	14	(b)	24	(d)	34	(b)	44	(b)	54	(c)	64	(d)	74	(a)	84	(a)	94	(c)
5	(c)	15	(a)	25	(c)	35	(a)	45	(d)	55	(d)	65	(d)	75	(d)	85	(c)	95	(d)
6	(b)	16	(c)	26	(c)	36	(a)	46	(d)	56	(c)	66	(c)	76	(c)	86	(c)	96	(b)
7	(c)	17	(b)	27	(b)	37	(b)	47	(d)	57	(c)	67	(c)	77	(b)	87	(d)	97	(c)
8	(a)	18	(a)	28	(b)	38	(a)	48	(c)	58	(c)	68	(b)	78	(c)	88	(b)	98	(d)
9	(a)	19	(c)	29	(b)	39	(c)	49	(d)	59	(a)	69	(a)	79	(b)	89	(c)	99	(d)
10	(a)	20	(c)	30	(a)	40	(c)	50	(a)	60	(c)	70	(d)	80	(a)	90	(d)	100	(a)

Hints & Solutions

1. (c) Potential at $C = V_C = 0$
Potential at $D = V_D$

$= k\left(\frac{-q}{L}\right) + \frac{kq}{3L}$

$= -\frac{2}{3}\frac{kq}{L}$

Potential difference

$V_D - V_C = \frac{-2}{3}\frac{kq}{L} = \frac{1}{4\pi\epsilon_0}\left(-\frac{2}{3}\cdot\frac{q}{L}\right)$

$\Rightarrow$ Work done $= Q(V_D - V_C)$

$= -\frac{2}{3} \times \frac{1}{4\pi\varepsilon_0}\frac{qQ}{L} = \frac{-qQ}{6\pi\varepsilon_0 L}$

2. (c) Potential energy of the system

$= 4 \times \frac{K}{2}\left(\frac{q^2}{L} + \frac{q^2}{L} + \frac{q^2}{L\sqrt{2}}\right) = 5.4\frac{Kq^2}{L}$

= Final kinetic energy of the system

$= 4 \times \frac{1}{2}mv^2 = 2mv^2$

$\therefore \quad v = \left[\frac{Kq^2}{mL}(2.7)\right]^{1/2}$

3. (c) $U = 2kq^2\left[-\frac{1}{a} + \frac{1}{2a} - \frac{1}{3a} + \frac{1}{4a} + ...\right]$

$= -\frac{2q^2}{4\pi\varepsilon_0 a}\left[1 - \frac{1}{2} + \frac{1}{3} - \frac{1}{4} + ...\right] = -\frac{2q^2 \log_e 2}{4\pi\varepsilon_0 a}$

4. (b) Since the given system is closed, the increase in KE is equal to decrease in P.E.

$\Rightarrow \frac{3}{2}mv^2 = \frac{2kq^2}{r_0} - \frac{3kq^2}{r}$

$\Rightarrow v = \sqrt{\frac{2kq^2(r - r_0)}{mrr_0}}$,

v will be max when $r \to \infty$

$\Rightarrow v_{max} = \sqrt{\frac{2kq^2}{mr_0}}$

The work performed by the interaction force during the variation of the system's configuration is equal to the decrease in the potential energy

$W = U_1 - U_2 = \frac{3kq^2}{r_0}$

$\therefore$ Work done per particle $= \frac{kq^2}{r_0}$

5. (c) We know that potential energy of discrete system of charges is given by

$U = \frac{1}{4\pi\epsilon_0}\left(\frac{q_1q_2}{r_{12}} + \frac{q_2q_3}{r_{23}} + \frac{q_3q_1}{r_{31}}\right)$

According to question,

$U_{initial} = \frac{1}{4\pi\epsilon_0}\left(\frac{q_1q_2}{0.3} + \frac{q_2q_3}{0.5} + \frac{q_3q_1}{0.4}\right)$

$U_{final} = \frac{1}{4\pi\epsilon_0}\left(\frac{q_1q_2}{0.3} + \frac{q_2q_3}{0.1} + \frac{q_3q_1}{0.4}\right)$

$U_{final} - U_{initial} = \frac{1}{4\pi\epsilon_0}\left(\frac{q_1q_2}{0.1} - \frac{q_2q_3}{0.5}\right)$

$= \frac{1}{4\pi\epsilon_0}[10q_2q_3 - 2q_2q_3] = \frac{q_3}{4\pi\epsilon_0}(8q_2)$

6. (b) $C_{medium} = K \times C_{air}$

7. (c) Work done $= \frac{1}{2}\frac{q^2}{C} = \frac{(8\times10^{-18})^2}{2\times100\times10^{-6}}$
$= 32 \times 10^{-32}$ J

8. (a) When S is closed, there will be no shifting of negative charge from plate A to B as the charge $-q$ is held by the charge $+q$. Neither there will be any shifting of charge from B to A.

9. (a) The molecules of liquid will convert into induced dipole, get oriented along the electric field produced between the plates and rise due to force of attraction.

10. (a) By symmetry $C_5 = 10\ \mu F$
Capacitor between B and C can be removed

so, now $C_{eq} = \frac{8\times8}{8+8} + \frac{8\times8}{8+8} = 8\ \mu F$

11. (b) Potential at any point inside the sphere = potential at the surface of the sphere = 10V.

12. (b) Potential should be same i.e. $V_A = V_B$

13. (d) Surface of metallic cube is an equipotential surface. Therefore, electric field is normal to the surface of the cube.

14. (b) $W_{el.} = q(V_i - V_f)$

or $6.4 \times 10^{-19} = -1.6 \times 10^{-19}(V_A - V_B)$

or $V_A - V_B = -4V$

or $V_A - V_C = -4V \quad (\because V_B = V_C)$

or $V_C - V_A = 4V$

15. (a) $\int_{\ell=\infty}^{\ell=0} -\vec{E}.d\vec{\ell} = V_0 - V_\infty$

$$= \frac{kq}{r} - 0$$

$$= \frac{9 \times 10^9 \times 1.11 \times 10^{-10}}{0.5}$$

$\approx 2V$

non uniform charge distribution on the ring

16. (c)

Let charge on each sphere $= q$ when they are connected together their potential will be equal.

Now let charge on $a = q_1$ and on $b = 2q - q_1$

$$\Rightarrow V_a = V_b \text{ or } \frac{1}{4\pi\varepsilon_o}\frac{q_1}{a} = \frac{1}{4\pi\varepsilon_o}\frac{2q - q_1}{b}$$

$$\Rightarrow \frac{q_1}{2q - q_1} = \frac{a}{b}$$

$$\frac{E_a}{E_b} = \frac{\frac{1}{4\pi\varepsilon_o}\frac{q_1}{a^2}}{\frac{1}{4\pi\varepsilon_o}\frac{q_2}{b^2}} = \left(\frac{q_1}{2q - q_1}\right)\frac{b^2}{a^2}$$

$$= \frac{a}{b}.\frac{b^2}{a^2} = \frac{b}{a} = b : a$$

17. (b) $E = -\nabla\phi = -\phi_0 2[x\hat{i} + y\hat{i} + z\hat{x}]$

$= \varepsilon_0 \nabla.E = -2\varepsilon_0\phi_0\nabla.(x\hat{i} + y\hat{i} + z\hat{x})$

$n = -6\phi_0\varepsilon_0$

18. (a) Potential at B, V_B is maximum

$V_B > V_C > V_A$

As in the direction of electric field potential decreases.

19. (c)

20. (c) Equipotential surfaces are normal to the electric field lines.

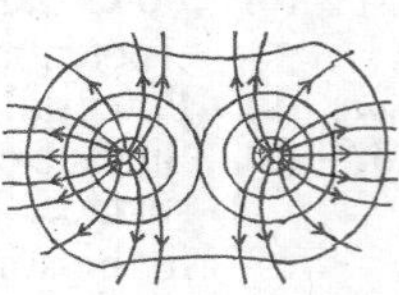

The following figure shows the equipotential surfaces along with electric field lines for a system of two positive charges.

21. (c) Using $dV = -\vec{E}.d\vec{r}$

$\Rightarrow \Delta V = -E\,\Delta r \cos\theta$

$$\Rightarrow E = \frac{-\Delta V}{\Delta r \cos\theta}$$

$$\Rightarrow E = \frac{-(20-10)}{10 \times 10^{-2} \cos 120^\circ}$$

$$= \frac{-10}{10 \times 10^{-2}(-\sin 30^\circ)} = 200 \text{ V/m}$$

Direction of E be perpendicular to the equipotential surface *i.e.* at 120° with X-axis.

22. (c) Charge on the outer sphere $= q_1 = 4\pi R^2\sigma$

Charge on the inner sphere $= q_2 = 4\pi r^2\sigma$

$$V = \frac{1}{4\pi\epsilon_0}\frac{q_1}{R} + \frac{1}{4\pi\epsilon_0}\frac{q_2}{r}$$

$$= \frac{1}{4\pi\epsilon_0}\left[\frac{4\pi R^2\sigma}{R} + \frac{4\pi r^2\sigma}{r}\right]$$

$$= \frac{4\pi\sigma}{4\pi\epsilon_0}(R + r) = \frac{\sigma}{\epsilon_0}(R + r)$$

23. (a) $E = 500$ V/m $\quad V = 3000$ V.

We know that electric field $(E) = 500 = \frac{V}{d}$

or $d = \frac{3000}{500} = 6$m

24. (d) In equilibrium, $F = qE = (ne)\frac{V}{d} = mg$

$$n = \frac{mgd}{eV} = \frac{1.96 \times 10^{-15} \times 9.8 \times 0.02}{1.6 \times 10^{-19} \times 800} = 3$$

25. (c) $q = 1\mu C = 1 \times 10^{-6}$C, $r = 4$ cm $= 4 \times 10^{-2}$ m

Potential $V = \frac{kq}{r} = \frac{9 \times 10^9 \times 10^{-6}}{4 \times 10^{-2}}$

$= 2.25 \times 10^5$ V.

Induced electric field $E = -\frac{kq}{r^2}$

$$= \frac{9 \times 10^9 \times 1 \times 10^{-6}}{16 \times 10^{-4}} = -5.625 \times 10^6 \text{ V/m}$$

26. (c) The potential at P due to whole disc is

$$V = \frac{\sigma}{2\epsilon_0}\left[\sqrt{R^2 + r^2} - r\right].$$

Now potential due to quarter disc,

$$V = \frac{V}{4} = \frac{\sigma}{8\epsilon_0}\left[\sqrt{R^2+r^2} - r\right].$$

27. (b) Due to conducting sphere
At centre, electric field $E = 0$

And electric potential $V = \dfrac{Q}{4\pi\epsilon_0 R}$

28. (b) $eE = m_e\omega^2 r$

$$\Rightarrow \int E\,dr = \frac{m_e\omega^2}{e}\int_0^R r\,dr$$

$$\Rightarrow V = \frac{m_e\omega^2 R^2}{2e}$$

29. (b)

30. (a) Here, $V(x) = \dfrac{20}{x^2-4}$ volt

We know that $E = -\dfrac{dV}{dx} = -\dfrac{d}{dx}\left(\dfrac{20}{x^2-4}\right)$

or, $E = +\dfrac{40x}{(x^2-4)^2}$

At $x = 4\ \mu m$,

$$E = +\frac{40\times 4}{(4^2-4)^2} = +\frac{160}{144} = +\frac{10}{9}\ \text{volt}/\mu m.$$

Positive sign indicates that $\vec{E}$ is in +ve x-direction.

31. (b) **32. (b)**

33. (c) Volume of big drop = 1000 × volume of each small drop

$$\frac{4}{3}\pi R^3 = 1000\times\frac{4}{3}\pi r^3 \Rightarrow R = 10r$$

$\because\ V = \dfrac{kq}{r}$ and $V' = \dfrac{kq}{R}\times 1000$

Total charge on one small droplet is q and on the big drop is 1000q.

$$\Rightarrow \frac{V'}{V} = \frac{1000r}{R} = \frac{1000}{10} = 100$$

$\therefore\ V' = 100V$

34. (b) $\vec{E} = \dfrac{\partial v}{\partial x}\hat{i} + \dfrac{\partial v}{\partial y}\hat{j}$

$\therefore |\vec{E}| = k(\sqrt{x^2+y^2}) = kr$

Given $v = -kxy$ $\qquad E \propto r$

$\therefore\ \vec{E} = ky\hat{i} + kx\hat{j}$

35. (a) Potential at the centre O, $V = k\dfrac{q}{r}$

$$V = k\left[\frac{2\times10^{-8}}{1} + \frac{-2\times10^{-8}}{1} + \frac{-3\times10^{-8}}{1} + \frac{6\times10^{-8}}{1}\right]$$

$$V = k\times 3\times10^{-8} = 9\times10^9\times3\times10^{-8}\ \text{volt} = 27\times10 = 270\ V$$

36. (a) $q_r + q_R = Q$ ... (1)

$$\frac{q_r}{q_R} = \frac{4\pi r^2\sigma}{4\pi R^2\sigma} = \frac{r^2}{R^2} \Rightarrow \frac{q_r}{q_R} = \frac{r^2}{R^2} \quad \text{... (2)}$$

From (1) and (2)

$$q_r = \frac{Qr^2}{R^2+r^2} \text{ and } q_R = \frac{QR^2}{R^2+r^2}$$

So, $V = \dfrac{q_r}{4\pi\varepsilon_0 r} + \dfrac{q_R}{4\pi\varepsilon_0 R}$

$$\Rightarrow V = \frac{Q}{4\pi\varepsilon_0}\left(\frac{r+R}{r^2+R^2}\right)$$

37. (b) Potential due to a charged shell at an outside point

$$= \frac{1}{4\pi\varepsilon_0}\times\frac{Q}{r}$$

while that inside the shell $= \dfrac{\sigma R}{\varepsilon_0}$, where R is the radius of the shell.

38. (a) $E = \dfrac{V}{d}$, $F = eE = eV/d$

$$a = \frac{F}{m} = \frac{eV}{md}$$

$$d = \frac{1}{2}at^2 \text{ or } t = \sqrt{\frac{2d}{a}}$$

$$\text{or}\quad t = \sqrt{\frac{2d\,md}{eV}} = \sqrt{\frac{2md^2}{eV}}$$

39. (c)

40. (c) Force on a charge in an electric field is $F = qE$

But $E = \dfrac{V}{d}$ $\qquad \therefore\ F = \dfrac{qV}{d}$

or $V = \dfrac{Fd}{q} = \dfrac{300N\times1\times10^{-2}m}{3C} = 10$ volt

41. (c) $\dfrac{kq_1}{a} + \dfrac{kq}{2d} = 0$

or $q_1 = \dfrac{-qa}{2d}$

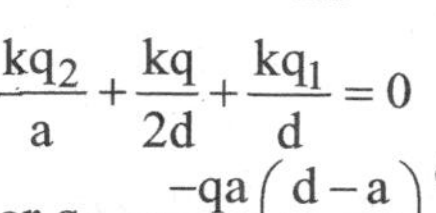

$$\frac{kq_2}{a} + \frac{kq}{2d} + \frac{kq_1}{d} = 0$$

$$\text{or } q_2 = \frac{-qa}{2d}\left(\frac{d-a}{d}\right)$$

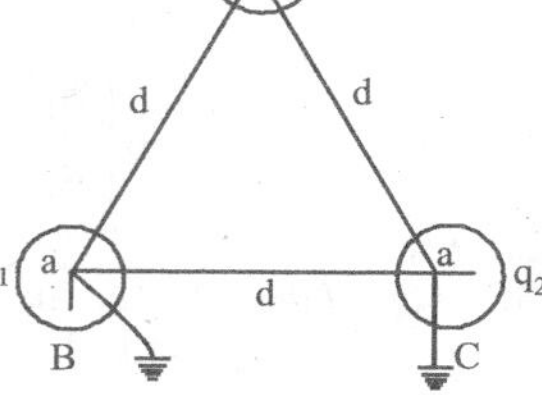

42. (d) $V = \frac{Kq.2a\cos\theta}{r^2} + \frac{Kq}{r} = \frac{Kq}{r}\left[\frac{2a\cos\theta}{r} + 1\right]$

43. (d)

44. (b) From given conditions, $V_A = V_C$ and $V_B = 0$

$\Rightarrow V_B = \frac{k(Q-q_1)}{3a} + \frac{kq_2}{2a} + \frac{kq_1}{2a} = 0$

$\Rightarrow 2Q + q_1 + 3q_2 = 0$ .... (i)

Using $V_A = V_C$

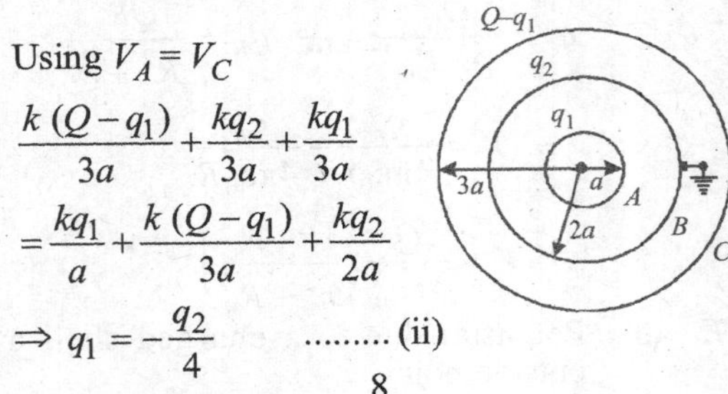

$\frac{k(Q-q_1)}{3a} + \frac{kq_2}{3a} + \frac{kq_1}{3a}$

$= \frac{kq_1}{a} + \frac{k(Q-q_1)}{3a} + \frac{kq_2}{2a}$

$\Rightarrow q_1 = -\frac{q_2}{4}$ .... (ii)

Using it in (1), $q_2 = -\frac{8}{11}Q$

45. (d) $V = \frac{1}{4\pi\varepsilon_0}\left[\frac{q}{x_0} + \frac{q}{3x_0} + \frac{q}{5x_0} + ...\right]$

$+ \frac{1}{4\pi\varepsilon_0}\left[\frac{(-q)}{2x_0} + \frac{(-q)}{4x_0} + \frac{(-q)}{6x_0} + ...\right]$

$= \frac{1}{4\pi\varepsilon_0}\frac{q}{x_0}\left[1 - \frac{1}{2} + \frac{1}{3} - \frac{1}{4} + \frac{1}{5} - \frac{1}{6} + ...\right]a$

$= \frac{1}{4\pi\varepsilon_0}\frac{q}{x_0}\log_e(1+1) = \frac{q\log_e 2}{4\pi\varepsilon_0 x_0}$

46. (d) 47. (d)

48. (c) Electric field is perpendicular to equipotential surfaces from high to low potential.

49. (d) The electric field at a point is equal to negative of potential gradient at that point.

$\vec{E} = -\frac{\partial V}{\partial r} = \left[-\frac{\partial V}{\partial x}\hat{i} - \frac{\partial V}{\partial y}\hat{j} - \frac{\partial V}{\partial z}\hat{k}\right]$

$= [(2xy + z^3)\,\hat{i} + \hat{j}\,x^2 + \hat{k}\,3xz^2]$

50. (a)

51. (a) $\phi_1 = \frac{kq_1}{a_1} + \frac{kq_2}{a_2}$

or $\phi_2 = \frac{kq_1}{a_2} + \frac{kq_2}{a_2}$

Solve to get

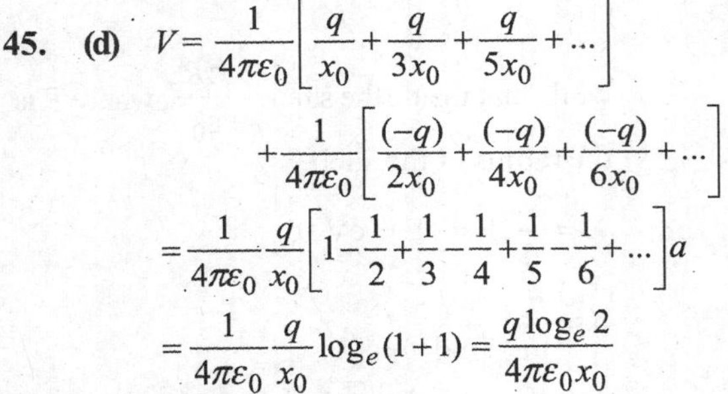

$q_1 = 4\pi\varepsilon_0\left[\frac{\phi_1 - \phi_2}{a_2 - a_1}\right]a_1 a_2$

52. (c)

53. (c) ABCDE is an equipotential surface, on equipotential surface no work is done in shifting a charge from one place to another.

54. (c) Potential energy decreases whenever there is attraction. A negative charge placed at centroid causes attraction.

55. (d) When $\frac{Q_1}{R_1} \neq \frac{Q_2}{R_2}$; current will flow in connecting wire so that energy decreases in the form of heat through the connecting wire.

56. (c) Since $W_{A\to B} = q(V_B - V_A)$

$\Rightarrow V_B - V_A = \frac{16}{4} = 4V$

57. (c) $W_{PQ} = q(V_Q - V_P)$
$= (-100 \times 1.6 \times 10^{-19})(-4-10)$
$= +2.24 \times 10^{-16}J$

58. (c) As work is done by the field, K.E. of the body increases by

K.E. = W = $q(V_A - V_B)$

$= 10^{-8}(600-0) = 6\times10^{-6}$ J

59. (a) We know that potential energy of two charge system is given by

$U = \frac{1}{4\pi\epsilon_0}\frac{q_1q_2}{r}$

According to question,

$U_A = \frac{1}{4\pi\epsilon_0}\frac{(+q)(-Q)}{a} = -\frac{1}{4\pi\varepsilon_0}\frac{Qq}{a}$

and $U_B = \frac{1}{4\pi\epsilon_0}\frac{(+q)(-Q)}{a} = -\frac{1}{4\pi\varepsilon_0}\frac{Qq}{a}$

$\Delta U = U_B - U_A = 0$

We know that for conservative force, $W = -\Delta U = 0$

60. (c) 61. (b)

62. (d) Length of body diagonal = $\sqrt{3}\ell$

∴ Distance of centre of cube from each corner $r = \frac{\sqrt{3}}{2}\ell$

P.E. at centre = 8 × Potential Energy due to $A = 8 \times \frac{Kq\times(-q)}{r}$

$= 8 \times \frac{1}{4\pi\varepsilon_0\sqrt{3}l} \times 2 \times q \times (-q) = \frac{-4q^2}{\sqrt{3}\pi\varepsilon_0 l}$

63. (b) Work done $W_{21} = (V_1 - V_2)q$

$V = \frac{1}{4\pi\epsilon_0}\left[\frac{Q_1}{R} + \frac{Q_2}{\sqrt{2}R}\right]$

and

$V_2 = \frac{1}{4\pi\epsilon_0}\left[\frac{Q_2}{R} + \frac{Q_1}{\sqrt{2}R}\right]$

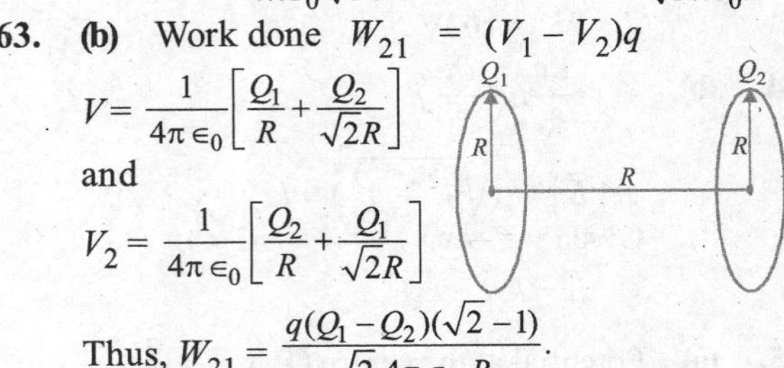

Thus, $W_{21} = \frac{q(Q_1 - Q_2)(\sqrt{2}-1)}{\sqrt{2}.4\pi\epsilon_0 R}$

64. (d) Since capacitance $C = \frac{\varepsilon_0 A}{d}$, as d decreases capacitance increases.

65. (d) 66. (c)

67. (c) Volume of 8 small drops = Volume of big drop

$$8\times\frac{4}{3}\pi R^3 = \frac{4}{3}\pi R^3 \Rightarrow R = 2r$$

As capacity is proportional to r, hence capacity becomes 2 times.

68. (b) The two capacitors are in parallel so

$$C = \frac{\varepsilon_0 A}{t\times 2}(k_1 + k_2)$$

69. (a)

70. (d) $C = 4\pi\varepsilon_0 R = \frac{1}{9\times10^9}\times 6400\times10^3 = 711\ \mu F$

71. (b) Area of plate $= \pi r^2 = \pi\times(8\times10^{-2})^2 = 0.0201\ m^2$

and $d = 1mm = 1\times10^{-3}$ m

Capacity of capacitor

$$C = \frac{\varepsilon_0\varepsilon_r A}{d} = \frac{8.85\times10^{-12}\times6\times0.0201}{1\times10^{-3}}$$

$= 1.068\times10^{-9}$ F

Potential difference, V = 150 volt

Energy stored,

$$U = \frac{1}{2}CV^2 = \frac{1}{2}\times(1.068\times10^{-9})\times(150)^2$$

$= 1.2\times10^{-5}$ J

72. (c) Capacitance are in series

$$\frac{1}{c} = \frac{1}{2}+\frac{1}{2}+\frac{1}{2} \Rightarrow c = \frac{2}{3}F$$

73. (a)

74. (a) In series $\frac{1}{C} = \frac{1}{C_1}+\frac{1}{C_2}+\frac{1}{C_3}$ and charge on each capacitor is same.

75. (d) Capacity of parallel plate capacitor

$$C = \frac{\varepsilon_r\varepsilon_0 A}{d}\qquad (\text{For air } \varepsilon_r = i)$$

So, $\frac{\varepsilon_0 A}{d} = 8\times10^{-12}$

If $d \to \frac{d}{2}$ and $\varepsilon_r \to 6$ then new capacitance

$$C' = 6\times\frac{\varepsilon_0 A}{d/2} = 12\frac{\varepsilon_0 A}{d} = 12\times 8\ pF = 96\ pF$$

76. (c) Capacitance will increase but not 5 times (because dielectric is not filled completely). Hence, new capacitance may be 200 μF.

77. (b)

78. (c) $V_0 = \frac{q}{C_0}$

$$V = \frac{q}{C} \Rightarrow \frac{V}{V_0} = \frac{C_0}{C} \Rightarrow \frac{C_0}{C} = \frac{500}{75} = \frac{20}{3}$$

By definition, $C = kC_0 \Rightarrow k = \frac{20}{3}$

79. (b)

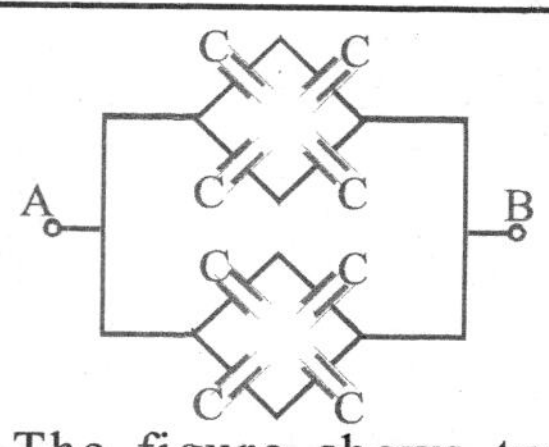

The figure shows two independent balanced wheatstone Brides connected in parallel each having a capacitance C. So, $C_{net} = C_{AB} = 2C$.

80. (a)

81. (c) $C = \frac{2\times2}{2+2}+2 = 3\ \mu F$

82. (a) Energy required to charge the capacitor is $W = U = QV$

$$\Rightarrow U = CV^2 = \frac{\varepsilon_0 A}{d}.V^2 = \frac{\varepsilon_0 Ad}{d^2}$$

$$V^2 = \varepsilon_0 E^2 Ad \qquad \left[\because E = \frac{V}{d}\right].$$

83. (b)

84. (a) Capacitance of spherical conductor $= 4\pi\varepsilon_0 a$ where a is radius of conductor.

Therefore, $C = \frac{1}{9\times10^9}\times1 = \frac{1}{9}\times10^{-9}$

$= 0.11\times10^{-9}$ F $= 1.1\times10^{-10}$ F

85. (c) $4\pi\varepsilon_0 r = C$

$\Rightarrow r = C/4\pi\varepsilon_0 = (1\times10^{-6}\times9\times10^9)$

$= 9\times10^3$ m $= 9$ km

86. (c) The network is equivalent to

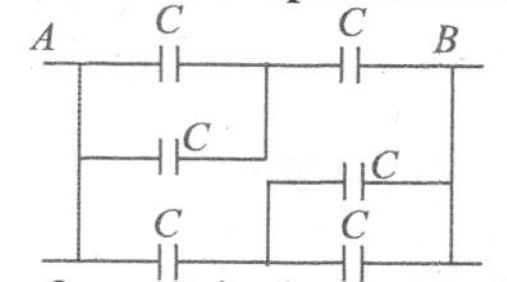

Therefore equivalent capacitance $= [2C \text{ series } C] // [C \text{ series } 2C]$

$$= 2\left(\frac{2C\times C}{2C+C}\right) = \frac{4C}{3}$$

87. (d)

88. (b) $C_1 < C_2$

$$\therefore \frac{C_1}{C_1+C_2} < \frac{1}{2} \text{ and } \frac{C_2}{C_1+C_2} > \frac{1}{2}$$

$$C = \frac{C_1C_2}{C_1+C_2} = C_1.\frac{C_2}{C_1+C_2} > \frac{C_1}{2}$$

Similarly, $C < \frac{C_2}{2}$

89. (c) Equivalent circuit is shown

$$\therefore C_{eq} = \frac{3\varepsilon_0 a^2}{5d}$$

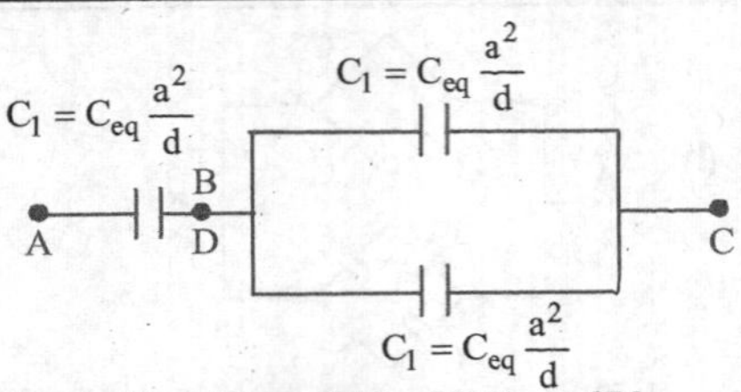

90. (d) $V_C = \frac{Q_1+Q_2}{C_1+C_2} = \frac{CV+2VC}{KC+2C} = \frac{3V}{K+2}$

91. (a)

92. (c) When key is on the position *BD*, the situation is shown in the figure.

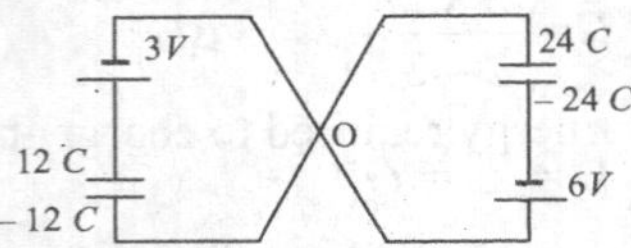

When key is on the position *AD*, the situation is shown in the figure.

– 24 C, 24 C, 6V, 3V, – 12 C, 12 C, O

The charge flows to each capacitor is 36 μ*C*, and so total charge flows through point O is 72 μ*C*.

93. (a) $C_1 = C, V_1 = V_0, C_2 = KC, V_2 = 0$

and $V_{common} = V$

We know that

$$V_{common} = \frac{C_1V_1 + C_2V_2}{C_1 + C_2}$$

$$\therefore V_{common} = \frac{CV_0 + KC \times O}{C + KC} \text{ or } K = \frac{V_0}{V} - 1$$

94. (c) Common potential $V' = \frac{C_1V + C_2 \times 0}{C_1 + C_2}$

$$= \frac{C_1}{C_1 + C_2}V$$

95. (d) $c_1 = \frac{(A/2)\varepsilon_0}{d/2} = \frac{A\varepsilon_0}{d}$,

$c_2 = K\frac{A\varepsilon_0}{d}, c_3 = K\frac{A\varepsilon_0}{2d}$

c_1, c_3, K, d, $\frac{d}{2}$, K c_2

$$\therefore c_{eq.} = \frac{c_1 \times c_2}{c_1 + c_2} + c_3 = \frac{(3+K)KA\varepsilon_0}{2d(K+1)}$$

($\because$ C_1 and C_2 are in series and resultant of these two in parallel with C_3)

96. (b)

97. (c) Before introducing a slab capacitance of plates

$$C_1 = \frac{\varepsilon_0 A}{3}$$

If a slab of dielectric constant K is introduced between plates then

$$C = \frac{K\varepsilon_0 A}{d} \text{ then } C_1' = \frac{\varepsilon_0 A}{2.4}$$

C_1 and C_1' are in series hence, C_1'

Slab

$$\frac{\varepsilon_0 A}{3} = \frac{k\frac{\varepsilon_0 A}{3}.\frac{\varepsilon_0 A}{2.4}}{k\frac{\varepsilon_0 A}{3} + \frac{\varepsilon_0 A}{2.4}}$$

$3k = 2.4k + 3 \quad 0.6k = 3$

Hence, the dielectric constant of slap is given by, $k = \frac{30}{6} = 5$

98. (d) $C_{eq} = \frac{\varepsilon_0}{\frac{d}{K_1} + \frac{d}{K_2} + \frac{d}{K_3}}$

Here, $K_1 = K_3 = 1, K_2 = \varepsilon/\varepsilon_0$

99. (d) Here, $V = 15 \times 10^6$ *V* dielectric strength $= 5 \times 10^7 \ V\,m^{-1}$

Maximum electric field,

E = 10% of dielectirc stength

$$\therefore \quad E = \frac{10}{100} \times 5 \times 10^7 = 5 \times 10^6 \ V\,m^{-1}$$

$$\text{As } E = \frac{V}{r} \quad \therefore r = \frac{V}{E} = \frac{15 \times 10^6}{5 \times 10^6} = 3 \ m$$

100. (a) If we thraw the charged particle just right of the center of the tunnel, the particle will cross the tunnel. Hence, applying conservation of momentum between start point and center of tunnel we get

$\Delta K + \Delta U = 0$

$$\text{or} \quad \left(0 + \frac{1}{2}mv^2\right) + q(V_f - V_i) = 0$$

$$\text{or} \quad V_f = \frac{V_s}{2}\left(3 - \frac{r^2}{R^2}\right) = \frac{\rho R^2}{6\varepsilon_0}\left(3 - \frac{r^2}{R^2}\right)$$

$$\text{Hence } r = \frac{R}{2}$$

$$V_f = \frac{\rho R^2}{6\varepsilon_0}\left(3 - \frac{R^2}{4R^2}\right) = \frac{11\rho R^2}{24\varepsilon_0}; V_i = \frac{\rho R^2}{3\varepsilon_0}$$

$$\frac{1}{2}mv^2 = 1\left[\frac{11\rho R^2}{24\varepsilon_0} - \frac{\rho R^2}{3\varepsilon_0}\right]$$

$$= \frac{\rho R^2}{\varepsilon_0}\left[\frac{11}{24} - \frac{1}{3}\right] = \frac{\rho R^2}{8\varepsilon_0} \text{ or } V = \left(\frac{\rho R^2}{4m\varepsilon_0}\right)^{1/2}$$

Hence velocity should be slightly greater than V.

17 Current Electricity

1. The total momentum of electrons in a straight wire of length 1000m carrying a current of 70A is closest to

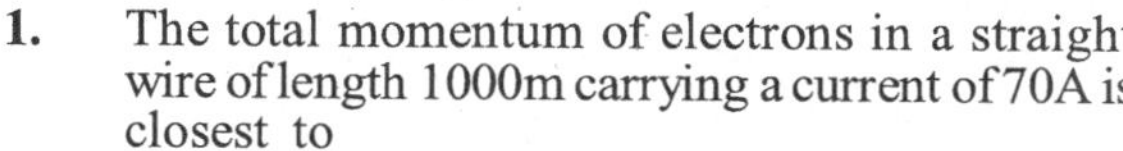

(a) 40×10^{-8} N-sec (b) 30×10^{-8} N-sec
(c) 50×10^{-8} N-sec (d) 70×10^{-8} N-sec

2. The resistance of a wire is R. It is bent at the middle by 180° and both the ends are twisted together to make a shorter wire. The resistance of the new wire is
(a) $2R$ (b) $R/2$ (c) $R/4$ (d) $R/8$

3. Suppose the drift velocity v_d in a material varied with the applied electric field E as $v_d \propto \sqrt{E}$. Then V – I graph for a wire made of such a material is best given by : **Tricky**

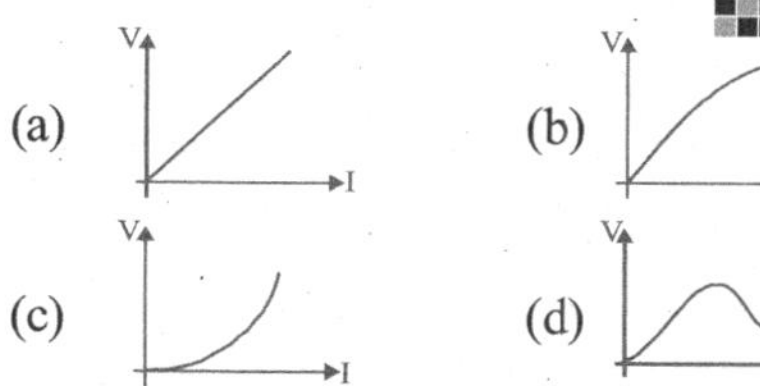

4. The current density varies with radial distance r as $J = a\,r^2$, in a cylindrical wire of radius R. The current passing through the wire between radial distance $R/3$ and $R/2$ is **Critical Thinking**
(a) $\dfrac{65\pi a R^4}{2592}$ (b) $\dfrac{25\pi a R^4}{72}$
(c) $\dfrac{65\pi a^2 R^3}{2938}$ (d) $\dfrac{81\pi a^2 R^4}{144}$

5. In a certain particle accelerator, electrons emerge in pulses at the rate of 250 pulses per second. Each pulse is of duration of 200ns and the electrons in the pulse constitute a current of 250mA. The number of electrons delivered by the accelerator per pulse is
(a) 8.00×10^{10} (b) 5.00×10^{8}
(c) 3.13×10^{11} (d) 9.60×10^{10}

6. Two wires of the same metal have same length, but their cross-sections are in the ratio 3:1. They are joined in series. The resistance of thicker wire is 10Ω. The total resistance of the combination will be
(a) 10 Ω (b) 20 Ω (c) 40 Ω (d) 100 Ω

7. In the network shown, each resistance is equal to R. The equivalent resistance between adjacent corners A and D is **Tricky**

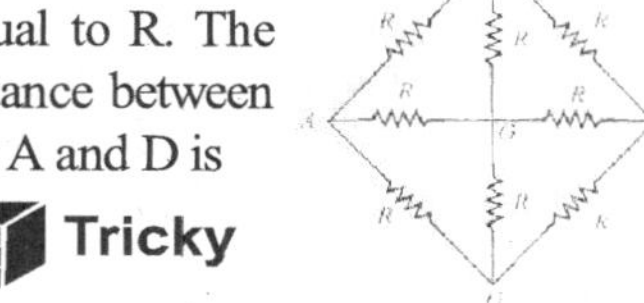

(a) R (b) $\dfrac{2}{3}$R
(c) $\dfrac{3}{7}$R (d) $\dfrac{8}{15}$R

8. Drift velocity of electrons is due to
(a) motion of conduction electrons due to random collisions.
(b) motion of conduction electrons due to electric field $\vec{E}$.
(c) repulsion to the conduction electrons due to inner electrons of ions.
(d) collision of conduction electrons with each other.

9. A metal wire is subjected to a constant potential difference. When the temperature of the metal wire increases, the drift velocity of the electron in it
(a) increases, thermal velocity of the electron increases
(b) decreases, thermal velocity of the electron increases
(c) increases, thermal velocity of the electron decreases
(d) decreases, thermal velocity of the electron decreases

10. If N, e, τ and m are representing electron density, charge, relaxation time and mass of an electron respectively, then the resistance of wire of length ℓ and cross-sectional area A is given by
(a) $\dfrac{m\ell}{Ne^2A^2\tau}$ (b) $\dfrac{2m\tau A}{Ne^2\ell}$
(c) $\dfrac{Ne^2\tau A}{2m\ell}$ (d) $\dfrac{Ne^2A}{2m\tau\ell}$

11. The electric resistance of a certain wire of iron is R. If its length and radius are both doubled, then
(a) the resistance and the specific resistance, will both remain unchanged
(b) the resistance will be doubled and the specific resistance will be halved
(c) the resistance will be halved and the specific resistance will remain unchanged
(d) the resistance will be halved and the specific resistance will be doubled

12. A wire X is half the diameter and half the length of a wire Y of similar material. The ratio of resistance of X to that of Y is
(a) 8 : 1 (b) 4 : 1 (c) 2 : 1 (d) 1 : 1

13. The voltage V and current I graphs for a conductor at two different temperatures T_1 and T_2 are shown in the figure. The relation between T_1 and T_2 is **Tricky**

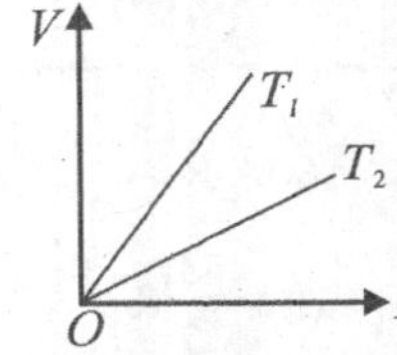

(a) $T_1 > T_2$
(b) $T_1 < T_2$
(c) $T_1 = T_2$
(d) $T_1 = \frac{1}{T_2}$

14. The amount of charge Q passed in time t through a cross-section of a wire is $Q = 5t^2 + 3t + 1$. The value of current at time t = 5 s is
(a) 9 A (b) 49 A
(c) 53 A (d) None of these

15. A conductor carries a current of 50 μ A. If the area of cross-section of the conductor is 50 mm^2, then value of the current density in Am^{-2} is
(a) 0.5 (b) 1 (c) 10^{-3} (d) 10^{-6}

16. The resistance of a wire is R. It is bent at the middle by 180° and both the ends are twisted together to make a shorter wire. The resistance of the new wire is **Critical Thinking**
(a) $2R$ (b) $R/2$ (c) $R/4$ (d) $R/8$

17. At what temperature will the resistance of a copper wire becomes three times its value at 0°C? (Temperature coefficient of resistance of copper is 4×10^{-3}/°C)
(a) 550°C (b) 500°C (c) 450°C (d) 400°C

18. When a current I is set up in a wire of radius r, the drift velocity is v_d. If the same current is set up through a wire of radius 2 r, the drift velocity will be
(a) $4 v_d$ (b) $2 v_d$ (c) $v_d/2$ (d) $v_d/4$

19. A straight conductor of uniform cross-section carries a current I. If s is the specific charge of an electron, the momentum of all the free electrons per unit length of the conductor, due to their drift velocity only is
(a) Is (b) $\sqrt{I/s}$ (c) I/s (d) $(I/s)^2$

20. If the resistance of a conductor is 5Ω at 50° C & 7Ω at 100° C, then mean temperature coefficient of resistance (of material) is
(a) 0.013/°C (b) 0.004/°C
(c) 0.006/°C (d) 0.008/°C

21. If negligibly small current is passed though a wire of length 15 m and resistance of 5Ω, having uniform cross section of 6×10^{-7} m^2, then coefficient of resistivity of material is
(a) 1×10^{-7} Ω–m (b) 2×10^{-7} Ω–m
(c) 3×10^{-7} Ω–m (d) 4×10^{-7} Ω–m

22. The resistance of a wire at room temperature 30°C is found to be 10 Ω. Now to increase the resistance by 10%, the temperature of the wire must be [The temperature coefficient of resistance of the material of the wire is 0.002 per °C]
(a) 36°C (b) 83°C (c) 63°C (d) 33°C

23. The number of free electrons per 100 mm of ordinary copper wire is 2×10^{21}. Average drift speed of electrons is 0.25 mm/s. The current flowing is
(a) 5 A (b) 80 A (c) 8 A (d) 0.8 A

24. At room temperature, copper has free electron density of 8.4×10^{28} per m^3. The copper conductor has a cross-section of 10^{-6} m^2 and carries a current of 5.4 A. The electron drift velocity in copper is **Critical Thinking**
(a) 400 m/s (b) 0.4 m/s (c) 0.4 mm/s (d) 72 m/s

25. An electron beam has an aperture 1.0 mm^2. A total of 6×10^{16} electrons go through any perpendicular cross section per second. Find the current density in the beam. (in A/m^2)
(a) 9.1×10^{13} (b) 9.6×10^{3}
(c) 6.6×10^{5} (d) 8.6×10^{11}

26. A wire of a certain material is stretched slowly by ten per cent. Its new resistance and specific resistance become respectively:
(a) 1.2 times, 1.3 times (b) 1.21 times, same
(c) both remain the same (d) 1.1 times, 1.1 times

27. A 4 ohm resistance wire is bent through 180° at its mid point and the two halves are twisted together. Then the resistance is
(a) 1 Ω (b) 2 Ω (c) 5 Ω (d) 8 Ω

28. The length of a given cylindrical wire is increased by 100%. Due to the consequent decrease in diameter the change in the resistance of the wire will be
(a) 200% (b) 100% (c) 50% (d) 300%

29. The masses of the three wires of copper are in the ratio of 1 : 3 : 5 and their lengths are in the ratio of 5 : 3 : 1. The ratio of their electrical resistance is **Tricky**
(a) 1 : 3 : 5 (b) 5 : 3 : 1
(c) 1 : 25 : 125 (d) 125 : 45 : 3

30. A uniform wire of length l and radius r has a resistance of 100 Ω. It is recast into a wire of radius $\frac{r}{2}$. The resistance of new wire will be :
(a) 1600 Ω (b) 400 Ω (c) 200 Ω (d) 100 Ω

31. A wire has a resistance 12 Ω. It is bent in the form of a circle. The effective resistance between two points on any diameter is
(a) 6 Ω (b) 3 Ω (c) 12 Ω (d) 24 Ω

32. In the network shown below, the ring has zero resistance. The equivalent resistance between the point A and B is **Tricky**

(a) 2R

(b) 4R

(c) 7R

(d) 10R

33. The equivalent resistance between points A and B is

(a) 2R

(b) (3/4) R

(c) (4/3) R

(d) (3/5) R

34. Six equal resistances are connected between points P, Q and R as shown in figure. Then net resistance will be maximum between :

(a) P and R

(b) P and Q

(c) Q and R

(d) Any two points

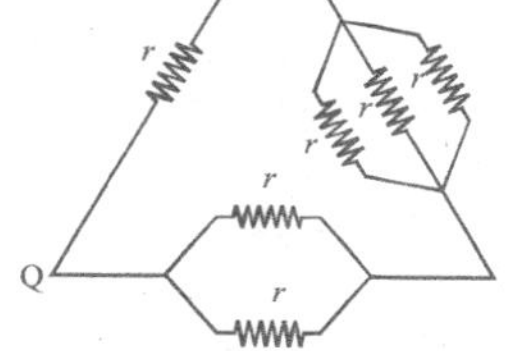

35. The current in the 1Ω resistor shown in the circuit is

(a) $\frac{2}{3}$ A

(b) 3 A

(c) 6 A

(d) 2 A

36. A wire of resistance 12 ohms per meter is bent to form a complete circle of radius 10 cm. The resistance between its two diametrically opposite points, A and B as shown in the figure, is

(a) 3Ω

(b) 6πΩ

(c) 6Ω

(d) 0.6πΩ

37. A wire has linear resistance ρ (in Ohm/m). Find the resistance R between points A and B if the side of the larger square is 'd'. **Critical Thinking**

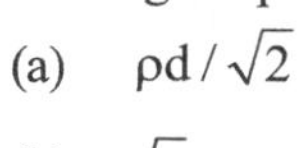

(a) $\rho d/\sqrt{2}$

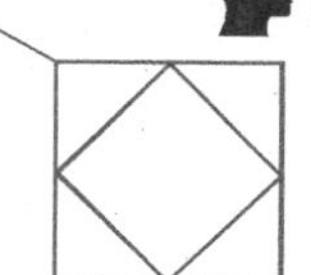

(b) $\sqrt{2}\rho d$

(c) 2 r d

(d) None of these

38. Six identical resistors are connected as shown in the figure. The equivalent resistance will be

(a) maximum between P and R

(b) maximum between Q and R

(c) maximum between P and Q

(d) All are equal

39. The circuit diagram shown in figure consists of a very large (infinite) number of elements. The resistances of the resistors in each subsequent element differ by a factor of k from the resistances of the resistors in the previous elements. Determine the resistance R_{AB} between points A and B if the resistances of the first element are R_1 and R_2. ($k = 1/2$) **Toughnut**

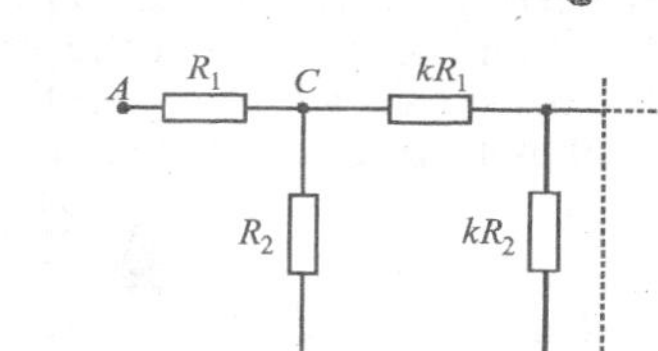

(a) $\frac{R_1 - R_2 + \sqrt{R_1^2 + R_2^2 + 6R_1R_2}}{2}$

(b) $\frac{R_1 + R_2 + \sqrt{R_1^2 + R_2^2 + 6R_1R_2}}{2}$

(c) $\frac{R_1 - R_2 - \sqrt{R_1^2 + R_2^2 + 6R_1R_2}}{2}$

(d) None of these

40. Each of the resistors shown in Fig. has resistance R. Find the equivalent resistance between A and B. **Tricky**

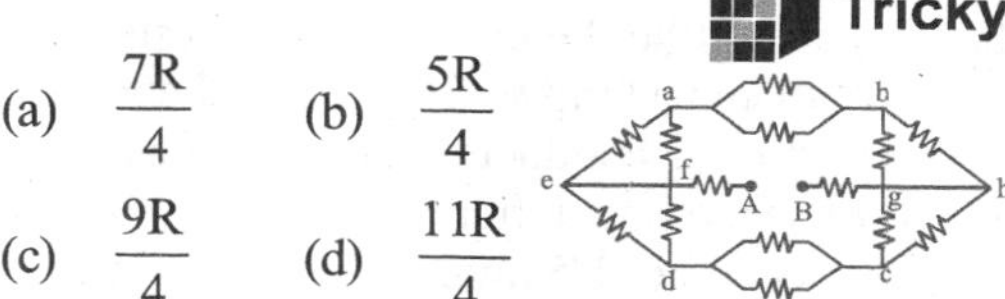

(a) $\frac{7R}{4}$ (b) $\frac{5R}{4}$

(c) $\frac{9R}{4}$ (d) $\frac{11R}{4}$

41. Find the equivalennt resistance between A and E (resistance of each resistor is R.) **Tricky**

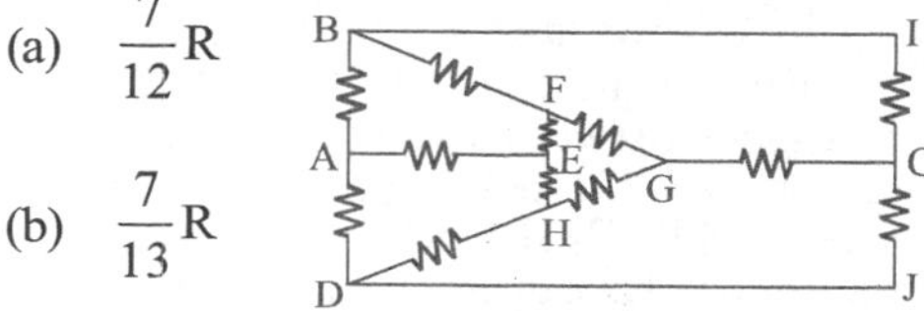

(a) $\frac{7}{12}$R

(b) $\frac{7}{13}$R

(c) $\frac{7}{15}$R (d) $\frac{8}{13}$R

42. In Fig. find the value of resistor to be connected between C and D,. so that the resistance of the entire circuit between A and B does not change wth the number of elementary sets. Tricky

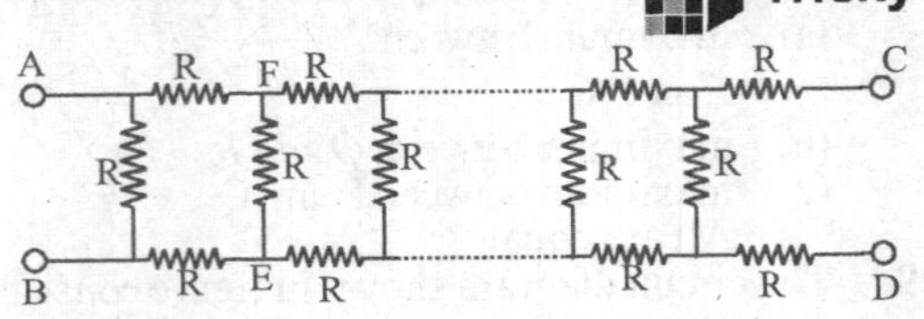

(a) R (b) $R(\sqrt{3}-1)$
(c) 3R (d) $R(\sqrt{3}+1)$

43. A hank of uninsulated wire consisting of seven and a half turns is stretched between two nails hammered into a board to which the ends of the wire are fixed. The resistance of the circuit between the nails is determined with the help of electrical measuring instruments. Determine the proportion in which the resistance will change if the wire is unwound so that the ends remain to be fixed to the nails. Critical Thinking

(a) 225 (b) 15 (c) 240 (d) 250

44. A 12 cm wire is given a shape of a right angled triangle *ABC* having sides 3 cm, 4 cm and 5 cm as shown in the figure. The resistance between two ends (*AB*, *BC*, *CA*) of the respective sides are measured one by one by a multi-meter. The resistances will be in the ratio of Critical Thinking

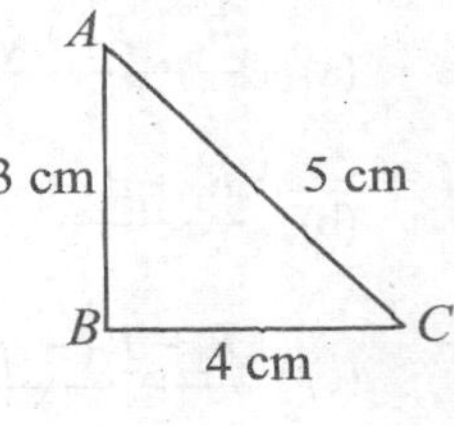

(a) 3:4:5 (b) 9:16:25
(c) 27:32:35 (d) 21:24:25

45. What will be the change in the resistance of a circuit consisting of five identical conductors if two similar conductors are added as shown by the dashed line in figure.

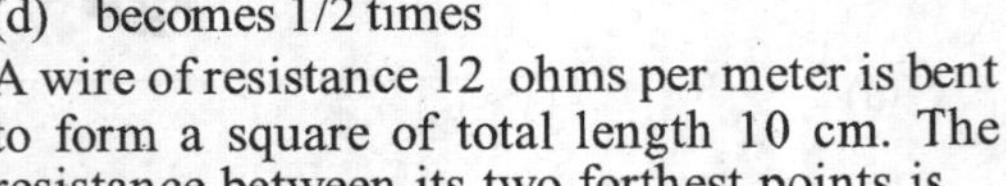

(a) becomes 1/5 times
(b) becomes 3/5 times
(c) becomes 2/5 times
(d) becomes 1/2 times

46. A wire of resistance 12 ohms per meter is bent to form a square of total length 10 cm. The resistance between its two forthest points is
(a) 1.2Ω (b) 6Ω (c) 1.5Ω (d) 2.8Ω

47. The Kirchhoff's second law ($\Sigma iR = \Sigma E$), where the symbols have their usual meanings, is based on
(a) conservation of momentum
(b) conservation of charge
(c) conservation of potential
(d) conservation of energy

48. A cell of internal resistance r is connected to an external resistance R. The current will be maximum in R, if
(a) R=r (b) R<r (c) R>r (d) R=r/2

49. If n cells each of emf ε and internal resistance *r* are connected in parallel, then the total emf and internal resistances will be
(a) $\varepsilon, \frac{r}{n}$ (b) ε, nr (c) $n\varepsilon, \frac{r}{n}$ (d) $n\varepsilon, nr$

50. Under what condition will the strength of current in a wire of resistance R be the same for connection is series and in parallel of n identical cells each of the internal resistance r?
(a) R=nr (b) R=r/n
(c) R=r (d) $R\to\infty, r\to 0$

51. A cell having an emf ε and internal resistance *r* is connected across a variable external resistance *R*. As the resistance *R* is increased, the plot of potential difference *V* across *R* is given by Tricky

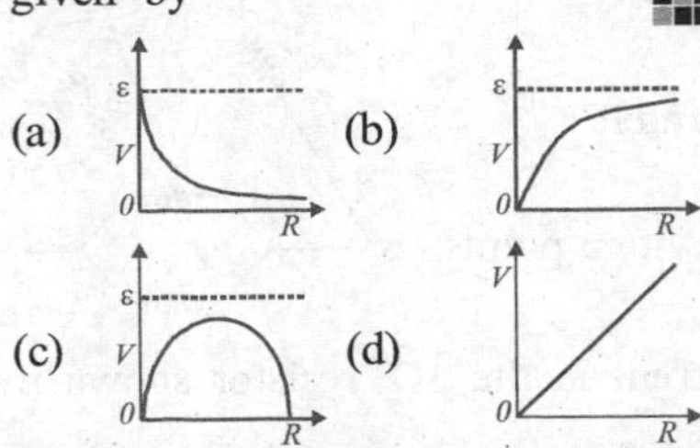

52. The internal resistance of a 2.1 V cell which gives a current of 0.2 A through a resistance of 10 Ω is
(a) 0.5 Ω (b) 0.8 Ω (c) 1.0 Ω (d) 0.2 Ω

53. A primary cell has an e.m.f. of 1.5 volt. When short-circuited it gives a current of 3 ampere. The internal resistance of the cell is
(a) 4.5 ohm (b) 2 ohm
(c) 0.5 ohm (d) (1/4.5) ohm

54. In the given network of four resistances, the equivalent resistance is :
(a) 20 Ω
(b) 5.4 Ω
(c) 12 Ω
(d) 4.5 Ω

55. In the given mesh, each resistor has resistance *R*. The effective resistance between the terminals *A* and *B* is

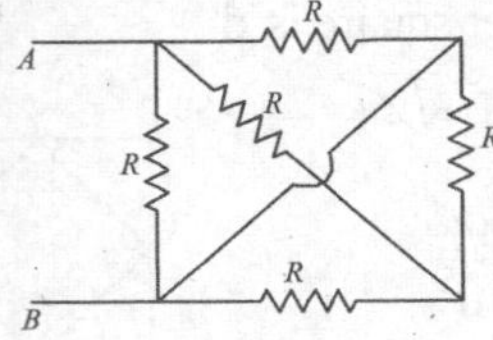

(a) $\frac{3R}{8}$ (b) $\frac{R}{2}$ (c) R (d) $2R$

56. All wires have same resistance and equivalent resistance between A and B is R_0. Now keys are closed, then the equivalent resistance will become **Tricky**

(a) $\frac{7R_0}{3}$

(b) $\frac{7R_0}{9}$

(c) $7R_0$

(d) $\frac{R_0}{3}$

57. Find equivalent resistance between A & B in the following circuit

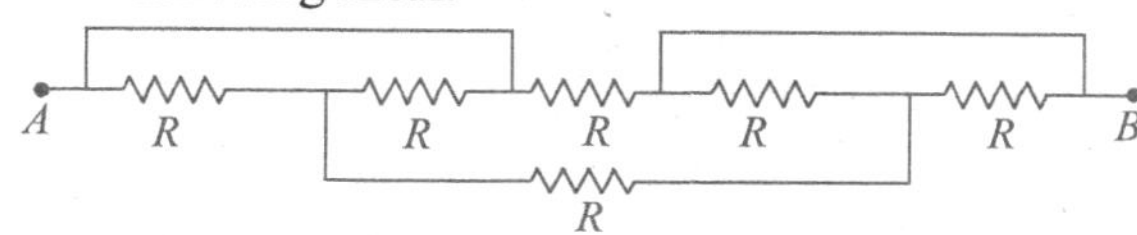

(a) $\frac{3R}{2}$ (b) $\frac{2R}{3}$ (c) 2R (d) 3R

58. The effective resistance between the terminals A and B is. **Tricky**

(a) 5 Ω

(b) 10 Ω

(c) 15 Ω

(d) 20 Ω

59. Determine the current in 2Ω resistor.

(a) 1 A

(b) 1.5 A

(c) 0.9 A

(d) 0.6 A

60. Two sources of equal emf are connected to an external resistance R. The internal resistance of the two sources are R_1 and R_2 ($R_2 > R_1$). If the potential difference across the source having internal resistance R_2 is zero, then

(a) $R = R_2 - R_1$

(b) $R = R_2 \times (R_1 + R_2)/(R_2 - R_1)$

(c) $R = R_1R_2/(R_2 - R_1)$

(d) $R = R_1R_2/(R_1 - R_2)$

61. In the series combination of n cells each cell having emf ε and internal resistance r. If three cells are wrongly connected, then total emf and internal resistance of this combination will be

(a) $n\varepsilon, (nr - 3r)$ (b) $(n\varepsilon - 2\varepsilon)\, nr$

(c) $(n\varepsilon - 4\varepsilon), nr$ (d) $(n\varepsilon - 6\varepsilon), nr$

62. A 3 volt battery with negligible internal resistance is connected in a circuit as shown in the figure. The current I, in the circuit will be

(a) 1 A

(b) 1.5 A

(c) 2 A

(d) 1/3 A

63. The four wires from a larger circuit intersect at junction A as shown. What is the magnitude and direction of the current between points A and B ?

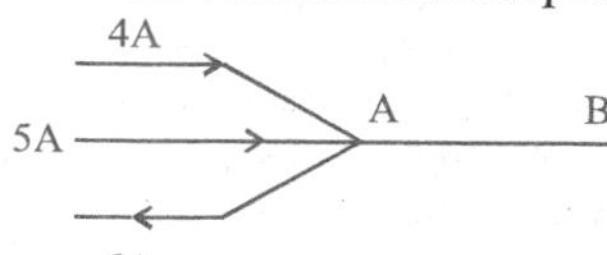

(a) 2 A from A to B (b) 2A from B to A

(c) 3A from A to B (d) 2A from B to A

64. A battery of emf 10V and internal resistance 3 0hm is conncted to a resister. The current in the circuit is 0.5 amp. The terminal voltage of the battery when the circuit is closed is

(a) 10V (b) zero (c) 1.5V (d) 8.5V

65. Five cells each of emf E and internal resistance r send the same amount of current through an external resistance R whether the cells are connected in parallel or in series. Then the ratio $\left(\frac{R}{r}\right)$ is **Critical Thinking**

(a) 2 (b) $\frac{1}{2}$ (c) $\frac{1}{5}$ (d) 1

66. Three batteries of emf 1V and internal resistance 1Ω each are connected as shown. Effective emf of combination between the points PQ is

(a) zero

(b) 1V

(c) 2V

(d) (2/3) V

67. In the circuit shown in Figure the current through

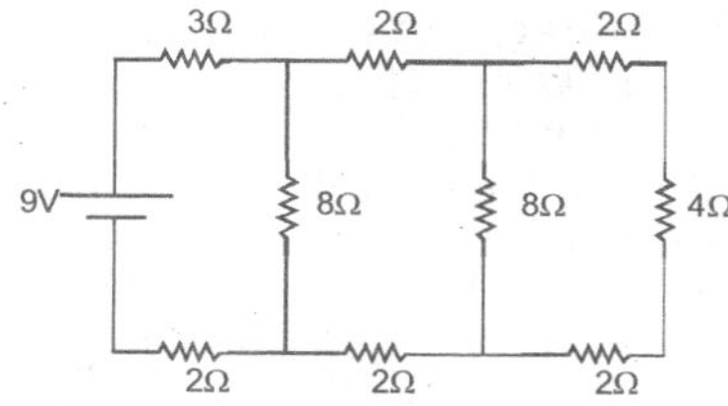

(a) the 3 Ω resistor is 0.50 A.

(b) the 3 Ω resistor is 0.25 A.

(c) the 4 Ω resistor is 0.50 A

(d) the 4 Ω resistor is 0.25 A.

68. In the circuit shown in the figure, find the current in 45 Ω. Tricky

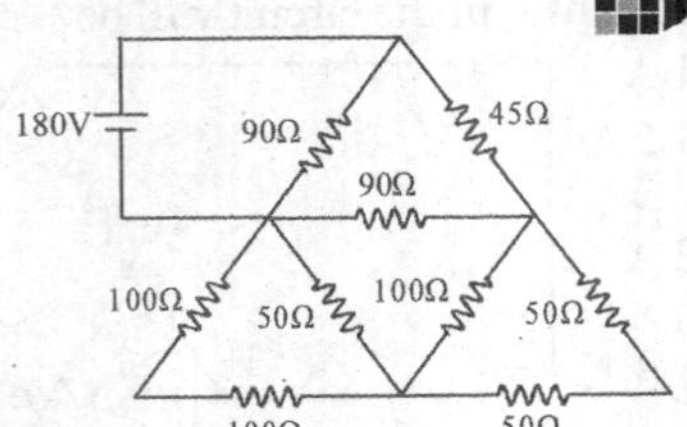

(a) 4 A (b) 2.5A
(c) 2 A (d) None of these

69. A, B and C are voltmeters of resistance R, 1.5 R and 3R respectively as shown in the figure. When some potential difference is applied between X and Y, the voltmeter readings are V_A, V_B and V_C respectively. Then

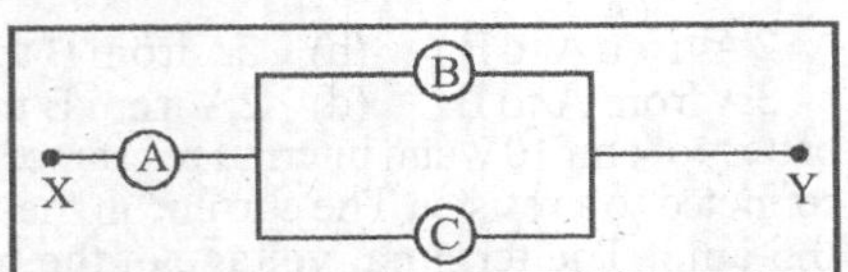

(a) $V_A \neq V_B = V_C$ (b) $V_A = V_B \neq V_C$
(c) $V_A \neq V_B \neq V_C$ (d) $V_A = V_B = V_C$

70.

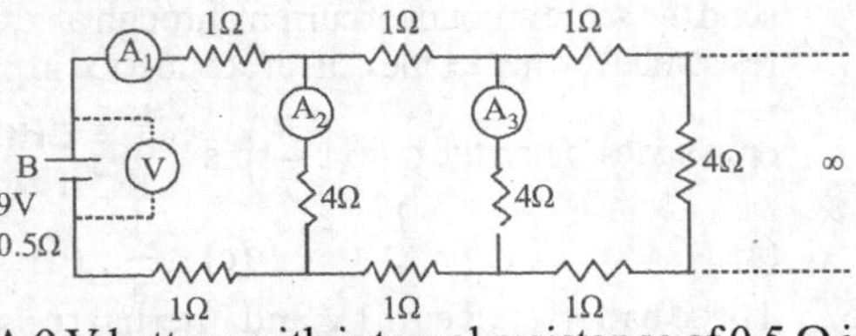

A 9 V battery with internal resistance of 0.5 Ω is connected across an infinite network as shown in the figure. All ammeters A_1, A_2, A_3 and voltmeter V are ideal.

Choose correct statement. Toughnut

(a) Reading of A_1 is 2 A
(b) Reading of A_1 is 18 A
(c) Reading of V is 9 V
(d) Reading of V is 7 V

71. Two batteries of emf 4 V and 8V with internal resistance 1 Ω and 2 Ω are connected in a circuit with a resistance of 9 Ω as shown in figure. The current and potential difference between the points *P* and *Q* are

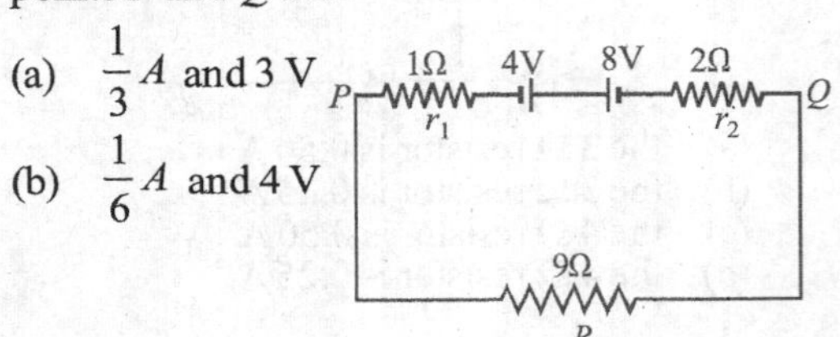

(a) $\frac{1}{3}A$ and 3 V
(b) $\frac{1}{6}A$ and 4 V
(c) $\frac{1}{9}A$ and 9 V
(d) $\frac{1}{12}A$ and 12 V

72. Find out the value of current through 2Ω resistance for the given circuit. Tricky

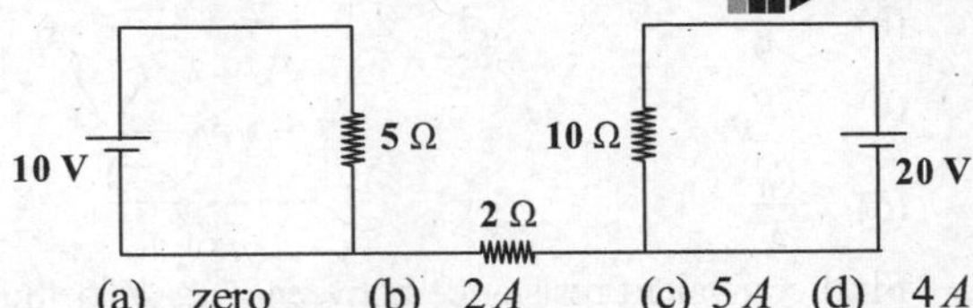

(a) zero (b) 2*A* (c) 5*A* (d) 4*A*

73. The circuit shown here has two batteries of 8.0 V and 16.0 V and three resistors 3 Ω, 9 Ω and 9 Ω and a capacitor of 5.0 μF.

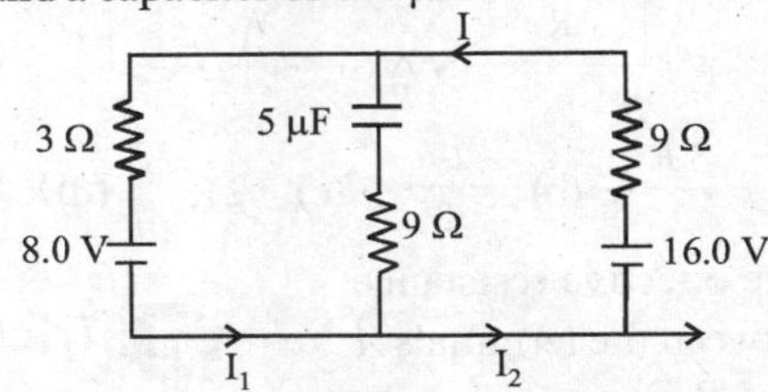

How much is the current I in the circuit in steady state? Critical Thinking

(a) 1.6A (b) 0.67A
(c) 2.5A (d) 0.25A

74. In the circuit shown, current (in A) through 50 V and 30 V batteries are, respectively.

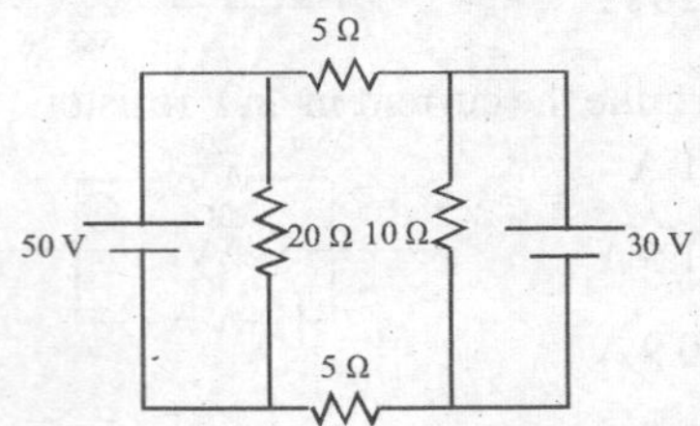

(a) 2.5 and 3 (b) 3.5 and 2
(c) 4.5 and 1 (d) 3 and 2.5

75. Cell having an emf ε and internal resistance *r* is connected across a variable external resistance *R*. As the resistance *R* is increased, the plot of potential difference *V* across *R* is given by : Tricky

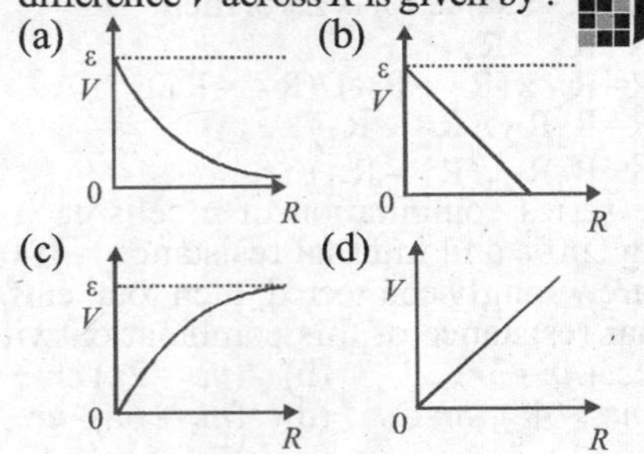

76. In the circuit shown in the figure, if potential at point A is taken to be zero, the potential at point B is

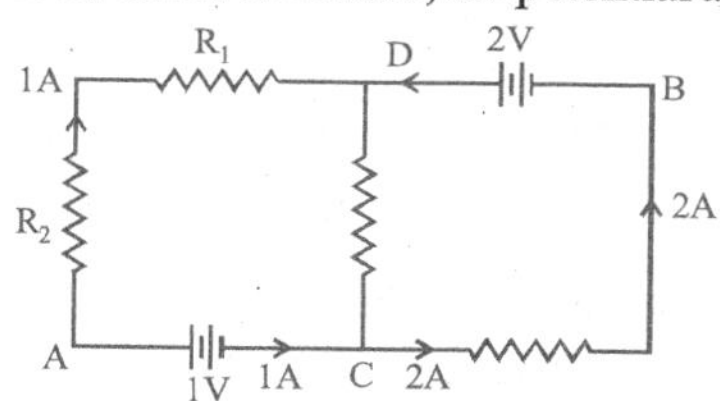

(a) –1V (b) +2V (c) –2V (d) +1V

77. Forty electric bulbs are connected in series across a 220 V supply. After one bulb is fused the remaining 39 are connected again in series across the same supply. The illumination will be
(a) more with 40 bulbs than with 39
(b) more with 39 bulbs than with 40
(c) equal in both the cases
(d) in the ratio $40^2 : 39^2$

78. A current of 3 amp flows through the 2Ω resistor shown in the circuit. The power dissipated in the 5-Ω resistor is:

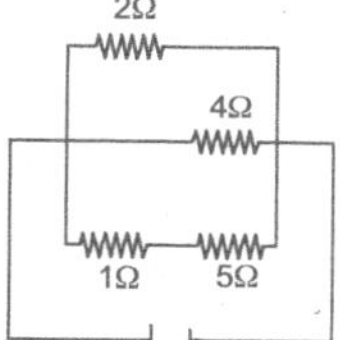

(a) 4 watt (b) 2 watt
(c) 1 watt (d) 5 watt

79. A electric tea kettle has two heating coils. When first coil of resistance R_1 is switched on, the kettle begins to boil tea in 6 minutes. When second coil of resistance R_2 is switched on, the boiling begins in 8 minutes. The value of R_1/R_2 is
(a) $\frac{7}{3}$ (b) $\frac{3}{7}$ (c) $\frac{3}{4}$ (d) $\frac{4}{3}$

80. An electric fan and a heater are marked as 100 W, 220 V and 1000 W, 220 V respectively. The resistance of heater is
(a) equal to that of fan (b) lesser than that of fan
(c) greater than that of fan (d) zero

81. Water boils in the electric kettle in 15 minutes after switching on. If the length of heating wire is decreased to 2/3 of its initial value, then the same amount of water will boil with the same supply voltage in **Critical Thinking**
(a) 8 minutes (b) 10 minutes
(c) 12 minutes (d) 15 minutes

82. A wire of radius r and another wire of radius 2r, both of same material and length are connected in series to each other. The combination is connected across a battery. The ratio of the heats produced in the two wires will be
(a) 4.00 (b) 2.00 (c) 0.50 (d) 0.25

83. A battery of e.m.f. 10 V and internal resistance 0.5 Ω is connected across a variable resistance R. The value of R for which the power delivered in it is maximum is given by
(a) 0.5 Ω (b) 1.0 Ω (c) 2.0 Ω (d) 0.25Ω

84. It takes 12 minutes to boil 1 litre of water in an electric kettle. Due to some defect it becomes necessary to remove 20% turns of heating coil of the kettle. After repair, how much time will it take to boil 1 litre of water? **Critical Thinking**
(a) 9.6 minute (b) 14.4 minute
(c) 16.8 minute (d) 18.2 minute

85. For maximum power from battery the internal resistance of battery r is

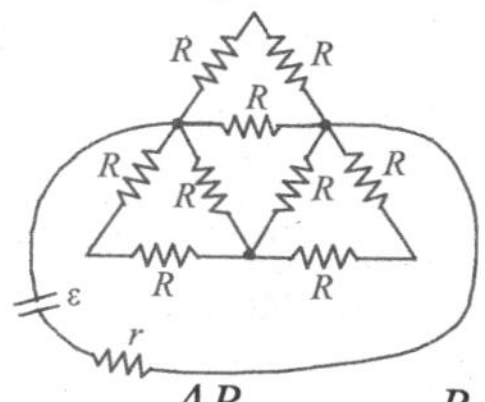

(a) $10R$ (b) $\frac{4R}{9}$ (c) $\frac{R}{8}$ (d) $\frac{10R}{9}$

86. Two cities are 150 km apart. Electric power is sent from one city to another city through copper wires. The fall of potential per km is 8 volt and the average resistance per km is 0.5 Ω. The power loss in the wires is: **Critical Thinking**
(a) 19.2 W (b) 19.2 kW
(c) 19.2 J (d) 12.2 kW

87. The charge flowing through a resistance R varies with time t as $Q = at - bt^2$. The total heat produced in R by the time current ceases is
(a) $a^3R/6b$ (b) $a^3R/3b$
(c) $\frac{a^3R}{2b}$ (d) $\frac{a^3R}{b}$

88. In a Wheatstone bridge all the four arms have equal resistance R. If the resistance of galvanometer arm is also R, the equivalent resistance of combination is
(a) 2R (b) R/4 (c) R/2 (d) R

89. The resistance of the four arms P, Q, R and S in a Wheatstone's bridge are 10 ohm, 30 ohm, 30 ohm and 90 ohm, respectively. The e.m.f. and internal resistance of the cell are 7 volt and 5 ohm respectively. If the galvanometer resistance is 50 ohm, the current drawn from the cell will be
(a) 0.2 A (b) 0.1 A (c) 2.0 A (d) 1.0 A

90. In a Wheatstone's bridge, three resistances P, Q and R connected in the three arms and the fourth arm is formed by two resistances S_1 and S_2 connected in parallel. The condition for the bridge to be balanced will be
(a) $\frac{P}{Q} = \frac{2R}{S_1 + S_2}$ (b) $\frac{P}{Q} = \frac{R(S_1 + S_2)}{S_1S_2}$
(c) $\frac{P}{Q} = \frac{R(S_1 + S_2)}{2S_1S_2}$ (d) $\frac{P}{Q} = \frac{R}{S_1 + S_2}$

91. 125 cm of potentiometer wire balances the emf. of a cell and 100 cm of the wire is required for balance, if the poles of the cell are joined by a 2Ω resitor. Then the internal resistance of the cell is **Tricky**

(a) 0.25 Ω (b) 0.5 Ω (c) 0.75 Ω (d) 1.25 Ω

92. The resistance of an ammeter is 13 Ω and its scale is graduated for a current upto 100 amps. After an additional shunt has been connected to this ammeter it becomes possible to measure currents upto 750 amperes by this meter. The value of shunt-resistance is

(a) 2 Ω (b) 0.2 Ω (c) 2 kΩ (d) 20 Ω

93. The resistances in the two arms of the meter bridge are 5Ω and RΩ, respectively. When the resistance R is shunted with an equal resistance, the new balance point is at 1.6 l_1. The resistance 'R' is: **Tricky**

(a) 10Ω

(b) 15Ω

(c) 20Ω

(d) 25Ω

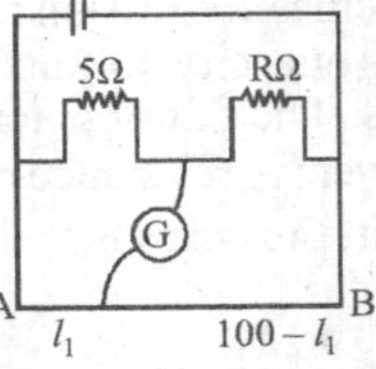

94. In a metre bridge, the balancing length from the left end (standard resistance of one ohm is in the right gap) is found to be 20 cm. The value of the unknown resistance is

(a) 0.8 Ω (b) 0.5 Ω (c) 0.4 Ω (d) 0.25 Ω

95. A potential difference of 220 V is maintained across a 12000 ohm rheostat, as shown in the figure. The voltmeter has a resistance of 6000 ohm and point *c* is at one-fourth of the distance from *a* to *b*. Therefore, the reading of the voltmeter will be

(a) 32 V

(b) 36 V

(c) 40 V

(d) 42 V

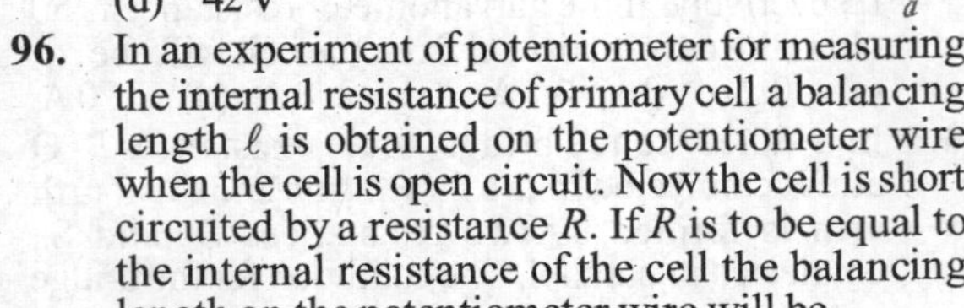

96. In an experiment of potentiometer for measuring the internal resistance of primary cell a balancing length ℓ is obtained on the potentiometer wire when the cell is open circuit. Now the cell is short circuited by a resistance *R*. If *R* is to be equal to the internal resistance of the cell the balancing length on the potentiometer wire will be

(a) ℓ (b) 2ℓ (c) $\ell/2$ (d) $\ell/4$

97. The incorrect option in the given circuit if the reading of ammeter is zero is

(a) The value of ε_1 will be $\frac{\varepsilon(R+r)}{R}$

(b) Current in *R* is $\frac{\varepsilon_1}{r+R}$

(c) Value of ε_1 will be ε

(d) Potential across 2*R* is zero.

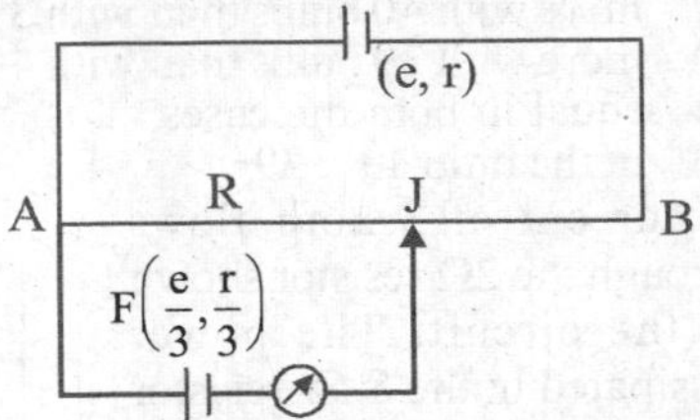

98. A potentiometer arrangement is shown in fig. The driver cell has emf e and internal resistance r. The resistance of potentiometer wire AB is R. F is the cell of em e/3 and internal resistance r/3. Balance point (J) can be obtained for all finite values of

(a) R > r/2 (b) R < r/2 (c) R > r/3 (d) R < r/3

99. In the figure ammeter A_1 reads a current of 10mA, while the voltmeter reads a potential difference of 3V. What does ammeter A_2 in *mA* read ? The ammeters are identical, the internal resistance of the battery is negligible. (Consider all ammeters and voltmeters as non-ideal.) **Critical Thinking**

(a) 6.67 mA

(b) 3.12 mA

(c) 1.12 mA

(d) 5.14 mA

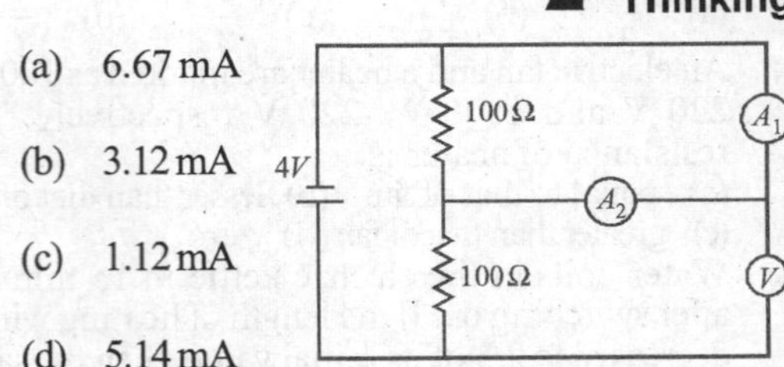

100. A battery of emf $E_0 = 6$ V is connected across a 2 m long uniform wire having resistance 4Ω/*m*. The cells of small emf $\varepsilon_1 = 2$ V and $\varepsilon_2 = 3$ V having internal resistance 2 Ω & 1 Ω respectively are connected as shown in the figure. The null point will be obtained at

(a) 0.10m

(b) 0.25m

(c) 0.50m

(d) 0.75m

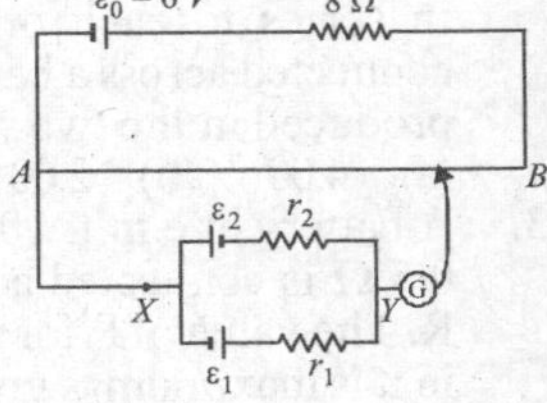

Answer KEYs

1	(a)	11	(c)	21	(b)	31	(b)	41	(a)	51	(b)	61	(d)	71	(a)	81	(b)	91	(b)
2	(c)	12	(c)	22	(b)	32	(a)	42	(b)	52	(a)	62	(b)	72	(a)	82	(a)	92	(a)
3	(c)	13	(a)	23	(d)	33	(d)	43	(a)	53	(c)	63	(c)	73	(b)	83	(a)	93	(b)
4	(a)	14	(c)	24	(c)	34	(b)	44	(c)	54	(d)	64	(d)	74	(a)	84	(a)	94	(d)
5	(c)	15	(b)	25	(b)	35	(d)	45	(b)	55	(a)	65	(d)	75	(c)	85	(b)	95	(c)
6	(c)	16	(c)	26	(b)	36	(a)	46	(a)	56	(b)	66	(a)	76	(d)	86	(b)	96	(c)
7	(d)	17	(b)	27	(a)	37	(a)	47	(d)	57	(b)	67	(d)	77	(b)	87	(a)	97	(c)
8	(b)	18	(d)	28	(d)	38	(c)	48	(a)	58	(a)	68	(c)	78	(d)	88	(d)	98	(a)
9	(b)	19	(c)	29	(d)	39	(a)	49	(a)	59	(c)	69	(d)	79	(c)	89	(a)	99	(a)
10	(a)	20	(a)	30	(a)	40	(d)	50	(c)	60	(c)	70	(a)	80	(b)	90	(b)	100	(d)

Hints & Solutions

1. **(a)** No. of electron in the wire = $nA\ell$ ($n = e^-$ density) and momentum $(nA\ell)\, m_e v_d$

$$= neAv_d \frac{\ell m_e}{e} = \frac{I\ell m_e}{e}$$

$$= \frac{70\times1000}{1.6\times10^{-19}}\times9.1\times10^{-31} = 40\times10^{-8}\text{ N-sec}$$

2. **(c)** Resistance of wire $(R) = \rho\frac{l}{A}$

If wire is bent in the middle then

$l' = \frac{l}{2}, A' = 2A$

$\therefore$ New resistance, $R' = \rho\frac{l'}{A'} = \frac{\rho\frac{l}{2}}{2A} = \frac{\rho l}{4A} = \frac{R}{4}$.

3. **(c)** $i = neAV_d$ and $V_d \propto \sqrt{E}$ (Given)

or, $i \propto \sqrt{E}$

$i^2 \propto E \quad i^2 \propto V$

Hence graph (c) correctly depicts the V-I graph for a wire made of such type of material.

4. **(a)** Given ; $J = ar^2$.

$$i = \int_1^2 J\times2\pi r\,dr = \int_{R/3}^{R/2} ar^2\times2\pi r\,dr$$

$$= 2\pi a\int_{R/3}^{R/2} r^3 dr = 2\pi a\left|\frac{r^4}{4}\right|_{R/3}^{R/2}$$

$$= \frac{\pi a}{2}\left[\left(\frac{R}{2}\right)^4 - \left(\frac{R}{3}\right)^4\right]$$

$$= \frac{\pi aR^4}{2}\times\frac{65}{81\times16} = \frac{65\pi aR^4}{2592}$$

5. **(c)** The charge in a pulse $dq = Idt$ $= 5.00\times10^{-8}$C.

$$n = \frac{q}{e} = \frac{5\times10^{-8}C}{1.6\times10^{-19}C} = 3.12\times10^{11}$$

6. **(c)** Length of each wire = ℓ; Area of thick wire (A_1) = 3A; Area of thin wire $(A_2) = A$ and resistance of thick wire (R_1) = 10 Ω.

Resistance $(R) = \rho\frac{\ell}{A} \propto \frac{1}{A}$ (if ℓ is constant)

$\therefore \frac{R_1}{R_2} = \frac{A_2}{A_1} = \frac{A}{3A} = \frac{1}{3}$

or, $R_2 = 3R_1 = 3\times10 = 30\,\Omega$

The equivalent resistance of these two resistors in series $= R_1 + R_2 = 30 + 10 = 40\Omega$.

7. **(d)** The equivalent circuit is as shown in figure. The resistance of arm AOD (= R + R) is in parallel to the resistance R of arm AD.

Their effective resistance

$$R_1 = \frac{2R\times R}{2R\times R} = \frac{2}{3}R$$

The resistance of arms AB, BC and CD is

$$R_2 = R + \frac{2}{3}R + R = \frac{8}{3}R$$

The resistance R_1 and R_2 are in parallel. The effective resistance between A and D is

$$R_3 = \frac{R_1 \times R_2}{R_1 + R_2} = \frac{\frac{2}{3}R \times \frac{8}{3}R}{\frac{2}{3}R + \frac{8}{3}R} = \frac{8}{15}R.$$

8. (b) Motion of conduction electrons due to random collisions has no preffered direction and average to zero. Drift velocity is caused due to motion of conduction electrons due to applied electric field $\vec{E}$.

9. (b) When the temperature increases, resistance increases. As the e.m.f. applied is the same, the current density decreases the drift velocity decreases. But the rms velocity of the electron due to thermal motion is proportional to $\sqrt{T}$. Therefore, the thermal velocity increases.

10. (a)

11. (c) $R = \frac{\rho \ell_1}{A_1}$, now $\ell_2 = 2\ell_1$

$A_2 = \pi (r_2)^2 = \pi (2r_1)^2 = 4\pi r_1^2 = 4A_1$

$\therefore \; R_2 = \frac{\rho(2\ell_1)}{4A_1} = \frac{\rho \ell}{2A} = \frac{R}{2}$

$\therefore$ Resistance is halved, but specific resistance remains the same.

12. (c) $R = \frac{\rho \ell}{(\pi D^2/4)}$ or $R \propto \frac{\ell}{D^2}$.

$$\frac{R_x}{R_y} = \frac{\ell_x}{D_x^2} \times \frac{D_y^2}{\ell_y} = \frac{\ell_y/2}{(D_y/2)^2} \times \frac{D_y^2}{\ell_y} = \frac{2}{1}$$

13. (a)

14. (c) $I = \frac{dQ}{dt} = 10t + 3$

At $t = 5s, I = 10 \times 5 + 3 = 53$ A

15. (b) Current density $J = I/A$

$= 50 \times 16^{-6} / 50 \times 10^{-6} = 1 \text{ Am}^{-2}$

16. (c) Resistance of wire $(R) = \rho \frac{l}{A}$

If wire is bent in the middle then

$l' = \frac{l}{2}, A' = 2A$

$\therefore$ New, $R' = \rho \frac{l'}{A'} = \frac{\rho \frac{l}{2}}{2A} = \frac{\rho l}{4A} = \frac{R}{4}$.

17. (b) $R_t = R_0(1 + \alpha t)$ at t°C $R_t = 3R_0$

$\alpha = 4 \times 10^{-3}$ /°C

$3R_0 = R_0 (1 + 4 \times 10^{-3} \times t)$

$\therefore 3 - 1 = 4 \times 10^{-3} t$

$\therefore \; t = \frac{2}{4 \times 10^{-3}} = 500°C$

18. (d) $I = nAev_d$ or $v_d \propto 1/\pi r^2$

19. (c)

20. (a) [Hint $\Rightarrow R_t = R_o(1 + \alpha t)$]

$5\Omega = R_0(1 + \alpha \times 50)$ and $7\Omega = R_0(1 + \alpha \times 100)$

or $\frac{5}{7} = \frac{1 + 50\alpha}{1 + 100\alpha}$ or $\alpha = \frac{2}{150} = 0.0133$/°C

21. (b) (Hint $\Rightarrow \rho = \frac{R.A}{\ell}$ = Coefficient of resistivity)

22. (b)

23. (d) $I = neAv_d$

$= 2 \times 10^{21} \times 1.6 \times 10^{-19} \times 10 \times 0.25 \times 10^{-3}$

$= 2 \times 1.6 \times 0.25 = \frac{8}{10} = 0.8$ A

24. (c) $v_d = \frac{I}{neA}$ Here, $I = 5.4$A, $n = 8.4 \times 10^{28}$, per m^3

$A = 10^{-6} m^2, e = 1.6 \times 10^{-19}$C

$\therefore \; v_d = \frac{5.4}{8.4 \times 10^{28} \times 1.6 \times 10^{-19} \times 10^{-6}} = 0.4$ mm/s

25. (b) $J = I/A = Q/tA = ne/tA$

$= 9.6 \times 10^3$ A/m^2

26. (b)

27. (a) $R = \rho \ell / A$; $R' = \frac{\rho \ell'}{A'} = \frac{\rho(\ell/2)}{2A} = \frac{1}{4}R$.

28. (d)

29. (d) $R = \frac{\rho l}{\pi r^2}$. But $m = \pi r^2 ld \; \therefore \pi r^2 = \frac{m}{ld}$

$\therefore \; R_1 = \frac{\rho l_1^2 d}{m_1}, \; R_2 = \frac{\rho l_2^2 d}{m_2}; \; R_3 = \frac{\rho l_3^2 d}{m_3}$

$R_1 : R_2 : R_3 = \frac{l_1^2}{m_1} : \frac{l_2^2}{m_2} : \frac{l_3^2}{m_3} = 125 : 45 : 3$

30. (a)

31. (b) Resistance of the wire of a semicircle = 12/2 = 6Ω

For equivalent resistance between two points on any diameter, 6Ω and 6Ω are in parallel.

32. (a) As the ring has no resistance, the three resistances of 3R each are in parallel.

$\Rightarrow \frac{1}{R'} = \frac{1}{3R} + \frac{1}{3R} + \frac{1}{3R} = \frac{1}{R} \Rightarrow R' = R$

$\therefore$ between point A and B equivalent resistance = R + R = 2R.

33. (d) $R_{eq} = \frac{\left(R + \frac{R}{2}\right) \times R}{\left(R + \frac{R}{2} + R\right)} = \frac{3}{5}R$

34. (b) **35.** (d)

36. (a)

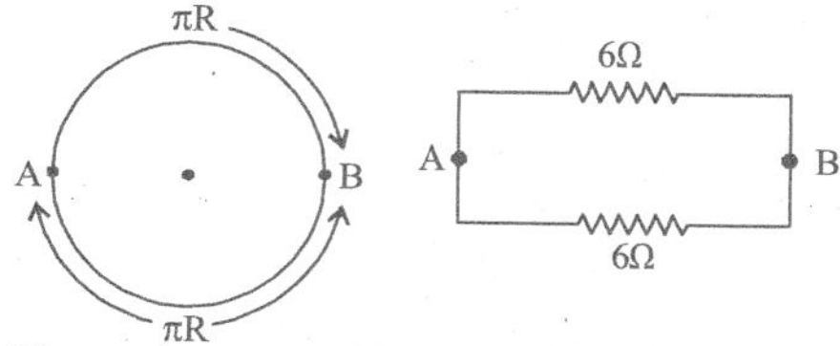

The resistance of length $2\pi R$ is 12Ω. Hence the resistance of length πR is 6Ω. Thus two resistances of 6Ω can be represented as shown in fig. 2.

$\therefore$ Equivalent resistance $R = \dfrac{6 \times 6}{12} = 3\Omega$

37. (a) Let each half side has resistance $r\,(= \rho d/2)$

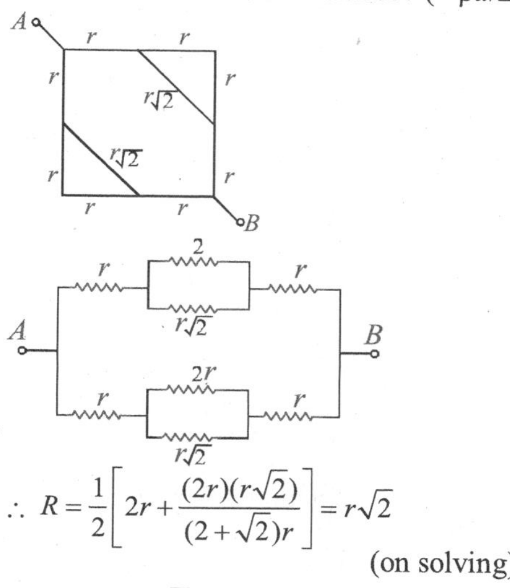

$$\therefore\ R = \frac{1}{2}\left[2r + \frac{(2r)(r\sqrt{2})}{(2+\sqrt{2})r}\right] = r\sqrt{2}$$

(on solving)

$\therefore\ R = \rho d/\sqrt{2}$

38. (c) For various combinations equivalent resistance is maximum between P and Q.

39. (a) It follows from symmetry considerations that if we remove the first element from the circuit, the resistance of the remaining circuit between points C and D will be $R_{CD} = kR_{AB}$. Therefore, the equivalent circuit of the infinite chain will have the form shown in figure.

Applying to this circuit the formulas for the resistance of series and parallel resistors, we obtain

$$R_{AB} = \frac{R_1 + R_2 kR_{AB}}{R_2 + kR_{AB}}$$

Solving the quadratic equation for R_{AB}, we obtain (in particular, for $k = 1/2$)

$$R_{AB} = \frac{R_1 - R_2 + \sqrt{R_1^2 + R_2^2 + 6R_1R_2}}{2}$$

40. (d) The figure can be redrawn as follows :

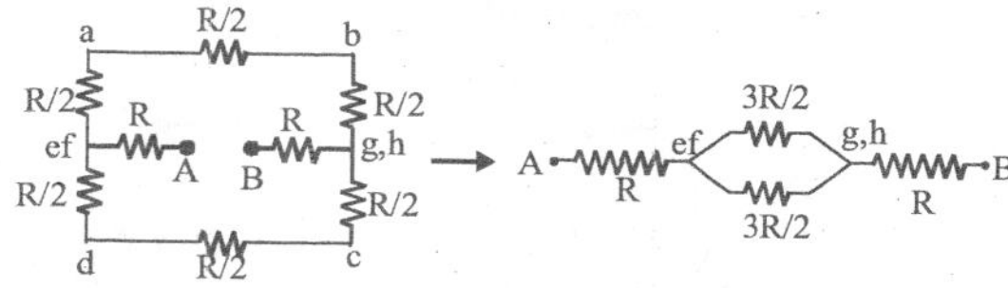

$$R_{AB} = R + \frac{3R}{4} + R = \frac{11R}{4}$$

41. (a) The figure can be redrawn as follows :

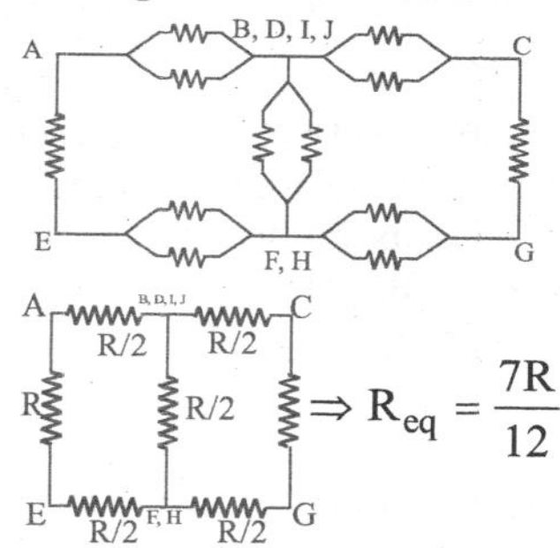

$$\Rightarrow R_{eq} = \frac{7R}{12}$$

42. (b) Let the resistor to be connected across CD be x. Then the equivalent resistance across EF should be x and also across AB should be x. So we get

$$\frac{(2R + x)R}{3R + x} = x$$

solve to get

$x = (\sqrt{3} - 1)R$

43. (a) Let the resistance of half the turn be R. Then in the former case, we have fifteen resistors of resistance R connected in parallel, the total resistance being $R/15$.

In the latter case, we have the same fifteen resistors connected in series, the total resistance being $15R$. Therefore, as a result of unwinding, the resistance of the wire will increase by a factor of 225.

44. (c) Resistance is directly proportional to length

$$\frac{1}{R_{AB}} = \frac{1}{3} + \frac{1}{4+5} = \frac{(4+5)+3}{(3)(4+5)}$$

$$R_{AB} = \frac{3 \times (4+5)}{3 + (4+5)} = \frac{27}{12}$$

Similarly,

$$R_{BC} = \frac{4 \times (3+5)}{4 + (3+5)} = \frac{32}{12}$$

$$R_{AC} = \frac{5 \times (3+4)}{5 + (3+4)} = \frac{35}{12}$$

$\therefore R_{AB} : R_{BC} : R_{AC} = 27 : 32 : 35$

45. **(b)** After adding two conductors, the circuit will acquire the form shown in figure. From symmetry considerations, we calculate that the central conductor will not participate in electric charge transfer.

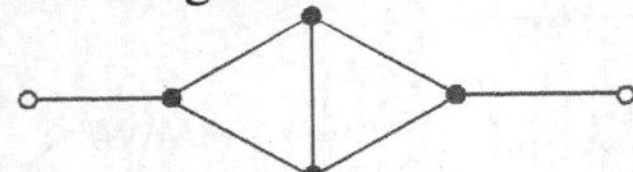

Therefore, if the initial resistance R_1 of the circuit was 5r, where r is the resistance of a conductor, the new resistance of the circuit will become

$$R_2 = 2r + \frac{2r}{2} = 3r \quad \text{Therefore,} \quad \frac{R_2}{R_1} = \frac{3}{5}$$

46. **(a)**

47. **(d)** Kirchhoff's first law is based on conservation of charge and Kirchhoff's second law is based on conservation of energy.

48. **(a)** **49. (a)** **50. (c)**

51. **(b)** Current in the circuit,

$$I = \frac{\varepsilon}{R+r}$$

Potential difference across R,

$$V = IR = \left(\frac{\varepsilon}{R+r}\right)R = \frac{\varepsilon}{1+\frac{r}{R}}$$

When $R = 0$, $V = 0$

$R = \infty$, $V = \varepsilon$

52. **(a)** Given : emf $\varepsilon = 2.1$ V

$I = 0.2$ A, $R = 10\Omega$

Internal resistance r = ?

From formula.

$\varepsilon - Ir = V = IR$

$2.1 - 0.2r = 0.2 \times 10$

$2.1 - 0.2r = 2$ or $0.2r = 0.1$

$$\Rightarrow r = \frac{0.1}{0.2} = 0.5\,\Omega$$

53. **(c)** $r = E/I = 1.5/3 = 0.5$ ohm.

54. **(d)** The equivalent circuit is shown in figure.

$$R = \frac{6\times4}{6+4} + \frac{7\times3}{7+3} = 4.5\,\Omega$$

6Ω 7Ω 4Ω 3Ω

55. **(a)** The effective circuit is shown in figure.

$$R_{AB} = \frac{3R}{8}$$

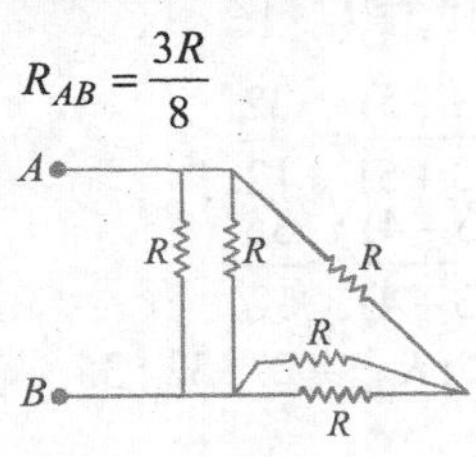

56. **(b)** If each resistor is r, then

$$R_0 = 3r \text{ or } r = \frac{R_0}{3}$$

A r r/3 r B

When keys are closed, then

$$R_{eq} = 7\frac{r}{3} = \frac{7R_0}{9}$$

57. **(b)**

58. **(a)** The effective circuit is shown in figure. Thus, $R_{AB} = 5\,\Omega$.

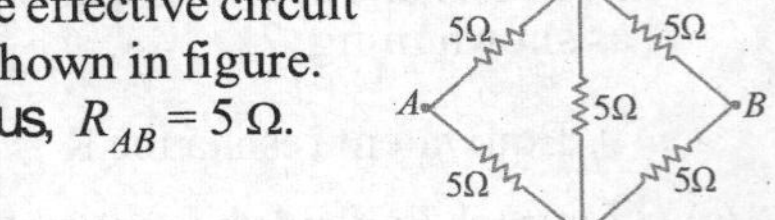

59. **(c)**

60. **(c)** $I = \dfrac{2\varepsilon}{R+R_1+R_2}$

Pot. difference across second cell

$= V = \varepsilon - IR_2 = 0$

$$\varepsilon = \frac{2\varepsilon}{R+R_1+R_2}.R_2 = 0$$

$R + R_1 + R_2 - 2R_2 = 0$

$R + R_1 - R_2 = 0 \quad \therefore\ R = R_2 - R_1$

61. **(d)** Since due to wrong connection of each cell the total emf reduced to 2ε then for wrong connection of three cells the total emf will reduced to $(n\varepsilon - 6\varepsilon)$ whereas the total or equivalent resistance of cell combination will be nr.

62. **(b)**

63. **(c)** Kirchhoff's junction rule states that the algebraic sum of all currents into and out of any branch point is zero : $\Sigma I = 0$. By convention, the sign of current entering a junction is positive and current leaving a junction is negative.

$4A + 5A - 6A + I_{AB} = 0$, therefore $I_{AB} = -3A$. The wire between points A and B carries a current of 3A away from the junction.

64. **(d)**

65. **(d)** Given : Number of cells, $n = 5$, emf of each cell $= E$

Internal resistance of each cell $= r$

In series, current through resistance R

$$I = \frac{nE}{nr+R} = \frac{5E}{5r+R}$$

In parallel, current through resistance R

$$I' = \frac{E}{\frac{r}{n}+R} = \frac{nE}{r+nR} = \frac{5E}{r+5R}$$

According to question, $I = I'$

$\therefore \dfrac{5E}{5r+5R} = \dfrac{5E}{r+5R} \Rightarrow 5r + R = r + 5R$

or $R = r \quad \therefore \dfrac{R}{r} = 1$

66. **(a)** $E_{net} = \dfrac{E_1 r_2 - E_2 r_1}{r_1 + r_2}$ or $E_{net} = \dfrac{2-2}{2+1} = 0$

67. **(d)**

68. **(c)**

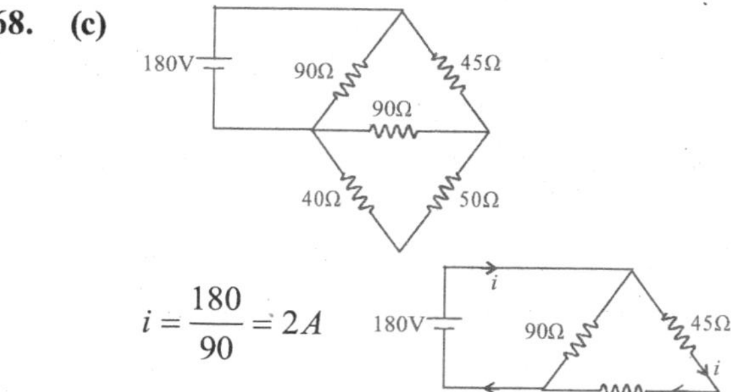

$i = \dfrac{180}{90} = 2A$

69. **(d)**

70. **(a)** The given circuit can be redrawn as,

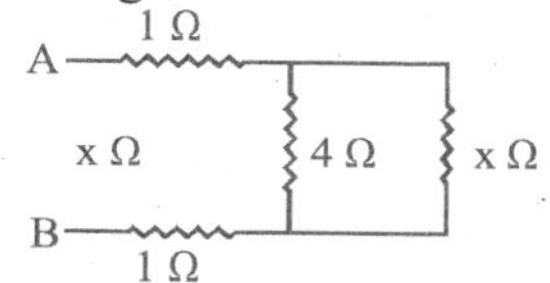

as 4 Ω and x Ω are parallel

$x' = \dfrac{1}{4} + \dfrac{1}{x} = \dfrac{(4+x)}{4x} \quad x' = \dfrac{4x}{4+x}$

& 1 Ω and 1 Ω are also parallel x" = 2 Ω

Now equivalent resistance of circuit

$x = \dfrac{4x}{4+x} + 2 = \dfrac{8+6x}{4+x}$

$4x + x^2 = 8 + 6x \Rightarrow x^2 - 2x - 8 = 0$

$x = \dfrac{2 \pm \sqrt{4 - 4(1)(-8)}}{2} = \dfrac{2 \pm \sqrt{36}}{2} = 4\Omega$

Reading of Ammeter $A_1 = \dfrac{V}{(R+r)}$

$A_1 = \dfrac{9}{4+0.5} = 2$ Ampere

71. **(a)** $I = \dfrac{8-4}{1+2+9} = \dfrac{4}{12} = \dfrac{1}{3}A;$

$V_P - V_Q = 4 - \dfrac{1}{3} \times 3 = 3$ volt

72. **(a)** The current in 2Ω resistor will be zero because it is not a part of any closed loop.

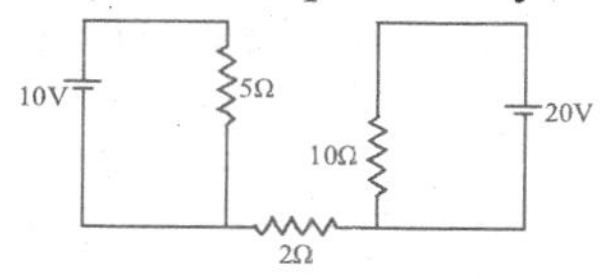

73. **(b)**

In steady state capacitor is fully charged hence no current will flow through line 2. By simplyfing the circuit

Hence resultant potential difference across resistances will be 8.0 V.

Thus current $I = \dfrac{V}{R} = \dfrac{8.0}{3+9} = \dfrac{8}{12}$

or, $I = \dfrac{2}{3} = 0.67$ A

74. **(a)** Current through 50 V and 30 V batteries are respectively 2.5 A and 3 A.

75. **(c)** The current through the resistance R

$I = \left(\dfrac{\varepsilon}{R+r}\right)$

The potential difference across R,

$V = IR = \left(\dfrac{\varepsilon}{R+r}\right)R$

$V = \dfrac{\varepsilon}{\left(1 + \dfrac{r}{R}\right)}$; when $R = 0, V = 0; R = \infty, V = \varepsilon$

Thus V increases as R increases upto certain limit, but it does not increase further.

76. **(d)** Current from D to C = 1A

$\therefore V_D - V_C = 2 \times 1 = 2V$

$V_A = 0 \qquad \therefore V_C = 1V, \therefore V_D - V_C = 2$

$\Rightarrow V_D - 1 = 2 \quad \therefore V_D = 3V$

$\therefore V_D - V_B = 2 \quad \therefore 3 - V_B = 2 \quad \therefore V_B = 1V$

77. **(b)** Since, the voltage is same for the two combinations, therefore $H \propto \dfrac{1}{R}$. Hence, the combination of 39 bulbs will glow more.

78. **(d)**

79. **(c)** Heat supplied $= \dfrac{V^2}{R} \times t$

$\Rightarrow \dfrac{t_1}{R_1} = \dfrac{t_2}{R_2} \Rightarrow \dfrac{6}{R_1} = \dfrac{8}{R_2} \Rightarrow \dfrac{R_1}{R_2} = \dfrac{3}{4}$

80. **(b)** As $R \propto V^2/P$ or $R \propto 1/P$, so resistance of heater is less than that of fan.

81. (b) $H=\frac{V^2}{R}\times 15\times 60=\frac{V^2}{(2/3)R}\times t$

or $t=\frac{2}{3}\times 15\times 60=600\,s=10$ minutes.

82. (a) $H=I^2Rt$. Here $R_1=\rho\frac{1}{\pi r^2}$ and

$R_2=\rho\frac{1}{\pi(2r)^2}$ That is, $R_1=4R_2$.

Hence, $\frac{H_1}{H_2}=4$

83. (a) According to maximum power theorem, the power in the circuit is maximum if the value of external resistance is equal to the internal resistance of battery.

84. (a) As voltage of appliance remains constant, the amount of heat produced is given by,

$$H=\frac{V^2}{R}t \quad \ldots (i)$$

when resistance is reduced by 20%, new resistance is R' = R – 0.2 R = 0.8 R

$$H'=\frac{V^2}{0.8R}t' \quad \ldots (ii)$$

Equating Eqs. (i) and (ii), we get

$$\frac{V^2}{R}t=\frac{V^2}{0.8R}t'$$

$\Rightarrow t'=0.8t=0.8\times 12=9.6$ min

85. (b)

86. (b) Total resistance R = (0.5 Ω/km) × (150 km) = 75 Ω

Total voltage drop = (8 V/km) × (150 km) = 1200 V

Power loss $=\frac{(\Delta V)^2}{R}=\frac{(1200)^2}{75}$ W $=19200$ W $=19.2$ kW

87. (a) $Q=at-bt^2$ $\quad i=\frac{dQ}{dt}=a-2bt$

$i=0$ for $t=t_0=a/2b$, i.e., current flow from $t=0$ to $t=t_0$.

The heat produced $=\int_0^{t_0} i^2R\,dt$. Putting the value of i we get heat produced $=\frac{a^3R}{6b}$.

88. (d)

89. (a) Given : V = 7 V and r = 5Ω

$R_{eq}=\frac{40\times 120}{40+120}\,\Omega$

$I=\frac{V}{R}=\frac{7}{5+\frac{40\times 120}{40+120}}=0.2$ A.

90. (b) $\frac{P}{Q}=\frac{R}{S}$ where $S=\frac{S_1S_2}{S_1+S_2}$

91. (b) $r=\frac{\ell_1-\ell_2}{\ell_2}\times R\Omega$

Here, $\ell_1=125$ cm, $\ell_2=100$ cm, $R=2\Omega$.

$\therefore\ r=0.5\Omega$

92. (a)

93. (b) This is a balanced wheatstone bridge condition,

$\frac{5}{R}=\frac{\ell_1}{100-\ell_1}$ and $\frac{5}{R/2}=\frac{1.6\ell_1}{100-1.6\ell_1}$

$\Rightarrow\ R=15\,\Omega$

94. (d) $\frac{P}{Q}=\frac{l}{(100-l)}$

or $P=\frac{l}{100-l}\times Q=\frac{20}{80}\times 1=0.25\Omega$.

95. (c) $\left(\frac{I}{4}\times 12000\right)/6000=2000$,

$I=\frac{220}{9000+2000}$ A

∴ Potential difference across voltmeter = 40 V.

96. (c) **97.** (c)

98. (a) Current in AB $=I=\frac{e}{R+r}$

potential difference across AB $=IR=\frac{eR}{R+r}$

To obtain balanced point,

$\frac{eR}{R+r}>\frac{e}{3}$ or $R>r/2$

99. (a) R_A = resistance of ammeter

$$\frac{4-V_1}{100}=\frac{V_1-1}{R_A}+\frac{V_1-0}{100} \quad \ldots\ldots\ldots (i)$$

$1V-0V=(10mA)R_A$

$$R_A=100\Omega \quad \ldots\ldots\ldots (ii)$$

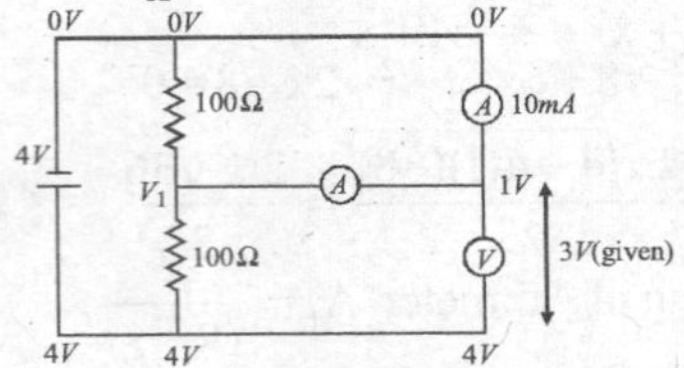

$$\frac{4-V_1}{100}=\frac{V_1-1}{100}+\frac{V_1-0}{100}$$

[By using eq. (1) and (2)]

$V_1=(5/3)$ V

$\frac{V_1-1}{R_A}$ = (Current in ammeter (II))

$\frac{(5/3)-1}{100}=6.67$ mA

100. (d) Effective emf of two cells in parallel

$$\varepsilon=\frac{\varepsilon_2r_1-\varepsilon_1r_2}{r_1+r_2}=\frac{3\times 2-2\times 1}{2+1}=\frac{4}{3}V$$

X terminal of the connection is positive. For neutral point, positive terminal of ε_0 must be connected to *X*.

18 Moving Charges and Magnetism

1. A charged particle of mass m and charge q travels on a circular path of radius r that is perpendicular to a magnetic field B. The time taken by the particle to complete one revolution is

(a) $\frac{2\pi q^2 B}{m}$ (b) $\frac{2\pi mq}{B}$ (c) $\frac{2\pi m}{qB}$ (d) $\frac{2\pi qB}{m}$

2. A charged particle enters in a magnetic field in a direction perpendicular to the magnetic field. Which of the following graphs show the correct variation of kinetic energy of the particle with time *t*?

Tricky

(a) (b)

(c) (d)

3. An electric charge $+q$ moves with velocity $\vec{v} = 3\hat{i} + 4\hat{j} + \hat{k}$ in an electromagnetic field given by $\vec{E} = 3\hat{i} + \hat{j} + 2\hat{k}$ and $\vec{B} = \hat{i} + \hat{j} - 3\hat{k}$ The *y*-component of the force experienced by $+q$ is :

(a) $11q$ (b) $5q$ (c) $3q$ (d) $2q$

4. A beam of electrons is moving with constant velocity in a region having simultaneous perpendicular electric and magnetic fields of strength 20 Vm^{-1} and 0.5 T respectively at right angles to the direction of motion of the electrons. Then the velocity of electrons must be

(a) 8 m/s (b) 20 m/s (c) 40 m/s (d) $\frac{1}{40}$ m/s

5. Two electron beams having their velocities in the ratio 1: 2 are subjected to identical magnetic fields acting at right angles to the direction of motion of electron beams. The ratio of deflection produced is:

(a) 2 : 1 (b) 1:2 (c) 4 : 1 (d) 1:4

6. Charge q is uniformly spread on a thin ring of radius R. The ring rotates about its axis with a uniform frequency f Hz. The magnitude of magnetic induction at the centre of the ring is

(a) $\frac{\mu_0 qf}{2R}$ (b) $\frac{\mu_0 q}{2fR}$ (c) $\frac{\mu_0 q}{2\pi fR}$ (d) $\frac{\mu_0 qf}{2\pi R}$

7. A charged particle moves insides a pipe which is bent as shown in fig. If $R > \frac{mv}{qB}$, then force exerted by the pipe on charged particle at P is (Neglect gravity)

Tricky

(a) toward center
(b) away from center
(c) zero
(d) none of these

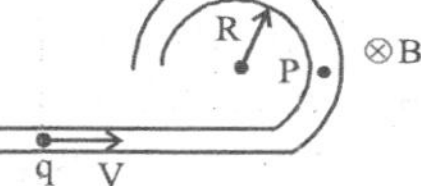

8. A proton carrying 1 MeV kinetic energy is moving in a circular path of radius *R* in uniform magnetic field. What should be the energy of an α-particle to describe a circle of same radius in the same field?

(a) 2 MeV (b) 1 MeV (c) 0.5 MeV (d) 4 MeV

9. Under the influence of a uniform magnetic field a charged particle is moving in a circle of radius R with constant speed v. The time period of the motion

(a) depends on both R and *v*
(b) is independent of both R and *v*
(c) depends on R and not on *v*
(d) depends on v and not on R

10. In a mass spectrometer used for measuring the masses of ions, the ions are initially accelerated by an electric potential V and then made to describe semicircular path of radius R using a magnetic field B. If V and B are kept constant, the ratio $\left(\frac{\text{charge on the ion}}{\text{mass of the ion}}\right)$ will be proportional to

Critical Thinking

(a) $1/R^2$ (b) R^2 (c) R (d) 1/R

11. An ionized gas contains both positive and negative ions. If it is subjected simultaneously to an electric field along the +x-direction and a magnetic field along the +*z*-direction, then

(a) positive ions deflect towards +*y*-direction and negative ions towards -*y* direction
(b) all ions deflect towards +*y*-direction
(c) all ions deflect towards –*y*-direction
(d) positive ions deflect towards –y-direction and negative ions towards + *y*-direction.

12. Three particles, an electron (e), a proton (p) and a helium atom (He) are moving in circular paths with constant speeds in the x – y plane in a region where a uniform magnetic field B exists along z-axis. The times taken by e, p and He inside the field to complete one revolution are t_e, t_p and t_{He} respectively. Then, **Tricky**

(a) $t_{He} > t_p = t_e$ (b) $t_{He} > t_p > t_e$
(c) $t_{He} = t_p = t_e$ (d) None of these

13. A particle of specific charge $\frac{q}{m} = \pi\ \text{Ckg}^{-1}$ is projected from the origin towards positive x-axis with a velocity of 10ms^{-1} in a uniform magnetic field $\vec{B} = -2\hat{k}$ T. The velocity $\vec{v}$ of particle after time t = 1/12 s will be (in ms^{-1})

(a) $5[\hat{i}+\sqrt{3}\hat{j}]$ (b) $5[\sqrt{3}\hat{i}+\hat{j}]$
(c) $5[\sqrt{3}\hat{i}-\hat{j}]$ (d) $5[\hat{i}+\hat{j}]$

14. A particle is projected in a plane perpendicular to a uniform magnetic field. The area bounded by the path described by the particle is proportional to

(a) the velocity (b) the momentum
(c) the kinetic energy (d) None of these

15. Consider a hypothetic spherical body. The body is cut into two parts about the diameter. One of hemispherical portion has mass distribution m whie the other portion has identical charge distribution q. The body is rotated about the axis with constant speed ω. Then, the ratio of magnetic moment to angular momentum is **Tricky**

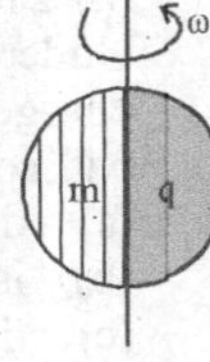

(a) $\frac{q}{2m}$ (b) $> \frac{q}{2m}$
(c) $< \frac{q}{2m}$ (d) cannot be calculated

16. There exist uniform magnetic and electric fields of magnitudes 1 T and 1 V m^{-1}, respectively, along positive y-axis. A charged particle of mass 1 kg and charge 1 C is having velocity 1 m s^{-1} along x-axis and is at origin at t = 0. Then, the coordinates of the particles at time πs will be

(a) (0, 1, 2) m (b) $(0, -\pi^2, -2)$ m
(c) $(2, \pi^2/2, 2)$m (d) $(0, \pi^2/2, 2)$ m

17. A particle of mass m and charge q moves with a constant velocity v along the positive x-direction. It enters a region containing a uniform magnetic field B directed along the negative z-direction, extending from $x = a$ to $x = b$. The minimum value of v required so that the particle can just enter the region $x > b$ is

(a) $\frac{qbB}{m}$ (b) $\frac{q(b-a)B}{m}$
(c) $\frac{qaB}{m}$ (d) $\frac{q(b+a)B}{2m}$

18. For a positively charged particle moving in a x-y plane initially along the x-axis, there is a sudden change in its path due to the presence of electric and/or magnetic fields beyond P. The curved path is shown in the x-y plane and is found to be non-circular. Which one of the following combinations is possible?

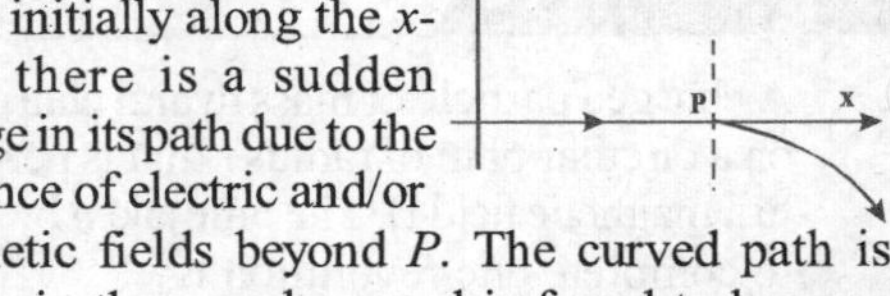

(a) $\vec{E} = 0; \vec{B} = b\hat{i} + c\hat{k}$ (b) $\vec{E} = a\hat{i}; \vec{B} = c\hat{k} + a\hat{i}$
(c) $\vec{E} = 0; \vec{B} = c\hat{j} + b\hat{k}$ (d) $\vec{E} = a\hat{i}; \vec{B} = c\hat{k} + b\hat{j}$

19. Two particles X and Y having equal charges, after being accelerated through the same potential difference, enter a region of uniform magnetic field and describe circular paths of radii R_1 and R_2 respectively. The ratio of the mass of X to that of Y is

(a) $(R_1/R_2)^{1/2}$ (b) R_2/R_1
(c) $(R_1/R_2)^2$ (d) R_1/R_2.

20. A uniform magnetic field of magnitude 1T exists in region $y \geq 0$ is along $\hat{k}$ direction as shown. A particle of charge 1C is projected from point $(-\sqrt{3}, -1)$ towards origin with speed 1 m/sec. If mass of particle is 1 kg, then co-ordinates of centre of circle in which particle moves are – **Tricky**

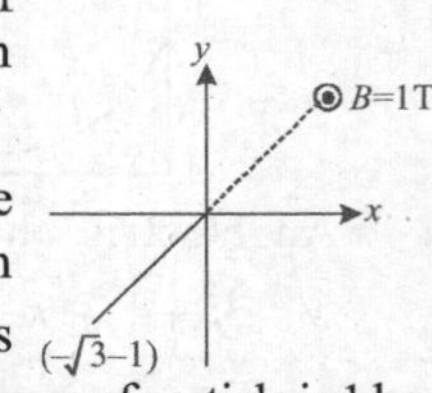

(a) $(1, \sqrt{3})$ (b) $(1, -\sqrt{3})$
(c) $\left(\frac{1}{2}, -\frac{\sqrt{3}}{2}\right)$ (d) $\left(\frac{\sqrt{3}}{2}, -\frac{1}{2}\right)$

21. An electron, charge –e, mass m, enters a uniform magnetic field $\vec{B} = B\vec{i}$ with an initial velocity $\vec{v} = v_x\vec{i} + v_y\vec{j}$. What is the velocity of the electron after a time interval of t second?

(a) $v_x\hat{i} + v_y\hat{j} + \frac{e}{m}v_yBt\hat{k}$
(b) $v_x\hat{i} + v_y\hat{j} - \frac{e}{m}v_yBt\hat{k}$
(c) $v_x\hat{i} + \left(v_y + \frac{e}{m}v_yBt\right)\hat{j}$
(d) $v_x\hat{i} + \left(v_y + \frac{e}{m}v_yBt\right)\hat{i} + v_y\hat{j}$

22. An alternating electric field, of frequency v, is applied across the dees (radius $= R$) of a cyclotron that is being used to accelerate protons (mass $= m$). The operating magnetic field (B) used in the cyclotron and the kinetic energy (K) of the proton beam, produced by it, are given by :

(a) $B = \frac{mv}{e}$ and $K = 2m\pi^2 v^2 R^2$

(b) $B = \frac{2\pi mv}{e}$ and $K = m^2 \pi v R^2$

(c) $B = \frac{2\pi mv}{e}$ and $K = 2m\pi^2 v^2 R^2$

(d) $B = \frac{mv}{e}$ and $K = m^2 \pi v R^2$

23. A uniform magnetic field is directed out of the page. A charged particle, moving in the plane of the page, follows a clockwise spiral of decreasing radius as shown. A reasonable explanation is

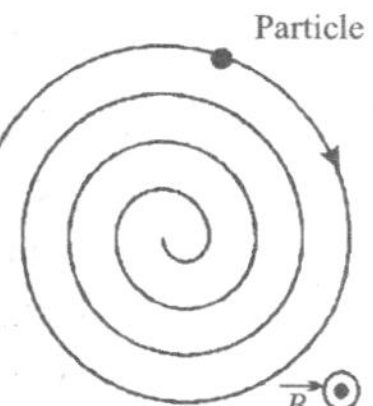

(a) the charge is positive and slowing down
(b) the charge is negative and slowing down
(c) the charge is positive and speeding up
(d) the charge is negative and speeding up

24. A 10 eV electron is circulating in a plane at right angles to a uniform field at magnetic induction 10^{-4} Wb/m^2 (= 1.0 gauss). The orbital radius of the electron is

(a) 12 cm (b) 16 cm (c) 11 cm (d) 18 cm

25. A deuteron of kinetic energy 50 keV is describing a circular orbit of radius 0.5 metre in a plane perpendicular to the magnetic field B. The kinetic energy of the proton that describes a circular orbit of radius 0.5 metre in the same plane with the same B is

(a) 25 keV (b) 50 keV (c) 200 keV (d) 100 keV

26. A particle of charge q and mass m starts moving from the origin under the action of an electric field $\vec{E} = E_0 \hat{i}$ and $\vec{B} = B_0 \hat{i}$ with velocity $\vec{v} = v_0 \hat{j}$. The speed of the particle will become $2v_0$ after a time

(a) $t = \frac{2mv_0}{qE}$ (b) $t = \frac{2Bq}{mv_0}$

(c) $t = \frac{\sqrt{3}Bq}{mv_0}$ (d) $t = \frac{\sqrt{3}mv_0}{qE}$

27. $OABC$ is current carrying square loop an electron is projected from the centre of loop along its diagonal AC as shown. Unit vector in the direction of initial acceleration will be

(a) $\hat{k}$

(b) $-\left(\frac{\hat{i}+\hat{j}}{\sqrt{2}}\right)$

(c) $-\hat{k}$

(d) $\frac{\hat{i}+\hat{j}}{\sqrt{2}}$

28. Two identical particles having the same mass m and charges +q and –q separated by a distance d enter a uniform magnetic field B directed perpendicular to paper inwards with in speeds v_1 and v_2 as shown in Fig. The particles will not collide if

Tricky

(a) $d > \frac{m}{Bq}(v_1 + v_2)$ (b) $d < \frac{m}{Bq}(v_1 + v_2)$

(c) $d > \frac{2m}{Bq}(v_1 + v_2)$ (d) $v_1 = v_2$

29. A charged sphere of mass m and charge $-q$ starts sliding along the surface of a smooth hemispherical bowl, at position P. The region has a transverse uniform magnetic field B. Normal force by the surface of bowl on the sphere at position Q is

Critical Thinking

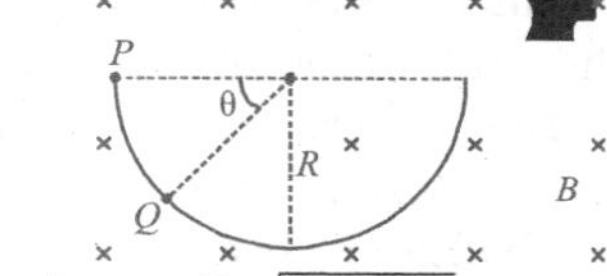

(a) $mg \sin\theta + qB\sqrt{2gR\sin\theta}$

(b) $3\,mg \sin\theta + qB\sqrt{2gR\sin\theta}$

(c) $mg \sin\theta - qB\sqrt{2gR\sin\theta}$

(d) $3\,mg \sin\theta - qB\sqrt{2gR\sin\theta}$

30. A positive charge 'q' of mass 'm' is moving along the + x axis. We wish to apply a uniform magnetic field B for time Δt so that the charge reverses its direction crossing the y axis at a distance d. Then:

(a) $B = \frac{mv}{qd}$ and $\Delta t = \frac{\pi d}{v}$

(b) $B = \frac{mv}{2qd}$ and $\Delta t = \frac{\pi d}{2v}$

(c) $B = \frac{2mv}{qd}$ and $\Delta t = \frac{\pi d}{2v}$

(d) $B = \frac{2mv}{qd}$ and $\Delta t = \frac{\pi d}{v}$

31. A cyclotron is operated at an oscillator frequency of 24 MHz and has a dee radius $R = 60$cm. What is magnitude of the magnetic field B (in Tesla) to accelerate deuterons (mass $= 3.34 \times 10^{-27}$) kg?

(a) 9.5 (b) 7.2 (c) 5.0 (d) 3.2

32. Two particles, each of mass m and charge q, are attached to the two ends of a light rigid rod of length $2R$. The rod is rotated at constant angular speed about a pependicular axis passing through its centre. The ratio of the magnitudes of the magnetic moment of the system and its angular momentum about the centre of the rod is

(a) $\frac{q}{2m}$ (b) $\frac{q}{m}$ (c) $\frac{2q}{m}$ (d) $\frac{q}{\pi m}$

33. A charged particle of specific charge (charge/mass) α is released from origin at time $t = 0$ with velocity $\vec{v} = v_0(\hat{i} + \hat{j})$ in uniform magnetic field $\vec{B} = B_0\hat{i}$. Coordinates of the particle at time $t = \pi/(B_0\alpha)$ are

Critical Thinking

(a) $\left(\frac{v_0}{2B_0\alpha}, \frac{\sqrt{2}v_0}{\alpha B_0}, \frac{-v_0}{B_0\alpha}\right)$ (b) $\left(\frac{-v_0}{2B_0\alpha}, 0, 0\right)$

(c) $\left(0, \frac{2v_0}{B_0\alpha}, \frac{v_0\pi}{2B_0\alpha}\right)$ (d) $\left(\frac{v_0\pi}{B_0\alpha}, 0, \frac{-2v_0}{B_0\alpha}\right)$

34. A particle of mass m and charge q enters a region of magnetic field (as shown) with speed v. There is a region in which the magnetic field is absent, as shown. The particle after entering the region collides elastically with a rigid wall. Time after which the velocity of particle becomes antiparallel to its initial velocity is

$\frac{mv}{\sqrt{2}qB}$ $\frac{mv}{\sqrt{2}qB}$ Wall

Critical Thinking

(a) $\frac{m}{2qB}(\pi + 4)$ (b) $\frac{m}{qB}(\pi + 2)$

(c) $\frac{m}{4qB}(\pi + 2)$ (d) $\frac{m}{4qB}(2\pi + 3)$

35. The figure shows a thin metalic rod whose one end is pivoted at point O. The rod rotates about the end O in a plane perpendicular to the uniform magnetic field with angular frequency ω in clockwise direction.

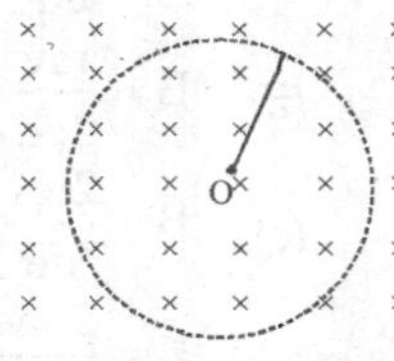

Which of the following is correct?

(a) The free electrons of the rod move towards the outer end

(b) The free electrons of the rod move towards the pivoted end.

(c) The free electrons of the rod move towards the mid-point of the rod.

(d) The free electrons of the rod do not move towards any end of the rod as rotation of rod has no effect on motion of free electrons.

36. A helium nucleus makes a full rotation in a circle of radius 0.8 meter in 2 sec. The value of the magnetic field induction B in tesla at the centre of circle will be

(a) $2 \times 10^{-19}\,\mu_0$ (b) $10^{-19}/\mu_0$

(c) $10^{-19}\,\mu_0$ (d) $2 \times 10^{-20}/\mu_0$

37. The correct plot of the magnitude of magnetic field $\vec{B}$ vs distance r from centre of the wire is, if the radius of wire is R

Tricky

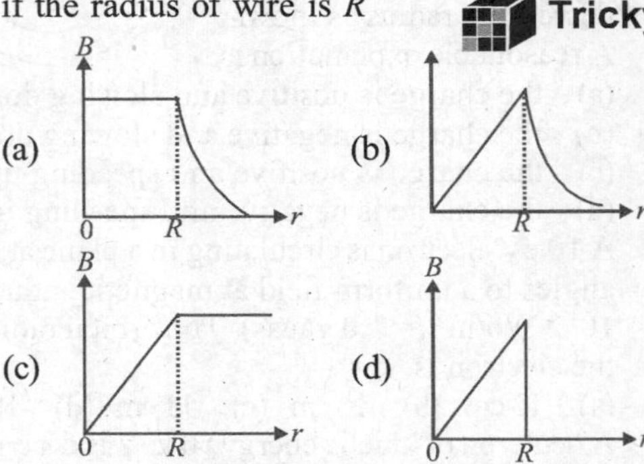

38. A current of I ampere flows in a wire forming a circular arc of radius r metres subtending an angle θ at the centre as shown. The magnetic field at the centre O in tesla is

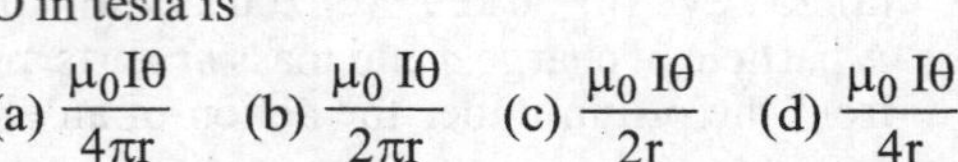

(a) $\frac{\mu_0 I\theta}{4\pi r}$ (b) $\frac{\mu_0 I\theta}{2\pi r}$ (c) $\frac{\mu_0 I\theta}{2r}$ (d) $\frac{\mu_0 I\theta}{4r}$

39. The magnetic field due to a current carrying circular loop of radius 3 cm at a point on the axis at a distance of 4 cm from the centre is 54 μT. What will be its value at the centre of loop ?

(a) 125 μT (b) 150 μT (c) 250 μT (d) 75 μT

40. If the magnetic field at P can be written as K tan $\left(\frac{\alpha}{2}\right)$, then K is

Tricky

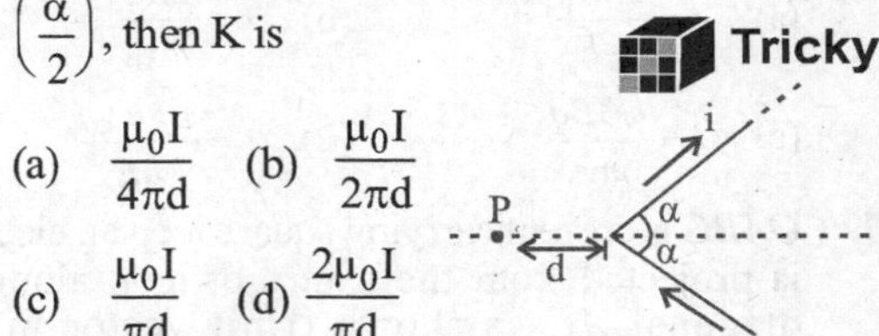

(a) $\frac{\mu_0 I}{4\pi d}$ (b) $\frac{\mu_0 I}{2\pi d}$

(c) $\frac{\mu_0 I}{\pi d}$ (d) $\frac{2\mu_0 I}{\pi d}$

41. A wire carries a current. Maintaining the same current it is bent first to form a circular plane coil of one turn which produces a magnetic field B at the centre of the coil. The same length is now bent more sharply to give a double loop of smaller radius. The magnetic field at the centre of the double loop, caused by the same current is **Critical Thinking**

(a) $4B$ (b) $B/4$ (c) $B/2$ (d) $2B$

42. Two long parallel wires P and Q are both perpendicular to the plane of the paper with distance of 5 m between them. If P and Q carry currents of 2.5 amp and 5 amp respectively in the same direction, then the magnetic field at a point half-way between the wires is

(a) $\frac{3\mu_0}{2\pi}$ (b) $\frac{\mu_0}{\pi}$ (c) $\frac{\sqrt{3}\mu_0}{2\pi}$ (d) $\frac{\mu_0}{2\pi}$

43. Consider two thin identical conducting wires covered with very thin insulating material. One of the wires is bent into a loop and produces magnetic field B_1, at its centre when a current I passes through it. The ratio $B_1 : B_2$ is: **Tricky**

(a) 1 : 1 (b) 1 : 3 (c) 1 : 9 (d) 9 : 1

44. Two wires are wrapped over a wooden cylinder to form two coaxial loops carrying current i_1 and i_2. If $i_2 = 8i_1$ the value of x for $B = 0$ at the origin O is :

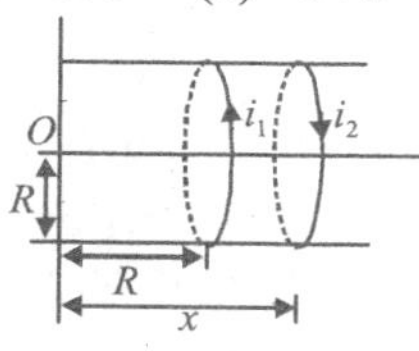

(a) $\sqrt{(\sqrt{7}-1)R}$ (b) $\sqrt{5}R$

(c) $\sqrt{3}R$ (d) $\sqrt{7}R$

45. Two long parallel wires carry equal current i flowing in the same direction are at a distance 2d apart. The magnetic field B at a point lying on the perpendicular line joining the wires and at a distance x from the midpoint is –

(a) $\frac{\mu_0 id}{\pi(d^2+x^2)}$ (b) $\frac{\mu_0 ix}{\pi(d^2-x^2)}$

(c) $\frac{\mu_0 ix}{(d^2+x^2)}$ (d) $\frac{\mu_0 id}{(d^2+x^2)}$

46. The current density $\vec{J}$ inside a long, solid, cylindrical wire of radius a = 12 mm is in the direction of the central axis, and its magnitude varies linearly with radial distance r from the axis according to $J = \frac{J_0 r}{a}$, where $J_0 = \frac{10^5}{4\pi}$ A/m^2. Find the magnitude of the magnetic field at $r = \frac{a}{2}$ in μT. **Toughnut**

(a) 10μT (b) 4μT (c) 5μT (d) 3μT

47. A long straight wire of radius a carries a steady current I. The current is uniformly distributed over its cross-section. The ratio of the magnetic fields B and B', at radial distances $\frac{a}{2}$ and 2a respectively, from the axis of the wire is :

(a) $\frac{1}{4}$ (b) $\frac{1}{2}$ (c) 1 (d) 4

48. An electron moving in a circular orbit of radius r makes n rotations per second. The magnetic field produced at the centre has magnitude:

(a) Zero (b) $\frac{\mu_0 n^2 e}{r}$ (c) $\frac{\mu_0 ne}{2r}$ (d) $\frac{\mu_0 ne}{2\pi r}$

49. A wire carrying current I has the shape as shown in adjoining figure. Linear parts of the wire are very long and parallel to X-axis while semicircular portion of radius R is lying in Y-Z plane. Magnetic field at point O is : **Critical Thinking**

(a) $\vec{B} = -\frac{\mu_0}{4\pi}\frac{I}{R}\left(\mu\hat{i} \times 2\hat{k}\right)$

(b) $\vec{B} = -\frac{\mu_0}{4\pi}\frac{I}{R}\left(\pi\hat{i} + 2\hat{k}\right)$

(c) $\vec{B} = \frac{\mu_0}{4\pi}\frac{I}{R}\left(\pi\hat{i} - 2\hat{k}\right)$

(d) $\vec{B} = \frac{\mu_0}{4\pi}\frac{I}{R}\left(\pi\hat{i} + 2\hat{k}\right)$

50. A small current element of length $d\ell$ and carrying current is placed at (1, 1, 0) and is carrying current in '+z' direction. If magnetic field at origin be $\vec{B}_1$ and at point (2, 2, 0) be $\vec{B}_2$ then:

(a) $\vec{B}_1 = \vec{B}_2$ (b) $|\vec{B}_1| = |2\vec{B}_2|$

(c) $\vec{B}_1 = -\vec{B}_2$ (d) $\vec{B}_1 = -2\vec{B}_2$

51. In the figure shown a coil of single turn is wound on a sphere of radius R and mass M. The plane of the coil is parallel to the plane and lies in the equatorial plane of the sphere. Current in the coil is i. The value of B if the sphere is in equilibrium is **Toughnut**

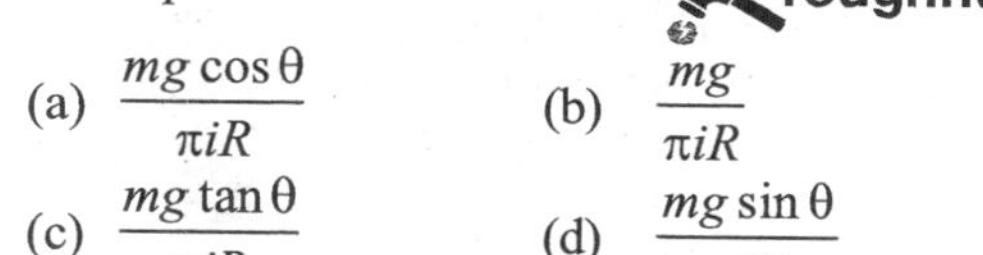

(a) $\frac{mg\cos\theta}{\pi iR}$ (b) $\frac{mg}{\pi iR}$

(c) $\frac{mg\tan\theta}{\pi iR}$ (d) $\frac{mg\sin\theta}{\pi iR}$

52. In the loops shown in figure. all curved sections are either semi-circles or quarter circles. All the loops carry same current. The magnetic fields at the centres have magnitudes B_1, B_2, B_3 and B_4. Then, which is correct? **Tricky**

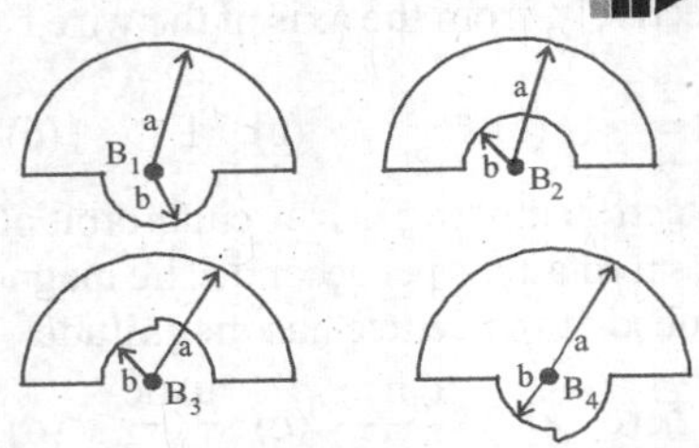

(a) $B_1 > B_2 > B_3 > B_4$ (b) $B_3 > B_4 > B_1 > B_1$
(c) $B_4 > B_1 > B_2 > B_3$ (d) $B_1 > B_4 > B_3 > B_2$

53. A current loop consists of two identical semicircular parts each of radius *R*, one lying in the *x-y* plane and the other in *x-z* plane. If the current in the loop is *i*., the resultant magnetic field due to the two semicircular parts at their common centre is

(a) $\frac{\mu_0 i}{\sqrt{2}R}$ (b) $\frac{\mu_0 i}{2\sqrt{2}R}$ (c) $\frac{\mu_0 i}{2R}$ (d) $\frac{\mu_0 i}{4R}$

54. Five very long, straight insulated wires are closely bound together to form a small cable. Currents carried by the wires are : $I_1 = 20$ A, $I_2 = -6$ A, $I_3 = 12$ A, $I_4 = -7$ A, $I_5 = 18$ A. (Negative currents are opposite in direction to the positive). The magnetic field induction at a distance of 10 cm from the cable is
(a) 5 μT (b) 15 μT (c) 74 μT (d) 128 μT

55. The magnetic field at O due to current in the infinite wire forming a loop as shown in Fig. is

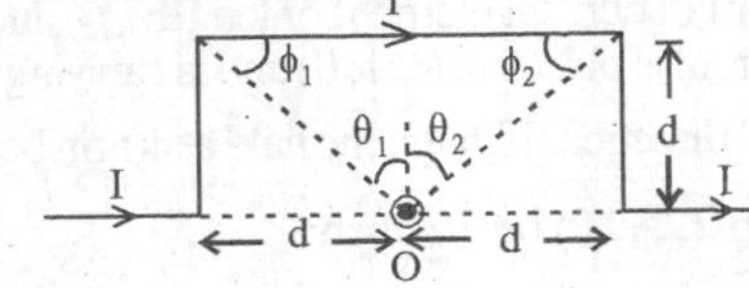

(a) $\frac{\mu_0 I}{2\pi d}(\cos\phi_1 + \cos\phi_2)$

(b) $\frac{\mu_0 I 2I}{4\pi d}(\tan\theta_1 + \tan\theta_2)$

(c) $\frac{\mu_0 I}{4\pi d}(\sin\phi_1 + \sin\phi_2)$

(d) $\frac{\mu_0 I}{4\pi d}(\cos\theta_1 + \sin\theta_2)$

56. A current I flows through a thin wire shaped as regular polygon of n sides which can be inscribed in a circle of radius R. The magnetic field induction at the center of polygon due to one side of the polygon is **Critical Thinking**

(a) $\frac{\mu_0 I}{\pi R}\left(\tan\frac{\pi}{n}\right)$ (b) $\frac{\mu_0 I}{4\pi R}\left(\tan\frac{\pi}{n}\right)$

(c) $\frac{\mu_0 I}{2\pi R}\left(\tan\frac{\pi}{n}\right)$ (d) $\frac{\mu_0 I}{2\pi R}\left(\cos\frac{\pi}{n}\right)$

57. A thin rod is bent in the shape of a small circle of radius *r*. If the charge per unit length of the rod is σ, and if the circle is rotated about its axis at a rate of *n* rotations per second, the magnetic induction at a point on the axis at a large distance *y* from the centre **Toughnut**

(a) $\mu_0 \pi r^3 n \frac{\sigma}{y^3}$ (b) $2\mu_0 \pi r^3 n \frac{\sigma}{y^3}$

(c) $\left(\frac{\mu_0}{4\pi}\right) r^3 n \frac{\sigma}{y^3}$ (d) $\left(\frac{\mu_0}{2\pi}\right) r^3 n \frac{\sigma}{y^3}$

58. Two identical long conducting wires AOB and COD are placed at right angle to each other, with one above other such that 'O' is their common point for the two. The wires carry I_1 and I_2 currents respectively. Point 'P' is lying at distance 'd' from 'O' along a direction perpendicular to the plane containing the wires. The magnetic field at the point 'P' will be :

(a) $\frac{\mu_0}{2\pi d}\left(\frac{I_1}{I_2}\right)$ (b) $\frac{\mu_0}{2\pi d}(I_1 + I_2)$

(c) $\frac{\mu_0}{2\pi d}(I_1^2 - I_2^2)$ (d) $\frac{\mu_0}{2\pi d}(I_1^2 \times I_2^2)^{1/2}$

59. Two similar coils of radius *R* are lying concentrically with their planes at right angles to each other. The currents flowing in them are *I* and 2 *I*, respectively. The resultant magnetic field induction at the centre will be: **Critical Thinking**

(a) $\frac{\sqrt{5}\mu_0 I}{2R}$ (b) $\frac{3\mu_0 I}{2R}$ (c) $\frac{\mu_0 I}{2R}$ (d) $\frac{\mu_0 I}{R}$

60. The magnetic field at point 'C' due to current flowing in 'M' shape figure is

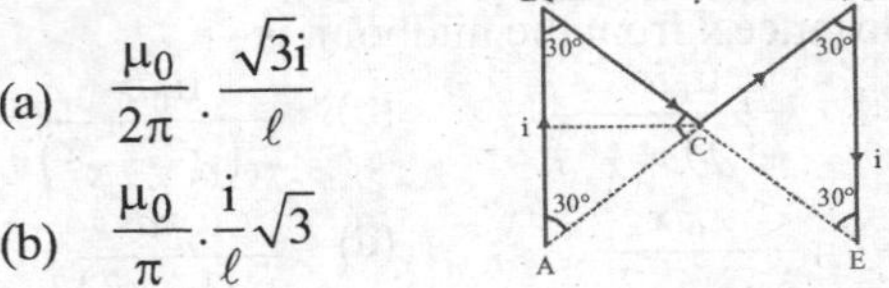

(a) $\frac{\mu_0}{2\pi}\cdot\frac{\sqrt{3}i}{\ell}$

(b) $\frac{\mu_0}{\pi}\cdot\frac{i}{\ell}\sqrt{3}$

(c) zero (d) $\frac{\mu_0}{4\pi}\cdot\frac{i}{\ell\sqrt{3}}$

61. A steady current is flowing in a circular coil of radius R, made up of a thin conducting wire. The magnetic field at the center of the loop is B_L. Now, a circular loop of radius R/n is made form the same wire without changing its length, by unfolding and refolding the loop, and the same current is passed through it. If new magnetic field at the center of the coil is B_C, then the ratio B_L/B_C is
(a) $1 : n^2$ (b) $n^{1/2}$
(c) n : 1 (d) None of these

62. A coaxial cable consists of a thin inner conductor fixed along the axis of a hollow outer conductor. The two conductors carry equal currents in opposites directions. Let B_1 and B_2 be the magnetic fields in the region between the conductors and outside the conductor, respectively Then, **Critical Thinking**

(a) $B_1 \neq 0, B_2 \neq 0$ (b) $B_1 = B_2 = 0$
(c) $B_1 \neq 0, B_2 = 0$ (d) $B_1 = 0, B_2 \neq 0$

63. Two very long, straight wires carrying, currents as shown in Fig. Find all locations where the net magnetic field is zero.

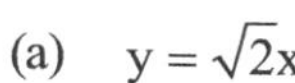

(a) $y = \sqrt{2}x$
(b) $y = x$
(c) $y = -x$
(d) $y = -(x/2)$

64. A current I flows in the anticlockwise direction through a square loop of side a lying in the XOY plane with its center at the origin. The magnetic induction at the center of the square loop is

(a) $\frac{2\sqrt{2}\mu_0 I}{\pi a}\hat{e}_x$ (b) $\frac{2\sqrt{2}\mu_0 I}{\pi a}\hat{e}_z$

(c) $\frac{2\sqrt{2}\mu_0 I}{\pi a^2}\hat{e}_z$ (d) $\frac{2\sqrt{2}\mu_0 I}{\pi a^2}\hat{e}_x$

65. Two equal electric currents are flowing perpendicular to each other as shown in the figure. *AB* and *CD* are perpendicular to each other and symmetrically placed with respect to the current flow. Where do we expect the resultant magnetic field to be zero?
(a) on *AB*
(b) on *CD*
(c) on both *AB* and *CD*
(d) on both *OD* and *BO*

66. A long straight wire of radius *R* carries current *i*. The magnetic field inside the wire at distance *r* from its centre is expressed as :

(a) $\left(\frac{\mu_0 i}{\pi R^2}\right).r$ (b) $\left(\frac{2\mu_0 i}{\pi R^2}\right).r$

(c) $\left(\frac{\mu_0 i}{2\pi R^2}\right).r$ (d) $\left(\frac{\mu_0 i}{2\pi R}\right).r$

67. Three infinitely long wires are placed equally apart on the circumference of a circle of radius a, perpendicular to its plane. Two of the wires carry current *I* each, in the same direction, while the third carries current 2*I* along the direction opposite to the other two. The magnitude of the magnetic indution $\vec{B}$ at a distance *r* from the centre of the circle, for $r > a$, is **Tricky**

(a) 0 (b) $\frac{2\mu_0}{\pi}\frac{I}{r}$

(c) $-\frac{2\mu_0}{\pi}\frac{I}{r}$ (d) $\frac{2\mu_0}{\pi}\frac{Ia}{r^2}$

68. The magnetic moment of a circular coil carrying current is
(a) directly proportional to the length of the wire in the coil
(b) inversely proportional to the length of the wire in the coil
(c) directly proportional to the square of the length of the wire in the coil
(d) inversely proportional to the square of the length of the wire in the coil

69. The figure shows two long straight current carrying wire separated by a fixed distance *d*. The magnitude of current, flowing in each wire varies with time but the magnitude of current in each wire is equal at all times.
Which of the following graphs shows the correct variation of force per unit length *f* between the two wires with current *i*? **Tricky**

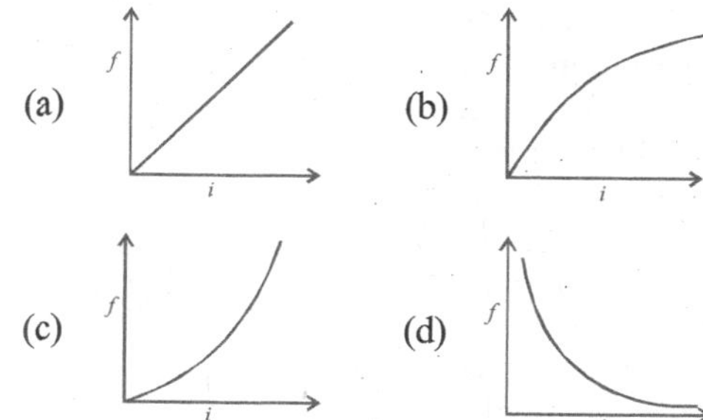

70. A square current carrying loop is suspended in a uniform magnetic field acting in the plane of the loop. If the force on one arm of the loop is $\vec{F}$, the net force on the remaining three arms of the loop is

(a) $3\vec{F}$ (b) $-\vec{F}$ (c) $-3\vec{F}$ (d) $\vec{F}$

71. A current carrying conductor placed in a magnetic field experiences maximum force when angle between current and magnetic field is
(a) $3\pi/4$ (b) $\pi/2$ (c) $\pi/4$ (d) zero

72. Through two parallel wires A and B, 10A and 2A of currents are passed respectively in opposite directions. If the wire A is infinitely long and the length of the wire B is 2m, then force on the conductor B, which is situated at 10 cm distance from A, will be
(a) 8×10^{-7} N (b) 8×10^{-5} N
(c) 4×10^{-7} N (d) 4×10^{-5} N

73. An 8 cm long wire carrying a current of 10 A is placed inside a solenoid perpendicular to its axis. If the magnetic field inside the solenoid is 0.3 T, then magnetic force on the wire is
(a) 0.14 N (b) 0.24 N (c) 0.34 N (d) 0.44 N

74. The orbital speed of electron orbiting around a nucleus in a circular orbit of radius 50 pm is 2.2×10^6 ms^{-1}. Then the magnetic dipole moment of an electron is
(a) 1.6×10^{-19} Am^2 (b) 5.3×10^{-21} Am^2
(c) 8.8×10^{-25} Am^2 (d) 8.8×10^{-26} Am^2

75. A conducting ring of mass 2 kg and radius 0.5 m is placed ring on a smooth horizonatal plane. The ring carries a current of i = 4A. A horizontal magnetic field B = 10 T is switched on at time t = 0 as shown in fig. The initial angular acceleration of the ring will be

Critical Thinking

(a) $40\ \pi$ rad s^{-2} (b) $20\ \pi$ rad s^{-2}
(c) $5\ \pi$ rad s^{-2} (d) $15\ \pi$ rad s^{-2}

76. A current carrying loop is placed in the non-uniform magnetic field whose variation in space is shown in fig. Direction of magnetic field is into the plane of paper. The magnetic force experienced by the loop is

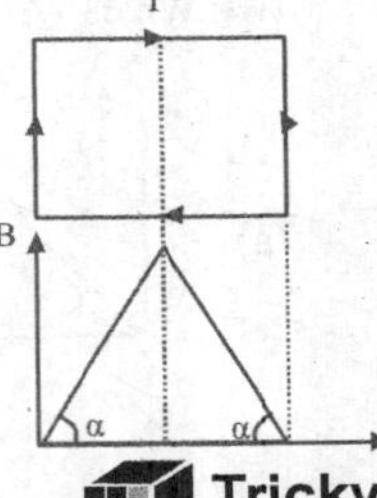

Tricky

(a) non-zero
(b) zero
(c) cannot say anything
(d) None of the above

77. An arrangement of three parallel straight wires placed perpendicular to plane of paper carrying same current 'I along the same direction is shown in fig. Magnitude of force per unit length on the middle wire 'B' is given by

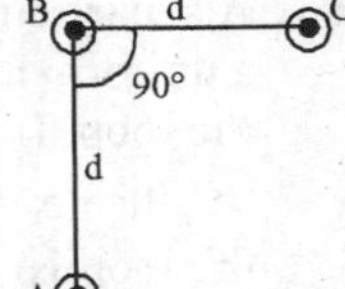

(a) $\dfrac{2\mu_0 i^2}{\pi d}$ (b) $\dfrac{\sqrt{2}\mu_0 i^2}{\pi d}$
(c) $\dfrac{\mu_0 i^2}{\sqrt{2}\pi d}$ (d) $\dfrac{\mu_0 i^2}{2\pi d}$

78. A square loop ABCD carrying a current i, is placed near and coplanar with a long straight conductor XY carrying a current I, the net force on the loop will be :

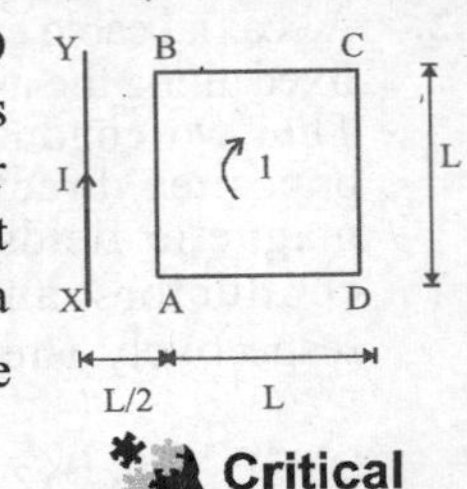

Critical Thinking

(a) $\dfrac{2\mu_0 Ii}{3\pi}$ (b) $\dfrac{\mu_0 Ii}{2\pi}$
(c) $\dfrac{2\mu_0 IiL}{3\pi}$ (d) $\dfrac{\mu_0 IiL}{2\pi}$

79. A long straight wire carries a certain current and produces a magnetic field of $2 \times 10^{-4}\ \dfrac{\text{weber}}{\text{m}^2}$ at a perpendicular distance of 5 cm from the wire. An electron situated at 5 cm from the wire moves with a velocity 10^7 m/s towards the wire along perpendicular to it. The force experienced by the electron will be (charge on electron $= 1.6 \times 10^{-19}$ C)
(a) Zero (b) 3.2 N
(c) 3.2×10^{-16} N (d) 1.6×10^{-16} N

80. A circular coil *ABCD* carrying a current *i* is placed in a uniform magnetic field. If the magnetic force on the segment *AB* is $\vec{F}$, the force on the remaining segment *BCDA* is

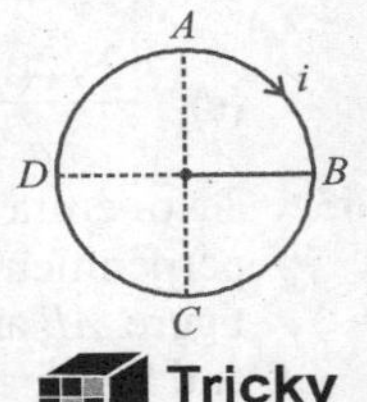

Tricky

(a) $\vec{F}$ (b) $-\vec{F}$ (c) $3\vec{F}$ (d) $-3\vec{F}$

81. A current carrying loop in the form of a right angle isosceles triangle ABC is placed in a uniform magnetic field acting along AB. If the magnetic force on the arm BC is F, what is the force on the arm AC?

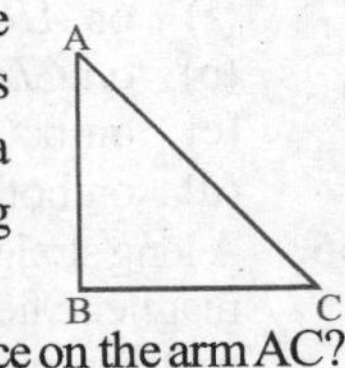

(a) $-\sqrt{2}\ \vec{F}$ (b) $-\vec{F}$
(c) $\vec{F}$ (d) $\sqrt{2}\ \vec{F}$

82. A closely wound solenoid of 2000 turns and area of cross-section 1.5×10^{-4} m^2 carries a current of 2.0 *A*. It suspended through its centre and perpendicular to its length, allowing it to turn in a horizontal plane in a uniform magnetic field 5×10^{-2} tesla making an angle of 30° with the axis of the solenoid. The torque on the solenoid will be:
(a) 3×10^{-2} N-m (b) 3×10^{-3} N-m
(c) 1.5×10^{-3} N-m (d) 1.5×10^{-2} N-m

83. The figure shows two infinite semi-cylindrical shells: shell-1 and shell-2. Shell-1 carries current i_1, in inward direction normal to the plane of paper, while shell-2 carries same current i_1, in opposite direction. A long straight conductor lying along the common axis of the shells is carrying current i_2 in direction same as that of current in shell-1. Force per unit length on the wire is **Critical Thinking**

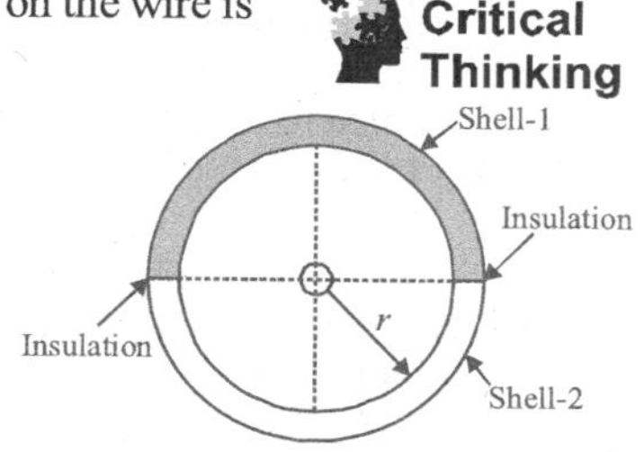

(a) zero

(b) $\frac{\mu_0 i_1 i_2}{2\pi r}$

(c) $\frac{2\mu_0 i_1 i_2}{\pi r}$

(d) $\frac{2\mu_0 i_1 i_2}{\pi^2 r}$

84. A circular arc *QTS* is kept in an external magnetic field $\vec{B}_0$ as shown in figure. The arc carries a current *I*. The magnetic field is directed normal and into the page. The force acting on the arc is

(a) $2IB_0R\hat{k}$

(b) $IB_0R\hat{k}$

(c) $-2IB_0R\hat{k}$

(d) $-IB_0R\hat{k}$

85. The magnetic force acting on the rod ABC in the presence of external magnetic field as shown in the figure is

(a) $BI\ell$

(b) $2BI\ell$

(c) $BI\ell\sqrt{3}$

(d) Zero

86. Two long straight parallel wires, carrying (adjustable) current I_1 and I_2, are kept at a distance d apart. If the force 'F' between the two wires is taken as 'positive' when the wires repel each other and 'negative' when the wires attract each other, the graph showing the dependence of 'F', on the product $I_1 I_2$, would be : **Tricky**

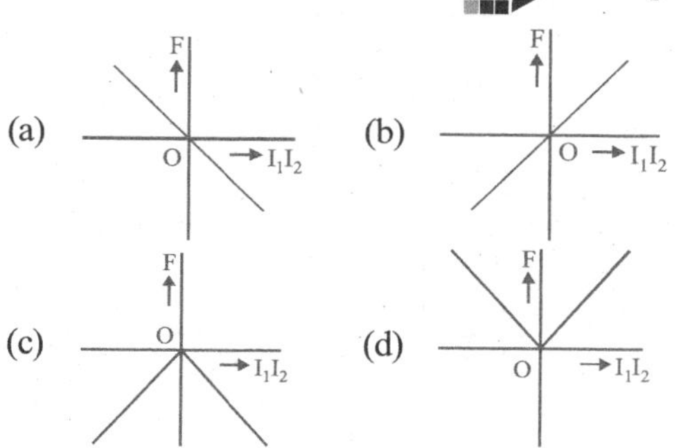

87. A closed loop *PQRS* carrying a current is placed in a uniform magnetic field. If the magnetic forces on segment *PS*, *SR* and *RQ* are F_1, F_2 and F_3 respectively and are in the plane of the paper and along the directions shown, the force on the segment *QP* is **Tricky**

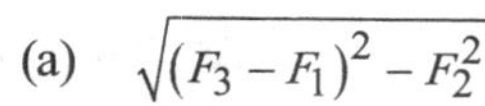

(a) $\sqrt{(F_3 - F_1)^2 - F_2^2}$

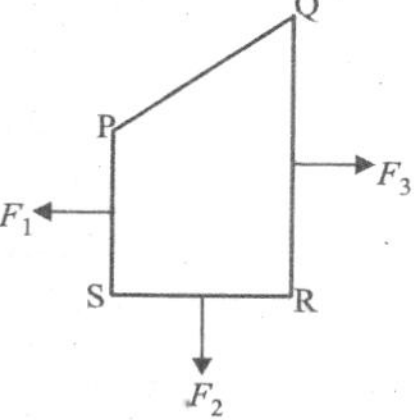

(b) $F_3 + F_1 - F_2$

(c) $F_3 - F_1 + F_2$

(d) $\sqrt{(F_3 - F_1)^2 + F_2^2}$

88. A conducting wire bent in the form of a parabola $y^2 = 2x$ carries a current $i = 2\,A$ as shown in figure. This wire is placed in a uniform magnetic field $\vec{B} = -4\hat{k}$ tesla. The magnetic force on the wire (in newton)

(a) $-16\hat{i}$ (b) $32\hat{i}$ (c) $-32\hat{i}$ (d) $16\hat{i}$

89. A conducting loop is placed in a magnetic field of strength *B* perpendicular to its plane. Radius of the loop is *r*, current in the loop is *i* and linear mass density of the wire of loop is *m*. Speed of any transverse wave in the loop will be

(a) $\sqrt{\frac{Bir}{m}}$ (b) $\sqrt{\frac{Bir}{2m}}$ (c) $\sqrt{\frac{2Bir}{m}}$ (d) $2\sqrt{\frac{Bir}{m}}$

90. A wire carrying current I is tied between points P and Q and is in the shape of a circular arc of radius R due to a uniform magnetic field B (perpendicular to the plane of the paper, shown by xxx) in the vicinity of the wire. If the wire subtends an angle $2\theta_0$ at the centre of the circle (of which it forms an arc) then the tension in the wire is :

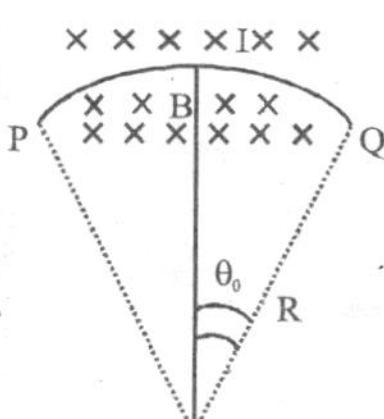

(a) $\frac{IBR}{2\sin\theta_0}$ (b) $\frac{IBR\theta_0}{\sin\theta_0}$ (c) IBR (d) $\frac{IBR}{\sin\theta_0}$

91. A moving coil galvanometer has 150 equal divisions. Its current sensitivity is 10-divisions per milliampere and voltage sensitivity is 2 divisions per millivolt. In order that each division reads 1 volt, the resistance in ohms needed to be connected in series with the coil will be

(a) 10^5 (b) 10^3 (c) 9995 (d) 99995

92. A 50 Ω resistance is connected to a battery of 5V. A galvanometer of resistance 100 Ω is to be used as an ammeter to measure current through the resistance, for this a resistance r_s is connected to the galvanometer. Which of the following connections should be employed if the measured current is within 1% of the current without the ammeter in the circuit? **Critical Thinking**

(a) $r_s = 0.5\ \Omega$ in series with the galvanometer
(b) $r_s = 1\ \Omega$ in series with galvanometer
(c) $r_s = 1\Omega$ in parallel with galvanometer
(d) $r_s = 0.5\ \Omega$ in parallel with the galvanometer

93. A galvanometer of resistance 50 Ω is connected to battery of 3V along with a resistance of 2950 Ω in series. A full scale deflection of 30 divisions is obtained in the galvanometer. In order to reduce this deflection to 20 divisions, the resistance in series should be

(a) 5050 Ω (b) 5550 Ω
(c) 6050 Ω (d) 4450 Ω

94. A circuit contains an ammeter, a battery of 30V and a resistance 40.8Ω all connected in series. If the ammeter has a coil of resistance 480Ω and a shunt of 20Ω, the reading in the ammeter will be:

(a) 0.25 A (b) 2A (c) 1 A (d) 0.5 A

95. In an ammeter 0.2% of main current passes through the galvanometer. If resistance of galvanometer is G, the resistance of ammeter will be:

(a) $\frac{1}{499}G$ (b) $\frac{499}{500}G$ (c) $\frac{1}{500}G$ (d) $\frac{500}{499}G$

96. A voltmeter has a range $0 - V$ with a series resistance R. With a series resistance $2R$, the range is $0 - V'$. The correct relation between V and V' is **Tricky**

(a) $V' = 2V$ (b) $V' > 2V$
(c) $V' >> 2V$ (d) $V' < 2V$

97. A microammeter has a resistance of 100 Ω and full scale range of 50 μA. It can be used as a voltmeter or as a higher range ammeter provided a resistance is added to it. Pick the correct range and resistance combination **Critical Thinking**

(a) 50 V range with 10 kΩ resistance in series
(b) 10 V range with 200 kΩ resistance in series
(c) 10 mA range with 1Ω resistance in parallel
(d) 10 mA range with 0.1 Ω resistance in parallel

98. A milli voltmeter of 25 milli volt range is to be converted into an ammeter of 25 ampere range. The value (in ohm) of necessary shunt will be

(a) 0.001 (b) 0.01 (c) 1 (d) 0.05

99. A galvanometer of resistance 5 ohms gives a full scale deflection for a potential difference of 10 mV. To convert the galvanometer into a voltmeter giving a full scale deflection for a potential difference of 1V, the size of the resistance that must be attached to the voltmeter is

(a) 0.495 ohm (b) 49.5 ohm
(c) 495 ohm (d) 4950 ohm

100. A moving coil galvanometer has a resistance of 900 Ω. In order to send only 10% of the main current through this galvanometer, the resistance of the required shunt is

(a) 0.9 Ω (b) 100 Ω (c) 405 Ω (d) 90 Ω

Answer KEYs

1	(c)	**11**	(c)	**21**	(a)	**31**	(d)	**41**	(a)	**51**	(b)	**61**	(a)	**71**	(b)	**81**	(b)	**91**	(c)
2	(d)	**12**	(b)	**22**	(c)	**32**	(a)	**42**	(d)	**52**	(c)	**62**	(c)	**72**	(b)	**82**	(d)	**92**	(d)
3	(a)	**13**	(b)	**23**	(a)	**33**	(d)	**43**	(b)	**53**	(b)	**63**	(c)	**73**	(b)	**83**	(d)	**93**	(d)
4	(c)	**14**	(c)	**24**	(c)	**34**	(a)	**44**	(d)	**54**	(c)	**64**	(b)	**74**	(c)	**84**	(b)	**94**	(d)
5	(b)	**15**	(a)	**25**	(d)	**35**	(b)	**45**	(b)	**55**	(a)	**65**	(a)	**75**	(a)	**85**	(a)	**95**	(c)
6	(a)	**16**	(d)	**26**	(d)	**36**	(c)	**46**	(a)	**56**	(c)	**66**	(c)	**76**	(b)	**86**	(a)	**96**	(d)
7	(a)	**17**	(b)	**27**	(d)	**37**	(b)	**47**	(c)	**57**	(a)	**67**	(a)	**77**	(c)	**87**	(d)	**97**	(b)
8	(b)	**18**	(b)	**28**	(c)	**38**	(a)	**48**	(c)	**58**	(d)	**68**	(c)	**78**	(a)	**88**	(b)	**98**	(a)
9	(b)	**19**	(c)	**29**	(b)	**39**	(c)	**49**	(b)	**59**	(a)	**69**	(c)	**79**	(c)	**89**	(a)	**99**	(c)
10	(a)	**20**	(c)	**30**	(c)	**40**	(b)	**50**	(c)	**60**	(b)	**70**	(b)	**80**	(b)	**90**	(c)	**100**	(b)

Hints & Solutions

1. **(c)** Equating magnetic force to centripetal force,
$$\frac{mv^2}{r} = qvB \sin 90°$$
Time to complete one revolution,
$$T = \frac{2\pi r}{v} = \frac{2\pi m}{qB}$$

2. **(d)** The change in K.E. is equal to work done by net force which is zero because the magnetic force is perpendicular to velocity. K.E. remains constant.

3. **(a)** Lorentz force acting on the particle
$$\vec{F} = q\left[\vec{E} + \vec{v} \times \vec{B}\right]$$
$$= q\left[3\hat{i} + \hat{j} + 2\hat{k} + \hat{i}(-12-1) - \hat{j}(-9-1) + k(3-4)\right]$$
$$F_y = 11q\hat{j}$$

4. **(c)** The electron moves with constant velocity without deflection. Hence, force due to magnetic field is equal and opposite to force due to electric field.

5. **(b)** Since, magnetic field is acting at right angles to the direction of motion of electron beams, so their paths will be circular.
$$r = \frac{mv}{qB} \text{ or } r \propto v \therefore \frac{r_1}{r_2} = \frac{v_1}{v_2} = 1/2 \text{ or } r_1 : r_2 = 1:2$$

6. **(a)**

7. **(a)** Radius of circular path of the charged particles $r = \frac{mV}{qB}$
Here R < r, hence the particle will press the outer wall of the pipe hence the force applied by the pipe on the particle should be towards the centre of the pipe.

8. **(b)**

9. **(b)** When a test charge q_0 enters a magnetic field $\vec{B}$ directed along z-axis, with a velocity $\vec{v}$ making angles d with the z-axis. The time period of the motion is independent of R and v.

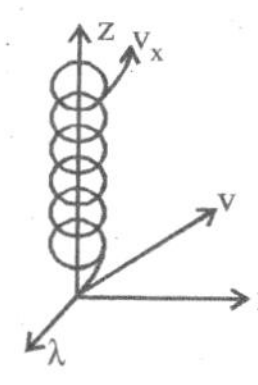

10. **(a)** In mass spectrometer, when ions are accelerated through potential V
$$\frac{1}{2}mv^2 = qV \quad \text{..........(i)}$$
As the magnetic field curves the path of the ions in a semicircular orbit
$$Bqv = \frac{mv^2}{R} \Rightarrow v = \frac{BqR}{m} \quad \text{..........(ii)}$$
Substituting (ii) in (i)
$$\frac{1}{2}m\left[\frac{BqR}{m}\right]^2 = qV \text{ or } \frac{q}{m} = \frac{2V}{B^2R^2}$$
Since V and B are constants,
$$\therefore \frac{q}{m} \propto \frac{1}{R^2}$$

11. **(c)**

12. **(b)** $T = 2\pi m / qB, \left(\frac{m}{q}\right)_{He} > \left(\frac{m}{q}\right)_p > \left(\frac{m}{q}\right)_e$
$\therefore \quad t_{He} > t_p > t_e$

13. **(b)** Time period,
$$T = \frac{2\pi m}{qB} = \frac{2\pi}{\pi \times 2} = 1s$$
Thus, particle will be at point P after $t = \frac{1}{12}$s
$$\vec{v} = 10\left[\cos 30\hat{i} + \sin 30\hat{j}\right]$$
$$\vec{v} = 10\left[\frac{\sqrt{3}}{2}\hat{i} + \frac{1}{2}\hat{j}\right] = 5\left[\sqrt{3}\hat{i} + \hat{j}\right]\text{ms}^{-1}$$

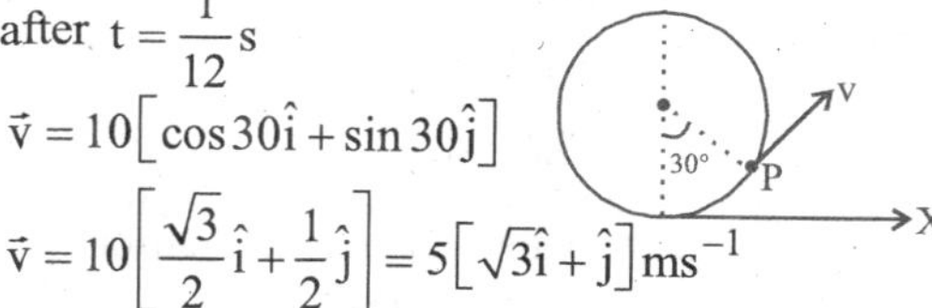

14. **(c)** As $r = \frac{mv}{qB} = \frac{P}{qB}$
$$\therefore \text{Area, } A = \pi r^2 = \pi\left(\frac{P}{qB}\right)^2 = \frac{\pi P^2}{qB} = \frac{2m\pi}{qB}K$$

15. **(a)** The ratio M/L is always $\frac{q}{2m}$

16. **(d)** 17. **(b)** 18. **(b)**

19. **(c)** K.E. of first particle $= \frac{1}{2}m_1v_1^2 = qV$ ... (i)
K.E. of second particle $= \frac{1}{2}m_2v_2^2 = qV$ ...(ii)
After entering the magnetic field, a magnetic force acts on the charged particle which moves the charged particle in circular path of radius
$$R = \frac{\sqrt{2mK}}{qB}$$
Here, K, q, B are equal
$$\therefore \quad R^2 \propto m \Rightarrow \frac{m_1}{m_2} = \frac{R_1^2}{R_2^2} \quad \text{From (i) and (ii)}$$

20. (c) The centre will be at C as shown :
Coordinates of the centre are
$(r\cos 60°, -r\sin 60°)$
where r = radius of circle

$$= \frac{mv}{Bq} = \frac{1\times 1}{1\times 1} = 1$$

i.e., $\left(\frac{1}{2}, -\frac{\sqrt{3}}{2}\right)$

21. (a)

22. (c) Time period of cyclotron is

$$T = \frac{1}{\upsilon} = \frac{2\pi m}{eB};\ B = \frac{2\pi m}{e}\upsilon;\ R = \frac{m\upsilon}{eB} = \frac{p}{eB}$$

$$\Rightarrow P = eBR = e\times\frac{2\pi m\upsilon}{e}R = 2\pi m\upsilon R$$

$$\text{K.E.} = \frac{p^2}{2m} = \frac{(2\pi m\upsilon R)^2}{2m} = 2\pi^2 m\upsilon^2 R^2$$

23. (a) $R = \frac{mv}{qB}$; $\vec{F} = q(\vec{v}\times\vec{B})$

As R decreases, v will decrease.
Clockwise motion for upward magnetic field $\rightarrow$ +ve

24. (c) K.E. of electron = $10\ eV \Rightarrow \frac{1}{2}mv^2 = 10\ eV$

$$\Rightarrow \frac{1}{2}(9.1\times 10^{-31})v^2 = 10\times 1.6\times 10^{-19}$$

$$\Rightarrow v^2 = \frac{2\times 10\times 1.6\times 10^{-19}}{9.1\times 10^{-31}}$$

$\Rightarrow v^2 = 3.52\times 10^{12} \Rightarrow v = 1.88\times 10^6$ m
Also, we know that for circular motion

$$\frac{mv^2}{r} = Bev \Rightarrow r = \frac{mv}{Be} = 11\text{ cm}$$

25. (d)

26. (d) Electric force on the particle, $F = Eq$, and

displacement $s = \frac{1}{2}at^2 = \frac{1}{2}\left(\frac{Eq}{m}\right)t^2$.

Now, $W = \Delta K$,

or $Fs = \frac{1}{2}m\left(v_f^2 - v_i^2\right)$ or $Eq\times\frac{1}{2}\left(\frac{Eq}{m}\right)t^2$

$= \frac{1}{2}m[(2v_0)^2 - v_0^2]$ $\quad\therefore\quad t = \frac{\sqrt{3}\,mv_0}{qE}$.

27. (d) $\vec{B} = -c\hat{k}$, and $\vec{v} = v\cos 45°\hat{i} - v\sin 45°\hat{j}$

$$= \frac{v}{\sqrt{2}}\hat{i} - \frac{v}{\sqrt{2}}\hat{j}.$$

Thus, $\vec{F} = q(\vec{v}\times\vec{B}) = q\left[\left(\frac{v}{\sqrt{2}}\hat{i} - \frac{v}{\sqrt{2}}\hat{j}\right)\times(-c\hat{k})\right]$

$= qcv\left[\frac{\hat{i}+\hat{j}}{\sqrt{2}}\right]$ $\quad\therefore\quad \vec{a} = \left[\frac{\hat{i}+\hat{j}}{\sqrt{2}}\right]$.

28. (c) The particle will not collide if
$d > 2(r_1 + r_2)$

or $d > 2\left(\frac{mv_1}{Bq} + \frac{mv_2}{Bq}\right)$ or $d > \frac{2m}{Bq}(v_1 + v_2)$

29. (b)

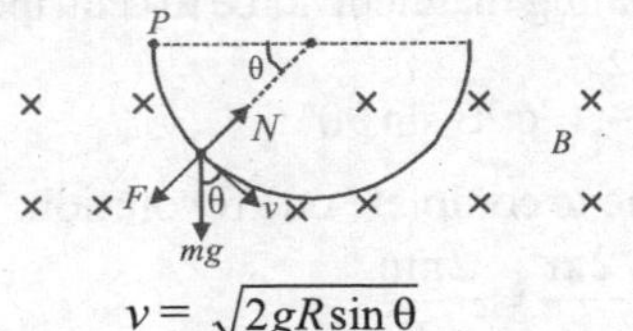

$$v = \sqrt{2gR\sin\theta}$$

and $F = qvB$

$$\text{Now } N - (F + mg\sin\theta) = \frac{mv^2}{R}$$

$$\therefore\quad N = F + mg\sin\theta + \frac{mv^2}{R}$$

$$= qB\sqrt{2gR\sin\theta} + mg\sin\theta + 2mg\sin\theta$$

$$\therefore\quad N = 3mg\sin\theta + qB\sqrt{2gR\sin\theta}$$

30. (c)

31. (d) Given: $f = 24\times 10^6$ Hz, $R = 0.60$ m

We know that, $R = \frac{mv}{qB}$

$$\therefore\quad B = \frac{mv}{qR} \quad ...(i)$$

where $v = \omega R = 2\pi fR$
$= 2\pi\times 24\times 10^6\times 0.60 = 9.04\times 10^7$ m/s
From (i),

$$B = \frac{(3.34\times 10^{-27})(9.04\times 10^7)}{1.6\times 10^{-19}\times 0.60} = 3.2\text{ T}.$$

32. (a)

33. (d) $\alpha = \frac{q}{m}$, path of the particle will be a helix of time period,

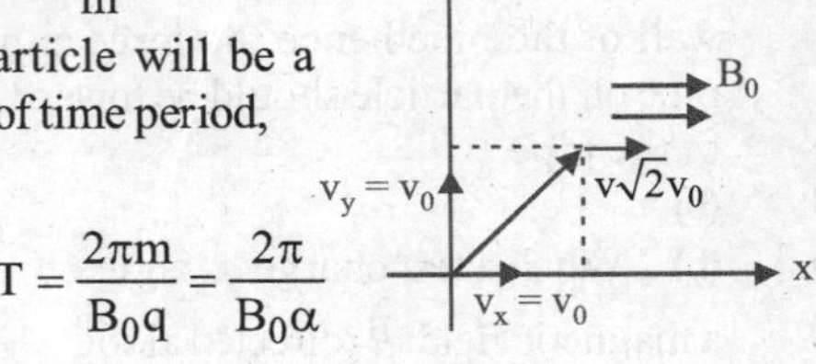

$$T = \frac{2\pi m}{B_0 q} = \frac{2\pi}{B_0\alpha}$$

The give time $t = \frac{\pi}{B_0\alpha} = \frac{T}{2}$

$\therefore$ Coordinates of particle at time $t = T/2$ would be $(vx\ T/2, 0, -2r)$

Here, $r = \frac{mv_0}{B_0 q} = \frac{v_0}{B_0\alpha}$

$\therefore$ The coordinate are $\left(\frac{v_0\pi}{B_0\alpha}, 0, \frac{-2v_0}{B_0\alpha}\right)$

34. **(a)** $r = \frac{mv}{qB}$

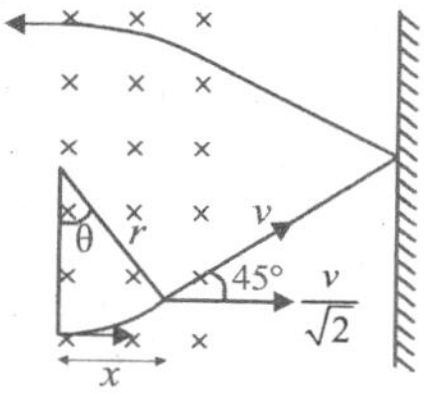

$\sin\theta = \frac{x}{r} = \frac{\frac{mv}{\sqrt{2}qB}}{\frac{mv}{qB}} = \frac{1}{\sqrt{2}}$

or $\theta = \frac{\pi}{4}$

Time to complete the circle (2π), $T = \frac{2\pi m}{qB}$

$\therefore$ time taken to traverses $\frac{\pi}{4}$, $t = \frac{\pi m}{4qB}$

Time taken to travel horizontal distance

$t_1 = \frac{\frac{mv}{\sqrt{2}qB}}{\frac{v}{\sqrt{2}}} = \frac{m}{qB}$

Total time taken $= 2t + 2t_1 = \frac{m}{2qB}(\pi + 4)$

35. **(b)**

36. **(c)** $B = \frac{\mu_0}{4\pi}\frac{2\pi i}{r}$ where

$i = \frac{2e}{t} = \frac{2\times1.6\times10^{-19}}{2} = 1.6\times10^{-19}$ A

$\therefore\ B = \frac{\mu_0 i}{2r} = \frac{\mu_0\times1.6\times10^{-19}}{2\times0.8} = \mu_0\times10^{-19}$ T

37. **(b)** The magnetic field from the centre of wire of radius R is given by

$B = \left(\frac{\mu_0 I}{2R^2}\right)r \quad (r < R) \Rightarrow B \propto r$

and $B = \frac{\mu_0 I}{2\pi r} \quad (r > R) \Rightarrow B \propto \frac{1}{r}$.

From the above descriptions, we can say that the graph (b) is a correct representation.

38. **(a)** $B = \frac{\mu_0 I}{2r}\times\frac{\theta}{2\pi} = \frac{\mu_0 I\theta}{4\pi r}$

39. **(c)** $B = \frac{\mu_0 i a^2}{2(x^2+a^2)^{3/2}}$

$B' = \frac{\mu_0 i}{2a} = \frac{\mu_0 i a^2}{2a(x^2+a^2)^{3/2}}\left(\frac{(x^2+a^2)^{3/2}}{a^2}\right)$

$B' = \frac{B.(x^2+a^2)^{3/2}}{a^3}$

Put $x = 4$ & $a = 3 \Rightarrow B' = \frac{54(5^3)}{3\times3\times3} = 250\mu$T

40. **(b)** Let us compute the magnetic field due to any one segment :

$B = \frac{\mu_0 i}{4\pi(d\sin\alpha)}(\cos 0° + \cos(180 - \alpha))$

$= \frac{\mu_0 I}{4\pi(d\sin\alpha)}(1 - \cos\alpha)$

$= \frac{\mu_0 I}{4\pi d}\tan\frac{\alpha}{2}$

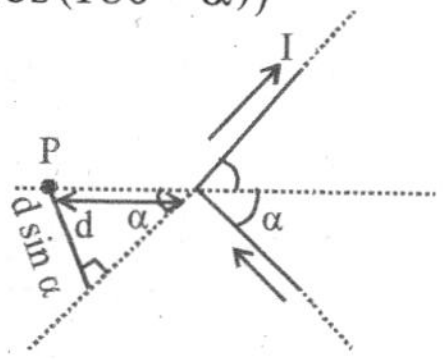

Resultant field will be

$B_{net} = 2B = \frac{\mu_0 I}{2\pi d}\tan\frac{\alpha}{2} \Rightarrow K = \frac{\mu_0 I}{2\pi d}$

41. **(a)** Let I be current and l be the length of the wire.

For Ist case : $B = \frac{\mu_0 In}{2r} = \frac{\mu_0 I\times\pi}{l}$ where $2\pi r = l$ and $n = 1$

For IInd Case : $l = 2(2\pi r') \Rightarrow r' = \frac{l}{4\pi}$

$B' = \frac{\mu_0 nI}{2r'} = \frac{\mu_0 2I}{2\frac{l}{4\pi}} = \frac{4\mu_0 I\pi}{l} = 4B$

42. **(d)**

43. **(b)** For loop $B = \frac{\mu_0 nI}{2a}$

where, a is the radius of loop.

Then, $B_1 = \frac{\mu_0 I}{2a}$

Now, for coil $B = \frac{\mu_0 I}{4\pi}.\frac{2nA}{x^3}$

at the centre x = radius of loop

$B_2 = \frac{\mu_0}{4\pi}.\frac{2\times3\times(I/3)\times\pi(a/3)^2}{(a/3)^3} = \frac{\mu_0.3I}{2a}$

$\therefore\ \frac{B_1}{B_2} = \frac{\mu_0 I/2a}{\mu_0.3I/2a} \qquad B_1 : B_2 = 1:3$

44. **(d)** Magnetic field on axis of circular coil is

$\frac{\mu_0 iR^2}{2(R^2+x^2)^{3/2}} = \frac{\mu_0 iR^2}{2(R^2+x^2)^{3/2}} = \frac{\mu_0 i_2 R^2}{2(R^2+x^2)^{3/2}}$

$x = \sqrt{7}R$

45. **(b)**

46. **(a)** Current in the element $= J(2\pi r.\,dr)$

Current enclosed by ampere loop of radius $a/2$

$I = \int_0^{a/2}\frac{J_0 r}{a}.2\pi r.dr$

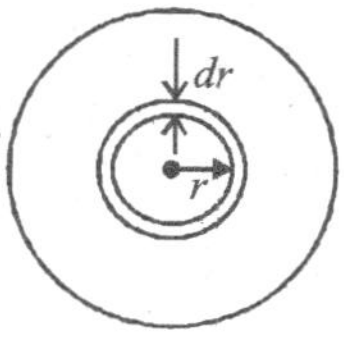

$= \frac{2\pi J_0}{3a}\left(\frac{a}{2}\right)^3 = \frac{\pi J_0 a^3}{12}$

Applying ampere's law

$B.2\pi.\frac{a}{2} = \mu_0.\frac{\pi J_0 a^2}{12} \Rightarrow B = \frac{\mu_0 J_0 a}{12}$

On putting values, $B = 10\mu$T

47. (c)

48. (c) Radius of circular orbit = r

No. of rotations per second = n

i.e., $T = \frac{1}{n}$

Magnetic field at its centre, $B_c = ?$

As we know, current

$i = \frac{e}{T} = \frac{e}{(1/n)} = en$ = equivalent current

Magnetic field at the centre of circular orbit,

$B_c = \frac{\mu_0 i}{2r} = \frac{\mu_0 ne}{2r}$.

49. (b) Magnetic field due to segment '1'

$\vec{B_1} = \frac{\mu_0 I}{4\pi R}[\sin 90° + \sin 0°](-\hat{k})$

$= \frac{-\mu_0 I}{4\pi R}(\hat{k}) = \vec{B_3}$

Magnetic field due to segment 2

$B_2 = \frac{\mu_0 I}{4R}(-\hat{i}) = \frac{-\mu_0 I}{4\pi R}(\pi\hat{i})$

$\therefore \vec{B}$ at centre

$\vec{B_c} = \vec{B_1} + \vec{B_2} + \vec{B_3}$

$= \frac{-\mu_0 I}{4\pi R}(\pi\hat{i} + 2\hat{k})$

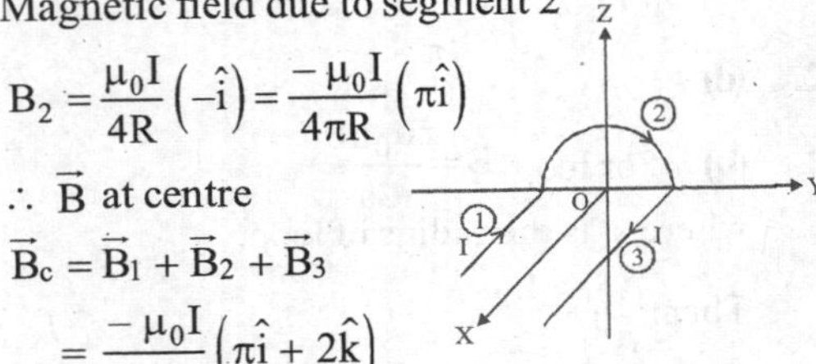

50. (c)

51. (b)

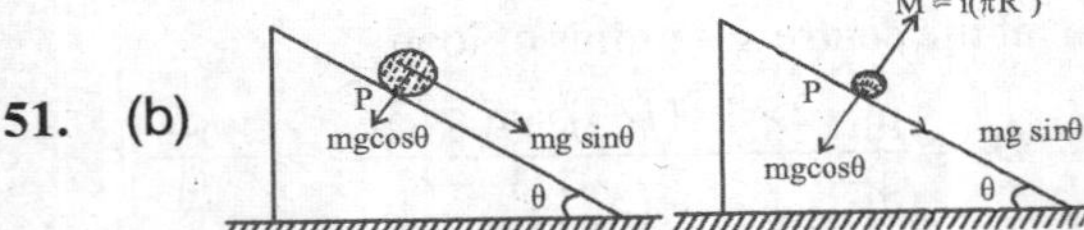

For rotational equilibrium of the sphere about point P.

$(mg \sin\theta) R = M(B \sin\theta)$

or, $mgR \sin\theta = i\pi R^2 B \sin\theta$

$B = mg/\pi iR$

52. (c) 53. (b)

54. (c) Net current is (20 – 6 + 12 – 7 + 18) A, i.e., 37 A

$r = \frac{10}{100} m = \frac{1}{10} m$

$B = \frac{\mu_0 I}{2\pi r} = \frac{4\pi \times 10^{-7} \times 37 \times 10}{2\pi \times 1} = 74 \times 10^{-6}$ T

$= 74\ \mu$ T.

55. (a) Using $B = \frac{\mu_0 I}{4\pi d}[\sin\theta_1 + \sin\theta_2]$

But $\theta_1 + \phi_1 = 90°$ or $\theta_1 = 90° - \phi$,

$\sin\theta_1 = \sin(90° - \phi_1) = \cos\phi_1$

Similarly, $\sin\theta_2 = \cos\phi_2$

$B_{net} = \frac{\mu_0 I}{2\pi d}(\cos\phi_1 + \cos\phi_2)$

56. (c) $B = \frac{\mu_0 I}{4\pi r}\left[2\sin\frac{\pi}{n}\right]$

$\cos\frac{\pi}{n} = \frac{r}{R}$

But or $r = R\cos\frac{\pi}{n}$

$\therefore B = \frac{\mu_0 I}{4\pi R\cos\frac{\pi}{n}}\left[2\sin\frac{\pi}{n}\right] = \frac{\mu_0 I}{2\pi R}\left[\tan\frac{\pi}{n}\right]$

57. (a) Current, $I = \frac{2\pi r\sigma}{\ell/n} = 2\pi r\sigma n$.

$B = \frac{\mu_0 I r^2}{2(y^2 + r^2)^{3/2}} = \frac{\mu_0 I r^2}{2y^3}$

If $y >> r$,

$B = \frac{\mu_0 r^2}{2y^2}(2\pi r\sigma n) = \frac{\mu_0 \pi r^3 \sigma n}{y^3}$

58. (d) Net magnetic field, $B = \sqrt{B_1^2 + B_2^2}$

$= \sqrt{\left(\frac{\mu_0 I_1}{2\pi d}\right)^2 + \left(\frac{\mu_0 I_2}{2\pi d}\right)^2} = \frac{\mu_0}{2\pi d}\sqrt{I_1^2 + I_2^2}$.

59. (a) The magnetic field, due the coil, carrying current I Ampere

$B_1 = \frac{\mu_0 I}{2R}$

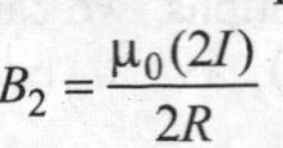

The magnetic field due to the coil, carrying current $2I$ Ampere

$B_2 = \frac{\mu_0(2I)}{2R}$

The resultant B

$B_{net} = \sqrt{B_1^2 + B_2^2 + 2B_1B_2\cos\theta}$, $\theta = 90°$

$B_{net} = \sqrt{B_1^2 + B_2^2} = \frac{\mu_0(2I)}{2R}\sqrt{1+4} = \frac{\sqrt{5}\mu_0 I}{2R}$

60. (b) B due to AB $= \frac{\mu_0}{4\pi}\frac{i}{d}\left[\sqrt{3}\right]$

$= \frac{\mu_0}{4\pi}\frac{i}{\ell/2}\left[\sqrt{3}\right] = \frac{\mu_0 i}{4\pi\ell}2\sqrt{3}$

Net magnetic field $= \frac{\mu_0}{2\pi}\frac{i}{\ell}\sqrt{3} \times 2 = \frac{\mu_0}{\pi}\frac{i}{\ell}\sqrt{3}$

61. (a) $B_L = \frac{\mu_0 I}{2\pi R}$

If the radius is R/n ; the number of turns will be n.

$B_C = \frac{n\mu_0 I}{2\pi(R/n)} = n^2\frac{\mu_0 I}{2\pi R}$ Hence, $\frac{B_L}{B_C} = \frac{1}{n^2}$

62. **(c)** Apply Ampere's circular law to the coaxial circular loops L_1 and L_2.
The magnetic field is B_1 at all points on L_1 and B_2 at all points on L_2. $\Sigma I \neq 0$ for L_1 and 0 for L_2. Hence, $B_1 \neq 0$ but $B_2 = 0$

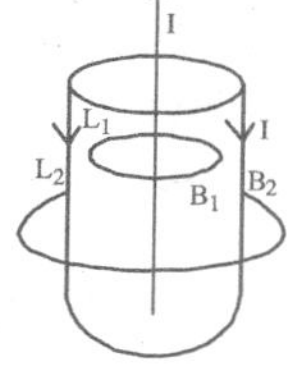

$\left[\text{As} \oint \vec{B}.d\vec{l} = \mu_0 \sum I\right]$

63. **(c)**

64. **(b)** Field due to one side of loop

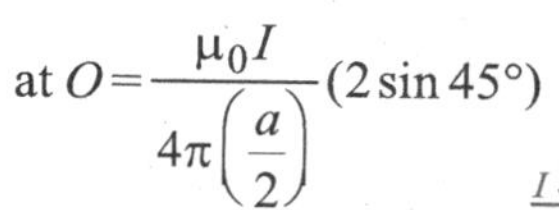

$$\text{at } O = \frac{\mu_0 I}{4\pi\left(\frac{a}{2}\right)}(2\sin 45°)$$

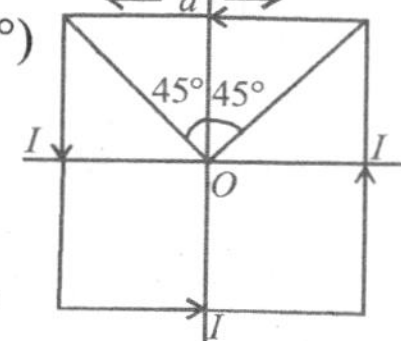

Field at O due to all four sides is along unit vector $\hat{e}_z$

$$\therefore \text{ Total field} = 4.\frac{\mu_0 I}{4\pi\left(\frac{a}{2}\right)}(2\sin 45°) = \frac{2\sqrt{2}\mu_0 I}{\pi a}$$

65. **(a)** 66. **(c)**

67. **(a)** From ampere circuital law
$\oint B.dl = \mu_0 i$
Where, i = Net current passing through the loop
Now $i = I + I - 2I = 0$
$\therefore \quad B.\,2\pi r = 0 \Rightarrow B = 0$

68. **(c)** $M = NiA \Rightarrow M \propto A \Rightarrow M \propto r^2$
$[\text{As } \ell = 2\pi r \Rightarrow \ell \propto r]$
$\Rightarrow M \propto \ell^2$

69. **(c)** The force per unit length is given by
$$f = \frac{\mu_0 i^2}{2\pi d}$$
i.e., $f \propto i^2$

70. **(b)**

71. **(b)** $F = iB l \sin\theta$. This is maximum when $\sin\theta = 1$ or $\theta = \pi/2$.

72. **(b)** $F = \frac{\mu_0}{4\pi}\frac{2I_1I_2}{r} \times \ell$
$$= \frac{10^{-7} \times 2 \times 10 \times 2}{0.1} \times 2 = 8 \times 10^{-5}\text{ N}$$

73. **(b)** $F = IlB$
Here, $\theta = 90°$, $I = 10\ A$
$l = 8\text{ cm} = 8 \times 10^{-2}\ m$, $B = 0.3\ T$
$\therefore \quad F = 10 \times 8 \times 10^{-2} \times 0.3 \times \sin 90° = 0.24\ N$

74. **(c)** Magnetic dipole moment
$$m = iA = \frac{e}{T} \times \pi r^2 = \frac{e}{(2\pi r/v)} \times \pi r^2 = \frac{erv}{2}.$$
$$= \frac{1.6 \times 10^{-19} \times 50 \times 10^{-12} \times 2.2 \times 10^6}{2}$$
$= 8.8 \times 10^{-25}\ \text{Am}^2$.

75. **(a)** Due to torque of magnetic field, ring will rotate about vertical diameter. $\tau = I\alpha \Rightarrow MB = I\alpha$
$$\Rightarrow \quad i\pi r^2 B = \frac{1}{2}mr^2\alpha$$
$$\Rightarrow \quad \alpha = \frac{2iB\pi}{m} = \frac{2 \times 4 \times 10\pi}{2} = 40\pi\text{ rad s}^{-2}$$

76. **(b)** Each and every pair of loop elements located symmetrically w.r.t. central line experience zero net force, so total magnetic force experienced by the loop is zero.

77. **(c)**

78. **(a)** The net force on the loop
$F_{net} = F_{AB} + F_{BC} + F_{CD} + F_{AD}$
$$\Rightarrow \quad F_{net} = \frac{\mu_o iI}{\pi} - \frac{\mu_o iI}{3\pi} = \frac{2\mu_o iI}{3\pi}$$

79. **(c)** Given:
Magnetic field $B = 2 \times 10^{-4}$ weber/m^2
Velocity of electron, $v = 10^7$ m/s
Lorentz force $F = qvB\sin\theta$
$= 1.6 \times 10^{-19} \times 10^7 \times 2 \times 10^{-4}\ (\because \theta = 90°)$
$= 3.2 \times 10^{-16}$ N

80. **(b)** Here, $\vec{F}_{AB} + \vec{F}_{BCDA} = \vec{0}$
$\Rightarrow \vec{F}_{BCDA} = -\vec{F}_{AB} = -\vec{F} \quad (\because F_{AB} = \vec{F})$

81. **(b)**

82. **(d)** Torque on the solenoid is given by
$\tau = MB\sin\theta$
where θ is the angle between the magnetic field and the axis of solenoid.
$M = $ niA
$\therefore \quad \tau = \text{niA B} \sin 30°$
$$= 2000 \times 2 \times 1.5 \times 10^{-4} \times 5 \times 10^{-2} \times \frac{1}{2}$$
$$= 1.5 \times 10^{-2}\ N - m$$

83. **(d)** Magnetic field at the centre of the shell $= 2\frac{\mu_0 i_1}{\pi^2 r}$

Force per unit length of the wire
$F = Bi_2 \times 1$
$$= \left[2\frac{\mu_0 i_1 i_2}{\pi^2 r}\right]$$

84. **(b)** We know, $\vec{F} = I(\vec{\ell} \times \vec{B})$ or $d\vec{F} = IB\,d\ell\sin\theta$
$$\therefore \quad F = \int_{60°}^{120°} IB_0\,d\ell\sin\theta\,d\ell \text{ where, } d\ell = rd\theta$$
$$F = \int_{60°}^{120°} IB_0 R\sin\theta\,d\theta = IBR\left[-\cos\theta\right]_{60°}^{120°}$$

$= IB_0R\left[-\left(-\frac{1}{2}\right)+\left(\frac{1}{2}\right)\right] = IB_0R$

Hence, force acting on the arc is $IB_0R\ \hat{k}$.

85. (a) Force $= BI\ell$

Here, equivalent length ℓ'

$= \sqrt{\ell^2+\ell^2-2\ell.\ell\cos 60°} = \sqrt{2\ell^2-\ell^2} = \ell$

86. (a) $I_1 I_2$ = Positive (attract) F = Negative

$I_1 I_2$ = Negative (repell) F = Positive

Hence, option (a) is the correct answer.

87. (d) If F_4 is the force on segment QP, then

$\vec{F}_1+\vec{F}_2+\vec{F}_3+\vec{F}_4=0$

and $\vec{F}_4 = -(\vec{F}_1+\vec{F}_2+\vec{F}_3)$

or $F_4 = \sqrt{(F_3-F_1)^2+F_2^2}$.

88. (b) At $x=2$, $y^2=2x=2\times 2$ or $y=2$.

Thus separation, $AB = 4$ cm

The force, $F = Bi\,(AB)$

$= 4\times 2\times 4 = 32$ N.

89. (a) $F = Bi\ell$

$2T\sin\frac{\theta}{2} = F$

or $2T\left(\frac{\theta}{2}\right) = Bi(r\theta)$

$\therefore \quad T = Bir$

Now speed, $v = \sqrt{\frac{T}{m}}$

$= \sqrt{\frac{Bir}{m}}$

90. (c) **91.** (c) **92.** (d)

93. (d) Total internal resistance $= (50+2950)\,\Omega = 3000\,\Omega$

Emf of the cell, $\varepsilon = 3$V

$\therefore$ Current $= \frac{\varepsilon}{R} = \frac{3}{3000} = 1\times 10^{-3}\text{ A} = 1.0\text{ mA}$

$\therefore$ Current for full scale deflection of 30 divisions is 1.0 mA.

$\therefore$ Current for a deflection of 20 divisions,

$I = \left(\frac{20}{30}\times 1\right)\text{mA}$ or $I = \frac{2}{3}\text{mA}$

Let the resistance be x Ω. Then

$x = \frac{\varepsilon}{I} = \frac{3V}{\left(\frac{2}{3}\times 10^{-3}\text{A}\right)} = \frac{3\times 3\times 10^3}{2}\Omega = 4500\,\Omega$

But the resistance of the galvanometer is 50Ω.

$\therefore$ Resistance to be added $= (4500-50)\,\Omega = 4450\,\Omega$

94. (d)

95. (c) As 0.2% of main current passes through the galvanometer hence $\frac{998}{1000}$I current through the shunt.

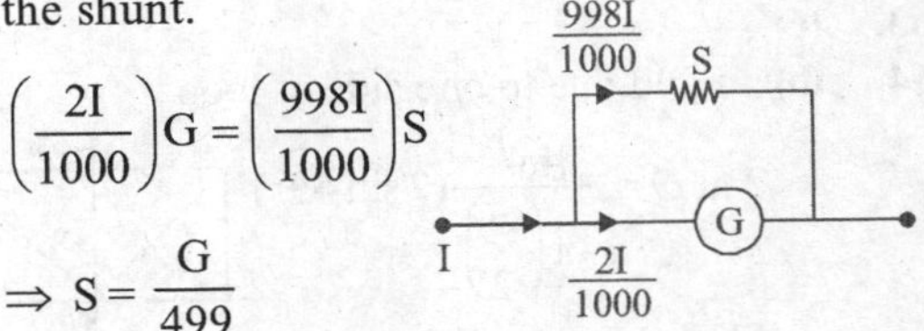

$\left(\frac{2I}{1000}\right)G = \left(\frac{998I}{1000}\right)S$

$\Rightarrow S = \frac{G}{499}$

Total resistance of Ammeter

$R = \frac{SG}{S+G} = \frac{\left(\frac{G}{499}\right)G}{\left(\frac{G}{499}\right)+G} = \frac{G}{500}$

96. (d) $V = i_g(G+R)$ and $V' = i_g(G+2R)$

It shows, $V < V' < 2V$

97. (b) $i_g = i\left(\frac{SG}{S+G}\right)$

or $50\times 10^{-6} = i\left(\frac{S}{S+100}\right)$ .... (i)

Also $V = i_g(G+R_0)$

or $V = 50\times 10^{-6}(100+R_0)$ .... (ii)

On simplifying above equations, we get

$V = 10\ V,\ R_0 = 200\times 10^{-3}\,\Omega$

98. (a) Galvanometer is converted into ammeter, by connected a shunt, in parallel with it.

$\frac{GS}{G+S} = \frac{V_G}{I} = \frac{25\times 10^{-3}}{25}$

$\frac{GS}{G+S} = 0.001\Omega$

Here $S << G$ so $S = 0.001\,\Omega$

99. (c) $I_g = 10\times 10^{-3}/5 = 2\times 10^{-3}$ A;

$R = \frac{V}{I_g} - R_g = \frac{1}{2\times 10^{-3}} - 5 = 495\Omega.$

100. (b) $I_g = 0.1\,I,\ I_s = 0.9\,I;\ S = I_gR_g/I_s$

$= 0.1\times 900/0.9 = 100\,\Omega$

19 Magnetism and Matter

1. If a bar magnet of pole strength m and magnetic moment M is cut perpendicular to its axis in two equal halves then its new pole strength m′ and magnetic moment M′ are respectively

(a) $m' = m$ and $M' = M$

(b) $m' = m$ and $M' = \frac{M}{2}$

(c) $m' = \frac{m}{2}$ and $M' = 2M$

(d) $m' = 2m$ and $M' = \frac{M}{2}$

2. A curve between magnetic moment and temperature of magnet is

Tricky

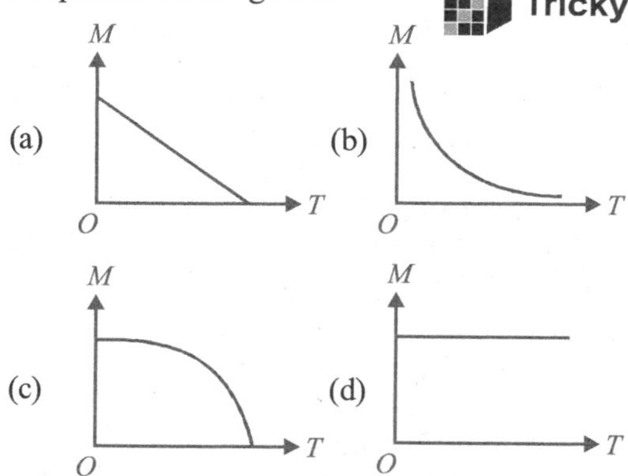

3. A bar magnet has a length 8 cm. The magnetic field at a point at a distance 3 cm from the centre in the broad side-on position is found to be $4\times10^{-6}\,T$. The pole strength of the magnet is.

(a) 6×10^{-5} Am (b) 5×10^{-5} Am
(c) 2×10^{-4} Am (d) 3×10^{-4} Am

4. A permanent magnet in the shape of a thin cylinder of length 10 cm has magnetisation $(M) = 10^6$ A m^{-1}. Its magnetization current I_M is

(a) 10^5 A (b) 10^6 A (c) 10^7 A (d) 10^8 A

5. Assume that each iron atom has a permanent magnetic moment equal to 2 Bohr magnetons (1 Bohr magneton $= 9.27\times10^{-24}$ A-m^2). The density of atoms in iron is 8.52×10^{28} atoms/m^3. Find the maximum magnetic moment per unit volume.

(a) 1.5×10^2 A|m (b) 1.58×10^6 A|m
(c) 1.2×10^5 A|m (d) 1.3×10^6 A|m

6. A bar magnet of magnetic moment 4.0 A-m^2 is free to rotate about a vertical axis through its center. The magnet is released from rest from east-west position. Kinetic energy of the magnet in north-south position will be ($H = 25\mu T$)

(a) 10^{-2}J (b) 10^{-4}J (c) 10^{-6}J (d) 0

7. A short bar magnet of magnetic moment 0.4J T^{-1} is placed in a uniform magnetic field of 0.16 T. The magnet is in stable equilibrium when the potential energy is

(a) –0.64 J (b) zero (c) –0.082 J (d) 0.064 J

8. A coil in the shape of an equilateral triangle of side l is suspended between the pole pieces of a permanent magnet such that $\vec{B}$ is in plane of the coil. If due to a current i in the triangle a torque τ acts on it, the side l of the triangle is

Critical Thinking

(a) $\frac{2}{\sqrt{3}}\left(\frac{\tau}{B.i}\right)^{\frac{1}{2}}$ (b) $2\left(\frac{\tau}{\sqrt{3}B.i}\right)^{\frac{1}{2}}$

(c) $\frac{2}{\sqrt{3}}\left(\frac{\tau}{B.i}\right)$ (d) $\frac{1}{\sqrt{3}}\frac{\tau}{B.i}$

9. A bar magnet of magnetic moment 3.0 A-m^2 is placed in a uniform magnetic field of 2×10^{-5} T. If each pole of the magnet experiences a force of 6×10^{-4} N, the length of the magnet is

(a) 0.5 m (b) 0.3 m (c) 0.2 m (d) 0.1 m

10. The net magnetic moment of two identical magnets each of magnetic moment M_0, inclined at 60° with each other is

Tricky

(a) M_0

(b) $\sqrt{2}\,M_0$

(c) $\sqrt{3}\,M_0$

(d) $2M_0$

11. Two identical magnetic dipoles of magnetic moments 1.0 A-m^2 each, placed at a separation of 2m with their axis perpendicular to each other. The resultant magnetic field at a point midway between the dipoles is

(a) $5\times10^{-7}\,T$ (b) $\sqrt{5}\times10^{-7}\,T$
(c) 10^{-7} T (d) None of these

12. A magnetic needle is kept in a non-uniform magnetic field. It experiences
(a) neither a force nor a torque
(b) a torque but not a force
(c) a force but not a torque
(d) a force and a torque

13. Force between two identical bar magnets whose centres are r metre apart is 4.8 N, when their axes are in the same line. If separation is increased to $2r$, the force between them is **Critical Thinking**
(a) 2.4N (b) 1.2N (c) 0.6N (d) 0.3N

14. A magnetic needle lying parallel to a magnetic field requires W units of work to turn it through 60°. The torque required to maintain the needle in this position will be
(a) $\sqrt{3}W$ (b) W (c) $\frac{\sqrt{3}}{2}W$ (d) $2W$

15. A uniform magnetic field parallel to the plane of the paper exists in space initially directed from left to right. When a bar of soft iron is placed in the field parallel to it, the lines of force passing through it will be represented by **Tricky**

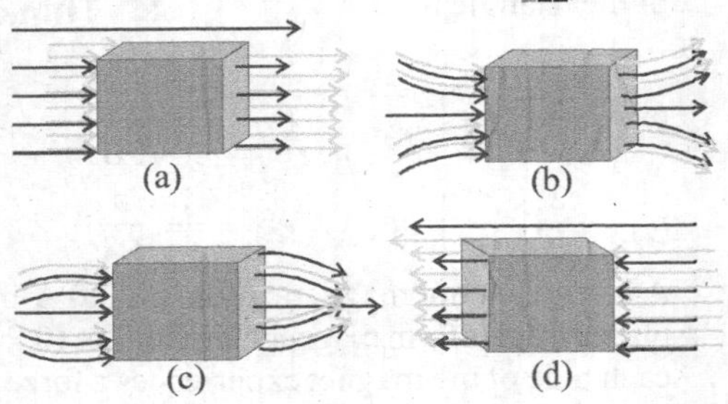

16. A bar magnet having centre O has a length of 4 cm. Point P_1 is in the broad side-on and P_2 is in the end side-on position with $OP_1 = OP_2 = 10$ metres. The ratio of magnetic intensities H at P_1 and P_2 is
(a) $H_1 : H_2 = 16 : 100$ (b) $H_1 : H_2 = 1 : 2$
(c) $H_1 : H_2 = 2 : 1$ (d) $H_1 : H_2 = 100 : 16$

17. The mid points of two small magnetic dipoles of length d in end-on positions, are separated by a distance x, (x > > d). The force between them is proportional to x^{-n} where n is:

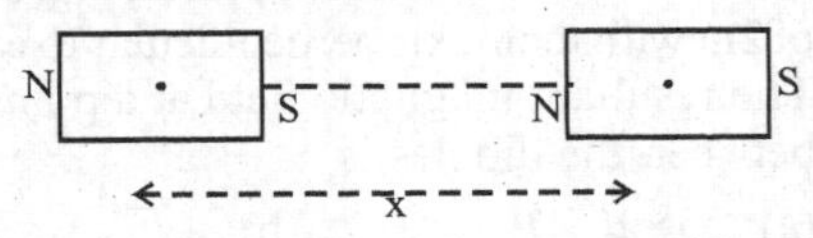

(a) 1 (b) 2 (c) 3 (d) 4

18. A bar magnet of length 'ℓ' and magnetic dipole moment 'M' is bent in the form of an arc as shown in figure. The new magnetic dipole moment will be **Critical Thinking**

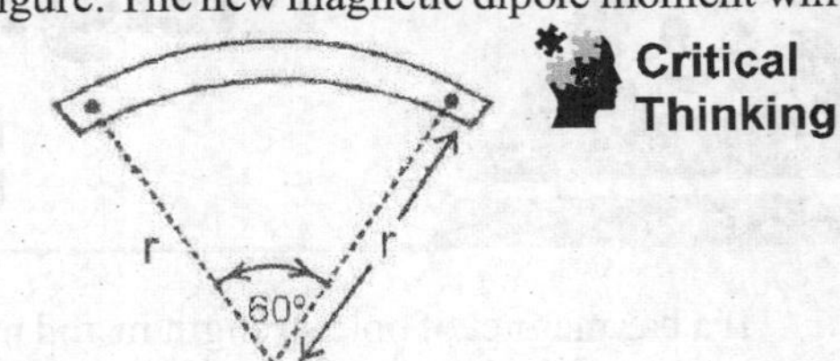

(a) $\frac{3}{\pi}$M (b) $\frac{2}{\pi}$M (c) $\frac{M}{2}$ (d) M

19. Two isolated point poles of strength 30 A-m and 60 A-m are placed at a distance of 0.3m. The force of repulsion is
(a) 2×10^{-3} N (b) 2×10^{-4} N
(c) 2×10^{5} N (d) 2×10^{-5} N

20. A magnet of magnetic moment $50\hat{i}$ A-m² is placed along the x-axis in a magnetic field $\vec{B} = (0.5\hat{i} + 3.0\hat{j})T$. The torque acting on the magnet is
(a) $175\,\hat{k}$ N-m (b) $150\,\hat{k}$ N-m
(c) $75\,\hat{k}$ N-m (d) $25\sqrt{37}\hat{k}$ N-m

21. A magnetic dipole is under the influence of two magnetic fields. The angle between the field directions is 60° and one of the fields has a magnitude of 1.2×10^{-2} T. If the dipole comes to stable equilibrium at an angle of 15° with this field, what is the magnitude of other field? **Toughnut**
(a) 4.4×10^{-3} tesla (b) 5.2×10^{-3} tesla
(c) 3.4×10^{-3} tesla (d) 7.8×10^{-3} tesla

22. Two short bar magnets P and Q are arranged such that their centres are on the X-axis and are separated by a large distance. The magnetic axes of P and Q are along X and Y axes respectively. At a point R, midway between their centres, if B is the magnitude of induction due to Q, the magnitude of total induction at R due to the both magnets is
(a) 3B (b) $\sqrt{5}$B (c) $\frac{\sqrt{5}}{2}$B (d) B

23. A circular coil of 16 turns and radius 10cm carries a current of 0.75 A and rest with its plane normal to an external magnetic field of 5.0×10^{-2} T. The coil is free to rotate about its stable equilibrium position with a frequency of 2.0 s^{-1} Compute the moment of inertia of the coil about its axis of rotation. **Toughnut**
(a) 3.4×10^{-5} kg m² (b) 1.2×10^{-4} kg m²
(c) 2.6×10^{-4} kg m² (d) 4.7×10^{-5} kg m²

24. The figure below shows the north and south poles of a permanent magnet in which n turn coil of area of cross-section A is resting, such that for a current i passed through the coil, the plane of the coil makes an angle θ with respect to the direction of magnetic field B. If the plane of the magnetic field and the coil are horizontal and vertical respectively, the torque on the coil will be

(a) $\tau = niAB\cos\theta$

(b) $\tau = niAB\sin\theta$

(c) $\tau = niAB$

(d) None of the above, since the magnetic field is radial

N Coil S

25. A magnetic dipole is acted upon by two magnetic fields which are inclined to each other at an angle of 75°. One of the fields has a magnitude of 15 mT. The dipole attains stable equilibrium at an angle of 30° with this field. The magnitude of the other field (in mT) is close to : **Critical Thinking**

(a) 1 (b) 11 (c) 36 (d) 1060

26. A 25 cm long solenoid has radius 2 cm and 500 total number of turns. It carries a current of 15 A. If it is equivalent to a magnet of the same size and magnetization $\vec{M}$ (magnetic moment/volume), then $|\vec{M}|$ is :

(a) $30000\pi\,Am^{-1}$ (b) $3\pi\,Am^{-1}$

(c) $30000\,Am^{-1}$ (d) $300\,Am^{-1}$

27. A magnet of magnetic moment *M* is rotated through 360° in a magnetic field *H*, the work done will be

(a) *MH* (b) 2*MH* (c) 2π*MH* (d) Zero

28. A steel wire of length ℓ has a magnetic moment M. It is bent in L-shape. The new magnetic moment is **Tricky**

(a) M (b) $\frac{M}{\sqrt{2}}$ (c) $\frac{M}{2}$ (d) 2M

29. A bar magnet having a magnetic moment of $2 \times 10^4\ JT^{-1}$ is free to rotate in a horizontal plane. A horizontal magnetic field $B = 6 \times 10^{-4}$ T exists in the space. The work done in taking the magnet slowly from a direction parallel to the field to a direction 60° from the field is

(a) 12 J (b) 6 J (c) 2 J (d) 0.6 J

30. Two identical short bar magnets, each having magnetic moment of 10 Am^2, are arranged such that their axial lines are perpendicular to each other and their centres be along the same straight line in a horizontal plane. If the distance between their centres is 0.2 m, the resultant magnetic induction at a point midway between them is ($\mu_0 = 4\pi \times 10^{-7}\ Hm^{-1}$) **Critical Thinking**

(a) $\sqrt{2}\times10^{-7}$ tesla (b) $\sqrt{5}\times10^{-7}$ tesla

(c) $\sqrt{2}\times10^{-3}$ tesla (d) $\sqrt{5}\times10^{-3}$ tesla

31. If the angular momentum of an electron is $\vec{J}$ then the magnitude of the magnetic moment will be

(a) $\frac{eJ}{m}$ (b) $\frac{eJ}{2m}$ (c) $eJ2m$ (d) $\frac{2m}{eJ}$

32. The strength of the earth's magnetic field is

(a) constant everywhere

(b) zero everywhere

(c) having very high value

(d) vary from place to place on the earth's surface

33. Which of the following is responsible for the earth's magnetic field?

(a) Convective currents in earth's core.

(b) Diversive current in earth's core.

(c) Rotational motion of earth.

(d) Translational motion of earth.

34. One can define ...*A*... of a place as the vertical plane which passes through the imaginary line joining the magnetic North and the South–poles.

Here, A refers to

(a) geographic meridian

(b) magnetic meridian

(c) magnetic declination

(d) magnetic inclination

35. At a place, if the earth's horizontal and vertical components of magnetic fields are equal, then the angle of dip will be

(a) 30° (b) 90° (c) 45° (d) 0°

36. The earth's magnetic field lines resemble that of a dipole at the centre of the earth. If the magnetic moment of this dipole is close to $8 \times 10^{22}\ Am^2$, the value of earth's magnetic field near the equator is close to **Critical Thinking**

(radius of the earth = 6.4×10^6 m)

(a) 0.6 Gauss (b) 1.2 Gauss

(c) 1.8 Gauss (d) 0.32 Gauss

37. The magnetic field of earth at the equator is approximately 4×10^{-5} T. The radius of earth is 6.4×10^6 m. Then the dipole moment of the earth will be nearly of the order of: Critical Thinking

(a) 10^{23} A m^2 (b) 10^{20} A m^2
(c) 10^{16} A m^2 (d) 10^{10} A m^2

38. Two magnets of same size and mass make respectively 10 and 15 oscillations per minute at certain place. The ratio of their magnetic moments is

(a) 4 : 9 (b) 9 : 4 (c) 2 : 3 (d) 3 : 2

39. Time periods of vibation of two bar magnets in sum and difference positions are 4 sec and 6 sec respectively. The ratio of their magnetic moments M_1 / M_2 is Tricky

(a) 6:4 (b) 30:16 (c) 2.6:1 (d) 1.5:1

40. A compass needle whose magnetic moment is 60 Am^2, is directed towards geographical north at any place experiencing moment of force of 1.2×10^{-3} Nm. At that place the horizontal component of earth field is 40 micro W/m^2. What is the value of dip angle at that place?

(a) 30° (b) 60° (c) 45° (d) 15°

41. A dip needle lies initially in the magnetic meridian when it shows an angle of dip θ at a place. The dip circle is rotated through an angle x in the horizontal plane and then it shows an angle of dip θ′.

Then $\frac{\tan \theta'}{\tan \theta}$ is

(a) $\frac{1}{\cos x}$ (b) $\frac{1}{\sin x}$ (c) $\frac{1}{\tan x}$ (d) cos x

42. A dip circle is so set that its needle moves freely in the magnetic meridian. In this position, the angle of dip is 40°. Now the dip circle is rotated so that the plane in which the needle moves makes an angle of 30° with the magnetic meridian. In this position, the needle will dip by an angle Critical Thinking

(a) 40° (b) 30°
(c) more than 40° (d) less than 40°

43. A compass needle placed at a distance r from a short magnet in Tan A position shows a deflection of 60°. If the distance is increased to r $(3)^{1/3}$, then deflection of compass needle is

(a) 30° (b) $60 \times 3^{\frac{1}{3}}$
(c) $60 \times 3^{\frac{2}{3}}$ (d) $60 \times 3^{\frac{3}{3}}$

44. The radius of the coil of a tangent galvanometer is 16cm. How many turns of the wire should be used if a current of 40m A is to produced at deflection of 45°. given, horizontal component of earth's field is 0.36×10^{-4} T.

(a) 458 (b) 229 (c) 200 (d) 115

45. A bar magnet 8 cms long is placed in the magnetic merdian with the N-pole pointing towards geographical north. Two netural points separated by a distance of 6 cms are obtained on the equatorial axis of the magnet. If horizontal component of earth's field = 3.2×10^{-5} T, then pole strength of magnet is Critical Thinking

(a) 5 ab-amp × cm (b) 10 ab-amp × cm
(c) 2.5 ab-amp × cm (d) 20 ab-amp × cm

46. The length of a magnet is large compared to its width and breadth. The time period of its oscillation in a vibration magnetometer is 2 s. The magnet is cut along its length into three equal parts and three parts are then placed on each other with their like poles together. The time period of this combination will be Toughnut

(a) 2 s (b) 2 / 3 s (c) $2\sqrt{3}\ s$ (d) $2/\sqrt{3}s$

47. The true value of angle of dip at a place is 60°, the apparent dip in a plane inclined at an angle of 30° with magnetic meridian is

(a) $\tan^{-1}\frac{1}{2}$ (b) $\tan^{-1}(2)$
(c) $\tan^{-1}\left(\frac{2}{3}\right)$ (d) $\tan^{-1}\left(\frac{\sqrt{3}}{4}\right)$

48. Let V and H be the vertical and horizontal components of earth's magnetic field at any point on earth. Near the north pole

(a) V>>H (b) V<<H
(c) V=H (d) V=H=0

49. A torque of 10^{-5} Nm is required to hold a magnet at 90° with the horizontal component H of the earth's magnetic field. The torque to hold it at 30° will be

(a) 5×10^{-6} Nm (b) 5Nm
(c) $\frac{1}{3} \times 10^{-5}$ Nm (d) $5\sqrt{3} \times 10^{-6}$ Nm

50. A tangent galvanometer is connected directly to an ideal battery. If the number of turns in the coil is doubled, the deflection will Tricky

(a) increase (b) decrease
(c) remain same (d) any of these

51. The magnetic needle of a tangent galvanometer is deflected at an angle 30° due to a magnet. The horizontal component of earth's magnetic field $0.34 \times 10^{-4} T$ is along the plane of the coil. The magnetic intensity is

(a) $1.96 \times 10^{-4} T$ (b) $1.96 \times 10^{-5} T$
(c) $1.96 \times 10^{4} T$ (d) $1.96 \times 10^{5} T$

52. A 10 cm long bar magnet of magnetic moment 1.34 Am^2 is placed in the magnetic meridian with its south pole pointing geographical south. The neutral point is obtained at a distance of 15 cm from the centre of the magnet. Calculate the horizontal component of earth's magnetic field.

Toughnut

(a) 0.12×10^{-4} T (b) 0.21×10^{-4} T
(c) 0.34×10^{-4} T (d) 0.87×10^{-7} T

53. A short bar magnet is placed in the magnetic meridian of the earth with north pole pointing north. Neutral points are found at a distance of 30 cm from the magnet on the East – West line, drawn through the middle point of the magnet. The magnetic moment of the magnet in Am^2 is close to: (Given $\frac{\mu_0}{4\pi} = 10^{-7}$ in SI units and B_H =Horizontal component of earth's magnetic field = 3.6×10^{-5} tesla)

(a) 14.6 (b) 19.4 (c) 9.7 (d) 4.9

54. At a place on earth, horizontal component of earth's magnetic field is B_1 and vertical component of earth's magnetic field is B_2. If a magnetic needle is kept vertical, in a plane making angle α with the horizontal component of magnetic field, then square of time period of oscillation of needle when slightly distributed is proportional to

(a) $\frac{1}{\sqrt{B_1 \cos\alpha}}$ (b) $\frac{1}{\sqrt{B_2}}$
(c) $\frac{1}{\sqrt{(B_1 \cos\alpha)^2 + B_2^2}}$ (d) infinite

55. A long straight horizontal cable carries a current of 2.5 A in the direction 10° south of west to 10° north of east. The magnetic meridian of the place happens to be 10° west of the geographic meridian. The earth's magnetic field at the location is 0.33 Gauss, and the angle of dip is zero. Locate the line of neutral points. (Ignore the thickness of the cable).

(a) 1.5 cm (b) 2.5 cm (c) 3.5 cm (d) 2.0 cm

56. The time period of oscillation of a bar magnet suspended horizontally along the magnetic meridian is T_0. If this magnet is replaced by another magnet of the same size and pole strength but with double the mass, the new time period will be

(a) $\frac{T_0}{2}$ (b) $\frac{T_0}{\sqrt{2}}$ (c) $\sqrt{2}T_0$ (d) $2T_0$

57. A magnet makes 40 oscillations per minute at a place having magnetic field intensity of $0.1 \times 10^{-5} T$. At another place, it takes 2.5 sec to complete one vibration. The value of earth's horizontal field at that place is

(a) $0.25 \times 10^{-6} T$ (b) $0.36 \times 10^{-6} T$
(c) $0.66 \times 10^{-8} T$ (d) $1.2 \times 10^{-6} T$

58. Two short magnets with their axes horizontal and perpendicular to the magnetic maridian are placed with their centres 40 cm east and 50 cm west of magnetic needle. If the needle remains undeflected, the ratio of their magnetic moments $M_1 : M_2$ is

Critical Thinking

(a) 4 : 5 (b) 16 : 25 (c) 64 : 125 (d) $2 : \sqrt{5}$

59. A current carrying coil is placed with its axis perpendicular to *N-S* direction. Let horizontal component of earth's magnetic field be H_0 and magnetic field inside the loop be H. If a magnet is suspended inside the loop, it makes angle θ with H. Then θ =

(a) $\tan^{-1}\left(\frac{H_0}{H}\right)$ (b) $\tan^{-1}\left(\frac{H}{H_0}\right)$
(c) $\operatorname{cosec}^{-1}\left(\frac{H}{H_0}\right)$ (d) $\cot^{-1}\left(\frac{H_0}{H}\right)$

60. A 30 cm long bar magnet is placed in the magnetic meridian with its north pole pointing south. The neutral point is obtained at a distance of 40cm from the center of the magnet. Find the pole strength of the magnet. The horizontal component of earth's magnetic field is 0.34 gauss.

(a) 25.7 Am (b) 23.7 Am
(c) 28.7 Am (d) 26.7 Am

Toughnut

61. A freely suspended magnet oscillates with period T in earth's horizontal magnetic field. When a bar magnet is brought near it, such that the magnetic field created by bar magnet is in same direction as earth's horizontal magnetic field, the period decreases to T/2. The ratio of the field of the magnet F to the earth's magnetic field (H) is

(a) 1 : 3 (b) 1 : 1 (c) 3 : 1 (d) 9 : 1

62. A compass which is allowed to move in a horizontal plane is taken to a geomagnetic pole. It
(a) will stay in north-south direction only
(b) will stay in east-west direction only
(c) will stay in any position
(d) None of these

63. A very small magnet is placed in the magnetic meridian with its south pole pointing north. The null point is obtained 20 cm away from the centre of the magnet. If the earth's magnetic field (horizontal component) at this point be 0.3 gauss, the magnetic moment of the magnet is **Critical Thinking**
(a) 8.0×10^2 *e.m.u.* (b) 1.2×10^3 *e.m.u.*
(c) 2.4×10^3 *e.m.u.* (d) 3.6×10^3 *e.m.u.*

64. In end on and broadside on position of a deflection magnetometer, if θ_1 and θ_2 are the deflections produced by short magnets at equal distances, then $\tan\theta_1/\tan\theta_2$ is
(a) 2 : 1 (b) 1 : 2 (c) 1 : 1 (d) 4 : 1

65. The magnetic needle has magnetic moment 8.7×10^{-2} Am^2 and moment of inertia 11.5×10^{-6} kgm^2. It performs 10 complete oscillations in 6.70 s, what is the magnitude of the magnetic field? **Critical Thinking**
(a) 0.012 T (b) 0.120 T
(c) 1.200 T (d) 2.10 T

66. If the period of oscillation of freely suspended bar magnet in earth's horizontal field H is 4 sec. When another magnet is brought near it, the period of oscillation is reduced to 2s. The magnetic field of second bar magnet is
(a) 4H (b) 3H (c) 2H (d) $\sqrt{3}$H

67. A vibration magnetometer consists of two identical bar magnets placed one over the other such that they are perpendicular and bisect each other. The time period of oscillation in a horizontal magnetic field is $2^{5/4}$ seconds. One of the magnets is removed and if the other magnet oscillates in the same field, then the time period in seconds is **Toughnut**
(a) $2^{1/4}$ (b) $2^{1/2}$ (c) 2 (d) $2^{3/4}$

68. The graph between χ and 1/T for paramagnetic material will be represented by **Tricky**

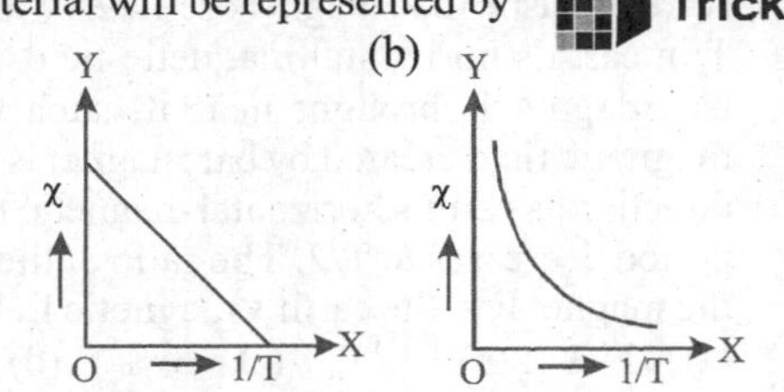

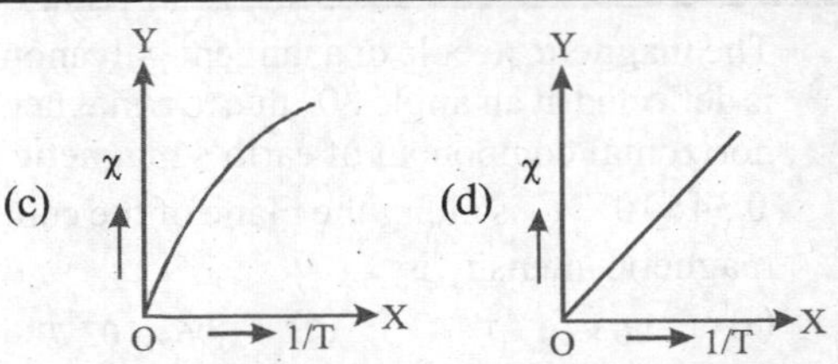

69. The B – H curve (i) and (ii) shown in fig associated with
(a) (i) diamagnetic and (ii) paramagnetic substance
(b) (i) paramagnetic and (ii) ferromagnetic substance
(c) (i) soft iron and (ii) steel
(d) (i) steel and (ii) soft iron

70. The magnetic susceptibility of a paramagnetic substances at –73°C is 0.0060, then its value at –173°C will be
(a) 0.0030 (b) 0.0120
(c) 0.0180 (d) 0.0045

71. Relative permeability of iron is 5500, then its mangetic susceptibility will be
(a) 5501 (b) 5499
(c) 5500×10^7 (d) None of these

72. Relative permittivity and permeability of a material ε_r and μ_r, respectively. Which of the following values of these quantities are allowed for a diamagnetic material?
(a) $\varepsilon_r = 0.5, \mu_r = 1.5$ (b) $\varepsilon_r = 1.5, \mu_r = 0.5$
(c) $\varepsilon_r = 0.5, \mu_r = 0.5$ (d) $\varepsilon_r = 1.5, \mu_r = 1.5$

73. A magnetising field of 2×10^3 amp/m produces a magnetic flux density of 8π Tesla in an iron rod. The relative permeability of the rod will be -
(a) 10^2 (b) 10^0 (c) 10^4 (d) 10^1

74. At a temperatur of 30°C, the susceptibility of a ferromagnetic material is found to be χ. Its susceptibility at 333°C is
(a) χ (b) 0.5χ (c) 2χ (d) 11.1χ

75. A watch glass containing some powdered substance is placed between the pole pieces of a magnet. Deep concavity is observed at the centre. The substance in the watch glass is
(a) iron (b) chromium
(c) carbon (d) wood

76. A permanent magnet in the shape of a thin cylinder of length 10 cm has magnetisation (M) = 10^6 A m^{-1}. Its magnetization current I_M is
(a) 10^5 A (b) 10^6 A (c) 10^7 A (d) 10^8 A

77. If the magnetic dipole moment of an atom of diamagnetic material, paramagnetic material and ferromagnetic material are denoted by μ_d, μ_p and μ_f respectively, then

(a) $\mu_d = 0$ and $\mu_p \neq 0$ (b) $\mu_d \neq 0$ and $\mu_p = 0$
(c) $\mu_p = 0$ and $\mu_f \neq 0$ (d) $\mu_d \neq 0$ and $\mu_f \neq 0$

78. The basic magnetization curve for a ferromagnetic material is shown in figure. Then, the value of relative permeability is highest for the point Tricky

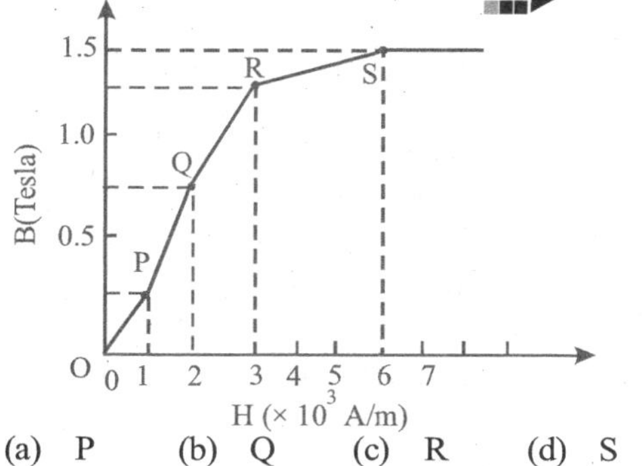

(a) P (b) Q (c) R (d) S

79. A domain in a ferromagnetic substance is in the form of a cube of side length 1 μm. If it contains 8×10^{10} atoms and each atomic dipole has a dipole moment of 9×10^{-24} A m^2, then the magnetization of the domain is Toughnut

(a) 7.2×10^5 A m^{-1} (b) 7.2×10^3 A m^{-1}
(c) 7.2×10^9 A m^{-1} (d) 7.2×10^{12} A m^{-1}

80. A Rowland ring of mean radius 15 cm has 3500 turns of wire wound on a ferromagnetic core of relative permeability 800. What is the magnetic field B in the core for a magnetising current of 1.2 amp ?

(a) 4.48 T (b) 2.48 T (c) 5.48 T (d) 3.12 T

81. An example of a perfect diamagnet is a superconductor. This implies that when a superconductor is put in a magnetic field of intensity B, the magnetic field B_s inside the superconductor will be such that: Tricky

(a) $B_s = -B$ (b) $B_s = 0$
(c) $B_s = B$ (d) $B_s < B$ but $Bs \neq 0$

82. If the susceptibility of dia, para and ferromagnetic materials are χ_d, χ_p, χ_f respectively, then

(a) $\chi_d < \chi_p < \chi_f$ (b) $\chi_d < \chi_f < \chi_p$
(c) $\chi_f < \chi_d < \chi_p$ (d) $\chi_f < \chi_p < \chi_d$

83. A bar magnet has coercivity 4×10^3 Am^{-1}. It is desired to demagnetise it by inserting it inside a solenoid 12 cm long and having 60 turns. The current that should be sent through the solenoid is

(a) 2 A (b) 4 A (c) 6 A (d) 8 A

84. The mass of a speciment of a ferromagnetic material is 0.6 kg. and its density is 7.8×10^3 kg/m^3. If the area of hysteresis loop of alternating magnetising field of frequency 50Hz is 0.722 MKS units then the hysteresis loss per second will be

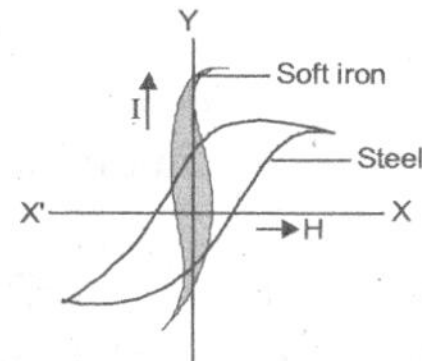

(a) 277.7×10^{-5} joule (b) 277.7×10^{-6} joule
(c) 277.7×10^{-4} joule (d) 27.77×10^{-4} joule

85. The coercivity of a small magnet where the ferromagnet gets demagnetized is 3×10^3 Am^{-1}. The current required to be passed in a solenoid of length 10 cm and number of turns 100, so that the magnet gets demagnetized when inside the solenoid, is: Toughnut

(a) 30 mA (b) 60 mA
(c) 3 A (d) 6 A

Answer KEYs

1	(b)	10	(c)	19	(a)	28	(b)	37	(a)	46	(b)	55	(a)	64	(a)	73	(c)	82	(a)
2	(c)	11	(b)	20	(b)	29	(b)	38	(a)	47	(b)	56	(c)	65	(a)	74	(b)	83	(d)
3	(a)	12	(d)	21	(a)	30	(d)	39	(c)	48	(a)	57	(b)	66	(a)	75	(a)	84	(a)
4	(a)	13	(d)	22	(b)	31	(b)	40	(a)	49	(a)	58	(c)	67	(c)	76	(a)	85	(c)
5	(b)	14	(a)	23	(b)	32	(d)	41	(a)	50	(c)	59	(a)	68	(d)	77	(a)		
6	(b)	15	(b)	24	(a)	33	(a)	42	(d)	51	(b)	60	(d)	69	(c)	78	(b)		
7	(d)	16	(b)	25	(b)	34	(b)	43	(a)	52	(c)	61	(c)	70	(b)	79	(a)		
8	(b)	17	(d)	26	(c)	35	(c)	44	(b)	53	(c)	62	(c)	71	(b)	80	(a)		
9	(d)	18	(a)	27	(d)	36	(a)	45	(a)	54	(c)	63	(b)	72	(b)	81	(b)		

Hints & Solutions

1. **(b)** When a bar magnet cut perpendicular to its axis into two equal parts then m′ = m and length $\ell' = \frac{\ell}{2}$

$\therefore M' = m\ell' = m\frac{\ell}{2} = \frac{M}{2}$

2. **(c)** Magnetism of a magnet falls with rise of temperatue and becomes practically zero above curie temperature.

3. **(a)** Magnetic field due to a bar magnet in the broad-side on position is given by

$$B = \frac{\mu_0}{4\pi}\frac{M}{\left[r^2 + \frac{\ell^2}{4}\right]^{3/2}};\ M = m\ell.$$

After substituting the values and simplifying we get

$$B = 6 \times 10^{-5}\ \text{A-m}$$

4. **(a)** $I_M = ML = 10^6 \times 0.1 = 10^{-5}$ A

5. **(b)** The total magnetic moment per unit volume.

$$= \frac{(8.52 \times 10^{28}) \times (2 \times 9.27 \times 10^{-24})}{1}$$

$= 1.58 \times 10^6$ A/m

6. **(b)** Loss in P.E. = gain in K.E.

$\therefore E_k = U_i - U_j = -$ MB cos 90° – (– MB cos0°) $= 4 \times 25 \times 10^{-6} = 10^{-4}$J

7. **(d)** For stable equilibrium

U = –MB = – (0.4) (0.16) = – 0.064 J

8. **(b)** τ = MB sinθ, τ = iAB sin90°

$\therefore A = \frac{\tau}{iB} = 1/2$ (BC) (AD)

But $\frac{1}{2}$(BC)(AD)

A, B, C, D, l

$$= \frac{1}{2}(l)\sqrt{l^2 - \left(\frac{l}{2}\right)^2} = \frac{\sqrt{3}}{4}l^2$$

$$\Rightarrow \frac{\sqrt{3}}{4}(l)^2 = \frac{\tau}{Bi} \Rightarrow \therefore l = 2\left(\frac{\tau}{\sqrt{3}\ Bi}\right)^{\frac{1}{2}}$$

9. **(d)** $F = mB$ or $6 \times 10^{-4} = m \times 2 \times 10^{-5}$

$\therefore$ $m = 30$ A-m

Now, $\ell = \frac{M}{m} = \frac{3}{30} = 0.1$ m

10. **(c)** $M_{net} = \sqrt{M_0^2 + M_0^2 + 2M_0^2 \cos 60°}$

$= \sqrt{3M_0^2} = \sqrt{3}M_0$

11. **(b)** $B_1 = \frac{\mu_0}{4\pi}.\frac{2M}{r^3} = 10^{-7} \times \frac{2 \times 1}{1^3} = 2 \times 10^{-7}\ T$

and $B_2 = \frac{\mu_0}{4\pi}.\frac{M}{r^3} = 10^{-7} \times \frac{1}{1^3} = 10^{-7}\ T$

Now, $B = \sqrt{B_1^2 + B_2^2} = \sqrt{5} \times 10^{-7}\ T.$

12. **(d)** A magnetic needle kept in non uniform magnetic field experience a force and torque due to unequal forces acting on poles.

13. **(d)** Force between two short bar magnets is given by

$$F = \frac{\mu_0}{4\pi}\frac{6M_1M_2}{r^4}$$

$$\therefore \frac{F_1}{F_2} = \frac{r_2^4}{r_1^4} = \left(\frac{2r}{r}\right)^4$$

or $F_2 = \frac{F_1}{16} = \frac{4.8}{16} = 0.3\ N.$

14. **(a)** $W = MB \cos 60°$

and $\tau = MB \sin 60°$

$\therefore \tau = \sqrt{3}W.$

15. **(b)** Permeability of soft iorn is maximum, so maximum lines of force tries to pass through the soft iron.

16. **(b)** Magnetic intensity on end side -on position is twice than broad side on position.

17. **(d)** In magnetic dipole

Force $\propto \frac{1}{r^4}$

In the given question, Force $\propto x^{-n}$ Hence, n = 4

18. **(a)** Magnetic dipole moment

$M = m \times \ell$ $\quad$ $M' = m \times r$

From figure

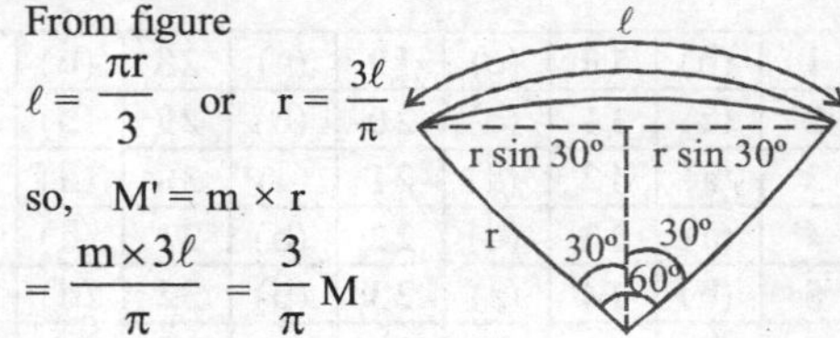

$\ell = \frac{\pi r}{3}$ or $r = \frac{3\ell}{\pi}$

so, $M' = m \times r$

$= \frac{m \times 3\ell}{\pi} = \frac{3}{\pi}M$

19. **(a)**

$$F = \frac{\mu_0}{4\pi}\frac{m_1 m_2}{r^2} = 10^{-7} \times \frac{30 \times 60}{(0.3)^2} = 2 \times 10^{-3}\ N.$$

20. (b) $\vec{\tau} = \vec{M} \times \vec{B} = 50\hat{i} \times (0.5\hat{i} + 3\hat{j}) = 150\,\hat{k}$ N-m

21. (a) Given that : $B_1 = 1.2 \times 10^{-2}$ T, orientation of dipole with the field B_1, $\theta_1 = 15°$
Hence, orientation of dipole with B_2, $\theta_2 = 60° - 15° = 45°$ (figure)
As the dipole is in equilibrium, therefore, the torque on the dipole due to the two fields must be equal and opposite.

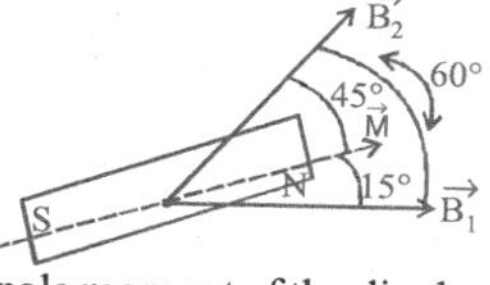

If M be the magnetic dipole moment of the dipole, then
$\tau_1 = \tau_2$ or $MB_1 \sin\theta_1 = MB_2 \sin\theta_2$

or, $$B_2 = \frac{B_1 \sin\theta_1}{\sin\theta_2} = \frac{1.2\times10^{-2}\sin 15°}{\sin 45°}$$

$$= \frac{1.2\times10^{-2}\times0.2588}{0.7071} = 4.4 \times 10^{-3} \text{ Tesla}$$

22. (b)

23. (b) The magnetic moment of the coil is
$M = NIA = 16 \times 0.75 \times \pi \times (0.1)^2 = 0.377\ Am^2$
If K be the moment of inertia of the coil about its axis of rotation, then its period of oscillation in a magnetic field B is given by

$$T = 2\pi\sqrt{\frac{K}{MB}}$$

or its frequency ν is $= \frac{1}{T} = \frac{1}{2\pi}\sqrt{\frac{MB}{K}}$

This gives $K = \frac{MB}{4\pi^2\nu^2}$

Given that $B = 5.0 \times 10^{-2}$ T, $M = 0.377$ A-m^2 and $\nu = 2s^{-1}$

$$\therefore\quad K = \frac{0.377\times5.0\times10^{-2}}{4\times(3.14)^2\times(2)^2} = 1.2 \times 10^{-4}\ kgm^2$$

24. (a) Plane of coil is having angle θ with the magnetic field.
$\therefore \tau = MB\sin(90° - \theta)$ or $\tau = niAB\cos\theta$ (As $M = niA$)

25. (b) We know that, magnetic dipole moment
$M = NiA\cos\theta$ i.e., $M \propto \cos\theta$
When two magnetic fields are inclined at an angle of 75° the equilibrium will be at 30°, so

$$\cos\theta = \cos(75° - 30°) = \cos 45° = \frac{1}{\sqrt{2}}$$

$$\frac{x}{\sqrt{2}} = \frac{15}{2} \therefore x \approx 11$$

26. (c) $\vec{M}$ (mag. moment/volume) $= \frac{NiA}{A\ell}$

$$= \frac{Ni}{\ell} = \frac{(500)15}{25\times10^{-2}} = 30000\ Am^{-1}$$

27. (d) $U_i = -MB\cos 0° = -MB$
and $U_f = -MB\cos 360° = -MB$
Now, $W = U_f - U_i = -MB - (-MB) = 0$.

28. (b) Magnetic moment, $M = m\ell$
$\frac{M}{\ell} = m$, where m is the polestrength.
Therefore distance between poles

$$= \sqrt{(\ell/2)^2 + (\ell/2)^2} = \frac{\ell}{\sqrt{2}}$$

So, $M' = \frac{m\ell}{\sqrt{2}} = \frac{M}{\sqrt{2}}$

29. (b) Work done $= MB(\cos\theta_1 - \cos\theta_2)$
$= MB(\cos 0° - \cos 60°)$

$$= MB\left(1 - \frac{1}{2}\right) = \frac{2\times10^4\times6\times10^{-4}}{2} = 6\text{ J}$$

30. (d) ① S N B_2 P B_1 S N ②
← 0.1m →← 0.1m →

From figure $B_{net} = \sqrt{B_a^2 + B_e^2}$

$$= \sqrt{\left(\frac{\mu_0}{4\pi}.\frac{2M}{d^3}\right)^2 + \left(\frac{\mu_0}{4\pi}.\frac{M}{d^3}\right)^2}$$

$$= \sqrt{5}.\frac{\mu_0}{4\pi}.\frac{M}{d^3} = \sqrt{5}\times10^{-7}\times\frac{10}{(0.1)^3} = \sqrt{5}\times10^{-3}\text{ tesla}$$

31. (b) We know that,
$$\frac{M}{J} = \frac{q}{2m} \qquad \therefore\ M = J\times\frac{e}{2m}.$$

32. (d) The strength of the earths magnetic field is not constant. It varies from one place to other place on the surface of earth. Its value being of the order of 10^{-5} T.

33. (a) The earth's core is hot and molten. Hence, convective current in earth's core is responsible for it's magnetic field.

34. (b)

35. (c) $\tan\theta = \frac{B_V}{B_H} = 1 \qquad \therefore\ \theta = 45°$

36. (a) Given $M = 8 \times 10^{22}\ Am^2$
$d = R_e = 6.4 \times 10^6 m$

Earth's magnetic field, $B=\frac{\mu_0}{4\pi}\cdot\frac{2M}{d^3}$

$$=\frac{4\pi\times10^{-7}}{4\pi}\times\frac{2\times8\times10^{22}}{(6.4\times10^6)^3}$$

$\cong 0.6$ Gauss

37. (a) Given, $B=4\times10^{-5}$ T

$R_E=6.4\times10^6$ m

Dipole moment of the earth M = ?

$$B=\frac{\mu_0}{4\pi}\frac{M}{d^3}$$

$$4\times10^{-5}=\frac{4\pi\times10^{-7}\times M}{4\pi\times\left(6.4\times10^6\right)^3}$$

$\therefore\ M\cong10^{23}\ Am^2$

38. (a) $T_1=\frac{60}{10}=6s$ and $T_2=\frac{60}{15}=4s$

For identical magnets,

$$\frac{M_1}{M_2}=\frac{T_2^2}{T_1^2}=\frac{4^2}{6^2}=\frac{16}{36}=\frac{4}{9}.$$

39. (c) $\frac{M_1}{M_2}=\frac{T_2^2+T_1^2}{T_2^2-T_1^2}=\frac{6^2+4^2}{6^2-4^2}=\frac{52}{20}=(2.6):1$

40. (a)

41. (a) $\tan\theta=\frac{V}{H},\tan\theta'=\frac{V}{H\cos x};\ \frac{\tan\theta'}{\tan\theta}=\frac{1}{\cos x}$

42. (d) $\delta_1=40°,\ \delta_2=30°,\ \delta=?$

$$\cot\delta=\sqrt{\cot^2\delta_1+\cot^2\delta_2}$$

$$=\sqrt{\cot^2 40°+\cot^2 30°}$$

$$\cot\delta=\sqrt{1.19^2+3}=2.1$$

$\therefore\ \delta=25°$ i.e. $\delta<40°$.

43. (a) $\frac{\tan\theta_2}{\tan\theta_1}=\frac{d_1^3}{d_2^3}=\frac{r^3}{[r(3)^{1/3}]^3}=\frac{1}{3}$

$$\tan\theta_2=\frac{1}{3}\tan\theta_1=\frac{\tan 60}{3}=\frac{\sqrt{3}}{3}=\frac{1}{\sqrt{3}}$$

$\therefore\ \theta_2=30°$

44. (b)

45. (a) Here, $2\ell=8$ cm, $\ell=4$ cm, $d=\frac{6}{2}=3$ cm.

At neutral point,

$$H=B=\frac{\mu_0}{4\pi}\frac{M}{(d^2+\ell^2)^{3/2}}$$

$$=10^{-7}\frac{M}{(5\times10^{-2})^3}=\frac{M}{1250}$$

$\therefore\ M=1250\,H=1250\times3.2\times10^{-5}\ Am^2$

$$m=\frac{M}{2\ell}=\frac{1250\times3.2\times10^{-5}}{8\times10^{-2}}\text{ Am.}=0.5$$

$$Am=0.5\times\frac{1}{10}\text{ ab amp}\times100\text{ cm}$$

$=5$ ab-amp cm.

46. (b) The moment of inertia of original magnet

$$I=\frac{m\ell^2}{12}$$

After dividing into three equal parts,

$$I'=3(m/3).\frac{(\ell/3)^2}{12}=\frac{I}{9}.$$

Also magnetic moment, $M'=M$

Now,

$$\frac{T'}{T}=\sqrt{\frac{I'}{I}\times\frac{M}{M'}}$$

$$\therefore\ T'=T\times\sqrt{\frac{I'}{I}}=2\times\sqrt{\frac{1}{9}}=\frac{2}{3}\text{s.}$$

47. (b) $\tan\theta'=\frac{\tan\theta}{\cos\alpha}=\frac{\tan 60°}{\cos 30°}=\frac{\sqrt{3}}{\sqrt{3}/2}=2$

$\therefore\ \theta'=\tan^{-1}(2).$

48. (a) $H=B\cos\theta,\ V=B\sin\theta$

Here B = earth's magnetic field

θ = angle of dip = 90° at north pole

$\Rightarrow H=B\cos 90°=0$

$V=B\sin 90°=B\Rightarrow V>>H$

49. (a)

50. (c) $i=\left(\frac{2RB_H}{\mu_0 N}\right)\tan\theta$

or $\frac{V}{R}=\left(\frac{2RB_H}{\mu_0 N}\right)\tan\theta$ ...(i)

When number of turns are doubled, resistance of the coil is also doubled, so

$\frac{V}{(2R)}=\left[\frac{2RB_H}{\mu_0(2N)}\right]\tan\theta'$ ...(ii)

From (i) and (ii), $\theta'=\theta$.

51. (b) We know that $\frac{B}{B_H}=\tan\theta$

or $B=B_H\tan\theta$

$=0.34\times10^{-4}\tan 30°=1.96\times10^{-5}$ T

52. (c) $B = \frac{\mu_0}{4\pi}\frac{M}{(r^2+\ell^2)^{3/2}} = B_H$

Given that $\mu_0 = 4\pi \times 10^{-7}$ T mA^{-1}, $M = 1.34$ Am^2, $r = 15$ cm $= 0.15$ m and $\ell = 5.0$ cm $= 0.05$ m

$\therefore\ B_H = 10^{-7} \times \frac{1.34}{[(0.15)^2+(0.5)^2]^{3/2}}$

$= 10^{-7} \times \frac{1.34}{0.025\sqrt{0.025}} = 0.34 \times 10^{-4}$ T

53. (c) Here, r = 30cm = 0.3cm

we know $\frac{\mu_0 M}{4\pi r^3} = B_H = 3.6\times10^{-5}$

$\Rightarrow M = \frac{3.6\times10^{-5}}{10^{-7}}(0.3)^3 = 9.7$ Am^2

54. (c) Resultant magnetic field in the plane,

$B = \sqrt{(B_1\cos\alpha)^2 + B_2^2}$

Time period,

$T = 2\pi\sqrt{\frac{I}{MB}}$

55. (a)

56. (c) $I_1 = \frac{m\ell^2}{12}$ and $I_2 = \frac{(2m)\ell^2}{12} = 2I_1$.

Now, $\frac{T_1}{T_2} = \sqrt{\frac{I_1}{I_2}}$ or $\frac{T_0}{T_2} = \sqrt{\frac{1}{2}}$

$\therefore\ T_2 = \sqrt{2}\,T_0$.

57. (b) $T_1 = \frac{60}{40} = 1.5$ s and $T_2 = 2.5$ s

We know that, $\frac{T_1}{T_2} = \sqrt{\frac{B_2}{B_1}}$

or $B_2 = \frac{T_1^2}{T_2^2}\times B_1 = \left(\frac{1.5}{2.5}\right)^2 \times (0.1\times10^{-5})$

$= 0.36 \times 10^{-6}$ T

58. (c) For null deflection

$\frac{M_1}{M_2} = \left(\frac{d_1}{d_2}\right)^3 = \left(\frac{40}{50}\right)^3 = \frac{64}{125}$

59. (a) $\tan\theta = \frac{H_0}{H}$

$\theta = \tan^{-1}\left(\frac{H_0}{H}\right)$

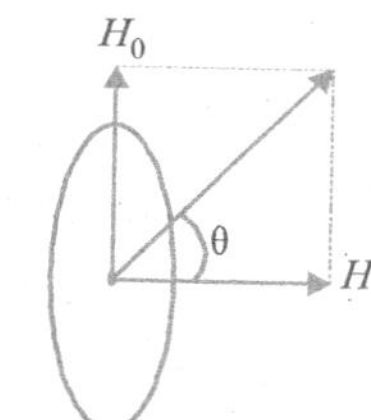

60. (d) $\because\ M = \frac{4\pi}{\mu_0}\frac{B_H(r^2-\ell^2)^2}{2r}$

$= 10^7 \times \frac{(0.34\times10^{-4}).[(0.40)^2-(0.15)^2]^2}{2\times0.40}$

$= 8.0$ Am^2

The pole strength of the magnet is,

$m = \frac{M}{2\ell} = \frac{8.0}{0.30} = 26.7$ Am

61. (c) $T_1 = 2\pi\sqrt{\frac{I}{HM}}$, and $T_2 = 2\pi\sqrt{\frac{I}{(H+F)M}}$

$\frac{T_1^2}{T_2^2} = \frac{H+F}{H} = \frac{T^2}{T^2/4} = \frac{4}{1} \Rightarrow H+F = 4H$

$\Rightarrow 3H = F$

62. (c) At geomagnetic poles, there is no horizontal component of earth field and so compass needle may stay at any position.

63. (b) $B = B_H$

or $\frac{\mu_0}{4\pi}\cdot\frac{2M}{r^3} = 0.3$

$\therefore\ M = 1.2 \times 10^3$ emu.

64. (a) $\frac{\tan\theta_1}{\tan\theta_2} = \frac{2}{1}$

65. (a) Magnetic moment, M = 8.7 × 10^{-2} Am^2 moment of inertia, I = 11.5 × 10^{-6} kg m^2 Time period of oscillation is

$T = \frac{6.70}{10} = 0.6755$ As, $T = 2\pi\sqrt{\frac{I}{MB}}$; $B = \frac{4\pi^2 I^2}{MT^2}$

$\therefore\ B = \frac{4\times(3.14)^2\times11.5\times10^{-6}}{8.7\times10^{-2}\times(0.67)^2} = 0.012$T

66. (a) Thus, $\frac{T}{T'} = \frac{2\pi\sqrt{\frac{I}{MH}}}{2\pi\sqrt{\frac{I}{MH'}}}$

Given, T = 4 sec, T' = 2 sec,

So, $\frac{4}{2} = \sqrt{\frac{H'}{H}} \Rightarrow H' = 4H$

67. (c) Initially magnetic moment of system

$$M_1 = \sqrt{M^2 + M^2} = \sqrt{2}M$$

and moment of inertia

$I_1 = I + I = 2I.$

Finally when one of the magnet is removed then

$M_2 = M$ and $I_2 = I$

So, $T = 2\pi\sqrt{\dfrac{I}{M B_H}}$

$$\frac{T_1}{T_2} = \sqrt{\frac{I_1}{I_2} \times \frac{M_2}{M_1}} = \sqrt{\frac{2I}{I} \times \frac{M}{\sqrt{2}M}}$$

$$\Rightarrow T_2 = \frac{2^{5/4}}{2^{1/4}} = 2\,\text{sec}$$

68. (d) Curi's Law $\chi_m = \dfrac{\mu_0 c}{T}$

69. (c) Soft iron → lowre tentivity and steel → high retentivity so option (c) is correct.

70. (b) As magnetic susceptibility $\chi_m \propto \dfrac{1}{T}$, therefore

$$\frac{\chi_2}{\chi_1} = \frac{T_1}{T_2} \Rightarrow \frac{\chi_2}{0.0060} = \frac{273-73}{273-173} = \frac{200}{100} = 2$$

$\chi_2 = 2 \times 0.0060 = 0.0120$

71. (b) We know that,

$\mu_r = 1 + \chi$

or $\chi = \mu_{r-1} = 5500 - 1 = 5499$

72. (b) For a diamagnetic material, the value of μ_r is less than one. For any material, the value of ϵ_r is always greater than 1.

73. (c) $\because \mu_r = \dfrac{\mu}{\mu_0} = \dfrac{B}{H\mu_0}$ or μ_r

$$= \frac{8\pi}{2 \times 10^3 \times 4\pi \times 10^{-7}} = 10^4$$

74. (b) $\dfrac{\chi_{m_1}}{\chi_{m_2}} = \dfrac{T_2}{T_1} = \dfrac{273+333}{273+30} = \dfrac{606}{303} = 2$

$\therefore \chi_{m_2} = \chi_{m_1}/2 = 0.5\chi_{m_1} = 0.5\chi \; (\because \chi_{m_1} = \chi)$

75. (a) Iron is ferromagnetic.

76. (a) As $BI = \mu_0 MI_M = \mu_0(I + I_M)$

Here, $I = 0$

Then $\mu_0 MI = \mu_0(I_M) \Rightarrow I_M = MI = 10^6 \times 0.1 = 10^{+5}$ A

77. (a)

78. (b) $B = \mu_0\mu_r H \Rightarrow \mu_r \propto \dfrac{B}{H}$ = slope of B-H curve

According to the given graph, slope of the graph is highest at point Q.

79. (a) The volume of the cube is

$V = (10^{-6}\,m)^3 = 10^{-18}\,m^3$

Net dipole moment m_{net}

$= 8 \times 10^{10} \times 9 \times 10^{-24}\,Am^2$

$= 72 \times 10^{-14}\,Am^2$

Intensity of magnetization is magnetic moment developed per unit volume.

$\therefore$ magnetization, $M = \dfrac{m_{net}}{\text{volume}}$

$$= \frac{72 \times 10^{-14}\,A\,m^2}{10^{-18}\,m^3} = 7.2 \times 10^5\,Am^{-1}$$

80. (a) Where μ is the permeability of the core and μ_r is the relative permeability

Now, $\mu_0 = 4\pi \times 10^{-7}\,TMA^{-1}$, $\mu_r = 800$,

$I = 1.2$ amp

and $n = \dfrac{N}{2\pi r} = \dfrac{3500}{2\pi r} = \dfrac{3500}{2\pi \times 0.15}\,m^{-1}$

$B = \mu n I = \mu_0\mu_r n I$

$$\therefore \quad B = \frac{(4\pi \times 10^{-7}) \times 800 \times 3500 \times 1.2}{(2\pi \times 0.15)} = 4.48\,T.$$

81. (b) Magnetic field inside the superconductor is zero. Diamagnetic substances are repelled in external magnetic field.

82. (a) $\chi_d < \chi_p < \chi_f$

For diamagnetic substance χ_d is small negative (10^{-5})

For paramagnetic substances χ_p is small and positive (10^{-3} to 10^{-5})

For ferromagnetic substanes χ_f is very large (10^3 to 10^5)

83. (d) $n = \dfrac{N}{l} = \dfrac{60}{0.12} = 500$ turn metre^{-1}

$$\Rightarrow i = \frac{H}{n} = \frac{4 \times 10^3}{500} = 8A$$

84. (a) $W_H = VAft = \dfrac{m}{d}Aft$

$$= \frac{0.6}{7.8 \times 10^3} \times 0.722 \times 50 = 277.7 \times 10^{-5}\text{ Joule}$$

85. (c) Magnetic field in solenoid $B = \mu_0 n i$

$\Rightarrow \dfrac{B}{\mu_0} = ni$ (Where n = number of turns per unit length)

$$\Rightarrow \frac{B}{\mu_0} = \frac{Ni}{L} \Rightarrow 3 \times 10^3 = \frac{100i}{10 \times 10^{-2}}$$

$\Rightarrow i = 3A$

20 Electromagnetic Induction

1. A conducting square loop is placed in a magnetic field B with its plane perpendicular to the field. The sides of the loop start shrinking at a constant rate α. The induced emf in the loop at an instant when its side is 'a' is
   (a) $2a\alpha B$ (b) $a^2\alpha B$ (c) $2a^2\alpha B$ (d) $a\alpha B$

2. A rectangular loop is present in the magnetic field region of an infinite long wire. Now the loop is being rotated as shown in the figure. Then the induced current in side AD will be

   Tricky

   (a) along DA

   (b) along AD
   (c) zero
   (d) None of these

3. A rod PQ of length L moves with a uniform velocity v parallel to a long straight wire carrying a current i, the end P remaining at a distance r from the wire. The emf induced across the rod is

   Critical Thinking

   (a) $\dfrac{\mu_0 iv^2}{2\pi}\ln\left(\dfrac{r+L}{r}\right)$ (b) $\dfrac{\mu_0 i^2v^2}{2\pi}\ln\left(\dfrac{r^2+L}{r}\right)$

   (c) $\dfrac{\mu_0 iv}{2\pi}\ln\left(\dfrac{r+L}{r}\right)$ (d) $\dfrac{\mu_0 iv}{2\pi}\ln\left(\dfrac{r^2+L^2}{L^2}\right)$

4. A coil having n turns and resistance $R\,\Omega$ is connected with a galvanometer of resistance $4R\,\Omega$. This combination is moved in time t second from a magnetic flux ϕ_1 weber to ϕ_2 weber. The induced current in the circuit is

   (a) $-\dfrac{\phi_2-\phi_1}{5\,Rnt}$ (b) $-\dfrac{n(\phi_2-\phi_1)}{5\,Rt}$

   (c) $-\dfrac{(\phi_2-\phi_1)}{Rnt}$ (d) $-\dfrac{n(\phi_2-\phi_1)}{Rt}$

5. A conducting disc of conductivity σ has a radius 'a' and thickness 't'. If the magnetic field B is applied in a direction perpendicular to the plane of the disc changes with time at the rate of $\dfrac{dB}{dt}=\alpha$. Calculate the power dissipated in the disc due to the induced current.

   (a) $\dfrac{\pi t\sigma a^4}{8}\alpha^2$ (b) $\dfrac{\pi t\sigma a^4}{4}\alpha^2$

   (c) $\dfrac{\pi t\sigma a^4}{2}\alpha^2$ (d) $\dfrac{2\pi t\sigma a^4}{3}\alpha^2$

6. Shown in the figure is a circular loop of radius r and resistance R. A variable magnetic field of induction $B=B_0e^{-t}$ is established inside the coil. If the key (K) is closed, the electrical power developed right after closing the switch is equal to

   Critical Thinking

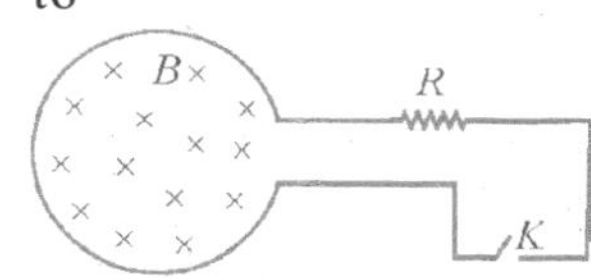

   (a) $\dfrac{B_0^2\pi r^2}{R}$ (b) $\dfrac{B_0 10r^3}{R}$

   (c) $\dfrac{B_0^2\pi^2r^4R}{5}$ (d) $\dfrac{B_0^2\pi^2r^4}{R}$

7. A conductor AB of length l moves in x – y plane with velocity $\vec{v}=v_0(\hat{i}-\hat{j})$. A magnetic field $\vec{B}=B_0(\hat{i}+\hat{j})$ exists in the region. The iduced emf is

   (a) zero (b) B_0lv_0
   (c) B_0lv_0 (d) $\sqrt{2}B_0lv_0$

8. A current carrying infinitely long wire is kept along the diameter of a circular wire loop, without touching it, the correct statement(s) is(are)

   I. The emf induced in the loop is zero if the current is constant.
   II. The emf induced in the loop is finite if the current is constant.
   III. The emf induced in the loop is zero if the current decreases at a steady rate.

   (a) I only (b) II only
   (c) I and II (d) I, II and III

9. Two different wire loops are concentric and lie in the same plane. The current in the outer loop (I) is clockwise and increases with time. The induced current in the inner loop

(a) is clockwise
(b) is zero
(c) is counter clockwise
(d) has a direction that depends on the ratio of the loop radii.

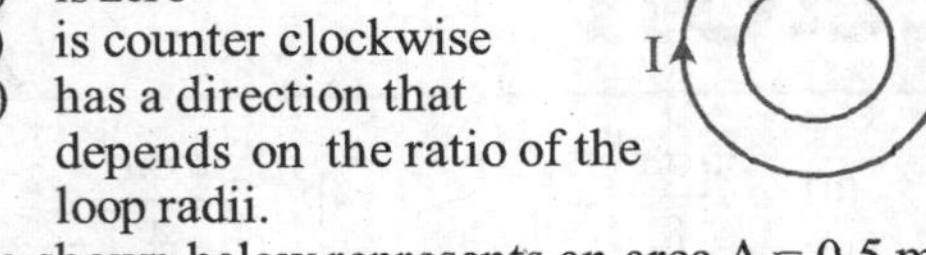

10. Fig shown below represents an area A = 0.5 m^2 situated in a uniform magnetic field B = 2.0 weber/m^2 and making an angle of 60° with respect to magnetic field. **Tricky**

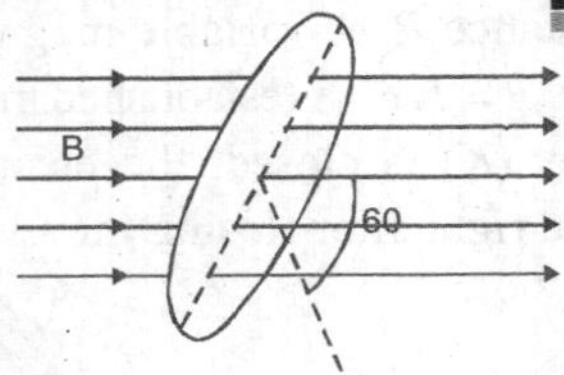

The value of the magnetic flux through the area would be equal to

(a) 2.0 weber (b) $\sqrt{3}$ weber
(c) $\sqrt{3}/2$ weber (d) 0.5 weber

11. In the figure the flux through the loop perpendicular to the plane of the coil and directed into the paper is varying according to the relation $\phi = 6t^2 + 7t + 1$ where ϕ is in milliweber and t is in second. The magnitude of the emf induced in the loop at t = 2 s and the direction of induce current through R are

(a) 39 mV; right to left
(b) 39 mV; left to right
(c) 31 mV; right to left
(d) 31 mV; left to right

R

12. Magnetic flux linked with a stationary loop of resistance R varies with respect to time during the time period T as follows:

$\phi = at(T - t)$

The amount of heat generated in the loop during that time (inductance of the coil is negligible) is

(a) $\frac{aT}{3R}$ (b) $\frac{a^2T^2}{3R}$ (c) $\frac{a^2T^2}{R}$ (d) $\frac{a^2T^3}{3R}$

13. A conducting circular loop is placed in a uniform magnetic field of 0.04 T with its plane perpendicular to the magnetic field. The radius of the loop starts shrinking at 2 mm/s. The induced emf in the loop when the radius is 2 cm is

(a) 4.8 π μV (b) 0.8 π μV
(c) 1.6 π μV (d) 3.2 π μV

Critical Thinking

14. In a uniform and constant magnetic field of induction B, two long conducting wires ab and cd are kept parallel to each other at distance ℓ with their plane perpendicular to B. The ends a and c are connected together by an ideal inductor of inductance L. A conducting slider wire PQ is imparted a speed v_0 at time $t = 0$. The situation is shown in the figure.

At time $t = \frac{\pi\sqrt{mL}}{4B\ell}$, the value of current I through the wire PQ is (ignore any resistance, electrical as well as mechanical)

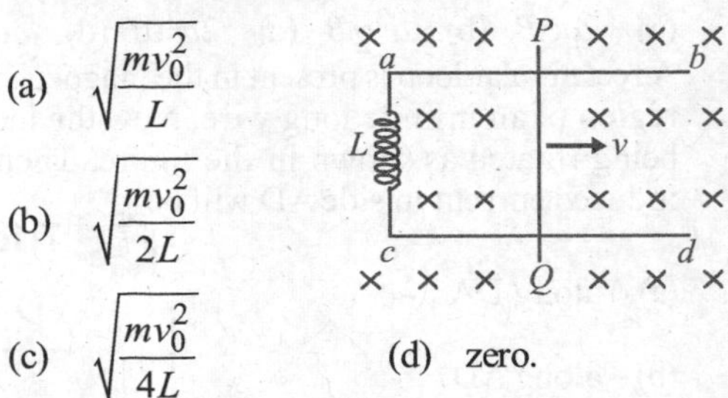

(a) $\sqrt{\frac{mv_0^2}{L}}$
(b) $\sqrt{\frac{mv_0^2}{2L}}$
(c) $\sqrt{\frac{mv_0^2}{4L}}$ (d) zero.

15. The figure shows a conducting loop consisting of half circle of area $A = 0.06$ m^2 and three straight segments. The half circle lies in a uniform changing magnetic field $B = 4t^2 + 2t + 5$ (SI unit), where t is the time in second. An ideal battery E = 2 V is connected as shown and the total resistance of the wire is 2Ω. The net current in the loop is at $t = 5$ second is:

(a) 1 A
(b) 1.5 A
(c) 0.26 A
(d) 0.10 A

B

16. A square loop with 2.0 m sides is perpendicular to a uniform magnetic field, with half the area of the loop in the field is shown in figure. The loop contains a 20.0 V battery with negligible internal resistance. If the magnitude of the field varies with time according to $B = 0.042 – 0.87\,t$, with B in tesla and t in second. The net emf of the circuit is: **Critical Thinking**

(a) 20.0 V
(b) 18.26 V
(c) 21.74 V
(d) None of these

20.0 V

17. A coil of circular cross-section having 1000 turns and 4 cm^2 face area is placed with its axis parallel to a magnetic field which decreases by 10^{-2} Wb m^{-2} in 0.01 s. The e.m.f. induced in the coil is:

(a) 400 mV (b) 200 mV
(c) 4 mV (d) 0.4 mV

18. The radius of the circular conducting loop shown in fig. is R. Magnetic field is decreasing at a constant rate α. Resistance per unit length of the loop is ρ.

Then, the current in wire AB is (AB is one of the diameters)

(a) $\frac{R\alpha}{2\rho}$ from A to B (b) $\frac{R\alpha}{2\rho}$ from B to A

(c) $\frac{R\alpha}{\rho}$ from A to B (d) 0

19. An equilateral triangular loop ABC made of uniform thin wires is being pulled out of a region with a uniform speed v, where a uniform magnetic field $\vec{B}$ perpendicular to the plane of the loop exists. At time $t = 0$, the point A is at the edge of the magnetic field. The induced current (I) vs time (t) graph will be as **Tricky**

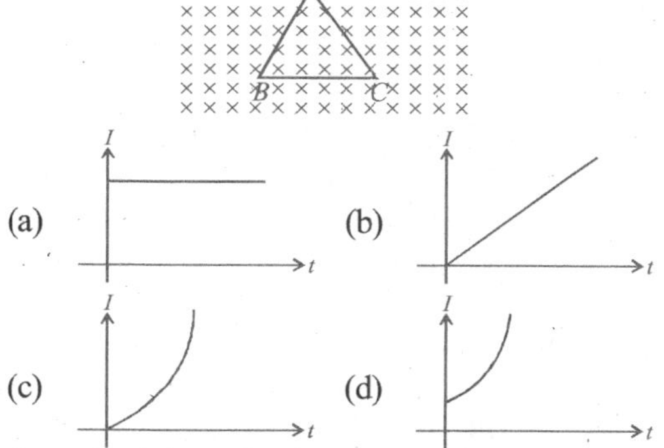

20. A vertical ring of radius r and resistance R falls vertically. It is in contact with two vertical rails which are joined at the top. The rails are without friction and resistance. There is a horizontal uniform magnetic field of magnitude B perpendicular to the plane of the ring and the rails. When the speed of the ring is v, the current in the top horizontal of the rail section is

(a) 0 (b) $\frac{2Brv}{R}$ (c) $\frac{4Brv}{R}$ (d) $\frac{8Brv}{R}$

21. A flexible wire loop in the shape of a circle has radius that grown linearly with time. There is a magnetic field perpendicular to the plane of the loop that has a magnitude inversely proportional to the distance from the center of the loop, $B(r) \propto \frac{1}{r}$. How does the emf E vary with time? **Tricky**

(a) $E \propto t^2$ (b) $E \propto t$

(c) $E \propto \sqrt{t}$ (d) E is constant

22. Charge Q is uniformly distributed on a thin insulating ring of mass m which is initially at rest. To what angular velocity will the ring be accelerated when a magnetic field B, perpendicular to the plane of the ring, is switched on?

(a) $\frac{QB}{2m}$ (b) $\frac{3QB}{2m}$ (c) $\frac{QB}{m}$ (d) $\frac{QB}{4m}$

23. A rectangular coil has a long straight wire passing through its centroid perpendicular to its plane as shown. If current through the wire varies as $i = i_0 \sin \omega t$, induced current in the coil will be (Given R = Resistance of the coil) **Tricky**

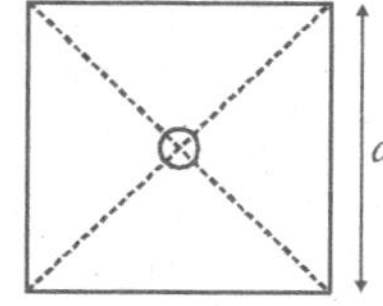

(a) 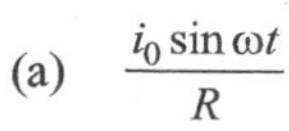 $\frac{i_0 \sin \omega t}{R}$

(b) $\frac{\pi a \sin \omega t}{bR}$

(c) zero

(d) $\frac{\pi a \cos \omega t}{bR}$

24. Two identical cycle wheels (geometrically) have different number of spokes connected from center to rim. One is having 20 spokes and the other having only 10 (the rim and the spokes are resistanceless). One resistance of value R is connected between center and rim. The current in R will be

(a) double in the first wheel than in the second wheel

(b) four times in the first wheel than in the second wheel

(c) will be double in the second wheel than that of the first wheel

(d) will be equal in both these wheels

25. The magnetic field in a region is given by $B = B_0\left(1+\frac{x}{a}\right)\hat{k}$. A square loop of edge-length d is placed with its edges along the x and y-axes. The loop is moved with a constant velocity $v = v_0\,\hat{i}$. The emf induced in the loop is :

(a) zero (b) v_0B_0d

(c) $\frac{v_0B_0d^3}{a^2}$ (d) $\frac{v_0B_0d^2}{a}$

26. A sliding wire of length 0.25 m and having a resistance of 0.5 Ω moves along conducting guiding rails AB and CD with a uniform speed of 4 m/s. A magnetic field of 0.5 T exists normal to the plane of $ABCD$ directed into the page. The guides are short -circuited with resistances of 4 and 2 Ω as shown. The current through the sliding wire is :

Critical Thinking

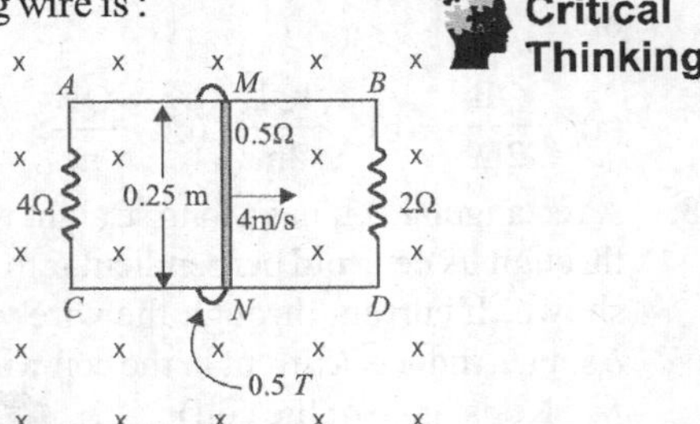

(a) 0.27 A (b) 0.37 A (c) 1.0 A (d) 0.72 A

27. An L-shaped conductor rod is moving in transverse magnetic field as shown in the figure. Potential difference between ends of the rod is maximum if the rod is moving with velocity

Tricky

(a) $4\hat{i}-6\hat{j}$ m/s

(b) $-4\hat{i}+6\hat{j}$ m/s

(c) $3\hat{i}+2\hat{j}$ m/s

(d) $\sqrt{13}\,\hat{i}$ m/s

28. A conducting rod of mass m and ℓ is placed on a smooth horizontal surface in a region where transverse uniform magnetic field B exists in the region. At $t = 0$, constant force F starts acting on the rod at its mid point as shown. Potential difference between ends of the rod, $V_P - V_Q$ at any time t is

(a) $\frac{BF\ell t}{2m}$

(b) $\frac{BF\ell t}{4m}$

(c) $\frac{5BF\ell t}{8m}$

(d) $\frac{7BF\ell t}{8m}$

29. A thin circular ring of area A is perpendicular to uniform magnetic field of induction B. A small cut is made in the ring and a galvanometer is connected across the ends such that the total resistance of circuit is R. When the ring is suddenly squeezed to zero area, the charge flowing through the galvanometer is

(a) $\frac{BR}{A}$ (b) $\frac{AB}{R}$ (c) ABR (d) B^2A/R^2

30. A 0.1 m long conductor carrying a current of 50 A is perpendicular to a magnetic field of 1.21 mT. The mechanical power to move the conductor with a speed of 1 ms^{-1} is

(a) 0.25 mW (b) 6.25 mW
(c) 0.625 W (d) 1 W

31. A wire is bent to form the double loop shown in Fig. There is a uniform magnetic field directed into the plane of the loop. If the magnitude of this field is decreasing, the current will flow from

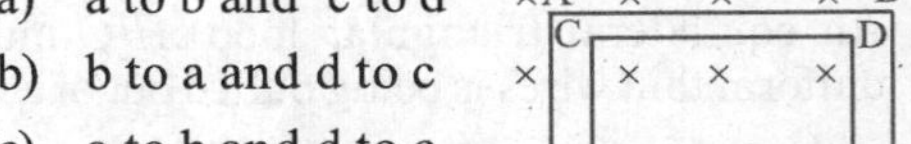

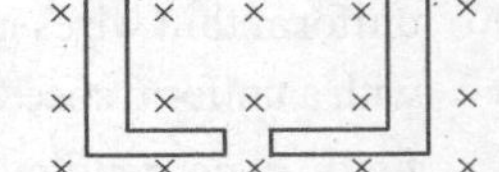

(a) a to b and c to d
(b) b to a and d to c
(c) a to b and d to c
(d) b to a and c to d

32. A conducting wire xy of length l and mass m is sliding without friction on vertical conduction rails ab and cd shown in Fig. A uniform magnetic field B exists perpendicular to the plane of the rails, x moves with a constant velocity of

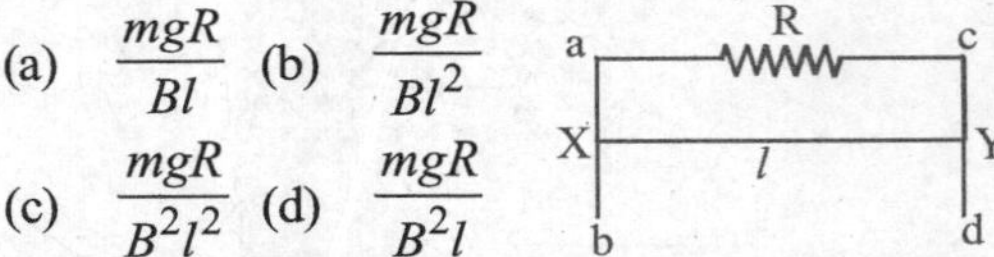

(a) $\frac{mgR}{Bl}$ (b) $\frac{mgR}{Bl^2}$

(c) $\frac{mgR}{B^2l^2}$ (d) $\frac{mgR}{B^2l}$

33. A wooden stick of length 3ℓ is rotated about an end with constant angular velocity ω in a uniform magnetic field B perpendicular to the plane of motion. If the upper one third of its length is coated with copper, the potential difference across the whole length of the stick is

Critical Thinking

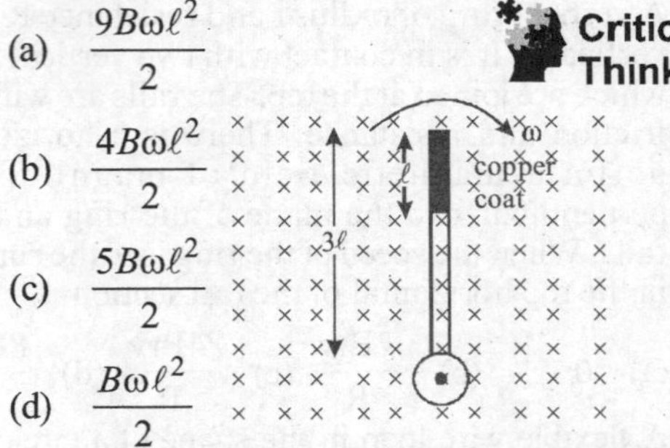

(a) $\frac{9B\omega\ell^2}{2}$

(b) $\frac{4B\omega\ell^2}{2}$

(c) $\frac{5B\omega\ell^2}{2}$

(d) $\frac{B\omega\ell^2}{2}$

34. A uniform magnetic field of induction B is confined to a cylindrical region of radius R. The magnetic field is increasing at a constant rate of $\frac{dB}{dt}$ (tesla/second). An electron of charge q, placed at the point P on the periphery of the field experiences an acceleration.

(a) $\frac{BR}{(\sqrt{2}+1)m}$ towards left

(b) $\frac{1}{2}\frac{eR}{m}\frac{dB}{dt}$ towards left

(c) $\frac{eR}{m}\frac{dB}{dt}$ towards left

(d) zero

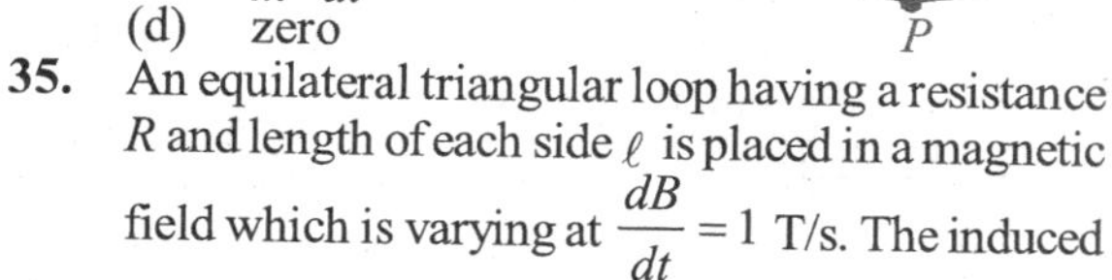

35. An equilateral triangular loop having a resistance R and length of each side ℓ is placed in a magnetic field which is varying at $\frac{dB}{dt} = 1$ T/s. The induced current in the loop will be

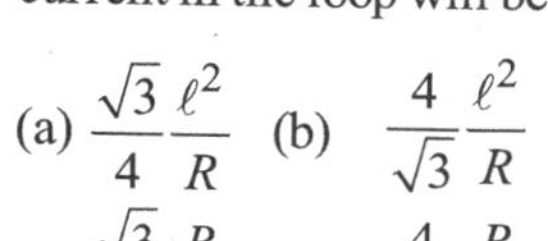

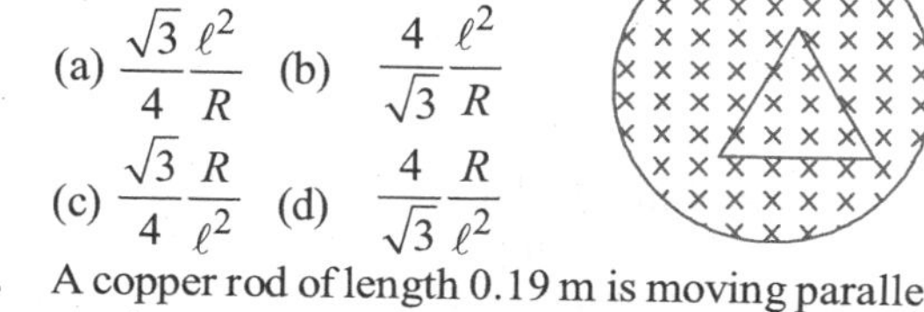

(a) $\frac{\sqrt{3}}{4}\frac{\ell^2}{R}$ (b) $\frac{4}{\sqrt{3}}\frac{\ell^2}{R}$

(c) $\frac{\sqrt{3}}{4}\frac{R}{\ell^2}$ (d) $\frac{4}{\sqrt{3}}\frac{R}{\ell^2}$

36. A copper rod of length 0.19 m is moving parallel to a long wire with a uniform velocity of 10 m/s. The long wire carries 5 ampere current and is perpendicular to the rod. The ends of the rod are at distances 0.01 m and 0.2 m from the wire. The emf induced in the rod will be- **Critical Thinking**

(a) 10 μV (b) 20 μV

(c) 30 μV (d) 40 μV

37. A metallic square loop $ABCD$ is moving in its own plane with velocity v in a uniform magnetic field perpendicular to its plane as shown in the figure. An electric field is induced

(a) in AD, but not in BC

(b) in BC, but not in AD

(c) neither in AD nor in BC

(d) in both AD and BC

38. A conducting wire of mass m slides down two smooth conducting bars, set at an angle θ to the horizintal as shown in Fig. The separation between the bars is l. The system is located in the magnetic field B, perpendicular to the plane of the sliding wire and bars. The constant velocity of the wire is **Toughnut**

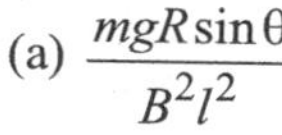

(a) $\frac{mgR\sin\theta}{B^2l^2}$

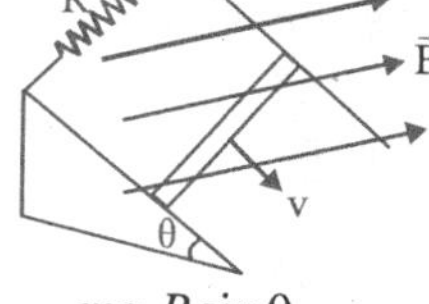

(b) $\frac{mgR\sin\theta}{Bl^3}$

(c) $\frac{mg\,R\sin\theta}{B^2l^5}$ (d) $\frac{mg\,R\sin\theta}{Bl^4}$

39. A plane loop, shaped as two squares of sides a = 1 m and b = 0.4 m is introduced into a uniform magnetic field ⊥ to the plane of loop. The magnetic field varies as B = 10^{-3} sin (100t) T. The amplitude of the current induced in the loop if its resistance per unit length is r = 5 mΩ m^{-1} is **Critical Thinking**

(a) 2 A

(b) 3 A

(c) 4 A

(d) 5 A

40. PQ is an infinite current carrying conductor. AB and CD are smooth conducting rods on which a conductor EF moves with constant velocity v as shown. The force needed to maintain constant speed of EF is

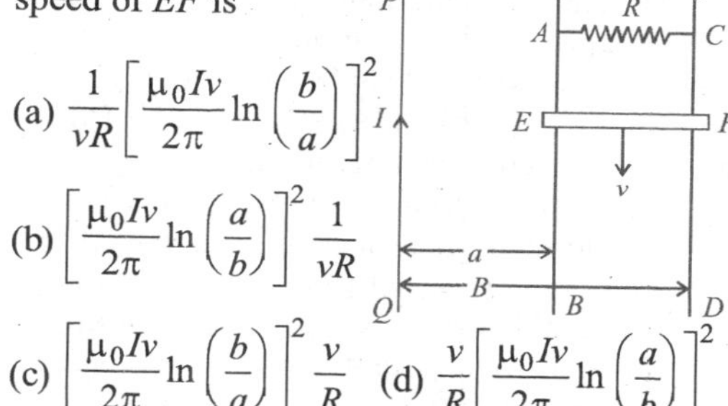

(a) $\frac{1}{vR}\left[\frac{\mu_0 Iv}{2\pi}\ln\left(\frac{b}{a}\right)\right]^2$

(b) $\left[\frac{\mu_0 Iv}{2\pi}\ln\left(\frac{a}{b}\right)\right]^2\frac{1}{vR}$

(c) $\left[\frac{\mu_0 Iv}{2\pi}\ln\left(\frac{b}{a}\right)\right]^2\frac{v}{R}$ (d) $\frac{v}{R}\left[\frac{\mu_0 Iv}{2\pi}\ln\left(\frac{a}{b}\right)\right]^2$

41. A coil of resistance 400Ω is placed in a magnetic field. If the magnetic flux ϕ (wb) linked with the coil varies with time t (sec) as $\phi = 50t^2 + 4$. The current in the coil at $t = 2$ sec is :

(a) 0.5 A (b) 0.1 A (c) 2 A (d) 1 A

42. A straight conducting metal wire is bent in the given shape and the loop is closed. Dimensions are as shown in the figure. Now the assembly is heated at a constant rate dT/dt = 1°C/s. The assembly is kept in a uniform magnetic field B = 1 T, perpendicular into the paper. Find the current in the loop at the moment, when the heating starts. Resistance of the loop is 10 Ω at any temperature. Coefficient of linear expansion α = 10^{-6}/°C.

(a) 1.5 × 10^{-6} A anticlockwise

(b) 1.5 × 10^{-6} A clockwise

(c) 0.75 × 10^{-6} A anticlockwise

(d) 0.75 × 10^{-6} A clockwise **Toughnut**

43. A rectangular coil of 20 turns and area of cross-section 25 sq. cm has a resistance of 100Ω. If a magnetic field which is perpendicular to the plane of coil changes at a rate of 1000 tesla per second, the current in the coil is

(a) 1 A (b) 50 A (c) 0.5 A (d) 5 A

44. A rectangular loop *PQRS*, is pulled with constant speed into a uniform transverse magnetic field by a force *F* (as shown). E.m.f. induced in side *PS* and potential difference between points *P* and *S* respectively are (Resistance of the loop = *r*)

(a) zero, $\frac{Fr}{B\ell}$

(b) zero, zero

(c) zero, $\frac{Fr}{6B\ell}$

(d) $\frac{Fr}{6B\ell}, \frac{Fr}{6B\ell}$

45. A horizontal ring of radius $r = \frac{1}{2}$ m is kept in a vertical constant magnetic field 1 T. The ring is collapsed from maximum area to zero area in 1s. Then the emf induced in the ring is

(a) 1 V (b) $(\pi/4)$ V (c) $(\pi/2)$ V (d) π V

46. A conducting rod of length l is hinged at point *O*. It is free to rotate in vertical plane. There exists a uniform magnetic field $\vec{B}$ in horizontal direction. The rod is released from position shown in the figure. When rod makes an angle θ from released position then potential difference between two ends of the rod is proportional to :

Critical Thinking

(a) $l^{1/2}$

(b) The lower end will be at a lower potential

(c) sin θ

(d) $(\sin\theta)^{1/2}$

47. An aluminium ring hangs vertically from a thread with its axis pointing east - west. A coil is fixed near to the ring and coaxial with it. What is the initial motion of the aluminium ring when the current in the coil is switched on?

(a) moves toward E (b) moves toward W

(c) moves toward N (d) moves toward S

48. A resistance less ring has 2 bulbs *A* and *B* rated at 2V, 10 W and 2V, 20 W respectively. The ring encloses an ideal solenoid whose magnetic field is as shown. The radius of solenoid is 1 m and the number of turns/length = 1000/m. The current changes at rate of 9 A/sec

Find the value of *P* if power dissipated in bulb *B* is $1.8\,P \times 10^{-4}$ watt.

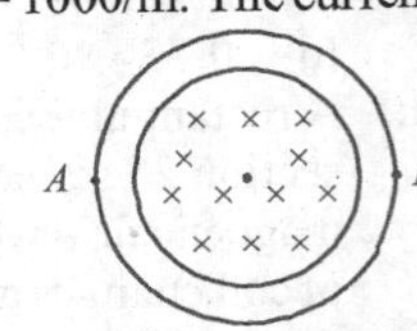

(a) 4 (b) 6

(c) 8 (d) 11

Toughnut

49. A cylindrical region of radius 1 m has instantaneous homogenous magnetic field of 5T and it is increasing at a rate of 2T/s. A regular hexagonal loop ABCDEFA of side 1 m is being drawn in to the region with a constant speed of 1 m/s as shown in the figure. What is the magnitude of emf developed in the loop just after the shown instant when the corner A of the hexagon is coinciding with the centre of the circle ?

Tricky

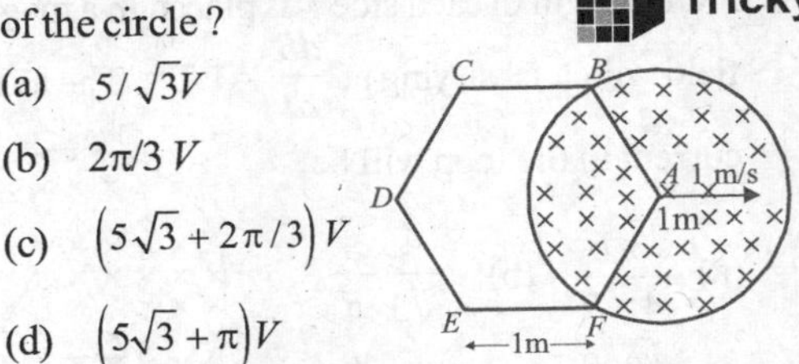

(a) $5/\sqrt{3}V$

(b) $2\pi/3\ V$

(c) $\left(5\sqrt{3} + 2\pi/3\right)V$

(d) $\left(5\sqrt{3} + \pi\right)V$

50. A thin semicircular conducting ring (PQR) of radius 'r' is falling with its plane vertical in a horizontal magnetic field B, as shown in figure. The potential difference developed across the ring when its speed is v, is :

(a) Zero

(b) $Bv\pi r^2/2$ and P is at higher potential

(c) πrBv and R is at higher potential

(d) 2rBv and R is at higher potential

51. A conducting square frame of side 'a' and a long staight wire carrying current I are located in the same plane as shown in the figure. The frame moves to the right with a constant velocity 'V'. The emf induced in the frame will be proportional to

Critical Thinking

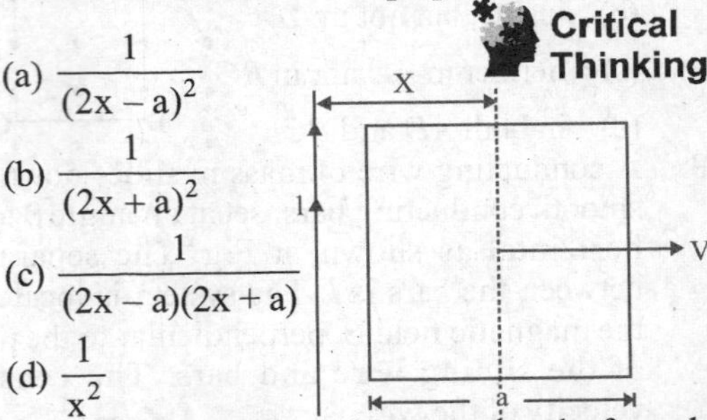

(a) $\frac{1}{(2x - a)^2}$

(b) $\frac{1}{(2x + a)^2}$

(c) $\frac{1}{(2x - a)(2x + a)}$

(d) $\frac{1}{x^2}$

52. The approximate formula expressing the formula of mutual inductance of two thin coaxial loops of the same radius a when their centers are separated by a distance l with $l >>$ a is

(a) $\frac{1}{2}\frac{\mu_0 \pi a^4}{l^3}$ (b) $\frac{1}{2}\frac{\mu_0 \pi a^4}{l^2}$

(c) $\frac{\mu_0}{4\pi}\frac{\pi a^4}{l^2}$ (d) $\frac{\mu_0}{\pi}\frac{a^4}{l^3}$

53. A bar magnet was pulled away from a hollow coil A as shown in fig. As the south pole came out of the coil, the bar magnet next to hollow coil B experienced a magnetic force

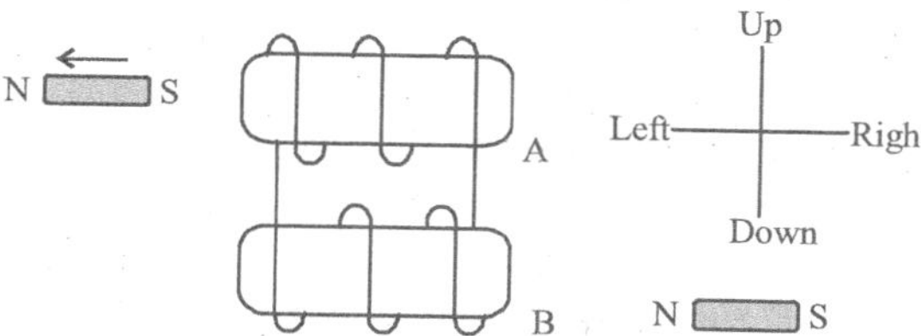

(a) to the right (b) to the left
(c) upward (d) equal to zero

54. A superconducting loop of radius R has self inductance L. A uniform and constant magnetic field B is applied perpendicular to the plane of the loop. Initially current in this loop is zero. The loop is rotated by 180°. The current in the loop after rotation is equal to

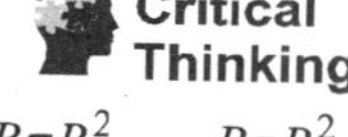

(a) zero (b) $\frac{B\pi R^2}{L}$ (c) $\frac{2B\pi R^2}{L}$ (d) $\frac{B\pi R^2}{2L}$

55. Two inductances L_1 and L_2 are placed closer and in parallel. Their combined inductance is

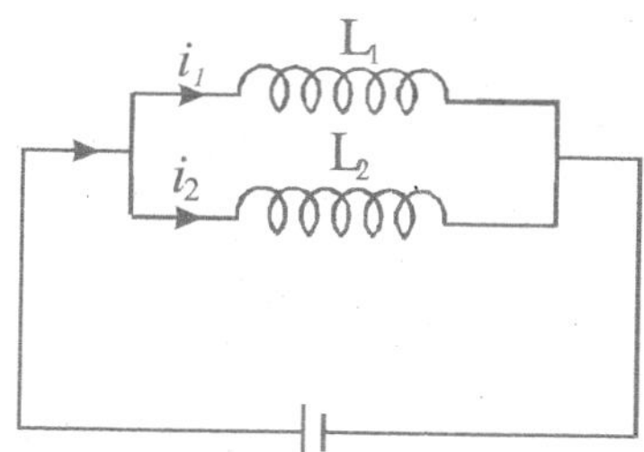

(a) $\frac{L_1L_2}{L_1+L_2}$ (b) (L_1+L_2)

(c) $(L_1+L_2)\frac{L_1}{L_2}$ (d) $(L_1+L_2)\frac{L_2}{L_1}$

56. Two coils are at fixed locations. Which coil 1 has no current and the current in coil 2 increases at the rate of 15.0 A s^{-1}, the emf in coil 1 is 25 mV, when coil 2 has no current and coil 1 has a current of 3.6 A, the flux linkage on coil 2 is

(a) 16 m Wb (b) 10 m Wb
(c) 4.00 m Wb (d) 6.00 mWb

57. What is the self-inductance of a coil which produces 5V when the current changes from 3 ampere to 2 ampere in one millisecond?

(a) 5000 henry (b) 5 milli-henry
(c) 50 henry (d) 5 henry

58. Figure shows a rectangular coil near a long wire. The mutual inductance of the combination is

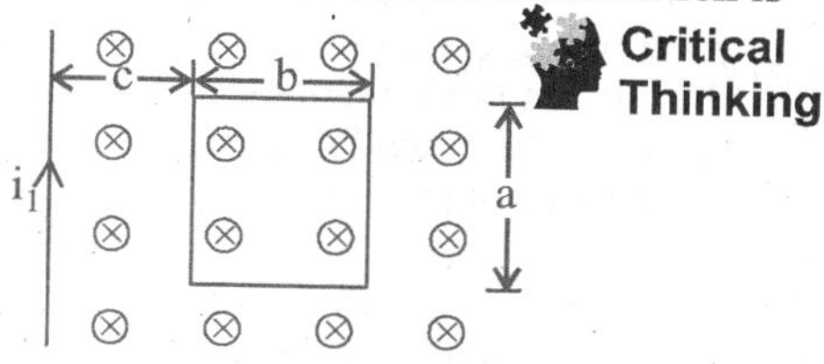

(a) $\frac{\mu_0 a}{2\pi} ln\left(1-\frac{b}{c}\right)$ (b) $\frac{\mu_0 a}{2\pi} ln\left(1+\frac{b}{c}\right)$

(c) $\frac{\mu_0 a}{\pi} ln\left(1+\frac{b}{c}\right)$ (d) $\frac{\mu_0 a}{\sqrt{2}\pi} ln\left(1+\frac{b}{c}\right)$

59. In an inductor of self-inductance L = 2 mH, current changes with time according to relation $i = t^2e^{-t}$. At what time emf is zero?

(a) 4s (b) 3s (c) 2s (d) 1s

60. A varying current in a coil changes from 10A to zero in 0.5 sec. If the average e.m.f induced in the coil is 220V, the self-inductance of the coil is

(a) 5 H (b) 6 H (c) 11 H (d) 12 H

61. A current of 1.5 A flows through a solenoid of length 20.0 cm, cross-section 20.0 cm^2 and 400 turns. The current is suddenly switched off in a short time of 1.0 millisecond. Ignoring the variation in the magnetic field the ends, the average back emf induced in the solenoid is:

(a) 0.3 V (b) 9.6 V (c) 30.0 V (d) 3.0 V

62. Two circular coils can be arranged in any of the three situations shown in the figure. Their mutual inductance will be

Tricky

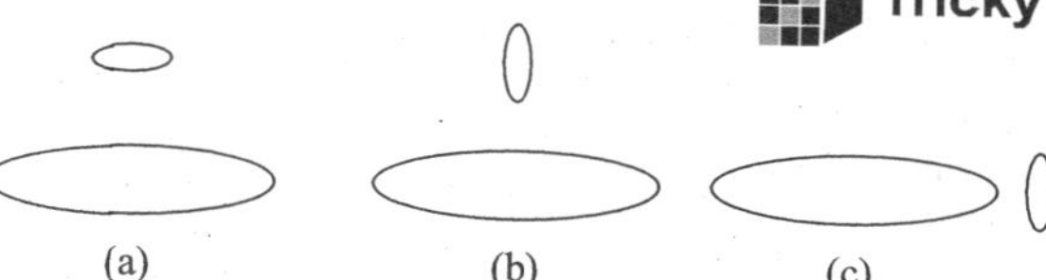

(a) (b) (c)

(a) maximum in situation (a)
(b) maximum in situation (b)
(c) maximum in situation (c)
(d) the same in all situations

63. A small square loop of wire of side ℓ is placed inside a large square loop of wire of side L $(L > \ell)$. The loop are coplanar and their centre coincide. The mutual inductance of the system is proportional to

(a) ℓ / L (b) ℓ^2 / L (c) L / ℓ (d) L^2 / ℓ

64. Two coils, one primary of 500 turns and one secondary of 25 turns, are wound on an iron ring of mean diameter 20 cm and cross-sectional area 12 cm^2. If the permeability of iron is 800, the mutual inductance is :

(a) 0.48 H (b) 2.4 H (c) 0.12 H (d) 0.24 H

65. When the current in a certain inductor coil is 5.0 A and is increasing at the rate of 10.0 A/s, the potential difference across the coil is 140V. When the current is 5.0 A and decreasing at the rate of 10.0 A/s, the potential difference is 60V. The self inductance of the coil is – **Critical Thinking**

(a) 2H (b) 4H (c) 8H (d) 12H

66. A wire of fixed lengths is wound on a solenoid of length ℓ and radius r. Its self inductance is found to be L. Now if same wire is wound on a solenoid of length $\ell/2$ and radius r/2, then the self inductance will be –

(a) 2L (b) L (c) 4L (d) 8L

67. A rectangular loop has a sliding connector PQ of length l and resistance R Ω and it is moving with a speed v as shown. The set-up is placed in a uniform magnetic field going into the plane of the paper. The three currents I_1, I_2 and I are **Tricky**

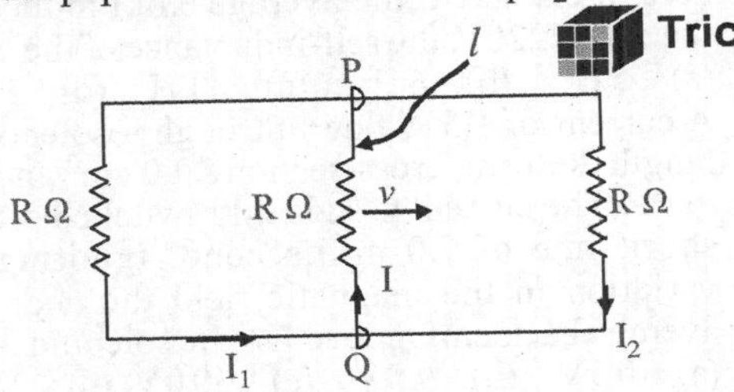

(a) $I_1 = -I_2 = \frac{Blv}{6R}, I = \frac{2Blv}{6R}$

(b) $I_1 = I_2 = \frac{Blv}{3R}, I = \frac{2Blv}{3R}$

(c) $I_1 = I_2 = I = \frac{Blv}{R}$

(d) $I_1 = I_2 = \frac{Blv}{6R}, I = \frac{Blv}{3R}$

68. A boat is moving due east in a region where the earth's magnetic field is 5.0×10^{-5} NA^{-1} m^{-1} due north and horizontal. The boat carries a vertical aerial 2 m long. If the speed of the boat is 1.50 ms^{-1}, the magnitude of the induced emf in the wire of aerial is **Tricky**

(a) 0.75 mV (b) 0.50 mV (c) 0.15 mV (d) 1mV

79. A horizontal straight wire 20 m long extending from east to west falling with a speed of 5.0 m/s, at right angles to the horizontal component of the earth's magnetic field 0.30×10^{-4} Wb/m^2. The instantaneous value of the e.m.f. induced in the wire will be

(a) 3mV (b) 4.5mV (c) 1.5mV (d) 6.0mV

70. A coil is suspended in a uniform magnetic field, with the plane of the coil parallel to the magnetic lines of force. When a current is passed through the coil it starts oscillating; It is very difficult to stop. But if an aluminium plate is placed near to the coil, it stops. This is due to :

(a) developement of air current when the plate is placed
(b) induction of electrical charge on the plate
(c) shielding of magnetic lines of force as aluminium is a paramagnetic material.
(d) electromagnetic induction in the aluminium plate giving rise to electromagnetic damping.

71. A long solenoid has 500 turns. When a current of 2 ampere is passed through it, the resulting magnetic flux linked with each turn of the solenoid is 4×10^{-3} Wb. The self- inductance of the solenoid is

(a) 2.5 henry (b) 2.0 henry
(c) 1.0 henry (d) 40 henry

72. A circular coil is radius 5 cm has 500 turns of a wire. The approximate value of the coefficient of self induction of the coil will be-

(a) 25mH (b) 25×10^{-3} mH
(c) 50×10^{-3} mH (d) 50×10^{-3} H

73. A uniform coil of self-inductance 1.8×10^{-4} H and resistance 6W is broken up into two identical coils. These two coils are then connected in parallel across a 12 V battery. The circuit time constant and steady state current through the battery respectively are :

(a) 30 μs, 8 A (b) 30 ms, 8 mA
(c) 30 s, 8 A (d) 300 s, 800 A

74. The back e.m.f. in a d.c. motor is maximum, when

(a) the motor has picked up max speed
(b) the motor has just started moving
(c) the speed of motor is still on the increase
(d) the motor has just been switched off

75. A motor having an armature of resistance 2Ω is designed to operate at 220 V mains. At full speed, it develops a back emf of 210V. When the motor is running at full speed, the current in the armature is:

(a) 3 A (b) 5 A (c) 7 A (d) 10 A

76. A simple electric motor has an armature resistance of 1 Ω and runs from a dc source of 12 volt. When running unloaded it draws a current of 2 amp. When a certain load is connected, its speed becomes one-half of its unloaded value. What is the new value of current drawn? **Critical Thinking**

(a) 7A (b) 3A (c) 5A (d) 4A

77. When a metallic plate swings between the poles of a magnet
(a) no effect on the plate
(b) eddy currents are set up inside the plate and the direction of the current is along the motion of the plate
(c) eddy currents are set up inside the plate and the direction of the current opposes the motion of the plate
(d) eddy currents are set up inside the plate

78. An electric potential difference will be induced between the ends of the conductor as shown in the diagram, when the conductor moves in the direction

M, L, N, Q, S, P

(a) P (b) Q
(c) L (d) M

79. A rectangular coil of 200 turns of wire 15×40 cm^2 makes 50 r.p.s. about an axis in its plane parallel to its longer side and perpendicular to a magnetic field of intensity 0.08 Wb/m^2. what is the instantaneous value of induced e.m.f. when the plane of the coil makes an angle with magnetic field of 45°
(a) 213.3 V (b) 301.7 V (c) 151.5 V (d) zero

80. A generator has an e.m.f. of 440 Volt and internal resistance of 4000 hm. Its terminals are connected to a load of 4000 ohm. The voltage across the load is
(a) 220 volt (b) 440 volt
(c) 200 volt (d) 400 volt

81. A metallic rod of length 'ℓ' is tied to a string of length 2ℓ and made to rotate with angular speed ω on a horizontal table with one end of the string fixed. If there is a vertical magnetic field 'B' in the region, the e.m.f. induced across the ends of the rod is **Toughnut**

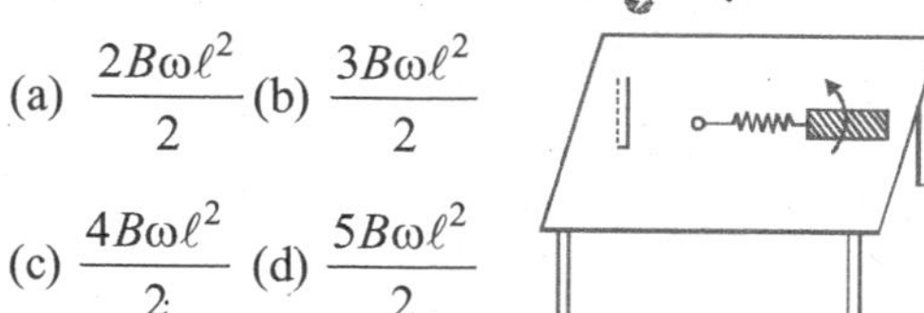

(a) $\frac{2B\omega\ell^2}{2}$ (b) $\frac{3B\omega\ell^2}{2}$

(c) $\frac{4B\omega\ell^2}{2}$ (d) $\frac{5B\omega\ell^2}{2}$

82. A circular loop of radius 0.3 cm lies parallel to amuch bigger circular loop of radius 20 cm. The centre of the small loop is on the axis of the bigger loop. The distance between their centres is 15 cm. If a current of 2.0 A flows through the smaller loop, then the flux linked with bigger loop is
(a) 9.1×10^{-11} weber (b) 6×10^{-11} weber
(c) 3.3×10^{-11} weber (d) 6.6×10^{-9} weber

83. In a coil of resistance 100 Ω, a current is induced by changing the magnetic flux through it as shown in the figure. The magnitude of change in flux through the coil is **Critical Thinking**

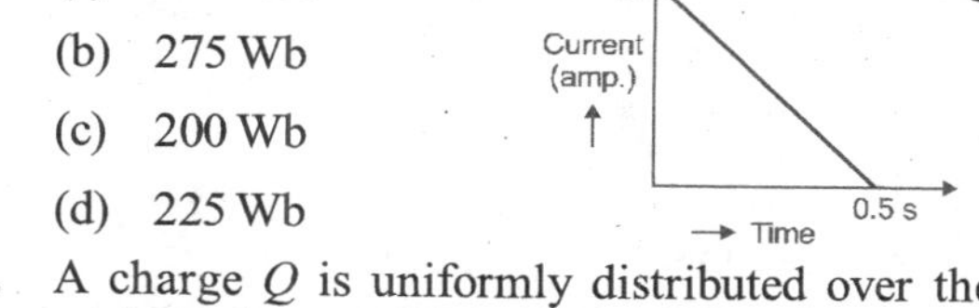

(a) 250 Wb
(b) 275 Wb
(c) 200 Wb
(d) 225 Wb

84. A charge Q is uniformly distributed over the surface of non-conducting disc of radius R. The disc rotates about an axis perpendicular to its plane and passing through its centre with an angular velocity ω. As a result of this rotation a magnetic field of induction B is obtained at the centre of the disc. If we keep both the amount of charge placed on the disc and its angular velocity to be constant and vary the radius of the disc then the variation of the magnetic induction at the centre of the disc will be represented by the figure : **Tricky**

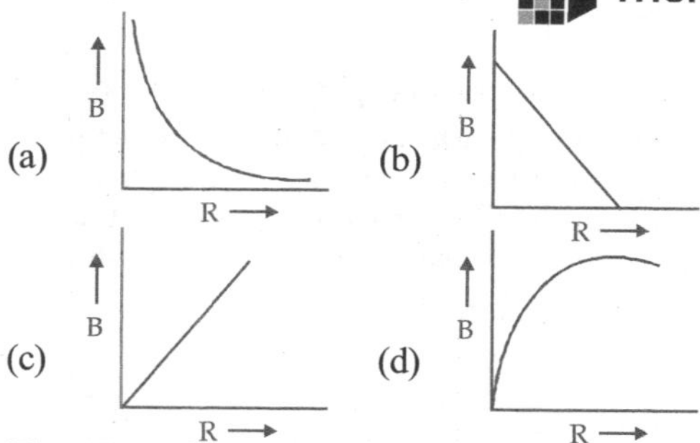

85. The two rails of a railway track, insulated from each other and the ground, are connected to a milli voltmeter. What is the reading of the milli voltmeter when a train travels at a speed of 180 km/hour along the track, given that the vertical component of earth's magnetic field is 0.2×10^{-4} weber/m^2 and the rails are separated by 1 m ?
(a) 1 (b) 3 (c) 5 (d) 7

86. A square metal wire loop of side 10 cm and resistance 1 ohm is moved with a constant velocity v_0 in a uniform magnetic field of induction $B = 2$ webers/m^2 as shown in the figure. The magnetic field lines are perpendicular to the plane of the loop (directed into the paper). The loop is connected to a network of resistors each of value 3 ohms. The resistances of the lead wires OS and PQ are negligible. What should be the speed of the loop so as to have a steady current of 1 milliampere in the loop ? Give the direction of current in the loop. **Toughnut**

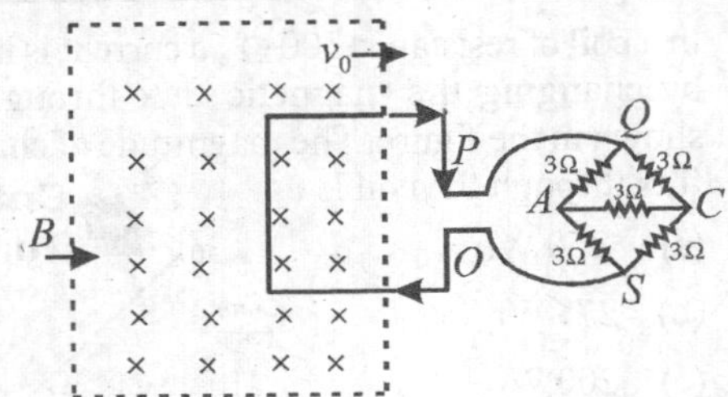

(a) 2.3 m/s (b) 1.2 m/s
(c) 0.3 m/s (d) 0.02 m/s

87. A conducting ring of radius r with a conducting spoke OA is in pure rolling on a horizontal surface in a region having a uniform magnetic field B as shown, v being the velocity of the centre of the ring. Then the potential difference $V_0 - V_A$ is:

(a) $\frac{Bvr}{2}$ (b) $\frac{3Bvr}{2}$

(c) $\frac{Bvr}{2}$ (d) $\frac{-3Bvr}{2}$

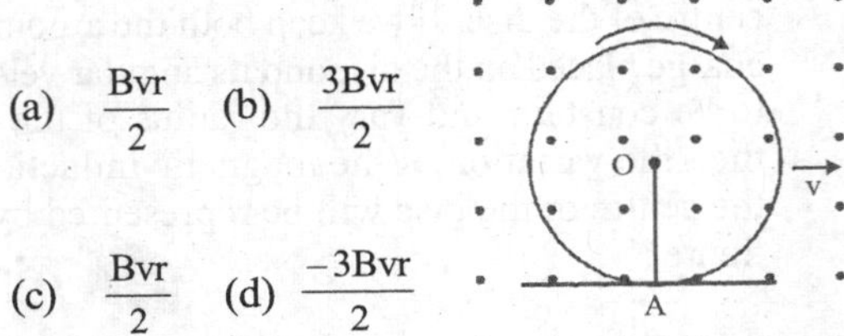

88. The figure below depicts a circular loop of radius R carrying a fixed current I. The upper half of the loop is placed in a uniform magnetic field of magnitude B, perpendicular to the plane of the paper, as shown. The magnitude of the force on the loop is (neglect gravity)

(a) BIR
(b) 2BIR
(c) πBIR
(d) 2πBIR

89. Two identical circular current carrying coils 1 and 2, each of radius R are placed adjacently a distance d apart ($d << R$). They carry equal currents I in opposite sense. Let L_1, L_2, M_{12}, and M_{21} denote the self and mutual inductances. Which of the following is at least approximately true?

(a) $|L_1| = |M_{21}|$
(b) $M_{21} > L_1$
(c) $L_1 = -L_2$
(d) $M_{12} = -M_{21}$

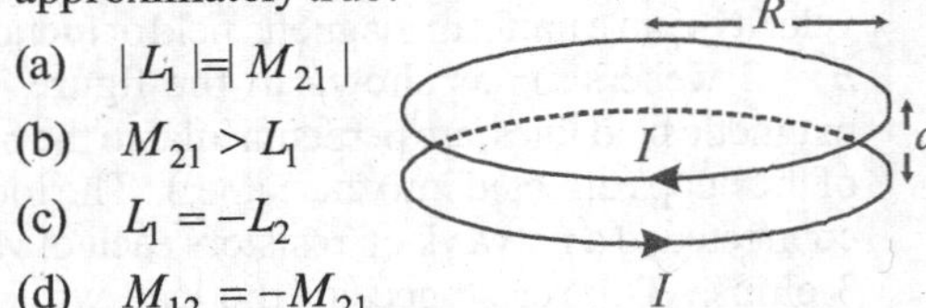

90. An infinitesimally small bar magnet of dipole moment $\vec{M}$ is pointing and moving with the speed v in the $\hat{x}$-direction. A small closed circular conducting loop of radius a and negligible self-inductance lies in the y-z plane with its center at $x = 0$, and its axis coinciding with the x-axis. Find the force opposing the motion of the magnet, if the resistance of the loop is R. Assume that the distance x of the magnet from the center of the loop is much greater than a.

(a) $\frac{21}{4}\frac{\mu_0^2 M^2 a^4 v}{Rx^8}$ (b) $\frac{16}{3}\frac{\mu_0 M^2 a^2 v^2}{Rx^3}$

(c) $\frac{3}{23}\frac{\mu_0 Mav^2}{Rx^3}$ (d) None of these

91. ABCD is a wire frame in the shape of an isosceles trapezium (i.e., length AB = length CD) enter a magnetic field with flux density B at t = 0 as shown in the figure. If the total resistance of wire frame is R. What is the value of the induced current in the wire frame after t seconds, assuming that the frame has to entered the field completely by then? [v = Velocity of frame]

(a) zero

(b) $\frac{Bv}{R}(2vt\sin\theta + \ell)$

(c) $\frac{Bv}{R}\left(\frac{2vt}{\tan\theta} + \ell\right)$ (d) $\frac{B}{v}$

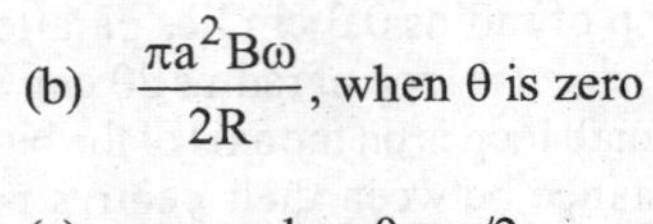

92. The magnetic flux φ linked with a conducting coil depends on time as $\phi = 4t^n + 6$, where n is a positive constant. The induced emf in the coil is e. Then which is wrong?

(a) If $0 < n < 1$, $e \neq 0$ and $|e|$ decreases with time
(b) If $n = 1$, e is constant
(c) If $n > 1$, $|e|$ increases with time
(d) If $n > 1$, $|e|$ decreases with time

93. A uniform circular loop of radius a and resistance R is placed perpendicular to a uniform magnetic field B. One half of the loop is rotated about the diameter with angular velocity ω as shown in Fig. Then, the current in the loop is

Critical Thinking

(a) $\frac{\pi a^2 B\omega}{4R}$, when θ is zero

(b) $\frac{\pi a^2 B\omega}{2R}$, when θ is zero

(c) zero, when θ = π/2

(d) $\frac{\pi a^2 B\omega}{2R}$, when θ = π/2

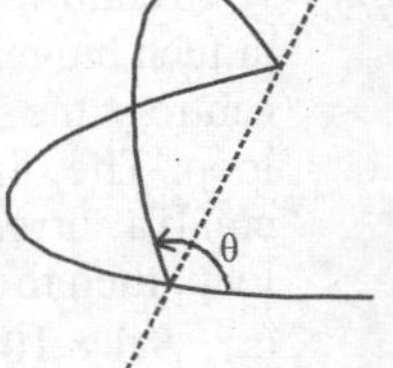

94. A metal rod moves at a constant velocity in a direction perpendicular to its length. A constant, uniform magnetic field exists in space in a direction perpendicular to the rod as well as its velocity. Select the correct statement from the following.
(a) The entire rod is at the same electric potential.
(b) There is an electric field in the rod.
(c) The electric potential is highest at the centre of the rod and decreases towards its ends.
(d) The electric potential is lowest at the centre of the rod, and increases towards its ends

95. A nonconducting ring of mass m and radius R has a charge Q uniformly distributed over its circumference. The ring is placed on a rough horizontal surface such that the plane of the ring is parallel to the surface. A vertical magnetic field $B = B_0t^2$ tesla is switched on. After 2 second from switching on the magnetic field the ring is just about to rotate about vertical axis through its centre. Then
(a) the induced electric field is quadratic in time t
(b) the force tangential to the ring is $9B_0QRt$
(c) until 2 seconds, the friction force does not come into play
(d) the friction coefficient between the ring and the surface is $\frac{2B_0RQ}{mg}$

96. There are three wire MO, NO and PQ, wires MO and NO are fixed and perpendicular to each other. Wire PQ moves with a constant velocity v as shown in the figure and resistance per unit length of each wire is λ and magnetic field exists perpendicular and inside the paper then

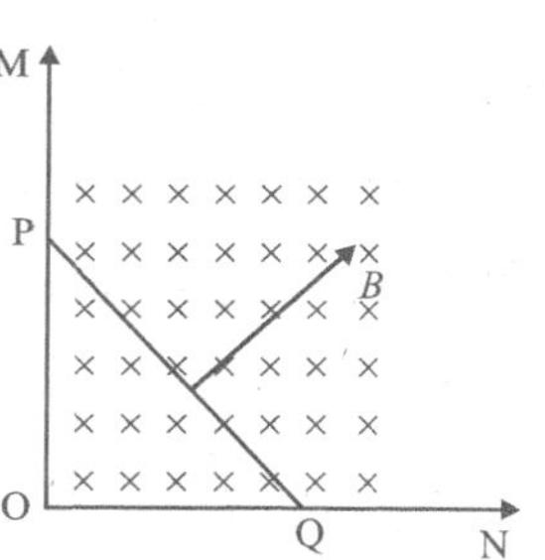

Which of the following is wrong?
(a) current in loop is anticlockwise
(b) magnitude of current in the loop is $\frac{Bv}{\lambda(\sqrt{2}+1)}$
(c) current in the loop is independent of time.
(d) magnitude of current decreases as time increases.

97. A (current versus time) graph of the current passing through a solenoid is shown in figure. If the back emf at t = 3s is e, find the back emf at t = 7 s

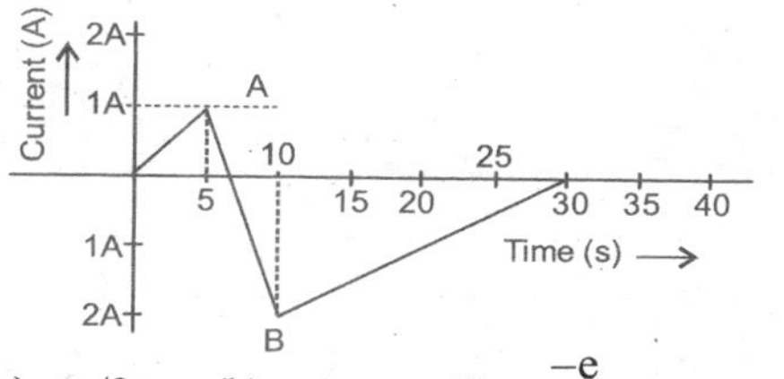

(a) e/2 (b) 0 (c) $\frac{-e}{2}$ (d) – 3e

98. The incorrect statements is/are
(a) The resistance offered by an inductor in a d.c circuit is always constant.
(b) The resistance of inductor in steady state is zero.
(c) An inductor is connected to a battery through a switch. The emf induced in the inductor is much larger when the switch is opened as compared to the emf induced when the switch is closed.
(d) To reduce the rate of increases of current through a solenoid should increase the time constant $\left(\frac{L}{R}\right)$.

99. In a conducting cycle wheel (of n spokes), each spoke of length ℓ is rotating with angular speed ω in uniform perpendicular magnetic field B. If due to flux cutting each metal spoke behaves as an identical cell of emf(e) then net emf of the system is **Critical Thinking**

(a) $nB\omega l^2$ (b) $\frac{1}{2}nB\omega l^2$ (c) $\frac{1}{2}B\omega l^2$ (d) $B\omega l^2$

100. A rod OA of length l is rotating (about end O) over a conducting ring in crossed magnetic field B with constant angular velocity ω as shown in figure

(a) Current flowing through the rod is $\frac{B\omega l^2}{R}$
(b) Magnetic force acting on the rod is $\frac{3B^2\omega l^3}{4R}$
(c) Torque due to magnetic force acting on the rod is $\frac{3B^2\omega l^4}{8R}$
(d) Magnitude of external force that acts perpendicularly at the end of the rod to maintain the constant angular speed is $\frac{3B^2\omega l^3}{5R}$

Answer KEYs

1	(a)	11	(d)	21	(d)	31	(c)	41	(a)	51	(c)	61	(a)	71	(c)	81	(d)	91	(c)
2	(a)	12	(d)	22	(a)	32	(c)	42	(a)	52	(a)	62	(a)	72	(a)	82	(a)	92	(d)
3	(c)	13	(d)	23	(c)	33	(c)	43	(c)	53	(a)	63	(b)	73	(a)	83	(a)	93	(d)
4	(b)	14	(b)	24	(d)	34	(b)	44	(c)	54	(c)	64	(d)	74	(a)	84	(a)	94	(b)
5	(a)	15	(c)	25	(d)	35	(a)	45	(b)	55	(a)	65	(b)	75	(b)	85	(a)	95	(d)
6	(d)	16	(c)	26	(a)	36	(c)	46	(d)	56	(d)	66	(a)	76	(a)	86	(d)	96	(d)
7	(a)	17	(a)	27	(c)	37	(d)	47	(a)	57	(b)	67	(b)	77	(c)	87	(c)	97	(d)
8	(a)	18	(d)	28	(a)	38	(a)	48	(a)	58	(b)	68	(c)	78	(d)	88	(b)	98	(a)
9	(c)	19	(b)	29	(b)	39	(b)	49	(c)	59	(c)	69	(a)	79	(a)	89	(a)	99	(c)
10	(d)	20	(d)	30	(b)	40	(a)	50	(d)	60	(c)	70	(d)	80	(d)	90	(a)	100	(c)

Hints & Solutions

1. **(a)** At any time t, the side of the square $a = (a_0 - \alpha t)$, where a_0 = side at $t = 0$.
At this instant, flux through the square :
$\phi = BA \cos 0° = B(a_0 - \alpha t)^2$
$\therefore$ emf induced $E = -\dfrac{d\phi}{dt}$
$\Rightarrow E = -B.2(a_0 - \alpha t)(0 - \alpha) = +2\alpha a B$

2. **(a)** Apply Lenz's law such that the net force on loop will be towards the infinite long wire.

3. **(c)**

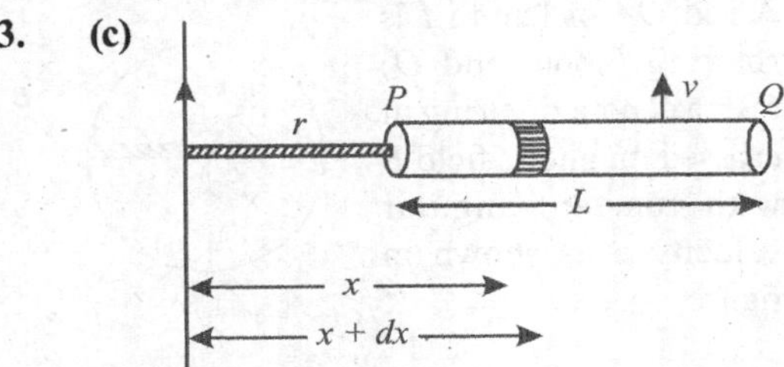

Consider a small element of length dx of the rod at a distance x and $(x + dx)$ from the wire.
The emf induced across the element

$$de = B\,v\,dx \quad \text{...(i)}$$

We know that magnetic field B at a distance x from a wire carrying a current i is given by

$$B = \frac{\mu_0}{2\pi} \cdot \frac{i}{x} \quad \text{...(ii)}$$

From eqs. (i) and (ii),

$$de = \frac{\mu_0 i}{2\pi x} v\,dx \quad \text{...(iii)}$$

The emf e induced in the entire length of the rod PQ is given by

$$e = \int de = \int_P^Q \frac{\mu_0}{2\pi}\frac{i}{x} v\,dx$$

$$= \int_r^{r+L} \frac{\mu_0}{2\pi}\frac{i}{x} v\,dx = \frac{\mu_0}{2\pi} i\,v \int_r^{r+L} \frac{dx}{x}$$

$$= \frac{\mu_0 i v}{2\pi} \log_e\left(\frac{r+L}{r}\right)$$

4. **(b)** The change in flux $= n(\phi_2 - \phi_1)$
Total resistance of the circuit $= 4R + R = 5R$
Thus induced current, $i = \dfrac{\Delta\phi/\Delta t}{R} = \dfrac{n(\phi_2 - \phi_1)}{5RT}$.

5. **(a)** Consider an elemental circle of thickness dr.
The induced emf in the circular path of radius r is

$$\varepsilon = \frac{d}{dt}(\pi r^2 B) = \pi r^2 \alpha$$

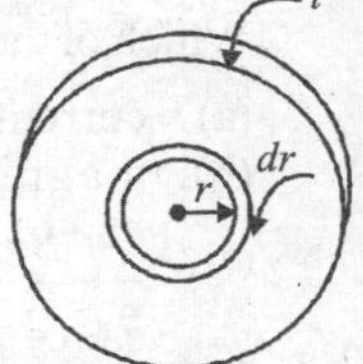

The resistance of circular path is

The length of the path being $2\pi r$ and tdr is the cross sectional area of current flow. For the element the power dissipated inside the path is

$$dP = \frac{\varepsilon^2}{R} = \frac{\pi t \sigma}{2} \alpha^2 r^3 dr$$

The total dissipated power P is

$$P = \frac{\pi t \sigma}{2}\alpha^2 \int_0^a r^3 dr = \frac{\pi t \sigma a^4}{8}\alpha^2$$

6. (d) $e = -\frac{d\phi}{dt} - -A\frac{dB}{dt} = -\pi r^2 \frac{d}{dt}(B_0 e^{-t}) = \pi r^2 B_0 e^{-t}$

At $t = 0$, $e = \pi r^2 B_0$, $P = \frac{e^2}{R} = \frac{(\pi r^2 B_0)^2}{R} = \frac{\pi^2 r^4 B_0^2}{R}$

7. (a) $\vec{\ell}, \vec{v}$, and $\vec{B}$ are coplanar.

8. (a)

9. (c) As I increases, ϕ increases
$\therefore$ I_i is such that it opposes the increases in ϕ. Hence, ϕ decreases (By Right Hand Rule). The induced current will be counter-clockwise.

10. (d) $\phi = BA\cos\theta = 2.0 \times 0.5 \times \cos 60^\circ$
$= \frac{2.0 \times 0.5}{2} = 0.5$ weber.

11. (d) $\phi = 6t^2 + 7t + 1 \Rightarrow \frac{d\phi}{dt} = 12t + 7$
At time, t = 2 sec.
$\frac{d\phi}{dt} = 24 + 7 = 31$ mv
Direction of current is from left to right according to Flemings right hand rule.

12. (d) Given that $\phi = at(T - t)$
Induced emf, $E = \frac{d\phi}{dt} = \frac{d}{dt}[at(T - t)]$
$= at(0 - 1) + a(T - t) = a(T - 2t)$
So, induced emf is also a function of time.
$\therefore$ Heat generated in time T is

$$H \int_0^T \frac{E^2}{R} dt = \frac{a^2}{R}\int_0^T (T - 2t)^2 dt = \frac{a^2 T^3}{3R}$$

13. (d) Induced emf in the loop is given by
$e = -B.\frac{dA}{dt}$ where A is the area of the loop.
$e = -B.\frac{d}{dt}(\pi r^2) = -B\pi 2r\frac{dr}{dt}$
$r = 2\text{cm} = 2 \times 10^{-2}$ m
$dr = 2$ mm $= 2 \times 10^{-3}$ m, dt = 1s

$$e = -0.04 \times 3.14 \times 2 \times 2 \times 10^{-2} \times \frac{2 \times 10^{-3}}{1} V$$
$$= 0.32\pi \times 10^{-5} V = 3.2\pi \times 10^{-6} V = 3.2\pi\mu V$$

14. (b) $I(t) = \sqrt{\frac{m}{L}} v_0 \sin\omega t$ where $\omega = \frac{B\ell}{\sqrt{mL}}$

$$\Rightarrow I = \sqrt{\frac{mv_0^2}{2L}}$$

15. (c) $e_{ind} = \frac{d\phi}{dt} = \frac{d}{dt}(BA) = A\frac{dB}{dt}$
$= A\frac{d}{dt}(4t^2 + 2t + 5) = A(8t + 2) = 0.06(8t + 2)$
$= 2.52$ at $t = 5$
$i = \frac{2 - e_{ind}}{2} = 0.26\,A$

16. (c) Induced emf. $|e| = \frac{d\phi}{dt} = A\left(\frac{dB}{dt}\right)$
$= \frac{2 \times 2}{2} \times \frac{d}{dt}(0.042 - 0.87t) = 2 \times 0.87 = 1.74$ V
The net induced emf = 20 + 1.74 = 21.74 V

17. (a) 18. (d)

19. (b) $I = \frac{\varepsilon}{R} = \frac{|d\phi/dt|}{R} = \frac{B.\frac{dA}{dt}}{R}$
$= \frac{B\frac{d}{dt}\left(\frac{1}{2}h.b\right)}{R} \propto \frac{Bb\frac{dh}{dt}}{R} \propto Bbv$
$\therefore b \propto t \Rightarrow I \propto t$

20. (d) $I = \frac{2Brv}{R/2} = \frac{4Brv}{R}$

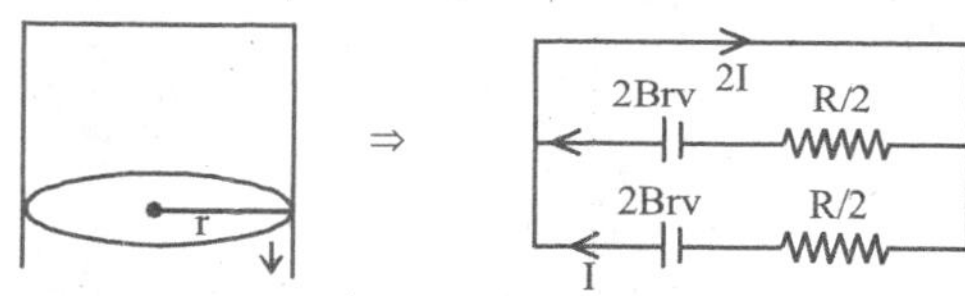

Current in the top horizontal $= 2I = \frac{8Brv}{R}$.

21. (d) Let radius of the loop is r at any time t and in further time dt, radius increases by dr.
The change in flux : $d\phi = (2\pi r dr)B$
$\Rightarrow e = \frac{d\phi}{dt} = 2\pi r\left(\frac{dr}{dt}\right)\frac{k}{r}$
$\Rightarrow e = 2\pi ck$ (constant)
$\left[\because \frac{dr}{dt} = c, B = \frac{k}{r}\right]$
The change in flux :
$d\phi = (2\pi r dr)B$

22. (a)

23. (c) The magnetic flux through the coil will be zero, and so induced emf in it will be zero.

24. (d) Since all the wires are connected between rim and axle, they will generate induced emf in parallel, hence it is same for any number of spokes.

25. (d) At $x = 0$, $B = B_0$, so $e_1 = B_0 v_0 d$

At, $x=d$, $B=B_0\left(1+\frac{d}{a}\right)$, so $e_2=B_0\left(1+\frac{d}{a}\right)v_0 d$

Now $e_{net}=e_2-e_1=\frac{B_0 v_0 d^2}{a}$

26. **(a)** The induced emf across the sliding wire $e=Bv\ell=0.5\times4\times0.25=0.5$ V

The effective circuit is shown in figure.

The equivalent resistance of the circuit

$r=\frac{4\times2}{4+2}+0.5=1.83\ \Omega$

Now, $i=\frac{V}{R}=\frac{0.5}{1.83}=0.27A$

27. **(c)** For maximum potential difference the velocity must be perpendicular to the line AB, and so

$\vec{v}=v\sin\theta\hat{i}+v\cos\theta\hat{j}$

$=v\left[\frac{3}{\ell}\hat{i}+\frac{2}{\ell}\hat{j}\right]=\frac{v}{\ell}[3\hat{i}+2\hat{j}]$

28. **(a)** Velocity at any time t,

$v=at=\frac{F}{m}t$

Now, $e=V_P-V_Q=Bv\ell\sin30°=B\left(\frac{Ft}{m}\right)\ell\times\frac{1}{2}$

$=\frac{BF\ell t}{2m}$

29. **(b)** $Q=\frac{\Delta\phi}{R}=\frac{\phi_2-\phi_1}{R}=\frac{BA-0}{R}=\frac{BA}{R}$

30. **(b)** $P=Fv=BIlv=1.25\times10^{-3}\times50\times0.1\times1W$
$=6.25\times10^{-3}\ W=6.25\ mW$

31. **(c)** By Lenz's law, clockwise current is induced in both loops. Greater the area, large will be the induced emf. Outer loop has greater area.

32. **(c)** $BI\ell=mg$ or $B\frac{Bv\ell}{R}\ell=mg$ or $v=\frac{mgR}{B^2\ell^2}$

33. **(c)** For A : $e_A=\frac{1}{2}B\omega(3\ell)^2$ and

$e_B=\frac{1}{2}B(-\omega)(2\ell)^2$

Resultant induced emf will be

$e=e_A+e_B=\frac{1}{2}B\omega\ell^2(9-4)$ or $e=\frac{5}{2}B\omega\ell^2$

34. **(b)**

35. **(a)** $\phi=\frac{\sqrt{3}}{4}\ell^2 B$

$\varepsilon=\left|\frac{d\phi}{dt}\right|=\frac{\sqrt{3}}{4}\ell^2\frac{dB}{dt}$ $\quad i=\frac{\varepsilon}{R}=\frac{\sqrt{3}\ell^2}{4R}$

36. **(c)** EMF induced in an element of length dx at a distance x from wire $=Bvdx$

$\therefore$ Total EMF induced in the rod

$E=\int_{0.01}^{0.2}Bv\,dx=\int_{0.01}^{0.2}\frac{\mu_0 iv}{2\pi x}dx=\frac{\mu_0 iv}{2\pi}\int_{0.01}^{0.2}\frac{1}{x}dx$

$E=\frac{\mu_0 iv}{2\pi}[\log_e x]_{0.01}^{0.2}=\frac{\mu_0 iv}{2\pi}[\log_{10}(0.2)-\log_{10}(0.01)]\times2.303$

$E=\frac{4\pi\times10^{-7}\times5\times10}{2\pi}[1.301]\times2.303$
$=2.99\times10^{-5}\ V\approx30\mu V$

37. **(d)**

38. **(a)** Component of weight along the inclied plane = mg sin θ

Again, $F=BI\ell=B\frac{B\ell v}{R}\ell=\frac{B^2\ell^2v}{R}$

Now, $\frac{B^2\ell^2v}{R}=mg\sin\theta$ or $v=\frac{mgR\sin\theta}{B^2\ell^2}$

39. **(b)** ϕ (flux linked) $=a^2B\cos0°-b^2B\cos180°$

$E=-\frac{d\phi}{dt}=-\left(a^2-b^2\right)\frac{dB}{dt}$
$=(a^2-b^2)B_0\,\omega\cos\omega t$

where $B=B_0$, sin ωt, $B_0=10^{-3}$ T, $\omega=100$

$\therefore\ I_{max}=\left(a^2-b^2\right)\frac{B_0\omega}{R}$

and $R=(4a+4b)r=4(a+b)r$

$\therefore I_{max}=\frac{(a-b)B_0\omega}{4r}=\frac{(1-0.4)\times10^{-3}\times100}{4\times5\times10^{-3}}=3A$

40. **(a)** Induced emf $=\int_a^b Bvdx=\int_a^b\frac{\mu_0 I}{2\pi x}vdx$

$\Rightarrow$ Induced emf $=\frac{\mu_0 Iv}{2\pi}\ln\left(\frac{b}{a}\right)$

$\Rightarrow$ Power dissipated $=\frac{E^2}{R}$

Also, power $=F.v\Rightarrow F=\frac{E^2}{vR}$

$\Rightarrow F=\frac{1}{vR}\left[\frac{\mu_0 Iv}{2\pi}\ln\left(\frac{b}{a}\right)\right]^2$

41. **(a)**

42. **(a)** Rate of change of area of the loop

$$\frac{dA}{dt} = A,\beta\frac{dT}{dt} = A.(2\alpha)\frac{dT}{dt} = \frac{3}{4}\times 2\times 10^{-6}\times 1$$

$= 11.5 \times 10^{-6} m^2/s$

$$\text{emf} = -\frac{d\phi}{dt} = -\frac{\beta.dA}{dt} = -1.5\times 10^{-6} V$$

current in the loop $= 1.5 \times 10^{-6}$A

The direction will be anticlockwise as the induced current will try to negate the increase in fluix due to increase in area.

43. **(c)** $i = \frac{e}{R} = \frac{\frac{nAdB}{dt}}{R}$

$$= \frac{20\times(25\times 10^{-4})\times 1000}{100} = 0.5\text{A}$$

44. **(c)** There is no charge in flux through PS and so induced emf will be zero. But

$V_P - V_B = V_Q - V_R$

Also, $F = Bil$ or $i = \frac{F}{B\ell}$

$\therefore V_P - V_B = V_Q - V_R = ir_{QR} = \frac{F}{B\ell}\times\frac{r}{6} = \frac{Fr}{6B\ell}$

45. **(b)** $|e| = B\frac{dA}{dt}$

$$|e| = B\frac{(A_2 - A_1)}{(t_2 - t_1)} \Rightarrow |e| = \frac{1\left(\frac{\pi}{4} - 0\right)}{(1-0)}$$

$e = \frac{\pi}{4}$V

46. **(d)** WET from A to B

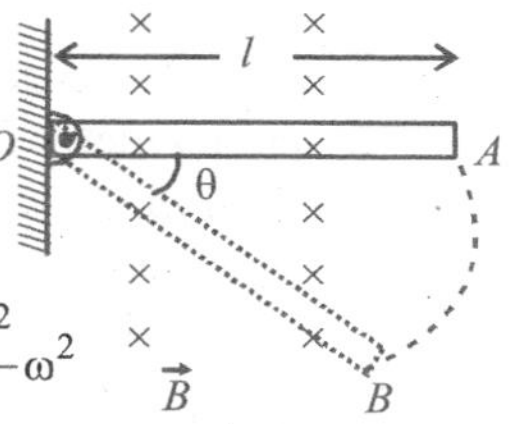

$\omega_{mg} = \Delta K.E.$

$$mg\frac{L}{2}\sin\theta = \frac{1}{2}I\omega^2$$

$$mg\frac{L}{2}\sin\theta = \frac{1}{2}\times\frac{mL^2}{3}\omega^2$$

$$\frac{3g}{L}\sin\theta = \omega^2$$

Induced emf produced in rod is

$$\varepsilon = \frac{1}{2}(B\omega L^2) = \frac{1}{2}B\times\sqrt{\frac{3g\sin\theta}{L}}\times L^2$$

$$\varepsilon = \frac{1}{2}\times B\times\sqrt{\frac{3g\sin\theta\times L^4}{L}} = \frac{1}{2}B\sqrt{3g\sin\theta L^3}$$

$\varepsilon \propto L^{3/2}$ $\varepsilon \propto \sqrt{\sin\theta}$

47. **(a)**

48. **(a)** Resistance of bulb $A = \frac{v^2}{P} = \frac{4}{10} = 0.4$

Resistance of bulb

$$B = \frac{v^2}{P} = 0.2$$

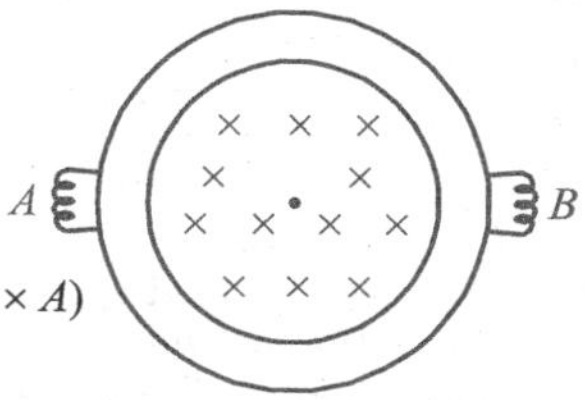

$$\text{emf} = \frac{d\phi}{dt} = \frac{d}{dt}(\mu_0 nI\times A)$$

$$= \mu_0 n\times A\times\frac{dI}{dt} = 10^{-7}\times 4\pi\times 1000\times\pi(1)^2\times 9$$

$v = 36\times 10^{-3}$

$$I = \frac{v}{R_{eq}} = \frac{36\times 10^{-3}}{0.6} = 6\times 10^{-2} A$$

Power dissipated through bulb $B = I^2R$

$= 36 \times 10^{-4} \times 0.2 = 7.2 \times 10^{-4}$ watt.

49. **(c)** The induced emf across the ends B and F due to motion of the loop,

$e_1 = Bv(BF) = 5 \times 1 \times 2 \sin 60° = 5\sqrt{3}\ V$.

The induced emf across the loop due to change in magnetic field

$$e_2 = A\frac{dB}{dt} = \frac{\pi R^2}{3}\left(\frac{dB}{dt}\right) = \frac{\pi(1)^2}{3}\times 2 = \frac{2\pi}{3}\ V.$$

So $e = e_1 + e_2 = \left(5\sqrt{3} + \frac{2\pi}{3}\right)\ V$.

50. **(d)**

51. **(c)** Emf induced in side 1 of frame $e_1 = B_1 V\ell$

$$B_1 = \frac{\mu_o I}{2\pi\,(x - a/2)}$$

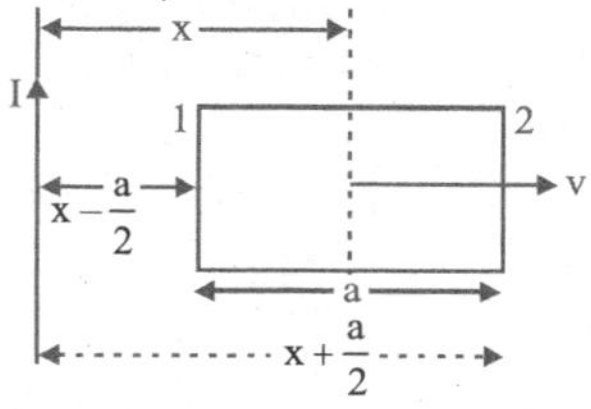

Emf induced in side 2 of frame $e_2 = B_2 V\ell$

$$B_2 = \frac{\mu_o I}{2\pi\,(x + a/2)}$$

Emf induced in square frame

$e = B_1V\ell - B_2V\ell$

$$= \frac{\mu_0 I}{2\pi\,(x - a/2)}\ell v - \frac{\mu_0 I}{2\pi\,(x + a/2)}\ell v$$

or, $e \propto \frac{1}{(2x - a)(2x + a)}$

52. **(a)** **53.** **(a)**

54. **(c)** Flux can't change in a superconducting loop.

$$\Delta\phi = 2\pi R^2.B$$

Initially current was zero, so self flux was zero.

$\therefore$ Finally $Li = 2\pi R^2 \times B$

$$i = \frac{2\pi R^2\times B}{L}$$

55. **(a)** Parallel $= \frac{L_1L_2}{L_1\times L_2}$

56. (d)

57. (b) $L = \frac{e}{\frac{di}{dt}} = \frac{e\,dt}{di} = \frac{5\times10^{-3}}{(3-2)}H = 5mH$

58. (b) Let current i_1 in straight wire be upward. Then the magnetic field due to the straight wire has magnitude $B_1 = \mu_0 i_1/2\pi r$ at distance r. In accrodance with right hand rule, B_1 points inward to the plane of page. We consider a differential strip of thickness dr, area dA_2 = a dr. Magnetic flux through area dA, $d\phi_B = B_1$ (a dr).
Total flux through the loop,

$$\phi_B = \int B_1 a dr = \int_c^{c+b} \frac{\mu_0 i_1}{2\pi r} a\, dr$$

$$= \frac{\mu_0 i_1 a}{2\pi}\int_c^{c+b}\frac{dr}{r} = \frac{\mu_0 i_1 a}{2\pi} ln\left(\frac{c+b}{c}\right)$$

Therefore mutual inductance,

$$M = \frac{\phi}{i_1} = \frac{\mu_0 a}{2\pi} ln\left(1+\frac{b}{c}\right)$$

59. (c) $L = 2mH, i = t^2e^{-t}$

$$E = -L\frac{di}{dt} = -L[-t^2e^{-t} + 2te^{-t}]$$

when $E = 0, -e^{-t}t^2 + 2te^{-t} = 0$
or, $2t\,e^{-t} = e^{-t}t^2 \Rightarrow t = 2$ sec.

60. (c) Initial current (I_1) = 10 A; Final current (I_2) = 0; Time (t) = 0.5 sec and induced e.m.f. (ε) = 220 V. Induced e.m.f. (ε)

$$= -L\frac{dI}{dt} = -L\frac{(I_2 - I_1)}{t} = -L\frac{(0-10)}{0.5} = 20L$$

or, $L = \frac{220}{20} = 11H$
[where L = Self inductance of coil]

61. (a) $e_{back} = L\frac{di}{dt}$ where $L = \frac{\mu_0 N^2 A}{l}$

$$\therefore e_{back} = \frac{\mu_0 N^2 A}{l}\left[\frac{1.5-0}{1\times10^3}\right]$$

$$= \frac{4\pi\times10^{-7}(400)^2\times20\times10^{-4}}{20\times10^{-2}}\times(1.5\times10^3)$$

$= 0.3\,V$

62. (a)

63. (b) $\phi_{total} = B_{large}A_{small} = \frac{\mu_0}{4\pi}\frac{i}{L/2}(2\sin 45°)\times\ell^2$

On comparing with $\phi_{total} = Mi$, we get $M \propto \frac{\ell^2}{L}$

64. (d) $M = \frac{\mu_0\mu_r N_1 N_2}{2R_1}A$

$$= \frac{4\pi\times10^{-4}\times800\times500\times25\times12\times10^{-4}}{2\times0.1} = 0.24H$$

65. (b)

Using $V_A - V_B = RI + L\frac{dI}{dt} \Rightarrow 140 = 5R + 10L$

$60 = 5R - 10L \quad \Rightarrow L = 4H$

66. (a) $L = \frac{\mu_0 N^2 \pi r^2}{\ell}$

Length of wire = N $2\pi r$ = constant (= C, suppose)

$$\therefore \quad L = \mu_0\left(\frac{C}{2\pi r}\right)^2\frac{\pi r^2}{\ell} \quad \therefore \; L \propto \frac{1}{\ell}$$

$\therefore$ Self inductance will become 2L.

67. (b)

68. (c) Induced emf $= vB_H l = 1.5\times5\times10^{-5}\times2$
$= 15\times10^{-5} = 0.15\,mV$

69. (a)

W ——→ E

$\varepsilon_{ind} = Bv\ell = 0.3\times10^{-4}\times5\times20$
$= 3\times10^{-3}V = 3\,mV.$

70. (d) Because of the Lenz's law of conservation of energy.

71. (c) Total number of turns in the solenoid, N = 500
Current, I = 2A.
Magnetic flux linked with each turn $= 4\times10^{-3}$ Wb
As, $\phi = LI$ or $N\phi = LI$

$$\Rightarrow L = \frac{N\phi}{1} = \frac{500\times4\times10^{-3}}{2}\text{ henry} = 1\,H.$$

72. (a) $\phi = Li \Rightarrow NBA = Li$
Since magnetic field at the centre of circular coil carrying current is given by

$$B = \frac{\mu_0}{4\pi}, \frac{2\pi Ni}{r}$$

$$\therefore N.\frac{\mu_0}{4\pi}.\frac{2\pi Ni}{r}.\pi r^2 = Li \Rightarrow L = \frac{\mu_0 N^2\pi r}{2}$$

Hence self inductance of a coil

$$\frac{4\pi\times10^{-7}\times500\times500\times\pi\times0.05}{2} = 25\,mH$$

73. (a)

74. (a) The back e.m.f. in a motor is induced e.m.f., which is maximum, when speed of rotation of the coil is maximum.

75. (b) $I = \frac{e_1 - e_2}{R} = \frac{220-210}{2} = 5A$

76. (a) Let initial e.m.f. induced = e.

$\therefore$ Initial current $i = \frac{E - e}{R}$ i.e., $2 = \frac{12 - e}{1}$

This gives $e = 12 - 2 = 10$ volt. As $e \propto \omega$. when speed is halved, the value of induced e.m.f. becomes

$\frac{e}{2} = \frac{10}{2} = 5$ volt

$\therefore$ New value of current

$i' = \frac{E - e}{R} = \frac{12 - 5}{1} = 7\text{ A}$

77. **(c)** Eddy currents are set up inside the plate and the direction of current opposes the motion of the metallic plate swings between the poles of a magnet.

78. **(d)** The velocity component of conductor must be perpendicular to $\vec{\ell}$ and $\vec{B}$.

79. **(a)**

80. **(d)** Total resistance of the circuit = 4000 + 400 = 4400 W

Current flowing $i = \frac{V}{R} = \frac{440}{4400} = 0.1$ amp.

Voltage across load = R i = 4000 × 0.1 = 400 volt.

81. **(d)** Here, induced e.m.f.

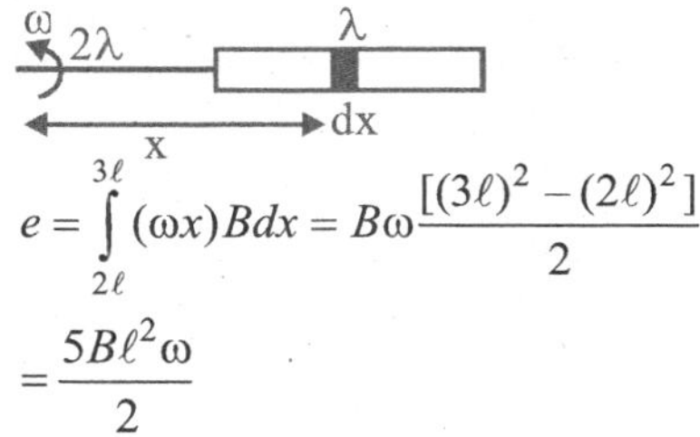

$$e = \int_{2\ell}^{3\ell} (\omega x) B dx = B\omega \frac{[(3\ell)^2 - (2\ell)^2]}{2}$$

$$= \frac{5B\ell^2\omega}{2}$$

82. **(a)** As we know, Magnetic flux, $\phi = B.A$

$$\frac{\mu_0 (2)(20 \times 10^{-2})^2}{2[(0.2)^2 + (0.15)^2]} \times \pi(0.3 \times 10^{-2})^2$$

On solving

$= 9.216 \times 10^{-11} = 9.2 \times 10^{-11}$ weber

83. **(a)** According to Faraday's law of electromagnetic induction, $\varepsilon = \frac{d\phi}{dt}$

Also, $\varepsilon = iR$ $\quad \therefore iR = \frac{d\phi}{dt} \Rightarrow \int d\phi = R\int i dt$

Magnitude of change in flux $(d\phi) = R \times$ area under current vs time graph

or, $d\phi = 100 \times \frac{1}{2} \times \frac{1}{2} \times 10 = 250$ Wb

84. **(a)** The magnetic field due to a disc is given as

$$B = \frac{\mu_0 \omega Q}{2\pi R} \quad \text{i.e., } B \propto \frac{1}{R}$$

85. **(a)**

86. **(d)**

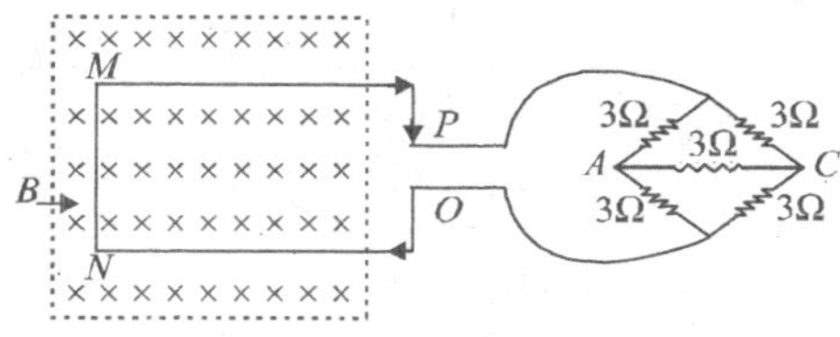

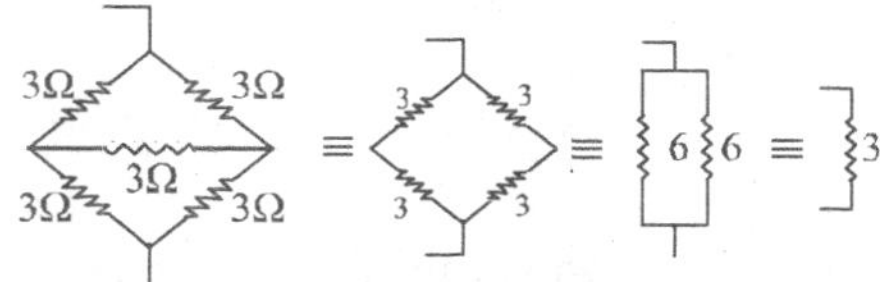

The network behaves like a balanced wheatstone bridge.

The induced emf developed is given by

$e = vB\ell = v \times 2 \times 0.1 = 0.2v$ ...(i)

Now, $e = IR$

$e = 10^{-3} \times 4 = 4 \times 10^{-3}$ amp ...(ii)

From (i) and (ii),

$0.2\,v = 4 \times 10^{-3} \quad \therefore v = \frac{4 \times 10^{-3}}{0.2} = 0.02\text{ m/s}$

87. **(c)** Considering pure rolling of OA about A: the induced emf across OA will be:

$|\vec{e}| = \frac{B\omega(r)^2}{2}$.

From lenz law, O will be the negative end, while A will be the positive end.

Hence $V_0 - V_A = -\frac{B\omega r^2}{2}$

88. **(b)** **89.** **(a)**

90. **(a)** $B = \frac{\mu_0}{2\pi}\frac{M}{x^3}$

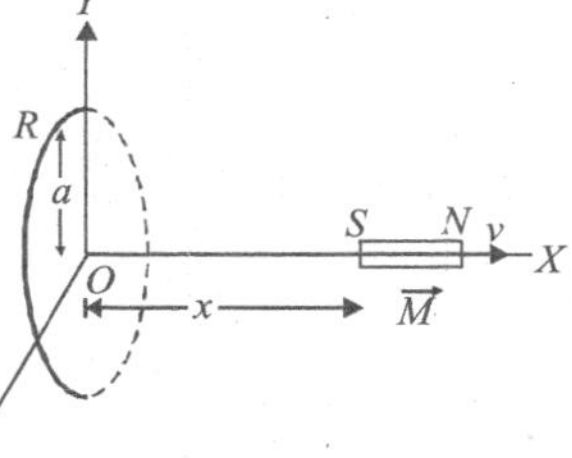

$\therefore$ Flux passing through the coil $= Ba^2$

$\therefore$ The induced emf in the coil

$$e = \frac{-d\phi}{dt} = -\frac{d}{dt}\left[\frac{\mu_0 M a^2}{2} \times \frac{1}{x^3}\right]$$

$$= \frac{3\mu_0 M a^2}{2x^4}\frac{dx}{dt} \left[\because v = \frac{dx}{dt}\right] = \frac{3\mu_0 M a^2}{2x^4} \times v$$

$\therefore$ Current in the coil $= I = \frac{e}{R} = \frac{3\mu_0 Ma^2 v}{2x^4 R}$

The magnetic moment of the loop

$$M' = I \times A = \frac{3\mu_0 Ma^2 v}{2x^4 R} \times (\pi a^2)$$

Now the potential energy of the loop placed in the magnetic field is

$$U = -M'B\cos 180° = \frac{3\mu_0 Ma^2 v \times \pi a^2}{2x^4 R} \times \frac{\mu_0 M}{2\pi x^3}$$

$$\therefore \quad U = \frac{3\mu_0^2 M^2 a^4 v}{4Rx^7}$$

Now, $|\vec{F}| = -\frac{dU}{dx}$ $\quad \therefore \quad F = \frac{21}{4}\frac{\mu_0^2 M^2 a^4 v}{Rx^8}$

Since by Newton's third law, action and reaction are equal. Therefore, the above calculated force is acting on the magnet. The direction of the force is in $-\hat{i}$ direction by Lenz's law.

91. **(c)** Total effective length crossing the flux lines

$$= \frac{2vt}{\tan\phi} + l \qquad \text{Current} = \frac{B\left(\frac{2vt}{\tan\phi} + l\right)v}{R}$$

92. **(d)** $\phi = 4t^n + 6$ $\quad \frac{d\phi}{dt} = 4nt^{n-1}$

$$|e| = 4nt^{n-1}, |e| = \frac{4n}{t^{1-n}}$$

93. **(d)** : $\theta = \omega t$. Only half circular part will be involved in inducing emf, so effective area $A = \frac{\pi a^2}{2}$

$\phi = BA\cos\theta$

$$e = -\frac{d\phi}{dt}$$

$$= +BA\sin\theta\left(\frac{d\theta}{dt}\right)$$

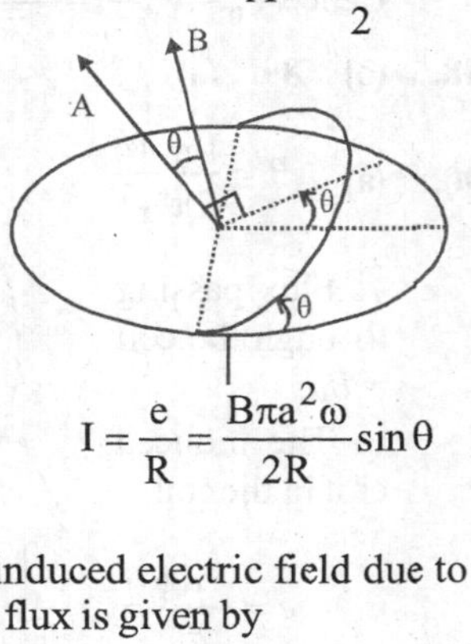

$$\Rightarrow e = \frac{B\pi a^2}{2}\omega\sin\theta \qquad I = \frac{e}{R} = \frac{B\pi a^2\omega}{2R}\sin\theta$$

94. **(b)**

95. **(d)** Magnitude of induced electric field due to change in magnetic flux is given by

$$\oint \vec{E}.d\vec{\ell} = \frac{d\phi}{dt} = A\frac{dB}{dt} \qquad (\because N = 1 \text{ and } \cos\theta = 1)$$

$$\text{or } E.\ell = \pi R^2(2B_0 t) \quad \left(\frac{dB}{dt} = 2B_0 t\right)$$

Here, E = induced electric field due to change in magnetic flux

$$E(2\pi R) = 2\pi R^2 B_0 t \text{ or } E = B_0 Rt$$

Hence, $F = QE = B_0 QRt$

This force is tangential to ring. Ring starts rotating when torque of this force is greater than the torque due to maximum friction ($f_{max} = \mu mg$)

or when $\tau_F \geq \tau_{f_{max}}$

Taking the limiting case, $\tau_F = \tau_{f_{max}}$ or

$F.R = (\mu mg)R$

It is given that ring starts rotating after 2 seconds.

So, putting $t = 2$ seconds, we get $\mu = \frac{2B_0 RQ}{mg}$

96. **(d)** $\phi = BA = \frac{B}{2}\frac{x}{\sqrt{2}}\frac{x}{\sqrt{2}}$

$$\varepsilon = \frac{d\phi}{dt} = \frac{2x}{4}\frac{dx}{dt}B = \frac{x}{2}Bv$$

$$\varepsilon = \frac{d\phi}{dt} = \frac{2x}{4}\frac{dx}{dt}B = \frac{x}{2}B\frac{d(2y)}{dt}$$

$$i = \frac{xBv}{2\lambda(x + \sqrt{2}x)} = \frac{Bv}{\lambda(1+\sqrt{2})}$$

97. **(d)** **98.** **(a)**

99. **(c)** each spoke will behave as a cell. induced emf of each spoke $= \frac{1}{2}Bl^2\omega$ But all (n) identical spoke behaving as a cell can connected in a parallel fashion, so the resultant emf = e $= \frac{1}{2}Bl^2\omega$ = same as single cell.

100. **(c)** $I = \frac{\varepsilon}{\frac{2R}{3}} = \frac{3\varepsilon}{2R}$

$$= \frac{3}{2R} \times \frac{1}{2}B\omega l^2$$

$$= \frac{3B\omega l^2}{4R}$$

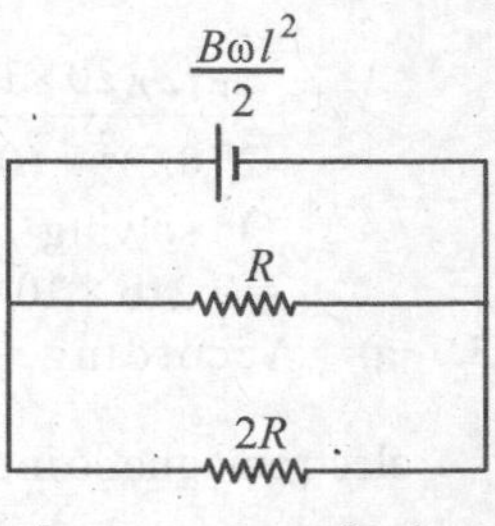

$\frac{B\omega l^2}{2}$ R 2R

Magnetic force $F = \frac{3B\omega l^2}{4R} \times l \times B = \frac{3B^2\omega l^3}{4R}$

$$\tau = \frac{3B^2\omega l^3}{4R} \times \frac{l}{2} = \frac{3B^2\omega l^4}{8R}$$

$\therefore$ Force to be applied at the end $= \frac{3B^2\omega l^3}{8R}$.

21 Alternating Current

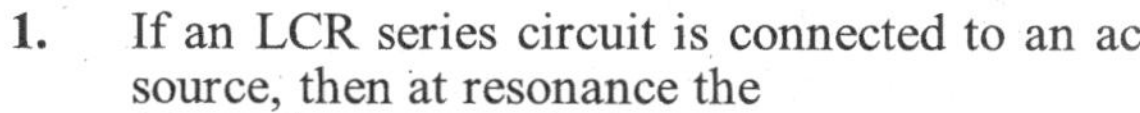

1. If an LCR series circuit is connected to an ac source, then at resonance the
   (a) voltage across R is zero
   (b) voltage across R equals the applied voltage
   (c) voltage across C is zero
   (d) voltage across L equals the applied voltage
2. Of the following about capacitive reactance which is correct?
   (a) The reactance of the capacitor is directly proportional to its ability to store charge
   (b) Capacitive reactance is inversely proportional to the frequency of the current
   (c) Capacitive reactance is measured in farad
   (d) The reactance of a capacitor in an A.C. circuit is similar to the resistance of a capacitor in a D.C. circuit
3. Which one of the following curves represents the variation of impedance (Z) with frequency f in series LCR circuit?

(a) 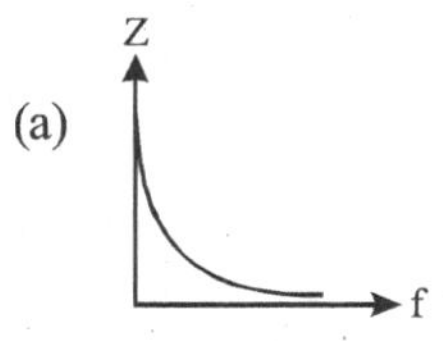

(b) 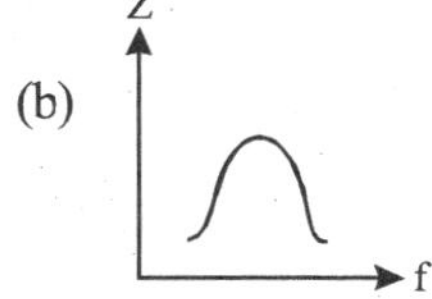

(c) 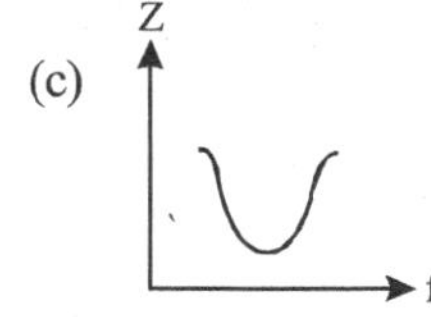

(d) 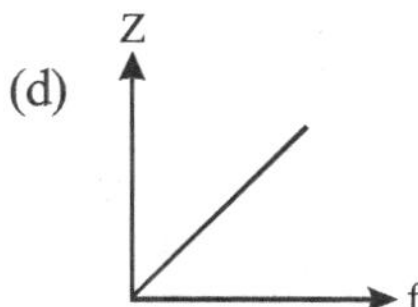

4. The power factor in a circuit connected to an A.C. The value of power factor is
   (a) unity when the circuit contains an ideal inductance only
   (b) unity when the circuit contains an ideal resistance only
   (c) zero when the circuit contains an ideal resistance only
   (d) unity when the circuit contains an ideal capacitance only
5. Which of the following graphs represents the correct variation of capacitive reactance X_C with frequency υ? **Tricky**

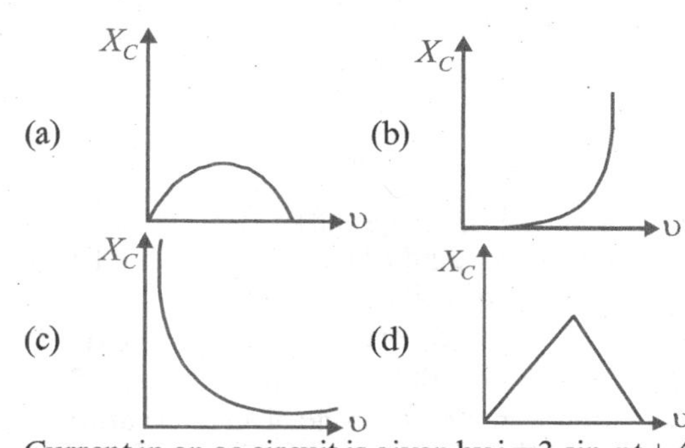

6. Current in an ac circuit is given by $i = 3 \sin \omega t + 4 \cos \omega t$ then
   (a) rms value of current is 5 A
   (b) mean value of this current in one half period will be $6/\pi$
   (c) if voltage applied is $V = V_m \sin \omega t$ then the circuit must be containing resistance and capacitance.
   (d) if voltage applied is $V = V_m \sin \omega t$, the circuit may contain resistance and inductance.
7. If we increase the driving frequency in a circuit with a purely capacitive load, then
   (a) amplitude V_C increases
   (b) amplitude V_C decreases
   (c) amplitude i_C increases
   (d) amplitude i_C decreases
8. If a direct current of value ampere is superimposed on an alternative current $I = b \sin \omega t$ flowing through a wire, what is the effective value of the resulting current in the circuit?

Critical Thinking

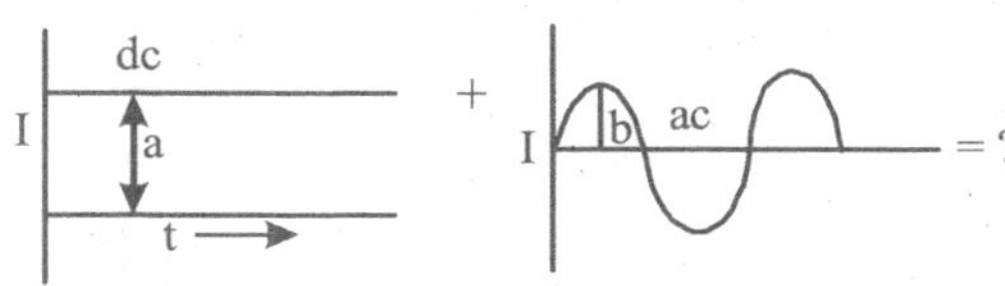

(a) $\left[a^2 - \frac{1}{2}b^2\right]^{1/2}$ (b) $[a^2 + b^2]^{1/2}$
(c) $\left[\frac{a^2}{2} + b^2\right]^{1/2}$ (d) $\left[a^2 + \frac{b^2}{2}\right]^{1/2}$

9. In a series LCR circuit, the difference of the frequencies at which current amplitude falls to $\frac{1}{\sqrt{2}}$ of the current amplitude at resonance is
(a) $\frac{R}{2\pi L}$ (b) $\frac{R}{\pi L}$ (c) $\frac{2R}{\pi L}$ (d) $\frac{3R}{2\pi L}$

10. If $i_1 = 3\sin\omega t$ and $i_2 = 4\cos\omega t$, then i_3 is
(a) $5\sin(\omega t + 53°)$
(b) $5\sin(\omega t + 37°)$
(c) $5\sin(\omega t + 45°)$
(d) $5\cos(\omega t + 53°)$

11. A sinusoidal voltage V(t) = 100 sin (500t) is applied across a pure inductance of L = 0.02 H. The current through the coil is:
(a) 10 cos (500 t) (b) – 10 cos (500t)
(c) 10 sin (500t) (d) – 10 sin (500t)

12. Figure shows one cycle of an alternating current with the segments AB, BC, CD, DE being symmetrical and parabolic. The root mean square value of this current over one cycle is x mA, find x.

Critical Thinking

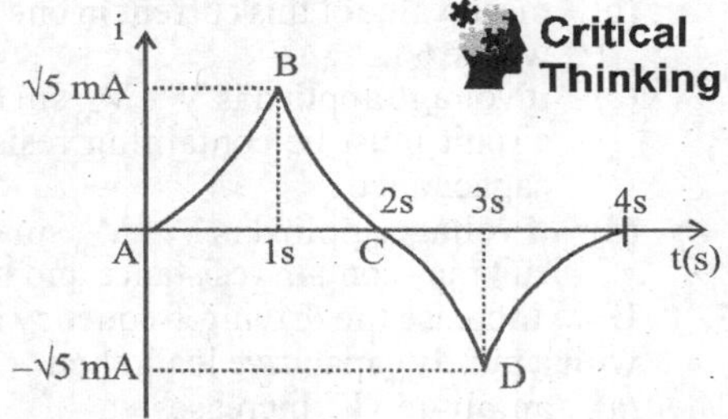

(a) 1 mA (b) 2 mA (c) 3 mA (d) 4 mA

13. In given circuit capacitor initially uncharged. Now at t = 0 switch S is closed then current given by source at any time t is

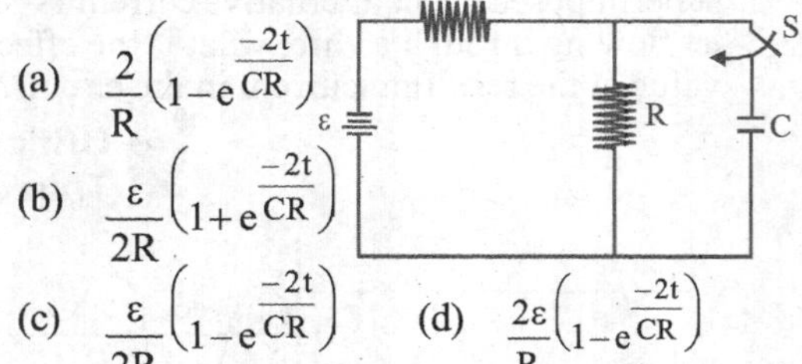

(a) $\frac{2}{R}\left(1 - e^{\frac{-2t}{CR}}\right)$
(b) $\frac{\varepsilon}{2R}\left(1 + e^{\frac{-2t}{CR}}\right)$
(c) $\frac{\varepsilon}{2R}\left(1 - e^{\frac{-2t}{CR}}\right)$ (d) $\frac{2\varepsilon}{R}\left(1 - e^{\frac{-2t}{CR}}\right)$

14. The transformer voltage induced in the secondary coil of a transformer is mainly due to
(a) a varying electric field
(b) a varying magnetic field
(c) the vibrations of the primary coil
(d) the iron core of the transformer

15. A coil of inductance 8.4 mH and resistance 6 Ω is connected to a 12 V battery. The current in the coil is 1.0 A at approximately the time
(a) 500 s (b) 25 s (c) 35 ms (d) 1 ms

16. Resonance frequency of a circuit is f. If the capacitance is made 4 times the initial value, then the resonance frequency will become :
(a) f/2 (b) 2f (c) f (d) f/4

17. The current (I) in the inductance is varying with time according to the plot shown in figure.

Tricky

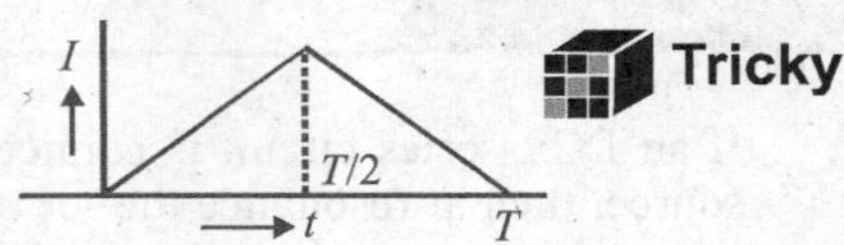

Which one of the following is the correct variation of voltage with time in the coil?

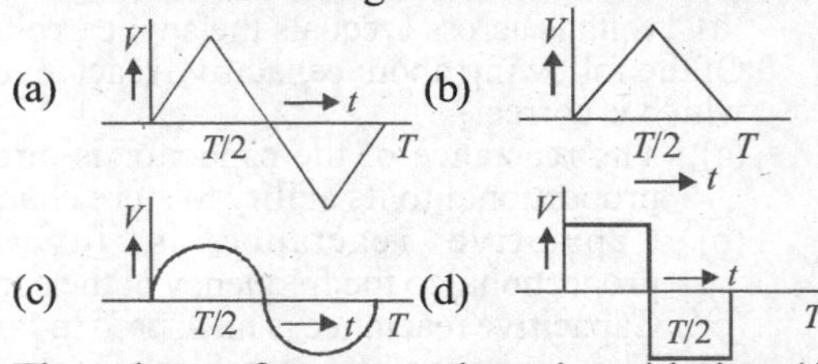

18. The voltage of an ac supply varies with time (t) as $V = 120 \sin 100\pi t \cos 100\pi t$. The maximum voltage and frequency respectively are
(a) 120 volts, 100 Hz (b) $\frac{120}{\sqrt{2}}$ volts, 100 Hz
(c) 60 volts, 200 Hz (d) 60 volts, 100 Hz

19. The equation of AC voltage is $E = 220 \sin(\omega t + \pi/6)$ and the A.C. current is $I = 10 \sin(\omega t + \pi/6)$. The average power dissipated is
(a) 150 W (b) 550 W (c) 250 W (d) 50 W

20. In an L-R circuit, the value of L is $(0.4/\pi)$ henry and the value of R is 30 ohm. If in the circuit, an alternating emf of 200 volt at 50 cycles per second is connected, the impedance of the circuit and current will be :
(a) 11.4 ohm, 17.5 ampere
(b) 30.7 ohm, 6.5 ampere
(c) 40.4 ohm, 5 ampere
(d) 50 ohm, 4 ampere.

21. In an ac circuit an alternating voltage $e = 200\sqrt{2} \sin 100t$ volts is connected to a capacitor of capacity 1 μF. The r.m.s. value of the current in the circuit is
(a) 10 mA (b) 100 mA (c) 200 mA (d) 20 mA

22. Determine the rms value of a semi - circular current wave which has a maximum value of a.

Critical Thinking

(a) $(1/\sqrt{2})a$
(b) $(\sqrt{3/2})a$
(c) $(\sqrt{2/3})a$
(d) $(\sqrt{1/3})a$

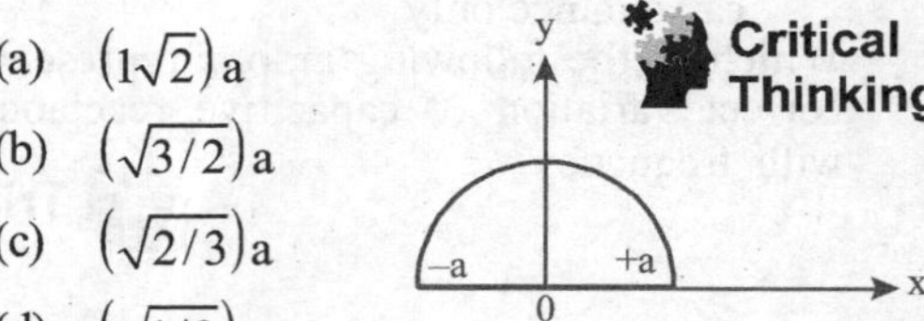

23. The r.m.s. value of potential difference V shown in the figure is

(a) V_0 (b) $V_0/\sqrt{2}$

(c) $V_0/2$ (d) $V_0/\sqrt{3}$

24. The equation of alternating current is : $I = 50\sqrt{2}\sin 400\pi t$ amp. Then the frequency and root mean square of current are respectively

(a) 200 Hz, 50 amp (b) 400π Hz, $50\sqrt{2}$ amp

(c) 200 Hz, $50\sqrt{2}$ amp (d) 50 Hz, 200 amp

25. A coil has resistance 30 ohm and inductive reactance 20 ohm at 50 Hz frequency. If an ac source, of 200 volt, 100 Hz, is connected across the coil, the current in the coil will be

(a) 4.0A (b) 8.0A (c) $\frac{20}{\sqrt{13}}$A (d) 2.0A

26. A small signal voltage $V(t) = V_0 \sin \omega t$ is applied across an ideal capacitor C :

(a) Current I (t), lags voltage V(t) by 90°.
(b) Over a full cycle the capacitor C does not consume any energy from the voltage source.
(c) Current I (t) is in phase with voltage V(t).
(d) Current I (t) leads voltage V(t) by 180°.

27. An ac voltage is applied to a resistance R and an inductor L in series. If R and the inductive reactance are both equal to 3Ω, the phase difference between the applied voltage and the current in the circuit is

(a) $\pi/6$ (b) $\pi/4$ (c) $\pi/2$ (d) zero

28. The voltage time (V-t) graph for triangular wave having peak value V_0 is as shown in figure. The rms value of V in time interval from t = 0 to T/4 is

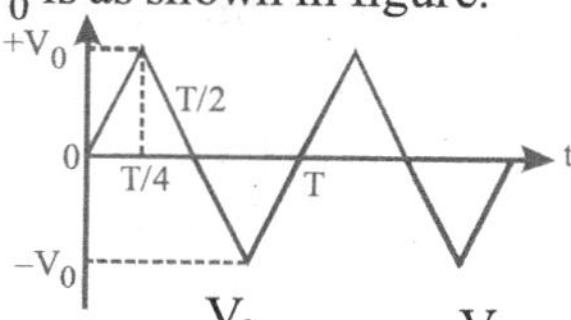

(a) $\frac{V_0}{\sqrt{3}}$ (b) $\frac{V_0}{2}$ (c) $\frac{V_0}{\sqrt{2}}$ (d) $\frac{V_0}{3}$

29. An inductor 20 mH, a capacitor 50 μF and a resistor 40Ω are connected in series across a source of emf V = 10 sin 340 t. The power loss in A.C. circuit is :

(a) 0.51 W (b) 0.67 W (c) 0.76 W (d) 0.89 W

30. A series R-C circuit is connected to an alternating voltage source. Consider two situations:

(A) When capacitor is air filled.
(B) When capacitor is mica filled.

Current through resistor is i and voltage across capacitor is V then :

(a) $V_a > V_b$ (b) $i_a > i_b$ (c) $V_a = V_b$ (d) $V_a < V_b$

31. Find the current passing through battery immediately after key (K) is closed. It is given that initially all the capacitors are uncharged. (given that R = 6 Ω and C = 4μF) **Tricky**

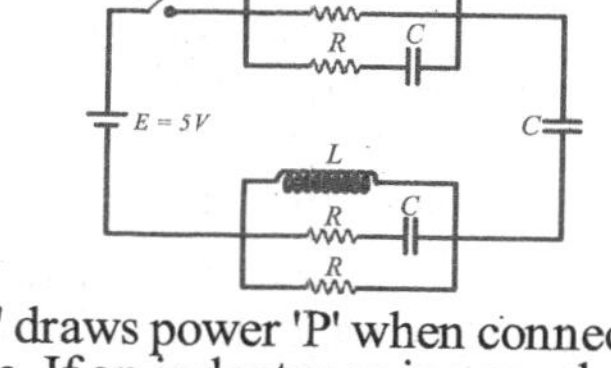

(a) 1 A
(b) 5 A
(c) 3 A
(d) 2 A

32. A resistance 'R' draws power 'P' when connected to an AC source. If an inductance is now placed in series with the resistance, such that the impedance of the circuit becomes 'Z', the power drawn will be

(a) $P\sqrt{\frac{R}{Z}}$ (b) $P\left(\frac{R}{Z}\right)$ (c) P (d) $P\left(\frac{R}{Z}\right)^2$

33. A condenser of capacity C is charged to a potential difference of V_1. The plates of the condenser are then connected to an ideal inductor of inductance L. The current through the inductor when the potential difference across the condenser reduces to V_2 is **Critical Thinking**

(a) $\left(\frac{C(V_1^2 - V_2^2)}{L}\right)^{1/2}$ (b) $\left(\frac{C(V_1 - V_2)^2}{L}\right)^{1/2}$

(c) $\frac{C(V_1^2 - V_2^2)}{L}$ (d) $\frac{C(V_1 - V_2)}{L}$

34. For the circuit as shown in figure; the applied current in A.C. circuit is zero ampere and $I_C = 10$A. Then the magnitude of current I_L is

$I_C = 10$ A; 2 μF = C; 4 mH = L

(a) 4 A
(b) 10 A
(c) 5 A
(d) undefined

35. What is the value of inductance L for which the current is maximum in a series LCR circuit with C = 10 μF and $\omega = 1000 s^{-1}$?

(a) 1 mH
(b) cannot be calculated unless R is known
(c) 10mH (d) 100mH

36. In the circuit shown, the symbols have their usual meanings. The cell has emf *E*. *X* is initially joined to *Y* for a long time. Then, *X* is joined to *Z*. The maximum charge on *C* at any later time will be

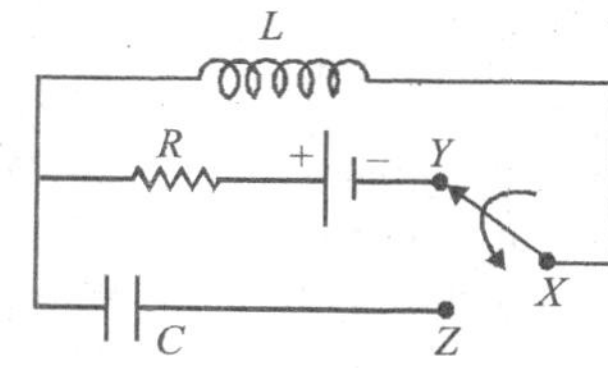

(a) $\frac{E}{R\sqrt{LC}}$ (b) $\frac{ER}{2\sqrt{LC}}$ (c) $\frac{E\sqrt{LC}}{2R}$ (d) $\frac{E\sqrt{LC}}{R}$

37. A coil of inductive reactance 31 Ω has a resistance of 8 Ω. It is placed in series with a condenser of capacitative reactance 25Ω. The combination is connected to an a.c. source of 110 volt. The power factor of the circuit is

(a) 0.64 (b) 0.80 (c) 0.33 (d) 0.56

38. The two capacitors, shown in the circuit, are initially uncharged and the cell is ideal. The switch S is closed at $t = 0$. Which of the following functions represents the current $i(t)$, through the cell as a function of time? Here i_0, i_1, i_2 are constants. Tricky

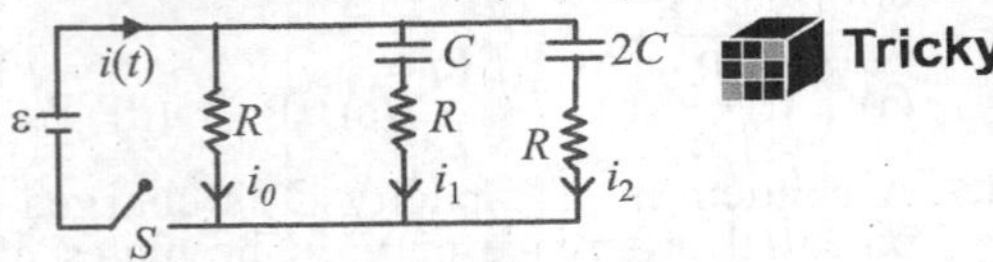

(a) $i(t) = i_0 + i_1 e^{-t/\tau}; \tau = 3C \times \frac{R}{3}$

(b) $i(t) = i_0 + i_1 e^{-t/\tau} + i_2 e^{-t/2\tau}; \tau = RC$

(c) $i(t) = i_1 + i_1 e^{-t/\tau}; \tau = 3C \times \frac{R}{3}$

(d) $i(t) = i_0 + i_1 e^{-t/\tau}; \tau = 3RC$

39. A coil of 40 henry inductance is connected in series with a resistance of 8 ohm and the combination is joined to the terminals of a 2 volt battery. The time constant of the circuit is

(a) 20 seconds (b) 5 seconds

(c) 1/5 seconds (d) 40 seconds

40. A resistor and an inductor are connected to an ac supply of 120 V and 50 Hz. The current in the circuit is 3 A. If the power consumed in the circuit is 108 W, then the resistance in the circuit is

(a) 12Ω (b) 40Ω

(c) $\sqrt{(52 \times 25)}$Ω (d) 360Ω

41. A short-circuited coil is placed in a time-varying magnetic field. Electrical power is dissipated due to the current induced in the coil. If the number of turns were to be quadrupled and the wire radius halved, the electrical power dissipated would be Critical Thinking

(b) the same

(c) doubled (d) quadrupled

42. An LCR series circuit with 100 Ω resistance is connected to an AC source of 200 V and angular frequency 300 radians per second. When only the capacitance is removed, the current lags behind the voltage by 60°. When only the inductance is removed, the current leads the voltage by 60°. Then the current and power dissipated in LCR circuit are respectively

(a) 1A, 200 watt (b) 1A, 400 watt

(c) 2A, 200 watt (d) 2A, 400 watt

43. In the circuits (a) and (b) switches S_1 and S_2 are closed at t = 0 and are kept closed for a long time. The variation of current in the two circuits for t ≥ 0 are roughly shown by figure (figures are schematic and not drawn to scale) : Tricky

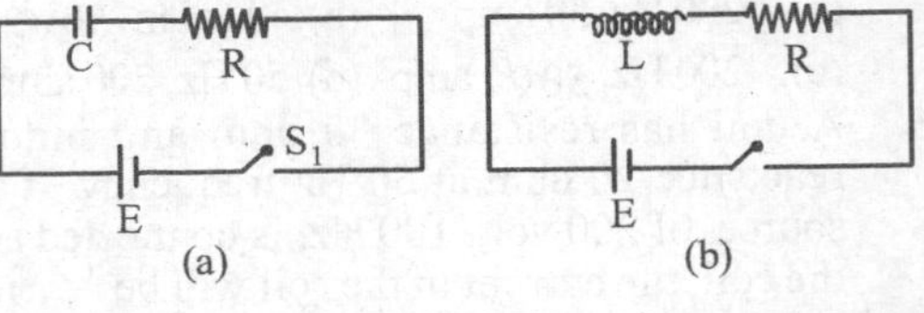

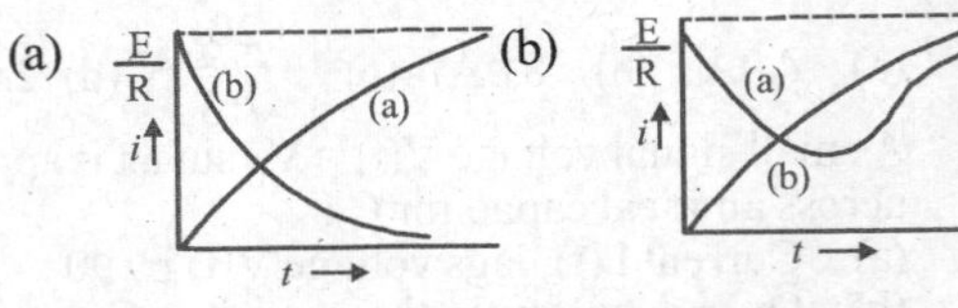

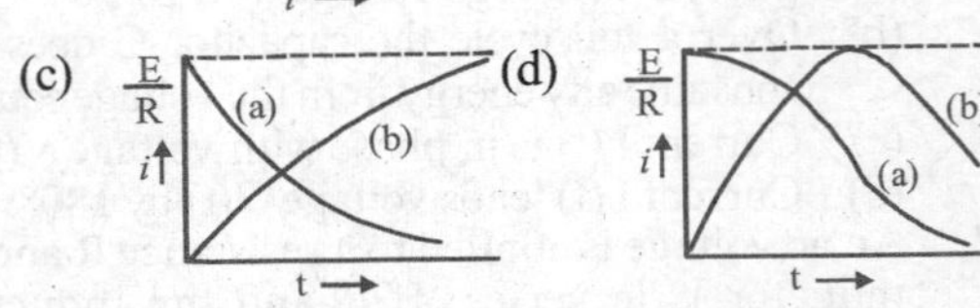

44. A coil has an inductance of 0.7 henry and is joined in series with a resistance of 220 Ω. When the alternating emf of 220 V at 50 Hz is applied to it then the phase through which current lags behind the applied emf and the wattless component of current in the circuit will be respectively

(a) 30°, 1 A (b) 45°, 0.5 A

(c) 60°, 1.5 A (d) None of these

45. In an L-C circuit shown in the figure, C = 1F, L = 4H. At time t = 0, charge in the capacitor is 4C and it is decreasing at a rate of $\sqrt{5}$ C/s. Choose the correct statements.

(a) maximum charge in the capacitor can be 6C

(b) maximum charge in the capacitor can be 8C

(c) charge in the capacitor will be maximum after time 2 $\sin^{-1}$ (2/3) sec

(d) None of these

46. Charges on the capacitors in four oscillating LC circuits vary as follows : (1) q = 2 cos 4t, (2) q = 4cos t, (3) q = 3 cos 4t, (4) q = 4 cos 2t, with q in coulomb and t in second. In which circuit(s) current amplitude is greatest?

(a) (1) (b) (2) (c) (3) (d) (4)

47. Figure shows a source of alternating voltage connected to a capacitor and resistor. Which of the following phasor diagrams correctly secrobes the phase relationship between I_c, the current between the source and the capacitor, and I_R, the current in the resistor? **Tricky**

(a) $I_C \rightarrow$, $I_R \rightarrow$

(b) $\leftarrow I_C$ $I_R \rightarrow$

(c) I_R (vertical), I_C (horizontal)

(d) I_R (inclined), I_C (horizontal)

48. A capacitor of 10 μF and an inductor of 1 H are joined in series. An ac of 50 Hz is applied to this combination. What is the impedance of the combination?

(a) $\frac{5(\pi^2-5)}{\pi}\Omega$ (b) $\frac{10^2(10-\pi^2)}{\pi}\Omega$

(c) $\frac{10(\pi^2-5)}{\pi}\Omega$ (d) $\frac{5^2(10-\pi^2)}{\pi}\Omega$

49. What is the amount of power delivered by the ac source in the circuit shown (in watts).

$X_C=12\Omega$, $R_1=5\Omega$, $X_L=8\Omega$, $R_2=6\Omega$, $E_{rms}=130V$

(a) 500 watt

(b) 1014 watt

(c) 1514 watt

(d) 2013 watt

50. The current in an $L-R$ circuit builds up to $(3/4)^{th}$ of its steady state value in 4 seconds. The time constant of this circuit is **Critical Thinking**

(a) $\frac{1}{\ln 2}$ sec (b) $\frac{2}{\ln 2}$ sec

(c) $\frac{3}{\ln 2}$ sec (d) $\frac{4}{\ln 2}$ sec

51. A resistor of resistance R, capacitor of capacitance C and inductor of inductance L are connected in parallel to AC power source of voltage $\varepsilon_0 \sin \omega t$. The maximum current through the resistance is half of the maximum current through the power source. Then value of R is

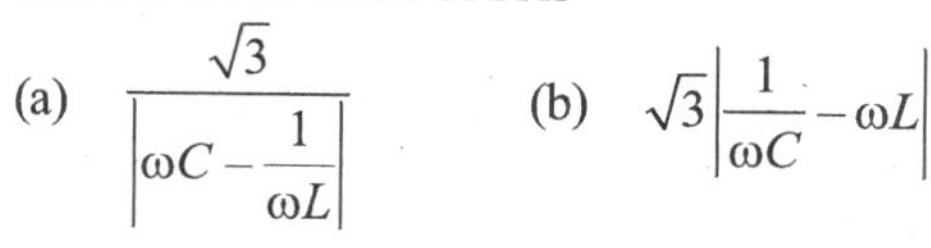

(a) $\frac{\sqrt{3}}{\left|\omega C-\frac{1}{\omega L}\right|}$ (b) $\sqrt{3}\left|\frac{1}{\omega C}-\omega L\right|$

(c) $\sqrt{5}\left|\frac{1}{\omega C}-\omega L\right|$ (d) None of these

52. When the rms voltages V_L, V_C and V_R are measured respectively across the inductor L, the capacitor C and the resistor R in a series LCR circuit connected to an AC source, it is found that the ratio $V_L : V_C : V_R = 1:2:3$. If the rms voltage of the AC sources is 100 V, the V_R is close to:

(a) 50 V (b) 70 V (c) 90 V (d) 100 V

53. Time constant of $L–R$ circuit will be

(a) $\frac{L}{R}$

(b) $\frac{2L}{R}$

(c) $\frac{L}{2R}$

(d) None of these

54. A series LR circuit is connected to an ac source of frequency ω and the inductive reactance is equal to 2R. A capacitance of capacitive reactance equal to R is added in series with L and R. The ratio of the new power factor to the old one is **Critical Thinking**

(a) $\sqrt{\frac{2}{3}}$ (b) $\sqrt{\frac{2}{5}}$ (c) $\sqrt{\frac{3}{2}}$ (d) $\sqrt{\frac{5}{2}}$

55. LC circuit contains a 20 mH inductor and a 50 μF capacitor with an initial charge of 10 mC. The resistance of the circuit is negligible. Let the instant the circuit is closed at t = 0. At what time is the energy stored completely magnetic ?

(a) t = 0 (b) t = 1.57 ms

(c) t = 3.14 ms (d) t = 6.28 ms

56. An inductor 20×10^{-3} Henry, a capacitor 100μF and a resistor 50Ω are connected in series across a source of EMF V = 10 sin 314t. If resistance is removed from the circuit and the value of inductance is doubled, then the variation of current with time in the new circuit is –

(a) 0.52 cos 314 t (b) 0.52 sin 314 t

(c) 0.52 sin (314 t + π/3) (d) None of these

57. In a LCR circuit capacitance is changed from C to $2C$. For the resonant frequency to remain unchanged, the inductance should be change from L to

(a) $4L$ (b) $2L$ (c) $L/2$ (d) $L/4$

58. As shown in figure, value of inductive reactance X_L will be if source voltage is 100 volt

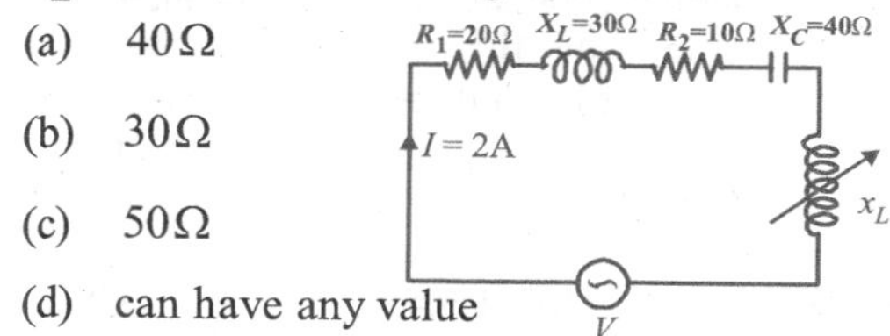

(a) 40Ω

(b) 30Ω

(c) 50Ω

(d) can have any value

59. In a series L-C-R circuit, C = 10^{-11} Farad, L = 10^{-5} Henry and R = 100 Ohm, when a constant D.C. voltage E is applied to the circuit, the capacitor acquires a charge 10^{-9} C. The D.C. source is replaced by a sinusoidal voltage source in which the peak voltage E_0 is equal to the constant D.C. voltage E. At resonance the peak value of the charge acquired by the capacitor will be :

(a) 10^{-15} C (b) 10^{-6} C
(c) 10^{-10} C (d) 10^{-8} C

60. In a uniform magnetic field of induction B a wire in the form of a semicircle of radius r rotates about the diameter of the circle with an angular frequency ω. The axis of rotation is perpendicular to the field. If the total resistance of the circuit is R, the mean power generated per period of rotation is **Toughnut**

(a) $\frac{(B\pi r\omega)^2}{2R}$ (b) $\frac{(B\pi r^2\omega)^2}{8R}$

(c) $\frac{B\pi r^2\omega}{2R}$ (d) $\frac{(B\pi r\omega^2)^2}{8R}$

61. An ac source of angular frequency ω is fed across a resistor R and a capacitor C in series. The current registered is I. If now the frequency of source is changed to ω/3 (but maintaining the same voltage), the current in the circuit is found to be halved. Then the ratio of reactance to resistance at the original frequency ω is **Toughnut**

(a) $\sqrt{3/5}$ (b) $\sqrt{5/3}$ (c) $\sqrt{2/3}$ (d) $\sqrt{3/2}$

62. In the LC circuit, the current is in the direction shown and the charges on the capacitor plates have the signs shown. At this time **Tricky**

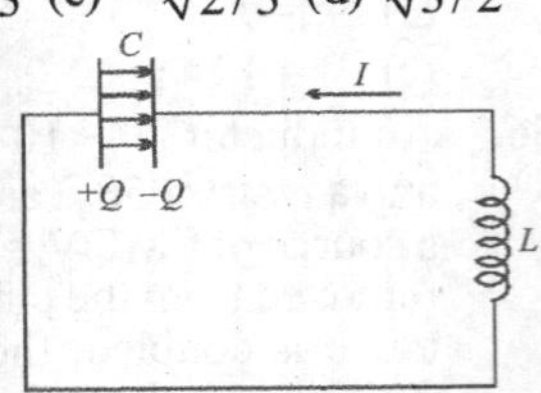

(a) I is increasing and Q is increasing
(b) I is increasing and Q is decreasing
(c) I is decreasing and Q is increasing
(d) I is decreasing and Q is decreasing

63. An LCR circuit contains resistance of 100 ohm and a supply of 200 volt at 300 radian angular frequency. If only capacitance is taken out from the circuit and the rest of the circuit is joined, current lags behind the voltage by 60°. If on the other hand, only inductor is taken out, the current leads by 60° with the applied voltage. The current flowing in the circuit is :

(a) 1 A (b) 1.5 A (c) 2 A (d) 2.5 A

64. In an electrical circuit R, L, C and an a.c. voltage source are all connected in series. When L is removed from the circuit, the phase difference between the voltage the current in the circuit is π/3. If instead, C is removed from the circuit, the phase difference is again π/3. The power factor of the circuit is : **Critical Thinking**

(a) 1/2 (b) $1/\sqrt{2}$ (c) 1 (d) $\sqrt{3}/2$

65. In given RC circuit, capacitance of capacitor C_1 = 3μF and C_2 = 1μF. It is given that time constant of circuit between A and B is 3 millisecond. Value of R will be **Tricky**

(a) 1Ω
(b) 10Ω
(c) 100Ω
(d) 1000Ω

66. The circuit shown has been operating for a long time. The instant after the switch in the circuit labeled S is opened, what is the voltage across the inductor V_L and which labeled point (A or B) of the inductor is at a higher potential ? Take R_1 = 4.0 Ω, R_2 = 8.0 Ω and L = 2.5 H.

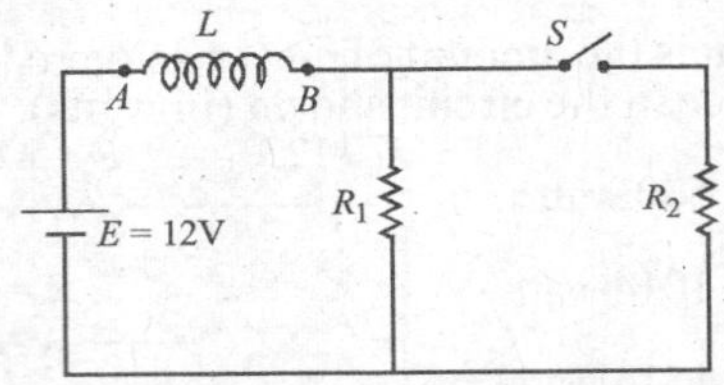

(a) V_L = 12V, point A is at the higher potential
(b) V_L = 12V, point B is at the higher potential
(c) V_L = 6V, point A is at the higher potential
(d) V_L = 6V, point B is at the higher potential

67. In the circuit shown below, the key K is closed at t = 0. The current through the battery is

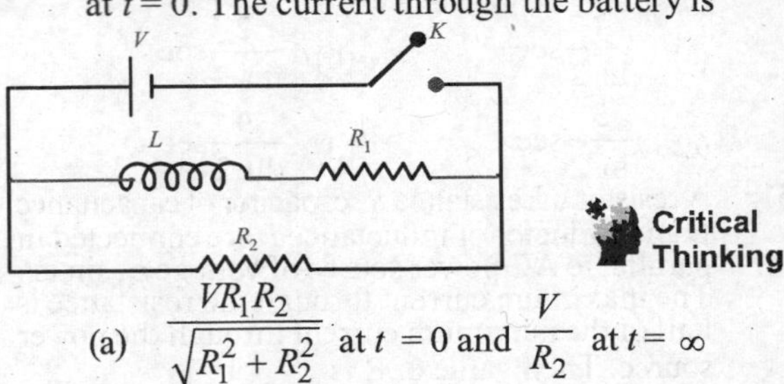

Critical Thinking

(a) $\frac{VR_1R_2}{\sqrt{R_1^2+R_2^2}}$ at t = 0 and $\frac{V}{R_2}$ at t = ∞

(b) $\frac{V}{R_2}$ at t = 0 and $\frac{V(R_1+R_2)}{R_1R_2}$ at t = ∞

(c) $\frac{V}{R_2}$ at t = 0 and $\frac{VR_1R_2}{\sqrt{R_1^2+R_2^2}}$ at t = ∞

(d) $\frac{V(R_1+R_2)}{R_1R_2}$ at t = 0 and $\frac{V}{R_2}$ at t = ∞

68. An arc lamp requires a direct current of 10 A at 80 V to function. If it is connected to a 220 V (rms), 50 Hz AC supply, the series inductor needed for it to work is close to :

(a) 0.044 H (b) 0.065 H
(c) 80 H (d) 0.08 H

69. In a series LCR circuit $R = 200\Omega$ and the voltage and the frequency of the main supply is 220V and 50 Hz respectively. On taking out the capacitance from the circuit the current lags behind the voltage by 30°. On taking out the inductor from the circuit the current leads the voltage by 30°. The power dissipated in the LCR circuit is

(a) 305 W (b) 210 W
(c) Zero W (d) 242 W

70. A fully charged capacitor C with initial charge q_0 is connected to a coil of self inductance L at $t = 0$. The time at which the energy is stored equally between the electric and the magnetic fields is:

(a) $\frac{\pi}{4}\sqrt{LC}$ (b) $2\pi\sqrt{LC}$ (c) $\sqrt{LC}$ (d) $\pi\sqrt{LC}$

71. An inductor (L = 0.03 H) and a resistor (R = 0.15 kΩ) are connected in series to a battery of 15V emf in a circuit shown below. The key K_1 has been kept closed for a long time. Then at t = 0, K_1 is opened and key K_2 is closed simultaneously. At t = 1 ms, the current in the circuit will be :

$(e^5 \cong 150)$ Critical Thinking

0.03 H
0.15 kΩ
K_2
15V
K_1

(a) 6.7 mA
(b) 0.67 mA
(c) 100 mA
(d) 67 mA

72. A resistor 'R' and 2μF capacitor in series is connected through a switch to 200 V direct supply. Across the capacitor is a neon bulb that lights up at 120 V. Calculate the value of R to make the bulb light up 5 s after the switch has been closed. ($\log_{10} 2.5 = 0.4$)

(a) $1.7 \times 10^5\,\Omega$ (b) $2.7 \times 10^6\,\Omega$
(c) $3.3 \times 10^7\,\Omega$ (d) $1.3 \times 10^4\,\Omega$

73. In the circuit shown here, the point 'C' is kept connected to point 'A' till the current flowing through the circuit becomes constant. Afterward, suddenly, point 'C' is disconnected from point 'A' and connected to point 'B' at time t = 0. Ratio of the voltage across resistance and the inductor at t = L/R will be equal to:

Tricky

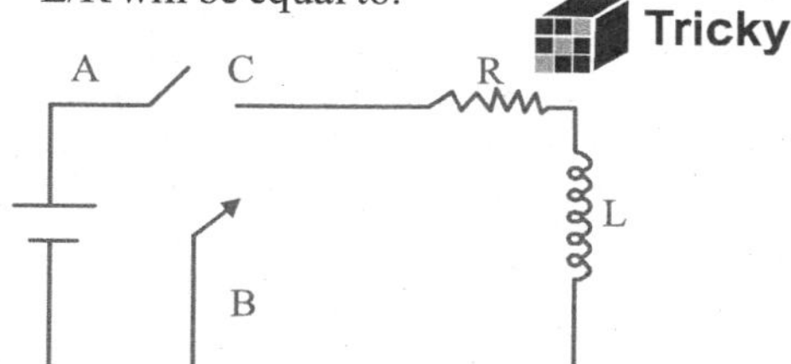

(a) $\frac{e}{1-e}$ (b) 1 (c) –1 (d) $\frac{1-e}{e}$

74. Combination of two identical capacitors, *a* resistor *R* and *a dc* voltage source of voltage 6V is used in an experiment on a (*C-R*) circuit. It is found that for a parallel combination of the capacitor the time in which the voltage of the fully charged combination reduces to half its original voltage is 10 second. For series combination the time needed for reducing the voltage of the fully charged series combination by half is

(a) 10 second (b) 5 second
(c) 2.5 second (d) 20 second

75. In an LCR circuit as shown below both switches are open initially. Now switch S_1 is closed, S_2 kept open. (q is charge on the capacitor and τ = RC is Capacitive time constant). Which of the following statements is correct ?

Critical Thinking

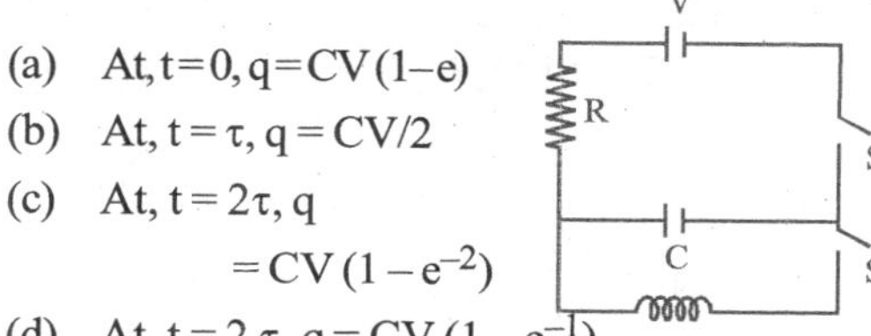

(a) At, t = 0, q = CV (1 – e)
(b) At, t = τ, q = CV/2
(c) At, t = 2τ, q = CV (1 – e^{-2})
(d) At, t = 2 τ, q = CV (1 – e^{-1})

76. Figure shows an iron - cored transformer assumed to be 100% efficient. The ratio olf the secondary turns to the primary turns is 1 : 20.

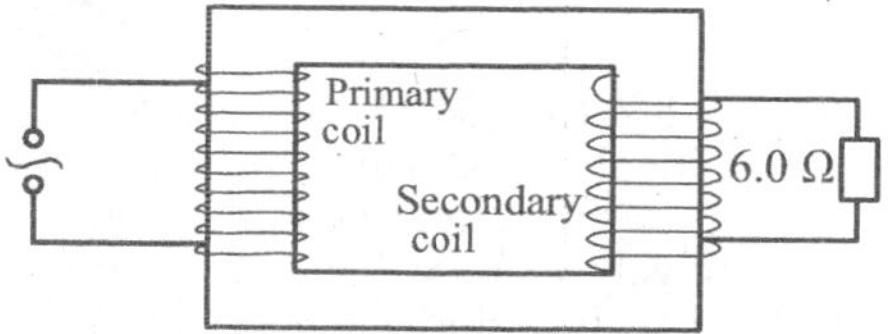

A 240 V ac supply is connected to the primary coil and a 6 W resistor is corrected to the secondary coil. What is the current in the primary coil?

(a) 0.10A (b) 0.14A (c) 2 A (d) 40 A

77. A step down transformer is connected to 2400 volts line and 80 amperes of current is found to flow in output load. The ratio of the turns in primary and secondary coil is 20 : 1. If transformer efficiency is 100%, then the current flowing in the primary coil will be

(a) 1600 amp (b) 20 amp
(c) 4 amp (d) 1.5 amp

78. At $t < 0$, the capacitor is charged and the switch is opened. At $t = 0$ the switch is closed. The shortest time T at which the charge on the capacitor will be zero is given by

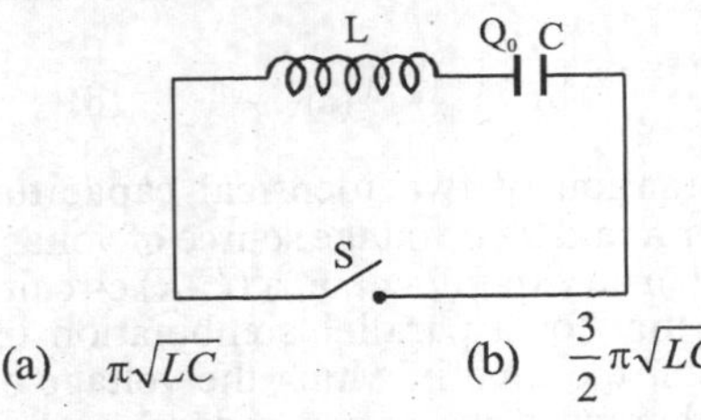

(a) $\pi\sqrt{LC}$ (b) $\frac{3}{2}\pi\sqrt{LC}$

(c) $\frac{\pi}{2}\sqrt{LC}$ (d) $2\pi\sqrt{LC}$

79. The primary and secondary coil of a transformer have 50 and 1500 turns respectively. If the magnetic flux ϕ linked with the primary coil is given by $\phi = \phi_0 + 4t$, where ϕ is in webers, t is time in seconds and ϕ_0 is a constant, the output voltage across the secondary coil is **Critical Thinking**

(a) 120 volts (b) 220 volts
(c) 30 volts (d) 90 volts

80. An ideal efficient transformer has a primary power input of 10 kW. The secondary current when the transformer is on load is 25 A. If the primary : secondary turns ratio is 8 : 1, then the potential difference applied to the primary coil is

(a) $\frac{10^4 \times 8^2}{25} V$ (b) $\frac{10^4 \times 8}{25} V$

(c) $\frac{10^4}{25 \times 8} V$ (d) $\frac{10^4}{25 \times 8^2} V$

81. A transformer is used to light a 140 W, 24 V bulb from a 240 V a.c. mains. The current in the main cable is 0.7 A. The efficiency of the transformer is **Tricky**

(a) 63.8 % (b) 83.3 % (c) 16.7 % (d) 36.2%

82. In an ideal transformer, the voltage and the current in the primary coil are 200 V and 2 A, respectively. If the voltage in the secondary coil is 2000 V, the value of current in the secondary coil will be

(a) 0.2 A (b) 2 A (c) 10 A (d) 20 A

83. In an oscillating LC circuit with L = 50 mH and C = 4.0 μF, the current is initially a maximum. How long will it take before the capacitor is fully discharged for the first time:

(a) 7×10^{-4} s (b) 14×10^{-4} s
(c) 28×10^{-4} s (d) none of these

84. The primary of a winding transformer has 400 turns while the secondary has 2000 turns. If the power output from the secondary at 1000 V is 12 kW, what is the primary voltage?

(a) 200 V (b) 300 V (c) 400 V (d) 500 V

85. Figure shows three oscillating LC circuit with identical inductors and capacitors. **Tricky** If t_1, t_2, t_3 are the time taken by the circuits I, II, III for fully discharge, then

(I) (II) (III)

(a) $t_1 > t_2 > t_3$
(b) $t_1 < t_2 < t_3$
(c) $t_2 < t_1 < t_3$ (d) $t_3 = \sqrt{t_1 t_2}$

86. The primary of a transformer when connected to a dc battery of 10 volt draws a current of 1 mA. The number of turns of the primary and secondary windings are 50 and 100 respectively. The voltage in the secondary and the current drawn by the circuit in the secondary are respectively **Tricky**

(a) 20 V and 0.5 mA
(b) 20 V and 2.0 mA
(c) 10 V and 0.5 mA
(d) Zero and therefore no current

87. A transformer having efficiency of 90% is working on 200V and 3kW power supply. If the current in the secondary coil is 6A, the voltage across the secondary coil and the current in the primary coil respectively are :

(a) 300 V, 15A (b) 450 V, 15A
(c) 450V, 13.5A (d) 600V, 15A

88. For long distance transmission, the A.C. is stepped up because of high voltage, the transmission is

(a) faster (b) economical
(c) undamped (d) less dangerous

89. A capacitor in an LC oscillator has a maximum potential difference of 17 V and a maximum energy of 160 μJ. When the capacitor has a potential difference of 5V and an energy of 10μJ, what is the energy stored in the magnetic field ?

(a) 10 μJ (b) 150 μJ (c) 160 μJ (d) 170 μJ

90. A transformer rated at 10 KW is used to connect a 5 KV transmission line to a 240 V circuit. The ratio of turns in the windings of the transformer is :

(a) 5 (b) 20.8 (c) 104 (d) 40

91. A step up transformer operates on a 230 V line and supplies a current of 2 ampere. The ratio of primary and secondary winding is 1:25. The current in primary is

(a) 25 A (b) 50 A (c) 15 A (d) 12.5 A

92. The tuning circuit of a radio receiver has a resistance of 50Ω, an inductor of 10 mH and a variable capacitor. A 1 MHz radio wave produces a potential difference of 0.1 mV. The values of the capacitor to produce resonance is (Take $\pi^2 = 10$) **Toughnut**

(a) 2.5 pF (b) 5.0 pF (c) 25 pF (d) 50 pF

93. A generator at a utility company produces 100 A of current at 4000 V. The voltage is stepped up to 2,40,000V by a transformer before it is sent on a high voltage transmission line. The current in transmission line is

(a) 3.67A (b) 2.67A (c) 1.67A (d) 2.40A

94. A transistor-oscillator using a resonant circuit with an inductor L (of negligible resistance) and a capacitor C in series produce oscillations of frequency f. If L is doubled and C is changed to 4C, the frequency will be

(a) 8f (b) $f/2\sqrt{2}$ (c) f/2 (d) f/4

95. A 220 volts input is supplied to a transformer. The output circuit draws a current of 2.0 ampere at 440 volts. If the efficiency of the transformer is 80%, the current drawn by the primary windings of the transformer is

(a) 3.6 ampere (b) 2.8 ampere
(c) 2.5 ampere (d) 5.0 ampere

Answer KEYs

1	(b)	11	(b)	21	(d)	31	(a)	41	(b)	51	(a)	61	(a)	71	(b)	81	(b)	91	(b)
2	(b)	12	(a)	22	(c)	32	(d)	42	(d)	52	(c)	62	(b)	72	(b)	82	(a)	92	(a)
3	(c)	13	(b)	23	(b)	33	(a)	43	(c)	53	(c)	63	(c)	73	(c)	83	(a)	93	(c)
4	(b)	14	(b)	24	(a)	34	(b)	44	(b)	54	(d)	64	(c)	74	(c)	84	(a)	94	(b)
5	(c)	15	(d)	25	(a)	35	(d)	45	(a)	55	(b)	65	(d)	75	(c)	85	(c)	95	(d)
6	(c)	16	(a)	26	(b)	36	(d)	46	(c)	56	(a)	66	(d)	76	(a)	86	(d)		
7	(c)	17	(d)	27	(b)	37	(b)	47	(a)	57	(c)	67	(c)	77	(c)	87	(b)		
8	(d)	18	(d)	28	(a)	38	(b)	48	(b)	58	(c)	68	(b)	78	(c)	88	(c)		
9	(a)	19	(b)	29	(a)	39	(b)	49	(c)	59	(d)	69	(d)	79	(a)	89	(b)		
10	(a)	20	(d)	30	(a)	40	(a)	50	(b)	60	(b)	70	(a)	80	(b)	90	(b)		

Hints & Solutions

1. **(b)** In series RLC circuit,

Voltage, $V = \sqrt{V_R^2 + (V_L - V_C)^2}$

And, at resonance, $V_L = V_C$

Hence, $V = V_R$

2. **(b)** $X_C = \frac{1}{\omega C} \Rightarrow X_C \alpha \frac{1}{\omega}$ for given C.

3. **(c)** Impedance at resonant frequency is minimum in series LCR circuit.

So, $Z = \sqrt{R^2 + \left(2\pi fL - \frac{1}{2\pi fC}\right)^2}$

When frequency is increased or decreased, Z increases.

4. **(b)** $\cos\phi = \frac{R}{Z}$, where Z is the impedance &

$Z = \sqrt{R^2 + (X_L - X_C)^2}$, if there is only resistance then $Z = R \Rightarrow \cos\phi = 1$

5. **(c)** Capacitive reactance, $X_C = \frac{1}{\omega C} = \frac{1}{2\pi\upsilon C}$

$\Rightarrow X_L \propto \frac{1}{\upsilon}$

With increases in frequency, X_C decreases. Hence, option (c) represents the corrent graph.

6. **(c)**

7. **(c)** As $X_C = \frac{1}{\omega C}$, so with the increase in frequency, X_C decreases and so i_C increases.

8. **(d)** as current any instant in the circuit will be $I = I_{dc} + I_{ac} = a + b\sin\omega t$

so,

$$I_{eff} = \left[\frac{\int_0^T I^2 dt}{\int_0^T dt}\right]^{1/2} = \left[\frac{1}{T}\int_0^T (a + b\sin\omega t)^2 dt\right]^{1/2}$$

i.e.

$$I_{eff} = \left[\frac{1}{T}\int_0^T \left(a^2 + 2ab\sin\omega t + b^2\sin^2\omega t\right) dt\right]^{1/2}$$

but as

$$\frac{1}{T}\int_0^T \sin\omega t dt = 0 \text{ and } \frac{1}{T}\int_0^T \sin^2\omega t\, dt = \frac{1}{2}$$

So, $I_{eff} = \left[a^2 + \frac{1}{2}b^2\right]^{1/2}$

9. **(a)** At resonance

$$I_R = \frac{E_0}{R} \Rightarrow \frac{I_R}{\sqrt{2}} = \frac{E_0}{\sqrt{2}R} = \frac{E_0}{\sqrt{R^2 + \left(\omega L - \frac{1}{\omega C}\right)^2}}$$

$\Rightarrow \omega_1 L - \frac{1}{\omega_1 C} = -R$ and $\omega_2 L - \frac{1}{\omega_2 C} = +R$

$\Rightarrow L(\omega_1 + \omega_2) = \left(\frac{\omega_1 + \omega_2}{\omega_1\omega_2}\right)\frac{1}{C} \Rightarrow \omega_1\omega_2 = \frac{1}{LC}$

and $L(\omega_2 - \omega_1) + \left(\frac{\omega_2 - \omega_1}{\omega_1\omega_2}\right)\frac{1}{C} = 2R$

Putting value of ω_1 and ω_2 we get

$$\omega_2 - \omega_1 = \frac{R}{L} \Rightarrow f_2 - f_1 = \frac{R}{2\pi L}$$

10. (a) $i_3 = i_1 + i_2 = 3\sin\omega t + 4\cos\omega t$
$= 5\sin(\omega t + \phi)$, where $\tan\phi = \frac{3}{4}$ or $\phi = 53°$.

11. (b) In a pure inductive circuit current always lags behind the emf by $\frac{\pi}{2}$.

If $v(t) = v_0 \sin\omega t$ then $I = I_0 \sin\left(\omega t - \frac{\pi}{2}\right)$

Now, given $v(t) = 100\sin(500t)$

and $I_0 = \frac{E_0}{\omega L} = \frac{100}{500 \times 0.02}$ [$\because$ L = 0.02 H]

$I_0 = 10\sin\left(500t - \frac{\pi}{2}\right) = -10\cos(500t)$

12. (a) RMS value over one cycle = RMS value over AB.

$0 \le t \le 15$

$i(t) = \sqrt{5}\, t^2$ c

$$i_{rms} = \sqrt{<i^2>} = \sqrt{\frac{5\int_0^1 t^4 dt}{\int_0^1 dt}} = 1\text{ mA}$$

13. (b) $Q = \frac{EC}{2}(1 - e^{-t/\tau})$

$I_2 = \frac{\varepsilon C}{2r} e^{-t/\tau} = \frac{\varepsilon}{R} e^{-t/\tau}$

$I_1 = \frac{Vc}{R} = \frac{\varepsilon/2}{R}(1 - e^{-t/\tau})$

$I_1 = \frac{\varepsilon}{R} - e^{-t/\tau} + \frac{\varepsilon}{2R} - \frac{\varepsilon}{2R}\varepsilon^{-t/\tau}$

$\frac{\varepsilon}{2R}(1 + e^{-t/\tau}) = \frac{\varepsilon}{2R}\left(1 + e^{\frac{-2t}{CR}}\right)$

[Circuit: battery ε, resistor R carrying I, branch I_1 through R, branch I_2 through C]

14. (b)

15. (d) Using $I = I_0(1 - e^{-t/\tau})$

But $I_0 = \frac{V}{R}$ and $\tau = \frac{L}{R}$

$\therefore\ I = \frac{V}{R}(1 - e^{-Rt/L}) = \frac{12}{6}\left[1 - e^{-6t/8.4\times10^{-3}}\right]$
$= 1$ (given)

$\therefore\ t = 0.97 \times 10^{-3}$ s ≈ 1ms

16. (a) $f = \frac{1}{2\pi\sqrt{LC}}$ i.e. $f \propto \frac{1}{\sqrt{C}} \rightarrow \frac{1}{\sqrt{4}} = \frac{1}{2}$ time

17. (d) $V = -L\frac{di}{dt}$ Here $\frac{di}{dt}$ is + ve for $\frac{T}{2}$ time and $\frac{di}{dt}$ is – ve for next $\frac{T}{2}$ time

18. (d) Given, $V = 120\sin(100\pi t)\cos(100\pi t)$
$= 60[2\sin(100\pi t)\cdot\cos(100\pi t)] = 60\sin 200\pi t$.

Thus, $V_0 = 60$ V, $\omega = 200\pi$ or $f = \frac{200\pi}{2\pi} = 100$Hz.

19. (b) We know that, $Z = \frac{E_0}{I_0}$

Given, $E_0 = 220$ and $I_0 = 10$

so $Z = \frac{220}{10} = 22$ ohm $\phi = \left[\frac{\pi}{6} - \left(-\frac{\pi}{6}\right)\right] = \frac{\pi}{3}$

$P_a = \frac{E_0}{\sqrt{2}} \times \frac{I_0}{\sqrt{2}} \times \cos\phi = \frac{220}{\sqrt{2}} \times \frac{10}{\sqrt{2}} \times \cos\frac{\pi}{3}$

20. (d) Here, $X_L = \omega L = 2\pi f L = 2\pi \times 50 \times \frac{0.4}{\pi} = 40\Omega$

$R = 30\Omega$

$\therefore\ Z = \sqrt{R^2 + X_L^2} = \sqrt{30^2 + 40^2} = 50\Omega$

$I_{rms} = \frac{V_{rms}}{Z} = \frac{200}{50} = 4$A

21. (d) $V_{rms} = \frac{200\sqrt{2}}{\sqrt{2}} = 200V$

$I_{rms} = \frac{V_{rms}}{X_C} = \frac{200}{\frac{1}{100 \times 10^{-6}}} = 2 \times 10^{-2} = 20$mA

22. (c) The equation of a semi - circular wave is $x^2 + y^2 = a^2$ or $y^2 = a^2 - x^2$

$$I_{rms} = \sqrt{\frac{1}{2a}\int_{-a}^{+a} y^2 dx}$$

$$I^2_{rms} = \frac{1}{2a}\int_{-a}^{+a}(a^2 - x^2)dx$$

$$= \frac{1}{2a}\int_{-a}^{+a}(a^2 - x^2)dx = \frac{1}{2a}\left|a^2 x - \frac{x^3}{3}\right|_{-a}^{+a}$$

$$= \frac{1}{2a}\left(a^3 - \frac{a^3}{3} + a^3 - \frac{a^3}{3}\right) = \frac{2a^2}{3}$$

$$I_{rms} = \sqrt{\frac{2a^2}{3}} = \sqrt{\frac{2}{3}}a$$

23. (b) $V_{rms} = \sqrt{\frac{(T/2)V_0^2 + 0}{T}} = \frac{V_0}{\sqrt{2}}$.

24. (a) $2\pi nt = 400\pi t$ $\therefore$ n = 200 $I_0 = 50\sqrt{2}$ amp.

r.m.s. current $= I_0/\sqrt{2} = 50$ amp.

25. (a) If $\omega = 50 \times 2\pi$ then $\omega L = 20\Omega$
If $\omega' = 100 \times 2\pi$ then $\omega' L = 40\Omega$
Current flowing in the coil is

$$I=\frac{200}{Z}=\frac{200}{\sqrt{R^2+(\omega' L)^2}}=\frac{200}{\sqrt{(30)^2+(40)^2}}$$

$I = 4A.$

26. (b) As we know, power $P = V_{rms} \cdot I_{rms} \cos\phi$

as $\cos\phi = 0$ $(\because \phi = 90°)$

$\therefore$ Power consumed $= 0$ (in one complete cycle)

27. (b) The phase difference ϕ is given by

$$\tan\phi=\frac{X_L}{R}=\frac{3}{3}=1 \Rightarrow \phi=\frac{\pi}{4}.$$

28. (a) $V=\frac{V_0}{T/4}t \Rightarrow V=\frac{4V_0}{T}t$

$$\Rightarrow V_{rms}=\sqrt{\langle V^2\rangle}=\frac{4V_0}{T}\sqrt{\langle t\rangle}$$

$$=\frac{4V_0}{T}\left\{\frac{\int_0^{T/4} t^2 dt}{\int_0^{T/4} dt}\right\}=\frac{V_0}{\sqrt{3}}$$

29. (a) Given: $L = 20$ mH; $C = 50\ \mu F$; $R = 40\ \Omega$

$V = 10 \sin 340t$ $\therefore V_{runs}=\frac{10}{\sqrt{2}}$

$$X_C=\frac{1}{\omega C}=\frac{1}{340\times 50\times 10^{-6}}=58.8\ \Omega$$

$X_L = \omega L = 340 \times 20 \times 10^{-3} = 6.8\ \Omega$

Impedance, $Z=\sqrt{R^2+(X_C-X_L)^2}$

$$=\sqrt{40^2+(58.8-6.8)^2}=\sqrt{4304}\ \Omega$$

Power loss in A.C. circuit,

$$P=i_{rms}^2 R=\left(\frac{V_{rms}}{Z}\right)^2 R=\left(\frac{10/\sqrt{2}}{\sqrt{4304}}\right)^2\times 40$$

$$=\frac{50\times 40}{4304}\simeq 0.51W$$

30. (a)

31. (a) $R_{eq}=\frac{5R}{6} \Rightarrow I=\frac{6E}{5R}=1A$

32. (d)

33. (a) $q = CV_1\cos\omega t \Rightarrow i=\frac{dq}{dt}=-\omega C v_1 \sin\omega t$

Also, $\omega^2=\frac{1}{LC}$ and $V = V_1\cos\omega t$

At $t = t_1$, $V = V_2$ and $i = -\omega C V_1 \sin\omega t_1$

$\therefore \cos\omega t_1=\frac{V_2}{V_1}$ (–ve sign gives direction)

$$\text{Hence, } i=V_1\sqrt{\frac{C}{L}}\left(1-\frac{V_2^2}{V_1^2}\right)^{1/2}=\left(\frac{C(V_1^2-V_2^2)}{L}\right)^{1/2}$$

34. (b) $I = I_C + (-I_L)$

Because I_C and I_L are 180° out of phase with each other.

35. (d) Condition for which the current is maximum in a series LCR circuit is,

$$\omega=\frac{1}{\sqrt{LC}}$$

$$1000=\frac{1}{\sqrt{L(10\times 10^{-6})}} \Rightarrow L=100\ \text{mH}$$

36. (d) Current in inductor $=\frac{E}{R}$

$\therefore$ its energy $=\frac{1}{2}\frac{LE^2}{R^2}$

Same energy is later stored in capacitor

$$\frac{Q^2}{2C}=\frac{1}{2}\frac{LE^2}{R^2}\Rightarrow Q=\sqrt{LC}\frac{E}{R}$$

37. (b) Power factor, $\phi=\frac{R}{\sqrt{\left(\omega L-\frac{1}{\omega C}\right)^2+R^2}}$

$$=\frac{8}{\sqrt{(31-25)^2+8^2}}=\frac{8}{\sqrt{6^2+8^2}}=\frac{8}{10}=0.8$$

38. (b) The three branches of the circuits carry currents

$i = i_0$, $i = i_1 e^{t/RC}$ and $i = i_2 2e^{-t/2RC}$ respectively.

The current through the cell, $i(t)$ can be found by using Kirchhoff's current law (or mode law).

39. (b) Time constant is L/R Given, $L = 40$H & $R = 8\Omega$ $\therefore \tau = 40/8 = 5$ sec.

40. (a) In an ac circuit, a pure indcutor does not consume any power. Therefore, power is consumed by the resistor only.

$\therefore$ $P = I_v^2 R$

or $108 = (3)^2$ R or $R = 12\ \Omega$

41. (b) $P=\frac{E^2}{R}=\frac{\pi r^2}{\rho\ell}\left(\frac{d\phi}{dt}\right)^2=\frac{\pi r^2}{\rho\ell}\left[\frac{d}{dt}(NBA)^2\right]$

$$=\frac{\pi r^2}{\rho\ell}N^2A^2\left(\frac{dB}{dt}\right)^2 \Rightarrow P\propto\frac{N^2r^2}{\ell}$$

Case 1 : $P_1\propto\frac{N^2r^2}{\ell}$

Case 2 : $P_2\propto\frac{(4N)^2(r/2)^2}{4\ell}$

Note : When we decrease the radius of the wire, its length increases but volume remains the same

$\Rightarrow \frac{P_1}{P_2}=\frac{1}{1}$ $\therefore$ Power remains the same.

42. (d)

43. (c) For capacitor circuit, $i = i_0 e^{-t/RC}$

For inductor circuit, $i=i_0\left(1-e^{-\frac{Rt}{L}}\right)$

Hence graph (c) correctly depicts i versus t graph.

44. (b)

45. (a) $i=\sqrt{5}A$

$$\frac{q_m^2}{2C}=\frac{q^2}{2C}+\frac{1}{2}Li^2 \Rightarrow q_{max}=6C$$

46. (c) $i_1=\frac{dq}{dt}=\frac{d}{dt}(2\cos 4t)=-8\sin 4t,$

$i_2=\frac{dq}{dt}=\frac{d}{dt}(4\cos t)=-4\sin t,$

$i_3=\frac{dq}{dt}=\frac{d}{dt}(3\cos 4t)=-12\sin 4t,$

$i_4=\frac{dq}{dt}=\frac{d(4\cos 2t)}{dt}=-8\sin 2t.$

Clearly, amplitude of i_3 is greatest.

47. (a) Since the cpacitor is connected in series to the resistor, current I_C from the supply and I_R through the resistor is in phase as represented by choice (a).

48. (b) Here, $X_L=\omega_L=2\pi fL=2\pi\times 50\times 1=100\pi\,\Omega$

$X_c=\frac{1}{\omega C}=\frac{1}{2\pi fC}=\frac{1}{2\pi\times 50\times 10\times 10^{-6}}=\frac{10^3}{\pi}\Omega$

So, X

$=|X_L-X_C|=\left|100\pi-\frac{10^3}{\pi}\right|=\left|10^2\left[\frac{\pi^2-10}{\pi}\right]\right|\Omega$

49. (c) $i_{1rms}=\frac{E_{rms}}{\sqrt{X_C^2+R_1^2}}=\frac{130}{13}=10A$

$i_{2rms}=\frac{E_{rms}}{\sqrt{X_L^2+R_2^2}}$

$=13A$

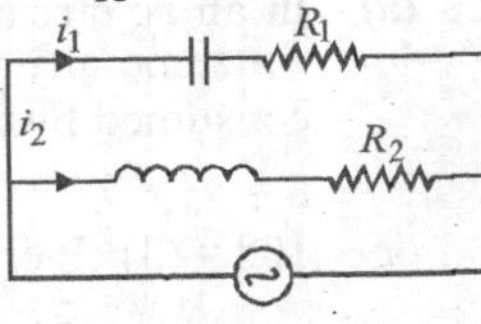

Power dissipated

$=i^2_{1rms}R_1+i^2_{2rms}R_2=10^2\times 5+13^2+6$

= power delivered by battery

$=500+169\times 6=1514$ watt

50. (b) $I=I_0(I-e^{-t/\tau})$

where $\tau\rightarrow$ time constant

$\therefore\quad \frac{3}{4}I_0=I_0(1-e^{-t/\tau})\Rightarrow\frac{3}{4}=l-e^{-t/\tau}$

$\Rightarrow\quad e^{-t/\tau}=\frac{1}{4}$

$\Rightarrow\quad \frac{-t}{\tau}\ln e=\ln\frac{1}{4}\Rightarrow\frac{-4}{\tau}=-2\ln 2\Rightarrow\tau=\frac{2}{\ln 2}$

51. (a)

52. (c) Given, $V_L:V_C:V_R=1:2:3$

V = 100 V V_R = ?

As we know, $V=\sqrt{V_R^2+(V_L-V_C)^2}$

Solving we get, $V_R\approx 90V$

53. (c) A circuit equivalent to the given circuit can be drawn as follows :

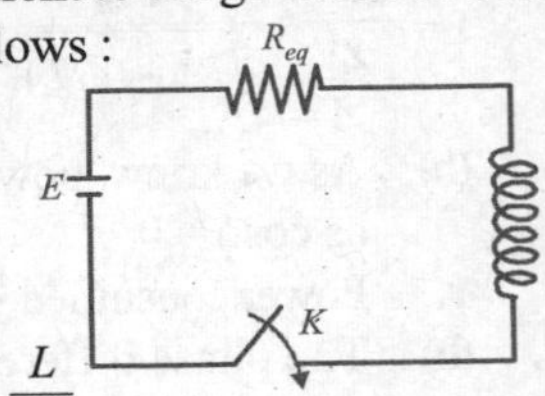

Where

$R_{eq}=R+\frac{2R\times 2R}{4R}$

$=R+R$

$R_{eq}=2R$

$\therefore$ Time constant $=\frac{L}{2R}$

54. (d) Power factor $_{(old)}$

$=\frac{R}{\sqrt{R^2+X_L^2}}=\frac{R}{\sqrt{R^2+(2R)^2}}=\frac{R}{\sqrt{5}R}$

Power factor$_{(new)}$

$=\frac{R}{\sqrt{R^2+(X_L-X_C)^2}}=\frac{R}{\sqrt{R^2+(2R-R)^2}}=\frac{R}{\sqrt{2}R}$

$\therefore\quad\frac{\text{New power factor}}{\text{Old power factor}}=\frac{\frac{R}{\sqrt{2}R}}{\frac{R}{\sqrt{5}R}}=\sqrt{\frac{5}{2}}$

55. (b)

56. (a) $X_C-X_L=31.85-2(6.28)=19.29$

$I_m=\frac{10}{19.29}=0.52$

Hence $I=0.52\sin(314t+\pi/2)=0.52\cos 314t$

57. (c) Resonance frequency,

$\omega_0=\frac{1}{\sqrt{LC}}=\frac{1}{\sqrt{L'(2C)}}\quad\therefore\quad L'=\frac{L}{2}.$

58. (c) $V=I\sqrt{(20+10)^2+(X_L+X_L-X_C)^2}$

$100=2\sqrt{(30)^2+(30+X_L-40)^2}$

$(50)^2=(30)^2+(X_L-10)^2$

$(X_L-10)^2=80\times 20:X_L-10=40;X_L=50\,\Omega$

59. (d)

60. (b) $\phi=\vec{B}\cdot\vec{A};\phi=BA\cos\omega t$

$\varepsilon=-\frac{d\phi}{dt}=\omega BA\sin\omega t\;;\;i=\frac{\omega BA}{R}\sin\omega t$

$P_{inst}=i^2R=\left(\frac{\omega BA}{R}\right)^2\times R\sin^2\omega t$

$$P_{avg}=\frac{\int_0^T P_{inst}\times dt}{\int_0^T dt}=\frac{(\omega BA)^2}{R}\frac{\int_0^T\sin^2\omega t dt}{\int_0^T dt}=\frac{1}{2}\frac{(\omega BA)^2}{R}$$

$$\therefore P_{avg}=\frac{(\omega B\pi r^2)^2}{8R}\qquad\left[A=\frac{\pi r^2}{2}\right]$$

61. **(a)** According to given problem,

$$I = \frac{V}{Z} = \frac{V}{[R^2 + (1/C\omega)^2]^{1/2}} \quad ....(i)$$

and $$\frac{I}{2} = \frac{V}{[R^2 + (3/C\omega)^2]^{1/2}} \quad ....(ii)$$

Substituting the value of I from eq. (i) in (ii),

$$4\left(R^2 + \frac{1}{C^2\omega^2}\right) = R^2 + \frac{9}{C^2\omega^2}$$

i.e., $\frac{1}{C^2\omega^2} = \frac{3}{5}R^2$ So that $\frac{X}{R}$

$$= \frac{(1/C\omega)}{R} = \frac{\left(\frac{3}{5}R^2\right)^{1/2}}{R} = \sqrt{\frac{3}{5}}$$

62. **(b)**

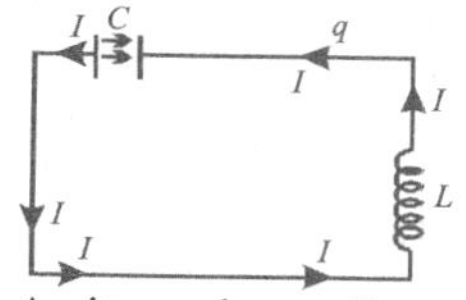

As it can be easily seen by the direction of I that Q is decreasing thus, energy of capacitor is decreasing and hence, energy of inductance is increasing or $\left(\frac{1}{2}LI^2\right)$ gives that I is increasing.

63. **(c)**

64. **(c)**

when L is removed from the circuit

$$\frac{X_C}{R} = \tan\frac{\pi}{3}$$

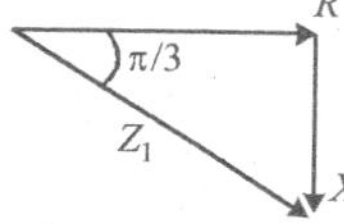

$X_C = R\tan\frac{\pi}{3}$ ....(i)

when C is remove from the circuit

$$\frac{X_L}{R} = \tan\frac{\pi}{3}$$

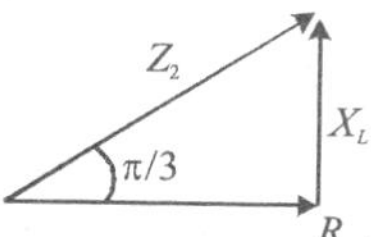

$X_C = R\tan\frac{\pi}{3}$ ...(ii)

net impedence $Z = \sqrt{R^2 + (X_L - X_C)^2} = R$

power factor $\cos\phi = \frac{R}{Z} = 1$

65. **(d)** Equivalent circuit can be drawn as below

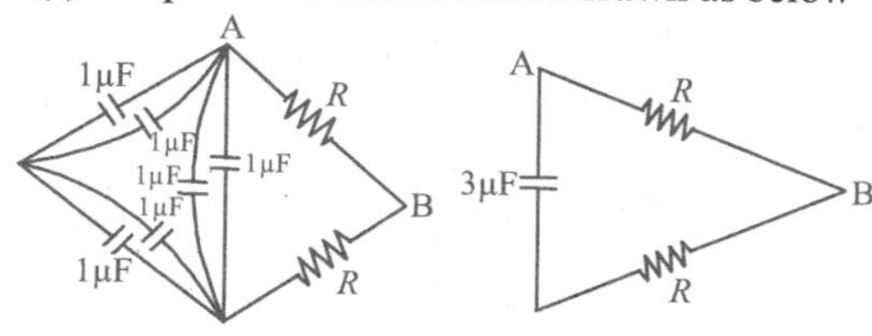

$RC = 3 \times 10^{-3}$ sec

$R \times 3 \times 10^{-6} = 3 \times 10^{-3}$

$R = 1000\,\Omega$

66. **(d)**

67. **(c)** At $t = 0$, no current will flow through L and R_1

$\therefore$ Current through battery $= \frac{V}{R_2}$

At $t = \infty$,

effective resistance, $R_{eff} = \frac{R_1R_2}{R_1 + R_2}$

$\therefore$ Current through battery $= \frac{V}{R_{eff}} = \frac{V(R_1 + R_2)}{R_1R_2}$

68. **(b)** Here

$$i = \frac{e}{\sqrt{R^2 + X_L^2}} = \frac{e}{\sqrt{R^2 + \omega^2L^2}} = \frac{e}{\sqrt{R^2 + 4\pi^2\nu^2L^2}}$$

$$10 = \frac{220}{\sqrt{64 + 4\pi^2(50)^2 L}} \quad \left[\because R = \frac{V}{I} = \frac{80}{10} = 8\right]$$

On solving we get $L = 0.065$ H

69. **(d)**

70. **(a)** Energy stored in magnetic field $= \frac{1}{2}Li^2$

Energy stored in electric field $= \frac{1}{2}\frac{q^2}{C}$

$\therefore \frac{1}{2}Li^2 = \frac{1}{2}\frac{q^2}{C}$

Also $q = q_0 \cos\omega t$ and $\omega = \frac{1}{\sqrt{LC}}$

On solving $t = \frac{\pi}{4}\sqrt{LC}$

71. **(b)** $I(0) = \frac{15 \times 100}{0.15 \times 10^3} = 0.1$ A

$I(\infty) = 0$

$I(t) = [I(0) - I(\infty)]\, e^{\frac{-t}{L/R}} + i(\infty)$

$I(t) = 0.1\, e^{\frac{-t}{L/R}} = 0.1\, e^{\frac{R}{L}}$

$I(t) = 0.1\, e^{\frac{0.15 \times 1000}{0.03}} = 0.67$ mA

72. **(b)** We have, $V = V_0(1 - e^{-t/RC})$

$\Rightarrow 120 = 200(1 - e^{-t/RC}) \Rightarrow t = RC \text{ in } (2.5) \Rightarrow R = 2.71 \times 10^6\,\Omega$

73. **(c)** **74.** **(c)**

75. **(c)** Charge on he capacitor at any time t is given by $q = CV\ (1 - e^{t/\tau})$ at $t = 2\tau$

$q = CV\,(1 - e^{-2})$

76. **(a)** The equivalent primary load is

$$R_1 = \left(\frac{N_1}{N_2}\right)^2 R_2 = \left(\frac{20}{1}\right)^2 (6.0) = 2400\,\Omega$$

Current in the primary coil $= \frac{240}{R_1} = \frac{240}{2400} = 0.1\,A$

77. **(c)** $\frac{I_s}{I_p} = \frac{n_p}{n_s};\ \frac{80}{I_p} = \frac{20}{1}$ or $I_p = 4$ amp.

78. **(c)**

79. **(a)** Since $\frac{V_s}{V_p} = \frac{N_s}{N_p}$

Where

N_s = No. of turns across primary coil = 50

N_p = No. of turns across secondary coil = 1500

and $V_p = \frac{d\phi}{dt} = \frac{d}{dt}(\phi_0 + 4t) = 4 \Rightarrow V_s = \frac{1500}{50} \times 4$

$= 120\,V$

80. **(b)** $P = V_1 i_1 = V_2 i_2$ $10 \times 10^3 = 25 \times V_2$

$V_2 = \frac{10^4}{25}$

Now $V_1 = \frac{n_1}{n_2} \times V_2 = \frac{8}{1} \times \frac{10^4}{25} V$

81. **(b)** Power of source = EI = 240 × 0.7 = 166

$\Rightarrow$ Efficiency $= \frac{140}{166} \Rightarrow \eta = 83.3\%$

82. **(a)**

83. **(a)** Time period,

$$T = 2\pi\sqrt{LC} = 2\pi\sqrt{(50\times10^{-3})\times4\times10^{-6}}$$
$$= 28 \times 10^{-4}\ s$$

Time taken by capacitor to charge fulley,

$t = \frac{T}{4} = 7\times10^{-4} s$.

84. **(a)** $N_P = 400, N_S = 2000$ and $V_S = 1000\,V$.

$\frac{V_P}{V_S} = \frac{N_P}{N_S}$ of, $V_P = \frac{V_S \times N_P}{N_S} = \frac{1000\times400}{2000} = 200V$.

85. **(c)** Time period, $T_1 = 2\pi\sqrt{LC}$, $T_2 = 2\pi\sqrt{\frac{LC}{2}}$,

$T_3 = 2\pi\sqrt{2LC}$.

Clearly $t_2 < t_1 < t_3$.

86. **(d)** A transformer is essentially an *AC* device. *DC* source so no mutual induction between coils

$\Rightarrow E_2 = 0$ and $I_2 = 0$

87. **(b)** Efficiency $\eta = \frac{V_s I_s}{V_p I_p} \Rightarrow 0.9 = \frac{V_s(6)}{3\times10^3} \Rightarrow V_s$

$= 450\,V$

As $V_p I_p = 3000$ so $I_p = \frac{3000}{V_p} = \frac{3000}{200} A = 15A$

88. **(c)** The transmission is undamped, because energy loss becomes negligible.

89. **(b)** $U_e + U_m = 160$ $\therefore U_m = 160 - U_e = 160 - 10$

$= 150\,\mu J$.

90. **(b)**

91. **(b)** $\frac{n_p}{n_s} = \frac{E_p}{E_s} = \frac{1}{25}$ $\therefore E_s = 25\,E_p$

But $E_s I_s = E_p I_p \Rightarrow I_p = \frac{E_S \times I_S}{E_P} \Rightarrow I_p = 50\,A$

92. **(a)** $L = 10\,mHz = 10^{-2}\,Hz$ $f = 1MHz = 10^6\,Hz$

$f = \frac{1}{2\pi\sqrt{LC}} \Rightarrow f^2 = \frac{1}{4\pi^2 LC}$

$\Rightarrow C = \frac{1}{4\pi^2 f^2 L} = \frac{1}{4\times10\times10^{-2}\times10^{12}}$

$= \frac{10^{-12}}{4} = 2.5\,pF$

93. **(c)** For a transformer,

$V_p I_p = V_s I_s \Rightarrow I_S = \frac{V_P I_P}{V_s} = \frac{4000\times100}{240000} A$

$= 1.67A$

94. **(b)** We know that frequency of electrical oscillation in L.C. circuit is

$f = \frac{1}{2\pi}\sqrt{\frac{1}{LC}}$

Now, L = 2L & C = 4C

$f' = \frac{1}{2\pi}\sqrt{\frac{1}{2L\cdot4C}} = \frac{1}{2\pi}\sqrt{\frac{1}{LC}} \times \frac{1}{2\sqrt{2}} \Rightarrow f'$

$= \frac{1}{2\sqrt{2}} \times f$

95. **(d)** $\frac{V_2}{V_1} = 0.8\frac{I_1}{I_2} \Rightarrow \frac{V_2 I_2}{V_1 I_1} = 0.8$

$V_1 = 220\,V, I_2 = 2.0\,A, V_2 = 440\,V$

$I_1 = \frac{V_2 I_2}{V_1} \times \frac{10}{8} = \frac{440\times2\times10}{220\times8} = 5A$

22 Electromagnetic Waves

1. The magnetic field in the plane electromagnetic field is given by:
$B_y = 2 \times 10^{-7} \sin(0.5 \times 10^3 z + 1.5 \times 10^{11} t)$ T
The expression for the electric field may be given by

(a) $E_y = 2 \times 10^{-7} \sin(0.5 \times 10^3 z + 1.5 \times 10^{11} t)$ V/m
(b) $E_x = 2 \times 10^{-7} \sin(0.5 \times 10^3 z + 1.5 \times 10^{11} t)$ V/m
(c) $E_y = 60 \sin(0.5 \times 10^3 z + 1.5 \times 10^{11} t)$ V/m
(d) $E_x = 60 \sin(0.5 \times 10^3 z + 1.5 \times 10^{11} t)$ V/m

2. The figure shows graphs of the electric field magnitude E versus time t for four uniform electric fields, all contained within identical circular regions. Which of them is according to the magnitudes of the magnetic field?

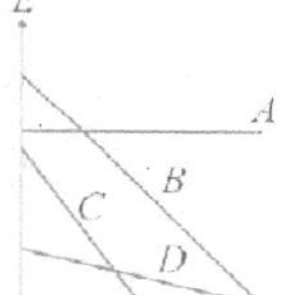

(a) A (b) B (c) C (d) D

3. In an apparatus, the electric field was found to oscillate with an amplitude of 24 V/m. The magnitude of the oscillating magnetic field will be
(a) 6×10^{-6} T (b) 2×10^{-8} T
(c) 8×10^{-8} T (d) 12×10^{-6} T

4. A circular ring of radius r is placed in a homogeneous magnetic field perpendicular to the plane of the ring. The field B changes with time according to equation B = kt, where, k is a constant and t is the time). The electric field in the ring is:

(a) $\frac{kr}{4}$ (b) $\frac{kr}{3}$ (c) $\frac{kr}{2}$ (d) $\frac{k}{2r}$

5. The rms value of the electric field of the light coming from the Sun is 720 N/C. The average total energy density of the electromagnetic wave is
(a) 4.58×10^{-6} J/m^3 (b) 6.37×10^{-9} J/m^3
(c) 81.35×10^{-12} J/m^3 (d) 3.3×10^{-3} J/m^3

6. For plane electromagnetic waves propagating in the z-direction, which one of the following combination gives the correct possible direction for $\vec{E}$ and $\vec{B}$ field respectively?

(a) $(2\hat{i} + 3\hat{j})$ and $(\hat{i} + 2\hat{j})$
(b) $(-2\hat{i} - 3\hat{j})$ and $(3\hat{i} - 2\hat{j})$
(c) $(3\hat{i} + 4\hat{j})$ and $(4\hat{i} - 3\hat{j})$
(d) $(\hat{i} + 2\hat{j})$ and $(2\hat{i} - \hat{j})$

7. A plane electromagnetic wave propagating in the X-direction has wavelength of 6.0 mm. The electric field is in the Y-direction and its maximum magnitude is 33 Vm^{-1}. The equation for the electric field as a function of x and t is
(a) $11 \sin \pi \left(t - \frac{x}{c}\right)$
(b) $33 \sin \left[\pi \times 10^{11} \left(t - \frac{x}{c}\right)\right]$
(c) $33 \sin \pi \left(t - \frac{x}{c}\right)$
(d) $11 \sin \left[\pi \times 10^{11} \left(t - \frac{x}{c}\right)\right]$

8. A new system of unit is evolved in which the values of μ_0 and ϵ_0 are 2 and 8 respectively. Then the speed of light in this system will be
(a) 0.25 (b) 0.5 (c) 0.75 (d) 1

9. An electromagnetic wave of frequency 1×10^{14} hertz is propagating along z-axis. The amplitude of electric field is 4 V/m. If $\varepsilon_0 = 8.8 \times 10^{-12}$ C^2/N-m^2, then average energy density of electric field will be:

Critical Thinking

(a) 35.2×10^{-10} J/m^3 (b) 35.2×10^{-11} J/m^3
(c) 35.2×10^{-12} J/m^3 (d) 35.2×10^{-13} J/m^3

10. In an electromagnetic wave, the electric and magnetic fields are 100 V m^{-1} and 0.265 A m^{-1}. The maximum energy flow is
(a) 26.5 W/m^2 (b) 36.5 W/m^2
(c) 46.7 W/m^2 (d) 765 W/m^2

11. If ε_0 and μ_0 are the electric permittivity and magnetic permeability in vacuum, ε and μ are corresponding quantities in medium, then refractive index of the medium is

(a) $\sqrt{\frac{\varepsilon}{\varepsilon_0}}$ (b) $\sqrt{\frac{\varepsilon_0\mu}{\varepsilon\mu_0}}$ (c) $\sqrt{\frac{\varepsilon_0\mu_0}{\varepsilon\mu}}$ (d) $\sqrt{\frac{\varepsilon\mu}{\varepsilon_0\mu_0}}$

12. For wave propagation of a 10 MHz signal, what should be the maximum electron density in ionosphere
(a) $\sim 1.2\times 10^{12} m^{-3}$ (b) $\sim 10^{6} m^{-3}$
(c) $\sim 10^{14} m^{-3}$ (d) $\sim 10^{22} m^{-3}$

13. A point source of electromagnetic radiation has an average power output of 800 W. The maximum value of electric field at a distance 4.0 m from the source is **Critical Thinking**
(a) 64.7 V/m (b) 57.8 V/m
(c) 56.72 V/m (d) 54.77 V/m

14. In a plane electromagnetic wave propagating in space has an electric field of amplitude 9×10^3 V/m, then the amplitude of the magnetic field is
(a) 2.7×10^{12} T (b) 9.0×10^{-3} T
(c) 3.0×10^{-4} T (d) 3.0×10^{-5} T

15. If c is the speed of electromagnetic waves in vacuum, its speed in a medium of dielectric constant K and relative permeability μ_r is

(a) $v=\frac{1}{\sqrt{\mu_r K}}$ (b) $v=c\sqrt{\mu_r K}$

(c) $v=\frac{c}{\sqrt{\mu_r K}}$ (d) $v=\frac{K}{\sqrt{\mu_r C}}$

16. An electromagnetic wave with frequency ω and wavelength λ travels in the $+y$ direction. Its magnetic field is along $+x$-axis. The vector equation for the associated electric field (of amplitude E_0) is

(a) $\vec{E}=-E_0\cos\left(\omega t+\frac{2\pi}{\lambda}y\right)\hat{x}$

(b) $\vec{E}=E_0\cos\left(\omega t-\frac{2\pi}{\lambda}y\right)\hat{x}$

(c) $\vec{E}=E_0\cos\left(\omega t-\frac{2\pi}{\lambda}y\right)\hat{z}$

(d) $\vec{E}=-E_0\cos\left(\omega t+\frac{2\pi}{\lambda}y\right)\hat{z}$

17. The magnetic field between the plates of radius 12 cm separated by distance of 4 mm of a parallel plate capacitor of capacitance 100 pF along the axis of plates having conduction current of 0.15 A is
(a) zero (b) 1.5 T (c) 15 T (d) 0.15 T

18. Consider an electromagnetic wave propagating in vacuum. Choose the correct statement: **Tricky**

(a) For an electromagnetic wave propagating in +y direction the electric field is $\vec{E}=\frac{1}{\sqrt{2}}E_{yz}(x,t)\hat{z}$ and the magnetic field is $\vec{B}=\frac{1}{\sqrt{2}}B_z(x,t)\hat{y}$

(b) For an electromagnetic wave propagating in +y direction the electric field is $\vec{E}=\frac{1}{\sqrt{2}}E_{yz}(x,t)\hat{y}$ and the magnetic field is $\vec{B}=\frac{1}{\sqrt{2}}B_{yz}(x,t)\hat{z}$

(c) For an electromagnetic wave propagating in +x direction the electric field is $\vec{E}=\frac{1}{\sqrt{2}}E_{yz}(y,z,t)(\hat{y}+\hat{z})$ and the magnetic field is $\vec{B}=\frac{1}{\sqrt{2}}B_{yz}(y,z,t)(\hat{y}+\hat{z})$

(d) For an electromagnetic wave propagating in +x direction the electric field is $\vec{E}=\frac{1}{\sqrt{2}}E_{yz}(x,t)(\hat{y}-\hat{z})$ and the magnetic field is $\vec{B}=\frac{1}{\sqrt{2}}B_{yz}(x,t)(\hat{y}+\hat{z})$

19. An electromagnetic wave of frequency $\nu = 3.0$ MHz passes from vacuum into a dielectric medium with permittivity $\epsilon = 4.0$. Then
(a) wavelength is halved and frequency remains unchanged
(b) wavelength is doubled and frequency becomes half
(c) wavelength is doubled and the frequency remains unchanged
(d) wavelength and frequency both remain unchanged.

20. A plane electromagnetic wave is incident on a plane surface of area A, normally and is perfectly reflected. If energy E strikes the surface in time t then force exerted on the surface is (c = speed of light) **Tricky**

(a) $\frac{2E}{Atc}$ (b) $\frac{E}{2c}$ (c) $\frac{2E}{ct}$ (d) zero

21. If a source is transmitting electromagnetic wave of frequency 8.2×10^6 Hz, then wavelength of the electromagnetic waves transmitted from the source will be
(a) 36.6m (b) 40.5m (c) 42.3m (d) 50.9m

22. In a wave $E_0 = 100\ Vm^{-1}$. Find the magnitude of Poynting's vector
(a) $13.25\ Wm^{-2}$ (b) $26.5\ Wm^{-2}$
(c) $18.25\ Wm^{-2}$ (d) $19.7\ Wm^{-2}$

23. The wave impendance of free space is
(a) zero (b) 376.6Ω
(c) 33.66Ω (d) $3.76\ \Omega$

24. In an apparatus, the electric field was found to oscillate with an amplitude of 18 V/m. The magnitude of the oscillating magnetic field will be
(a) $4 \times 10^{-6}T$ (b) $6 \times 10^{-8}T$
(c) $9 \times 10^{-9}T$ (d) $11 \times 10^{-11}T$

25. The average electric field of electromagnetic waves in certain region of free space is $9 \times 10^{-4}\ NC^{-1}$. Then the average magnetic field in the same region is of the order of
(a) 27×10^{-4} T (b) 3×10^{-12} T
(c) $\left(\frac{1}{3}\right) \times 10^{-12}$ T (d) 3×10^{12} T

26. A plane electromagnetic wave travels in free space along X-direction. If the value of $\vec{B}$ (in tesla) at a particular point in space and time is $1.2 \times 10^{-8}\ \hat{k}$. The value of $\vec{E}$ (in Vm^{-1}) at that point is
(a) $1.2\hat{j}$ (b) $3.6\hat{k}$ (c) $1.2\hat{k}$ (d) $3.6\hat{j}$

27. A plane electromagnetic wave is incident on a material surface. If the wave delivers momentum p and energy E, then
(a) $p = 0, E = 0$ (b) $p \neq 0, E \neq 0$
(c) $p \neq 0, E = 0$ (d) $p = 0, E \neq 0$

28. In order to establish an instantaneous displacemet current of 1 mA in the space between the plates of $2\mu F$ parallel plate capacitor, the potential difference need to apply is
(a) $100\ Vs^{-1}$ (b) $200\ Vs^{-1}$
(c) $300\ Vs^{-1}$ (d) $500\ Vs^{-1}$

29. In an electromagnetic wave, power is
(a) transmitted along the magnetic field
(b) transmitted along the electric field
(c) equally transferred along the electric and magnetic fields
(d) transmitted in a direction perpendicular to both electric and magnetic fields

30. The electric field component of a monochromatic radiation is given by

$$\vec{E} = 2E_0\ \hat{i}\ \cos kz \cos \omega t$$

Its magnetic field $\vec{B}$ is then given by :
(a) $\frac{2E_0}{c}\hat{j} \sin kz \cos \omega t$
(b) $-\frac{2E_0}{c}\hat{j} \sin kz \sin \omega t$
(c) $\frac{2E_0}{c}\hat{j} \sin kz \sin \omega t$
(d) $\frac{2E_0}{c}\hat{j} \cos kz \cos \omega t$

31. Which of the following has/have zero average value in a plane electromagnetic wave?
(a) Both magnetic and electric fields
(b) Electric field only
(c) Magnetic field only
(d) Magnetic energy

32. The energy density of electromagnetic wave in vacuum is given by the relation
(a) $\frac{E^2}{2\varepsilon_0} + \frac{B^2}{2\mu_0}$ (b) $\frac{1}{2}\varepsilon_0 E^2 + \frac{1}{2}\mu_0 B^2$
(c) $\frac{E^2 + B^2}{c}$ (d) $\frac{1}{2}\varepsilon_0 E^2 + \frac{B^2}{2\mu_0}$

33. The waves which are electromagnetic in nature are
(a) sound waves and light waves
(b) water waves and radio waves
(c) light waves and X-rays
(d) sound waves and water waves

34. We consider the radiation emitted by the human body. Which of the following statements is true?
(a) the radiation emitted lies in the ultraviolet region and hence is not visible.
(b) the radiation emitted is in the infra-red region.
(c) the radiation is emitted only during the day.
(d) the radiation is emitted during the summers and absorbed during the winters.

35. An electromagnetic radiation has an energy 11 keV. To which region of electromagnetic spectrum does it belong ?
(a) X-ray region (b) Ultra violet region
(c) Infrared region (d) Visible region

36. Electromagnetic radiation of highest frequency is
(a) infrared radiations (b) visible radiation
(c) radio waves (d) γ-rays

37. In which one of the following regions of the electromagnetic spectrum will the vibrational motion of molecules give rise to absorption?
(a) Ultraviolet (b) Microwaves
(c) Infrared (d) Radio waves

38. Which of the following statements is true?
(a) The wavelength of microwaves is greater than that of UV-rays
(b) The wavelength of IR rays is lesser than that of UV-rays
(c) The wavelength of microwaves is lesser than that of IR rays
(d) Gamma rays has largest wavelength in the electromagnetic spectrum.

39. Which of the following electromagnetic waves has minimum frequency?
(a) Microwaves (b) Audible waves
(c) Ultrasonic wave (d) Radiowaves

40. All components of the electromagnetic spectrum in vacuum have the same
(a) energy (b) velocity
(c) wavelength (d) frequency

41. The range of wavelength of the visible light is
(a) 10 Å to 100 Å (b) 4,000 Å to 8,000 Å
(c) 8,000 Å to 10,000Å (d) 10,000Å to 15,000 Å

42. Given below is a list of electromagnetic spectrum and its mode of production. Which one does not match?
(a) Gamma rays – Radioactive decay of the nucleus
(b) Ultraviolet rays – Magnetron valve
(c) Radio wave – Rapid acceleration and deceleration of electrons in conducting wires
(d) X-rays – coolidge tube

43. Which of the following shows green house effect?
(a) Ultraviolet rays (b) Infrared rays
(c) X-rays (d) None of these

44. Given below is a list of E.M spectrum and its use. Which one does not match?
(a) U.V. rays — finger prints detection
(b) I.R.. rays — Secret writing on ancient walls
(c) X- rays — Atomic structure
(d) Microwaves — forged document detection

45. Which radiation in sunlight, causes heating effect?
(a) Ultraviolet (b) Infrared
(c) Visible light (d) All of these

46. A capacitor, made of two parallel plates each of plate area A and separation *d*, is being charged by an external ac source. If I_1 and I_2 be the displacement current inside the capacitor and the instanteneous current charging the capacitor, then
(a) $I_1 = I_2$ (b) $I_1 > I_2$
(c) $I_1 < I_2$ (d) None of these

47. An EM wave radiates outwards from a dipole antenna, with E_0 as the amplitude of its electric field vector. The electric field E_0 which transports significant energy from the source falls off as [r = distance from dipole antenna]
(a) $\frac{1}{r^3}$ (b) $\frac{1}{r^2}$
(c) $\frac{1}{r}$ (d) remains constant

48. Find the average energy density corresponding to maximum electric field, if magnetic field in a plane electromagnetic wave is given by $B = 200 \times 10^{-6} \sin[(4 \times 10^{15})(t - x/c)]$
(a) $1.6\ J\ m^{-3}$ (b) $0.16\ J\ m^{-3}$
(c) $0.016\ J\ m^{-3}$ (d) $0.0016\ J\ m^{-3}$

49. The amplitude of the electric field if the intensity of a plane electromagnetic wave is given as 2.0 Wm^{-2} is
(a) $38.8\ NC^{-1}$ (b) $48.3\ NC^{-1}$
(c) $50.2\ NC^{-1}$ (d) $68.8\ NC^{-1}$

50. Which of the following is the correct equation for magnetic field as a function of *x*, and *t* where a plane electromagnetic wave propagating in the *x*-direction has a wavelength of 5.0 mm? The electric field is in the y-direction and its maximum value is 30 Vm^{-1}. Critical Thinking
(a) $B = 10^{-3} \sin\left[\frac{2\pi}{5\times10^{-3}}(ct - x)\right]$
(b) $B = 10^{-7} \sin\left[\frac{2\pi}{5\times10^{-3}}(ct - x)\right]$
(c) $B = 10^{-5} \sin\left[\frac{\pi}{1.5\times10^{-3}}(ct - x)\right]$
(d) None of these

51. The ratio of contributions made by the electric field and magnetic field components to the intensity of an EM wave is
(a) c : 1 (b) $c^2 : 1$
(c) 1 : 1 (d) $\sqrt{c} : 1$

52. Which of the following waves have the maximum wavelength
(a) X-rays (b) I.R. rays
(c) UV rays (d) Radio waves

53. If microwaves, X rays, infrared, gamma rays, ultra-violet, radio waves and visible parts of the electromagnetic spectrum are denoted by M, X, I, G, U, R and V then which of the following is the arrangement in ascending order of wavelength ?
(a) R, M, I, V, U, X and G
(b) M, R, V, X, U, G and I
(c) G, X, U, V, I, M and R
(d) I, M, R, U, V, X and G

54. Microwave oven acts on the principle of :
(a) giving rotational energy to water molecules
(b) giving translational energy to water molecules
(c) giving vibrational energy to water molecules
(d) transferring electrons from lower to higher energy levels in water molecule

55. Beyond which frequency, the ionosphere bends any incident E.M. radiation but do not reflect it back towards the earth.
(a) 80 MHz (b) 40 MHz
(c) 20 MHz (d) 10 MHz

56. Radio waves and visible light in vacuum have
(a) same velocity but different wavelength
(b) continuous emission spectrum
(c) band absorption spectrum
(d) line emission spectrum

57. Which of the following is the infrared ray wavelength?
(a) 10^{-4} cm (b) 10^{-5} cm
(c) 10^{-6} cm (d) 10^{-7} cm

58. A parallel plate capacitor with *a* and seperation *d*, is charged by a constant current *I*. Consider a plane surface of area $\frac{a}{2}$ parallel to the plates and drawn symmetrically between the plates. What is the displacement current through this area?
(a) $I/2$ (b) $I/3$ (c) $I/5$ (d) $I/7$

59. The magnetic field of a beam emerging from a filter facing a floodlight is given by
$B_0 = 12 \times 10^{-8} \sin(1.20\times10^7 z - 3.60 \times 10^{15} t)$ T.
What is the average intensity of the beam?
(a) 1.72×10^2 W/m^2 (b) 1.72 W/m^2
(c) 2.31 W/m^2 (d) 2.31×10^2 W/m^2

60. The charge on a parallel plate capacitor varies as $q = q_0 \cos 2\pi\nu t$. The plates are very large and close together (area = A, separation = d). Neglecting the edge effects, the displacement current through the capacitor is **Critical Thinking**
(a) $2\pi\nu q_0 \sin 2\pi\nu t$ (b) $3/5\pi\nu q_0 \sin \pi\nu t$
(c) $4\pi\nu q_0 \sin 2\pi\nu t$ (d) None of these

61. The energy contained in a cylinder of cross-section 10cm^2 and length 50cm along the x-axis if electric field in an electromagnetic wave is given by
$E = 50 \sin \omega(t - x/c)$ is **Critical Thinking**
(a) 1.2×10^{-10} J (b) 5.5×10^{-12} J
(c) 2.3×10^{-10} J (d) 3.6×10^{-12} J

62. Match List-I (Electromagnetic wave type) with List-II (Its association/application) and select the correct option from the choices given below the lists:

	Column I		Column II
1.	Infrared waves	(i)	To treat muscular strain
2.	Radio waves	(ii)	For broadcasting
3.	X-rays	(iii)	To detect fracture of bones
4.	Ultraviolet rays	(iv)	Absorbed by the ozone layer of the atmosphere

	1	2	3	4
(a)	(iv)	(iii)	(ii)	(i)
(b)	(i)	(ii)	(iv)	(iii)
(c)	(iii)	(ii)	(i)	(iv)
(d)	(i)	(ii)	(iii)	(iv)

63. Arrange the following electromagnetic radiations per quantum in the order of increasing energy :
A : Blue light B : Yellow light
C : X-ray D : Radiowave.
(a) C, A, B, D (b) B, A, D, C
(c) D, B, A, C (d) A, B, D, C

64. During the propagation of electromagnetic waves in a medium
(a) electric energy density is double of the magnetic energy density.
(b) electric energy density is half of the magnetic energy density.
(c) electric energy density is equal to the magnetic energy density.
(d) both electric and magnetic energy densities are zero.

65. An EM wave of intensity I falls on a surface kept in vacuum and exerts radiation pressure p on it. Which of the following is wrong?

(a) Radiation pressure is $\frac{I}{c}$ if the wave is totally absorbed

(b) Radiation pressure is $\frac{I}{c}$ if the wave is totally reflected

(c) Radiation pressure is $\frac{2I}{c}$ if the wave is totally reflected

(d) Radiation pressure is in the range $\frac{I}{c} < p < \frac{2I}{c}$ for real surfaces

66. Which of the following statements is and correct.

(a) When variable frequency a.c. source is connected to a capacitor, displacement current increases with increase in frequency

(b) As frequency increases conduction current also increases

(c) Short wave bands are used for transmission of radio waves to a large distance

(d) Ultraviolet radiation are absorbed by the atmosphere and converts Ozone to Oxygen

67. A charged particle oscillates about its mean equilibrium position with a frequency of 10^9 Hz. The electromagnetic waves produced

(a) will have frequency of 10^6 Hz

(b) will have frequency of 2×10^3 Hz

(c) will have wavelength of 0.3 m

(d) fall in the region of U.V. waves

68. The magnetic field in a travelling electromagnetic wave has a peak value of 20 nT. The peak value of electric field strength is

(a) 3 V/m (b) 6 V/m (c) 9 V/m (d) 12 V/m

69. In a plane electromagnetic wave, the electric field oscillates sinusoidally at a frequency of 2.0×10^{10} Hz and amplitude 48 V m^{-1}. Then

(a) the wavelength of the wave is 1.5×10^{-5} m

(b) the amplitude of the oscillating magnetic field is 16×10^{-3}T

(c) the average energy density of the E is equal to the average energy density of the B. [$c = 3 \times 10^8$ m s^{-1}.]

(d) None of these

70. An LC resonant circuit contains a 4 pF capacitor and a 10 mH inductor. It is set into oscillation coupled to an instrument. The wavelength of the radiated electromagnetic waves is: **Toughnut**

(a) 355 cm (b) 377 m

(c) 400 mm (d) 600 mm

Answer KEYs

1	(d)	8	(a)	15	(c)	22	(b)	29	(d)	36	(d)	43	(b)	50	(b)	57	(a)	64	(c)
2	(c)	9	(c)	16	(c)	23	(b)	30	(c)	37	(b)	44	(d)	51	(c)	58	(a)	65	(b)
3	(c)	10	(a)	17	(a)	24	(b)	31	(a)	38	(d)	45	(b)	52	(d)	59	(b)	66	(d)
4	(c)	11	(d)	18	(d)	25	(b)	32	(d)	39	(b)	46	(a)	53	(c)	60	(a)	67	(c)
5	(a)	12	(a)	19	(a)	26	(d)	33	(c)	40	(b)	47	(c)	54	(c)	61	(b)	68	(b)
6	(b)	13	(d)	20	(c)	27	(b)	34	(b)	41	(b)	48	(c)	55	(b)	62	(d)	69	(c)
7	(b)	14	(d)	21	(a)	28	(d)	35	(a)	42	(d)	49	(a)	56	(a)	63	(c)	70	(b)

Hints & Solutions

1. (d) $B_y = 2 \times 10^{-7} \sin(0.5 \times 10^3 z + 1.5 \times 10^{11} t)$ T
The electric vector is perpendicular to B as well as direction of propagation of electromagnetic wave.
Therefore E_x has to be taken.
Further, $E_0 = B_0 \times c$
$= 2 \times 10^{-7} \times 3 \times 10^8$ V/m
$E_0 = 2 \times 10^{-7} \times 3 \times 10^8 = 60$ V/m
$\therefore$ The corresponding value of the electric field is
$Ex = 60 \sin(0.5 \times 10^3 z + 1.5 \times 10^{11} t)$ V/m

2. (c) $\oint \vec{B}.d\vec{\ell}\sqrt{b^2-4ac} = \mu_0 \in_0 \frac{d\phi}{dt}$
or $B \times 2\pi r = \mu_0 \in_0 A\left(\frac{dE}{dt}\right) \quad \therefore B \propto \left(\frac{dE}{dt}\right)$

3. (c) $\frac{E}{B} = C \Rightarrow B = \frac{E}{C} = \frac{24}{3\times10^8} = 8\times10^{-8}$ T

4. (c) $\oint \vec{E}\cdot d\vec{\ell} = -\frac{d\phi_B}{dt}$
or $E(2\pi r) = \frac{d}{dt}(kt \times \pi r^2) \quad \therefore \quad E = \frac{kr}{2}$

5. (a) $E_{rms} = 720$
The average total energy density
$= \frac{1}{2}\in_0 E_0^2 = \frac{1}{2}\in_0 [\sqrt{2}E_{rms}]^2 = \in_0 E_{rms}^2$
$= 8.85 \times 10^{-12} \times (720)^2 = 4.58 \times 10^{-6}$ J/m^3

6. (b) As we know, $\vec{E}\cdot\vec{B} = 0 \quad \because [\vec{E} \perp \vec{B}]$
and $\vec{E}\times\vec{B}$ should be along Z direction
As $(-2\hat{i}-3\hat{j})\times(3\hat{i}-2\hat{j}) = 5\hat{k}$
Hence option (b) is the correct answer.

7. (b) $\omega = 2\pi v = \frac{2\pi c}{\lambda} = \frac{2\pi\times3\times10^8}{6\times10^{-3}}$
$= \pi \times 10^{11}$ rad/sec
The equation for the electric field, along y-axis in the electromagnetic wave is
$$Ey = E_0 \sin\omega\left(t-\frac{x}{c}\right) = 33\sin\left[\pi\times10^{11}\left(t-\frac{x}{c}\right)\right]$$

8. (a) The speed of light
$C = \frac{1}{\sqrt{\mu_0\varepsilon_0}} = \frac{1}{\sqrt{2\times8}} = \frac{1}{4} = 0.25$

9. (c) **Given:** Amplitude of electric field,
$E_0 = 4$ v/m
Absolute permitivity,
$\varepsilon_0 = 8.8 \times 10^{-12}$ c^2/N-m^2
Average energy density $u_E = ?$
Applying formula,
Average energy density $u_E = \frac{1}{4}\varepsilon_0 E^2$
$\Rightarrow u_E = \frac{1}{4}\times8.8\times10^{-12}\times(4)^2 = 35.2\times10^{-12}$ J/m^3

10. (a) Here, amplitude of electric field, $E_0 = 100$ V/m; amplitude of magnetic field, $H_0 = 0.265$ A/m. We know that the maximum rate of energy flow,
$S = E_0 \times H_0 = 100 \times 0.265 = 26.5$ W/m^2.

11. (d)

12. (a) If maximum electron density of the ionosphere is N_{max} per m^3 then the critical frequency f_c is given by $f_c = 9(N_{max})^{1/2}$
$\Rightarrow 10\times10^6 = 9(N)^{1/2} \Rightarrow N = 1.2\times10^{12} m^{-3}$

13. (d) Intensity of EM wave is given by
$$I = \frac{P}{4\pi R^2} = v_{av}\cdot c = \frac{1}{2}\varepsilon_0 E_0^2 \times c$$
$$\Rightarrow E_0 = \sqrt{\frac{P}{2\pi R^2\varepsilon_0 c}}$$
$$= \sqrt{\frac{800}{2\times3.14\times(4)^2\times8.85\times10^{-12}\times3\times10^8}} = 54.77\frac{V}{m}$$

14. (d) $B_0 = \frac{E_0}{c} = \frac{9\times10^3}{3\times10^8} = 3\times10^{-5}T.$

15. (c) Speed of light of vacuum $c = \frac{1}{\sqrt{\mu_0\varepsilon_0}}$ and in another medium $v = \frac{1}{\sqrt{\mu\varepsilon}}$
$\therefore \frac{c}{v} = \sqrt{\frac{\mu\varepsilon}{\mu_0\varepsilon_0}} = \sqrt{\mu_r K} \Rightarrow v = \frac{c}{\sqrt{\mu_r K}}.$

16. (c) In an electromagnetic wave electric field and magnetic field are perpendicular to the direction of propagation of wave. The vector equation for the electric field is

$$\vec{E} = E_0 \cos\left(\omega t - \frac{2\pi}{\lambda} y\right)\hat{z}$$

17. (a) As B $\propto$ r, since the point is on the axis where r = 0, So B = 0

18. (d) Wave in X-direction means E and B should be function of x and t.

$\hat{y} - \hat{z} \perp \hat{y} + \hat{z}$

19. (a) Frequency remains constant during refraction

$$v_{med} = \frac{1}{\sqrt{\mu_0 \in_0 \times 4}} = \frac{c}{2}$$

$$\frac{\lambda_{med}}{\lambda_{air}} = \frac{v_{med}}{v_{air}} = \frac{c/2}{c} = \frac{1}{2}$$

$\therefore$ wavelength is halved and frequency remains unchanged.

20. (c) Incident momentum, $p = \frac{E}{c}$

For perfectly reflecting surface with normal incidence

$$\Delta p = 2p = \frac{2E}{c}; \quad F = \frac{\Delta p}{\Delta t} = \frac{2E}{ct};$$

21. (a) $\lambda = \frac{c}{\nu} = \frac{3\times10^8}{8.2\times10^6} = 36.5\text{m}$

22. (b) $\vec{s} = \frac{E^2}{C\mu_0} = 26.5\text{Wm}^{-2}$

23. (b) Wave impedance $= Z = \sqrt{\frac{\mu_0}{\varepsilon_0}} = 376.6\ \Omega$

24. (b) $c = \frac{E}{B} \Rightarrow B = \frac{E}{c} = \frac{18}{3\times10^8} = 6\times10^{-8}T$.

25. (b) For electromagnetic waves we know that,

$$\frac{E}{B} = c \quad \therefore \quad \frac{9\times10^{-4}}{B} = 3\times10^8\ \text{ms}^{-1}$$

$\Rightarrow B = 3\times10^{-12}$ T.

26. (d) Given: $\vec{B} = 1.2\times10^{-8}\hat{k}T$

$\vec{E} = ?$

From formula,

$E = Bc = (1.2\times10^{-8}\text{ T})(3\times10^8\text{ ms}^{-1}) = 3.6\ \text{Vm}^{-1}$

$\vec{B}$ is along Z-direction and the wave propagates along X-direction. Therefore $\vec{E}$ should be along Y-direction.

Thus, $\vec{E} = 3.6\hat{j}\ \text{Vm}^{-1}$

27. (b) EM waves carry momentum and hence can exert pressure on surfaces. They also transfer energy to the surface so $p \neq 0$ and $E \neq 0$.

28. (d) $I_d = 1\ mA = 10^{-3}$ A
$C = 2\mu F = 2\times10^{-6}$ F

$$I_D = I_C = \frac{d}{dt}(CV) = C\frac{dV}{dt}$$

Therefore, $\frac{dV}{dt} = \frac{I_D}{C} = \frac{10^{-3}}{2\times10^{-6}} = 500\ \text{Vs}^{-1}$

Therefore, applying a varying potential difference of 500 V s^{-1} would produce a displacement current of desired value.

29. (d) The power per unit area carried by an E.M. wave i.e., energy transported per unit time across a unit crossection area is perpendicular to the direction in which the wave is travelling.

30. (c) Magnetic field vector, $\vec{B} = \frac{2E_0}{c}\hat{j}\sin kz \sin \omega t$

31. (a) Both magnetic and electric fields have zero average value in a plane e.m. wave.

32. (d)

33. (c) Light waves are electromagnetic waves.

34. (b) Depends on the magnitude of frequency

35. (a) $E = \frac{hc}{\lambda} \Rightarrow \lambda = \frac{hc}{E}$

$$\Rightarrow \lambda = \frac{6.6\times10^{-34}\times3\times10^8}{11\times1000\times1.6\times10^{-19}} = 12.4\ \text{Å}$$

Increasing order of frequency →

x-rays | u-v rays | visible | Infrared

wavelength range of visible region is 4000Å to 7800Å.

36. (d) $\nu_{\gamma\text{-rays}} > \nu_{\text{visible radiation}} > \nu_{\text{infrared}} > \nu_{\text{Radio waves}}$

37. (b) Molecular spectra due to vibrational motion lie in the microwave region of EM-spectrum. Due to Kirchhoff's law in spectroscopy the same will be absorbed.

38. (d) G > X > U > V > I > M > R [frequency order]

39. (b) Audible waves are not electromagnetic wave.

40. (b) All components of electromagnetic spectrum travel in vacuum with velocity 3×10^8 m/s.

41. (b) Wavelength of visible spectrum is 3900Å – 7800 Å.

42. (d) X-rays – coolidge tube

43. (b) Infrared radiations reflected by low lying clouds and keeps the earth warm.

44. (d) Ultravoilet radiations are used in the detection of invisible writing, forged documents, finger prints in forensic lab. While microwaves are used in microwave oven.

45. (b) Infrared causes heating effect.

46. (a) 47. (c)

48. (c) Average energy density corresponding to maximum electric field,

$$u_{av} = \frac{B_0^2}{2\mu_0} = \frac{(200 \times 10^{-6})^2}{2 \times 4\pi \times 10^{-7}} \cong 0.016 \text{ Jm}^{-3}$$

49. (a) Intersity of a plane electromagnetic wave

$$= \frac{1}{2}\varepsilon_0 E_0^2 c \text{, so } E = \sqrt{\frac{2I}{\varepsilon_0 c}} = 38.8 \text{ NC}^{-1}$$

50. (b) $E = E_0 \sin \omega\left(t - \frac{x}{c}\right)$; $B = B_0 \sin \omega\left(t - \frac{x}{c}\right)$

$$I_d = I_0 \sin\left(\omega t + \frac{\pi}{2}\right) = I \text{ and } \omega = 2\pi\nu = \frac{2\pi}{\lambda}c$$

$$E_0 = 30\sin\left[\frac{2\pi}{5 \times 10^{-3}}(ct - x)\right] \text{ and } B_0 = \frac{E_0}{C} = 10^{-7}\text{ T},$$

$$B = B_0 \sin\left[\frac{2\pi}{\lambda}(ct - x)\right] = 10^{-7} \sin\left[\frac{2\pi}{5 \times 10^{-3}}(ct - x)\right]$$

51. (c)

52. (d) $\lambda_{\text{Radiowave}} > \lambda_{\text{IR rays}} > \lambda_{\text{UV rays}} > \lambda_{\text{x-rays}}$

53. (c) Gamma rays < X-rays < Ultra violet < Visible rays < Infrared rays < Microwaves < Radio waves.

54 (c) Microwave oven acts on the principle of giving vibrational energy to water molecules.

55. (b) Wave is uv rays.

56. (a) In vacuum velocity of all EM waves are same but their wavelengths are different.

57. (a) The wavelength of infrared region is 8×10^{-5} cm to 3×10^{-3} cm. So maximum wavelength of infrared region $= 8 \times 10^{-5}$ cm $\approx 10^{-4}$ cm.

58. (a)

59. (b) As we know that,
The standard equation of magnetic field

$$B = B_0 \sin(kx - \omega t)$$

And, the given equation is
$B = 12 \times 10^{-8} \sin(1.20 \times 10^7 z - 3.60 \times 10^{15} t)$T.
On comparing this equation with standard equation (i), we have.

$$B_0 = 12 \times 10^{-8}$$

So, the average intensity of the beam

$$I_{av} = \frac{1}{2}\frac{B_0^2}{\mu_0} \cdot c = \frac{1}{2} \times \frac{(12 \times 10^{-8})^2 \times 3 \times 10^8}{4\pi \times 10^{-7}}$$

$$= 1.72 \text{ W/m}^2$$

60. (a) As we know that,
As the displacement current through the capacitor is,

$$I_d = I_c = \frac{dq}{dt} \quad \text{...(i)}$$

As given that, $q = q_0 \cos 2\pi\nu t$
Putting this value of q in Eq (i), we get

So, $$I_d = I_c = \frac{d}{dt}(q_0 \cos^2 \pi\nu t)$$

$$I_d = I_c = -q_0 \sin 2\pi\nu t \times 2\pi\nu$$

$$I_d = I_c = 2\pi\nu q_0 \sin 2\pi\nu t$$

61. (b) Energy density $= \frac{1}{2}\varepsilon_0 E_0^2 = 1.1 \times 10^{-8}$ J m^{-3}

Volume of cylinder $= A \times L = 5 \times 10^{-4}$ m^3
So, required energy $= 1.1 \times 10^{-8} \times 5 \times 10^{-4}$
$= 5.5 \times 10^{-12}$ J

62. (d)

63. (c) E, Decreases →

γ-rays X-rays uv-rays Visible rays (VIBGYOR) IR rays Microwaves Radio waves

Radio wave < yellow light < blue light < X-rays

(Increasing order of energy)

64. (c) $E_0 = CB_0$ and $C = \frac{1}{\sqrt{\mu_0 \varepsilon_0}}$

Electric energy density $= \frac{1}{2}\varepsilon_0 E_0^2 = \mu_E$

Magnetic energy density $= \frac{1}{2}\frac{Bo^2}{\mu_0} = \mu_B$

Thus, $\mu_E = \mu_B$

Energy is equally divided between electric and magnetic field

65. (b)

66. (d) Short wave (wavelength 30 km to 30 cm). These waves are used for radio transmission and for general communication purpose to a longer distance from ionosphere.

The wavelength of these wave ranges between 4000 Å to 100 Å that is smaller wavelength and higher frequency. They are absorbed by atmosphere and convert oxygen into ozone. They cause skin diseases and they are harmful to eye and cause permanent blindness.

67. (c) The frequency of the charged particles oscillates about its mean equilibrium position

$= 10^9$ Hz.

Vibrating particle produces electric and magnetic field.

So, frequency of electromagnetic waves produced by the charged particle is $\nu = 10^9$ Hz.

Wavelength $\lambda = \frac{c}{\nu} = \frac{3\times10^8}{10^9} = 0.3$ m

Since the range of radiowaves is between 10 Hz to 10^{12} Hz and hence 10^9 Hz lies in region of radio waves.

68. (b) From question, $B_0 = 20\text{ nT} = 20 \times 10^{-9}\text{T}$

($\because$ velocity of light in vacuum $C = 3 \times 10^8\text{ ms}^{-1}$)

$\vec{E}_0 = \vec{B}_0 \times \vec{C}$

$|\vec{E}_0| = |\vec{B}|\cdot|\vec{C}| = 20\times10^{-9}\times3\times10^8 = 6\text{ V/m}.$

69. (c) Here $c = 3\times10^8$ m/s, $\nu = 2\times10^{10}$ Hz, $E_0 = 48$ V/m

$\lambda = ?, B_0 = ?, \overrightarrow{u_E} = ?, \overrightarrow{u_B} = ?$ $\quad \because c = \nu\lambda$

$\Rightarrow \lambda = \frac{c}{\nu} = \frac{3\times10^8}{2\times10^{10}} = 1.5\times10^{-2}\text{m}$

$\frac{E_0}{B_0} = c \Rightarrow B_0 = \frac{48}{3\times10^8} = 16\times10^{-8}T.$

Energy density due to electric field and magnetic field

$u_E = \frac{1}{2}\varepsilon_o E^2$ & $u_B = \frac{1}{2}\mu_o B^2$

We have $E = cB$ and $C^2 = \frac{1}{\mu_o\varepsilon_o}$ so,

$U_E = \frac{1}{2}\varepsilon_o(cB)^2 = \frac{1}{2}\varepsilon_o\left(\frac{1}{\mu_o\varepsilon_o}\right)B^2 = \frac{1}{2\mu_o}B^2 = U_B.$

70. (b) $\upsilon = \frac{1}{2\pi\sqrt{LC}}$

$\lambda = \frac{C}{\upsilon} \Rightarrow \lambda = C\times2\pi\sqrt{LC}$

$\lambda = 3\times10^8\times2\times3.14\times\sqrt{10\times10^{-3}\times4\times10^{-12}}$

$= 377$ m

23 Ray Optics and Optical Instruments

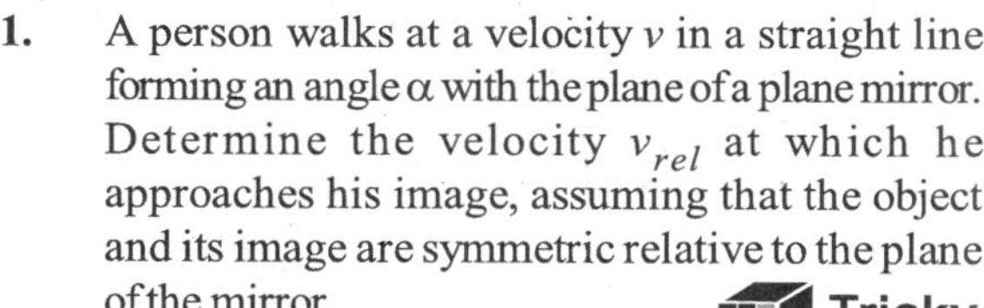

1. A person walks at a velocity v in a straight line forming an angle α with the plane of a plane mirror. Determine the velocity v_{rel} at which he approaches his image, assuming that the object and its image are symmetric relative to the plane of the mirror. **Tricky**

(a) $2v \sin \alpha$ (b) $2v \cos \alpha$
(c) $v \sin \alpha$ (d) $v \cos \alpha$

2. A point object is located at a distance 15cm. from the pole of a concave mirror of focal length 10cm on its principal axis is moving with a velocity $(8\hat{i} + 11\hat{j})$ cm/s and velocity of mirror is $(4\hat{i} + 2\hat{j})$ cm/s as shown. If $\vec{v}$ is the velocity of image. Then find the value of $|\vec{v}|$ in (cm/s).

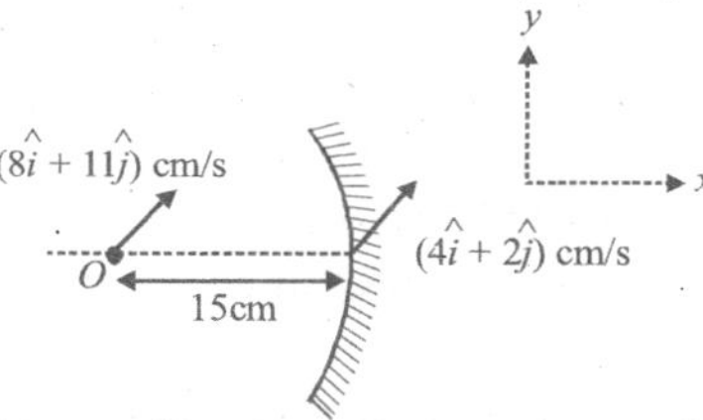

(a) 20 (b) 30 (c) 10 (d) 40

3. Two mirrors, one concave and the other convex, are placed 60 cm apart with their reflecting surfaces facing each other. An object is placed 30 cm from the pole of either of them on their axis. If the focal lengths of both the mirrors are 15 cm, the position of the image formed by reflection, first at the convex and then at the concave mirror, is : **Critical Thinking**

(a) 19.09 cm from the pole of the concave mirror
(b) 19.09 cm from the pole of the convex mirror
(c) 11.09 cm from the pole of the concave mirror
(d) 11.09 cm from the pole of the convex mirror

4. Two plane mirrors A and B are aligned parallel to each other, as shown in the figure. A light ray is incident at an angle 30° at a point just inside one end of A. The plane of incidence coincides with the plane of the figure. The maximum number of times the ray undergoes reflections (including the first one) before it emerges out is **Tricky**

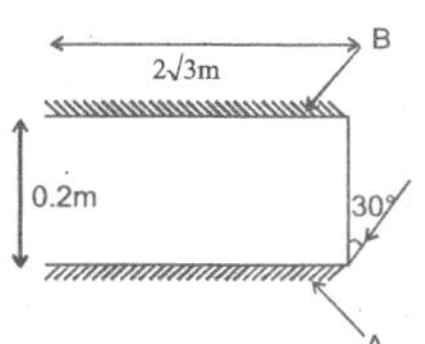

(a) 28 (b) 30 (c) 32 (d) 34

5. The inner surface of a cone coated by a reflecting layer forms a conical mirror. A thin incandescent filament is stretched in the cone along its axis. Determine the minimum angle α of the cone for which the rays emitted by the filament will be reflected from the conical surface not more than once

(a) 120° (b) 90° (c) 150° (d) 135°

6. I is the image of a point object O formed by spherical mirror, then which of the following statement is incorrect? (Take real or virtual objects at finite distances from pole)

(a) If O and I are on the same side of the principal axis, then they have to be on opposite sides of the mirror
(b) If O and I are on opposite sides of the principal axis, then they have to be on same side of the mirror
(c) If O and I are on opposite sides of the principal axis, then they have to be on opposite sides of the mirror as well
(d) If O is on principal axis, then I has to lie on principal axis only

7. When an object is placed at a distance of 25 cm from a mirror, the magnification is m_1. The object is moved 15cm further away with respect to the earlier position, and the magnification becomes m_2. If $m_1/m_2 = 4$, the focal length of the mirror is:

(a) 10 cm (b) 30 cm (c) 15 cm (d) 20 cm

8. In an experiment to determine the focal length (f) of a concave mirror by the u - v method, a student places the object pin A on the principal axis at a distance x from the pole P. The student looks at the pin and its inverted image from a distance keeping his/her eye in line with PA. When the student shifts his/her eye towards left, the image appears to the right of the object pin. Then,

(a) $x < f$ (b) $f < x < 2f$
(c) $x = 2f$ (d) $x > 2f$

9. A plane mirror is kept parallel to y-axis. A point object is approaching the mirror with velocity $\vec{u} = (10\hat{i} + 10\hat{j})$ m/s. The magnitude of relative velocity of objective w.r.t image is equal to
(a) $20\sqrt{2}$ m/s (b) 20 m/s
(c) $10\sqrt{2}$ m/s (d) 10 m/s

10. The width of man's face is 10cm. The distance between the eyes of the man is 4cm. Then the minimum width of plane mirror to see his full face, is
(a) 5 cm (b) 4 cm (c) 3 cm (d) 10 cm

11. A point source has been placed as shown in the figure. What is the length on the screen that will receive reflected light from the mirror?
(a) $2H$
(b) $3H$
(c) H
(d) None of these

12. A point object is kept in front of a plane mirror. The plane mirror is doing SHM of amplitude 2 cm. The plane mirror moves along the x - axis which is normal to the mirror. The amplitude of the mirror is such that the object is always in front of the mirror. The amplitude of SHM of the image is
(a) 0 (b) 2 cm (c) 4 cm (d) 1 cm

13. An object is moving with speed v_0 towards a spherical mirror with radius of curvature R, along the central axis of the mirror. The speed of the image with respect to the mirror is (U is the distance of the object from mirror at any given time t)

Critical Thinking

(a) $+\left(\frac{R}{U-2R}\right)v_0^2$ (b) $-\left(\frac{R}{R-2U}\right)^2 v_0$
(c) $-\left(\frac{R}{2U-2R}\right)^2 v_0$ (d) $+\left(\frac{R}{2U-2}\right)v_0^2$

14. A point object is moving with velocity $\vec{u} = 2\hat{i} + \hat{j} - \hat{k}$ m/s in front of a stationary plane mirror. The magnitude of relative velocity of the image with respect to object is maximum if the normal of the plane mirror will be along
(a) $2\hat{i} + \hat{j} + \hat{k}$ (b) $-2\hat{i} + \hat{j} - \hat{k}$
(c) $2\hat{i} + \hat{j} - \hat{k}$ (d) $2\hat{i} - \hat{j} - \hat{k}$

15. A ray of light is incident at 50° on the middle of one of the two plane mirrors arranged at an angle of 60° between them. The ray then touches the second mirror, get reflected back to the first mirror, making an angle of incidence of

Tricky

(a) 50° (b) 60° (c) 70° (d) 80°

16. A ray parallel to principal axis is incident at 30° from normal on concave mirror having radius of curvature R. The point on principal axis where rays are focussed is Q such that PQ is

Critical Thinking

(a) $\frac{R}{2}$
(b) $\frac{R}{\sqrt{3}}$
(c) $\frac{2\sqrt{R}-R}{\sqrt{2}}$
(d) $R\left(1-\frac{1}{\sqrt{3}}\right)$

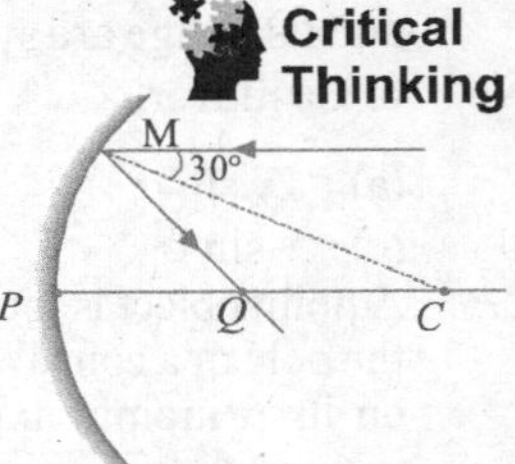

17. A concave mirror is placed on a horizontal table, with its axis directed vertically upwards. Let O be the pole of the mirror and C its centre of curvature. A point object is placed at C. It has a real image, also located at C. If the mirror is now filled with water, the image will be.
(a) real, and will remain at C.
(b) real, and located at a point between C and ∞.
(c) virtual, and located at a point between C and O.
(d) real, and located at a point between C and O

18. A point object P moves towards a stationary convex mirror with a constant speed v, along the optical axis. The speed of the image
(a) is always less than v
(b) may be greater than, equal to or less than v, depending upon the position of P
(c) is always greater than v
(d) None of these

19. A short linear object of length ℓ lies along the axis of a concave mirror of focal length f at a distance u from the pole of the mirror. The size of the image is approximately equal to
(a) $\ell\left(\frac{u-f}{f}\right)^{1/2}$ (b) $\ell\left(\frac{u-f}{f}\right)^2$
(c) $\ell\left(\frac{f}{u-f}\right)^{1/2}$ (d) $\ell\left(\frac{f}{u-f}\right)^2$

20. A point object ' O' is at the centre of curvature of a concave mirror. The mirror starts to move at a speed u, in a direction perpendicular to the principal axis. Then, the initial velocity of the image is
    (a) 2u, in the direction opposite to that of mirror's velocity
    (b) 2u, in the direction same as that of mirror's velocity
    (c) zero
    (d) u, in the direction same as that of mirror's velocity

21. You are asked to design a shaving mirror assuming that a person keeps it 10 cm from his face and views the magnified image of the face at the closest comfortable distance of 25 cm. The radius of curvature of the mirror would then be :
    (a) 60 cm (b) –24 cm (c) – 60 cm (d) 24 cm

22. Two rays are incident on a spherical mirror of radius R = 5 cm. parallel to its optical axis at distances $h_1 = 0.5$ cm. and $h_2 = 3$ cm. Determine the distance Δx (approximately) between the points at which these rays intersect the optical axis after being reflected at the mirror.
    (a) 0.2 cm. (b) 1.5 cm.
    (c) 0.6 cm. (d) 1.0 cm.

23. A cube of side 2 m is placed in front of a concave mirror of focal length 1m with its face P at a distance of 3 m and face Q at a distance of 5 m from the mirror. The distance between the image of face P and Q is

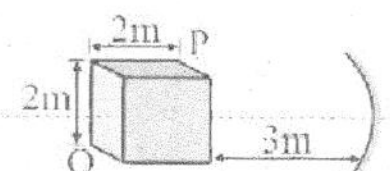

Critical Thinking

(a) 1m (b) 0.5m (c). 0.5m (d) 0.25m

24. In figure find the total magnification after two successive reflections first on M_1 and then on M_2.
    (a) +1 (b) –2
    (c) +2 (d) –1

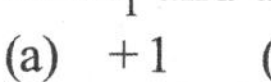

25. A boy of height h is walking away from a street lamp with a constant speed v. The height of the street lamp is 3 h. The rate at which the length of the boy's shadow is increasing when he is at a distance of 10 h from the base of the street lamp is :

Tricky

(a) 2v (b) v (c) v/2 (d) v/3

26. A point source of light B is placed at a distance L in front of the centre of a mirror of width 'd' hung vertically on a wall. A man walks in front of the mirror along a line parallel to the mirror at a distance 2L from it as shown in fig. The greatest distance over which he can see the image of the light source in the mirror is
    (a) d/2 (b) d (c) 2d (d) 3d

27. An object kept on the principle axis is moving in the same directions as that of mirror as shown in figure. Speed of object and mirror is 10m/s and 40/12 m/s. Radius of the curvature of the mirror is 20 cm. What should be the distance of object from the mirror at this instant so that the image is statonary?
    (a) 25 cm
    (b) 45 cm
    (c) 37.5 cm
    (d) 15 cm

28. A ray of light is incident at the glass-water interface at an angle i, it emerges finally parallel to the surface of water, then the value of μ_g would be

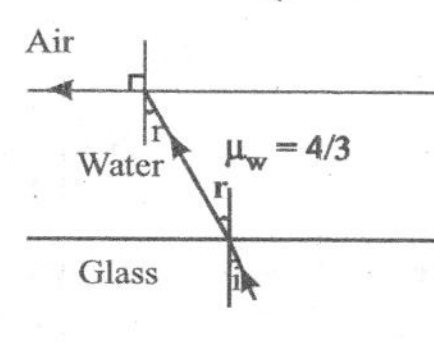

(a) (4/3)sin$i$ (b) 1/sin$i$
(c) 4/3 (d) 1

29. A point light source S is outside a cylinder on its axis near the end face (base). Determine the minimum refractive index n of the cylinder material for which none of the rays entering the base will emerge from the lateral surface.
    (a) $1/\sqrt{2}$ (b) $\sqrt{2}$ (c) 1/2 (d) 1

30. A hemispherical glass body of radius 10 cm and refractive index 1.5 is silvered on its curved surface. A small air bubble is 6 cm below the flat surface inside it along the axis. The position of the iamge of the air bubble made by the mirror is seen :

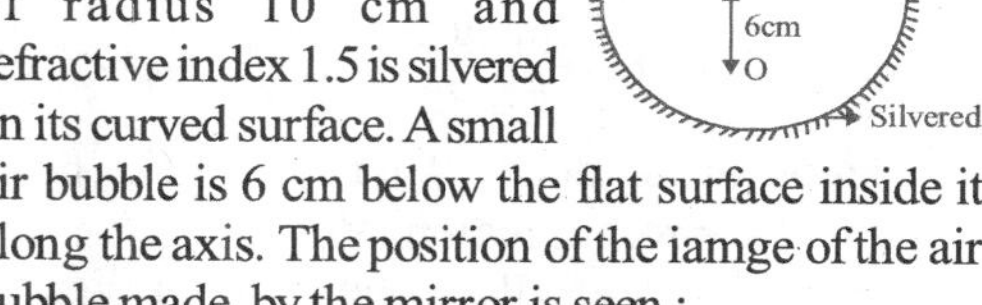

(a) 14 cm below flat surface
(b) 20 cm below flat surface
(c) 16 cm below flat surface
(d) 30 cm below flat surface

Toughnut

31. A point object is placed at the centre of a glass sphere of radius 6 cm and refractive index 1.5. The distance of virtual image from the surface is
    (a) 6 cm (b) 4 cm (c) 12 cm (d) 9 cm

32. A plane mirror is held at a height h above the bottom of an empty beaker. The beaker is now filled with water up to depth d. The general expression for the distance from a scratch at the bottom of the beaker to its image in terms of h and the depth d of water in the beaker is

Toughnut

(a) $2h - d\left(\frac{\mu}{\mu - 1}\right)$

(b) $2h - \frac{d}{2}\left(\frac{\mu - 1}{\mu}\right)$

(c) $2h - d\left(\frac{\mu - 1}{\mu}\right)$

(d) $2h - d\left(\frac{2\mu - 1}{\mu}\right)$

33. A diver looking up through the water sees the outside world contained in a circular horizon. The refractive index of water is $\frac{4}{3}$, and the diver's eyes are 15 cm below the surface of water. Then the radius of the circle is:

(a) $15 \times 3 \times \sqrt{5}$ cm (b) $15 \times 3\sqrt{7}$ cm

(c) $\frac{15 \times \sqrt{7}}{3}$ cm (d) $\frac{15 \times 3}{\sqrt{7}}$ cm

34. A ball is dropped from a height of 20 m above the surface of water in a lake. The refractive index of water is 4.3. A fish inside the lake, in the line of fall of the ball, is looking at the ball. At an instant, when the ball is 12.8 m above the water surface, the fish sees the speed of ball as [Take $g = 10\ m/s^2$.

Critical Thinking

(a) 9 m/s (b) 12 m/s

(c) 16 m/s (d) 21.33 m/s

35. A light beam is travelling from Region I to IV (figure). The refractive index in regionals I, II, III and IV are $n_0, \frac{n_0}{2}, \frac{n_0}{6}$ and $\frac{n_0}{8}$ respectively. The angle of incidence θ for which the beam just misses entering region IV is –

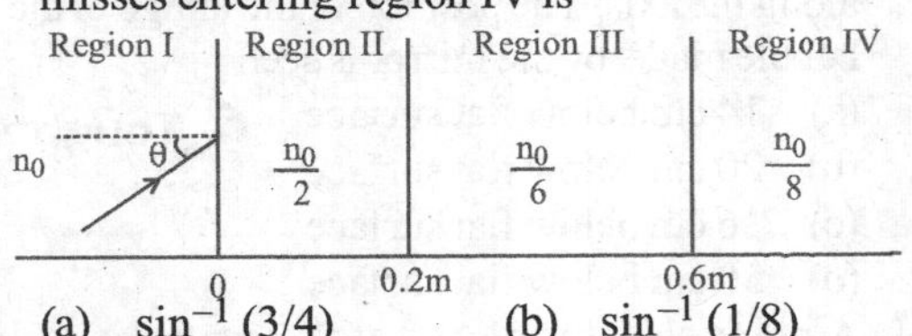

(a) $\sin^{-1}(3/4)$ (b) $\sin^{-1}(1/8)$

(c) $\sin^{-1}(1/4)$ (d) $\sin^{-1}(1/3)$

36. A glass prism of refractive index 1.5 is immersed in water (refractive index 4/3). A light beam incident normally on the face AB is totally reflected to reach on the face BC if

Critical Thinking

(a) $\sin\theta \geq \frac{8}{9}$

(b) $\frac{2}{3} < \sin\theta < \frac{8}{9}$

(c) $\sin\theta \leq \frac{2}{3}$

(d) $\frac{1}{2} < \sin\theta < 1$

37. A diverging beam of light from a point source S having divergence angle α, falls symmetrically on a glass slab as shown. The angles of incidence of the two extreme rays are equal. If the thickness of the glass slab is t and the refractive index n, then the divergence angle of the emergent beam is

(a) zero (b) α

(c) $\sin^{-1}\left(\frac{1}{n}\right)$ (d) $2\sin^{-1}\left(\frac{1}{n}\right)$

38. A mango tree is at the bank of river and one of the branch of tree extends over the river. A tortoise lives in river. A mango falls just above the tortoise. The acceleration of the mango falling from tree appearing to the tortoise is (Refractive index of water is 4/3 and the tortoise is stationary)

Critical Thinking

(a) g (b) 3g/4

(c) 4g/3 (d) None of these

39. A container is filled with water ($\mu = 1.33$) upto a height of 33.25 cm. A concave mirror is placed 15 cm above the water level and the image of an object placed at the bottom is formed 25 cm below the water level. Focal length of the mirror is

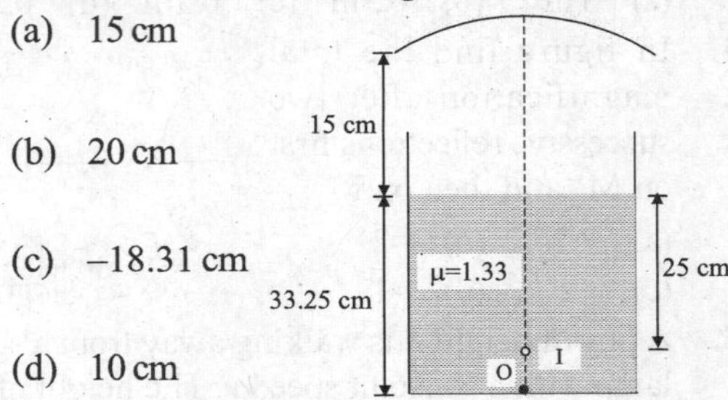

(a) 15 cm

(b) 20 cm

(c) –18.31 cm

(d) 10 cm

40. A ray of light passes through four transparent media with refractive indices μ_1, μ_2, μ_3 and μ_4 as shown in the figure. The surfaces of all media are parallel. If the emergent ray CD is parallel to the incident ray AB, we must have

Tricky

(a) $\mu_1 = \mu_2$

(b) $\mu_2 = \mu_3$

(c) $\mu_3 = \mu_4$

(d) $\mu_4 = \mu_1$

41. A point light source is moving with a constant velocity v inside a transparent thin spherical shell of radius R, which is filled with a transparent liquid. If at $t = 0$ light source is at the centre of the sphere, then at what time a thin dark ring will be visible for an observer outside the sphere. The refractive index of liquid with respect to that of shell is $\sqrt{2}$. **Toughnut**

(a) $\frac{R}{\sqrt{2}V}$ (b) $\frac{R}{2V}$ (c) $\frac{R}{3V}$ (d) $\frac{R}{\sqrt{3}V}$

42. A ray of light is incident on a thick slab of glass (thickness t) as shown below. The emergent ray is parallel to the incident ray but displaced sideways by a distance d. If the angles are small then d is:

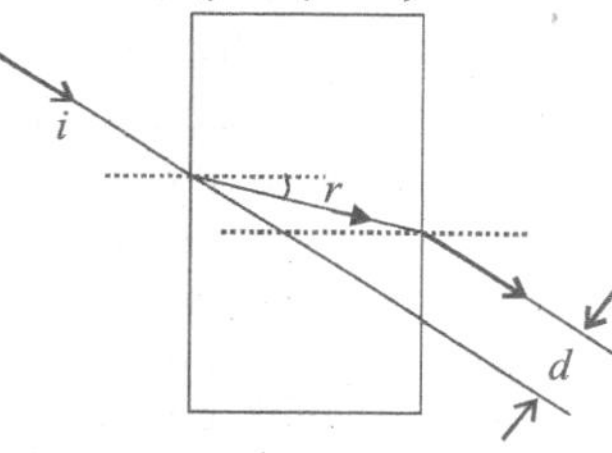

(a) $t\,(1 - i/r)$

(b) $r\,t\,(1 - i/r)$

(c) $i\,t\left(1 - \frac{r}{i}\right)$

(d) $t\left(1 - \frac{r}{i}\right)$

43. Light is incident from a medium into air at two possible angles of incidence (A) 20° and (B) 40°. In the medium light travels 3.0 cm in 0.2 ns. The ray will :

(a) suffer total internal reflection in both cases (A) and (B)

(b) suffer total internal reflection in case (B) only

(c) have partial reflection and partial transmission in case (B)

(d) have 100% transmission in case (A)

44. An observer can see through a pin-hole the top end of a thin rod of height h, placed as shown in the figure. The beaker height is 3h and its radius h. When the beaker is filled with a liquid up to a height 2h, he can see the lower end of the rod. Then the refractive index of the liquid is **Critical Thinking**

(a) $\frac{5}{2}$ (b) $\sqrt{\frac{5}{2}}$ (c) $\sqrt{\frac{3}{2}}$ (d) $\frac{3}{2}$

45. A point source of light is placed at a depth of h below the surface of water of refractive index μ. A floating opaque disc is placed on the surface of water so that light from the sources is not visible from the surface. The minimum diameter of the disc is **Toughnut**

(a) $2h/(\mu^2 - 1)^{1/2}$ (b) $2h(\mu^2 - 1)^{1/2}$

(c) $h/[2(\mu^2 - 1)^{1/2}]$ (d) $h(\mu^2 - 1)^{1/2}$

46. A ray of light is incident from a denser to a rarer medium. The critical angle for total internal reflection is θ_{iC} and Brewster's angle of incidence is θ_{iB}, such that $\sin\theta_{iC}/\sin\theta_{iB} = \eta = 1.28$. The relative refractive index of the two media is:

(a) 0.2 (b) 0.4 (c) 0.8 (d) 0.9

47. Let the refractive index of a denser medium with respect to a rarer medium be n_{12} and its critical angle be θ_C. At an angle of incidence A when light is travelling from denser medium to rarer medium, a part of the light is reflected and the rest is refracted and the angle between reflected and refracted rays is 90°. Angle A is given by :

(a) $\frac{1}{\cos^{-1}(\sin\theta_C)}$ (b) $\frac{1}{\tan^{-1}(\sin\theta_C)}$

(c) $\cos^{-1}(\sin\theta_C)$ (d) $\tan^{-1}(\sin\theta_C)$

48. In a thick glass slab of thickness ℓ and refractive index n_1 a cuboidal cavity of thickness m is carved as shown in the figure and is filled with liquid of R.I. n_2 ($n_1 > n_2$). The ratio of ℓ/m, so that shift produced by this slab is zero when an observer A observes an object B with paraxial rays is **Tricky**

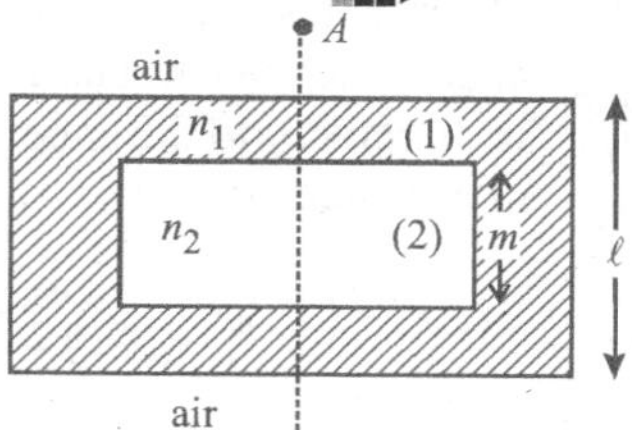

(a) $\frac{n_1 - n_2}{n_2 - 1}$

(b) $\frac{n_1 - n_2}{n_2(n_1 - 1)}$

(c) $\frac{n_1 - n_2}{n_1 - 1}$

(d) $\frac{n_1 - n_2}{n_1(n_2 - 1)}$

49. A printed page is pressed by a glass of water. The refractive index of the glass and water is 1.5 and 1.33, respectively. If the thickness of the bottom of glass is 1 cm and depth of water is 5 cm, how much the page will appear to be shifted if viewed from the top ?

(a) 1.033 cm (b) 3.581 cm

(c) 1.533 cm (d) 1.90 cm

50. A rectangular glass slab $ABCD$ of refractive index n_1 is immersed in water of refractive index $n_2(n_1 > n_2)$. A ray of light is incident at the surface AB of the slab as shown. The maximum value of the angle of incidence α_{max} such that the ray comes out only from the other surface CD is given by

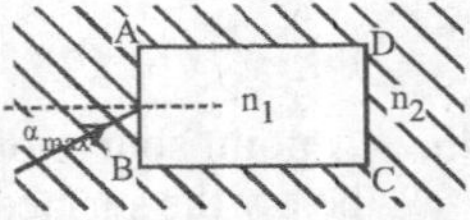

(a) $\sin^{-1}\left[\frac{n_1}{n_2}\cos\left(\sin^{-1}\left(\frac{n_2}{n_1}\right)\right)\right]$

(b) $\sin^{-1}\left[n_1\cos\left(\sin^{-1}\left(\frac{1}{n_2}\right)\right)\right]$

(c) $\sin^{-1}\left(\frac{n_1}{n_2}\right)$ (d) $\sin^{-1}\left(\frac{n_2}{n_1}\right)$

51. Two point source S_1 and S_2 are 24 cm apart. Where should a convex lens of focal length 9 cm be placed in between them so that the images of both sources are formed at the same place? **Critical Thinking**

(a) 6 cm from S_1 (b) 15 cm from S_1
(c) 10 cm from S_1 (d) 12 cm from S_1

52. Two thin symmetrical lenses of different nature and of different material have equal radii of curcature R = 15 cm. The lenses are put close together and immersed in water ($\mu_w = 4/3$). The focal length of the system in water is 30 cm. The difference between refractive indices of the two lenses is

(a) 1/2 (b) 1/4 (c) 1/3 (d) 3/4

53. A glass sphere, refractive index 1.5 and radius 10 cm, has a sphrical cavity of radius 5 cm concentric with it. A narrow beam of parallel light is directed into the sphere. Find the final image and its nature. **Toughnut**

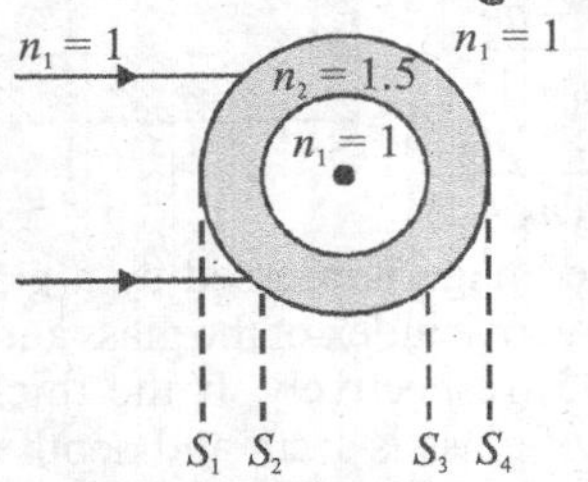

(a) 25 cm left of S_4, virtual
(b) 25 cm right of S_4, real
(c) 15 cm left of S_4, virtual
(d) 20 cm right of S_4, virtual

54. A concave lens of glass, refractive index 1.5 has both surfaces of same radius of curvature R. On immersion in a medium of refractive index 1.75, it will behave as a

(a) convergent lens of focal length 3.5 R
(b) convergent lens of focal length 3.0 R
(c) divergent lens of focal length 3.5 R
(d) divergent lens of focal length 3.0 R

55. Choose the correct ray diagram of a thin equi-convex lens which is cut as shown in the figure.

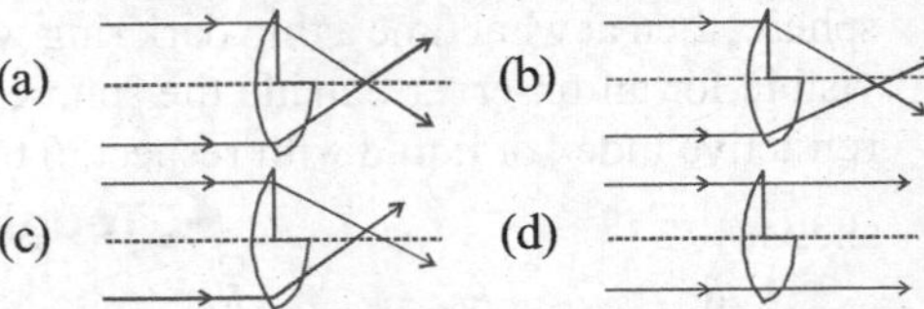

56. A scuba diver in an empty swimming pool uses a magnifier ($n = 1.25$) to enlarge the print on a plastic instruction sheet. If the pool is filled with water ($n = 1.33$), what happens to the magnification of the print? **Tricky**

(a) It increases and is greater than one.
(b) It stays the same.
(c) It decreases, but is still greater than one.
(d) It decreases and is less than one.

57. A convex lens, of focal length 30 cm, a concave lens of focal length 120 cm, and a plane mirror are arranged as shown. For an object kept at a distance of 60 cm from the convex lens, the final image, formed by the combination, is a real image, at a distance of : **Critical Thinking**

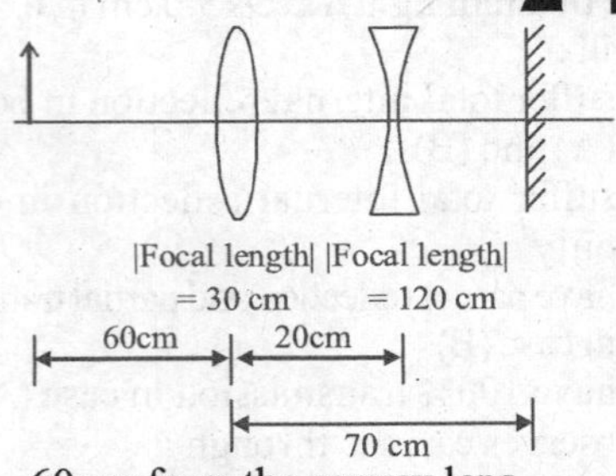

(a) 60 cm from the convex lens
(b) 60 cm from the concave lens
(c) 70 cm from the convex lens
(d) 70 cm from the concave lens

58. The image of an illuminated square is obtained on a screen with the help of a converging lens. The distance of the square from the lens is 40 cm. The area of the image is 9 times that of the square. The focal length of the lens is :

(a) 36 cm (b) 27 cm (c) 60 cm (d) 30 cm

59. Convex surface of thin concavo – convex lens of refractive index 1.5 is silvered as shown. A small object is kept in air at 30 cm left of the lens on its principal axis. The distance of the final image is
(a) 20 cm (b) 30 cm (c) 10 cm (d) 15 cm

60. Rays of light from Sun falls on a biconvex lens of focal length f and the circular image of Sun of radius r is formed on the focal plane of the lens. Then
(a) area of image is πr^2 and area is directly proportional of f
(b) area of image is πr^2 and area is directly proportional to f^2
(c) intensity of image increases if f is increased
(d) if lower half of the lens is covered with black paper area will become half

61. A plastic hemisphere has a radius of curvature of 8 cm and an index of refraction of 1.6. On the axis halfway between the plane surface and the spherical one (4 cm from each) is a small object O. The distance between the two images when viewed along the axis from the two sides of the hemisphere is approximately

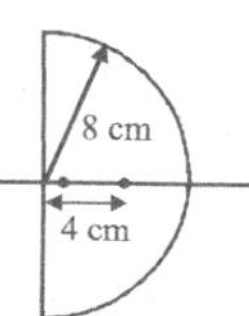

(a) 1.0 cm (b) 1.5 cm
(c) 3.75 cm (d) 2.5 cm

62. A small straight rod is placed at an inclination with the optical axis of a thin lens as shown in the figure. The base of the rod is on the optical axis and at a distance $2f$ (f = focal length of the lens) from the lens. The image of the rod would be

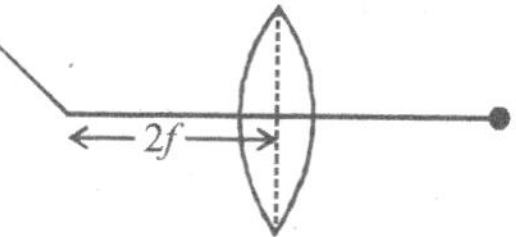

(a) a straight line leaning towards the lens
(b) a straight line leaning away from the lens
(c) a curve leaning towards the lens
(d) a curve leaning away from the lens

63. A luminous object and a screen are at a fixed distance D apart. A converging lens of focal length f is placed between the object and screen. A real image of the object in formed on the screen for two lens positions of they are separated by a distance d equal to
(a) $\sqrt{D(D+4f)}$ (b) $\sqrt{D(D-4f)}$
(c) $\sqrt{2D(D-4f)}$ (d) $\sqrt{D^2+4f}$

64. A thin convex lens of focal length 'f' is put on a plane mirror as shown in the figure. When an object is kept at a distance 'a' from the lens-mirror combination, its image is formed at a distance $\frac{a}{3}$ in front of the combination. The value of 'a' is :

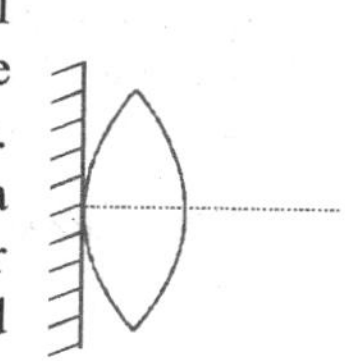

(a) 3f (b) $\frac{3}{2}$f (c) f (d) 2f

65. A plano convex lens fits exactly into a plano concave lens. Their plane surface are parallel to each other. If the lenses are made of different materials of refractive indices μ_1 & μ_2 and R is the radius of curvature of the curved surface of the lenses, then focal length of combination is

Critical Thinking

(a) $\frac{R}{\mu_1-\mu_2}$ (b) $\frac{2R}{\mu_1-\mu_2}$
(c) $\frac{R}{2(\mu_1-\mu_2)}$ (d) $\frac{R}{2-(\mu_1+\mu_2)}$

66. A convex lens is in contact with concave lens. The magnitude of the ratio of their focal length is 2/3. Their equivalent focal length is 30 cm. What are their individual focal lengths?
(a) –15, 10 (b) –10, 15
(c) 75, 50 (d) –75, 50

Critical Thinking

67. A diminished image of an object is to be obtained on a screen 1.0 m from it. This can be achieved by appropriately placing
(a) a concave mirror of suitable focal length
(b) a convex mirror of suitable focal length
(c) a convex lens of focal length less than 0.25 m
(d) a concave lens of suitable focal length

68. Two lenses of focal length f_1 = 10 cm and f_2 = – 20 cm are kept as shown. The resultant power of combination will be

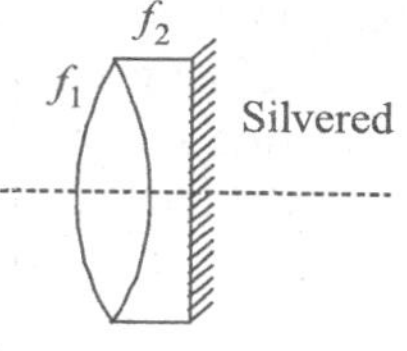

(a) –10D (b) 5D (c) 0 (d) 10D

69. A hollow double concave lens is made of very thin transparent material. It can be filled with air or either of two liquids L_1 or L_2 having refractive indices μ_1 and μ_2 respectively ($\mu_2 > \mu_1 > 1$). The lens will diverge a parallel beam of light if it is filled with

(a) air and placed in air
(b) air and immersed in L_1
(c) L_1 and immersed in L_2
(d) L_2 and immersed in L_1

70. On silvering the convex side of a plano convex lens, focal length of the resulting mirror is: [where R = radius of convex lens, μ = refractive index of glass]

(a) $\frac{R}{2\mu}$ (b) $\frac{2R}{\mu}$ (c) $\frac{R}{\mu}$ (d) $\frac{3R}{2\mu}$

71. A thin convex lens of focal length 10 cm and refractive index 1.5 is cut vertically into two equal pieces. They are placed as shown with a liquid of refractive index 3 between them. What is the focal length of the combination?

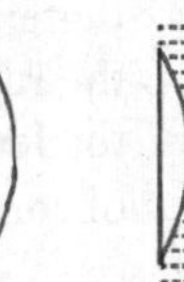

(a) –10 cm. (b) –10/4 cm.
(c) –10/3 cm. (d) None of these

72. An eye specialist prescribes spectacles having combination of convex lens of focal length 40 cm in contact with a concave lens of focal length 25 cm. The power of this lens combination in diopters is

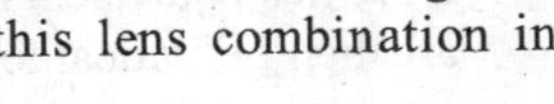

(a) +1.5 (b) – 1.5 (c) +6.67 (d) –6.67

73. Figure shows a concavo-convex lens μ_2. What is the condition on the reflective indices so that the lens is diverging?

Tricky

(a) $2\mu_3 < \mu_1 + \mu_2$
(b) $2\mu_3 > \mu_1 + \mu_2$
(c) $\mu_3 > 2(\mu_1 - \mu_2)$
(d) None of these

2R, μ_1, μ_3, μ_2, R

74. Two identical thin planoconvex lenses of refractive index n are silvered, one on the plane side and the other on the convex side. The ratio of their focal lengths is

Critical Thinking

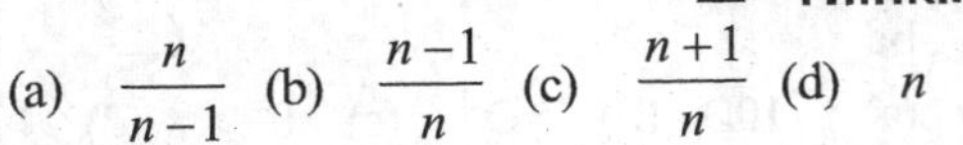

(a) $\frac{n}{n-1}$ (b) $\frac{n-1}{n}$ (c) $\frac{n+1}{n}$ (d) n

75. The size of the image of an object, which is at infinity, as formed by a convex lens of focal length 30 cm is 2 cm. If a concave lens of focal length 20 cm is placed between the convex lens and the image at a distance of 26 cm from the convex lens, calculate the new size of the image.

(a) 1.25 cm (b) 2.5 cm (c) 1.05 cm (d) 2 cm

76. The given lens is broken into four parts and rearranged as shown. If the initial focal length is f then after rearrangement the equivalent focal length is –

1 2 3 4 — in air

(a) f (b) $f/2$ (c) $f/4$ (d) $4f$

77. A transparent sphere of radius R has a cavity of radius $R/2$ as shown in figure. Find the refractive index of the sphere if a parallel beam of light falling on left surface focuses at point P.

O, P

Toughnut

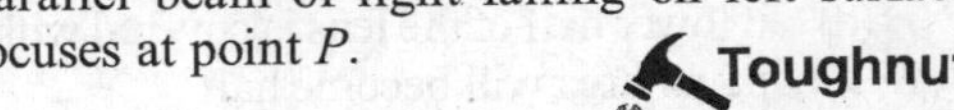

(a) $\mu = \frac{3+\sqrt{5}}{2}$ (b) $\mu = \frac{3-\sqrt{5}}{2}$
(c) $\mu = 3+\sqrt{5}$ (d) $\mu = \frac{1+\sqrt{5}}{2}$

78. A triangular prism of glass is inside water. A ray, incident normally, on one of the faces, is totally reflected from face BC. Then the minimum refractive index of glass is –

A, C, 45°, B, $n_w = 4/3$

(a) $\frac{\sqrt{3}}{2}$ (b) $\frac{5}{3}$
(c) $\frac{2\sqrt{2}}{5}$ (d) $\frac{4\sqrt{2}}{3}$

79. ACB is right - angle prism with other angles as 60° and 30°. Refractive index of the prism is 1.5. AB has thin layer of liquid on it as shown. Light falls normally on the face AC. For total internal reflections, maximum refractive index of the liquid is

A, 60°, 60°, 30°, B, C

(a) 1.4
(b) 1.3
(c) 1.2
(d) 1.6

80. A equilateral prism is made of a transparent material of refractive index $\sqrt{2}$. A ray of light AB is incident at 45° as shown. The net deviation in the path of ray when it comes out of prism is

60°, Silvered, 45°, B, A, 60°, 60°

(a) 135°
(b) 120°
(c) 30°
(d) 150°

81. A thin prism P_1 with angle 4° and made from glass of refractive index 1.54 is combined with another thin prism P_2 made from glass of refractive index 1.72 to produce dispersion without deviation. The angle of the prism P_2 is

(a) 5.33° (b) 4° (c) 3° (d) 2.6°

82. A ray is incident normally on a right angle prism whose refractive index is $\sqrt{3}$ and prism angle α = 30°, after crossing prism ray passes through glass sphere. It strikes the glass sphere at $\frac{R}{\sqrt{3}}$ distance from principal axis, as shown in figure sphere is half polished. Find the net angle of deviation of incident ray. Toughnut

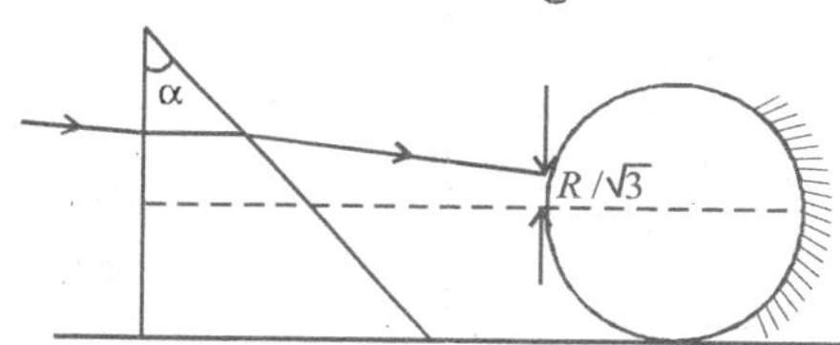

(a) 120° (b) 150° (c) 90° (d) 180°

83. A prism placed in air made up of flint glass is such that any incident ray on one surface does not emerge from the second surface. Critical angle for flint glass is 36° in air. Then, refracting angle A may be

(a) 37° (b) 54° (c) 71° (d) 73°

84. The following data are given for a crown glass prism ;

refractive index for violet light $n_v = 1.521$

refractive index for red light $n_r = 1.510$

refractive index for yellow light $n_y = 1.515$

Then the dispersive power of a parallel glass slab made of the same material is

(a) 0.01 (b) 0.03 (c) 0 (d) 0.02

85. Two beams of red and violet colours are made to pass separately through a prism (angle of the prism is 60°). In the position of minimum deviation, the angle of refraction will be

(a) 30° for both the colours

(b) greater for the violet colour

(c) greater for the red colour

(d) equal but not 30° for both the colours

86. r and r' denote the angles inside an equilateral prism, as usual, in degrees. Consider that during some time interval from $t = 0$ to $t = t$, r' varies with time as $r' = 10 + t^2$. During this time r will vary as (assume that r and r' are in degree)

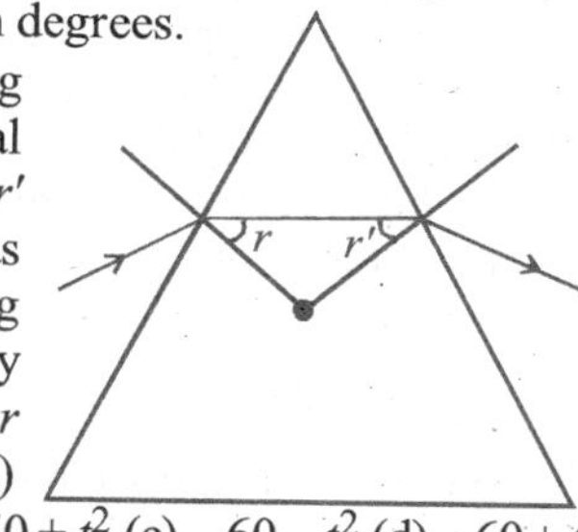

(a) $50 - t^2$ (b) $50 + t^2$ (c) $60 - t^2$ (d) $60 + t^2$

87. For an optical fibre consists of a core (refractive index = μ_1) and cladding (refractive index = μ_2):

(a) $\mu_1 > \mu_2$ (b) $\mu_1 < \mu_2$ (c) $\mu_1 = \mu_2$ (d) $\frac{\mu_1}{\mu_2} = \frac{1}{2}$

88. An isosceles prism of angle 120° has a refractive index 1.44. Two parallel monochromatic rays enter the prism parallel to each other in air as shown. The rays emerge from the opposite faces Tricky

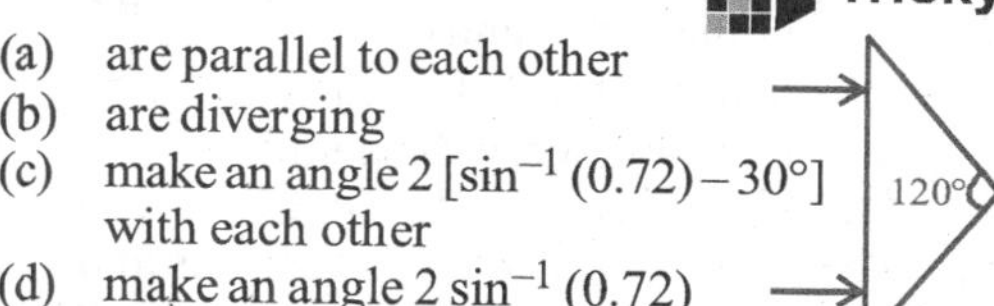

(a) are parallel to each other

(b) are diverging

(c) make an angle $2[\sin^{-1}(0.72) - 30°]$ with each other

(d) make an angle $2\sin^{-1}(0.72)$ with each other

89. In Fig. ABC is the cross section of a right - angled prism and ACDE is the cross section of a glass slab. The value of θ so that incident normally on the face AB does not cross the face AC is (given $\sin^{-1}(3/5) = 37°$). Tricky

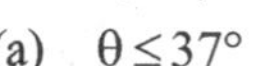

(a) $\theta \le 37°$

(b) $\theta < 37°$

(c) $\theta \le 53°$

(d) $\theta < 53°$

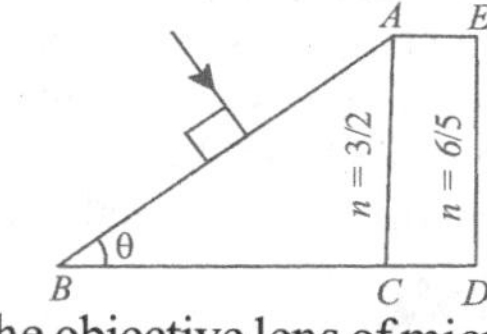

90. The diameter of the objective lens of microscope makes an angle β at the focus of the microscope. Further, the medium between the object and the lens is an oil of refractive index n. Then the resolving power of the microscope

(a) increases with decreasing value of n

(b) increases with decreasing value of β

(c) increases with increasing value of $n \sin 2\beta$

(d) increases with increasing value of $\frac{1}{n \sin 2\beta}$

91. A telescope has an objective lens of focal length 150 cm and an eyepiece of focal length 5 cm. If a 50 m tall tower at a distance of 1 km is observed through this telescope in normal setting, the angle formed by the image of the tower is θ, then θ is close to :

(a) 6.1 rad (b) 3.2 rad (c) 1.5 rad (d) 0.2 rad

92. An object is located in a fixed position in front of a screen. Sharp image is obtained on the screen for two positions of a thin lens separated by 10 cm. The size of the images in two situations are in the ratio 3 : 3. What is the distance between the screen and the object?
(a) 124.5 cm (b) 144.5 cm
(c) 65.0 cm (d) 99.0 cm

93. The focal lengths of the objective and the eye piece of a compound microscope are 2.0 cm and 3.0 cm, respectively. The distance between the objective and the eye piece is 15.0 cm. The final image formed by the eye piece is at infinity. The two lenses are thin. The distance in cm of the object and the image produced by the objective, measured from the objective lens, are respectively
(a) 2.4 and 12.0 (b) 2.4 and 15.0
(c) 2.0 and 12.0 (d) 2.0 and 3.0

94. In a compound microscope, the intermediate image is **Tricky**
(a) virtual, erect and magnified
(b) real, erect and magnified
(c) real, inverted and magnified
(d) virtual, erect and reduced

95. The focal lengths of objective lens and eye lens of a Galilean telescope are respectively 30 cm and 3.0 cm. telescope produces virtual, erect image of an object situated far away from it at least distance of distinct vision from the eye lens. In this condition, the magnifying power of the Galilean telescope should be:
(a) +11.2 (b) –11.2 (c) –8.8 (d) +8.8

96. In a compound microscope, the focal length of objective lens is 1.2 cm and focal length of eye piece is 3.0 cm. When object is kept at 1.25 cm in front of objective, final image is formed at infinity. Magnifying power of the compound microscope should be:
(a) 200 (b) 100 (c) 400 (d) 150

97. The focal length of the objective and the eyepiece of a telescope are 50 cm and 5 cm respectively. If the telescope is focussed for distinct vision on a scale distant 2 m from its objective, then its magnifying power will be : **Critical Thinking**
(a) –4 (b) – 8
(c) +8 (d) – 2

98. When a ray of light enters a glass slab from air,
(a) its wavelength decreases.
(b) its wavelength increases.
(c) Its frequency decreases.
(d) neither its wavelength nor its frequency changes.

99. A fish looking up through the water sees the outside world contained in a circular horizon. If the refractive index of water is 4/3 and the fish is 12 cm below the surface, the radius of this circle (in cm) is – **Critical Thinking**
(a) $36\sqrt{5}$ (b) $4\sqrt{5}$ (c) $36\sqrt{7}$ (d) $36/\sqrt{7}$

100. In a vessel, as shown in Fig. point P is just visible when no liquid is filled in vessel through a telescope in the air. When liquid is filled in the vessel completely, point Q is visible without moving the vessel or telescope. Find the refractive index of the liquid.
(a) $\frac{\sqrt{14}}{3}$ (b) $\frac{\sqrt{85}}{5}$
(c) $\sqrt{2}$ (d) $\sqrt{3}$

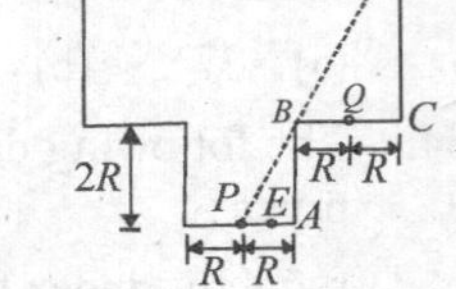

Answer KEYs

1	(a)	11	(a)	21	(c)	31	(a)	41	(a)	51	(a)	61	(d)	71	(c)	81	(c)	91	(c)
2	(a)	12	(c)	22	(c)	32	(c)	42	(c)	52	(c)	62	(c)	72	(b)	82	(d)	92	(d)
3	(a)	13	(b)	23	(d)	33	(d)	43	(b)	53	(a)	63	(b)	73	(b)	83	(d)	93	(a)
4	(b)	14	(c)	24	(c)	34	(c)	44	(b)	54	(a)	64	(d)	74	(a)	84	(d)	94	(c)
5	(a)	15	(c)	25	(c)	35	(b)	45	(a)	55	(c)	65	(a)	75	(b)	85	(a)	95	(d)
6	(c)	16	(d)	26	(d)	36	(a)	46	(c)	56	(d)	66	(a)	76	(b)	86	(a)	96	(a)
7	(d)	17	(d)	27	(a)	37	(b)	47	(d)	57	(a)	67	(c)	77	(a)	87	(a)	97	(d)
8	(b)	18	(a)	28	(b)	38	(c)	48	(b)	58	(d)	68	(d)	78	(d)	88	(c)	98	(a)
9	(b)	19	(d)	29	(b)	39	(c)	49	(c)	59	(c)	69	(d)	79	(b)	89	(b)	99	(d)
10	(c)	20	(b)	30	(b)	40	(d)	50	(a)	60	(b)	70	(a)	80	(d)	90	(c)	100	(b)

Hints & Solutions

1. (a) We resolve the velocity vector $\vec{v}$ of the person into two components, one parallel to the mirror, $\vec{v}_{\|}$ and the other perpendicular to the mirror, $\vec{v}_{\perp}$, i.e. $\vec{v} = \vec{v}_{\|} + \vec{v}_{\perp}$ (figure). The velocity of the image will obviously be $\vec{v}' = \vec{v}_{\|} - \vec{v}_{\perp}$. Therefore, the velocity at which the person approaches his image is defined as his velocity relative to the image. From the formula $v_{rel} = 2v_{\perp} = 2v \sin\alpha$.

2. (a) $\vec{v}_{obj,\, mirror} = 4\hat{i} + 9\hat{j}$ $\quad \frac{dx}{dt} = 4, \; \frac{dy}{dt} = 9; \; u = -x$

$$-\frac{1}{10} = \frac{1}{V} + \frac{1}{-x}; \quad V = \frac{-10x}{x-10}$$

$$v_{1x} = \frac{dV}{dt} = \left(\frac{10}{x-10}\right)^2 \left(\frac{dx}{dt}\right); \quad v_{1x} = -16$$

$$m = -\frac{V}{-x} = \frac{-10}{x-10} = \frac{y_1^0}{y}$$

$$v_{1y} = -\left(\frac{10}{x-10}\right)\frac{dy}{dt}; \quad v_{1y} = -18$$

$$\vec{v}_{image,\, mirror} = -16\hat{i} - 18\hat{j}; \quad \vec{v}_{image} = -12\hat{i} - 16\hat{j}$$

$|v| = 20$ cm/s

3. (a) For convex mirror :

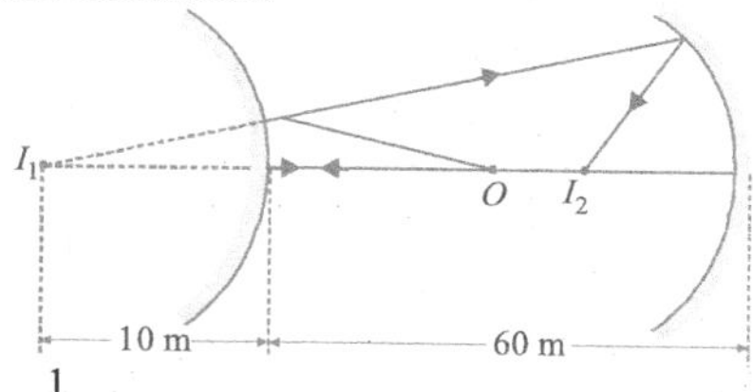

$$\frac{1}{v} + \frac{1}{-30} = \frac{1}{+15} \text{ or } v = 10 \text{ m}$$

For concave mirror : $\frac{1}{v} + \frac{1}{-70} = \frac{1}{-15}$

or $v = -19.09$ cm

4. (b) Maximum number of reflection $= \frac{2\sqrt{3}}{x}$

where $x = 0.2 \tan 30° = 0.2/\sqrt{3}$.

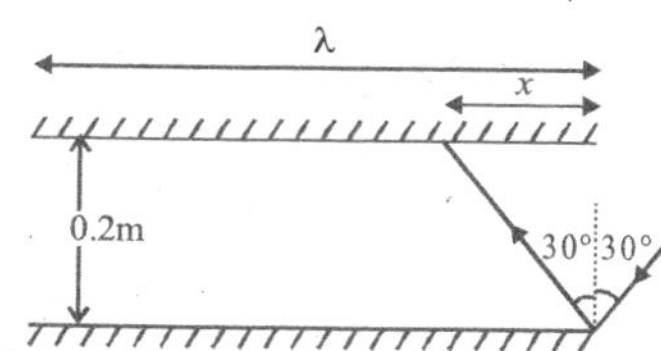

5. (a)

6. (c) Use the concepts related to image formation by spherical mirrors.

7. (d) $m = -\frac{v}{u} = -\left(\frac{f}{u-f}\right)$

Now $m_1 = -\left(\frac{f}{25-f}\right)$ ...(i)

and $m_2 = -\left(\frac{f}{40-f}\right)$ ...(ii)

$\therefore \frac{m_1}{m_2} = \frac{40-f}{25-f}$ or $4 = \frac{40-f}{25-f}$ or $f = 20$ cm.

8. (b)

9. (b) Velocity of image in mirror

$\vec{v} = -10\hat{i} + 10\hat{j} \qquad \vec{v}_{rel} = \vec{u} - \vec{v} = 20\hat{i}.$

10. (c) Min. width of plane mirror to see full face

$= \frac{D-d}{2} = \frac{10-4}{2} = 3$cm.

11. (a) In ΔABD, $\frac{BD}{3H} = \tan 45°$ or $BD = 3H$

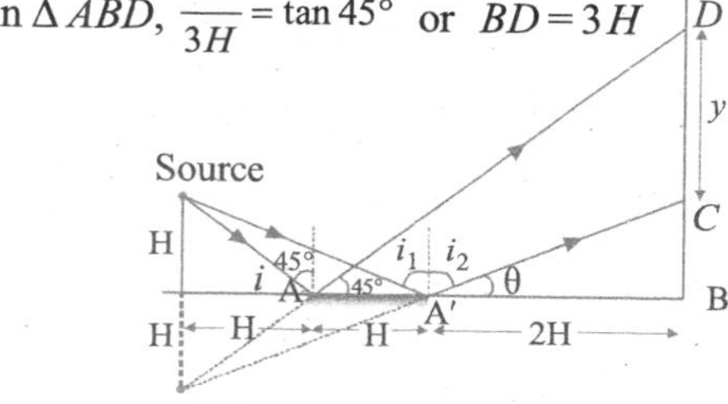

And in $\Delta A'BC$, $\frac{BC}{2H} = \tan\theta = \frac{1}{2}$ or $BC = H$

Now, $y = BD - BC = 3H - H = 2H$.

12. (c)

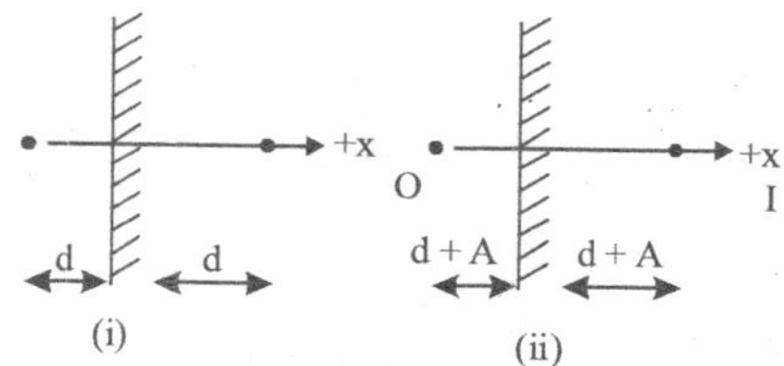

From Fig. (i) and (ii), its is clear that if the mirror moves distance 'A', then the image moves a distance '2A'.

13. (b) For concave mirror

$$\frac{2}{R} = \frac{1}{v} + \frac{1}{u} \text{ or } \frac{2}{-R} = \frac{1}{v} + \frac{1}{-u}$$

$$\therefore \frac{1}{v} = \frac{1}{U} - \frac{2}{R} = \frac{R-2U}{UR} \quad \text{or } v = \left[\frac{RU}{R-2U}\right]$$

In spherical mirror, image velocity

$$v_i = -\left[\frac{v^2}{u^2}\right]v_0 = -\left[\frac{RU}{R-2U}\right]^2\frac{v_0}{U^2} = -\left[\frac{R}{R-2U}\right]^2 v_0$$

14. (c)

15. (c) See geometry of the figure.

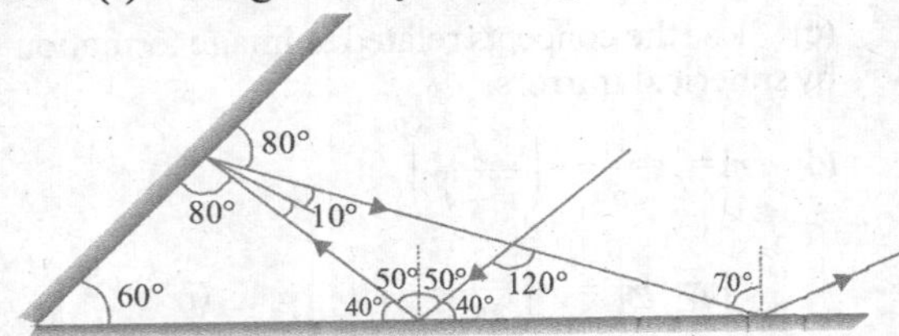

16. (d) From similar triangles,

$$\frac{QC}{\sin 30°} = \frac{R}{\sin 120°}$$

or $QC = R \times \frac{\sin 30°}{\sin 120°} = \frac{R}{\sqrt{3}}$

Thus $PQ = PC - QC = R - \frac{R}{\sqrt{3}} = R\left(1 - \frac{1}{\sqrt{3}}\right)$

17. (d)

18. (a) Image speed $= -\left(\frac{v^2}{u^2}\right)\frac{du}{dt}$

For convex mirror $\left|\frac{v}{u}\right| < 1$

$\therefore$ Image speed is always less than v.

19. (d) The axial magnification $\delta v = -\frac{v^2}{u^2}(\delta u) = \left(\frac{f}{u-f}\right)^2 \ell$.

20. (b) Image velocity (w.r.t mirror)
= –m × object velocity (w.r.t mirror)
Here m = 1

21. (c) **22.** (c)

23. (d) For P: $\frac{1}{v} + \frac{1}{u} = \frac{1}{f}$

or $\frac{1}{v_P} + \frac{1}{-3} = \frac{1}{-1}$ or $v_P = -\frac{3}{2}\,m$

For Q: $\frac{1}{v_Q} + \frac{1}{-5} = \frac{1}{-1}$ or $v_Q = \frac{-5}{4}\,m$

So horizontal distance between image of P and Q is 0.25 m

24. (c) For M_1: V = – 60, m_1 = –2
For M_1: u = +20, F = 10

$\frac{1}{V} - \frac{1}{20} = \frac{1}{10} \Rightarrow V = 20 \quad \therefore M_2 = -\frac{20}{20} = -1$

$\therefore$ $M = m_1 \times m_2 = +2$

25. (c) Let x be the length of the shadow of the boy, when he is at a distance y from the lamp.

From similar triangles, we have

$\frac{h}{x} = \frac{3h}{x+y}$ or $x = y/2$.

Thus $\frac{dx}{dt} = \frac{dy/dt}{2}$

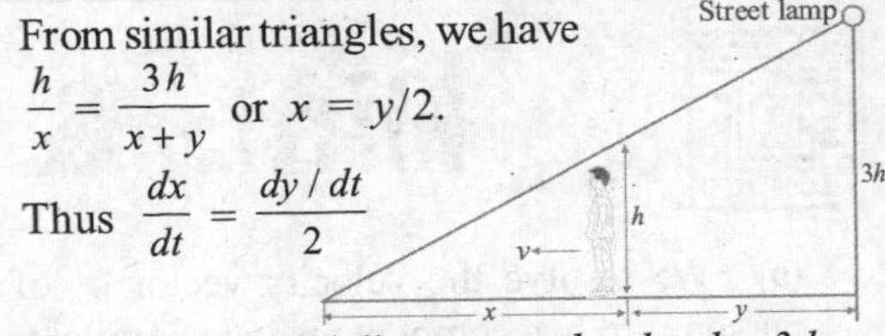

26. (d) So, required distance $= d + d + d = 3d$.

27. (a) $\vec{V}_{im} = -\frac{V^2}{u^2}\vec{V}_{om}$

$$0 - \frac{40}{13} = -\left(\frac{-10}{-10+x}\right)^2\left(10 - \frac{40}{13}\right) \qquad \frac{-10}{-10+x} = \pm\frac{2}{3}$$

x = 25, – 5cm

Object is infront of mirror Hence 25 cm.

28. (b) $\mu g \sin i = \mu_{air} \sin 90°$ $\mu_g = \frac{1}{\sin i}$

29. (b)

30. (b) Given, radius of hemispherical glass R = 10 cm

$\therefore$ Focal length $f = \frac{10}{2} = -5$ cm

u = (10 – 6) = –4 cm.

By using mirror formula,

$\frac{1}{v} + \frac{1}{u} = \frac{1}{f} \Rightarrow \frac{1}{v} + \frac{1}{-4} = \frac{1}{-5} \Rightarrow v = 20$ cm.

Apparent height, $h_a = h_r\frac{\mu_1}{\mu_2} = 30 \times \frac{1}{1.5} = 20$ cm below flat surface.

31. (a)

32. (c) The distance of bottom of the beaker from mirror $= h - d\left(1 - \frac{1}{\mu}\right)$

So it will be at a distance $= h - d\left(1 - \frac{1}{\mu}\right)$ from mirror. Now distance between bottom of beaker and image

$= h + h - d\left(1 - \frac{1}{\mu}\right) = 2h - d\left(\frac{\mu - 1}{\mu}\right)$.

33. (d) Given, $\mu = \frac{4}{3}$ h = 15 cm

$\frac{\sin 90°}{\sin C} = \mu \Rightarrow \sin C = \frac{1}{\mu} = \frac{R}{\sqrt{R^2 + h^2}} = \frac{3}{4}$

$\Rightarrow 16R^2 = 9R^2 + 9h^2$ or, $7R^2 = 9h^2$

or, $R = \frac{3}{\sqrt{7}}h = \frac{3}{\sqrt{7}} \times 15$ cm

34. (c) Consider the activity A to B

Applying $v^2 - u^2 = 2as$

$v^2 - 0^2 = 2 \times 10 \times 7.2$

$\Rightarrow v = 12$ m/s

The velocity of ball as perceived by fish is

$v' = {}_w\mu \times v = \frac{4}{3} \times 12 = 16$ m/s

35. (b)

36. (a) The phenomenon of total internal reflection takes place during reflection at P.

$\sin\theta = \frac{1}{{}_g^w\mu}$ ...(i)

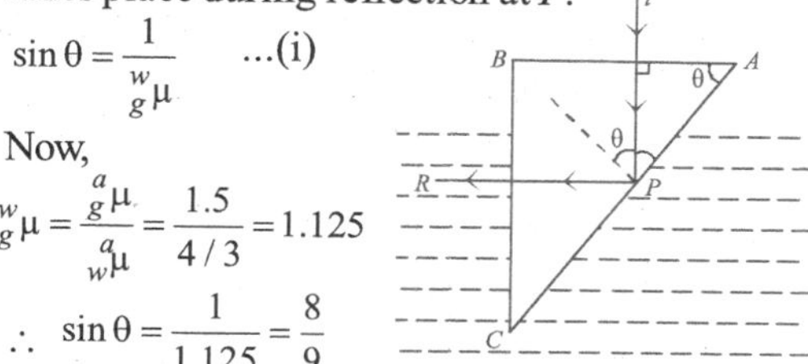

Now,

$${}_g^w\mu = \frac{{}_g^a\mu}{{}_w^a\mu} = \frac{1.5}{4/3} = 1.125$$

$\therefore \sin\theta = \frac{1}{1.125} = \frac{8}{9}$

$\therefore$ $\sin\theta$ should be greater than $\frac{8}{9}$.

37. (b)

38. (c) $\frac{x}{1} = \frac{x_{rel}}{\mu} \Rightarrow x_{rel} = \mu x$

$$\Rightarrow \frac{d^2x_{rel}}{dt^2} = \mu^2\frac{d^2x}{dt^2} \Rightarrow a_{rel} = \mu g$$

39. (c) The image I' for first refraction (i.e., when the ray comes out of liquid) is at a depth of

$$= \frac{33.25}{1.33} = 25\,\text{cm} \quad \left[\because \text{Apparent depth} = \frac{\text{Real depth}}{\mu}\right]$$

Now, reflection will occur at concave mirror. For this I' behaves as an object.

$\therefore \quad u = -(15+25) = -40$ cm and

$$v = -\left[15 + \frac{25}{1.33}\right]$$

Where $\frac{25}{1.33}$ is the real depth of the image.

Using mirror formula we get

$\frac{1}{f} = \frac{1}{v} + \frac{1}{u}$, $\quad f = -18.31$ cm

40. (d)

41. (a) This dark ring will be visible if ray from source gets total internal reflection from the spherical shell.

Let the source at any instant be at point P then at point Q ray will be totally reflected if θ is equal to or greater than critical angle. If QP is equal to x, then

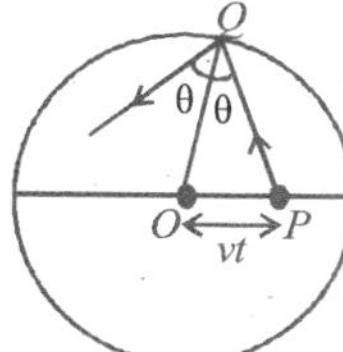

$$z = \cos\theta = \frac{R^2 + x^2 - v^2t^2}{2Rx}$$

For θ to be minimum

$$\frac{dz}{dx} = \frac{2x(2Rx) - 2R(R^2 + x^2 - v^2t^2)}{4R^2x^2} = 0$$

$$\Rightarrow x = \sqrt{R^2 - v^2t^2}$$

So, $\cos\theta = \frac{2(R^2 - v^2t^2)}{2R\sqrt{R^2 - v^2t^2}} = \frac{\sqrt{R^2 - v^2t^2}}{R}$

For no light come out, $\sin\theta \geq \frac{1}{\sqrt{2}}$ or $\theta \geq 45°$

$$\frac{\sqrt{R^2 - v^2t^2}}{R} = \frac{1}{\sqrt{2}}; \quad t = \frac{R}{\sqrt{2}V}$$

42. (c) The lateral displacement is given by

$$d = \frac{t}{\cos r}\sin(i - r)$$

For small angle i, angle r is also small, and so

$$d = t(i - r) = t\,i\,(1 - r/i)$$

43. (b)

44. (b) For the image of point P to be seen by the observer, it should be formed at point Q.

In ΔQNS, $NS = QS = 2h$

$\therefore \quad \angle NQS = 45°$

$\therefore \quad r = 45°$

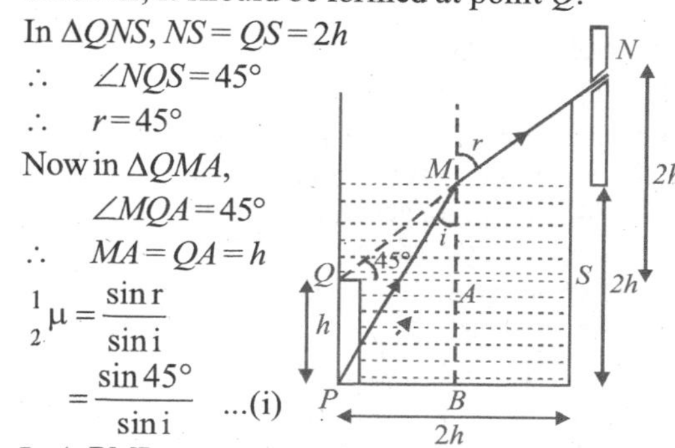

Now in ΔQMA,

$\angle MQA = 45°$

$\therefore \quad MA = QA = h$

$${}_2^1\mu = \frac{\sin r}{\sin i} = \frac{\sin 45°}{\sin i} \quad \text{...(i)}$$

In ΔPMB,

$PM^2 = 4h^2 + h^2 = 5h^2$

$\therefore \quad \sin i = \frac{h}{\sqrt{5}h} = \frac{1}{\sqrt{5}}$ ...(ii)

From (i) and (ii) ${}_2^1\mu = \sqrt{\frac{5}{2}}$

45. (a) The figure shows incidence from water at critical angle θ_c for the limiting case.

Now, $\sin\theta_c = 1/\mu$ so that $\tan\theta_c = 1/(\mu^2 - 1)^{1/2}$ which is also equal to r/h where r is the radius of the disc.

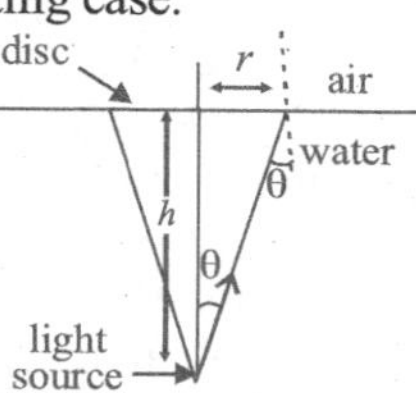

Therefore, diameter of the disc is $2r = 2h\tan\theta_c$.

46. (c) 47. (d)

48. (b) Shift $= (\ell - m)\left(1 - \frac{1}{n_1}\right) + m\left(1 - \frac{1}{n_2}\right) = 0$

49. (c) Real depth = 5 cm + 1 cm = 6 cm

Apparent depth $= \frac{d_1}{\mu_1} + \frac{d_2}{\mu_2} + ...$

$= \frac{5}{1.33} + \frac{1}{1.5} \simeq 3.8 + 0.7$

$= 4.5$ cm

$\therefore$ Shift = 6 cm − 4.5 cm

$\simeq 1.5$ cm

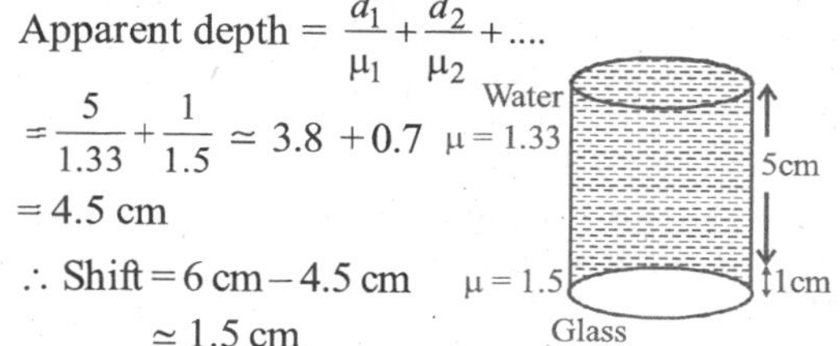

50. (a)

51. (a) In this case, one of the image will be real and the other virtual. Let us assume that image of S_1 is real and that of S_2 is virtual.

Applying $\frac{1}{v}-\frac{1}{u}=\frac{1}{f}$

For $S_1 : \frac{1}{y}+\frac{1}{x}=\frac{1}{9}$ ...(i)

For $S_2 : \frac{1}{y}+\frac{1}{24-x}$

$=\frac{1}{9}$ ...(ii)

Solving eqs. (i) and (ii), we get $x = 6$ cm

52. (c)

53. (a) We will have single surface refractions successively at the four surfaces S1, S2, S3 and S4. Do not forget to shift origin to the vertex of respective surface.

Refractive at first surface S1 : Light travels from air to glass.

$$\frac{1.5}{v_1}-\frac{1}{\infty}=\frac{(1.5-1)}{(+10)} \qquad v_1 = 30 \text{ cm}$$

First image is object for the refraction at second surface. For refraction at surafce S_2 : Light travels from glass to air

$$\frac{1}{v_2}-\frac{1.5}{(+25)}=\frac{1-1.5}{(+)5} \qquad v_2 = -25 \text{ cm}$$

For refraction at surface S_3 : Light travels from air to glass.

$$\frac{1.5}{v_3}-\frac{1}{(-35)}=\frac{(1.5-1)}{(-5)} \qquad v_3 = -35/3 \text{ cm}$$

For refractio at surface S_4: Light travels from glass to air

simillarly, $v_4 = -25$ cm

The final image is virtual, formal at 25 cm to the left of the vertex of surface S_4.

54. (a) Use lens maker's formula

$$\frac{1}{f}=\left({}_{g}^{m}\mu-1\right)\left(\frac{1}{R_1}-\frac{1}{R_2}\right)$$

Now, ${}_{g}^{m}\mu=\frac{{}_{g}\mu}{{}_{m}\mu}=\frac{1.5}{1.75}$

$$\therefore \frac{1}{f}=\left(\frac{1.5}{1.75}-1\right)\left(-\frac{1}{R}-\frac{1}{R}\right)=+\frac{0.25\times 2}{1.75R}$$

$\Rightarrow f = +3.5R$

The positive sign shows that the lens behaves as convergent lens.

55. (c)

56. (d) Since the water has a greater index of refraction than the lens, the magnifier now acts like a diverging lens, resulting in an image that is smaller than the object.

57. (a) For convex lens,

$$\frac{1}{30}=\frac{1}{v}+\frac{1}{60}\Rightarrow\frac{1}{60}=\frac{1}{v}$$

Similarly for concave lens

$$\frac{1}{-120}=\frac{1}{v}-\frac{1}{40}\Rightarrow\frac{1}{v}=\frac{1}{60}$$

Virtual object 10 cm behind plane mirror.

Hence real image 10 cm infront of mirror or, 60 cm from convex lens.

58. (d) If side of object square $= \ell$ and side of image square $= \ell'$

From question, $\frac{\ell'^2}{\ell}=9$ or $\frac{\ell'}{\ell}=3$

i.e., magnification $m = 3$ $\quad u = -40$ cm

$v = 3\times 40 = 120$ cm $f = ?$

From formula, $\frac{1}{v}-\frac{1}{u}=\frac{1}{f}$

$$\frac{1}{120}-\frac{1}{-40}=\frac{1}{f} \text{ or, } \frac{1}{f}=\frac{1}{120}+\frac{1}{40}=\frac{1+3}{120}$$

$\therefore f = 30$ cm

59. (c) Power of combinations

$$P=2(\mu-1)\left(\frac{1}{-60}-\frac{1}{-20}\right)+\left(\frac{-1}{-20/2}\right)$$

$$P=\frac{2}{15}=\frac{-1}{f}$$

$\therefore$ Resultant focal length $f=-\frac{15}{2}$

$\therefore$ From mirror formula $\frac{1}{-30}+\frac{1}{v}=\frac{1}{f}$

Solving $v = -10$ cm.

60. (b)

61. (d) Distance of image from the plane surface is as follows :

$$x_1=\frac{4}{1.6}=2.5\text{ cm}\left(d_{app}=\frac{d_{actual}}{\mu}\right)$$

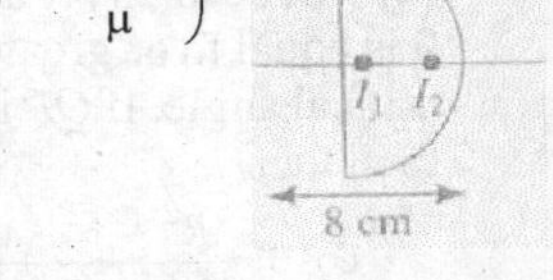

For the curved side

$$\frac{1.6}{4}+\frac{1}{x_2}=\frac{1-1.6}{8}$$

$\therefore x_2 \approx -3.0$ cm

The minus sign means the image is on the object side.

$\therefore I_1I_2 = (8-2.5-3.0)$ cm $= 2.5$ cm

62. (c)

63. (b) Let the object distance be x. Then, the image distance is $D-x$.

From lens equation, $\frac{1}{x}+\frac{1}{D-x}=\frac{1}{f}$

On algebraic rearrangement, we get

$$x2 - Dx + Df = 0$$

On solving for x, we get

$$x_1 = \frac{D - \sqrt{D(D-4f)}}{2} \quad x_2 = \frac{D + \sqrt{D(D-4f)}}{2}$$

The distance between the two object positions is

$$d = x_2 - x_1 = \sqrt{D(D-4f)}$$

64. **(d)**

65. **(a)** $\frac{1}{F} = \frac{1}{f_1} + \frac{1}{f_2}$

$$\frac{1}{F} = (\mu_1 - 1)\left(\frac{1}{\infty} + \frac{1}{R}\right) + (\mu_2 - 1)\left(\frac{1}{-R} - \frac{1}{\infty}\right) = \frac{\mu_1 - \mu_2}{R}$$

$$F = \frac{R}{\mu_1 - \mu_2}$$

66. **(a)** $\frac{|f_1|}{|f_2|} = \frac{2}{3}$

f_1 : focal length of convex lens. $\frac{1}{f} = \frac{1}{f_1} - \frac{1}{f_2}$

$$\Rightarrow \frac{1}{30} = \frac{1}{f_1} - \frac{2}{3f_1}$$

$f_1 = 10$ cm, $f_2 = -15$ cm

67. **(c)**

68. **(d)** $P = 2\left[\frac{100}{10} + \frac{100}{-20} + 0\right]$ $\quad P = 10$ dioptre.

69. **(d)** **70.** **(a)**

71. **(c)** $\frac{1}{f} = (\mu - 1)\left(\frac{1}{R_1} - \frac{1}{R_2}\right)$;

$R = 10$ cm.

$$\frac{1}{f'} = (3-1)\left(\frac{1}{-10} - \frac{1}{10}\right),$$

$$f' = -\frac{10}{4}$$

f=20 f' f=20

$$\frac{1}{f_{eq}} = \frac{1}{20} - \frac{4}{10} + \frac{1}{20} = \frac{2}{20} - \frac{4}{10} \Rightarrow f_{eq} = -\frac{10}{3}$$

72. **(b)** $f_1 = +40$ cm $= 0.4$ m

(for convex lens)

$f_2 = -25$ cm $= -0.25$ m

(for concave lens)

$\therefore$ Focal length (f) of the combination

$$\frac{1}{f} = \frac{1}{f_1} + \frac{1}{f_2} = \frac{1}{0.40} - \frac{1}{0.25} = \frac{0.25 - 0.4}{0.40 \times 0.25}$$

$$= -\frac{0.15}{0.1} = -1.5 \text{ dioptre.} \Rightarrow P = \frac{1}{f} = -1.5 \text{ dioptre}$$

73. **(b)** $\frac{\mu_2}{V} - \frac{\mu_1}{-u_0} = \frac{\mu_2 - \mu_1}{-2R}$

$$\frac{\mu_3}{V_f} - \frac{\mu_2}{V} = \frac{\mu_3 - \mu_2}{-R}$$

$$\frac{\mu_3}{V_f} + \frac{\mu_1}{u_0} = \frac{\mu_1 - \mu_2}{2R} + \frac{\mu_2 - \mu_3}{R}$$

$$\therefore \frac{\mu_1 - \mu_2}{2} < \mu_3 - \mu_2 \Rightarrow \mu_1 - \mu_2 < 2\mu_3 - 2\mu_2$$

$$\Rightarrow \mu_1 + \mu_2 < 2\mu_3$$

74. **(a)** $\frac{1}{f} = \frac{2}{f_e} + \frac{2}{f_m}$

$f_m = \frac{R}{2}$ $f_m = 0$

$\frac{1}{f_e} = \frac{n-1}{R}$ $\quad f_e = \frac{n-1}{R}$ $\quad f_2 = \frac{R}{2n}$

$f_1 = \frac{R}{2(n-1)}$ $\quad$ Ratio $\frac{f_1}{f_2} = \frac{n}{n-1}$

75. **(b)**

76. **(b)** Cutting a lens in transverse direction doubles their focal length i.e. $2f$.

Using the formula of equivalent focal length,

$$\frac{1}{f} = \frac{1}{f_1} + \frac{1}{f_2} + \frac{1}{f_3} + \frac{1}{f_4}$$

We get equivalent focal length as $f/2$.

77. **(a)** Let refractive index of glass be μ.

Let after first refraction, image distance be v then

$$\frac{\mu}{v} - \frac{1}{\infty} = \frac{\mu - 1}{R} \Rightarrow v = \frac{\mu R}{\mu - 1}$$

Now second refraction will take place.

So distance of first image from O is

$$u_1 = \frac{\mu R}{\mu - 1} - R = \frac{R}{\mu - 1}$$ and image is formed at R

$$\therefore \frac{1}{R} - \frac{\mu(\mu - 1)}{R} = \frac{2(1-\mu)}{R}$$

$$\Rightarrow \mu^2 - 3\mu + 1 = 0 \text{ So, } \mu = \frac{3 + \sqrt{5}}{2}$$

78. **(d)** For T.I.R. $45° > C$

$\therefore \sin 45° > \sin C$

$\therefore \frac{1}{\sqrt{2}} > \frac{4/3}{n}$

$\therefore n > \frac{4\sqrt{2}}{3}$

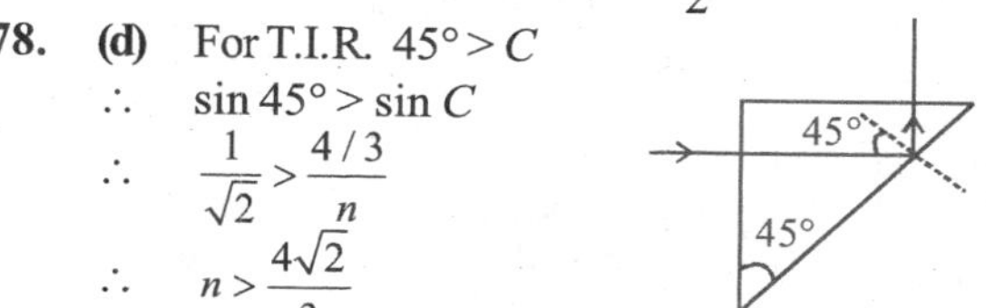

79. **(b)**

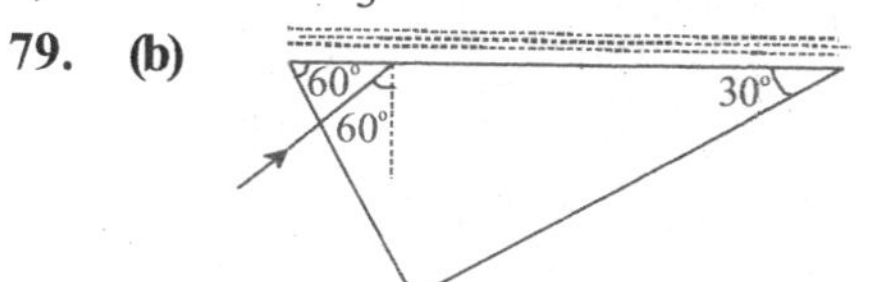

Clearly, $i_c \le 60°$

So, maximum possible value of i_c is 60°.

Now, $_1\mu_g = \dfrac{1}{\sin i_c}$

$\dfrac{\mu_g}{\mu_1} = \dfrac{1}{\sin i_c}$ or $\mu_1 = \mu_g \sin i_c = 1.5 \sin 60° = 1.5 \times \dfrac{\sqrt{3}}{2}$

$= 1.5 \times 0.866 = 1.299 = 1.3$

80. (d) $\delta = (45° - 30°) + (180° - 60°) + (45° - 30°)$
$= 150°$ clockwise.

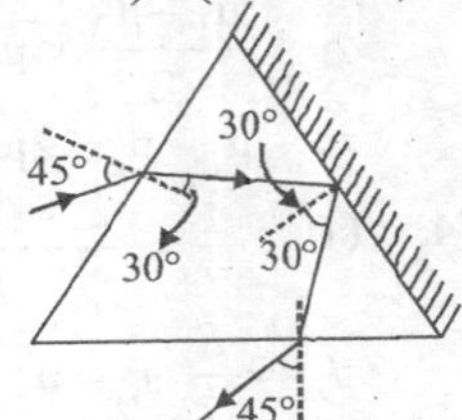

81. (c) **82.** (d) **83.** (d)

84. (d) Dispersive power (ω)

$= \dfrac{n_v - n_r}{n_y - 1} = \dfrac{1.521 - 1.510}{1.515 - 1} = 0.02$

85. (a) For minimum deviation the ray in the prism is parallel to the base of the prism. This condition does not depend on the colour (or wave length) of incident radiation. So in both the cases, by geometry, r = 30°. So (a) is correct option.

86. (a) In a prism : $r + r' = A \Rightarrow r = A - r'$

$\therefore\ r = 60° - (10 + t^2) = 50 - t^2$

87. (a)

88. (c) Applying Snell's law at *P*,

$\mu = \dfrac{\sin r}{\sin 30°}$

$\sin r = \dfrac{1.44}{2} = 0.72$

$\therefore\ \delta = r - 30° = \sin^{-1}(0.72) - 30°$

$\therefore$ The rays make an angle of

$2\delta = 2[\sin^{-1}(0.72) - 30°]$ with each other.

89. (b) $A = 90° - \theta$

$\Rightarrow r_2 = A = 90° - \theta > \theta_c$

$\cos\theta > \sin\theta_c = \dfrac{6/5}{2/3} = \dfrac{4}{5}$

(θ_c is critical angle)

$\theta < \cos^{-1}\dfrac{4}{5} = 37°$

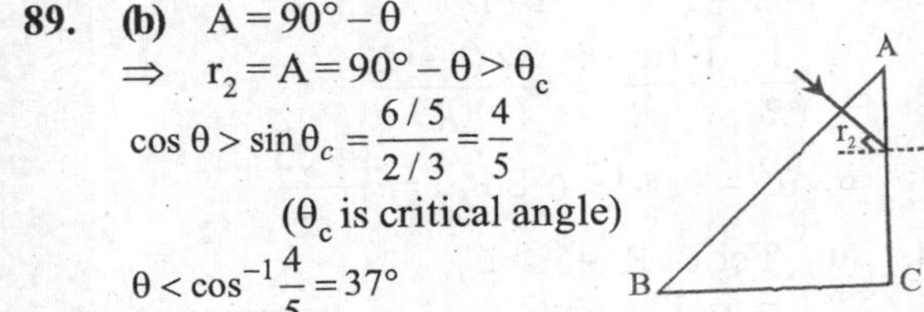

90. (c)

91. (c) Magnifying power of telescope,

$MP = \dfrac{\beta \text{ (angle subtended by image at eye piece)}}{\alpha \text{ (angle subtended by object on objective)}}$

Also, $MP = \dfrac{f_o}{f_e} = \dfrac{150}{5} = 30 \quad \alpha = \dfrac{50}{1000} = \dfrac{1}{20}$ rad

$\therefore\ \beta = \theta = MP \times \alpha = 30 \times \dfrac{1}{20} = \dfrac{3}{2} = 1.5$ rad

92. (d) **93.** (a)

94. (c) The intermediate image in compound microscope is real, inverted and magnified.

95. (d) Given, Focal length of objective, $f_0 = 30$ cm
focal length of eye lens, $f_e = 3.0$ cm
Magnifying power, M = ?
Magnifying power of the Galilean telescope,

$M_D = \dfrac{f_0}{f_e}\left(1 - \dfrac{f_e}{D}\right) = \dfrac{30}{3}\left(1 - \dfrac{3}{25}\right)$ [$\because$ D = 25 cm]

$= 10 \times \dfrac{22}{25} = 8.8$ cm

96. (a)

97. (d) Given : $f_0 = 50$ cm, $f_e = 5$ cm
$d = 25$ cm, $u_0 = -200$ cm

Magnification M = ?

As $\dfrac{1}{v_0} - \dfrac{1}{u_0} = \dfrac{1}{f_0}$

$\Rightarrow \dfrac{1}{v_0} = \dfrac{1}{f_0} + \dfrac{1}{u_0} = \dfrac{1}{50} - \dfrac{1}{200} = \dfrac{4-1}{200} = \dfrac{3}{200}$

or $v_0 = \dfrac{200}{3}$ cm Now $v_e = d = -25$ cm

From, $\dfrac{1}{v_e} - \dfrac{1}{u_e} = \dfrac{1}{f_e}$

$\dfrac{1}{u_e} = \dfrac{1}{f_e} - \dfrac{1}{v_e} = \dfrac{1}{5} + \dfrac{1}{25} = \dfrac{6}{25}$ or, $v_e = \dfrac{-25}{6}$ cm

Magnification $M = M_0 \times M_e$

$= \dfrac{v_0}{u_0} \times \dfrac{v_e}{u_e} = \dfrac{-200/3}{200} \times \dfrac{-25}{-25/6} = -\dfrac{1}{3} \times 6 = -2$

98. (a) When the ray enters a glass slab from air, its frequency remains unchanged.
Since glass slab in an optically denser medium, the velocity of light decreases and therefore we can conclude that the wavelength decreases. ($\because v = \nu\lambda$)

99. (d) $\sin i_c = \dfrac{1}{\mu} = \dfrac{r}{\sqrt{r^2 + h^2}}$

Using $h = 12$ cm, $\mu = 4/3$

We get $r = \dfrac{36}{\sqrt{7}}$ cm.

100. (b) From similar triangles APB and CBD, point P will be just visible if CD = 4 R

$\mu \sin i = 1 \times \sin r$

$\mu \sin(\angle CDQ) = 1 \times \sin(\angle CDB)$

$\mu \times \dfrac{R}{R\sqrt{17}} = 1 \times \dfrac{2R}{R\sqrt{20}} \Rightarrow \mu = \dfrac{2\sqrt{17}}{\sqrt{20}} = \dfrac{\sqrt{85}}{5}$

24 Wave Optics

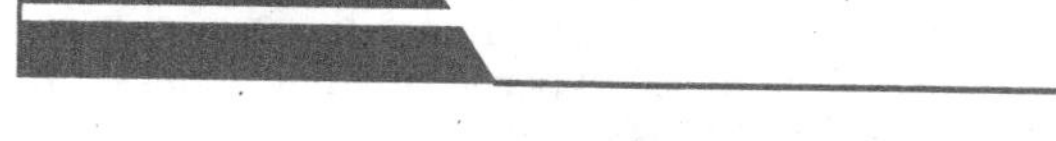

1. Huygen's concept of secondary wave
   (a) allows us to find the focal length of a thick lens
   (b) is a geometrical method to find a wavefront
   (c) is used to determine the velocity of light
   (d) is used to explain polarisation

2. A wavefront AB passing through a system C emerges as DE. The system C could be
   (a) a slit
   (b) a biprism
   (c) a prism
   (d) a glass slab

3. The phase difference between incident wave and reflected wave is 180° when light ray
   Tricky
   (a) enters into glass from air
   (b) enters into air from glass
   (c) enters into glass from diamond
   (d) enters into water from glass

4. A beam of monochromatic light is refracted from vacuum into a medium of refractive index 1.5. The wavelength of refracted light will be
   (a) dependent on intensity of refracted light
   (b) same (c) smaller (d) larger

5. The correct curve between refractive index μ and wavelength λ will be
   (a) A
   (b) D
   (c) B
   (d) C

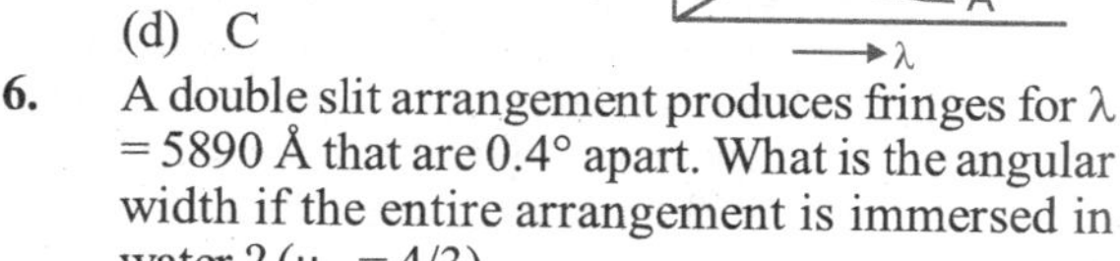

6. A double slit arrangement produces fringes for $\lambda = 5890$ Å that are 0.4° apart. What is the angular width if the entire arrangement is immersed in water ? ($\mu_w = 4/3$)
   (a) 0.3° (b) 2.3° (c) 0.8° (d) 1.3°

7. Young's double slit experiment is made in a liquid. The 10th bright fringe lies in liquid where 6th dark fringe lies in vacuum. The refractive index of the liquid is approximately
   Critical Thinking
   (a) 1.8 (b) 1.5 (c) 1.3 (d) 1.6

8. A possible means for making an airplane invisible to radar is to coat the plane with an anti reflective polymer. Radar waves have a wavelength of 3.00 cm and the index of refraction of the polymer is $n = 1.50$. How thick would you make the coating?
   Critical Thinking
   (a) 1.50 cm (b) 3.00 cm
   (c) 0.50 cm (d) None of these

9. In the adjacent diagram, CP represents a wavefront and AO & BP, the corresponding two rays. Find the condition on θ for constructive interference at P between the ray BP and reflected ray OP.

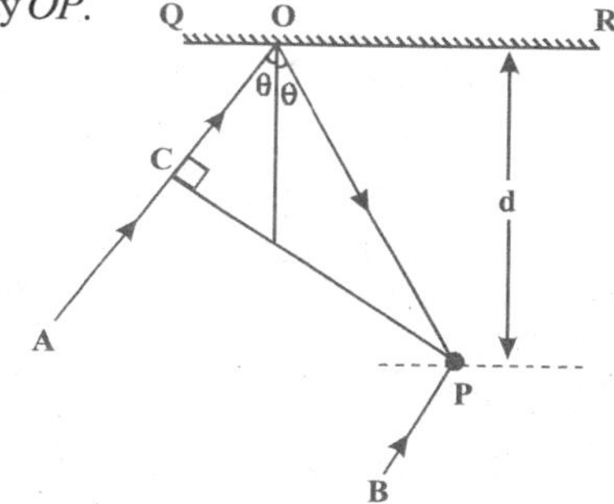

   (a) $\cos\theta = 3\lambda/2d$ (b) $\cos\theta = \lambda/4d$
   (c) $\sec\theta - \cos\theta = \lambda/d$ (d) $\sec\theta - \cos\theta = 4\lambda/d$

10. From a medium of index of refraction n_1, monochromatic light of wavelength λ is incident normally on a thin film of uniform thickness L (where $L > 0.1\lambda$) and index of refraction n_2. The light transmitted by the film travels into a medium with refractive index n_3. The value of minimum film thickness when maximum light is transmitted if ($n_1 < n_2 < n_3$) is
    (a) $\frac{n_1\lambda}{2n_2}$ (b) $\frac{n_1\lambda}{4n_2}$ (c) $\frac{\lambda}{4n_2}$ (d) $\frac{\lambda}{2n_2}$

11. Two slits separated by a distance of 1 mm are illuminated with red light of wavelength 6.5×10^{-7} m. The interference fringes are observed on a screen placed 1 m from the silts. The distance of the third dark fringe from the central fringe will be equal to :
    Critical Thinking
    (a) 0.65 mm (b) 1.30 mm (c) 1.62 mm (d) 1.95 mm

12. Interference fringes were produced in Young's double slit experiment using light of wavelength 5000 Å. When a film of material 2.5×10^{-3} cm thick was placed over one of the slits, the fringe pattern shifted by a distance equal to 20 fringe width. The refractive index of the material of the film is
    (a) 1.25 (b) 1.33 (c) 1.4 (d) 1.5

13. In an interference arrangement similar to Young's double-slit experiment, the slits S_1 and S_2 are illuminated with coherent microwave sources, each of frequency 10^6 Hz. The sources are synchronized to have zero phase difference. The slits are separated by a distance d = 150.0 m. The intensity I (θ) is measured as a function of θ, where θ is defined as shown. If I_0 is the maximum intensity, then I (θ) for $0 \le \theta \le 90°$ is given by

(a) $I(\theta) = I_0/2$ *for* $\theta = 30°$

(b) $I(\theta) = I_0/4$ *for* $\theta = 90°$

(c) $I(\theta) = I_0$ *for* $\theta = 0°$

(d) $I(\theta)$ is constant for all values of θ.

14. If white light is used in the Newton's rings experiment, the colour observed in the reflected light is complementary to that observed in the transmitted light through the same point. This is due to

(a) 90° change of phase in one of the reflected waves

(b) 180° change of phase in one of the reflected waves

(c) 145° change of phase in one of the reflected waves

(d) 45° change of phase in one of the reflected waves

15. In Fresnel's biprism experiment the width of 10 fringes is 2cm which are formed at a distance of 2 meter from the slit. If the wavelength of light is 5100 Å then the distance between two coherent sources will be

(a) 5.1×10^{-4} m (b) 5.1×10^{4} cm

(c) 5.1×10^{-4} mm (d) 10.1×10^{-4} cm

16. Two beams of light having intensities I and $4I$ interfere to produce a fringe pattern on a screen. The phase difference between the beams is $\pi/2$ at point A and π at point B. Then the difference between the resultant intensities at A and B is

(a) $2I$ (b) $4I$ (c) $5I$ (d) $7I$

17. In the ideal double-slit experiment, when a glass-plate (refractive index 1.5) of thickness t is introduced in the path of one of the interfering beams (wave-lenght λ), the intensity at the position where the central maximum occurred previously remains unchanged. The minimum thickness of the glass-plate is

(a) 2λ (b) 2λ/3 (c) λ/3 (d) λ

18. Monochromatic light of wavelength 400 nm and 560 nm are incident simultaneously and normally on double slits apparatus whose slits separation is 0.1 mm and screen distance is 1m. Distance between areas of total darkness will be

(a) 4mm (b) 5.6mm (c) 14mm (d) 28mm

19. White light used to illuminate the two slits in Young's double slit experiment. The separation between the slits is d and the distance between the screen and the slit is D (>>d). At a point on the screen in front of one of the slits, certain wave-lengths are missing. The missing wavelengths are

Critical Thinking

(a) $\lambda = \dfrac{d^2}{(2n+1)D}$ (b) $\lambda = \dfrac{(2n+1)d^2}{D}$

(c) $\lambda = \dfrac{d^2}{(n+1)D}$ (d) $\lambda = \dfrac{(n+1)D}{d^2}$

20. A thin slice is cut out of a glass cylinder along a plane parallel to its axis. The slice is placed on a flat glass plate as shown in Figure. The observed interference fringes from this combination shall be

Tricky

(a) straight

(b) circular

(c) equally spaced

(d) having fringe spacing which increases as we go outwards

21. Interference fringes were produced using white light in a double slit arrangement. When a mica sheet of uniform thickness of refractive index 1.6 (relative to air) is placed in the path of light from one of the slits, the central fringe moves through some a distance. This distance is equal to the width of 30 interference bands if light of wavelength 4800 Å is used. The thickness (in μm) of mica is

(a) 90 (b) 12 (c) 14 (d) 24

22. In a Young's double slit experiment, 12 fringes are observed to be formed in a certain segment of the screen when light of wavelength 600 nm is used. If the wavelength of light is changed to 400 nm, number of fringes observed in the same segment of the screen is given by

(a) 12 (b) 18 (c) 24 (d) 30

23. Two monochromatic light beams of intensity 16 and 9 units are interfering. The ratio of intensities of bright and dark parts of the resultant pattern is:

(a) $\dfrac{16}{9}$ (b) $\dfrac{4}{3}$ (c) $\dfrac{7}{1}$ (d) $\dfrac{49}{1}$

24. In Young's double slit experiment intensity at a point is (1/4) of the maximum intensity. Angular position of this point is

(a) $\sin^{-1}(\lambda/d)$ (b) $\sin^{-1}(\lambda/2d)$
(c) $\sin^{-1}(\lambda/3d)$ (d) $\sin^{-1}(\lambda/4d)$

25. A thin sheet of mica (12×10^{-7} m thick) is placed in the path of one of the interfering beams in a biprism arrangement. It is found that the central bright band shifts a distance equal to the width of a bright fringe. The refractive index of mica (Given $\lambda = 6 \times 10^{-7}$ m) is

(a) 1.0 (b) 1.5 (c) 1.75 (d) 1.25

26. Two sources S_1 and S_2 emitting coherent light waves of wavelength λ in the same phase are situated as shown. The distance *OP*, so that the light intensity detected at *P* is equal to that at *O* is

Critical Thinking

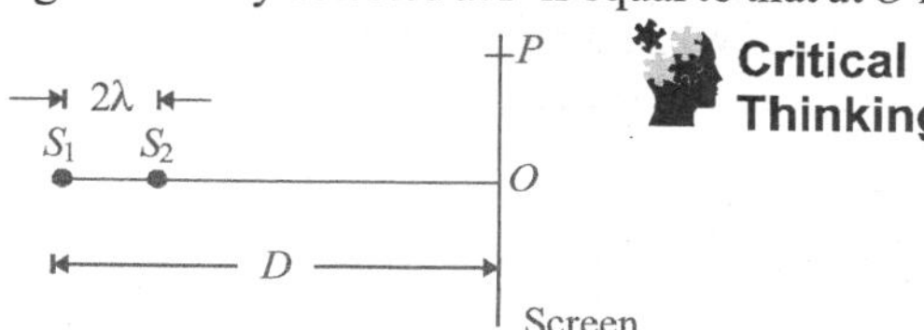

(a) $D\sqrt{2}$ (b) $D/2$ (c) $D\sqrt{3}$ (d) $D/\sqrt{3}$

27. When a plastic thin film of refractive index 1.45 is placed in the path of one of the interfering waves then the central fringe is displaced through width of five fringes. The thickness of the film, if the wavelength of light is 5890Å, will be

(a) 6.544×10^{-4} cm (b) 6.544×10^{-4} m
(c) 6.54×10^{-4} cm (d) 6.5×10^{-4} cm

28. A plastic sheet (refractive index = 1.6) covers one slit of a double slit arrangement meant for the Young's experiment. When the double slit is illuminated by monochromatic light (wavelength in air = 6600 Å), the centre of the screen appears dark rather than bright. The minimum thickness of the plastic sheet to be used for this to happen is:

(a) 3300 Å (b) 6600 Å (c) 2062 Å (d) 5500 Å

29. The YDSE apparatus is as shown in Fig. The condition for point P to be a dark fringe is

Tricky

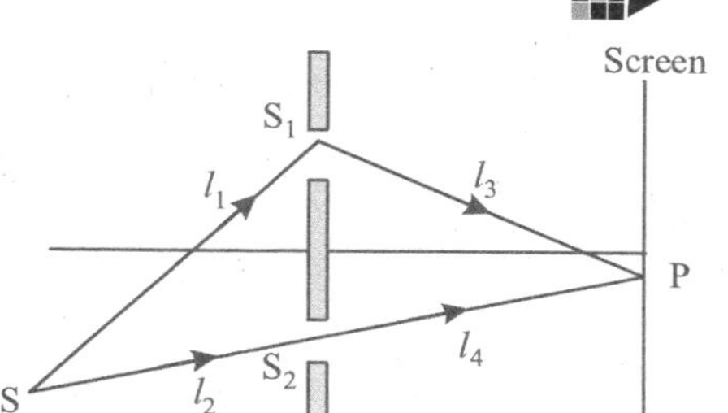

(a) $(l_1 - l_3) + (l_2 - l_4) = n\lambda$

(b) $(l_1 - l_2) + (l_3 - l_4) = \frac{(2n-1)}{2}\lambda$

(c) $(l_1 - l_3) + (l_2 - l_4) = \frac{(2n-1)\lambda}{2}$

(d) $(l_1 - l_2) + (l_4 - l_3) = \frac{(2n-1)\lambda}{2}$

30. A plane wave of monochromatic light falls normally on a uniform thin layer of oil which covers a glass plate. The wavelength of source can be varies continuosly. Complete destructive interference is observed for λ = 5000 Å and λ = 1000 Å and for no other wavelength in between. If μ of oil is 1.3 and that of glass is 1.5, the thickness of the film will be Toughnut

(a) 6.738×10^{-5} cm (b) 5.7×10^{-5} cm
(c) 4×10^{-5} cm (d) 2.8×10^{-5} cm

31. A broad source of light (I = 680 nm) illuminates normally two glass plates 120 mm long that touch at one end and are separated by a wire 0.034 mm in diameter at the other end. The total number of bright fringes that appear over the 120 mm distance is –

incident light

t

0.034mm

120mm

(a) 50
(b) 100
(c) 200
(d) 400

32. Two light waves having the same wavelength λ in vacuum are in phase initially. Then the first ray travels a path of length L_1 through a medium of refractive index μ_1. The second ray travels a path of length L_2 through a medium of refractive index μ_2. The two waves are then combined to observe interference effects. The phase difference between the two, when they interfere, is

(a) $\frac{2\pi}{\lambda}(L_1 - L_2)$ (b) $\frac{2\pi}{\lambda}(\mu_1 L_1 - \mu_2 L_2)$

(c) $\frac{2\pi}{\lambda}(\mu_2 L_1 - \mu_1 L_2)$ (d) $\frac{2\pi}{\lambda}\left[\frac{L_1}{\mu_1} - \frac{L_2}{\mu_2}\right]$

33. In YSDE, both slits are covered by transparent slab. Upper slit is covered by slab of R.I. 1.5 and thickness t and lower is covered by R.I. $\frac{4}{3}$ and thickness $2t$, then central maxima Tricky

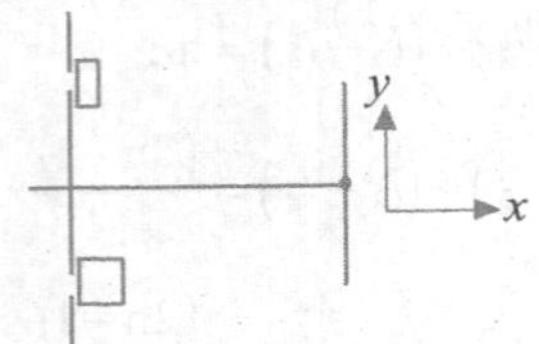

(a) shifts in +ve y-axis direction
(b) shifts in –ve y-axis direction
(c) remains at same position
(d) may shift in upward or downward depending upon wavelength of light

34. A light of wavelength 6000 Å shines on two narrow slits separated by a distance 1.0 mm and illuminates a screen at a distance 1.5 m away. When one slit is covered by a thin glass plate of refractive index 1.8 and other slit by a thin glass plate of refractive index μ, the central maxima shifts by 0.1 rad. Both plates have the same thickness of 0.5 mm. The value of refractive index μ of the glass is **Toughnut**

(a) 1.4 (b) 1.5
(c) 1.6 (d) None of these

35. In Young's double slit experiment intensity at a point is how much times of the maximum intensity if angular position of this point is $\sin^{-1}(\lambda/6d)$?

(a) 1/2 (b) 2/5
(c) 3/4 (d) 3/5

36. A monochromatic light is used in Young's double slit experiment when one of the slits is covered by a transparent sheet of thickness 1.8 mm, made of material of refractive index μ_1 number of fringes which shift is 18, when another sheet of thickness 3.6 mm, made of material of refractive index μ_2 is used, number of fringes which shift is 9. Relation between μ_1 and μ_2 is given by **Critical Thinking**

(a) $4\mu_2 - \mu_1 = 3$ (b) $4\mu_1 - \mu_2 = 3$
(c) $3\mu_2 - \mu_1 = 4$ (d) $2\mu_1 - \mu_2 = 4$

37. Two coherent sources separated by distance d are radiating in phase having wavelength λ. A detector moves in a big circle around the two sources in the plane of the two sources. The angular position of $n = 4$ interference maxima is given as

(a) $\sin^{-1}\dfrac{n\lambda}{d}$

(b) $\cos^{-1}\dfrac{4\lambda}{d}$

(c) $\tan^{-1}\dfrac{d}{4\lambda}$ (d) $\cos^{-1}\dfrac{\lambda}{4d}$

38. A glass plate 0.40 micron thick is illuminated by a beam of white light normal to the plate. The refractive index of glass is 1.50 and the limits of the visible spectrum are $\lambda_v = 4000$ Å and $\lambda_R = 7000$ Å. The wavelengths that get intensified in the reflected beam are

(a) 4800 Å and 5200 Å (b) 4800 Å and 6700 Å
(c) 4800 Å only (d) 5200 Å only

39. In a *YDSE* experiment if a slab whose refraction index can be varied is placed in front of one of the slits then the variation of resultant intensity at mid-point of screen with 'μ' will be best represented by (μ ≥ 1). [Assume slits of equal width and there is no absorption by slab; mid point of screen is the point where waves interfere with zero phase difference in absence of slab] **Tricky**

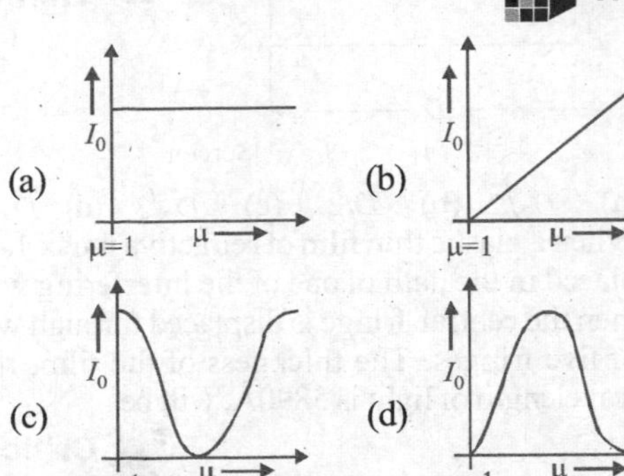

40. In an experiment, sodium light (λ = 5890 A°) is employed and interference fringes are obtained in which 20 fringes equally spaced occupy 2.30 cm on the screen. When sodium light is replaced by blue light, the setup remaining the same otherwise, 30 fringes occupy 2.80 cm. The wavelength of blue light is

(a) 4780 Å (b) 5760 Å (c) 9720 Å (d) 6390 Å

41. A ray of light of intensity I is incident on a parallel glass slab at point A as shown in diagram. It undergoes partial reflection and refraction. At each reflection, 25% of incident energy is reflected. The rays AB and A'B' undergo interference. The ratio of I_{max} and I_{min} is :

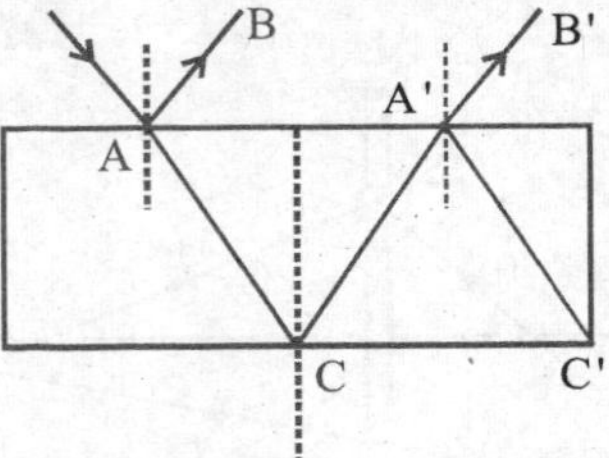

(a) 49 : 1 (b) 7 : 1 (c) 4 : 1 (d) 8 : 1

42. A wedged shaped air film having an angle of 40 second is illuminated by a monochromatic light and the fringes are observed vertically down through a microscope. The fringe separation between two consecutive bright fringes is 0.12 cm. The wavelength of light is :
(a) 5545 Å (b) 6025 Å (c) 4925 Å (d) 4655 Å

43. Figure shows two coherent sources $S_1 - S_2$ vibrating in same phase. *AB* is an irregular wire lying at a far distance from the sources S_1 and S_2. Let $\frac{\lambda}{d} = 10^{-3}$. $\angle BOA = 0.12°$. How many bright spots will be seen on the wire, including points *A* and *B*.
(a) 2
(b) 3
(c) 4
(d) more than 4

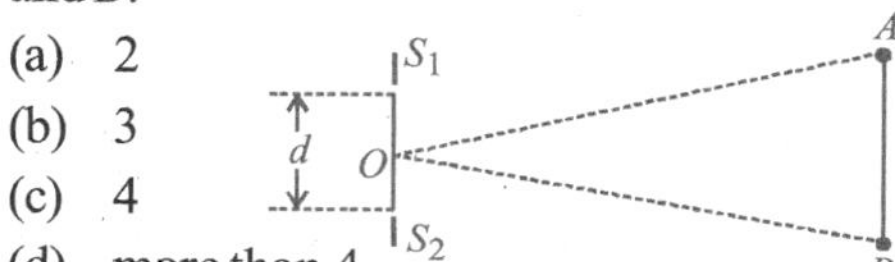

44. In young's double - slit experiment, the separation between the slits is d, distance between the slit and screen is D (D >> d). In the interference pattern, there is a maxima exactly in front of each slit. Then the possible wavelength(s) used in the experiment are **Critical Thinking**
(a) $d^2/D, d^2/2D, d^2/3D$ (b) $d^2/D, d^2/3D, d^2/5D$
(c) $d^2/2D, d^2/4D, d^2/6D$ (d) None of these

45. A YDSE is conducted in water (μ_1) as shown in figure. A glass plate of thickness *t* and refractive index μ_2 is placed in the path of S_2. The optical path difference at *O* is
(a) $(\mu_2 - 1)t$
(b) $(\mu_1 - 1)t$
(c) $\left(\frac{\mu_2}{\mu_1} - 1\right)t$
(d) $(\mu_2 - \mu_1)t$

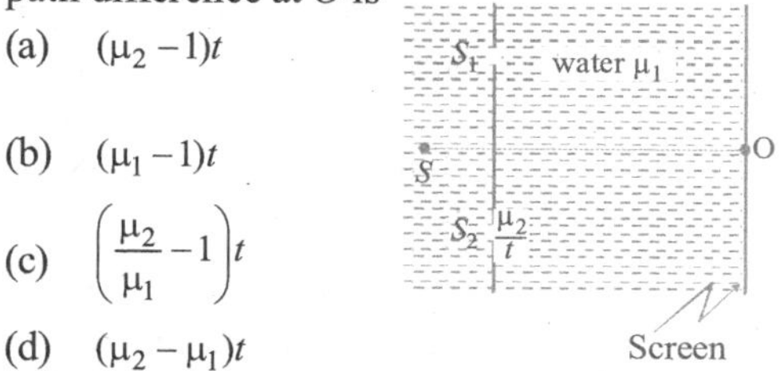

46. In a Young's double slit experiment with light of wavelength λ, fringe pattern on the screen has fringe width β. When two thin transparent glass (refractive index μ) plates of thickness t_1 and t_2 ($t_1 > t_2$) are placed in the path of the two beams respectively, the fringe pattern will shift by a distance
(a) $\frac{\beta(\mu-1)}{\lambda}\left(\frac{t_1}{t_2}\right)$ (b) $\frac{\mu\beta}{\lambda}\frac{t_1}{t_2}$
(c) $\frac{\beta(\mu-1)}{\lambda}(t_1 - t_2)$ (d) $(\mu-1)\frac{\lambda}{\beta}(t_1 + t_2)$

47. Two identical coherent sources are placed on a diameter of a circle of radius R at separation x (<< R) symmetrical about the center of the circle. The sources emit identical wavelength λ each. The number of points on the circle of maximum intersity is (x = 5λ) **Toughnut**
(a) 20 (b) 22 (c) 24 (d) 26

48. To produce a minimum reflection of wavelengths near the middle of visible spectrum (550 nm), how thick should a coating of MgF_2 (μ = 1.38) coated on a glass surface?
(a) 10^{-7} (b) 10^{-10} (c) 10^{-9} m (d) 10^{-8} m

49. In YDSE distance between the S_1 and S_2 is d. P_1 and P_2 are two points equidistance from O at an angular position β as shown. A parallel beam of monochromatic light is incident at an angle α on the slits. Then the ratio of path difference at P_1 and P_2 is:

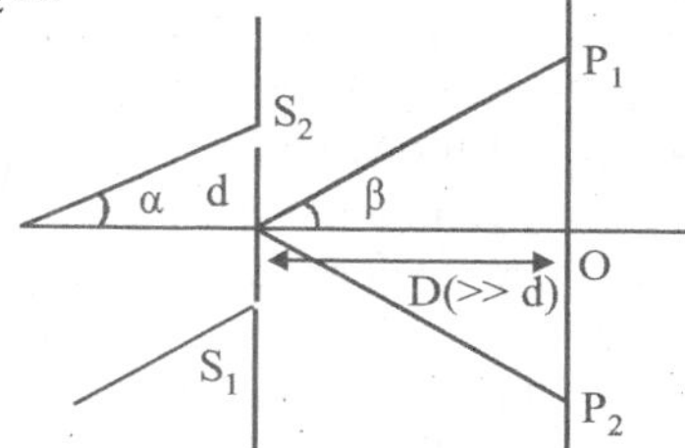

(a) $\cot\frac{\alpha-\beta}{2}\cot\frac{\alpha+\beta}{2}$ (b) $\tan\frac{\alpha+\beta}{2}\cot\frac{\alpha-\beta}{2}$
(c) $\sin\frac{\alpha+\beta}{2}\cos\frac{\alpha-\beta}{2}$ (d) $\tan\frac{\alpha-\beta}{2}\cot\frac{\alpha+\beta}{2}$

50. A physics professor wants to find the diameter of a human hair by placing it between two flat glass plates, illuminating the plates with light of vaccum wavelength λ = 552nm and counting the number of bright fringes produced along the plates. The Professor find 125 bright fringes between the edge of the plates and the hair. What is the diameter of the hair? **Toughnut**
(a) 525×10^{-9}m (b) 344×10^{-3}m
(c) 3.44×10^{-5}m (d) None of the above

51. In the figure shown in a *YDSE*, a parallel beam of light is incident on the slits from a medium of refractive index n_1. The wavelength of light in this medium is λ_1. A transparent slab of thickness t and refractive index is put infront of one slit. The medium between the screen and the plane of the slits is n_2. The phase difference between the light waves reaching point *O* (symmetrical, relative to the slit) is

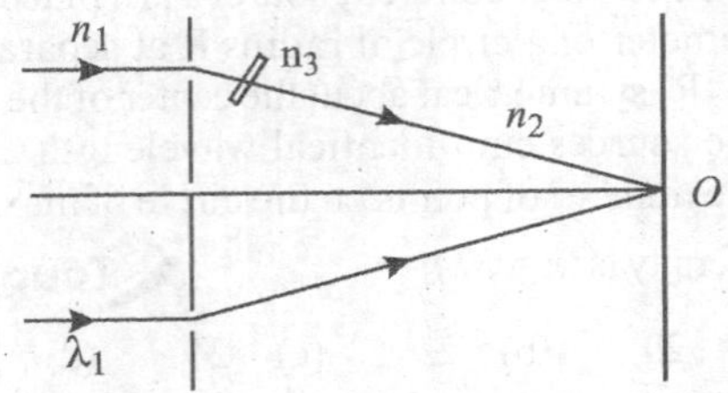

(a) $\frac{2\pi}{n_1\lambda_1}(n_3 - n_2)\,t$ (b) $\frac{2\pi}{\lambda_1}(n_3 - n_2)\,t$

(c) $\frac{2\pi n_1}{n_2\lambda_1}\left(\frac{n_3}{n_2} - 1\right)t$ (d) $\frac{2\pi n_1}{\lambda_1}(n_3 - n_2)\,t$

52. A beam of light consisting of two wavelength 6500A° and 5200A° is used to obtain interference fringes in a young's double slit experiment. The distance between the slits is 2.0 mm and the distance between the plane of the slits and the screen is 120 cm. What is the least distance from the central maximum where the bright fringes due to both the wave length coincide ?

(a) 0.156 cm (b) 0.152 cm

(c) 0.17 cm (d) 0.16 cm

53. A double slit, $S_1 - S_2$ is illuminated by a light source S emitting light of wavelength λ. The slits are separated by a distance d. A plane mirror is placed at a distance D in front of the slits and a screen is placed at a distance 2D behind the slits. The screen receives light reflected only by the plane mirror. The fringe-width of the interference pattern on the screen is

Tricky

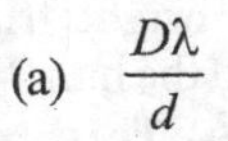

(a) $\frac{D\lambda}{d}$

(b) $\frac{2D\lambda}{d}$

(c) $\frac{3D\lambda}{d}$

(d) $\frac{4D\lambda}{d}$

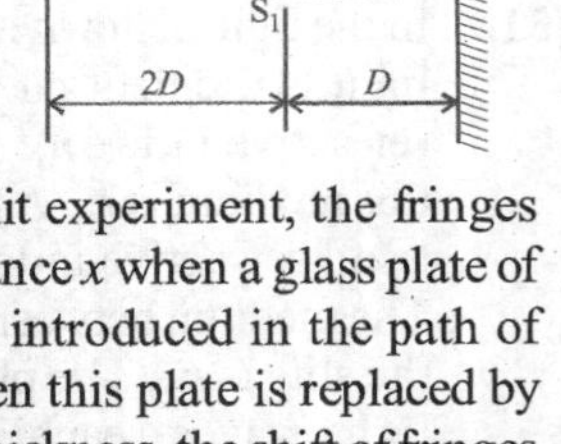

54. In a Young's double slit experiment, the fringes are displaced by a distance x when a glass plate of refractive index 1.5 is introduced in the path of one of the beams. When this plate is replaced by another plate of same thickness, the shift of fringes is $(3/2)\,x$. The refractive index of second plate is

(a) 1.75 (b) 1.50 (c) 1.25 (d) 1.00

55. A certain region of a soap bubble reflects red light of wavelength $\lambda = 650$nm. What is the minimum thickness that this region of the soap bubble could have? Take the index of reflection of the soap film to be 1.41.

Toughnut

(a) 1.2×10^{-7}m (b) 650×10^{-9}m

(c) 120×10^{7}m (d) 650×10^{5}m

56. In figure, Young's double slit experiment Q is the position of the first bright fringe on the right side of O. P is the 11th fringe on the other side, as measured from Q. If $\lambda = 6000$ Å, then S_1B will be equal to

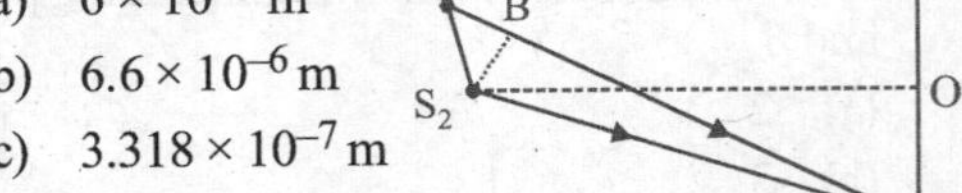

(a) 6×10^{-6} m

(b) 6.6×10^{-6} m

(c) 3.318×10^{-7} m

(d) 3.144×10^{-7} m

57. The intensity of a point source of light, S, placed at a distance d in front of a screen A, is I_0 at the center of the screen. Find the light intensity at the center of the screen if a completely reflecting plane mirror M is placed at a distance d behind the source, as shown in figure.

Critical Thinking

(a) $\frac{27I_0}{9}$ (b) $\frac{25I_0}{9}$

(c) $\frac{17I_0}{9}$ (d) $\frac{10I_0}{9}$

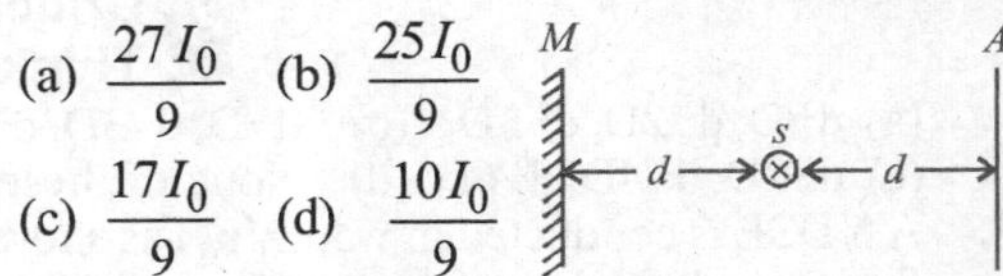

58. In Young's double slit interference experiment, the slit widths are in the ratio 1 : 25. Then the ratio of intensity at the maxima and minima in the interference pattern is

(a) 3:2 (b) 1:25 (c) 9:4 (d) 1:5

59. The maximum number of possible interference maxima for slit separation equal to 1.8λ, where λ is the wavelength of light used, in a Young's double slit experiment is

(a) zero (b) 3 (c) infinite (d) 5

60. In a Young's double slit experiment, if the incident light consists of two wavelengths λ_1 and λ_2, the slit separation is d, and the distance between the slit and the screen is D, the maxima due to the two wavelengths will coincide at a distance from the central maxima, given by :

Tricky

(a) $\frac{\lambda_1\lambda_2}{2\,Dd}$ (b) $(\lambda_1 - \lambda_2)\cdot\frac{2d}{D}$

(c) LCM of $\lambda_1 \cdot \frac{D}{d}$ and $\lambda_2 \cdot \frac{D}{d}$

(d) HCF of $\frac{\lambda_1 D}{d}$ and $\frac{\lambda_2 D}{d}$

61. Consider the YDSE arrangement shown in figure. If $d = 10\lambda$ then position of 8th maxima is

(a) $y = \frac{D}{10}$

(b) $y = \frac{D}{3}$

(c) $y = \frac{4}{5}D$ (d) $y = \frac{4D}{3}$

62. Two rectangular glass plates are in contact at one edge while the other edges are separated by a space of some suitable thickness so as to form a low angle wedge. The spacer is placed parallel to the line of contact and is at a distance of 10 cm from it. When viewed normally in light of wavelength 5500 Å, a series of evenly spaced dark bands 0.5 mm apart are seen. The thickness of the spacer is : **Toughnut**

(a) 0.0425 cm (b) 0.0036 cm
(c) 0.0055 cm (d) 0.0254 cm

63. At two points P and Q on screen in Young's double slit experiment, waves from slits S_1 and S_2 have a path difference of 0 and $\frac{\lambda}{4}$, respectively. The ratio of intensities at P and Q will be :

(a) 2 : 1 (b) $\sqrt{2} : 1$ (c) 4 : 1 (d) 3 : 2

64. In a Young's double slit experiment, the two slits act as coherent sources of wave of equal amplitude A and wavelength λ. In another experiment with the same arrangement the two slits are made to act as incoherent sources of waves of same amplitude and wavelength. If the intensity at the middle point of the screen in the first case is I_1 and in the second case is I_2, then the ratio $\frac{I_1}{I_2}$ is

(a) 2 (b) 1 (c) 0.5 (d) 4

65. On a hot summer night, the refractive index of air is smallest near the ground and increases with height from the ground. When a light beam is directed horizontally, the Huygens' principle leads us to conclude that as it travels, the light beam :
(a) bends downwards
(b) bends upwards
(c) becomes narrower
(d) goes horizontally without any deflection

66. Assuming human pupil to have a radius of 0.25 cm and a comfortable viewing distance of 25 cm, the minimum separation between two objects that human eye can resolve at 500 nm wavelength is :
(a) 100 μm (b) 300 μm (c) 1 μm (d) 30 μm

67. In Young's double slit experiment, one of the slit is wider than other, so that amplitude of the light from one slit is double of that other slit. If I_m be the maximum intensity, the resultant intensity I when they interfere at phase difference ϕ is given by

(a) $\frac{I_m}{9}(4+5\cos\phi)$ (b) $\frac{I_m}{3}\left(1+2\cos^2\frac{\phi}{2}\right)$
(c) $\frac{I_m}{5}\left(1+4\cos^2\frac{\phi}{2}\right)$ (d) $\frac{I_m}{9}\left(1+8\cos^2\frac{\phi}{2}\right)$

68. A beam of unpolarised light of intensity I_0 is passed through a polaroid A and then through another polaroid B which is oriented so that its principal plane makes an angle of 45° relative to that of A. The intensity of the emergent light is
(a) I_0 (b) $I_0/2$ (c) $I_0/4$ (d) $I_0/8$

69. Two beams, A and B, of plane polarized light with mutually perpendicular planes of polarization are seen through a polaroid. From the position when the beam A has maximum intensity (and beam B has zero intensity), a rotation of polaroid through 30° makes the two beams appear equally bright. If the initial intensities of the two beams are I_A and I_B respectively, then $\frac{I_A}{I_B}$ equals **Tricky**

(a) 3 (b) $\frac{3}{2}$ (c) 1 (d) $\frac{1}{3}$

70. In a Young's double slit experiment, slits are separated by 0.5 mm, and the screen is placed 150 cm away. A beam of light consisting of two wavelengths, 650 nm and 520 nm, is used to obtain interference fringes on the screen. The least distance from the common central maximum to the point where the bright fringes due to both the wavelengths coincide is :
(a) 9.75 mm (b) 15.6 mm (c) 1.56 mm (d) 7.8 mm

71. The box of a pin hole camera, of length L, has a hole of radius a. It is assumed that when the hole is illuminated by a parallel beam of light of wavelength λ the spread of the spot (obtained on the opposite wall of the camera) is the sum of its geometrical spread and the spread due to diffraction. The spot would then have its minimum size (say b_{min}) when : **Critical Thinking**

(a) $a = \sqrt{\lambda L}$ and $b_{min} = \sqrt{4\lambda L}$

(b) $a = \frac{\lambda^2}{L}$ and $b_{min} = \sqrt{4\lambda L}$

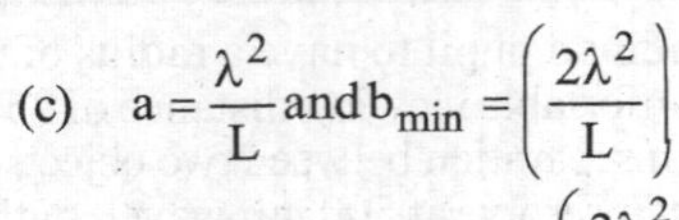

(c) $a = \frac{\lambda^2}{L}$ and $b_{min} = \left(\frac{2\lambda^2}{L}\right)$

(d) $a = \sqrt{\lambda l}$ and $b_{min} = \left(\frac{2\lambda^2}{L}\right)$

72. Two polaroids are placed in the path of unpolarized beam of intensity I_0 such that no light is emitted from the second polaroid. If a third polarioid whose polarization axis makes an angle θ with the polarization axis of first polaroid, is placed between these polaroids then the intensity of light emerging from the last polaroid will be

(a) $\left(\frac{I_0}{8}\right)\sin^2 2\theta$ (b) $\left(\frac{I_0}{4}\right)\sin^2 2\theta$

(c) $\left(\frac{I_0}{2}\right)\cos^4\theta$ (d) $I_0\cos^4\theta$

73. The width of the diffraction band varies
(a) inversely as the wavelength
(b) directly as the width of the slit
(c) directly as the distance between the slit and the screen
(d) inversely as the size of the source from which the slit is illuminated

74. The first diffraction minimum due to the single slit diffraction is seen at θ = 30° for a light of wavelength 5000 Å falling perpendicularly on the slit. The width of the slit is
(a) 2.5×10^{-5} cm (b) 1.25×10^{-5} cm
(c) 10×10^{-5} cm (d) 5×10^{-5} cm

75. An unpolarised beam of intensity I_0 is incident on a pair of nicols making an angle of 60° with each other. The intensity of light emerging from the pair is
(a) I_0 (b $I_0/2$ (c) $I_0/4$ (d) $I_0/8$

76. The angle substanded by the first diffraction minimum for a point source viewed in the hydrogen line at 1420 MHz with a radio telescope having an aperture of 25 m is : Tricky
(a) 0.8° (b) 0.64° (c) 1.2° (d) 2.2°

77. The angle of incidence at which reflected light is totally polarized for reflection from air to glass (refractive index *n*), is
(a) $\tan^{-1}(1/n)$ (b) $\sin^{-1}(1/n)$
(c) $\sin^{-1}(n)$ (d) $\tan^{-1}(n)$

78. A polaroid is placed at 45° to an incoming light of intensity I. Now the intensity of light after polarisation would be
(a) I (b) I/2 (c) $I/\sqrt{2}$ (d) zero

79. Unpolarised light is incident on a dielectric of refractive index $\sqrt{3}$. What is the angle of incidence if the reflected beam is completely polarised?
(a) 30° (b) 45° (c) 60° (d) 75°

80. When the diffraction pattern from a certain slit illuminated with laser light (λ = 6330 A°) is projected on a screen 150 cm from the slit, the second minima on each side are separated by 8 cm. This tells us that : Critical Thinking
(a) the slit is approximately 0.005 cm wide
(b) the slit is approximately 0.05 cm wide
(c) a / λ is approximately 7.5 (a is the slit width)
(d) a / λ is approximately 750

81. When an unpolarized light of intensity I_0 is incident on a polarizing sheet, the intensity of the light which does not get transmitted is
(a) $\frac{1}{4}I_0$ (b) $\frac{1}{2}I_0$ (c) I_0 (d) zero

82. The angle of polarisation for any medium is 60°. The critical angle for this is
(a) $\sin^{-1}\frac{1}{\sqrt{3}}$ (b) $\cos^{-1}\sqrt{3}$
(c) $\sin^{-1}\sqrt{3}$ (d) $\tan^{-1}\sqrt{3}$

83. If I_0 is the intensity of the principal maximum in the single slit diffraction pattern, then what will be its intensity when the slit width is doubled?
(a) $4I_0$ (b) $2I_0$ (c) $\frac{I_0}{2}$ (d) I_0

84. Two polaroids have their polarizing directions parallel so that the intensity of a transmitted light is maximum. The angle through which either polaroid must be turned if the intensity is to drop by one-half is Tricky
(a) 135° (b) 90° (c) 120° (d) 180°

85. In an experiment of single slit diffraction pattern, first minimum for red light coincides with first maximum of some other wavelength. If wavelength of red light is 6600 Å, then wavelength of first maximum will be:
(a) 3300Å (b) 4400Å (c) 5500Å (d) 6600Å

86. A mixture of polarised and unpolarised light is incident on a polariser. As the polariser is rotated through 360°, it is found that the minimum and the maximum of the transmitted intensity is in the ratio 1 : q. The ratio of intensities of polarised to unpolarised light in the incident beam is
(a) $\frac{1}{2}(q-1):1$ (b) $1:\frac{1}{2}(q-1)$
(c) $\frac{1}{2}q:1$ (d) $1:\frac{1}{2}q$

87. The intensity of the central maximum in Young's double-slit experiment is 4I. The intensity at the first minimum is zero and the distance between two consecutive maxima is w. The distance from the central maximum to the position where the intensity falls to I is

(a) $\frac{2}{3}w$ (b) $\frac{1}{4}w$ (c) $\frac{1}{2}w$ (d) $\frac{1}{3}w$

88. In Young's double slit experiment using monochromatic light the fringe pattern shifts by a certain distance on the screen when a mica sheet of refractive index 1.6 and thickness 1.964 microns is introduced in the path of one of the interfering waves. The mica sheet is then removed and the distance between the slits and the screen is doubled. It is found that the distance between successive maxima (or minima) now is the same as the observed fringe shift upon the introduction of the mica sheet. Calculate the wavelength of the monochromatic light used in the experiment. **Toughnut**

(a) 38.2A° (b) 68.32A°
(c) 5892A° (d) 528.32A°

89. In a Young's double slit experiment 10 fringes are observed in a given segment of the screen when light of wavelength 500.0 *nm* is used. When the experiment is performed in a medium of refractive index 1.2, the number of fringes observed in the same segment remains 10. The wavelength of light used in this case is

(a) 400.0 nm (b) 417.7 nm
(c) 500.0 nm (d) 600.0 nm

90. A narrow monochromatic beam of light of intensity I is incident on a glass plate as shown in figure. Another identical glass plate is kept close to the first one and parallel to it. Each glass plate reflects 25 per cent of the light incident on it and transmits the remaining. Find the ratio of the minimum and the maximum intensities in the interference pattern formed by the two beams obtained after one reflection at each plate.

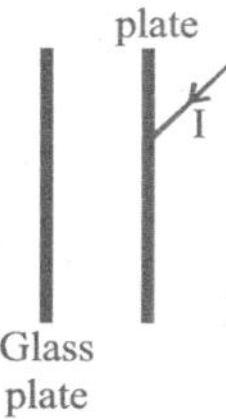

(a) 1 : 30 (b) 1 : 49
(c) 1 : 51 (d) 1 : 73

91. In *YDSE* a light containing two wavelengths 500 nm and 700 nm are used. Find the minimum distance where maxima of two wavelengths coincide. Given $D/d = 10^3$, where D is the distance between the slits and the screen and d is the distance between the slits.

(a) 1.2m (b) 3.5mm (c) 2.8cm (d) 8.1mm

92. In a double slit experiment slits S_1, S_2 is illuminated by a coherent light of wavelength λ. The slits are separated by a distance d. The experimental set up is modified by using plane mirrors as shown in figure. Find the fringe width of interference pattern on the screen. **Critical Thinking**

(a) $\frac{(3D_1 + 2D_2)\lambda}{d}$ (b) $\frac{(2D_1 + 3D_2)\lambda}{d}$

(c) $\frac{(3D_2 - 3D_1)\lambda}{d}$ (d) $\frac{(3D_1 - 2D_2)\lambda}{2d}$

93. A radio transmitting station operating at a frequency of 120 MHz has two identical antennas that radiate in phase. Antenna B is 9 m to the right of antenna A. Consider point P at a horizontal distance x to the right of antenna A as shown in Fig. The value of x and order for which the constructive interference will occur at point P is **Critical Thinking**

(a) x = 14.95 m, n = 2
(b) x = 5.6 m, n = 2
(c) x = 1.65 m, n = 3
(d) x = 0, n = 3.6

94. If one of the slits of a standard Young's double slit experiment is covered by a thin parallel sided glass slab so that it transmits only one half the light intensity of the other, then–

(a) the fringe pattern will get shifted towards the covered slit
(b) the fringe pattern will get shifted away from the covered slit
(c) the bright fringes will become more bright and the dark ones will become less bright
(d) the fringe width will changed

95. If screen is shifted in *x* direction away from source, then which of the following is incorrect?

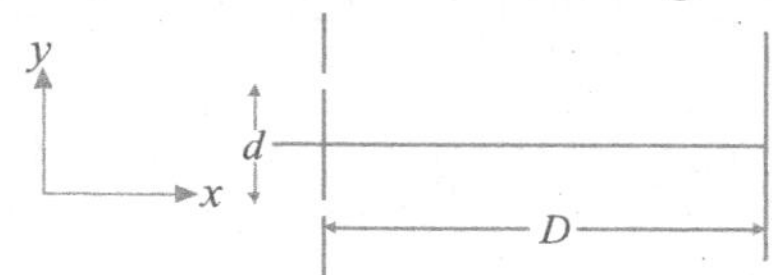

(a) Central maxima is shifted along *x*-axis
(b) Position of all maximas except the central maxima change
(c) Fringe width remains constant
(d) Angular width changes due to shifting

96. A light wave of wavelength λ_0 propagates from point A to point B. We introduce in its path a glass plate of refractive index n and thickness l. The introduction of the plate alters the phase of the plate at B by an angle ϕ. If λ is the wavelength of lights on emerging from the plate, then

(a) $\Delta\phi = 0$ (b) $\Delta\phi = \dfrac{2\pi l}{\lambda_0}$

(c) $\Delta\phi = 2\pi l n^2\left(\dfrac{1}{\lambda} - \dfrac{1}{\lambda_0}\right)$ (d) $\Delta\phi = \dfrac{2\pi l}{\lambda_0}(n-1)$

97. A ray of light of wavelength λ_0 and frequency v_0 enters a glass slab of refractive index μ from air. Then
(a) its wavelength increases, frequency decreases
(b) its wavelength decreases, frequency remain same
(c) its wavelength increases, frequency remain same
(d) bothe remains contant

(distance between slit and screen D = 12cm and distance between slits d = 5cm.) then the wavelength of the radiation used can be–
(a) 3/4 cm (b) 4 cm, 5cm
(c) 2/3 cm, 2/5 cm (d) 4/3 cm, 5/3 cm

99. In Young's double-slit experiment, let A and B be the two slits. A thin film of thickness t and refractive index μ is placed in front of A. Let β = frings width. Then the central maxima will shift
(a) toward A (b) toward B
(c) by $t(\mu-1)\dfrac{\beta}{\lambda}$ (d) by $\mu t\dfrac{\beta}{\lambda}$

100. A parallel beam of light (λ = 5000Å) is incident at an angle α = 30° with the normal to the slit plane in YDSE. Assume that the intensity due to each slit at any point on the screen is I_0. Point O is equidistant from S_1 and S_2. The distance between slit is 1 mm, then the intensity at

Toughnut

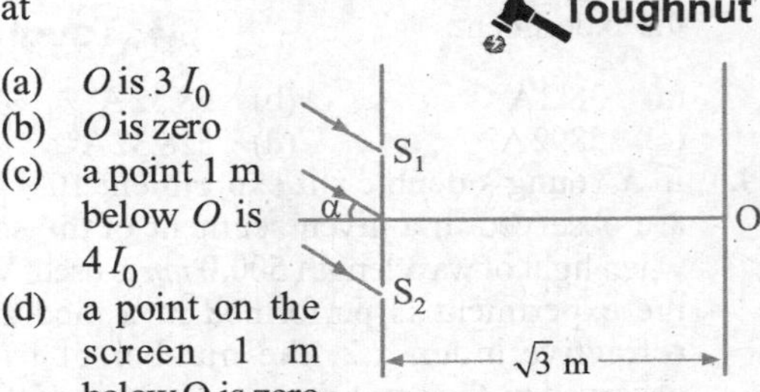

(a) O is $3\,I_0$
(b) O is zero
(c) a point 1 m below O is $4\,I_0$
(d) a point on the screen 1 m below O is zero

Answer KEYs

1	(b)	**11**	(c)	**21**	(d)	**31**	(b)	**41**	(a)	**51**	(a)	**61**	(d)	**71**	(a)	**81**	(b)	**91**	(b)
2	(c)	**12**	(c)	**22**	(b)	**32**	(b)	**42**	(d)	**52**	(a)	**62**	(c)	**72**	(a)	**82**	(a)	**92**	(a)
3	(a)	**13**	(c)	**23**	(d)	**33**	(b)	**43**	(a)	**53**	(d)	**63**	(a)	**73**	(c)	**83**	(a)	**93**	(b)
4	(c)	**14**	(b)	**24**	(c)	**34**	(c)	**44**	(c)	**54**	(a)	**64**	(a)	**74**	(c)	**84**	(a)	**94**	(a)
5	(a)	**15**	(a)	**25**	(b)	**35**	(c)	**45**	(d)	**55**	(a)	**65**	(b)	**75**	(c)	**85**	(b)	**95**	(c)
6	(a)	**16**	(b)	**26**	(c)	**36**	(a)	**46**	(c)	**56**	(a)	**66**	(d)	**76**	(c)	**86**	(a)	**96**	(d)
7	(a)	**17**	(a)	**27**	(a)	**37**	(b)	**47**	(a)	**57**	(d)	**67**	(d)	**77**	(d)	**87**	(d)	**97**	(b)
8	(c)	**18**	(d)	**28**	(d)	**38**	(c)	**48**	(a)	**58**	(c)	**68**	(c)	**78**	(b)	**88**	(c)	**98**	(c)
9	(b)	**19**	(a)	**29**	(b)	**39**	(c)	**49**	(d)	**59**	(b)	**69**	(d)	**79**	(c)	**89**	(d)	**99**	(a)
10	(b)	**20**	(a)	**30**	(a)	**40**	(a)	**50**	(c)	**60**	(c)	**70**	(d)	**80**	(a)	**90**	(b)	**100**	(c)

Hints & Solutions

1. **(b)** Huygen's principle gives us a geometrical method of tracing a wavefront.
2. **(c)** A slit would give divergent; a biprism would give double; a glass slab would give a parallel wavefront. Edge is downward.
3. **(a)** When light reflects from denser surface phase change of π occurs.
4. **(c)** From $\mu = \frac{c}{v} = \frac{n\lambda_v}{n\lambda_m}, \lambda_m = \frac{\lambda_v}{\mu}$

   Here, c = velocity of light in vacuum and v = velocity of light in medium
   m = refractive index of the medium.
   Hence, wavelength in medium $(\lambda_m) < \lambda_v$ ($\because$ $\mu > 1$, given)
   So, the required wavelength decreases.
5. **(a)** According to Cauchy's formula

   $\mu = A + \frac{B}{\lambda^2} + ...$ Therefore, as λ increases, μ decreases. Curve A is correct.
6. **(a)**
7. **(a)** Fringe width $\omega = \frac{\lambda D}{d}$.

   When the apparatus is immersed in a liquid, then λ will decrease μ times and hence ω is reduced μ (refractive index) times. $10\omega' = (5.5)\omega$

   or $10\lambda'\left(\frac{D}{d}\right) = (5.5)\frac{\lambda D}{d}$ or $\frac{\lambda}{\lambda'} = \frac{10}{5.5} = \mu$

   or $\mu = 1.8$
8. **(c)** Treating the anti-reflectance coating like a camera-lens coating, one can obtain for its thickness t:

   $$2t = \left(m + \frac{1}{2}\right)\frac{\lambda}{n}, m = 0, \pm 1, \pm 2, ... \quad ......(i)$$

   Condition (1) is the condition for the destructive interference for the normally incident and reflected electromagnetic wave. The minimum thickness of the coat refers to $m = 0$. This gives for the thickness:

   $$t = \frac{\lambda}{4n} = \frac{3.00}{4(1.5)} \text{cm} = 0.500 \text{ cm}.$$

   This anti-reflectance coating could be easily countered by changing the wavelength of the radar to 1.50 cm now creating maximum reflection.
9. **(b)**
10. **(b)** Equation of path difference form maxima in transmission (or weak reflection)

    $$\Delta P_{opt} = 2n_2 L = \frac{\lambda_{vacuum}}{2}, \frac{3\lambda_{vacuum}}{2} ....$$

    $$\Rightarrow 2\left(\frac{n_2}{n_1}\right)L = \frac{\lambda}{2}, \frac{3\lambda}{2}, ... \Rightarrow L = \frac{\lambda}{4n_2}$$

    (notice that λ = wavelength in medium is related to λ_{vacuum} as, $\lambda_{vacuum} = n_1\lambda$)
11. **(c)** Given $d = 1$mm, $\lambda = 6.5 \times 10^{-7}$m, $D = 1$m.

    For n^{th} dark fringe $y_n = \left(\frac{2n-1}{2}\right)\frac{D\lambda}{d}$

    For third fringe,

    $$y_3 = \left(\frac{2 \times 3 - 1}{2}\right) \times \frac{1 \times 6.5 \times 10^{-7}}{1 \times 10^{-3}} = 1.625 \text{ mm}.$$
12. **(c)** $n = \frac{(\mu - 1)tD}{d}$ but $\beta = \frac{\lambda D}{d} \Rightarrow \frac{D}{d} = \frac{\beta}{\lambda}$

    $n = (\mu - 1)t\beta/\lambda$

    $20\beta = (\mu - 1)\, 2.5 \times 10^{-3} (\beta/5000 \times 10^{-8})$

    $$\mu - 1 = \frac{20 \times 5000 \times 10^{-8}}{2.5 \times 10^{-3}} \Rightarrow \mu = 1.4$$
13. **(c)** We know that

    $$I(\theta) = I_0 \cos^2\frac{\delta}{2} \text{ where } \delta = \frac{2\pi d \tan\theta}{\lambda}$$

    $$I(\theta) = I_0 \cos^2\left(\frac{\pi d \tan\theta}{\lambda}\right) = I_0 \cos^2\left(\frac{\pi \times 150 \times \tan\theta}{3 \times 10^8 / 10^6}\right)$$

    $$= I_0 \cos^2\left(\frac{\pi}{2}\tan\theta\right)$$

    For $\theta = 30°$;

    $$I(\theta) = I_o \cos^2\left(\frac{\pi}{2\sqrt{3}}\right)$$

    For $\theta = 90°$; $I(\theta) = I_o \cos^2(\infty)$
    For $\theta = 0°$ $I(\theta) = I_0$
    $I(\theta)$ is not constant.
    Alternatively, when θ is zero the path difference between wave originating from S_1 and that from S_2 will be zero. This corresponds to a maxima.
14. **(b)**
15. **(a)** $d = \frac{D\lambda}{\beta}$ ...(i)

    According to question,
    $\lambda = 5100 \times 10^{-10}$m

    $\beta = \frac{2}{10} \times 10^{-2}$m ...(ii)

    $D = 2$m, $d = ?$
    From eqs. (i) and (ii)

    $$d = \frac{2 \times 51 \times 10^{-8}}{2 \times 10^{-3}} = 5.1 \times 10^{-4} \text{m}$$

16. (b)

17. (a) Path difference $= (\mu - 1)t = n\lambda$;
For minimum $t, n = 1$; $\therefore t = 2\lambda$

18. (d) At the area of total darkness minima will occur for both the wavelengths.

$$\therefore \frac{(2n+1)}{2}\lambda_1 = \frac{(2m+1)}{2}\lambda_2$$

$$\Rightarrow (2n+1)\lambda_1 = (2m+1)\lambda_2$$

or $\frac{(2n+1)}{(2m+1)} = \frac{560}{400} = \frac{7}{5}$ or $10n = 14m + 2$

by inspection for $m = 2, n = 3$ and for $m = 7, n = 10$, the distance between them will be the distance between such points.

i.e., $\Delta s = \frac{D\lambda_1}{d}\left\{\frac{(2n_2+1)-(2n_1+1)}{2}\right\}$

put $n_2 = 10, n_1 = 3$
On solving we get, $\Delta s = 28$mm.

19. (a) n^{th} minimum has a distance from the centre

$= x = (2n+1)\frac{1}{2}\frac{\lambda D}{d}$

For a point on the screen directly in front of one of the slits, $x = d/2$

$\therefore$ for minimum intensity in front of one of the slits

$$= \frac{d}{2} = (2n+1)\frac{\lambda}{2}\frac{D}{d} \therefore \lambda = \frac{d^2}{(2n+1)D}$$

20. (a) Locus of equal path difference are lines running parallel to axis of the cylinder. Hence straight fringes will be observed.

21. (d) Shift of fringe pattern $= (\mu - 1)\frac{tD}{d}$

$$\therefore \frac{30D\,(4800\times10^{-10})}{d} = (0.6)t\frac{D}{d}$$

$$30\times4800\times10^{-10} = 0.6t$$

$$t = \frac{30\times4800\times10^{-10}}{0.6} = \frac{1.44\times10^{-5}}{0.6} = 24\times10^{-6}$$

22. (b) $\frac{12\lambda_1 D}{d} = \frac{k\lambda_2 D}{d}$, $k = \frac{12\times600}{400} = 18$

23. (d) Intensity $\propto$ (amplitude)^2

$$\frac{I_1}{I_2} = \frac{16}{9} = \frac{a_1^2}{a_2^2} \Rightarrow a_1 = 4;\ a_2 = 3$$

Therefore the ratio of intensities of bright and dark parts

$$\frac{I_{Bright}}{I_{Dark}} = \frac{(a_1+a_2)^2}{(a_1-a_2)^2} = \frac{(4+3)^2}{(4-3)^2} = \frac{49}{1}$$

24. (c) 25. (b)

26. (c) Referring to the figure, the path difference between the two waves starting from S_1 and S_2 turns out to be $(2\lambda\cos\theta) = n\lambda$ where n is taken as 1 to get the point of maximum intensity which is the same as a point O.

Therefore, the above relation gives $\cos\theta = 1/2$ so that $\theta = 60°$ and $\tan\theta = PO/D$

$= \sqrt{3}$, giving $PO = D\sqrt{3}$.

27. (a) $\therefore X_0 = \frac{\beta}{\lambda}(\mu-1)t \Rightarrow 5\beta = \frac{\beta(0.45)t}{5890\times10^{-10}}$

$$\therefore t = \frac{5\times5890\times10^{-10}}{0.45} = 6.544\times10^{-4}\text{ cm}$$

28. (d) The path difference produced by a sheet
$\Delta = (\mu - 1)t$
According to the given condition (for minimum thickness)

$$(\mu-1)t = \frac{\lambda}{2}$$

$$\therefore t = \frac{\lambda}{2(\mu-1)} = \frac{6600\times10^{-10}}{2(1.6-1)} = 5500\text{ Å}.$$

29. (b) For dark fringe,

Path difference $= (2n-1)\frac{\lambda}{2}$

30. (a) In this case, both the rays suffere a phase change of 180° and the conditions for destructive interference is

$$2nd = \left(m+\frac{1}{2}\right)\lambda_1 \text{ and } 2nd = \left(m+\frac{3}{2}\right)\lambda_2$$

$$\therefore \frac{m+\frac{1}{2}}{m+\frac{3}{2}} = \frac{\lambda_2}{\lambda_1} = \frac{5000}{700} = \frac{5}{7}$$

and $d = \frac{\left(m+\frac{1}{2}\right)\lambda_1}{2n} = \frac{2.5\times7000}{2\times1.3}$
$= 6.738\times10^{-5}$ cm

31. (b) Constructive interference happens when $2t = (m-1/2)\lambda$. The minimum value for $m =$ is $m = 1$, the maximum value is the integer portion of

$$\frac{2d}{\lambda} + \frac{1}{2} = \frac{2\times0.034\times10^{-3}}{680\times10^{-9}} + \frac{1}{2} = 100.5$$

$m_{max} = 100$

32. (b) Effective path difference is $\mu_1 L_1 - \mu_2 L_2$.

33. **(b)** $\Delta x_1 = (\mu_1 - 1)t = (1.5 - 1)t = 0.5t$

and $\Delta x_2 = (\mu_2 - 1) \times 2t = \left(\frac{4}{3} - 1\right) \times 2t = \frac{2}{3}t$.

As $\Delta x_2 > \Delta x_1$, so shift will be along –ve y-axis.

34. **(c)** Change in path difference for any point on screen is $|\mu - 1.8|$ t.
For central maxima, phase difference = 0.
Hence,

$d \sin\theta - |\mu - 1.8|\, t = 0$

$\Rightarrow$ $d\theta = |\mu - 1.8|\, t$ [q is very small and is in radian]

$$\Rightarrow \quad |\mu - 1.8| = \frac{10^{-3} \times 0.1}{0.5 \times 10^{-3}} = 0.2 \Rightarrow \mu = 2 \quad \text{or } 1.6$$

35. **(c)**

36. **(a)** $(\mu - 1)t = n\beta$

$$\frac{(\mu_1 - 1) \times 1.8 \times 10^{-5}}{(\mu_2 - 1) \times 3.6 \times 10^{5}} = \frac{18\beta}{9\beta}$$

$(\mu_1 - 1) = 4(\mu_2 - 1) \qquad 4\mu_2 - \mu_1 = 3$

37. **(b)** Path difference,

$\Delta x = S_1P - S_2P = d\cos\theta$.

$\therefore d\cos\theta = 4\lambda$

or $\cos\theta = (4\lambda/d)$

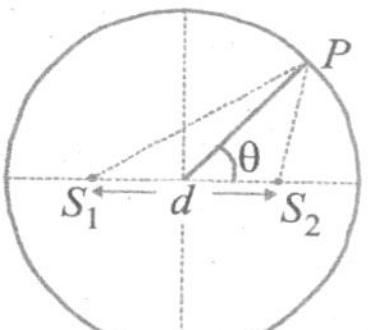

38. **(c)**

39. **(c)** In absence of film or for $m = 0$ intensity is maximum at screen. As the value of m is increased, intensity shall decrease and then increase alternately. Hence the correct variation is (c).

40. **(a)**

$$\beta_1 = \frac{2.30}{20} = \frac{D\lambda_1}{d}$$

and $$\beta_2 = \frac{2.80}{30} = \frac{D\lambda_2}{d}$$

or $$\frac{\beta_1}{\beta_2} = \frac{2.30 \times 30}{20 \times 2.80} = \frac{\lambda_1}{\lambda_2}$$

$\therefore \lambda_2 = 0.81\ \lambda_1 = 0.81 \times 5890 = 4780$ Å.

41. **(a)**

42. **(d)** $$\beta = \frac{\lambda}{2\mu \tan\alpha} \simeq \frac{\lambda}{2\mu\alpha}$$

$\therefore \lambda = \beta \times 2\mu\alpha$

$$= 0.12 \times 10^{-2} \times 2 \times 1 \times \left(\frac{40}{60 \times 60} \times \frac{\pi}{180}\right)$$

= 4655Å.

43. **(a)** Angular width $= \frac{\lambda}{d} = 10^{-3}$ (given)

$\therefore$ No. of fringes within 0.12° will be

$$n = \frac{0.12 \times 2\pi}{360 \times 10^{-3}} \cong [2.09]$$

$\therefore$ The number of bright spots will be two

44. **(c)** $$S_2P - S_1P = \frac{dy}{D} = \frac{d \times (d/2)}{D} = \frac{d^2}{2D}$$

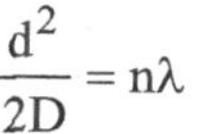

$$\frac{d^2}{2D} = n\lambda$$

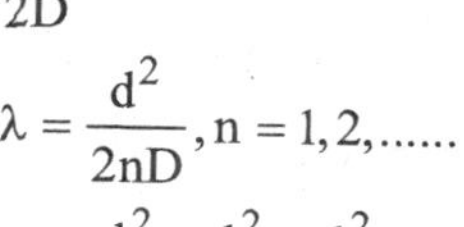

$$\lambda = \frac{d^2}{2nD}, n = 1, 2, \ldots\ldots$$

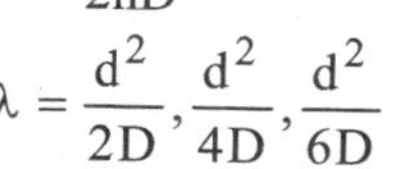

$$\lambda = \frac{d^2}{2D}, \frac{d^2}{4D}, \frac{d^2}{6D}$$

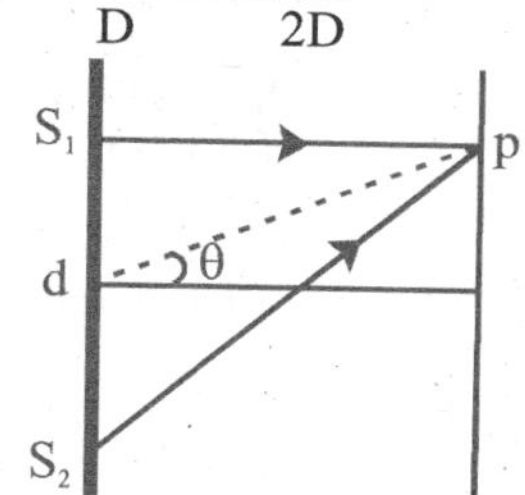

45. **(d)** Optical path difference

$\Delta x = (\mu_2 - \mu_1)t$.

46. **(c)** $$\text{Shift} = \frac{\beta(\mu - 1)}{\lambda}t_1 - \frac{\beta(\mu - 1)}{\lambda}t_2$$

$$= \frac{\beta(\mu - 1)}{\lambda}(t_1 - t_2)$$

47. **(a)** Path difference at P is

$$\Delta x = 2\left(\frac{x}{2}\cos\theta\right) = x\cos\theta$$

For intensity to be maximum,

$\Delta x = n\lambda$

(n = 0, 1, 2, 3,....)

or $x\cos\theta = n\lambda$

or $$\cos\theta = \frac{n\lambda}{x} \geq 1$$

$$\therefore \quad n \geq \frac{x}{\lambda}$$

Subsituting x = 51, we get

n ≥ 5 or n = 1, 2, 3, 4, 5,......

Therefore in all four quadrants there can be 20 maxima. There are more maxima at θ = 0° and θ = 180°. But n = 5 corresponds to θ = 90° and θ = 270° which are coming only twice while we have multuplies it four times. Therefore, total number of maxima are still 20, i.e., n = 1 to 4 in four quadrants (total 16) plus more at θ = 0°, 90°, 180° and 270°.

48. **(a)** The optical path difference needed for destructive interference is

$$2\mu d = (2n + 1)\frac{\lambda}{2}, n = 0, 1, 2, \ldots.$$

Note that 2μd is the total optical path length that the rays traverse when n = 0.

$\therefore \quad d = \frac{\lambda/2}{2\mu} = \frac{\lambda}{4\mu} = \frac{350\times10^{-9}}{4\times1.38}$

$= 100 \text{ nm} = 1\times10^{-7} \text{ m}$

49. (d) $\frac{\Delta P_1}{\Delta P_2} = \frac{\sin\alpha - \sin\beta}{\sin\alpha + \sin\beta} = \tan\frac{\alpha-\beta}{2}\cot\frac{\alpha+\beta}{2}$

50. (c) The reflections from the boundaries will cause a net 180° phase shift.
The condition for bright fringes is $2t = (m + \frac{1}{2})\lambda_{film}$
Now, m = 124 since there is a bright fringe for m = 0 and $\lambda_{film} = \frac{\lambda}{n}$

$$t = \frac{\left(m+\frac{1}{2}\right)\lambda_{film}}{2} = \frac{\left(m+\frac{1}{2}\right)\lambda}{2n}$$

$$= \frac{\left(124+\frac{1}{2}\right)(552\times10^{-9} m)}{2\times(1.00)} = 3.44\times10^{-5} m$$

51. (a) Optical path difference between the waves $= (n_3 - n_2)t$

$\therefore$ Phase difference

$$= 2\pi\frac{(n_3-n_2)t}{\lambda_{(Vacuum)}} = 2\pi\frac{(n_3-n_2)t}{n_1\lambda_1}$$

52. (a)

53. (d) Due to reflection virtual source will be formed at distance D from mirror.
The effective distance of the screen = 2D + 2D = 4D

$\therefore$ Fringe width $= \frac{4D\lambda}{d}$

54. (a) Using relation, $y_0 = \frac{D}{d}(\mu - 1)t$
We have,

$\frac{z}{3} = \frac{(1.5-1)}{(\mu-1)} \Rightarrow \frac{2}{3} = \frac{1}{2(\mu-1)}$

$\frac{\frac{3}{2}z}{\mu-1}$ ... $\frac{1}{\mu-1} = \frac{4}{3} \Rightarrow \mu - 1 = \frac{3}{4}$

$\mu = \frac{7}{4} = 1.75$

55. (a) There is air on both sides of the soap film.
$\therefore$ the reflections of the light produce a net 180° phase shift.
The condition for bright fringes is
$2t = (m + \frac{1}{2})\lambda_{film}$

$$t = \frac{(m+\frac{1}{2})\lambda_{film}}{2} = \frac{(m+\frac{1}{2})\lambda}{2n}$$

$$= \frac{(\frac{1}{2})(650\times10^{-9} m)}{2(1.41)} = 1.2\times10^{-7} m$$

56. (a) Path difference between first and 11th bright fringe
$= S_1B$ (10 bright fringes) $= 10\lambda$
$= 10 \times (6000\times10^{-7})\text{m} = 6\times10^{-6}$ m

57. (d) $E_1 = I_0 = \frac{L}{d^2}$, without mirror

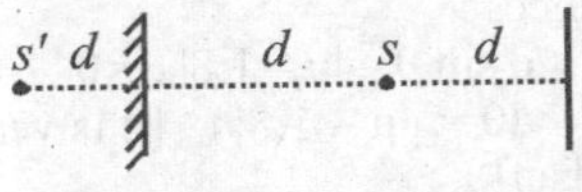

s′ d | d s d

and $E_2 = \frac{L}{d^2} + \frac{L}{(3d)^2} = \frac{L}{d^2}\left[1+\frac{1}{9}\right] = \frac{10I_0}{9}$

with mirror.

58. (c)

59. (b) As $\sin\theta = \frac{n\lambda}{d}$ and $\sin\theta$ cannot be $\not>1$

$\therefore 1 = \frac{n\lambda}{1.8\lambda}$ or $n = 1.8$

Hence maximum number of possible interference maximas, 0, ±1 i.e. 3

60. (c) $y_n = \frac{nD\lambda_1}{d} = \frac{(n+1)D\lambda_2}{d}$

or $n = \left(\frac{\lambda_2}{\lambda_2 - \lambda_1}\right)$

and $y_n = \left(\frac{\lambda_1\lambda_2}{\lambda_2-\lambda_1}\right)\left(\frac{D}{d}\right)$.

61. (d) For maxima $d\sin\theta = n\lambda$

$\sin\theta = \frac{n\lambda}{d} = \frac{8\lambda}{10\lambda} \Rightarrow \sin\theta = \frac{4}{5} \Rightarrow \tan\theta = \frac{4}{3}$

Also $\tan\theta = \frac{y}{D} \quad \therefore \; y = \frac{4D}{3}$

62. (c) $\beta = \frac{\lambda}{2\mu\alpha} = \frac{\lambda}{2\mu\left(\frac{t}{x}\right)}$

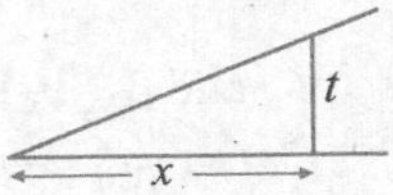

$= \frac{\lambda x}{2\mu t}$ or $t = \frac{\lambda x}{2\beta\mu}$

63. (a)

64. (a) For coherent sources :
$I_1 = 4I_0$
For incoherent sources
$I_2 = 2I_0 \quad \therefore \; \frac{I_1}{I_2} = \frac{2}{1}$

65. (b)

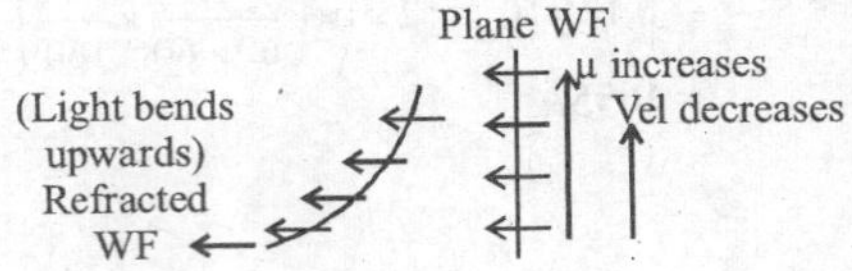

Plane WF
μ increases
Vel decreases
(Light bends upwards)
Refracted WF

66. (d) $\sin\theta = \frac{0.25}{25} = \frac{1}{100}$

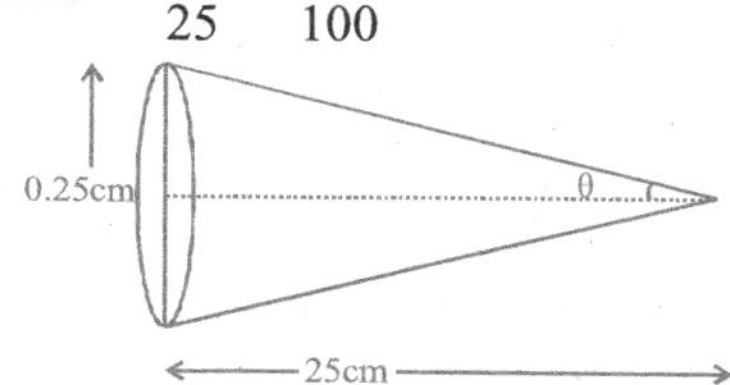

Resolving power $= \frac{1.22\lambda}{2\mu\sin\theta} = 30\ \mu m.$

67. (d) **68.** (c)

69. (d) According to malus law, intensity of emerging beam is given by,

$I = I_0 \cos^2\theta$

Now, $I_{A'} = I_A \cos^2 30°$

$I_{B'} = I_B \cos^2 60°$ As $I_{A'} = I_{B'}$

$\Rightarrow I_A \times \frac{3}{4} = I_B \times \frac{1}{4} = \frac{I_A}{I_B} = \frac{1}{3}$

70. (d) For common maxima, $n_1\lambda_1 = n_2\lambda_2$

$\Rightarrow \frac{n_1}{n_2} = \frac{\lambda_2}{\lambda_1} = \frac{520\times10^{-9}}{650\times10^{-9}} = \frac{4}{5}$

For λ_1

$y = \frac{n_1\lambda_1 D}{d}, \lambda_1 = 650\,\text{nm}$

$y = \frac{4\times650\times10^{-9}\times1.5}{0.5\times10^{-3}}$ or, $y = 7.8$ mn

71. (a) Given geometrical spread = a

Diffraction spread $= \frac{\lambda}{a}\times L = \frac{\lambda L}{a}$

The sum $b = a + \frac{\lambda L}{a}$

For b to be minimum

$\frac{db}{da} = 0 \Rightarrow \frac{d}{da}\left(a + \frac{\lambda L}{a}\right) = 0 \Rightarrow a = \sqrt{\lambda L}$

$b_{min} = \sqrt{\lambda L} + \sqrt{\lambda L} = 2\sqrt{\lambda L} = \sqrt{4\lambda L}$

72. (a)

73. (c) The width of the diffraction band is given by

$\beta = \frac{\lambda D}{d} \Rightarrow \beta \propto \lambda$

$\beta \propto D$ and $\beta \propto \frac{1}{d}$.

74. (c) For first minimum,

$d\sin\theta = \lambda$

$\Rightarrow d = \frac{\lambda}{\sin\theta} = \frac{5000\times10^{-8}\text{ cm}}{\sin 30°}$

$= \frac{5000\times10^{-8}\text{ cm}}{1/2} = 10\times10^{-5}\text{ cm}$

75. (c) According to Malus' law

$I = I_0\cos^2\theta = I_0(\cos^2 60°) = I_0\times\left(\frac{1}{2}\right)^2 = \frac{I_0}{4}.$

76. (c) $\lambda = \frac{c}{f} = \frac{3\times10^8}{1420\times10^6} = 0.214\,\text{m}$

$\sin\theta = \frac{1.22\lambda}{d} = \frac{1.22\times0.219}{25} = 0.010$

or $\theta = 0.6°$, and $2\theta = 1.2°$

77. (d)

78. (b) $I' = I\cos^2\theta = I\cos^2(45°) = I\left(\frac{1}{\sqrt{2}}\right)^2 = \frac{I}{2}.$

79. (c) $\mu = \tan i$

$\Rightarrow i = \tan^{-1}(\mu) = \tan^{-1}(\sqrt{3}) = 60°.$

80. (a) $\sin\theta_2 = \frac{2\lambda}{d}$ or $\theta_2 = \sin^{-1}\left(\frac{2\lambda}{d}\right)$

Given $y = D(2\theta_2) = 8\times10^{-2}$

or $1.5\times2\times\sin^{-1}\left(\frac{2\lambda}{d}\right) = 8\times10^{-2}$

$\Rightarrow d \simeq 0.005$ cm

81. (b)

82. (a) $\mu = \tan i_p = \tan 60° = \sqrt{3}$

$\sin i_C = \frac{1}{\mu} = \frac{1}{\sqrt{3}} \Rightarrow i_C = \sin^{-1}\frac{1}{\sqrt{3}}$

83. (a)

84. (a) For $I = \frac{I_0}{2}$ and $I = I_0\cos^2\theta = \frac{I_0}{2}$

$\therefore \theta = 45°$

Therefore the angle through which either polaroids turned is 135° (= 180° − 45°)

85. (b) In a single slit experiment,

For diffraction maxima, $a\sin\theta = (2n+1)\frac{\lambda}{2}$

and for diffraction minima, $a\sin\theta = n\lambda$

According to question,

$(2\times1+1)\frac{\lambda}{2} = 1\times6600 \quad (\because \lambda_R = 6600\text{Å})$

$\lambda = \frac{6600\times2}{3} = 4400\text{Å}$

86. (a) **87.** (d)

88. (c) Fringe shift when a sheet of thickness t and refractive index μ is introduced in path of one of interfering waves is

$\Delta x = \frac{Dt(\mu-1)}{d} = \frac{D\times1.964\times10^{-6}(1.6-1)}{d}$ ...(i)

The distance between two maxima where mica

sheet is remove and the distance between the slits and the screen is doubled $= \frac{\lambda(2D)}{d}$ ...(ii)

Given that the value in eq. (i) and eq. (ii) are equal

$\therefore \frac{D \times 1.964 \times 10^{-6} 0.6}{d} = \frac{\lambda \times 2D}{d}$

$\Rightarrow \lambda = \frac{1.964 \times 10^{-6} \times 0.6}{2}$

$= 0.5892 \times 10^{-6} \text{m} = 5892 \text{Å}$

89. **(d)**

90. **(b)** As shown in the figure, the interference will be between

$0.25I = I_1$ and $0.14I = I_2$

$$\frac{I_{max}}{I_{min}} = \frac{(\sqrt{I_1} + \sqrt{I_2})^2}{(\sqrt{I_1} - \sqrt{I_2})^2} = \frac{[\sqrt{0.25I} + \sqrt{0.14I}]^2}{[\sqrt{0.25I} - \sqrt{0.14I}]^2} = \frac{49}{1}.$$

91. **(b)** At the place where maxima for both the wavelengths coincide, y will be same for both the maxima, i.e.,

$$\frac{n_1\lambda_1 D}{d} = \frac{n_2\lambda_2 D}{d} \Rightarrow \frac{n_1}{n_2} = \frac{\lambda_1}{\lambda_2} = \frac{700}{500} = \frac{7}{5}$$

Minimum integral value of n_2 is 5.

$\therefore$ Minimum distance of maxima of the two wavelengths fr1om central fringe

$$= \frac{n_2\lambda_2 D}{d} = 5 \times 700 \times 10^{-9} \times 10^3 = 3.5 \text{ mm}.$$

92. **(a)** The fringe width β is given by, $\beta = \frac{D\lambda}{d}$

where $D = D_1 + (D_1 + D_2) + (D_1 + D_2) = 3D_1 + 2D_2$

$\therefore \beta = \frac{(3D_1 + 2D_2)\lambda}{d}$

93. **(b)** Path difference, $\delta = BP - AP$

$\sqrt{x^2 + 9^2} - x = n\lambda \Rightarrow x^2 + 9^2 = n^2\lambda^2 + x^2 + 2n\lambda x$

$\Rightarrow x = \frac{9^2 - n^2\lambda^2}{2n\lambda}$ also $\lambda = \frac{c}{v} = \frac{3 \times 10^8}{120 \times 10^6} = 2.5 \text{ m}$

n = 1, x = 14.95 m; n = 3, x = 1.65 m;

n = 2, x = 5.6 m; n = 4 not possible

94. **(a)** $I_{max} = (\sqrt{I_1} + \sqrt{I_2})^2 = \left(\sqrt{I_1} + \sqrt{\frac{I}{2}}\right)^2 < 4I$

$I_{min} = \left(\sqrt{I_1} - \sqrt{\frac{I}{2}}\right)^2 > 0$

95. **(c)**

96. **(d)** $\phi_i = \frac{2\pi}{\lambda_0}\ell \quad \phi_f = \frac{2\pi}{\lambda}\ell$

$$\Delta\phi = 2\pi\ell\left(\frac{1}{\lambda} - \frac{1}{\lambda_0}\right)$$

Further, by Snell's law,

$$n\lambda = (1)\lambda_0 \Rightarrow \lambda = \frac{\lambda_0}{n} \Rightarrow \Delta\phi = \frac{2\pi l}{\lambda_0}(n-1)$$

97. **(b)**

98. **(c)** Path difference $= \sqrt{D^2 + d^2} - D = 1$ cm. Also,

$[\sqrt{D^2 + d^2} - D] = (2n-1)\frac{\lambda}{2} \Rightarrow \lambda = \frac{2(1)}{2n-1}$

For n = 1, 2, 3...................

$\lambda = 2\text{cm}, \frac{2}{3}\text{cm}, \frac{2}{5}\text{cm}, ...$

99. **(a)** For central maximum at P,

$$\frac{xd}{D} = t(\mu - 1) \text{ or } x = t(\mu - 1)\frac{D}{d}$$

Now, $\beta = \frac{D\lambda}{d} \quad \therefore \frac{D}{d} = \frac{\beta}{\lambda}$

Hence, $x = t(\mu - 1)\frac{\beta}{\lambda}$

100. **(c)**

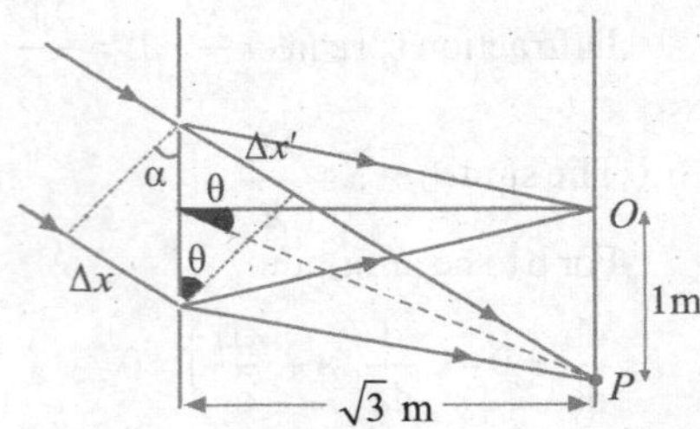

The path difference at O,

$\Delta x = d\sin\alpha = d\sin 30° = \frac{10^{-3}}{2}$ m

Now $\phi = \frac{2\pi}{\lambda}.\Delta x = \frac{2\pi}{5000 \times 10^{-10}} \times \frac{10^{-3}}{2} = 2\pi \times 10^3$

So $I = I_o + I_o + 2\sqrt{I_o I_o}\cos(2\pi \times 10^3) = 4I_o$

The angular position of P, $\tan\theta = \frac{1}{\sqrt{3}}$;

or $\theta = 30°$. It means path diff^n at P is also Δx and hence $I = 4I_o$

25 Dual Nature of Radiation and Matter

1. If the momentum of electron is changed by P, then the de Broglie wavelength associated with it changes by 0.5%. The initial momentum of electron will be :

(a) $200P$ (b) $400P$ (c) $\frac{P}{200}$ (d) $100P$

2. Which of the following figures represents the variation of particle momentum and the associated de-Broglie wavelength? **Tricky**

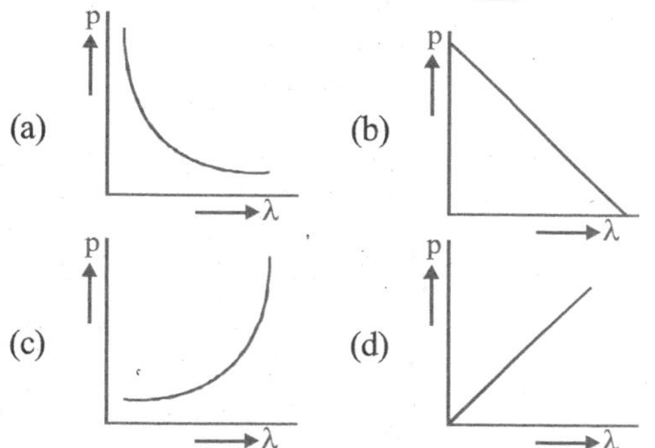

3. Electrons are accelerated through a potential difference V and protons are accelerated through a potential difference 4 V. The de-Broglie wavelengths are λ_e and λ_p for electrons and protons respectively. The ratio of $\frac{\lambda_e}{\lambda_p}$ is given by : (given m_e is mass of electron and m_p is mass of proton)

(a) $\sqrt{\frac{m_p}{m_e}}$ (b) $\sqrt{\frac{m_e}{m_p}}$ (c) $\frac{1}{2}\sqrt{\frac{m_e}{m_p}}$ (d) $2\sqrt{\frac{m_p}{m_e}}$

4. An electron of mass 'm' and charge 'e' initially at rest gets accelerated by a constant electric field E. The rate of change of de-Broglie wavelength of this electron at time t, ignoring relativistic effects is : **Critical Thinking**

(a) $\frac{-h}{eEt^2}$ (b) $\frac{-eht}{E}$ (c) $\frac{-mh}{eEt^2}$ (d) $\frac{-h}{eE}$

5. A proton and a deuteron are accelerated through the same accelerating potential. Which one of the two has greater value of de-Broglie wavelength associated with it, and less momentum? **Tricky**

(a) Proton
(b) Deutron
(c) Both have equal values
(d) None of these

6. A particle of mass M at rest decays into two particles of masses m_1 and m_2, having non-zero velocities. The ratio of the de Broglie wavelengths of the particles, λ_1/λ_2, is

(a) m_1/m_2 (b) m_2/m_1 (c) 1.0 (d) $\sqrt{m_2}/\sqrt{m_1}$

7. A light of wavelength λ and intensity I is incident normally on the surface. If reflection coefficient of surface is r, the pressure exerted by light on the surface is equal to

(a) $\frac{2I}{c}$ (b) $\frac{I}{c}r$

(c) $\frac{I}{c}(1+r)$ (d) $\frac{I}{c}(1-r)$

8. An electron microscope uses electrons accelerated by a voltage of 50 kV. If λ be the de-Broglie wavelength associated with the electrons. Also taking other factors, such as numerical aperture etc. to be same, the resolving power of an electron microscope be Re & that of an optical microscope which used yellow light is R_o, then

(a) $\lambda = 0.05$Å, $R_e >> R_o$
(b) $\lambda = 0.15$Å, $R_e << R_o$
(c) $\lambda = 0.11$Å, $R_e < R_o$
(d) None of these

9. A proton has kinetic energy $E = 100$ keV which is equal to that of a photon. The wavelength of photon is λ_2 and that of proton is λ_1. The ration of λ_2/λ_1 is proportional to

(a) E^2 (b) $E^{1/2}$ (c) E^{-1} (d) $E^{-1/2}$

10. Electrons with de-Broglie wavelength λ fall on the target in an X-ray tube. The cut-off wavelength of the emitted X-rays is

(a) $\lambda_0 = \frac{2mc\lambda^2}{h}$ (b) $\lambda_0 = \frac{2h}{mc}$

(c) $\lambda_0 = \frac{2m^2c^2\lambda^3}{h^2}$ (d) $\lambda_0 = \lambda$

11. The de-Broglie wavelength of a neutron in thermal equilibrium with heavy water at a temperature T (Kelvin) and mass m, is :-

(a) $\frac{h}{\sqrt{3mkT}}$ (b) $\frac{2h}{\sqrt{3mkT}}$ (c) $\frac{2h}{\sqrt{mkT}}$ (d) $\frac{h}{\sqrt{mkT}}$

12. A homogeneous ball (mass = m) of ideal black material at rest is illuminated with a radiation having a set of photons (wavelength = λ), each with the same momentum and the same energy. The rate at which photons fall on the ball is n. The linear acceleration of the ball is

(a) mλ/nh (b) nh/mλ (c) nh/mλ (d) 2mλ/nh

13. A particle of mass m is projected from ground with velocity u making angle θ with the vertical. The de-Broglie wavelength of the particle at the highest point is –

Tricky

(a) ∞ (b) $\frac{h}{mu\sin\theta}$

(c) $\frac{h}{mu\cos\theta}$ (d) $\frac{h}{mu}$

14. An electron of mass m and a photon have same energy E. The ratio of de-Broglie wavelengths associated with them is :

(a) $\frac{1}{c}\left(\frac{E}{2m}\right)^{\frac{1}{2}}$ (b) $\left(\frac{E}{2m}\right)^{\frac{1}{2}}$

(c) $c(2mE)^{\frac{1}{2}}$ (d) $\frac{1}{xc}\left(\frac{2m}{E}\right)^{\frac{1}{2}}$

15. An electron is accelerated through a potential difference of V volt. It has a wavelength λ associated with it. Through what potential difference an electron must be accelerated so that its de Broglie wavelength is the same as that of a proton? Take mass of proton to be 1837 times larger than the mass of electron.

Critical Thinking

(a) V volt (b) 1837 V volt

(c) V/1837 volt (d) $\sqrt{1837}$ V volt

16. Assume that the de Broglie wave associated with an electron can form a standing wave between the atoms arranged in a one dimensional array with nodes at each of the atomic sites. It is found that one such standing wave is formed if the distance d between the atoms of the array is 2Å. A similar standing wave is again formed if d is increased to 2.5 Å but not for any intermediate value of d. The energy of the electrons in electron volt and the least value of d for which the standing wave of the type described above can form, respectively are

(a) 112 eV, 1.5 A° (b) 151 eV, 0.5 A°

(c) 211 eV, 1.5 A° (d) None of these

17. When a beam of 10.6 eV photons of intensity 2.0 W/m^2 falls on a platinum surface of area 1.0 × 10^{-4} m^2 and work function 5.6 eV, 0.53% of the incident photons eject photoelectrons, then the number of photoelectrons emitted per second and their minimum & maximum energies (in eV) [Take 1eV = 1.6 × 10^{-19} J] are respectively.

Toughnut

(a) 1.18 × 10^{10}, 2 eV, 5 eV

(b) 1.18 × 10^{14}, 0 eV, 5 eV

(c) 2.18 × 10^{13}, 0 eV, 5 eV

(d) 3.11 × 10^{11}, 1 eV, 6 eV

18. When a surface is irradiated with light of wavelength 4950 Å, a photocurrent appears which vanishes if a retarding potential greater than 0.6 V is applied across the photo tube. When a different source of light is used, it is found that the critical retarding potential is changed to 1.1 V. Find the wavelength of second source.

Critical Thinking

(a) 4077 Å (b) 992 Å (c) 628 Å (d) 238 Å

19. Find the frequency of light which ejects electrons from a metal surface fully stopped by a retarding potential of 3V. The photoelectric effect begin in this metal at frequency of 6 × 10^{14} per second.

(a) 1.32 × 10^{15} Hz (b) 3.28 × 10^{14} Hz

(c) 6.22 × 10^{15} Hz (d) 2.22 × 10^{11} Hz

20. The potential energy of a particle of mass *m* is given by $V(x) = \begin{cases} E_0; & 0 \le x \le 1 \\ 0; & x > 1 \end{cases}$

λ_1 and λ_2 are the de-Broglie wavelengths of the particle, when $0 \le x \le 1$ and $x > 1$ respectively. If the total energy of particle is $2E_0$, find λ_1 / λ_2.

(a) $\sqrt{2}$ (b) $\sqrt{3}$ (c) $\sqrt{5}$ (d) $2\sqrt{2}$

21. Light of wavelength 180 nm ejects photoelectron from a plate of a metal whose work function is 2 eV. If a uniform magnetic field of 5×10^{-5} T is applied parallel to plate, what would be the radius of the path followed by electrons ejected normally from the plate with maximum energy?

(a) 1.239m (b) 0.149m
(c) 3.182m (d) 2.33m

22. A cylindrical rod of some laser material 5×10^{-2} m long and 10^{-2} m in diameter contains 2×10^{25} ions per m^3. If on excitation all the ions are in the upper energy level and de-excite simultaneously emitting photons in the same direction, calculate the maximum energy contained in a pulse of radiation of wavelength 6.6×10^{-7} m. If the pulse lasts for 10^{-7}s. the average power of the laser during the pulse is **Toughnut**

(a) 532 MW (b) 352 MW
(c) 235MW (d) 325 MW

23. Radiation of wavelength λ, is incident on a photocell. The fastest emitted electron has speed v. If the wavelength is changed to $\frac{3\lambda}{4}$, the speed of the fastest emitted electron will be:

(a) $= v\left(\frac{4}{3}\right)^{\frac{1}{2}}$ (b) $= v\left(\frac{3}{4}\right)^{\frac{1}{2}}$
(c) $> v\left(\frac{4}{3}\right)^{\frac{1}{2}}$ (d) $< v\left(\frac{4}{3}\right)^{\frac{1}{2}}$

24. An electron beam is accelerated by a potential difference V to hit a metallic target to produce X-rays. It produces continuous as well as characteristic X-rays. If λ_{min} is the smallest possible wavelength of X-ray in the spectrum, the variation of log λ_{min} with log V is correctly represented in : **Tricky**

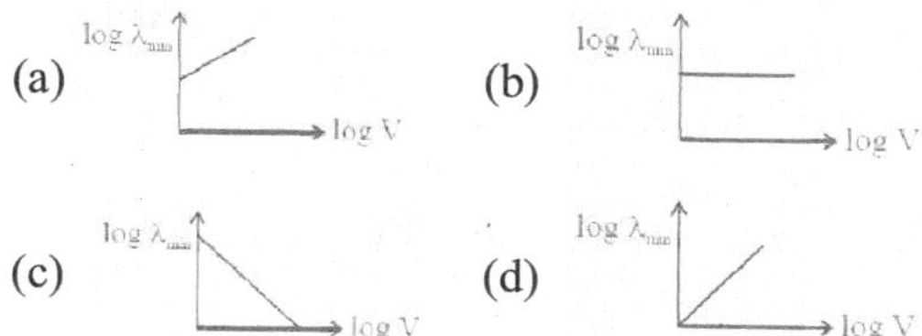

25. A particle A of mass m and initial velocity v collides with a particle B of mass $\frac{m}{2}$ which is at rest. The collision is head on, and elastic. The ratio of the de-Broglie wavelengths λ_A to λ_B after the collision is **Toughnut**

(a) $\frac{2}{3}$ (b) $\frac{1}{2}$ (c) $\frac{1}{3}$ (d) 2

26. A mono chromatic source of light operating at 400 w emits 8×10^{20} photons per second, the wavelength of light is

(a) 100 nm (b) 200 nm (c) 300 nm (d) 400 nm

27. A point source of light is placed at the centre of curvature of a hemispherical surface. The radius of curvature is R and the inner surface is completly reflecting The force on the hemisphere due to the light falling on it if the source emits a power P [c is the speed of ligtht]

(a) $\frac{2P}{C}$ (b) $\frac{P}{2C}$ (c) $\frac{5P}{2C}$ (d) $\frac{4P}{3C}$

28. An α-particle having a de-Broglie wavelength λ_i collides with a stationary carbon nucleus. The α-particle moves off in a different direction as shown below.

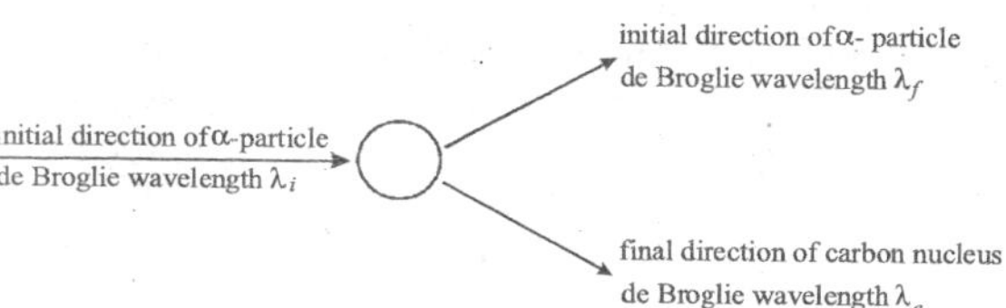

After the collision, the de Broglie wavelengths of the α-particle and the carbon nucleus are λ_f and λ_e respectively. Which of the following relations about de Broglie wavelengths is correct

(a) $\lambda_i < \lambda_f$ (b) $\lambda_i > \lambda_f$ (c) $\lambda_f = \lambda_e$ (d) $\lambda_i = \lambda_e$

29. A photon of frequency f causes the emission of a photoelectron of maximum kinetic energy E_k from a metal. If a photon of frequency $3f$ is incident on the same metal, the maximum kinetic energy of the emitted photoelectron

(a) equals $3E_k$
(b) is greater than $3E_k$
(c) is less than $3E_k$
(d) may be equal to, less than or, greater than $3E_k$

30. A 5 watt source emits monochromatic light of wavelength 5000 Å. When placed 0.5 m away, it liberates photoelectrons from a photosensitive metallic surface. When the source is moved to a distance of 1.0 m, the number of photoelectrons liberated will be reduced by a factor of

(a) 8 (b) 16 (c) 2 (d) 4

31. The work functions of metals A and B are in the raio 1 : 2. If light of frequencies f and 2f are incident on the surfaces of A and B respectively, the ratio of the maximum kinetic energies of photoelectrons emitted is (f is greater than threshold frequency of A, 2f is greater than threshold frequency of B) **Critical Thinking**

(a) 1 : 1 (b) 1 : 2 (c) 1 : 3 (d) 1 : 4

32. The fastest photoelectrons emitted from a metallic surface during the photoelectric effect have an energy of 8eV. The same photons when incident on atomic hydrogen, are strongly absorbed, exciting the atoms from the ground state to the first excited state. The work function of the metal is

(a) 5.6 eV (b) 2.2 eV (c) 4.6 eV (d) 11.4 eV

33. An ultraviolet light bulb, emitting 400 nm and an infrared light bulb, emitting at 700nm, each are rated at 130 W. Then the ratio of the number of photons emitted per second by the UV and IR sources is –

(a) 0.57 (b) 1.75 (c) 28 (d) 0.04

34. A source of light is placed at a distance of 50 cm from a photocell and the stopping potential is found to be V_0. If the distance between the light source and photocell is made 25 cm, the new stopping potential will be

(a) $2V_0$ (b) $V_0/2$ (c) V_0 (d) $4V_0$

35. A sensor is exposed for time t to a lamp of power P placed at a distance l. The sensor has an opening that is 4d in diameter. Assuming all energy of the lamp is given off as light, the number of photons entering the sensor if the wavelength of light is λ is **Critical Thinking**

(a) $N = P\lambda d^2 t/hcl^2$ (b) $N = 4P\lambda d^2 t/hcl^2$
(c) $N = P\lambda d^2 t/4hcl^2$ (d) $N = P\lambda d^2 t/16hcl^2$

36. The cathode of a photoelectric cell is changed such that the work function changes from W_1 to W_2 ($W_2 > W_1$). If the current before and after changes are I_1 and I_2, all other conditions remaining unchanged, then (assuming $h\nu > W_2$) **Tricky**

(a) $I_1 = I_2$ (b) $I_1 < I_2$
(c) $I_1 > I_2$ (d) $I_1 < I_2 < 2I_1$

37. In a photo-emissive cell, with exciting wavelength λ, the maximum kinetic energy of electron is K. If the exciting wavelength is changed to $3\lambda/4$ the kinetic energy of the fastest emitted electron will be

(a) 3K/4 (b) 4K/3
(c) less than 4K/3 (d) greater than 4K/3

38. A photoelectric cell is illuminated by a point source of light 1m away. When the source is shifted to 2m then

(a) number of electrons emitted is a quarter of the initial number
(b) each emitted electron carries one quarter of the initial energy
(c) number of electrons emitted is half the initial number
(d) each emitted electron carries half the initial energy

39. The maximum kinetic energy of the electrons hitting a target so as to produce X-ray of wavelength 1 Å is

(a) 1.24 keV (b) 12.4 keV
(c) 124 keV (d) None of these

40. When the X-ray tube is operated at 1kV, then X-rays of minimum wavelength 6.22 Å are produced. If the tube is operated at 10 kV, then the minimum wavelength of x-rays will be

(a) 0.622 Å (b) 6.22 Å
(c) 3.11 Å (d) zero

41. Which one of the following statements is WRONG in the context of X-rays generated from a X-ray tube?

(a) Wavelength of characteristic X-rays decreases when the atomic number of the target increases.
(b) Cut-off wavelength of the continuous X-rays depends on the atomic number of the target
(c) Intensity of the characteristic X-rays depends on the electrical power given to the X-ray tube
(d) Cut-off wavelength of the continuous X-rays depends on the energy of the electrons in the X-ray tube

42. X-rays are produced in an X-ray tube operating at a given accelerating voltage. The wavelength of the continuous X-rays has values from **Tricky**

(a) 0 to ∞
(b) λ_{min} to ∞ where $\lambda_{min} > 0$
(c) 0 to λ_{max} where $\lambda_{max} < \infty$
(d) λ_{min} to λ_{max} where $0 < \lambda_{min} < \lambda_{max} < \infty$

43. The ratio of energies of X-rays of the wavelength 0.01Å and 0.5 Å will be

(a) 1 : 1 (b) 1 : 2 (c) 1 : 5 (d) 50 : 1

44. An x-ray tube is operating at 30 KV then the minimum wavelength of the x-rays coming out of the tube is –

(a) 1.24 Å (b) 0.413 Å (c) 0.124 Å (d) 0.13 Å

45. A parallel beam of electrons travelling in x-direction falls on a slit of width d (see figure). If after passing the slit, an electron acquires momentum p_y in the y-direction then for a majority of electrons passing through the slit (h is Planck's constant): **Tricky**

(a) $|P_y| d > h$

(b) $|P_y| d < h$

(c) $|P_y| d \simeq h$

(d) $|P_y| d >> h$

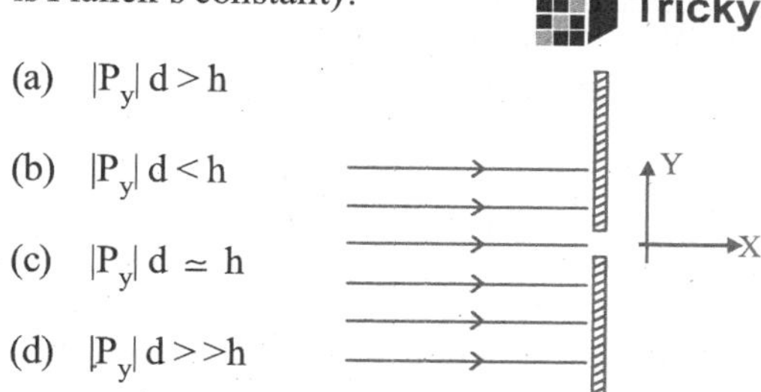

46. The potential difference applied to an X-ray tube is 5kV and the current through it is 3.2mA. Then the number of electrons striking the target per second is

(a) 2×10^{16} (b) 5×10^{6} (c) 1×10^{17} (d) 4×10^{15}

47. When the minimum wavelength of X-rays is 2Å then the applied potential difference between cathode and anticathode will be

(a) 6.2kV (b) 2.48 kV (c) 24.8 kV (d) 62 kV

48. A horizontal cesium plate (work function = 1.9 eV) is moved vertically downward at a constant speed V in a room full of radiation of wavelength 250 nm and above. The minimum value of v so that the vertically upward component of velocity is nonpositive for each photoelectron.

(a) $1.04 \times 10^6 ms^{-1}$ (b) $2.03 \times 10^7 ms^{-1}$
(c) $4.03 \times 10^8 ms^{-1}$ (d) $5.11 \times 10^9 ms^{-1}$

49. The energy in monochromatic X-rays of wavelength 1 Å is roughly equal to

(a) 2×10^{-15} J (b) 2×10^{-16} J
(c) 2×10^{-17} J (d) 2×10^{-18} J

50. The short wavelength limit of continuous X-radiation emitted by an X-ray tube operating at 30 kV is 0.414 Å. Calculate Planck's constant.

(a) 6.22×10^{-34} erg-sec (b) 6.624×10^{-24} erg-sec
(c) 6.624×10^{-27} J-sec (d) 6.624×10^{-34} J-sec

51. A particle having mass 1.67×10^{-27}kg moving at a speed v absorbs a photon of wavelength 122 nm and stops. The value of v is

(a) 3.25 ms^{-1} (b) 1.28 ms^{-1}
(c) 0.82 ms^{-1} (d) 0.33 ms^{-1}

52. Electrons with energy 80 keV are incident on the tungsten target of an X-ray tube. K-shell electrons of tungsten have 72.5 keV energy. X-rays emitted by the tube contain only **Tricky**

(a) a continuous X-ray spectrum (Bremsstrahlung) with a minimum wavelength of 0.155Å
(b) a continuous X-ray spectrum (Bremsstrahlung) with all wavelengths
(c) the characteristic X-ray spectrum of tungsten
(d) a continuous X-ray spectrum (Bremsstrahlung) with a minimum wavelength of 0.155Å and the characteristic X-ray spectrum of tungsten.

53. An X-ray tube is operated at 15 kV. Calculate the upper limit of the speed of the electrons striking the target. **Critical Thinking**

(a) 7.26×10^7 m/s (b) 7.62×10^7 m/s
(c) 7.62×10^7 cm/s (d) 7.26×10^9 m/s

54. Thickness of the medium (μ) at which intensity of emergent x-rays becomes half is:

(a) $\frac{1.23}{\mu}$ (b) $\frac{3.318}{\mu}$ (c) $\frac{0.126}{\mu}$ (d) $\frac{0.693}{\mu}$

55. When X-rays of wavelength 0.5 Å would be transmitted by an aluminium tube of thickness 7 mm, its intensity remains one-fourth. The absorption coefficient of aluminium for these X-rays is

(a) 0.188 mm^{-1} (b) 0.189 mm^{-1}
(c) 0.198 mm^{-1} (d) None of these

56. An X-ray tube with Cu target is operated at 25 kV. The glancing angle for a NaCl. Crystal for the Cu k_α line is 15.8°. Find the wavelength of this line. (d for NaCl = 2.82 Å, h = 6.62×10^{-27} erg-sec)

(a) 3.06 Å (b) 1.53 Å
(c) 0.75 Å (d) None of these

57. The threshold wavelength of the tungsten is 2300 Å. If ultraviolet light of wavelength 1800 Å is incident on it, then the maximum kinetic energy of photoelectrons would be about –

(a) 1.49 eV (b) 2.2 eV (c) 3.0 eV (d) 5.0 eV

58. Two identical photocathodes receive light of frequencies f_1 and f_2. If the velocites of the photo electrons (of mass m) coming out are respectively v_1 and v_2, then **Critical Thinking**

(a) $v_1^2 - v_2^2 = \frac{2h}{m}(f_1 - f_2)$

(b) $v_1 + v_2 = \left[\frac{2h}{m}(f_1 + f_2)\right]^{1/2}$

(c) $v_1^2 + v_2^2 = \frac{2h}{m}(f_1 + f_2)$

(d) $v_1 - v_2 = \left[\frac{2h}{m}(f_1 - f_2)\right]^{1/2}$

59. When photons of wavelength λ_1 are incident on an isolated sphere, the corresponding stopping potential is found to be V. When photons of wavelength λ_2 are used, the corresponding stopping potential was thrice that of the above value. If light of wavelength λ_3 is used then find the stopping potential for this case :

(a) $\frac{hc}{e}\left[\frac{1}{\lambda_3}+\frac{1}{\lambda_2}-\frac{1}{\lambda_1}\right]$

(b) $\frac{hc}{e}\left[\frac{1}{\lambda_3}+\frac{1}{2\lambda_2}-\frac{1}{\lambda_1}\right]$

(c) $\frac{hc}{e}\left[\frac{1}{\lambda_3}-\frac{1}{\lambda_2}-\frac{1}{\lambda_1}\right]$

(d) $\frac{hc}{e}\left[\frac{1}{\lambda_3}+\frac{1}{2\lambda_2}-\frac{3}{2\lambda_1}\right]$

60. Photons with energy 5 eV are incident on a cathode C, on a photoelectric cell. The maximum energy of the emitted photoelectrons is 2 eV. When photons of energy 6 eV are incident on C, no photoelectrons will reach the anode A if the stopping potential of A relative to C is

(a) 3 V (b) –3 V (c) –1 V (d) 4 V

61. According to Einstein's photoelectric equation, the plot of the kinetic energy of the emitted photo electrons from a metal Versus the frequency, of the incident radiation gives a straight line whose slope

(a) depends both on the intensity of the radiation and the metal used
(b) depends on the intensity of the radiation
(c) depends on the nature of the metal used
(d) is the same for the all metals and independent of the intensity of the radiation

62. 1.5 mW of 400 nm light is directed at a photoelectric cell. If 0.10 per cent of the incident photons produce photoelectrons, then find the current in the cell. **Critical Thinking**

(a) 4.8 μA (b) 48 μA (c) 1.8 μA (d) 0.48 μA

63. Photoelectrons are ejected from a metal when light of frequency υ falls on it. Pick out the wrong statement from the following.

(a) No electrons are emitted if υ is less than *W*/*h*, where *W* is the work function of the metal
(b) The ejection of the photoelectrons is instantaneous.
(c) The maximum energy of the photoelectrons is hυ.
(d) The maximum energy of the photoelectrons is independent of the intensity of the light.

64. If an electron is accelerated through a potential difference 150 volt, its de-Broglie wavelength is:

(a) 1 A° (b) 100A° (c) 5000A° (d) 90000A°

65. The photoelectric threshold wavelength of silver is 3250×10^{-10}m. The velocity of the electron ejected from a silver surface by ultraviolet light of wavelength 2536×10^{-10} m is
(Given $h = 4.14 \times 10^{-15}$ eVs and $c = 3 \times 10^8$ ms^{-1})

(a) $\approx 0.6 \times 10^6$ ms^{-1} (b) $\approx 61 \times 10^3$ ms^{-1}
(c) $\approx 0.3 \times 10^6$ ms^{-1} (d) $\approx 6 \times 10^6$ ms^{-1}

66. An atom emits a photon of wavelength λ = 600 nm by transition from an excited state of life time 8×10^{-9} s. If $\Delta\nu$ represents the minimum uncertainty in the frequency of the photon, the fractional width $\frac{\Delta\nu}{\nu}$ of the spectral line is of the order of

(a) 10^{-4} (b) 10^{-6} (c) 10^{-8} (d) 10^{-10}

67. Find the number of photon emitted per second by a 25 watt source of monochromatic light of wavelength 6600 Å. What is the photoelectric current assuming 3% efficiency for photoelectric effect ? **Toughnut**

(a) $\frac{25}{3}\times10^{19}$ J, 0.4 amp (b) $\frac{25}{4}\times10^{19}$ J, 6.2 amp
(c) $\frac{25}{2}\times10^{19}$ J, 0.8 amp (d) None of these

68. Radiation of two photon energies twice and five times the work function of metal are incident successively on the metal surface. The ratio of the maximum velocity of photoelectrons emitted is the two cases will be

(a) 1 : 2 (b) 2 : 1 (c) 1 : 4 (d) 4 : 1

69. In the photoelectric experiment, if we use a monochromatic light, the I-V curve is as shown. If work function of the metal is 2eV, estimate the power of light used. (Assume efficiency of photo emission = 10^{-3}%, i.e., number of photoelectrons emitted are 10^{-3}% of number of photons incident on metal)

I
10μA
V
–5 volt

(a) 2W
(b) 5W
(c) 7W
(d) 10W

70. Light of wavelength 200 Å fall on aluminium surface. Work function of aluminium is 4.2 eV. What is the kinetic energy of the fastest emitted photoelectrons?

(a) 2 eV (b) 1 eV (c) 4 eV (d) 0.2 eV

71. When a metallic surface is illuminated with radiation of wavelength λ, the stopping potential is V. If the same surface is illuminated with radiation of wavelength 2λ, the stopping potential is $\frac{V}{4}$. The threshold wavelength for the metallic surface is :

(a) 4λ (b) 5λ (c) $\frac{5}{2}\lambda$ (d) 3λ

72. A beam of light has two wavelengths of 4972Å and 6216Å with a total intensity of 3.6×10^{-3} Wm^{-2} equally distributed among the two wavelengths. The beam falls normally on an area of 1 cm^2 of a clean metallic surface of work function 2.3 eV. Assume that there is no loss of light by reflection and that each capable photon ejects one electron. The number of photoelectrons liberated in 2s is approximately:

(a) 6×10^{11} (b) 9×10^{11}
(c) 11×10^{11} (d) 15×10^{11}

73. At $t = 0$, light of intensity 10^{12} photons/s–m^2 of energy $6eV$ per photon start falling on a plate with work function 2.5 eV. If area of the plate is 2×10^{-4} m^2 and for every 10^5 photons one photoelectron is emitted, charge on the plate at $t = 25$s is

(a) 8×10^{-15}C (b) 4×10^{-14}C
(c) 12×10^{-14}C (d) 16×10^{-14}C

74. The work function for sodium surface is 2.0 eV and that for aluminium surface is 4.2 eV. The two metal are lluminated with appropriate radiation so as to cause photoemission. Then **Tricky**

(a) the threshold frequency for sodium will be less than that for aluminium
(b) the threshold frequency for sodium will be more than that of aluminium
(c) both sodium and aluminium will have the same threshold frequency
(d) none of the above

75. Light coming from a discharge tube filled with hydrogen falls on the cathode of the photoelectric cell. The work function of the surface of cathode is 4eV. Which of the following values of the anode voltage (in volts) with respect to the cathode will likely to make the photo current zero?

(a) – 4 (b) – 6 (c) – 8 (d) – 10

76. In an experiment on photoelectric effect, a student plots stopping potential V_0 against reciprocal of the wavelength λ of the incident light for two different metals A and B. These are shown in the figure. **Tricky**

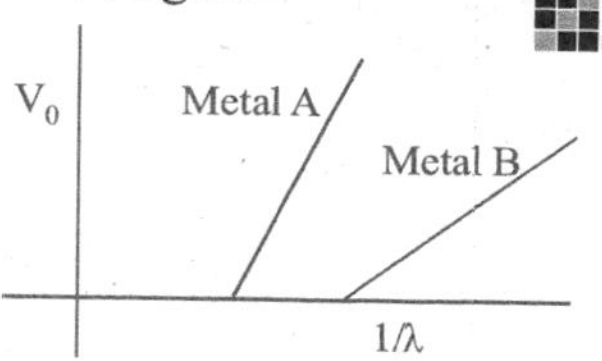

Looking at the graphs, you can most appropriately say that:

(a) Work function of metal B is greater than that of metal A
(b) For light of certain wavelength falling on both metal, maximum kinetic energy of electrons emitted from A will be greater than those emitted from B.
(c) Work function of metal A is greater than that of metal B
(d) Students data is not correct

77. A metal surface is illuminated by a light of given intensity and frequency to cause photoemission. If the intensity of illumination is reduced to one - fourth of its original value, then the maximum KE, of emitted photoelectrons will become

(a) $(1/16)^{th}$ of original value
(b) unchanged
(c) twice the original value
(d) four times the original value

78. The threshold frequency for a metallic surface corresponds to an energy of 6.2 eV and the stopping potential for a radiation incident on this surface is 5 V. The incident radiation lies in

(a) ultra-violet region (b) infra-red region
(c) visible region (d) *X*-ray region

79. Light of wavelength 500 nm is incident on a metal with work function 2.28 eV. The wavelength of the emitted electron is: **Critical Thinking**

(a) $< 2.8 \times 10^{-9}\,m$ (b) $\geq 2.8 \times 10^{-9}\,m$
(c) $\leq 2.8 \times 10^{-12}\,m$ (d) $< 2.8 \times 10^{-10}\,m$

80. 10^{-3} W of 5000Å light is directed on a photoelectric cell. If the current in the cell is 0.16 μA, the percentage of incident photons which produce photoelectrons, is

(a) 0.4% (b) 0.04% (c) 20% (d) 10%

81. The surface of a metal is illuminted with the light of 400 nm. The kinetic energy of the ejected photoelectrons was found to be 1.68 eV. The work function of the metal is:
(hc = 1240 eV.nm)

(a) 1.41 eV (b) 1.51 eV (c) 1.68 eV (d) 3.09 eV

82. For photoelectric emission from certain metal the cut-off frequency is ν. If radiation of frequency 2ν impinges on the metal plate, the maximum possible velocity of the emitted electron will be (m is the electron mass)

(a) $\sqrt{h\nu/m}$ (b) $\sqrt{2h\nu/m}$

(c) $2\sqrt{h\nu/m}$ (d) $\sqrt{h\nu/(2m)}$

83. Electromagnetic radiation falls on a metallic body whose work function is 2eV. For a particular radiation of frequency ν, the maximum kinetic energy of the photoelectron is found to be 4 eV. What would be the maximum kinetic energy of photoelectron for the radiation of frequency $\frac{5\nu}{3}$?

(a) $\frac{8}{3}$eV (b) 8 eV (c) $\frac{10}{3}$eV (d) $\frac{20}{3}$eV

84. In a photo-emissive cell, with exciting wavelength λ, the fastest electron has speed v. If the exciting wavelength is changed to $\frac{5\lambda}{4}$, the speed of the fastest emitted electron will be

(a) $(3/4)^{1/2}$. v (b) $(4/5)^{1/2}$. v

(c) less than $(4/5)^{1/2}$. v

(d) greater than $(4/5)^{1/2}$. v

85. In a photoelectric emission, electrons are ejected from metal X and Y by light of frequency f. The potential difference V required to stop the electrons is measured for various frequencies. If Y has a greater work function than X, which graph illustrates the expected results? Tricky

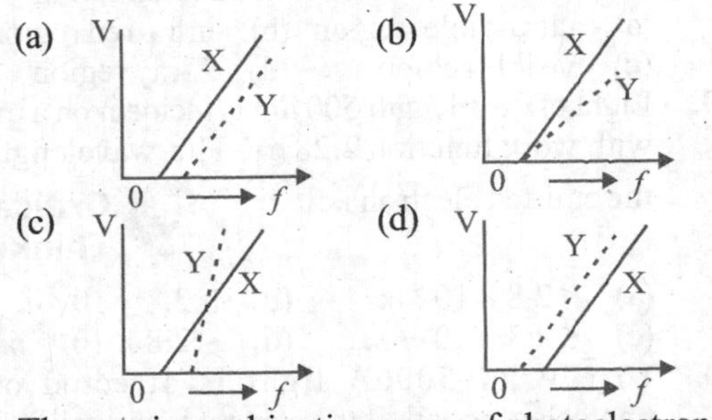

86. The maximum kinetic energy of photoelectrons emitted from a surface when photons of energy 6 eV fall on it is 4 eV. The stopping potential, in volt, is

(a) 2 (b) 4 (c) 6 (d) 10

87. A perfectly relflecting solid sphere of radius x is kept in the path of a paralled beam of light of large aperture. If the beam carries an intensity I, find the force exerted by the beam on the sphere.

(a) $\frac{\pi x^2 i}{C}$ (b) $\frac{\pi x^3 i^2}{2C}$ (c) $\frac{\pi x i^2}{2C^2}$ (d) $\frac{3\pi x^2 i}{C}$

88. A parallel beam of light of wavelength 600 nm and intensity 100 w m^{-2}. How many photons cross 1 cm^2 area perpendicular to the beam in 1 sec?

(a) 3.0×10^{16} (b) 4.0×10^{15}

(c) 2.0×10^{14} (d) 1.0×10^{13}

89. A point source causes photoelectric effect from a small metal plate. Which of the curves in Fig may represent the saturation photo - current as a function of the distance between the source and the metal?

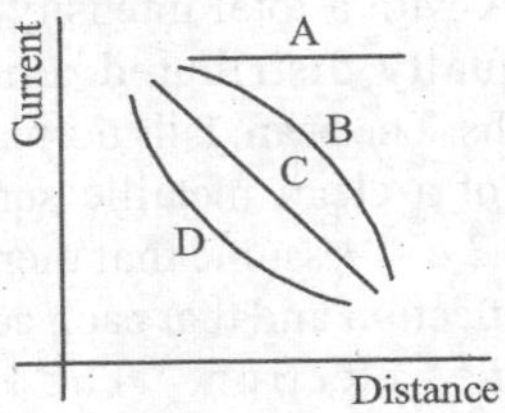

(a) A (b) B (c) C (d) D

90. A photoelectric surface is illuminated successively by monochromatic light of wavelengths λ and $\frac{\lambda}{2}$. If the maximum kinetic energy of the emitted photoelectrons in the second case is 3 times that in the first case, the work function of the surface is : Critical Thinking

(a) $\frac{hc}{2\lambda}$ (b) $\frac{hc}{\lambda}$ (c) $\frac{hc}{3\lambda}$ (d) $\frac{3hc}{\lambda}$

91. The de Broglie wavelength of an electron is the same as that of a 50 KeV X-ray photon. The ratio of the energy of the photon to the kinetic energy of the electron is: (the energy equivalent of electron mass is 0.5 MeV)

(a) 1:50 (b) 1:20 (c) 20:1 (d) 50:1

92. A 200 W sodium street lamp emits yellow light of wavelength 0.6 μm. Assuming it to be 25% efficient in converting electrical energy to light, the number of photons of yellow light it emits per second is

(a) 1.5×10^{20} (b) 6×10^{18}

(c) 62×10^{20} (d) 3×10^{19}

93. Photoelectric emission is observed from a metallic surface for frequencies v_1 and v_2 of the incident light rays ($v_1 > v_2$). If the maximum values of kinetic energy of the photoelectrons emitted in the two cases are in the ratio of 1 : k, then the threshold frequency of the metallic surface is

(a) $\frac{v_1 - v_2}{k-1}$ (b) $\frac{kv_1 - v_2}{k-1}$

(c) $\frac{kv_2 - v_1}{k-1}$ (d) $\frac{v_2 - v_1}{k}$

94. A material particle with a rest mass m_0 is moving with a velocity of light c. Then the wavelength of the de Broglie wave associated with it is :

(a) (h/m_0c) (b) zero (c) ∞ (d) (m_0c/h)

95. To decrease the cut-off wavelength of continuous X-rays by 25%, the potential difference across X-ray tube **Critical Thinking**

(a) must be increased by $\frac{100}{3}\%$

(b) must be decreased by $\frac{100}{3}\%$

(c) must be increased by 25%

(d) must be decreased by 25%

96. The shortest wavelength of X-ray emitted from an X-ray tube operated at 2×10^6 volt, is of the order of

(a) 10^{-5} Å (b) 10^{-2} Å (c) 0.15 Å (d) 1 Å

97. De–broglie wavelength associated with the hydrogen atom moving with most probable velocity at 27 °C is 1.26 Å . De–broglie wavelength associated with helium atom moving with r.m.s. velocity at 51 °C is :

(a) 2.268 Å (b) 1.852 Å
(c) 0.5 Å (d) 0.7 Å

98. Two identical non – relative particles A and B move at right angles to each other, processing de Broglie wavelengths λ_1 and λ_2, respectively. The de Broglie wavelength of each particle in their centre of mass frame of reference is **Toughnut**

(a) $\lambda_1 + \lambda_2$ (b) $2\lambda_1\lambda_2 / \left(\sqrt{\lambda_1^2 + \lambda_2^2}\right)$

(c) $\lambda_1\lambda_2 / \left(\sqrt{|\lambda_1^2 + \lambda_2^2|}\right)$ (d) $(\lambda_1 + \lambda_2)/2$

99. The ratio of the λ_{min} in a Coolidge tube to $\lambda_{deBroglie}$ of the electrons striking the target depends on accelerating potential V as

(a) $\frac{\lambda_{min}}{\lambda_{de\,Broglie}} \propto \sqrt{V}$ (b) $\frac{\lambda_{min}}{\lambda_{de\,Broglie}} \propto V$

(c) $\frac{\lambda_{min}}{\lambda_{de\,Broglie}} \propto \frac{1}{\sqrt{V}}$ (d) $\frac{\lambda_{min}}{\lambda_{de\,Broglie}} \propto \frac{1}{V}$.

100. In a photoelectric experiment, anode potential is plotted against plate current in figure. then

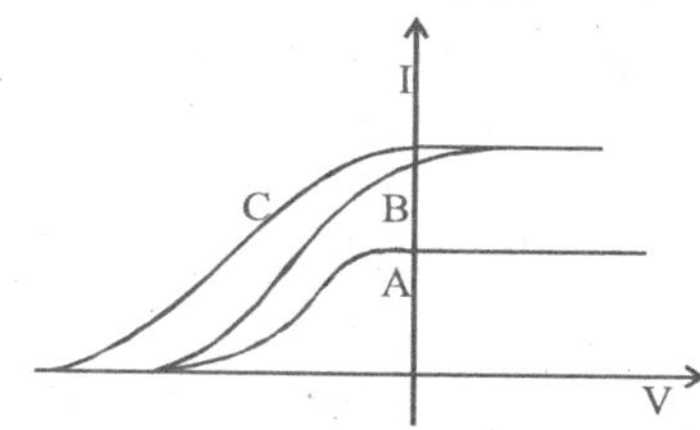

(a) A and B will have different intensities while B and C will have different frequencies

(b) B and C will have different intensities while A and C will have different frequencies

(c) A and B will have different intensities while A and C will have equal frequencies

(d) B and C will have equal intensities while A and B will have same frequencies

Answer KEYs

1	(a)	11	(a)	21	(b)	31	(b)	41	(b)	51	(a)	61	(d)	71	(d)	81	(a)	91	(c)
2	(a)	12	(b)	22	(c)	32	(b)	42	(b)	52	(d)	62	(d)	72	(b)	82	(b)	92	(a)
3	(d)	13	(b)	23	(c)	33	(a)	43	(d)	53	(a)	63	(c)	73	(a)	83	(b)	93	(b)
4	(a)	14	(a)	24	(c)	34	(c)	44	(b)	54	(d)	64	(a)	74	(a)	84	(d)	94	(b)
5	(a)	15	(c)	25	(d)	35	(a)	45	(a)	55	(c)	65	(a)	75	(d)	85	(a)	95	(a)
6	(c)	16	(b)	26	(d)	36	(a)	46	(a)	56	(b)	66	(b)	76	(d)	86	(b)	96	(b)
7	(c)	17	(b)	27	(a)	37	(d)	47	(a)	57	(a)	67	(a)	77	(b)	87	(a)	97	(c)
8	(a)	18	(a)	28	(a)	38	(a)	48	(a)	58	(a)	68	(a)	78	(a)	88	(a)	98	(b)
9	(d)	19	(a)	29	(b)	39	(b)	49	(a)	59	(d)	69	(c)	79	(b)	89	(d)	99	(c)
10	(a)	20	(a)	30	(d)	40	(a)	50	(d)	60	(b)	70	(a)	80	(b)	90	(a)	100	(d)

Hints & Solutions

1. (a) The de-Broglie's wavelength associated with the moving electron $\lambda = \frac{h}{P}$
Now, according to problem
$\frac{d\lambda}{\lambda} = -\frac{dp}{P}$; $\frac{0.5}{100} = \frac{P}{P'} \Rightarrow P' = 200\,P$

2. (a) According to De-broglie $p = \frac{h}{\lambda}$ or $P \propto \frac{1}{\lambda}$
$P \propto \frac{1}{\lambda}$ represents rectangular hyperbola.

3. (d) Energy in joule (E) = charge × potential diff. in volt
$E_{electron} = q_e V$ and $E_{proton} = q_p\, 4V$
de-Broglie wavelength $\lambda = \frac{h}{P} = \frac{h}{\sqrt{2mE}}$
$\lambda_e = \frac{h}{\sqrt{2m_e eV}}$ and $\lambda_P = \frac{h}{\sqrt{2m_P e4V}}$
$(\because q_e = q_P)$
$$\therefore \frac{\lambda_e}{\lambda_P} = \frac{\frac{h}{\sqrt{2m_e eV}}}{\frac{h}{\sqrt{2m_P e4V}}} = \sqrt{\frac{2m_P e4V}{2m_e eV}} = 2\sqrt{\frac{m_P}{m_e}}$$

4. (a) Acceleration, $a = \frac{Ee}{m}$
Velocity of electron, $v = at = \frac{Eet}{m}$.
Wavelength, $\lambda = \frac{h}{mv} = \frac{h}{Eet}$.
Now $\frac{d\lambda}{dt} = -\frac{h}{eEt^2}$.

5. (a) de-Broglie wavelength of a charged particle is given by, $\lambda \propto \frac{1}{\sqrt{mq}}$
If m_p and e are mass and charge of a proton respectively,
and, m_D and e are mass and charge of a deutron respectively, then
$$\frac{\lambda_p}{\lambda_D} = \sqrt{\frac{m_D q_D}{m_p q_p}} = \sqrt{\frac{(2m_p)(e)}{(m_p)(e)}} = \sqrt{2}$$
$\lambda_p = \sqrt{2}\lambda_D$

Momentum, is given by, $P = \frac{h}{\lambda}$ or, $p \propto \frac{1}{\lambda}$
where, h = plank's constant
Since the wavelength of a proton is more than that of deutron thus, the momentum of a proton is lesser than that of deutron. Hence, the momentum of proton is less.

6. (c) Applying conservation of linear momentum,
Initial momentum = Final momentum
$0 = m_1v_1 - m_2v_2 \Rightarrow m_1v_1 = m_2v_2$
Now, $\frac{\lambda_1}{\lambda_2} = \frac{h/m_1v_1}{h/m_2v_2} = 1$

7. (c) $F = \frac{dP}{dt} = \left(\frac{IA}{hc/\lambda}\right)(1+r)\frac{h}{\lambda}$
$\Rightarrow P = \frac{F}{A} = \frac{I}{c}(1+r)$

8. (a)

9. (d) For photon,
$\lambda_2 = \frac{hc}{E}$ ...(i)
For proton, $p = \sqrt{2mE}$
$\lambda_1 = \frac{h}{p} = \frac{h}{\sqrt{2mE}}$ ...(ii)
$$\frac{\lambda_2}{\lambda_1} = \frac{hc}{E \times \frac{h}{\sqrt{2mE}}} \propto E^{-1/2}$$

10. (a) The cut off wavelength is given by
$\lambda_0 = \frac{hc}{eV}$ ...(i)
According to de Broglie equation
$\lambda = \frac{h}{p} = \frac{h}{\sqrt{2meV}}$
$\Rightarrow \lambda^2 = \frac{h^2}{2meV} \Rightarrow V = \frac{h^2}{2me\lambda^2}$ ...(ii)
From (i) and (ii),
$\lambda_0 = \frac{hc \times 2me\lambda^2}{eh^2} = \frac{2mc\lambda^2}{h}$

11. (a) We know that,
de-Broglie wavelength $\lambda = \frac{\hbar}{P} = \frac{h}{\sqrt{2m(KE)}}$

K.E. of thermal neutron $= \frac{3}{2}kT$

$= \frac{h}{\sqrt{2m\left(\frac{3}{2}kT\right)}} \Rightarrow \lambda = \frac{h}{\sqrt{3mkT}}$

12. (b) Momentum imparted per unit time = np

$\Rightarrow \quad F = \frac{nh}{\lambda} \quad \therefore$ Acceleration $= \frac{nh}{m\lambda}$

13. (b) Velocity at highest point = u sin θ.

$\therefore \lambda_D = \frac{h}{mu\sin\theta}$ (Since θ is velocity wrt vertical)

14. (a) For electron De-Broglie wavelength,

$\lambda_e = \frac{h}{\sqrt{2mE}}$; For photon E = pc

$\Rightarrow$ De-Broglie wavelength, $\lambda_{Ph} = \frac{hc}{E}$

$\therefore \quad \frac{\lambda_e}{\lambda_{Ph}} = \frac{h}{\sqrt{2mE}} \times \frac{E}{hc} = \left(\frac{E}{2m}\right)^{1/2} \frac{1}{c}$

15. (c) $\lambda = \frac{h}{\sqrt{2mqV}}$ and mV = constant

$1837\,V' = 1 \times V$ or $V' = \frac{V}{1837}$ volt

16. (b)

17. (b) No. of photons/sec

$= \frac{\text{Energy incident on platinum surface per second}}{\text{Energy of one photon}}$

No. of photons incident per second

$= \frac{2 \times 10 \times 10^{-4}}{10.6 \times 1.6 \times 10^{-19}} = 1.18 \times 10^{14}$

As 0.53% of incident photon can eject photoelectrons

∴ No. of photoelectrons ejected per second

$= 1.18 \times 10^{14} \times \frac{0.53}{100} = 6.25 \times 10^{11}$

Minimum energy = 0 eV,

Maximum energy = (10.6 – 5.6) eV = 5 eV

18. (a) By Einstein equation

$hf = W_0 + eV_0.$

$W_0 = \frac{hc}{\lambda} - eV_0$

$= \frac{(6.63 \times 10^{-34}) \times (3 \times 10^8)}{4950 \times 10^{-10}} - (1.6 \times 10^{-19}) \times 0.6$

$= 3.12 \times 10^{-19}$ J

For the second surface

$hf' = W_0 + eV_0'$

or $\quad \frac{hc}{\lambda'} = 3.12 \times 10^{-19} + 1.6 \times 10^{-19} \times 1.1$

$\therefore \quad \lambda' = 4077$ Å

19. (a) The work function of the metal

$W_0 = hf_0 = (6.63 \times 10^{-34}) \times (6 \times 10^{14})$

$= 3.98 \times 10^{19}$ J

If f is the frequency of the light, then

$hf = W_0 + eV_0$

$\therefore \quad f = \frac{W_0 + eV_0}{h}$

$= \frac{3.98 \times 10^{-19} + 1.6 \times 10^{-19} \times 3}{6.63 \times 10^{-34}} = 1.32 \times 10^{15}\ Hz$

20. (a)

21. (b) If v_{max} is the speed of the fastest electron emitted from the metal surface, then

$\frac{hc}{\lambda} = W_0 + \frac{1}{2}mv_{max}^2$

$\frac{(6.63 \times 10^{-34}) \times (3 \times 10^8)}{(180 \times 10^{-9})}$

$= 2 \times (1.6 \times 10^{-19}) + \frac{1}{2}(9.1 \times 10^{-31})v_{max}^2$

$\therefore \quad v = 1.31 \times 10^6$ m/s

The radius of the electron is given by

$r = \frac{mv}{qB} = \frac{(9.1 \times 10^{-31}) \times (1.31 \times 10^6)}{(1.6 \times 10^{-19}) \times (5 \times 10^{-9})} = 0.149$ m

22. (c) Total number of ions in the rod

$N = (2 \times 10^{25}) \times \left[\frac{\pi}{4} \times (10^{-2})^2\right] \times (5 \times 10^{-2})$

$= 7.85 \times 10^{19}$

The energy of excitation

$E = N \times \frac{hc}{\lambda}$

$= (7.85 \times 10^{19}) \times \frac{(6.6 \times 10^{-34}) \times (3 \times 10^8)}{6.6 \times 10^{-7}}$

$= 23.55$ J

Average power P

$= \frac{E}{t} = \frac{23.55}{10^{-7}} = 23.55 \times 10^7 W = 235$ MW

23. (c) $h\nu_0{}^2 - h\nu_0 = \frac{1}{2}mv^2$

$\therefore \quad \frac{4}{3}h\nu_0 - h\nu_0 = \frac{1}{2}mv'^2$

$\therefore \frac{v'^2}{v^2} = \frac{\frac{4}{3}v - v_0}{v - v_0}$ $\therefore v' = v\sqrt{\frac{\frac{4}{3}v - v_0}{v - v_0}}$.

$\therefore v' > v\sqrt{\frac{4}{3}}$

24. (c) In X-ray tube, $\lambda_{min} = \frac{hc}{eV}$

$\ln \lambda_{min} = \ln\left(\frac{hc}{e}\right) - \ln V$

Clearly, log λ_{min} versus log V graph slope is negative hence option (c) correctly depicts.

25. (d) From question, $m_A = M; m_B = \frac{m}{2}$

$u_A = V \quad u_B = 0$

Let after collision velocity of $A = V_1$ and velocity of $B = V_2$

Applying law of conservation of momentum,

$mu = mv_1 + \left(\frac{m}{2}\right)v_2$ or, $2\mu = 2v_1 + v_2$ ....(*i*)

By law of collision

$e = \frac{v_2 - v_1}{u - 0}$ or, $u = v_2 - v_1$ ....(*ii*)

[$\because$ collision is elastic, $e = 1$]

using eqns (*i*) and (*ii*)

$v_1 = \frac{1}{3}\mu$ and $v_2 = \frac{4}{3}u$

de-Broglie wavelength $\lambda = \frac{h}{p}$

$\therefore \frac{\lambda_A}{\lambda_B} = \frac{P_B}{P_A} = \frac{\frac{m}{2} \times \frac{4}{3}u}{m \times \frac{1}{3}} = 2$

26. (d) **27.** (a)

28. (a) Final speed of α-particle will be less than the initial speed therefore,

$\left(\frac{h}{\lambda_f}\right) < \left(\frac{h}{\lambda_i}\right)$ or $\lambda_f > \lambda_i$

29. (b) We write Einstein's equation for the photoelectric effect,

$hf - W = E_K$ ... (i)

$3hf - W = E'_K$ ... (ii)

From (i) and (ii), we get,

$E'_K - 3E_K = 2W > 0.$

30. (d) Number of emitted electrons $N_E \propto$ Intensity

$\propto \frac{1}{(\text{Distance})^2}$

Therefore, as distance is doubled, N_E decreases by (1/4) times.

31. (b) $E = W_0 + K_{max}$ ...(i)

$\Rightarrow hf = W_A + K_A$ ...(ii)

and $2hf = W_B + K_B = 2W_A + K_B$

$\left(\because \frac{W_A}{W_B} = \frac{1}{2}\right)$

Dividing equation (i) by (ii)

$\frac{1}{2} = \frac{W_A + K_A}{2W_A + K_B} \Rightarrow \frac{K_A}{K_B} = \frac{1}{2}$

32. (b)

33. (a) Use $n = \frac{I}{h\nu} = \frac{I\lambda}{hc}$

$\therefore \frac{n_{UV}}{n_{IR}} = \frac{\lambda_{UV}}{\lambda_{IR}} = \frac{400}{700} = \frac{4}{7} = 0.57$

34. (c) Since, stopping potential is independent of distance hence new stopping potential will remain unchanged i.e., new stopping potential = V_0.

35. (a) $E = \frac{hc}{\lambda}$

Number of photons emitted is

$\frac{Pt}{\left(\frac{hc}{\lambda}\right)} = n_0 \qquad n_0 = \frac{P\lambda t}{hc}$

Since the radiation is spherically symmetric, so total number of photons entring the sensor is n_0 times the ratio of aperture area to the area of a sphere of radius ℓ.

$N = n_0 \frac{\pi(2d)^2}{4\pi\ell^2} = \frac{P\lambda t}{hc}\frac{d^2}{\ell^2}$

36. (a) The work function has no effect on photoelectric current so long as $h\nu > W_0$. The photoelectric current is proportional to the intensity of incident light. Since there is no change in the intensity of light, hence $I_1 = I_2$.

37. (d)

38. (a) Power $\propto$ No. of electrons emitted (N)

$P \propto \frac{1}{r^2} \Rightarrow N \propto \frac{1}{r^2}$

39. (b) $\lambda_{min} = 1$ Å (given)

$\because \lambda_{min} = \frac{1240}{E}$ (eV) (nm)

Thus, $E = \frac{1240(eV)(nm)}{0.01(nm)} = 12400$ eV

E = 12.4 KeV

40. (a) $\frac{\lambda_{m_2}}{\lambda_{m_1}} = \frac{V_1}{V_2} \Rightarrow \lambda_{m_2} = \frac{6.22\times10^3}{10^4} = 0.622$ Å

41. (b)

42. (b) The continuous X-ray spectrum is shown in figure.

All wavelengths $> \lambda_{min}$ are found,

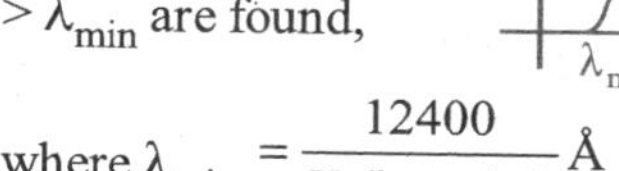

where $\lambda_{min} = \frac{12400}{V \text{ (in volts)}}$ Å

Here V is the applied voltage.

43. (d) $\frac{E_1}{E_2} = \frac{hc}{\lambda_1}\frac{\lambda_2}{hc} = \frac{\lambda_2}{\lambda_1} \Rightarrow \frac{E_1}{E_2} = \frac{50}{1}$

44. (b) $\lambda_{min} = \frac{12400}{V}$ Å $= \frac{12400}{30000} = 0.413$Å

45. (a) From Bragg's equation

$d \sin\theta = \lambda \qquad \sin\theta = \frac{\lambda}{d} < 1$

$\therefore \quad \lambda < d$

$\frac{h}{|p_y|} < d \qquad \left[\because \lambda = \frac{h}{|p_y|}\right]$

$\therefore \quad h < |p_y| d$

46. (a) $I = \frac{q}{t} = \frac{ne}{t}$

No. of electrons striking the target per second

$= \frac{I}{e} = 2\times10^{16}$

47. (a) $V = \frac{12400}{\lambda_{min}(Å)}$ Volt $\Rightarrow V = \frac{12400}{2} = 6.2$ KV

48. (a)

49. (a) $E = hc/\lambda = 6.6 \times 10^{-34} \times 3 \times 10^8/1 \times 10^{-10}$
$= 2 \times 10^{-15}$ J

50. (d) $\lambda_{min} = \frac{ch}{eV} \;\therefore\; h = \left(\frac{eV}{c}\right)\lambda_{min}$

$h = \frac{1.6\times10^{-19}\times30000\times0.141\times10^{-10}}{3\times10^8}$

$\Rightarrow h = 6.624 \times 10^{-34}$ J-sec

51. (a)

52. (d) $\lambda_{min} = \frac{hc}{E} \quad \therefore \lambda_{min} = \frac{12400}{80\times10^3}$ Å $= 0.155$Å

Energy of incident electrons is greater than the ionization energy of electrons in K-shell, the K-shell electrons will be knocked off. Hence, characteristic X-ray spectrum will be obtained.

53. (a) The maximum kinetic energy of an electron accelerated through a potential difference of V volt is $\frac{1}{2}mv^2 = eV$

$\therefore$ maximum velocity $v = \sqrt{\frac{2eV}{m}}$

$v = \sqrt{\frac{2\times1.6\times10^{-19}\times15000}{9.1\times10^{-31}}}$,

$v = 7.26 \times 10^7$ m/s

54. (d)

55. (c) $\frac{I_0}{I} = 4$, x = 7mm (given)

$\Rightarrow \mu = \frac{2.303 \log_{10}\frac{I_0}{I}}{x}$

$\mu = \frac{2.303 \log_{10} 4}{7} = \frac{2.303\times0.6023}{7}$

$\Rightarrow \mu = 0.198$ mm^{-1}

56. (b) According to Bragg's law, $\Rightarrow 2d \sin\theta = n\lambda$, n = 1 for first order

$\Rightarrow 2 \times 2.82 \sin 15.8 = \lambda$

$\Rightarrow \lambda = 5.64 \times 0.2723 = 1.53$ Å

57. (a) $K_{max} = h\nu - h\nu_0 = hc\left(\frac{1}{\lambda} - \frac{1}{\lambda_0}\right)$

Use $hc = 1.24\times10^{-6}$(eV) m

$hc = 1.24\times10^{-6}\left(\frac{10^8}{18} - \frac{10^8}{23}\right)$

$= \frac{1.24\times100\times(23-18)}{18\times23} = 1.49$ eV

58. (a) For one photocathode

$$hf_1 - W = \frac{1}{2}mv_1^2 \quad ....(i)$$

For another photo cathode

$$hf_2 - W = \frac{1}{2}mv_2^2 \quad ....(ii)$$

Subtracting (ii) from (i) we get

$$(hf_1 - W) - (hf_2 - W) = \frac{1}{2}mv_1^2 - \frac{1}{2}mv_2^2$$

$$\therefore h(f_1 - f_2) = \frac{m}{2}(v_1^2 - v_2^2)$$

$$\therefore v_1^2 - v_2^2 = \frac{2h}{m}(f_1 - f_2)$$

59. (d) **60.** (b)

61. (d) From Equation $K.E = h\nu - \phi$ slope of graph of K.E & ν is h (Plank's constant) which is same for all metals

62. (d) The number of photons per second directed at the cell is

$$n = \frac{Power}{h\nu} = \frac{P\lambda}{hc} = \frac{1.5\times10^{-3}\times400\times10^{-9}}{6.62\times10^{-34}\times3\times10^{8}}$$

$$n = \frac{1}{3.31}\times10^{16}\ (\text{photons / s})$$

But it is given that only 0.1% of these photons produce photoelectrons. Therefore number of photoelectrons produced per second is

$$n_e = \frac{0.1}{100}\times\frac{10^{16}}{3.31} = \frac{10^{13}}{3.31}$$

Therefore, the current is

$$I = n_e e$$

$$= \frac{10^{13}\times1.6\times10^{-19}}{3.31} = 0.48\times10^{-6} = 0.48\ \mu A$$

63. (c) According to photo-electric equation : $K.E_{max} = h\nu - h\nu_0$ (Work function) Some sort of energy is used in ejecting the photoelectrons.

64. (a) $\lambda = \frac{h}{\sqrt{2meV}} = \left(\frac{150}{V}\right)^{\frac{1}{2}} A^o = \left(\frac{150}{150}\right)^{\frac{1}{2}} A^o = 1A^o$

65. (a) Given, $\lambda_0 = 3250\times10^{-10}$ m

$\lambda = 2536\times10^{-10}$ m

$$\phi = \frac{hc}{\lambda_0} = \frac{4.14\times10^{-15}\times3\times10^{8}}{3250\times10^{-10}} = 3.82\text{eV}$$

$$h\nu = \frac{hc}{\lambda} = \frac{4.14\times10^{-15}\times3\times10^{8}}{2536\times10^{-10}} = 4.89\text{eV}$$

According to Einstein's photoelectric equation,

$K_{max} = h\nu - \phi$

$KE_{max} = (4.89 - 3.82)\text{eV} = 1.077\text{ eV}$

$$\frac{1}{2}mv^2 = 1.077\times1.6\times10^{-19}$$

$$\Rightarrow v = \sqrt{\frac{2\times1.077\times1.6\times10^{-19}}{9.1\times10^{-31}}}$$

or, $v = 0.6\times10^{6}$ m/s or 6×10^{5} m/s

66. (b) $\Delta E.\Delta t \sim h$

$$E = \frac{hc}{\lambda} \qquad \Delta E = \frac{hc}{\lambda}\frac{\Delta\lambda}{\lambda}$$

$$\therefore \frac{hc}{\lambda}\frac{\Delta\lambda}{\lambda}.\Delta t \sim h$$

Now, $c = \nu\lambda$

$\nu\Delta\lambda + \lambda\Delta\nu = 0$

$\nu D\lambda = -\lambda\Delta\nu$

$$\therefore \frac{\Delta\lambda}{\lambda} = -\frac{\Delta\nu}{\nu} \qquad \therefore \frac{ch}{\lambda}\frac{\Delta\nu}{\nu}\Delta t \sim h$$

$$\therefore \frac{\Delta\nu}{\nu} \sim \frac{h}{\Delta t}.\frac{\lambda}{hc} \sim \frac{\lambda}{\Delta tc} = \frac{600\times10^{-9}}{8\times10^{-9}\times3\times10^{8}}$$

$$\frac{\Delta\nu}{\nu} \sim 10^{-6}$$

67. (a) $P_{in} = 25\text{W}, \lambda = 6600\ \text{Å} = 6600\times10^{-10}$ m

$nh\nu = P$

$\Rightarrow$ Number of photons emitted/sec,

$$n = \frac{P}{hc/\lambda} = \frac{P\lambda}{hc} = \frac{25\times6600\times10^{-10}}{6.64\times10^{-34}\times3\times10^{8}}$$

$$= 8.28\times10^{19} = \frac{25}{3}\times10^{19}$$

3% of emitted photons are producing current

$$\therefore I = \frac{3}{100}\times ne = \frac{3}{100}\times\frac{25}{3}\times10^{19}\times1.6\times10^{-19} = 0.4\text{A}$$

68. (a)

69. (c) The energy of incident photons is given by

$h\nu = eV_s + \phi_0 = 2 + 5 = 7\text{eV}$

(V_s is stopping potential and ϕ_0 is work function)

Saturation current $= 10^{-5}$ A

$= \frac{\eta P}{h\nu}e = \frac{10^{-5}P}{7\times e}e$ (η is photon emission efficiency)

$\therefore P = 7W$

70. **(a)** By Einstein's equation of photo-electric effect, the maximum kinetic energy of emiited photo-electrons is given by

$E_k = h\nu - W$ or $E_k = \frac{hc}{\lambda} - W$

Where, h = Planck's constant

ν = frequency of incident light

W = work function of metal

λ = wavelength of incident light

$$E_k = \frac{6.6\times10^{-34}\times3\times10^{8}}{2000\times10^{-10}\times1.6\times10^{-19}}\text{eV} - 4.2\text{eV}$$

So, $E_k = 2$eV

71. **(d)** According to Einstein's photoelectric effect,

$$eV = \frac{hc}{\lambda} - \frac{hc}{\lambda_0} \quad \text{...(i)}$$

$$eV/4 = \frac{hc}{2\lambda} - \frac{hc}{\lambda_0} \quad \text{...(ii)}$$

Dividing equation (i) by (ii) by

$$\Rightarrow \quad 4 = \frac{\frac{1}{\lambda} - \frac{1}{\lambda_0}}{\frac{1}{2\lambda} - \frac{1}{\lambda_0}} \text{ on solving we get,}$$

$\lambda_0 = 3\lambda$

72. **(b)**

73. **(a)** Number of emitted electrons

$$= \frac{10^{12}\times2\times10^{-4}\times25}{10^{5}} = 5\times10^{4} \text{ electrons}$$

$q = +ne = 8\times10^{-15}$ C

74. **(a)** As work function $W = h\nu_0$, where ν_0 is the threshold frequency.

Greater the work function, greater is the threshold frequency. Therefore, the threshold frequency of sodium will be lesser than that for aluminium.

75. **(d)** **76.** **(d)**

77. **(b)** Maximum KE depends on the frequency of incident radiation, not on intensity.

78. **(a)** $\phi = 6.2$ eV $= 6.2\times1.6\times10^{-19}$ J

$V = 5$ volt

$$\frac{hc}{\lambda} - \phi = eV_0 \Rightarrow \lambda = \frac{hc}{\phi + eV_0}$$

$$= \frac{6.6\times10^{-34}\times3\times10^{8}}{1.6\times10^{-19}(6.2+5)} \approx 10^{-7}\text{ m}$$

This range lies in ultra violet range.

79. **(b)** Work function ϕ of metal $= 2.28$ eV

Wavelength of light $\lambda = 500$ nm $= 500\times10^{-9}$m

$$KE_{max} = \frac{hc}{\lambda} - \phi$$

$$KE_{max} = \frac{6.6\times10^{-34}\times3\times10^{8}}{5\times10^{-7}\times1.6\times10^{-19}} - 2.28$$

$KE_{max} = 2.48 - 2.28 = 0.2$ ev

$$\lambda_{min} = \frac{h}{p} = \frac{h}{\sqrt{2m(KE)_{max}}}$$

$$\lambda_{min} = \frac{25}{9}\times10^{-9} = 2.80\times10^{-9}\text{ nm}$$

$\therefore \lambda \geq 2.8\times10^{-9}$ m

80. **(b)**

81. **(a)** $\lambda = 400$ nm, $hc = 1240$ eV.nm, K.E. $= 1.68$ eV

We know that,

$$\frac{hc}{\lambda} - W = K.E \Rightarrow W = \frac{hc}{\lambda} - K.E$$

$$\Rightarrow W = \frac{1240}{400} - 1.68 = 3.1 - 1.68 = 1.42\text{ eV}$$

82. **(b)** From photo-electric equation,

$h\nu' = h\nu + K_{max}$ ...(i)

$h.2\nu = h\nu + \frac{1}{2}mV^2_{max}$ $[\therefore \nu' = 2\nu]$

$$\Rightarrow h\nu = \frac{1}{2}mV^2_{max} \Rightarrow V_{max} = \sqrt{\frac{2h\nu}{m}}$$

83. **(b)** $h\nu = \phi + K_{max}$

$h\nu = 2 + 4 = 6$ eV

Now, $h\left(\frac{5}{3}\nu\right) = 2 + K_{max}$

$K_{max} = \frac{5}{3}\times6 - 2 \Rightarrow K_{max} = 8$ eV.

84. **(d)**

85. **(a)** V versus f has a constant slope of h/e, so both lines must be parallel. Also, work function is equal to intercept on f - axis.

86. **(b)** Stopping potential is the negative potential applied to stop the electrons having maximum kinetic energy. Therefore, stopping potential will be 4 volt.

87. **(a)**

88. **(a)** number of photon $= \frac{P \times t \times \pi r^2}{hc/\lambda}$

89. **(d)** Saturation current is inversely proportional to the square of distance of cathode from point source.

90. **(a)** From Einstein's photoelectric equation

$$K.E._{\lambda} = \frac{hc}{\lambda} - \phi \quad ...(i)$$

(for monochromatic light of wavelength λ)
where ϕ is work function

$$K.E._{\lambda/2} = \frac{hc}{\lambda/2} - \phi \quad ...(ii)$$

(for monochromatic light of wavelength $\lambda/2$)
From question,

$$K.E._{\lambda/2} = 3(K.E._{\lambda}) \Rightarrow \frac{hc}{\lambda/2} - \phi = 3\left(\frac{hc}{\lambda} - \phi\right)$$

$$\frac{2hc}{\lambda} - \phi = 3\frac{hc}{\lambda} - 3\phi \Rightarrow 2\phi = \frac{hc}{\lambda} \therefore \phi = \frac{hc}{2\lambda}$$

91. **(c)**

92. **(a)** Give that, only 25% of 200W converter electrical energy into light of yellow colour

$$\left(\frac{hc}{\lambda}\right) \times N = 200 \times \frac{25}{100}$$

Where N is the No. of photons emitted per second,
h = plank's constant, c, speed of light.

$$N = \frac{200 \times 25}{100} \times \frac{\lambda}{hc}$$

$$= \frac{200 \times 25 \times 0.6 \times 10^{-6}}{100 \times 6.2 \times 10^{-34} \times 3 \times 10^{8}} = 1.5 \times 10^{20}$$

93. **(b)** By using $h\nu - h\nu_0 = K_{max}$

$\Rightarrow h(\nu_1 - \nu_0) = K_1$ ..... (i)

And $h(\nu_2 - \nu_0) = K_2$ ..... (ii)

$$\Rightarrow \frac{\nu_1 - \nu_0}{\nu_2 - \nu_0} = \frac{K_1}{K_2} = \frac{1}{K},$$

Hence $\nu_0 = \frac{k\nu_1 - \nu_2}{K - 1}$.

94. **(b)**

95. **(a)** $\lambda_1 = \frac{hc}{ev_1}$; $\lambda_2 = \frac{hc}{ev_2} \Rightarrow \frac{\lambda_2}{\lambda_1} = \frac{v_1}{v_2}$

$$\Rightarrow \frac{3}{4} = \frac{v_1}{v_2} \Rightarrow \left(\frac{v_2 - v_1}{v_1}\right) \times 100 = \frac{100}{3}$$

Hence, P.D. must be increased by $\frac{100}{3}\%$.

96. **(b)** $E = eV = 2 \times 10^{6} \times 1.6 \times 10^{-19}$ J

$$\Rightarrow \lambda = \frac{hc}{E} = \frac{6.6 \times 10^{-34} \times 3 \times 10^{8}}{3.2 \times 10^{-13}} \approx 10^{-2}\ \text{Å}$$

97. **(c)**

98. **(b)** Let m be the mass of each particle, then $\lambda_1 = (h/mv_1)$ and $\lambda_2 = (h/mv_2)$, where v_1 and v_2 are the velocities of two particles as shown in the figure.

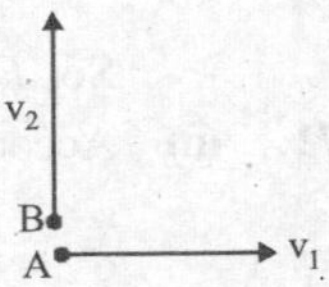

$$\vec{v}_{CM} = \frac{m\vec{v}_1 + m\vec{v}_2}{2m} = \frac{\vec{v}_1 + \vec{v}_2}{2}$$

Velocity of A w.r.t. C frame is

$$\vec{v}_{1c} = \vec{v}_1 - \vec{v}_{CM} = \frac{\vec{v}_1 - \vec{v}_2}{2}$$

$$|\vec{v}_{1c}| = \frac{\sqrt{\vec{v}_1 - \vec{v}_2}}{2} = |\vec{v}_{2c}|$$

So, required wavelength is

$$\lambda = \frac{h}{m|\vec{v}_{1c}|} = \frac{h}{m} \times \frac{2}{\frac{h}{m}\sqrt{\frac{1}{\lambda_1^2} + \frac{1}{\lambda_2^2}}}$$

$$= \frac{2\lambda_1\lambda_2}{\sqrt{\lambda_1^2 + \lambda_2^2}}$$

99. **(c)** $\lambda_{min} = \frac{hc}{eV}$

and $\lambda_{de\text{-}Broglie} = \frac{h}{\rho} = \frac{h}{\sqrt{2meV}}$.

100. **(d)** From the graph it is clear that A and B have the same stopping potential and therefore, the same frequency. Also, B and C have the same intensity.

26 Atoms

1. The energy of He^+ in the ground state is – 54.4 eV, then the energy of Li^{++} in the first excited state will be
   (a) –30.6 eV (b) 27.2 eV
   (c) –13.6 eV (d) –27.2 eV

2. The wavelength K_α of X-rays for two metals 'A' and 'B' are $\frac{4}{1875R}$ and $\frac{1}{675R}$ respectively, where 'R' is Rydberg constant. Find the number of elements lying between A and B according to their atomic numbers
   (a) 3 (b) 1 (c) 4 (d) 5

3. If in hydrogen atom, radius of nth Bohr orbit is r_n, frequency of revolution of electron in n^{th} orbit is f_n, choose the correct option. Tricky

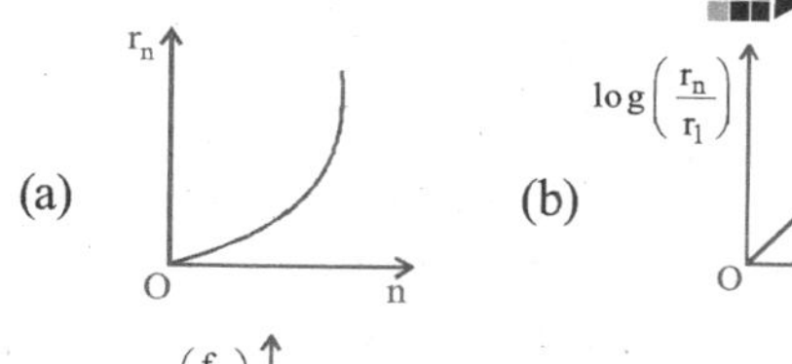

(a) (b)

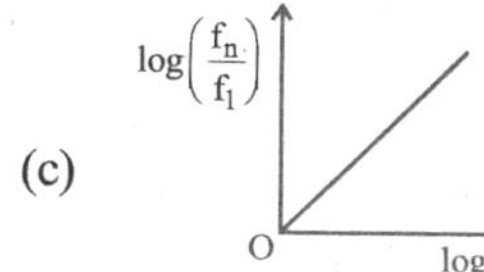

(c) (d) Both (a) and (b)

4. The de–Broglie wavelength associated with the electron in the n = 4 level is :
   (a) $\frac{1}{4}$ th of the de–Broglie wavelength of the electron in the ground state.
   (b) four times the de–Broglie wavelength of the electron in the ground state
   (c) two times the de–Broglie wavelength of the electron in the ground state
   (d) half of the de–Broglie wavelength of the electron in the ground state

5. Excitation energy of a hydrogen like ion in its excitation state is 40.8 eV. Energy needed to remove the electron from the ion in ground state is
   (a) 54.4 eV (b) 13.6 eV
   (c) 40.8 eV (d) 27.2 eV

6. The diagram shows the energy levels for an electron in a certain atom. Which transition shown represents the emission of a photon with the most energy ? Tricky

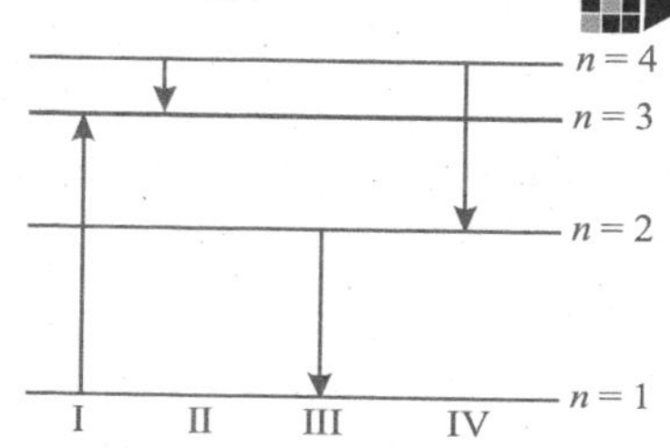

   (a) I (b) II (c) III (d) IV

7. If one were to apply Bohr model to a particle of mass 'm' and charge 'q' moving in a plane under the influence of a magnetic field 'B', the energy of the charged particle in the nth level will be :

   (a) $n\left(\frac{hqB}{2\pi m}\right)$ (b) $n\left(\frac{hqB}{8\pi m}\right)$
   (c) $n\left(\frac{hqB}{4\pi m}\right)$ (d) $n\left(\frac{hqB}{\pi m}\right)$

8. The acceleration of an electron in the first orbit of the hydrogen atom (z = 1) is :
   (a) $\frac{h^2}{\pi^2 m^2 r^3}$ (b) $\frac{h^2}{8\pi^2 m^2 r^3}$
   (c) $\frac{h^2}{4\pi^2 m^2 r^3}$ (d) $\frac{h^2}{4\pi m^2 r^3}$

9. In Rutherford's experiment, the number of α-particles scattered through an angle of 60° by a silverfoil is 200 per minute. When the silver foil is replaced by a copper foil of the same thickness, the number of α-particles scattered through an angle of 60° per minute is:
   (a) $\frac{200 \times Z_{Cu}}{Z_{Ag}}$ (b) $200 \times \left(\frac{Z_{Cu}}{Z_{Ag}}\right)^2$
   (c) $200 \times \frac{Z_{Ag}}{Z_{Cu}}$ (d) $200 \times \left(\frac{Z_{Ag}}{Z_{Cu}}\right)^2$

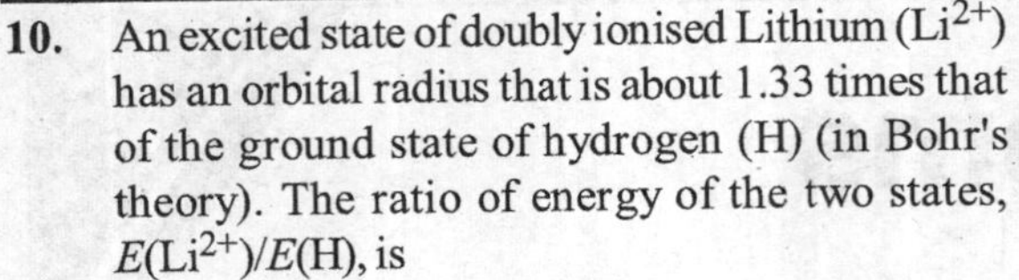

10. An excited state of doubly ionised Lithium (Li^{2+}) has an orbital radius that is about 1.33 times that of the ground state of hydrogen (H) (in Bohr's theory). The ratio of energy of the two states, $E(Li^{2+})/E(H)$, is
(a) 2.25 (b) 4.5 (c) 1 (d) 9

11. If elements with principal quantum number $n > 4$ were not allowed in nature, the number of possible elements would be
(a) 60 (b) 32 (c) 4 (d) 64

12. In the Bohr's model of hydrogen-like atom the force between the nucleus and the electron is modified as $F = \frac{e^2}{4\pi\varepsilon_0}\left(\frac{1}{r^2} + \frac{\beta}{r^3}\right)$, where β is a constant. For this atom, the radius of the nth orbit in terms of the Bohr radius $\left(a_0 = \frac{\varepsilon_0 h^2}{m\pi e^2}\right)$ is :

Critical Thinking

(a) $r_n = a_0 n - \beta$ (b) $r_n = a_0 n^2 + \beta$
(c) $r_n = a_0 n^2 - \beta$ (d) $r_n = a_0 n + \beta$

13. Consider the spectral line resulting from the transition $n = 2 \rightarrow n = 1$ in the atoms and ions given below. The shortest wavelength is produced by
(a) Hydrogen atom
(b) Deuterium atom
(c) Singly ionized Helium
(d) Doubly ionised Lithium

14. In the Bohr model of a hydrogen atom, the centripetal force is furnished by the coulomb attraction between the proton and the electron. If a_0 is the radius of the ground state orbit, m is the mass, e is the charge on the electron and ε_0 is the vacuum permittivity, the speed of the electron is

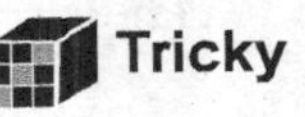

Tricky

(a) 0 (b) $\frac{e}{\sqrt{\varepsilon_0 a_0 m}}$
(c) $\frac{e}{\sqrt{4\pi\varepsilon_0 a_0 m}}$ (d) $\frac{\sqrt{4\pi\varepsilon_0 a_0 m}}{e}$

15. A hydrogen atom and a doubly ionized lithium atom are both in the second excited state. If L_H and L_{Li} respectively represent their electronic angular momenta and E_H and E_{Li} their energies, then
(a) $L_H > L_{Li}$ and $|E_H| > |E_{Li}|$
(b) $L_H = L_{Li}$ and $|E_H| < |E_{Li}|$
(c) $L_H = L_{Li}$ and $|E_H| > |E_{Li}|$
(d) $L_H < L_{Li}$ and $|E_H| < |E_{Li}|$

16. As per Bohr model, the minimum energy (in eV) required to remove an electron from the ground state of doubly ionized Li atom ($Z = 3$) is
(a) 1.51 (b) 13.6 (c) 40.8 (d) 122.4

17. Ionization energy of a hydrogen-like ion A is greater than that of another hydrogen-like ion B. If r, u, E and L represent the radius of the orbit, speed of the electron, energy of the atom and orbital angular momentum of the electron respectively then in ground state
(a) $r_A > r_B$ (b) $u_A > u_B$
(c) $E_A > E_B$ (d) $L_A > L_B$

18. An electron in a hydrogen atom makes a transition from first excited state to ground state. The equivalent current due to circulating electron –
(a) increases 2 times (b) increases 4 times
(c) increases 8 times (d) remains the same

19. According to Bohr's theory, the time averaged magnetic field at the centre (i.e. nucleus) of a hydrogen atom due to the motion of electrons in the nth orbit is proportional to : (n = principal quantum number)
(a) n^{-4} (b) n^{-5} (c) n^{-3} (d) n^{-2}

20. The momentum of hydrogen atom when a photon is emitted in a transition from $n_i = 10$ to $n_f = 1$ is –
(a) 7×10^{-27} kg/ms (b) 7×10^{27} kg/ms
(c) 3.5×10^{-27} kg/ms (d) 3.5×10^{27} kg/ms

21. An α-particle of energy 5 MeV is scattered through 180° by a fixed uranium nucleus. The distance of closest approach is of the order of
(a) 10^{-12} cm (b) 10^{-10} cm
(c) 10^{-14} cm (d) 10^{-15} cm

22. In a hydrogen atom following the Bohr's postulates the product of linear momentum and angular momentum is proportional to $(n)^x$ where 'n' is the orbit number. Then 'x' is-
(a) 0 (b) 2 (c) –2 (d) 1

23. A hydrogen atom makes a transition from n = 2 to n = 1 and emits a photon. This photon strikes a doubly ionized lithium atom (z = 3) in excited state and completely removes the orbiting electron. The least quantum number for the excited state of the ion for the process is :
(a) 2 (b) 4 (c) 5 (d) 3

24. In the hydrogen atom, an electron makes a transition from n = 2 to n = 1. The magnetic field produced by the circulating electron at the nucleus –

Critical Thinking

(a) decreases 16 times (b) increases 4 times
(c) decreases 4 times (d) increases 32 times

25. Suppose that a material emits X–rays of wavelengths $\lambda_{K_\alpha}, \lambda_{K_\beta}, \lambda_{L_\alpha}$, when it is excited by fast moving electrons; the wavelengths corresponding to K_α, K_β, L_α X– rays of the material respectively. Then we can write
(a) $\lambda_{K_\beta} = \lambda_{K_\alpha} + \lambda_{L_\alpha}$
(b) $\sqrt{\lambda_{K_\beta}} = \sqrt{\lambda_{K_\alpha}} + \sqrt{\lambda_{L_\alpha}}$
(c) $\frac{1}{\lambda_{K_\beta}} = \frac{1}{\lambda_{K_\alpha}} + \frac{1}{\lambda_{L_\alpha}}$
(d) $\frac{1}{\sqrt{\lambda_{K_\beta}}} = \frac{1}{\sqrt{\lambda_{K_\alpha}}} + \frac{1}{\sqrt{\lambda_{L_\alpha}}}$

26. Which of the following statements are true regarding Bohr's model of hydrogen atom?
(I) Orbiting speed of electron decreases as it shifts to discrete orbits away from the nucleus
(II) Radii of allowed orbits of electron are proportional to the principal quantum number
(III) Frequency with which electrons orbit around the nucleus in discrete orbits is inversely proportional to the cube of principal quantum number
(IV) Binding force with which the electron is bound to the nucleus increases as it shifts to outer orbits
Select correct answer using the codes given below.
Codes :
(a) I and II (b) II and IV
(c) I, II and III (d) II, III and IV

27. Suppose an electron is attracted towards the origin by a force $\frac{k}{r}$ where k is a constant and r is the distance of the electron from the origin. By applying Bohr model to this system, the radius of the nth orbital of the electorn is found to be 'r_n' and the kinetic energy of the electron to be 'K_n'. Then which of the following is true
(a) K_n independent of n, $r_n \propto n$
(b) $K_n \propto \frac{1}{n}$, $r_n \propto n$
(c) $K_n \propto \frac{1}{n}$, $r_n \propto n^2$
(d) $K_n \propto \frac{1}{n^2}$, $r_n \propto n^2$

28. Find the maximum angular speed (in $\times 10^{16}$ rad/sec) of the electron of a hydrogen atom in a stationary orbit.
(a) 2.1 (b) 4.1 (c) 8.5 (d) 11.6

29. Assume in specific conditions only those transitions are allowed to hydrogen atoms in which the principal quantum number n changes by 2. Find the smallest wavelength emitted by hydrogen and wavelenght emitted by hydrogen in the visible range (380 nm to 780 nm). respectively (in nm)
(a) 103,487 (b) 123,532
(c) 211,410 (c) 320,435

30. An electron in a hydrogen atom makes a transition from $n = n_1$ to $n = n_2$. The time period of electron in the initial state is eight times that in the final state. Then which of the following statement is true ? **Tricky**
(a) $n_1 = 3n_2$ (b) $n_1 = 4n_2$
(c) $n_1 = 2n_2$ (d) $n_1 = 5n_2$

31. If the angular momentum of an electron in an orbit is J then the K.E. of the electron in that orbit is
(a) $\frac{J^2}{2mr^2}$ (b) $\frac{Jv}{r}$ (c) $\frac{J^2}{2m}$ (d) $\frac{J^2}{2\pi}$

32. A neutron travelling with a velocity v and kinetic energy E has a perfectly elastic head-on collision with a nucleus of an atom of mass number A at rest. The fraction of total energy retained by the neutron is approximately
(a) $[(A-1)/(A+1)]^2$ (b) $[(A+1)/(A-1)]^2$
(c) $[(A-1)/A]^2$ (d) $[(A+1)/A]^2$

33. In a Rutherford experiment, the number of particles scattered at 90° angle are 28 per minute then number of scattered particles at an angle 60° and 120° will be **Critical Thinking**
(a) 117 per minute, 25 per minute
(b) 50 per minute, 12.5 per minute
(c) 100 per minute, 200 per minute
(d) 112 per minute, 12.4 per minute

34. The energy difference between the first two levels of hydrogen atom is 10.2 eV for another element of atomic number 10 and mass number 20, this will be
(a) 2040 eV (b) 0.201 eV
(c) 510 eV (d) 1020 eV

35. If a proton had a radius R and the charge was uniformly distributed, calculate using Bohr theory, the total ground state energy of a H–atom when R = 0.1Å.
(a) –13.6 eV (b) –5.6 eV
(c) –3.67 eV (d) –2.67 eV

36. An unknown hot gas emits radiation of wavelengths 46.0 nm, 82.8 nm and 103.5 nm only. Assume that the atoms have only two excited states and the difference between consecutive energy levels decreases as energy is increased. Taking

the energy of the highest energy state to be zero, find the energies of the ground state and the first excited state.
(a) –27 eV, – 12 eV (b) –6 eV, – 3 eV
(c) –11 eV, – 8 eV (d) –9 eV, –3 eV

37. Consider 3rd orbit of He^+ (Helium), using non-relativistic approach, the speed of electron in this orbit will be [given $K = 9 \times 10^9$ constant, Z = 2 and h (Plank's Constant) $= 6.6 \times 10^{-34}$ J s]
(a) 1.46×10^6 m/s (b) 0.73×10^6 m/s
(c) 3.0×10^8 m/s (d) 2.92×10^6 m/s

38. A hydrogen atom in state n = 6 makes two successive transitions and reaches the ground state. In the first transition a photon of 1.13 eV is emitted. Find the energy of the photon emitted in the second transition and the value of n in the intermediate state.
(a) 12.1 eV, n = 3 (b) 6.1 eV, n = 4
(c) 2.1 e V, n = 3 (d) 1.1 eV, n = 5

39. Electrons in a certain energy level $n = n_1$, can emit 3 spectral lines. When they are in another energy level, $n = n_2$. They can emit 6 spectral lines. The orbital speed of the electrons in the two orbits are in the ratio of
(a) 4 : 3 (b) 3 : 4 (c) 2 : 1 (d) 1 : 2

40. An energy of 24.6 eV is required to remove one of the electrons from a neutral helium atom. The energy (in eV) required to remove both the electrons from a neutral helium atom is
(a) 38.2 (b) 49.2 (c) 51.8 (d) 79.0

41. The wavelength of K_a X-rays produced by an X-ray tube is 0.76 Å. Find the atomic number of the anode material of the tube ?
(a) 40 (b) 30 (c) 20 (d) 10

42. A hydrogen atom in its ground state absorbs 10.2 eV of energy. The orbital angular momentum is increased by
(a) 1.05×10^{-34} J-s (b) 3.16×10^{-34} J-s
(c) 2.11×10^{-34} J-s (d) 4.22×10^{-34} J-s

43. An α particle passes rapidly through the exact centre of a hydrogen molecule, moving on a line perpendicular to the internuclear axis. The distance between the nuclei is *b*. Where on its path does the α particle experience the greatest force? (Assume that the nuclei do not move much during the passage of the α particle. Also neglect the electric field of the electrons in the molecule.)

(a) $\frac{b}{2}$
(b) $\frac{b}{2\sqrt{2}}$
(c) $\frac{b}{\sqrt{2}}$
(d) None of these

+e
+2e
b
α -particle
+e
H_2 molecule

44. The element which has a K_α x-rays line of wavelength 1.8 Å is ($R = 1.1 \times 10^{-7}$ m^{-1}, $b = 1$ and $\sqrt{5/33} = 0.39$)
(a) Co, Z = 27 (b) Iron, Z = 26
(c) Mn, Z = 25 (d) Ni, Z = 28

45. Which of the following transitions in hydrogen atoms emit photons of highest frequency?
(a) $n = 1$ to $n = 2$ (b) $n = 2$ to $n = 6$
(c) $n = 6$ to $n = 2$ (d) $n = 2$ to $n = 1$

46. In the Rutherford experiment, α-particles are scattered from a nucleus as shown. Out of the four paths, which path is not possible?

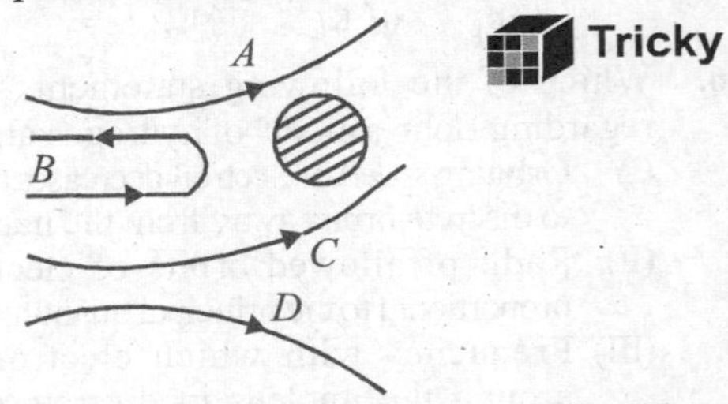

(a) *D* (b) *B* (c) *C* (d) *A*

47. Electrons are bombarded to excite hydrogen atoms and six spectral lines are observed. If E_g is the ground state energy of hydrogen, the minimum energy the bombarding electrons should posses is

Critical Thinking

(a) $\frac{8E_g}{9}$ (b) $\frac{15E_g}{16}$ (c) $\frac{35E_g}{36}$ (d) $\frac{48E_g}{49}$

48. In Rutherford scattering experiment, what will be the correct angle for α-scattering for an impact parameter, $b = 0$?
(a) 90° (b) 270° (c) 0° (d) 180°

49. If the atom ${}_{100}Fm^{257}$ follows the Bohr model and the radius of ${}_{100}Fm^{257}$ is *n* times the Bohr radius, then find *n*.
(a) 100 (b) 200 (c) 4 (d) 1/4

50. In Bohr theory of hydrogen atom, let *r*, *v* and *E* be the radius of orbit, speed of electron and the total energy of the electron respectively. Which of the following quantities is proportional to the quantum number n?
(a) *vr* (b) *rE* (c) *r/E* (d) *r/v*

51. The ionization energy of a hydrogen-like Bohr atom is 4 Rydbergs. Find the wavelength of radiation emitted when the electron jumps from the first excited state to the ground state:
[1 Rydberg = 2.2×10^{-18}, $h = 6.6 \times 10^{-34}$ Js, $c = 3 \times 10^8$ m/s.
Bohr radius of hydrogen atom = 5×10^{-11}m]
(a) 400 Å (b) 300 Å (c) 500 Å (d) 600 Å

52. An α-particle of 10 MeV collides head-on with a copper nucleus (Z = 29) and is deflected back. Then, the minimum distance of approach between the centres of the two is:
(a) 8.4×10^{-15} cm (b) 8.4×10^{-15} m
(c) 4.2×10^{-15} m (d) 4.2×10^{-15} cm

53. In Rutherford scattering experiment, the number of α-particles scattered at 60° is 5×10^6. The number of α-particles scattered at 120° will be
(a) 15×10^6 (b) $\frac{3}{5} \times 10^6$
(c) $\frac{5}{9} \times 10^6$ (d) None of these

54. Suppose potential energy between electron and proton at separation r is given by $U = K \ln(r)$, where K is a constant. For such a hypothetical hydrogen atom, the ratio of energy difference between energy levels ($n = 1$ and $n = 2$) and ($n = 2$ and $n = 4$) is

Critical Thinking

(a) 1 (b) 2 (c) 3 (d) 4

55. The energy of a hydrogen atom in the first excited state if the potential energy is taken to be zero in the ground state.
(a) 23.8 eV (b) 50.2 eV (c) 10.3 eV (d) 6.3 eV

56. If potential energy between a proton and an electron is given by $|U| = ke^2/2R^3$, where e is the charge of electron and R is the radius of atom, then radius of Bohr's orbit is given by (h = Planck's constant, k = constant)

Critical Thinking

(a) $\frac{ke^2m}{h^2}$ (b) $\frac{6\pi^2}{n^2}\frac{ke^2m}{h^2}$
(c) $\frac{2\pi}{n}\frac{ke^2m}{h^2}$ (d) $\frac{4\pi^2ke^2m}{n^2h^2}$

57. In a hypothetical system, a particle of mass m and charge –3q is moving around a very heavy particle charge q. Assume that Bohr's model is applicable to this system, then velocity of mass m in the first orbit is
(a) $\frac{3q^2}{2\varepsilon_0 h}$ (b) $\frac{3q^2}{4\varepsilon_0 h}$
(c) $\frac{3q}{2\pi\varepsilon_0 h}$ (d) $\frac{3q}{4\pi\varepsilon_0 h}$

58. The energy of an electron in an excited hydrogen atom is –3.4 eV. Then, according to Bohr's theory, the angular momentum of this electron, in Js, is
(a) 2.11×10^{-34} (b) 3×10^{-34}
(c) 1.055×10^{-34} (d) 0.5×10^{-34}

59. Energy required for the electron excitation in Li^{++} from the first to the third Bohr orbit is :
(a) 36.3 eV (b) 108.8 eV
(c) 122.4 eV (d) 12.1 eV

60. Hydrogen ($_1H^1$), Deuterium ($_1H^2$), singly ionised Helium $(_2He^4)^+$ and doubly ionised lithium $(_3Li^6)^{++}$ all have one electron around the nucleus. Consider an electron transition from n = 2 to n = 1. If the wavelengths of emitted radiation are λ_1, λ_2, λ_3 and λ_4 respectively then approximately which one of the following is correct?
(a) $4\lambda_1 = 2\lambda_2 = 2\lambda_3 = \lambda_4$
(b) $\lambda_1 = 2\lambda_2 = 2\lambda_3 = \lambda_4$
(c) $\lambda_1 = \lambda_2 = 4\lambda_3 = 9\lambda_4$
(d) $\lambda_1 = 2\lambda_2 = 3\lambda_3 = 4\lambda_4$

61. Hydrogen atom is excited from ground state to another state with principal quantum number equal to 4. Then the number of spectral lines in the emission spectra will be :
(a) 2 (b) 3 (c) 5 (d) 6

62. As an electron makes a transition from an excited state to the ground state of a hydrogen - like atom/ion :
(a) kinetic energy decreases, potential energy increases but total energy remains same
(b) kinetic energy and total energy decrease but potential energy increases
(c) its kinetic energy increases but potential energy and total energy decrease
(d) kinetic energy, potential energy and total energy decrease

63. In a hydrogen like atom electron make transition from an energy level with quantum number n to another with quantum number (n – 1). If n>>1, the frequency of radiation emitted is proportional to :
(a) $\frac{1}{n}$ (b) $\frac{1}{n^2}$ (c) $\frac{1}{n^{3/2}}$ (d) $\frac{1}{n^3}$

64. A diatomic molecule is made of two masses m_1 and m_2 which are separated by a distance r. If we calculate its rotational energy by applying Bohr's rule of angular momentum quantization, its energy will be given by : (n is an integer)

Toughnut

(a) $\frac{(m_1+m_2)^2n^2h^2}{2m_1^2m_2^2r^2}$ (b) $\frac{n^2h^2}{2(m_1+m_2)r^2}$
(c) $\frac{2n^2h^2}{(m_1+m_2)r^2}$ (d) $\frac{(m_1+m_2)n^2\hbar^2}{2m_1m_2r^2}$

65. Some energy levels of a molecule are shown in the figure. The ratio of the wavelengths $r = \lambda_1 / \lambda_2$, is given by

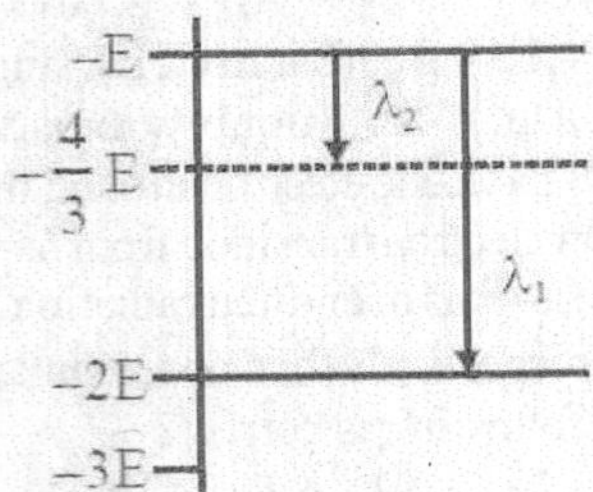

(a) $r = \frac{3}{4}$ (b) $r = \frac{1}{3}$ (c) $r = \frac{4}{3}$ (d) $r = \frac{2}{3}$

66. Consider a 3-level system with energies E_1, E_2 and E_3 in ascending order. λ_1, λ_2 and λ_3 are the wavelengths of radiation corresponding to the transitions $E_2 \to E_1$, $E_3 \to E_2$ and $E_3 \to E_1$ respectively. The wavelengths are related by

(a) $\lambda_1 = \lambda_2\lambda_3 / (\lambda_3 - \lambda_2)$
(b) $\lambda_2 = \lambda_3 - \lambda_1$
(c) $\lambda_2 = \lambda_1\lambda_3 / (\lambda_1 + \lambda_3)$
(d) $\lambda_3 = \lambda_1\lambda_2 / (\lambda_1 + \lambda_2)$

67. Ultraviolet light of wavelengths λ_1 and λ_2 when allowed to fall on hydrogen atoms in their ground state is found to liberate electrons with kinetic energy $K.E_1$ and $K.E_2$ respectively. Find the value of Planck's constant. **Tricky**

(a) $h = \left|\frac{(KE_2 - KE_1)(\lambda_1 + \lambda_2)}{C(\lambda_1 - \lambda_2)}\right|$

(b) $h = \left|\frac{(KE_1 - KE_2)(\lambda_2 - \lambda_1)}{C\lambda_1\lambda_2}\right|$

(c) $h = \left|\frac{(KE_1 - KE_2)\lambda_1\lambda_2}{C(\lambda_2 - \lambda_1)}\right|$

(d) None of these

68. In an experiment on photoelectric effect photons of wavelength 300 nm eject electrons from a metal of work function 2.25eV. A photon of energy equal to that of the most energetic electron corresponds to the following transition in the hydrogen atom:

(a) n = 2 to n = 1 state. (b) n = 3 to n = 1 state.
(c) n = 3 to n = 2 state. (d) n = 4 to n = 3 state.

69. Electrons in hydrogen like atom ($Z = 3$) make transitions from the fifth to the fourth orbit and from the fourth to the third orbit. The resulting radiations are incident normally on a metal plate and eject photoelectrons. The stopping potential for the photoelectrons ejected by the shorter wavelength is 3.95 volts. Find the stopping potential for the photoelectrons ejected by the longer wavelength, then. (Rydberg constant $= 1.094 \times 10^7 \, m^{-1}$) **Toughnut**

(a) 5V (b) 2 V (c) 0.754 V (d) 2.99 V

70. The ionization energy of a hydrogen like Bohr atom is 4 Rydbergs. Find the wavelength of the radiation emitted when the electron jumps from the first excited state to the ground state

(a) 300 A° (b) 2.5×10^{-11} m
(c) 100 A° (d) 1.5×10^{-11} m

71. A neutron of kinetic energy 65eV collides inelastically with a singly ionized helium atom at rest. It is scattered at an angle of 90° with respect of its original direction. If the atom get de-excited subsequently by emitting radiation, then the impossible frequency of the emitted radiation is [Given : Mass of He atom = 4 × (mass of neutron), Ionization energy of H atom = 13.6eV]

(a) 1.82×10^{15} Hz (b) 6.11×10^{15} Hz
(c) 11.67×10^{15} Hz (d) 9.84×10^{15} Hz

72. An electron, in a hydrogen-like atom, is in an excited state. It has a total energy of –3.4 eV. The kinetic energy and the de-Broglie wavelength of the electron are respectively

(a) +3.4 eV, 0.66×10^{-9} m
(b) –3.4 eV, 1.99×10^{-9} m
(c) 2.8 eV, 2.38×10^{-10} m
(d) 1.1 eV, 1.28×10^{-9} m

73. The wavelength of radiation is λ_0 when an electron jumps from third to second orbit of hydrogen atom. For the electron to jump from the fourth to the second orbit of the hydrogen atom, the wavelength of radiation emitted will be

(a) $\frac{16}{25}\lambda_0$ (b) $\frac{20}{27}\lambda_0$ (c) $\frac{27}{20}\lambda_0$ (d) $\frac{25}{16}\lambda_0$

74. In the hydrogen atom spectrum λ_{3-1} and λ_{2-1} represent wavelengths emitted due to transition from second and first excited states to the ground state respectively. The value of $\frac{\lambda_{3-1}}{\lambda_{2-1}}$ is

(a) 27/32 (b) 32/27 (c) 4/9 (d) 9/4

75. The transition from the state $n = 4$ to $n = 3$ in a hydrogen-like atom results in ultraviolet radiation. Infrared radiation will be obtained in the transition

(a) $2 \rightarrow 1$ (b) $3 \rightarrow 2$ (c) $4 \rightarrow 2$ (d) $5 \rightarrow 4$

76. One of the lines in the emission spectrum of Li^{2+} has the same wavelength as that of the Ist line of Lyman series in hydrogen spectrum. The electronic transition corresponding to this line is $n = 6 \rightarrow 3 = x$. Find the value of x.

(a) 8 (b) 3 (c) 7 (d) 5

77. The ionisation potential of H-atom is 13.6 V. When it is excited from ground state by monochromatic radiations of 970.6 Å, the number of emission lines will be (according to Bohr's theory)

(a) 10 (b) 8 (c) 6 (d) 4

78. The ratio of frequencies of the shortest wavelengths of Balmer and Lyman series of hydrogen atom is

(a) 4 : 1 (b) 1 : 4 (c) 27 : 5 (d) 5 : 27

79. A Spectroscopic instrument can resolve two nearby wavelength λ and $\lambda + \Delta\lambda$ if $\lambda/\Delta\lambda$ is smaller than 8000. This is used to study the spectral lines of the Balmer series of hydrogen. Approximately how many lines will be resolved by the instrument ?

(a) 60 (b) 43 (c) 38 (d) 21

80. The difference between the longest wavelength line of the Balmer series and shortest wavelength line of the Lyman series for a hydrogenic atom (atomic number Z) equal to $\Delta\lambda$. The value of the Rydberg constant for the given atom is :

(a) $\frac{5}{31}\frac{1}{\Delta\lambda . Z^2}$ (b) $\frac{5}{36}\frac{Z^2}{\Delta\lambda}$

(c) $\frac{31}{5}\frac{1}{\Delta\lambda . Z^2}$ (d) none of these

81. A hydrogen atom is in ground state. Then to get six lines in emission spectrum, wavelength of incident radiation should be

(a) 800 Å (b) 825 Å (c) 975 Å (d) 01025 Å

82. The ratio of minimum to maximum wavelengths of radiation that an excited electron in a hydrogen atom can emit while going to the ground state is

(a) 1/2 (b) Zero (c) 3/4 (d) 27/32

83. Whenever a photon is emitted by hydrogen in Balmer series, it is followed by another photon in Lyman series What wavelength does this latter photon correspond to?

(a) 122 nm (b) 221 A°

(c) 321 nm (d) 111 A°

84. In hydrogen spectrum the wavelength of H_α line is 656 nm, whereas in the spectrum of a distant galaxy, H_α line wavelength is 706 nm. Estimated speed of the galaxy with respect to earth is,

(a) 2×10^8 m/s (b) 2×10^7 m/s

(c) 2×10^6 m/s (d) 2×10^5 m/s

85. The first member of Balmer series of hydrogen has a wavelength of 6563 Å the wavelength of its second member will be

(a) 4861 Å (b) 6563 Å (c) 3561 Å (d) 1215Å

86. When a gas of hydrogen-like ions is prepared in a particular excited state X. It emits photons having wavelength equal to the wavelength of the first line of the Lyman series together with photons of five other wavelengths Identify the gas and find the principal quantum number of the state X respectively.

(a) $He^+, 4$ (b) $Li^{++}, 3$

(c) $H^+, 2$ (d) None of these

87. Taking Rydberg's constant $R_H = 1.097 \times 10^7$m, first and second wavelength of Balmer series in hydrogen spectrum is

(a) 2000 Å, 3000 Å (b) 1575 Å, 2960 Å

(c) 6529 Å, 4280 Å (d) 6552 Å, 4863 Å

88. The energy of electron in the nth orbit of hydrogen atom is expressed as $E_n = \frac{-13.6}{n^2}$ eV.

The shortest and longest wavelength of Lyman series will be

(a) 910 Å, 1213 Å (b) 5463 Å, 7858 Å

(c) 1315 Å, 1530 Å (d) None of these

89. A 12.5 eV electron beam is used to bombard gaseous hydrogen at room temperature. Hydrogen atoms would be excited upto nth energy level the hence the wavelength obtained first member of Balmer series is λ_B, then.

(a) $\lambda_B = 1415$Å (b) $\lambda_B = 1215$Å

(c) $\lambda_B = 6563$Å (d) $\lambda_B = 8523$Å

90. If the wavelength of the first line of the Balmer series in the hydrogen spectrum is λ, then the wavelength of the first line of the Lyman series is

(a) $(27/5)\lambda$ (b) $(5/27)\lambda$

(c) $(32/27)\lambda$ (d) $(27/32)\lambda$

91. One of the lines in the emission spectrum of Li^{2+} has the same wavelength as that of the 2nd line of Balmer series in hydrogen spectrum. The electronic transition corresponding to this line is

(a) $n = 4 \rightarrow n = 2$ (b) $n = 8 \rightarrow n = 2$

(c) $n = 8 \rightarrow n = 4$ (d) $n = 12 \rightarrow n = 6$

92. Hydrogen atom excites energy level from fundamental state to n = 3. Number of spectral lines according to Bohr, is **Tricky**

(a) 4 (b) 3 (c) 1 (d) 2

93. When an electron in a hydrogen atom makes a transition from 2nd excited stated to ground state it emits a photon of frequency f. The frequency of photon emitted when an electron of Li^{++} makes a transition from Ist excited state to ground state is

(a) $\frac{243}{32}f$ (b) $\frac{141}{32}f$ (b) $\frac{81}{32}f$ (d) $\frac{63}{32}f$

94. The photon radiated from hydrogen corresponding to 2nd line of Lyman series is absorbed by a hydrogen like atom X in 2nd excited state. As a result the hydrogen like atom X makes a transition to nth orbit. Then –

(a) $X = He^+, n = 4$ (b) $X = Li^{++}, n = 6$

(c) $X = He^+, n = 6$ (d) $X = Li^{++}, n = 9$

95. A Spectroscopic instrument can resolve two nearby wavelength λ and $\lambda + \Delta\lambda$ if $\lambda/\Delta\lambda$ is smaller than 8000. This is used to study the spectral lines of the Balmer series of hydrogen. Approximately how many lines will be resolved by the instrument?

(a) 60 (b) 43 (c) 38 (d) 21

96. If the series limit wavelength of Lyman series for the hydrogen atom is 912 Å, then the series limit wavelength for Balmer series of hydrogen atoms is

(a) 912 Å (b) 912 × 2 Å

(c) 912 × 4 Å (d) $\frac{912}{2}$ Å

97. The third line of the Balmer series spectrum of a hydrogen like ion of atomic number Z equals to 108.5 nm. Then Z is

(a) 2 (b) 5 (c) 3 (d) 6

98. The ground state energy of hydrogen atom is–13.6 eV. If an electron makes a transition from an energy level – 0.85 eV to – 1.51 eV, calculate the wavelength λ of the spectral line emitted.

(a) 1.88×10^{-6} m (b) 1.87×10^{-10} m

(c) 1.66×10^{-9} m (d) 2.36×10^{-9} m

99. If the electron revolving around the nucleus in a radius 'r' with orbital speed 'v' has magnetic moment $evr/2$. Hence, using Bohr's postulate of the quantization of angular momentum obtain the magnetic moment (M) of hydrogen atom in its ground state and current (I) due to revolution of electron.

(a) $M = \frac{eh}{4\pi m}, I = \frac{eV}{2\pi r}$ (b) $M = \frac{2eh}{5\pi m}, I = \frac{eV}{4\pi r}$

(c) $M = \frac{h}{\pi m}, I = \frac{e}{\pi r}$ (d) $M = \frac{eh}{\pi m}, I = \frac{eV}{\pi r}$

100. If radiation corresponding to first line of "Balmer series" of He^+ ion knocked out electron from 1st excited state of H atom, the kinetic energy of ejected electron from H atom would be (eV) –

[Given $E_n = -\frac{Z^2}{n^2}(13.6 \text{ eV})$]

(a) 4.155 eV (b) 8.310 eV

(c) 2.515 eV (d) 5.550 eV

Answer KEYs

1	(a)	11	(a)	21	(a)	31	(a)	41	(a)	51	(b)	61	(d)	71	(b)	81	(c)	91	(d)
2	(c)	12	(c)	22	(a)	32	(a)	42	(a)	52	(b)	62	(c)	72	(a)	82	(c)	92	(b)
3	(d)	13	(d)	23	(b)	33	(d)	43	(b)	53	(c)	63	(d)	73	(b)	83	(a)	93	(a)
4	(b)	14	(c)	24	(d)	34	(d)	44	(a)	54	(a)	64	(d)	74	(a)	84	(b)	94	(d)
5	(a)	15	(b)	25	(c)	35	(a)	45	(d)	55	(a)	65	(b)	75	(d)	85	(a)	95	(c)
6	(c)	16	(d)	26	(a)	36	(a)	46	(b)	56	(b)	66	(d)	76	(b)	86	(a)	96	(c)
7	(c)	17	(b)	27	(a)	37	(a)	47	(d)	57	(a)	67	(c)	77	(c)	87	(d)	97	(a)
8	(c)	18	(c)	28	(b)	38	(a)	48	(a)	58	(a)	68	(c)	78	(a)	88	(a)	98	(a)
9	(b)	19	(d)	29	(a)	39	(a)	49	(a)	59	(b)	69	(c)	79	(c)	89	(c)	99	(a)
10	(a)	20	(a)	30	(c)	40	(d)	50	(a)	60	(c)	70	(a)	80	(c)	90	(b)	100	(a)

Hints & Solutions

1. (a) Energy of electron in n^{th} orbit is

$$E_n = -(Rch)\frac{Z^2}{n^2} = -54.4 \text{ eV}$$

For He^+ is ground state

$$E_1 = -(Rch)\frac{(2)^2}{(1)^2} = -54.4 \Rightarrow Rch = 13.6$$

$\therefore$ For Li^{++} in first excited state (n = 2)

$$E' = -13.6 \times \frac{(3)^2}{(2)^2} = -30.6 \text{ eV}$$

2. (c) Using $\frac{1}{\lambda} = R(Z-1)^2\left[\frac{1}{n_2^2} - \frac{1}{n_1^2}\right]$

For α particle, $n_1 = 2,\ n_2 = 1$

For metal A :

$$\frac{1875\ R}{4} = R\ (Z_1 - 1)^2\left(\frac{3}{4}\right) \Rightarrow Z_1 = 26$$

For metal B :

$$675\ R = R\ (Z_2 - 1)^2\left(\frac{3}{4}\right) \Rightarrow Z_2 = 31$$

Therefore, 4 elements lie between A and B.

3. (d) Radius of nth orbit $r_n \propto n^2$, graph between r_n and n is a parabola. Also,

$$\frac{r_n}{r_1} = \left(\frac{n}{1}\right)^2 \Rightarrow \log_e\left(\frac{r_n}{r_1}\right) = 2\log_e(n)$$

Comparing this equation with y = mx + c,

Graph between $\log_e\left(\frac{r_n}{r_1}\right)$ and $\log_e(n)$ will be a straight line, passing from origin.

Similarly it can be proved that graph between $\log_e\left(\frac{f_n}{f_1}\right)$ and $\log_e n$ is a straight line. But with negative slops.

4. (b)

5. (a) Excitation energy $\Delta E = E_2 - E_1 = 13.6\ Z^2 \left[\frac{1}{1^2} - \frac{1}{2^2}\right]$

$$\Rightarrow 40.8 = 13.6 \times \frac{3}{4} \times Z^2 \Rightarrow Z = 2.$$

Now required energy to remove the electron from ground state

$$= \frac{+13.6Z^2}{(1)^2} = 13.6(Z)^2 = 54.4 \text{ eV.}$$

6. (c) $\Delta E = E_0\left[\frac{1}{n_1^2} - \frac{1}{n_2^2}\right]$

here (I^{st}) transition represent absorption not emission. So, (c) is correct option.

7. (c) $qVB = \frac{mv^2}{r}$ and $\frac{nh}{2\pi} = mvr$

Multiplying equation (i) and (ii),

$$\frac{qBnh}{2\pi} = m^2v^2$$

Now multiplying both sides by $\frac{1}{2m}$,

$$n\frac{qBh}{4\pi m} = \frac{1}{2}mv^2 \quad \text{i.e. KE} = n\left[\frac{qBh}{4\pi m}\right]$$

8. (c) Speed of electron in first orbit (n = 1) of hydrogen atom (z = 1),

$$v = \frac{e^2}{2\varepsilon_0 h} \qquad r = \frac{h^2\varepsilon_0}{\pi me^2} \Rightarrow \varepsilon_0 = \frac{r\pi me^2}{h^2}$$

Acceleration of electron,

$$\frac{v^2}{r} = \frac{e^4}{4\varepsilon_0^2 h^2} \times \frac{\pi me^2}{h^2\varepsilon_0}$$

$$= \frac{e^4 \times \pi me^2}{4h^4\varepsilon_0^3} \qquad \text{.......(ii)}$$

eliminating ε_0

$$= \frac{e^4\ \pi me^2 h^6}{4h^4 r^3 \pi^3 m^3 e^6} = \frac{h^2}{4\pi^2 m^2 r^3}$$

9. (b) 10. (a)

11. (a) The maximum number of electrons in an orbit is $2n^2$. $n > 4$ is not allowed.

Therefore the number of maximum electron that can be in first four orbits are

$$2\,(1)^2 + 2\,(2)^2 + 2\,(3)^2 + 2\,(4)^2$$

$$= 2 + 8 + 18 + 32 = 60$$

Therefore, possible element are 60.

12. (c) As $F = \frac{mv^2}{r} = \frac{e^2}{4\pi\varepsilon_0}\left(\frac{1}{r^2} + \frac{\beta}{r^3}\right)$

and $mvr = \frac{nh}{2\pi} \Rightarrow v = \frac{nh}{2\pi mr}$

$$\therefore\ m\left(\frac{nh}{2\pi mr}\right)^2 \times \frac{1}{r} = \frac{e^2}{4\pi\epsilon_0}\left(\frac{1}{r^2}+\frac{\beta}{r^3}\right)$$

or, $\frac{1}{r^2}+\frac{\beta}{r^3} = \frac{mn^2h^2 4\pi\epsilon_0}{4\pi^2 m^2 e^2 r^3}$

or, $\frac{a_0 n^2}{r^3} = \frac{1}{r^2}+\frac{\beta}{r^3}$ $\left(\because a_0 = \frac{\epsilon_0 h^2}{m\pi e^2}\ \text{Given}\right)$

$\therefore\ r = a_0 n^2 - \beta$

13. (d)

14. (c) Centripetal force = Coulombian force

$$\frac{mv^2}{a_0} = \frac{1}{4\pi\varepsilon_0}\cdot\frac{e\times e}{a_0^2}$$

$$\Rightarrow v^2 = \frac{e^2}{4\pi\varepsilon_0 . a_0 . m} \Rightarrow v = \frac{e}{\sqrt{4\pi\varepsilon_0 . a_0 . m}}$$

15. (b)

16. (d) $E_n = -13.6\frac{(Z^2)}{(n^2)}$ eV

Therefore, ground state energy of doubly ionized lithium atom ($Z = 3, n = 1$) will be

$$E_1 = (-13.6)\frac{(3)^2}{(1)^2} = -122.4 \text{ eV}$$

$\therefore$ Ionization energy of an electron in ground state of doubly ionized lithium atom will be 122.4eV.

17. (b)

18. (c) $i = \frac{q}{T}$ Now, $T^2 \propto r^3 \propto n^6 \Rightarrow i \propto \frac{1}{n^3}$

19 (d) Magnetic field at the centre of nucleus of H-atom,

$$B = \frac{\mu_0 I}{2r} \quad \text{..... (i)}$$

According to Bohr's model, radius of orbit $r \propto n^2$ from eq. (i) we can also write as $B \propto n^{-2}$

20. (a)

21. (a) Distance of closest approach

$$r_0 = \frac{Ze(2e)}{4\pi\varepsilon_0\left(\frac{1}{2}mv^2\right)}$$

Energy, $E = 5\times 10^6 \times 1.6\times 10^{-19}$ J

$$\therefore r_0 = \frac{9\times10^9 \times (92\times1.6\times10^{-19})(2\times1.6\times10^{-19})}{5\times10^6\times1.6\times10^{-19}}$$

$\Rightarrow r = 5.2\times 10^{-14}\ m = 5.3\times 10^{-12}$ cm.

22. (a) Linear momentum $\Rightarrow mv\ \alpha\ 1/n$

angular momentum $\Rightarrow mv\ \alpha n$

$\therefore$ product of linear momentum and angular momentum αn^0.

23. (b)

24. (d) $\because$ $B = \frac{\mu_0 I}{2r}$ and $I = \frac{e}{T}$

$B = \frac{\mu_0 e}{2rT}$ $[r \propto n^2, T \propto n^5]$;

$B \propto \frac{1}{n^5}$

25. (c) The energy level diagram of the atom is shown in the figure. It is clear that

$E_{K\beta} = E_{K\alpha} + E_{L\alpha}$

or, $\nu_{K\beta} = \nu_{K\alpha} + \nu_{L\alpha}$

or, $\frac{1}{\lambda_{K\beta}} = \frac{1}{\lambda_{K\alpha}} + \frac{1}{\lambda_{L\alpha}}$

K_α K_β L_α

26. (a) Orbital speed varies inversely as the radius of the orbit.

$v \propto \frac{1}{n}$

27. (a) $\frac{mv^2}{r} = \frac{K}{r}$ or $v = \sqrt{\frac{K}{m}} = cr^\circ$

Now $mvr = \frac{nh}{2\pi}$ or $r = \frac{nh}{2\pi mv}$

$\therefore$ $r \propto n.$

Also kinetic energy

$K = \frac{mv^2}{2} = \frac{m}{2}\times\frac{K}{m} = \frac{K}{2}.$

28. (b) **29.** (a)

30. (c) In the nth orbit, let r_n be the radius and v_n be the speed of electron.

Time period, $T_n = \frac{2\pi r_n}{v_n} \propto \frac{r_n}{v_n}$

Now $r_n \propto n^2$; $v_n \propto \frac{1}{n}$ $\therefore \frac{r_n}{v_n} \propto n^3$ or $T_n \propto n^3$

Here $8 = \left(\frac{n_1}{n_2}\right)^3$ or $\frac{n_1}{n_2} = 2$ i.e., $n_1 = 2n_2$

31. (a) Angular momentum = mrv = J $\therefore v = \frac{J}{mr}$

K. E. of electron $= \frac{1}{2}mv^2 = \frac{1}{2}m\left(\frac{J}{mr}\right)^2 = \frac{J^2}{2mr^2}$

32. (a) $v'_1 = \frac{(m_1 - m_2)v_1 + 2m_2v_2}{m_1+m_2}$

As v_2 is zero, $m_2 > m_1$, v'_1 is in the opposite direction.

$m_1 = 1, m_2 = A.$ $\therefore |v'_1| = \frac{(A-1)}{(A+1)}v_1$

The fraction of total energy retained is

$$\frac{1/2mv'^2_1}{1/2v_1^2} = \frac{(A-1)^2}{(A+1)^2}$$

33. (d) No. of particles scattered through an angle

$$\theta = N(\theta) = \frac{k\cancel{Z}^2}{\sin^4\left(\frac{\theta}{2}\right)(\text{K.E.})^2}$$

$$\therefore 28 = \frac{4kcz^2}{(\text{K.E.})^2} \text{ for } \theta = 90°$$

$$\therefore \frac{kz^2}{(\text{K.E.})^2} = \frac{28}{4} = 7$$

$$\therefore N(60°) = \frac{7}{\sin^4\left(\frac{60°}{2}\right)} = 16 \times 7 = 112\,/\text{min.}$$

$$N(120°) = \frac{7}{\sin^4\left(\frac{120°}{2}\right)} = 12.4\,/\,\text{min}$$

34. (d) For H atom

$E_2 - E_1 = 10.2$ eV

For $Z = 10$,

$$E_1 = \frac{-13.6 \times 10^2}{1^2} = -1360 \text{ eV}$$

$$E_2 = \frac{-13.6 \times 10^2}{2^2} = -3.4 \times 100 = -340 \text{ eV}$$

$$\therefore \quad E_2 - E_1 = -340 - (-1360) = 1020 \text{ eV}$$

35. (a) **36. (a)**

37. (a) Speed of electron in nth orbit

$$V_n = \frac{2\pi\, KZe^2}{nh}$$

$$V = (2.19 \times 10^6 \text{ m/s})\ \frac{Z}{n}$$

$$V = (2.19 \times 10^6)\ \frac{2}{3}\ (Z = 2\ \&\ n = 3)$$

$V = 1.46 \times 10^6$ m/s

38. (a) **39. (a)**

40. (d) For helium ion, $Z = 2$ and for first orbit $n = 1$.

$$\therefore \quad E_1 = \frac{-13.6}{(1)^2} \times 2^2 = -54.4 \text{ eV}$$

$\therefore$ Energy required to remove this $e^- = +54.4$ eV

$\therefore$ Total energy required $= 54.4 + 24.6 = 79$ eV

41. (a) $$\frac{1}{\lambda} = RZ^2\left(\frac{1}{1^2} - \frac{1}{2^2}\right) = \frac{3}{4}RZ^2$$

or $$Z^2 = \frac{4}{3R\lambda}$$

$$= \frac{4}{3\text{x}(1.097\text{x}10^7)\text{x}(0.76\text{x}10^{-10})} = 1599.25$$

or $Z^2 = 1600 \quad \therefore \quad Z = 40.$

42. (a) Electron after absorbing 10.2 eV energy goes to its first excited state (n = 2) from ground state (n = 1).

$$\therefore \text{ Increase in momentum } = \frac{h}{2\pi}$$

$$= \frac{6.6 \times 10^{-34}}{6.28} = 1.05 \times 10^{-34} \text{ J-s.}$$

43. (b) $$E_p = \frac{2K_{ex}}{\left(x^2 + \frac{b^2}{4}\right)^{3/2}}$$

For maximum E_P

$$\frac{dE_P}{dx} = 0$$

$$\Rightarrow x = \frac{b}{2\sqrt{2}}$$

44. (a) $$\frac{1}{\lambda_\alpha} = \frac{3R}{4}(Z-1)^2$$

$$\Rightarrow \quad (Z-1) = \sqrt{\frac{4}{3R\lambda_\alpha}} = \sqrt{\frac{4}{3 \times 1.1 \times 10^7 \times 1.8 \times 10^{-10}}}$$

$$= \frac{200}{3}\sqrt{\frac{5}{33}} = \frac{78}{3} = 26$$

$\Rightarrow \quad Z = 26 + 1 = 27$

45. (d) Frequency is given by

$$h\nu = -13.6\left(\frac{1}{n_2^2} - \frac{1}{n_1^2}\right)$$

For transition from $n = 6$ to $n = 2$,

$$\nu_1 = \frac{-13.6}{h}\left(\frac{1}{6^2} - \frac{1}{2^2}\right) = \frac{2}{9} \times \left(\frac{13.6}{h}\right)$$

For transition from n = 2 to n = 1,

$$\nu_2 = \frac{-13.6}{h}\left(\frac{1}{2^2} - \frac{1}{1^2}\right) = \frac{3}{4} \times \left(\frac{13.6}{h}\right).$$

$\therefore \nu_1 > \nu_2$

46. (c) As α-particles are doubly ionised helium He^{++} i.e. Positively charged and nucleus is also positively charged and we know that like charges repel each other.

47. (b) To obtain 6 spectral line, as electron must be excited to fourth orbit with energy $\frac{E_g}{16}$, so that the difference is $\frac{15E_g}{16}$.

48. (d) $$b = \frac{Ze^2 \cot\left(\frac{\theta}{2}\right)}{4\pi \epsilon_0 k_i} = 0 \Rightarrow \cot\left(\frac{\theta}{2}\right) = 0$$

$$\Rightarrow \frac{\theta}{2} = 90° \text{ or } \theta = 180°$$

49. (d)

50. (a) Note that v is inversely proportional to n whereas r is directly to n^2.

51. (b) The energy in the ground state

$E_1 = -4$ Rydberg $E_1 = -4 \times 2.2 \times 10^{-18}$J

The energy of the first excited state ($n = 2$)

$$E_2 = \frac{E_1}{4} = -2.2 \times 10^{-18} J$$

The energy difference

$\Delta E = E_2 - E_1 = 3 \times 2.2 \times 10^{-18}$J

Now, the wavelength of radiation emitted is

$$\lambda = \frac{hc}{\Delta E}$$

$$\lambda = \frac{6.6 \times 10^{-34} \times 3 \times 10^8}{3 \times 2.2 \times 10^{-18}} = 300 Å$$

52. (b)

53. (c) $N \propto \frac{1}{\sin^4 \theta/2}$; $\frac{N_2}{N_1} = \frac{\sin^4(\theta_1/2)}{\sin^4(\theta_2/2)}$

or $N_2 = 5 \times 10^6 \times \left(\frac{1}{2}\right)^4 \left(\frac{2}{\sqrt{3}}\right)^4 = \frac{5}{9} \times 10^6$

54. (a) $-\frac{dU}{dr} = F$ (conservative force field)

$\Rightarrow F = \frac{-K}{r}$ provides the centrifugal force for circular motion of electron.

$$\frac{mv^2}{r} = \frac{K}{r} \Rightarrow r = \frac{nh}{2\pi\sqrt{mK}}$$

K.E. of electron $= \frac{1}{2}mv^2 = \frac{1}{2}K$

P.E. of electron $= K \ln r$

$E(n) =$ Total energy $=$ K.E. + P.E.

$$= \frac{1}{2}K + K \ln r = \frac{K}{2}\left[1 + \log \frac{n^2h^2}{4\pi^3 mk}\right]$$

$$\text{Required ratio} = \frac{E(2) - E(1)}{E(4) - E(2)} = 1$$

55. (a)

56. (b) $U = -\frac{ke^2}{2R^2}$, $F = -\frac{dU}{dR} = \frac{3ke^2}{2R^4}$

But, $F = \frac{mv^2}{R} \Rightarrow \frac{mv^2}{R} = \frac{3ke^2}{2R^4}$

Also, $mvR = \frac{nh}{2\pi}$

Solve to get : $R = \frac{6\pi^2 ke^2 m}{n^2h^2}$

57. (a) $\frac{mv^2}{r} = \frac{3q^2}{4\pi\varepsilon_0 r^2} \Rightarrow mvr = \frac{3q^2}{4\pi\varepsilon_0 v}$ ...(i)

and $\frac{nh}{2\pi} = mvr$ ...(ii)

Using (i) and (ii) and putting n = 1

$$\frac{h}{2\pi} = \frac{3q^2}{4\pi\varepsilon_0 v} \Rightarrow v = \frac{3q^2}{2\varepsilon_0 h}$$

58. (a)

59. (b) Energy of excitation,

$$\Delta E = 13.6\, \pi^2 \left(\frac{1}{\eta_1} - \frac{1}{\eta_2}\right) eV$$

$$\Rightarrow \Delta E = 13.6\,(3)^2 \left(\frac{1}{1^2} - \frac{1}{3^2}\right) = 108.8\, eV$$

60. (c)

61. (d) The possible number of the spectral lines is given

$$= \frac{n(n-1)}{2} = \frac{4(4-1)}{2} = 6$$

62. (c) $U = -K\frac{ze^2}{r}$; T.E $= -\frac{k}{2}\frac{ze^2}{r}$

K.E $= \frac{k}{2}\frac{ze^2}{r}$. Here r decreases

63. (d) $\Delta E = hv$

$$v = \frac{\Delta E}{h} = k\left[\frac{1}{(n-1)^2} - \frac{1}{n^2}\right] = \frac{k(2n-1)}{n^2(n-1)^2}$$

$$\approx \frac{2k}{n^3} \quad \text{or,} \quad v \propto \frac{1}{n^3}$$

64. (d) The energy of the system of two atoms of diatomic molecule $E = \frac{1}{2}I\omega^2$ where I = moment of inertia

ω = Angular velocity $= \frac{L}{I}$,

L = Angular momentum

$$I = \frac{1}{2}(m_1 r_1^2 + m^2 r_2^2)$$

Thus, $E = \frac{1}{2}(m_1 r_1^2 + m_2 r_2^2)\omega^2$ ...(i)

$$E = \frac{1}{2}(m_1 r_1^2 + m_2 r_2^2)\frac{L^2}{I^2}$$

$L = n\hbar$ (According Bohr's Hypothesis)

$$E = \frac{1}{2}(m_1 r_1^2 + m_2 r_2^2)\frac{L^2}{(m_1 r_1^2 + m_2 r_2^2)^2}$$

$$E=\frac{1}{2}\frac{L^2}{(m_1r_1^2+m_2r_2^2)}=\frac{n^2h^2}{8\pi^2(m_1r_1^2+m_2r_2^2)}$$

$$E=\frac{(m_1+m_2)n^2h^2}{8\pi^2r^2m_1m_2}$$

65. (b) **66.** (d)

67. (c) $\frac{hc}{\lambda_1}-\frac{hc}{\lambda_0}=\text{K.E.}_1$ and $\frac{hc}{\lambda_2}-\frac{hc}{\lambda_0}=\text{K.E.}_2$

$$\Rightarrow \frac{hc}{\lambda_1}-\frac{hc}{\lambda_2}=\text{K.E.}_1-\text{K.E}_2$$

$$\Rightarrow hc\left[\frac{\lambda_2-\lambda_1}{\lambda_1\lambda_2}\right]=\text{K.E.}_1-\text{K.E}_2$$

$$\therefore h=\frac{(\text{K.E.}_1-\text{K.E.}_2)\lambda_1\lambda_2}{c(\lambda_2-\lambda_1)}$$

68. (c)

69. (c) For transition from $n=5$ to $n=4$,

$$h\nu=13.6\times9\left[\frac{1}{16}-\frac{1}{25}\right]=\frac{13.6\times9\times9}{16\times25}=2.754\,\text{eV}$$

For transition from $n=4$ to $n=3$,

$$h\nu'=13.6\times9\left[\frac{1}{9}-\frac{1}{16}\right]=\frac{13.6\times9\times7}{9\times16}=5.95\text{eV}$$

For transition $n=4$ to $n=3$, the frequency is high and hence wavelength is short.

For photoelectric effect, $h\nu'-W=eV_0$, where W = work function

$5.95\times1.6\times10^{-19}-W=1.6\times10^{-19}\times3.95$

$\Rightarrow W=2\times1.6\times10^{-19}=2\text{ eV}$

Again applying $h\nu-W=eV'_0$

We get, $2.754\times1.6\times10^{-19}-2\times1.6\times10^{-19}=1.6\times10^{-19}V'_0$

$\Rightarrow V'_0=0.754\text{ V}$

70. (a) $E_n=-\frac{\text{I.E.}}{n^2}$ for Bohr's hydrogen atom.

Here, I.E. $=4R$ $\therefore E_n=\frac{-4R}{n^2}$

$$\therefore E_2-E_1=\frac{-4R}{2^2}-\left(-\frac{4R}{1^2}\right)=3R \quad ..(i)$$

$$E_2-E_1=h\nu=\frac{hc}{\lambda} \quad ...(ii)$$

From (i) and (ii)

$$\frac{hc}{\lambda}=3R$$

$$\therefore \lambda=\frac{hc}{3R}=\frac{6.6\times10^{-34}\times3\times10^8}{2.2\times10^{-18}\times3}=300\text{Å}$$

71. (b)

72. (a) $E_n=-3.4\text{ eV}$

The kinet2ic energy is equal to the magnitude of total energy in this case.

$\therefore$ K.E. $=+3.4$ eV

The de Broglie wavelength of electron

$$\lambda=\frac{h}{\sqrt{2mK}}=\frac{6.64\times10^{-34}}{\sqrt{2\times9.1\times10^{-31}\times3.4\times1.6\times10^{-19}}}$$

$=0.66\times10^{-9}$ m

73. (b)

74. (a) $\frac{1}{\lambda_{3-1}}=R\left(\frac{1}{1^2}-\frac{1}{3^2}\right)=\frac{8R}{9}$

$$\frac{1}{\lambda_{2-1}}=R\left(\frac{1}{1^2}-\frac{1}{2^2}\right)=\frac{3R}{4}\Rightarrow\frac{\lambda_{3-1}}{\lambda_{2-1}}=\frac{27}{32}$$

75. (d) In Lyman series energy is released in U.V. region.

In Balmer series energy is released in Visible region.

In Paschen/Bracket/P-fund series energy is released in I.R. region.

76. (b) For I[st] line of Lyman series in hydrogen spectrum

$$\frac{1}{\lambda}=\text{R (1)}\left(\frac{1}{1^2}-\frac{1}{2^2}\right)=\frac{3}{4}\text{R}$$

For Li^{2+} $\left[\frac{1}{\lambda}=\text{R}\times9\left(\frac{1}{3^2}-\frac{1}{6^2}\right)=\frac{3\text{R}}{4}\right]$

which is satisfied by $n=6\rightarrow n=3$.

77. (c) $\frac{1}{\lambda}=\text{R}\left[\frac{1}{n_1^2}-\frac{1}{n_2^2}\right]$

$$\Rightarrow\frac{1}{970.6\times10^{-10}}=1.097\times10^7\left[\frac{1}{1^2}-\frac{1}{n_2^2}\right]\Rightarrow n_2=4$$

$\therefore$ Number of emission line $N=\frac{n(n-1)}{2}=\frac{4\times3}{2}=6$

78. (a) **79.** (c)

80. (c) $\Delta\lambda=\lambda_1-\lambda_2$

$$\frac{1}{R\left[\frac{1}{(2)^2}-\frac{1}{(3)^2}\right]Z^2}=-\frac{1}{R\left[1-0\right]Z^2}$$

On solving, $R=\frac{31}{5}\frac{1}{\Delta\lambda\,.\,Z^2}$

81. (c) Number of possible spectral lines emitted when an electron jumps back to ground state from n[th] orbit $=\frac{n(n-1)}{2}$

Here, $\frac{n(n-1)}{2}=6\Rightarrow n=4$

Wavelength λ from transition from n = 1 to n = 4 is given by,

$$\frac{1}{\lambda} = R\left(\frac{1}{1} - \frac{1}{4^2}\right) \Rightarrow \lambda = \frac{16}{15R} = 975\ \text{Å}$$

82. (c) E is minimum when $n_1 = 1$ & $n_2 = 2$

$$\Rightarrow E_{min} = 13.6\left(\frac{1}{1} - \frac{1}{4}\right) eV = 13.6 \times \frac{3}{4} eV$$

E is maximum when $n_1 = 1$ & $n_2 = \infty$ (the atom is ionized, that is known as ionization energy)

$$\Rightarrow E_{max} = 13.6\left(1 - \frac{1}{\infty}\right) = 13.6 eV$$

$$\therefore \frac{E_{min}}{E_{max}} = \frac{3}{4} \Rightarrow \frac{hc/\lambda_{max}}{hc/\lambda_{min}} = \frac{3}{4} \Rightarrow \frac{\lambda_{min}}{\lambda_{max}} = \frac{3}{4}$$

83. (a) 84. (b) 85. (a) 86. (a)

87. (d) $\frac{1}{\lambda} = R\left[\frac{1}{n_1^2} - \frac{1}{n_2^2}\right]$. For first wavelength, $n_1 = 2, n_2 = 3$

$\Rightarrow \lambda_1 = 6563$ Å. For second wavelength, $n_1 = 2, n_2 = 4$

$\Rightarrow \lambda_2 = 4861$ Å

88. (a)

$$\frac{1}{\lambda_{max}} = R\left[\frac{1}{(1)^2} - \frac{1}{(2)^2}\right] \Rightarrow \lambda_{max} = \frac{4}{3R} \approx 1213\text{Å}$$

and $\frac{1}{\lambda_{min}} = R\left[\frac{1}{(1)^2} - \frac{1}{\infty}\right] \Rightarrow \lambda_{min} = \frac{1}{R} \approx 910\text{Å}.$

89. (c)

90. (b) For first line of Balmer series

$$\frac{1}{\lambda} = R\left(\frac{1}{4} - \frac{1}{9}\right) \Rightarrow R = \frac{36}{5\lambda}$$

∴ Wavelength of the first line, λ_L of the Lyman series is given by

$$\frac{1}{\lambda_L} = R\left(1 - \frac{1}{4}\right) = \frac{36}{5\lambda} \times \frac{3}{4} = \frac{27}{5\lambda}$$

$$\Rightarrow \lambda_L = \frac{5\lambda}{27}$$

91. (d)

92. (b) No. of lines $N_E = \frac{n(n-1)}{2} = \frac{3(3-1)}{2} = 3$

93. (a) $E = hf = 13.6 Z^2\left(\frac{1}{n_2^2} - \frac{1}{n_1^2}\right)$

$$\frac{f_2}{f_1} = \frac{3^2\left(\frac{1}{1^2} - \frac{1}{2^2}\right)}{1^2\left(\frac{1}{1^2} - \frac{1}{3^2}\right)} \Rightarrow f_2 = \frac{243}{32} f.$$

94. (d) Energy of nth state in hydrogen is same as energy of $3n^{th}$ state in Li^{++}.

∴ 3 → 1 transition in H would give same energy as the 3 × 3 → 1 × 3 transitor in Li^{++}.

95. (c)

96. (c) $\frac{1}{\lambda} = R\left[\frac{1}{n_1^2} - \frac{1}{n_2^2}\right]$

For limiting wavelength of Lyman series

$n_1 = 1, n_2 = \infty \qquad \frac{1}{\lambda_L} = R$

For limiting wavelength of Balmer series

$n_1 = 2, n_2 = \infty$

$$\frac{1}{\lambda_B} = R\left(\frac{1}{4}\right) \Rightarrow \lambda_B = \frac{4}{R}$$

$\therefore \lambda_B = 4\lambda_L = 4 \times 912$ Å.

97. (a) For the third line of Balmer series, $n_1 = 2, n_2 = 5$

$$\therefore \frac{1}{\lambda} = RZ^2\left(\frac{1}{n_1^2} - \frac{1}{n_2^2}\right) = RZ^2\left(\frac{1}{2^2} - \frac{1}{5^2}\right) = \frac{21 RZ^2}{100}$$

E = – 13.6 eV

$$Z^2 \times \frac{21}{100} = \frac{hc}{\lambda} = \frac{1242\ eVnm}{108.5\ nm}$$

$$Z^2 = \frac{1242 \times 100}{108.5 \times 21 \times 13.6} = 4 \Rightarrow Z = 2$$

98. (a)

99. (a) $I = \frac{e}{T} = \frac{eV}{2\pi r}$ so, $M = \frac{ev}{2\pi r} \times \pi r^2 = \frac{evr}{2}$

According to Bohr's theory angular momentum

$mvr = \frac{nh}{2\pi}$ or $vr = \frac{nh}{2\pi m}$ so, $M = \frac{neh}{4\pi m}$

For the ground state n = 1 so, $M = \frac{eh}{4\pi m}$

100. (a) Energy of photon corresponding to first line of Balmer series = $(13.6)(2)^2\left[\frac{1}{4} - \frac{1}{9}\right]$

Energy need to eject electron from n = 2 level in H atom

$= (13.6)\left(\frac{1}{4}\right)$

So, required kinetic energy

$$= (13.6)\left[\left(\frac{1}{4} - \frac{1}{9}\right) - \left(\frac{1}{4}\right)\right] eV$$

$$= 13.6 \times \left(\frac{11}{36}\right) = 4.155\ eV$$

27 Nuclei

1. The nuclear radius of $_8O^{16}$ is 3×10^{-15} m. If an atomic mass unit is 1.67×10^{-27} kg, then the nuclear density is approximately
(a) 2.35×10^{17} g cm^{-3} (b) 2.35×10^{17} kg m^{-3}
(c) 2.35×10^{17} gm^{-3} (d) 2.35×10^{17} kg mm^{-3}

2. If the nuclear radius of ^{27}Al is 3.6 Fermi, the approximate nuclear radius of ^{64}Cu in Fermi is
(a) 4.8 (b) 3.6 (c) 2.4 (d) 1.2

3. The nuclear radius of a nucleus with nucleon number 16 is 3×10^{-15} m. Then, the nuclear radius of a nucleus with nucleon number 128 is
(a) 3×10^{-15} m (b) 1.5×10^{-15} m
(c) 6×10^{-15} m (d) 4.5×10^{-15} m

4. Atomic weight of boron is 10.81 and it has two isotopes $_5B^{10}$ and $_5B^{11}$. Then ratio of $_5B^{10}$: $_5B^{11}$ in nature would be
(a) 19 : 81 (b) 10 : 11 (c) 15 : 16 (d) 81 : 19

5. If the nucleus $^{27}_{13}Al$ has nuclear radius of about 3.6 fm, then $^{125}_{32}Te$ would have its radius approximately as
(a) 9.6 fm (b) 12.0 fm
(c) 4.8 fm (d) 6.0 fm.

6. The radius of germanium (Ge) nuclide is measured to be twice the radius of $^{9}_{4}$Be. The number of nucleons in Ge are
(a) 74 (b) 75 (c) 72 (d) 73

7. Calculate binding energy of $_{92}U^{238}$.
Given $M(U^{238}) = 238.050783$ amu, $m_n = 1.008665$ amu and $m_p = 1.007825$ amu
(a) 801.7 MeV (b) 18.7 MeV
(c) 0.7 MeV (d) 1801.7 MeV

8. The binding energy of deuteron ($^{2}_{1}$H) is 1.15 MeV per nucleon and an alpha particle ($^{4}_{2}$He) has a binding energy of 7.1 MeV per nucleon. Then in the reaction

$$^{2}_{1}H + ^{2}_{1}H \longrightarrow ^{4}_{2}He + Q$$

the energy released Q is :
(a) 5.95 MeV (b) 26.1 MeV
(c) 23.8 MeV (d) 289.4 MeV

9. Binding energy per nucleon vs mass number curve for nuclei is shown in the Figure. W, X, Y and Z are four nuclei indicated on the curve. The process that would release energy is **Tricky**

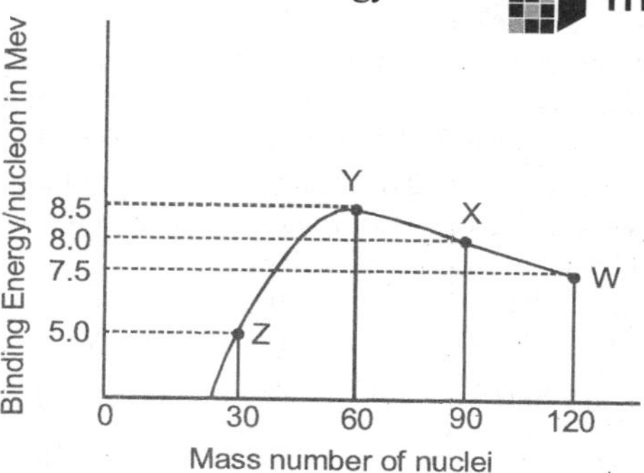

(a) $Y \to 2Z$ (b) $W \to X + Z$
(c) $W \to 2Y$ (d) $X \to Y + Z$

10. A proton is bombarded on a stationary Lithium nucleus. As a result of collision two α-particles are produced. The direction of motion of the α-particles with the initial direction of motion makes an angle $\cos^{-1}\frac{1}{4}$. If B.E/Nucleon for Li^7 and He^4 are 5.60 MeV and 7.06 MeV respectively, then :

(a) Kinetic energy of striking proton is 17.28 MeV
(b) Kinetic energy of striking proton is 8.64 MeV
(c) Kinetic energy of striking proton is 4.32 MeV
(d) Kinetic energy of striking proton is 2.16 MeV

11. The binding energy per nucleon of ^{10}X is 9 MeV and that of ^{11}X is 7.5 MeV where X represents an element. The minimum energy required to remove a neutron from ^{11}X is
(a) 7.5 MeV (b) 2.5 MeV
(c) 8 MeV (d) 0.5 MeV

12. The binding energies per nucleon for a deuteron and an α-particle are x_1 and x_2 respectively. What will be the energy Q released in the reaction $_1H^2 + _1H^2 \to _2He^4 + Q$
(a) $4(x_1 + x_2)$ (b) $4(x_2 - x_1)$
(c) $2(x_1 + x_2)$ (d) $2(x_2 - x_1)$

13. The masses of neutron and proton are 1.0087 a.m.u. and 1.0073 a.m.u. respectively. If the neutrons and protons combine to form a helium nucleus (alpha particles) of mass 4.0015 a.m.u the binding energy fo the helium nucleus will be (1 a.m.u. = 931 MeV)
(a) 28.4 MeV (b) 20.8 MeV
(c) 27.3 MeV (d) 14.2 MeV

14. The mass of $_7N^{15}$ is 15.00011 amu, mass of $_8O^{16}$ is 15.99492 amu and m_p = 1.00783 amu. Determine binding energy of last proton of $_8O^{16}$.
(a) 2.13 MeV (b) 0.13 MeV
(c) 10 MeV (d) 12.13 MeV

15. A heavy nucleus having mass number 200 gets disintegrated into two small fragments of mass number 80 and 120. If binding energy per nucleon for parent atom is 6.5 MeV and for daughter nuclei is 7 MeV and 8 MeV respectively, then the energy released in the decay will be –
(a) 200 MeV (b) –220 MeV
(c) 220 MeV (d) 180 MeV

16. If the binding energy per nucleon in ${}_3^7Li$ and ${}_2^4He$ nuclei are 5.60 MeV and 7.06 MeV respectively, then in the reaction

$$p + {}_3^7Li \longrightarrow 2\ {}_2^4He$$

energy of proton must be
(a) 28.24 MeV (b) 17.28 MeV
(c) 1.46 MeV (d) 39.2 MeV

17. A gamma ray creates an electron-positron pair. If the rest mass energy of an electron is 0.5 MeV and the total kinetic energy of the electron-positron pair is 0.78 MeV, then the energy of the gamma ray must be
(a) 0.78 MeV (b) 1.78 MeV
(c) 1.28 MeV (d) 0.28 MeV

18. If the total binding energies of ${}_1^2H$, ${}_2^4He$, ${}_{26}^{56}Fe$, & ${}_{92}^{235}U$ nuclei are 2.22, 28.3, 492 and 1786 MeV respectively, identify the most stable nucleus of the following
(a) ${}_{26}^{56}Fe$ (b) ${}_1^2H$ (c) ${}_{92}^{235}U$ (d) ${}_2^4He$

19. The nuclear fusion reaction $2\,{}_1H^2 \rightarrow {}_2He^4$ + Energy, is proposed to be used for the production of industrial power. Assuming the efficiency of process for production of power is 20%, find the mass of the deuterium required approximately for a duration of 1 year. Given mass of ${}_1H^2$ nucleus = 2.0141 a.m.u and mass of ${}_2He^4$ nuclei = 4.0026 a.m.u and 1 a.m.u. = 31 MeV
(a) 165 kg (b) 138 kg (c) 180 kg (d) 60 kg

20. Two deuterons undergo nuclear fusion to form a Helium nucleus. Energy released in this process is : (given binding energy per nucleon for deuteron=1.1 MeV and for helium=7.0 MeV)

Toughnut

(a) 30.2 MeV (b) 32.4 MeV
(c) 23.6 MeV (d) 25.8 MeV

21. In the nuclear fusion reaction

$${}_1^2H + {}_1^3H \rightarrow {}_2^4He + n$$

given that the repulsive potential energy between the two nuclei is ~ 7.7×10^{-14} J, the temperature at which the gases must be heated to initiate the reaction is nearly
[Boltzmann's Constant $k = 1.38 \times 10^{-23}$ J/K]
(a) 10^7 K (b) 10^5 K
(c) 10^3 K (d) 10^9 K

22. When Uranium is bombarded with neutrons, it undergoes fission. The fission reaction can be written as :

$${}_{92}U^{235} + {}_0n^1 \rightarrow {}_{56}Ba^{141} + {}_{36}Kr^{92} + 3x + Q$$

where three particles named x are produced and energy Q is released. What is the name of the particle x?
(a) electron (b) α-particle
(c) neutron (d) neutrino

23. A reactor converts 100% of given mass into energy and that it operates at a power of 9×10^7 watt. The mass of the fuel consumed in 30 min in the reactor will be :
(a) 12×10^{-3} kg (b) 25×10^{-8} kg
(c) 18×10^{-7} kg (d) 11×10^{-4} kg

24. If a star can convert all the He nuclei completely into oxygen nuclei, the energy released per oxygen nuclei is [Mass of He nucleus is 4.0026 amu and mass of Oxygen nucleus is 15.9994 amu]
(a) 7.6 MeV (b) 56.12 MeV
(c) 10.24 MeV (d) 23.9 MeV

25. Determine the power output of a ${}_{92}U^{235}$ reactor if it takes 30 days to use 2kg of fuel. Energy released per fission is 200 MeV and $N = 6.023 \times 10^{26}$ per kilomole.

Toughnut

(a) 63.28 MW (b) 3.28 MW
(c) 0.6 MW (d) 50.12 MW

26. In the sun about 4 billion kg of matter is converted to energy each second. The power output of the sun in watt is
(a) 3.6×10^{26} (b) 0.36×10^{26}
(c) 36×10^{26} (d) 0.036×10^{26}

27. For fission of uranium $^{235}_{92}U$, one needs
 (a) very fast neutrons (a few electrons volts)
 (b) low energy neutrons (≈ 0.04 eV at 300K)
 (c) a moderator making neutrons slow
 (d) both (b) and (c)

28. Imagine that a reactor converts all given mass into energy and that it operates at a power level of 10^9 watt. The mass of the fuel consumed per hour in the reactor will be : (velocity of light, c is 3×10^8 m/s)
 (a) 0.96 gm (b) 0.8 gm
 (c) 4×10^{-2} gm (d) 6.6×10^{-5} gm

29. Consider the following reaction
 $_1H^2 +_1 H^2 \rightarrow_2 He^4 + Q$. If
 $m(_1H^2) = 2.0141amu$;
 $m(_2He^4) = 4.0024amu$. The energy Q released (in MeV) in this fusion reaction is
 (a) 12 (b) 6 (c) 24 (d) 48

30. If 200 MeV energy is released in the fission of a single U^{235} nucleus, the number of fissions required per second to produce 1 kilowatt power shall be (Given $1eV = 1.6 \times 10^{-19}$ J)
 (a) 3.125×10^{13} (b) 3.125×10^{14}
 (c) 3.125×10^{15} (d) 3.125×10^{16}

31. A radioactive sample S_1 having an activity 5μCi has twice the number of nuclei as another sample S_2 which has an activity of 10 μCi. The half lives of S_1 and S_2 can be
 (a) 20 years and 5 years, respectively
 (b) 20 years and 10 years, respectively
 (c) 10 years each
 (d) 5 years each

32. A mixture consists of two radioactive materials A_1 and A_2 with half lives of 20 s and 10 s respectively. Initially the mixture has 40 g of A_1 and 160 g of A_2. The amount of the two in the mixture will become equal after
 (a) 60 s (b) 80 s (c) 20 s (d) 40 s

33. Two radioactive materials A and B have decay constant $\frac{13}{7}$ units and $\frac{19}{14}$ units respectively. Directally both have the same number of nuclei. The time (in same system of units), after which the ratio of their remaining nuclei will be 1/e is
 (a) 2 (b) 3 (c) 4 (d) 5

34. A radioactive sample contains 10^{-3} kg each of two nuclear species A and B with half-life 4 days and 8 days respectively. The ratio of the amounts of A and B after a period of 16 days is
 (a) 1 : 2 (b) 4 : 1 (c) 1 : 4 (d) 2 : 1

35. A radioactive nucleus X decays to a stable nucleus 'Y'. Then the graph of rate of formation of 'Y' against time 't' will be 

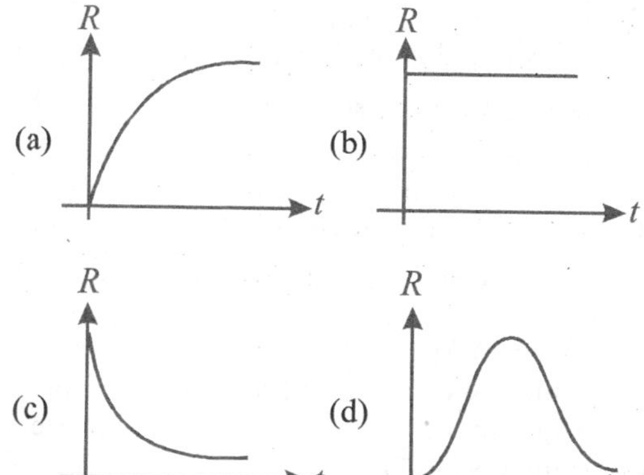

36. A radioactive material of half life $\ell n2$ was produced in a nuclear reactor. Consider two different instants A and B. The number of undecayed nuclei at instant B was twice of that of instant A. If the activities at instants A and B are A_1 and A_2 respectively then the difference in the age of he sample at these instants equals.
 (a) $\left|\ell n\left(\frac{2A_1}{A_2}\right)\right|$ (b) $\ell n2\left|\ell n\left(\frac{A_1}{A_2}\right)\right|$
 (c) $\left|\ell n\left(\frac{A_2}{2A_1}\right)\right|$ (d) $\ell n2\left|\ell n\left(\frac{A_2}{A_1}\right)\right|$

37. An archaeologist analyses the wood in a prehistoric structure and finds that C^{14} (Half life = 5700 years) to C^{12} is only one-fourth of that found in the cells of buried plants. The age of the wood is about
 (a) 5700 years (b) 2850 years
 (c) 11,400 years (d) 22,800 years

38. Half lives for α and β emission of a radioactive material are 16 years and 48 years respectively. When material decays giving α and β emission simultaneously, time in which 3/4th material decays is
 (a) 29 years (b) 24 years
 (c) 64 years (d) 12 years

39. Actinium 231, $^{231}AC_{89}$, emit in succession two β particles, four α-particles, one β and one α plus several γ rays. What is the resultant isotope?
 (a) $^{221}Au_{79}$ (b) $^{211}Au_{79}$
 (c) $^{221}Pb_{82}$ (d) $^{211}Pb_{82}$

40. A radioactive source in the form of metal sphere of diameter 10^{-3} m emits beta particle at a constant rate of 6.25×10^{10} particles per second. If the source is electrically insulated, how long will it take for its potential to rise by 1.0 volt, assuming that 80% of the emitted beta particles escape from the source? **Toughnut**

(a) 6.95 μ sec (b) 0.95 μ sec
(c) 1.95 μ sec (d) 2.15 μ sec

41. A radioactive nucleus undergoes a series of decay according to the scheme **Tricky**

$$A \xrightarrow{a} A_1 \xrightarrow{\beta} A_2 \xrightarrow{\alpha} A_3 \xrightarrow{\gamma} A_4$$

If the mass number and atomic number of 'A' are 180 and 72 respectively, then what are these numbers for A_4

(a) 172 and 69 (b) 174 and 70
(c) 176 and 69 (d) 176 and 70

42. Half lives of two isotopes X and Y of a material are known to be 2×10^9 years and 4×10^9 years respectively. If a planet was formed with equal number of these isotopes, then the current age of planet, given that currently the material has 20% of X and 80% of Y by number, will be –

(a) 2×10^9 years (b) 4×10^9 years
(c) 6×10^9 years (d) 8×10^9 years

43. At radioactive equilibrium, the ratio between the atoms of two radioactive elements (*X*) and (*Y*) was found to be $3.2 \times 10^9 : 1$ respectively. If half-life of the element (*X*) is 1.6×10^{10} years, then half-life of the element (*Y*) would be

(a) 3.2×10^9 years (b) 5×10^9 years
(c) 1.6×10^{10} years (d) 5 years

44. There are two radioactive substances *A* and *B*. Decay constant of *B* is two times that of *A*. Initially, both have equal number of nuclei. After n half lives of *A*, rate of disintegration of both are equal. The value of *n* is

(a) 4 (b) 2 (c) 1 (d) 5

45. A 280 days old radioactive substance shows an activity of 6000 dps, 140 days later its activity becomes 3000 dps. What was its initial activity?

(a) 20000 dps (b) 24000 dps
(c) 12000 dps (d) 6000 dps

46. A radioactive element X converts into another stable element Y. Half life of X is 2 hrs. Initially only X is present. After time t, the ratio of atoms of X and Y is found to be 1 : 4, then t in hours is

(a) 2 (b) 4
(c) between 4 and 6 (d) 6

47. For a radioactive sample the counting rate changes from 6520 counts/minute to 3260 counts/minute in 2 minutes. Determine the decay constant.

(a) 1.78 per sec (b) 0.78 per sec
(c) 2.78 per sec (d) 5.78 per sec

48. Consider a radioactive material of half-life 1.0 minute. If one of the nuclei decays now, the next one will decay

(a) after 1 minute
(b) after $\frac{1}{\log_e 2}$ minute
(c) after $\frac{1}{N}$ minute, where N is the number of nuclei present at that moment
(d) after any time

49. After 150 days, the activity of a radioactive sample is 5000 dps. The activity becomes 2500 dps after another 75 days. The initial activity of the sample is

(a) 20000 dps (b) 40000 dps
(c) 7500 dps (d) 10000 dps

50. If a radioactive material contains 0.1 mg of Th^{234} how much of it will remain unchanged after 120 days. Given Half life is 24 days

(a) 1.12 μg (b) 0.1 μg
(c) 3.125 μg (d) 125 μg

51. Consider α particles, β particles and γ - rays, each having an energy of 0.5 MeV. In increasing order of penetrating powers, the radiations are:

(a) α,β,γ (b) α,γ,β (c) β,γ,α (d) γ,β,α

52. The activity of a radioactive sample is measured as N_0 counts per minute at $t = 0$ and N_0/e counts per minute at $t = 5$ minutes. The time (in minutes) at which the activity reduces to half its value is

(a) $\log_e 2/5$ (b) $\frac{5}{\log_e 2}$
(c) $5 \log_{10} 2$ (d) $5 \log_e 2$

53. A sample of radioactive material decays simultaneously by two processes A and B with half lives 1/2 and 1/4 hr. respectively. For first half hour it decays with the process A, next one hour with the process B and for further half an hour with both A and B. If originally there were N_0 nuclei, the number of nuclei after 2 hour of such decay is

$N_0\left(\frac{1}{2}\right)^x$ then find the value of x.

(a) 5 (b) 4 (c) 3 (d) 8

54. Beta rays emitted by a radioactive material are
(a) electromagnetic radiations
(b) the electrons orbiting around the nucleus
(c) charged particles emitted by the nucleus
(d) neutral particles

55. Two radioactive materials X_1 and X_2 have decay constants 10λ and λ respectively. If initially they have the same number of nuclei, then the ratio of the number of nuclei of X_1 to that of X_2 will be $1/e$ after a time

(a) $\frac{1}{10\lambda}$ (b) $\frac{1}{11\lambda}$ (c) $\frac{11}{10\lambda}$ (d) $\frac{1}{9\lambda}$

56. The half life of a radioactive isotope 'X' is 50 years. It decays to another element 'Y' which is stable. The two elements 'X' and 'Y' were found to be in the ratio of 1 : 15 in a sample of a given rock. The age of the rock was estimated to be

(a) 150 years (b) 200 years
(c) 250 years (d) 100 years

57. Let N_β be the number of β particles emitted by 1 gram of Na^{24} radioactive nuclei (half life = 15 hrs) in 7.5 hours, N_β is close to (Avogadro number = 6.023×10^{23}/g mole):
(a) 6.2×10^{21} (b) 7.5×10^{21}
(c) 1.25×10^{22} (d) 1.75×10^{22}

58. The activity of a radioactive sample is A_1 at time t_1 and A_2 at time t_2. If τ is average life of sample then the number of nuclei decayed in time $(t_2 - t_1)$ is

(a) $A_1t_1 - A_2t_2$ (b) $\frac{(A_1 - A_2)}{2}\tau$
(c) $(A_1 - A_2)(t_2 - t_1)$ (d) $(A_1 - A_2)\tau$.

59. One gram of a radioactive sample of half-life 10 min is sealed in a capsule at time $t = 0$. Amount of sample decayed upto 5 min is
(a) 0.293g (b) 0.5g (c) 0.25g (d) 0.707g

60. Two radioactive substances A and B have decay constants 5λ and λ respectively. At t = 0 they have the same number of nuclei. The ratio of number of nuclei of A to those of B will be $(1/e)^2$ after a time interval
(a) 4λ (b) 2λ (c) $1/2\lambda$ (d) $1/4\lambda$

61. The half life of radioactive Radon is 3.8 days. The time at the end of which $\frac{1}{20}$th of the radon sample will remain undecayed is (given $\log_{10} e = 0.4343$)
(a) 3.8 days (b) 16.5 days
(c) 33 days (d) 76 days.

62. Average life of a radioactive sample is 4 ms. Initially the total number of nuclei is N_0. A charged capacitor of capacity 20 μF is connected across a resistor R. The value of R such that ratio of number of nuclei remaining to charge on capacitor remains constant with time is
(a) 100 Ω (b) 200 Ω (c) 300 Ω (d) 400 Ω

63. A radioactive nucleus can decay by two different processes. The half life for the first process is t_1 and that for the second process is t_2. If effective half life is t, then

(a) $t = t_1 + t_2$ (b) $\frac{1}{t} = \frac{1}{t_1} + \frac{1}{t_2}$

(c) $t = \frac{2t_1t_2}{t_1 + t_2}$ (d) $t = \frac{t_1 + t_2}{2}$

64. In a sample of rock, the ratio of ^{206}Pb to ^{238}U nuclei is found to be 0.5. The age in year of the rock is (given half–life of U^{238} is 4.5×10^9 years)

Critical Thinking

(a) 2.25×10^9
(b) $4.5 \times 10^9 \ln 3$
(c) $4.5 \times 10^9 \frac{\ln\left(\frac{3}{2}\right)}{\ln 2}$
(d) $2.25 \times 10^9 \ln\left(\frac{3}{2}\right)$

65. In an α-decay the kinetic energy of α-particle is 48 MeV and Q-value of the reaction is 50 MeV. The mass number of the mother nucleus is (Assume that daughter nucleus is in ground state)
(a) 96 (b) 100 (c) 104 (d) 110

66. If nuclei of a radioactive element is produced at constant rate α and they decays with decay constant λ. At $t = 0$, number of nuclei is zero than the number of nuclei at time t is

(a) $\frac{\alpha}{\lambda}(1 - e^{-\lambda t})$ (b) $\alpha - \frac{\alpha}{\lambda}e^{-\lambda t}$

(c) $\frac{\alpha}{\lambda}e^{-\lambda t}$ (d) $\alpha(1 - e^{-\lambda t})$

67. Two radioactive substances X and Y emit α and β particles respectively. Their disintegration constants are in the ratio 2 : 3. To have equal probabilities of getting emission of α and β particles, the ratio of number of atoms X to that of Y at any time instant is

(a) 2:3 (b) 3:2 (c) e:1 (d) $(e-1)$:1

68. A container is filled with a radioactive substance for which the half – life is 2 days. A week later, when the container is opened, it contains 5 g of the substance. Approximately how many grams of the substance were initially placed in the container?

(a) 40 (b) 60 (c) 80 (d) 100

69. Which sample, A or B shown in Fig. has shorter mean-life? **Tricky**

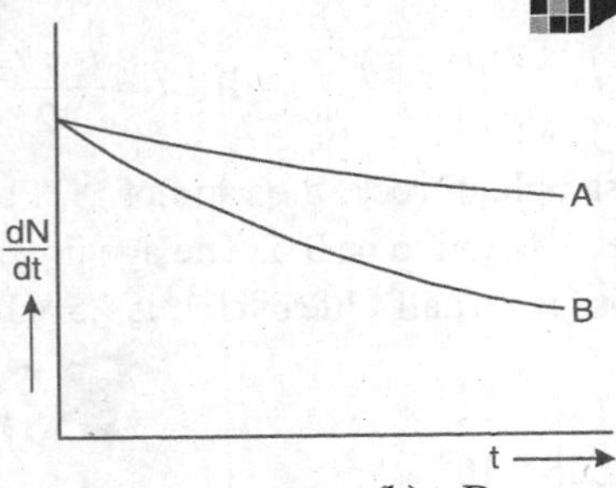

(a) A (b) B
(c) both A and B have same mean-life
(d) Can not be predicted

70. The ratio of number of atoms of ^{14}C to ^{12}C in living matter is measured to be 1.3×10^{-12} at the present time. A 12 g sample of carbon produces 180 decays/min due to the small amount of ^{14}C in it. The half-life of ^{14}C is nearly [1 year $= 3.15 \times 10^7$ s]

(a) 574 years (b) 5740 years
(c) 2870 years (d) 287 years

71. A radioactive substance with decay constant of $0.5 s^{-1}$ is being produced at a constant rate of 50 nuclei per second. If there are no nuclei present initially, the time (in second) after which 25 nuclei will be present is

(a) 1 (b) ln 2 (c) ln (4/3) (d) 2 ln (4/3)

72. When the nucleus of a radium-226, which is at rest, decays, an α particle and the nucleus of radon are created. The released energy during the decay is 4.87 MeV, which appears as the kinetic energy of the two resulted particles. Calculate the kinetic energy of α particle and radon nucleus.

(a) 4.78 MeV and 0.09 MeV
(b) 4.67 MeV and 0.2 MeV
(c) 4.84 MeV and 0.03 MeV
(d) 4.81 MeV and 0.06 MeV

73. When a U^{238} nucleus originally at rest, decays by emitting an alpha particle having a speed 'u', the recoil speed of the residual nucleus is

(a) $\frac{4u}{238}$ (b) $-\frac{5u}{234}$ (c) $\frac{4u}{234}$ (d) $-\frac{4u}{238}$

74. There are n number of radioactive nuclei in a sample that undergoes beta decay. If from the sample, n' number of β- particles are emitted every 2 s, then half - life of nuclei is

(a) n'/2 (b) 0.693 × (2n/n')
(c) 0.693 ln (2n/n') (d) 0.693 × n/n'

75. A radioactive element decays to $\frac{1}{4}$th of its initial value in time t. The ratio of its half life to mean life is

(a) $\frac{\ln 2}{\lambda^2}$ (b) $\frac{\ln 2}{\lambda}$ (c) 2 ln 2 (d) ln 2

76. Starting with a sample of pure ^{66}Cu, $\frac{7}{8}$ of it decays into Zn in 15 minutes. The corresponding half life is

(a) 15 minutes (b) 10 minutes
(c) $7\frac{1}{2}$ minutes (d) 5 minutes

77. The luminous dials watches are usually made by mixing a zinc sulphide phosphor with an a-particle emitter. The mass of radium (mass number 226, half - life 1620 years) that is needed to produce an average of 10 a-particles per second for this purpose is **Toughnut**

(a) 2.77 mg (b) 2.77 g
(c) 2.77×10^{-23} g (d) 2.77×10^{-13} kg

78. A sample of radioactive substance has 10^6 nuclei. If half life is 20 seconds, the number of nuclei left in the sample after 10 second is

(a) 10^4 (b) 2×10^5 (c) 7×10^5 (d) 11×10^5

79. The half-life of a radioactive element A is the same as the mean-life of another radioactive element B. Initially both substances have the same number of atoms, then : **Tricky**

(a) A and B decay at the same rate always.
(b) A and B decay at the same rate initially.
(c) A will decay at a faster rate than B.
(d) B will decay at a faster rate than A.

80. A piece of bone of an animal from a ruin is found to have ^{14}C activity of 12 disintegrations per minute per gm of its carbon content. The ^{14}C activity of a living animal is 16 disintegrations per minute per gm. How long ago nearly did the animal die? (Given half life of ^{14}C is $t_{1/2} = 5760$ years)

(a) 1672 years (b) 2391 years
(c) 3291 years (d) 4453 years

81. An element A decays into an element C by a two step process: $A \to B + He_2^4$ and $B \to C + 2e_0^{-1}$. Then

(a) A and C are isotopes
(b) A and C are isobars
(c) B and C are isotopes
(d) A and B are isobars

82. Radium ^{226}Ra, spontaneously decays to radon with the emission of an α-particle and a γ ray. If the speed of the α particle upon emission from an initially stationary radium nucleus is 1.5×10^7 m/s, what is the recoil speed of the resultant radon nucleus? Assume the momentum of γ ray is negligible compared to that of α particle.

(a) 1.6×10^4 m/s (b) 2.7×10^5 m/s
(c) 1.3×10^6 m/s (d) 2.6×10^3 m/s

83. In a certain hypothetical radioactive decay process, species A decays into species B and species B decays into species C according to the reactions :

$$A \longrightarrow 2B + \text{particles} + \text{energy}$$
$$B \longrightarrow 3C + \text{particles} + \text{energy}$$

The decay constant for species is $\lambda_1 = 1$ sec^{-1} and that for the species B is $\lambda_2 = 100$ sec^{-1}. Initially 10^4 moles of species of A were present while there was none of B and C. It was found that species B reaches its maximum number at a time $t_0 = 2 \ln(10)$ sec. Calculate the value of maximum number of moles of B.

(a) 1 (b) 2 (c) 5 (d) 4

84. A radioactive isotope has a half life of 100 years. How long will it take the activity to reduce to 1% of its original valve?

(a) 330 years (b) 460 years
(c) 660 years (d) 920 years

85. A radioactive nucleus (initial mass number A and atomic number Z emits 3 α - particles and 2 positrons. The ratio of number of neutrons to that of protons in the final nucleus will be

(a) $\frac{A-Z-8}{Z-4}$ (b) $\frac{A-Z-4}{Z-8}$
(c) $\frac{A-Z-12}{Z-4}$ (d) $\frac{A-Z-4}{Z-2}$

86. The half life of a radioactive substance is 20 minutes. The approximate time interval $(t_2 - t_1)$ between the time t_2 when $\frac{2}{3}$ of it had decayed and time t_1 when $\frac{1}{3}$ of it had decayed is :

(a) 14 min (b) 20 min (c) 28 min (d) 7 min

87. After absorbing a slowly moving neutron of mass m_N (momentum ≈ 0) a nucleus of mass M breaks into two nuclei of masses m_1 and $5m_1$ ($6m_1 = M + m_N$) respectively. If the de Broglie wavelength of the nucleus with mass m_1 is λ, the de Broglie wavelength of the nucleus will be

(a) 5λ (b) λ/5 (c) λ (d) 25λ

88. Assume that a neutron breaks into a proton and an electron. The energy released during this process is : (mass of neutron = 1.6725×10^{-27} kg, mass of proton = 1.6725×10^{-27} kg, mass of electron = 9×10^{-31} kg).

(a) 0.51 MeV (b) 7.10 MeV
(c) 6.30 MeV (d) 5.4 MeV

89. Half-lives of two radioactive elements A and B are 20 minutes and 40 minutes, respectively. Initially, the samples have equal number of nuclei. After 80 minutes, the ratio of decayed number of A and B nuclei will be :

(a) 1 : 4 (b) 5 : 4 (c) 1 : 16 (d) 4 : 1

90. A radioactive nucleus A with a half life T, decays into a nucleus B. At t = 0, there is no nucleus B. At sometime t, the ratio of the number of B to that of A is 0.3. Then, t is given by

(a) $t = T\log(1.3)$ (b) $t = \frac{T}{\log(1.3)}$
(c) $t = T\frac{\log 2}{\log 1.3}$ (d) $t = \frac{\log 1.3}{\log 2}T$

91. A ^7Li target is bombarded with a proton beam current of 10^{-4} A for 1 hour to produce ^7Be of activity 1.8×10^8 disintegrations per second. Assuming that one ^7Be radioactive nucleus is produced by bombarding 1000 protons, determine its half life.

Toughnut

(a) 8.6×10^6 s (b) 4.2×10^5 s
(c) 3.1×10^5 s (d) 1.1×10^6 s

92. The disintegration rate of a certain radioactive sample at any instant is 4750 disintegrations per minute. Five minutes later the rate becomes 2700 disintegrations per minute. Calculate half life of the sample.
(a) 9.1 min (b) 6.1 min (c) 2.3 min (d) 1.1 min

93. A nucleus at rest undergoes a decay emitting an α-particle of de-Broglie wavelength $\lambda = 5.76 \times 10^{-15} m$. If the mass of the daughter nucleus is 223.610 amu and that of the α-particle is 4.002 amu, determine the mass of the parent nucleus inamu. (1 amu = 931.470 MeV/c^2].

Toughnut

(a) 227.62 amu (b) 112.11 amu
(c) 90.3 amu (d) 23.8 amu

94. In an ore containing uranium, the ratio of U^{238} to Pb^{206} nuclei is 3. Calculate the age of the ore, assuming that all the lead present in the ore in the final stable product of U^{238}. Take the half life of U^{238} to be 4.5×10^9 year.
(a) 1.8×10^9 year (b) 2.3×10^{10} year
(c) 5.1×10^7 year (d) 6.2×10^6 year

95. The proton -proton mechanism that accounts for energy production in the sun releases 26.7 MeV energy for each event. In this process, protons fuse to form an alpha particle (^4He). At what rate $\frac{dm}{dt}$ is hydrogen being consumed in the core of the sun by the p-p cycle? Power of sun is 3.90×10^{26} W.
(a) 1.6×10^{10} kg/s (b) 2.3×10^9 kg/s
(c) 6.2×10^{11} kg/s (d) 5.5×10^{10} kg/s

96. Consider the fusion in helium plasma. Find the temperature at which the average thermal energy 1.5kT equals the coulomb potential energy at 2 fm.

Toughnut

(a) 1.2×10^5 K (b) 2.23×10^{10} K
(c) 5.4×10^3 K (d) 1.1×10^5 K

97. The element curium $_{96}Cm^{248}$ has a mean life of 10^{13} s. Its primary decay modes are spontaneous fission and α- decay, the former with a probability of 8% and the later with a probability of 92 %. Each fission releases 200 MeV of energy. The masses involved in α decay are as follows:
$_{96}Cm^{248} = 248.072220u$, $_{94}Pu^{244} = 244.064100u$ and $_2He^4 = 4.002603u$ Calculate the power output from a sample of 10^{20} Cm atoms.
(1u = 931 MeV/c^2).

Toughnut

(a) 1.6×10^{-5}W (b) 2.6×10^{-3}W
(c) 3.3×10^{-5}W (d) 5.1×10^{-3}W

98. A radioactive element decays by β emission. A detector records n beta particles in 2 second and in next 2 seconds it records 0.75 n beta particles. Find mean life corrected to nearest whole number.
Given $\ell n2 = 0.6931$ and $\ell n3 = 1.0986$.
(a) 6.9 sec (b) 9.9 sec
(c) 10.1 sec (d) 12.2 sec

99. The count rate from a radioactive sample falls from 4.0×10^6 per second to 1×10^6 per second in 20 hour. What will be the count rate, 100 hour after the beginning?
(a) 3.91×10^3 sec^{-1} (b) 5.81×10^4 sec^{-1}
(c) 6.22×10^5 sec^{-1} (d) None of these

100. The ratio of half-life times of two elements A and B is $\frac{T_A}{T_B}$. The ratio of respective decay constant $\frac{\lambda_A}{\lambda_B}$, is
(a) T_B / T_A (b) T_A / T_B
(c) $\frac{T_A + T_B}{T_A}$ (d) $\frac{T_A - T_B}{T_A}$

Answer KEYs

1	(b)	11	(b)	21	(d)	31	(a)	41	(a)	51	(a)	61	(b)	71	(d)	81	(a)	91	(a)
2	(a)	12	(b)	22	(c)	32	(d)	42	(d)	52	(d)	62	(b)	72	(a)	82	(b)	92	(b)
3	(c)	13	(a)	23	(c)	33	(a)	43	(d)	53	(d)	63	(b)	73	(c)	83	(b)	93	(a)
4	(a)	14	(d)	24	(c)	34	(c)	44	(c)	54	(c)	64	(c)	74	(b)	84	(c)	94	(a)
5	(d)	15	(c)	25	(a)	35	(c)	45	(b)	55	(d)	65	(b)	75	(d)	85	(b)	95	(c)
6	(c)	16	(b)	26	(b)	36	(c)	46	(c)	56	(b)	66	(a)	76	(d)	86	(b)	96	(b)
7	(d)	17	(b)	27	(d)	37	(c)	47	(d)	57	(b)	67	(b)	77	(d)	87	(c)	97	(c)
8	(c)	18	(a)	28	(c)	38	(b)	48	(d)	58	(d)	68	(c)	78	(c)	88	(a)	98	(a)
9	(c)	19	(b)	29	(c)	39	(d)	49	(a)	59	(a)	69	(b)	79	(d)	89	(b)	99	(a)
10	(a)	20	(c)	30	(a)	40	(a)	50	(c)	60	(c)	70	(b)	80	(b)	90	(d)	100	(a)

Hints & Solutions

1. (b) For nucleus of $_8O^{16}$
Mass = (16) (1.67×10^{-27}) kg

Volume $= \frac{4}{3}\pi R^3$

$= \frac{4}{3}\pi(3\times10^{-15})^3 m^3 = 36\pi \times 10^{-45} m^3$

Density $= \frac{\text{mass}}{\text{volume}} = \frac{16\times1.67\times10^{-27}kg}{36\pi\times10^{-45}m^3}$

$= 2.35 \times 10^{17}$ kgm^{-3}

2. (a) Nuclear radius, $r \propto A^{1/3}$
where A is mass number

$\frac{r_{Cu}}{r_{Al}} = \left(\frac{A_{Cu}}{A_{Al}}\right)^{1/3} = \left(\frac{64}{27}\right)^{1/3}$

$\Rightarrow r_{Cu} = 3.6\left(\frac{4}{3}\right) = 4.8$ Fermi

3. (c) Radius R os a nucleus changes with the nucleon number A of the nucleus as
$R = 1.3 \times 10^{-15} \times A^{1/3}$ m
Hence,

$\frac{R_2}{R_1} = \left(\frac{A_2}{A_1}\right)^{1/3} = \left(\frac{128}{16}\right)^{1/3} = (8)^{1/3} = 2$

$\therefore R_2 = 2R_1 = 2(3\times10^{-15})m = 6\times10^{-15}$ m

4. (a) Let the percentage of B^{10} atoms be x, then average atomic weight

$= \frac{10x + 11(100-x)}{100} = 10.81 \Rightarrow x = 19$

$\therefore \frac{N_{B^{10}}}{N_{B^{11}}} = \frac{19}{81}$

5. (d) It has been known that a nucleus of mass number A has radius
$R = R_0 A^{1/3}$,
where $R_0 = 1.2 \times 10^{-15}$m and A = mass number

In case of $^{27}_{13}A\ell$, let nuclear radius be R_1

and for $^{125}_{32}Te$, nuclear radius be R_2

For $^{27}_{13}Al$, $R_1 = R_0(27)^{1/3} = 3R_0$

For $^{125}_{32}Te$, $R_2 = R_0(125)^{1/3} = 5R_0$

$\frac{R_2}{R_1} = \frac{5R_0}{3R_0} = \frac{5}{3}R_1 = \frac{5}{3}\times3.6 = 6$fm

6. (c) We use the formula,
$R = R_0 A^{1/3}$
This represents relation between atomic mass and radius of the nucleus.
For berillium, $R_1 = R_0(9)^{1/3}$
For germanium, $R_2 = R_0 A^{1/3}$

$\frac{R_1}{R_2} = \frac{(9)^{1/3}}{(A)^{1/3}} \Rightarrow \frac{1}{2} = \frac{(9)^{1/3}}{(A)^{1/3}}$

$\Rightarrow \frac{1}{8} = \frac{9}{A} \Rightarrow A = 8\times9 = 72.$

7. (d) Mass defect
$\Delta m = [Zm_p + (A-Z)m_n] - M(U^{238})$

$= [92 \times 1.007825\ (238-92) \times 1.008665] - 238.050783$
$\Delta m = 1.93421$ amu
$BE = 1.93421 \times 931.5$MeV $= 1801.7$ MeV

8. (c) Given, $_1H^2 + {_1H^2} \rightarrow {_2H^4} + Q$
The total binding energy of the deutrons $= 4\times1.15 = 4.60$ MeV
The total binding energy of alpha particle $= 4\times7.1 = 28.4$ MeV
The energy released in the process $= 28.4 - 4.60 = 23.8$ MeV.

9. (c) Energy is released when stability increases. This will happen when binding energy per nucleon increases.

	Reactant	Product
Reaction (a)	60×8.5MeV $= 510$MeV	$2\times30\times5 = 300$ MeV
Reaction (b)	$120\times7.5 = 900$ MeV	$(90\times8+30\times5) = 870$ MeV
Reaction (c)	$120\times7.5 = 900$ MeV	$2\times60\times8.5 = 1020$ MeV
Reaction (d)	$90\times8 = 720$ MeV	$(60\times8.5+30\times5) = 600$ MeV

10. **(a)** Q value of the reaction
$Q = (2 \times 4 \times 7.06 - 7 \times 5.6) = 17.28$ MeV....(i)
$K_P + Q = 2K_\alpha$ ....(ii)
$\sqrt{2m_P K_P} = 2\sqrt{2m_\alpha K_\alpha}\cos\alpha$
$K_P = K_\alpha$ ...(iii)
So $K_P = 17.28$ MeV.

11. **(b)**

12. **(b)** $Q = 4(x_2 - x_1)$

13. **(a)** B.E. $= \Delta mc^2 = \Delta \times 931$ MeV
$= [2(1.0087 + 1.0073) - 4.0015] \times 931$
$= 28.4$ MeV

14. **(d)** $M({}_8O^{16}) = M({}_7N^{15}) + 1m_P$
binding energy of last proton
$= M(N^{15}) + m_P - M({}_1O^{16})$
$= 15.00011 + 1.00783 - 15.99492$
$= 0.01302$ amu $= 12.13$ MeV

15. **(c)** Energy released $= (80 \times 7 + 120 \times 8 - 200 \times 6.5)$
$= 220$ MeV.

16. **(b)** Let E be the energy of proton, then
$E + 7 \times 5.6 = 2 \times [4 \times 7.06]$
$\Rightarrow E = 56.48 - 39.2 = 17.28$MeV

17. **(b)** Energy of γ-ray photon = Rest mass energy + K.E.
$= 2(0.5) + 0.78 = 1.78$ MeV

18. **(a)**

19. **(b)** Mass defect $\Delta m = 2 \times 2.014 - 4.0026 = 0.0256$ a.m.u.
Energy released when two ${}_1H^2$ nuclei fuse
$= 0.0256 \times 931 = 23.8$ MeV
Total energy required to be produced by nuclear reaction in 1 year
$= 2500 \times 10^6 \times 3.15 \times 10^7 = 7.88 \times 10^{16}$ J
No. of nuclei of ${}_1H^2$ required

$$= \frac{7.88 \times 10^{16}\text{ J}}{23.8 \times 1.6 \times 10^{-13}} \times 2 = 4.14 \times 10^{28}$$

Mass of Deuterium required

$$= \frac{4.14 \times 10^{28}}{6.02 \times 10^{23}} \times 2 \times 10^{-3}\text{ kg} = 138\text{kg}$$

20. **(c)** ${}_1H^2 + {}_1H^2 \rightarrow {}_2He^4$
Total binding energy of two deuterium nuclei $= 1.1 \times 4 = 4.4$ MeV
Binding energy of a $({}_2He^4)$ nuclei $= 4 \times 7 = 28$ MeV
Energy released in this process $= 28 - 4.4 = 23.6$ MeV

21. **(d)** average kinetic energy per molecule $= \frac{3}{2}kT$

This kinetic energy should be able to provide the repulsive potential energy

$$\therefore \frac{3}{2}kT = 7.7 \times 10^{-14}$$

$$\Rightarrow T = \frac{2 \times 7.7 \times 10^{-14}}{3 \times 1.38 \times 10^{-23}} = 3.7 \times 10^9 \text{ K}$$

22. **(c)**

23. **(c)** the reactor will conume $\frac{9 \times 10^7}{9 \times 10^{16}} \times 1800$kg
$= 18 \times 10^{-7}$ kg

24. **(c)** $4\,{}_2^4He \longrightarrow {}_8^{16}O$
B.E. $= \Delta m \times 931.5$ MeV
$= (4 \times 4.0026 - 15.9994) \times 931.5 = 10.24$MeV

25. **(a)** Number of atoms in 2kg fuel

$$= \frac{2}{235} \times 6.023 \times 10^{26} = 5.12 \times 10^{24}$$

number of fission per second

$$= \frac{5.12 \times 10^{24}}{30 \times 24 \times 60 \times 60} = 1.978 \times 10^{18}$$

Energy released per fission
$= 200$ MeV $= 200 \times 1.6 \times 10^{-13} = 3.2 \times 10^{-11}$ J
Power output $= 3.2 \times 10^{-11} \times 1.978 \times 10^{18}$
$= 63.28 \times 10^6$ W $= 63.28$ MW

26. **(b)** $\frac{m}{t} = 4 \times 10^8$ kgs^{-1}

$E = mc^2$

$$\Rightarrow \frac{E}{t} = \left(\frac{m}{t}\right)c^2 \Rightarrow \frac{E}{t} = 4 \times 10^8 \times 9 \times 10^{16}$$

$$\Rightarrow \frac{E}{t} = 3.6 \times 10^{25}\text{ Js}^{-1} \Rightarrow \frac{E}{t} = 3.6 \times 10^{25}\text{ W}$$

27. **(d)** Fission of ${}_{92}^{235}U$ is possible by only slow neutrons. To produce slow neutrons, one has to use moderators so that fission is possible.

28. **(c)** Power level of reactor, $P = \frac{E}{\Delta t} = \frac{\Delta mc^2}{\Delta t}$
mass of the fuel consumed per hour in the reactor,

$$\frac{\Delta m}{\Delta t} = \frac{P}{c^2} = \frac{10^9}{(3\times10^8)^2}\text{kg/sec} = \frac{3600\times10^9}{9\times10^{16}}\times10^3\ \frac{\text{gm}}{\text{hr}}$$

29. **(c)** ${}_1H^2 + {}_1H^2 \rightarrow {}_2He^4 + Q$

$\Rightarrow \Delta m = m({}_2He^4) - 2m({}_1H^2)$

$\Rightarrow \Delta m = 4.0024 - 2(2.0141)$

$\Rightarrow \Delta m = -0.0258$ amu

Since, $Q = c^2\Delta m$

$\Rightarrow Q = (0.0258)(931.5)$ MeV $\Rightarrow Q = 24$ MeV.

30. **(a)** $P = n\left(\frac{E}{t}\right) \Rightarrow 1000 = \frac{n\times200\times10^6\times1.6\times10^{-19}}{t}$

$\Rightarrow \frac{n}{t} = 3.125\times10^{13}.$

31. **(a)**

32. **(d)**

33. **(a)** $\frac{N_1}{N_2} = e^{(\lambda_2-\lambda_1)} = \frac{1}{e}$

$\Rightarrow \quad t = \frac{1}{\lambda_2-\lambda_1} = 2$

34. **(c)** Ratio of number of half life taken is given as:

After 16 days

$T_{A1/2} = \frac{16}{4} = 4;\ T_{B1/2} = \frac{16}{8} = 2$

$N = N_0\left(\frac{1}{2}\right)^n \Rightarrow \frac{N_A}{N_B} = \frac{1}{2^4} : \frac{1}{2^2} = 2^2 : 2^4$

$= 4 : 16, = 1 : 4$

35. **(c)** $N = N_0e^{-\lambda t} \Rightarrow N_y = N_0(1-e^{-\lambda t})$

Rate of formation of $Y = \frac{dN}{dt} = +\lambda N_0e^{-\lambda t}$

$X \longrightarrow Y$

At $t=0,\ R = \lambda N_0 \Rightarrow t=0 \quad N_0$

$t=\infty,\ R=0 \quad \Rightarrow t=t \quad N$

$N_y = N_0(1-e^{-\lambda t})$

36. **(c)** $A_1 = (\lambda N_0)e^{-\lambda t_1}$ .....(i)

$A_2 = (\lambda 2N_0)e^{-\lambda t_2}$ .... (ii)

$t_1 - t_2 = \frac{1}{\lambda}\ell n\left(\frac{A_2}{2A_1}\right) = \ell n\left(\frac{A_2}{2A_1}\right)$

37. **(c)** $\frac{C_{14}}{C_{12}} = \frac{1}{4} = \left(\frac{1}{2}\right)^{t/5700}$

$\Rightarrow \frac{t}{5700} = 2 \Rightarrow t = 11400$ years

38. **(b)** Effective half life is calculated as

$\frac{1}{T} = \frac{1}{T_1} + \frac{1}{T_2}$

$\frac{1}{T} = \frac{1}{16} + \frac{1}{48} \Rightarrow T = 12$ years

Time in which $\frac{3}{4}$ will decay is 2 half lives $= 24$ years

39. **(d)**

40. **(a)** Let $t =$ time for the potential of metal sphere to rise by one volt.

Now $\beta-$ particles emitted in this time

$= (6.25\times10^{11})\times t$

Number of β-particles escaped in this time

$= (80/100)\times(6.25\times10^{10})t = 5\times10^{10}t$

$\therefore$ Charge acquired by the sphere in t sec.

$Q = (5\times10^{10}t)\times(1.6\times10^{-19}) = 8\times10^{-19}t$ ...(i)

($\because$ emission of β-particle lends to a charge e on metal sphere)

The capacitance C of a metal sphere is given by

$C = 4\pi\varepsilon_0\times r$

$= \left(\frac{1}{9\times10^9}\right)\times\left(\frac{10^{-3}}{2}\right) = \frac{10^{-12}}{18}$ farad ...(ii)

we know that $Q = C\times V$ {Here $V = 1$ volt}

$\therefore \quad (8\times10^{-9})t = \left(\frac{10^{-12}}{18}\right)\times1$

Solving it for t, we get $t = 6.95$ µsec.

41. **(a)** ${}_{72}A^{180} \xrightarrow{\alpha} {}_{70}A_1^{176} \xrightarrow{\beta} {}_{71}A_2^{176}$

$\xrightarrow{\alpha} {}_{69}A_3^{172} \xrightarrow{\gamma} {}_{69}A_4^{172}$

42. (d)

$$N_x = \frac{N_0}{2}e^{-\lambda_1 t} = 0.2N_0\ N_y = \frac{N_0}{2}e^{-\lambda_2 t} = 0.8N_0$$

$e^{(\lambda_1-\lambda_2)} = 4 \Rightarrow t = 8\times10^9$ years

43. (d) $X \xrightarrow{\lambda_1} Y \xrightarrow{\lambda_2} Z$

At radioactive equilibrium,

$(\lambda_X)\times(N_X) = (\lambda_Y)\times(N_Y)$

$$\frac{\lambda_X}{\lambda_Y} = \frac{N_Y}{N_X}\frac{(t_{1/2})_Y}{(t_{1/2})_X} = \frac{N_Y}{N_X} \text{ or } \frac{(t_{1/2})_Y}{1.6\times10^{10}}$$

$$= \frac{1}{3.2\times10^9}(t_{1/2})_Y = 5 \text{ years}$$

44. (c)

45. (b) $\lambda = \frac{2.303}{280}\log\frac{A_0}{6000} = \frac{2.303}{420}\log\frac{A_0}{3000}$

On solving, we get, $A_0 = 24000$ dps.

46. (c) Let N_0 be the number of atoms of X at time t = 0.

Then at t = 4 hrs (two half lives)

$$N_x = \frac{N_0}{4} \text{ and } N_y = \frac{3N_0}{4}$$

$\therefore N_x/N_y = 1/3$

and at t = 6 hrs (three half lives)

$$N_x = \frac{N_0}{8} \text{ and } N_y = \frac{7N_0}{8} \text{ or } \frac{N_x}{N_y} = \frac{1}{7}$$

The given ratio $\frac{1}{4}$ lies between $\frac{1}{3}$ and $\frac{1}{7}$.

Therefore, t lies between 4 hrs and 6 hrs.

47. (d) at time t = 0 ; $A_0 = \frac{dN_0}{dt}$

$$\frac{A}{A_0} = \frac{dN/dt}{dN_0/dt} = \frac{3260}{6520}$$

$$\text{or } \lambda = \frac{2.303}{t}\log\frac{A_0}{A} = \frac{2.303}{2\times60}\log 2$$

= 5.78 per sec

48. (d) Because radioactivity is a spontaneous phenomenon.

49. (a) Activity of sample becomes 2500 from 5000 in 75 days therefore its half life is 75 days, so

$$R = \frac{R_0}{2^{\frac{150}{75}}} = 5000 \Rightarrow R_0 = 5000\times4 = 20,000$$

50. (c) Using

$$\frac{M}{M_0} = \left(\frac{1}{2}\right)^{t/T} = \left(\frac{1}{2}\right)^{120/24} = \left(\frac{1}{2}\right)^5 = \frac{1}{32}$$

$$\text{So } M = \frac{M_0}{32} = \frac{0.1 \text{ mg}}{32} = 3.125\ \mu\text{g}.$$

51. (a) The penetrating power is dependent on velocity.

For a given energy, the velocity of γ radiation is highest and α-particle is least.

52. (d) $N = N_0 e^{-\lambda t}$

Here, $t = 5$ minutes

$$\frac{N_0}{e} = N_0\cdot e^{-5\lambda} \Rightarrow 5\lambda = 1,\ \lambda = \frac{1}{5},$$

$$\text{Now, } T_{1/2} = \frac{\ell n2}{\lambda} = 5\ \ell n2$$

53. (d)

54. (c) β-particles are charged particles emitted by the nucleus.

55. (d) $N_1 = N_0 e^{-10\lambda t}$ and $N_2 = N_0 e^{-\lambda t}$.

$$\therefore \frac{N_1}{N_2} = \frac{e^{-10\lambda t}}{e^{-\lambda t}} = \frac{1}{e^{9\lambda t}}$$

$$\text{Given } \frac{N_1}{N_2} = \frac{1}{e};\ \therefore \frac{1}{e^{9\lambda t}} = \frac{1}{e}$$

$$\text{or, } 9\lambda t = 1 \text{ or } t = \left(\frac{1}{9\lambda}\right)$$

56. (b) Let number of atoms in X = N_x

Number of atoms in Y = N_y

By question

$$\frac{N_x}{N_y} = \frac{1}{15}$$

$$\therefore \text{Part of } N_x = \frac{1}{16}(N_x + N_y) = \frac{1}{2^4}(N_x + N_y)$$

So, total 4 half lives are passed, so, age of rock is $4\times50 = 200$ years

57. (b) We know that $N_\beta = N_0(1 - e^{-\lambda t})$

$$N_\beta = \frac{6.023\times10^{23}}{24}\left[1 - e^{\frac{\ell n 2}{15}\times7.5}\right]$$

on solving we get,

$N_\beta = 7.4\times10^{21}$

If α and B are emitted simultaneously.

58. **(d)** Let N_0 be the initial number of nuclei, then

$N_1 = N_0 e^{-\lambda t_1}$ and $N_2 = N_0 e^{-\lambda t_2}$

$\therefore$ Number of nuclei decayed $= N_1 - N_2$

$$= N_0(e^{-\lambda t_1} - e^{-\lambda t_2}) = \frac{A_0}{\lambda}(e^{-\lambda t_1} - e^{-\lambda t_2})$$

$$= \frac{A_1 - A_2}{\lambda} = (A_1 - A_2)\tau.$$

59. **(a)** Use the relation $N = N_0 e^{-\lambda t}$ where the decay constant $\lambda = (0.693 / T_{half})$. The half life period T_{half} is given to be 10min. The amount N of radioactive sample remaining at the end of 5 min comes out to be about 0.707g, so that the amount of sample decayed is $(1 - 0.707) = 0.293$g.

60. **(c)** $\lambda_A = 5\lambda$ and $\lambda_B = \lambda$

At $t = 0$, $(N_0)_A = (N_0)_B$

Given, $\frac{N_A}{N_B} = \left(\frac{1}{e}\right)^2$

According to radioactive decay, $\frac{N}{N_0} = e^{-\lambda t}$

$$\therefore \frac{N_A}{(N_0)_A} = e^{-\lambda_A t} \text{ and } \frac{N_B}{(N_0)_B} = e^{-\lambda_B t}$$

From (1) and (2), $\frac{N_A}{N_B} = e^{-(5\lambda - \lambda)t}$

$$\Rightarrow \left(\frac{1}{e}\right)^2 = e^{-4\lambda t} = \left(\frac{1}{e}\right)^{4\lambda t} \Rightarrow 4\lambda t = 2 \therefore t = \frac{1}{2\lambda}.$$

61. **(b)** $t_{1/2} = 3.8$ day

$$\therefore \quad \lambda = \frac{0.693}{t_{1/2}} = \frac{0.693}{3.8} = 0.182$$

If the initial number of atom is $a = A_0$ then after time t the number of atoms is $a/20 = A$. We have to find t.

$$t = \frac{2.303}{\lambda}\log\frac{A_0}{A} = \frac{2.303}{0.182}\log\frac{a}{a/20}$$

$$= \frac{2.303}{0.182}\log 20 = 16.46 \text{ days}$$

62. **(b)** $N = N_0 e^{-\lambda t}$ $q = q_0 e^{-t/RC}$ and $\frac{N}{q} =$ constant

when $e^{-\lambda t} = e^{t/RC} \Rightarrow \lambda = \frac{1}{RC}$ $R = 200\,\Omega$.

63. **(b)** $\frac{dN}{dt} = -N(\lambda_1 + \lambda_2) = -N\lambda$

$$\Rightarrow \lambda = \lambda_1 + \lambda_2 \text{ so } \frac{1}{t} = \frac{1}{t_1} + \frac{1}{t_2} \left[\text{as } \lambda = \frac{0.693}{t}\right]$$

64. **(c)** Suppose an initial radionuclide I decays to a final product F w ith a half – life $T_{1/2}$.

At any time, $N_I = N_0 e^{-\lambda t}$

Number of product nuclei $= N_F = N_0 - N_I$

$$\frac{N_F}{N_I} = \frac{N_0 - N_I}{N_I} = \left(\frac{N_0}{N_I} - I\right)$$

$$\frac{N_0}{N_I} = \left(1 + \frac{N_F}{N_I}\right) = 1 + 0.5 = 1.5$$

$$\therefore \quad \frac{T_{1/2}\ln(1.5)}{\ln 2} = 4.5 \times 10^9 \frac{\ell n\left(\frac{3}{2}\right)}{\ell n 2} \text{ year}$$

65. **(b)** We have $K_\alpha = \frac{m_y}{m_y + m_\alpha}.Q$

$$\Rightarrow K_\alpha = \frac{A-4}{A}.Q \Rightarrow 48 = \frac{A-4}{4}.50 \Rightarrow A = 100$$

66. **(a)** $\frac{dN}{dt} = \alpha - \lambda N$ $\quad \int_0^N \frac{dN}{\alpha - \lambda N} = \int_0^t dt$

Solving $N = \frac{\alpha}{\lambda}(1 - e^{-\lambda t})$.

67. **(b)** $N_1\lambda_1 = N_2\lambda_2$ $\frac{\lambda_1}{\lambda_2} = \frac{2}{3} \Rightarrow \frac{N_2}{N_1} = \frac{2}{3}$

$\Rightarrow$ ratio of no. of nuclei of X to that of $Y = 3/2$

68. **(c)** For this substance 7 days correspond to 3.5 half - lives. Over 3 half - lifes the sample reduces to $\frac{1}{2^3} = \frac{1}{8}$ of its initial mass. After 4 half - lifes, the sample has only $\frac{1}{2^4} = \frac{1}{16}$ of its initial mass. Hence, after 3.5 half - lives the sample must contain somewhere between 1/8 and 1/16 of its initial mass. Hence 5 g is somewhere between 1/8 and 1/16 of the initial mass.

So, the initial mass is somewhere between $8 \times 5 = 40$ g and $16 \times 5 = 80$ g.

69. **(b)**

70. **(b)** $\frac{^{14}C}{^{12}C} = 1.3 \times 10^{-12}$

12 g contain 6.022×10^{23} atoms

No. atoms of $^{14}C = 6.022 \times 10^{23} \times 1.3 \times 10^{-12}$

$= 7.8286 \times 10^{11} \Rightarrow N = N_0 e^{-\lambda t}$

$$-\frac{dN}{dt} = N_0 e^{-\lambda t} \times \lambda \Rightarrow -\frac{dN}{dt} = N \times \lambda$$

$$\frac{180}{60} = 7.8286 \times 10^{11} \times \lambda$$

$$\Rightarrow \frac{1}{\lambda} = \frac{7.8286 \times 10^{11}}{3} \Rightarrow \lambda = 0.3832 \times 10^{-11}$$

$$t_{1/2} = \frac{0691}{\lambda} = \frac{0.692}{0.3832 \times 10^{-11}} = 1.80 \times 10^{11} \text{ sec}$$

Half life $= 1.8032 \times 10^{11}$ sec $= 0.5740 \times 10^{4}$ year $= 5740$ years.

71. **(d)** $\frac{dN}{dt} = 50 - \frac{N}{0.5}$ $\int_0^N \frac{dN}{50 - 2N} = \int_0^t dt$

$N = (100(1 - e^{-t/2})) = 25$

$t = 2 \ln(4/3)$

72. **(a)** $k_\alpha = \frac{M}{M+4} Q = \frac{222}{226} \times 4.87 = 4.78$ MeV

73. **(c)** **74.** **(b)**

75. **(d)** Half life, $t_{1/2} = \frac{\ln 2}{\lambda}$

Mean life, $\tau = \frac{1}{\lambda}$ $\therefore \frac{t_{1/2}}{\tau} = \ln 2.$

76. **(d)** $\frac{7}{8}$ of Cu decays in 15 minutes.

$$\therefore \text{Cu undecayed} = N = 1 - \frac{7}{8} = \frac{1}{8} = \left(\frac{1}{2}\right)^3$$

$\therefore$ No. of half lifes $= 3$

$$n = \frac{t}{T} \text{ or } 3 = \frac{15}{T}$$

$$\Rightarrow T = \text{half life period} = \frac{15}{3} = 5 \text{ minutes}$$

77. **(d)** $\left|\frac{d/N}{dt}\right| = \lambda N$

Number of radius nuclei in mg $= \frac{N_A m}{226}$

Decay constant,

$$\lambda = \frac{0.693}{t_{1/2}} = \frac{0.693}{1620 \times 3.16 \times 10^7}$$

$$\left|\frac{dN}{dt}\right| = 10 = \frac{6.02 \times 10^{23} m}{226} \times \frac{0.693}{1620 \times 3.16 \times 10^7}$$

$$\therefore \quad m = \frac{10 \times 226 \times 162 \times 3.16 \times 10^7}{6.02 \times 10^{23} \times 0.693}$$

$= 2.77 \times 10^{-10}$ g $= 2.77 \times 10^{-13}$ kg

78. **(c)** $N = N_0 e^{-\lambda/t} = 10^6 e^{-\left[\frac{\ln 2}{20} \times 10\right]}$

$$= 10^6 e^{-\ln\sqrt{2}} = \frac{10^6}{\sqrt{2}} = \approx 7.07 \times 10^5$$

79. **(d)** $(T_{1/2})_A = (t_{mean})_B$

$$\Rightarrow \frac{0.693}{\lambda_A} = \frac{1}{\lambda_B} \Rightarrow \lambda_A = 0.693 \lambda_B$$

or $\lambda_A < \lambda_B$

Also rate of decay $= \lambda N$

Initially number of atoms (N) of both are equal but since $\lambda_B > \lambda_A$, therefore B will decay at a faster rate than A

80. **(b)** Given, for ^{14}C

$A_0 = 16$ dis min^{-1} g^{-1}

$A = 12$ dis min^{-1} g^{-1}

$t_{1/2} = 5760$ years

Now, $\lambda = \frac{0.693}{t_{1/2}} = \frac{0.693}{5760}$ per year

Then, from, $t = \frac{2.303}{\lambda} \log_{10} \frac{A_0}{A}$

$$= \frac{2.303 \times 5760}{0.693} \log_{10} \frac{16}{12} \approx 2391 \text{ years}$$

81. **(a)** Isotopes : A & C have same number of protons.

82. **(b)**

83. **(b)** $\frac{dN_A}{dt} = -\lambda_1 N_A$, $\frac{dN_B}{dt} = 2\lambda_1 N_A - \lambda_2 N_B$,

$N_B = \text{maximum} \Rightarrow \frac{dN_B}{dt} = 0$

$\Rightarrow 2\lambda_1 N_A = \lambda_2 N_{B_{max}} \Rightarrow N_{B_{max}} = \frac{2\lambda_1}{\lambda_2} N_A$

$\Rightarrow N_{B_{max}} = \frac{2\lambda_1}{\lambda_2} N_0 e^{-\lambda_1 t} = 2$

84. **(c)** $N = N_0 \left(\frac{1}{2}\right)^n$ so, for 1% decay

$\frac{1}{100} N_0 = N_0 \left(\frac{1}{2}\right)^n \Rightarrow \frac{1}{100} = \left(\frac{1}{2}\right)^n$

$\Rightarrow n = \frac{2}{\log 2} \Rightarrow \frac{t}{T} = \frac{2}{\log 2} \Rightarrow t = 6.6T$

$= 6.6 \times 100 = 660$ years

If α and B are emitted simultaneously.

85. **(b)** As a result of emission of 1 α-particle, the mass number decreases by 4 units and atomic number decreases by 2 units. And by the emission of 1 positron the atomic number decreases by 1 unit but mass number remains constant.

$\therefore$ Mass number of final nucleus $= A - 12$

Atomic number of final nucleus $= Z - 8$

$\therefore$ Number of neutrons $= (A - 12) - (Z - 8)$
$= A - Z - 4$

Number of protons $= Z - 8$

$\therefore$ Required ratio $= \frac{A - Z - 4}{Z - 8}$

86. **(b)** **87.** **(c)**

88. **(a)** ${}_0^1n \longrightarrow {}_1^1H + {}_{-1}e^0 + \bar{\nu} + Q$

The mass defect during the process

$\Delta m = m_n - m_H - m_e = 1.6725 \times 10^{-27}$
$- (1.6725 \times 10^{-27} + 9 \times 10^{-31} \text{kg})$
$= -9 \times 10^{-31}$ kg

The energy released during the process

$E = \Delta mc^2$

$E = 9 \times 10^{-31} \times 9 \times 10^{16} = 81 \times 10^{-15}$ Joules

$$E = \frac{81 \times 10^{-15}}{1.6 \times 10^{-19}} = 0.511\text{MeV}$$

89. **(b)**

90. **(d)** Let initially there are total N_0 number of nuclei

At time t $\frac{N_B}{N_A} = 0.3\text{(given)} \Rightarrow N_B = 0.3N_A$

$N_0 = N_A + N_B = N_A + 0.3N_A \therefore N_A = \frac{N_0}{1.3}$

As we know $N_t = N_0 e^{-\lambda t}$ or, $\frac{N_0}{1.3} = N_0 e^{-\lambda t}$

$\frac{1}{1.3} = e^{-\lambda t} \Rightarrow ln(1.3) = \lambda t$

or, $t = \frac{ln(1.3)}{\lambda} \Rightarrow t = \frac{ln(1.3)}{\frac{ln(2)}{T}} = \frac{ln(1.3)}{ln(2)} T$

91. **(a)** The total number of protons bombarded

$= \frac{it}{\lambda} = \frac{10^{-4} \times 3600}{1.6 \times 10^{-19}} = 22.5 \times 10^{17}$

Number of ^{7}Be produced

$N = \frac{22.5 \times 10^{17}}{1000} = 22.5 \times 10^{14}$

We know that activity

$A = \lambda N$ or $A = \left(\frac{0.693}{t_{1/2}}\right) N$

$\therefore t_{1/2} = 0.693 \frac{N}{A} = 0.693 \times \frac{22.5 \times 10^{14}}{1.8 \times 10^8} = 8.63 \times 10^6$ s

92. **(b)** We know that the rate of integration $\left|-\frac{dN}{dt}\right| = A$

$\therefore A = A_0 e^{-\lambda t}$ or $2700 = 4750 e^{-\lambda \times 5}$

or $\lambda = 0.1131$ per minute

Half life $t_{1/2} = \frac{0.693}{\lambda} = \frac{0.693}{0.1131} = 6.1$ minute

93. **(a)** By conservation of momentum, we have

$0 = \vec{P}_\alpha + \vec{P}_d \Rightarrow \vec{P}_\alpha = -\vec{P}_d$ or $P_\alpha = P_d = P$

The kinetic energy released in the process

$K = K_\alpha + K_p = \frac{P^2}{2m_\alpha} + \frac{P^2}{2m_d}$

$= \frac{P^2}{2m_\alpha}\left(1 + \frac{m_\alpha}{m_d}\right) = \frac{(h/\lambda)^2}{2m_\alpha}\left(1 + \frac{m_\alpha}{m_d}\right)$

After substituting the given values, we get

$K = 6.25$ MeV

If m_p is the mass of the parent nucleus, then

$$K + (m_\alpha + m_d)c^2 = m_p c^2$$

or $\quad 6.25 + (223.61 + 4.002)c^2 = m_p c^2$

After simplifying, we get $\quad m_p = 227.62$ amu.

94. (a) Suppose x is the number of Pb^{206} nulei. The number of U^{238} nuclei will be 3x, Thus

$$3x + x = N_0$$

We know that $N = N_0 e^{-\lambda t}$ or $3x = 4xe^{-\lambda t}$

$$\therefore e^{\lambda t} = \frac{4}{3} \text{ or } t = \frac{\ln 4/3}{\lambda} = \frac{\ln 4/3}{(0.693/t_{1/2})}$$

$$= \frac{\ln 4/3}{(0.693/4.5\times 10^9)} = 1.868\times 10^9 \text{ year.}$$

95. (c) The rate dm/dt can be calculate as ;

$$\text{Power, } P = \frac{dE}{dt} = \frac{dE}{dm}\times\frac{dm}{dt} = \frac{\Delta E}{\Delta m}\times\frac{dm}{dt}$$

$$\therefore \frac{dm}{dt} = \frac{\Delta m}{\Delta E}P \qquad \ldots(i)$$

We known that 26.2 MeV $= 4.20 \times 10^{-12}$ J of thermal energy is produced when four protons are consumed. This is $\Delta E = 4.20\times 10^{-12}$ J for $\Delta m = 4 \times (1.67 \times 10^{-27}$ kg).

Substituting these values in equation (i), we have

$$\frac{dm}{dt} = \frac{\Delta m}{\Delta E}P = \frac{4(1.67\times 10^{-27})}{4.20\times 10^{-12}}\times(3.90\times 10^{26})$$

$= 6.2 \times 10^{11}$ kg/s

96. (b)

97. (c) The total energy released

$E =$ Energy released in fission process + energy released in α- decay process

$$= N_F \times 200 + N_\alpha \times (0.005517 \times 931)$$

$$= \left(\frac{8}{100}\times 10^{20}\right)\times 200 + \left(\frac{92}{100}\times 10^{20}\right)\times(0.005517\times 931)$$

$$= 20.725\times 10^{20} \text{ MeV}$$

Power output $P = E/t$

$$= \frac{20.725\times 10^{20}\times 1.6\times 10^{-13}}{10^{13}}$$

$= 3.3 \times 10^{-5}$ W.

98. (a) We know that, $N = N_0 e^{-\lambda t}$

After 2 second, $N_1 = N_0 e^{-\lambda\times 2}$

After (2 + 2) second, $N_2 = N_0 e^{-\lambda\times 4}$

According to given conditions,

$$N_0 - N_1 = n$$

or $\quad N_0 - N_0 e^{-2\lambda} = n \qquad \ldots(i)$

and $\quad N_0 e^{-2\lambda} - N_0 e^{-4\lambda} = 0.75\,n \ldots(ii)$

After solving above equations, we get

$$\lambda = 0.145 \text{ s}$$

Mean life $\qquad T = 1/\lambda = 6.9$ s.

99. (a) If A_0 is the initial activity of radioactive sample, then activity at any time

$$A = A_0 e^{-\lambda t}$$

or $1 \times 10^6 = 4 \times 10^6 e^{-\lambda\times 20}$

or $e^{-20\lambda} = \frac{1}{4}$

The count rate after 100 hour is given by

$$A' = A_0 e^{-\lambda\times 100} = A_0 e^{-100\lambda} = A_0[e^{-20\lambda}]^5$$

$$= 4\times 10^6\left[\frac{1}{4}\right]^5 = 3.91 \times 10^3 \text{ per second}$$

100. (a) $T_{1/2} = \frac{\ln 2}{\lambda} \therefore \lambda = \frac{\ln 2}{T_{1/2}}$

$$\Rightarrow \lambda_A = \frac{\ln 2}{T_A}, \lambda_B = \frac{\ln 2}{T_B} \Rightarrow \frac{\lambda_A}{\lambda_B} = \frac{T_B}{T_A}.$$

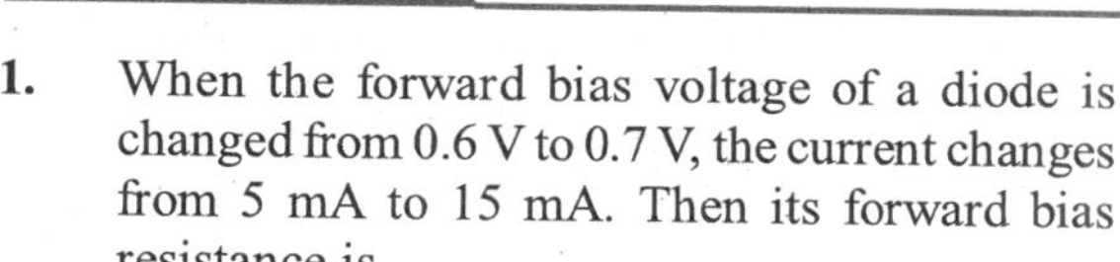

28 Semiconductor Electronics: Materials, Devices and Simple Circuits

1. When the forward bias voltage of a diode is changed from 0.6 V to 0.7 V, the current changes from 5 mA to 15 mA. Then its forward bias resistance is
(a) 0.01 Ω (b) 0.1 Ω (c) 10 Ω (d) 100 Ω

2. The I-V characteristic of a P-N junction diode is shown below. The approximate dynamic resistance of the p-n junction when a forward bias of 2 volt is applied is **Tricky**

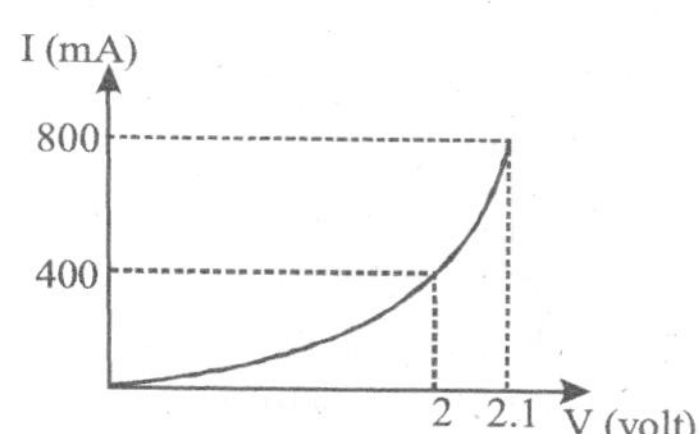

(a) 1 Ω (b) 0.25 Ω (c) 0.5 Ω (d) 5 Ω

3. Zener breakdown in a semi-conductor diode occurs when
(a) forward current exceeds certain value
(b) reverse bias exceeds certain value
(c) forward bias is equal to certain value
(d) potential barrier is reduced to zero

4. The electric conductivity of an intrinsic semiconductor increases when the electromagnetic waves of wavelength equal or shorter than 3895 nm is incident on it. The band gap of the semiconductor is
(a) 0.2 eV (b) 0.3 eV (c) 1.0 eV (d) 1.5 eV

5. A silicon diode has a threshold voltage of 0.7 V. If an input voltage given by 2 sin (πt) is supplied to a half wave rectifier circuit using this diode, the rectified output has a peak value of
(a) 2V (b) 1.4V (c) 1.3V (d) 0.7V

6. A potential barrier of 0.50 V exists across a p-n junction. If the depletion region is 5.0×10^{-7} m wide, the intensity of the electric field in this region is
(a) 1.0×10^6 V/m (b) 1.0×10^5 V/m
(c) 2.0×10^5 V/m (d) 2.0×10^6 V/m

7. The concentration of hole - electron pairs in pure silicon at T = 300 K is 7×10^{15} per cubic meter. Antimony is doped into silicon in a proportion of 1 atom in 10^7 Si atoms. Assuming that half of the impurity atoms contribute electron in the conduction band, calculate the factor by which the number of charge carriers increases due to doping. The number of silicon atoms per cubic meter is 5×10^{28} **Critical Thinking**
(a) 2.8×10^5 (b) 3.1×10^2
(c) 4.2×10^5 (d) 1.8×10^5

8. A zener diode of voltage V_Z (= 6V) is used to maintain a constant voltage across a load resistance R_L (= 1000 Ω) by using a series resistance R_s (= 100Ω). If the e.m.f. of source is E (= 9 V), what is the power being dissipated in Zener diode ?
(a) 0.144 watt (b) 0.324 watt
(c) 0.244 watt (d) 0.544 watt

9. If the ratio of the concentration of electrons to that of holes in a semiconductor is $\frac{7}{5}$ and the ratio of currents is $\frac{7}{4}$, then what is the ratio of their drift velocities?
(a) $\frac{5}{8}$ (b) $\frac{4}{5}$ (c) $\frac{5}{4}$ (d) $\frac{4}{7}$

10. The circuit has two oppositively connected ideal diodes in parallel. What is the current flowing in the circuit?

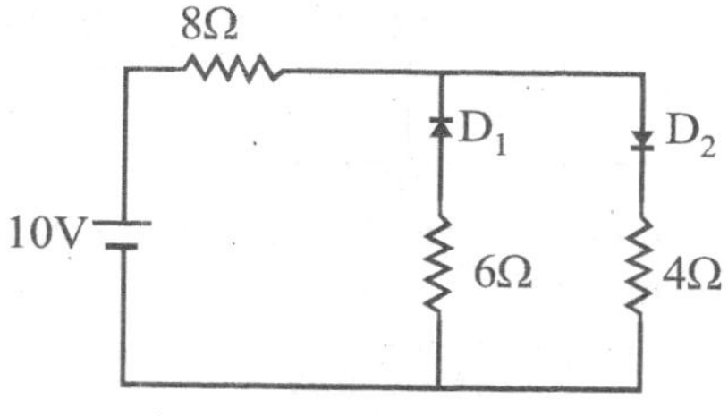

(a) 0.4A (b) 0.8A (c) 0.6A (d) 0.5A

11. A diode having potential difference 0.5 V across its junction which does not depend on current, is connected in series with resistance of 20Ω across source. If 0.1 A current passes through resistance then what is the voltage of the source?
(a) 1.5V (b) 2.0V (c) 2.5V (d) 5V

12. If the lattice constant of this semiconductor is decreased, then which of the following is correct?

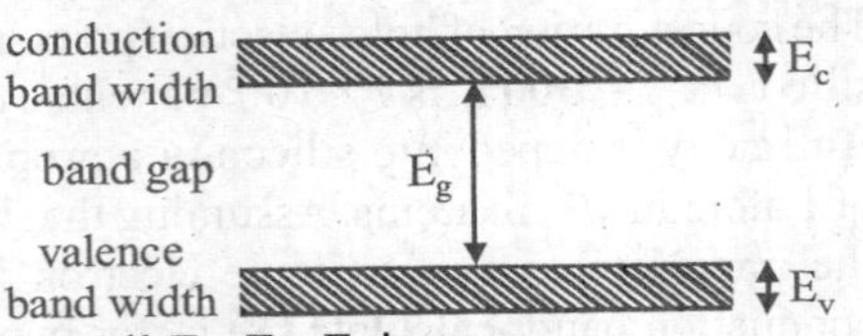

(a) All E_c, E_g, E_v increase
(b) E_c and E_v increase, but E_g decreases
(c) E_c and E_v decrease, but E_g increases
(d) All E_c, E_g, E_v decrease

13. In the following, which one of the diodes reverse biased?

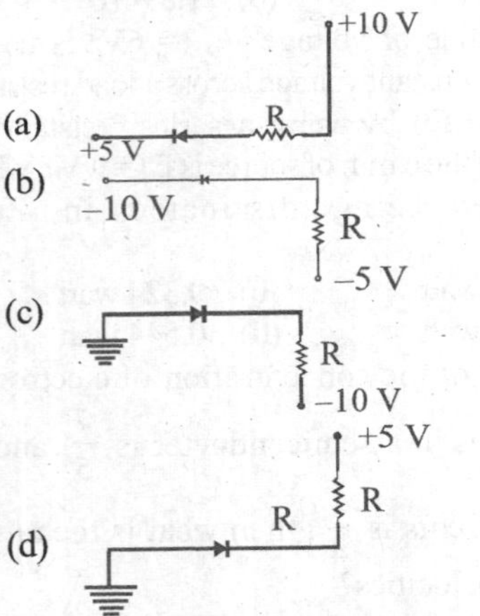

14. A Zener diode is connected to a battery and a load as show below:

Critical Thinking

4 kΩ A I_L
I_Z
I
60 V
10 V = V_Z
R_L = 2kΩ
B

The currents, I, I_Z and I_L are respectively.
(a) 15 mA, 5 mA, 10 mA
(b) 15 mA, 7.5 mA, 7.5 mA
(c) 12.5 mA, 5 mA, 7.5 mA
(d) 12.5 mA, 7.5 mA, 5 mA

15. Suppose a 'n'-type wafer is creatd by doping Si crystal having 5×10^{28} atoms/m^3 with 1ppm concentration of As. On the surface 200 ppm Boron is added to create 'P' region in this wafer. Considering $n_i = 1.5 \times 10^{16}$ m^{-3} calculate the density of the minority charge carriers in the p regions. Toughnut
(a) 2.25×10^7/m^3 (b) 1.12×10^3/m^3
(c) 3.11×10^6/m^3 (d) 2.11×10^5/m^3

16. Which of the following statements is incorrect? Tricky
(a) The drift current in a p-n junction is from the n-side to the p-side.
(b) Drift current is due to free electrons only.
(c) Silicon is preferred over germanium for making semiconductor devices.
(d) The energy gap in germanium is more than the energy gap in silicon.

17. For LED's to emit light in visible region of electromagnetic light, it should have energy band gap in the range of:
(a) 0.1 eV to 0.4 eV (b) 0.5 eV to 0.8 eV
(c) 0.9 eV to 1.6 eV (d) 1.7 eV to 3.0 eV

18. When a potential difference is applied across, the current passing through
(a) an insulator at 0K is maximum
(b) a semiconductor at 0K is zero
(c) a metal at 0K is finite
(d) a P-N diode at 300K is zero, if it is reverse biased

19. The circuit shown in following figure contains two diode D_1 and D_2 each with a forward resistance of 50 ohms and with infinite backward resistance. If the battery voltage is 6 V, the current through the 100 ohm resistance (in amperes) is

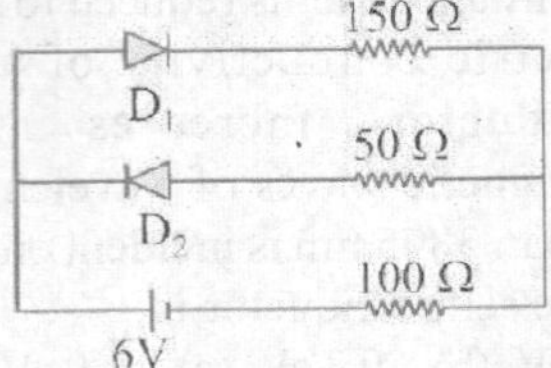

(a) zero (b) 0.02 (c) 0.03 (d) 0.036

20. Pure Si at 500K has equal number of electron (n_e) and hole (n_h) concentrations of 1.5×10^{16} m^{-3}. Doping by indium increases n_h to 4.5×10^{22} m^{-3}. The doped semiconductor is of
(a) n–type with electron concentration $n_e = 5 \times 10^{22}$ m^{-3}
(b) p–type with electron concentration $n_e = 2.5 \times 10^{10}$ m^{-3}
(c) n–type with electron concentration $n_e = 2.5 \times 10^{23}$ m^{-3}
(d) p–type having electron concentration $n_e = 5 \times 10^9$ m^{-3}

21. When p-n junction diode is forward biased then
(a) both the depletion region and barrier height are reduced
(b) the depletion region is widened and barrier height is reduced
(c) the depletion region is reduced and barrier height is increased
(d) Both the depletion region and barrier height are increased

22. If in a *p-n* junction diode, a square input signal of 10 V is applied as shown

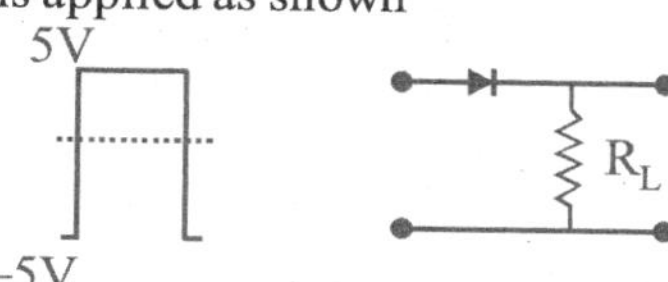

Then the output signal across R_L will be

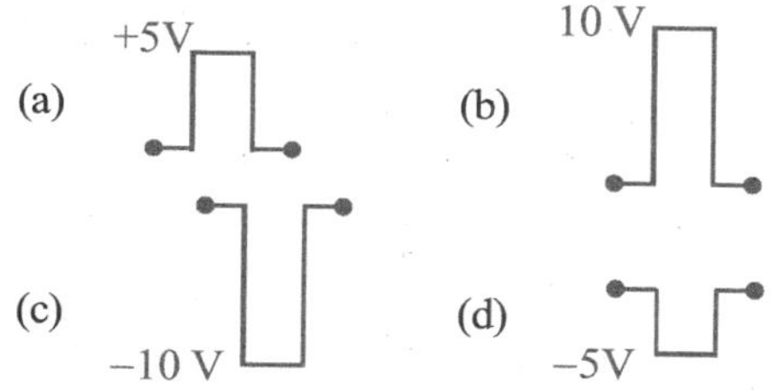

23. In a p-type semiconductor the acceptor level is situated 60 meV above the valence band. The maximum wavelength of light required to produce a hole will be
(a) 0.207×10^{-5} m (b) 2.07×10^{-5} m
(c) 20.7×10^{-5} m (d) 2075×10^{-5} m

24. In a full wave rectifier circuit operating from 150 Hz mains frequency, the fundamental frequency in the ripple would be
(a) 25 Hz (b) 50 Hz (c) 70.7 Hz (d) 300 Hz

25. Carbon, silicon and germanium have four valence electrons each. At room temperature which one of the following statements is most appropriate?
(a) The number of free electrons for conduction is significant only in Si and Ge but small in C.
(b) The number of free conduction electrons is significant in C but small in Si and Ge.
(c) The number of free conduction electrons is negligibly small in all the three.
(d) The number of free electrons for conduction is significant in all the three.

26. A 2-V battery is connected across the points A and B as shown in the figure given below. Assuming that the resistance of each diode is zero in forward bias, and infinity in reverse bias, the current supplied by the battery when its positive terminal is connected to A, is

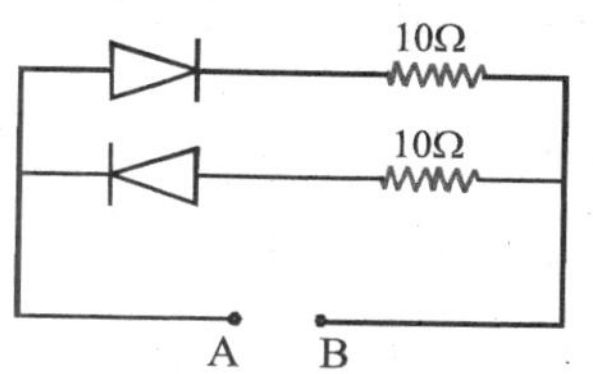

(a) 0.2 A (b) 0.4 A (c) 0.3A (d) 0.1 A

27. The correct statement is
(a) A p-type semiconductors is a negative type crystal.
(b) A p- type semiconductor is a charged crystal.
(c) The diffusion current in a p-n junction is from the p-side to the n-side.
(d) The diffusion current in a p-n junction is smaller than the drift current when the junction is in forward biased.

28. Different voltages are applied across a P-N junction and the currents are measured for each value. Which of the following graphs is obtained between voltage and current

Tricky

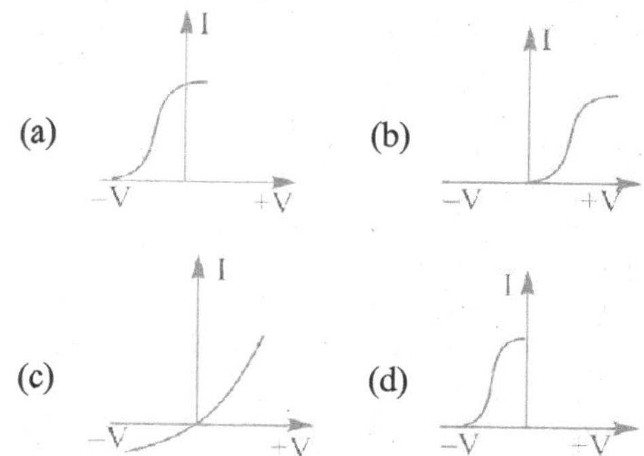

29. On doping germanium with donor atoms of density 10^{17} cm^{-3} its conductivity in mho/cm will be [Given $\mu_e = 3800$ cm^2/V–s and $n_i = 2.5 \times 10^{13}$ cm^{-13}]
(a) 30.4 (b) 60.8 (c) 91.2 (d) 121.6

30. In a p-n junction having depletion layer of thickness 10^{-6} m the potential across it is 0.1 V. The electric field is
(a) 10^7 V/m (b) 10^{-6} V/m
(c) 10^5 V/m (d) 10^{-5} V/m

31. The intrinsic conductivity of germanium at 27° is 2.13 mho m^{-1} and mobilities of electrons and holes are 0.38 and 0.18 m^{2}V^{-1}s^{-1} respectively. The density of charge carriers is
(a) 2.37×10^{19} m^{-3} (b) 3.28×10^{19} m^{-3}
(c) 7.83×10^{19} m^{-3} (d) 8.47×10^{19} m^{-3}

32. Currents flowing in each of the circuits A and B respectively are **Critical Thinking**

(Circuit A) (Circuit B)

(a) 1 A, 2A (b) 2A, 1A
(c) 4A, 2A (d) 2A, 4A

33. By increasing the temperature, the specific resistance of a conductor and a semiconductor
(a) increases for both
(b) decreases for both
(c) increases, decreases
(d) decreases, increases

34. A zener diode, having breakdown voltage equal to 15V, is used in a voltage regulator circuit shown in figure. The current through the diode is

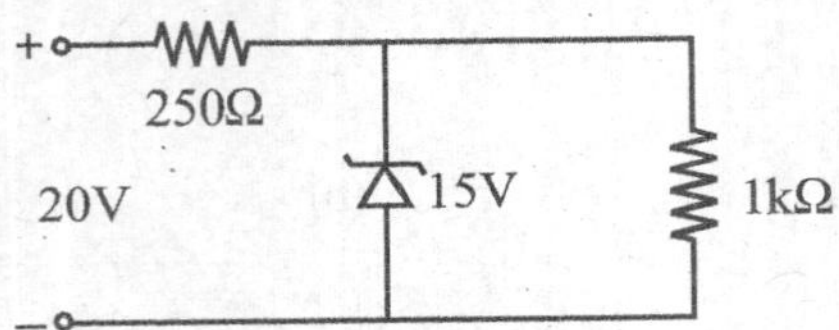

(a) 10 mA (b) 15 mA (c) 20 mA (d) 5 mA

35. The ratio of electron and hole currents in a semiconductor is 7/4 and the ratio of drift velocities of electrons and holes is 5/4, then ratio of concentrations of electrons and holes will be
(a) 5/7 (b) 7/5 (c) 25/49 (d) 49/25

36. A d.c. battery of V volt is connected to a series combination of a resistor R and an ideal diode D as shown in the figure below. The potential difference across R will be

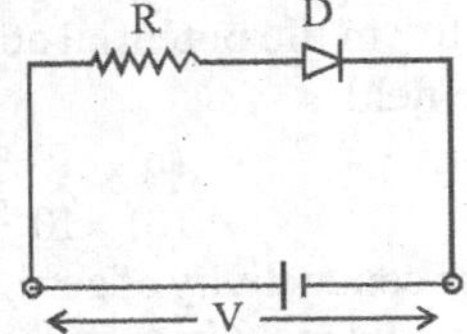

(a) 2V when diode is forward biased
(b) zero when diode is forward biased
(c) V when diode is reverse biased
(d) V when diode is forward biased

37. In fig., the input is across the terminals A and C and the output is across B and D. Then the output is

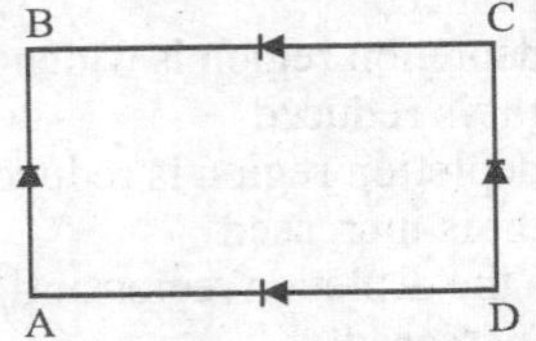

(a) zero (b) same as the input
(c) half wave rectified (d) full wave rectified

38. Assuming the diodes to be of silicon with forward resistance zero, the current I in the following circuit is **Tricky**

(a) 0 (b) 9.65 mA
(c) 10 mA (d) 10.36 mA

39. An experiment is performed to determine the 1–V characteristics of a Zener diode, which has a protective resistance of R = 100 Ω, and a maximum power of dissipation rating of 1W. The minimum voltage range of the DC source in the circuit is:
(a) 0 – 5V (b) 0 – 24 V
(c) 0 – 12 V (d) 0 – 8V

40. In germanium the energy gap is about 0.75 eV. The wavelength of light which germanium starts absorbing is
(a) 5000 Å (b) 1650 Å
(c) 16500 Å (d) 165000 Å

41. The grid voltage of any triode valve is changed from –1 volt to –3 volt and the mutual conductance is 3×10^{-4} mho. The change in plate circuit current will be
(a) 0.8 mA (b) 0.6 mA (c) 0.4 mA (d) 1 mA

42. Which one of the following statement is false ?
(a) Pure Si doped with trivalent impurities gives a p-type semiconductor
(b) Majority carriers in a n-type semiconductor are holes
(c) Minority carriers in a p-type semiconductor are electrons
(d) The resistance of intrinsic semiconductor decreases with increase of temperature

43. If a battery of 5V is connected across a plate of the semiconductor of area 2×10^{-4} m^2 and of thickness 1.2×10^{-3} m, then the current flowing through the plate is
(a) 0.32 mA (b) 0.64 mA
(c) 1.28 mA (d) 2.56 mA

44. In a triode, $g_m = 2 \times 10^{-3}$ ohm^{-1} ; $\mu = 42$; resistance of load, R = 50 kilo ohm. The voltage amplification obtained from this triode will be

(a) 30.42 (b) 29.57 (c) 28.18 (d) 27.15

45. In the energy band diagram of a material shown below, the open circles and filled circles denote holes and electrons respectively. The material is

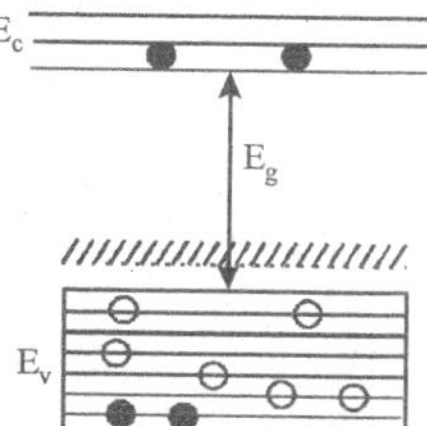

(a) an insulator
(b) a metal
(c) an n-type semiconductor
(d) a p-type semiconductor

46. The electric conductivity of an intrinsic semiconductor increases when the electromagnetic waves of wavelength equal or shorter than 2475 nm is incident on it. The charge carrier concentration of this semiconductor at room temperature is 1.6×10^{16} m^{-3} and the mobilities of electrons and holes at the same temperature are 0.4 m^2 V^{-1} S^{-1} and 0.2 m^2 V^{-1}S^{-1} respectively. Regarding these information the band gap B is

(a) B = 0.5 eV (b) B = 1.3 eV
(c) B = 1.54×10^{-3} eV (d) None of these

47. A transistor is connected in common -emitter configuration. The collector supply is 8V and the voltage drop across a resistor of 800 W in the collector is 0.5 V. If the current gain factor α is 0.96. Find the base current.

(a) 26 μA (b) 16 μA
(c) 10 μA (d) 5 μA

48. Copper, a monovalent, has molar mass 63.54 g/mol and density 8.96 g/cm^3. What is the number density n of conduction electron in copper?

(a) 3.2×10^{20} m^{-3} (b) 8.49×10^{26} m^{-3}
(c) 6.2×10^{31} m^{-3} (a) None

49. A pnp transistor is used in common-emitter mode in an amplifier circuit. A change of 40 μA in the base current brings a change of 2 mA in collector current and 0.04 V in base emitter voltage. Which of the following is the correct value of the input resistor R_{in}?

(a) R_{in} = 9 kΩ (b) R_{in} = 75 kΩ
(c) R_{in} = 1 kΩ (d) None of these

50. The Si diode and the Ge diode conduct at 0.7V and 0.3V respectively, in the circuit given below:

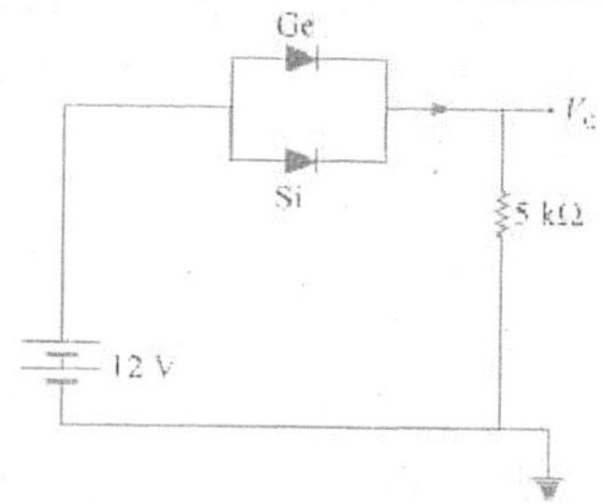

If now the Ge diode connection are reversed, what will be the new value of i′

(a) i′ = 15.8 mA (b) i′ = 6.2 mA
(c) i′ = 1.17 mA (d) i′ = 2.26 mA

51. A 5V battery may be connected across the points A and B as shown in figure. Assume that the resistance of each diode is zero in forward bias and infinity in reverse bias. Find the current i_2 by the battery if the positive terminal of the battery is connected to the point B.

(a) i_2 = 1.3 A
(b) i_2 = 2.8 A
(c) i_2 = 0.25 A
(d) i_2 = 0.75 A

52. A potential barrier of 0.3 Volt exists across a p-n junction. The depletion region is 1 μm wide. The intensity of electric field in this region is E. If electron with speed 5×10^5 m/s approaches this p-n junction from n-side, what will be its speed V on entering the p-side? **Toughnut**

(a) V = 12×10^5 m/s (b) V = 1×10^4 m/s
(c) V = 3.8×10^5 m/s (d) V = 1.6×10^4 m/s

53. A npn transistor in a common emitter mode is used as a simple voltage amplifier with a collector current of 4mA. the terminal of a 8V battery is connected to the collector through a load resistance R_L and to the base through a resistance R_B. The collector emitter voltage V_{CE} = 4V, base-emitter voltage V_{BE} = 0.6 and the base current amplification factor β_{dc} = 100, calculate the value of R_B. **Toughnut**

(a) R_B = 15 kΩ (b) R_B = 200 kΩ
(c) R_B = 1 kΩ (d) R_B = 185 kΩ

54. In a common emitter (CE) amplifier having a voltage gain G, the transistor used has transconductance 0.03 mho and current gain 25. If the above transistor is replaced with another one with transconductance 0.02 mho and current gain 20, the voltage gain will be

(a) 1.5G (b) $\frac{1}{3}$G (c) $\frac{5}{4}$G (d) $\frac{2}{3}$G

55. The current gain in the common emitter mode of a transistor is 10. The input impedance is $20\,k\Omega$ and load of resistance is $100\,k\Omega$. The power gain is
(a) 300 (b) 500 (c) 200 (d) 100

56. A transistor connected in common emitter configuration has input resistance $R_B = 2\ k\Omega$ and load resistance of $5\ k\Omega$. If $\beta = 60$ and an input signal 12 mV is applied, calculate the voltage gain, the power gain and the value of output voltage

Critical Thinking

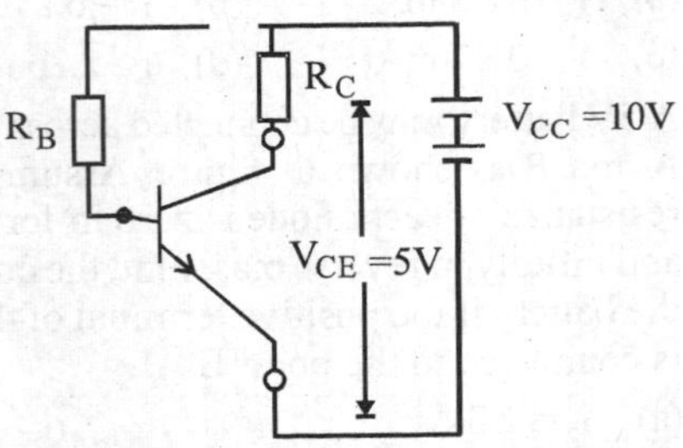

(a) $A_v = 150, V_{out} = 1.8V$, and power gain $= 9000$
(b) $A_v = 20, V_{out} = 1V$, and power gain $= 2000$
(c) $A_v = 150, V_{out} = 1.5V$, and power gain $= 8500$
(d) $A_v = 20, V_{out} = 1.5V$, and power gain $= 2000$

57. If the given transistor is used as an amplifier then for input resistance of 80Ω and load resistance of $16k\Omega$, the output voltage corresponding to the input voltage of 12mV will be
(a) 37.5mV (b) 37500V
(c) 300V (d) 300mV

58. An NPN-transistor circuit is arranged as shown in figure. It is

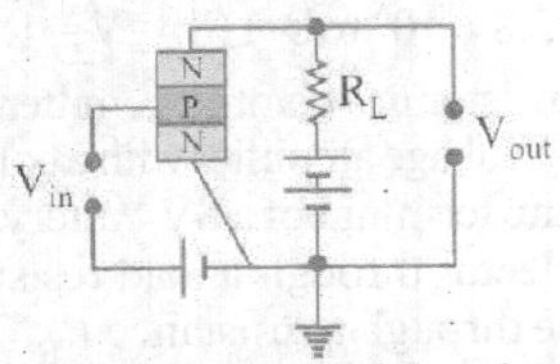

(a) a common base amplifier circuit
(b) a common emitter amplifier circuit
(c) a common collector amplifier circuit
(d) neither of the above

59. A npn transistor is connected in common emitter configuration in a given amplifier. A load resistance of $800\ \Omega$ is connected in the collector circuit and the voltage drop across it is 0.8 V. If the current amplification factor is 0.96 and the input resistance of the circuit is 192Ω, the voltage gain and the power gain of the amplifier will respectively be :
(a) 4, 3.84 (b) 3.69, 3.84
(c) 4, 4 (d) 4, 3.69

60. A transistor can be used as an amplifier when
(a) emitter is forward biased and collector is reverse biased
(b) both are reverse biased
(c) both emitter and collector are forward
(d) None of these

61. In the case of a common emitter transistor amplifier the ratio of the collector current to the emitter current I_c/I_e is 0.96. The current gain of the amplifier is
(a) 6 (b) 48 (c) 24 (d) 12

62. In a npn transistor 10^{10} electrons enter the emitter in 10^{-6} s. 4% of the electrons are lost in the base. The current transfer ratio will be
(a) 0.98 (b) 0.97 (c) 0.96 (d) 0.94

63. The input resistance of a silicon transistor is 100 W. Base current is changed by 40 μA which results in a change in collector current by 2 mA. This transistor is used as a common emitter amplifier with a load resistance of $4\ K\Omega$. The voltage gain of the amplifier is :
(a) 2000 (b) 3000 (c) 4000 (d) 1000

64. For transistor action **Tricky**
(1) Base, emitter and collector regions should have similar size and doping concentrations.
(2) The base region must be very thin and lightly doped.
(3) The eimtter-base junction is forward biased and base-collector junction is reverse based.
(4) Both the emitter-base junction as well as the base-collector junction are forward biased.
(a) (3) and (4) (b) (4) and (1)
(c) (1) and (2) (d) (2) and (3)

65. In the following common emitter configuration an NPN transistor with current gain $\beta = 100$ is used. The output voltage of the amplifier will be

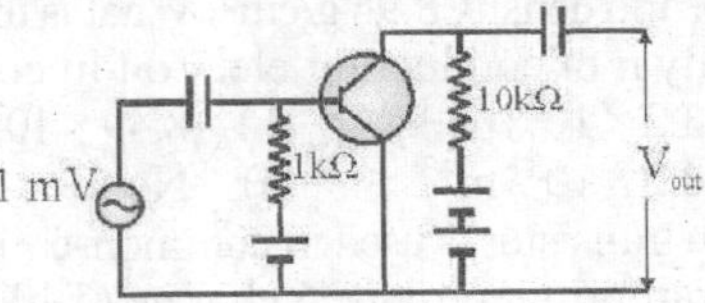

(a) 10 mV (b) 0.1 V (c) 1.0 V (d) 10 V

66. The transfer ratio β of a transistor is 50. The input resistance of the transistor when used in the common emitter configuration is $1\ k\Omega$. The peak value of the collector A.C. current for an A.C. input voltage of 0.01 V peak is
(a) 100 μA (b) 0.01 mA
(c) 0.25 mA (d) 500 μA

67. In a common base mode of a transistor, the collector current is 5.488 mA for an emitter current of 5.60 mA. The value of the base current amplification factor (β) will be
(a) 49 (b) 50 (c) 51 (d) 48

68. A transistor has $\beta = 40$. A change in base current of 100 μ A, produces change in collector current
(a) 40×100 microampere
(b) $(100 - 40)$ microampere
(c) $(100 + 40)$ microampere
(d) 100/40 microampere

69. Transfer characteristics [output voltage (V_0) vs input voltage (V_1)] for a base biased transistor in CE configuration is as shown in the figure. For using transistor as a switch, it is used **Tricky**
(a) in region (III)
(b) both in region (I) and (III)
(c) in region (II)
(d) in region (I)

70. When NPN transistor is used as an amplifier
(a) electrons move from base to emitter
(b) holes move from emitter to base
(c) electrons move from collector to base
(d) holes move from base to emitter

71. The ratio of work function and temperature of two emitters are 1 : 2, then the ratio of current densities obtained by them will be
(a) 4 : 1 (b) 2 : 1 (c) 1 : 2 (d) 1 : 4

72. In a common emitter transistor amplifier the audio signal voltage across the collector is 3V. The resistance of collector is 3 kΩ. If current gain is 100 and the base resistance is 2 kΩ, the voltage and power gain of the amplifier is **Critical Thinking**
(a) 15 and 200 (b) 150 and 15000
(c) 20 and 2000 (d) 200 and 1000

73. A transistor has a base current of 1 mA and emitter current 90 mA. The collector current will be
(a) 90 mA (b) 1 mA (c) 89 mA (d) 91 mA

74. In an experiment with npn transistor amplifier in common emitter configuration, the current gain of the transistor is 100. If the collector current changes by 1mA, what will be the change in emitter current?
(a) 1.1 mA (b) 1.01 mA
(c) 0.01 mA (d) 10 mA

75. In common emitter amplifier the $\frac{I_c}{I_e}$ is 0.98. The current gain will be
(a) 4.9 (b) 7.8 (c) 49 (d) 78

76. For a common base amplifier, the values of resistance gain and voltage gain are 3000 and 2800 respectively. The current gain will be
(a) 1.1 (b) 0.98 (c) 0.93 (d) 0.83

77. In a transistor, the change in base current from 100 μA to 125 μA causes a change in collector current from 5 mA to 7.5 mA, keeping collector-to-emitter voltage constant at 10 V. What is the current gain of the transistor? **Critical Thinking**
(a) 200 (b) 100 (c) 50 (d) 25.

78. A common emitter amplifier has a voltage gain of 50, an input impedance of 100Ω and an output impedance of 200Ω. The power gain of the amplifier is
(a) 1000 (b) 1250 (c) 100 (d) 500

79. Current gain of a transistor in common base mode is 0.95. Its value in common emitter mode is
(a) 0.95 (b) 1.5 (c) 19 (d) $(19)^{-1}$

80. The current gain in transistor in common base mode is 0.99. To change the emitter current by 5 mA, the necessary change in collector will be
(a) 0.196 mA (b) 2.45 mA
(c) 4.95 mA (d) 5.1 mA

81. In common emitter amplifier, the current gain is 62. The collector resistance and input resistance are 5 kΩ an 500Ω respectively. If the input voltage is 0.01 V, the output voltage is **Critical Thinking**
(a) 0.62 V (b) 6.2 V (c) 62 V (d) 620 V

82. What is the voltage gain in a common emitter amplifier, where input resistance is 3 Ω and load resistance 24 Ω, $\beta = 0.6$?
(a) 8.4 (b) 4.8 (c) 2.4 (d) 480

83. A transistor is operated in common emitter configuration at $V_C = 2$V such that a change in the base current from 100 μA to 300 μA produces a change in the collector current from 10mA to 20 mA. The current gain is
(a) 50 (b) 75 (c) 100 (d) 25

84. The figure shows a logic circuit with two inputs A and B and the output C. The voltage wave forms across A, B and C are as given. The logic gate circuit is :

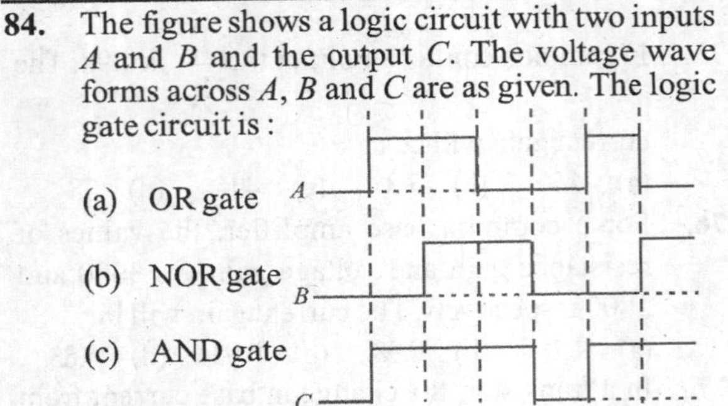

(a) OR gate
(b) NOR gate
(c) AND gate
(d) NAND gate

85. For the given combination of gates, if the logic states of inputs A, B, C are as follows A = B = C = 0 and A = B = 1, C = 0 then the logic states of output D are

(a) 0, 0 (b) 0, 1
(c) 1, 0 (d) 1, 1

86. The following circuit represents

(a) OR gate
(b) XOR gate
(c) AND gate
(d) NAND gate

87. In the circuit below, A and B represent two inputs and C represents the output.
The circuit represents

(a) NOR gate
(b) AND gate
(c) NAND gate
(d) OR gate

88. What will be the input of A and B for the Boolean expression $\overline{(A+B)}.\overline{(A.B)}=1$

(a) 0, 0 (b) 0, 1 (c) 1, 0 (d) 1, 1

89. The logic circuit shown below has the input waveforms 'A' and 'B' as shown. Pick out the correct output waveform.

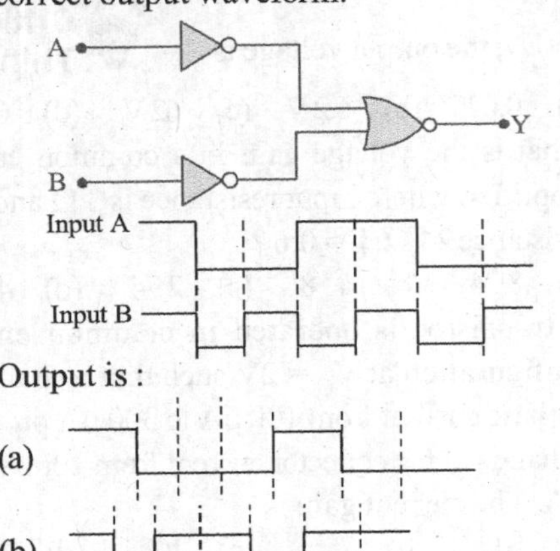

Output is :

(a)

(b)

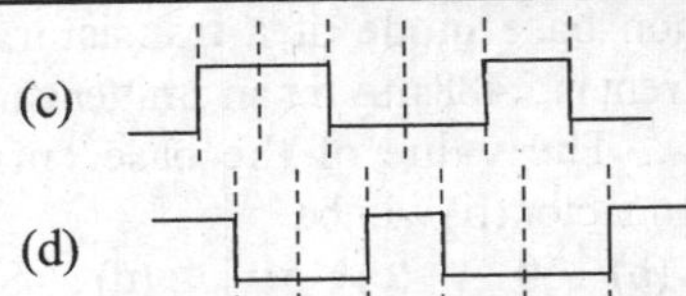

(c)

(d)

90. A system of four gates is set up as shown. The 'truth table' corresponding to this system is : **Tricky**

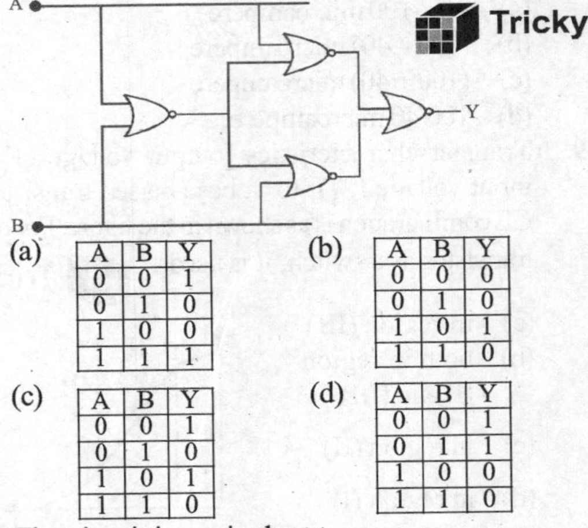

(a)

A	B	Y
0	0	1
0	1	0
1	0	0
1	1	1

(b)

A	B	Y
0	0	0
0	1	0
1	0	1
1	1	0

(c)

A	B	Y
0	0	1
0	1	0
1	0	1
1	1	0

(d)

A	B	Y
0	0	1
0	1	1
1	0	0
1	1	0

91. The circuit is equivalent to

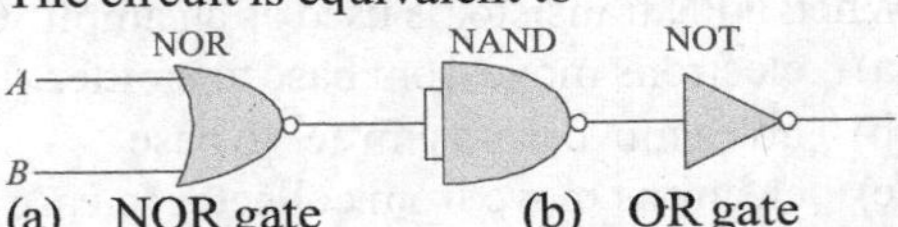

(a) NOR gate (b) OR gate
(c) AND gate (d) NAND gate

92. The combination of gates shown below yields

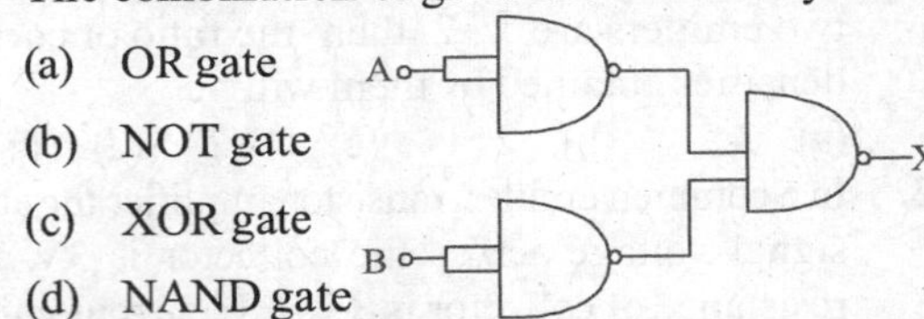

(a) OR gate
(b) NOT gate
(c) XOR gate
(d) NAND gate

93. The output of an OR gate is connected to both the inputs of a NAND gate. The combination will serve as a:

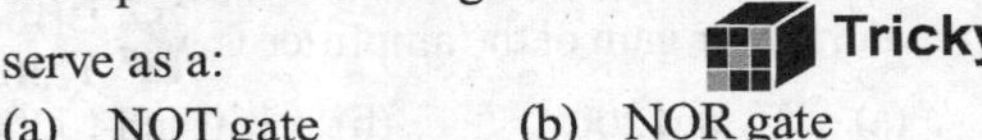

(a) NOT gate (b) NOR gate
(c) AND gate (d) OR gate

94. The forward biased diode connection is:

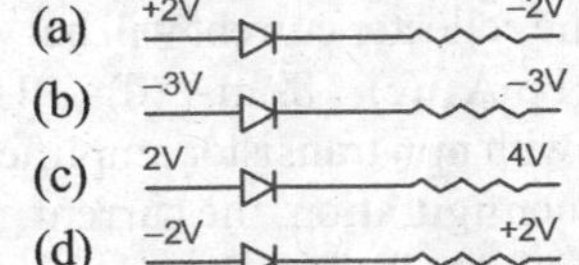

(a)
(b)
(c)
(d)

95. A red LED emits light at 0.1 watt uniformly around it. The amplitude of the electric field of the light at a distance of 1 m from the diode is : **Toughnut**

(a) 5.48 V/m (b) 7.75 V/m
(c) 1.73 V/m (d) 2.45 V/m

Answer KEYs

1	(c)	11	(c)	21	(a)	31	(a)	41	(b)	51	(c)	61	(c)	71	(d)	81	(b)	91	(a)
2	(b)	12	(c)	22	(a)	32	(c)	42	(b)	52	(c)	62	(c)	72	(b)	82	(b)	92	(a)
3	(b)	13	(d)	23	(b)	33	(c)	43	(c)	53	(d)	63	(a)	73	(c)	83	(a)	93	(b)
4	(b)	14	(d)	24	(d)	34	(d)	44	(b)	54	(d)	64	(d)	74	(b)	84	(a)	94	(a)
5	(c)	15	(a)	25	(a)	35	(b)	45	(d)	55	(b)	65	(c)	75	(c)	85	(d)	95	(d)
6	(a)	16	(d)	26	(a)	36	(d)	46	(a)	56	(a)	66	(d)	76	(c)	86	(b)		
7	(d)	17	(d)	27	(c)	37	(d)	47	(a)	57	(c)	67	(a)	77	(b)	87	(d)		
8	(a)	18	(b)	28	(c)	38	(c)	48	(b)	58	(b)	68	(a)	78	(b)	88	(a)		
9	(c)	19	(b)	29	(b)	39	(c)	49	(c)	59	(a)	69	(b)	79	(c)	89	(a)		
10	(b)	20	(d)	30	(c)	40	(c)	50	(d)	60	(a)	70	(d)	80	(c)	90	(a)		

Hints & Solutions

1. **(c)** Forward bias resistance $= \dfrac{\Delta V}{\Delta I}$

$$= \frac{(0.7-0.6)V}{(15-5)mA} = \frac{0.1}{10\times10^{-3}} = 10\Omega.$$

2. **(b)** $R = \dfrac{\Delta V}{\Delta I} = \dfrac{2.1-2}{(800-400)\times10^{-3}} = \dfrac{1}{4} = 0.25\Omega$

3. **(b)** At the zener breakdown voltage the electric field of the junction is high (10^6 V/m) enough to pull valence electrons from the host atoms. The emission of electrons from the host atoms accounts for high current at the breakdown.

4. **(b)** Forbidden energy gap $= \dfrac{12400\ (eV-Å)}{\lambda\ (Å)}$

$$= \frac{12400}{38950} eV = 0.3\ eV.$$

5. **(c)** Peak value of rectified output voltage = peak value of input voltage – barrier voltage
$= 2 - 0.7 = 1.3$ V.

6. **(a)** $E = \dfrac{V}{d} = \dfrac{0.50}{5\times10^{-7}} V/m = 1.0\times10^{6} V/m$

7. **(d)** In pure semiconductor electron-hole pair $= 7\times10^{15}/m^3$
$n_{initial} = n_h + n_e = 14\times10^{15}$ after doping donor Impurity

$$N_D = \frac{5\times10^{28}}{10^7} = 5\times10^{21} \text{ and } n_e = \frac{N_D}{2}$$

$= 2.5\times10^{21}$

So, $n_{final} = n_h + n_e$

$\Rightarrow n_{final} \approx n_e \approx 2.5\times10^{21}$ ($\because n_e >> n_h$)

$$\text{Factor} = \frac{n_{final} - n_{initial}}{n_{initial}}$$

$$= \frac{2.5\times10^{21} - 14\times10^{15}}{14\times10^{15}} \approx \frac{2.5\times10^{21}}{14\times10^{15}} = 1.8\times10^{5}$$

8. **(a)** Here, $E = 9V$; $V_z = 6$; $R_L = 1000\Omega$ and $R_s = 100\Omega$,
Potential drop across series resistor
$V = E - V_Z = 9 - 6 = 3V$
Current through series resistance R_S is

$$I = \frac{V}{R} = \frac{3}{100} = 0.03\ A$$

Current through load resistance R_L is

$$I_L = \frac{V_Z}{R_L} = \frac{6}{1000} = 0.006\ A$$

Current through Zener diode is
$I_Z = I - I_L = 0.03 - 0.006 = 0.024$ amp.
Power dissipated in Zener diode is
$P_Z = V_Z I_Z = 6\times0.024 = 0.144$ Watt

9. **(c)** $\dfrac{I_e}{I_h} = \dfrac{n_e eAv_e}{n_h eAv_h} \Rightarrow \dfrac{7}{4} = \dfrac{7}{5}\times\dfrac{v_e}{v_h} \Rightarrow \dfrac{v_e}{v_h} = \dfrac{5}{4}$

10. **(b)** D_2 is forward biased whereas D_1 is reversed biased.
So effective resistance of the circuit

$R = 8 + 4 = 12\Omega$ $\qquad \therefore i = \dfrac{10}{12} = 0.8$ A

11. **(c)** 12. **(c)**

13. **(d)** p-side connected to low potential and n-side is connected to high potential.

14. **(d)** Here, $R = 4\,k\Omega = 4\times10^3\,\Omega$ $V_i = 60$ V
Zener voltage $V_z = 10$ V $R_L = 2\,k\Omega = 2\times10^3\,\Omega$

Load current, $I_L = \frac{V_Z}{R_L} = \frac{10}{2\times10^3} = 5$ mA

Current through R, $I = \frac{V_i - V_Z}{R} = \frac{60-10}{4\times10^3}$

$= 12.5$ mA

Fom circuit diagram,

$I = I_Z + I_L$

$\Rightarrow 12.5 = I_Z + 5 \Rightarrow I_Z = 12.5 - 5 = 7.5$ mA

15. (a) In 'n' region; number of e^- is due to As:

$n_e = N_D = 1\times10^{-6}\times5\times10^{28}$ atoms/m^3

$n_e = 5\times10^{22}$/m^3

The minority carriers (hole) is

$$n_h = \frac{n_i^2}{n_e} = \frac{(1.5\times10^{16})^2}{5\times10^{22}} = \frac{2.25\times10^{32}}{5\times10^{22}}$$

$n_h = 0.45\times10$/m^3

Similarly, when Boron is implanted a 'p' type is created with holes

$n_h = N_A = 200\times10^{-6}\times5\times10^{28} = 1\times10^{25}$/m^3

This is far greater than e^- that existed in 'n' type wafer on which Boron was diffused.

Therefore, minority carriers in created 'p' region

$$n_e = \frac{n_i^2}{n_h} = \frac{2.25\times10^{32}}{1\times10^{25}} = 2.25\times10^7/\text{m}^3$$

16. (d) Silicon is cheaper than germanium, so it is preferred over germanium. But energy gap in germanium is smaller than silicon.

17. (d)

18. (b) At 0 K, the motion of electrons cease and so electric current becomes zero.

19. (b) The current will flow through diode D_1 and so

$$i = \frac{6}{(150+50+100)} = 0.02A$$

20. (d) $n_i^2 = n_e n_h$

$(1.5\times10^{16})^2 = n_e(4.5\times10^{22})$

$\Rightarrow n_e = 0.5\times10^{10}$ or $n_e = 5\times10^9$

Given $n_h = 4.5\times10^{22} \Rightarrow n_h >> n_e$

$\therefore$ Semiconductor is p-type and

$n_e = 5\times10^9$ m^{-3}.

21. (a) Both the depletion region and barrier height is reduced.

22. (a) The current will flow through R_L when the diode is forward biased.

23. (b) $\lambda = \frac{hc}{E}$

$$= \frac{6.62\times10^{-34}\times3\times10^8}{(60\times10^{-3}\times1.6\times10^{-19})} = 2.07\times10^{-5}\text{ m}$$

24. (d) Input frequency, $f = 150$ Hz $\Rightarrow T = \frac{1}{150}$

For full wave rectifier, $T_1 = \frac{T}{2} = \frac{1}{300}$

$\Rightarrow f_1 = 300$ Hz.

25. (a)

26. (a) As one of the diodes (the lower one) will be in reverse bias, no current will pass through it, so the effective resistance of the circuit $= 10\Omega$.

27. (c) **28.** (c)

29. (b) Conductivity

$\sigma = n_i e\mu_e = 10^{17}\times(1.6\times10^{-19})\times3800$

$= 60.8$ mho/cm

30. (c) $E = \frac{V}{d} = \frac{0.1}{10^{-6}} = 10^5$ V/m

31. (a)

32. (c) In circuit A, both the diodes are in forward bias and in parallel, so

$R = 2\,\Omega$

Current, $i = \frac{V}{R} = \frac{8}{2} = 4A$.

In circuit B, only one diode is in forward bias,

so current, $i = \frac{8}{4} = 2$A.

33. (c) The resistivity of conductor increases with increase in temperature. The resistivity of semiconductor decreases as the temperature increases

34. (d)

35. (b) $I = nAev_d$ or $I \propto nv_d$

$$\therefore \frac{I_e}{I_h} = \frac{n_e v_e}{n_h v_h} \text{ or } \frac{n_e}{n_h} = \frac{I_e}{I_h}\times\frac{v_h}{v_e} = \frac{7}{4}\times\frac{4}{5} = \frac{7}{5}$$

36. (d)

37. (d) It is the circuit of full wave rectifier.

38. (c) $i = \frac{V}{R} = \frac{20}{2\times10^3} = 10$ mA.

39. (c) The minimum voltage range of DC source is given by

$V^2 = PR$ $\because$ $P = 1$ watt, $R = 100\Omega = 1\times100$

$\therefore V = 10$ volt.

40. (c) $\lambda = \frac{hc}{E} = \frac{6.63\times10^{-34}\times3\times10^8}{0.75\times1.6\times10^{-19}} = 16500$Å

41. (b) $g_m = \frac{\Delta I_p}{\Delta V_g}$

or $\Delta I_p = g_m\times\Delta V_g = 3\times10^{-4}\times[-3-(-1)]$

$= -0.6\times10^{-3}$ A $=$ shortage of 0.6×10^{-3} A

42. (b) Majority carriers in an n-type semiconductor are electrons.

43. (c)

44. (b) $A_v = \frac{\mu R}{r_p + R} = \frac{\mu R}{(\mu/g_m)+R}$

$$= \frac{42\times(50\times10^3)}{42/(2\times10^{-3})+50\times10^3} = 29.57$$

45. **(d)**

46. **(a)** Forbidden energy gap $= \dfrac{12400\ (\text{eV}-\text{Å})}{\lambda\ (\text{Å})}$

$$= \frac{12400}{24750}\text{eV} = 0.5 \text{ eV}$$

47. **(a)**

48. **(b)** If M is the molar mass and ρ is the density then volume of one mole

$$V = \frac{M}{\rho}.$$

The number of atoms per unit volume

$$= \frac{N_A}{V} = \frac{N_A}{M/\rho} = \frac{N_A\rho}{M} = \frac{(6.02\times10^{23})\times(8.96)}{63.54}$$

$= 8.49\times10^{22}\ \text{cm}^{-3} = 8.49\times10^{28}\ \text{m}^{-3}$

As each copper (monovalent) atom has one electron, so number of electrons per unit volume $= 8.49\times10^{26}\ \text{m}^{-3}$

49. **(c)** Input resistance

$$R_{in} = \frac{\Delta V_{BE}}{\Delta i_B} = \frac{0.04}{40\times10^{-6}} = 1k\Omega$$

50. **(d)** **51.** **(c)**

52. **(c)** The intensity of electric field

$$E = \frac{V}{d} = \frac{0.3}{1\times10^{-6}} = 3\times10^5\ V/m$$

The electric field retarded the electron. The retardation

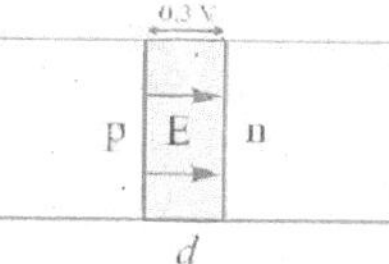

$$a = \frac{Ee}{m}$$

The speed of the electron on entering p-side

$$v^2 = u^2 - 2as = u^2 - 2\frac{eE}{m}.d$$

$$= (5\times10^5)^2 - \frac{2\times1.6\times10^{-19}\times3\times10^5}{9.1\times10^{-31}}\times(1\times10^{-6})$$

$$= 3.8\times10^5\ \text{m/s}$$

53. **(d)** By the definition

$$\beta_{dc} = \frac{i_C}{i_B}$$

$$\therefore\quad i_B = \frac{i_C}{\beta_{dc}} = \frac{4mA}{100} = 0.04\ \text{mA}$$

From figure

$$i_B R_B + V_{BE} = V_{CC}$$

$$\therefore R_B = \frac{V_{CC}-V_{BE}}{i_B} = \frac{(8-0.6)}{0.04\times10^{-3}} = 185\ \text{k}\Omega$$

Also $R_L i_C + V_{CE} = V_{CC}$

$$\therefore\ R_L = \frac{V_{CC}-V_{CE}}{i_C} = \frac{8-4}{8\times10^{-3}} = 1k\Omega$$

54. **(d)**

55. **(b)** The power gain in case of CE amplifier,

Power gain $= \beta^2\times$ Resistance gain

$$= \beta^2\times\frac{R_o}{R_i} = (10)^2\times5 = 500.$$

56. **(a)** Voltage gain

$$A_v = \beta\frac{R_C}{R_B} = 60\times\frac{5\times10^3}{2\times10^3} = 150$$

Power gain $= \beta^2\dfrac{R_C}{R_B} = 60\times60\times2.5 = 9000$

Base current $= I_B = \dfrac{12\times10^{-3}}{2\times10^3} = 6\times10^{-6}\,\text{A}$

$I_C = \beta I_B = 60\times6\times10^{-6} = 3.6\times10^{-4}\,\text{A}$

Output $= I_C R_L = 3.6\times10^{-4}\times5\times10^3\Omega = 1.8\text{V}$

57. **(c)**

58. **(b)** In the circuit emitter is common.

59. **(a)** $P_v = A_v\alpha = 4\times0.96 = 3.84$

60. **(a)** The emitter produces majority charge carriers, so in order that these reach collector, emitter is forward biased, while collector is reverse biased.

61. **(c)** $\dfrac{I_c}{I_e} = 0.96 \Rightarrow I_c = 0.96 I_e$

But $I_e = I_c + I_b = 0.96 I_e + I_b$

$\Rightarrow I_b = 0.04 I_e$

$\therefore$ Current gain $\beta = \dfrac{I_c}{I_b} = \dfrac{0.96 I_e}{0.04 I_e} = 24$

62. **(c)**

63. **(a)** Voltage gain $(A_V) = \dfrac{V_{out}}{V_{in}} = \dfrac{I_{out}}{I_{in}}\times\dfrac{R_{out}}{R_{in}}$

$$A_V = \frac{2\times10^{-3}}{40\times10^{-6}}\times\frac{4\times10^3}{100} = 2\times100 = 2000$$

64. **(d)**

65. **(c)** Voltage gain,

$$\frac{V_0}{V_i} = \beta\frac{R_2}{R_1}\ \text{ or }\ V_0 = \left[\beta\frac{R_2}{R_1}\right]\times V_i$$

$$= \left[100\times\frac{10}{1}\right]\times1\times10^{-3} = 1\text{V}.$$

66. **(d)** $i_c = \beta\cdot\dfrac{V_s}{R_{in}} = 50\times\dfrac{0.01}{1000} = 500\times10^{-6}\,\text{A}$

$= 500\ \mu\text{A}$

67. (a) $I_C = 5.488\text{ mA}, I_e = 5.6\text{ mA}$

$$\alpha = \frac{5.488}{5.6}, \beta = \frac{\alpha}{1-\alpha} = 49$$

68. (a) $\beta = \frac{\Delta I_C}{\Delta I_B}$ or $\Delta I_C = \beta \Delta I_B = 40 \times 100\ \mu A$

69. (b) I $\rightarrow$ ON II $\rightarrow$ OFF
In IInd state it is used as a amplifier it is active region.

70. (d) holes will diffuse from base to emitter

71. (d) $\frac{J_1}{J_2} = \frac{AT_1^2 e^{-W_1/kT_1}}{AT_2^2 e^{-W_2/kT_2}}$

$$= \left(\frac{T_1}{T_2}\right)^2 e^{-\frac{W_1}{kT_1} + \frac{W_2}{kT_2}} = \left(\frac{1}{2}\right)^2 e^0 = \frac{1}{4}$$

72. (b) Given, current gain $\beta = 100, R_c = 3k\Omega, R_b = 2k\Omega$

Voltage gain $(A_v) = \beta \frac{R_c}{R_b} = 100\left(\frac{3}{2}\right) = 150$

Power gain $= A_v \beta = 150\ (100) = 15000$

73. (c) $I_C = I_E - I_B = 90 - 1 = 89\text{ mA}$

74. (b) $\beta = \frac{\Delta i_c}{\Delta i_b}$, so $\Delta i_b = \frac{1 \times 10^{-3}}{100} = 0.01\text{ mA}$

$$\Delta i_e = \Delta i_b + \Delta i_c = 0.01 + 1 = 1.01\text{ mA}$$

75. (c)

76. (c) Current gain, $\alpha = \frac{A_V}{A_R} = \frac{2800}{3000} = 0.93$

77. (b) Current gain $= \frac{\Delta I_C}{\Delta I_B}$ when V_{CE} is constant.

$$= \frac{2.5 \times 10^{-3}}{25 \times 10^{-6}} = 0.1 \times 10^3 = 100$$

$[\Delta I_B = 125\ \mu A - 100\ \mu A = 25\ \mu A$
$\Delta I_C = 7.5\text{ mA} - 5\text{ mA} = 2.5\text{ mA}]$

78. (b) Voltage gain $= \beta \times$ Impedance gain

$$50 = \beta \times \frac{200}{100} = 2\beta \Rightarrow \beta = 25$$

and power gain $= \beta^2 \times \frac{200}{100} = 1250.$

79. (c) $\beta = \frac{\alpha}{1-\alpha} = \frac{0.95}{1-0.95} = \frac{0.95}{0.05} = 19$

80. (c) $\Delta I_C = \alpha\, \Delta I_E = 0.99 \times 5 = 4.95\text{ mA}$

81. (b) $\frac{V_o}{V_{in}} = \frac{R_o}{R_{in}} \times \beta = \frac{5 \times 10^3 \times 62}{500} = 10 \times 62 = 620$

$V_o = 620 \times V_{in} = 620 \times 0.01 = 6.2\text{ V}$

$\therefore V_o = 6.2$ volt.

82. (b) Voltage gain, $A_v = \beta \frac{R_L}{R_i} = 0.6 \times \frac{24}{3} = 4.8$

83. (a) The current gain

$$\beta = \frac{\Delta I_C}{\Delta I_B} = \frac{10\text{mA}}{200\mu A} = \frac{10 \times 10^3}{200} = 50$$

84. (a)

A	0	1	1	0
B	0	0	1	1
C	1	1	1	1

OR gate

85. (d) Truth table of the gate is as follows :

A	B	C	$A+B$	$Y = (A+B).C$	$\overline{Y}$
0	0	0	0	0	1
1	1	0	1	0	1

86. (b) Output of upper AND gate $= \overline{A}B$
Output of lower AND gate $= A\overline{B}$
$\therefore$ Output of OR gate, $Y = A\overline{B} + B\overline{A}$
This is boolean expression for XOR gate.

87. (d) The truth table for the above logic gate is :

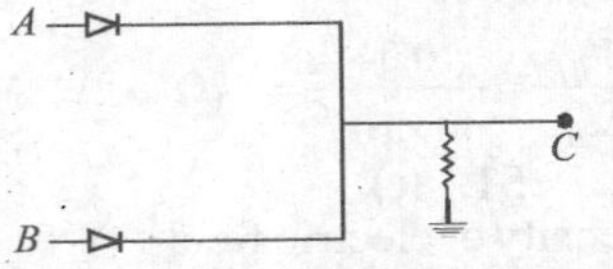

A	B	C
1	1	1
1	0	1
0	1	1
0	0	0

This truth table follows the boolean algebra $C = A + B$ which is for OR gate

88. (a) For $A = 0, B = 0$, $\overline{A+B} = 1$ and $\overline{A.B} = 1$

So $\overline{(A+B).(A.B)} = 1.$

89. (a) **90.** (a)

91. (a) NOR + (NAND = NOT) + NOT $\Rightarrow$ NOR.

92. (a) The final boolean expression is,

$$X = \overline{(\overline{A}.\overline{B})} = \overline{\overline{A}} + \overline{\overline{B}} = A + B \Rightarrow \text{OR gate}$$

93. (b) $\overline{(A+B)}$ = NOR gate
When both inputs of NAND gate are connected, it behaves as NOT gate
OR + NOT = NOR.

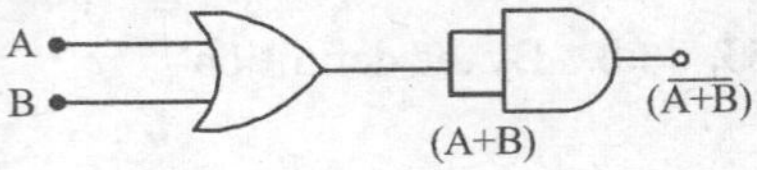

94. (a) P —▷|— n
For forward bias, p-side must be at higher potential than n-side. $\Delta V = (+)Ve$

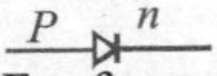

95. (d) Using $U_{av} = \frac{1}{2}\varepsilon_0 E^2$ But $U_{av} = \frac{P}{4\pi r^2 \times c}$

$$\therefore \frac{P}{4\pi r^2} = \frac{1}{2}\varepsilon_0 E^2 \times c \Rightarrow E = \left(\frac{2P}{4\pi\varepsilon_0} \times \frac{1}{cr^2}\right)^{\frac{1}{2}}$$

$$\Rightarrow E = \left(2 \times 9 \times 10^9 \times \frac{0.1}{3 \times 10^8 \times 1 \times 1}\right)^{\frac{1}{2}}$$

$$= \sqrt{6} = 2.45\text{ V/m}$$

29 Communication Systems

1. Which mode of communication is not employed for the transmission of T.V. signals?
   (a) Ground wave propagation
   (b) sky wave propagation
   (c) space wave propagation
   (d) None of these
2. Audio signal cannot be transmitted because

   Critical Thinking

   (a) the signal has more noise
   (b) the signal cannot be amplified for distance communication
   (c) the transmitting antenna length is very small to design
   (d) the transmitting antenna length is very large and impracticable
3. An antenna is a device
   (a) that converts electromagnetic energy into radio frequency signal
   (b) that converts radio frequency signal into electromagnetic energy
   (c) that converts guided electromagnetic waves into free space electromagnetic waves and vice-versa
   (d) None of these
4. Modulation is the process of superposing
   (a) low frequency audio signal on high frequency waves
   (b) low frequency radio signal on low frequency audio waves
   (c) high frequency audio signal on low frequency radio waves
   (d) low frequency audio signal on low frequency radio waves
5. The maximum peak to peak voltage of an AM wire is 24 mV and the minimum peak to peak voltage is 8 mV. The modulation factor in percentage is
   (a) 10% (b) 20% (c) 25% (d) 50%
6. A laser beam is used for carrying out surgery because it

   Tricky

   (a) is highly monochromatic
   (b) is highly coherent
   (c) is highly directional
   (d) can be sharply focussed
7. Which of the following frequencies will be suitable for beyond-the horizon communication using sky waves?
   (a) 10 kHz (b) 10 MHz
   (c) 1 GHz (d) 1000 GHz
8. The sky wave propagation is suitable for radio-waves of frequency
   (a) upto 2 MHz
   (b) from 2 MHz to 20 MHz
   (c) from 2 MHz to 30 MHz
   (d) from 2 MHz to 50 MHz
9. The maximum range d_{max}, of radar is
   (a) proportional to the cube root of the peak transmitted power
   (b) proportional to the fourth root of the peak transmitted power
   (c) proportional to the square root of the peak transmitted power
   (d) Not related to the peak transmitted power at all
10. Frequencies in the UHF range normally propagate by means of –
    (a) Ground waves (b) Sky waves.
    (c) Surface waves (d) Space waves
11. A T.V. tower has a height 150 m. What is the population density around the T.V. tower if the total population covered is 50 lakh? The radius of earth is 6.4×10^6 m.

    Critical Thinking

    (a) 800 km^{-2} (b) 829 km^{-2}
    (c) 82.6 km^{-2} (d) 826.6 km^{-2}
12. What will be the image frequency of an FM radio receiver that is tuned to 98.6 MHz broadcast station?
    (a) 111.8 MHz (b) 108 MHz
    (c) 121.6 MHz (d) 132 MHz
13. The temperature of earth's atmosphere increases with height in
    (a) troposphere
    (b) stratosphere and ionosphere
    (c) mesosphere and tonosphere
    (d) None of these
14. Critical frequency ν_c for the ionosphere is –
    (a) $\nu_c = 9.N_{max}$ (b) $\nu_c = 9\sqrt{N_{max}}$
    (c) $\nu_c = 6.IN max$ (d) $\nu_c = 6.N_{max}$

15. The phenomenon by which light travels in an optical fibres is
(a) Reflection
(b) Refraction
(c) Total internal reflection
(d) Transmission

16. The attenuation in optical fibre is mainly due to
(a) absorption
(b) scattering
(c) neither absorption nor scattering
(d) both (a) and (b)

17. The fundamental radio antenna is a metal rod which has a length equal to
(a) λ in free space at the frequency of operation
(b) $\lambda/2$ in free space at the frequency of operation
(c) $\lambda/4$ in free space at the frequency of operation
(d) $3\lambda/4$ in free space at the frequency of operation

18. Long range radio transmission is possible when the radio waves are reflected from the ionosphere. For this to happen the frequency of the radio waves must be in the range:
(a) 80 - 150 MHz (b) 8 - 25 MHz
(c) 1 - 3 MHz (d) 150 - 1500 kHz

19. The TV telecast is to cover a radius of 120 km (given the radius of the Earth = 6400 km), the height of the transmitting antenna is
(a) 1280 m (b) 1125 m (c) 1560 m (d) 79 m

20. The radio waves of frequency 300 MHz to 3000 MHz belong to
(a) high frequency band
(b) very high frequency band
(c) ultra high frequency band
(d) super high frequency band

21. Electromagnetic waves with frequencies greater than the critical frequency of ionosphere cannot be used for communication using sky wave propagation because: **Tricky**
(a) the refractive index of the ionosphere becomes very high for $f > f_c$
(b) the refractive index of the ionosphere becomes very low for $f > f_c$
(c) the refractive index of the ionosphere very high for $f < f_c$
(d) none of the above

22. In an amplitude modulated wave for audio frequency of 500 cycle/second, the appropriate carrier frquency will be
(a) 50 cycles/s (b) 100 cycles/s
(c) 500 cycles/s (d) 50, 000 cycles/s

23. Long distance short-wave radio broadcasting uses
(a) ground wave (b) ionospheric wave
(c) direct wave (d) sky wave

24. What is the modulation index of an over modulated wave
(a) 1 (b) Zero (c) <1 (d) >1

25. Which of the following statements is correct?
(a) Short waves are completely absorbed by ionosphere.
(b) Radio waves are unpolarised electromagnetic waves.
(c) Optical interference between fibres is zero.
(d) Shorter the wavelength, the smaller is the velocity of wave propagation.

26. Broadcasting antennas are generally
(a) omnidirectional type
(b) vertical type
(c) horizontal type (d) None of these

27. Which range of frequencies is suitable for sky wave propagation?
(a) 1 kHz to 500 kHz (b) 1 MHz to 2 MHz
(c) 2 MHz to 20 MHz (d) Above 30 MHz

28. In earth's atmosphere, for F_2-layer, the virtual height and critical frequency in night time are
(a) 210 km and 5 MHz (b) 250 km and 6 MHz
(c) 280 km and 7 MHz (d) 350 km and 6 MHz

29. In ground or surface wave propagation, the loss of power in a signal wave is due to* **Critical Thinking**
(a) induced charges in earth which travel in ground along with the wave, resulting the alternating currents in the earth's surface
(b) interference of waves
(c) superposition of waves
(d) high frequency of signal wave.

30. A transducer used at the transmitting end, serves the purpose of converting
(a) electrical signal to sound form
(b) sound signal to electrical form
(c) electrical signal to magnetic form
(d) sound signal to magnetic form

31. In which frequency range, space waves are normally propagated
(a) HF (b) VHF (c) UHF (d) SHF

32. In an optical fibre, the diameter of the core is nearly
(a) 10^{-2} m (b) 10^{-4} m
(c) 10^{-8} m (d) 10^{-10} m

33. Communication satellite is used when the distance of transmission is greater than –
(a) 100km (b) 500km
(c) 10km (d) 1000km

34. The electron density of a layer of ionosphere at a height 150 km from the earth's surface is 9×10^9 per m^3. For the sky transmission from this layer up to a range of 250 km,
The critical frequency of the layer is
(a) 2 Hz (b) 2.7 Hz
(c) 2.78 kHz (d) 2.7 MHz

35. What should be the maximum acceptance angle at the aircore interface of an optical fibre if n_1 and n_2 are the refractive indices of the core and the cladding, respectively?
(a) $\sin^{-1}(n_2/n_1)$ (b) $\sin^{-1}\sqrt{n_1^2 - n_2^2}$
(c) $\left[\tan^{-1}\frac{n_2}{n_1}\right]$ (d) $\left[\tan^{-1}\frac{n_1}{n_2}\right]$

36. Space wave propagation is useful for –
(a) low frequency waves and large distances
(b) low frequency waves and small distances
(c) high frequency waves
(d) ultra-high frequency waves

37. Which one of the following is used in line-transmission –
(a) coaxial cable (b) optical fibre cable
(c) twisted pair (d) all the above

38. A TV tower has a height of 100m. By how much the height of tower be increased to triple its coverage range?
(a) 800 m (b) 700m (c) 600m (d) 500m

39. In a radio receiver, noise is generally developed at –
(a) IF stage (b) audio stage
(c) RF stage (d) receiving antenna

40. Range of frequencies allotted for commercial FM radio broadcast is –
(a) 88 to 108 MHz (b) 88 to 108 kHz
(c) 8 to 88 MHz (d) 88 to 108 GHz

41. A super heterodyne receiver is designed to receive transmitted signals between 5 and 10 MHz. The tuning range of the local oscillate for IF frequency 600 kHz for high-side tuning would be

Critical Thinking

(a) 4.6 to 9.6 MHz (b) 5.6 to 10.6 MHz
(c) 4.6 to 10.6 MHz (d) 5.6 to 9.6 MHz

42. For television broadcasting, the frequency employed is normally –
(a) 30-300 MHz (b) 3-30 MHz
(c) 30-300 KHz (d) 30-300 Hz

43. If the heights of transmitting and the receiving antennas are each equal to h, the maximum line-of-sight distance between them is (R is the radius of earth)
(a) $\sqrt{2Rh}$ (b) $\sqrt{4Rh}$ (c) $\sqrt{6Rh}$ (d) $\sqrt{8Rh}$

44. A signal of peak voltage 5 V at 10 kHz is modulated using a carrier wave of peak voltage 10 V at 2 MHz. The side band frequencies are
(a) 2010 kHz, 1990 kHz (b) 2000 kHz, 1990 kHz
(c) 2010 kHz, 2000 kHz (d) 2100 kHz, 1900 kHz

45. If the maximum amplitude of an amplitude modulated wave is 25 V and the minimum amplitude is 5 V, the modulation index is :
(a) $\frac{1}{5}$ (b) $\frac{1}{3}$ (c) $\frac{3}{2}$ (d) $\frac{2}{3}$

46. The electromagnetic waves of frequency 2 MHz to 30 MHz are used
(a) in ground wave propagation
(b) in sky wave propagation
(c) in microwave propagation
(d) in satellite communication

47. 100% modulation in FM means

Tricky

(a) actual frequency deviation > maximum allowed frequency deviation
(b) actual frequency deviation = maximum allowed frequency deviation
(c) actual frequency deviation ≥ maximum allowed frequency deviation
(d) actual frequency deviation < maximum allowed frequency deviation

48. The antenna current of an AM transmitter is 8 A when only the carrier is sent, but it increases to 893 A when the carrier is sent, but it increases to 893 A when the carrier is modulated by a single sine wave. Find the percentage modulation.
(a) 60.1% (b) 70.1% (c) 80.1% (d) 50.1%

49. When radio waves passes through ionosphere, phase difference between space current and capacitive displacement current is
(a) 0 rad (b) $(3\pi/2)$ rad
(c) $(\pi/2)$ rad (d) π rad

50. The rms value of a carrier voltage is 100 volts. Compute its rms value when it has been amplitude modulated by a sinusoidal audio voltage to a depth of 30%.

Toughnut

(a) 94 V (b) 104.5V (c) 114.4 V (d) 124 V

51. A carrier frequency of 1 MHz and peak value of 10 V is amplitude modulated with a signal frequency of 10 kHz with peak value of 0.5 V. Then, the modulation index and the side band frequencies respectively are **Critical Thinking**
(a) 0.05 and 1 ± 0.010 MHz
(b) 0.5 and 1 ± 0.010 MHz
(c) 0.5 and 1 ± 0.005 MHz
(d) 0.05 and 1 ± 0.005 MHz

52. The worng statement is
(a) Electromagnetic waves with frequencies more than the critical frequency of ionosphere cannot be used for communication using sky wave propagation.
(b) The refractive index of the ionosphere becomes very high for frequencies higher than the critical frequency.
(c) Diode lasers are used as optical sources in optical communication.
(d) Diode lasers consume more energy.

53. A sinusoidal carrier voltage of frequency 10 MHz and amplitude 200 volts is amplitude modulated by a sinusoidal voltage of frequency 10 kHz producing 40% modulation. Calculate the frequency of upper and lower sidebands.
(a) 10010 kHz, 9990 kHz
(b) 1010 kHz, 990 kHz
(c) 10100 Hz, 9990 Hz
(d) 1010 MHz, 990 MHz

54. The most useful transmission for low frequency and small distance is through –
(a) sky waves (b) ground waves
(c) surface waves (d) space waves

55. A 1000 kHz carrier is simultaneously modulated with 300 Hz, 800 Hz and 2 kHz audio waves. The frequencies present in the output is
(a) 999.7 kHz, 100.3 kHz, 999.2 kHz
(b) 1000.8 kHz, 998 kHz, 1002 kHz
(c) 1002.8 kHz, 996 kHz, 1106 kHz
(d) both (a) and (b)

56. The maximum distance upto which TV transmission from a TV tower of height h can be received is proportional to –
(a) $h^{1/2}$ (b) h (c) $h^{3/2}$ (d) h^2

57. For good demodulation of AM signal of carrier frequency f, the value of RC should be
(a) $RC = \frac{1}{f}$ (b) $RC < \frac{1}{f}$
(c) $RC \geq \frac{1}{f}$ (d) $RC >> \frac{1}{f}$

58. An audio frequency of 10 kHz is transmitted by SSB after AM with carrier waves of frequency 1MHz. The frequency of current in output load is **Tricky**
(a) 1010 kHz or 990 kHz
(b) 1010 MHz or 1010 MHz
(c) 110 kHz or 990 kHz
(d) 110 MHz or 990 MHz

59. At maximum modulation (i.e. m = 1), the power in each sideband is .... of that of carrier.
(a) 25% (b) 50% (c) 40% (d) 60%

60. In an FM system a 7 kHz signal modulates 108 MHz carrier so that frequency deviation is 50 kHz. The carrier swing is
(a) 7.1 (b) 2.8 (c) 5.8 (d) 3.2

61. An FM signal has a resting frequency of 105 MHz and highest frequency of 105.03 MHz when modulated by a signal of frequency 5 kHz. The carrier swing is
(a) 25 kHz (b) 54 kHz
(c) 60 kHz (d) 75 kHz

62. A detector circuit must –
(a) rectify the signal
(b) deliver an audio output signal
(c) perform the above two functions
(d) None of the above

63. An oscillator is producing FM waves of frequency 2 kHz with a variation of 10 kHz. What is the modulating index?
(a) 5 (b) 4 (c) 3 (d) 2

64. An AM- signal is given as
$x_{AM}(t) = 100\,[p(t) + 0.5g(t)] \cos \omega_c t$
in interval $0 \leq t \leq 1$. One set of possible values of the modulating signal and modulation index would be **Critical Thinking**
(a) t, 0.5 (b) t, 1.0 (c) t, 1.5 (d) t^2, 2.0

65. In frequency modulated wave:
(a) frequency varies with time
(b) amplitude varies with time
(c) both frequency and amplitude vary with time
(d) both frequency and amplitude are constant

66. In light modulation, which characteristic of the carrier light wave is varied?
(a) Amplitude (b) Frequency
(c) Phase (d) Intensity

67. An FM radio station operating at 630 kHz is permitted to broadcast audio frequencies up to 6 kHz. The band pass filter in its modulation circuit can retain the frequencies
(a) 636 kHz, 630 kHz (b) 12 kHz, 6 kHz
(c) 1260 kHz, 6 kHz (d) 1260 kHz, 630 kHz

68. What is the band width in amplitude modulation?
(a) Equal to signal frequency
(b) Two times the signal frequency
(c) Half the signal frequency
(d) One fourth the signal frequency

69. A 10 kW carrier is sinusoidally modulated by two carriers corresponding to a modulation index of 30% and 40% respectively then total power radiated by the modulator is Toughnut
(a) 10.25 kW (b) 11.25 kW
(c) 12.75 kW (d) 17 kW

70. As the modulation level is increased, the carrier power –
(a) is increased (b) is decreased
(c) remains the same (d) None of the above

71. A device with input x(t) and output y(t) is characterized by: $y(t) = x^2(t)$.
An FM signal with frequency deviation of 90 kHz and modulating signal bandwidth of 5 kHz is applid to this device. The bandwidth of the output signal is Critical Thinking
(a) 370 kHz (b) 190 kHz
(c) 380 kHz (d) 95 kHz.

72. 12 signals each band limited to 5 kHz are to be transmitted by frequency-division multiplexer. If AM-SSB modulation guard band of 1 kHz is used then the bandwidth of multiplexed signal is
(a) 101 kHz (b) 99 kHz
(c) 84 kHz (d) 71 kHz

73. If level of modulation is increased .... power is increased.
(a) carrier
(b) sideband
(c) carrier as well as sideband
(d) none of the above

74. Determine the modulation index and carrier swing of an FM carrier having a frequency deviation of 25 kHz and a modulating signal of 5 kHz.
(a) 7,70 kHz (b) 8,80 kHz
(c) 5,50 kHz (d) 10,100 kHz

75. A radio station has two channels. One is AM at 1020 kHz and the other FM at 89.5 MHz. For good results you will use
(a) longer antenna for the AM channel and shorter for the FM
(b) shorter antenna for the AM channel and longer for the FM
(c) same length antenna will work for both
(d) information given is not enough to say which one to use for which

76. Band width is dependent on –
(a) the frequency of carrier
(b) the frequency of modulated signal
(c) the thickness of the cable
(d) the resistance of the circuit.

77. An audio signal represented as 25 sin 2π (2000 t) amplitude modulated by a carrier wave : 60 sin 2π(100, 000)t. The modulation index of the modulated signal is Tricky
(a) 25% (b) 41.6% (c) 50% (d) 75%

78. If a carrier wave $c(t) = A \sin \omega_c t$ is amplitude modulated by a modulator signal $m(t) = A \sin \omega_m t$ then the equation of modulated signal $[C_m(t)]$ and its modulation index are respectively
(a) $C_m(t) = A(1 + \sin \omega_m t) \sin \omega_c t$ and 2
(b) $C_m(t) = A(1 + \sin \omega_m t) \sin \omega_m t$ and 1
(c) $C_m(t) = A(1 + \sin \omega_m t) \sin \omega_c t$ and 1
(d) $C_m(t) = A(1 + \sin \omega_c t) \sin \omega_m t$ and 2

79. When the modulating frequency is doubled, the modulation index is halved and the modulating voltage remains constant. The modulation system is
(a) Amplitude modulation
(b) Phase modulation
(c) Frequency modulation
(d) All of the above

80. Pre-emphasis in FM system is done to
(a) compress modulating signal
(b) expand modulating signal
(c) amplify lower frequency component of the modulating signal
(d) amplify higher frequency component of the modulating signal

81. A broadcast radio transmitter radiates 12 kW when percentage of modulation is 50%, then the unmodulated carrier power is

(a) 5.67 kW (b) 7.15 kW
(c) 9.6 kW (d) 12 kW

82. AM is used for broadcasting because

(a) it is more noise immune than other modulation systems
(b) it requires less transmitting power compared with other systems
(c) its use avoids receiver complexity
(d) no other modulation system can provide the necessary bandwidth faithful transmission.

83. The frequency deviation in a FM transmission is 18.75 KHz. If it broadcasts in 88-108 MHz band, then the percent modulation is

(a) 10% (b) 25% (c) 50% (d) 75%

84. A transmitter radiates 10 kW of power with the carrier unmodulated and 11.8 kW with the carrier sinusoidally modulated. The modulation factor is

(a) 56% (b) 60% (c) 72% (d) 84%

85. A modulated signal $C_m(t)$ has the form $C_m(t) = 30 \sin 300\pi t + 10 (\cos 200\pi t - \cos 400\pi t)$. The carrier frequency f_c, the modulating frequency (message frequency) f_ω and the modulation index μ are respectively given by : * Critical Thinking

(a) $f_c = 200$ Hz; $f_w = 50$ Hz; $\mu = \frac{1}{2}$

(b) $f_c = 150$ Hz; $f_w = 50$ Hz; $\mu = \frac{2}{3}$

(c) $f_c = 150$ Hz; $f_w = 30$ Hz; $\mu = \frac{1}{3}$

(d) $f_c = 200$ Hz; $f_w = 30$ Hz; $\mu = \frac{1}{2}$

86. The T.V. transmission tower in Delhi has a height of 240 m. The distance upto which the broadcast can be received. (Radius of earth is 6.4×10^6 m)

(a) 60 km (b) 100 km (c) 50 km (d) 55 km

87. An audio signal consists of two distinct sounds: one a human speech signal in the frequency band of 200 Hz to 2700 Hz, while the other is a high frequency music signal in the frequency band of 10200 Hz to 15200 Hz. The ratio of the AM signal bandwidth required to send both the signals together to the AM signal bandwidth requried to send just the human speech is: * Critical Thinking

(a) 2 (b) 5 (c) 6 (d) 3

88. A signal of 5 kHz frequency modulates a carrier of frequency 1 MHz and peak voltage 25V. If the amplitude at the side bands of the amplitude modulated signal is 5V, then the modulation index is :

(a) 0.8 (b) 0.6 (c) 0.5 (d) 0.4

89. Television signals on earth cannot be received at distances greater than 100 km from the transmission station. The reason behind this is that

(a) the receiver antenna is unable to detect the signal at a disance greater than 100 km
(b) the TV programme consists of both audio and video signals
(c) the TV signals are less powerful than radio signals
(d) the surface of earth is curved like a sphere

90. A carrier frequency of 10 kV at 1 MHz is amplitude modulated by a 1 kHz signal of 6 kV peak voltage. When the modulation pattern is observed on a calibrated CRO, the voltage indicated by the CRO will be

(a) 16 (b) 32 (c) 864 (d) 84

91. Determine the modulation index and carrier swing of an FM carrier having a frequency deviation of 56 kHz and a modulating signal of 8 kHz.

(a) 9, 580 kHz (b) 3, 101 kHz
(c) 7, 112 kHz (d) 8, 80 kHz

92. Given the electric field of a complete amplitude modulated wave as

$$\vec{E} = \hat{i}E_c\left(1 + \frac{E_m}{E_c}\cos\omega_m t\right)\cos\omega_c t.$$

Where the subscript c stands for the carrier wave and m for the modulating signal. The frequencies present in the modulated wave are

(a) ω_c and $\sqrt{\omega_c^2 + \omega_m^2}$

(b) $\omega_c, \omega_c + \omega_m$ and $\omega_c - \omega_m$

(c) ω_c and ω_m

(d) ω_c and $\sqrt{\omega_c \omega_m}$

93. In optical communication system operating at 1200 nm, only 2% of the source frequency is available for TV transmission having a bandwidth of 5 MHz. What is the number of TV channels that can be transmitted?

Toughnut

(a) 1 million (b) 50 million
(c) 1 thousand (d) 50 thousand

94. The velocity of all radio waves in free space is 3×10^8 m/s. What is the frequency of a radio wave of wavelength 150m.

(a) 5 MHz (a) 2 MHz
(c) 0.5 MHz (d) 0 MHz

95. Maximum usable frequency (MUF) in F-region layer is x, when the critical frequency is 60 MHz and the angle of incidence is 70°. Then calculate x.

(a) 122 MHz (b) 216 Hz
(c) 175 MHz (d) 75 Hz

96. If $E_c = 20 \sin 10^5 \pi t$ and $E_m = 10 \sin 400\pi t$ are carrier and modulating signals, the modulation index is:

(a) 56% (b) 30% (c) 50% (d) 48%

97. The area of the region covered by the TV broadcast by a TV tower of 100 m height is (Radius of the earth = 6.4×10^6 m).

(a) $1.28\,\pi \times 10^3$ km^2 (b) $2\,\pi \times 10^4$ km^2
(c) 3.1×10^4 km^2 (d) $6\,\pi \times 10^4$ km^2

98. Determine the modulation index (m_f) and carrier swing (S) of an FM carrier having a frequency deviation of 25 kHz and a modulating signal of 5 kHz.

(a) $m_f = 3$ (b) $m_f = 5$
(c) S = 80 kHz (d) S = 60 kHz

99. Calculate the power developed by an amplitude modulated wave in a load resistance of 100 Ω, if the peak voltage of carrier wave is 100 V and modulation index is 0.4.

(a) 50 watt (b) 54 watt
(c) 104 watt (d) 4 watt

100. The maximum range for the tropospheric transmission of radio wave of wavelength 3m using the transmitting antenna and receiving antenna of hights 100m and 60m respectively is

(a) 8m (b) 800m (c) 8km (d) 80km

Answer KEYs

1	(a)	11	(b)	21	(a)	31	(c)	41	(b)	51	(a)	61	(c)	71	(c)	81	(c)	91	(c)
2	(b)	12	(c)	22	(d)	32	(b)	42	(a)	52	(d)	62	(c)	72	(d)	82	(c)	92	(b)
3	(c)	13	(b)	23	(c)	33	(d)	43	(d)	53	(a)	63	(a)	73	(b)	83	(b)	93	(a)
4	(a)	14	(b)	24	(d)	34	(d)	44	(a)	54	(b)	64	(a)	74	(c)	84	(b)	94	(b)
5	(d)	15	(c)	25	(c)	35	(b)	45	(d)	55	(d)	65	(a)	75	(b)	85	(b)	95	(c)
6	(d)	16	(d)	26	(b)	36	(d)	46	(b)	56	(a)	66	(a)	76	(b)	86	(d)	96	(c)
7	(b)	17	(c)	27	(c)	37	(d)	47	(b)	57	(d)	67	(a)	77	(b)	87	(c)	97	(a)
8	(c)	18	(b)	28	(d)	38	(a)	48	(b)	58	(a)	68	(b)	78	(c)	88	(d)	98	(b)
9	(b)	19	(b)	29	(a)	39	(c)	49	(a)	59	(a)	69	(b)	79	(c)	89	(d)	99	(b)
10	(c)	20	(c)	30	(b)	40	(a)	50	(b)	60	(a)	70	(c)	80	(d)	90	(b)	100	(c)

Hints & Solutions

1. (a) The mode of communication used for T.V. signals transmission is either space wave propagation, *i.e.*, line of sight propagation or satellite communication.
2. (b) As the audio frequencies range from 20 Hz to 20 kHz, therefore if they are transmitted directly into space, the length of the transmitting antenna required would be extremely large. Hence audio signal can not be transmitted directly into space.
3. (c) An antenna is a metallic structure used to radiate or receive EM waves.
4. (a) Modulation is a process of superposing a low frequency audio signals (called modulating signal) on a high frequency radio wave called carrier wave.
5. (d) Here $V_{max} = \frac{24}{2} = 12\ mV$

   and $V_{min} = \frac{8}{2} = 4\ mV$

   Now,

   $$m = \frac{V_{max} - V_{min}}{V_{max} + V_{min}} = \frac{12-4}{12+4} = \frac{8}{16} = \frac{1}{2} = 0.5 = 50\%$$
6. (d) Surgery needs sharply focused beam of light and laser can be sharply focussed.
7. (b) 10 kHz cannot be radiated (antenna size), 1 GHz and 1000 GHz will penetrate.
8. (c) The sky waves are the radiowaves of frequency between 2 MHz to 30 MHz. These waves can propagate through atmosphere and are reflected back the ionosphere of earth's atmosphere.
9. (b) Maximum Range of the radar is given by

   $$R_{max} = \left(\frac{P_t A^2 S}{4\pi\lambda^2 P_{min}}\right)^{1/4}$$

   Where P_t : peak value of transmitted power
   A : caputre area of the receiving antenna
   S : radar cross-sectional area
   λ : wavelength of RADAR wave
   P_{min} : minimum receivable power of the receiver
10. (c) 11. (b)
12. (c) FM (I.F)=11.5MHz

    $F(\text{image}) = f_s + 2 f_{IF}$

    $= 98.6 + 2\times 11.5 = 121.6$ MHz
13. (b) The temperature of earth's atmosphere increases with height in stratosphere, from 220 K to 280 K and in ionosphere, from 180 K to 700 K.
14. (b)
15. (c) In optical fibre, light travels inside it, due to total internal reflection.
16. (d) A very small part of light energy is lost from an optical fibre due to absorption or due to light leaving the fibre as a result of scattering of light sideways by impurities in the glass fibre.
17. (c)
18. (b) Frequency of radio waves for sky wave propagation is 2 MHZ to 30 MHZ.
19. (b) Height of transmitting antena is given by :

    $$d = \sqrt{2Rh_T},\ h_T = \frac{d^2}{2R}$$

    $$h_T = \frac{(120\times 10^3\ m)^2}{2\times 6400\times 10^3 m} = 1125\ m$$
20. (c)
21. (a) For total internal reflection of sky waves, the refractive index of the ionosphere should be less. If $f > f_c$, the refractive index will increase. By this waves will just pass without reflection taking place.
22. (d) Carrier frequency > audio frequency
23. (c)
24. (d) When $m_a > 1$ then carrier is said to be over modulated.
25. (c) It is true the radiowaves are polarised electromagnetic waves. The antenna of portable AM radio is sensitive to only magnetic components of electromagnetic waves. On account of this the set should be put horizontal and in proper situation so that the signals are received properly from radio station.

Optical communication is a system by which we transfer the informations over any distance from one location to other through optical range of frequency using optical fibre. The optical interference between fibres is zero. Hence, optical fibre communication has immunity to cross talk.

The electromagnetic waves of shorter wavelength do not suffer much diffraction from the obstacles of earth's atmosphere so they can travel long distance.

Also, shorter the wavelength, shorter is the velocity of wave propagation.

26. (b)

27. (c) Sky wave propagation is suited for frequency range 2 MHz to 20 MHz.

28. (d)

29. (a) In ground or surface wave propagation, the loss of power in a signal wave is due to attenuation, *i.e.,* the absorption of energy by the earth which inturn is due to induced charges in earth which travel in ground along with the wave, resulting in the alternating currents in the earth's surface, as well as due to tilting effect, *i.e.,* diffraction of waves.

30. (b) The message from the information source may not be in electrical form so to convert these information i.e. in form of sound to electrical form a transducer like a microphone is used.

31. (c)

32. (b) The diameter of the core in an optical fibre is nearly 10^{-4} m.

33. (d)

34. (d) $f_c = 9\sqrt{N_m} = 9\times\sqrt{9\times10^{10}}$

$= 2.7 \times 10^6$ Hz $= 2.7$ MHz

35. (b) Core of acceptance angle

$\theta = \sin^{-1}\sqrt{n_1^2 - n_2^2}$

36. (d) **37.** (d)

38. (a) $h = 100m,\ d \propto \sqrt{h}$

To triple coverage range, d is to be increased to 3d

$\Rightarrow h' = 9h = 900m$

Increase in height $= h' - h = 800m$

39. (c) **40.** (a)

41. (b) For high-side tunning

$f_{LO} = f_M + f_{IF},\ P_{IF} = 0.6$ MHz

$f_{LOL} = 5 + 0.6 = 5.6$ MHz

$f_{LOU} = 10 + 0.6 = 10.6$ MHz

42. (a)

43. (d) The maximum line-of-sight distance between the transmitting and receiving antennas is

$$d_M = \sqrt{2Rh_T} + \sqrt{2Rh_R}$$

where h_T and h_R are the heights of transmitting and receiving antennas respectively.

$$\therefore d_M = \sqrt{2Rh} + \sqrt{2Rh} = 2\sqrt{2Rh} = \sqrt{8Rh}$$

$$(\because h_T = h_R = h)$$

44. (a) Given, $\upsilon_m = 10$ kHz

$\upsilon_c = 2$ MHz $= 2000$ kHz

Side band frequencies are

$\upsilon_{USB} = \upsilon_c + \upsilon_m = (2000 + 10)$kHz $= 2010$ kHz

$\upsilon_{LSB} = \upsilon_c - \upsilon_m = (2000 - 10)$kHz $= 1990$ kHz

45. (d) **46.** (b)

47. (b)

$$m = \frac{\text{actual frequency deviation}}{\text{max. allowed frequency deviation}} \times 100\%$$

$$= \frac{(\Delta f)\text{actual}}{(\Delta f)\text{max}} \times 100\%$$

if (Δf) actual $= (\Delta f)$ max

$m = 100\%$

48. (b) **49.** (a)

50. (b) $P_t = P_c\left[1 + \frac{m_a^2}{2}\right]$

$$\Rightarrow \frac{V^2_{rms}}{2} = \frac{V^2_c}{2}\left[1 + \frac{m_a^2}{2}\right]$$

$$V^2_{rms} = V^2_c\left[1 + \frac{m_a^2}{2}\right]$$

$$\Rightarrow V_{rms} = V_c\sqrt{1 + \frac{m_a^2}{2}} \Rightarrow V_{rms} = 100\sqrt{1 + \frac{(0.3)^2}{2}}$$

$= 104.5$ volts.

51. (a) Modulation index, $M = \frac{A_m}{A_C} = \frac{0.5V}{10V} = 0.05$

The side band frequencies are

$V_{SB} = V_C \pm V_m = 1 \pm 0.010$ MHz

52. (d)

53. (a) Modulating signal frequency → 10kHz

Carrier signal frequency → 10 MHz

∴ Side band frequency are

USB = 10 MHz + 10 kHz = 10010 kHz

LSB = 10MHz – 10 kHz = 9990 kHz

54. (b)

55. (d) Frequencies present in the output

$= 100 \pm 0.3$

1000 ± 21 i.e. →

999.7 KHz, 100.3 KHz, 999.2 KHz

1000.8 KHz, 998 KHz, 1002 KHz

56. (a)

57. (d) For good demodulation,

$\frac{1}{f} << RC$ or, $RC >> \frac{1}{f}$

58. (a) SSB transmission to signal are possible at load

$\omega_c + \omega_m$ or $\omega_c - \omega_m$

⇒ (1000 + 10) kHz or (100 – 10) kHz.

⇒ 1010 kHz or 990 kHz

59. (a)

60. (a) Carrier Swing $= \frac{\text{Frequency deviation}}{\text{Modulating frequency}}$

$= \frac{50}{7} = 7.143$

61. (c) Carrier Swing $= 2 \times \Delta f$

$= 2 \times 105.03 - 105 = 2 \times 0.03$ MHz

$= 60$ kHz

62. (c)

63. (a) The formula for modulating index is given by

$$m_f = \frac{\delta}{v_m} = \frac{\text{Frequency deviation}}{\text{Modulating frequency}} = \frac{10 \times 10^3}{2 \times 10^3} = 5$$

64. (a) Comparing $(x_{AM})t = 100\,[1 + 0.5\,t]\cos\omega_c t$ for $0 < t < 1$

with standard AM signal $x_{AM} = E_c\,[1 + m_a \cos \omega_m t] \cos \omega_c t$

We have modulating signal t and $m_a = 0.5$.

65. (a)

66. (a) In light modulation, amplitude is varied.

67. (a) Low pass filter allows 624 kHz to pass and retains 630 kHz and 636 kHz frequencies.

68. (b) (Band width)$_{AM}$ or Channel width

69. (b) $P_c = \frac{E_c^2}{2} = 10$ kW

$m_a = \sqrt{m_1^2 + m_2^2} = \sqrt{0.30^2 + 0.40^2} = 0.50$

$$\therefore\ P_t = P_c\left(1 + \frac{m_a^2}{2}\right) = 10\left(1 + \frac{(0.5)^2}{2}\right) = 11.25 \text{ kW}$$

70. (c)

71. (c) For x(t), BW = 2(Δω + ω)

Δω is deviation and ω is the band width of modulating signal.

∴ BW = 2(90 + 5) = 190

For x^2 (t), BW = 2 × 190 = 380

72. (d) Total signal B.W = 12 × 5 = 60 kHz

11 guard band are required between 12 signal

∴ guard bandwidth = 11 × 1kHz = 11kHz

∴ total bandwidth = 60 + 11 = 71kHz

73. (b)

74. (c) Frequency deviation, Δf = 25 kHz

Frequency of modulating signal, $f_m = 5$ kHz

Modulation index, $m_f = \frac{\Delta f}{f_m} = \frac{25}{5} = 5$

75. (b) The frequency of AM channel is 1020 kHz whereas for the FM it is 89.5 MHz (given). For higher frequencies (MHz), space wave communication is needed. Very tall towers are used as antennas.

76. (b)

77. (b) Modulation index $= \frac{B}{A}$

B = 25, A = 60

$\Rightarrow$ M.I. $= \frac{25}{60} = 0.416 \Rightarrow m\% = 41.6\%$

78. **(c)** Modulation index

$$m_a = \frac{E_m}{E_c} = \frac{A}{A} = 1$$

Equation of modulated signal $[C_m(t)]$

$= E_{(C)} + m_a E_{(C)} \sin \omega_m t = A(1 + \sin \omega_m t) \sin \omega_c t$

(As $E_{(C)} = A \sin\omega_c t$)

79. **(c)**

80. **(d)** Pre-emphasis of higher frequency component is required in FM-system because high frequency terms of modulating signal have small amplitude and therefore small power relative to those of low frequency term.

In the reproduced program at the o/p, these high frequency terms have poor S/N ratio and at time noise may completely mask the signal at these high frequencies, so it is necessary to provide pre-emphasis of high frequencies.

81. **(c)** $P_c = \dfrac{P_t}{1+\dfrac{m_a^2}{2}} = \dfrac{12}{1+\dfrac{(0.5)^2}{2}} = \dfrac{12}{1.25} = 9.6 \text{ kW}$

82. **(c)**

83. **(b)** For given transmission band 88-108 MHz

$(\Delta f)_{max} = 75 \text{ kHz}$

given $(\Delta f)_{actual} = 18.75 \text{ kHz}$

$\therefore$ % modulation $m = \dfrac{(\Delta f)_{actual}}{(\Delta f)_{max}} \times 100 = \dfrac{18.75}{75} = 25\%$

84. **(b)** $P_c = P_t\left(1+\dfrac{m^2}{2}\right)$

$\Rightarrow \quad 11.8 = 10\left(1+\dfrac{m^2}{2}\right)$

$\Rightarrow \quad m = 0.6 \Rightarrow$ % modulation $= 60\%$.

85. **(b)** Comparing the given equation with standard modulated signal wave equation,

$$m = A_c \sin \omega_c t + \frac{\mu A_c}{2} \cos(\omega_c - \omega_s)t - \frac{\mu A_c}{2} \cos(\omega_c + \omega_s)t$$

$\mu \dfrac{A_c}{2} = 10 \Rightarrow \mu = \dfrac{2}{3}$ (modulation index)

$A_c = 30$

$\omega_c - \omega_s = 200\pi$

$\omega_c + \omega_s = 400\pi$

$\Rightarrow f_c = 150,\ f_s = 50 \text{ Hz}$.

86. **(d)** Here $R = 6.4 \times 10^6$ m,

$h = 240$ m,

so, required distance

$d = \sqrt{2Rh} = \sqrt{2 \times 6.4 \times 10^6 \times 240}$

$= 55 \times 10^3$ m

$= 55$ km

87. **(c)** Ratio of AM signal Bandwidths

$$= \frac{15200-200}{2700-200} = \frac{15000}{2500} = 6.$$

88. **(d)**

89. **(d)** Due to the finite curvature of the earth, the waves can not be received beyond the tangent points on the curve.

90. **(b)** Voltage observed on CRO of modulated signal (peak to peak)

= (carrier voltage + modulating signal voltage) × 2

$= (10+6) \times 2 = 16 \times 2 = 32 \text{ kV}$

91. **(c)**

92. **(b)** The frequencies present in amplitude modulated wave are:

Carrier frequency $= \omega_c$

Upper side band frequency $= \omega_c + \omega_m$

Lower side band frequency $= \omega_c - \omega_m$.

93. **(a)** Optical source frequency $v = \dfrac{c}{\lambda}$

$$\Rightarrow v = \frac{3 \times 10^8 \text{ ms}^{-1}}{1200 \times 10^6 m} = 2.5 \times 10^{14} Hz$$

Bandwidth of channel (2% of the source frequency) $= 5 \times 10^{12}$ Hz

Number of channels

$$= \frac{\text{Total bandwidth}}{\text{Bandwidth needed per channel}}$$

$$= \frac{5 \times 10^{12} \text{Hz}}{5 \times 10^6 \text{Hz}} = 10^6 = 1 \text{ million.}$$

94. (b) Here : Velocity of electromagnetic waves in free space and wavelength

$v = 3 \times 10^8$ m/s and $\lambda = 150\, m$

The frequency of radio waves is given by

$$= \frac{v}{\lambda} = \frac{3\times10^8}{150} = 2\times10^6\, Hz = 2\text{ MHz}.$$

95. (c) $MUF = \frac{f_c}{\cos\theta} = \frac{60}{\cos 70°} = 175\, MHz$

96. (c) Modulation index $= \frac{10}{20} = 0.5 = 50\%$

97. (a) Area covered by T.V. signals

$A = 2\pi hR = 2 \times 3.14 \times 100 \times 6.4 \times 10^6$

$= 128\pi \times 10^8$

$\Rightarrow A = 1.28\pi \times 10^3\, km^2$

98. (b) Frequency deviation, $\Delta f = 25$ kHz

Frequency of modulating signal, $f_m = 5$ kHz

Modulation index, $m_f = \frac{\Delta f}{f_m} = \frac{25}{5} = 5$

Carrier swing $= 2 \times \Delta f = 2 \times 25\text{ kHz} = 50\text{ kHz}$

99. (b) $E_c = 100\,V$, $m_a = 0.4$, $R = 100\,\Omega$,

$$P_c = \frac{E_c^{\,2}}{2R} = \frac{(100)^2}{2\times100} = 50 \text{ watt}$$

$$P = \left(1 + \frac{m_a^{\,2}}{2}\right)P_c = \left[1 + \frac{(0.4)^2}{2}\right]\times 50 = 54 \text{ watt}$$

100. (c) $\lambda = 3m, H = 100m, h = 60m$

Path difference between direct wave and wave obtained ofter reflection from earth surface is

$$\frac{2\pi}{\lambda}\left(\frac{2Hh}{d}\right) + \pi = 2\pi$$

$$\text{or } \left(\frac{4Hh}{\lambda d}\right) = 1 \text{ or } d = \frac{4Hh}{\lambda d}$$

$$\Rightarrow d = \frac{4\times100\times60}{3} = 8 \times 10^3 \text{ or } 8 \text{ km}$$